Neural Frontiers

Neural Frontiers

Selected Articles Published by MDPI

MDPI

Basel • Beijing • Wuhan • Barcelona • Belgrade • Novi Sad • Cluj • Manchester

This is a reprint of articles published online by the open access publisher MDPI (available at: www.mdpi.com). The responsibility for the book's title and preface lies with Alexandru Vladimir Ciurea, who compiled this selection.

For citation purposes, cite each article independently as indicated on the article page online and as indicated below:

Lastname, A.A.; Lastname, B.B. Article Title. *Journal Name* **Year**, *Volume Number*, Page Range.

ISBN 978-3-0365-9764-5 (Hbk)
ISBN 978-3-0365-9765-2 (PDF)
doi.org/10.3390/books978-3-0365-9765-2

Contents

Preface

In the dynamic and ever-evolving landscape of neurological sciences, the intersection of knowledge between neurology, neurodegeneration, neuroscience, and neurosurgery is both vast and deep. *Neural Frontiers* is an ambitious endeavor that seeks to encapsulate the essence of these interconnected disciplines through a tapestry of articles and research findings.

The primary aim of this yearbook is to provide a comprehensive overview of the latest developments and breakthroughs in these fields. Our purpose is not only to showcase groundbreaking research and innovative practices but also to foster a deeper understanding of the complexities of the human brain and nervous system. This compilation serves as both a reflection of the current state of knowledge and a beacon guiding future explorations.

The motivation for this work stems from a collective desire to bridge gaps between theory and practice, between academics and clinicians, and between the diverse specialties within neurological sciences.

By bringing together a wide array of perspectives, we hope to offer a holistic view of the subject matter that is both informative and inspiring.

This yearbook is primarily intended for students, educators, practitioners, and researchers in the fields of neurology, neuroscience, and neurosurgery. However, its comprehensive nature makes it a valuable resource for anyone interested in the intricacies of brain science, from medical professionals to curious lay readers.

The contributions within these pages are the result of the tireless efforts of a diverse group of authors, including seasoned professors and practicing doctors. Each author brings a unique perspective, shaped by their individual experiences and areas of expertise, which collectively enriches the content of this yearbook.

We extend our deepest gratitude to all those who have played a role in bringing this project to fruition. This includes not only our esteemed authors but also the mentors, colleagues, and peers who have provided invaluable feedback, support, and inspiration. A special acknowledgment is due to the core students that worked tirelessly on the production of the articles reunited in this book:

- Razvan-Adrian Covache-Busuioc;
- Bogdan-Gabriel Bratu;
- Horia Petre Costin;
- Antonio-Daniel Corlatescu;
- Luca-Andrei Glavan;
- David-Ioan Dumitrascu;
- Andrei Adrian Popa;
- Matei Serban.

In conclusion, *Neural Frontiers* stands as a testament to the collaborative spirit and intellectual curiosity that drives progress in neurological sciences. We hope that this yearbook will enlighten, inspire, and contribute significantly to the ongoing dialogue in these vital and fascinating fields.

Alexandru Vladimir Ciurea

International Journal of
Molecular Sciences

MDPI

Review

Unraveling Molecular and Genetic Insights into Neurodegenerative Diseases: Advances in Understanding Alzheimer's, Parkinson's, and Huntington's Diseases and Amyotrophic Lateral Sclerosis

Alexandru Vlad Ciurea [1,2], Aurel George Mohan [3,4,*], Razvan-Adrian Covache-Busuioc [1], Horia-Petre Costin [1], Luca-Andrei Glavan [1], Antonio-Daniel Corlatescu [1] and Vicentiu Mircea Saceleanu [5,6]

1 Department of Neurosurgery, "Carol Davila" University of Medicine and Pharmacy, 020021 Bucharest, Romania; prof.avciurea@gmail.com (A.V.C.); razvan-adrian.covache-busuioc0720@stud.umfcd.ro (R.-A.C.-B.); horiacostin2001@yahoo.com (H.-P.C.); glavan.luca@gmail.com (L.-A.G.); antonio.corlatescu@gmail.com (A.-D.C.)
2 Neurosurgery Department, Sanador Clinical Hospital, 010991 Bucharest, Romania
3 Department of Neurosurgery, Bihor County Emergency Clinical Hospital, 410167 Oradea, Romania
4 Department of Neurosurgery, Faculty of Medicine, Oradea University, 410610 Oradea, Romania
5 Neurosurgery Department, Sibiu County Emergency Hospital, 550245 Sibiu, Romania; vicentiu.saceleanu@gmail.com
6 Neurosurgery Department, "Lucian Blaga" University of Medicine, 550024 Sibiu, Romania
* Correspondence: mohanaurel@uoradea.ro

Abstract: Neurodegenerative diseases are, according to recent studies, one of the main causes of disability and death worldwide. Interest in molecular genetics has started to experience exponential growth thanks to numerous advancements in technology, shifts in the understanding of the disease as a phenomenon, and the change in the perspective regarding gene editing and the advantages of this action. The aim of this paper is to analyze the newest approaches in genetics and molecular sciences regarding four of the most important neurodegenerative disorders: Alzheimer's disease, Parkinson's disease, Huntington's disease, and amyotrophic lateral sclerosis. We intend through this review to focus on the newest treatment, diagnosis, and predictions regarding this large group of diseases, in order to obtain a more accurate analysis and to identify the emerging signs that could lead to a better outcome in order to increase both the quality and the life span of the patient. Moreover, this review could provide evidence of future possible novel therapies that target the specific genes and that could be useful to be taken into consideration when the classical approaches fail to shed light.

Keywords: molecular genetics; neurodegenerative disease; molecular pathology; epigenetics; gene expression; therapeutic targets; biomarkers; Alzheimer's disease; Parkinson's disease; Huntington's disease; amyotrophic lateral sclerosis

Citation: Ciurea, A.V.; Mohan, A.G.; Covache-Busuioc, R.-A.; Costin, H.-P.; Glavan, L.-A.; Corlatescu, A.-D.; Saceleanu, V.M. Unraveling Molecular and Genetic Insights into Neurodegenerative Diseases: Advances in Understanding Alzheimer's, Parkinson's, and Huntington's Diseases and Amyotrophic Lateral Sclerosis. *Int. J. Mol. Sci.* **2023**, *24*, 10809. https://doi.org/10.3390/ijms241310809

Academic Editors: Orietta Pansarasa and Matteo Bordoni

Received: 13 May 2023
Revised: 26 June 2023
Accepted: 26 June 2023
Published: 28 June 2023

1. Introduction

Neurodegenerative diseases are, according to recent studies, one of the main causes of disability and death worldwide. Interest in molecular genetics has started to experience exponential growth thanks to numerous advancements in technology, shifts in the understanding of the disease as a phenomenon, and the change in the perspective regarding gene editing and the advantages of this action. However, the concept of genes is not, as one might consider, a late 20th century notion. Aristotle predicted the existence of genes through postulating that the mother had her characteristics encoded inside the menstrual blood, while the father had his inside the semen, In addition, Hippocrates' theory resembled what Charles Darwin later described as "pangenesis" [1]. However, two breakthroughs came a few centuries later. The first was when the Czech scientist Johann Gregor Mendel coined the terms "recessive, discrete and dominant factors" by observing his hybridization experiments performed on peas [2]. Later that century, Wilhelm von Waldeyer familiarized the scientific world with the term "Chromosomen", derived from the work of Theodor Boveri, who coined the notion of "Chromatinelemente" [3]. The Nobel prize for Physiology or Medicine was awarded, in 1962, to Francis Crick and James D. Watson, for discovering the key to understanding not only molecular genetics, but also the fundament of life itself—the molecular structure of nucleic acids. This discovery was a huge milestone that led to understanding the base structure of life—DNA, RNA, and the creation of proteins; however, it also led to the comprehension of how a small misplacement of some nucleotide subunit can lead to a plethora of changes in the created proteins, further developing the disease.

Neurodegenerative diseases are represented by a group of disorders that are usually associated with protein deposits or misfoldings leading to chemical changes, loss of function, and apoptosis inside the neurons of the brain and spinal cord [4]. They are chronic and progressive, and, most importantly, there are many treatments that slow the development of the diseases and help manage the symptoms but do not cure the actual problem. Molecular genetics play a very important role in understanding the mechanics of the neurodegenerative disorders as it leads to identifying certain genes that are associated with this type of pathology and can also be a way of finding more efficient treatments [5]. In addition, molecular genetics has played an important role in identifying the specific proteins implicated in forming aggregates inside the cells that lead to the apparition of neurodegenerative disorders. The most common types of proteins that are implicated in forming these aggregates are amyloid-β, tau protein, α-synuclein, and prion protein [4,6]. The neurodegenerative disorders have different types of mechanisms behind each one of them, presenting a unique symptomatology. The most common neurodegenerative disorders are Alzheimer's disease, Parkinson's disease, Huntington's disease, and amyotrophic lateral sclerosis, each one of which is associated with different genes, and where the common element is the formation of protein aggregates that lead to changing the physical and chemical properties of the nervous cell [4,7].

2. Alzheimer's Disease (AD)

Alzheimer's disease is the most common neurodegenerative disease that occurs in humans, representing 70% of the total dementia cases [7]. At first, Alois Alzheimer observed, in 1901, the interesting behavior of a 51-year-old woman who suffered from sleep disorders, memory loss, and progressive confusion. During the autopsy, Alois Alzheimer discovered the presence of neurofibrillary tangles and neuritic plaques, concluding that the disease was caused by the agglomeration of these structures. Neurofibrillary tangles—hyperphosphorylated tau proteins and neuritic plaques—aggregate beta amyloids [8]. A study conducted by Gatz et al. found that environmental factors can influence to some degree the chance of developing AD in humans that have the predisposing genes [9]. The perspective of Alzheimer's disease was governed for more than 30 years by the amyloid cascade formation, which finished with the formation of the beta amyloids. However, newer studies tend to identify the AD amyloid cascade as a simplified view of the pathophysiology involved in the disease, emphasizing the glymphatic system, as well as the Lipoprotein receptor-related protein-1; RAGE [10].

2.1. *Amyloid Precursor Protein*

The genetic landscape of AD is dominated by mutations of, firstly, the amyloid precursor protein (APP); these mutations generate autosomal dominant alleles leading to the development of early onset AD [11]. APP can follow either the amyloidogenic or the non-amyloidogenic path, being cleaved by three different secretases: alpha, beta, or gamma [12]. Physiologically, the APP is cleaved by the alpha or gamma secretases [13] during the non-amyloidogenic path. During the amyloidogenic path, the APP is cleaved by the beta and gamma secretases, resulting in *Aβ38*, *Aβ40*, and *Aβ42* [14]. Subsequently, Aβ plaques are formed in a process that starts from Aβ monomers and ends with the development of amyloid plaques. Of the three beta amyloids previously mentioned, *Aβ42* is the least soluble [15].

Naturally, studies have shown that APP, which is coded on chromosome number 21, is directly correlated with a higher risk of developing early-onset AD in trisomy 21 individuals [16].

The relevance of amyloid pathogenicity is explainable by taking into consideration a study conducted in 2012 by Jonsson, T. et al. [17] on a population of 1795 Icelandic subjects. This study proved that the *A673T* (rs63750847) (or A2T) substitution, also known as the Icelandic mutation, decreases the risk of developing Alzheimer's (and cognitive decline associated with aging) by lowering the overall aggregation and production of *Aβ40* and *Aβ42* by approximately 40% [17,18]. However, a study conducted in 2014 on a population of 2641 Chinese subjects showed that this gene does not explain the longevity of the old Chinese subjects who participated in this study because A2T was not included in all their genomes [19]. Another study performed in 2015 on 3487 Danish subjects showed that the *A673T* mutation is present in only one subject (0.014%) [20], in contrast with 0.43% in the Nordic population [17]. On the other hand, the mutation *A673V*, a mutation which manifests itself in a homozygous state, is a gene that is linked with early-onset AD [21]. However, this mutation leads to a distinctive manifestation of the disease in comparison to dominant genes that determine AD: familial-inherited AD usually develops amyloid deposits inside the striatum [22]. In comparison, this mutation tends to avoid the striatum in the incipient phase, focusing the amyloid deposits inside the cerebellum [21].

There are genes that protect against AD, and conversely there are genes that lead to early-onset AD (Osaka and Arctic mutations) or Cerebral Amyloid Angiopathy (Dutch and Italian mutations). For example, there are four mutations that happen on the *E693* position on the APP coding gene, exon 17: the Dutch mutation (*E693Q*), the Osaka mutation (*E693del*), the Italian mutation (*E693K*), and the Arctic mutation (E693G), all resulting in a change of the 22nd amino acid in Aβ, thus developing a peptide called E22. *E693Q* (rs63750579) mutation leads to a peptide called *E22Q*, and *E693Q* mouse models identify features shared with human Alzheimer's brain pathology [23]. This mutation's clinical phenotype is called hereditary cerebral hemorrhage with amyloidosis (HCHWA-D), and occurs in patients suffering from recurrent strokes and dementia. Aβ accumulates in the cerebral vessel walls because its overall production increases with this mutation, resulting in the formation of deposits in the meningo cortical vessels that lead to cerebral amyloid angiopathy [24,25]. *E693del* (dbSNP ID: NA) creates a mutant peptide known as *E22Δ*, which has been shown in a study conducted by Uddin, M.S., Tewari, D., Sharma, G. et al. to lead to increased endoplasmic reticulum stress by increasing the oligomerization of Aβ [26]. Further stress is created, as determined by an overexpressed chaperone *GRP78* and by the expression of the *GFAP* (glial fibrillary acidic protein), by *E22Δ*, thus correlating AD with the glymphatic system [27]. More properties of *E22Δ* were derived in a 2008 study that discovered that this mutant peptide creates fewer overall amyloid deposits; these deposits, however, are more resistant to proteolysis [28]. Recent studies performed by McKnelly et al. show the destabilizing and cytotoxic effect E22 peptides have on in vitro cell membranes by disturbing the phosphatidylcholine and phosphatidylserine constituting the cell membrane. This analysis showcases that these peptides tend to have a disturbing effect on the cell membrane proportional with the quantity of positive charges in the molecule. Thus, the peptides are, in order from least charged to most charged: *E22Δ* (two positive charges), *E22G*, *E22Q* (both have three positive charges), and *E22K* (four positive charges), meaning that the Italian mutation generates the most cytotoxic Aβ [29] (Table 1).

Table 1. Most prominent mutations in APP gene.

Mutation	Pathogenicity	Type of Mutation	Biological Effect	Citation
A673T (Icelandic)	Alzheimer's Disease—Protective	Substitution	This particular type is linked to limited build-up of amyloid and is believed to guard against amyloid-related issues. It results in a decrease of approximately 40 percent in the production of amyloidogenic Aβ peptides, and the Aβ that is produced has a reduced tendency to form clumps.	[17–20]
A673V	Not Classified	Substitution	According to the CERAD criteria, a clear diagnosis of AD was made, as evidenced by substantial Aβ and tau pathology deposits (Braak stage VI) along with cerebral amyloid angiopathy. The deposits found contained elevated levels of Aβ40 and were notably larger, with fewer preamyloid deposits. Perivascular localization was frequently observed. In laboratory studies, it was discovered that A673V caused a shift in β-secretase processing of APP toward the amyloidogenic pathway and amplified Aβ aggregation.	[21,22]
E693Q (Dutch)	Hereditary Cerebral Hemorrhage with Amyloidosis—Pathogenic	Substitution	There is a substantial accumulation of amyloid in the cerebral blood vessels, accompanied by hemorrhages and some diffuse plaques in the brain tissue. In laboratory experiments, it was observed that this condition speeds up Aβ aggregation in vitro, leading to greater fibril formation, and may also modify APP processing.	[24,25]
E693del (Osaka, E693Δ, *E693delta*	Alzheimer's Disease—Pathogenic	Deletion	This variant led to an increased oligomerization and nucleation of Aβ aggregates in vitro. Furthermore, it was found that there was no alteration in the Aβ42/Aβ40 ratio, but there was a decrease in both Aβ42 and Aβ40. This variant was also discovered to be more resistant to degradation by neprilysin and insulin-degrading enzyme. Additionally, this variant had a greater inhibitory effect on long-term potentiation (LTP) compared to wild-type Aβ, which suggests a potential negative impact on synaptic plasticity.	[26–29]
E693K (Italian)	Hereditary Cerebral Hemorrhage with Amyloidosis—Pathogenic	Substitution	The observed symptoms include small to large hematomas, subarachnoid bleeding, scars with hemosiderin deposits, small infarcts, and cortical calcifications. Aβ immunoreactivity was observed in vessel walls and neuropil, but there was an absence of neurofibrillary changes and neuritic plaques. Despite a reduction in the Aβ42/Aβ40 ratio and a decrease in Aβ42 levels, the mutant peptide was found to be toxic in cells and aggregates at a faster rate.	[23]
E693G (Arctic, E22G)	Alzheimer's Disease—Pathogenic	Substitution	Several carriers displayed neuropathology that was indicative of AD. Plaques were observed to have a "targetoid" shape, containing heterogeneous truncated Aβ peptides in the center and surrounded by Aβ42. Cell-based assays revealed a reduction in the production of both Aβ40 and Aβ42. Additionally, there was a decrease in proteolytic degradation of Aβ by neprilysin, a type of enzyme that breaks down proteins.	[23,25]
c.-488C>A (rs532314089)	Alzheimer's Disease	Substitution	Predicted to disrupt binding of transcription factor EGR1. PHRED-scaled CADD = 0.26. Negative regulator in multiple cell types including PC12 neuronal-like rat chromaffin cells, SK-N-SH neuroblastoma cells, C6 glial cells and U373 astroctyoma cells among others	[30–32]
c.24+38G>A (rs373985746)	Alzheimer's Disease	Substitution	Predicted benign in silico (PHRED-scaled CADD = 10).	[32]
c.24+288G>A (rs192348494)	Alzheimer's Disease	Substitution	Predicted benign in silico (PHRED-scaled CADD = 12).	[32]
c.-23-377A>G (rs150375400)	Alzheimer's Disease	Substitution	Predicted benign in silico (PHRED-scaled CADD = 10).	[32]
A18T	Alzheimer's Disease, Cardiovascular Disease	Substitution	Predicted to disrupt signal peptide cleavage and affect APOE secretion. PHRED-scaled CADD = 22.	[32,33]

CADD—Combined Annotation-Dependent Depletion.

2.2. Presenilin1

Presenilin1, part of the γ-secretase, is encoded by *PSEN1*, which is located on chromosome *14q24.3* [34]. The change in the nucleotides of this gene is responsible for approximately 70 to 80% of manifestations of autosomal dominant Alzheimer's disease [35,36].

Naturally, mutations to this gene will determine an unnatural development of Aβ and, in general, *PSEN1* mutations have a negative effect on the cleavage activity of the γ-secretase, resulting in an increased production in *Aβ42* rather than *Aβ40* [37]. However, it is important to note that the overall amplification of amyloid production might not be the right answer to explaining why a change in nucleotides in the presenilin gene determines AD. Besides the amyloid cascade theory, there is another theory that derives from mouse studies, which indicate that presenilin is truly important in the processes of memorizing, learning, and nervous cell survivability [38]. Mutations that can happen on exon 4 of the *PSEN1* gene include *A79V*, *M84V*, and *L85P*. *A79V* is an autosomal dominant mutation that determines an amplified proportion of *Aβ42* in comparison to *Aβ40*, by decreasing the latter's production [39,40]. In a 2018 study, Koriath et al. discuss the fact that four alleles have been identified with a frequency of 0.00014% in the gnomAD database, deducing that this mutation has a low penetrance [41]. *M84V* is another autosomal dominant mutation that determines multiple types of atrophies—temporal and frontal lobe and also cerebellar and cortical atrophy [42]. It determines a greater *Aβ42* to *Aβ40* ratio, this time by increasing the overall quantity of *Aβ42* [43]. Moreover, mutation *M84T* has also been linked to Alzheimer's disease. The *L85P* mutant determines a greater ratio of *Aβ42* to *Aβ40*, but is also shown to increase the production of *Aβ43* [44]. Autopsies of people who suffered from this mutation showed an aggregation of amyloid inside the basal ganglia and cortex, but there is not yet enough evidence to support theories regarding the pathophysiology of this mutation [45]. Up to this moment, only one mutation on the intron 4 of the *PSEN1* gene has been identified, namely *Int4del* (also referred to as L113_I114insT), regarding the deletion of a G nucleotide in the splice region of the *PSEN1* gene, just after the exon4, and therefore transcribing into three altered transcripts, one of which codes a protein with an extra threonine in addition to the *PSEN1* protein; as well as two smaller transcripts due to the apparition of premature stop codons [46]. Studies have shown that this mutant leads to an increase in the *Aβ42* to *Aβ40* ratio, by decreasing the quantity of *Aβ40* and *Aβ38* [47,48]. The *M139V* mutation happens on exon 5 of the presenilin 1 gene. Subjects with this mutation develop Alzheimer's disease without many distinctive clinical and morphological features [49]. Regarding amyloid formation, this mutant will lead to increased *Aβ42* and *Aβ43*, and an overall decrease in *Aβ40*, *Aβ38*, and *Aβ37* [50]. Another mutation of the presenilin 1 gene is seen on exon 7 and is called *S212Y*. Neuropathologically, this mutation resembles the typical AD neurofibrillary tangles and neuritic plaques [51]. Liu et al. determined that in this mutation, production of Aβ42 is increased [47]. Another mutation of the presenilin 1 gene associated with early-onset and late-onset AD is known as *A434C*. This change in nucleotides leads to accumulation of plaques and amyloids inside the neocortex, accompanied by neurofibrillary tangles, and hippocampal and amygdala gliosis [52]. A 2022 study suggests that this mutant leads to a different conformation that results in a particular mode of interaction with the γ-secretase, creating a larger quantity of *Aβ42* than normal [53].

2.3. *Presenilin2*

Presenilin2, part of the γ-secretase, is encoded by *PSEN2*. The *PSEN2* gene is found on chromosome lq42.13 and has a total of 12 exons [53]. Mutations in presenilin genes usually lead, as stated before, to early-onset Alzheimer's disease. However, not all changes in the nucleotides that compose the *PSEN2* gene determine AD. For example, a 2018 screening discovered a mutation in a Belgian subject, now known as mutation *K82fs* (situated on exon 5), who suffered from frontotemporal dementia. This mutation seems to determine a drop in presenilin 2 production in the hippocampal region and in the frontal cortex; further analysis of this pathology revealed Pick's disease [54]. Two other studies performed in 2020 and 2021 revealed a new mutation on this gene: *c.*71C>A* [55], which happens on exon 13 3'UTR. The fact that this mutation is in the untranslated region is showcased by the fact that it lies where miR-183-5p attaches to the gene, inhibiting the suppressing activity of miR-183-5p [56]. Moreover, pathology associated with this mutation is relevant in the diagnosis of AD, for example, a greater ratio of Aβ42/Aβ40 and hippocampal atrophy [55,56]. Furthermore, *M239V* is a mutation, discovered in an 1995 study, that lies on exon 8, and linked with early-onset AD [57]. Autopsies performed on subjects carrying this substitution showcased beta amyloid aggregations and tau NFTs, the golden standard for diagnosing Alzheimer's disease (Figure 1, Table 2).

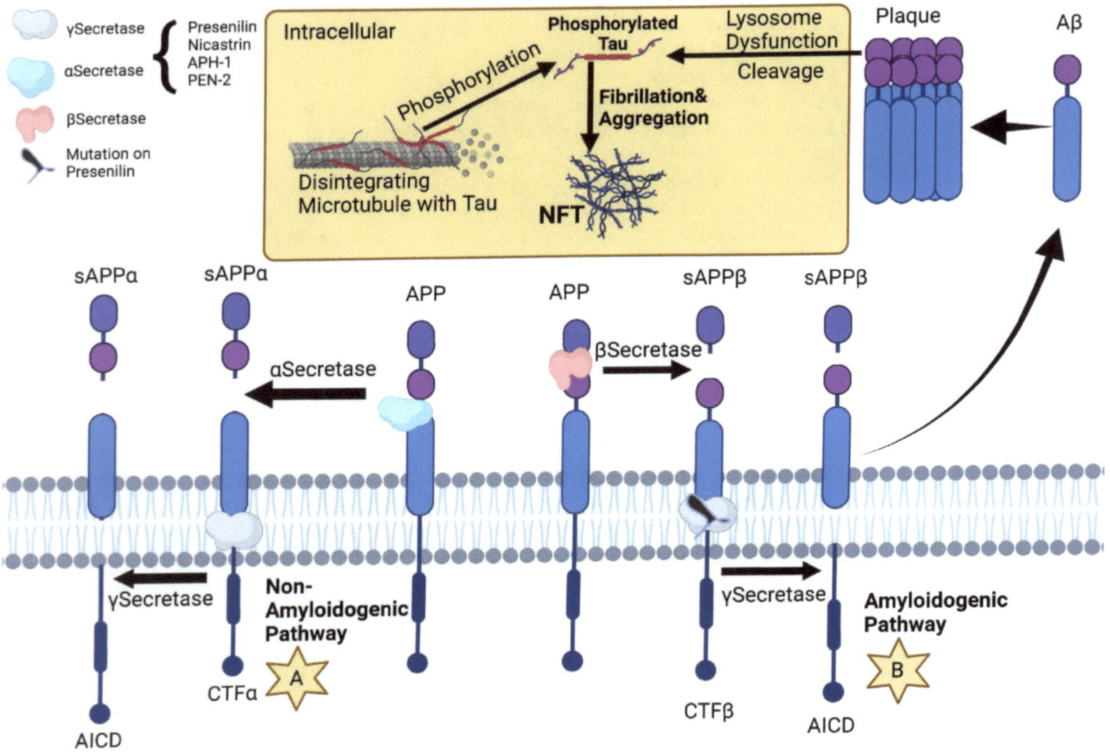

Figure 1. Amyloid precursor protein (APP) is a protein that spans the cell membrane. Its processing can occur through two pathways: the non-amyloidogenic (**A**) and amyloidogenic (**B**) pathways. In the non-amyloidogenic pathway, APP is cleaved in the middle of Aβ by α-secretase, resulting in the production of soluble APPα (sAPPα) and C-terminal fragment α (CTFα), which is then hydrolyzed by γ-secretase to generate the APP intracellular domain (AICD). In the amyloidogenic pathway, APP is cleaved by β-secretase, resulting in the release of N-terminal soluble APPβ (sAPPβ) and the C-terminal fragment β (CTFβ), which is then hydrolyzed by γ-secretase to produce Aβ and AICD. γ-secretase is composed of several parts, including presenilin, nicastrin, anterior pharynx-defective 1 (APH-1), and presenilin enhancer 2 (*PEN-2*). Mutations in the PSEN gene may increase the activity of γ-secretase, leading to the formation of plaques. Moreover, the plaques lead to lysosomal dysfunction (interacting with Caspaze 2/3), promoting cleavage of Tau and NFT (neurofibrillary tangles) formation. Created with BioRender.com.

Table 2. Recent prominent mutations in the *PSEN2* gene.

Mutation	Pathogenicity	Type of Mutation	Biological Effect	Citation
A79V	Alzheimer's Disease —Pathogenic	Substitution	The observed neuropathology was in line with that of AD. It was observed that this variant led to an increase in the Aβ42/Aβ40 ratio and a decrease in the Aβ37/Aβ42 ratios in cells.	[39–41]
M84V	Alzheimer's Disease —Pathogenic	Substitution	In two cases, the observed neuropathology was consistent with AD. Additionally, MRI scans revealed cortical and cerebellar atrophy in these two cases. In the third case, frontal and temporal lobe atrophy was observed. Cell studies revealed an increase in both Aβ42 and the Aβ42/Aβ40 ratio.	[42,43]
L85P	Alzheimer's Disease —Pathogenic	Substitution	SPECT and PET scans showed bilateral hypoperfusion and hypometabolism in the occipital and temporal lobes. Cell studies revealed an increase in the Aβ42/Aβ40 ratio as well as increased Aβ42 levels in transfected cells. In vitro studies indicated a decrease in Aβ42 production and the complete absence of Aβ40 production.	[44,45]
L113_I114insT (int4del)	Alzheimer's Disease —Pathogenic	Substitution	The observed neuropathology was consistent with AD, and included neuron loss in the hippocampus and entorhinal cortex, the presence of neuritic plaques and neurofibrillary tangles in the hippocampus, and amyloid angiopathy, which was particularly evident in the cerebellum. The identified mutation involved a deletion of a G in the splice donor site of intron 4, resulting in the production of three aberrant transcripts. Further investigations indicated an increase in both Aβ42 and the Aβ42/Aβ40 ratio, as well as a reduction in Aβ40 and Aβ38 production in patient brain membranes.	[46–48]
M139V	Alzheimer's Disease —Pathogenic	Substitution	Decrease in the levels of Aβ40, Aβ38, and Aβ37, and an increase in the levels of Aβ42 and Aβ43. In iPSC-derived neurons, the levels of mutant protein were found to be variable, suggesting protein instability.	[49–51]

Thus, correlating the anatomopathological features discovered during autopsies with the familial information and heritage of these patients determined the classification of this gene as a gene that is linked with early-onset Alzheimer's disease [58].

2.4. Apolipoprotein E

However, early-onset Alzheimer's disease accounts for approximately 1–2% of the total cases of AD. The majority of reported genetic cases of AD are caused by mutations of *APOE* (apolipoprotein E). Apolipoprotein E is a glycoprotein that consists of 299 amino acids, and is created inside the central nervous system by a great number of glial cells, including microglias, astrocytes, the cells of the choroid plexus, mural vascular cells, and neurons that undergo stress. *APOE* is abundantly expressed both peripherally and centrally; however, due to the BBB, it exists as separate pools. Therefore, it is crucial to understand what roles each pool may play in AD pathogenesis as well as therapeutic opportunities they present. Peripherally, *APOE* is produced primarily by the liver and plays an essential role in redistribution and metabolism of lipids such as triglycerides, cholesterol, cholesteryl esters, and phospholipids through lipoprotein particles [59]. *APOE* isoforms are associated with different lipoprotein particles in peripheral circulation; for instance, *APOE4* tends to preferentially associate itself with triglyceride-rich particles, while *APOE2* and *APOE3* prefer high density lipoproteins (HDLs). APOE-mediated cholesterol and lipid transport is critical for proper CNS formation and repair. *APOE3* shows a greater effect than *APOE4* at stimulating neurite outgrowth after injury, hence its prevalence among astrocytes. *APOE4* alters structural reorganization of neurons, decreases expression of key synaptic proteins, and inhibits glutamatergic signaling critical for neuronal plasticity and network maintenance. APOE's effects vary by cell type; for instance, it is expressed by astrocytes, microglia, pericytes, and oligodendrocytes under various circumstances. Therefore, in order to gain a comprehensive understanding of its role within the brain it is crucial to examine its structure, lipidation status, and biochemical properties among different cell types that express it [60]. Research has demonstrated that *APOE*, a protein found in the brain, plays an integral role in Alzheimer's disease by its interaction with amyloid-beta protein. *APOE* was discovered co-depositing with Aβ in amyloid plaques, thus contributing directly to AD risk. Knocking out *APOE* in amyloid model mice alters their Aβ plaque morphology significantly, indicating it plays a crucial role in fibrilization and amyloid deposition processes.

Studies of APOE's effects on amyloid pathology have demonstrated isoform-dependent effects, with *APOE4* having the greatest impact, followed by *APOE3* and then *APOE2*. Studies indicate that those carrying an increased level Aβ compared to *APOE3* carriers and earlier deposition, with greater overall deposition, wider cortical distribution, and earlier deposition onset than its *APOE2* counterpart, while delaying deposition, less severe pathology, and protecting cognitive function, were seen with these carriers [61].

Studies have also demonstrated that *APOE4* stabilizes soluble, cytotoxic Aβ fragments and enhances fibrillogenesis to accelerate early amyloid pathology seeding. Thus, their interaction may serve as a potential target for therapeutic intervention at early stages of amyloid disease progression [62,63].

APOE plays an essential role in clearing away antibodies via several mechanisms, including receptor-mediated clearance and proteolytic degradation. Neurons utilize LRP1 receptors to absorb Aβ /APOE complexes from neurons via LRP1; this process is impaired for carriers of APOE4 due to reduced complex stability between *APOE4* and Aβ. Soluble Aβ can also be removed by proteolytic enzymes; however *APOE4* proves less effective at this than *APOE2* or *APOE3*, leading to reduced clearance overall [64].

2.5. APOE and Tau

One defining characteristic of Alzheimer's disease (AD) pathology is the formation of neurofibrillary tangles (NFTs). NFTs consist of hyperphosphorylated tau aggregates as well as Aβ plaques. Studies have demonstrated that carrying the *APOE4* allele increases tau phosphorylation more than either *APOE2* or *APOE3*, particularly when exposed to Aβ oligomers. Furthermore, PET imaging studies on humans with this allele reveal greater tau deposition regardless of plaque presence. Additionally, neuronal *APOE4* was found to promote tau phosphorylation and cell death more effectively than *APOE3* in induced pluripotent stem cell cultures; animal models indicate that this genotype was also associated with higher total tau and phospho-tau levels, exacerbating tau-mediated neurodegeneration through modulating microglial activation [65]. Recently published research has demonstrated that deletion of Astrocytic *APOE4* can significantly decrease tau-related synaptic degeneration and disease-associated gene signatures, protecting against microglial phagocytosis as well as providing protection from tau. One study using AAV-tau delivery found that *APOE2* may cause tau phosphorylation and aggregation to increase, potentially due to formation of tau/APOE complexes primarily produced when non-lipidated *APOE2* was present. Recent genome-wide association study (GWAS) results indicate that *APOE2* may offer protection from AD risk by differentially regulating protein phosphatase 2A (PP2A), an important tau phosphatase in the human brain, unlike the detrimental impact of *APOE4* [66,67]. Taken together, these results show how the impact of *APOE* on tauopathy pathogenesis and tau-mediated neurotoxicity depends on which isoform is chosen. APOE's role in tau pathology has drawn much interest both within AD research and among researchers studying related tau-related diseases such as FTD (frontotemporal dementia), CTE (chronic traumatic encephalopathy), and CBD (corticobasal degeneration). For instance, FTD patients carrying the *APOE4* genotype display earlier onset tau pathology, more severe neurodegeneration, and greater cognitive decline than non-APOE4 carriers, suggesting *APOE* may influence tau pathology independently of Aβ pathology. Therefore, understanding its molecular mechanisms within tauopathy may provide an important insight for developing strategies against AD and related neurodegenerative conditions such as FTD or CTE/CBD [68].

Studies conducted to date have demonstrated that *APOE* binds to regions of tau that contribute to pathogenic NFT formation, and one potential mechanism is that *APOE* may bind tau and block its phosphorylation sites. This interaction has been shown to be isoform-specific, with *APOE3* showing stronger binding affinity to tau's microtubule-binding region than *APOE4*. According to research, reduced binding affinity of *APOE4* to tau may increase GSK3-mediated tau hyperphosphorylation and subsequent formation of NFTs. Alternately, some experts hypothesize that *APOE4* inhibits the Wnt signaling pathway through LRP5/6 receptors by increasing GSK3 activity and leading to tau phosphorylation. Current research is further exploring these potential mechanisms so as to understand APOE's contributions in tau pathogenesis [69].

2.6. *APOE and Neuroinflammation*

The recent literature indicates that inflammation is an integral component of neurodegeneration, with its modulation by *APOE* gaining increasing attention. *APOE* may contribute to AD pathogenesis through various pathways; however, evidence is mounting that suggests they converge into neuroinflammation. Microglia cells often surround plaques found in postmortem brain tissue and play an active role in orchestrating an inflammatory response and clearing out amyloid plaques from memory cells. Studies conducted on mice lacking *APOE* have demonstrated decreased microglial response to plaques, suggesting it may be necessary for proper microglial activation in response to amyloid aggregation. Emerging research has also demonstrated that disease-associated microglia (DAM) or microglial neurodegenerative (MGnD) phenotypes exhibit a consistent transcriptional signature across Alzheimer's mouse models, with APOE serving as a central regulator. *APOE*'s effect on microglial function appears to vary depending on its isoform, with recent research showing that APOE3 induces more effective microglial responses to Aβ injection than its isoform APOE4. This observation could be explained by Triggering Receptor Expressed on Myeloid Cells 2 (TREM2), which interacts with *APOE* with high affinity to modulate microglial responses. Evidence indicates that binding of *APOE* to TREM2 depends on both its isoform and lipidation status, potentially explaining differences in microglial function between isoforms. *APOE4* may impair homeostatic microglial functions due to reduced lipidation or affinity with TREM2, possibly accounting for its less potent homeostatic microglial responses compared with other isoforms [70,71]. Studies have demonstrated that C-reactive protein (CRP), produced by hepatocytes and released into plasma or serum, can be modulated by an individual's *APOE* genotype in their peripheral immune system. CRP is an inflammatory protein produced in response to inflammation or injury and its levels vary accordingly. Proteomic analysis of cerebrospinal fluid has demonstrated lower CRP levels among *APOE4* carriers compared with individuals carrying either *APOE3* or *APOE2*. CSF samples also reveal reduced concentrations of CRP and complement cascade proteins among these carriers, in comparison with individuals who carry either *APOE2* or *APOE3* [72]. However, in spite of this tendency in the genotype, AD prevalence increases sharply with increasing serum CRP levels, with its greatest impact seen among *APOE4* carriers. However, in a longitudinal cohort, the *APOE* haplotype, but not the CRP haplotype, was associated with life-long cognitive decline, thus disproving any association between CRP and cognitive decline. Therefore, those carrying the *APOE4* allele may experience abnormal immune reactions, in response to pathological development, that lead to injury responses and cognitive deficits [73]. Therefore, targeting APOE-mediated inflammatory responses as part of therapeutic approaches for Alzheimer's disease or neurodegeneration could prove useful and should be explored further as a potential solution.

2.7. *Important APOE Mutations Involved in AD Onset*

2.7.1. *c.-488C>A*

The biological impact of this variant found within the *APOE* promoter remains unknown [30]. It falls within the functional domain of the HuD protein that spans nucleotides

−651 to −366 [31] which has been shown to act as a negative regulator in multiple cell types including PC12 neuronal-like rat chromaffin cells, SK-N-SH neuroblastoma cells, C6 glial cells, and U373 astrocytoma cells. Furthermore, substitution of nucleotide 488C will remove potential binding sites used by transcription factor EGR1, which would interact directly with this transcription factor [32].

2.7.2. *c.-24+38G>A*

Yee et al. conducted a study that sequenced the *APOE* genes of 257 Southern Chinese individuals spanning 69 AD patients, 83 subjects with mild cognitive impairment (MCI), and 105 cognitively healthy controls in South China. Two AD patients (1.4%), one control (0.5%), and no MCI patients (0%) carried this variant; it was detected globally at an incidence frequency of 0.00033 in the gnomAD variant database, with most carriers having East Asian heritage (0.0037 frequency; 43 heterozygotes) [32].

2.7.3. *c.-24+288G>A*

Yee et al. conducted a study where this variant was identified in 257 Southern Chinese individuals spanning Alzheimer's disease (AD), mild cognitive impairment (MCI), and cognitively healthy controls from Southern China. One AD patient (0.7%), one MCI patient (0.6%), and one control (0.5%) carried it [32]; gnomAD reported the variant *c.-24+288G>A* as having a worldwide frequency of 0.00016 with only five heterozygote carriers of East Asian origin worldwide. Conversely it was significantly more prevalent among East Asians with an East Asian ancestry population, with an incidence rate of 0.015 according to gnomAD v2.1.1 (Oct 2022); out of 22 carriers listed there, all were had an East Asian ancestry with at least one homozygote from this region.

2.7.4. *c.-23-377A>G*

This variant was identified in a study which involved sequencing the APOE genes of 257 individuals of Southern Chinese origin, comprising 69 AD patients, 83 subjects with MCI, and 105 cognitively healthy individuals—including six AD patients (4.3%), three MCI patients (1.8%), and three controls (1.4%) [32].

In the gnomAD variant database, this variant was reported at an overall frequency of 0.00073 and at A much higher frequency among individuals of East Asian ancestry; 22.2 carriers were identified from East Asia alone with one being homozygous for it.

2.7.5. *A18T*

Yee et al. conducted an in-depth analysis of 257 Southern Chinese individuals' APOE genes, comprising AD 69 patients, 83 subjects WITH mild cognitive impairment, and 105 healthy controls, and found one AD patient (0.7%), three MCI patients (18%), and four controls (1.9%) [32].

Zhou et al. identified this variant as one of six *APOE* variants with potential clinical relevance and functional consequences due to its high prevalence in at least one population and predicted deleterious effects by in silico algorithms. They performed whole genome and exome sequencing analyses from 138,632 individuals from different populations and discovered that this variation alters the *APOE* signal peptide sequence at its cleavage site, potentially hindering secretory efficiency by disrupting recognition at this spot [33] (Table 3).

Table 3. Recently discovered new APOE gene mutations.

Mutation	Pathogenicity	Type of Mutation	Biological Effect	Citation
K82fs	Tauopathy and Pick's Disease	Deletion	The neuropathological findings were consistent with Pick's disease. A frameshift was identified to start at K82, and the mutant protein was found to be reduced in the frontal cortex and hippocampus.	[54]
*c.*71C>A*	Alzheimer's Disease —Pathogenic	Substitution	In one case, an MRI scan revealed widening of the sulcus, fissure, and temporal horn, along with a decrease in hippocampal volume. Additionally, FDG-PET showed hypometabolism in the bilateral frontal, parietal, and temporal lobes. Among the five affected carriers, CSF analysis showed Aβ42, total tau, and phospho-tau levels consistent with AD. The study suggests a possible reduction in the binding of PSEN2 expression suppressor miR-183-5p, which may lead to an increased Aβ42/Aβ40 ratio.	[55,56]
M239V	Alzheimer's Disease —Pathogenic	Substitution	The brain pathology showed diffuse cerebral atrophy, senile plaques, neurofibrillary tangles (Braak and Braak stage VI), ectopic neurons in subcortical white matter, and extracellular "ghost" neurofibrillary tangles. In cell-based assays, there was an increase in the Aβ42/Aβ40 ratio and an increase in Aβ42 levels. However, there was no change in the proteolytic products PSEN2-CTF and PSEN2-NTF.	[57,58]

2.8. Microtubule-Associated Protein Tau

The discovery that microtubule-associated protein tau (MAPT) gene mutations caused frontotemporal dementia with parkinsonism linked to chromosome 17 (FTLD-17) was a historic moment, providing genetic evidence of dysfunction within tau alone as sufficient cause of neurodegeneration independent of Aβ. Abnormal accumulation of tau is seen across various central nervous system disorders such as AD, Pick's disease (PiD), progressive supranuclear palsy (PSP), corticobasal degeneration (CBD), and argyrophilic grain disease; thus, targeting tau offers the possibility not only of treating AD itself but also of treating many other tauopathies associated with Aβ. Human brains produce six distinct isoforms of tau through alternative splicing of the *MAPT* gene located on chromosome 17q21. The different isoforms result from alternatively splicing exons 2 and 3, leading to variants with zero (0N), one (1N), or two (2N) N-terminus inserts [74]. Exon 10 can also affect protein production, leading to three (3R) or four (4R) microtubule-binding domains residing on C-terminal tau proteins based on whether they contain three (3R) or four (4R). 3R tau binds less tightly than 4R tau, so that six tau isoforms could exist: *3R0N*, *3R1N*, *3R2N*, and *4R0N* are all possible isoforms. An average brain contains equal levels of 3R and 4R tau. However, in certain tauopathies—for instance, those linked to frontotemporal dementia with parkinsonism linked to mutations near exon 10 on chromosome 17 (*FTDP-17*)—there may be an increase in 4R tau, increasing interaction with microtubules [75,76]. Tau undergoes various post-translational modifications during both normal physiological processes and stress-induced responses, such as glycosylation, ubiquitination, glycation, nitration, and oxidation processes, with phosphorylation being the most widely studied. When exposed to healthy brain environments such as Alzheimer's disease or other tauopathies such as multiple myeloma or parkinsonism, the levels of tau phosphorylation vary; in healthy brain tissue there are around two or three residues, while neurodegenerative conditions such as Alzheimer's or tauopathies involve much higher phosphorylation, with nine or ten phosphates per molecule being created by imbalanced activity between tau kinases and phosphatases. This results in hyperphosphorylated tau being localized within its environment, in turn resulting in multiple serine/threonine/tyrosine residues on different places on its protein structure due to an imbalance between its kinase/phosphatase activity [77]. Glycosylation, ubiquitination, glycation, nitration, and oxidation are among the many post-translational modifications that play a role in controlling tau during both normal and stress-induced responses. Of these modifications, phosphorylation has been widely studied. Tau is typically phosphorylated on two to three residues in healthy brains. However, in AD and other tauopathies, its hyperphosphorylation occurs at nine phosphates per molecule. This imbalance results from disruptions in the activity of tau kinases and tau phosphatases, leading to decreased affinity of tau for microtubules as well as resistance against degradation by both ubiquitin-proteasome pathway degradation and calcium-activated neutral proteases. Hyperphosphorylated tau forms fibrils and aggregates into NFTs over time. Major tau kinases include GSK-3b, CDK5, PKA, and MAPK CaMK II MARK [78,79], while PP2A has been identified as the primary dephosphorylation enzyme for abnormal tau. Changes in tau kinases and phosphatases have long been documented as markers of AD and related conditions, with expression and activation rates of tau kinases and phosphatases often increasing over time [80]. A variety of processes, including Aβ, impaired brain glucose metabolism, inflammation, and infection all play a part in abnormal tau hyperphosphorylation; therefore, identifying pathways governing post-translational modifications of tau may prove extremely valuable when searching for therapeutic targets.

Research has demonstrated that an abnormal hyperphosphorylation of tau occurs prior to its accumulation in Alzheimer's disease-affected neurons. This hyperphosphorylated tau has been identified both within neurofibrillary tangles as well as within the cytosols of AD brains. Utilizing mAβ Tau-1 for immunocytochemical studies has demonstrated that abnormally phosphorylated tau (not normal tau) accumulates in neurons without tangles (stage "0" tangles) in Alzheimer's and aged hippocampi [81]. At present, tau found in neurofibrillary tangles is known to be ubiquitinated while abnormally hyperphosphorylated tau isolated from AD brain cytosol does not display this property, indicating abnormal hyperphosphorylation occurs prior to its accumulation into neurofibrillary tangles. Davies et al. [82] demonstrated that tau phosphorylation occurs prior to PHF formation in the AD brain by employing monoclonal antibodies targeting mitotic phosphor epitopes. One possible explanation for abnormal hyperphosphorylation of tau is conformational changes occurring within diseased brains that make it an even more favorable substrate for phosphorylation and/or dephosphorylation, respectively. Moreover, they have developed monoclonal antibodies to detect conformational changes in tau, and have demonstrated that tau indeed undergoes conformational changes both in AD patients and transgenic mice that overexpress human tau [82]. As in *FTDP-17*, which is caused by certain missense mutations of tau, these mutations make tau more susceptible to hyperphosphorylation by brain protein kinases and lead to its hyperphosphorylation. However, in AD it is less likely that tau mutations alone are responsible for hyperphosphorylation; several neuronal proteins become over-phosphorylated due to an imbalance between protein phosphorylation and dephosphorylation processes. Biochemical analyses have demonstrated that Alzheimer's brain tissue contains excessive levels of tubulin and neurofilaments that have become hyperphosphorylated; immunocytochemical analysis shows neurofilaments and *MAP1B* to also be affected. Furthermore, both PHF-abnormally hyperphosphorylated tau and its cytosolic counterpart are readily dephosphorylated by in vitro phosphatases [83].

2.9. *Important MAPT Mutations Involved in AD Onset*

2.9.1. *MAPT IVS10+12 C>T*

This mutation was identified as the causative mutation in the Kumamoto pedigree, a Japanese kindred with frontotemporal dementia [84]. Primary clinical symptoms included parkinsonism and dementia manifesting during their fifth decade, with an average onset age of 53 years and length of illness lasting seven years (n = 6).

Brain tissue from affected individuals was observed to contain elevated exon 10 tau transcripts and 4R tau isoforms, with elevated exon 10 tau aggregates observed both in neurons and glial cells; isolated tau filaments displayed twisted ribbon-like morphologies made up of hyperphosphorylated 4R tau. Yasuda et al. reported neuropathological findings for one member of this pedigree while Takamatsu et al. discussed neuropathological findings for one individual within this pedigree [84,85].

2.9.2. *MAPT A152T*

In a large series of American and European people, the *A152T* variant was found to be associated with an increased risk of DLB, but not PD [86]. The variant was found in 10 out of 2456 controls (minor allele frequency 0.20 percent). Among PD patients, 18 out of 3229 carried the variant (MAF 0.28 percent), and among DLB patients, six out of 442 patients carried the variant (MAF 0.68 percent). In addition, two out of 181 patients with multiple system atrophy carried the variant (MAF 0.55 percent), a non-significant increase in frequency.

Consistent with the variable clinical presentations associated with this variant, neuropathological reports are similarly diverse. Abnormal tau accumulation appears to be the unifying feature in all cases for which postmortem findings are available. In some cases, prominent Lewy body pathology is seen [87]. In other cases, the pathology is indicative of PNLA, as indicated by the prominent neuronal loss and tau deposition in the globus pallidus, subthalamic nucleus, and substantia nigra, with lower levels of pathology in the motor cortex, striatum, pontine nuclei, and cerebellum.

This variant has been shown to impair tau's ability to bind microtubules, resulting in less efficient microtubule assembly and impaired microtubule stability. In addition, although the mutant protein appears to aggregate with lower efficiency than wild-type protein overall, it is more prone to oligomer formation [88]. Isogenic human iPSCs generated from fibroblasts of an *A152T* carrier showed that the mutant tau is predisposed to proteolysis by caspases and other proteases and leads to greater tau pathology.

2.9.3. *MAPT K257T*

Autopsy analysis revealed Pick's disease, a subtype of FTD characterized by severe frontotemporal atrophy, particularly in the temporal lobes. The neocortex, hippocampus, and some subcortical regions displayed numerous tau-positive Pick bodies while diffuse hyperphosphorylated tau was detected in certain cell bodies [89]. Recombinant tau protein with the *K257T* mutation displayed reduced capacity to facilitate microtubule assembly [89].

2.9.4. *MAPT L266V*

Kobayashi et al. reported one case who underwent autopsy that revealed severe frontotemporal atrophy with Pick-like pathology, evident by prominent atrophy of both frontal and temporal cortices as well as caudate nucleus and substantia nigra, prominent neuronal threads, coiled bodies, and ballooned neurons throughout all layers of the cortex and the brainstem; abundant tau-positive inclusions within neurons and astrocytes with a high concentration of tau-positive inclusions were present throughout all cortical layers and the brainstem; there were abundant tau-positive inclusions found among neurons and astrocytes, with a high concentration found throughout all cortical layers as well as a high concentration found within the caudate nucleus; and numerous neuronal threads, coiled bodies, and ballooned neurons were observed. Additionally, neuronal threads and ballooned neurons were observed—all were present and notable for its severity [90].

Hogg et al. reported another case characterized by severe frontotemporal atrophy with Pick-like pathology, as well as significant atrophy of the hippocampus and parietal lobe. Neuronal loss was extreme across the cortex and substantia nigra with severe atrophy in these regions; there was also significant gliosis present, while tau-positive inclusions were widely distributed, including within the hippocampus, striatum, and substantia nigra. Pick bodies were Gallyas silver positive and contained straight filaments distributed randomly throughout layers of cortex [91].

In vitro, this mutation alters exon 10's splicing, leading to higher levels of tau transcripts, with four microtubule binding repeat domains (4R tau). This leads to decreased rates of microtubule assembly induced by tau and lower tubulin polymerization levels; more specifically, with 3R tau isoforms being more likely to assemble than their 4R counterparts (Table 4).

Table 4. Most relevant mutations in the MAPT gene leading to AD.

Mutation	Pathogenicity	Type of Mutation	Biological Effect	Citation
IVS10+12 C>T	Familial Danish Dementia—Pathogenic	Substitution	The mutant protein leads to the formation of tau aggregates in both neurons and glia, and isolated tau filaments exhibit a twisted, ribbon-like morphology and consist of hyperphosphorylated 4-repeat (4R) tau isoforms. The mutation also causes a destabilization of a stem-loop structure that regulates the alternative splicing of exon 10, resulting in a higher frequency of inclusion of exon 10 and an increased proportion of 4R tau isoforms.	[84,85]
A152T	Alzheimer's Disease—Risk	Substitution	The presence of tau pathology is a common feature, often accompanied by Lewy bodies, amyloid plaques, or TDP-43 pathology. The mutant tau has a decreased ability to bind to microtubules, leading to less efficient microtubule assembly and impaired microtubule stability. Additionally, it has an increased propensity to form tau oligomers and is more susceptible to proteolysis by caspases.	[86–88]
K257T	Tauopathy and Frontotemporal —Pathogenic	Substitution	The patient exhibited frontotemporal atrophy with significant temporal lobe involvement. Tau-positive Pick bodies were found in the neocortex, hippocampus, and subcortical regions similar to those seen in sporadic Pick's disease. Some cell bodies showed diffuse hyperphosphorylated tau. In vitro analysis showed that recombinant tau protein with the K257T mutation had a decreased ability to promote microtubule assembly.	[89]
L266V	Frontotemporal —Pathogenic	Substitution	The patient had severe atrophy of the frontal and temporal lobes, with extensive neuronal loss and gliosis. Tau-positive inclusions, including Pick bodies, and tau-positive argyrophilic astrocytes with stout filaments and round or irregular argyrophilic inclusions were also observed. In molecular studies, there were increased levels of exon 10+ tau mRNA and soluble four-repeat (4R) tau. The patient showed a decreased rate and extent of tau-induced microtubule assembly, as well as a specific increase in tau self-assembly for the 3R isoform.	[90,91]

2.10. The Evolving Landscape of Alzheimer's Disease Donanemab Treatment: Exploring Current and Future Perspectives

Donanemab, a humanized antibody that targets the N truncated pyroglutamate-amyloid-b peptide (*pGlu3Aβ, AβpE3*), has shown potential to reduce cerebral amyloid depositions in Alzheimer's disease, constituting a promising treatment option for AD. This therapy aims either to reduce *pGlu3* Aβ formation at glutamyl cyclase, or to clear *pGluAβ* after formation and/or block aggregation [92–94]. Donanemab was found to be highly active against amyloid, especially cored plaques within the CNS. However, its efficacy as a treatment of AD is still uncertain. The binding properties of antibodies targeting *AβpE3* are different against the soluble and aggregated forms of *AβpE3-42* [95,96].

Lowe et. al. [97] proved that Donanemab shows good tolerance to dosages up to 10 mg/kg, with a terminal half-life mean of four days following a single dose of 0.1–3 mg/kg. The half-life was increased to 10 days with a dosage of 10 mg/kg. A 40–50% decrease in amyloid levels was seen at 24 weeks with a Standardized Uptake Ratio (SUVR), which decreased from 1.65 to 0.36, and a change in Centiloids (CLs), which decreased from −44.4 (SD 14.2) from baseline. Moreover, 90% of the subjects developed antidrug antibodies 3 months after a single dose.

In a separate report by Lowe et al. [97], it was shown that Donanemab caused rapid amyloid decreases even after just one dose [98]. The mean reduction in PET amyloid was −16.5 CL at 10 mg/kg, −40.0 CL at 20 mg/kg, and −50.6% at 40 mg/kg. The multiple-dosage groups at week 24 showed a mean reduction in amyloid levels in the 10 mg/kg Q4weekly arm, a 50.2 CL for the 10 mg/kg Q2weekly arm, and a 58.4 CL for the 20 mg/kg Q4weekly arm. In both the single-dose and multi-dose cohorts, some patients had an amyloid clearance level below 24.2. Donanemab was effective in treating all but one patient (97.8%) [98].

In another cohort, nearly 40% of participants (46 of 115) receiving Donanemab reached the full amyloid clearing threshold (24.1 CL), and their baseline amyloid levels were lower than those of the group as a whole. In the first 24-week period, there was a moderately negative correlation (r = −0.54) between the baseline amyloid level and that of the plaques removed. The amyloid clearing was sustained, with a very low rate of re-accumulation (0.02 mean rate over one year). Participants who had an amyloid concentration of =11 CL by week 24 but discontinued treatment, would require about 3.9 to accumulate amyloid up to 24.1 CL. The overall tau accumulation was reduced by 34% in the Donanemab groups compared to placebos at week 76 [99].

In another study by Lowe et al., two patients experienced asymptomatic amyloid imaging abnormalities as a result of cerebral microhemorrhages. One patient discontinued treatment because of these reactions. In a second study by the same authors, seven serious adverse effects were reported in six patients. Only one patient died from a non-drug-related myocardial ischemia. In another study by the same authors [98], seven serious adverse events were reported among six patients, with only one patient dying due to a non-drug-related myocardial infarction. Two patients (4%) from the interventional arms stopped taking Donanemab because of adverse reactions.

Another investigation relates that the rate of mortality from all causes was lower for participants who received Donanemab (0.76%) compared to those who received placebo (1.6%). The study also found that there were no differences between the groups in terms of the rates of serious adverse reactions, which were respectively 19.85% and 20.0% in the Donanemab group and the placebo group [99].

As a conclusion, Donanemab is now being tested as an alternative therapy for Alzheimer's. It has been a long time since the urgent need to slow the disease's progression was met. Alzheimer's disease treatments have been approved by the FDA, but there has been controversy over this approval. There is also a need for better and more effective treatments. According to a systematic review of phase III trials for preclinical Alzheimer's, the first anti-amyloid therapies were performed. It is important to plan carefully, to perform longitudinal assessments, and to store and manage data effectively as clinical trials and new therapeutics are developed. The research conducted will determine Donanemab's effectiveness for a diverse population, increasing the retention of the treatment and improving referrals to clinical trials.

2.11. Brief Reflection Point

The section offers an engaging panorama of Alzheimer's research progress and challenges, and advocates for an interdisciplinary global collaboration to overcome them. With each step we take towards unravelling Alzheimer's complex web, the promise of effective interventions—perhaps eventually leading to cures—becomes more tangible, bearing immense significance for millions across the globe who suffer.

3. Parkinson's Disease (PD)

Parkinson's disease (PD) is a progressive neurodegenerative condition, most often seen among elderly individuals worldwide. It is estimated to affect between 0.3% of the general population and 1–3% of those over the age of 65. By 2030, its numbers are expected to climb from 8.7 million to 9.3 million. James Parkinson first described PD symptoms in 1817, and they typically include dysfunctions of the somatomotor system, including rigidity, bradykinesia, postural instability, gait dysfunction, and tremors.

Disease progression leads to progressive degeneration of the nigrostriatal dopaminergic pathway, leading to significant neuron loss in substantia nigra pars compacta (SNpc) neurons and depletion of dopamine (DA). Non-motor dysfunctions such as dementia, hyposmia, and gastrointestinal abnormalities often accompany disease progression.

Pathological hallmarks of Parkinson disease (PD) include accumulations of a-synuclein aggregates known as Lewy bodies or neurites in certain areas of the central nervous system, such as the basal ganglia, dorsal motor nucleus of vagus (DMV), olfactory bulb (OB), locus coeruleus (LC), intermediolateral nucleus in spinal cord (IML), celiac ganglia, and enteric nervous system (ENS) [100].

New research indicates that Parkinson's disease (PD) neuropathology could be caused by environmental stressors and the natural process of aging itself. Exposure to environmental toxins, drugs of abuse, or the stress of aging may lead to chronic low-level inflammation in the brain, leading to something known as "inflammageing," and thus to neuron cellular senescence.

Pathologically, Parkinson's patients typically display damage in the substantia nigra pars compacta and pontine locus coeruleus regions of their brains characterized by depigmentation, neuronal loss, and gliosis. By the time symptoms manifest themselves, approximately 60–70% of neurons from this region have already been lost [101].

Genetic factors have been estimated to account for roughly 25% of the risk associated with Parkinson's disease, and genetic variants associated with it vary both in terms of frequency and risk. While rare mutations within individual genes (known as monogenic causes) may contribute to its development (known as monogenic causes), these were generally discovered through linkage analysis in affected families using linkage analysis; some common genetic variants that only contribute a small amount to risk were also discovered via genome-wide association studies (*GWASs*), including many common genetic variants that contribute an intermediate risk, such as *GBA* or *LRRK2* variants.

Genetic classification of Parkinson's can lead to various treatment approaches and prognoses for each subgroup, often depending on age of onset, family history, and pathogenic variant presence; age at onset, family history and presence of pathogenic variants are frequently used as criteria for stratifying this form of PD. Monogenic forms may or may not represent typical forms of idiopathic PD. Importantly, some genes involved in monogenic PD have also been identified through *GWAS* studies as common variants. One such gene, *SNCA*, which was discovered through these analyses to have common variants, is also implicated in monogenic PD pathogenesis, supporting the role of a-synuclein. Other pathways may also play a part in its pathogenesis, such as tau aggregation, which is linked with other neurodegenerative conditions such as Alzheimer's and frontotemporal dementia [102].

Familial Parkinson's, also referred to as Mendelian or monogenic PD, is characterized by rare yet high-penetrance genetic variants that increase risk. Autosomal dominant (e.g., *SNCAA53T* and *VPS35D620N*) and recessive forms of familial Parkinson disease have been identified using linkage analysis in families with the help of next-generation sequencing technologies, though only 5–10% of cases fall under these single gene variants. Conversely, low-penetrance genetic variants with more frequent associations with sporadic Parkinson's disease have been identified through genome-wide association studies (*GWASs*). At first glance, distinguishing familial from sporadic disease may help with diagnosis, prognosis, and genetic counseling for at-risk family members; however, such classification may obscure shared genetic or biological mechanisms that underlie them both.

An example is that both rare and common genetic variants associated with *SNCA* have been shown to increase Parkinson's risk, underscoring its role as an aSyn-mediated disease mechanism. Missense variants in *SNCA* such as *p.A53T*, *p.A30P* and *p.E46K* cause autosomal dominant familial Parkinson disease, while the common risk variant *SNCArs356168* occurs in approximately 40% of European-ancestry populations and has only modest effects on disease risk [103].

SNCA, or synuclein complex A, is a 14.5 kDa protein consisting of 140 amino acids encoded by 5 exons and having a transcript length of 3041bps. Located on *4q21.3-q22* of human chromosome 4, this synuclein protein family also includes *SNCB* (*5q35*) and *SNCG* (*10q23.2-q23.3*). The structure of *SNCA* protein comprises an N-terminal region with incomplete KXKEGV motifs, an extremely hydrophobic NAC domain, and an acidic C-terminal domain. Under physiological conditions, it appears as either an intrinsically disordered monomer or helically folded tetramer structure. Although it was previously thought to be toxic in this form, recent observations have refuted this idea.

Over the last two decades, various hypotheses have been put forward regarding the toxic structural form of *SNCA*; none has yet been unanimously agreed upon. What is known is that its neurotoxic form accumulates within neurons before disseminating throughout anatomically interconnected regions in the Parkinson's disease brain through interneural transmission using various mechanisms.

Although *SNCA* is most abundantly expressed in the brain, it also appears in heart, skeletal, muscle, and pancreas cells. While its exact function remains undetermined, several hypotheses have been proposed based on its structure, physical properties, and interactions with interacting partners. *SNCA* may play an essential role in regulating dopamine release and transport, inducing microtubule-associated protein tau fibrillization and exerting a neuroprotective phenotype in non-dopaminergic neurons by modulating both p53 expression and transactivation of proapoptotic genes leading to decreased caspase-3 activation.

Given *SNCA*'s central role in neurodegenerative processes, its essentiality may suggest that selective forces among sarcopterygians play a vital role in modulating its molecular and cellular mechanisms. Fine-tuning of these mechanisms through minute changes to protein activity could have contributed to evolutionary adaptations that meet different environmental and ecological needs. Current evidence indicates that amino acids 32 to 58 of *SNCA*'s N-terminal lipid binding domain are critical to its normal cellular functioning and disease pathogenesis. Lineage-specific substitutions could have led to structural remodeling and functional adaptation in *SNCA* over generations, and any mutation affecting its critical regions is likely to be harmful. These discoveries provide the framework for investigating their critical roles through various interaction studies as well as targeting them with drug discovery efforts to treat FPD [104].

Mutations in *LRRK2* account for 5–12% of familial parkinsonism cases and 1–5% of sporadic cases. So far, seven missense *LRRK2* mutations have been identified as pathogenic: *R1441G*, *R1441C*, and *R1441H* were all found to be pathogenic; these variants can be found within different functional domains of *LRRK2*, including *R1441G* located on *R1441C*, which affects *R1441H*; *Y1699C* was also involved, as well *G2019S*, *R1628P*, *G2385R*, and *I2020T* variants specific to certain populations. The *G2019S* mutation, which leads to constitutive activation of the kinase, is one of the most prevalent. It accounts for an estimated 36% of familial and sporadic Parkinson's cases among North African Arabs; approximately 30% among Ashkenazi Jewish populations; up to 6% among familial cases in Europe and North America; and up to 3% among apparently sporadic cases; however, it does not occur among Asian populations. Various other *LRRK2* mutations such as *G2385R, R1628P, S1647T, R1398H,* and *N551K* have also been associated with parkinsonism within certain Asian populations. Studies conducted among Asian populations spanning Singapore, Taiwan, and mainland China have established that *LRRK2* variants *G2385R* or *R1628P* may increase risk for Parkinson disease. Furthermore, the *G2385R* variant has been found to increase risk for Parkinson's disease among Japanese and Korean populations; these variants were not seen among Indians and Caucasians. Although *LRRK2* mutations exist in familial PD, no differences exist in clinical features or neurochemical differentiation between idiopathic and familial forms of parkinsonism. Both forms of Parkinson disease (PD) involve profound dopaminergic neuronal degeneration and gliosis in the SNpc, decreased dopamine levels in the caudate putamen, and Lewy body pathology in the brainstem; therefore, understanding *LRRK2* plays an essential role for all forms of PD [105].

Mutations in the *PINK1* gene are an important cause of early-onset Parkinson's disease (EOPD), accounting for 1–9% of genetic cases and 15% of early-onset cases—second only to Parkin mutations. First identified by Unoki and Nakamura in 2001, its 18 Kb span contains 8 exonic regions that encode for an essential serine/threonine protein kinase essential for mitochondrial functioning and metabolism.

As reported by the MDSGene database, worldwide there have been 151 *PINK1* mutation carriers who carry 62 different disease-causing sequence variants involved with both sporadic and familial Parkinson disease cases; 13 definitely pathogenic mutations exist alongside 44 possibly pathogenic variants (13 definitely pathogenic mutations and 44 possibly pathogenic variants).

PINK1, an encoded protein from the *PINK1* gene, primarily localizes to mitochondria where it serves as a serine/threonine-type protein kinase that regulates mitochondrial quality control (mitoQC). MitoQC involves maintaining respiring mitochondrial networks while selectively eliminating damaged ones through mitophagy, an essential process critical for cell homeostasis. Furthermore, in addition to mitoQC functions, *PINK1* also plays an anti-death, pro-survival role under various forms of stress conditions, preventing neuronal cell death under various stress conditions. Additionally, its protein contains an N-terminal mitochondrial targeting sequence (MTS or TMD), transmembrane sequence (TMS or TMD), and C-terminal domain [106].

3.1. *Perspectives of Treatment*

The metal-based hypothesis of neurodegeneration is an attractive explanation for the pathophysiology behind Parkinson's disease. This hypothesis proposes that reactive oxygen species are generated by redox-active metals, particularly iron. ROS (reactive oxygen species) cause membrane phospholipids to be peroxidized, resulting in the production of reactive aldehydes. Both ROS and reactive aldehydes modify α-synuclein, causing it to aggregate. Aggregated α-synuclein causes mitochondrial dysfunction, resulting in a vicious cycle of increased ROS production and decreased ATP synthesis. In order to provide a more effective treatment of PD, a multi-task strategy targeting these events is needed [107].

Coenzyme Q10 is a vital antioxidant that is important in reducing oxidative stresses, a factor implicated in Parkinson's disease and other neurodegenerative diseases. In order to establish their potential as a marker of disease, a number of studies have investigated the levels of CoQ10 found in different tissues of people with PD or other parkinsonian disorders. Several studies have also explored the therapeutic potential of CoQ10 for the treatment of PD or PS. Several clinical studies have examined the ability of ubiquinol (the antioxidant form of Coenzyme 10 or CoQ10) to reduce oxidative damages observed in PD. CoQ10 can restore mitochondrial function by bypassing Complex I dysfunction, which is a feature of sporadic PD. A meta-analysis consisting of 8 controlled trials involving 899 patients found that CoQ10 is well-tolerated, safe, and does not improve motor symptoms compared to placebo. The study authors do not recommend CoQ10 as a routine treatment for PD except in cases where levodopa is wearing off [108].

Recent clinical trials have shown that iron chelation therapy is a promising approach to treating Parkinson's disease. Due to the multifactorial nature of PD, targeting a specific factor, such as iron, may not be enough for complete neuroprotection. It may be necessary to develop and test multifunctional drugs that combine the iron chelation process with other protective properties.

A growing global population is aging, and central nervous system disorders such as Parkinson's and Alzheimer's diseases are becoming more prevalent. These disorders are linked to iron accumulation in certain areas of the mind. Finding effective treatments for these conditions is therefore crucial to improving the longevity and quality of life of elderly people [109]. DFO (deferrioxamine) was administered intramuscularly in early studies of Alzheimer's patients. DFP (deferiprone) was the first oral chelator used to treat Friedreich's Ataxia [110]. This condition is characterized by frataxin deficiency, which is the mitochondrial chaperone for iron. Animal studies have shown that DFO or DFP can reduce iron in different brain regions, and also provide neuroprotection for an animal model of Parkinson's disease. In two clinical trials, oral DFP was administered to PD patients in cohorts. MRI measurements showed that the iron content of the substantia nigra, as measured by DFP, decreased. UPDRS scores also improved. Iron chelation was not effective in patients who had high levels of inflammatory marker IL-6. DFP has a major problem with agranulocytosis, and neutropenia. This requires testing of white blood counts every week and complicates logistics. DFP is currently being tested in phase II clinical trials on early-stage PD [107,111].

α-Synuclein is another point of interest regarding the treatment of PD. Although clinical trials using monoclonal antibodies to treat α-synuclein aggregates have failed to show any improvement in Parkinson's symptoms, other studies in progress or recruiting participants may prove that targeting α-synucleinopathies as a therapeutic option is possible.

3.2. *Brief Reflection Point*

This section leaves us with a clear message: while we have made significant strides toward understanding Parkinson's disease, much remains unsaid. To truly make progress against it possible for millions worldwide living with Parkinson's, interdisciplinary research must continue as we unravel its complexities. Continuing the search for knowledge is not simply essential scientifically; it also enhances quality of life for millions affected by this condition worldwide.

4. Huntington's Disease (HD)

Huntington's disease represents a neurodegenerative disorder that bears the name of the American physiologist George Huntington (1850–1916). He was the first one to observe the main manifestation of this disease, characterized by uncontrolled motor activities that tend to be compared to dance-like movements of the body, named chorea, as well as abnormalities regarding both the personality and patient's way of thinking. The motor disturbance can be split into two different stages: the incipient stage, characterized by a hyperkinetic syndrome, due to the loss of the medium spiny neurons (MSNs) of the indirect pathway, followed in the later stages of the disease by the loss of the MSNs of the direct pathway, which lead to a hypokinetic syndrome [112]. Moreover, Huntington was the one to observe that this pathology has both a hereditary nature as well as a progressive onset, saying "Once it begins it clings to the bitter end" [113]

However, Huntington's disease represents a rare pathology, with an incidence of 10.6 to 13.7 out of 100,000 individuals. The statistics vary across ethnic groups due to the differences in the HTT gene. According to a study conducted by Bates et al. in 2015 [114], the average length of repeated CAG trinucleotide sequences varies from 18.4–18.7 in the European population and 17.5–17.7 in the East Asian population.

As it was increasingly studied, it was shown that Huntington's disease represents an autosomal dominant progressive neurodegenerative disorder, consisting of a repetitive set of $(CAG)_n$ trinucleotide sequences in a gene found on the chromosome 4p16.3, between *D4S10* and *D4S98* [115] (Gusella et al., 1983), called huntingtin (HTT), leading to a polyglutamine expansion. This repetitive sequence of trinucleotides leads to a mutation of the huntingtin gene (mHTT).

However, the number of CAG units repeated in an allele has a strong significance when it comes to predicting whether the allele is a disease generating one or not. The normal range for the healthy population is between 6 and 35 units. Between 36 and 39 units, the disease is not guaranteed to occur, but there are also chances of developing it. Over 40 units, the mutation is regarded as highly penetrant and it will generate a phenotype along the adult population [116].

Conversely, the length of the CAG repeated sequence is not only correlated with the chances of developing the Huntington's disease, but also with the onset age of this pathology. The study conducted by Persichetti et al. (1994) showed that the bigger the sequence of trinucleotides, the earlier the beginning of the neurological symptoms [117].

Additionally, other variations have been identified in the HTT gene beyond its polymorphic/expanded CAG repeat. These include modifications in both its coding sequence (such as an expanded CCG repeat after CAG repeat and deletion polymorphism at codon 2642), as well as untranslated sequences, intron sequences, and those flanking its centromeric and telomeric ends. These variations have been used to define HTT haplotypes, which represent groups of sequence variants on specific chromosomes that tend to remain relatively unchanged between generations due to limited recombination events in this relatively small segment of genome [118].

These haplotypes, carrying expanded alleles in HD patients, have revealed that approximately 50% of Europeans with HD have one common ancestor, while multiple independent mutations on different chromosomal backbones account for the rest. However, none of the most frequent haplotypes found either on HD chromosomes or among HD heterozygotes appear to significantly alter motor diagnosis age. Therefore, while natural sequence variation at HTT might occasionally serve as a source of disease modification in HD, its contribution is not significant enough [119].

The mHTT gene leads to the formation of an abnormal huntingtin protein with an extended polyglutamine tail at the NH2-terminal end.

The toxicity of the mHTT gene is generated through the formation of two kinds of mRNA. The first is represented by the HTT mRNA, while the second is a HTT mRNA exon 1, which encodes only the first exon due to the CAG repeated sequence [120]. Therefore, from the first type of mRNA will result a full-length huntingtin protein made of huntingtin fragments, which are loops that are used for proteolytic cleavage and will later represent the site for complex post-translational modifications that will lead to the HTT exon 1 fragment, as well as some individual boxes. The HTT mRNA exon 1 on the other side translates itself with the help of the ribosome just in a huntingtin fragment, which is arranged in three parts. The first is described as a mixed sequence of 17 amino acids, the second part is the polyglutamine sequence, also called polyQ, while the last part of the exon 1 fragment is characterized by a proline-rich domain (PRD) (Figure 2).

Upon translation, either the full-length huntingtin protein or the HTT exon1 protein is generated. The HTT exon1 fragment comprises the HTTNT sequence, the polyglutamine sequence encoded by the CAG repeat, and a proline-rich domain. On the other hand, the full-length huntingtin protein includes the HTT exon1 sequence, as well as ordered and disordered protein segments represented by boxes and loops, respectively.

Proteolytic cleavage, which occurs at recognition sequences located in the disordered segments, leads to the formation of various products, including HTT exon1-like fragments. Fragments with expanded polyQ segments play an important role in the development of Huntington's disease through molecular mechanisms that have yet to be fully understood.

Huntington's disease begins early, before any symptoms have emerged, with transcriptional dysregulation taking place due to mutations of HTT that disrupt transcriptional machinery via interactions with transcription factors and molecular mediators such as CBP (cAMP response element-binding protein).

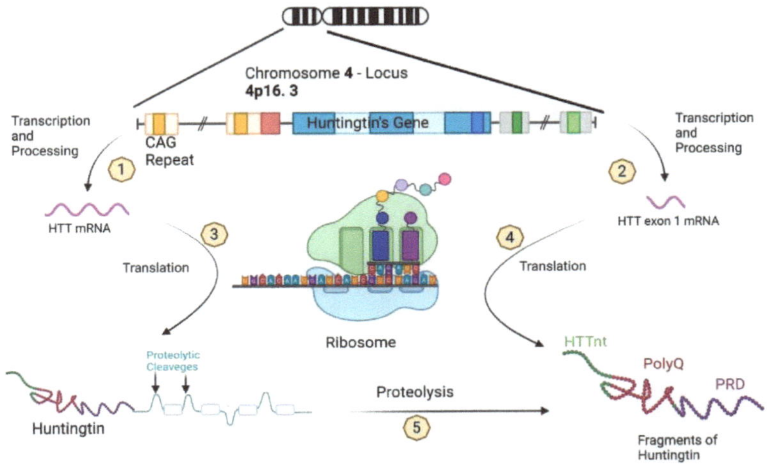

Figure 2. The huntingtin protein can take on various forms and undergo changes depending on the expression of the HTT gene. Normally, the expression of HTT results in the production of an RNA transcript that codes for the complete huntingtin protein. However, if the gene has an expanded CAG repeat, the RNA transcript can be processed abnormally, producing an mRNA that encodes only the HTT exon1 protein. 1 & 2—Transcription and Processing; 3 & 4—Translation; 5—Proteolysis. Created with BioRender.com.

Recent research has demonstrated changes to chromatin remodeling through impaired histone activity, and reductions in mitogen-activated and stress-activated protein kinase 1 (*MSK-1*) activity among striatal neurons of Huntington's disease patients and animal models. Overexpression of *MSK-1* resulted in increased expression of peroxisome proliferator-activated receptor gamma coactivator alpha (PGC-1a), a transcriptional coactivator involved in mitochondrial biogenesis that may also protect against neuronal death [121].

Histone deacetylase (HDAC) inhibitors have shown promising results in animal models of Huntington's disease, and may represent a viable therapeutic target. Class III HDACs (sirtuins) have demonstrated promise as neuroprotective targets, with one study revealing an improvement in motor function and reduced brain atrophy in a Huntington's disease mouse model by overexpressing *Sirt1*, an NAD-dependent protein deacetylase. Notably, *Sirt1* also restored brain-derived neurotrophic factor (BDNF) [122].

The following actions that happen after the translation are represented by a condensation and oligomerization of the protein fragments in cytoplasm, which will lead to a dysfunctional proteostasis of the cell. Moreover, these fragments will go through an aggregation process inside the nucleus of the cell, binding to the DNA as inclusions and therefore altering the entire process of transcription. All of these pathological processes will alter both the axonal transport and inter-synaptic transmission, as well as the mitochondria of the cell, causing a decreased energy output [121].

According to a study conducted by G. Vonsatellel et al. in 1998 [123], the mutated huntingtin protein can be found in both the dystrophic neurites and in nuclear inclusions of the neuron, with a higher prevalence in the cortex and neostriatum in comparison with the globus pallidus and cerebellum, where this type of mutated protein cannot be found. Moreover, this study attested that mHTT protein was found in 38% to 52% of the neurons of the patients with juvenile HD (with an age under 20 years) and in 3% to 6% of the neurons regarding the adult onset of the HD.

A study conducted by the GWAS (Genome-Wide Association Study) in 2017 discovered the existence of a correlation between the HD onset and a gene called *MSH3*, that together with the *MSH2* gene, will lead to a heterodimer called MutSβ, whose main goal is to repair the possible mismatches of the DNA after replication. However, a variant of *MSH3* seems to be involved in the somatic expansion of the CAG repetitive sequence, leading to an increased risk of developing HD, affecting mostly the brain striatum, which is the most affected by HD. This test was conducted using the post hoc analysis [124].

However, even if today we understand the pathogeny of HD better than at any previous time, the therapeutic directions are quite limited, and it is difficult to achieve a conclusive result. According to a study conducted by Travessa et al. [125], of 99 trials that proposed to study 41 compounds to treat HD, only 2 trials made it to phase 4 (2%), with a success rate of only 3.5%.

4.1. Treatment

Tetrabenazine is the only drug approved to treat Huntington's chorea in North America and some European countries; however, this could increase depression as a potential side effect. Post hoc analysis showed that advanced Huntington's disease patients who already took antidepressant drugs did not experience worsening depression after starting treatment with tetrabenazine. Antipsychotic drugs are most often employed for chorea treatment in Europe, while both tetrabenazine and antipsychotics are applied at an equal level in North America and Australia.

Amantadine's efficacy for chorea treatment remains inconclusive, while pridopidine, a dopaminergic stabilizer, has been assessed as a symptomatic therapy in Huntington's disease; trials showed mild stimulatory and inhibitory effects depending on dopaminergic tone levels.

However, a recent phase III trial involving 437 patients failed to demonstrate significant improvements in modified motor score after six months of treatment with doses up to 90 mg/day; further analysis suggested potential benefits in UHDRS total motor score as an endpoint. Huntington's disease offers few treatment options for cognitive dysfunction and behavioral abnormalities, and trials of cholinesterase inhibitors have proven unsuccessful in improving cognitive dysfunction. Moreover, addressing the behavioral abnormalities remains challenging.

Therapies designed to lower levels of mutant huntingtin (*mHTT*) represent one of the most promising approaches for disease modification. These emerging therapies target either DNA or RNA of the *mHTT* gene; targeting can take the form of antisense oligonucleotides (ASOs), RNA interference (RNAi), or small molecule splicing inhibitors. Currently, ASOs are being investigated in a human phase 1b/2a study delivered intrathecally that catalyze degradation of *HTT* mRNA via RNAse H; in animal models, this approach resulted in a decrease of up to 80% in *HTT* mRNA levels over time [126].

Attaining lower levels of *mHTT* may prove particularly effective for Huntington's disease modification. One approach aimed at this goal involves targeting either DNA or RNA of the *mHTT* gene; targeting its expression using ASOs, RNAi, or small molecule splicing inhibitors are possible solutions.

RNAi-based approaches use RNA molecules that bind to mRNA in the cytoplasm and prompt its removal by Argonaute 2, an RNAse enzyme found within an RNA-induced silencing complex. While therapeutic strategies using this approach are still in their preclinical stage, one treatment could potentially provide permanent *HTT* reduction through intracranial injection into the striatum of a small molecule splicing modifier that has shown promising results in animal models with muscular atrophy; screening is underway to identify small molecule modulators of *mHTT*.

Targeting the DNA of *mHTT* can be done using two approaches: zinc finger proteins and the CRISPR/Cas9 system. Zinc finger proteins are structural motifs that bind directly to DNA; synthetic zinc finger transcription factors targeting CAG have been used successfully to lower levels of *mHTT* protein in animal models. However, as they create non-native proteins, they could potentially trigger immune reactions, and further research is required before conclusively targeting them.

CRISPR/Cas9 is a groundbreaking gene editing technology, allowing scientists to make precise changes to an organism's DNA. CRISPR stands for Clustered Regularly Interspaced Short Palindromic Repeats found in bacteria and archaea genomes; the repeats are connected by "Cas" genes which encode for proteins such as the Cas9 enzyme.

CRISPR/Cas9 works by employing a small RNA molecule called a guide RNA (gRNA), designed to target specific DNA sequences with high specificity. Engineered specifically to bind to its target sequence, gRNA allows for highly targeted binding between the DNA molecule and itself—with Cas9 enzyme acting like "molecular scissors" at this location, cutting its DNA at target locations. More recently it has also become a genome editing tool with various applications in human disease treatment. Huntington's disease patients were treated using this technology to excise promoter regions, the transcription start site, and CAG mutation expansion of the *mHTT* gene found in their fibroblasts, leading to permanent allele-specific inactivation of this gene. This approach has also been successfully tested in an HD rodent model, providing proof of concept. Further preclinical work needs to be completed before CRISPR/Cas9 gene editing technology can reach clinical application. Recently raised concerns over unexpected off-target mutations have necessitated more study [127].

No treatment has yet been shown to stop or slow Huntington's disease from progressing, although various clinical trials have investigated its potential effectiveness, including of minocycline, riluzole, and remacemide. However, none has shown significant effects.

Coenzyme Q10 (CoQ10) at 600 mg/day was observed to slow functional decline, although this did not reach statistical significance. Additional research is evaluating effects of doses up to 2400 mg/day as well as testing on preHD gene carriers. Safety and tolerability data showed that doses of up to 3600 mg/day were well-tolerated by most study participants without experiencing serious adverse events [128].

Creatine is widely acknowledged for its antioxidant properties and potential to improve mitochondrial function and cell bioenergetics. Unfortunately, an earlier study conducted using 10 g/day did not demonstrate significant improvements in total motor score, functional capacity, or neuropsychological testing scores compared with controls. Therefore, phase III studies are now taking place with higher dosages up to 40 g/day as part of CREST-E (Creatine Safety Tolerability Efficacy in Huntington's disease) [129].

4.2. Brief Reflection Point

In essence, this section reaffirms the significance of continued and interdisciplinary research in unraveling Huntington's disease. As our understanding deepens, so does the potential to enhance diagnosis, develop effective treatments, and ultimately improve the lives of those affected by this disease. Our journey in decoding Huntington's is far from over; each step forward holds promise and hope for millions globally.

5. Amyotrophic Lateral Sclerosis (ALS)

Amyotrophic lateral sclerosis is a neurodegenerative disease usually manifesting with an onset of focal muscular weakness and fatigue, having the tendency to spread selectively among upper and lower motor neurons; in some cases dysarthria, dysphonia, and dysphagia can also occur [130]. It affects the health state of the patient as it spreads typically from distal muscles to more proximal ones. The survival rate is about 2 to 5 years from the date of diagnosis, with the cause of death usually being respiratory failure. ALS can be classified as being familial (10% of the cases) and sporadic (90% of the cases), both of which show similarities between symptomatology. This shows how important epigenetic studies are in discovering new treatments for ND and how both epigenetics and environmental factors can determine the apparition of ALS and neurodegenerative disorders in general [131]. Various studies have shown that ALS has a slightly higher incidence in the female population. Furthermore, great geographical discrepancies show that there is a prevalence in the European region, while in Asia and the Middle East, far fewer ALS cases have been reported [132–134]. In addition, it has been proven that environmental exposure to different agents such as infectious agents or heavy metals can lead to genetic alterations and can also facilitate the development of degenerative diseases. The most common genetic genes that cause ALS are *C9orf72, SOD1, TARDBP, FUS,* and *TBK1* [130,131] (Figure 3).

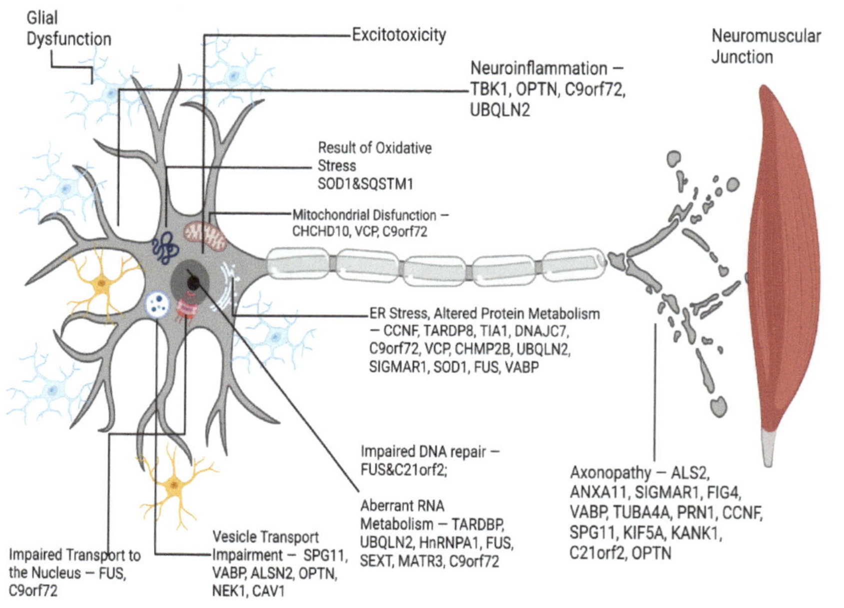

Figure 3. Advances in large-scale genomic analysis have uncovered a variety of causative genes and risk factors for amyotrophic lateral sclerosis (ALS). These gene variants map onto key pathogenic mechanisms relevant to all motor neuron cellular compartments as well as neighboring cells such as glia and interneurons. In this way, these mechanisms are genetically validated, enabling a greater confidence in their targeting for therapeutic benefit. Some of these mechanisms have emerged only in recent years due to new genetic information, including gene changes highlighting dysregulation of RNA processing and metabolism. There is significant overlap of some genes with those found in closely related disorders such as frontotemporal dementia (for example, *C9orf72, CHCHD10, SQSTM1, TBK1, CCNF, FUS, TARDBP, OPTN, UBQLN2, TUBA4A, ATAXN2, VCP,* and *CHMP2B*). This suggests a closer relationship with broader neurodegenerative disorders, and indeed many of the pathways depicted are relevant in, for example, Alzheimer's disease. ER, endoplasmic reticulum. Created with BioRender.com.

Numerous studies have confirmed to date that *C9orf72* mutations are the most common in familial ALS disease, which only represents 10% of the ALS cases ([135]), compared to other gene mutations that lead to the apparition of this pathology. *C9orf72* is a gene that is formed out of 11 exons and it is implicated in splicing mechanisms that produce transcripts and 2 isoforms. The mutation that is widely spread throughout ALS cases especially in

Europe and North America is the GGGGCC hexanucleotide expansion, located in the first intron of variants 1 and 3 and in the promoter region of variant 2 [132,134]. More than 22 repeats were found in neurodegenerative disorders and it has been stated that the patients that present the GGGGCC hexanucleotide repeat might share the Finnish founder risk haplotype; this statement that can be supported by the fact that numerous studies have shown that the expansion is most common in the Finnish population compared to other regions [136,137]. The mechanisms behind this expansion can be explained by its capacity to be transcribed into repetitive RNA, which then forms both sense and antisense RNA foci and five dipeptide repeat proteins (DPRs). This can explain possible mechanisms of pathology represented by dysfunction of *C9orf72* protein or toxic accumulations of either RNA foci, which can sequester RNA binding proteins, creating inclusions or DPRs inside the nucleus of the nervous cells [135,138].

Mutations of the *SOD1* gene represent one of the most common factors of apparition for familial ALS and represent about 10–20% of FALS (familial ALS) cases, being far less common in sporadic ALS cases. At the moment, there are over 155 gene mutations of the *SOD1* that can have implications in ND diseases, including ALS. These can result in different modifications to the *SOD1* protein regarding its level, structure, or enzymatic activity, which can eventually lead to toxic protein accumulations due to misfolding of the *SOD1* proteins [139]. It was first stated that the *SOD1* mutations are implicated in the apparition of ALS by a mechanism based on the loss of function of this protein, but this hypothesis was soon abandoned after some experimental studies [139–141]. The *SOD1* protein, also known as superoxide dismutase 1, is a very well-conserved gene located inside the nucleus, cytoplasm, and mitochondria's membrane. It is formed by 5 exons that encode 153 amino acid metalloenzyme, binding Cu and Zn ions, forming a dismutase that removes radicals from the cells and metabolizes them into oxygen and hydrogen peroxide ([139]). The most common mutations of SOD1 associated with ALS are usually based on changes in the structure of the protein, specifically, the amino acid positions, and might include *A4V* [142], *G93A* [143], and *L84F* [144,145].

A4V mutation changes the alanine from codon 4 to valine in exon 1 and it is associated with aggressive forms of ALS representing 50% of the SOD1 mutation cases reported in North America [139,146]. *G93A* is a heavily studied gene mutation of the *SOD1* protein that represents a substitution of glycine with alanine from the codon 93 of the *SOD1* protein, changing its conformation, and is also responsible for approximately 20% of the familial ALS cases. According to a study conducted by the Chinese Pharmaceutical Association in 2023 [147], oxidative stress has a major impact on ALS pathology; patients showed different oxidative markers, such as glutamate excitotoxicity, and dysfunctions at several levels, such as mitochondria, due to calcium influx, and axon as well as protein oxidation. These modifications observed were at *SOD1-G93A* in mice. This mutation is relatively rare in the general population but it is very common in familial ALS, and multiple studies on animal models have also shown that having the *SOD1-G93A* mutation is enough to cause motor-neuron degeneration [147–149]. Understanding the mechanisms as well as the specific effects of *SOD1* mutations on protein structure and functions is an important area of research for developing effective treatments for ALS and neurodegenerative diseases in general.

The *L84F* mutation changes the amino acid at position 84 from leucine to phenylalanine, and is associated with a slightly mild form of ALS compared to other mutations such as *G93A*, resulting in both protein instability and misfolding that can lead to forming protein accumulations [144]. According to a study on the population of central Italy [145], the *L84F* mutation is a rare mutation that has been identified in several families as a result of a common ancestor who carried the mutation. In addition, the article reported several pathological characteristics of the patients with ALS who had this mutation, including young age of onset, a relatively slow progression of the disease, and the predominance of upper limbs' motor neuron involvement, suggesting that the clinical features might be caused by the specific mutation (Table 5).

Table 5. Important SOD1 mutations associated with ALS onset.

Mutation	Pathogenicity	Type of Mutation	Biological Effect	Citation
A4V	ALS Pathogenic	Substitution	This mutation is responsible for a rapidly progressive dominant form of amyotrophic lateral sclerosis (ALS) that exclusively affects lower motor neurons, and it accounts for 50% of SOD1 mutations associated with familial ALS in North America. However, it is a rare mutation in Europe.	[142,146]
G93A	ALS Pathogenic	Substitution	Patients showing different oxidative markers, such as glutamate excitotoxicity, and dysfunctions at several levels, such as mitochondria, due to calcium influx, and axon as well as protein oxidation; modifications were observed at SOD1-G93A in mice. This mutation is relatively rare in the general population but it is very common in familial ALS, and multiple studies on animal models have also shown that having the SOD1-G93A mutation is enough to cause motor-neuron degeneration.	[147–149]
L84F	ALS Pathogenic	Substitution	Protein instability and misfolding that can lead to forming protein accumulations.	[144,145]

Mutations of the TARDBP gene are associated with neurodegenerative diseases, especially with amyotrophic lateral sclerosis; this gene encodes the *TDP-43* protein, which is implicated in the apparition of this ND. The exact mechanisms of how this mutation leads to motor neuron degeneration are not fully understood, but one of the reasons might be that it affects the structure of the function of *TDP-43*, thus leading to the formation of toxic protein aggregates [150,151]. According to an article published in *Nature Genetics* [152] that investigated the frequency of *TARDBP* mutations by sampling 93 familial ALS cases and 109 sporadic ALS cases, the mutations that were identified were mostly missense mutations, with *A382T* being the most common one found. In addition, regarding the clinical symptoms, the patients found with *TARDBP* mutations had an early onset age and usually bulbar onset, as well as a shorter survival rate compared to that of other ALS cases. The *TARDBP* gene was present in 5% of the familial ALS cases and 1.9% of the sporadic cases. TARDBP mutations interfere with the normal function of *TDP-43*, leading to abnormal protein clumps in motor neurons, which contribute to death and degeneration of these motor neurons. While its exact mechanisms for inducing ALS development remain enigmatic, speculation points toward changes in RNA splicing, protein translation, or degradation processes being involved. *TARDBP* mutations are an important genetic risk factor for ALS and studying them can provide insights into its mechanisms of development, potentially providing potential therapies targeted specifically towards these mutations [153]. *TARDBP* gene mutations have long been linked with amyotrophic lateral sclerosis. One of the more often-seen *TARDBP* mutations linked to ALS is *M337V*; this mutation can be found in both familial and sporadic cases of the condition. Other *TARDBP* variants that have also been frequently connected with this form of neurological disease include *A315T, G348C,* and *A382T* mutations [150,154,155].

The TDP-43 *M337V* mutation has been associated with familial amyotrophic lateral sclerosis (ALS) in Japan. A study conducted on 41 families living with familial ALS revealed that 11 of those families contained *TDP-43 M337V* mutations. The *TDP-43 M337V* mutation results in the production of an abnormal *TDP-43* protein that aggregates abnormally and accumulates in motor neurons, ultimately leading to their degeneration and death, and thus contributing to ALS development. It is thought that this accumulation may play a key role in its manifestation [154,155].

The *TDP-43 A315T* mutation has been linked with familial motor neuron disease. This genetic alteration occurs within the *TARDBP* gene, which contains instructions for producing the *TDP-43* protein; mutations disrupting this production can lead to abnormal protein accumulation within motor neurons and eventually death. According to [155], attaining abnormal levels of *TDP-43* through mutations such as *A315T* has been linked with degeneration and death of motor neurons, eventually leading to motor neuron diseases such as ALS. Although this mutation only appears in a small percentage of familial motor neuron disease cases, it is a pathogenic one responsible for inducing symptoms in those who carry it. Undertaking research into the *TDP-43 A315T* mutation and its impact on motor neuron disease could provide key insights into its underlying mechanisms and possible therapeutic targets, and genetic testing may play a pivotal role in identifying individuals carrying this mutation who could be at an increased risk of familial motor neuron disease. Another study shows that the *TDP-43* (A315T) transgenic mouse model provides an opportunity to investigate how mutations of the TDP-43 gene *A315T* affect amyotrophic lateral sclerosis and other motor neuron diseases, leading to abnormal accumulations of protein aggregates, which eventually cause degeneration and death of motor neurons. It reveals that transgenic mice carrying the *A315T* mutation of *TDP-43* may succumb to early death due to digestive complications before fully manifesting neurological signs associated with ALS, suggesting it also influences their digestive systems and may contribute to their early demise. Although the exact mechanisms underlying gastrointestinal complications remain poorly understood, experts speculate that abnormal *TDP-43* protein build-up in intestinal cells may lead to dysfunction and damage. The early death of these transgenic mice highlights the necessity of conducting further studies into the effects of *A315T* mutation on non-neuronal systems within the body, and possible therapeutic approaches for managing them. The *TDP-43* (A315T) transgenic mouse model provides an essential way of studying its underlying mechanisms as well as devising potential treatments against it [156].

The *TARDBP* gene mutation, *A382T*, has been associated with an increased susceptibility for motor neuron diseases such as amyotrophic lateral sclerosis. Studies have demonstrated that this mutation results in the production of an abnormal *TDP-43* protein that accumulates and damages motor neurons, ultimately leading to their decline and eventual demise. The *A382T* mutation is considered an uncommon pathogenic mutation found only in a small proportion of familial ALS cases. Further research is necessary to fully understand how this mutation leads to disease development and potential therapeutic approaches that target it. Genetic testing could prove invaluable in identifying individuals carrying this variant who may be at greater risk of ALS and related motor neuron diseases [157,158] (Table 6).

Table 6. Important TARDBP mutations associated with ALS onset.

Mutation	Pathogenicity	Type of Mutation	Biological Effect	Citation
M337V	ALS Pathogenic	Substitution	Production of an abnormal TDP-43 protein that aggregates abnormally and accumulates in motor neurons, ultimately leading to their degeneration and death, and thus contributing to ALS development.	[150,154,155]
A315T	ALS Pathogenic	Substitution	Transgenic mice carrying the A315T mutation of TDP-43 may succumb to early death due to digestive complications before fully manifesting neurological signs associated with ALS, suggesting it also influences their digestive systems and may contribute to their early demise. Although the exact mechanisms underlying gastrointestinal complications remain poorly understood, experts speculate that abnormal TDP-43 protein build-up in intestinal cells may lead to dysfunction and damage.	[156]
A382T	Possible ALS Pathogenic	Substitution	Unknown mechanism.	[157,158]

The Fused in Sarcoma (FUS) gene represents another common gene that can cause ALS apparition and encodes a protein involved in controlling RNA processing and transport within cells. Mutations of this gene have been linked with neurodegenerative diseases. Mutations that alter FUS protein accumulation in neurons may result in their dysfunction and apoptosis. Both missense and truncating mutations of FUS may alter protein function and disease development. Some mutations are inherited in an autosomal dominant manner, while others may occur spontaneously. Further research must be conducted in order to understand how FUS mutations work and to develop therapeutic approaches; genetic testing could help identify individuals at increased risk of FUS-related diseases [159,160].

Regarding amyotrophic lateral sclerosis (ALS), researchers have discovered several mutations of the FUS gene that may contribute to its progression. *R521C*, *R521H*, and *P525L* mutations are among the most frequently encountered, occurring in approximately 5–10% of familial cases and a smaller proportion of sporadic cases of ALS.

Mutations result in the production of a mutated FUS protein that accumulates in motor neuron cytoplasm, eventually leading to their degeneration and death. While the exact mechanisms underlying ALS development due to such mutations remain enigmatic, experts speculate that they disrupt RNA metabolism and transport processes that are vitally important in maintaining proper motor nerve functionality [159].

A previous study investigated the prevalence and clinical features of FUS mutations among Italian patients with familial ALS. From 54 cases studied, 2 individuals (3.7%) carried *FUS R521C* mutations that showed early disease onset, as well as more severe disease progression, than patients without this mutation, suggesting FUS mutations may play an integral role in familial ALS prevalence among Italian populations. This suggests FUS mutations may play a vital part in creating familial ALS within Italy's population [161,162].

The *TBK1* gene is also implicated in ALS apparition and plays an essential role in regulating various cellular processes, such as autophagy, inflammation, and immune response. Mutations have been identified as one cause of ALS and FTD; such mutations result in abnormal protein aggregate accumulations within motor neurons, causing degeneration and death. Interference with regular cellular processes may play a part in contributing to this condition and to its progression. Mutations in *TBK1* have been identified as an uncommon cause of ALS and FTD. These mutations cause abnormal protein aggregates to build up in motor neurons, eventually leading to their degeneration and death, possibly interfering with normal cellular processes that contribute to disease development in this way, contributing further to ALS or FTD development [163,164]. Mutations in the *TBK1* gene have been identified in patients suffering from amyotrophic lateral sclerosis (ALS), including missense, frameshift, and truncating mutations. Some of the more prevalent mutations include the frameshift mutation *TBK1* p.Ile383Thr, which results in truncated protein expression [164]. Missense mutations such as *p.Arg357Ser* and *p.Gly290Val* have also been found; these latter two also appear in familial cases as these affect highly conserved amino acids found within its protein. Additionally, missense mutations impair its ability to bind its target proteins, allowing greater protein expression from within.

These mutations are thought to disrupt *TBK1*'s role in regulating various cellular processes, such as autophagy and immune responses, potentially contributing to motor neuron degeneration in ALS patients [164,165].

According to [164], various studies have demonstrated *TBK1* gene mutations among familial and sporadic cases of ALS, suggesting *TBK1* dysfunction may play an integral part in its progression. Furthermore, this piece delves into potential pathways through which these mutations could contribute to ALS progression, such as disrupted autophagy, neuroinflammation, and mitochondrial malfunction.

This article highlights the various functions of *TBK1* in terms of its involvement with innate immunity, autophagy, protein aggregate clearance, and neuroinflammation; these processes have been implicated as being causal in ALS. Studies have identified mutations of the *TBK1* gene among ALS patients, indicating its involvement. Furthermore, dysfunction caused by mutations may contribute to neuroinflammation, which results in protein aggregate accumulation as well as motor neuron death, resulting in the progression of the condition.

This article suggests that *TBK1* could be an effective therapeutic target for treating ALS, specifically by increasing autophagy and decreasing neuroinflammation. Furthermore, further research needs to be conducted in order to fully comprehend how this gene contributes to ALS, as well as to create effective treatments that target it [164–166].

In recent years, epigenetic studies have led to major discoveries regarding neurodegenerative diseases and appear to be the key element in finding a treatment that can cure the condition rather than just alleviate it. The term epigenetic refers to any genetic alteration that occurs at any molecular level, associated with several factors such as environmental agents or stress, that does not implicate the modification of the DNA directly [167]. The most common mechanisms regarding epigenetics are DNA methylation, miRNAs, and PTM (post-translational modification) of histones [168].

ALS is a disease characterized by the loss of neuromuscular junctions, and is also associated with apoptosis of UMN (upper motor neuron) and LMN (lower motor neuron) cells, as well as the surrounding astrocytes and microglial cells. It is also associated with inclusions in which the main protein involved is *TDP-43*, which can cause protein aggregation by its mislocalization between the nucleus and the cytoplasm [169]. In addition, there are multiple pathological causes of ALS; the most relevant are the failure of protein degradation, changes in RNA metabolism, and axonal transport.

The failure of protein degradation can be the leading cause implicated in the formation of protein aggregates that can disturb the homeostasis of the cells, and it is also one of the similarities between neurodegenerative diseases. Protein chaperones can aid the cells to the point of saturation by refolding the misfolded protein. This can explain the formation of protein aggregates; however, if the chaperones are overloaded, ubiquitin inclusions are formed inside the nucleus. The genes that lead to protein accumulation relating to ALS pathologies are *UBQLN2* which is associated with the formation of the inclusions, *SQSTM1, TBK1, VCP*, and *C9orf72* protein, which interacts with the inclusions previously formed [131,170–172].

Another mechanism that contributes to the pathogenesis of ALS is abnormal RNA metabolism, which includes abnormalities in RNA processing and miRNA expression that lead to misfolding proteins and formation of aggregates. RNA processing includes many mechanisms, such as transcription, splicing, editing, transport, and degradation. The key elements implicated in the deviations are RNA-binding proteins such as *TDP-43* and *FUS*, but there are additional targeted proteins that can present mutations, such as *ANG, STX, ATXN2, MATR3, hnRNPA1*, and *hnRNPA2B1*. This demonstrates that any mutations occurring at this level have an important role in ALS, although we cannot state that these mutations lead to the apparition of ALS [173].

Specific gene mutations that may contribute to ALS are *PFN1, SOD1*, and *TUBA4A*. The mutations of the C-terminus of kinesin-1 may also contribute, with the implicated mechanism affecting the integrity of the cytoarchitecture and the transport throughout the axon by destabilizing the tubulin networks, which may also affect the molecules carried inside the axon. The binding of the molecules transported inside are stabilized by *DCTN1*. It is mentioned that mutations at this level are not very common and do not play an important role in familial ALS apparition, but they rather contribute to diagnosing several ND diseases, especially in late-onset Parkinson's, ALS, and FTD [131,174,175].

The different motor phenotypes of ALS can be classified based on whether upper or lower motor neurons are affected, and which regions they impact. Each subtype may present with differing life expectancies; some may even present with cognitive and behavioral deficits. Classic ALS is the most prevalent subtype, marked by signs of both upper and lower motor neuron loss in various body areas. By contrast, primary lateral sclerosis (PLS) typically presents as progressive spasticity with slowing movements, as well as isolated upper motor neuron signs. PLS patients exhibit no muscle atrophy, visible fasciculations, or denervation on EMG four years post-symptom onset. Their median survival is over 20 years, while PLS can transition into ALS three to four years after disease onset. A subtype called UMN-predominant ALS shows some LMN involvement, but less prominently than classic ALS. Patients in this subtype progress more slowly but live shorter lives than PLS patients [176,177].

There are multiple motor phenotypes of ALS, defined by their degree of involvement with upper versus lower motor neurons and regional distribution of symptoms. Understanding each subtype's life expectancy and degree of cognitive and behavioral impairment is imperative.

One form of ALS, called LMN predominant ALS, involves limited UMN involvement with variable rates of progression. PMA (progressive muscular atrophy), on the other hand, involves progressive isolated LMN signs without evidence of UMN dysfunction. Up to 30% of PMA patients may develop these signs during follow-up.

Bulbar ALS is a devastating form of the disease characterized by rapid decline and an average lifespan of only two years from disease onset. It manifests as spastic dysarthria due to bulbar UMN dysfunction and tongue wasting and fasciculation due to bulbar LMN dysfunction. While only 30% initially exhibit bulbar symptoms, most ultimately experience difficulty speaking or swallowing due to this form of the condition.

Pseudobulbar Palsy (PBP) is another subtype, distinguished by absent facial expressions, spastic dysarthria, difficulty chewing, dysphagia, and tongue protrusion due to spasticity, but no fasciculation or wasting. This disorder originates in the upper motor neurons (UMNs), distinguishing it from progressive bulbar Palsy, which only affects LMNs but may not be universally recognized [178,179].

5.1. Perspectives for Treatment

The complexity of ALS has been demonstrated by its failure in over 40 randomized controlled trials that attempted to find effective disease-modifying medications. Riluzole is currently approved in most European countries as the sole disease-modifying medication; administered twice daily at 50 mg dose, its antiglutamatergic properties increase mean patient survival by approximately six months. The most frequently experienced side effects include nausea, diarrhea, fatigue dizziness, and liver issues.

More recently, edaravone (a free radical scavenger) has been studied in ALS patients. A phase III, double-blind study administering 60 mg/day of edaravone intravenously over 2 weeks per month resulted in significantly less decline in scores on the revised ALS Functional Rating Scale (ALSFRS-R). After six months, significantly fewer scores had decreased [180–182].

Masitinib, an oral tyrosine kinase inhibitor, is currently under study as a possible treatment for ALS. A randomized controlled trial administered masitinib as an add-on therapy with riluzole at 4.5 mg/kg/day, and showed positive effects in decreasing ALSFRS-R scores, especially among those experiencing typical disease progression. Its effectiveness will be further examined in a confirmatory study [183].

Honokiol was found to provide neuroprotective benefits by mitigating oxidative stress and improving mitochondrial function in affected neurons. Researchers also observed increased expression of key antioxidant enzymes and reduced expression of pro-inflammatory cytokines in those neurons treated with honokiol.

Overall, studies suggest that honokiol could serve as an effective therapy to treat ALS by decreasing oxidative stress and improving mitochondrial function in affected neurons. More research should be conducted in order to confirm these results, as well as to determine optimal dosing and treatment regimens for honokiol in ALS patients [147].

5.2. Brief Reflection Point

This section highlights the fact that ALS is a multifactorial disorder affected by genetic, environmental, and lifestyle influences; thus, integrating genetic/molecular knowledge with larger biopsychosocial models is vital in order to gain an overall understanding of this illness. At its core, ALS research is about helping those living with this devastating condition improve their quality of life, and eventually finding a cure. While great strides have been taken in understanding it thus far, much more work remains. Ongoing multidisciplinary research efforts must continue; each step forward in research contributes toward making patients' lives better while ultimately finding an endpoint cure for this devastating illness.

6. Conclusions and Future Directions

Over the last five years, tremendous advances have been made in understanding both the pathophysiology and genetic basis of AD. Revamp of the amyloid-β cascade hypothesis and greater insight into AD preclinical phases has enabled improved comprehension. Genetic investigations have progressed from identifying three causal and one risk gene, to discovering multiple genetic markers that can be used to create a polygenic risk score for AD. Biomarker diagnosis has fundamentally transformed how AD is classified, making it possible to enroll patients earlier in studies when blood biomarkers become accessible. Molecular imaging will enhance diagnostic categorization and pathophysiology of any disease by providing visual evidence of co-pathology or regional protein aggregated deposits. At this rate, insights into risk reduction, primary and secondary prevention, and non-pharmacological and pharmacological treatments could become available and in parallel earlier than ever before. The early identification and multimodal treatment of patients could soon become a reality [184].

PD has been recognized for over 200 years and presents serious healthcare challenges worldwide. However, Parkinson's is treatable when interventions are tailored specifically for each patient by trained healthcare providers. Here we highlight several promising advancements in Parkinson's research and treatments which offer hope that services will continue to evolve to make a meaningful difference to those living with the disorder worldwide [185].

Understanding of Huntington's disease (HD) has advanced substantially due to advances in genetic technology and large cohorts of individuals living with HD. Thus, new genetic modifiers of the disease have been identified. Somatic instability of CAG repeats is most prevalent in tissues most susceptible to HD pathology, and its degree is negatively correlated with age at disease onset. DNA repair components involved with mismatch repair may act to control somatic instability and the disease course. MutSb's attempts at repair may induce loop-outs within CAG tracts targeted by MutSb, leading to potentially significant expansion. Reducing *MSH3*, *PMS2*, and *LIG1* pro-instability factors or hindering their function is thought to reduce somatic instability while being well-tolerated. *FAN1* expression decreases somatic instability while postponing disease onset; its upregulation could serve as protection from HD. Modulating these DNA repair components may also reduce instability in other pathogenic repeat sequences, suggesting these potential therapeutic options could also prove effective against repeat expansion diseases. Additionally, *mHTT* sequesters components of the NPC in aggregates, disrupting nucleocytoplasmic transport. Modulating nuclear transport pathways has proven protective in cell models of HD, opening new avenues for therapeutic intervention. Cerebrospinal fluid (CSF) offers an ideal source of CNS proteins for clinical trials due to its accessibility throughout. Notably, neurofilament light chain (NfL) is one of the proteins released into CSF and plasma after neuronal damage to be used as a biomarker or surrogate endpoint for clinical trials. Its levels seem to correlate strongly with disease progression; thus, its levels could serve both functions simultaneously. Furthermore, *mHTT* could also be released by damaged neurons, with CSF concentration changes representing early changes to be found prior to manifestation of HD [112,186].

In recent years we have witnessed an increased focus on developing precision medicine approaches for ALS subtypes with known genetic causes. One promising therapeutic avenue involves antisense oligonucleotides (ASOs), which are short sequences of nucleotides that can modulate gene expression and splicing. ASOs have proven effective against preclinical models of ALS characterized by *SOD1* mutations and *C9orf72* repeat expansions;

clinical studies utilizing intrathecal administration of ASOs targeting these genes are currently ongoing. Stem cell therapies such as granulocyte-colony stimulating factor (G-CSF) are being explored as possible therapies for ALS. Peripheral blood stem cells, bone marrow mesenchymal stem cells, and non-neural progenitor cells induced by granulocyte-colony stimulating factor have all shown safety and tolerability when administered to ALS patients. However, their efficacy against disease progression remains undetermined. Current phase II and III clinical trials offer hope that more precise classification of ALS cases based on pathogenic mechanisms will lead to targeted therapies with favorable results in particular ALS subgroups, eventually making ALS disease manageable [187,188].

Author Contributions: Conceptualization, A.V.C. and A.G.M.; methodology, R.-A.C.-B. and H.-P.C.; software, L.-A.G. and A.-D.C.; validation, A.V.C., V.M.S. and A.G.M.; formal analysis, R.-A.C.-B. and H.-P.C.; investigation, L.-A.G. and A.-D.C.; resources, A.V.C. and A.G.M.; data curation, V.M.S.; writing—original draft preparation, R.-A.C.-B. and H.-P.C.; writing—review and editing, L.-A.G. and A.-D.C.; visualization, A.G.M.; supervision, A.V.C.; project administration, V.M.S.; funding acquisition, A.V.C. All authors have read and agreed to the published version of the manuscript.

Funding: The APC was funded by the "Carol Davila" University of Medicine and Pharmacy, Bucharest, through an internal project. The authors would like to express their gratitude to the "Carol Davila" University of Medicine and Pharmacy, Bucharest, which supported the APC.

Institutional Review Board Statement: Not applicable.

Informed Consent Statement: Not applicable.

Data Availability Statement: All data is available on PubMed and on ALZFORUM database.

Conflicts of Interest: The authors declare no conflict of interest.

Abbreviations

ND	Neurodegenerative disease
AD	Alzheimer's disease
APP	Amyloid precursor protein
Aβ	Amyloid beta peptide
HCHWA-D	Hereditary cerebral hemorrhage with amyloidosis
PiD	Pick's disease
GFAP	Glial fibrillary acidic protein
SPECT	Single-photon emission computed tomography
PET	Positron emission tomography
APOE	Apolipoprotein E
BBB	Brain Blood Barrier
HDL	High density lipoprotein
CNS	Central Nervous System
NFTs	Neurofibrillary Tangles
GWAS	Genome-wide association study
PP2A	Protein phosphatase 2A
FTD	Frontotemporal Dementia
CTD	Chronic traumatic encephalopathy
CBD	Corticobasal degeneration
DAM	Disease-associated microglia
MGnD	Microglial neurodegenerative
TREM2	Triggering receptor expressed on myeloid cells 2
CRP	C-reactive protein
MCI	Mild cognitive impairment
MAPT	Microtubule-associated protein tau
PSP	Progressive supranuclear palsy
PHF	Abnormally hyperphosphorylated taus

CL	Centiloids
SUVR	Standardized Uptake Ration
PD	Parkinson's disease
SNpc	Substantia nigra pars compacta
DA	Dopamine
DMV	Dorsal motor nucleus of vagus
OB	Olfactory bulb
LC	Locus coeruleus
IML	Intermediolateral nucleus in spinal cord
ENS	Enteric nervous system
EOPD	Early-onset Parkinson's disease
mitoQC	Mitochondrial quality control
DFO	Deferrioxamine
DFP	Deferiprone
HD	Huntington's disease
MSNs	Medium spiny neurons
mHTT	Mutation of the huntingtin gene
HTT	Huntingtin gene
CBP	cAMP response element-binding protein
MSK1	Mitogen-activated and stress-activated protein kinase 1
PGC-1a	Proliferator-activated receptor gamma coactivator alpha
HDAC	Histone deacetylase
BDNF	Brain-derived neurotrophic factor
ASOs	Antisense oligonucleotides
RNAi	RNA interference
CoQ10	Coenzyme Q10
CREST-E	Creatine safety tolerability efficacy in Huntington's disease
ALS	Amyotrophic lateral sclerosis
FALS	Familial amyotrophic lateral sclerosis
DPR	Dipeptide repeat proteins
FUS	Fused in Sarcoma
miRNA	Micro RNA
PTM	Post-translational modification
UMN	Upper motor neuron
LMN	Lower motor neuron
PMA	Progressive muscular atrophy
PBP	Pseudobulbar Palsy

References

1. Berry, R.M. The genetic revolution and the physician's duty of confidentiality: The role of the old Hippocratic virtues in the regulation of the new genetic intimacy. *BMJ* **1997**, *18*, 401–441. [CrossRef] [PubMed]
2. De Castro, M. Johann Gregor Mendel: Paragon of experimental science. *Mol. Genet. Genom. Med.* **2016**, *4*, 3–8. [CrossRef] [PubMed]
3. Saceleanu, V.M.; Mohan, A.G.; Covache-Busuioc, R.A.; Costin, H.P.; Ciurea, A.V. Wilhelm von Waldeyer: Important Steps in Neural Theory, Anatomy and Citology. *Brain Sci.* **2022**, *12*, 224. [CrossRef] [PubMed]
4. Kovacs, G.G. Concepts and classification of neurodegenerative diseases. In *Handbook of Clinical Neurology*; Elsevier: Amsterdam, The Netherlands, 2018; pp. 301–307. [CrossRef]
5. Brettschneider, J.; Del Tredici, K.; Lee, V.M.-Y.; Trojanowski, J.Q. Spreading of pathology in neurodegenerative diseases: A focus on human studies. *Nat. Rev. Neurosci.* **2015**, *16*, 109–120. [CrossRef] [PubMed]
6. Guo, J.L.; Lee, V.M.Y. Cell-to-cell transmission of pathogenic proteins in neurodegenerative diseases. *Nat. Med.* **2014**, *20*, 130–138. [CrossRef] [PubMed]
7. Castellani, R.J.; Rolston, R.K.; Smith, M.A. Alzheimer Disease. *Disease-a-Month* **2010**, *56*, 484–546. [CrossRef]
8. Ciurea, V.A.; Covache-Busuioc, R.-A.; Mohan, A.G.; Costin, H.P.; Voicu, V.; Academy, B.R. Alzheimer's disease: 120 years of research and progress. *J. Med. Life* **2023**, *16*, 173–177. [CrossRef]
9. Gatz, M.; Reynolds, C.A.; Fratiglioni, L.; Johansson, B.; Mortimer, J.A.; Berg, S.; Fiske, A.; Pedersen, N.L. Role of Genes and Environments for Explaining Alzheimer Disease. *Arch. Gen. Psychiatry* **2006**, *63*, 168–174. [CrossRef]

10. An, F.; Zhao, R.; Xuan, X.; Xuan, T.; Zhang, G.; Wei, C. Calycosin ameliorates advanced glycation end product-induced neurodegenerative changes in cellular and rat models of diabetes-related Alzheimer's disease. *Chem. Interact.* **2022**, *368*, 110206. [CrossRef]
11. Van Cauwenberghe, C.; Van Broeckhoven, C.; Sleegers, K. The genetic landscape of Alzheimer disease: Clinical implications and perspectives. *Anesth. Analg.* **2016**, *18*, 421–430. [CrossRef]
12. Roberts, S.; Ripellino, J.; Ingalls, K.; Robakis, N.; Felsenstein, K. Non-amyloidogenic cleavage of the beta-amyloid precursor protein by an integral membrane metalloendopeptidase. *J. Biol. Chem.* **1994**, *269*, 3111–3116. [CrossRef]
13. Gao, C.-X.; Wu, Q.; Sun, J.-X.; Song, X.-H.; Wang, J.; Xiong, C.-Q.; Teng, F.-X. Blocking beta 2-adrenergic receptor inhibits dendrite ramification in a mouse model of Alzheimer's disease. *Neural Regen. Res.* **2017**, *12*, 1499–1506. [CrossRef]
14. Okochi, M.; Tagami, S.; Yanagida, K.; Takami, M.; Kodama, T.S.; Mori, K.; Nakayama, T.; Ihara, Y.; Takeda, M. γ-Secretase Modulators and Presenilin 1 Mutants Act Differently on Presenilin/γ-Secretase Function to Cleave Aβ42 and Aβ43. *Cell Rep.* **2013**, *3*, 42–51. [CrossRef]
15. Hampel, H.; Hu, Y.; Hardy, J.; Blennow, K.; Chen, C.; Perry, G.; Kim, S.H.; Villemagne, V.L.; Aisen, P.; Vendruscolo, M.; et al. The amyloid-β pathway in Alzheimer's disease: A plain language summary. *Neurodegener. Dis. Manag.* **2023**, *13*, 141–201. [CrossRef]
16. Tosh, J.L.; Rhymes, E.R.; Mumford, P.; Whittaker, H.T.; Pulford, L.J.; Noy, S.J.; Cleverley, K.; Strydom, A.; Fisher, E.M.C.; Wiseman, F.K.; et al. Genetic dissection of down syndrome-associated alterations in APP/amyloid-β biology using mouse models. *Sci. Rep.* **2021**, *11*, 5736. [CrossRef] [PubMed]
17. Jonsson, T.; Atwal, J.K.; Steinberg, S.; Snaedal, J.; Jonsson, P.V.; Bjornsson, S.; Stefansson, H.; Sulem, P.; Gudbjartsson, D.F.; Maloney, J.; et al. A mutation in APP protects against Alzheimer's disease and age-related cognitive decline. *Nature* **2012**, *488*, 96–99. [CrossRef]
18. Benilova, I.; Gallardo, R.; Ungureanu, A.-A.; Cano, V.C.; Snellinx, A.; Ramakers, M.; Bartic, C.; Rousseau, F.; Schymkowitz, J.; De Strooper, B. The Alzheimer Disease Protective Mutation A2T Modulates Kinetic and Thermodynamic Properties of Amyloid-β (Aβ) Aggregation. *J. Biol. Chem.* **2014**, *289*, 30977–30989. [CrossRef]
19. Liu, Y.-W.; He, Y.-H.; Zhang, Y.-X.; Cai, W.-W.; Yang, L.-Q.; Xu, L.-Y.; Kong, Q.-P. Absence of A673T variant in APP gene indicates an alternative protective mechanism contributing to longevity in Chinese individuals. *Neurobiol. Aging* **2014**, *35*, 935.e11–935.e12. [CrossRef] [PubMed]
20. Mengel-From, J.; Jeune, B.; Pentti, T.; McGue, M.; Christensen, K.; Christiansen, L. The APP A673T frequency differs between Nordic countries. *Neurobiol. Aging* **2015**, *36*, 2909.e1–2909.e4. [CrossRef] [PubMed]
21. Giaccone, G.; Morbin, M.; Moda, F.; Botta, M.; Mazzoleni, G.; Uggetti, A.; Catania, M.; Moro, M.L.; Redaelli, V.; Spagnoli, A.; et al. Neuropathology of the recessive A673V APP mutation: Alzheimer disease with distinctive features. *Acta Neuropathol.* **2010**, *120*, 803–812. [CrossRef] [PubMed]
22. Villemagne, V.L.; Ataka, S.; Mizuno, T.; Brooks, W.S.; Wada, Y.; Kondo, M.; Jones, G.; Watanabe, Y.; Mulligan, R.S.; Nakagawa, M.; et al. High Striatal Amyloid β-Peptide Deposition Across Different Autosomal Alzheimer Disease Mutation Types. *Arch. Neurol.* **2009**, *66*, 1537–1544. [CrossRef] [PubMed]
23. Readhead, B.; Haure-Mirande, J.-V.; Zhang, B.; Haroutunian, V.; Gandy, S.; Schadt, E.E.; Dudley, J.T.; Ehrlich, M.E. Molecular systems evaluation of oligomerogenic APPE693Q and fibrillogenic APPKM670/671NL/PSEN1Δexon9 mouse models identifies shared features with human Alzheimer's brain molecular pathology. *Mol. Psychiatry* **2016**, *21*, 1099–1111. [CrossRef] [PubMed]
24. Natté, R.; Maat-Schieman, M.L.C.; Haan, J.; Bornebroek, M.; Roos, R.A.; Van Duinen, S.G. Dementia in hereditary cerebral hemorrhage with amyloidosis-Dutch type is associated with cerebral amyloid angiopathy but is independent of plaques and neurofibrillary tangles. *Ann. Neurol.* **2001**, *50*, 765–772. [CrossRef] [PubMed]
25. Nishitsuji, K.; Tomiyama, T.; Ishibashi, K.; Kametani, F.; Ozawa, K.; Okada, R.; Maat-Schieman, M.L.; Roos, R.A.; Iwai, K.; Mori, H. Cerebral vascular accumulation of Dutch-type Aβ42, but not wild-type Aβ42, in hereditary cerebral hemorrhage with amyloidosis, Dutch type. *J. Neurosci. Res.* **2007**, *85*, 2917–2923. [CrossRef]
26. Uddin, M.S.; Tewari, D.; Sharma, G.; Kabir, M.T.; Barreto, G.E.; Bin-Jumah, M.N.; Perveen, A.; Abdel-Daim, M.M.; Ashraf, G.M. Molecular Mechanisms of ER Stress and UPR in the Pathogenesis of Alzheimer's Disease. *Mol. Neurobiol.* **2020**, *57*, 2902–2919. [CrossRef] [PubMed]
27. Alberdi, E.; Wyssenbach, A.; Alberdi, M.; Sánchez-Gómez, M.V.; Cavaliere, F.; Rodríguez, J.J.; Verkhratsky, A.; Matute, C. Ca^{2+}-dependent endoplasmic reticulum stress correlates with astrogliosis in oligomeric amyloid β-treated astrocytes and in a model of Alzheimer's disease. *Aging Cell* **2013**, *12*, 292–302. [CrossRef]
28. Tomiyama, T.; Nagata, T.; Shimada, H.; Teraoka, R.; Fukushima, A.; Kanemitsu, H.; Takuma, H.; Kuwano, R.; Imagawa, M.; Ataka, S.; et al. A new amyloid β variant favoring oligomerization in Alzheimer's-type dementia. *Ann. Neurol.* **2008**, *63*, 377–387. [CrossRef]
29. McKnelly, K.J.; Kreutzer, A.G.; Howitz, W.J.; Haduong, K.; Yoo, S.; Hart, C.; Nowick, J.S. Effects of Familial Alzheimer's Disease Mutations on the Assembly of a β-Hairpin Peptide Derived from $A\beta_{16-36}$. *Biochemistry* **2022**, *61*, 446–454. [CrossRef]
30. Paik, Y.K.; Chang, D.J.; Reardon, C.A.; Walker, M.D.; Taxman, E.; Taylor, J.M. Identification and characterization of transcriptional regulatory regions associated with expression of the human apolipoprotein E gene. *J. Biol. Chem.* **1988**, *263*, 13340–13349. [CrossRef]
31. Maloney, B.; Ge, Y.-W.; Alley, G.M.; Lahiri, D.K. Important differences between human and mouse APOE gene promoters: Limitation of mouse APOE model in studying Alzheimer's disease. *J. Neurochem.* **2007**, *103*, 1237–1257. [CrossRef]

32. Yee, A.; Tsui, N.B.; Kwan, R.Y.; Leung, A.Y.; Lai, C.K.; Chung, T.; Lau, J.Y.; Fok, M.; Dai, D.L.; Lau, L.-T. Apolipoprotein E Gene Revisited: Contribution of Rare Variants to Alzheimer's Disease Susceptibility in Southern Chinese. *Curr. Alzheimer Res.* **2021**, *18*, 67–79. [CrossRef] [PubMed]
33. Zhou, Y.; Mägi, R.; Milani, L.; Lauschke, V.M. Global genetic diversity of human apolipoproteins and effects on cardiovascular disease risk. *J. Lipid Res.* **2018**, *59*, 1987–2000. [CrossRef] [PubMed]
34. Saunders, A.M. Gene identification in Alzheimer's disease. *Pharmacogenomics* **2001**, *2*, 239–249. [CrossRef] [PubMed]
35. Theuns, J.; Del-Favero, J.; Dermaut, B.; van Duijn, C.M.; Backhovens, H.; Broeck, M.V.D.; Serneels, S.; Corsmit, E.; Van Broeckhoven, C.; Cruts, M. Genetic variability in the regulatory region of presenilin 1 associated with risk for Alzheimer's disease and variable expression. *Hum. Mol. Genet.* **2000**, *9*, 325–331. [CrossRef] [PubMed]
36. Cruts, M.; Theuns, J.; Van Broeckhoven, C. Locus-specific mutation databases for neurodegenerative brain diseases. *Hum. Mutat.* **2012**, *33*, 1340–1344. [CrossRef] [PubMed]
37. Sun, L.; Zhou, R.; Yang, G.; Shi, Y. Analysis of 138 pathogenic mutations in presenilin-1 on the in vitro production of Aβ42 and Aβ40 peptides by γ-secretase. *Proc. Natl. Acad. Sci. USA* **2016**, *114*, E476–E485. [CrossRef]
38. Wines-Samuelson, M.; Schulte, E.C.; Smith, M.J.; Aoki, C.; Liu, X.; Kelleher, R.J.; Shen, J. Characterization of Age-Dependent and Progressive Cortical Neuronal Degeneration in Presenilin Conditional Mutant Mice. *PLoS ONE* **2010**, *5*, e10195. [CrossRef]
39. Kauwe, J.S.K.; Jacquart, S.; Chakraverty, S.; Wang, J.; Mayo, K.; Fagan, A.M.; Holtzman, D.M.; Morris, J.C.; Goate, A.M. Extreme cerebrospinal fluid amyloid β levels identify family with late-onset Alzheimer's disease presenilin 1 mutation. *Ann. Neurol.* **2007**, *61*, 446–453. [CrossRef]
40. Kumar-Singh, S.; Theuns, J.; Van Broeck, B.; Pirici, D.; Vennekens, K.; Corsmit, E.; Cruts, M.; Dermaut, B.; Wang, R.; Van Broeckhoven, C. Mean age-of-onset of familial alzheimer disease caused by presenilin mutations correlates with both increased Aβ42 and decreased Aβ40. *Hum. Mutat.* **2006**, *27*, 686–695. [CrossRef]
41. Koriath, C.; Kenny, J.; Adamson, G.; Druyeh, R.; Taylor, W.; Beck, J.; Quinn, L.; Mok, T.H.; Dimitriadis, A.; Norsworthy, P.; et al. Predictors for a dementia gene mutation based on gene-panel next-generation sequencing of a large dementia referral series. *Mol. Psychiatry* **2020**, *25*, 3399–3412. [CrossRef]
42. Gallo, M.; Frangipane, F.; Cupidi, C.; De Bartolo, M.; Turone, S.; Ferrari, C.; Nacmias, B.; Grimaldi, G.; Laganà, V.; Colao, R.; et al. The novel PSEN1 M84V mutation associated to frontal dysexecutive syndrome, spastic paraparesis, and cerebellar atrophy in a dominant Alzheimer's disease family. *Neurobiol. Aging* **2017**, *56*, 213.e7–213.e12. [CrossRef] [PubMed]
43. Hsu, S.; Gordon, B.A.; Hornbeck, R.; Norton, J.B.; Levitch, D.; Louden, A.; Ziegemeier, E.; Laforce, R., Jr.; Chhatwal, J.; Day, G.S.; et al. Discovery and validation of autosomal dominant Alzheimer's disease mutations. *Alzheimers Res. Ther.* **2018**, *10*, 67. [CrossRef] [PubMed]
44. Kakuda, N.; Takami, M.; Okochi, M.; Kasuga, K.; Ihara, Y.; Ikeuchi, T. Switched Aβ43 generation in familial Alzheimer's disease with presenilin 1 mutation. *Transl. Psychiatry* **2021**, *11*, 558. [CrossRef]
45. López-García, S.; Jiménez-Bonilla, J.; Delgado, A.L.; Balaguer, P.O.; Ceberio, J.I.; Marraco, I.B.; Rodríguez, E.R.; Sánchez-Juan, P. A Rare PSEN1 (Leu85Pro) Mutation Causing Alzheimer's Disease in a 29-Year-Old Woman Presenting as Corticobasal Syndrome. *J. Alzheimers Dis.* **2019**, *70*, 655–658. [CrossRef] [PubMed]
46. Arber, C.; Llerena, C.V.; Toombs, J.; Pocock, J.M.; Ryan, N.S.; Fox, N.C.; Zetterberg, H.; Hardy, J.; Wray, S. Amyloid precursor protein processing in human neurons with an allelic series of the PSEN1 intron 4 deletion mutation and total presenilin-1 knockout. *Brain Commun.* **2019**, *1*, fcz024. [CrossRef] [PubMed]
47. De Jonghe, C.; Cruts, M.; Rogaeva, E.A.; Tysoe, C.; Singleton, A.; Vanderstichele, H.; Meschino, W.; Dermaut, B.; Vanderhoeven, I.; Backhovens, H.; et al. Aberrant Splicing in the Presenilin-1 Intron 4 Mutation Causes Presenile Alzheimer's Disease by Increased A 42 Secretion. *Hum. Mol. Genet.* **1999**, *8*, 1529–1540. [CrossRef] [PubMed]
48. Szaruga-Bracke, M.; Veugelen, S.; Benurwar, M.; Lismont, S.; Sepulveda-Falla, D.; Lleo, A.; Ryan, N.S.; Lashley, T.; Fox, N.; Murayama, S.; et al. Qualitative changes in human γ-secretase underlie familial Alzheimer's disease. *J. Exp. Med.* **2015**, *212*, 2003–2013. [CrossRef]
49. Fox, N.C.; Kennedy, A.M.; Harvey, R.J.; Lantos, P.L.; Roques, P.K.; Collinge, J.; Hardy, J.; Hutton, M.; Stevens, J.M.; Warrington, E.K.; et al. Clinicopathological features of familial Alzheimer's disease associated with the M139V mutation in the presenilin 1 gene. Pedigree but not mutation specific age at onset provides evidence for a further genetic factor. *Brain* **1997**, *120*, 491–501. [CrossRef]
50. Chávez-Gutiérrez, L.; Bammens, L.; Benilova, I.; Vandersteen, A.; Benurwar, M.; Borgers, M.; Lismont, S.; Zhou, L.; Van Cleynenbreugel, S.; Esselmann, H.; et al. The mechanism of γ-Secretase dysfunction in familial Alzheimer disease. *EMBO J.* **2012**, *31*, 2261–2274. [CrossRef]
51. Ringman, J.M.; Gylys, K.H.; Medina, L.D.; Fox, M.; Kepe, V.; Flores, D.L.; Apostolova, L.G.; Barrio, J.R.; Small, G.; Silverman, D.H.; et al. Biochemical, neuropathological, and neuroimaging characteristics of early-onset Alzheimer's disease due to a novel PSEN1 mutation. *Neurosci. Lett.* **2011**, *487*, 287–292. [CrossRef]
52. Devi, G.; Fotiou, A.; Jyrinji, D.; Tycko, B.; DeArmand, S.; Rogaeva, E.; Song, Y.-Q.; Medieros, H.; Liang, Y.; Orlacchio, A.; et al. Novel presenilin 1 mutations associated with early onset of dementia in a family with both early-onset and late-onset Alzheimer disease. *Arch. Neurol.* **2000**, *57*, 1454–1457. [CrossRef] [PubMed]

53. Levy-Lahad, E.; Wasco, W.; Poorkaj, P.; Romano, D.M.; Oshima, J.; Pettingell, W.H.; Yu, C.E.; Jondro, P.D.; Schmidt, S.D.; Wang, K.; et al. Candidate gene for the chromosome 1 familial Alzheimer's disease locus. *Science* **1995**, *269*, 973–977. [CrossRef] [PubMed]
54. Perrone, F.; Cacace, R.; Van Mossevelde, S.; Bossche, T.V.D.; De Deyn, P.P.; Cras, P.; Engelborghs, S.; van der Zee, J.; Van Broeckhoven, C. Genetic screening in early-onset dementia patients with unclear phenotype: Relevance for clinical diagnosis. *Neurobiol. Aging* **2018**, *69*, 292.e7–292.e14. [CrossRef] [PubMed]
55. Jia, L.; Fu, Y.; Shen, L.; Zhang, H.; Zhu, M.; Qiu, Q.; Wang, Q.; Yan, X.; Kong, C.; Hao, J.; et al. *PSEN1*, *PSEN2*, and *APP* mutations in 404 Chinese pedigrees with familial Alzheimer's disease. *Alzheimers Dement.* **2020**, *16*, 178–191. [CrossRef]
56. Pang, Y.; Li, T.; Wang, Q.; Qin, W.; Li, Y.; Wei, Y.; Jia, L. A Rare Variation in the 3′ Untranslated Region of the Presenilin 2 Gene Is Linked to Alzheimer's Disease. *Mol. Neurobiol.* **2021**, *58*, 4337–4347. [CrossRef]
57. Wallon, D.; Rousseau, S.; Rovelet-Lecrux, A.; Quillard-Muraine, M.; Guyant-Maréchal, L.; Martinaud, O.; Pariente, J.; Puel, M.; Rollin-Sillaire, A.; Pasquier, F.; et al. The French Series of Autosomal Dominant Early Onset Alzheimer's Disease Cases: Mutation Spectrum and Cerebrospinal Fluid Biomarkers. *J. Alzheimers Dis.* **2012**, *30*, 847–856. [CrossRef]
58. Marcon, G.; Giaccone, G.; Cupidi, C.; Balestrieri, M.; Beltrami, C.A.; Finato, N.; Bergonzi, P.; Sorbi, S.; Bugiani, O.; Tagliavini, F. Neuropathological and Clinical Phenotype of an Italian Alzheimer Family with M239V Mutation of Presenilin 2 Gene. *J. Neuropathol. Exp. Neurol.* **2004**, *63*, 199–209. [CrossRef]
59. Huang, Y.; Weisgraber, K.H.; Mucke, L.; Mahley, R.W. Apolipoprotein E: Diversity of Cellular Origins, Structural and Biophysical Properties, and Effects in Alzheimer's Disease. *J. Mol. Neurosci.* **2004**, *23*, 189–204. [CrossRef]
60. Liu, C.-C.; Kanekiyo, T.; Xu, H.; Bu, G. Apolipoprotein E and Alzheimer disease: Risk, mechanisms and therapy. *Nat. Rev. Neurol.* **2013**, *9*, 106–118. [CrossRef]
61. Kara, E.; Marks, J.D.; Fan, Z.; Klickstein, J.A.; Roe, A.D.; Krogh, K.A.; Wegmann, S.; Maesako, M.; Luo, C.C.; Mylvaganam, R.; et al. Isoform- and cell type-specific structure of apolipoprotein E lipoparticles as revealed by a novel Forster resonance energy transfer assay. *J. Biol. Chem.* **2017**, *292*, 14720–14729. [CrossRef]
62. Lin, Y.-T.; Seo, J.; Gao, F.; Feldman, H.M.; Wen, H.-L.; Penney, J.; Cam, H.P.; Gjoneska, E.; Raja, W.K.; Cheng, J.; et al. APOE4 Causes Widespread Molecular and Cellular Alterations Associated with Alzheimer's Disease Phenotypes in Human iPSC-Derived Brain Cell Types. *Neuron* **2018**, *98*, 1141–1154.e7. [CrossRef] [PubMed]
63. Kang, S.S.; Ebbert, M.T.W.; Baker, K.E.; Cook, C.; Wang, X.; Sens, J.P.; Kocher, J.-P.; Petrucelli, L.; Fryer, J.D. Microglial translational profiling reveals a convergent APOE pathway from aging, amyloid, and tau. *J. Exp. Med.* **2018**, *215*, 2235–2245. [CrossRef] [PubMed]
64. Mathys, H.; Davila-Velderrain, J.; Peng, Z.; Gao, F.; Mohammadi, S.; Young, J.Z.; Menon, M.; He, L.; Abdurrob, F.; Jiang, X.; et al. Single-cell transcriptomic analysis of Alzheimer's disease. *Nature* **2019**, *570*, 332–337. [CrossRef]
65. Therriault, J.; Benedet, A.L.; Pascoal, T.A.; Mathotaarachchi, S.; Chamoun, M.; Savard, M.; Thomas, E.; Kang, M.S.; Lussier, F.; Tissot, C.; et al. Association of Apolipoprotein E ε4 With Medial Temporal Tau Independent of Amyloid-β. *JAMA Neurol.* **2020**, *77*, 470–479. [CrossRef] [PubMed]
66. Shi, Y.; Alzheimer's Disease Neuroimaging Initiative; Yamada, K.; Liddelow, S.A.; Smith, S.T.; Zhao, L.; Luo, W.; Tsai, R.M.; Spina, S.; Grinberg, L.T.; et al. ApoE4 markedly exacerbates tau-mediated neurodegeneration in a mouse model of tauopathy. *Nature* **2017**, *549*, 523–527. [CrossRef] [PubMed]
67. Harris, F.M.; Brecht, W.J.; Xu, Q.; Mahley, R.W.; Huang, Y. Increased tau Phosphorylation in Apolipoprotein E4 Transgenic Mice Is Associated with Activation of Extracellular Signal-regulated Kinase. *J. Biol. Chem.* **2004**, *279*, 44795–44801. [CrossRef]
68. Wadhwani, A.R.; Affaneh, A.; Van Gulden, S.; Kessler, J.A. Neuronal apolipoprotein E4 increases cell death and phosphorylated tau release in alzheimer disease. *Ann. Neurol.* **2019**, *85*, 726–739. [CrossRef]
69. Gibb, G.M.; Pearce, J.; Betts, J.C.; Lovestone, S.; Hoffmann, M.M.; Maerz, W.; Blackstock, W.P.; Anderton, B.H. Differential effects of apolipoprotein E isoforms on phosphorylation at specific sites on tau by glycogen synthase kinase-3β identified by nano-electrospray mass spectrometry. *FEBS Lett.* **2000**, *485*, 99–103. [CrossRef]
70. Park, J.-S.; Ji-Seon, P.; Sang-Wook, K.; Kang, M.-J.; Kang, S.-W.; Kim, D.-H.; Yoon, S.-Y. Disease-Associated Mutations of TREM2 Alter the Processing of N-Linked Oligosaccharides in the Golgi Apparatus. *Traffic* **2015**, *16*, 510–518. [CrossRef]
71. Saceleanu, V.M.; Covache-Busuioc, R.-A.; Costin, H.-P.; Glavan, L.-A.; Ciurea, A.V. An Important Step in Neuroscience: Camillo Golgi and His Discoveries. *Cells* **2022**, *11*, 4112. [CrossRef]
72. Ulrich, J.D.; Ulland, T.K.; Mahan, T.E.; Nyström, S.; Nilsson, K.P.; Song, W.M.; Zhou, Y.; Reinartz, M.; Choi, S.; Jiang, H.; et al. ApoE facilitates the microglial response to amyloid plaque pathology. *J. Exp. Med.* **2018**, *215*, 1047–1058. [CrossRef] [PubMed]
73. Keren-Shaul, H.; Spinrad, A.; Weiner, A.; Matcovitch-Natan, O.; Dvir-Szternfeld, R.; Ulland, T.K.; David, E.; Baruch, K.; Lara-Astaiso, D.; Toth, B.; et al. A Unique Microglia Type Associated with Restricting Development of Alzheimer's Disease. *Cell* **2017**, *169*, 1276–1290.e17. [CrossRef] [PubMed]
74. Hutton, M.; Lendon, C.L.; Rizzu, P.; Baker, M.; Froelich, S.; Houlden, H.; Pickering-Brown, S.; Chakraverty, S.; Isaacs, A.; Grover, A.; et al. Association of missense and 5′-splice-site mutations in tau with the inherited dementia FTDP-17. *Nature* **1998**, *393*, 702–705. [CrossRef] [PubMed]
75. Spillantini, M.G.; Bird, T.D.; Ghetti, B. Frontotemporal Dementia and Parkinsonism Linked to Chromosome 17: A New Group of Tauopathies. *Brain Pathol.* **2006**, *8*, 387–402. [CrossRef] [PubMed]

76. Poorkaj, P.; Bird, T.D.; Wijsman, E.; Nemens, E.; Garruto, R.M.; Anderson, L.; Andreadis, A.; Wiederholt, W.C.; Raskind, M.; Schellenberg, G.D. Tau is a candidate gene for chromosome 17 frontotemporal dementia. *Ann. Neurol.* **1998**, *43*, 815–825. [CrossRef] [PubMed]
77. Goedert, M.; Spillantini, M.G.; Jakes, R.; Rutherford, D.; Crowther, R.A. Multiple isoforms of human microtubule-associated protein tau: Sequences and localization in neurofibrillary tangles of Alzheimer's disease. *Neuron* **1989**, *3*, 519–526. [CrossRef]
78. Baudier, J.; Cole, R. D Interactions between the microtubule-associated tau proteins and S100b regulate tau phosphorylation by the Ca2+/calmodulin-dependent protein kinase II. *J. Biol. Chem.* **1988**, *263*, 5876–5883. [CrossRef]
79. Drewes, G.; Ebneth, A.; Preuss, U.; Mandelkow, E.-M.; Mandelkow, E. MARK, a Novel Family of Protein Kinases That Phosphorylate Microtubule-Associated Proteins and Trigger Microtubule Disruption. *Cell* **1997**, *89*, 297–308. [CrossRef]
80. Drewes, G.; Lichtenberg-Kraag, B.; Döring, F.; Mandelkow, E.; Biernat, J.; Goris, J.; Dorée, M.; Mandelkow, E. Mitogen activated protein (MAP) kinase transforms tau protein into an Alzheimer-like state. *EMBO J.* **1992**, *11*, 2131–2138. [CrossRef]
81. Bancher, C.; Brunner, C.; Lassmann, H.; Budka, H.; Jellinger, K.; Wiche, G.; Seitelberger, F.; Grundke-Iqbal, I.; Iqbal, K.; Wisniewski, H. Accumulation of abnormally phosphorylated τ precedes the formation of neurofibrillary tangles in Alzheimer's disease. *Brain Res.* **1989**, *477*, 90–99. [CrossRef]
82. Davies, P.; Vincent, I.J. On the role of tau phosphorylation in the formation of PHF. *Neurobiol. Aging* **1995**, *16*, 420–422. [CrossRef]
83. Wang, J.-Z.; Wu, Q.; Smith, A.; Grundke-Iqbal, I.; Iqbal, K. τ is phosphorylated by GSK-3 at several sites found in Alzheimer disease and its biological activity markedly inhibited only after it is prephosphorylated by A-kinase. *FEBS Lett.* **1998**, *436*, 28–34. [CrossRef]
84. Yasuda, M.; Takamatsu, J.; D'Souza, I.; Crowther, R.A.; Kawamata, T.; Hasegawa, M.; Hasegawa, H.; Spillantini, M.G.; Tanimukai, S.; Poorkaj, P.; et al. A novel mutation at position +12 in the intron following exon 10 of the tau gene in familial frontotemporal dementia (FTD-Kumamoto). *Ann. Neurol.* **2000**, *47*, 422–429. [CrossRef] [PubMed]
85. Takamatsu, J.; Kondo, A.; Ikegami, K.; Kimura, T.; Fujii, H.; Mitsuyama, Y.; Hashizume, Y. Selective Expression of Ser 199/202 Phosphorylated Tau in a Case of Frontotemporal Dementia. *Dement. Geriatr. Cogn. Disord.* **1998**, *9*, 82–89. [CrossRef]
86. Labbé, C.; Ogaki, K.; Lorenzo-Betancor, O.; Soto-Ortolaza, A.I.; Walton, R.L.; Rayaprolu, S.; Fujioka, S.; Murray, M.E.; Heckman, M.G.; Puschmann, A.; et al. Role for the microtubule-associated protein tau variant p.A152T in risk of α-synucleinopathies. *Neurology* **2015**, *85*, 1680–1686. [CrossRef]
87. Jin, S.C.; Pastor, P.; Cooper, B.; Cervantes, S.; Benitez, B.A.; Razquin, C.; Goate, A.; Cruchaga, C.; Ibero-American Alzheimer Disease Genetics Group Researchers. Pooled-DNA sequencing identifies novel causative variants in PSEN1, GRN and MAPT in a clinical early-onset and familial Alzheimer's disease Ibero-American cohort. *Alzheimers Res. Ther.* **2012**, *4*, 34. [CrossRef] [PubMed]
88. Coppola, G.; Chinnathambi, S.; Lee, J.J.; Dombroski, B.A.; Baker, M.C.; Soto-Ortolaza, A.I.; Lee, S.E.; Klein, E.; Huang, A.Y.; Sears, R.; et al. Evidence for a role of the rare p.A152T variant in MAPT in increasing the risk for FTD-spectrum and Alzheimer's diseases. *Hum. Mol. Genet.* **2012**, *21*, 3500–3512. [CrossRef]
89. Rizzini, C.; Goedert, M.; Hodges, J.R.; Smith, M.J.; Jakes, R.; Hills, R.; Xuereb, J.H.; Crowther, R.A.; Spillantini, M.G. *Tau* Gene Mutation K257T Causes a Tauopathy Similar to Pick's Disease. *J. Neuropathol. Exp. Neurol.* **2000**, *59*, 990–1001. [CrossRef]
90. Kobayashi, T.; Ota, S.; Tanaka, K.; Ito, Y.; Hasegawa, M.; Umeda, Y.; Motoi, Y.; Takanashi, M.; Yasuhara, M.; Anno, M.; et al. A novel L266V mutation of the tau gene causes frontotemporal dementia with a unique tau pathology. *Ann. Neurol.* **2003**, *53*, 133–137. [CrossRef]
91. Hogg, M.; Grujic, Z.M.; Baker, M.; Demirci, S.; Guillozet, A.L.; Sweet, A.P.; Herzog, L.L.; Weintraub, S.; Mesulam, M.-M.; LaPointe, N.E.; et al. The L266V tau mutation is associated with frontotemporal dementia and Pick-like 3R and 4R tauopathy. *Acta Neuropathol.* **2003**, *106*, 323–336. [CrossRef]
92. Schilling, S.; Zeitschel, U.; Hoffmann, T.; Heiser, U.; Francke, M.; Kehlen, A.; Holzer, M.; Hutter-Paier, B.; Prokesch, M.; Windisch, M.; et al. Glutaminyl cyclase inhibition attenuates pyroglutamate Aβ and Alzheimer's disease–like pathology. *Nat. Med.* **2008**, *14*, 1106–1111. [CrossRef] [PubMed]
93. Frost, J.L.; Liu, B.; Kleinschmidt, M.; Schilling, S.; Demuth, H.-U.; Lemere, C.A. Passive Immunization against Pyroglutamate-3 Amyloid-β Reduces Plaque Burden in Alzheimer-Like Transgenic Mice: A Pilot Study. *Neurodegener. Dis.* **2012**, *10*, 265–270. [CrossRef] [PubMed]
94. Frost, J.L.; Liu, B.; Rahfeld, J.-U.; Kleinschmidt, M.; O'Nuallain, B.; Le, K.X.; Lues, I.; Caldarone, B.J.; Schilling, S.; Demuth, H.-U.; et al. An anti-pyroglutamate-3 Aβ vaccine reduces plaques and improves cognition in APPswe/PS1ΔE9 mice. *Neurobiol. Aging* **2015**, *36*, 3187–3199. [CrossRef] [PubMed]
95. Hoffmann, T.; Meyer, A.; Heiser, U.; Kurat, S.; Böhme, L.; Kleinschmidt, M.; Bühring, K.-U.; Hutter-Paier, B.; Farcher, M.; DeMuth, H.-U.; et al. Glutaminyl Cyclase Inhibitor PQ912 Improves Cognition in Mouse Models of Alzheimer's Disease—Studies on Relation to Effective Target Occupancy. *Experiment* **2017**, *362*, 119–130. [CrossRef] [PubMed]
96. DeMattos, R.B.; Lu, J.; Tang, Y.; Racke, M.M.; DeLong, C.A.; Tzaferis, J.A.; Hole, J.T.; Forster, B.M.; McDonnell, P.C.; Liu, F.; et al. A Plaque-Specific Antibody Clears Existing β-amyloid Plaques in Alzheimer's Disease Mice. *Neuron* **2012**, *76*, 908–920. [CrossRef]
97. Lowe, S.L.; Willis, B.A.; Hawdon, A.; Natanegara, F.; Chua, L.; Foster, J.; Shcherbinin, S.; Ardayfio, P.; Sims, J.R. Donanemab (LY3002813) dose-escalation study in Alzheimer's disease. *Alzheimers Dement. Transl. Res. Clin. Interv.* **2021**, *7*, e12112. [CrossRef]

98. Lowe, S.L.; Evans, C.D.; Shcherbinin, S.; Cheng, Y.-J.; Willis, B.A.; Gueorguieva, I.; Lo, A.C.; Fleisher, A.S.; Dage, J.L.; Ardayfio, P.; et al. Donanemab (LY3002813) Phase 1b Study in Alzheimer's Disease: Rapid and Sustained Reduction of Brain Amyloid Measured by Florbetapir F18 Imaging. *J. Prev. Alzheimers Dis.* **2021**, *8*, 414–424. [CrossRef]
99. Shcherbinin, S.; Evans, C.D.; Lu, M.; Andersen, S.W.; Pontecorvo, M.J.; Willis, B.A.; Gueorguieva, I.; Hauck, P.M.; Brooks, D.A.; Mintun, M.A.; et al. Association of Amyloid Reduction After Donanemab Treatment With Tau Pathology and Clinical Outcomes: Association of amyloid reduction after donanemab treatment with tau pathology and clinical outcomes: The TRAILBLAZER-ALZ randomized clinical trial. *JAMA Neurol.* **2022**, *79*, 1015. [CrossRef]
100. Raza, C.; Anjum, R.; Shakeel, N.U.A. Parkinson's disease: Mechanisms, translational models and management strategies. *Life Sci.* **2019**, *226*, 77–90. [CrossRef]
101. Beitz, J.M. Parkinson s disease: A review. *Front. Biosci.* **2014**, *S6*, 65–74. [CrossRef]
102. Day, J.O.; Mullin, S. The Genetics of Parkinson's Disease and Implications for Clinical Practice. *Genes* **2021**, *12*, 1006. [CrossRef] [PubMed]
103. Ye, H.; Robak, L.A.; Yu, M.; Cykowski, M.; Shulman, J.M. Genetics and Pathogenesis of Parkinson's Syndrome. *Annu. Rev. Pathol. Mech. Dis.* **2023**, *18*, 95–121. [CrossRef] [PubMed]
104. Siddiqui, I.J.; Pervaiz, N.; Abbasi, A.A. The Parkinson Disease gene SNCA: Evolutionary and structural insights with pathological implication. *Sci. Rep.* **2016**, *6*, 24475. [CrossRef] [PubMed]
105. Rui, Q.; Ni, H.; Li, D.; Gao, R.; Chen, G. The Role of LRRK2 in Neurodegeneration of Parkinson Disease. *Curr. Neuropharmacol.* **2018**, *16*, 1348–1357. [CrossRef] [PubMed]
106. Vizziello, M.; Borellini, L.; Franco, G.; Ardolino, G. Disruption of Mitochondrial Homeostasis: The Role of PINK1 in Parkinson's Disease. *Cells* **2021**, *10*, 3022. [CrossRef]
107. Crichton, R.R.; Dexter, D.T.; Ward, R.J. Brain iron metabolism and its perturbation in neurological diseases. In *Metal Ions in Neurological Systems*; Linert, W., Kozlowski, H., Eds.; Springer Vienna: Vienna, Austria, 2012; pp. 1–15. [CrossRef]
108. Chandra, G.; Shenoi, R.; Anand, R.; Rajamma, U.; Mohanakumar, K. Reinforcing mitochondrial functions in aging brain: An insight into Parkinson's disease therapeutics. *J. Chem. Neuroanat.* **2019**, *95*, 29–42. [CrossRef]
109. Borgna-Pignatti, C.; Rugolotto, S.; De Stefano, P.; Zhao, H.; Cappellini, M.D.; Del Vecchio, G.C.; Romeo, M.A.; Forni, G.L.; Gamberini, M.R.; Ghilardi, R.; et al. Survival and complications in patients with thalassemia major treated with transfusion and deferoxamine. *Haematologica* **2004**, *89*, 1187–1193.
110. Boddaert, N.; Sang, K.H.L.Q.; Rötig, A.; Leroy-Willig, A.; Gallet, S.; Brunelle, F.; Sidi, D.; Thalabard, J.-C.; Munnich, A.; Cabantchik, Z.I. Selective iron chelation in Friedreich ataxia: Biologic and clinical implications. *Blood* **2007**, *110*, 401–408. [CrossRef]
111. Devos, D.; Moreau, C.; Devedjian, J.C.; Kluza, J.; Petrault, M.; Laloux, C.; Jonneaux, A.; Ryckewaert, G.; Garçon, G.; Rouaix, N.; et al. Targeting Chelatable Iron as a Therapeutic Modality in Parkinson's Disease. *Antioxid. Redox Signal.* **2014**, *21*, 195–210. [CrossRef]
112. McColgan, P.; Tabrizi, S.J. Huntington's disease: A clinical review. *Eur. J. Neurol.* **2017**, *25*, 24–34. [CrossRef]
113. Wexler, A.; Wild, E.; Tabrizi, S.J. George Huntington: A legacy of inquiry, empathy and hope. *Brain* **2016**, *139*, 2326–2333. [CrossRef]
114. Bates, G.P.; Dorsey, R.; Gusella, J.F.; Hayden, M.R.; Kay, C.; Leavitt, B.R.; Nance, M.; Ross, C.A.; Scahill, R.I.; Wetzel, R.; et al. Huntington Disease. *Nat. Rev. Dis. Prim.* **2015**, *1*, 15005. [CrossRef]
115. MacDonald, M.E.; Ambrose, C.M.; Duyao, M.P.; Myers, R.H.; Lin, C.; Srinidhi, L.; Barnes, G.; Taylor, S.A.; James, M.; Groot, N.; et al. The Huntington's Disease Collaborative Research Group: A novel gene containing a trinucleotide repeat that is expanded and unstable on Huntington's disease chromosomes. *Cell* **1993**, *72*, 971–983. [CrossRef]
116. Squitieri, F.; Andrew, S.; Goldberg, Y.; Kremer, B.; Spence, N.; Zelsler, J.; Nichol, K.; Theilmann, J.; Greenberg, J.; Goto, J.; et al. DNA haplotype analysis of Huntington disease reveals clues to the origins and mechanisms of CAG expansion and reasons for geographic variations of prevalence. *Hum. Mol. Genet.* **1994**, *3*, 2103–2114. [CrossRef] [PubMed]
117. Persichetti, F.; Srinidhi, J.; Kanaley, L.; Ge, P.; Myers, R.H.; D'Arrigo, K.; Barnes, G.T.; MacDonald, M.E.; Vonsattel, J.-P.; Gusella, J.F.; et al. Huntington's disease CAG trinucleotide repeats in pathologically confirmed post-mortem brains. *Neurobiol. Dis.* **1994**, *1*, 159–166. [CrossRef] [PubMed]
118. Gusella, J.F.; MacDonald, M.E.; Lee, J.-M. Genetic modifiers of Huntington's disease. *Mov. Disord.* **2014**, *29*, 1359–1365. [CrossRef] [PubMed]
119. Lee, J.-M.; Gillis, T.; Mysore, J.S.; Ramos, E.M.; Myers, R.H.; Hayden, M.R.; Morrison, P.J.; Nance, M.; Ross, C.A.; Margolis, R.L.; et al. Common SNP-Based Haplotype Analysis of the 4p16.3 Huntington Disease Gene Region. *Am. J. Hum. Genet.* **2012**, *90*, 434–444. [CrossRef]
120. Landles, C.; Sathasivam, K.; Weiss, A.; Woodman, B.; Moffitt, H.; Finkbeiner, S.; Sun, B.; Gafni, J.; Ellerby, L.M.; Trottier, Y.; et al. Proteolysis of Mutant Huntingtin Produces an Exon 1 Fragment That Accumulates as an Aggregated Protein in Neuronal Nuclei in Huntington Disease. *J. Biol. Chem.* **2010**, *285*, 8808–8823. [CrossRef]
121. Ha, A.D.; Fung, V.S. Huntington's disease. *Curr. Opin. Neurol.* **2012**, *25*, 491–498. [CrossRef]
122. Ross, C.A.; Tabrizi, S.J. Huntington's disease: From molecular pathogenesis to clinical treatment. *Lancet Neurol.* **2011**, *10*, 83–98. [CrossRef]
123. Vonsattel, J.P.G.; DiFiglia, M. Huntington Disease. *J. Neuropathol. Exp. Neurol.* **1998**, *57*, 369–384. [CrossRef]

124. Moss, D.J.H.; Pardiñas, A.F.; Langbehn, D.; Lo, K.; Leavitt, B.R.; Roos, R.; Durr, A.; Mead, S.; Holmans, P.; Jones, L.; et al. Identification of genetic variants associated with Huntington's disease progression: A genome-wide association study. *Lancet Neurol.* **2017**, *16*, 701–711. [CrossRef]
125. Travessa, A.M.; Rodrigues, F.B.; Mestre, T.A.; Ferreira, J.J. Fifteen Years of Clinical Trials in Huntington's Disease: A Very Low Clinical Drug Development Success Rate. *J. Huntingtons Dis.* **2017**, *6*, 157–163. [CrossRef]
126. Kordasiewicz, H.B.; Stanek, L.M.; Wancewicz, E.V.; Mazur, C.; McAlonis, M.M.; Pytel, K.A.; Artates, J.W.; Weiss, A.; Cheng, S.H.; Shihabuddin, L.S.; et al. Sustained Therapeutic Reversal of Huntington's Disease by Transient Repression of Huntingtin Synthesis. *Neuron* **2012**, *74*, 1031–1044. [CrossRef]
127. Agustín-Pavón, C.; Mielcarek, M.; Garriga-Canut, M.; Isalan, M. Deimmunization for gene therapy: Host matching of synthetic zinc finger constructs enables long-term mutant Huntingtin repression in mice. *Mol. Neurodegener.* **2016**, *11*, 64. [CrossRef]
128. Yang, L.; Calingasan, N.Y.; Wille, E.J.; Cormier, K.; Smith, K.; Ferrante, R.J.; Beal, M.F. Combination therapy with Coenzyme Q_{10} and creatine produces additive neuroprotective effects in models of Parkinson's and Huntington's Diseases. *J. Neurochem.* **2009**, *109*, 1427–1439. [CrossRef]
129. McGarry, A.; Auinger, P.; Kieburtz, K.D.; Bredlau, A.-L.; Hersch, S.M.; Rosas, H.D. Suicidality Risk Factors Across the CARE-HD, 2CARE, and CREST-E Clinical Trials in Huntington Disease. *Neurol. Clin. Pract.* **2022**, *12*, 131–138. [CrossRef]
130. Basavarajappa, B.S.; Subbanna, S. Histone Methylation Regulation in Neurodegenerative Disorders. *Int. J. Mol. Sci.* **2021**, *22*, 4654. [CrossRef] [PubMed]
131. Masrori, P.; Van Damme, P. Amyotrophic lateral sclerosis: A clinical review. *Eur. J. Neurol.* **2020**, *27*, 1918–1929. [CrossRef] [PubMed]
132. Renton, A.E.; Majounie, E.; Waite, A.; Simon-Saánchez, J.; Rollinson, S.; Gibbs, J.R.; Schymick, J.C.; Laaksovirta, H.; van Swieten, J.C.; Myllykangas, L.; et al. A Hexanucleotide Repeat Expansion in C9ORF72 Is the Cause of Chromosome 9p21-Linked ALS-FTD. *Neuron* **2011**, *72*, 257–268. [CrossRef] [PubMed]
133. Woollacott, I.O.C.; Mead, S. The C9ORF72 expansion mutation: Gene structure, phenotypic and diagnostic issues. *Acta Neuropathol.* **2014**, *127*, 319–332. [CrossRef] [PubMed]
134. Balendra, R.; Isaacs, A.M. C9orf72-mediated ALS and FTD: Multiple pathways to disease. *Nat. Rev. Neurol.* **2018**, *14*, 544–558. [CrossRef] [PubMed]
135. Herdewyn, S.; Zhao, H.; Moisse, M.; Race, V.; Matthijs, G.; Reumers, J.; Kusters, B.; Schelhaas, H.J.; Berg, L.H.v.D.; Goris, A.; et al. Whole-genome sequencing reveals a coding non-pathogenic variant tagging a non-coding pathogenic hexanucleotide repeat expansion in C9orf72 as cause of amyotrophic lateral sclerosis. *Hum. Mol. Genet.* **2012**, *21*, 2412–2419. [CrossRef]
136. Majounie, E.; Renton, A.E.; Mok, K.; Dopper, E.G.; Waite, A.; Rollinson, S.; Chiò, A.; Restagno, G.; Nicolaou, N.; Simon-Sanchez, J.; et al. Frequency of the C9orf72 hexanucleotide repeat expansion in patients with amyotrophic lateral sclerosis and frontotemporal dementia: A cross-sectional study. *Lancet Neurol.* **2012**, *11*, 323–330. [CrossRef]
137. Xi, Z.; Zinman, L.; Grinberg, Y.; Moreno, D.; Sato, C.; Bilbao, J.M.; Ghani, M.; Hernández, I.; Ruiz, A.; Boada, M.; et al. Investigation of C9orf72 in 4 Neurodegenerative Disorders. *Arch. Neurol.* **2012**, *69*, 1583–1590. [CrossRef] [PubMed]
138. Česnik, A.B.; Darovic, S.; Mihevc, S.P.; Štalekar, M.; Malnar, M.; Motaln, H.; Lee, Y.-B.; Mazej, J.; Pohleven, J.; Grosch, M.; et al. Nuclear RNA foci from *C9ORF72* expansion mutation form paraspeckle-like bodies. *J. Cell Sci.* **2019**, *132*, jcs.224303. [CrossRef]
139. Saccon, R.A.; Bunton-Stasyshyn, R.K.A.; Fisher, E.M.; Fratta, P. Is SOD1 loss of function involved in amyotrophic lateral sclerosis? *Brain* **2013**, *136*, 2342–2358. [CrossRef]
140. Bosco, D.A.; Morfini, G.; Karabacak, N.M.; Song, Y.; Gros-Louis, F.; Pasinelli, P.; Goolsby, H.; Fontaine, B.A.; Lemay, N.; McKenna-Yasek, D.; et al. Wild-type and mutant SOD1 share an aberrant conformation and a common pathogenic pathway in ALS. *Nat. Neurosci.* **2010**, *13*, 1396–1403. [CrossRef]
141. Bruijn, L.I.; Houseweart, M.K.; Kato, S.; Anderson, K.L.; Anderson, S.D.; Ohama, E.; Reaume, A.G.; Scott, R.W.; Cleveland, D.W. Aggregation and Motor Neuron Toxicity of an ALS-Linked SOD1 Mutant Independent from Wild-Type SOD1. *Science* **1998**, *281*, 1851–1854. [CrossRef]
142. Saeed, M.; Yang, Y.; Deng, H.-X.; Hung, W.-Y.; Siddique, N.; Dellefave, L.; Gellera, C.; Andersen, P.M.; Siddique, T. Age and founder effect of SOD1 A4V mutation causing ALS. *Neurology* **2009**, *72*, 1634–1639. [CrossRef]
143. Hensley, K.; Mhatre, M.; Mou, S.; Pye, Q.N.; Stewart, C.; West, M.; Williamson, K.S. On the Relation of Oxidative Stress to Neuroinflammation: Lessons Learned from the G93A-SOD1 Mouse Model of Amyotrophic Lateral Sclerosis. *Antioxid. Redox Signal.* **2006**, *8*, 2075–2087. [CrossRef] [PubMed]
144. Curti, D.; Rognoni, F.; Alimonti, D.; Malaspina, A.; Feletti, F.; Tessera, S.; Finotti, N.; Rehak, L.; Mazzini, L.; Zerbi, F.; et al. SOD1 activity and protective factors in familial ALS patients with L84F SOD1 mutation. *Lateral Scler. Other Motor Neuron Disord.* **2002**, *3*, 115–122. [CrossRef] [PubMed]
145. Ceroni, M.; Malaspina, A.; Poloni, T.E.; Alimonti, D.; Rognoni, F.; Habgood, J.; Imbesi, F.; Antonelli, P.; Alfonsi, E.; Curti, D.; et al. Clustering of ALS patients in central Italy due to the occurrence of the L84F SOD1 gene mutation. *Neurology* **1999**, *53*, 1064. [CrossRef] [PubMed]
146. Juneja, T.; Pericak-Vance, M.A.; Laing, N.G.; Dave, S.; Siddique, T. Prognosis in Familial Amyotrophic Lateral Sclerosis: Progression and Survival in Patients with Glu100gly and Ala4val Mutations in Cu,Zn Superoxide Dismutase. *Neurology* **1997**, *48*, 55–57. [CrossRef]

147. Zhou, Y.; Tang, J.; Lan, J.; Zhang, Y.; Wang, H.; Chen, Q.; Kang, Y.; Sun, Y.; Feng, X.; Wu, L.; et al. Honokiol alleviated neurodegeneration by reducing oxidative stress and improving mitochondrial function in mutant SOD1 cellular and mouse models of amyotrophic lateral sclerosis. *Acta Pharm. Sin. B* **2023**, *13*, 577–597. [CrossRef]
148. Cassina, P.; Cassina, A.; Pehar, M.; Castellanos, R.; Gandelman, M.; de León, A.; Robinson, K.M.; Mason, R.P.; Beckman, J.S.; Barbeito, L.; et al. Mitochondrial Dysfunction in SOD1^{G93A}-Bearing Astrocytes Promotes Motor Neuron Degeneration: Prevention by Mitochondrial-Targeted Antioxidants. *J. Neurosci.* **2008**, *28*, 4115–4122. [CrossRef]
149. Zhang, X.; Li, L.; Chen, S.; Yang, D.; Wang, Y.; Zhang, X.; Wang, Z.; Le, W. Rapamycin treatment augments motor neuron degeneration in SOD1^{G93A}mouse model of amyotrophic lateral sclerosis. *Autophagy* **2011**, *7*, 412–425. [CrossRef]
150. Daoud, H.; Valdmanis, P.N.; Kabashi, E.; Dion, P.; Dupre, N.; Camu, W.; Meininger, V.; Rouleau, G.A. Contribution of TARDBP mutations to sporadic amyotrophic lateral sclerosis. *J. Med. Genet.* **2008**, *46*, 112–114. [CrossRef]
151. Lattante, S.; Rouleau, G.A.; Kabashi, E. *TARDBP* and *FUS* Mutations Associated with Amyotrophic Lateral Sclerosis: Summary and Update. *Hum. Mutat.* **2013**, *34*, 812–826. [CrossRef]
152. Kabashi, E.; Valdmanis, P.; Dion, P.; Spiegelman, D.; McConkey, B.J.; Velde, C.V.; Bouchard, J.-P.; Lacomblez, L.; Pochigaeva, K.; Salachas, F.; et al. TARDBP mutations in individuals with sporadic and familial amyotrophic lateral sclerosis. *Nat. Genet.* **2008**, *40*, 572–574. [CrossRef]
153. Corcia, P.; Valdmanis, P.; Millecamps, S.; Lionnet, C.; Blasco, H.; Mouzat, K.; Daoud, H.; Belzil, V.; Morales, R.; Pageot, N.; et al. Phenotype and genotype analysis in amyotrophic lateral sclerosis with TARDBP gene mutations. *Neurology* **2012**, *78*, 1519–1526. [CrossRef] [PubMed]
154. Tamaoka, A.; Arai, M.; Itokawa, M.; Arai, T.; Hasegawa, M.; Tsuchiya, K.; Takuma, H.; Tsuji, H.; Ishii, A.; Watanabe, M.; et al. TDP-43 M337V Mutation in Familial Amyotrophic Lateral Sclerosis in Japan. *Intern. Med.* **2010**, *49*, 331–334. [CrossRef] [PubMed]
155. Watanabe, S.; Oiwa, K.; Murata, Y.; Komine, O.; Sobue, A.; Endo, F.; Takahashi, E.; Yamanaka, K. ALS-linked TDP-43M337V knock-in mice exhibit splicing deregulation without neurodegeneration. *Mol. Brain* **2020**, *13*, 8. [CrossRef] [PubMed]
156. Esmaeili, M.A.; Panahi, M.; Yadav, S.; Hennings, L.; Kiaei, M. Premature death of TDP-43 (A315T) transgenic mice due to gastrointestinal complications prior to development of full neurological symptoms of amyotrophic lateral sclerosis. *Int. J. Exp. Pathol.* **2013**, *94*, 56–64. [CrossRef] [PubMed]
157. Orrù, S.; Coni, P.; Floris, A.; Littera, R.; Carcassi, C.; Sogos, V.; Brancia, C. Reduced stress granule formation and cell death in fibroblasts with the A382T mutation of TARDBP gene: Evidence for loss of TDP-43 nuclear function. *Hum. Mol. Genet.* **2016**, *25*, 4473–4483. [CrossRef]
158. Zanini, G.; Selleri, V.; Nasi, M.; De Gaetano, A.; Martinelli, I.; Gianferrari, G.; Lofaro, F.D.; Boraldi, F.; Mandrioli, J.; Pinti, M. Mitochondrial and Endoplasmic Reticulum Alterations in a Case of Amyotrophic Lateral Sclerosis Caused by TDP-43 A382T Mutation. *Int. J. Mol. Sci.* **2022**, *23*, 11881. [CrossRef]
159. Deng, H.; Gao, K.; Jankovic, J. The role of FUS gene variants in neurodegenerative diseases. *Nat. Rev. Neurol.* **2014**, *10*, 337–348. [CrossRef]
160. Corrado, L.; Del Bo, R.; Castellotti, B.; Ratti, A.; Cereda, C.; Penco, S.; Sorarù, G.; Carlomagno, Y.; Ghezzi, S.; Pensato, V.; et al. Mutations of FUS gene in sporadic amyotrophic lateral sclerosis. *J. Med. Genet.* **2010**, *47*, 190–194. [CrossRef]
161. Rademakers, R.; Stewart, H.; Dejesus-Hernandez, M.; Krieger, C.; Graff-Radford, N.; Fabros, M.; Briemberg, H.; Cashman, N.; Eisen, A.; Mackenzie, I.R. *Fus* gene mutations in familial and sporadic amyotrophic lateral sclerosis. *Muscle Nerve* **2010**, *42*, 170–176. [CrossRef]
162. Ticozzi, N.; Silani, V.; LeClerc, A.L.; Keagle, P.; Gellera, C.; Ratti, A.; Taroni, F.; Kwiatkowski, T.J.; McKenna-Yasek, D.M.; Sapp, P.C.; et al. Analysis of FUS gene mutation in familial amyotrophic lateral sclerosis within an Italian cohort. *Neurology* **2009**, *73*, 1180–1185. [CrossRef]
163. Weinreich, M.; Shepheard, S.R.; Verber, N.; Wyles, M.; Heath, P.R.; Highley, J.R.; Kirby, J.; Shaw, P.J. Neuropathological characterization of a novel TANK binding kinase (*TBK1*) gene loss of function mutation associated with amyotrophic lateral sclerosis. *Neuropathol. Appl. Neurobiol.* **2019**, *46*, 279–291. [CrossRef]
164. Oakes, J.A.; Davies, M.C.; Collins, M.O. TBK1: A new player in ALS linking autophagy and neuroinflammation. *Mol. Brain* **2017**, *10*, 5. [CrossRef]
165. Gurfinkel, Y.; Polain, N.; Sonar, K.; Nice, P.; Mancera, R.L.; Rea, S.L. Functional and structural consequences of TBK1 missense variants in frontotemporal lobar degeneration and amyotrophic lateral sclerosis. *Neurobiol. Dis.* **2022**, *174*, 105859. [CrossRef]
166. Zarei, S.; Carr, K.; Reiley, L.; Diaz, K.; Guerra, O.; Altamirano, P.; Pagani, W.; Lodin, D.; Orozco, G.; Chinea, A. A comprehensive review of amyotrophic lateral sclerosis. *Surg. Neurol. Int.* **2015**, *6*, 171. [CrossRef]
167. Deans, C.; Maggert, K.A. What Do You Mean, "Epigenetic"? *Genetics* **2015**, *199*, 887–896. [CrossRef] [PubMed]
168. Burggren, W. Epigenetic Inheritance and Its Role in Evolutionary Biology: Re-Evaluation and New Perspectives. *Biology* **2016**, *5*, 24. [CrossRef]
169. Neumann, M.; Sampathu, D.M.; Kwong, L.K.; Truax, A.C.; Micsenyi, M.C.; Chou, T.T.; Bruce, J.; Schuck, T.; Grossman, M.; Clark, C.M.; et al. Ubiquitinated TDP-43 in Frontotemporal Lobar Degeneration and Amyotrophic Lateral Sclerosis. *Science* **2006**, *314*, 130–133. [CrossRef] [PubMed]
170. Fecto, F.; Yan, J.; Vemula, S.P.; Liu, E.; Yang, Y.; Chen, W.; Zheng, J.G.; Shi, Y.; Siddique, N.; Arrat, H.; et al. SQSTM1 Mutations in Familial and Sporadic Amyotrophic Lateral Sclerosis. *Arch. Neurol.* **2011**, *68*, 1440–1446. [CrossRef] [PubMed]

171. Maruyama, H.; Morino, H.; Ito, H.; Izumi, Y.; Kato, H.; Watanabe, Y.; Kinoshita, Y.; Kamada, M.; Nodera, H.; Suzuki, H.; et al. Mutations of optineurin in amyotrophic lateral sclerosis. *Nature* **2010**, *465*, 223–226. [CrossRef] [PubMed]
172. Johnson, J.O.; Mandrioli, J.; Benatar, M.; Abramzon, Y.; Van Deerlin, V.M.; Trojanowski, J.Q.; Gibbs, J.R.; Brunetti, M.; Gronka, S.; Wuu, J.; et al. Exome Sequencing Reveals VCP Mutations as a Cause of Familial ALS. *Neuron* **2010**, *68*, 857–864. [CrossRef]
173. Boeynaems, S.; Bogaert, E.; Van Damme, P.; Bosch, L.V.D. Inside out: The role of nucleocytoplasmic transport in ALS and FTLD. *Acta Neuropathol.* **2016**, *132*, 159–173. [CrossRef]
174. Vilarino-Guell, C.; Wider, C.; Soto-Ortolaza, A.I.; Cobb, S.A.; Kachergus, J.M.; Keeling, B.H.; Dachsel, J.C.; Hulihan, M.M.; Dickson, D.W.; Wszolek, Z.K.; et al. Characterization of DCTN1 genetic variability in neurodegeneration. *Neurology* **2009**, *72*, 2024–2028. [CrossRef] [PubMed]
175. Araki, E.; Tsuboi, Y.; Daechsel, J.; Milnerwood, A.; Vilarino-Guell, C.; Fujii, N.; Mishima, T.; Oka, T.; Hara, H.; Fukae, J.; et al. A Novel *DCTN1* mutation with late-onset parkinsonism and frontotemporal atrophy. *Mov. Disord.* **2014**, *29*, 1201–1204. [CrossRef] [PubMed]
176. Al-Chalabi, A.; Hardiman, O.; Kiernan, M.C.; Chiò, A.; Rix-Brooks, B.; van den Berg, L.H. Amyotrophic lateral sclerosis: Moving towards a new classification system. *Lancet Neurol.* **2016**, *15*, 1182–1194. [CrossRef] [PubMed]
177. Chiò, A.; Calvo, A.; Moglia, C.; Mazzini, L.; Mora, G.; PARALS Study Group. Phenotypic heterogeneity of amyotrophic lateral sclerosis: A population based study. *J. Neurol. Neurosurg. Psychiatry* **2011**, *82*, 740–746. [CrossRef] [PubMed]
178. Finegan, E.; Chipika, R.H.; Shing, S.L.H.; Hardiman, O.; Bede, P. Pathological Crying and Laughing in Motor Neuron Disease: Pathobiology, Screening, Intervention. *Front. Neurol.* **2019**, *10*, 260. [CrossRef]
179. Wijesekera, L.C.; Mathers, S.; Talman, P.; Galtrey, C.; Parkinson, M.H.; Ganesalingam, J.; Willey, E.; Ampong, M.A.; Ellis, C.M.; Shaw, C.E.; et al. Natural history and clinical features of the flail arm and flail leg ALS variants. *Neurology* **2009**, *72*, 1087–1094. [CrossRef]
180. Doble, A. The pharmacology and mechanism of action of riluzole. *Neurology* **1996**, *47* (Suppl. S4), 233S–241S. [CrossRef]
181. Lacomblez, L.; Amyotrophic Lateral Sclerosis/Riluzole Study Group II; Bensimon, G.; Meininger, V.; Leigh, P.; Guillet, P. Dose-ranging study of riluzole in amyotrophic lateral sclerosis. Amyotrophic Lateral Sclerosis/Riluzole Study Group II. *Lancet* **1996**, *347*, 1425–1431. [CrossRef]
182. Hinchcliffe, M.; Smith, A. Riluzole: Real-world evidence supports significant extension of median survival times in patients with amyotrophic lateral sclerosis. *Degener. Neurol. Neuromuscul. Dis.* **2017**, *7*, 61–70. [CrossRef]
183. Mora, J.S.; Genge, A.; Chio, A.; Estol, C.J.; Chaverri, D.; Hernández, M.; Marín, S.; Mascias, J.; Rodriguez, G.E.; Povedano, M.; et al. Masitinib as an add-on therapy to riluzole in patients with amyotrophic lateral sclerosis: A randomized clinical trial. *Amyotroph. Lateral Scler. Front. Degener.* **2019**, *21*, 5–14. [CrossRef] [PubMed]
184. Scheltens, P.; De Strooper, B.; Kivipelto, M.; Holstege, H.; Chételat, G.; Teunissen, C.E.; Cummings, J.; van der Flier, W.M. Alzheimer's disease. *Lancet* **2021**, *397*, 1577–1590. [CrossRef] [PubMed]
185. Bloem, B.R.; Henderson, E.J.; Dorsey, E.R.; Okun, M.S.; Okubadejo, N.; Chan, P.; Andrejack, J.; Darweesh, S.K.L.; Munneke, M. Integrated and patient-centred management of Parkinson's disease: A network model for reshaping chronic neurological care. *Lancet Neurol.* **2020**, *19*, 623–634. [CrossRef]
186. Tabrizi, S.J.; Flower, M.; Ross, C.A.; Wild, E.J. Huntington disease: New insights into molecular pathogenesis and therapeutic opportunities. *Nat. Rev. Neurol.* **2020**, *16*, 529–546. [CrossRef] [PubMed]
187. Deda, H.; Inci, M.; Kürekçi, A.; Sav, A.; Kayıhan, K.; Özgün, E.; Üstünsoy, G.; Kocabay, S. Treatment of amyotrophic lateral sclerosis patients by autologous bone marrow-derived hematopoietic stem cell transplantation: A 1-year follow-up. *Cytotherapy* **2009**, *11*, 18–25. [CrossRef]
188. Moviglia, G.A.; Moviglia-Brandolino, M.T.; Varela, G.S.; Albanese, G.; Piccone, S.; Echegaray, G.; Martinez, G.; Blasseti, N.; Farias, J.; Farina, P.; et al. Feasibility, Safety, and Preliminary Proof of Principles of Autologous Neural Stem Cell Treatment Combined with T-Cell Vaccination for ALS Patients. *Cell Transplant.* **2012**, *21*, 57–63. [CrossRef]

International Journal of
Molecular Sciences

MDPI

Review

Decoding Neurodegeneration: A Comprehensive Review of Molecular Mechanisms, Genetic Influences, and Therapeutic Innovations

Victor Voicu [1,2], Calin Petre Tataru [3,4,*], Corneliu Toader [5,6,*], Razvan-Adrian Covache-Busuioc [5], Luca Andrei Glavan [5], Bogdan-Gabriel Bratu [5], Horia Petre Costin [5], Antonio Daniel Corlatescu [5] and Alexandru Vlad Ciurea [5,7]

1 Pharmacology, Toxicology and Clinical Psychopharmacology, "Carol Davila" University of Medicine and Pharmacy in Bucharest, 020021 Bucharest, Romania; victor.voicu@yahoo.com
2 Medical Section within the Romanian Academy, 010071 Bucharest, Romania
3 Department of Opthamology, "Carol Davila" University of Medicine and Pharmacy, 020021 Bucharest, Romania
4 Central Military Emergency Hospital "Dr. Carol Davila", 010825 Bucharest, Romania
5 Department of Neurosurgery, "Carol Davila" University of Medicine and Pharmacy, 020021 Bucharest, Romania; razvan-adrian.covache-busuioc0720@stud.umfcd.ro (R.-A.C.-B.); luca-andrei.glavan0720@stud.umfcd.ro (L.A.G.); bogdan.bratu@stud.umfcd.ro (B.-G.B.); horia-petre.costin0720@stud.umfcd.ro (H.P.C.); antonio.corlatescu0920@stud.umfcd.ro (A.D.C.); prof.avciurea@gmail.com (A.V.C.)
6 Department of Vascular Neurosurgery, National Institute of Neurology and Neurovascular Diseases, 077160 Bucharest, Romania
7 Neurosurgery Department, Sanador Clinical Hospital, 010991 Bucharest, Romania
* Correspondence: calin.tataru@umfcd.ro (C.P.T.); corneliu.toader@umfcd.ro (C.T.)

check for updates

Citation: Voicu, V.; Tataru, C.P.; Toader, C.; Covache-Busuioc, R.-A.; Glavan, L.A.; Bratu, B.-G.; Costin, H.P.; Corlatescu, A.D.; Ciurea, A.V. Decoding Neurodegeneration: A Comprehensive Review of Molecular Mechanisms, Genetic Influences, and Therapeutic Innovations. *Int. J. Mol. Sci.* **2023**, *24*, 13006. https://doi.org/10.3390/ijms241613006

Academic Editor: Hongjun (Harry) Fu

Received: 1 August 2023
Revised: 17 August 2023
Accepted: 18 August 2023
Published: 21 August 2023

Abstract: Neurodegenerative disorders often acquire due to genetic predispositions and genomic alterations after exposure to multiple risk factors. The most commonly found pathologies are variations of dementia, such as frontotemporal dementia and Lewy body dementia, as well as rare subtypes of cerebral and cerebellar atrophy-based syndromes. In an emerging era of biomedical advances, molecular–cellular studies offer an essential avenue for a thorough recognition of the underlying mechanisms and their possible implications in the patient's symptomatology. This comprehensive review is focused on deciphering molecular mechanisms and the implications regarding those pathologies' clinical advancement and provides an analytical overview of genetic mutations in the case of neurodegenerative disorders. With the help of well-developed modern genetic investigations, these clinically complex disturbances are highly understood nowadays, being an important step in establishing molecularly targeted therapies and implementing those approaches in the physician's practice.

Keywords: neurodegenerative disorders; molecular mechanisms; genetic mutations; frontotemporal dementia; tauopathies; protein-encoding gene

1. Frontotemporal Demetia (FTD)

Frontotemporal lobar degeneration (FTLD) syndromes exhibit a complex neuropathology characterized by heterogeneity. Two key forms are seen, known as FTLD-tau and TDP, distinguished by the presence of tau or TDP-43-positive inclusions, respectively [1]. Recently, FUS-positive inclusions have also been detected in some FTLD cases. Two rare neuropathological subtypes of FTLD exist. FTLD-UPS is characterized by inclusions positive for ubiquitin but negative for tau, TDP-43, and FUS. CHMP2B mutation cases tend to display this form, while FTLD-ni lacks discernable inclusions [2]. CHMP2B gene mutations were first identified in Danish kindred who suffered from frontotemporal dementia linked to chromosome 3 (FTD-3) [3]. This mutation alters the splice acceptor site for CHMP2B's final exon, leading to the production of two novel transcripts known as CHMP2BIntron5 and CHMP2BDelta10. Furthermore, an autosomal dominant Belgian FTLD pedigree revealed another mutation known as CHMP2BQ165X, which causes premature stop codons, resulting in the protein lacking its final 49 amino acids, leading to its premature degradation [4].

Limited available data indicate that FTD occurs in approximately 11 cases per 100,000 individuals, with an incidence rate of 1.6 cases per 100,000 individuals. However, these figures appear to significantly increase from the fifth to seventh decades of life and likely underestimate actual prevalence due to misdiagnosis among older individuals. FTD accounts for 40% of dementia cases with early onset that have been confirmed through postmortem examination, although its onset typically begins between middle age and the ninth decade [5].

As serotonergic function is reduced in FTD, the use of serotonergic modulators is more apposite a priori, and modest behavioral benefit has been shown for trazodone and citalopram, though not for paroxetine [6]. Due to decreased serotonergic activity observed in FTD, serotonergic modulators appear more suitable as treatment options. Trazodone and citalopram have both demonstrated modest behavioral improvement; paroxetine did not achieve such benefits. If psychosis, aggression, or intrusive compulsions require management, then neuroleptic medications may be required, though care must be taken as individuals living with FTD can experience significant extrapyramidal side effects from even newer-generation antipsychotics. Therefore, in such instances, we suggest low-dose risperidone or quetiapine, with a strict clinical monitoring to avoid extrapyramidal side effects from antipsychotics [7].

Numerous potential strategies for treating FTD have been proposed. These include antitau antibodies and agents that stabilize microtubules, methods to increase the expression and release of progranulin, modulating autoimmunity and neuroinflammation (particularly applicable for GRN mutations), as well as using antisense oligonucleotides to silence toxic C9orf72 messenger RNA messengers—strategies which hold promise in targeting the root mechanisms while making advancements in FTD treatment [8].

GRN belongs to a family of growth factors with cysteine-rich polypeptides and can be found throughout various tissues. The GRN gene can be found 1.7 Mb upstream from MAPT's microtubule-associated protein tau-encoding gene MAPT. GRN comprises 12 exons that code for its precursor glycoprotein of 68.5 kDa, divided into seven distinct granulins of 6 kDa each for secretion purposes by cells. GRN plays its role through cell signaling and signal transduction pathways [9].

Recent studies have demonstrated that mutations of the GRN gene are responsible for frontotemporal lobar degeneration, known as FTLD-U. This form is characterized by neuronal inclusions positive for ubiquitin but negative for tau. Mutations affecting GRN may lead to functional loss via an NMD process based on analysis of its cDNA both in brain cells and lymphoblastoid cells [10].

Frontotemporal dementia was initially clinically described by Arnold Pick in 1892. Later, Alois Alzheimer identified neuropathological lesions characteristic of Pick's disease in 1911 [11]; these lesions, now known as Pick bodies, were later found in the 1960s to contain abnormal filaments made up of hyperphosphorylated microtubule-associated protein tau, and these neurofibrillary lesions closely resemble those described by Alzheimer in 1907, hence its naming after him [12].

In 1994, an autosomal dominant familial form of frontotemporal dementia with Parkinsonism was identified that was linked with chromosome 17q21.2. Missense mutations found in this form were thought to negatively impact how effectively tau protein interacts with microtubules; reduced interactions can be seen as partial loss of function, leading to destabilized microtubules that disrupt crucial cellular processes, such as rapid axonal transport [13].

Variants of FTD forms in other neurodegenerative disorders are as follows:

- PiD (Pick's disease) typically presents as behavioral variant frontotemporal dementia (bvFTD) or nonfluent variant primary progressive aphasia (nfvPPA), with motor deficits being rare. Histological characteristics of PiD include neuronal loss and swelling known as Pick cells; distinct large spherical neuronal cytoplasmic inclusions called Pick bodies may also be observed in some individuals with the disorder [14].
- Progressive supranuclear palsy (PSP) typically presents as a movement disturbance characterized by early postural instability, axial rigidity, bradykinesia, and ophthalmoplegia. Cognitive impairment may be mild; however, some cases with PSP pathology show dementia similar to bvFTD or nfvPPA. PSP pathology also involves degeneration in multiple subcortical regions, including the striatum, globus pallidus, subthalamic nucleus, midbrain, tectum/tegmentum, substantia nigra, basis, pontis, cerebellar dentate nucleus, and cerebellar peduncles [15].
- Patients diagnosed with corticobasal degeneration (CBD) often display corticobasal syndrome (CBS), characterized by bradykinesia, rigidity, dystonia, apraxia, cortical sensory signs, alien limb phenomenon, and bradykinesia [16].
- They may also show features of FTD (bvFTD or nfvPPA), displaying depigmentation in the substantia nigra, atrophy of globus pallidus, as well as focal and asymmetric cerebral cortical atrophy—histopathological features that overlap between PSP and CBD such as tau-immunoreactive glial cells as well as NCI histopathological features of both syndromes [17].
- FTD caused by mutations of the MAPT gene is an autosomal dominant form linked to chromosome 17 (FTDP-17T) that accounts for roughly 10% of familial FTD cases. MAPT pathogenic mutations include missense or deletions in exons 1 and 9–13 or mutations after exon 10 that mainly manifest themselves through behavior changes, personality shifts, cognitive dysfunction, and atypical Parkinsonism, typically seen through behavior and personality alterations, cognitive deficits, and Parkinsonian-like symptoms as well as neuropathology featuring hyperphosphorylated tau deposits within gray and white matter structures [18].
- Mutations to the VCP gene (valosin-containing protein) cause a rare familial syndrome known as inclusion body myopathy, Paget's disease of bone, and FTD with variable penetrance (IBMFD). VCP belongs to the AAA-ATPase gene superfamily and serves as a molecular chaperone involved in various cellular activities that are either directly or indirectly controlled by UPS (ubiquitin–proteasome system) [19].

This histological pattern has been observed frequently among cases associated with tau mutations in various exons, including exons 9 (K257T, L266V, and G272V), 10 and 11 (L315R and S320F), 12 (E342V K369I), and 13 (G389R). Notably, in cases of FTDP-17 linked to I260V mutation in exon 9, brain scans did not reveal typical pathological tau species such as Pick-like bodies and neurofibrillary tangles, whereas S352L mutation also did not lead to the deposition of insoluble tau adopting pathological forms [20].

Patients who carry mutations in exon 10 or intron 10 typically display neuronal and glial tau pathology with ribbon-like filaments primarily composed of 4R tau. On the other hand, missense mutations outside exon 10 tend to present selective neuronal pathology that includes all six isoforms deposited as either paired helical filaments (PHFs) or straight filaments (SFs) [21].

Studies of FTD linked to chromosome 17 have been complicated by genetic heterogeneity. Some families that clearly link with 17q21 have been found not positive for MAPT mutations despite extensive analysis of their coding regions. Those families without tau pathology might suggest the presence of another defective gene at 17q21; however, MAPT mutations or complex forms, such as chromosomal rearrangements, cannot be entirely excluded as causes. Investigating the disease mechanism in these tau-negative FTD families is of vital importance, given that most FTD patients do not exhibit tau pathology according to immunohistochemistry tests. Investigating both tau-positive and tau-negative families should contribute to our knowledge about tau and related tauopathies, leading to more effective therapeutic approaches for this devastating disorder.

Targeting TDP-43 aggregates: One therapeutic strategy involves decreasing TDP-43 clearance or inhibiting their formation to help remove accumulations of toxic aggregates that accumulate soluble TDP-43 or toxic aggregates [22].

Targeting specific mutants: For specific mutations such as C9orf72, methods designed to decrease its transcript have shown promising results. Antisense oligonucleotides (ASOs) have been extensively researched and have proven successful at decreasing this mutated transcript.

FTD-FUS (FTD fused in sarcoma) and therapeutic strategies: Studies have demonstrated that treating cell cultures with methylation inhibitors may help decrease cytoplasmic mislocalization and aggregates associated with FUS mutants in FTD cases.

Mutations associated with FTD-UPS (ubiquitin–proteasome system): For therapeutic intervention, silencing the pathologically involved genes such as mutant CHMP2B by administering siRNA treatments has been observed to reverse cellular pathology in patient fibroblasts. FTD has several pathological conditions associated with it, yet no definitive mapping exists between these conditions and particular clinical presentations. One underlying condition often associated with FTD is "frontal lobe degeneration of the non-Alzheimer type", or dementia lacking distinctive histopathology (DLDH), which helps distinguish its clinical presentation from any potential associated pathologies [23].

Human-induced pluripotent stem cells (iPSCs) are extensively employed in studies related to neurological and neurodegenerative conditions (Table 1). These iPSCs originate from specialized somatic cells, commonly from fibroblasts or peripheral blood mononuclear cells, through the heightened expression of the reprogramming agents Oct4, Klf4, Sox2, and c-Myc [24].

Table 1. Overview of targeted molecular therapies in frontotemporal dementia.

Study Focus	Key Findings and Techniques	Citation
Origin and employment of iPSCs	Derived from somatic cells like fibroblasts. Uses agents Oct4, Klf4, Sox2, and c-Myc.	[22]
iPSC models for FTLD-tau	Focuses on fibroblasts from MAPT alterations.	[25]
Patient-sourced iPSCs with 10 + 16 mutation	Noted surge in 4R tau expression, leading to increased 4R:3R tau isoform ratio.	[26]
Tau pathology in N279K iPSC-derived neurons	Observed shifts in 4R:3R tau isoform balance, with increased 4R tau levels.	[27]
iPSCs from patients with N279K mutation	Differentiated into NPCs. Used CRISPR/Cas9 for isogenic controls. Differentiated into astrocytes using Sox10 and growth factors.	[27]

A distinct study noted marked tau pathology in N279K iPSC-derived neurons. Here, the neurons displayed shifts in the 4R:3R tau isoform balance, showcasing increased 4R tau levels and augmented tau fragmentation as early as 28 days after the maturation process of iPSC-derived neural progenitor cells (NPCs) [25].

In an initial analysis of patient-sourced iPSCs possessing the 10 + 16 mutation, a significant surge in 4R tau expression was documented during neuronal development. This led to a heightened 4R:3R tau isoform proportion in these cells [26].

iPSCs from patients with the N279K mutation were differentiated into NPCs. Subsequently, CRISPR/Cas9 genome editing was utilized to produce isogenic control cells. These patient-derived and isogenic control cells underwent differentiation into astrocytes by amplifying the expression of Sox10 and introducing pro-astrocytic growth and neurotrophic factors, including ciliary neurotrophic factor [27].

2. Spinocerebellar Ataxias (SCA)

Autosomal dominant spinocerebellar ataxias (SCAs, or ADCAs) comprise an expansive group of inherited ataxias with symptoms usually appearing between 30 and 50 years of age, although specific subtypes of SCA can appear earlier or even after 60 years [28].

Recent epidemiological evidence indicates that SCAs may be more prevalent than previously estimated. Prevalence rates exceeding 5–7 in 100,000 have been documented across various geographical regions; Japan, in particular, has reported 18.5 cases per 100,000 when considering dominant, recessive, and sporadic SCAs plus familial spastic paraplegia [29]. These numbers are comparable with other less frequent motor neurodegenerative conditions like Huntington's disease or motor neuron diseases [30].

At present, over 30 SCA genes or loci have been identified, each one contributing differently to spinocerebellar neurodegeneration. Polyglutamine expansions in specific genes (SCAs 1, 2, 3, 6, 7, 17, and DRPLA) cause abnormally long polyQ tracts in encoded proteins. Noncoding expansions contribute to SCA 10 and 12 cases. Conventional mutations affect genes encoding cytoskeletal proteins (bIII spectrin in SCA5), voltage-gated potassium channel Kv3.3 in SCA13), protein kinases (tau Tubulin Kinase 2, SCA11; protein Kinase C Gamma 14), intracellular calcium channels (Inositol 1,4,5-triphosphate receptor 1), and fibroblast growth factors (FGF14 and ATPases AFG3L2 in SCA27 and 28) [31] (Table 2).

Table 2. Molecular pathways useful in spinocerebellar ataxias treatment.

Molecular Target	Pathway	SCA Subtype	Possible Medication
PP2	PP2	1, 12	PP2-mediated regulators
PRKC	PRCK	1, 14	PRKC-mediated regulators
Gene transcriptors	Multiple	1–3, 6, 7	HDACis
Aggregation	Autophagy Transglutaminase	1–3, 6, 7, 17, DRLPA	Rapamycin Cystamine
Chaperons	HSR, UPR	1–3, 6, 7, 17, DRLPA	Arimoclomol
Ubiquitin	UPS	1–3, 6, 7, 17, DRLPA	UPS derivates
Mitochondrial approach	Multiple	Any	Coenzyme Q10
Calcium activity	Calcium mechanisms	1, 6	Ca^{2+} blockers
Dopamine pathway	Dopamine	1–3, 6, 17, 27	Levodopa, anticholinergic, and dopaminergic pharmaceutical therapies
Neurotransmitters	GABA	Any	Glutamate inhibitors
Ataxins	RNAi	Any	RNAi
Caspases	Caspases	Any	Cystamine

PP2—serine/threonine protein phosphatase 2; PRKC—protein kinase C; HDACis—histone deacetylase inhibitors; DRPLA—dentatorubral pallidoluysian atrophy; HSR—heat shock response; UPR—unfolded protein response; UPS—ubiquitin/proteasome system; GABA—gamma aminobutyric acid; RNAi—RNA interference.

Dysregulation of transcription and gene expression is seen in SCA1, where ataxin 1 interacts with various transcription factors involved in transcriptional regulation, such as Lanp/Anp32a, PQBP-1, Silencing Mediator of Retinoid and Thyroid Hormone Receptors (SMRT), Boat, Gfi-1/senseless, Capicua, and Tip60. Mutant forms of ataxin 1 disrupt these transcriptional regulators' activity by disrupting gene expression levels, leading to changes in Wnt-receptor signaling and Retinoic Acid/Thyroid Hormone signaling, as well as nucleic acid binding [32].

Other SCAs such as SCA2, SCA3, SCA7, SCA17, and DRPLA, as well as atrophin 1 gene product, directly participate in transcription as components of transcriptional regulatory complexes. For example, ataxins 2, 3, and 7 interact with basal transcription factor TATA-binding protein while atrophin 1 interacts with various transcriptional regulators—such interactions interfere with CREB-dependent transcription, which has an effect on gene expression [33].

The balance between protein acetylation and deacetylation, essential to optimal cell functioning, can be disturbed when mutant proteins contain expanded polyglutamine repeats that cause pathogenesis of neurodegenerative diseases. Restoring this equilibrium through a genetic or pharmacological adjustment in histone deacetylases (HDACs) potentially offers therapeutic strategies using HDAC inhibitors against neurodegeneration [34].

Alterations of synaptic neurotransmission play an important part in the neurodegenerative mechanisms underlying SCAs. Motor dysfunction precedes neuronal death in SCA1 transgenic mice, leading to Purkinje cell dysfunction that compromises Purkinje cell functions as well as changes to synaptic plasticity [35].

SCA8 mice share similar symptoms to human patients of this disease, specifically a loss of GABAergic inhibition in the cerebellum and intranuclear inclusions with expanded polyglutamine content in Purkinje cells and other neurons [36]. This finding could provide one explanation for SCA8's lack of disease penetrance; it has been associated with large CTG repeat expansion in an antisense RNA for KLHL1 gene antisense RNA, as well as neuropathological analysis showing degeneration of Purkinje cells, inferior olivary neurons, and nigral neurons [37].

Alterations of calcium homeostasis play an integral part in SCAs. For instance, SCA6 is caused by polyglutamine expansions in the CACNA1A gene that encodes the alpha (2.1) subunit of CaV2.1 voltage-dependent P/Q-type calcium channel [38], while mutations of the protein kinase C gamma (PKC gamma) gene affect the C1 domain, essential for translocation and regulation of PKC gamma kinase activity [39]. Mutations in the PRKCG gene, which codes PKC gamma, result in the dysfunction of calcium homeostasis [40]. Finally, SCA15 causes deletions or missense mutations within the ITPR1 gene that are involved with intracellular calcium release from the endoplasmic reticulum (Table 2).

These changes in neurotransmission and calcium homeostasis play a key role in the pathogenesis of SCAs, providing therapeutic targets for intervention.

Mitochondrial Stress and Apoptosis: Polyglutamine-expanded cellular death of cerebellar neurons by polyglutamine-expanded containing proteins is preceded by the recruitment of caspases into polyQ aggregates. This is followed by the activation of caspases 3 and 9 and of mitochondrial apoptotic pathways mediated by members of the Bcl-2 family, such as Bax and Bcl-x(L). Both factors are known key components of neuronal apoptosis by regulating the mitochondrial release of cytochrome-c and Smac/DIABLO [41].

A method that has proven effective in diminishing the expression of mutant ATXN1 in vivo involves introducing AAV vectors that carry an shRNA targeting ATXN1 to the cerebellum [42].

This results in a significant reduction in mutant ATXN1 expression, leading to considerable enhancement in motor function and normalization of the structure of Purkinje cells in the transformed cells. Even though this SCA1 research utilized an shRNA targeting both the wild-type and mutant ATXN1, strategies focusing specifically on the mutant allele are being devised for numerous polyQ disorders, SCA3 included [43] (Table 3).

In a sequence of research projects using a transgenic mouse model for SCA17, Xioa-Jiang Li and his team discovered indications suggesting that the polyQ expansion in TBP modifies its DNA-binding capacity and interaction with transcription regulators. This could potentially lead to the decreased expression of certain genes, such as HSPB1 and TrkA. Hence, for SCA7 and SCA17, the evidence suggests that polyQ expansion particularly affects these transcription regulators' competency to manage the expression of certain key genes crucial for neuronal function, while not altering their regulation of the majority of other genes [44] (Table 3).

Table 3. Overview of targeted molecular therapies in spinocerebellar ataxias.

Treatment/Research Focus	Research Focus	Citation
AAV vector method targeting ATXN1	Introducing AAV vectors with shRNA targeting ATXN1 led to reduced mutant ATXN1 expression. This improved motor function and normalized Purkinje cells. Research now focuses on specifically targeting mutant alleles in polyQ disorders, including SCA3.	[42,43]
Research on polyQ expansion in SCA17	PolyQ expansion in TBP affects its DNA binding and interactions with transcription regulators. This might reduce the expression of specific genes, impacting neuronal function. However, the majority of genes remain unaffected.	[44]

3. Lewy Body Dementia (DLB)

Amyloid-beta protein has been implicated as an inducer of Lewy-type pathology, while mutations of alpha-synuclein can often result in cortical Lewy bodies as well as brainstem Lewy bodies; patients carrying these mutations frequently exhibit dementia symptoms. Other contributing factors of Lewy-type pathology include male gender, having late-onset Parkinson's disease (PD), and carrying the CYP2D6*4 allele and specific alpha-synuclein haplotypes such as L478 and Rep1, among others. Mutations in the Parkin gene cause an early-onset autosomal recessive form of Parkinson's disease characterized by changes to the substantia nigra without Lewy pathology development, suggesting that related metabolic pathways may influence vulnerable cell populations without leading to Lewy pathology development [45].

There may be an association between alpha-synuclein aggregation and Parkin mutations and proteasomal dysfunction and cell death pathways [46]; additionally, dementia found among DLB patients may be due to Alzheimer-type pathologies. Alpha-synuclein interacts with amyloid-beta protein through beta-amyloid's ability to increase fibrillization and aggregation, leading to DLB pathology characterized by both Alzheimer-type pathology as well as alpha-synuclein pathology. Notably, alpha-synuclein aggregates may contribute to Parkinsonian features while amyloid-beta protein aggregates have been associated with Alzheimer's disease (AD); hyperphosphorylated tau abnormalities may contribute to frontotemporal lobar degeneration [47].

Synucleinopathies are neurodegenerative conditions characterized by an accumulation of aggregated forms of the protein a-synuclein (a-syn) within various brain cells. Synucleinopathies, often associated with aging, are becoming increasingly prevalent due to extended life expectancy. Of all neurodegenerative conditions resulting in dementia, synucleinopathies rank second only to AD [48]. Most synucleinopathies fall under the category of Lewy body diseases (LBD), as they involve the build-up of aggregated a-syn in Lewy bodies within vulnerable neurons and Lewy neurites on neuronal processes. Parkinson's disease, Parkinson's disease dementia (PDD), and DLB are three of the more recognizable forms of Lewy body dementia; there may also be fewer common conditions [49].

Discovering a-syn as an essential component of LBD was made possible through findings of mutations of the SNCA gene (which encodes for a-syn) in familial forms of Parkinson's disease and subsequent identification as one of the major components of Lewy bodies. Studies have demonstrated a strong link between mutations of SNCA and DLB occurrence sporadically. This association can be anticipated given that Lewy bodies contain a-syn, which has been implicated as playing an essential part in the pathophysiology of DLB, PD, and PDD. Unsurprisingly, research indicates a correlation between certain regions of the SNCA gene and different phenotypes of Parkinson's disease and DLB; specifically, the 3′ region was linked with Parkinsonism while the 5′ region was linked with DLB pathology [50].

This may have an impact on the gene expression as well as brain distribution of Lewy body pathology. DLB has been linked with several genes, including SNCA, LRRK2, PSEN1, PSEN2, APP, SNCB MAPT SCARB2 GBA, and APOE. Noteworthy is the possibility that rare variants in AD-related genes (PSEN1, PSEN2, and APP) found in dementia cases could be misdiagnosed due to inadequate neuropathological assessment. Lewy body pathology is a relatively common feature of AD and may contribute to disease phenotype, leading to DLB. Recent genome-wide association studies have confirmed previously reported associations (APOE, SNCA, and GBA), and identified CNTN1 as an additional likely locus, providing an unbiased investigation of the genetics behind DLB [51].

APOE e4 allele and glucocerebrosidase (GBA) have emerged as two of the strongest genetic risk factors for DLB. APOE e4 allele is associated with an increased risk of DLB and is frequently found among individuals who exhibit mixed DLB-AD pathology, but is also overrepresented among cases of pure DLB and Parkinson's disease dementia. Studies have established a correlation between APOE e4 and more severe Lewy body pathology, particularly among individuals with lower AD pathology [52].

GBA mutations are significantly more frequent among DLB patients compared to individuals without this condition. GBA mutations have been associated with early age of onset, higher disease severity, and faster disease progression in DLB cases. Similar to APOE, GBA may play an integral role in the formation and spread of Lewy body pathology, although its precise mechanisms have yet to be identified. Furthermore, an association was recently observed between DLB and SCARB2, a gene linked with Parkinson's disease; this highlights lysosomal pathways as possible mediators of DLB development [53].

Studies conducted on DLB cases confirmed pathologically have revealed an association between GBA1 mutations and the condition. Initial investigations identified GBA1 mutations in an impressive percentage of DLB cases ranging from 3.5% to 28% depending on the specific research study conducted [54]. Subsequent multisite studies demonstrated GBA1 mutations in approximately 7.6% of DLB patients and 3.6% of individuals with both Lewy body disease and Alzheimer's neuropathology [55]. These findings demonstrate that GBA1 mutations may be an influential risk factor for DLB and may impact disease development and progression. Furthermore, in a separate clinical study focused on Parkinson's disease, GBA1 mutations had an enormous effect on its phenotype; specifically, the development of dementia symptoms in patients [56].

GBA1 mutations have been associated with earlier disease onset and death ages in DLB compared to noncarriers. DLB patients carrying GBA1 mutations generally experience disease onset approximately five years earlier compared to noncarriers, their median ages of disease onset being 63.5 and 68.9, respectively. The duration between diagnosis and death in cases carrying GBA1 mutations remains similar to noncarriers. Furthermore, they may demonstrate higher scores on the Hoehn Yahr scale and Unified Parkinson's Disease Rating Scale than noncarriers, while E326K variant but not T369M has been associated with DLB and Parkinson's disease dementia [57].

GBA1-associated Parkinsonism remains incompletely understood. Emerging evidence reveals that impaired lysosomal function, caused by deficient or mutant glucocerebrosidase enzyme, may impede alpha-synuclein degradation, an integral protein involved in DLB pathology [58].

Studies conducted using murine models and human neuronal cells demonstrate that dysfunctional glucocerebrosidase leads to accumulation and aggregation of alpha-synuclein, which, in turn, impairs trafficking and lysosomal activity of glucocerebrosidase activity, further worsening disease progression. Furthermore, autophagy disruption may contribute to its pathological mechanisms as part of DLB progression [59].

Recent biochemical studies have demonstrated similar levels of Ab40 and Ab42 in plaques between classic DLB and DLB with rapid progression (rpDLB), as well as abnormal solubility/aggregation of a-synuclein aggregates and increased binding to membranes of b-amyloid proteins in frontal cortex areas of both DLB and rpDLB cases [60].

Protein expression levels have been found to decrease significantly in DLB, with levels of NDUFA7, NDUFA10, NDUFB8, SDHB, UQCRC2, MTCO1, ATP5A, and ATP50 all showing significant reduction. No significant variations exist between rpDLB and DLB with respect to protein expression, with the exception of NDUFA7, which exhibits reduced expression even when normalized against VDAC expression levels, suggesting that rapid progression may have less of an impact on mitochondria compared to DLB [59].

Even though gene and protein expression vary significantly between DLB and rpDLB patients, the mitochondrial enzymatic activity of complexes I, II, III, and IV was significantly reduced in frontal cortex area 8 in both conditions, suggesting altered mitochondrial function as one major contributor to DLB and rpDLB pathogenesis in the frontal cortex [59].

Compared to DLB, three genes associated with purine metabolism (ENTPD2, NME3, and PRUNE) are significantly upregulated in rpDLB [59].

Additionally, reduced expression of NPM1 in the frontal cortex of DLB may indicate nucleolar stress linked to altered ribosomal biogenesis and protein expression of several transcription initiation factors at the ribosome, more so for rpDLB than DLB [61], while expression levels for elongation factors eEF1A and eEF2 were preserved across both types [62].

Notably, no significant variations were observed between DLB and controls in terms of gene expression for various cytokines and inflammatory mediators [63].

Genetic studies have identified variations in genes linked to other neurodegenerative disorders, including Parkinson's disease (SNCA) and AD (APP, PSEN1, and PSEN2) among unrelated and sporadic DLB patients [64].

DLB was only recently recognized as a distinct disease entity, and research in molecular genetics of DLB lags behind that of Alzheimer's and Parkinson's diseases.

Rivastigmine, administered orally up to 12 mg/day, was the first to undergo a comprehensive evaluation in a large-scale (n = 120) randomized, double-blind, placebo-controlled international trial [65].

Remarkably, 63% of those treated with rivastigmine exhibited a 30% or greater enhancement on a digital cognitive assessment, contrasting with the placebo group, where only 30% exhibited improvement after 23 weeks. Rivastigmine-treated patients were observed to be less anxious and experienced fewer hallucinations. A comparable trend was detected with galantamine, given orally up to 24 mg/day, in a smaller (n = 50) open-label multi-center investigation on DLB patients. Here, a notable improvement from the starting point was seen on the Clinician's Global Impression of Change (CGIC) scale (an increase of +0.5 out of 7 points; $p = 0.01$) after 24 weeks [66] Table 4.

Research using animal models for Parkinsonism and other neurodegenerative conditions has pointed towards excessive glutaminergic activity at cortical synapses. Memantine, an N-methyl d-aspartate (NMDA) receptor antagonist, benefits dementia patients by mitigating the harmful impacts of glutamate in their brains [67].

During the mild to intermediate stages of DLB, cholinesterase inhibitors such as rivastigmine, galantamine, and donepezil typically succeed in diminishing the frequency and intensity of hallucinations and delusions. Parkinsonian symptoms in DLB tend to be alleviated by the dopamine precursor levodopa, but its dosage must be moderated to prevent exacerbating visual hallucinations or the onset of unrest or excessive daytime drowsiness [68] Table 4.

Zonisamide, an anticonvulsant medication, obstructs sodium and T-type calcium channels and curtails the release of glutamate and carbonic anhydrase activity. It has secured approval in Japan for PD treatment. Its prevalent side effects involve weight reduction and diminished appetite, although these adverse reactions are infrequent [69].

Table 4. Overview of targeted molecular therapies in Lewy body dementia.

Treatment/Drug	Key Details and Outcomes	Citation
Rivastigmine	- Administered orally up to 12 mg/day. - 63% of treated individuals showed a 30% or greater cognitive improvement vs. 30% in the placebo group after 23 weeks. - Reduced anxiety and hallucinations	[65,66]
Galantamine	- Administered orally up to 24 mg/day. - Notable improvement on the CGIC scale after 24 weeks.	[66,68]
Memantine	- NMDA-receptor antagonist. - Benefits dementia patients by reducing harmful impacts of glutamate.	[67]
Levodopa	- Alleviates Parkinsonian symptoms in DLB. - Dosage moderation is vital to prevent increased hallucinations or unrest.	[68]
Zonisamide	- Anticonvulsant. - Approved in Japan for PD treatment. - Side effects include weight loss and reduced appetite, though infrequent.	[69]

4. Friedreich's Ataxia (FA)

Friedreich's ataxia is an autosomal recessive trait. At first, only one locus on chromosome 9ql3 had been identified and mapped; however, more recently, FRDA2 has been proposed for certain families unrelated to chromosome 9 [70].

Friedreich's ataxia affects approximately 2 to 4 cases per 100,000 individuals among Caucasians [71].

Nicolaus Friedreich, a German doctor, first described this condition between 1863 and 1877 in various reports. Two females and seven males from three sibships showed symptoms such as ataxia, dysarthria, sensory loss, muscle weakness, scoliosis, foot deformity, and cardiac symptoms. Onset typically occurred around puberty.

Friedreich's ataxia patients most commonly exhibit an expansion of a GAA trinucleotide repeat within an Alu element within the first intron of the frataxin gene. Roughly 96% are homozygous for this GAA expansion while 4% possess both expanded allele and point mutation in their genetic coding sequence. Many missense mutations of frataxin occur in its carboxy-terminal half, an important functional region. Such missense mutations include L106S, D122Y, G130V, R165P/C, and L182F, which have been linked with milder clinical presentations among individuals heterozygous for such mutations [72].

The FRDA gene spans 80 kb of genomic DNA and comprises seven exons. Exons 1–5a encode a major transcript, 1.3 kb mRNA, which translates to frataxin protein with a 210-amino-acid sequence. Frataxin mRNA expression has been detected most prominently in the spinal cord and heart tissue as well as liver, skeletal muscle, and pancreas tissues; its distribution correlates with pathological features observed among those suffering from this condition [73].

Friedreich's ataxia is an autosomal recessive disorder with variable symptoms, unlike many other recessive traits. Deviant variants have been mapped back to one locus on chromosome 9, including late-onset Friedreich's ataxia (LOFA), with symptoms beginning 25 years after diagnosis; Friedreich's ataxia with retained reflexes (FARR); and the Acadian form of Friedreich ataxia, which progresses more slowly without cardiomyopathy or diabetes mellitus [74].

Friedreich's ataxia can be diagnosed through analysis of its GAA repeat. Molecular testing has uncovered unexpected presentations of Friedreich's ataxia, such as pure sensory ataxia, spastic paraparesis, and chorea [75]. Although Geoffrey et al. proposed highly specific criteria for diagnosis, molecular testing may help avoid underdiagnosis in cases with very late-onset symptoms [76].

There have been various medications developed to alleviate oxidative stress caused by intracellular iron imbalance, which is either the result of frataxin deficiency directly or secondary effects thereof. Iron chelators such as desferoxamine and antioxidants like ascorbic acid or coenzyme Q10 analogs may be effective at decreasing iron overload in mitochondria; however, desferrioxamine only chelates iron from extracellular fluid and cytosol sources, not necessarily those in mitochondria. Therefore, its use should be limited strictly to controlled therapeutic trials only [77].

Studies have revealed an increase in urinary 8-hydroxy-2′-deoxyguanosine (8OH2′dG), an indicator of DNA damage, in those suffering from Friedreich's ataxia. This finding indicates that reactive oxygen species production could play a part in its pathogenesis. Treatment with antioxidant idebenone produced a significant drop in normalized urinary concentrations of 8OH2′dG after just one dose in one small group of patients; further longitudinal research is needed to establish its efficacy as a biological marker and therapeutic benefits of other antioxidants or similar agents against this form of neurodegeneration [78].

Friedreich's ataxia most often strikes young individuals of both genders. A systematic study with 115 patients from 90 families determined the mean age of onset as 10.52 ± 7.4 years, and death occurred 37.54 ± 14.35 years after initial symptoms appeared. Harding et al. reported these findings in their clinical and genetic investigation of 90 families; additionally, they investigated early diagnostic criteria as well as intrafamilial clustering of clinical features [79].

Not all patients of Friedreich's ataxia possess homozygous GAA expansions in their frataxin gene. Two to four percent are compound heterozygotes containing one GAA expansion with either a point mutation or deletion on another allele; these individuals may exhibit some unusual symptoms, including less dysarthria than with homozygotes but a greater tendency toward optic pallor. Cossee et al. provided these data when investigating point mutations and clinical presentation among Friedreich's ataxia patients [80].

Frataxin is a highly conserved protein essential for maintaining iron homeostasis and mitigating intracellular oxidative stress by modulating levels of reactive oxygen species (ROS). Cells lacking frataxin experience mitochondrial iron accumulation, an excess of reactive oxygen species, impaired antioxidant enzyme activity, and an increase in their susceptibility to oxidative stress. Marmolino et al. conducted research that demonstrated how an azelaoyl PAF PPAR-g agonist, known as APAF (azelaoyl PAF), could increase expression of frataxin mRNA and protein in primary fibroblasts from both healthy individuals and Friedreich's ataxia patients; similar effects were also observed with neuroblastoma cell line called SKNBE, suggesting that APAF might regulate the expression of the FXN gene in tissues relevant to disease [81].

Past studies have investigated the efficacy and safety of BML-210 as a histone deacetylase inhibitor compound to treat Friedreich's ataxia in mice carrying GAA230 repeats, using human lymphoblastoid cells and 3-nitropropionic acid as examples of such compounds with potential. Their results suggested promise as potential therapies; however, some have demonstrated mitochondrial toxicity, so further investigation should take place to evaluate their safety and effectiveness for humans [82].

Based on previous observations, fibroblasts of patients with Friedreich's ataxia (FRDA) exhibited decreased cytoskeletal organization and elevated levels of glutathione bound to cytoskeletal proteins. Furthermore, abnormal immunoreactivity for glutathionylated proteins such as GS-Pro (glutathione bound to proteins) was seen both in gray matter neurons as well as white matter cells and axons of patients, suggesting abnormal protein glutathionylation attributed to FRDA. These results suggest the presence of oxidative stress due to reduced frataxin expression; glutathionylated proteins serve as biomarkers of such effects of stress [83].

Studies involving mitochondria isolated from adipocyte-like cells that overexpress frataxin demonstrated an association between frataxin and mitochondrial calcium (Ca^{2+}).

Frataxin expression led to activation of the respiratory chain, increased membrane potential and production, elevated ATP production, as well as the uptake of mitochondrial Ca^{2+}. This increase may have stimulated tricarboxylic acid cycle activity, indicating frataxin's connection with energy conversion through mitochondria [84].

Frataxin-deficient dorsal root ganglia sensory neurons produced an increase in intracellular free Ca^{2+} following its reduction, leading to caspase activation, increased nitrosation, and activation of pro-apoptotic gene Bax activation—events observed in FRDA patients as fragmentation of alpha-fodrin, axonal degeneration, and eventual apoptosis of these neurons, as seen through altered calcium homeostasis and cell death. Furthermore, it was determined that depletion could be prevented via its BH4 domain of Bcl-xL protein, which plays an active role in both cell death as well as altered calcium homeostasis regulation [85].

Alterations of mitochondrial bioenergetics and Ca^{2+} homeostasis of FRDA neurons could have profound ramifications on axonal transport and distribution along axons and synapses, which is particularly noteworthy given the extensive axonal network. Studies on Drosophila larvae focusing on mitochondrial transport showed that decreased frataxin expression led to early decreases in membrane potential, defects in retrograde transport (primarily affecting retrograde direction), abnormal accumulation of mitochondria at synapses, and dying-back neuropathy. These findings suggest that mitochondria may contribute significantly to the impairment of axonal transport in FRDA patients [86].

5. Progressive Supranuclear Palsy (PSP)

PSP is an uncommon neurodegenerative disease with an estimated prevalence of about 7–10 individuals per 100,000, considered one of the more frequent atypical Parkinsonian syndromes and marked by motor, behavioral, and language abnormalities [87].

PSP displays clinical heterogeneity and has many different phenotypes, with Richardson's syndrome (PSP-RS), also known as Steele–Richardson–Olszewski syndrome, being the classic example, first described in 1964 and first identified by its symptoms of vertical gaze palsy, pseudobulbar palsy, nuchal dystonia, and dementia [88].

Pathologically, PSP can be described by the presence of 4-repeat (4R) tau inclusions such as neurofibrillary tangles, neuropil threads, tufted astrocytes, and oligodendroglial coiled bodies, predominantly found in basal ganglia, diencephalon, and brainstem, with prominent involvement of globus pallidus subthalamic nucleus and substantia nigra. PSP is the most prevalent primary 4R-tauopathy, with abnormal accumulation beginning prior to the presymptomatic phase. Although most commonly seen sporadically, there have been familial forms as well as a few pedigrees showing characteristics similar to PSP, which suggest an autosomal dominant inheritance pattern [89].

Mutations in the microtubule-associated protein tau (MAPT) gene have been implicated in PSP, with approximately 0.6% to 14.3% of cases carrying MAPT mutations. Individuals diagnosed with PSP often carry multiple mutations that contribute to its development and worsen its disease phenotype [90].

One mutation, known as the S305S mutation of the tau gene, has been linked with mild cortical atrophy and extensive subcortical neurofibrillary tangles, consistent with the neuropathological diagnosis of PSP. Moreover, neuropathology is distinguished by a prevalence of subcortical neurofibrillary and glial tangles, neuropil threads composed mainly of hyperphosphorylated tau filaments, and thread-like neuropils that often appear. Pathological changes occur predominantly in the basal ganglia, brainstem, and cerebellum without significant amyloid deposition. Neuronal loss has been noted in both output nuclei of the basal ganglia as well as the focal frontal and temporal cortical regions; frontotemporal atrophy remains minimal while specific pathological features, including ballooned neurons and tufted astrocytes, can be observed within affected brain regions [91].

PSP neuropathology serves as an authoritative benchmark, offering insight into its underlying mechanisms and providing a definitive diagnosis.

Transactivation-responsive DNA-binding protein of Mr 43 kDa (TDP-43) is a nuclear protein that plays an integral part in transcriptional repression and alternative splicing. It was initially identified as one of the major contributors to abnormal protein accumulations found in the frontotemporal cortex and motor neurons of people diagnosed with frontotemporal lobar degeneration-U (FTLD-U) or amyotrophic lateral sclerosis (ALS), marked by both ubiquitin-positive and tau-negative inclusions [92].

TDP-43 accumulation was initially believed to be limited to FTLD-U and ALS, but subsequent studies have demonstrated abnormal accumulation in certain cases of other neurodegenerative diseases, including AD, Parkinson's disease with and without dementia, DLB, etc.

Recent research demonstrated that a significant proportion (26% to be exact) of patients with PSP, typically classified as tauopathy, had abnormal accumulations of TDP-43 in the limbic system, suggesting that pathological TDP-43 accumulation can also occur within this form of tauopathy and could contribute to hippocampal sclerosis within this form of PSP cases [92].

Early studies involving TDP-43 immunohistochemistry and immunoblot analysis with nonspecific antibodies suggested that PSP cases do not exhibit abnormal accumulation of TDP-43 protein. However, more specific antibodies and biochemical analysis revealed alterations in TDP-43 pathways similar to those seen in FTLD-TDP and ALS patients [93].

However, previous studies of AD suggest that the presence of TDP-43 pathology is associated with later age of onset and death as well as diminished cognitive function [94].

Some cases of early-onset autosomal dominant PSP have been linked to mutations (G303V) in the tau gene that result in the overexpression of 4R isoform tau and its hyperphosphorylation, producing specific protein bands on analysis. Mutations in this gene are relatively uncommon but certain polymorphisms and haplotypes have been found associated with PSP, while individuals carrying mutations may exhibit atypical features of PSP [95].

Levodopa, combined with a peripheral decarboxylase inhibitor like carbidopa or benserazide, stands as the primary drug used for dopaminergic replacement therapy in PSP. Typically, the PSP–Parkinsonism (PSP-P) subtype shows a more pronounced response. Symptoms like bradykinesia, rigidity, and tremor in any PSP phenotype might respond comparably to those in PSP-P, but improvements in postural stability are less likely. Retrospective research suggests that 20–30% of pathologically confirmed PSP patients and 20–40% of clinically diagnosed PSP patients have observed positive effects from levodopa, either as a standalone or in tandem with another dopaminergic drug like amantadine [96] Table 5.

Table 5. Overview of targeted molecular therapies in Progressive Supranuclear Palsy.

Treatment/Drug	Key Details and Outcomes	Citation
Levodopa with carbidopa or benserazide	- Primary drug for dopaminergic replacement in PSP. - PSP-P subtype shows a better response. - 20–30% of pathologically confirmed and 20–40% of clinically diagnosed PSP patients have positive effects when combined with another dopaminergic drug like amantadine.	[96]
Donepezil	- Can exacerbate motor symptoms while improving cognitive abilities in PSP patients.	[97]
Amitriptyline	- Anecdotal claims of improvement in motor functions, but limited data on its effects on gait issues in PSP.	[98]
Amantadine	- NMDA-receptor antagonist with inconsistent results in PSP.	[98]
Carbidopa-levodopa	- Indicated as the most effective agent in a study of 147 patients, improving Parkinsonism in 20% of the treated group.	[98]

Despite anecdotal claims of marked improvement in motor functions on low doses of amitriptyline, data on pharmacological treatments for gait issues in PSP remain scanty. The ineffectiveness or lack of response of gait problems or freezing to levodopa and other dopaminergic drugs is a common observation. Amantadine, an NMDA-receptor antagonist, has shown inconsistent results [97] Table 5.

A retrospective analysis involving 147 clinically diagnosed patients indicated that carbidopa–levodopa was the most potent agent, enhancing Parkinsonism in 20% of the treated group. Dopamine agonists, amantadine, and selegiline did not yield favorable outcomes [98].

Fluctuations related to dosage in dopaminergic response have not been distinctly reported for PSP/CBS. Thus, there is no requirement for controlled-release or extended-release versions or for monoamine oxidase B inhibitors or catecholamine O-methyltransferase inhibitors in such contexts [99].

6. Corticobasal Degeneration (CBD)

Rebeiz, Kolodny, and Richardson first introduced CBD as an entity between 1967 and 1968. Three patients displayed symptoms including slow and awkward voluntary limb movement, tremor, dystonic posturing, stiffness, lack of dexterity, and the sensation of "numbness or deadness" in one or both affected limbs. Over time, these symptoms intensified and included gait disorder, limb rigidity, impaired position sense, and other sensory deficits. Pathological examination revealed an asymmetric frontoparietal cortical atrophy and neuronal loss with associated gliosis. Nissl substance was absent in affected neurons, thus giving rise to the term "achromatic." Swollen eosinophilic and hyaline appearing pyramidal neurons were observed in the third and fifth cortical layers; there was significant loss of pigmented neurons from substantia nigra, variable involvement among subcortical neurons, as well as secondary degeneration of corticospinal tract [100].

CBD usually presents itself in middle to late adulthood, with the average age of onset being 63 years old [101]; however, cases have been reported as early as 40 years old and confirmed through pathology at 45 years. Both males and females can be affected, although some studies have noted a greater number of women [102].

Studies show that tau protein in CBD comes from transcripts of exon 10 on chromosome 17. Pathological tau displays abnormal phosphorylation and solubility properties as well as abnormal filamentous structures. Ballooned neurons, neurofibrillary tangles, neuropil threads, grains, glia, and neuronal inclusions all display tau immunoreactivity; neurofibrillary lesions showing tau immunoreactivity can often be found in locations like locus ceruleus, raphe nuclei, tegmental gray matter, or even the substantia nigra [103].

CBD, with its characteristic tau-immunoreactive astrocytic lesion that resembles AD, shares several traits with Alzheimer's. These tau-positive astrocytic plaques are argyrophilic structures often observed in CBD, although they have also been identified in PSP. Unlike AD, however, these plaques in CBD originate from glial cells rather than being amyloidogenic in origin [103]. CBD is associated with specific tau-immunoreactive lesions in neurons and glial cell processes. Minimal pathological features for diagnosing CBD may include cortical and striatal tau-positive neuronal and glial lesions, particularly astrocytic plaques and thread-like structures in white matter; neuronal loss in specific cortical regions and the substantia nigra may also indicate its presence [104].

CBD and PSP share characteristics, as evidenced by identical H1 and H1/H1 tau haplotype statuses seen across both disorders. This shared genetic predisposition suggests that both disorders share a common genetic factor contributing to their development; however, these may still represent separate nosological entities caused by different etiological factors occurring simultaneously or extreme forms of the same disorder arising due to variations in genetic background or specific trigger agents [105].

Studies have uncovered decreased complex III activity and an increase in markers of lipid oxidative damage among PSP cybrid cell lines containing mitochondrial DNA from PSP patients [106], but contradictory reports regarding aconitase activity across both types of cell lines suggest that they do not always accurately reflect changes occurring in human brain tissue postmortem [107].

Patients presenting initially with primary progressive aphasia (PPA) or frontotemporal dementia may later develop movement disorders or symptoms characteristic of CBD, typically two to seven years after the initial manifestation occurred.

The treatment of Parkinsonism in CBD/CBS often involves levodopa, though it generally yields minimal effects. Some reports suggest a brief mild to moderate improvement [101] Table 6.

Table 6. Overview of targeted molecular therapies in corticobasal dementia.

Treatment/Drug	Application/Condition	Key Outcomes	Citation
Dopaminergic treatments, benzodiazepines, anticholinergics	Parkinsonism	Typically ineffective and might lead to side effects such as cognitive decline.	[98]
Levodopa	Parkinsonism in CBD/CBS	Minimal effects, with some reports indicating brief mild to moderate improvement.	[101]
Botulinum toxin injections	Dystonia in CBS	Effective for pain relief, enhancing hygiene, countering secondary contractures, and occasionally improving limb functionality in early disease stages.	[108]
Benzodiazepines (e.g., Clonazepam), anticholinergic drugs, muscle relaxants	Dystonia	Oral medications like these are often tried but seldom prove effective.	[109]
Clonazepam	Myoclonus	Shows a favorable response, especially effective.	[109]
Levetiracetam, gabapentin, valproic acid	Myoclonus	Can be beneficial, with some accounts of use.	[109]

Other therapeutic approaches for Parkinsonism, encompassing other dopaminergic treatments, benzodiazepines, and anticholinergics, typically do not offer much assistance and can sometimes lead to side effects, like cognitive decline [98] Table 6.

Dystonia tends to respond best to precise botulinum toxin injections. In the context of CBS, botulinum toxin shots can be administered to alleviate pain, enhance hygiene, counteract secondary contractures, and occasionally bolster limb functionality during the early disease stages. Although some practitioners might experiment with oral medications like benzodiazepines, anticholinergic drugs, or muscle relaxants, they seldom prove effective. The principal line of treatment remains botulinum toxin injections [108] Table 6.

7. Wilson's Disease

Wilson's disease, also referred to as hepatolenticular degeneration, is an autosomal recessive genetic condition caused by mutations in the ATP7B gene. Reports of individuals suffering from symptoms including liver cirrhosis and neuronal degeneration date back as far as 1850; however, Kinnear Wilson first formalized this clinical entity in 1912 by identifying seven familial patients suffering from progressive degeneration of their lenticular nuclei and liver cirrhosis upon autopsy. Wilson speculated that there may have been toxic chemicals affecting liver cirrhosis, but copper's role in Wilson's disease pathogenesis was only recognized 35 years after his initial observations [110].

Scheinberg and Gitlin first described Wilson's disease in 1952 after noting a deficiency of ceruloplasmin in affected individuals' serum, a diagnostic test that remains important today. Wilson's disease can occur anywhere worldwide and affects individuals aged 3 to 80; children and adolescents seem more prone than adults, due to differences in estrogen levels and iron metabolism [111].

Wilson's disease affects approximately 1 out of every 30,000 individuals globally, with an approximate carrier rate of approximately 0.011 and a gene frequency of 0.56. Recent clinical studies have reassessed its prevalence [112].

Hepatocytes, the primary target for copper uptake and accumulation in the liver, possess a remarkable ability to assess intracellular copper status and control its excretion through bile [113]. This regulation is enabled by an ATPase encoded at Wilson's disease locus that transports copper out of cells into trans-Golgi networks [114]. Hepatocytes express this enzyme widely. As copper levels in hepatocytes increase, ATPase moves from its original location within the trans-Golgi network to a vesicular compartment near the canalicular membrane. Copper accumulates within this vesicular compartment, leading to decreased cytoplasmic copper levels and prompting ATPase redistribution back towards the trans-Golgi network [115]. Ultimately, copper excretes into bile for excretion from cells, effectively maintaining intracellular copper homeostasis while simultaneously rapidly eliminating excess cytosolic copper levels [116] Figure 1.

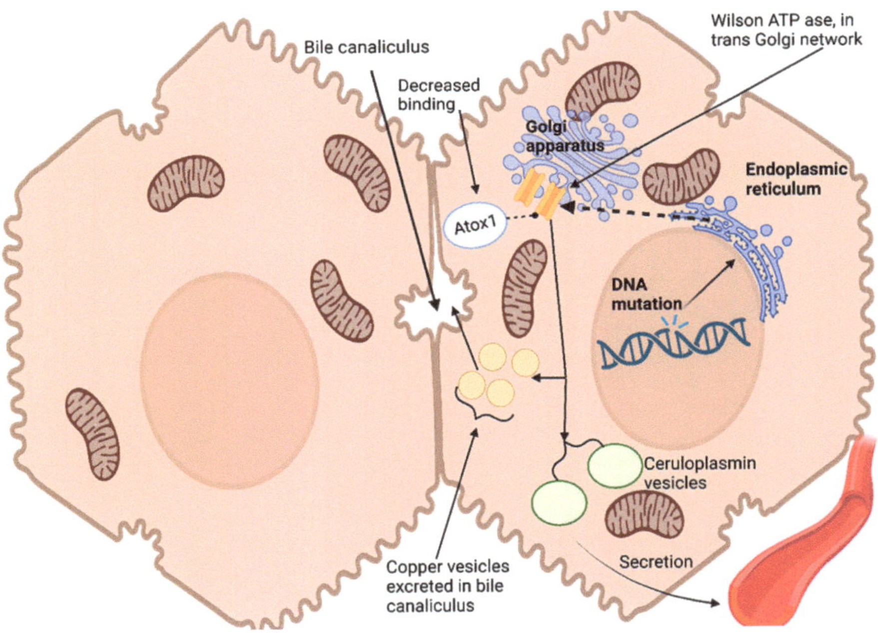

Figure 1. The proposed model depicts various potential molecular mechanisms responsible for the disturbed copper homeostasis observed in hepatocytes expressing Wilson's disease ATPase mutations. Under normal circumstances, this ATPase normally resides in the trans-Golgi network, but when exposed to high copper levels, it relocates into a vesicular compartment of its cell's cytoplasm, where it accumulates copper before returning to its trans-Golgi network and eventually excreting copper into its bile.

Copper has no direct effect on the production or secretion rate of ceruloplasmin, an essential protein responsible for copper transport in blood plasma, yet failure to incorporate copper results in secreting non-functional forms that quickly degrade in plasma. Wilson ATPase plays an essential role in transporting copper to secretory pathways of hepatocytes; impaired movement into these compartments results in the disruption of ceruloplasmin production, leading to significant drops in serum ceruloplasmin levels due to Wilson's disease [117].

Genetic studies conducted with yeast have provided insight into the role of atox1 protein as a copper chaperone for delivery to the secretory pathway. Atox1 interacts with Wilson ATPase in an atox1-dependent fashion, and mutations that conserve copper-binding domains (an amino terminus of Wilson ATPase) lead to decreased atox1 binding; these observations suggest that impaired delivery could be the source of Wilson's disease in patients carrying such mutations, thus making atox1 an essential copper chaperone that plays critical roles in copper homeostasis regulation and homeostasis regulation [113].

Wilson's disease (WD) patients do not all carry causative mutations. One such gene, located on 13q14.3, spans 80 kb of genomic space and contains 21 exons encoding ATP7B or Wilson ATPase, which transport copper via P-type ATPase transporters. This ATPase contains 1465 amino acids and is synthesized in the endoplasmic reticulum before localizing to the trans-Golgi network (TGN) of hepatocytes. However, its expression has also been detected in various other tissues including the brain, kidney, lung, and placenta. Over 500 mutations of the ATP7B gene have been identified, most being extremely rare [118]. Mutations include missense or nonsense single-nucleotide variants (60%), insertions/deletions (26%), and splice site mutations (9%). Compound heterozygotes are often observed among WD patients where multiple different mutations exist on both gene copies [119].

ATP7B gene mutations vary significantly across geographical regions. Some mutations, like H1069Q and R778L, are particularly prevalent among European and Asian populations, while most other reported mutations have an incidence rate below 10%. Hotspots for WD gene mutations in Europe can be found between exons 8 and 18. In India, however, mutations within exons 2–5 have been linked with severe phenotypes [119].

H1069Q mutation in exon 14 is the most frequently occurring type, occurring in 30–70% of WD patients. This mutation involves replacing histidine from the SEHPL motif in the N domain with glutamic acid, leading to misfolding and abnormal phosphorylation of the P domain, resulting in reduced ATP binding affinity of half the normal level; additionally, it shows decreased thermal stability as endoplasmic reticulum abnormally migrates toward TGN localization [120].

Modifier genes are known to exacerbate or alleviate phenotypes caused by other disease-causing genes. One such modifier, apolipoprotein E (ApoE), found on 19q13.2, has been strongly linked with Western dry eye disease. ApoE exists as three alleles: e2, e3, and e4. Of these alleles and genotypes, ApoEe3 and e3/3 are most often found together and play an essential role in lipid metabolism, while ApoEe4 is linked with neurodegenerative disorders like AD as well. Evidence shows that patients with the e3/3 genotype may delay their onset age due to the neuroprotective and antioxidative properties of ApoE3, while ApoEe4 has been associated with earlier-onset symptoms of WD [121].

Penicillamine was one of the first successful oral drugs designed to treat Wilson's disease, providing many patients with lifesaving therapy. It has become widely prescribed, but nowadays, this can become an issue when physicians prescribe it while being unaware of safer pharmacological options. Penicillamine works by chelating copper from liver cells. Unfortunately, due to its toxicity, it also acts as a known teratogen, which may lead to neurological symptoms in unborn babies if administered as initial treatment in pregnant patients with neurological symptoms [122].

Among patients treated with penicillamine, 50–75% experience further neurological deterioration and 25–35% do not return to pre-penicillamine levels of functioning; one out of every four treated may suffer permanent additional disturbances and severe disability as a result of taking penicillamine due to copper mobilization from their livers leading to further elevation in brain copper levels. This worsening of symptoms could be attributed to its mobilization, leading to increases in brain copper levels as a result.

Trientine was first introduced as an alternative to penicillamine in 1982 as an option for patients who are sensitive to it, specifically those intolerant of it. Like penicillamine, trientine also acts as a chelator, inducing urinary excretion of copper; however, trientine is generally considered safer as its dose and administration method remain similar. Another advantage is that trientine does not trigger hypersensitivity reactions associated with penicillamine. Moreover, trientine has been extensively studied as an initial therapy instead of penicillamine. Nevertheless, trientine is generally seen as a possible alternative when administered instead of penicillamine for those intolerant of penicillamine treatment plans [123].

Tetrathiomolybdate (TM) is an experimental treatment for Wilson's disease. When taken with food, TM works by binding copper and protein together and producing an imbalance that prevents food absorption as well as endogenously secreted copper from reaching your system, leading to reduced overall copper levels in your body. When taken between meals, TM absorbs into the bloodstream, where it forms complexes with blood copper and albumin, which makes this complexed copper non-available for uptake by cells, making this non-toxic [124].

Dysarthria is one of the primary neurologic manifestations of Wilson's disease, affecting approximately 89% to 97% of patients who exhibit neurological involvement. Dysarthria typically takes on a mixed form characterized by spastic, ataxic, hypokinetic, dystonic, and spastic components, often congruent with dystonia as it displays dystonic qualities with strained or harsh voice qualities; conversely, those suffering from Parkinsonism may exhibit hypokinetic properties when speaking [125].

Dystonia is a characteristic of Wilson's disease that is found in approximately 11–65% of patients, often manifesting itself through focal, segmental, multifocal, or generalized dystonia ranging in severity from mild to severe. One notable manifestation is "risus sardonicus", an exaggerated and forced smile caused by dystonic facial muscles [126].

Wilsonian tremor can be observed in 22–25% to 55% of individuals diagnosed with neurologic Wilson's disease. This tremor may occur at rest, when taking certain postures, or while performing actions. Parkinsonism, marked by bradykinesia, cogwheel rigidity, or imbalance, affects 19–62% of patients living with Wilson's disease [125].

Wilson's disease patients experience seizures at a rate that is ten times higher than in the general population; however, seizures rarely present themselves initially and have often been associated with initiating chelating therapy [127].

Kayser–Fleischer rings, copper deposits in the limbic region of the cornea, can be found in nearly 100% of those suffering from neurologic Wilson's disease and 50% with presymptomatic and hepatic forms of it. Sunflower cataracts do not impair vision but must be detected with an ophthalmoscope; their visibility cannot be observed through naked-eye viewing alone [128].

Distinguishing Wilson's disease from other common and rare neurologic conditions is crucial, given that Wilson's is treatable. Young individuals experiencing movement issues should consider it as a possible diagnosis. Wilson's disease can often be confused with essential tremor, young-onset Parkinson's disease, and generalized dystonia, all common neurologic conditions that often present themselves similarly. Rare juvenile genetic extrapyramidal disorders, including Huntington's disease, Hallervorden–Spatz's disease, idiopathic torsion dystonia, chorea-acanthocytosis, and benign familial chorea can all present with symptoms similar to Wilson's disease. Wilson's disease should certainly be considered in any attempt at differential diagnosis; however, other potential causes must first be explored. Psychological abnormalities associated with Wilson's disease could easily be misinterpreted as affective disorders, early schizophrenia, or drug abuse, requiring careful differential diagnosis to ensure appropriate diagnosis [128].

Dimercaptosuccinic acid (DMSA) is an alternative maintenance therapy for Wilson's disease, with lower urinary copper excretion and more severe adverse effects than dimercaptopropanol. Na-DMPS (sodium dimercaptosuccinate) has low toxicity but may not be appropriate for advanced Wilson's disease or critically ill patients; calcium disodium edetate (CaNa2-EDTA) also chelates copper less effectively compared with zinc or iron. Therefore, DMSA/NaDMPS is recommended as alternative maintenance therapy when D-penicillamine cannot be tolerated [129].

Zinc has long been recognized as an effective adjuvant therapy to decrease copper absorption in Wilson's disease patients. Studies have demonstrated that oral zinc sulfate or zinc gluconate significantly increased urinary copper excretion, and clinical symptoms improved dramatically for those using zinc sulfate as part of maintenance therapy during follow-up. Zinc also has shown great efficacy with both preclinical patients as well as those using copper chelators; they even show promising clinical efficacy among asymptomatic and preclinical patients and those in maintenance after starting with copper chelators treatment [130].

Studies on Wilson's disease using magnetic resonance imaging (MRI) have extensively explored its structural changes, which correspond with abnormal postsynaptic dopaminergic function as measured by [11C] raclopride PET/IBZM-SPECT scans [131]. Wilson's disease patients typically do not respond well to levodopa treatment due to abnormal dopaminergic function, possibly explaining their subpar response to it. PET and SPECT studies have also indicated the presence of presynaptic dopaminergic damage comparable to that seen in Parkinson's patients; this damage may contribute to occasional levodopa responsiveness. Such presynaptic damage could have an impactful influence on neurological symptoms in particular patients [132].

Wilson's disease causes presynaptic damage that varies widely among its patients, leading to inconsistent reports of levodopa responsiveness and no correlation between extrapyramidal symptoms and MRI findings [133]. There is a notable relationship between extrapyramidal symptoms and olfactory dysfunction. Those suffering from marked deficits could potentially benefit from dopaminergic medications; however, clinical experience with dopamine-antagonistic medication has not proven very fruitful, further disproving any essential role played by this pathway in neurological-type Wilson's disease patients [134].

An intriguing possibility is that therapy-induced increases in dopamine D2-receptor binding could signal functional recovery of the striatal dopaminergic system, possibly linked to olfactory function. Therefore, metabolically induced damage of nigrostriatal structures could contribute to Wilson's disease as olfactory dysfunction is caused by metabolic damage to specific structures within basal ganglia that process odorous stimuli; further research is necessary in this area of dopaminergic function, structural changes, and deficits within Wilson's disease [135].

Wilson's disease stands out from its counterparts in that its usual progression pattern does not apply; rather, there appears to be an apparent incongruence [136]. Pathological studies have revealed no damage in the substantia nigra region commonly affected by Parkinson's disease [137]; instead, research shows nigrostriatal dopaminergic damage occurs more commonly at nerve terminals rather than cell bodies, which corresponds to the functional nature of lesions found in Wilson's disease [137].

Wilson's disease usually results in lower total serum copper levels due to reduced ceruloplasmin levels; however, under certain circumstances, such as severe liver injury or acute hepatic failure due to Wilson's disease, these may rise significantly due to copper released from damaged liver tissue stores and stored reserves [134].

Copper can be detoxified when it binds covalently to high-affinity copper-binding proteins that act as chaperones or endogenous chelators within cells, acting as chaperones or endogenous chelators within cells. Copper bound to ceruloplasmin in the bloodstream is non-toxic, while remaining serum copper, known as free copper, may become toxic through less-tight binding to proteins and being mobilized more readily, potentially leading to copper toxicity and contributing to reactive oxygen species formation. Additional research implicates acid sphingomyelinase and ceramide in liver cell death due to mitochondria being the prime target in liver cell death; more recent research has indicated acid sphingomyelinase and ceramide in liver cell death [138].

Wilson's disease does not typically involve changes to either the olfactory bulb or piriform cortex; instead, their symptoms could be related to direct, metabolically damaged dopaminergic nerve terminals projecting to olfactory areas in the brain that project these scents [139].

Although chelator-induced neurologic worsening and free copper toxicity appear to have an apparent relationship in Wilson's disease, further clinical and experimental data must be collected to fully establish this concept. Reacting to biochemical readings before clinical worsening could potentially benefit patients; however, certain limitations must also be considered [140].

Calculating non-ceruloplasmin-bound copper involves precise measurements of both copper and ceruloplasmin levels; however, inflammation or immunologic assays that detect nonfunctional protein precursors (e.g., apo-ceruloplasmin) can lead to overestimation of ceruloplasmin levels, leading to inaccurate values for non-ceruloplasmin-bound copper and making their calculations useless [141].

To address this challenge, direct measurement of the lower-affinity or non-covalently bound copper pool has been proposed as an approach to assess copper toxicity in Wilson's disease more easily and reliably. Ultrafiltration or other techniques could be employed for this purpose. Measuring serum-free copper may offer a simpler and more reliable means of doing so [141].

Exchangeable copper (CuEXC) and its relative exchangeable copper (REC), both promising approaches that could offer valuable insights into copper dynamics and disease progression, are also promising approaches. But for these strategies to become truly useful clinical tools, further research and evidence must be collected and utilized effectively in clinical practice [142].

A model incorporating the R778L mutation was effectively differentiated into hepatocyte-like cells (HLCs), suggesting the successful creation of a Wilson's disease model from hESCs without needing a patient sample. Notably, HLCs with the R778L mutation displayed a heightened susceptibility to excessive copper, making them more sensitive to copper-induced cytotoxicity than the standard HLCs. This newly developed model, which does not require a patient sample, offers a fresh avenue for testing the efficacy of drugs before their clinical use [143] Table 7.

Table 7. Overview of targeted molecular therapies in Wilson's disease.

Model/Component	Description/Effect	Citation
R778L mutation model	Differentiated into hepatocyte-like cells (HLCs)	[143]
HLCs with R778L mutation	Showed heightened susceptibility to excessive copper, more sensitive to copper-induced cytotoxicity	[143]

8. Niemann–Pick Disease

Since the early 1980s, Niemann–Pick disease has been divided into two distinct entities. Acid sphingomyelinase deficiencies encompass types A and B, and Niemann–Pick disease type C includes types C and D. This illness results from deficiencies in either NPC1 or NPC2 transport proteins, leading to lipid storage diseases [144].

Acid sphingomyelinase-deficient Niemann–Pick disease, more commonly referred to as ASM-deficient NPD or ASM deficiency, is an autosomal recessive disorder caused by mutations of the SMPD1 gene. A deficiency of ASM lysosomal enzyme leads to progressive accumulation of sphingomyelin throughout body organs in all forms and neuronopathic forms, with secondary accumulations of other lipids as well [145].

History shows us that ASM deficiencies can generally be divided into two broad types, known as types A and B. Although intermediate forms have also been observed, ASM deficiency represents an entire continuum that ranges from severe neuronopathic forms such as type A, more prevalent among Ashkenazi Jews than other populations, to non-neuronopathic forms like type B, with estimated incidence rates estimated at one out of every 200,000 births in France and intermediate forms more frequent throughout central Europe [144].

Classical Niemann–Pick Disease Type A: Neonatal life usually begins normally, with symptoms like vomiting or diarrhea typically manifesting within months of life. Poor weight gain and an enlarged liver and spleen (hepatosplenomegaly) may also signal classical Niemann–Pick disease type A. Nearly 70% of cases may exhibit mild physical abnormalities and brownish pigmentation of their skin, with neurological signs typically appearing between 5 and 10 months of age, beginning with decreased muscle tone (hypotonia) before eventually progressing to progressive loss of acquired motor skills and interest in their surroundings. Nerve conduction typically slows, and macular cherry-red spots may develop at an advanced stage of the disease. Hypotonia then progresses to bilateral pyramidal signs, spasticity increases over time, deep tendon reflexes become absent, and seizures may occur but usually do not pose major threats. Death typically occurs between 1.5 and 3 years, although in some instances, symptoms may develop over an extended period of time [146].

Niemann–Pick Disease Type B: Niemann–Pick disease type B is a non-neuronopathic form, with variable degrees of systemic involvement and diagnosis typically in late infancy or childhood; it can sometimes also appear later in adulthood. Most patients living with type B will reach late adulthood, while some children may develop severe systemic disease, leading to their premature demise [147].

Niemann–Pick Disease Type C (NP-C) is an autosomal recessive atypical lysosomal lipid storage disorder caused by mutations of either of two genes: NPC1 (95% of cases) or NPC2. Both genetic variants lead to similar metabolic impairment in terms of processing and using endocytosed cholesterol, leading to excess storage in extraneural tissues. Diseased cells tend to rely on cholesterol trafficking pathways, while neurons mainly accumulate GM2 and GM3 gangliosides instead of cholesterol. Unfortunately, NPC1 and NPC2 proteins remain poorly understood. Neuropathological features may include neuronal storage, significant loss (especially of Purkinje cells), ectopic dendrites, neuroaxonal dystrophy, and similar changes that mirror AD. Clinical and biochemical features do not distinguish between those carrying NPC1 mutations (the dominant group) and those carrying rarer NPC2 mutations (rare group), though both groups can be present. Furthermore, this disease affects people of all ethnicities, with an estimated incidence rate estimated at 1 in 100,000 live births [148].

Changed cholesterol efflux from lysosomes disrupts neuronal firing patterns. The mechanism involves the upregulation of ABCA1 transporter, which decreases plasma membrane PtdIns(4,5)P2 content via PtdIns(4,5)P2-floppase activity, leading to decreased PtdIns(4,5)P2-floppase activity and therefore decreasing voltage-gated KCNQ2/3 potassium channel activity that regulates neuronal excitability when neurons affected by NPC1 disease are present, resulting in hyperexcitability. Lysosomal cholesterol efflux plays an essential role in shaping electrical and functional characteristics of neuronal plasma membranes under healthy as well as disease conditions [145] Figure 2.

NPC1 disease cases typically result from mutations of the NPC1 gene, coding for an integral membrane protein called NPC1. NPC1 protein resides in late endosomal vesicles within cells and plays an essential role in maintaining cholesterol and sphingolipid balance within them. Unlike enzyme deficiencies found in traditional lysosomal storage diseases (LSDs), however, NPC1 protein does not secrete itself from cells, making cross-correction unlikely through transduced cells alone; therefore, to achieve clinical benefits over CNS storage diseases with enzyme deficiencies alone, technical challenges would need to be met in order to achieve clinical benefits in comparison to CNS storage diseases caused by enzyme deficiencies alone [149].

Cells containing NPC2 mutations may be cross-corrected using an NPC2-enriched medium, making these patients suitable candidates for gene therapy or gene product transduction. Unfortunately, this approach does not address cholesterol accumulation caused by cells carrying NPC1 mutations [150].

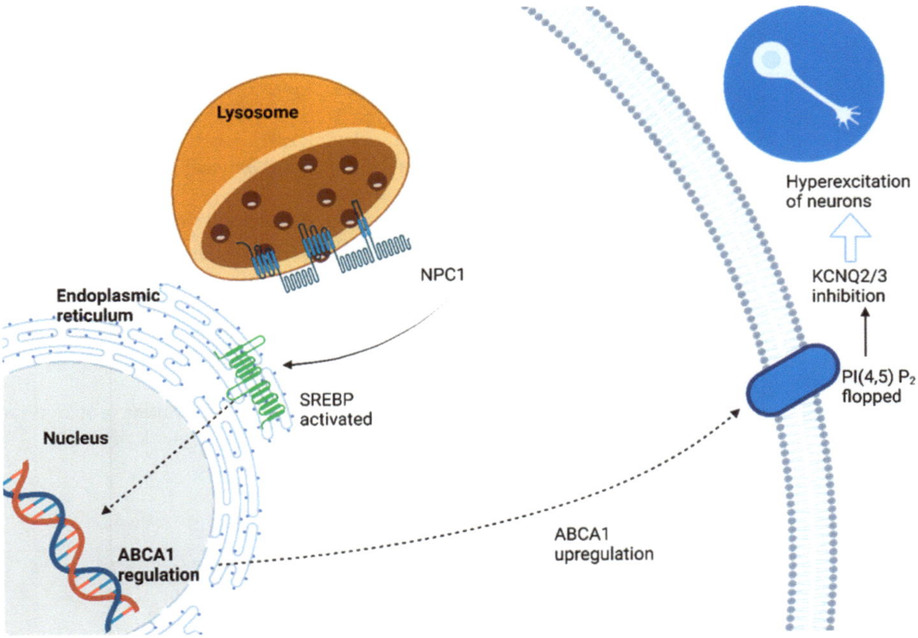

Figure 2. NPC1 disease is a neurodegenerative disorder caused by genetic mutations of the lysosomal cholesterol transporter NPC1. Cholesterol accumulates in these organelles and begins its pathological process, ultimately resulting in greater excitability of neurons affected.

An investigation of the role of GM2 ganglioside in NPC disease was undertaken by crossbreeding NPC mice with those lacking the ability to synthesize GM2 and GD2 gangliosides. Ganglioside-deficient mice demonstrated no neuronal GM2 ganglioside, supporting its role in NPC pathogenesis [149].

N-butyldeoxynojirimycin (NB-DNJ), an inhibitor of GSL biosynthesis, provided another approach to studying their roles in NPC mouse models by decreasing levels of all GlcCer-derived GSLs; substrate reduction therapy has proven useful against other glycosphingolipid storage diseases resulting from lysosomal enzyme deficiencies [151].

Erratic expression and distribution of TDP-43 were observed in an NPC mouse model as well as an in vitro human NPC neuronal model system. Cytoplasmic TDP43 sequestered in stress granules could indicate mutations, disease processes, or environmental stress, leading to neurodegeneration. Furthermore, an NPC neuronal model displayed abnormal gene expression under TDP-43 control at the RNA processing level, which suggests possible connections with the observed neuropathology of NPC [152].

NPCD, like other lysosomal storage disorders, is characterized by an accumulation of various lipids such as di- and triacylglycerols, phosphoinositides, sphingosine, sphingomyelin, GlcCer, and gangliosides. This accumulation may be affected by factors like increased pH causing decreased acid lipase activity, favorable physical/chemical interactions between stored lipids, and reduced enzyme activity due to stored lipids [153].

Surprisingly, although NPCD cells accumulate sphingomyelin, mice with the lysosomal hydrolase of this lipid knocked out (mimicking Niemann–Pick disease type A) do not accumulate cholesterol at the whole cell level. This observation points out the need for subcellular lipidomics to accurately quantify lipid accumulation due to anisotropic distribution and the potential significance of minor populations of lipids. Furthermore, suborganellar lipidomics may help shed more insight into this disease; specifically, by identifying whether cholesterol accumulates within intralysosomal membranes or on its limits, or both [153].

Despite the disease's association with cholesterol trafficking anomalies, drugs aiming to lower cholesterol have not demonstrated effectiveness in altering the disease trajectory [154].

Arimoclomol, a derivative of bimoclomol, is known to bolster heat shock protein (HSP) gene expression. It aids in triggering HSPs, thus enhancing the inherent cellular protective mechanisms during cellular distress situations [155].

Treatment strategies for NPC have concentrated on several areas [156] (Table 8):

- Lessening the volume of intra-lysosomal free cholesterol;
- Curbing the production of glucosylceramide by inhibiting its synthase;
- Regulating inflammatory reactions and immune responses;
- Augmenting the transfer of free cholesterol from the lysosomal section into the cytosol;
- Modulating the expression of genes vital for initiating cell differentiation by hindering histone deacetylases (HDAC);
- Employing pharmacological chaperones to promote cellular protein repair pathways via the activation of molecular chaperones, like heat shock proteins;
- Exploring the potential of gene therapy.

Table 8. Overview of targeted molecular therapies in Niemann–Pick disease.

Treatment/Therapeutical Approaches	Description/Effect	Citations
Cholesterol-lowering drugs	Not effective in altering disease trajectory	[154]
Regulating inflammatory reactions		[154]
Curbing glucosylceramide production	By inhibiting its synthase	[155]
Augmenting free cholesterol transfer	Stimulating the transfer from lysosomal section into the cytosol	[155]
Modulating gene expression	By hindering histone deacetylases (HDAC)	[155]
Pharmacological chaperones	Promote cellular protein repair pathways via activation of molecular chaperones like heat shock proteins	[155]
HP-β-CD	Cyclic oligosaccharide; limited by inability to penetrate blood–brain barrier	[154]
Arimoclomol	Enhances heat shock protein gene expression, aiding cellular protective mechanisms	[155]
Miglustat	Sole officially approved treatment in EU for neurological symptoms; may slow or mitigate disease progression	[157]

9. Tay–Sachs Disease

Tay–Sachs disease is an autosomal recessive condition caused by an enzyme deficiency called beta-hexosaminidase A, an important lysosomal enzyme responsible for breaking down GM2 ganglioside. Without enough of it being broken down by nerve cell lysosomes, an accumulation of this GM2 ganglioside occurs, which leads to nerve cell dysfunction. This condition falls under GM2 gangliosides, a group of lysosomal storage disorders where various defective peptides (alpha and beta subunits of beta-hexosaminidase A and the GM2 activator protein) contribute to the degradation of GM2 ganglioside. Infantile Tay–Sachs disease, also known as type 1 GM2 gangliosidosis or infantile Tay–Sachs disease, is typically characterized by almost-zero beta-hexosaminidase A activity; however, juvenile and late-onset forms exist with residual enzyme activity contributing to milder presentations of disease [158].

Dr. Warren Tay, a British ophthalmologist and physician, presented the case of an infant suffering from progressive weakness and changes to her eye's yellow spot at one year of age before reporting similar cases within her family. Meanwhile, an American physician named Bernard Sachs independently described clinical manifestations and pathological features associated with infantile Tay–Sachs disease despite Dr. Tay's reports. Since Sachs first identified autosomal recessive inheritance and increased Ashkenazic Jewish child occurrence in his initial observations of Tay–Sachs disease, further research confirmed this pattern and increased incidence. Robinson and Stirling identified two hexosaminidase isoenzymes A and B in the human spleen, which allowed Okada and O'Brien in California as well as Sandhoff in Germany to observe an absence of hexosaminidase-A activity among frozen tissue samples from patients suffering from Tay–Sachs disease [159].

Tay–Sachs disease is an autosomal recessive genetic condition caused by an insufficient supply of beta-hexosaminidase A (Hex A). This enzyme has two polypeptide chains encoded on chromosomes 15 and 5, known as alpha and beta subunits, respectively. These subunits combine to form an active enzyme responsible for breaking down specific substrates found within lysosomes, such as GM2 ganglioside. Hydrolysis of this substrate requires activator protein activation within its specific environment in the lysosome. Additionally, beta subunits can dimerize to form beta-hexosaminidase B (Hex B), an active enzyme that acts on water-soluble neutral substrates. Mutations in this beta subunit lead to Sandhoff's disease, with deficient amounts of both Hex A and Hex B being present in its composition [160].

Mutations in the alpha-chain gene disrupt its catalytic function and consequently lead to the accumulation of GM2 gangliosides in neuronal tissue, leading to neurodegeneration and mental and motor retardation. Tay–Sachs disease's clinical course depends on how much residual Hex A activity remains due to mutation severity. Mutations that completely disrupt Hex A activity cause the severe infantile form of Tay–Sachs disease, which presents early and rapidly, leading to death by early childhood. By contrast, the late infantile form presents later and has a later age of demise than its infantile form counterpart, with certain mutations associated with this variant of Tay–Sachs disease [160].

Adult Tay–Sachs disease is distinguished by its clinical variability, typically appearing during the second or third decade of life. Usually, there is less motor and neurological deterioration compared to other variants; psychosis often appears first. Missense mutations are usually responsible for this form of Tay–Sachs disease in adult individuals; Gly269Ser and 805G->A mutations found frequently within exon 7 are typically linked with it [161].

Tay–Sachs disease (TSD) was initially linked to Jewish families of Eastern European ancestry who resided within Sachs' initial study area in New York City. TSD has now come to be recognized as not being limited to any one ethnic group but is widespread, impacting Blacks and Asians alike [162]. DNA sequencing technology has shown that TSD is caused by numerous mutations, disproving the myth that its causes were sole. French Canadians and Ashkenazi Jews, for instance, have shown evidence of genetic heterogeneity when it comes to TSD within these populations [163]. Furthermore, Ashkenazi Jews can have multiple mutations leading to TSD while Moroccan Jews with TSD can display distinct mutations as causes [164].

TSD and other severe forms of GM2 gangliosidosis, including its variants, are associated with an almost total deficiency in Hex A activity; however, total Hex activity remains within normal levels due to active Hex B production. On the other hand, in Sandhoff disease (SD), another form of GM2 gangliosidosis, only a minimal amount (2–4% of normal) of Hex activity can be detected despite normal levels of a-subunit mRNA. This is likely attributable to the low affinity between a-subunits, leading to detectable but limited amounts of Hex S. The less severe juvenile and adult-onset forms of TSD can often be traced back to mutations that allow some preservation of Hex A function (approximately 1–5% of normal) due to improper folding or dimerization of enzymes. Most missense mutations result in misfolded proteins, which are detected by the endoplasmic reticulum quality control system and degraded through its endoplasmic reticulum-associated degradation pathway (ERAD) before finally being processed through proteasome degradation [165].

Enzyme-Enhancing Therapy: When treating LSDs such as GM2 gangliosidosis or other lysosomal storage diseases, missense mutations often destabilize native folded proteins before reaching the lysosome for degradation. To address this, researchers have explored using pharmacological chaperones, or molecules that bind and stabilize enzymes and prevent their degradation; commonly employed are enzyme-specific competitive inhibitors. Another approach involves manipulating the endoplasmic Reticulum-Associated Degradation (ERAD) system to block the recognition of defective proteins or improve endogenous chaperone functions [166].

Gene Therapy: Neuropathic LSDs are an ideal candidate for gene therapy as their root causes lie within single genes with naturally secreted products that require corrective enzyme to restore deficient tissues; transducing just a small number of cells can deliver sufficient corrective enzyme to make up for deficiencies across an extensive region. Studies have demonstrated that even modest levels (around 10% of normal activity) in healthy individuals with pseudo-deficiencies are sufficient to avoid disease progression [167].

Enzyme Replacement Therapy and Cell Transplantation: This strategy has proven successful for some LSDs with systemic manifestations, including MPS I, Fabry disease, Pompe disease, MPS VI, and MPS II. Although enzyme therapy is currently being investigated in clinical trials for certain neuropathic LSDs, it may not provide an effective solution for GM2 gangliosidosis. B-hexosaminidase has a lower abundance in cultured microglia when compared with other lysosomal glycosidases, making it less amenable to enzyme replacement therapy [166]. Bone marrow transplantation has shown some promise for treating certain LSDs, such as human a-mannosidosis, but its effectiveness varies between diseases; for instance, in MPS IIIA (Sanfilippo type A), this therapy can extend life without providing neurological improvements [168].

Substrate Reduction Therapy: Miglustat, an iminosugar known as N-butyldeoxynojirimycin, belongs to a class of molecules known as glucosylceramide synthase inhibitors, which block glycolipid biosynthesis. While Miglustat showed promising results when tested on SD mice by improving lifespan and clinical features, its efficacy endpoint target in late-onset TSD patients was not satisfactory. Regardless, research into inhibitors to reduce substrate accumulation remains an active area [167,168].

One such success in this field is Cerdelga eliglustat, approved as Cerdelga, which has become a first-line therapy for type 1 Gaucher disease. Substrate reduction therapy using orally active small molecules such as Eliglustat has demonstrated therapeutic advantages and efficacy comparable to enzyme therapy for non-neuronopathic Gaucher disease patients. In an initial clinical trial involving patients with late-onset GM2 gangliosidosis, oral administration of pyrimethamine (PYR) at a lower dosage than typically used to treat parasitic diseases showed significant enhancement of Hex A activity in peripheral blood leukocytes; the degree of this enhancement correlated to plasma PYR concentration. The results also confirmed this improvement across Sandhoff and Tay–Sachs variants of GM2 gangliosidosis patients alike. Notably, PYR treatment's increase of Hex A activity was limited to mutant forms of Hex A and not due to any non-specific lysosomotropic effect, as indicated by levels of Glcr and ss-galactosidase activity in plasma and leukocytes, which remained unchanged following PYR administration [169].

For Tay–Sachs disease, current data support a cellular pathology model whereby the alpha subunit (a) of an enzyme is translocated to the endoplasmic reticulum (ER), where it undergoes cotranslational glycosylation and folding via the calreticulin/calnexin cycle. When properly folded, this dimer forms with beta subunits (b). From there it travels to the Golgi apparatus, where mannose-6-phosphate signals direct it back toward lysosomes for its intended function [170] Figure 3.

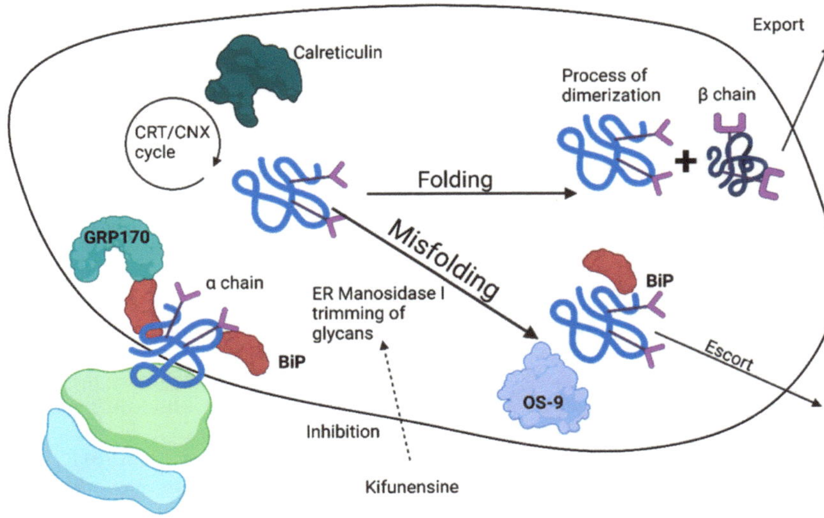

Figure 3. This diagram depicts the complex cellular model of Tay–Sachs disease. After alpha subunit undergoes glycosylation into the ER and structural stabilization from calreticulin/calnexin cycle, it can properly fold, forming the beta subunit, and both will be exported to the Golgi. If alpha subunit undergoes improper folding, glycans are removed before proteasomal degradation. A possibility to avoid this issue is Kifunensine, an inhibitor for ER Manosidase I trimming of glycans.

However, in cases where ER folding of the alpha subunit becomes impaired due to mutations like G269S and E482K, its glycan trimming process—often in conjunction with OS-9 association—can initiate a disposal pathway. When this occurs, retrotranslocon protein channels allow access to the cytosol. PNGase removes any remaining glycans before degradation occurs at 26S proteasome and ultimately results in the removal of improperly folded enzyme molecules, helping explain the pathology of Tay–Sachs disease [171].

10. Fahr's Syndrome

Basal ganglia calcification, also referred to as Fahr's syndrome or Fahr's disease, is an extremely rare neurological condition with an incidence rate of less than 1 out of 1 million individuals, usually affecting those in their third and fourth decades of life [170]. First described by German neurologist Karl Theodor Fahr in 1930, Fahr's disease manifests itself by abnormal calcium deposition within certain brain regions responsible for controlling movements, such as basal ganglia, thalamus, dentate nucleus, cerebral cortex, cerebellum, and hippocampus [172].

Patients often present with extrapyramidal symptoms. Furthermore, they may also exhibit cerebellar dysfunction, speech difficulties, dementia, and neuropsychiatric signs and symptoms. Genetic studies have implicated various loci in Fahr's disease development. One commonly implicated locus (IBGC1) lies at 14q, while two other loci have been discovered on chromosome 8 and two more on 2. Additionally, loss-of-function mutations found within SLC20A2 on chromosome 8 have been suggested as contributing to its pathophysiology [173].

As part of an appropriate diagnostic approach for Fahr's syndrome, it is advisable to conduct sequencing of the SLC20A2 gene first. If no mutations or deletions are identified within SLC20A2, deletion/duplication analysis could then be explored further. In the absence of either identifiable mutations or deletions within SLC20A2, sequence analysis on PDGFRB would then be necessary; should no disease-causing mutations or deletions be identified through molecular genetic tests, then other genetic disorders linked with brain calcifications should be explored further [173].

Fahr's syndrome may arise due to various underlying conditions, including endocrine disorders (such as idiopathic hypoparathyroidism, secondary hypoparathyroidism, and hyperparathyroidism), adult-onset neurodegenerative conditions (e.g., neurodegeneration with brain iron accumulation disease), infectious diseases such as intrauterine or perinatal infections, as well as inherited syndromes like Aicardi–Goutieres syndrome or tuberous sclerosis complex [174].

Seizures and movement disorders related to Fahr's syndrome may be improved by correcting phosphate and calcium levels. Treatment with alpha hydroxy vitamin D3 and corticosteroids has also been shown to reverse neurological deficits in some instances [174]. Clonazepam and atypical antipsychotics may also help with managing symptoms for those living with Fahr's syndrome, though lithium use should be used with caution as it may increase seizure risk. Furthermore, carbamazepine, benzodiazepines, and barbiturates may worsen underlying gait disorders; therefore, these strategies should also be employed carefully when treating patients living with Fahr's syndrome [175].

Genes for Basal Ganglia Calcification:

Mutations of the SLC20A2 gene found on 8p11.21 can be passed down autosomally dominantly. Proper inorganic phosphate transport is essential in keeping the calcium and phosphate balance within cells intact, with impaired PiT2 function leading to the accumulation of calcium phosphate deposits in extracellular matrix tissues of vessels. Research using Slc20a2-knockout mice (KOs) has shed light on pathophysiology associated with SLC20A2, showing increased cerebrospinal fluid, suggesting potential therapeutic implications as a therapeutic option.

Located on 1q25.3, the XPR1 gene is tightly associated with PiT2. It encodes for the xenotropic and polytropic retrovirus receptor that plays an essential role in exporting phosphate from cells, contributing directly to intracellular phosphate homeostasis as well as calcium deposition in some instances. In families, mutations of this gene may be passed down as autosomal dominant traits [176].

PDGFRB is a gene linked to familial primary brain calcification (PFBC). It encodes one of two receptors for platelet-derived growth factor (PDGF), with subunit ss being its main ligand and playing an essential role in maintaining the integrity of the blood–brain barrier. Loss-of-function mutations could disrupt pericyte permeability surrounding brain blood vessels, resulting in the deposition of calcium deposits within brain tissues. Mutations in both PDGFB and PDGFRB genes exhibit autosomal dominant inheritance patterns. PDGFB plays an essential role in pericyte recruitment, blood–brain barrier regulation, angiogenesis, and calcium regulation through BBB disruption, leading to progressive calcinosis and disruption. Furthermore, PDGF proteins may regulate phosphate transporters such as XPR1 and PiT in the brain [177].

PTH resistance disorders may play a part in some cases of brain calcification. When that is the case, loss of function of GNAS on the maternal allele can cause basal ganglia calcification—although not considered primary familial brain calcification (PFBC). GNAS is a complex imprinted locus responsible for producing multiple transcripts via alternative splicing and promoters [178].

Keller et al. conducted genetic analysis on six families to discover that mutations in the gene encoding PDGF-B (located at 22q13.1) could be responsible for familial IBGC in patients who previously tested negative for SLC20A2 and PDGFRB mutations. Furthermore, these experiments also showed that mice that expressed 50% less endothelial PDGF-B developed significant brain calcification within one year [179].

However, it should be remembered that mutations of PDGF-B and SLC20A2 each have different pathophysiological mechanisms. Loss of endothelial PDGF-B has been associated with pericyte deficiency as well as blood–brain barrier deficiency; on the other hand, loss of SLC20A2, as a member of the type III sodium-dependent phosphate transporters family, has been linked with accumulations of inorganic phosphate in the brain, leading to calcium phosphate deposition [180].

To date, multiple treatment modalities have been explored in Fahr's patients, striving for remission or stabilization. These treatments span a range of biological hypotheses and derive from limited clinical observations. Medications are primarily prescribed to manage symptoms like anxiety, depression, obsessive–compulsive disorder, and dystonia. For urinary incontinence, oxybutynin is preferred, while antiepileptics are chosen for seizures. Interestingly, seizures and movement disorders associated with Fahr's syndrome, especially those linked to parathyroid imbalances, can be addressed by regulating phosphate and calcium levels. As an illustration, employing alpha-hydroxy vitamin D3 combined with corticosteroids has reversed some neurological deficits [181] Table 9.

Table 9. Overview of targeted molecular therapies in Fahr's syndrome.

General Medications	Manage Symptoms: Anxiety, Depression, OCD, Dystonia	Citations
Oxybutynin	Treat urinary incontinence	[181]
Antiepileptics	Address seizures	
Phosphate and calcium level regulation	Address seizures and movement disorders, especially those linked to parathyroid imbalances	
Alpha hydroxy vitamin D3 + corticosteroids	Reversed some neurological deficits	
Clonazepam and atypical antipsychotics	Provide therapeutic benefits	
Lithium	Prescribed with caution due to increased seizure risk	
Carbamazepine, benzipenes, barbiturates	Might intensify gait disturbances	
Antidepressants and anxiolytics	Prescribed with caution due to potential side effects at lower thresholds in Fahr's syndrome patients	
iPSC models (SLC20A2 mutations)	Research tool for studying Fahr's disease mechanisms, potential for drug- and gene-based intervention studies	[182]

Research has pointed toward various SLC20A2 mutations as potential culprits behind Fahr's disease, which is a subset of primary familial brain calcification (PFBC). Yet, there have been only a handful of patient-derived induced pluripotent stem cell (iPSC) models to study this condition. Notably, Zhang and colleagues discovered a novel SLC20A2 mutation in a Fahr's disease-affected family and managed to secure dermal fibroblasts from a member of this family. They successfully reprogrammed these fibroblasts into iPSCs using episomal plasmids carrying genes like OCT3/4, SOX2, KLF4, LIN28, and L-MYC. Such efforts present valuable resources and platforms, paving the way for more detailed investigations into Fahr's disease mechanisms. This could potentially enhance the development and assessment of both drug and gene-based interventions [182] Table 9.

11. Conclusions and Future Perspectives

Neurodegenerative ailments have grave implications, and conventional medications often fall short. Presently, stem cell therapy is gaining traction as a potentially transformative solution, but much of the current research is rooted in animal studies, creating a knowledge void regarding its long-term impacts and results in humans. Positive outcomes in animals demand more rigorous investigation before being translated into human therapies. Among the various neurodegenerative conditions, Parkinson's disease and amyotrophic lateral sclerosis have received more attention compared to Huntington's disease and Alzheimer's disease. Before stem cell therapies can become a mainstream treatment for these conditions, numerous hurdles, such as cost, safety concerns, the expertise involved, and post-procedure monitoring, need to be addressed [183].

Traditional treatments for neurodegenerative diseases, such as small-molecule medications and immunotherapy aimed at harmful proteins, largely alleviate symptoms without actually stopping or reversing the disease's advance. New therapeutic techniques like autophagy therapy, miRNA therapy, and stem cell therapy have been introduced. However, they face challenges such as difficulty in crossing the blood–brain barrier, unexpected side effects, and challenges in addressing intracellular proteins. The emergence of protein-targeted degradation technologies brings hope by potentially targeting proteins previously considered unreachable for drug interventions. These novel degradation methods each have their distinctive mechanisms and objectives, accompanied by their advantages and drawbacks. This examination explores their present roles in neurodegenerative diseases, weighs their strengths and weaknesses, and foresees their future trajectory in the domain [184].

Gene therapy offers a hopeful avenue for addressing neurodegenerative diseases. Transitioning these therapies from theoretical frameworks to practical clinical applications has, however, been difficult. Preliminary studies indicate that delivering gene therapies to the central nervous system is safe and typically well-tolerated. To improve delivery, new vectors, such as AAV9, are being researched. AAV9 stands out for its ability to cross the blood–brain barrier and target neurons predominantly. Yet, its efficiency might fluctuate based on factors like the recipient's age at the time of treatment. Also, the high production costs of AAV9 for human testing make its broad application challenging. Another exciting advancement in this domain is the tricistronic lentiviral vector, highlighted in the Prosavin trial. Efforts are also ongoing to refine gene therapy delivery methods to the CNS, including intracerebroventricular, intrathecal, and direct brain and spinal cord injections. However, these delivery advancements would be moot if the genes delivered are not efficacious. Therefore, a key priority is the discovery and assessment of new therapeutic genes, underpinned by a richer understanding of the origins and evolution of neurodegenerative conditions. As our knowledge deepens, we expect improved early diagnosis, facilitating interventions before significant cellular damage [185].

Author Contributions: Conceptualization, V.V. and C.T.; methodology, C.P.T. and R.-A.C.-B.; software, L.A.G.; validation, A.V.C., B.-G.B. and H.P.C.; formal analysis, A.D.C.; investigation, C.P.T. and R.-A.C.-B.; resources, L.A.G.; data curation, H.P.C.; writing—original draft preparation, B.-G.B.; writing—review and editing, A.D.C.; visualization, C.T.; supervision, V.V.; project administration, A.V.C.; funding acquisition, A.V.C. All authors have read and agreed to the published version of the manuscript.

Funding: This research received no external funding.

Institutional Review Board Statement: Not applicable.

Informed Consent Statement: Not applicable.

Data Availability Statement: All data is available online on libraries such as PubMed.

Conflicts of Interest: The authors declare no conflict of interest.

References

1. Cairns, N.J.; Bigio, E.H.; Mackenzie, I.R.A.; Neumann, M.; Lee, V.M.-Y.; Hatanpaa, K.J.; White, C.L.; Schneider, J.A.; Grinberg, L.T.; Halliday, G.; et al. Neuropathologic Diagnostic and Nosologic Criteria for Frontotemporal Lobar Degeneration: Consensus of the Consortium for Frontotemporal Lobar Degeneration. *Acta Neuropathol.* **2007**, *114*, 5–22. [CrossRef] [PubMed]
2. Mackenzie, I.R.A.; Neumann, M.; Bigio, E.H.; Cairns, N.J.; Alafuzoff, I.; Kril, J.; Kovacs, G.G.; Ghetti, B.; Halliday, G.; Holm, I.E.; et al. Nomenclature and Nosology for Neuropathologic Subtypes of Frontotemporal Lobar Degeneration: An Update. *Acta Neuropathol.* **2010**, *119*, 1–4. [CrossRef] [PubMed]
3. Skibinski, G.; Parkinson, N.J.; Brown, J.M.; Chakrabarti, L.; Lloyd, S.L.; Hummerich, H.; Nielsen, J.E.; Hodges, J.R.; Spillantini, M.G.; Thusgaard, T.; et al. Mutations in the Endosomal ESCRTIII-Complex Subunit CHMP2B in Frontotemporal Dementia. *Nat. Genet.* **2005**, *37*, 806–808. [CrossRef] [PubMed]

4. Isaacs, A.M.; Johannsen, P.; Holm, I.; Nielsen, J.E. Consortium, Fr. Frontotemporal Dementia Caused by CHMP2B Mutations. *Curr. Alzheimer Res.* **2011**, *8*, 246–251. [CrossRef] [PubMed]
5. Snowden, J.S.; Thompson, J.C.; Stopford, C.L.; Richardson, A.M.T.; Gerhard, A.; Neary, D.; Mann, D.M.A. The Clinical Diagnosis of Early-Onset Dementias: Diagnostic Accuracy and Clinicopathological Relationships. *Brain* **2011**, *134*, 2478–2492. [CrossRef]
6. Herrmann, N.; Black, S.E.; Chow, T.; Cappell, J.; Tang-Wai, D.F.; Lanctôt, K.L. Serotonergic Function and Treatment of Behavioral and Psychological Symptoms of Frontotemporal Dementia. *Am. J. Geriatr. Psychiatry* **2012**, *20*, 789–797. [CrossRef]
7. Czarnecki, K.; Kumar, N.; Josephs, K.A. Parkinsonism and Tardive Antecollis in Frontotemporal Dementia—Increased Sensitivity to Newer Antipsychotics?: Parkinsonism and Tardive Antecollis in Frontotemporal Dementia. *Eur. J. Neurol.* **2008**, *15*, 199–201. [CrossRef]
8. Sivasathiaseelan, H.; Marshall, C.; Agustus, J.; Benhamou, E.; Bond, R.; Van Leeuwen, J.; Hardy, C.; Rohrer, J.; Warren, J. Frontotemporal Dementia: A Clinical Review. *Semin. Neurol.* **2019**, *39*, 251–263. [CrossRef]
9. Mukherjee, O.; Wang, J.; Gitcho, M.; Chakraverty, S.; Taylor-Reinwald, L.; Shears, S.; Kauwe, J.S.K.; Norton, J.; Levitch, D.; Bigio, E.H.; et al. Molecular Characterization of Novel Progranulin (*GRN*) Mutations in Frontotemporal Dementia. *Hum. Mutat.* **2008**, *29*, 512–521. [CrossRef]
10. Terryn, J.; Verfaillie, C.M.; Van Damme, P. Tweaking Progranulin Expression: Therapeutic Avenues and Opportunities. *Front. Mol. Neurosci.* **2021**, *14*, 713031. [CrossRef]
11. Spillantini, M.G.; Van Swieten, J.C.; Goedert, M. Tau Gene Mutations in Frontotemporal Dementia and Parkinsonism Linked to Chromosome 17 (FTDP-17). *Neurogenetics* **2000**, *2*, 193–205. [CrossRef] [PubMed]
12. Pollock, N.J.; Mirra, S.S.; Binder, L.I.; Hansen, L.A.; Wood, J.G. Filamentous Aggregates in Pick's Disease, Progressive Supranuclear Palsy, And Alzheimer's Disease Share Antigenic Determinants with Microtubule-Associated Protein, Tau. *Lancet* **1986**, *328*, 1211. [CrossRef] [PubMed]
13. Wilhelmsen, K.C.; Lynch, T.; Pavlou, E.; Higgins, M.; Nygaard, T.G. Localization of Disinhibition-Dementia-Parkinsonism-Amyotrophy Complex to 17q21-22. *Am. J. Hum. Genet.* **1994**, *55*, 1159–1165. [PubMed]
14. Mackenzie, I.R.A.; Neumann, M. Molecular Neuropathology of Frontotemporal Dementia: Insights into Disease Mechanisms from Postmortem Studies. *J. Neurochem.* **2016**, *138*, 54–70. [CrossRef]
15. Sha, S.; Hou, C.; Viskontas, I.V.; Miller, B.L. Are Frontotemporal Lobar Degeneration, Progressive Supranuclear Palsy and Corticobasal Degeneration Distinct Diseases? *Nat. Clin. Pract. Neurol.* **2006**, *2*, 658–665. [CrossRef]
16. Parmera, J.B.; Rodriguez, R.D.; Studart Neto, A.; Nitrini, R.; Brucki, S.M.D. Corticobasal Syndrome: A Diagnostic Conundrum. *Dement. Neuropsychol.* **2016**, *10*, 267–275. [CrossRef] [PubMed]
17. Parmera, J.B.; Oliveira, M.C.B.D.; Rodrigues, R.D.; Coutinho, A.M. Progressive Supranuclear Palsy and Corticobasal Degeneration: Novel Clinical Concepts and Advances in Biomarkers. *Arq. Neuropsiquiatr.* **2022**, *80* (5 Suppl. S1), 126–136. [CrossRef]
18. Wszolek, Z.K.; Tsuboi, Y.; Ghetti, B.; Pickering-Brown, S.; Baba, Y.; Cheshire, W.P. Frontotemporal Dementia and Parkinsonism Linked to Chromosome 17 (FTDP-17). *Orphanet J. Rare Dis.* **2006**, *1*, 30. [CrossRef] [PubMed]
19. Kimonis, V.E.; Fulchiero, E.; Vesa, J.; Watts, G. VCP Disease Associated with Myopathy, Paget Disease of Bone and Frontotemporal Dementia: Review of a Unique Disorder. *Biochim. Biophys. Acta BBA Mol. Basis Dis.* **2008**, *1782*, 744–748. [CrossRef] [PubMed]
20. Pickering-Brown, S.M.; Baker, M.; Nonaka, T.; Ikeda, K.; Sharma, S.; Mackenzie, J.; Simpson, S.A.; Moore, J.W.; Snowden, J.S.; De Silva, R.; et al. Frontotemporal Dementia with Pick-type Histology Associated with Q336R Mutation in the Tau Gene. *Brain* **2004**, *127*, 1415–1426. [CrossRef]
21. Rademakers, R.; Cruts, M.; Van Broeckhoven, C. The Role of Tau (MAPT) in Frontotemporal Dementia and Related Tauopathies. *Hum. Mutat.* **2004**, *24*, 277–295. [CrossRef] [PubMed]
22. De Conti, L.; Borroni, B.; Baralle, M. New Routes in Frontotemporal Dementia Drug Discovery. *Expert Opin. Drug Discov.* **2017**, *12*, 659–671. [CrossRef] [PubMed]
23. Grossman, M. Frontotemporal Dementia: A Review. *J. Int. Neuropsychol. Soc.* **2002**, *8*, 566–583. [CrossRef] [PubMed]
24. Takahashi, K.; Yamanaka, S. Induction of Pluripotent Stem Cells from Mouse Embryonic and Adult Fibroblast Cultures by Defined Factors. *Cell* **2006**, *126*, 663–676. [CrossRef] [PubMed]
25. Ehrlich, M.; Hallmann, A.-L.; Reinhardt, P.; Araúzo-Bravo, M.J.; Korr, S.; Röpke, A.; Psathaki, O.E.; Ehling, P.; Meuth, S.G.; Oblak, A.L.; et al. Distinct Neurodegenerative Changes in an Induced Pluripotent Stem Cell Model of Frontotemporal Dementia Linked to Mutant TAU Protein. *Stem Cell Rep.* **2015**, *5*, 83–96. [CrossRef] [PubMed]
26. Sposito, T.; Preza, E.; Mahoney, C.J.; Setó-Salvia, N.; Ryan, N.S.; Morris, H.R.; Arber, C.; Devine, M.J.; Houlden, H.; Warner, T.T.; et al. Developmental Regulation of Tau Splicing Is Disrupted in Stem Cell-Derived Neurons from Frontotemporal Dementia Patients with the 10 + 16 Splice-Site Mutation in MAPT. *Hum. Mol. Genet.* **2015**, *24*, 5260–5269. [CrossRef]
27. Hallmann, A.-L.; Araúzo-Bravo, M.J.; Mavrommatis, L.; Ehrlich, M.; Röpke, A.; Brockhaus, J.; Missler, M.; Sterneckert, J.; Schöler, H.R.; Kuhlmann, T.; et al. Astrocyte Pathology in a Human Neural Stem Cell Model of Frontotemporal Dementia Caused by Mutant TAU Protein. *Sci. Rep.* **2017**, *7*, 42991. [CrossRef]
28. Duenas, A.M. Molecular Pathogenesis of Spinocerebellar Ataxias. *Brain* **2006**, *129*, 1357–1370. [CrossRef]
29. Tsuji, S.; Onodera, O.; Goto, J.; Nishizawa, M.; On Behalf of the Study Group on Ataxic Diseases*. Sporadic Ataxias in Japan—A Population-Based Epidemiological Study. *Cerebellum* **2008**, *7*, 189–197. [CrossRef]
30. Matilla-Dueñas, A.; Sánchez, I.; Corral-Juan, M.; Dávalos, A.; Alvarez, R.; Latorre, P. Cellular and Molecular Pathways Triggering Neurodegeneration in the Spinocerebellar Ataxias. *Cerebellum* **2010**, *9*, 148–166. [CrossRef]

31. Matilla-Dueñas, A. The Highly Heterogeneous Spinocerebellar Ataxias: From Genes to Targets for Therapeutic Intervention. *Cerebellum* **2008**, *7*, 97–100. [CrossRef] [PubMed]
32. Goold, R.; Hubank, M.; Hunt, A.; Holton, J.; Menon, R.P.; Revesz, T.; Pandolfo, M.; Matilla-Duenas, A. Down-Regulation of the Dopamine Receptor D2 in Mice Lacking Ataxin 1. *Hum. Mol. Genet.* **2007**, *16*, 2122–2134. [CrossRef] [PubMed]
33. Shimohata, T.; Nakajima, T.; Yamada, M.; Uchida, C.; Onodera, O.; Naruse, S.; Kimura, T.; Koide, R.; Nozaki, K.; Sano, Y.; et al. Expanded Polyglutamine Stretches Interact with TAFII130, Interfering with CREB-Dependent Transcription. *Nat. Genet.* **2000**, *26*, 29–36. [CrossRef] [PubMed]
34. Steffan, J.S.; Bodai, L.; Pallos, J.; Poelman, M.; McCampbell, A.; Apostol, B.L.; Kazantsev, A.; Schmidt, E.; Zhu, Y.-Z.; Greenwald, M.; et al. Histone Deacetylase Inhibitors Arrest Polyglutamine-Dependent Neurodegeneration in Drosophila. *Nature* **2001**, *413*, 739–743. [CrossRef]
35. Clark, H.B.; Burright, E.N.; Yunis, W.S.; Larson, S.; Wilcox, C.; Hartman, B.; Matilla, A.; Zoghbi, H.Y.; Orr, H.T. Purkinje Cell Expression of a Mutant Allele of *SCA1* in Transgenic Mice Leads to Disparate Effects on Motor Behaviors, Followed by a Progressive Cerebellar Dysfunction and Histological Alterations. *J. Neurosci.* **1997**, *17*, 7385–7395. [CrossRef]
36. Corral, J.; Genís, D.; Banchs, I.; San Nicolás, H.; Armstrong, J.; Volpini, V. Giant SCA8 Alleles in Nine Children Whose Mother Has Two Moderately Large Ones: Inheriting Giant SCA8 Alleles. *Ann. Neurol.* **2005**, *57*, 549–553. [CrossRef]
37. Koob, M.D.; Moseley, M.L.; Schut, L.J.; Benzow, K.A.; Bird, T.D.; Day, J.W.; Ranum, L.P.W. An Untranslated CTG Expansion Causes a Novel Form of Spinocerebellar Ataxia (SCA8). *Nat. Genet.* **1999**, *21*, 379–384. [CrossRef]
38. Liu, J.; Tang, T.-S.; Tu, H.; Nelson, O.; Herndon, E.; Huynh, D.P.; Pulst, S.M.; Bezprozvanny, I. Deranged Calcium Signaling and Neurodegeneration in Spinocerebellar Ataxia Type 2. *J. Neurosci.* **2009**, *29*, 9148–9162. [CrossRef]
39. Adachi, N.; Kobayashi, T.; Takahashi, H.; Kawasaki, T.; Shirai, Y.; Ueyama, T.; Matsuda, T.; Seki, T.; Sakai, N.; Saito, N. Enzymological Analysis of Mutant Protein Kinase Cγ Causing Spinocerebellar Ataxia Type 14 and Dysfunction in Ca2+ Homeostasis. *J. Biol. Chem.* **2008**, *283*, 19854–19863. [CrossRef]
40. Chen, D.-H.; Brkanac, Z.; Christophe Verlinde, L.M.J.; Tan, X.-J.; Bylenok, L.; Nochlin, D.; Matsushita, M.; Lipe, H.; Wolff, J.; Fernandez, M.; et al. Missense Mutations in the Regulatory Domain of PKCγ: A New Mechanism for Dominant Nonepisodic Cerebellar Ataxia. *Am. J. Hum. Genet.* **2003**, *72*, 839–849. [CrossRef]
41. Chou, A.-H.; Yeh, T.-H.; Kuo, Y.-L.; Kao, Y.-C.; Jou, M.-J.; Hsu, C.-Y.; Tsai, S.-R.; Kakizuka, A.; Wang, H.-L. Polyglutamine-Expanded Ataxin-3 Activates Mitochondrial Apoptotic Pathway by Upregulating Bax and Downregulating Bcl-XL. *Neurobiol. Dis.* **2006**, *21*, 333–345. [CrossRef] [PubMed]
42. Xia, H.; Mao, Q.; Eliason, S.L.; Harper, S.Q.; Martins, I.H.; Orr, H.T.; Paulson, H.L.; Yang, L.; Kotin, R.M.; Davidson, B.L. RNAi Suppresses Polyglutamine-Induced Neurodegeneration in a Model of Spinocerebellar Ataxia. *Nat. Med.* **2004**, *10*, 816–820. [CrossRef]
43. Alves, S.; Nascimento-Ferreira, I.; Auregan, G.; Hassig, R.; Dufour, N.; Brouillet, E.; Pedroso De Lima, M.C.; Hantraye, P.; Pereira De Almeida, L.; Déglon, N. Allele-Specific RNA Silencing of Mutant Ataxin-3 Mediates Neuroprotection in a Rat Model of Machado-Joseph Disease. *PLoS ONE* **2008**, *3*, e3341. [CrossRef]
44. Shah, A.G.; Friedman, M.J.; Huang, S.; Roberts, M.; Li, X.-J.; Li, S. Transcriptional Dysregulation of TrkA Associates with Neurodegeneration in Spinocerebellar Ataxia Type 17. *Hum. Mol. Genet.* **2009**, *18*, 4141–4152. [CrossRef] [PubMed]
45. Cummings, J.L. Dementia with Lewy Bodies: Molecular Pathogenesis and Implications for Classification. *J. Geriatr. Psychiatry Neurol.* **2004**, *17*, 112–119. [CrossRef] [PubMed]
46. Hardy, J.; Cookson, M.R.; Singleton, A. Genes and Parkinsonism. *Lancet Neurol.* **2003**, *2*, 221–228. [CrossRef]
47. Masliah, E.; Rockenstein, E.; Veinbergs, I.; Sagara, Y.; Mallory, M.; Hashimoto, M.; Mucke, L. β-Amyloid Peptides Enhance α-Synuclein Accumulation and Neuronal Deficits in a Transgenic Mouse Model Linking Alzheimer's Disease and Parkinson's Disease. *Proc. Natl. Acad. Sci. USA* **2001**, *98*, 12245–12250. [CrossRef]
48. Van Den Eeden, S.K. Incidence of Parkinson's Disease: Variation by Age, Gender, and Race/Ethnicity. *Am. J. Epidemiol.* **2003**, *157*, 1015–1022. [CrossRef]
49. Outeiro, T.F.; Koss, D.J.; Erskine, D.; Walker, L.; Kurzawa-Akanbi, M.; Burn, D.; Donaghy, P.; Morris, C.; Taylor, J.-P.; Thomas, A.; et al. Dementia with Lewy Bodies: An Update and Outlook. *Mol. Neurodegener.* **2019**, *14*, 5. [CrossRef]
50. Guerreiro, R.; Ross, O.A.; Kun-Rodrigues, C.; Hernandez, D.G.; Orme, T.; Eicher, J.D.; Shepherd, C.E.; Parkkinen, L.; Darwent, L.; Heckman, M.G.; et al. Investigating the Genetic Architecture of Dementia with Lewy Bodies: A Two-Stage Genome-Wide Association Study. *Lancet Neurol.* **2018**, *17*, 64–74. [CrossRef]
51. Leverenz, J.B.; Fishel, M.A.; Peskind, E.R.; Montine, T.J.; Nochlin, D.; Steinbart, E.; Raskind, M.A.; Schellenberg, G.D.; Bird, T.D.; Tsuang, D. Lewy Body Pathology in Familial Alzheimer Disease: Evidence for Disease- and Mutation-Specific Pathologic Phenotype. *Arch. Neurol.* **2006**, *63*, 370. [CrossRef] [PubMed]
52. Dickson, D.W.; Heckman, M.G.; Murray, M.E.; Soto, A.I.; Walton, R.L.; Diehl, N.N.; Van Gerpen, J.A.; Uitti, R.J.; Wszolek, Z.K.; Ertekin-Taner, N.; et al. *APOE* E4 Is Associated with Severity of Lewy Body Pathology Independent of Alzheimer Pathology. *Neurology* **2018**, *91*, e1182–e1195. [CrossRef] [PubMed]
53. Nalls, M.A.; Duran, R.; Lopez, G.; Kurzawa-Akanbi, M.; McKeith, I.G.; Chinnery, P.F.; Morris, C.M.; Theuns, J.; Crosiers, D.; Cras, P.; et al. A Multicenter Study of Glucocerebrosidase Mutations in Dementia with Lewy Bodies. *JAMA Neurol.* **2013**, *70*, 727. [CrossRef]

54. Sidransky, E.; Nalls, M.A.; Aasly, J.O.; Aharon-Peretz, J.; Annesi, G.; Barbosa, E.R.; Bar-Shira, A.; Berg, D.; Bras, J.; Brice, A.; et al. Multicenter Analysis of Glucocerebrosidase Mutations in Parkinson's Disease. *N. Engl. J. Med.* **2009**, *361*, 1651–1661. [CrossRef]
55. Clark, L.N.; Kartsaklis, L.A.; Wolf Gilbert, R.; Dorado, B.; Ross, B.M.; Kisselev, S.; Verbitsky, M.; Mejia-Santana, H.; Cote, L.J.; Andrews, H.; et al. Association of Glucocerebrosidase Mutations with Dementia with Lewy Bodies. *Arch. Neurol.* **2009**, *66*. [CrossRef] [PubMed]
56. Tsuang, D.; Leverenz, J.B.; Lopez, O.L.; Hamilton, R.L.; Bennett, D.A.; Schneider, J.A.; Buchman, A.S.; Larson, E.B.; Crane, P.K.; Kaye, J.A.; et al. GBA Mutations Increase Risk for Lewy Body Disease with and without Alzheimer Disease Pathology. *Neurology* **2012**, *79*, 1944–1950. [CrossRef]
57. Setó-Salvia, N.; Pagonabarraga, J.; Houlden, H.; Pascual-Sedano, B.; Dols-Icardo, O.; Tucci, A.; Paisán-Ruiz, C.; Campolongo, A.; Antón-Aguirre, S.; Martín, I.; et al. Glucocerebrosidase Mutations Confer a Greater Risk of Dementia during Parkinson's Disease Course. *Mov. Disord.* **2012**, *27*, 393–399. [CrossRef]
58. Granek, Z.; Barczuk, J.; Siwecka, N.; Rozpędek-Kamińska, W.; Kucharska, E.; Majsterek, I. GBA1 Gene Mutations in α-Synucleinopathies—Molecular Mechanisms Underlying Pathology and Their Clinical Significance. *Int. J. Mol. Sci.* **2023**, *24*, 2044. [CrossRef]
59. Mazzulli, J.R.; Xu, Y.-H.; Sun, Y.; Knight, A.L.; McLean, P.J.; Caldwell, G.A.; Sidransky, E.; Grabowski, G.A.; Krainc, D. Gaucher Disease Glucocerebrosidase and α-Synuclein Form a Bidirectional Pathogenic Loop in Synucleinopathies. *Cell* **2011**, *146*, 37–52. [CrossRef]
60. Ferman, T.J.; Boeve, B.F. Dementia with Lewy Bodies. *Neurol. Clin.* **2007**, *25*, 741–760. [CrossRef]
61. Garcia-Esparcia, P.; López-González, I.; Grau-Rivera, O.; García-Garrido, M.F.; Konetti, A.; Llorens, F.; Zafar, S.; Carmona, M.; Del Rio, J.A.; Zerr, I.; et al. Dementia with Lewy Bodies: Molecular Pathology in the Frontal Cortex in Typical and Rapidly Progressive Forms. *Front. Neurol.* **2017**, *8*. [CrossRef] [PubMed]
62. Olson, M.O.J. Sensing Cellular Stress: Another New Function for the Nucleolus? *Sci. STKE* **2004**, *2004*. [CrossRef]
63. Kapp, L.D.; Lorsch, J.R. The Molecular Mechanics of Eukaryotic Translation. *Annu. Rev. Biochem.* **2004**, *73*, 657–704. [CrossRef]
64. Shepherd, C.E.; Thiel, E.; McCann, H.; Harding, A.J.; Halliday, G.M. Cortical Inflammation in Alzheimer Disease but Not Dementia with Lewy Bodies. *Arch. Neurol.* **2000**, *57*, 817. [CrossRef]
65. McKeith, I.; Del Ser, T.; Spano, P.; Emre, M.; Wesnes, K.; Anand, R.; Cicin-Sain, A.; Ferrara, R.; Spiegel, R. Efficacy of Rivastigmine in Dementia with Lewy Bodies: A Randomised, Double-Blind, Placebo-Controlled International Study. *Lancet* **2000**, *356*, 2031–2036. [CrossRef] [PubMed]
66. Edwards, K.; Royall, D.; Hershey, L.; Lichter, D.; Hake, A.; Farlow, M.; Pasquier, F.; Johnson, S. Efficacy and Safety of Galantamine in Patients with Dementia with Lewy Bodies: A 24-Week Open-Label Study. *Dement. Geriatr. Cogn. Disord.* **2007**, *23*, 401–405. [CrossRef] [PubMed]
67. Warre, R.; Thiele, S.; Talwar, S.; Kamal, M.; Johnston, T.H.; Wang, S.; Lam, D.; Lo, C.; Khademullah, C.S.; Perera, G.; et al. Altered Function of Glutamatergic Cortico-Striatal Synapses Causes Output Pathway Abnormalities in a Chronic Model of Parkinsonism. *Neurobiol. Dis.* **2011**, *41*, 591–604. [CrossRef]
68. Fernandez, H.H.; Wu, C.-K.; Ott, B.R. Pharmacotherapy of Dementia with Lewy Bodies. *Expert Opin. Pharmacother.* **2003**, *4*, 2027–2037. [CrossRef]
69. Murata, M.; Odawara, T.; Hasegawa, K.; Iiyama, S.; Nakamura, M.; Tagawa, M.; Kosaka, K. Adjunct Zonisamide to Levodopa for DLB Parkinsonism: A Randomized Double-Blind Phase 2 Study. *Neurology* **2018**, *90*, e664–e672. [CrossRef]
70. Meeus, B.; Theuns, J.; Van Broeckhoven, C. The Genetics of Dementia with Lewy Bodies: What Are We Missing? *Arch. Neurol.* **2012**, *69*, 1113–1118. [CrossRef]
71. Smeyers, P.; Monrós, E.; Vílchez, J.; Lopez-Arlandis, J.; Prieto, F.; Palau, F. A Family Segregating a Friedreich Ataxia Phenotype That Is Not Linked to the FRDA Locus. *Hum. Genet.* **1996**, *97*, 824–828. [CrossRef] [PubMed]
72. Romeo, G.; Menozzi, P.; Ferlini, A.; Fadda, S.; Di Donato, S.; Uziel, G.; Lucci, B.; Capodaglio, L.; Filla, A.; Campanella, G. Incidence of Friedreich Ataxia in Italy Estimated from Consanguineous Marriages. *Am. J. Hum. Genet.* **1983**, *35*, 523–529. [PubMed]
73. Palau, F. Friedreich's Ataxia and Frataxin: Molecular Genetics, Evolution and Pathogenesis (Review). *Int. J. Mol. Med.* **2001**, *7*, 581–589. [CrossRef] [PubMed]
74. Campuzano, V.; Montermini, L.; Moltò, M.D.; Pianese, L.; Cossée, M.; Cavalcanti, F.; Monros, E.; Rodius, F.; Duclos, F.; Monticelli, A.; et al. Friedreich's Ataxia: Autosomal Recessive Disease Caused by an Intronic GAA Triplet Repeat Expansion. *Science* **1996**, *271*, 1423–1427. [CrossRef]
75. De Michele, G.; Filla, A.; Cavalcanti, F.; Di Maio, L.; Pianese, L.; Castaldo, I.; Calabrese, O.; Monticelli, A.; Varrone, S.; Campanella, G. Late Onset Friedreich's Disease: Clinical Features and Mapping of Mutation to the FRDA Locus. *J. Neurol. Neurosurg. Psychiatry* **1994**, *57*, 977–979. [CrossRef]
76. Berciano, J.; Combarros, O.; Castro, M.D.; Palau, F. Intronic GAA Triplet Repeat Expansion in Friedreich's Ataxia Presenting with Pure Sensory Ataxia. *J. Neurol.* **1997**, *244*, 390–391. [CrossRef]
77. Geoffroy, G.; Barbeau, A.; Breton, G.; Lemieux, B.; Aube, M.; Leger, C.; Bouchard, J.P. Clinical Description and Roentgenologic Evaluation of Patients with Friedreich's Ataxia. *Can. J. Neurol. Sci. J. Can. Sci. Neurol.* **1976**, *3*, 279–286. [CrossRef]
78. Wilson, R.B.; Lynch, D.R.; Fischbeck, K.H. Normal Serum Iron and Ferritin Concentrations in Patients with Friedreich's Ataxia. *Ann. Neurol.* **1998**, *44*, 132–134. [CrossRef]

79. Schulz, J.B.; Dehmer, T.; Schols, L.; Mende, H.; Hardt, C.; Vorgerd, M.; Burk, K.; Matson, W.; Dichgans, J.; Beal, M.F.; et al. Oxidative Stress in Patients with Friedreich Ataxia. *Neurology* **2000**, *55*, 1719–1721. [CrossRef]
80. Harding, A.E. Friedreich's Ataxia: A Clinical and Genetic Study of 90 Families with an Analysis of Early Diagnostic Criteria and Intrafamilial Clustering of Clinical Features. *Brain J. Neurol.* **1981**, *104*, 589–620. [CrossRef]
81. Cossée, M.; Dürr, A.; Schmitt, M.; Dahl, N.; Trouillas, P.; Allinson, P.; Kostrzewa, M.; Nivelon-Chevallier, A.; Gustavson, K.-H.; Kohlschütter, A.; et al. Friedreich's Ataxia: Point Mutations and Clinical Presentation of Compound Heterozygotes. *Ann. Neurol.* **1999**, *45*, 200–206. [CrossRef] [PubMed]
82. Marmolino, D.; Acquaviva, F.; Pinelli, M.; Monticelli, A.; Castaldo, I.; Filla, A.; Cocozza, S. PPAR-γ Agonist Azelaoyl PAF Increases Frataxin Protein and MRNA Expression. New Implications for the Friedreich's Ataxia Therapy. *Cerebellum* **2009**, *8*, 98–103. [CrossRef] [PubMed]
83. Rai, M.; Soragni, E.; Jenssen, K.; Burnett, R.; Herman, D.; Coppola, G.; Geschwind, D.H.; Gottesfeld, J.M.; Pandolfo, M. HDAC Inhibitors Correct Frataxin Deficiency in a Friedreich Ataxia Mouse Model. *PLoS ONE* **2008**, *3*, e1958. [CrossRef] [PubMed]
84. Sparaco, M.; Gaeta, L.M.; Santorelli, F.M.; Passarelli, C.; Tozzi, G.; Bertini, E.; Simonati, A.; Scaravilli, F.; Taroni, F.; Duyckaerts, C.; et al. Friedreich's Ataxia: Oxidative Stress and Cytoskeletal Abnormalities. *J. Neurol. Sci.* **2009**, *287*, 111–118. [CrossRef]
85. Ristow, M.; Pfister, M.F.; Yee, A.J.; Schubert, M.; Michael, L.; Zhang, C.-Y.; Ueki, K.; Michael, M.D.; Lowell, B.B.; Kahn, C.R. Frataxin Activates Mitochondrial Energy Conversion and Oxidative Phosphorylation. *Proc. Natl. Acad. Sci. USA* **2000**, *97*, 12239–12243. [CrossRef]
86. Mincheva-Tasheva, S.; Obis, E.; Tamarit, J.; Ros, J. Apoptotic Cell Death and Altered Calcium Homeostasis Caused by Frataxin Depletion in Dorsal Root Ganglia Neurons Can Be Prevented by BH4 Domain of Bcl-XL Protein. *Hum. Mol. Genet.* **2014**, *23*, 1829–1841. [CrossRef]
87. Shidara, Y.; Hollenbeck, P.J. Defects in Mitochondrial Axonal Transport and Membrane Potential without Increased Reactive Oxygen Species Production in a *Drosophila* Model of Friedreich Ataxia. *J. Neurosci.* **2010**, *30*, 11369–11378. [CrossRef]
88. Kawashima, M.; Miyake, M.; Kusumi, M.; Adachi, Y.; Nakashima, K. Prevalence of Progressive Supranuclear Palsy in Yonago, Japan. *Mov. Disord.* **2004**, *19*, 1239–1240. [CrossRef]
89. Steele, J.C. Progressive Supranuclear Palsy: A Heterogeneous Degeneration Involving the Brain Stem, Basal Ganglia and Cerebellum with Vertical Gaze and Pseudobulbar Palsy, Nuchal Dystonia and Dementia. *Arch. Neurol.* **1964**, *10*, 333. [CrossRef]
90. Wen, Y.; Zhou, Y.; Jiao, B.; Shen, L. Genetics of Progressive Supranuclear Palsy: A Review. *J. Park. Dis.* **2021**, *11*, 93–105. [CrossRef]
91. Yabe, I.; Yaguchi, H.; Kato, Y.; Miki, Y.; Takahashi, H.; Tanikawa, S.; Shirai, S.; Takahashi, I.; Kimura, M.; Hama, Y.; et al. Mutations in Bassoon in Individuals with Familial and Sporadic Progressive Supranuclear Palsy-like Syndrome. *Sci. Rep.* **2018**, *8*, 819. [CrossRef] [PubMed]
92. Stanford, P.M.; Halliday, G.M.; Brooks, W.S.; Kwok, J.B.J.; Storey, C.E.; Creasey, H.; Morris, J.G.L.; Fulham, M.J.; Schofield, P.R. Progressive Supranuclear Palsy Pathology Caused by a Novel Silent Mutation in Exon 10 of the Tau Gene. *Brain* **2000**, *123*, 880–893. [CrossRef] [PubMed]
93. Yokota, O.; Davidson, Y.; Bigio, E.H.; Ishizu, H.; Terada, S.; Arai, T.; Hasegawa, M.; Akiyama, H.; Sikkink, S.; Pickering-Brown, S.; et al. Phosphorylated TDP-43 Pathology and Hippocampal Sclerosis in Progressive Supranuclear Palsy. *Acta Neuropathol.* **2010**, *120*, 55–66. [CrossRef] [PubMed]
94. Arai, T.; Hasegawa, M.; Akiyama, H.; Ikeda, K.; Nonaka, T.; Mori, H.; Mann, D.; Tsuchiya, K.; Yoshida, M.; Hashizume, Y.; et al. TDP-43 Is a Component of Ubiquitin-Positive Tau-Negative Inclusions in Frontotemporal Lobar Degeneration and Amyotrophic Lateral Sclerosis. *Biochem. Biophys. Res. Commun.* **2006**, *351*, 602–611. [CrossRef] [PubMed]
95. Josephs, K.A.; Whitwell, J.L.; Knopman, D.S.; Hu, W.T.; Stroh, D.A.; Baker, M.; Rademakers, R.; Boeve, B.F.; Parisi, J.E.; Smith, G.E.; et al. Abnormal TDP-43 Immunoreactivity in AD Modifies Clinicopathologic and Radiologic Phenotype. *Neurology* **2008**, *70*, 1850–1857. [CrossRef]
96. Xie, T.; Kang, U.J.; Kuo, S.-H.; Poulopoulos, M.; Greene, P.; Fahn, S. Comparison of Clinical Features in Pathologically Confirmed PSP and MSA Patients Followed at a Tertiary Center. *Npj Park. Dis.* **2015**, *1*, 15007. [CrossRef]
97. Nieforth, K.A.; Golbe, L.I. Retrospective Study of Drug Response in 87 Patients with Progressive Supranuclear Palsy. *Clin. Neuropharmacol.* **1993**, *16*, 338–346. [CrossRef] [PubMed]
98. Kompoliti, K.; Goetz, C.G.; Boeve, B.F.; Maraganore, D.M.; Ahlskog, J.E.; Marsden, C.D.; Bhatia, K.P.; Greene, P.E.; Przedborski, S.; Seal, E.C.; et al. Clinical Presentation and Pharmacological Therapy in Corticobasal Degeneration. *Arch. Neurol.* **1998**, *55*, 957. [CrossRef] [PubMed]
99. Bluett, B.; Pantelyat, A.Y.; Litvan, I.; Ali, F.; Apetauerova, D.; Bega, D.; Bloom, L.; Bower, J.; Boxer, A.L.; Dale, M.L.; et al. Best Practices in the Clinical Management of Progressive Supranuclear Palsy and Corticobasal Syndrome: A Consensus Statement of the CurePSP Centers of Care. *Front. Neurol.* **2021**, *12*, 694872. [CrossRef]
100. Saranza, G.M.; Whitwell, J.L.; Kovacs, G.G.; Lang, A.E. Corticobasal Degeneration. *Int. Rev. Neurobiol.* **2019**, *149*, 87–136. [CrossRef]
101. Wenning, G.K.; Litvan, I.; Jankovic, J.; Granata, R.; Mangone, C.A.; McKee, A.; Poewe, W.; Jellinger, K.; Chaudhuri, K.R.; D'Olhaberriague, L.; et al. Natural History and Survival of 14 Patients with Corticobasal Degeneration Confirmed at Postmortem Examination. *J. Neurol. Neurosurg. Psychiatry* **1998**, *64*, 184–189. [CrossRef]
102. Rinne, J.O.; Lee, M.S.; Thompson, P.D.; Marsden, C.D. Corticobasal Degeneration: A Clinical Study of 36 Cases. *Brain* **1994**, *117*, 1183–1196. [CrossRef]

103. Feany, M.B.; Dickson, D.W. Neurodegenerative Disorders with Extensive Tau Pathology: A Comparative Study and Review. *Ann. Neurol.* **1996**, *40*, 139–148. [CrossRef]
104. Lang, A.E. Corticobasal Degeneration: Selected Developments. *Mov. Disord.* **2003**, *18*, 51–56. [CrossRef]
105. Houlden, H.; Baker, M.; Morris, H.R.; MacDonald, N.; Pickering-Brown, S.; Adamson, J.; Lees, A.J.; Rossor, M.N.; Quinn, N.P.; Kertesz, A.; et al. Corticobasal Degeneration and Progressive Supranuclear Palsy Share a Common Tau Haplotype. *Neurology* **2001**, *56*, 1702–1706. [CrossRef]
106. Chirichigno, J.W.; Manfredi, G.; Beal, M.F.; Albers, D.S. Stress-Induced Mitochondrial Depolarization and Oxidative Damage in PSP Cybrids. *Brain Res.* **2002**, *951*, 31–35. [CrossRef]
107. Fitzmaurice, P.S.; Bamsey, C.L.; Ang, L.; Guttman, M.; Rajput, A.H.; Furukawa, Y.; Kish, S.J. Brain Aconitase Activity Is Not Decreased in Progressive Supranuclear Palsy. *Neurology* **2002**, *59*, 137–138. [CrossRef]
108. Vanek, Z.; Jankovic, J. Dystonia in Corticobasal Degeneration. *Mov. Disord.* **2001**, *16*, 252–257. [CrossRef]
109. Boeve, B.F.; Josephs, K.A.; Drubach, D.A. Current and Future Management of the Corticobasal Syndrome and Corticobasal Degeneration. *Handb. Clin. Neurol.* **2008**, *89*, 533–548. [CrossRef]
110. Kertesz, A.; Martinez-Lage, P.; Davidson, W.; Munoz, D.G. The Corticobasal Degeneration Syndrome Overlaps Progressive Aphasia and Frontotemporal Dementia. *Neurology* **2000**, *55*, 1368–1375. [CrossRef]
111. Cumings, J.N. The Copper and Iron Content of Brain and Liver in the Normal and in Hepato-Lenticular Degeneration. *Brain* **1948**, *71*, 410–415. [CrossRef]
112. Litwin, T.; Gromadzka, G.; Członkowska, A. Gender Differences in Wilson's Disease. *J. Neurol. Sci.* **2012**, *312*, 31–35. [CrossRef]
113. Loudianos, G.; Gitlin, J.D. Wilson's Disease. *Semin. Liver Dis.* **2000**, *20*, 353–364. [CrossRef]
114. Hung, I.H.; Suzuki, M.; Yamaguchi, Y.; Yuan, D.S.; Klausner, R.D.; Gitlin, J.D. Biochemical Characterization of the Wilson Disease Protein and Functional Expression in the Yeast Saccharomyces Cerevisiae. *J. Biol. Chem.* **1997**, *272*, 21461–21466. [CrossRef]
115. Schaefer, M.; Hopkins, R.G.; Failla, M.L.; Gitlin, J.D. Hepatocyte-Specific Localization and Copper-Dependent Trafficking of the Wilson's Disease Protein in the Liver. *Am. J. Physiol. Gastrointest. Liver Physiol.* **1999**, *276*, G639–G646. [CrossRef] [PubMed]
116. Schaefer, M.; Roelofsen, H.; Wolters, H.; Hofmann, W.J.; Müller, M.; Kuipers, F.; Stremmel, W.; Vonk, R.J. Localization of the Wilson's Disease Protein in Human Liver. *Gastroenterology* **1999**, *117*, 1380–1385. [CrossRef] [PubMed]
117. Frydman, M. Genetic Aspects of Wilson's Disease. *J. Gastroenterol. Hepatol.* **1990**, *5*, 483–490. [CrossRef]
118. Larin, D.; Mekios, C.; Das, K.; Ross, B.; Yang, A.-S.; Gilliam, T.C. Characterization of the Interaction between the Wilson and Menkes Disease Proteins and the Cytoplasmic Copper Chaperone, HAH1p. *J. Biol. Chem.* **1999**, *274*, 28497–28504. [CrossRef]
119. Cocoş, R.; Şendroiu, A.; Schipor, S.; Bohîlţea, L.C.; Şendroiu, I.; Raicu, F. Genotype-Phenotype Correlations in a Mountain Population Community with High Prevalence of Wilson's Disease: Genetic and Clinical Homogeneity. *PLoS ONE* **2014**, *9*, e98520. [CrossRef]
120. Chuang, L.M.; Wu, H.P.; Jang, M.H.; Wang, T.R.; Sue, W.C.; Lin, B.J.; Cox, D.W.; Tai, T.Y. High Frequency of Two Mutations in Codon 778 in Exon 8 of the ATP7B Gene in Taiwanese Families with Wilson Disease. *J. Med. Genet.* **1996**, *33*, 521–523. [CrossRef]
121. Wang, X.P.; Wang, X.H.; Bao, Y.C.; Zhou, J.N. Apolipoprotein E Genotypes in Chinese Patients with Wilson's Disease. *QJM* **2003**, *96*, 541–542. [CrossRef]
122. Brewer, G.J.; Dick, R.D.; Johnson, V.D.; Brunberg, J.A.; Kluin, K.J.; Fink, J.K. Treatment of Wilson's Disease with Zinc: XV Long-Term Follow-up Studies. *J. Lab. Clin. Med.* **1998**, *132*, 264–278. [CrossRef]
123. Brewer, G.J.; Terry, C.A.; Aisen, A.M.; Hill, G.M. Worsening of Neurologic Syndrome in Patients With Wilson's Disease with Initial Penicillamine Therapy. *Arch. Neurol.* **1987**, *44*, 490–493. [CrossRef]
124. Brewer, G.J.; Fink, J.K.; Hedera, P. Diagnosis and Treatment of Wilson's Disease. *Semin. Neurol.* **1999**, *19*, 261–270. [CrossRef]
125. Gooneratne, S.R.; Howell, J.M.; Gawthorne, J.M. An Investigation of the Effects of Intravenous Administration of Thiomolybdate on Copper Metabolism in Chronic Cu-Poisoned Sheep. *Br. J. Nutr.* **1981**, *46*, 469–480. [CrossRef]
126. Machado, A.; Fen Chien, H.; Mitiko Deguti, M.; Cançado, E.; Soares Azevedo, R.; Scaff, M.; Reis Barbosa, E. Neurological Manifestations in Wilson's Disease: Report of 119 Cases. *Mov. Disord.* **2006**, *21*, 2192–2196. [CrossRef]
127. Lorincz, M.T. Neurologic Wilson's Disease: Wilson's Disease. *Ann. N. Y. Acad. Sci.* **2010**, *1184*, 173–187. [CrossRef]
128. Dening, T.R.; Berrios, G.E.; Walshe, J.M. Wilson's Disease and Epilepsy. *Brain* **1988**, *111*, 1139–1155. [CrossRef]
129. Lorincz, M. Recognition and Treatment of Neurologic Wilson's Disease. *Semin. Neurol.* **2013**, *32*, 538–543. [CrossRef]
130. Liu, J.; Luan, J.; Zhou, X.; Cui, Y.; Han, J. Epidemiology, Diagnosis, and Treatment of Wilson's Disease. *Intractable Rare Dis. Res.* **2017**, *6*, 249–255. [CrossRef]
131. Wang, X.-P.; Zhang, W.-F.; Huang, H.-Y.; Preter, M. Neurology in the People's Republic of China—An Update. *Eur. Neurol.* **2010**, *64*, 320–324. [CrossRef] [PubMed]
132. Westermark, K.; Tedroff, J.; Thuomas, K.-Å.; Hartvig, P.; Långström, B.; Andersson, Y.; Hörnfeldt, K.; Aquilonius, S.-M. Neurological Wilson's Disease Studied with Magnetic Resonance Imaging and with Positron Emission Tomography Using Dopaminergic Markers: PET and MRI in Neurological Wilson's Disease. *Mov. Disord.* **1995**, *10*, 596–603. [CrossRef] [PubMed]
133. Berio, A.; Vento, R.; Di Stefano, A. Favorable results with an association of L-dopa and amantadine added to penicillamine in the treatment of Wilson's disease. *Minerva Pediatr.* **1973**, *25*, 807–813. [PubMed]
134. Jeon, B.; Kim, J.M.; Jeong, J.M.; Kim, K.M.; Chang, Y.S.; Lee, D.S.; Lee, M.C. Dopamine Transporter Imaging with [123I]-Beta -CIT Demonstrates Presynaptic Nigrostriatal Dopaminergic Damage in Wilson's Disease. *J. Neurol. Neurosurg. Psychiatry* **1998**, *65*, 60–64. [CrossRef] [PubMed]

135. Mueller, A.; Reuner, U.; Landis, B.; Kitzler, H.; Reichmann, H.; Hummel, T. Extrapyramidal Symptoms in Wilson's Disease Are Associated with Olfactory Dysfunction. *Mov. Disord.* **2006**, *21*, 1311–1316. [CrossRef]
136. Savic, I.; Gulyas, B.; Larsson, M.; Roland, P. Olfactory Functions Are Mediated by Parallel and Hierarchical Processing. *Neuron* **2000**, *26*, 735–745. [CrossRef]
137. Nyberg, P.; Gottfries, C.-G.; Holmgren, G.; Persson, S.; Roos, B.-E.; Winblad, B. Advanced Catecholaminergic Disturbances in the Brain in a Case of Wilson's Disease. *Acta Neurol. Scand.* **1982**, *65*, 71–75. [CrossRef]
138. Weiss, K.H.; Stremmel, W. Clinical Considerations for an Effective Medical Therapy in Wilson's Disease: Treatment Efficacy in Wilson Disease. *Ann. N. Y. Acad. Sci.* **2014**, *1315*, 81–85. [CrossRef]
139. Lang, P.A.; Schenck, M.; Nicolay, J.P.; Becker, J.U.; Kempe, D.S.; Lupescu, A.; Koka, S.; Eisele, K.; Klarl, B.A.; Rübben, H.; et al. Liver Cell Death and Anemia in Wilson Disease Involve Acid Sphingomyelinase and Ceramide. *Nat. Med.* **2007**, *13*, 164–170. [CrossRef]
140. Zischka, H.; Lichtmannegger, J.; Schmitt, S.; Jägemann, N.; Schulz, S.; Wartini, D.; Jennen, L.; Rust, C.; Larochette, N.; Galluzzi, L.; et al. Liver Mitochondrial Membrane Crosslinking and Destruction in a Rat Model of Wilson Disease. *J. Clin. Investig.* **2011**, *121*, 1508–1518. [CrossRef]
141. Walshe, J.M. Wilson's Disease: The Importance of Measuring Serum Caeruloplasmin Non-Immunologically. *Ann. Clin. Biochem. Int. J. Lab. Med.* **2003**, *40*, 115–121. [CrossRef]
142. Walshe, J.M. Serum "free" Copper in Wilson Disease. *QJM* **2012**, *105*, 419–423. [CrossRef] [PubMed]
143. Kim, D.; Kim, S.-B.; Ryu, J.L.; Hong, H.; Chang, J.-H.; Yoo, T.-J.; Jin, X.; Park, H.-J.; Han, C.; Lee, B.H.; et al. Human Embryonic Stem Cell-Derived Wilson's Disease Model for Screening Drug Efficacy. *Cells* **2020**, *9*, 872. [CrossRef] [PubMed]
144. El Balkhi, S.; Poupon, J.; Trocello, J.-M.; Leyendecker, A.; Massicot, F.; Galliot-Guilley, M.; Woimant, F. Determination of Ultrafiltrable and Exchangeable Copper in Plasma: Stability and Reference Values in Healthy Subjects. *Anal. Bioanal. Chem.* **2009**, *394*, 1477–1484. [CrossRef] [PubMed]
145. Vanier, M.T. Niemann–Pick Diseases. *Handb. Clin. Neurol.* **2013**, *113*, 1717–1721. [CrossRef]
146. Schuchman, E.H. The Pathogenesis and Treatment of Acid Sphingomyelinase-deficient Niemann–Pick Disease. *J. Inherit. Metab. Dis.* **2007**, *30*, 654–663. [CrossRef]
147. McGovern, M.M.; Aron, A.; Brodie, S.E.; Desnick, R.J.; Wasserstein, M.P. Natural History of Type A Niemann-Pick Disease: Possible Endpoints for Therapeutic Trials. *Neurology* **2006**, *66*, 228–232. [CrossRef]
148. Labrune, P.; Bedossa, P.; Huguet, P.; Roset, F.; Vanier, M.T.; Odievre, M. Fatal Liver Failure in Two Children with Niemann-Pick Disease Type B. *J. Pediatr. Gastroenterol. Nutr.* **1991**, *13*, 104–109. [CrossRef]
149. Vivas, O.; Tiscione, S.A.; Dixon, R.E.; Ory, D.S.; Dickson, E.J. Niemann-Pick Type C Disease Reveals a Link between Lysosomal Cholesterol and PtdIns(4,5)P2 That Regulates Neuronal Excitability. *Cell Rep.* **2019**, *27*, 2636–2648.e4. [CrossRef]
150. Carstea, E.D.; Morris, J.A.; Coleman, K.G.; Loftus, S.K.; Zhang, D.; Cummings, C.; Gu, J.; Rosenfeld, M.A.; Pavan, W.J.; Krizman, D.B.; et al. Niemann-Pick C1 Disease Gene: Homology to Mediators of Cholesterol Homeostasis. *Science* **1997**, *277*, 228–231. [CrossRef]
151. Naureckiene, S.; Sleat, D.E.; Lackland, H.; Fensom, A.; Vanier, M.T.; Wattiaux, R.; Jadot, M.; Lobel, P. Identification of *HE1* as the Second Gene of Niemann-Pick C Disease. *Science* **2000**, *290*, 2298–2301. [CrossRef] [PubMed]
152. Butters, T.D.; Dwek, R.A.; Platt, F.M. Inhibition of Glycosphingolipid Biosynthesis: Application to Lysosomal Storage Disorders. *Chem. Rev.* **2000**, *100*, 4683–4696. [CrossRef] [PubMed]
153. Dardis, A.; Zampieri, S.; Canterini, S.; Newell, K.L.; Stuani, C.; Murrell, J.R.; Ghetti, B.; Fiorenza, M.T.; Bembi, B.; Buratti, E. Altered Localization and Functionality of TAR DNA Binding Protein 43 (TDP-43) in Niemann-Pick Disease Type C. *Acta Neuropathol. Commun.* **2016**, *4*, 52. [CrossRef] [PubMed]
154. Patterson, M.C.; Di Bisceglie, A.M.; Higgins, J.J.; Abel, R.B.; Schiffmann, R.; Parker, C.C.; Argoff, C.E.; Grewal, R.P.; Yu, K.; Pentchev, P.G.; et al. The Effect of Cholesterol-lowering Agents on Hepatic and Plasma Cholesterol in Niemann-Pick Disease Type C. *Neurology* **1993**, *43*, 61. [CrossRef] [PubMed]
155. Hargitai, J.; Lewis, H.; Boros, I.; Rácz, T.; Fiser, A.; Kurucz, I.; Benjamin, I.; Vígh, L.; Pénzes, Z.; Csermely, P.; et al. Bimoclomol, a Heat Shock Protein Co-Inducer, Acts by the Prolonged Activation of Heat Shock Factor-1. *Biochem. Biophys. Res. Commun.* **2003**, *307*, 689–695. [CrossRef]
156. Sitarska, D.; Tylki-Szymańska, A.; Ługowska, A. Treatment Trials in Niemann-Pick Type C Disease. *Metab. Brain Dis.* **2021**, *36*, 2215–2221. [CrossRef]
157. On behalf of the NPC Registry investigators; Patterson, M.C.; Mengel, E.; Vanier, M.T.; Schwierin, B.; Muller, A.; Cornelisse, P.; Pineda, M. Stable or Improved Neurological Manifestations during Miglustat Therapy in Patients from the International Disease Registry for Niemann-Pick Disease Type C: An Observational Cohort Study. *Orphanet J. Rare Dis.* **2015**, *10*, 65. [CrossRef]
158. Wheeler, S.; Sillence, D.J. Niemann–Pick Type C Disease: Cellular Pathology and Pharmacotherapy. *J. Neurochem.* **2020**, *153*, 674–692. [CrossRef]
159. Rosenberg, R.N.; Pascual, J.M. *Rosenberg's Molecular and Genetic Basis of Neurological and Psychiatric Disease*, 5th ed.; Elsevier Academic Press: Amsterdam, The Netherlands, 2015.
160. Fernandes Filho, J.A.; Shapiro, B.E. Tay-Sachs Disease. *Arch. Neurol.* **2004**, *61*, 1466. [CrossRef]
161. Myerowitz, R. Tay-Sachs Disease-Causing Mutations and Neutral Polymorphisms in the Hex A Gene. *Hum. Mutat.* **1997**, *9*, 195–208. [CrossRef]

162. Navon, R.; Kolodny, E.H.; Mitsumoto, H.; Thomas, G.H.; Proia, R.L. Ashkenazi-Jewish and Non-Jewish Adult GM2 Gangliosidosis Patients Share a Common Genetic Defect. *Am. J. Hum. Genet.* **1990**, *46*, 817–821. [PubMed]
163. Akalin, N.; Shi, H.-P.; Vavougios, G.; Hechtman, P.; Lo, W.; Scriver, C.R.; Mahuran, D.; Kaplan, F. Novel Tay-Sachs Disease Mutations from China. *Hum. Mutat.* **1992**, *1*, 40–46. [CrossRef] [PubMed]
164. Myerowitz, R. Splice Junction Mutation in Some Ashkenazi Jews with Tay-Sachs Disease: Evidence against a Single Defect within This Ethnic Group. *Proc. Natl. Acad. Sci. USA* **1988**, *85*, 3955–3959. [CrossRef] [PubMed]
165. Kaufman, M.; Grinshpun-Cohen, J.; Karpati, M.; Peleg, L.; Goldman, B.; Akstein, E.; Adam, A.; Navon, R. Tay-Sachs Disease AndHEXA Mutations among Moroccan Jews. *Hum. Mutat.* **1997**, *10*, 295–300. [CrossRef]
166. Cachon-Gonzalez, M.B.; Zaccariotto, E.; Cox, T.M. Genetics and Therapies for GM2 Gangliosidosis. *Curr. Gene Ther.* **2018**, *18*, 68–89. [CrossRef] [PubMed]
167. Maegawa, G.H.B.; Tropak, M.; Buttner, J.; Stockley, T.; Kok, F.; Clarke, J.T.R.; Mahuran, D.J. Pyrimethamine as a Potential Pharmacological Chaperone for Late-Onset Forms of GM2 Gangliosidosis. *J. Biol. Chem.* **2007**, *282*, 9150–9161. [CrossRef]
168. Danielsen, E.R.; Lund, A.M.; Thomsen, C. Cerebral Magnetic Resonance Spectroscopy Demonstrates Long-Term Effect of Bone Marrow Transplantation in α-Mannosidosis. In *JIMD Reports—Volume 11*; Zschocke, J., Gibson, K.M., Brown, G., Morava, E., Peters, V., Eds.; JIMD Reports; Springer: Berlin/Heidelberg, Germany, 2013; Volume 11, pp. 49–52. [CrossRef]
169. Maegawa, G.H.B.; Banwell, B.L.; Blaser, S.; Sorge, G.; Toplak, M.; Ackerley, C.; Hawkins, C.; Hayes, J.; Clarke, J.T.R. Substrate Reduction Therapy in Juvenile GM2 Gangliosidosis. *Mol. Genet. Metab.* **2009**, *98*, 215–224. [CrossRef]
170. Dersh, D.; Iwamoto, Y.; Argon, Y. Tay–Sachs Disease Mutations in HEXA Target the α Chain of Hexosaminidase A to Endoplasmic Reticulum–Associated Degradation. *Mol. Biol. Cell* **2016**, *27*, 3813–3827. [CrossRef]
171. Clarke, J.T.R.; Mahuran, D.J.; Sathe, S.; Kolodny, E.H.; Rigat, B.A.; Raiman, J.A.; Tropak, M.B. An Open-Label Phase I/II Clinical Trial of Pyrimethamine for the Treatment of Patients Affected with Chronic GM2 Gangliosidosis (Tay–Sachs or Sandhoff Variants). *Mol. Genet. Metab.* **2011**, *102*, 6–12. [CrossRef]
172. Manyam, B.V.; Walters, A.S.; Narla, K.R. Bilateral Striopallidodentate Calcinosis: Clinical Characteristics of Patients Seen in a Registry. *Mov. Disord.* **2001**, *16*, 258–264. [CrossRef]
173. Chiu, H.F.K.; Lam, L.C.W.; Shum, P.P.S.; Li, K.W. Idiopathic Calcification of the Basal Ganglia. *Postgrad. Med. J.* **1993**, *69*, 68–70. [CrossRef]
174. Saleem, S.; Aslam, H.M.; Anwar, M.; Anwar, S.; Saleem, M.; Saleem, A.; Rehmani, M.A.K. Fahr's Syndrome: Literature Review of Current Evidence. *Orphanet J. Rare Dis.* **2013**, *8*, 156. [CrossRef]
175. el Maghraoui, A.; Birouk, N.; Zaim, A.; Slassi, I.; Yahyaoui, M.; Chkili, T. [Fahr syndrome and dysparathyroidism. 3 cases]. *Presse Medicale Paris Fr. 1983* **1995**, *24*, 1301–1304.
176. Jensen, N.; Autzen, J.K.; Pedersen, L. Slc20a2 Is Critical for Maintaining a Physiologic Inorganic Phosphate Level in Cerebrospinal Fluid. *Neurogenetics* **2016**, *17*, 125–130. [CrossRef] [PubMed]
177. Keller, A.; Westenberger, A.; Sobrido, M.J.; García-Murias, M.; Domingo, A.; Sears, R.L.; Lemos, R.R.; Ordoñez-Ugalde, A.; Nicolas, G.; Da Cunha, J.E.G.; et al. Mutations in the Gene Encoding PDGF-B Cause Brain Calcifications in Humans and Mice. *Nat. Genet.* **2013**, *45*, 1077–1082. [CrossRef] [PubMed]
178. Lemos, R.R.; Ferreira, J.B.M.M.; Keasey, M.P.; Oliveira, J.R.M. An Update on Primary Familial Brain Calcification. *Int. Rev. Neurobiol.* **2013**, *110*, 349–371. [CrossRef]
179. Turan, S.; Bastepe, M. GNAS Spectrum of Disorders. *Curr. Osteoporos. Rep.* **2015**, *13*, 146–158. [CrossRef]
180. Batla, A.; Tai, X.Y.; Schottlaender, L.; Erro, R.; Balint, B.; Bhatia, K.P. Deconstructing Fahr's Disease/Syndrome of Brain Calcification in the Era of New Genes. *Parkinsonism Relat. Disord.* **2017**, *37*, 1–10. [CrossRef]
181. Abe, S.; Tojo, K.; Ichida, K.; Shigematsu, T.; Hasegawa, T.; Morita, M.; Sakai, O. A Rare Case of Idiopathic Hypoparathyroidism with Varied Neurological Manifestations. *Intern. Med.* **1996**, *35*, 129–134. [CrossRef]
182. Zhang, Q.; Li, Z.; Sun, H.; Zhang, S.; Zhang, J.; Wang, Y.; Fang, H.; Xu, Y. Generation of Induced Pluripotent Stem Cell Line (ZZUi0012-A) from a Patient with Fahr's Disease Caused by a Novel Mutation in SLC20A2 Gene. *Stem Cell Res.* **2019**, *35*, 101395. [CrossRef]
183. Sakthiswary, R.; Raymond, A.A. Stem Cell Therapy in Neurodegenerative Diseases: From Principles to Practice. *Neural Regen. Res.* **2012**, *7*, 1822–1831. [CrossRef] [PubMed]
184. Fang, Y.; Wang, J.; Zhao, M.; Zheng, Q.; Ren, C.; Wang, Y.; Zhang, J. Progress and Challenges in Targeted Protein Degradation for Neurodegenerative Disease Therapy. *J. Med. Chem.* **2022**, *65*, 11454–11477. [CrossRef] [PubMed]
185. O'Connor, D.M.; Boulis, N.M. Gene Therapy for Neurodegenerative Diseases. *Trends Mol. Med.* **2015**, *21*, 504–512. [CrossRef] [PubMed]

International Journal of *Molecular Sciences*

MDPI

Review

From Recognition to Remedy: The Significance of Biomarkers in Neurodegenerative Disease Pathology

Corneliu Toader [1,2,†], Nicolaie Dobrin [3,*], Felix-Mircea Brehar [1,4,*], Constantin Popa [5,6,7,†], Razvan-Adrian Covache-Busuioc [1], Luca Andrei Glavan [1], Horia Petre Costin [1], Bogdan-Gabriel Bratu [1], Antonio Daniel Corlatescu [1], Andrei Adrian Popa [1] and Alexandru Vlad Ciurea [1,7,8]

1 Department of Neurosurgery, "Carol Davila" University of Medicine and Pharmacy, 020021 Bucharest, Romania; corneliu.toader@umfcd.ro (C.T.); luca-andrei.glavan0720@stud.umfcd.ro (L.A.G.); horia-petre.costin0720@stud.umfcd.ro (H.P.C.); bogdan.bratu@stud.umfcd.ro (B.-G.B.); antonio.corlatescu0920@stud.umfcd.ro (A.D.C.); prof.avciurea@gmail.com (A.V.C.)
2 Department of Vascular Neurosurgery, National Institute of Neurology and Neurovascular Diseases, 077160 Bucharest, Romania
3 Department of Neurosurgery, Clinical Emergency Hospital "Prof. Dr. Nicolae Oblu", 700309 Iasi, Romania
4 Department of Neurosurgery, Clinical Emergency Hospital "Bagdasar-Arseni", 041915 Bucharest, Romania
5 Department of Neurology, "Carol Davila" University of Medicine and Pharmacy, 020021 Bucharest, Romania
6 Department of Neurology, National Institute of Neurology and Neurovascular Diseases, 077160 Bucharest, Romania
7 Medical Science Section, Romanian Academy, 060021 Bucharest, Romania
8 Neurosurgery Department, Sanador Clinical Hospital, 010991 Bucharest, Romania
* Correspondence: dobrin_nicolaie@icloud.com (N.D.); felix.brehar@umfcd.ro (F.-M.B.)
† These authors contributed equally to this work.

Abstract: With the inexorable aging of the global populace, neurodegenerative diseases (NDs) like Alzheimer's disease (AD), Parkinson's disease (PD), and amyotrophic lateral sclerosis (ALS) pose escalating challenges, which are underscored by their socioeconomic repercussions. A pivotal aspect in addressing these challenges lies in the elucidation and application of biomarkers for timely diagnosis, vigilant monitoring, and effective treatment modalities. This review delineates the quintessence of biomarkers in the realm of NDs, elucidating various classifications and their indispensable roles. Particularly, the quest for novel biomarkers in AD, transcending traditional markers in PD, and the frontier of biomarker research in ALS are scrutinized. Emergent susceptibility and trait markers herald a new era of personalized medicine, promising enhanced treatment initiation especially in cases of SOD1-ALS. The discourse extends to diagnostic and state markers, revolutionizing early detection and monitoring, alongside progression markers that unveil the trajectory of NDs, propelling forward the potential for tailored interventions. The synergy between burgeoning technologies and innovative techniques like -omics, histologic assessments, and imaging is spotlighted, underscoring their pivotal roles in biomarker discovery. Reflecting on the progress hitherto, the review underscores the exigent need for multidisciplinary collaborations to surmount the challenges ahead, accelerate biomarker discovery, and herald a new epoch of understanding and managing NDs. Through a panoramic lens, this article endeavors to provide a comprehensive insight into the burgeoning field of biomarkers in NDs, spotlighting the promise they hold in transforming the diagnostic landscape, enhancing disease management, and illuminating the pathway toward efficacious therapeutic interventions.

Keywords: neurodegenerative diseases; aging population; socio-economic implications; Alzheimer's disease; Parkinson's disease; amyotrophic lateral sclerosis; biomarker identification; wet and dry markers; early diagnosis; disease monitoring; therapeutic efficacy; susceptibility markers; trait markers; personalized medicine; diagnostic accuracy; disease progression; innovative techniques; -omics technologies; histologic assessments; imaging technology

Citation: Toader, C.; Dobrin, N.; Brehar, F.-M.; Popa, C.; Covache-Busuioc, R.-A.; Glavan, L.A.; Costin, H.P.; Bratu, B.-G.; Corlatescu, A.D.; Popa, A.A.; et al. From Recognition to Remedy: The Significance of Biomarkers in Neurodegenerative Disease Pathology. *Int. J. Mol. Sci.* **2023**, *24*, 16119. https://doi.org/10.3390/ijms242216119

Academic Editor: Hideaki Matsui

Received: 10 October 2023
Revised: 28 October 2023
Accepted: 6 November 2023
Published: 9 November 2023

1. Introduction

1.1. The Growing Challenge of Neurodegenerative Diseases Due to the Aging Population

Given the recent shifts in global demography, there has been a significant rise in the older adult cohort, specifically those above 65 years. Contemporary demographic studies indicate that this age group is expanding more rapidly than its counterparts. Data from 2019 show that 703 million individuals globally were aged 65 or older, which is a figure that is projected to surge to 1.5 billion by 2050 [1]. In a 2018 study by the American Academy of Neurology, it was inferred that between 15 and 20% of those aged 65 and above exhibited symptoms of mild cognitive impairment (MCI), which is a syndrome characterized by subtle cognitive decline with negligible impact on daily functional activities [2]. The escalating aging trend worldwide is concomitant with the rise of age-related health challenges. Notably, recent analyses have emphasized that neurological disorders stand as the primary contributors to DALYs (disability-adjusted life-years), accounting for 276 million cases, and are the second predominant cause of mortality, with 90 million cases [3]. Corroborating this, there is compelling epidemiological data suggesting a potential link between physical frailty and cognitive setbacks in older age, including the onset of Alzheimer's disease (AD), MCI, vascular dementia (VaD), non-AD dementias, and the presence of AD pathology even in older individuals not diagnosed with dementia [4]. However, the current epidemiological landscape presents a scarcity of comprehensive data on cognitive frailty, particularly its prevalence and implications [5]. The trajectory from cognitive frailty to full-blown dementia remains ambiguous at present.

1.2. Understanding the Socioeconomic Implications of Increasing Prevalence of Conditions like AD, PD, and ALS

Addressing the economic repercussions due to the upsurge in Alzheimer's disease is paramount. A proactive approach involving early detection and intervention is vital not only to mitigate the prevalence of AD but also to elevate the life quality of both the affected individuals and their caregivers. The institution of robust social support mechanisms is integral to this strategy. Non-pharmacological measures emerge as the most preferred modalities in both the prevention and management of AD [6]. There is a prevailing hypothesis linking socioeconomic standing to AD, although the underlying cause for this association has not been unequivocally elucidated by prior research. A study by Wang et al. employed Mendelian randomization to delve into the potential influence of socioeconomic strata on AD and probed if elevated income exerted a protective effect against the disease's onset [7]. From a health economics perspective, evaluations bifurcate into comparative analyses, assessing the cost–benefit ratio of varied therapeutic avenues, and cost-of-illness (COI) evaluations, which ascertain the economic strain of an ailment from a defined standpoint. Parkinson's disease has been the subject of numerous COI studies across diverse global regions [8]. The protracted nature of PD, characterized by escalating disability and increasing dependence in activities of daily living (ADLs), imposes a substantial socioeconomic load. Advanced stages necessitate specialized institutional care, entailing significant resources and expenditures. Moreover, the familial impact of PD is profound with most ADL-dependent patients relying on home-based care provided by family members [9]. A subsequent exploration hypothesizes that there is a potential correlation between ALS risk and dietary habits, specifically the frequent intake of expensive, high-trophic level fish species known for elevated mercury content. This led to a detailed examination of the interplay between ALS, socioeconomic status, and mercury exposure via fish consumption [10]. Furthermore, the CDC's NIOSH National Occupational Mortality Surveillance (NOMS) research discerned that professions linked to a superior socioeconomic tier, such as computer-related fields, engineering, legal practices, and business operations, manifested augmented ALS mortality rates after adjusting for demographic variables like age, gender, and ethnicity [11].

1.3. The Critical Need for New Biomarkers in the Context of Disease Diagnosis, Monitoring, and Treatment

The discernment of consistent biomarkers holds promise in advancing the early detection of neurodegenerative diseases, paving the way for the initiation of tailored therapeutic regimens. At present, the realm of epigenetics lacks robust and dependable biomarkers conducive to the diagnosis, categorization, or tracking of neurodegenerative disease progression [12]. In the context of available diagnostic modalities for neurodegenerative ailments, while pathological evaluations are held in high esteem across diverse afflictions, their applicability is limited in discerning neurodegenerative diseases during a patient's lifetime. Alternatives like positron emission tomography (PET) scans or emergent biomarkers (encompassing genomics and proteomics) present potential breakthroughs and are being integrated into refined diagnostic parameters [13]. However, it is noteworthy that parameters such as DNA methylation levels, SIRT activity, and BDNF expression witness a marked decline in individuals diagnosed with dementia or Parkinson's disease. Hence, the concurrent assessment of these epibiomarkers might enhance the diagnostic accuracy for neurodegenerative diseases. Given the reversibility of epigenetic alterations, gauging parameters like DNA methylation levels, SIRT activity, and BDNF expression could equip medical practitioners with insights to evaluate the efficacy of therapeutic interventions [14].

2. The Significance of Biomarkers in Neurodegenerative Diseases

2.1. What Are Biomarkers and Why Are They Important?

The concept of a "biomarker", derived from the amalgamation of "biological" and "marker", encompasses a broad range of medical signs. These signs provide objective evidence of a patient's health condition and can be consistently and accurately quantified. This is distinct from medical symptoms, which are subjective sensations or complaints reported by the patient [15]. Biomarkers serve as pivotal tools in the methodical evolution of pharmaceuticals and medical apparatuses [16]. Yet, despite their immense significance, there exists a pronounced ambiguity surrounding their foundational definitions and the intricacies of their application in both research and clinical settings [17]. The spectrum of biomarkers ranges from elementary metrics like pulse and blood pressure to intricate laboratory assessments of blood and other biological specimens. Historically, medical signs have always been integral to clinical practice, with biomarkers representing the pinnacle of objective and quantifiable indicators that contemporary lab sciences can consistently measure. In the realm of drug innovation and broader biomedical investigations, biomarkers hold a transformative role. Deciphering the interplay between quantifiable biological mechanisms and clinical results is paramount for bolstering our repertoire of disease interventions and for a profound comprehension of standard physiological processes [18]. For biomarkers to be genuinely efficacious as replacements for clinically relevant endpoints, there is a prerequisite to thoroughly grasp the standard biological mechanisms, the alterations in disease conditions, and the impacts of varied interventions, be they drug-induced, device-based, or other [15]. The imperative for the prompt and precise identification of neurodegenerative conditions in clinical environments cannot be overstated. Beyond furnishing diagnostic and future insights, this need also encompasses the fine tuning of therapeutic approaches, ensuring apt care and support, and offering patients avenues to participate in clinical therapeutic studies [19].

2.2. Differentiating between Risk, Prodromal, Clinical, Wet, Dry Markers and Surrogate Endpoints

The methodology of risk assessment finds its application across diverse clinical spheres and for a variety of clinical outcomes. Regardless of the specific clinical domain or outcome in question, the foundational principles and techniques for evaluating risk markers and risk assessment remain consistent. Risk is typically gauged by counting the number of outcome incidents over a specified time span. This is traditionally encapsulated either via a survival curve or by denoting the fraction of incidents within a designated time frame, such as 30 days or a year [20]. As a result, there is often a strong interrelation among multiple biomarkers, complicating the process of pinpointing a singular prominent marker. Within the field of periodontology, the quest for risk biomarkers that can predict potential disease onset in individuals devoid of clinical symptoms is ongoing [21]. For Parkinson's disease in its prodromal phase, while markers can facilitate diagnosis, it is imperative to understand four central characteristics of these markers, especially if they are to guide the selection of neuroprotective treatments. Among these, understanding the specificity or predictive accuracy of the marker is crucial, given the notable variances in specificity and positive predictive value (PPV) among different prodromal markers [22]. In this context, a "wet biomarker" is delineated as a prospective biomarker that can be objectively ascertained within a body fluid [23]. Biomarkers have been categorized into two main types: "dry" markers, which encompass imaging parameters, and "wet" markers, which refer to genetic and biochemical elements detectable in fluids such as blood, serum, urine, and tissue samples [24]. There are also surrogate markers (or surrogate endpoint), which are markers that are used as a distant relationship between an action and a clinical endpoint. An example of this would be the easy-to-understand relationship between smoking and lung cancer [25]. A surrogate endpoint of smoking would be death. Therefore, smoking is a surrogate marker of death via lung cancer. The utility of these endpoints would be of great value because it would clarify more easily the barrier between the general population and disease. It is a challenging task to pick surrogate endpoints and demonstrate their efficacy, because this action requires an extraordinary understanding of the disease's pathophysiology. Several studies in the current literature have clarified the important yet difficult task to create these surrogate endpoints, and they have demonstrate the failure of this viewpoint in numerous studies, including neurodegenerative disease [26–30].

2.3. Overview of Their Roles in Early Diagnosis, Monitoring Disease Progression, and Evaluating Therapeutic Efficacy

At present, the categorization of most biomarkers hinges on the pathogenic processes they signify. For conditions like Alzheimer's disease and frontotemporal lobar degeneration (FTLD) spectrum, the primary focus is on biomarkers indicative of pathology, such as those for amyloid-β (Aβ) and tau pathologies. These biomarkers are predominantly evaluated through CSF examinations, blood tests, and positron emission tomography scans [31]. In the preclinical stages of AD, while there are detectable biomarkers signaling brain alterations, clinical manifestations remain absent [32]. Conversely, in Parkinson's disease (PD), the onset of classic motor symptoms is observed only after a significant proportion, over half, of neurons in the substantia nigra (SN) have already degenerated [33]. Consequently, pinpointing these conditions early is imperative for implementing strategies geared toward preventing neuronal loss. Over recent years, there has been a concerted effort by researchers to bolster the advancement of reliable biomarkers for neurodegenerative ailments. Despite these endeavors, results have often been inconsistent and not always meeting optimal standards. The trajectory of medical practice is increasingly leaning toward precision medicine, underscoring the pressing need to seamlessly incorporate disease-specific biomarkers in clinical routines and to engineer potent disease-altering treatments [31]. A double approach regarding neurodegenerative disease could be, firstly, neuroinflammation, which is a key factor that is both result and cause of neurodegeneration [34]. Secondly, in the last decade, research has pinpointed another key factor of neurodegeneration: cIMT (carotid intima media thickness). cIMT has been long debated as a surrogate endpoint of neurodegenerative disease; however, nowadays, it is a relevant influence in neurodegenerative disease [35–37].

2.4. Overview of Biomarkers in Huntington's Disease, Multiple Sclerosis, Frontotemporal Dementia and Essential Tremor

2.4.1. Huntington's Disease

Increasing emphasis has been placed on the significance of white matter in the degenerative process [38], as widespread alterations can be detected over a decade prior to anticipated disease onset [39]. A comprehensive study amalgamated clinical and morphometric imaging data from 1082 participants, sourced from the IMAGE-HD, TRACK-HD, and PREDICT-HD studies, with longitudinal observations spanning 1–10 years. The findings from this research indicate that imaging might be a viable endpoint in clinical trials due to its potential heightened sensitivity [40].

Regarding the wet biomarkers, a study indicates that mutant HTT levels exhibit correlations with clinical scores both cross-sectionally and in relation to CSF tau and neurofilament light chain (NfL) [41], both being indicators of neuronal damage [42]. This suggests that mHTT is likely released from compromised or deteriorating neurons. Given the pivotal role of mHTT in HD pathogenesis, it emerges as a salient potential biomarker. Not only is it the pathogenic agent in itself, but in the context of Huntington-lowering, it stands as a crucial gauge of pharmacodynamics, signifying whether the therapeutic agent has effectively engaged its target and manifested the anticipated immediate biological effect [43] (See Table 1).

One study indicated that the accumulated data suggest a discernible segment of mHTT in the CSF is derived from striatal cells. These results advocate for the application of CSF mHTT as a PD biomarker in evaluating the engagement of therapeutic interventions tailored to decrease mHTT levels in the striatum [44].

A subsequent study explored the feasibility of utilizing noninvasive positron emission tomography (PET) for direct assessment of therapeutic efficacy and monitoring disease evolution in relation to mHTT. In this context, the novel radioligand [11C]CHDI-626 was characterized and examined longitudinally for mHTT PET imaging within the zQ175DN mouse model of HD. Notwithstanding its rapid metabolism and kinetics, the radioligand proved efficacious for mHTT PET imaging [45].

Table 1. Biomarkers used for Huntington's Disease.

References	Biomarker(s) Type/Tool	Key Findings	Implication/Significance
[38,39]	White Matter (Imaging)	Emphasis on its significance; alterations seen over a decade before anticipated disease onset.	Crucial for understanding disease progression.
[40]	Imaging Data	Data amalgamated from 1082 participants over 1–10 years from various studies.	Indicates imaging as a viable endpoint in clinical trials due to heightened sensitivity.
[41,42]	mHTT (Wet Biomarker)	Correlates with clinical scores, CSF tau, and NfL.	mHTT potentially released from compromised neurons; possible biomarker for HD.
[43]	mHTT (Wet Biomarker)	Significant in HD pathogenesis.	Crucial gauge of pharmacodynamics for huntingtin-lowering therapies.
[44]	CSF mHTT (Wet Biomarker)	Derived from striatal cells.	Suggested as PD biomarker for therapeutic engagement evaluations.
[45]	[11C]CHDI-626 (PET)	Examined for mHTT PET imaging in zQ175DN mouse model.	Proves effective for mHTT PET imaging despite rapid metabolism.
[46]	mHTT and CBVa	Early HTT-lowering treatment defers onset and decelerates progression in mHTT mouse model; CBVa alteration influenced by mHTT on neural activity.	Indicates potential therapeutic interventions and understanding neuronal dysfunction mechanisms.

Liu et al.'s study furnishes initial evidence indicating that the early introduction of HTT-lowering treatment, prior to the manifestation of motor symptoms and striatal atrophy, can defer the onset and decelerate the progression of pathology and phenotype in a mouse model expressing full-length mHTT [46]. Concurrently, the research findings posit that the observed alteration in CBVa in premanifest zQ175 mice is a subsequent effect stemming from the influence of mHTT on neural activity/metabolism. Furthermore, the study suggests that a diminished rate of oxygen/nutrient delivery, attributed to a reduced cerebral blood volume and a decline in glucose transporter GLUT1 across a jeopardized neurovascular network during the manifest stage, may eventually instigate neuronal dysfunction and degeneration [46] (See Table 1).

2.4.2. Multiple Sclerosis

In multiple sclerosis (MS), magnetic resonance imaging (MRI) elucidates the dimensions, quantity, chronology, and evolution of lesions within the central nervous system (CNS). Consequently, MRI is integral to the diagnostic process and therapeutic surveillance [47–49]. A study by Huang et al. demonstrates an up-regulation of MIP-1a and CXCL10 in the cerebrospinal fluid (CSF) of patients diagnosed with multiple sclerosis. Collectively, these cytokine biomarkers serve as a significant indicator of T cell activity, offering a measure that is both independent and complementary to the previously documented CXCL13, which is a chemokine targeting B lymphocytes [50]. To date, the singular cerebrospinal fluid (CSF) biomarker of clinical significance for MS is the presence of immunoglobulin G (IgG) oligoclonal bands (OCBs). These OCBs signify the intrathecal production of IgG, acting as a broader indicator of adaptive immunity activation within the CNS. It is pertinent to note that OCBs are not exclusive to MS; they have been identified in various inflammatory neurological disorders. Additionally, approximately 5% of MS instances do not exhibit CSF OCBs based on conventional assays [51–55] (See Table 2).

Blood-based serum neurofilament light chain (sNfL) is a potential and easily accessible prognostic and treatment response biomarker for patients diagnosed with multiple sclerosis. It is important to note that without the inclusion of supplementary clinical context, sNfL on its own does not suffice for diagnosing multiple sclerosis or distinguishing it from other neuroinflammatory conditions characterized by neuroaxonal damage and elevated sNfL levels, such as neuromyelitis optica spectrum disorders or myelin oligodendrocyte glycoprotein (MOG) encephalomyelitis [56–59].

Table 2. Biomarkers used for Multiple Sclerosis.

References	Biomarker(s)	Description/Function	Sample Origin
[47–49]	MRI	Used to elucidate the dimensions, quantity, chronology, and evolution of lesions in the CNS	Central Nervous System (CNS)
[50]	MIP-1a	Cytokine biomarker indicative of T cell activity	Cerebrospinal Fluid (CSF)
[50]	CXCL10	Cytokine biomarker indicative of T cell activity	Cerebrospinal Fluid (CSF)
[50]	CXCL13	Chemokine targeting B lymphocytes	Not Specified
[51–55]	IgG Oligoclonal Bands (OCBs)	Indicator of adaptive immunity activation in the CNS	Cerebrospinal Fluid (CSF)
[56–59]	sNfL (serum neurofilament light chain)	Potential prognostic and treatment response biomarker	Blood Serum

2.4.3. Frontotemporal Dementia

Over the past decade, neurofilament light chain (NfL) has garnered attention as a potential biomarker for FTLD due to its sensitivity in detecting neurodegeneration. Moreover, its levels demonstrate a correlation with the pace of clinical progression, providing prognostic insights. Recent scholarly investigations underscore the utility of NfL as a discriminative biomarker between bvFTD and primary psychiatric disorders, exhibiting areas under the curve ranging from 0.84 to 0.94 [60–64].

Progranulin (GRN) can be quantified in both blood and CSF, although the preponderance of research has been conducted on blood samples. Preliminary investigations reported remarkable sensitivity and specificity (both exceeding 95%) with a threshold of 61.5 ng/mL (ascertained in plasma using the Adipogen assay). However, subsequent research has proposed an elevated threshold of 71.0 ng/mL, boasting a sensitivity of 98.1% and specificity of 98.5%. It is posited that these levels are diminished from birth, as they appear to be low even when first assessed during late adolescence. Furthermore, these levels manifest consistent stability over extended periods, remaining relatively unaltered for up to four years as evidenced in one study [65–67] (See Table 3).

Table 3. Biomarkers used for Frontotemporal Dementia.

References	Biomarker(s)	Sample Type	Key Findings	Clinical Implications
[60–64]	Neurofilament light chain (NfL)	Not specified	Correlation with the pace of clinical progression. Discriminative potential between bvFTD and primary psychiatric disorders; areas under the curve: 0.84 to 0.94	Prognostic insights and discriminative utility between bvFTD and psychiatric disorders
[65–67]	Progranulin (GRN)	Blood, CSF	Remarkable sensitivity and specificity exceeding 95% at a threshold of 61.5 ng/mL (using AdipoGen assay). Elevated threshold proposed: 71.0 ng/mL, with a sensitivity of 98.1% and specificity of 98.5%. Levels show stability over extended periods	Potentially discriminative and diagnostic for FTLD. Stability over time suggests reliable biomarker potential

2.4.4. Essential Tremor

In a forward-looking study that distinguished between sporadic-ET and hereditary-ET cases, the levels of uric acid were juxtaposed with those of controls. The results did not indicate significant deviations, thereby not affirming a neuroprotective function of uric acid in ET. Nonetheless, it is noteworthy that a correlation emerged between reduced uric acid levels and a later age of onset in sporadic cases, suggesting its potential significance as an indicator of neurodegeneration in such patients [68].

A study by Wang et al. introduced a methodologically sound consensus-based approach to scrutinize cerebellar involvement in ET, leveraging an augmented cohort for enhanced statistical power and taking into account the implications of MRI processing pipelines and statistical frameworks. This examination did not identify cerebellar involvement for advanced ET when synthesizing findings from three MRI biomarkers: voxel-based morphometry, cerebellar gray matter and white matter volumetry, and cerebellar lobular volumetry. The hypothesis was further assessed using ten prevalent statistical models based on biomarkers from Freesurfer, SUIT, and MAGeT. Notably, no cerebellar ROI derived from these three pipelines exhibited a consistent significant discrepancy [69] (See Table 4).

Another study performed by Yu et al. revealed that erythrocytic total and aggregated α-syn concentrations were significantly elevated in PD and ET patients in comparison to HCs. Notably, erythrocytic total α-syn levels were observed to be markedly higher in the ET cohort than in the PD group. Additionally, the ratios of erythrocytic aggregated to total α-syn levels in the ET group were discernibly reduced relative to those in the PD and HC groups. A significant correlation was also identified between erythrocytic aggregated α-syn levels and the disease duration in ET patients [70].

Table 4. Biomarkers used for Frontotemporal Dementia.

References	Biomarker(s)	Sample Type	Key Findings	Clinical Implications
[68]	Uric acid	Not specified	No significant deviations between sporadic ET, hereditary ET, and controls. A correlation between reduced uric acid levels and later age of onset in sporadic ET.	Uric acid levels may have significance as an indicator of neurodegeneration in sporadic ET patients.
[69]	Cerebellar MRI biomarkers (voxel-based morphometry, cerebellar gray and white matter volumetry, cerebellar lobular volumetry)	MRI	No cerebellar involvement identified for advanced ET across multiple MRI biomarkers and statistical models.	No significant cerebellar alterations in advanced ET.

Table 4. *Cont.*

References	Biomarker(s)	Sample Type	Key Findings	Clinical Implications
[70]	Erythrocytic total and aggregated α-syn	Blood (erythrocyte)	Erythrocytic total and aggregated α-syn levels significantly elevated in PD and ET vs. HCs. Erythrocytic total α-syn levels higher in ET than PD. Reduced ratios of erythrocytic aggregated to total α-syn in ET vs. PD and HCs. Correlation between erythrocytic aggregated α-syn levels and disease duration in ET.	Erythrocytic α-syn concentrations might have diagnostic potential for distinguishing ET, PD, and HCs. The biomarker also correlates with ET progression.

3. Alzheimer's Disease (AD): The Quest for Novel Biomarkers

3.1. Current State of AD Biomarker Identification

Lecanemab (BAN2401), an IgG1 monoclonal antibody, is designed to target soluble aggregated forms of amyloid beta (Aβ), spanning oligomers, protofibrils, and insoluble fibrils. The BAN2401-G000-201 clinical trial, structured as a randomized double-blind study with a Bayesian design, evaluated three doses of lecanemab against a placebo in the early stages of Alzheimer's disease, covering both mild cognitive impairment due to AD and mild AD dementia [31]. The primary evaluation criterion was the change from the outset at 12 months based on ADCOMS [71]. Essential secondary criteria encompassed changes in brain amyloid through PET Standard Uptake Value ratio (SUVr), ADCOMS, Clinical Dementia Rating-Sum-of-Boxes (CDR-SB), Alzheimer Disease Assessment Scale-Cognitive Subscale (ADAS-Cog14), CSF biomarkers, and total hippocampal volume as discerned by volumetric magnetic resonance imaging (vMRI). An additional focal point was the assessment of lecanemab's efficacy in comparison to a placebo at 18 months based on ADCOMS, CDR-SB, and ADAS-Cog14 within specific clinical subgroups [72].

Given the lack of therapeutic solutions for AD, physical activity has emerged as a pivotal lifestyle determinant that might mitigate or delay the disease's onset [73]. Delving into the impacts of exercise on systemic biomarkers linked to AD risk and correlating them with pivotal metabolomic shifts can propel preventive, monitoring, and therapeutic endeavors. A study evaluated systemic biomarkers, namely CTSB, BDNF, and klotho, and conducted a metabolomics analysis after a 26-week aerobic regimen [74]. In terms of CSF biomarkers, specific dietary patterns manifested varying effects on Aβ40 and Aβ42/40 ratios among different participant groups [75]. Another investigative endeavor probed into plasma biomarkers tied to neuroinflammation in relation to AD among a preclinical AD cohort. Only GFAP was found to be significantly elevated in the preclinical AD group compared to the healthy elderly [76]. See Figure 1.

Another study incorporated both subjective and objective cognitive performance metrics, in addition to parameters like sleep, stress, mood, and quality of life, facilitating a comprehensive evaluation of cognitive function and psychosocial well-being in relation to AD biomarker shifts [77]. An exploration into the relationship between certain plasma biomarkers and clinical efficacy endpoints underscored the predictive capacity of these markers in gauging cognitive decline [78]. The study also highlighted the potential utility of plasma biomarkers in monitoring lecanemab's therapeutic effects and possibly individual patient responses. These insights are formative and will be delved into further in upcoming phase 3 lecanemab clinical trials [78]. Additionally, gantenerumab treatment showcased a dose-dependent impact on CSF biomarkers indicative of AD's core pathological processes [79], including synaptic dysfunction [80]. An exploratory analysis from the TRAILBLAZER-ALZ study indicated that donanemab treatment modulates plasma levels of specific biomarkers relative to a placebo with these changes correlating with amyloid plaque shifts as identified by amyloid PET imaging [81].

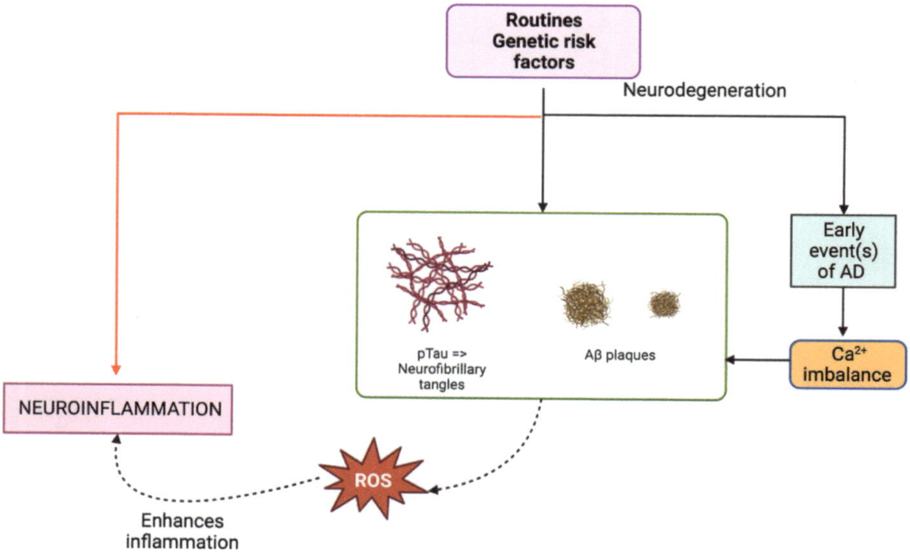

Figure 1. In examining the underpinnings and evolution of Alzheimer's disease (AD), one observes substantial influence from lifestyle and risk-associated genes. A notable precursor event to AD is calcium dyshomeostasis, which plays a crucial role in the synthesis of amyloid beta (Aβ) and phosphorylated Tau (pTau). The formation of toxic oligomers from Aβ can subsequently amalgamate into amyloid plaques. Similarly, pTau oligomers can lead to the establishment of neurofibrillary tangles. Such occurrences are widely regarded as pivotal to the process of neurodegeneration. Concurrently, these and other associated phenomena generate reactive oxygen species (ROS), which not only underscore neuroinflammation, a hallmark of neurodegeneration, but also potentially exacerbate or catalyze further pathological events pertinent to AD.

3.2. Novel "Wet" and "Dry" Markers in the Horizon

The current understanding of biomarkers hinges on their capability to provide insights into the underlying mechanisms of AD pathology. Established CSF analytes have shown varying degrees of accuracy across different studies, emphasizing the urgency for innovative biomarkers that can enhance diagnostic precision. There is a growing demand for biomarkers that can elucidate additional aspects of AD pathogenesis, highlighting areas like neuroinflammation and early neuronal dysfunction preceding overt cell death [82]. An extended study, named ALFA+, seeks to conduct the in-depth phenotyping of a subset of participants from the ALFA parent cohort. The research will incorporate both wet (CSF, blood, urine) and imaging (MRI and PET) biomarkers for a holistic evaluation [83].

The REST protein has emerged as a potential novel biomarker for AD, albeit its exclusive detection in the central nervous system and in vitro models limits its application in translational research [84]. A trend in REST levels has been observed with declining levels corresponding to increasing clinical severity of the disease. Preliminary findings suggest certain biomarkers, such as NPTXR, might be indicative of AD progression [85]. The APOE ε4 allele stands out as a significant genetic determinant for AD susceptibility with carriers exhibiting distinct pathological traits, including a higher prevalence of amyloid plaques [86]. A comprehensive analysis of 30 brain-centric proteins as potential CSF biomarkers for AD progression revealed NPTXR as a promising candidate with levels decreasing commensurately as AD advances [87].

CSF samples, pivotal in these evaluations, were diligently processed and analyzed at the Leonard Wolfson Biomarker Laboratory, University College London. Standardized protocols were employed for the assessment of Aβ42, T-tau, and P-tau [88]. The VIVIAD trial, a phase 2b study, is evaluating varoglutamstat's potential as a disease-altering treatment for AD with an emphasis on its correlation with both novel and traditional biomarkers [89]. Biomarker alterations serve as crucial outcome metrics in phase 2b AD trials. For instance, the SAPHIR trial observed a decline in YKL-40, a marker for AD-related neuroinflammation, following varoglutamstat administration [90]. As the field of biomarkers evolves, developing advanced nanotechnologies to monitor neuronal activity within networks, especially focusing on [Ca^{2+}] changes in living organisms, will be a pivotal challenge [91].

3.3. How Emerging Technologies Are Aiding in the Identification of New Markers for AD

Neuroinflammatory markers, pivotal in understanding neurodegenerative processes, have been evaluated using the Luminex xMAP technology in significant cohort studies. However, these investigations have produced inconsistent results, which poses challenges, especially for academic entities and small biotechnology companies striving to develop treatments targeting neuroinflammation [92]. Although there have been substantial advancements in technology and methodologies for target identification and evaluation, the journey from recognizing promising targets to early drug discovery remains intricate and uncertain [93].

Metabolomics, with its diverse analytical platforms, offers powerful diagnostic tools and insights into disease mechanisms. These technologies have been employed in both animal and human studies, encompassing plasma and CSF evaluations. They have identified metabolic pathways that are disrupted in conditions like AD and MCI [94]. A notable discovery is the reduced plasma levels of desmosterol, a cholesterol precursor, in AD patients. This decrease correlates with cognitive changes, suggesting its potential as an AD diagnostic biomarker. Merging metabolomic signatures with other biomarkers could further enhance diagnostic specificity [95] (See Table 5).

DNA microarray techniques have also been employed to delve into neurobiology and neurodegeneration. Recent publications have emphasized the significance and appropriate utilization of this technology while exploring neurodegenerative mechanisms [96]. Innovative imaging technologies, such as PET, hold promise for enhancing early diagnostic precision in AD's prodromal states, especially in patients with MCI, potentially fast tracking the evolution of disease-altering treatments [97]. The emergence of high-throughput DNA genotyping and sequencing has facilitated numerous genome-wide association studies (GWASs) in AD [98].

The arena of stem cell technology is witnessing rapid advancements. Many patients are opting to have their stem cells collected and reprogrammed. One advantage is the creation of cellular models representing "aged" cells, but there is caution to exercise, as reprogrammed cells may not perfectly replicate native neurons [91].

Table 5. Emerging Biomarkers for Alzheimer's Disease.

References	Biomarker(s)	Sample Type/Method	Key Findings/Notes
[71]	ADCOMS	Not specified	Primary evaluation criterion for lecanemab trial
[72]	Amyloid through PET SUVr, ADCOMS, CDR-SB, ADAS-Cog14, CSF biomarkers, Hippocampal volume (vMRI)	PET, CSF, vMRI	Secondary criteria for lecanemab trial
[74]	CTSB, BDNF, klotho	Blood (systemic)	Evaluated after a 26-week aerobic regimen

Table 5. *Cont.*

References	Biomarker(s)	Sample Type/Method	Key Findings/Notes
[75]	Aβ40, Aβ42/40	CSF	Dietary patterns influenced these biomarkers
[76]	GFAP	Plasma	Elevated in preclinical AD vs. healthy elderly
[84]	REST protein	CNS and in vitro	Potential novel biomarker for AD
[85]	NPTXR	Not specified	Indicative of AD progression
[86]	APOE ε4 allele	Genetic	Major determinant for AD susceptibility
[87]	NPTXR and 29 other brain-centric proteins	CSF	NPTXR levels decrease as AD progresses
[88]	Aβ42, T-tau, P-tau	CSF	Standard protocols used for assessment
[90]	YKL-40	Not specified	Marker for AD-related neuroinflammation
[95]	Desmosterol	Plasma	Reduced levels in AD patients; potential diagnostic biomarker

4. Parkinson's Disease: Beyond Traditional Markers

4.1. Challenges in Early Diagnosis and Monitoring of PD

A family history showcasing a similar tremor pattern may point toward essential tremors, particularly given that this condition often exhibits an autosomal dominant inheritance pattern. Conversely, indicators such as a classic rest tremor, primarily unilateral tremor presentation, leg tremor, associated rigidity, and a response to levodopa are suggestive of Parkinson's disease [99]. The diagnosis of idiopathic Parkinson's disease is still largely clinical despite technological advancements in radiological assessments. Distinct clinical signs required for diagnosis include a distal resting tremor ranging between 3 and 6 Hz, rigidity, bradykinesia, and an asymmetrical onset [100]. Other hallmark signs encompass late-onset postural instability, olfactory deficits, and micrographia.

Machine learning (ML) has been harnessed by researchers aiming for early Parkinson's disease diagnosis, utilizing motion data gathered from individuals' upper limbs [101]. Experiments had participants, both those diagnosed with PD and healthy individuals, wear a device on their upper limbs while performing specific tasks [102]. To determine the optimal model for PD diagnosis, numerous experiments were conducted. The selected network topology comprised a single hidden layer with eight neurons. Tanh, Relu, and sigmoid functions were designated as activation functions for input, hidden, and output layers, respectively [102].

Early clinical diagnosis of PD is intricate, as overt differences in motor and cognitive features are elusive. Comprehensive understanding of clinical symptoms, pathological alterations, and neural dysfunction is imperative for a definitive disease diagnosis [103]. While many ML-based models have been proposed for ESPD diagnosis [104], the BNA neuromarker, derived from easily obtainable EEG data, stands out for its clinical utility and repeatability [105]. Alongside therapeutic interventions like gene therapy, neuroprotection, and pharmacology, the search for PD's biological markers is relentless, aiming at early diagnosis [103].

Balance training, in particular, suffers from the lack of standardized approaches in monitoring training programs, making incomplete descriptions problematic [106]. During the trial, the influence of both study treatments, CBT and clinical monitoring, on depression in Parkinson's patients remained uncertain. However, factors such as the chronic depression experienced by the sample, the progressive nature of PD, and the durable gains from CBT over 14 weeks suggest that the benefits of CBT might surpass mere placebo effects [107].

Digital biomarkers have shown potential in passive monitoring, indicating decreased mobility in PD participants relative to controls. These biomarkers could detect significant irregularities even when traditional exams did not, hinting at their heightened sensitivity, making them suitable for long-term clinical trials and treatment monitoring [108]. In scenarios where therapy response is subpar and alternative explanations are absent, more advanced methods like electronic compliance monitoring may prove beneficial [109].

4.2. Innovative "Wet" and "Dry" Biomarkers for PD

NTK stands as a previously identified biomarker panel, which was validated through a comprehensive, longitudinal study involving 2743 early AD patients. During this study, multiple CSF biomarkers exhibited notable alterations [110]. In relation to Parkinson's disease, there has been a documented decline of 10–15% in CSF αSyn in comparison to healthy controls (HCs) [111]. This discovery was further confirmed using an independent methodology. However, this research stands as the inaugural longitudinal CSF study that focused on PD and HC using this specific biomarker panel. Notably, apart from αSyn, the study found no significant variations in other evaluated biomarkers [112] (See Table 6).

The T1w/T2w ratio within the midbrain is considered to embody a culmination of multiple PD-associated changes. These include modifications in neurons, dendrites, microglia, and iron content. Such data might produce a pronounced contrast that could be more effective than alternative MRI sequences in detecting PD-associated pathology. This ratio could potentially serve as an early detection biomarker for PD. To further this hypothesis, a subsequent MRI–pathology correlation study is recommended [113].

There have been indications that platelet CoQ10 redox ratios are considerably reduced in PD patients [114]. However, this test has not transitioned into clinical applications yet. The identification of a peripheral biomarker that can recognize decreased coenzyme Q10 activity may expedite research and improve clinical outcomes concerning PD [115].

Sargramostim, when administered in low doses, has shown the potential to modify immune functions, influence T cell phenotypes, and amplify treatment-induced biomarker levels. These changes have been associated with improved MDS–UPDRS Part III scores. The treatment also amplified Treg-mediated immunosuppressive functions, which remained consistent throughout the study. It is noteworthy that Tregs from PD patients previously exhibited a hindered ability to suppress Teff proliferation, which was linked to heightened disease severity [116].

Additional biomarkers like α-synuclein, neurofilament light chain, tau, phospho-tau, and beta-amyloid were assessed as potential exploratory endpoints over a 4-week treatment period. However, this duration might have been insufficient to detect significant clinical changes in these parameters. It is important to mention that there are not any validated biomarkers for PD presently. Future research endeavors might investigate the influence of venglustat on biomarkers and the progression of the disease over extended treatment periods [117] (See Table 6).

The effects of nilotinib on CSF biomarkers suggest that reducing oligomeric α-synuclein and p-tau could enhance dopamine metabolism in PD patients [118]. The data from both clinical and biomarker perspectives indicate that pioglitazone may not be a promising neuroprotective agent for PD. An intriguing point is that even though an epidemiological study pinpointed a reduced PD risk among individuals exposed to glitazone drugs, this association was not validated in a subanalysis that was specific to pioglitazone [119]. In this study, DaT-SPECT was employed as a PD enrichment biomarker, unveiling a SWEDD incident rate (3.8%). This rate was considerably lower than what is typically observed in multiple large multicenter studies with analogous PD populations [120,121].

Table 6. The wide variety of biomarkers used in Parkinson Disease and their clinical association.

References	Biomarker(s) or Indicator	Association/Significance
[99]	Classic rest tremor, Unilateral tremor presentation, Leg tremor, Associated rigidity, Response to levodopa	Indicators suggestive of Parkinson's disease
[100]	Distal resting tremor (3 to 6 Hz), Bradykinesia, Asymmetrical onset, Late-onset postural instability, Olfactory deficits, Micrographia	Clinical signs required for Parkinson's disease diagnosis
[105]	BNA neuromarker (from EEG data)	Stands out for its clinical utility and repeatability in ESPD diagnosis
[108]	Digital biomarkers (mobility)	Potential in passive monitoring indicative of decreased mobility in PD participants
[111]	CSF αSyn decline	Documented decline in Parkinson's patients compared to healthy controls
[113]	T1w/T2w ratio within the midbrain	Could serve as an early detection biomarker for PD due to various PD-associated changes
[114]	Platelet CoQ10 redox ratios	Indicative of reduced platelet CoQ10 redox in PD patients
[116]	Treatment-induced biomarker levels (Sargramostim)	Association with improved MDS-UPDRS Part III scores and modified immune functions
[117]	α-Synuclein, neurofilament light chain, tau, etc.	Assessed as potential exploratory endpoints, but duration was potentially insufficient for significant changes
[118]	Oligomeric α-synuclein and p-tau (effects of nilotinib)	Suggest that reducing these could enhance dopamine metabolism in PD patients
[119]	DaT-SPECT	Employed as a PD enrichment biomarker

4.3. Potential for Tailored Therapies and Improved Diagnostic Accuracy

Currently, Alzheimer's disease treatment employs only three AChE inhibitors: donepezil, rivastigmine, and galantamine. These medications serve primarily to offer symptomatic relief and are predominantly prescribed for mild to moderate dementia cases [122]. Art therapy has demonstrated notable benefits for patients, such as enhanced visual exploration patterns that begin to align with those of a control group. This suggests that art-centric visual training can foster the adoption of efficient visual exploration techniques [123]. To elaborate, art therapy has been shown to yield significant enhancements in visuospatial abilities, visual exploration strategies, and motor functions in PD patients with mild to moderate impairment. These improvements coincide with functional connectivity (FC) changes, pointing to a functional reorganization within primary and associative visual networks. This indicates that art therapy may serve as a valuable supplementary treatment to existing pharmacological interventions [123].

The core objective of a specific study was to ascertain if a customized tai chi program could bolster postural stability in Parkinson's disease patients [124]. The results revealed that practicing tai chi twice weekly for 24 weeks, in comparison to resistance training or stretching programs, effectively enhanced postural stability and other functional aspects in patients with mild-to-moderate Parkinson's disease. Moreover, tai chi training led to a marked reduction in fall incidents compared to the stretching routine. These positive outcomes persisted three months post-intervention, aligning with prior studies focused on individuals aged 70 and above [124]. Given the chronic and progressive nature of PD, it is recommended that the visual feedback VR technique be adopted as a long-term treatment strategy, complementing physical therapy, to sustain gait and postural performance in PD patients [125].

A study used a classification method which was validated using LOOCV and achieved an impressive classification accuracy of 93.62%. Most of the altered functional connections that exhibited high discriminative power were predominantly found within or across specific networks and the cerebellum [126]. Some studies managed to obtain a high classification accuracy of 94.4%, but the employed imaging method was invasive, making it unsuitable for routine diagnostics [127]. In contrast, certain noninvasive techniques have attained commendable classification accuracy using multi-type feature combinations. However, none have reached the high accuracy levels of this classification results [128]. The classification model, incorporating the basic SVM model and FG III, surpassed other ensemble classification models in performance. The final ensemble model was assessed using independent test data, achieving a 75.8% accuracy in distinguishing between early-stage PD and ET [129]. While the early-stage PD and ET classification model showcased good feasibility and potential, it was not exceptional [129].

The reliability of assessments using wearable sensors is influenced by factors such as sensor positioning, sensor-to-segment alignment, and frequently, the total number of sensors. This often results in increased costs and obtrusiveness [130]. In contrast, devices like the Wii Balance Board (WBB) and the Kinect for Windows v2 provide potential solutions for human motion tracking, circumventing the challenges posed by wearable sensors [131].

5. Amyotrophic Lateral Sclerosis (ALS): The Frontier of Biomarker Research

5.1. Overview of the Unique Challenges Posed by ALS

While the fundamental definition of ALS appears clear-cut, emerging insights suggest that ALS is not a singular disease but encompasses a diverse array of conditions with shared clinical characteristics [132]. People diagnosed with ALS face unique challenges compared to other patient groups where expressive disclosure has been employed as a therapeutic strategy, such as those with cancer, rheumatoid arthritis, or asthma. This uniqueness stems from the rapid progression of ALS, leading to paralysis, loss of independence, communication barriers, and the inevitable fatal prognosis. Given the swift and dynamic nature of ALS, the physical and emotional hurdles faced by patients may evolve significantly within a span of six months post-intervention. As such, emotional expression interventions tailored for ALS and similar rapidly progressing diseases might benefit from periodic 'booster' sessions. These sessions can address the evolving challenges and emotional shifts that patients encounter as the disease progresses [133]. See Figure 2.

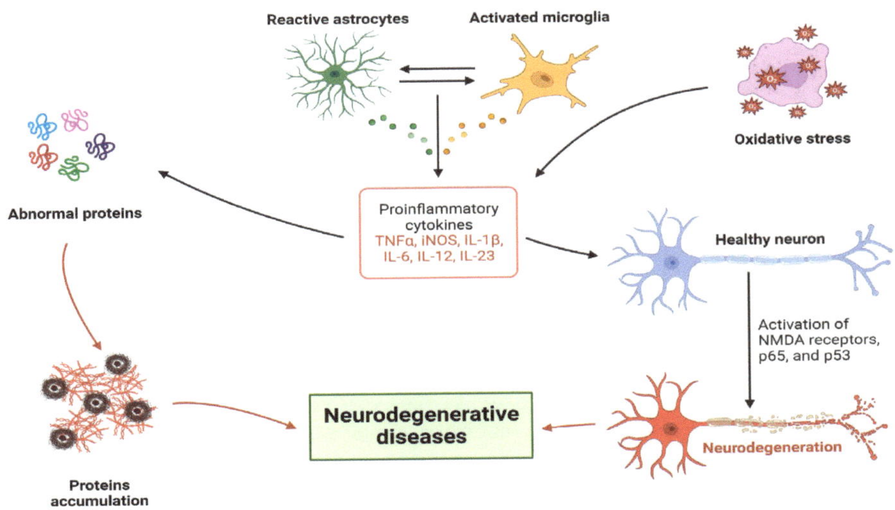

Figure 2. General pathophysiological mechanisms of neurodegenerative diseases.

Caregivers attending to ALS or progressive muscular atrophy (PMA) patients navigate numerous challenges as they witness the relentless progression and fatal trajectory of the disease. They grapple with the physical decline of the patient and potential cognitive and behavioral changes, escalating the caregiver's responsibilities and emotional strain [134]. Prolonged clinical research has identified systemic metabolic irregularities in ALS patients. While some discrepancies remain, a significant portion of studies highlight disturbances in functional metrics, like diminished glucose tolerance, insulin resistance, and abnormal fatty acid utilization [135]. However, the evident shortcomings of existing models in replicating ALS-specific conditions and associated pathologies cast doubt on their applicability for ALS research. The absence of accurate TDP-43 and FUS disease models represents a significant hurdle in ALS research. There is an urgent need for alternative models that can faithfully capture all dimensions of the disease [136].

5.2. Newly Identified Biomarkers and Their Potential Implications

The strategy of focusing on easily obtainable biofluids and evaluating markers directly linked to ALS pathogenesis stands as a cornerstone for the effective development of biomarkers. The occurrence of ferroptosis in motor neurons is increasingly acknowledged as a vital aspect of ALS with markers like lipid and iron accumulation signaling this specific type of programmed cell death [137,138]. Neurofilament light chain (NfL) and phosphorylated heavy chain (pNfH) are renowned indicators of neural integrity specific to ALS [139]. Within the Mitotarget/TRO19622 study, a cohort comprising 512 ALS patients from 15 European centers engaged in a phase III trial of olesoxime, and these biomarkers were assessed [140]. Notably, higher baseline levels of NfL, 4-HNE, 8-oxo-dG, and FT were linked with a steeper decline in ALSFRS-r during an 18-month monitoring period. Intriguingly, alterations in these markers outpaced functional deterioration with discernible differences between rapid and slow disease progressors observable at the 6-month mark [141].

Analyzing both the MCP-1 and FOXP3 mRNA, distinct effects were observed within the PP population. Among other findings, a significant change over time was solely detected in the PP population for actin-NT with no other notable effects identified for other biomarkers in the studied populations. To detect a 44% decrease in the progression rate of PPIA with 80% power, a total of 142 patients was deemed necessary. Likewise, the study aimed to discern a 43% reduction in ALSFRS-R progression over 24 weeks and a 25% absolute decrease in patients becoming non-self-sufficient at the 24-week mark [142].

Incepted in 2007, the Pre-Symptomatic Familial ALS (Pre-fALS) study is a longitudinal examination of unaffected individuals with a heightened genetic susceptibility to ALS. Its objectives encompass characterizing the pre-symptomatic disease phase, pinpointing biomarkers indicative of the imminent clinical manifestation, and collating essential data to pave the way for early intervention or preventive trials [143]. The ATLAS initiative holds promise in unearthing early markers of disease activity that extend beyond NfL. The periodic collection of CSF and urine/blood samples will facilitate the discovery of other potential fluid markers indicative of disease activity. Additionally, thorough electromyography (EMG) could provide insights into the temporal relationship between NfL elevation and the appearance of EMG anomalies, shedding light on the comparative sensitivity of these biomarkers [144]. Previous research by Keizman et al. pinpointed a notable correlation between clinical disability in ALS patients and inflammatory biomarkers, including CRP. This emphasizes the critical role of inflammation in ALS, underscoring CRP as a readily obtainable biomarker from blood samples irrespective of the patient's clinical status. Elevated CRP levels have also been detected in the cerebrospinal fluid of ALS patients, accentuating the importance of neuroinflammation in the disease progression [145].

5.3. Technological and Methodological Advancements in ALS Biomarker Discovery

The Mitotarget/TRO19622 study was a phase III trial focused on olesoxime, which was carried out as a negative, randomized, double-blinded, and placebo-controlled trial.

This trial incorporated 512 ALS patients drawn from 15 European centers [140]. This research strictly adhered to the guidelines and regulations set by both French and European authorities. The objective of this study was to evaluate the influence of RNS60 treatment on potential markers indicative of inflammation and neurodegeneration in the peripheral blood of ALS patients. The markers under scrutiny included MCP-1, PPIA, actin-NT, 3-NT, IL-17, NfL, and Tregs, which were identified through FOXP3 and CD25 mRNA [142] (See Table 7).

Part A of the study served as the natural history run-in phase. Throughout this period, participants underwent monthly monitoring to identify changes in their plasma NfL levels or the onset of clinically evident ALS. The design of Part A, which prioritized feedback from the ALS community, aimed to ensure minimal inconvenience for the participants. As such, most assessments, such as monthly blood draws for NfL monitoring, could be conducted within the confines of participants' homes. Part B of the study was a randomized phase, which was double-blind and placebo-controlled. Here, pre-symptomatic participants exhibiting elevated NfL levels were randomized to either receive tofersen or a placebo. This randomization process was dynamic, factoring in aspects like SOD1 variant type, the last recorded plasma NfL level before randomization, and age [144].

To delve into the prognostic potential of CRP, serum levels were gauged at the inception of the study. These levels were then correlated with various clinical demographics of ALS patients, such as age at the time of diagnosis, gender, disease duration by the time of evaluation, onset site, ALSFRS-R total score, body mass index, smoking habits, and overall survival [145]. Following intrathecal infusions, participants' cells were chased using CSF drawn prior to the transplantation process. After this process, participants were advised to maintain a specific position, the Trendelenburg position, for a duration of up to two hours. Throughout the study, participants, trial investigators, and personnel from the sponsor remained blind to treatment allocations. These allocations were assigned at the cell culture manufacturing facility once the clinical site informed them of participant eligibility [146].

In another study, patients were given either a placebo or increasing doses of NP001. The main endpoints for monitoring were safety, shifts in clinical status, and the reactions of blood monocyte immune activation markers CD16 and HLA-DR to NP001. These values were sourced from an independent flow cytometry laboratory at UCSF, which employed validated procedures for determinations. The statistical analysis for these values was conducted independently for CD16, while Neuraltus scientists managed the analysis for HLA-DR values [147] (See Table 7).

Table 7. New avenues in biomarker development of Amyotrophic Lateral Sclerosis.

References	Biomarker(s) or Indicator	Association/Significance
[94,95]	Ferroptosis markers (lipid and iron accumulation)	Linked to ALS-associated programmed cell death
[139]	Neurofilament light chain (NfL)	Indicator of neural integrity specific to ALS
[139]	Phosphorylated heavy chain (pNfH)	Indicator of neural integrity specific to ALS
[141]	NfL	Higher baseline levels linked with steeper decline in ALSFRS-r
[141]	4-HNE	Linked with steeper decline in ALSFRS-r
[141]	8-oxo-dG	Linked with steeper decline in ALSFRS-r
[141]	FT	Linked with steeper decline in ALSFRS-r
[142]	MCP-1	Observed distinct effects in PP population
[142]	FOXP3 mRNA	Observed distinct effects in PP population
[142]	Actin-NT	Significant change over time detected in the PP population
[142]	PPIA	Associated with progression rate

Table 7. *Cont.*

References	Biomarker(s) or Indicator	Association/Significance
[145]	CRP (inflammation marker)	Correlated with clinical disability; role in inflammation
[142]	MCP-1	Indicator of inflammation and neurodegeneration
[142]	PPIA	Indicator of inflammation and neurodegeneration
[142]	Actin-NT	Indicator of inflammation and neurodegeneration
[142]	3-NT	Indicator of inflammation and neurodegeneration
[142]	IL-17	Indicator of inflammation and neurodegeneration
[142]	NfL	Indicator of inflammation and neurodegeneration
[142]	Tregs (identified through FOXP3 and CD25 mRNA)	Indicator of inflammation and neurodegeneration
[144]	Plasma NfL levels	Used for monitoring onset of clinically evident ALS
[147]	Blood monocyte immune activation markers CD16	Monitored for reactions to NP001
[147]	HLA-DR	Monitored for reactions to NP001

6. Emerging Susceptibility and Trait Markers for Neurodegenerative Diseases

6.1. Introducing the Importance of Susceptibility and Trait Markers

The influence of cumulative lead exposure on cognitive functions may be mediated by the APOE genotype. Specifically, the E4 allele of the APOE gene is a recognized risk factor for Alzheimer's disease. Research has shown that individuals carrying at least one E4 allele, as opposed to those without the E4 allele, experience a more pronounced negative effect of bone lead on their neurobehavioral test scores, especially if they have been occupationally exposed to lead [148].

Certain studies indicate that the susceptibility to Parkinson's disease from pesticide exposure might be influenced by alterations in genes responsible for detoxifying enzymes. For instance, a recent investigation revealed that changes in neuronal aldehyde dehydrogenase enzymes correlate with an elevated risk of developing PD [149].

Genetic susceptibility testing, which is currently witnessing rapid advancements, presents risk information that can be described as a "moving target." The field of genetic testing is evolving quickly both in the variety of tests available and our growing comprehension of the intricate relationships between genes, environment, and behavior [150].

Quantitative Susceptibility Mapping (QSM) studies generally indicate heightened susceptibility, hinting at increased iron content, in brain regions linked to the pathophysiology of several neurodegenerative diseases. For instance, the substantia nigra in PD, the basal ganglia in Huntington's disease (HD), the amygdala and caudate nucleus (CN) in AD, the motor cortex in amyotrophic lateral sclerosis, and the cerebellar dentate nucleus (DN) in Friedreich's ataxia (FRDA) all show these changes [151].

Several studies have documented persistent sleep changes, such as shortened Rapid Eye Movement (REM) latencies, even during periods of remission from depression. Longitudinal research has consistently observed stable REM latencies, which suggests potential trait markers for some of these sleep alterations. This notion of a trait marker is further bolstered by the discovery of similar REM sleep changes in individuals with a pronounced family history of depression even if they were asymptomatic at the study's time [152,153].

The parkin gene is expansive, spanning over 1.5 Mb with approximately 12 exons. It is located on chromosome 6q25.2-27. A specific mutation in this gene, specifically a homozygous exon depletion, was initially identified as a trait responsible for early-onset autosomal recessive Parkinson's in a Japanese family [154].

6.2. Exploration of New Markers Identified through Genetic, Epidemiologic, and Epigenetic Studies

Biomarker data from the present research offer valuable insights into the disease's mechanism of action, targeting several pathways, including neuroprotection, neuroinflammation, and neurodegeneration. Notably, all participants treated with MSC-NTF exhibited significant, consistent, and lasting changes in numerous neuroinflammatory and neurodegenerative biomarkers such as MCP-1 and NfL. These findings align with prior trials [146] and underscore the potential of a treatment associated with slowing disease progression [155].

In initial experiments with two cell lines, the accumulation of TDP43 fragments was diminished, and TDP-43 nuclear localization was reinstated when mTOR was inhibited by Rapamycin [156]. Furthermore, in both mouse and human stem cell-derived neurons and astrocytes containing mutant TDP43, enhancing autophagy led to improved TDP43 clearance and localization, emphasizing that autophagy induction counteracts neurodegeneration via TDP43 clearance [157].

The single nucleotide polymorphism (SNP) rs75932628-T has been associated with genetically higher sTREM2 levels in CSF and an increased risk of Alzheimer's disease onset. Comparable results were observed for sTREM2 in CSF with a notable relative change from the baseline in the high exercise group versus the control group [158].

Evidence suggests a causal relationship between the LC and disease-modifying processes. Both genetic and neurotoxin-induced LC lesions exacerbate neuropathology and cognitive impairments in mouse models of AD, highlighting the LC's pivotal role in regulating neuroinflammation [159]. Moreover, advanced techniques like DREADD chemogenetics and traditional pharmacological enhancement of NE neurotransmission can reverse AD's pathophysiological features, boost microglial phagocytosis, and improve cognitive functions [160].

The detected levels correspond with those documented in extensive metabolizers. Anticipations for such outcomes were based on a pre-screening process designed to exclude potential carriers of CYP2D6 genetic variants, which are present in approximately 10% of the general population and are known to decelerate atomoxetine metabolism [118]. Nevertheless, it is essential to consider that CSF levels may be influenced by the permeability of the blood–brain barrier, factors correlated with aging, AD [119], and LC degeneration [120].

In the adopted reference-free methodology, all quantitative data post-assembly are disregarded. Under optimal circumstances, sequences deriving from identical genetic sources should culminate in a unique contig per sample. Within the framework of the BusyBee methodology, such a contig would emerge as a singular point, minimally impacting the comprehensive density distribution. The conspicuous signal emanating from the high-density cluster within the PD + RS group suggests the presence of multiple contigs. These contigs are sufficiently distinct to resist merging during assembly, yet they exhibit qualitative properties suggesting an association with Rhodococcus [121].

6.3. The Future Potential of These Markers in Personalized Medicine

Recent research has unveiled potential biomarkers that could prove instrumental in monitoring the progression of Parkinson's disease. These findings also open up novel avenues for deeper investigation into the underlying mechanisms of PD. A study set out to validate these preliminary findings with additional sample sets. Furthermore, the study aimed to explore if the identified compounds that seem to predict the progression of PD can also distinguish between PD patients, healthy individuals, and those diagnosed with other neurodegenerative diseases [161]. Despite the consistent epidemiological associations between elevated urate levels and a decreased risk and progression rate of PD, the trial's outcomes do not advocate for a protective role of urate [162]. Adding to the complexity, recent Mendelian randomization research challenges the protective nature of high urate levels against PD [163], while another study indicates its potential protective effect in slowing the progression of established PD [164]. The focus of the ATLAS study (NCT04856982) is to ascertain the effects of early administration of tofersen in individuals who are pre-symptomatic carriers of certain SOD1 mutations, which are known for their association with aggressive disease progression and increased plasma NfL levels. Given the potential benefits of early intervention in ALS, insights from ATLAS, combined with data from the VALOR study and its subsequent open-label extension, aim to provide clarity on the ideal timeframe for initiating treatment in cases of SOD1-ALS [144].

7. Diagnostic and State Markers: Revolutionizing Early Detection and Monitoring

7.1. Delving into Novel Diagnostic Markers for AD, PD, and ALS

There is growing emphasis on discovering new biomarkers that can bridge the gap between psychological risk factors and Alzheimer's disease to foster a deeper understanding of the illness. Recent studies have highlighted the dysregulation of REST in depression, which is a psychological disorder linked with stress that elevates the risk for AD [165]. Given REST's role in stress responses, it might serve as a pivotal biological link between psychological risks and AD. In exploration of the relationship between REST and previously pinpointed plasma protein markers associated with mild cognitive impairment (MCI) transitioning to AD and cortical atrophy [166], significant correlations were identified with BDNF, RANTES, PAI-1, and NSE. These correlations were independently validated in the Intervention cohort. BDNF, akin to REST, is believed to play a neuroprotective role under pathological conditions [167]. A cutting-edge PET method employing 11C-labeled AA was implemented, granting the first-ever visualization of in vivo dopaminergic neurotransmission in a resting state. Echoing findings from animal research [168], we observed significant increases in the incorporation coefficient K^* for AA when exposed to apomorphine across several brain regions. This is believed to reflect neuronal signaling events associated with activated D2 receptors connected to cPLA2 [169]. NP001 is a specialized, pH-balanced stabilized sodium chlorite variant and presents a groundbreaking effector molecule that introduces a fresh drug category targeting inflammatory macrophages and modulating their function in vitro and in vivo [170]. Chlorite's anti-inflammatory influence in macrophages is attributed to the elevated intracellular presence of taurine chloramine, which is known to suppress NF-κB triggered inflammatory pathways [171]. Prior clinical investigations with an alternate chlorite form have showcased its ability to counter inflammation and reset systemic macrophages to their natural wound-healing phagocytic state [172]. Contemporary studies indicate a direct correlation between the progression of the G93A strain of ALS mice and the infiltration of inflammatory monocytes into the spinal cord [147]. In this research, 30 brain-centric proteins were assessed as potential CSF biomarkers indicative of AD severity using multiplex mass spectrometry-based quantification. NPTXR emerged as a prime candidate for tracking disease progression. Intriguingly, two prior studies also flagged NPTXR as a promising progression biomarker for AD. As AD intensifies, CSF NPTXR levels proportionally decrease. This observation requires further validation in an expanded cohort observed longitudinally. It is hypothesized that NPTXR could be a pivotal CSF biomarker for gauging the effectiveness of emerging AD therapies [87].

7.2. Implications for Improved Diagnostic Accuracy

Studies of this domain unveil an intriguing observation: among older adults at a heightened risk of dementia, an 8-week stress reduction regimen led to a notable surge in REST levels, positioning REST as a potential adjustable target. The intricate role of REST in managing cortisol levels, primarily through the modulation of the CYP11B1 gene, [173] offers insight into the possibility that the intervention may have influenced cortisol concentrations, subsequently impacting REST levels. In our study, the selection criteria for the participant inclusion across the two cohorts either strictly involved individuals devoid of psychiatric ailments (like depression, anxiety, ANM) or those diagnosed with one (intervention cohort). Given this setup, the research could not draw a direct comparison of REST levels between mentally healthy seniors and those grappling with depression or anxiety. Consequently, a direct exploration between cognitive debt and this newfound biological indicator remains pending [84]. The innovative use of [1-11C]arachidonate PET in assessing healthy human participants showcased tangible impacts on the regional brain AA integration and rCBF following a pharmacological nudge with apomorphine, which is a D1/D2 receptor stimulant. The findings underscore the potential of this approach in capturing real-time signal transduction events tied to dopaminergic neurotransmission in a living brain. This paves the way for subsequent explorations into the efficacy of this technique in evaluating disruptions in cerebral dopaminergic functionality in disorders such as Parkinson's disease and schizophrenia [174].

7.3. The Promise of These Markers in Disease Monitoring and Evaluating Drug Efficacy

Spatially normalized images were used to conduct the T1w/T2w ratio comparisons both for VBA and ROI-centric studies. A potential concern might be that these results could have been influenced by volumetric or morphometric data. However, this seems improbable, considering that the atlas-based segmentation approach (essentially the inverse of normalization) did not reveal any significant volume discrepancies between the PD patient group and the control group [113]. Building upon these insights, future endeavors might consider a phase II study that utilizes a seemingly immune-regulatory dosage of NP001 chlorite (2 mg/kg). This could be juxtaposed against a minimal effective dose (1 mg/kg) and a placebo, with the study span extended to discern if modulating inflammation influences the pace of ALS disease progression. Thus, research has laid the groundwork for deploying specific NP001 dosages targeting inflammation markers in ALS patients. This aims to explore the hypothesis that inflammation might play a pivotal role in the onset and development of ALS [147].

8. Progression Markers: Tracking Disease Evolution

8.1. Importance of Progression Markers in Neurodegenerative Diseases

Findings indicate that the diversity observed in Parkinson's disease, especially concerning different ages of onset, might manifest through distinct deviations in both imaging and non-imaging biomarkers. When planning future clinical trials aiming to assess neuroprotective medications, it is crucial to factor in this biomarker variability associated with different PD onset ages. Opting for participants with a consistent age of onset could mitigate this variability, enhancing statistical power even with fewer participants. Continuous monitoring of the PPMI cohort will further elucidate the influence of onset age on the progression of PD and its potential interaction with these biomarkers [175]. The Parkinson Progression Markers Initiative (PPMI) operates as a global, multicenter cohort study, spanning 21 US and 12 international locations. This study focuses on patients newly diagnosed with PD who have not received any treatment at the point of enrollment, and it also includes healthy controls [176]. When it comes to understanding UPSIT scores, especially among individuals prone to or diagnosed with neurological conditions like PD, it is pivotal to rely on refreshed normative data. These data should ideally stem from a sizable sample that mirrors the demographic profile of PD patients. The UPSIT, given its adaptability for mail distribution and at-home self-administration, is perfectly aligned for large-scale investigations. It is worth noting that UPSIT has been employed in comprehensive studies such as Parkinson Associated Risk Syndrome (PARS) and the aforementioned PPMI. Our current research was tailored to offer normative data for UPSIT, segmented by age and gender, drawing insights from percentiles derived from the extensive, forward-looking cohorts of both PARS and PPMI [177].

8.2. Newly Identified Markers and Their Potential Role in Understanding Disease Trajectory

The observed lack of a significant statistical variation between the individuals undergoing DRT and those not undergoing DRT in terms of incident ICD symptoms may be attributed to multiple factors. The sample size being a limited one, the ability of distinguishment between the impact of dopamine agonists and other DRTs was diminished. Additionally, the potential variances in DRT and DAT availability among the four primary ICDs and behaviors such as punding, hobbyism, and walkabout were not thoroughly evaluated. The QUIP, which was primarily developed as a high-sensitivity (94%) screening tool but with a reduced specificity (72%) [178], might have resulted in certain participants displaying ICD symptoms that were either false positives or clinically non-pertinent [179]. Moreover, the study analyzed the expression levels of established apoptotic markers. These markers encompass pro-caspase 3, the p17 subunit of active caspase 3, cleaved PARP, and the anti-apoptotic protein BCL2. Notably, pro-caspase 3 is activated to form caspase 3 at the onset of the apoptotic process, leading to the proteolytic cleavage of the DNA repair enzyme, PARP, producing an 89 kDa apoptosis-specific PARP fragment [180]. In phase 2A of the study, the plasma concentrations of wr-CRP as potential biomarkers were quantified, and for phase 2B, hs-CRP plasma values were incorporated as a part of the enrollment criteria. To amalgamate CRP data from phase 2A and 2B for comprehensive analysis, a calibration equation derived from Ziv-Baran et al. [181] was employed to calibrate the phase 2A wr-CRP values. Baseline clinical and demographic data were analyzed according to their respective treatment groups [182]. The concluding evaluation of NP001 efficacy revealed a notably elevated percentage of non-progressors in the NP001 treatment group over a 6-month duration, in comparison to the placebo group. In patients displaying clinically relevant plasma CRP concentrations exceeding 3 mg/L, a 10:1 response favorability was observed for NP001-treated individuals versus placebo controls. Aligning with NP001′s anti-inflammatory properties, individuals displaying higher inflammation levels, as demarcated by blood CRP concentrations, exhibited a greater likelihood of benefiting from the treatment.

8.3. The Promise for Better Disease Management and Tailored Interventions

A reduction in DAT availability, especially a continuous decline over a period, might serve as an indicator for the likelihood of forthcoming ICD manifestations in early-stage PD patients post-initiation of DRT. Both neurobiological determinants and clinical attributes act as predisposing elements for the emergence of ICD symptoms in the context of DRT administration. Such understanding will aid in mitigating patient risks and devising innovative treatment strategies [179]. The potential influence of non-response bias merits attention. In the PARS study, 53% of qualifying participants submitted a completed UPSIT. Compared to non-participants, those who responded tended to be younger, female, Caucasian, have a familial history of PD, and did not indicate a diminished olfaction [183]. In the PPMI study, around 60% of the eligible cohort returned an UPSIT, but direct comparisons between participants and non-participants were not feasible. Regarding this specific study, interpretations concerning the prevalence of smoking in this amalgamated group are constrained as the data are exclusive to PARS participants. Such a limitation impedes the capacity to examine if smoking, linked with elevated olfactory dysfunction risk but reduced PD risk [184], might be a significant factor influencing the findings. An intriguing avenue for subsequent research would be an in-depth exploration of the interrelationships among smoking habits, olfactory function, and PD susceptibility [177]. Riluzole's administration was largely well-received, with side effects being comparable to placebo. Riluzole has been a staple in ALS treatment for numerous years. Yet, to achieve a holistic assessment of riluzole's safety and effectiveness within the Alzheimer's demographic, extensive and prolonged studies are imperative before its administration to Alzheimer's patients outside controlled clinical environments [185]. To sum up recent findings, riluzole-administered Alzheimer's patients exhibited a more gradual decline in cerebral glucose metabolism compared to their placebo counterparts across various Alzheimer's-relevant brain sectors. This decline was in correlation with their cognitive functionality. Such observations bolster the necessity for subsequent extensive clinical studies to further appraise riluzole's potential as a prospective medicinal treatment for Alzheimer's [186].

9. Innovative Techniques and Technologies in Biomarker Discovery

9.1. Highlighting the Role of -Omics, Histologic Assessments, and Imaging in Biomarker Identification

While numerous clinico-pathological and molecular biomarkers have been assessed for their potential benefits to FTD/TPI, their translation to clinical practice remains elusive [187]. Studies indicate that the KRAS mutational assessment, a globally recognized standard-of-care test, can discern patients with KRASG12 mutant mCRC who are less likely to derive benefits from FTD/TPI treatment. This identification aids in circumventing unnecessary patient side effects and optimizing healthcare resources. Consequently, this study presents the inaugural evidence of a genomics-driven precision approach for chemotherapy in mCRC, holding significant promise to enhance patient selection criteria for FTD/TPI therapeutic interventions [188].

9.2. The Synergy between Technology and Biomarker Discovery

Resveratrol appears to play a potential role in preserving the blood–brain barrier integrity primarily through the mitigation of MMP9 levels. Furthermore, resveratrol may stimulate adaptive immune mechanisms, potentially bolstering the brain's resilience against amyloid accumulation. The compound's potential to decelerate cognitive regression in Alzheimer's disease may be attributed to a synchronized immune response, both peripheral and central, that could potentially halt neuronal apoptosis. Summarizing, the preliminary observations from the investigation underscore the need for a more extensive study to validate the supposition that resveratrol can fortify a compromised BBB, subsequently leading to cognitive and functional enhancements in a broader AD patient cohort [189].

10. Conclusions and Future Outlook

10.1. A Reflection on the Advancements Made and the Challenges Ahead

Exenatide, a therapeutic agent classified as a glucagon-like peptide 1 agonist and traditionally employed for type 2 diabetes, has recently demonstrated potential positive impacts on motor functionalities in a controlled trial involving Parkinson's disease patients. Prevailing research posits that dysfunctional brain insulin and protein kinase B (Akt) signaling might be implicated in PD development. Nonetheless, comprehensively assessing the degree of drug interaction with these potential mechanisms in a live setting presents considerable difficulties [190]. Neurodegeneration, an intrinsic aging phenomenon, manifests in all aging populations. Given the rise in longevity, confronting neurodegeneration has emerged as a significant concern for healthcare frameworks, predominantly in developed nations. The rate of neuronal death progression serves as a pivotal metric for neurodegeneration. Predominant theories suggest the existence of multiple external and internal factors that can either accelerate or retard this process. Such determinants could be inherent, like an individual's unique metabolic processes, or they might be external and associated with environmental conditions [191].

10.2. Potential Challenges and Areas of Unmet Need

The intricate endeavor of extrapolating specific therapeutic interventions in human-centric randomized controlled trials (RCTs) can be exemplified by neurofibromatosis type 1 (NF1). NF1 is a genetically inherited condition linked with cognitive anomalies impacting a vast majority, approximately 80%, of pediatric patients [192]. Preliminary trials have pinpointed multiple therapeutic prospects. Lovastatin rectifies synaptic functionality and ameliorates learning anomalies in Nf1+/− mice by targeting RAS activity. In contrast, compounds like methylphenidate and L-dopa bolster attention by restoring dopamine equilibrium in specific Nf1+/− variants with bi-allelic deactivation in neuroglial progenitor cells [193]. Recent reports indicate that atrophy of the basal forebrain cholinergic system (BFCS) often precedes both entorhinal cortex degeneration and memory dysfunctions in Alzheimer's disease (AD). This revelation challenges established paradigms concerning the chronological progression of AD-associated topographical pathology [194]. Over the past decade, BACE1 has been a focal point for the development of potential AD treatments. However, formulating such compounds has proven arduous, with challenges like cellular penetration, oral absorption/metabolism, and brain accessibility. Utilizing a fragment-based chemical approach, LY2811376 ((S)-4-(2,4-difluoro-5-pyrimidin-5-yl-phenyl)-4-methyl-5,6-dihydro-4H-[1,3]thiazin-2-ylamine), the inaugural non-peptidic BACE1 inhibitor with oral bioavailability, was synthesized. This compound has demonstrated significant reductions in Aβ levels in experimental models [195].

10.3. Multidisciplinary Collaboration for Accelerated Biomarker Discovery

To encapsulate, a holistic methodology has been showcased, a methodology that embodies the seamless amalgamation of foundational, translational, and clinical research factions within a unified, adaptive structure centered around the NBGB. This framework consolidates clinical data, biomarkers, and post-mortem samples, coupled with comprehensive information from well-documented subjects. This integration facilitates various research teams to efficiently collate expansive and enriched data repositories, furthering investigations across multiple neurodegenerative diseases [196]. Advancing our comprehension of the intricate interplay between oxidation, antioxidants, and neurodegenerative maladies will necessitate a multidisciplinary approach [197]. Among the array of neuroimaging modalities, three techniques stand out in specialized clinical contexts due to their advanced validation stages relative to other biomarkers. These are structural MRI for atrophy detection, FDG-PET for hypometabolism assessment, and amyloid-PET for amyloid deposition quantification. In addition to these three types of MRI, neuromelanin-sensitive MRI can offer valuable information in patients with Parkinson's disease [198,199]. The sequential application of these tools, as recommended by a consortium of multidisciplinary experts [200], draws upon their individual merits and limitations, as succinctly delineated in subsequent literature [201].

Author Contributions: Conceptualization, C.T. and N.D.; methodology, C.T. and F.-M.B.; software, R.-A.C.-B. and L.A.G.; validation, H.P.C.; formal analysis, B.-G.B.; investigation, N.D.; resources, A.D.C.; data curation, A.A.P.; writing—original draft preparation, R.-A.C.-B.; writing—review and editing, L.A.G. and B.-G.B.; visualization, N.D. and C.P.; supervision, F.-M.B. and C.P.; project administration, A.V.C.; funding acquisition, A.V.C. All authors have read and agreed to the published version of the manuscript.

Funding: This research received no external funding.

Institutional Review Board Statement: Not applicable.

Informed Consent Statement: Not applicable.

Data Availability Statement: All data are available online in libraries such as PubMed.

Acknowledgments: Publication of this paper was supported by the "Carol Davila" University of Medicine and Pharmacy, through the institutional program Publish not Perish.

Conflicts of Interest: The authors declare no conflict of interest.

References

1. García-Cordero, J.; Pino, A.; Cuevas, C.; Puertas-Martín, V.; San Román, R.; De Pascual-Teresa, S. Neurocognitive Effects of Cocoa and Red-Berries Consumption in Healthy Adults. *Nutrients* **2021**, *14*, 1. [CrossRef] [PubMed]
2. Petersen, R.C.; Lopez, O.; Armstrong, M.J.; Getchius, T.S.D.; Ganguli, M.; Gloss, D.; Gronseth, G.S.; Marson, D.; Pringsheim, T.; Day, G.S.; et al. Practice guideline update summary: Mild cognitive impairment: Report of the Guideline Development, Dissemination, and Implementation Subcommittee of the American Academy of Neurology. *Neurology* **2018**, *90*, 126–135. [CrossRef] [PubMed]
3. Feigin, V.L.; Nichols, E.; Alam, T.; Bannick, M.S.; Beghi, E.; Blake, N.; Culpepper, W.J.; Dorsey, E.R.; Elbaz, A.; Ellenbogen, R.G.; et al. Global, regional, and national burden of neurological disorders, 1990–2016: A systematic analysis for the Global Burden of Disease Study 2016. *Lancet Neurol.* **2019**, *18*, 459–480. [CrossRef]
4. Panza, F.; Solfrizzi, V.; Giannini, M.; Seripa, D.; Pilotto, A.; Logroscino, G. Nutrition, frailty, and Alzheimer's disease. *Front. Aging Neurosci.* **2014**, *6*, 221. [CrossRef] [PubMed]

5. Sugimoto, T.; Sakurai, T.; Ono, R.; Kimura, A.; Saji, N.; Niida, S.; Toba, K.; Chen, L.-K.; Arai, H. Epidemiological and clinical significance of cognitive frailty: A mini review. *Ageing Res. Rev.* **2018**, *44*, 1–7. [CrossRef]
6. Kim, D. The Effects of a Recollection-Based Occupational Therapy Program of Alzheimer's Disease: A Randomized Controlled Trial. *Occup. Ther. Int.* **2020**, *2020*, 6305727. [CrossRef]
7. Bayraktar, Y.; Isik, E.; Isik, I.; Ozyilmaz, A.; Toprak, M.; Kahraman Guloglu, F.; Aydin, S. Analyzing of Alzheimer's Disease Based on Biomedical and Socio-Economic Approach Using Molecular Communication, Artificial Neural Network, and Random Forest Models. *Sustainability* **2022**, *14*, 7901. [CrossRef]
8. Findley, L.; Aujla, M.; Bain, P.G.; Baker, M.; Beech, C.; Bowman, C.; Holmes, J.; Kingdom, W.K.; MacMahon, D.G.; Peto, V.; et al. Direct economic impact of Parkinson's disease: A research survey in the United Kingdom. *Mov. Disord.* **2003**, *18*, 1139–1145. [CrossRef]
9. Winter, Y.; Balzer-Geldsetzer, M.; Spottke, A.; Reese, J.P.; Baum, E.; Klotsche, J.; Rieke, J.; Simonow, A.; Eggert, K.; Oertel, W.H.; et al. Longitudinal study of the socioeconomic burden of Parkinson's disease in Germany. *Eur. J. Neurol.* **2010**, *17*, 1156–1163. [CrossRef]
10. Hoffman, H.I.; Bradley, W.G.; Chen, C.Y.; Pioro, E.P.; Stommel, E.W.; Andrew, A.S. Amyotrophic Lateral Sclerosis Risk, Family Income, and Fish Consumption Estimates of Mercury and Omega-3 PUFAs in the United States. *Int. J. Environ. Res. Public Health* **2021**, *18*, 4528. [CrossRef]
11. Beard, J.D.; Steege, A.L.; Ju, J.; Lu, J.; Luckhaupt, S.E.; Schubauer-Berigan, M.K. Mortality from Amyotrophic Lateral Sclerosis and Parkinson's Disease among Different Occupation Groups—United States, 1985–2011. *MMWR Morb. Mortal. Wkly. Rep.* **2017**, *66*, 718–722. [CrossRef] [PubMed]
12. Ciceri, F.; Rotllant, D.; Maes, T. Understanding Epigenetic Alterations in Alzheimer's and Parkinson's Disease: Towards Targeted Biomarkers and Therapies. *Curr. Pharm. Des.* **2017**, *23*, 839–857. [CrossRef] [PubMed]
13. Turner, R.S.; Stubbs, T.; Davies, D.A.; Albensi, B.C. Potential New Approaches for Diagnosis of Alzheimer's Disease and Related Dementias. *Front. Neurol.* **2020**, *11*, 496. [CrossRef] [PubMed]
14. Martínez-Iglesias, O.; Naidoo, V.; Cacabelos, N.; Cacabelos, R. Epigenetic Biomarkers as Diagnostic Tools for Neurodegenerative Disorders. *Int. J. Mol. Sci.* **2021**, *23*, 13. [CrossRef] [PubMed]
15. Strimbu, K.; Tavel, J.A. What are biomarkers? *Curr. Opin. HIV AIDS* **2010**, *5*, 463–466. [CrossRef] [PubMed]
16. Robb, M.A.; McInnes, P.M.; Califf, R.M. Biomarkers and Surrogate Endpoints: Developing Common Terminology and Definitions. *JAMA* **2016**, *315*, 1107. [CrossRef]
17. Califf, R.M. Biomarker definitions and their applications. *Exp. Biol. Med.* **2018**, *243*, 213–221. [CrossRef]
18. Ellenberg, S.S.; Hamilton, J.M. Surrogate endpoints in clinical trials: Cancer. *Stat. Med.* **1989**, *8*, 405–413. [CrossRef]
19. Hansson, O. Biomarkers for neurodegenerative diseases. *Nat. Med.* **2021**, *27*, 954–963. [CrossRef]
20. Hlatky, M.A.; Greenland, P.; Arnett, D.K.; Ballantyne, C.M.; Criqui, M.H.; Elkind, M.S.V.; Go, A.S.; Harrell, F.E.; Hong, Y.; Howard, B.V.; et al. Criteria for Evaluation of Novel Markers of Cardiovascular Risk: A Scientific Statement from the American Heart Association. *Circulation* **2009**, *119*, 2408–2416. [CrossRef]
21. Bouchard, P.; Carra, M.C.; Boillot, A.; Mora, F.; Rangé, H. Risk factors in periodontology: A conceptual framework. *J. Clin. Periodontol.* **2017**, *44*, 125–131. [CrossRef] [PubMed]
22. Postuma, R.B.; Berg, D. Advances in markers of prodromal Parkinson disease. *Nat. Rev. Neurol.* **2016**, *12*, 622–634. [CrossRef] [PubMed]
23. Silajdžić, E.; Björkqvist, M. A Critical Evaluation of Wet Biomarkers for Huntington's Disease: Current Status and Ways Forward. *J. Huntingt. Dis.* **2018**, *7*, 109–135. [CrossRef]
24. Kraus, V.B.; Burnett, B.; Coindreau, J.; Cottrell, S.; Eyre, D.; Gendreau, M.; Gardiner, J.; Garnero, P.; Hardin, J.; Henrotin, Y.; et al. Application of biomarkers in the development of drugs intended for the treatment of osteoarthritis. *Osteoarthr. Cartil.* **2011**, *19*, 515–542. [CrossRef]
25. Aronson, J.K. Biomarkers and surrogate endpoints. *Br. J. Clin. Pharmacol.* **2005**, *59*, 491–494. [CrossRef]
26. Beach, T.G. A Review of Biomarkers for Neurodegenerative Disease: Will They Swing Us across the Valley? *Neurol. Ther.* **2017**, *6*, 5–13. [CrossRef]
27. Coley, N.; Andrieu, S.; Delrieu, J.; Voisin, T.; Vellas, B. Biomarkers in Alzheimer's disease: Not yet surrogate endpoints. *Ann. N. Y. Acad. Sci.* **2009**, *1180*, 119–124. [CrossRef]
28. Li, D.; Mielke, M.M.; Bell, W.R.; Reilly, C.; Zhang, L.; Lin, F.V.; Yu, F. Blood biomarkers as surrogate endpoints of treatment responses to aerobic exercise and cognitive training (ACT) in amnestic mild cognitive impairment: The blood biomarkers study protocol of a randomized controlled trial (the ACT Trial). *Trials* **2020**, *21*, 19. [CrossRef]
29. Boone, C.W.; Kelloff, G.J. Intraepithelial neoplasia, surrogate endpoint biomarkers, and cancer chemoprevention. *J. Cell. Biochem. Suppl.* **1993**, *17F*, 37–48. [CrossRef]
30. Weintraub, W.S.; Lüscher, T.F.; Pocock, S. The perils of surrogate endpoints. *Eur. Heart J.* **2015**, *36*, 2212–2218. [CrossRef]
31. Tondo, G.; De Marchi, F. From Biomarkers to Precision Medicine in Neurodegenerative Diseases: Where Are We? *J. Clin. Med.* **2022**, *11*, 4515. [CrossRef] [PubMed]
32. Dubois, B.; Hampel, H.; Feldman, H.H.; Scheltens, P.; Aisen, P.; Andrieu, S.; Bakardjian, H.; Benali, H.; Bertram, L.; Blennow, K.; et al. Preclinical Alzheimer's disease: Definition, natural history, and diagnostic criteria. *Alzheimers Dement.* **2016**, *12*, 292–323. [CrossRef] [PubMed]

33. Cheng, H.-C.; Ulane, C.M.; Burke, R.E. Clinical progression in Parkinson disease and the neurobiology of axons. *Ann. Neurol.* **2010**, *67*, 715–725. [CrossRef] [PubMed]
34. Zhang, W.; Xiao, D.; Mao, Q.; Xia, H. Role of neuroinflammation in neurodegeneration development. *Signal Transduct. Target. Ther.* **2023**, *8*, 1–32. [CrossRef]
35. Wang, W.; Norby, F.L.; Alonso, A.; Gottesman, R.F.; Jack, C.R.; Meyer, M.L.; Knopman, D.S.; Sullivan, K.J.; Hughes, T.M.; Lakshminarayan, K.; et al. Association of Carotid Intima-Media Thickness with Brain MRI Markers in the Atherosclerosis Risk in Communities Neurocognitive Study (ARIC-NCS). *J. Stroke Cerebrovasc. Dis. Off. J. Natl. Stroke Assoc.* **2022**, *31*, 106388. [CrossRef]
36. Willeit, P.; Tschiderer, L.; Allara, E.; Reuber, K.; Seekircher, L.; Gao, L.; Liao, X.; Lonn, E.; Gerstein, H.C.; Yusuf, S.; et al. Carotid Intima-Media Thickness Progression as Surrogate Marker for Cardiovascular Risk. *Circulation* **2020**, *142*, 621–642. [CrossRef]
37. Saleh, C. Carotid artery intima media thickness: A predictor of cognitive impairment? *Front. Biosci. Elite Ed.* **2010**, *2*, 980–990. [CrossRef]
38. Paulsen, J.S.; Langbehn, D.R.; Stout, J.C.; Aylward, E.; Ross, C.A.; Nance, M.; Guttman, M.; Johnson, S.; MacDonald, M.; Beglinger, L.J.; et al. Detection of Huntington's disease decades before diagnosis: The Predict-HD study. *J. Neurol. Neurosurg. Psychiatry* **2008**, *79*, 874–880. [CrossRef]
39. Tabrizi, S.J.; Langbehn, D.R.; Leavitt, B.R.; Roos, R.A.; Durr, A.; Craufurd, D.; Kennard, C.; Hicks, S.L.; Fox, N.C.; Scahill, R.I.; et al. Biological and clinical manifestations of Huntington's disease in the longitudinal TRACK-HD study: Cross-sectional analysis of baseline data. *Lancet Neurol.* **2009**, *8*, 791–801. [CrossRef]
40. Abeyasinghe, P.M.; Long, J.D.; Razi, A.; Pustina, D.; Paulsen, J.S.; Tabrizi, S.J.; Poudel, G.R.; Georgiou-Karistianis, N. Tracking Huntington's disease progression using motor, functional, cognitive and imaging markers. *Mov. Disord. Off. J. Mov. Disord. Soc.* **2021**, *36*, 2282–2292. [CrossRef]
41. Byrne, L.M.; Rodrigues, F.B.; Blennow, K.; Durr, A.; Leavitt, B.R.; Roos, R.A.C.; Scahill, R.I.; Tabrizi, S.J.; Zetterberg, H.; Langbehn, D.; et al. Neurofilament light protein in blood as a potential biomarker of neurodegeneration in Huntington's disease: A retrospective cohort analysis. *Lancet Neurol.* **2017**, *16*, 601–609. [CrossRef] [PubMed]
42. Wild, E.J.; Boggio, R.; Langbehn, D.; Robertson, N.; Haider, S.; Miller, J.R.C.; Zetterberg, H.; Leavitt, B.R.; Kuhn, R.; Tabrizi, S.J.; et al. Quantification of mutant huntingtin protein in cerebrospinal fluid from Huntington's disease patients. *J. Clin. Investig.* **2015**, *125*, 1979–1986. [CrossRef] [PubMed]
43. Rodrigues, F.B.; Wild, E.J. Huntington's Disease Clinical Trials Corner: February 2018. *J. Huntingt. Dis.* **2018**, *7*, 89–98. [CrossRef] [PubMed]
44. Caron, N.S.; Banos, R.; Aly, A.E.; Xie, Y.; Ko, S.; Potluri, N.; Anderson, C.; Black, H.F.; Anderson, L.M.; Gordon, B.; et al. Cerebrospinal fluid mutant huntingtin is a biomarker for huntingtin lowering in the striatum of Huntington disease mice. *Neurobiol. Dis.* **2022**, *166*, 105652. [CrossRef] [PubMed]
45. Bertoglio, D.; Verhaeghe, J.; Miranda, A.; Wyffels, L.; Stroobants, S.; Mrzljak, L.; Khetarpal, V.; Skinbjerg, M.; Liu, L.; Dominguez, C.; et al. Longitudinal preclinical evaluation of the novel radioligand [11C]CHDI-626 for PET imaging of mutant huntingtin aggregates in Huntington's disease. *Eur. J. Nucl. Med. Mol. Imaging* **2022**, *49*, 1166–1175. [CrossRef] [PubMed]
46. Liu, H.; Zhang, C.; Xu, J.; Jin, J.; Cheng, L.; Miao, X.; Wu, Q.; Wei, Z.; Liu, P.; Lu, H.; et al. Huntingtin silencing delays onset and slows progression of Huntington's disease: A biomarker study. *Brain* **2021**, *144*, 3101–3113. [CrossRef]
47. Cook, S.D.; Dhib-Jalbut, S.; Dowling, P.; Durelli, L.; Ford, C.; Giovannoni, G.; Halper, J.; Harris, C.; Herbert, J.; Li, D.; et al. Use of Magnetic Resonance Imaging as Well as Clinical Disease Activity in the Clinical Classification of Multiple Sclerosis and Assessment of Its Course: A Report from an International CMSC Consensus Conference, March 5–7, 2010. *Int. J. MS Care* **2012**, *14*, 105–114. [CrossRef]
48. van Munster, C.E.P.; Uitdehaag, B.M.J. Outcome Measures in Clinical Trials for Multiple Sclerosis. *CNS Drugs* **2017**, *31*, 217–236. [CrossRef]
49. Filippi, M.; Brück, W.; Chard, D.; Fazekas, F.; Geurts, J.J.G.; Enzinger, C.; Hametner, S.; Kuhlmann, T.; Preziosa, P.; Rovira, À.; et al. Association between pathological and MRI findings in multiple sclerosis. *Lancet Neurol.* **2019**, *18*, 198–210. [CrossRef]
50. Huang, J.; Khademi, M.; Fugger, L.; Lindhe, Ö.; Novakova, L.; Axelsson, M.; Malmeström, C.; Constantinescu, C.; Lycke, J.; Piehl, F.; et al. Inflammation-related plasma and CSF biomarkers for multiple sclerosis. *Proc. Natl. Acad. Sci. USA* **2020**, *117*, 12952–12960. [CrossRef]
51. McNicholas, N.; Lockhart, A.; Yap, S.M.; O'Connell, K.; Tubridy, N.; Hutchinson, M.; McGuigan, C. New versus old: Implications of evolving diagnostic criteria for relapsing-remitting multiple sclerosis. *Mult. Scler. Houndmills Basingstoke Engl.* **2019**, *25*, 867–870. [CrossRef] [PubMed]
52. Bankoti, J.; Apeltsin, L.; Hauser, S.L.; Allen, S.; Albertolle, M.E.; Witkowska, H.E.; von Büdingen, H.C. In multiple sclerosis, oligoclonal bands connect to peripheral B-cell responses. *Ann. Neurol.* **2014**, *75*, 266–276. [CrossRef] [PubMed]
53. Petzold, A. Intrathecal oligoclonal IgG synthesis in multiple sclerosis. *J. Neuroimmunol.* **2013**, *262*, 1–10. [CrossRef] [PubMed]
54. Freedman, M.S.; Thompson, E.J.; Deisenhammer, F.; Giovannoni, G.; Grimsley, G.; Keir, G.; Ohman, S.; Racke, M.K.; Sharief, M.; Sindic, C.J.; et al. Recommended standard of cerebrospinal fluid analysis in the diagnosis of multiple sclerosis: A consensus statement. *Arch. Neurol.* **2005**, *62*, 865–870. [CrossRef] [PubMed]
55. Link, H.; Kostulas, V. Utility of isoelectric focusing of cerebrospinal fluid and serum on agarose evaluated for neurological patients. *Clin. Chem.* **1983**, *29*, 810–815. [CrossRef] [PubMed]

56. Chitnis, T. Serum NFL levels should be used to monitor multiple sclerosis evolution—Commentary. *Mult. Scler. Houndmills Basingstoke Engl.* **2020**, *26*, 21–22. [CrossRef]
57. Kim, H.; Lee, E.J.; Kim, S.; Choi, L.K.; Kim, K.; Kim, H.W.; Kim, K.K.; Lim, Y.M. Serum biomarkers in myelin oligodendrocyte glycoprotein antibody-associated disease. *Neurol. Neuroimmunol. Neuroinflamm.* **2020**, *7*, e708. [CrossRef]
58. Watanabe, M.; Nakamura, Y.; Michalak, Z.; Isobe, N.; Barro, C.; Leppert, D.; Matsushita, T.; Hayashi, F.; Yamasaki, R.; Kuhle, J.; et al. Serum GFAP and neurofilament light as biomarkers of disease activity and disability in NMOSD. *Neurology* **2019**, *93*, e1299–e1311. [CrossRef]
59. Peng, L.; Bi, C.; Xia, D.; Mao, L.; Qian, H. Increased cerebrospinal fluid neurofilament light chain in central nervous system inflammatory demyelinating disease. *Mult. Scler. Relat. Disord.* **2019**, *30*, 123–128. [CrossRef]
60. Meeter, L.H.; Dopper, E.G.; Jiskoot, L.C.; Sanchez-Valle, R.; Graff, C.; Benussi, L.; Ghidoni, R.; Pijnenburg, Y.A.; Borroni, B.; Galimberti, D.; et al. Neurofilament light chain: A biomarker for genetic frontotemporal dementia. *Ann. Clin. Transl. Neurol.* **2016**, *3*, 623–636. [CrossRef]
61. van der Ende, E.L.; Meeter, L.H.; Poos, J.M.; Panman, J.L.; Jiskoot, L.C.; Dopper, E.G.P.; Papma, J.M.; de Jong, F.J.; Verberk, I.M.W.; Teunissen, C.; et al. Serum neurofilament light chain in genetic frontotemporal dementia: A longitudinal, multicentre cohort study. *Lancet Neurol.* **2019**, *18*, 1103–1111. [CrossRef] [PubMed]
62. Thijssen, E.H.; La Joie, R.; Wolf, A.; Strom, A.; Wang, P.; Iaccarino, L.; Bourakova, V.; Cobigo, Y.; Heuer, H.; Spina, S.; et al. Diagnostic value of plasma phosphorylated tau181 in Alzheimer's disease and frontotemporal lobar degeneration. *Nat. Med.* **2020**, *26*, 387–397. [CrossRef] [PubMed]
63. Illán-Gala, I.; Lleo, A.; Karydas, A.; Staffaroni, A.M.; Zetterberg, H.; Sivasankaran, R.; Grinberg, L.T.; Spina, S.; Kramer, J.H.; Ramos, E.M.; et al. Plasma Tau and Neurofilament Light in Frontotemporal Lobar Degeneration and Alzheimer Disease. *Neurology* **2021**, *96*, e671–e683. [CrossRef] [PubMed]
64. Rojas, J.C.; Wang, P.; Staffaroni, A.M.; Heller, C.; Cobigo, Y.; Wolf, A.; Goh, S.-Y.M.; Ljubenkov, P.A.; Heuer, H.W.; Fong, J.C.; et al. Plasma Neurofilament Light for Prediction of Disease Progression in Familial Frontotemporal Lobar Degeneration. *Neurology* **2021**, *96*, e2296–e2312. [CrossRef] [PubMed]
65. Ghidoni, R.; Stoppani, E.; Rossi, G.; Piccoli, E.; Albertini, V.; Paterlini, A.; Glionna, M.; Pegoiani, E.; Agnati, L.F.; Fenoglio, C.; et al. Optimal Plasma Progranulin Cutoff Value for Predicting Null Progranulin Mutations in Neurodegenerative Diseases: A Multicenter Italian Study. *Neurodegener. Dis.* **2011**, *9*, 121–127. [CrossRef]
66. Sellami, L.; Rucheton, B.; Ben Younes, I.; Camuzat, A.; Saracino, D.; Rinaldi, D.; Epelbaum, S.; Azuar, C.; Levy, R.; Auriacombe, S.; et al. Plasma progranulin levels for frontotemporal dementia in clinical practice: A 10-year French experience. *Neurobiol. Aging* **2020**, *91*, 167.e1–167.e9. [CrossRef]
67. Bird, T.D. Progranulin plasma levels in the diagnosis of frontotemporal dementia. *Brain* **2009**, *132*, 568–569. [CrossRef]
68. Lee, C.Y.; Yoon, J.H.; Park, J.-H.; Hong, J.Y.; Sunwoo, M.K.; Kang, S.Y. Uric acid level may not be reduced in essential tremor. *Int. J. Neurosci.* **2018**, *128*, 1163–1167. [CrossRef]
69. Wang, Q.; Aljassar, M.; Bhagwat, N.; Zeighami, Y.; Evans, A.C.; Dagher, A.; Pike, G.B.; Sadikot, A.F.; Poline, J.-B. Reproducibility of cerebellar involvement as quantified by consensus structural MRI biomarkers in advanced essential tremor. *Sci. Rep.* **2023**, *13*, 581. [CrossRef]
70. Yu, Z.; Liu, G.; Zheng, Y.; Huang, G.; Feng, T. Erythrocytic alpha-synuclein as potential biomarker for the differentiation between essential tremor and Parkinson's disease. *Front. Neurol.* **2023**, *14*, 1173074. [CrossRef]
71. Satlin, A.; Wang, J.; Logovinsky, V.; Berry, S.; Swanson, C.; Dhadda, S.; Berry, D.A. Design of a Bayesian adaptive phase 2 proof-of-concept trial for BAN2401, a putative disease-modifying monoclonal antibody for the treatment of Alzheimer's disease. *Alzheimers Dement. Transl. Res. Clin. Interv.* **2016**, *2*, 1–12. [CrossRef]
72. Swanson, C.J.; Zhang, Y.; Dhadda, S.; Wang, J.; Kaplow, J.; Lai, R.Y.K.; Lannfelt, L.; Bradley, H.; Rabe, M.; Koyama, A.; et al. A randomized, double-blind, phase 2b proof-of-concept clinical trial in early Alzheimer's disease with lecanemab, an anti-Aβ protofibril antibody. *Alzheimers Res. Ther.* **2021**, *13*, 80. [CrossRef] [PubMed]
73. Voss, M.W.; Soto, C.; Yoo, S.; Sodoma, M.; Vivar, C.; Van Praag, H. Exercise and Hippocampal Memory Systems. *Trends Cogn. Sci.* **2019**, *23*, 318–333. [CrossRef] [PubMed]
74. Gaitán, J.M.; Moon, H.Y.; Stremlau, M.; Dubal, D.B.; Cook, D.B.; Okonkwo, O.C.; Van Praag, H. Effects of Aerobic Exercise Training on Systemic Biomarkers and Cognition in Late Middle-Aged Adults at Risk for Alzheimer's Disease. *Front. Endocrinol.* **2021**, *12*, 660181. [CrossRef]
75. Hoscheidt, S.; Sanderlin, A.H.; Baker, L.D.; Jung, Y.; Lockhart, S.; Kellar, D.; Whitlow, C.T.; Hanson, A.J.; Friedman, S.; Register, T.; et al. Mediterranean and Western diet effects on Alzheimer's disease biomarkers, cerebral perfusion, and cognition in mid-life: A randomized trial. *Alzheimers Dement.* **2022**, *18*, 457–468. [CrossRef] [PubMed]
76. Prins, S.; De Kam, M.L.; Teunissen, C.E.; Groeneveld, G.J. Inflammatory plasma biomarkers in subjects with preclinical Alzheimer's disease. *Alzheimers Res. Ther.* **2022**, *14*, 106. [CrossRef]
77. Innes, K.E.; Selfe, T.K.; Brundage, K.; Montgomery, C.; Wen, S.; Kandati, S.; Bowles, H.; Khalsa, D.S.; Huysmans, Z. Effects of Meditation and Music-Listening on Blood Biomarkers of Cellular Aging and Alzheimer's Disease in Adults with Subjective Cognitive Decline: An Exploratory Randomized Clinical Trial. *J. Alzheimers Dis.* **2018**, *66*, 947–970. [CrossRef]

78. McDade, E.; Cummings, J.L.; Dhadda, S.; Swanson, C.J.; Reyderman, L.; Kanekiyo, M.; Koyama, A.; Irizarry, M.; Kramer, L.D.; Bateman, R.J. Lecanemab in patients with early Alzheimer's disease: Detailed results on biomarker, cognitive, and clinical effects from the randomized and open-label extension of the phase 2 proof-of-concept study. *Alzheimers Res. Ther.* **2022**, *14*, 191. [CrossRef]
79. Blennow, K.; Hampel, H.; Weiner, M.; Zetterberg, H. Cerebrospinal fluid and plasma biomarkers in Alzheimer disease. *Nat. Rev. Neurol.* **2010**, *6*, 131–144. [CrossRef]
80. Kester, M.I.; Teunissen, C.E.; Crimmins, D.L.; Herries, E.M.; Ladenson, J.H.; Scheltens, P.; Van Der Flier, W.M.; Morris, J.C.; Holtzman, D.M.; Fagan, A.M. Neurogranin as a Cerebrospinal Fluid Biomarker for Synaptic Loss in Symptomatic Alzheimer Disease. *JAMA Neurol.* **2015**, *72*, 1275. [CrossRef]
81. Pontecorvo, M.J.; Lu, M.; Burnham, S.C.; Schade, A.E.; Dage, J.L.; Shcherbinin, S.; Collins, E.C.; Sims, J.R.; Mintun, M.A. Association of Donanemab Treatment With Exploratory Plasma Biomarkers in Early Symptomatic Alzheimer Disease: A Secondary Analysis of the TRAILBLAZER-ALZ Randomized Clinical Trial. *JAMA Neurol.* **2022**, *79*, 1250. [CrossRef] [PubMed]
82. Fagan, A.M.; Perrin, R.J. Upcoming candidate cerebrospinal fluid biomarkers of Alzheimer's disease. *Biomark. Med.* **2012**, *6*, 455–476. [CrossRef] [PubMed]
83. Molinuevo, J.L.; Gramunt, N.; Gispert, J.D.; Fauria, K.; Esteller, M.; Minguillon, C.; Sánchez-Benavides, G.; Huesa, G.; Morán, S.; Dal-Ré, R.; et al. The ALFA project: A research platform to identify early pathophysiological features of Alzheimer's disease. *Alzheimers Dement. Transl. Res. Clin. Interv.* **2016**, *2*, 82–92. [CrossRef] [PubMed]
84. Ashton, N.J.; Hye, A.; Leckey, C.A.; Jones, A.R.; Gardner, A.; Elliott, C.; Wetherell, J.L.; Lenze, E.J.; Killick, R.; Marchant, N.L. Plasma REST: A novel candidate biomarker of Alzheimer's disease is modified by psychological intervention in an at-risk population. *Transl. Psychiatry* **2017**, *7*, e1148. [CrossRef]
85. Hendrickson, R.C.; Lee, A.Y.H.; Song, Q.; Liaw, A.; Wiener, M.; Paweletz, C.P.; Seeburger, J.L.; Li, J.; Meng, F.; Deyanova, E.G.; et al. High Resolution Discovery Proteomics Reveals Candidate Disease Progression Markers of Alzheimer's Disease in Human Cerebrospinal Fluid. *PLoS ONE* **2015**, *10*, e0135365. [CrossRef]
86. Schmechel, D.E.; Saunders, A.M.; Strittmatter, W.J.; Crain, B.J.; Hulette, C.M.; Joo, S.H.; Pericak-Vance, M.A.; Goldgaber, D.; Roses, A.D. Increased amyloid beta-peptide deposition in cerebral cortex as a consequence of apolipoprotein E genotype in late-onset Alzheimer disease. *Proc. Natl. Acad. Sci. USA* **1993**, *90*, 9649–9653. [CrossRef]
87. Begcevic, I.; Tsolaki, M.; Brinc, D.; Brown, M.; Martinez-Morillo, E.; Lazarou, I.; Kozori, M.; Tagaraki, F.; Nenopoulou, S.; Gkioka, M.; et al. Neuronal pentraxin receptor-1 is a new cerebrospinal fluid biomarker of Alzheimer's disease progression. *F1000Research* **2018**, *7*, 1012. [CrossRef]
88. Pan, X.; Cunningham, E.L.; Passmore, A.P.; McGuinness, B.; McAuley, D.F.; Beverland, D.; O'Brien, S.; Mawhinney, T.; Schott, J.M.; Zetterberg, H.; et al. Cerebrospinal Fluid Spermidine, Glutamine and Putrescine Predict Postoperative Delirium Following Elective Orthopaedic Surgery. *Sci. Rep.* **2019**, *9*, 4191. [CrossRef]
89. Vijverberg, E.G.B.; Axelsen, T.M.; Bihlet, A.R.; Henriksen, K.; Weber, F.; Fuchs, K.; Harrison, J.E.; Kühn-Wache, K.; Alexandersen, P.; Prins, N.D.; et al. Rationale and study design of a randomized, placebo-controlled, double-blind phase 2b trial to evaluate efficacy, safety, and tolerability of an oral glutaminyl cyclase inhibitor varoglutamstat (PQ912) in study participants with MCI and mild AD—VIVIAD. *Alzheimers Res. Ther.* **2021**, *13*, 142. [CrossRef]
90. Scheltens, P.; Hallikainen, M.; Grimmer, T.; Duning, T.; Gouw, A.A.; Teunissen, C.E.; Wink, A.M.; Maruff, P.; Harrison, J.; Van Baal, C.M.; et al. Safety, tolerability and efficacy of the glutaminyl cyclase inhibitor PQ912 in Alzheimer's disease: Results of a randomized, double-blind, placebo-controlled phase 2a study. *Alzheimers Res. Ther.* **2018**, *10*, 107. [CrossRef]
91. Alzheimer's Association Calcium Hypothesis Workgroup; Khachaturian, Z.S. Calcium Hypothesis of Alzheimer's disease and brain aging: A framework for integrating new evidence into a comprehensive theory of pathogenesis. *Alzheimers Dement.* **2017**, *13*, 178. [CrossRef]
92. Hu, W.T.; Holtzman, D.M.; Fagan, A.M.; Shaw, L.M.; Perrin, R.; Arnold, S.E.; Grossman, M.; Xiong, C.; Craig-Schapiro, R.; Clark, C.M.; et al. Plasma multianalyte profiling in mild cognitive impairment and Alzheimer disease. *Neurology* **2012**, *79*, 897–905. [CrossRef] [PubMed]
93. Van Eldik, L.J.; Carrillo, M.C.; Cole, P.E.; Feuerbach, D.; Greenberg, B.D.; Hendrix, J.A.; Kennedy, M.; Kozauer, N.; Margolin, R.A.; Molinuevo, J.L.; et al. The roles of inflammation and immune mechanisms in Alzheimer's disease. *Alzheimers Dement. Transl. Res. Clin. Interv.* **2016**, *2*, 99–109. [CrossRef] [PubMed]
94. Snyder, H.M.; Carrillo, M.C.; Grodstein, F.; Henriksen, K.; Jeromin, A.; Lovestone, S.; Mielke, M.M.; O'Bryant, S.; Sarasa, M.; Sjøgren, M.; et al. Developing novel blood-based biomarkers for Alzheimer's disease. *Alzheimers Dement.* **2014**, *10*, 109–114. [CrossRef] [PubMed]
95. Sato, Y.; Suzuki, I.; Nakamura, T.; Bernier, F.; Aoshima, K.; Oda, Y. Identification of a new plasma biomarker of Alzheimer's disease using metabolomics technology. *J. Lipid Res.* **2012**, *53*, 567–576. [CrossRef]
96. Pasinetti, G.M. Use of cDNA microarray in the search for molecular markers involved in the onset of Alzheimer's disease dementia. *J. Neurosci. Res.* **2001**, *65*, 471–476. [CrossRef]
97. Nordberg, A.; Rinne, J.O.; Kadir, A.; Långström, B. The use of PET in Alzheimer disease. *Nat. Rev. Neurol.* **2010**, *6*, 78–87. [CrossRef]

98. Hampel, H.; Lista, S.; Teipel, S.J.; Garaci, F.; Nisticò, R.; Blennow, K.; Zetterberg, H.; Bertram, L.; Duyckaerts, C.; Bakardjian, H.; et al. Perspective on future role of biological markers in clinical therapy trials of Alzheimer's disease: A long-range point of view beyond 2020. *Biochem. Pharmacol.* **2014**, *88*, 426–449. [CrossRef]
99. Tolosa, E.; Wenning, G.; Poewe, W. The diagnosis of Parkinson's disease. *Lancet Neurol.* **2006**, *5*, 75–86. [CrossRef]
100. Italian Neurological Society; Italian Society of Clinical Neurophysiology; Guidelines for the Treatment of Parkinson's Disease 2002. The diagnosis of Parkinson's disease. *Neurol. Sci. Off. J. Ital. Neurol. Soc. Ital. Soc. Clin. Neurophysiol.* **2003**, *24* (Suppl. S3), S157–S164.
101. Cavallo, F.; Moschetti, A.; Esposito, D.; Maremmani, C.; Rovini, E. Upper limb motor pre-clinical assessment in Parkinson's disease using machine learning. *Parkinsonism Relat. Disord.* **2019**, *63*, 111–116. [CrossRef] [PubMed]
102. Karapinar Senturk, Z. Early diagnosis of Parkinson's disease using machine learning algorithms. *Med. Hypotheses* **2020**, *138*, 109603. [CrossRef] [PubMed]
103. Gao, L.; Wu, T. The study of brain functional connectivity in Parkinson's disease. *Transl. Neurodegener.* **2016**, *5*, 18. [CrossRef] [PubMed]
104. Bhat, S.; Acharya, U.R.; Hagiwara, Y.; Dadmehr, N.; Adeli, H. Parkinson's disease: Cause factors, measurable indicators, and early diagnosis. *Comput. Biol. Med.* **2018**, *102*, 234–241. [CrossRef] [PubMed]
105. Hassin-Baer, S.; Cohen, O.S.; Israeli-Korn, S.; Yahalom, G.; Benizri, S.; Sand, D.; Issachar, G.; Geva, A.B.; Shani-Hershkovich, R.; Peremen, Z. Identification of an early-stage Parkinson's disease neuromarker using event-related potentials, brain network analytics and machine-learning. *PLoS ONE* **2022**, *17*, e0261947. [CrossRef] [PubMed]
106. Farlie, M.K.; Molloy, E.; Keating, J.L.; Haines, T.P. Clinical Markers of the Intensity of Balance Challenge: Observational Study of Older Adult Responses to Balance Tasks. *Phys. Ther.* **2016**, *96*, 313–323. [CrossRef]
107. Dobkin, R.D.; Menza, M.; Allen, L.A.; Gara, M.A.; Mark, M.H.; Tiu, J.; Bienfait, K.L.; Friedman, J. Cognitive-Behavioral Therapy for Depression in Parkinson's Disease: A Randomized, Controlled Trial. *Am. J. Psychiatry* **2011**, *168*, 1066–1074. [CrossRef]
108. Lipsmeier, F.; Taylor, K.I.; Kilchenmann, T.; Wolf, D.; Scotland, A.; Schjodt-Eriksen, J.; Cheng, W.; Fernandez-Garcia, I.; Siebourg-Polster, J.; Jin, L.; et al. Evaluation of smartphone-based testing to generate exploratory outcome measures in a phase 1 Parkinson's disease clinical trial. *Mov. Disord.* **2018**, *33*, 1287–1297. [CrossRef]
109. Grosset, K.A. Measuring therapy adherence in Parkinson's disease: A comparison of methods. *J. Neurol. Neurosurg. Psychiatry* **2006**, *77*, 249–251. [CrossRef]
110. Milà-Alomà, M.; Salvadó, G.; Gispert, J.D.; Vilor-Tejedor, N.; Grau-Rivera, O.; Sala-Vila, A.; Sánchez-Benavides, G.; Arenaza-Urquijo, E.M.; Crous-Bou, M.; González-de-Echávarri, J.M.; et al. Amyloid beta, tau, synaptic, neurodegeneration, and glial biomarkers in the preclinical stage of the Alzheimer's continuum. *Alzheimers Dement.* **2020**, *16*, 1358–1371. [CrossRef]
111. Tokuda, T.; Salem, S.A.; Allsop, D.; Mizuno, T.; Nakagawa, M.; Qureshi, M.M.; Locascio, J.J.; Schlossmacher, M.G.; El-Agnaf, O.M.A. Decreased α-synuclein in cerebrospinal fluid of aged individuals and subjects with Parkinson's disease. *Biochem. Biophys. Res. Commun.* **2006**, *349*, 162–166. [CrossRef] [PubMed]
112. Bartl, M.; Dakna, M.; Galasko, D.; Hutten, S.J.; Foroud, T.; Quan, M.; Marek, K.; Siderowf, A.; Franz, J.; Trenkwalder, C.; et al. Biomarkers of neurodegeneration and glial activation validated in Alzheimer's disease assessed in longitudinal cerebrospinal fluid samples of Parkinson's disease. *PLoS ONE* **2021**, *16*, e0257372. [CrossRef] [PubMed]
113. Du, G.; Lewis, M.M.; Sica, C.; Kong, L.; Huang, X. Magnetic resonance T 1w/T 2w ratio: A parsimonious marker for P arkinson disease. *Ann. Neurol.* **2019**, *85*, 96–104. [CrossRef] [PubMed]
114. Götz, M.E.; Gerstner, A.; Harth, R.; Dirr, A.; Janetzky, B.; Kuhn, W.; Riederer, P.; Gerlach, M. Altered redox state of platelet coenzyme Q 10 in Parkinson's disease. *J. Neural Transm.* **2000**, *107*, 41–48. [CrossRef]
115. Mischley, L.K.; Allen, J.; Bradley, R. Coenzyme Q10 deficiency in patients with Parkinson's disease. *J. Neurol. Sci.* **2012**, *318*, 72–75. [CrossRef]
116. Saunders, J.A.H.; Estes, K.A.; Kosloski, L.M.; Allen, H.E.; Dempsey, K.M.; Torres-Russotto, D.R.; Meza, J.L.; Santamaria, P.M.; Bertoni, J.M.; Murman, D.L.; et al. CD4+ Regulatory and Effector/Memory T Cell Subsets Profile Motor Dysfunction in Parkinson's Disease. *J. Neuroimmune Pharmacol.* **2012**, *7*, 927–938. [CrossRef]
117. on behalf of the MOVES-PD Investigators; Peterschmitt, M.J.; Saiki, H.; Hatano, T.; Gasser, T.; Isaacson, S.H.; Gaemers, S.J.M.; Minini, P.; Saubadu, S.; Sharma, J.; et al. Safety, Pharmacokinetics, and Pharmacodynamics of Oral Venglustat in Patients with Parkinson's Disease and a GBA Mutation: Results from Part 1 of the Randomized, Double-Blinded, Placebo-Controlled MOVES-PD Trial. *J. Park. Dis.* **2022**, *12*, 557–570. [CrossRef]
118. Pagan, F.L.; Hebron, M.L.; Wilmarth, B.; Torres-Yaghi, Y.; Lawler, A.; Mundel, E.E.; Yusuf, N.; Starr, N.J.; Anjum, M.; Arellano, J.; et al. Nilotinib Effects on Safety, Tolerability, and Potential Biomarkers in Parkinson Disease: A Phase 2 Randomized Clinical Trial. *JAMA Neurol.* **2020**, *77*, 309. [CrossRef]
119. Brauer, R.; Bhaskaran, K.; Chaturvedi, N.; Dexter, D.T.; Smeeth, L.; Douglas, I. Glitazone Treatment and Incidence of Parkinson's Disease among People with Diabetes: A Retrospective Cohort Study. *PLoS Med.* **2015**, *12*, e1001854. [CrossRef]
120. The Parkinson Study Group PRECEPT Investigators. Mixed lineage kinase inhibitor CEP-1347 fails to delay disability in early Parkinson disease. *Neurology* **2007**, *69*, 1480–1490. [CrossRef]
121. Whone, A.L.; Watts, R.L.; Stoessl, A.J.; Davis, M.; Reske, S.; Nahmias, C.; Lang, A.E.; Rascol, O.; Ribeiro, M.J.; Remy, P.; et al. Slower progression of Parkinson's disease with ropinirole versus levodopa: The REAL-PET study. *Ann. Neurol.* **2003**, *54*, 93–101. [CrossRef] [PubMed]

122. Cummings, J.; Lee, G.; Ritter, A.; Sabbagh, M.; Zhong, K. Alzheimer's disease drug development pipeline: 2020. *Alzheimers Dement. Transl. Res. Clin. Interv.* **2020**, *6*, e12050. [CrossRef]
123. Cucca, A.; Di Rocco, A.; Acosta, I.; Beheshti, M.; Berberian, M.; Bertisch, H.C.; Droby, A.; Ettinger, T.; Hudson, T.E.; Inglese, M.; et al. Art therapy for Parkinson's disease. *Parkinsonism Relat. Disord.* **2021**, *84*, 148–154. [CrossRef] [PubMed]
124. Li, F.; Harmer, P.; Fitzgerald, K.; Eckstrom, E.; Stock, R.; Galver, J.; Maddalozzo, G.; Batya, S.S. Tai Chi and Postural Stability in Patients with Parkinson's Disease. *N. Engl. J. Med.* **2012**, *366*, 511–519. [CrossRef]
125. Feng, H.; Li, C.; Liu, J.; Wang, L.; Ma, J.; Li, G.; Gan, L.; Shang, X.; Wu, Z. Virtual Reality Rehabilitation Versus Conventional Physical Therapy for Improving Balance and Gait in Parkinson's Disease Patients: A Randomized Controlled Trial. *Med. Sci. Monit.* **2019**, *25*, 4186–4192. [CrossRef] [PubMed]
126. Chen, Y.; Yang, W.; Long, J.; Zhang, Y.; Feng, J.; Li, Y.; Huang, B. Discriminative Analysis of Parkinson's Disease Based on Whole-Brain Functional Connectivity. *PLoS ONE* **2015**, *10*, e0124153. [CrossRef]
127. Acton, P.D.; Newberg, A. Artificial neural network classifier for the diagnosis of Parkinson's disease using [^{99m}Tc]TRODAT-1 and SPECT. *Phys. Med. Biol.* **2006**, *51*, 3057–3066. [CrossRef]
128. Long, D.; Wang, J.; Xuan, M.; Gu, Q.; Xu, X.; Kong, D.; Zhang, M. Automatic Classification of Early Parkinson's Disease with Multi-Modal MR Imaging. *PLoS ONE* **2012**, *7*, e47714. [CrossRef]
129. Lin, S.; Gao, C.; Li, H.; Huang, P.; Ling, Y.; Chen, Z.; Ren, K.; Chen, S. Wearable sensor-based gait analysis to discriminate early Parkinson's disease from essential tremor. *J. Neurol.* **2023**, *270*, 2283–2301. [CrossRef]
130. Díaz, S.; Stephenson, J.B.; Labrador, M.A. Use of Wearable Sensor Technology in Gait, Balance, and Range of Motion Analysis. *Appl. Sci.* **2019**, *10*, 234. [CrossRef]
131. Álvarez, I.; Latorre, J.; Aguilar, M.; Pastor, P.; Llorens, R. Validity and sensitivity of instrumented postural and gait assessment using low-cost devices in Parkinson's disease. *J. NeuroEng. Rehabil.* **2020**, *17*, 149. [CrossRef] [PubMed]
132. Marangi, G.; Traynor, B.J. Genetic causes of amyotrophic lateral sclerosis: New genetic analysis methodologies entailing new opportunities and challenges. *Brain Res.* **2015**, *1607*, 75–93. [CrossRef] [PubMed]
133. Averill, A.J.; Kasarskis, E.J.; Segerstrom, S.C. Expressive disclosure to improve well-being in patients with amyotrophic lateral sclerosis: A randomised, controlled trial. *Psychol. Health* **2013**, *28*, 701–713. [CrossRef]
134. De Wit, J.; Vervoort, S.C.J.M.; Van Eerden, E.; Van Den Berg, L.H.; Visser-Meily, J.M.A.; Beelen, A.; Schröder, C.D. User perspectives on a psychosocial blended support program for partners of patients with amyotrophic lateral sclerosis and progressive muscular atrophy: A qualitative study. *BMC Psychol.* **2019**, *7*, 35. [CrossRef]
135. Joardar, A.; Manzo, E.; Zarnescu, D.C. Metabolic Dysregulation in Amyotrophic Lateral Sclerosis: Challenges and Opportunities. *Curr. Genet. Med. Rep.* **2017**, *5*, 108–114. [CrossRef] [PubMed]
136. Laferriere, F.; Polymenidou, M. Advances and challenges in understanding the multifaceted pathogenesis of amyotrophic lateral sclerosis. *Swiss Med. Wkly.* **2015**, *145*, w14054. [CrossRef]
137. Stockwell, B.R.; Friedmann Angeli, J.P.; Bayir, H.; Bush, A.I.; Conrad, M.; Dixon, S.J.; Fulda, S.; Gascón, S.; Hatzios, S.K.; Kagan, V.E.; et al. Ferroptosis: A Regulated Cell Death Nexus Linking Metabolism, Redox Biology, and Disease. *Cell* **2017**, *171*, 273–285. [CrossRef]
138. Chen, L.; Hambright, W.S.; Na, R.; Ran, Q. Ablation of the Ferroptosis Inhibitor Glutathione Peroxidase 4 in Neurons Results in Rapid Motor Neuron Degeneration and Paralysis. *J. Biol. Chem.* **2015**, *290*, 28097–28106. [CrossRef]
139. Moreau, C.; Danel, V.; Devedjian, J.C.; Grolez, G.; Timmerman, K.; Laloux, C.; Petrault, M.; Gouel, F.; Jonneaux, A.; Dutheil, M.; et al. Could Conservative Iron Chelation Lead to Neuroprotection in Amyotrophic Lateral Sclerosis? *Antioxid. Redox Signal.* **2018**, *29*, 742–748. [CrossRef]
140. Lenglet, T.; Lacomblez, L.; Abitbol, J.L.; Ludolph, A.; Mora, J.S.; Robberecht, W.; Shaw, P.J.; Pruss, R.M.; Cuvier, V.; Meininger, V.; et al. A phase II–III trial of olesoxime in subjects with amyotrophic lateral sclerosis. *Eur. J. Neurol.* **2014**, *21*, 529–536. [CrossRef]
141. Devos, D.; Moreau, C.; Kyheng, M.; Garçon, G.; Rolland, A.S.; Blasco, H.; Gelé, P.; Timothée Lenglet, T.; Veyrat-Durebex, C.; Corcia, P.; et al. A ferroptosis–based panel of prognostic biomarkers for Amyotrophic Lateral Sclerosis. *Sci. Rep.* **2019**, *9*, 2918. [CrossRef] [PubMed]
142. Beghi, E.; Pupillo, E.; Bianchi, E.; Bonetto, V.; Luotti, S.; Pasetto, L.; Bendotti, C.; Tortarolo, M.; Sironi, F.; Camporeale, L.; et al. Effect of RNS60 in amyotrophic lateral sclerosis: A phase II multicentre, randomized, double-blind, placebo-controlled trial. *Eur. J. Neurol.* **2023**, *30*, 69–86. [CrossRef] [PubMed]
143. Benatar, M.; Wuu, J. Presymptomatic studies in ALS: Rationale, challenges, and approach. *Neurology* **2012**, *79*, 1732–1739. [CrossRef] [PubMed]
144. Benatar, M.; Wuu, J.; Andersen, P.M.; Bucelli, R.C.; Andrews, J.A.; Otto, M.; Farahany, N.A.; Harrington, E.A.; Chen, W.; Mitchell, A.A.; et al. Design of a Randomized, Placebo-Controlled, Phase 3 Trial of Tofersen Initiated in Clinically Presymptomatic SOD1 Variant Carriers: The ATLAS Study. *Neurotherapeutics* **2022**, *19*, 1248–1258. [CrossRef] [PubMed]
145. Lunetta, C.; Lizio, A.; Maestri, E.; Sansone, V.A.; Mora, G.; Miller, R.G.; Appel, S.H.; Chiò, A. Serum C-Reactive Protein as a Prognostic Biomarker in Amyotrophic Lateral Sclerosis. *JAMA Neurol.* **2017**, *74*, 660. [CrossRef]
146. Berry, J.D.; Cudkowicz, M.E.; Windebank, A.J.; Staff, N.P.; Owegi, M.; Nicholson, K.; McKenna-Yasek, D.; Levy, Y.S.; Abramov, N.; Kaspi, H.; et al. NurOwn, phase 2, randomized, clinical trial in patients with ALS: Safety, clinical, and biomarker results. *Neurology* **2019**, *93*, e2294–e2305. [CrossRef]

147. Miller, R.G.; Zhang, R.; Block, G.; Katz, J.; Barohn, R.; Kasarskis, E.; Forshew, D.; Gopalakrishnan, V.; McGrath, M.S. NP001 regulation of macrophage activation markers in ALS: A phase I clinical and biomarker study. *Amyotroph. Lateral Scler. Front. Degener.* **2014**, *15*, 601–609. [CrossRef]
148. Stewart, W.F.; Schwartz, B.S.; Simon, D.; Kelsey, K.; Todd, A.C. ApoE genotype, past adult lead exposure, and neurobehavioral function. *Environ. Health Perspect.* **2002**, *110*, 501–505. [CrossRef]
149. Fitzmaurice, A.G.; Rhodes, S.L.; Lulla, A.; Murphy, N.P.; Lam, H.A.; O'Donnell, K.C.; Barnhill, L.; Casida, J.E.; Cockburn, M.; Sagasti, A.; et al. Aldehyde dehydrogenase inhibition as a pathogenic mechanism in Parkinson disease. *Proc. Natl. Acad. Sci. USA* **2013**, *110*, 636–641. [CrossRef]
150. Roberts, J.S.; Uhlmann, W.R. Genetic susceptibility testing for neurodegenerative diseases: Ethical and practice issues. *Prog. Neurobiol.* **2013**, *110*, 89–101. [CrossRef]
151. Ravanfar, P.; Loi, S.M.; Syeda, W.T.; Van Rheenen, T.E.; Bush, A.I.; Desmond, P.; Cropley, V.L.; Lane, D.J.R.; Opazo, C.M.; Moffat, B.A.; et al. Systematic Review: Quantitative Susceptibility Mapping (QSM) of Brain Iron Profile in Neurodegenerative Diseases. *Front. Neurosci.* **2021**, *15*, 618435. [CrossRef] [PubMed]
152. Rush, A.J. Polysomnographic Findings in Recently Drug-Free and Clinically Remitted Depressed Patients. *Arch. Gen. Psychiatry* **1986**, *43*, 878. [CrossRef] [PubMed]
153. Lauer, C.J. In Quest of Identifying Vulnerability Markers for Psychiatric Disorders by All-Night Polysomnography. *Arch. Gen. Psychiatry* **1995**, *52*, 145. [CrossRef]
154. Kitada, T.; Asakawa, S.; Hattori, N.; Matsumine, H.; Yamamura, Y.; Minoshima, S.; Yokochi, M.; Mizuno, Y.; Shimizu, N. Mutations in the parkin gene cause autosomal recessive juvenile parkinsonism. *Nature* **1998**, *392*, 605–608. [CrossRef]
155. Cudkowicz, M.E.; Lindborg, S.R.; Goyal, N.A.; Miller, R.G.; Burford, M.J.; Berry, J.D.; Nicholson, K.A.; Mozaffar, T.; Katz, J.S.; Jenkins, L.J.; et al. A randomized PLACEBO-CONTROLLED phase 3 study of mesenchymal stem cells induced to secrete high levels of neurotrophic factors in amyotrophic lateral sclerosis. *Muscle Nerve* **2022**, *65*, 291–302. [CrossRef]
156. Caccamo, A.; Majumder, S.; Deng, J.J.; Bai, Y.; Thornton, F.B.; Oddo, S. Rapamycin Rescues TDP-43 Mislocalization and the Associated Low Molecular Mass Neurofilament Instability. *J. Biol. Chem.* **2009**, *284*, 27416–27424. [CrossRef] [PubMed]
157. Barmada, S.J.; Serio, A.; Arjun, A.; Bilican, B.; Daub, A.; Ando, D.M.; Tsvetkov, A.; Pleiss, M.; Li, X.; Peisach, D.; et al. Autophagy induction enhances TDP43 turnover and survival in neuronal ALS models. *Nat. Chem. Biol.* **2014**, *10*, 677–685. [CrossRef]
158. Jensen, C.S.; Bahl, J.M.; Østergaard, L.B.; Høgh, P.; Wermuth, L.; Heslegrave, A.; Zetterberg, H.; Heegaard, N.H.H.; Hasselbalch, S.G.; Simonsen, A.H. Exercise as a potential modulator of inflammation in patients with Alzheimer's disease measured in cerebrospinal fluid and plasma. *Exp. Gerontol.* **2019**, *121*, 91–98. [CrossRef]
159. Weinshenker, D. Long Road to Ruin: Noradrenergic Dysfunction in Neurodegenerative Disease. *Trends Neurosci.* **2018**, *41*, 211–223. [CrossRef]
160. Heneka, M.T.; Nadrigny, F.; Regen, T.; Martinez-Hernandez, A.; Dumitrescu-Ozimek, L.; Terwel, D.; Jardanhazi-Kurutz, D.; Walter, J.; Kirchhoff, F.; Hanisch, U.-K.; et al. Locus ceruleus controls Alzheimer's disease pathology by modulating microglial functions through norepinephrine. *Proc. Natl. Acad. Sci. USA* **2010**, *107*, 6058–6063. [CrossRef]
161. LeWitt, P.A.; Li, J.; Lu, M.; Guo, L.; Auinger, P. Metabolomic biomarkers as strong correlates of Parkinson disease progression. *Neurology* **2017**, *88*, 862–869. [CrossRef] [PubMed]
162. Weisskopf, M.; O'Reilly, E.; Chen, H.; Schwarzschild, M.; Ascherio, A. Plasma Urate and Risk of Parkinson's Disease. *Am. J. Epidemiol.* **2007**, *166*, 561–567. [CrossRef]
163. Kobylecki, C.J.; Nordestgaard, B.G.; Afzal, S. Plasma urate and risk of Parkinson's disease: A mendelian randomization study. *Ann. Neurol.* **2018**, *84*, 178–190. [CrossRef] [PubMed]
164. Simon, K.C.; Eberly, S.; Gao, X.; Oakes, D.; Tanner, C.M.; Shoulson, I.; Fahn, S.; Schwarzschild, M.A.; Ascherio, A.; on behalf of the Parkinson Study Group. Mendelian randomization of serum urate and parkinson disease progression. *Ann. Neurol.* **2014**, *76*, 862–868. [CrossRef] [PubMed]
165. Otsuki, K.; Uchida, S.; Wakabayashi, Y.; Matsubara, T.; Hobara, T.; Funato, H.; Watanabe, Y. Aberrant REST-mediated transcriptional regulation in major depressive disorder. *J. Psychiatr. Res.* **2010**, *44*, 378–384. [CrossRef]
166. Hye, A.; Riddoch-Contreras, J.; Baird, A.L.; Ashton, N.J.; Bazenet, C.; Leung, R.; Westman, E.; Simmons, A.; Dobson, R.; Sattlecker, M.; et al. Plasma proteins predict conversion to dementia from prodromal disease. *Alzheimers Dement.* **2014**, *10*, 799. [CrossRef] [PubMed]
167. Arancibia, S.; Silhol, M.; Moulière, F.; Meffre, J.; Höllinger, I.; Maurice, T.; Tapia-Arancibia, L. Protective effect of BDNF against beta-amyloid induced neurotoxicity in vitro and in vivo in rats. *Neurobiol. Dis.* **2008**, *31*, 316–326. [CrossRef]
168. Bhattacharjee, A.K.; Chang, L.; White, L.; Bazinet, R.P.; Rapoport, S.I. Imaging apomorphine stimulation of brain arachidonic acid signaling via D2-like receptors in unanesthetized rats. *Psychopharmacology* **2008**, *197*, 557–566. [CrossRef]
169. Bhattacharjee, A.K.; Chang, L.; Lee, H.-J.; Bazinet, R.P.; Seemann, R.; Rapoport, S.I. D2 but not D1 dopamine receptor stimulation augments brain signaling involving arachidonic acid in unanesthetized rats. *Psychopharmacology* **2005**, *180*, 735–742. [CrossRef]
170. McGrath, M.S.; Kahn, J.O.; Herndier, B.G. Development of WF10, a novel macrophage-regulating agent. *Curr. Opin. Investig. Drugs Lond. Engl. 2000* **2002**, *3*, 365–373.
171. Joo, K.; Lee, Y.; Choi, D.; Han, J.; Hong, S.; Kim, Y.M.; Jung, Y. An anti-inflammatory mechanism of taurine conjugated 5-aminosalicylic acid against experimental colitis: Taurine chloramine potentiates inhibitory effect of 5-aminosalicylic acid on IL-1β-mediated NFκB activation. *Eur. J. Pharmacol.* **2009**, *618*, 91–97. [CrossRef] [PubMed]

172. Butovsky, O.; Siddiqui, S.; Gabriely, G.; Lanser, A.J.; Dake, B.; Murugaiyan, G.; Doykan, C.E.; Wu, P.M.; Gali, R.R.; Iyer, L.K.; et al. Modulating inflammatory monocytes with a unique microRNA gene signature ameliorates murine ALS. *J. Clin. Investig.* **2012**, *122*, 3063–3087. [CrossRef] [PubMed]
173. Somekawa, S.; Imagawa, K.; Naya, N.; Takemoto, Y.; Onoue, K.; Okayama, S.; Takeda, Y.; Kawata, H.; Horii, M.; Nakajima, T.; et al. Regulation of Aldosterone and Cortisol Production by the Transcriptional Repressor Neuron Restrictive Silencer Factor. *Endocrinology* **2009**, *150*, 3110–3117. [CrossRef]
174. Thambisetty, M.; Gallardo, K.A.; Liow, J.-S.; Beason-Held, L.L.; Umhau, J.C.; Bhattacharjee, A.K.; Der, M.; Herscovitch, P.; Rapoport, J.L.; Rapoport, S.I. The Utility of ^{11}C-Arachidonate PET to Study in vivo Dopaminergic Neurotransmission in Humans. *J. Cereb. Blood Flow Metab.* **2012**, *32*, 676–684. [CrossRef]
175. Pagano, G.; Ferrara, N.; Brooks, D.J.; Pavese, N. Age at onset and Parkinson disease phenotype. *Neurology* **2016**, *86*, 1400–1407. [CrossRef]
176. Marek, K.; Jennings, D.; Lasch, S.; Siderowf, A.; Tanner, C.; Simuni, T.; Coffey, C.; Kieburtz, K.; Flagg, E.; Chowdhury, S.; et al. The Parkinson Progression Marker Initiative (PPMI). *Prog. Neurobiol.* **2011**, *95*, 629–635. [CrossRef] [PubMed]
177. Brumm, M.C.; Pierz, K.A.; Lafontant, D.-E.; Caspell-Garcia, C.; Coffey, C.S.; Siderowf, A.; Marek, K. Updated Percentiles for the University of Pennsylvania Smell Identification Test in Adults 50 Years of Age and Older. *Neurology* **2023**, *100*, e1691–e1701. [CrossRef]
178. Weintraub, D.; Hoops, S.; Shea, J.A.; Lyons, K.E.; Pahwa, R.; Driver-Dunckley, E.D.; Adler, C.H.; Potenza, M.N.; Miyasaki, J.; Siderowf, A.D.; et al. Validation of the questionnaire for impulsive-compulsive disorders in Parkinson's disease. *Mov. Disord.* **2009**, *24*, 1461–1467. [CrossRef]
179. Smith, K.M.; Xie, S.X.; Weintraub, D. Incident impulse control disorder symptoms and dopamine transporter imaging in Parkinson disease. *J. Neurol. Neurosurg. Psychiatry* **2016**, *87*, 864–870. [CrossRef]
180. Datta, A.; Qian, J.; Chong, R.; Kalaria, R.N.; Francis, P.; Lai, M.K.P.; Chen, C.P.; Sze, S.K. Novel pathophysiological markers are revealed by iTRAQ-based quantitative clinical proteomics approach in vascular dementia. *J. Proteom.* **2014**, *99*, 54–67. [CrossRef]
181. Ziv-Baran, T.; Shenhar-Tsarfaty, S.; Etz-Hadar, I.; Goldiner, I.; Gottreich, A.; Alcalay, Y.; Zeltser, D.; Shapira, I.; Angel, Y.; Friedensohn, L.; et al. The ability of the wide range CRP assay to classify individuals with low grade inflammation into cardiovascular risk groups. *Clin. Chim. Acta* **2017**, *471*, 185–190. [CrossRef] [PubMed]
182. Miller, R.G.; Zhang, R.; Bracci, P.M.; Azhir, A.; Barohn, R.; Bedlack, R.; Benatar, M.; Berry, J.D.; Cudkowicz, M.; Kasarskis, E.J.; et al. Phase 2B randomized controlled trial of NP001 in amyotrophic lateral sclerosis: Pre-specified and post hoc analyses. *Muscle Nerve* **2022**, *66*, 39–49. [CrossRef] [PubMed]
183. Siderowf, A.; Jennings, D.; Eberly, S.; Oakes, D.; Hawkins, K.A.; Ascherio, A.; Stern, M.B.; Marek, K.; the PARS Investigators. Impaired olfaction and other prodromal features in the Parkinson At-Risk Syndrome study. *Mov. Disord.* **2012**, *27*, 406–412. [CrossRef] [PubMed]
184. Berg, D.; Postuma, R.B.; Adler, C.H.; Bloem, B.R.; Chan, P.; Dubois, B.; Gasser, T.; Goetz, C.G.; Halliday, G.; Joseph, L.; et al. MDS research criteria for prodromal Parkinson's disease: MDS Criteria for Prodromal PD. *Mov. Disord.* **2015**, *30*, 1600–1611. [CrossRef] [PubMed]
185. Matthews, D.C.; Ritter, A.; Thomas, R.G.; Andrews, R.D.; Lukic, A.S.; Revta, C.; Kinney, J.W.; Tousi, B.; Leverenz, J.B.; Fillit, H.; et al. Rasagiline effects on glucose metabolism, cognition, and tau in Alzheimer's dementia. *Alzheimers Dement. Transl. Res. Clin. Interv.* **2021**, *7*, e12106. [CrossRef] [PubMed]
186. Matthews, D.C.; Mao, X.; Dowd, K.; Tsakanikas, D.; Jiang, C.S.; Meuser, C.; Andrews, R.D.; Lukic, A.S.; Lee, J.; Hampilos, N.; et al. Riluzole, a glutamate modulator, slows cerebral glucose metabolism decline in patients with Alzheimer's disease. *Brain* **2021**, *144*, 3742–3755. [CrossRef]
187. Voutsadakis, I.A. Biomarkers of Trifluridine-Tipiracil Efficacy. *J. Clin. Med.* **2021**, *10*, 5568. [CrossRef]
188. Van De Haar, J.; Ma, X.; Ooft, S.N.; Van Der Helm, P.W.; Hoes, L.R.; Mainardi, S.; Pinato, D.J.; Sun, K.; Salvatore, L.; Tortora, G.; et al. Codon-specific KRAS mutations predict survival benefit of trifluridine/tipiracil in metastatic colorectal cancer. *Nat. Med.* **2023**, *29*, 605–614. [CrossRef]
189. Moussa, C.; Hebron, M.; Huang, X.; Ahn, J.; Rissman, R.A.; Aisen, P.S.; Turner, R.S. Resveratrol regulates neuro-inflammation and induces adaptive immunity in Alzheimer's disease. *J. Neuroinflamm.* **2017**, *14*, 1. [CrossRef]
190. Athauda, D.; Gulyani, S.; Karnati, H.K.; Li, Y.; Tweedie, D.; Mustapic, M.; Chawla, S.; Chowdhury, K.; Skene, S.S.; Greig, N.H.; et al. Utility of Neuronal-Derived Exosomes to Examine Molecular Mechanisms That Affect Motor Function in Patients With Parkinson Disease: A Secondary Analysis of the Exenatide-PD Trial. *JAMA Neurol.* **2019**, *76*, 420. [CrossRef]
191. Carlini, V.; Verduci, I.; Cianci, F.; Cannavale, G.; Fenoglio, C.; Galimberti, D.; Mazzanti, M. CLIC1 Protein Accumulates in Circulating Monocyte Membrane during Neurodegeneration. *Int. J. Mol. Sci.* **2020**, *21*, 1484. [CrossRef] [PubMed]
192. Hyman, S.L.; Shores, A.; North, K.N. The nature and frequency of cognitive deficits in children with neurofibromatosis type 1. *Neurology* **2005**, *65*, 1037–1044. [CrossRef] [PubMed]
193. Payne, J.M.; Hearps, S.J.C.; Walsh, K.S.; Paltin, I.; Barton, B.; Ullrich, N.J.; Haebich, K.M.; Coghill, D.; Gioia, G.A.; Cantor, A.; et al. Reproducibility of cognitive endpoints in clinical trials: Lessons from neurofibromatosis type 1. *Ann. Clin. Transl. Neurol.* **2019**, *6*, 2555–2565. [CrossRef] [PubMed]

194. Cavedo, E.; Grothe, M.J.; Colliot, O.; Lista, S.; Chupin, M.; Dormont, D.; Houot, M.; Lehéricy, S.; Teipel, S.; Dubois, B.; et al. Reduced basal forebrain atrophy progression in a randomized Donepezil trial in prodromal Alzheimer's disease. *Sci. Rep.* **2017**, *7*, 11706. [CrossRef] [PubMed]
195. May, P.C.; Dean, R.A.; Lowe, S.L.; Martenyi, F.; Sheehan, S.M.; Boggs, L.N.; Monk, S.A.; Mathes, B.M.; Mergott, D.J.; Watson, B.M.; et al. Robust Central Reduction of Amyloid-β in Humans with an Orally Available, Non-Peptidic β-Secretase Inhibitor. *J. Neurosci.* **2011**, *31*, 16507–16516. [CrossRef]
196. Toledo, J.B.; Van Deerlin, V.M.; Lee, E.B.; Suh, E.; Baek, Y.; Robinson, J.L.; Xie, S.X.; McBride, J.; Wood, E.M.; Schuck, T.; et al. A platform for discovery: The University of Pennsylvania Integrated Neurodegenerative Disease Biobank. *Alzheimers Dement.* **2014**, *10*, 477. [CrossRef]
197. Chang, K.-H.; Cheng, M.-L.; Chiang, M.-C.; Chen, C.-M. Lipophilic antioxidants in neurodegenerative diseases. *Clin. Chim. Acta* **2018**, *485*, 79–87. [CrossRef]
198. Trujillo, P.; Aumann, M.A.; Claassen, D.O. Neuromelanin-sensitive MRI as a promising biomarker of catecholamine function. *Brain J. Neurol.* **2023**, awad300. [CrossRef]
199. Huddleston, D.E.; Chen, X.; Hwang, K.; Langley, J.; Tripathi, R.; Tucker, K.; McKay, J.L.; Hu, X.; Factor, S.A. Neuromelanin-sensitive MRI correlates of cognitive and motor function in Parkinson's disease with freezing of gait. *Front. Dement.* **2023**, *2*, 1215505. [CrossRef]
200. Chételat, G. How to use neuroimaging biomarkers in the diagnosis framework of neurodegenerative diseases? *Rev. Neurol.* **2022**, *178*, 490–497. [CrossRef]
201. Chételat, G.; Arbizu, J.; Barthel, H.; Garibotto, V.; Law, I.; Morbelli, S.; Van De Giessen, E.; Agosta, F.; Barkhof, F.; Brooks, D.J.; et al. Amyloid-PET and 18F-FDG-PET in the diagnostic investigation of Alzheimer's disease and other dementias. *Lancet Neurol.* **2020**, *19*, 951–962. [CrossRef] [PubMed]

brain sciences

MDPI

Opinion

The Brain's Glymphatic System: Drawing New Perspectives in Neuroscience

Alexandru Vlad Ciurea [1,2], Aurel George Mohan [3,4,*], Razvan-Adrian Covache-Busuioc [1], Horia Petre Costin [1] and Vicentiu Mircea Saceleanu [5,6]

1 Neurosurgery Department, "Carol Davila" University of Medicine and Pharmacy, 020021 Bucharest, Romania; prof.avciurea@gmail.com (A.V.C.); razvan-adrian.covache-busuioc0720@stud.umfcd.ro (R.-A.C.-B.); horiacostin2001@yahoo.com (H.P.C.)
2 Neurosurgery Department, Sanador Clinical Hospital, 010991 Bucharest, Romania
3 Department of Neurosurgery, Bihor County Emergency Clinical Hospital, 410167 Oradea, Romania
4 Department of Neurosurgery, Faculty of Medicine, Oradea University, 410610 Oradea, Romania
5 Neurosurgery Department, Sibiu County Emergency Hospital, 550245 Sibiu, Romania; vicentiu.saceleanu@gmail.com
6 Department of Neurosurgery, "Lucian Blaga" University of Medicine, 550024 Sibiu, Romania
* Correspondence: mohanaurel@yahoo.com

Abstract: This paper delves into the intricate structure and functionality of the brain's glymphatic system, bringing forth new dimensions in its neuroscientific understanding. This paper commences by exploring the cerebrospinal fluid (CSF)—its localization, production, and pivotal role within the central nervous system, acting as a cushion and vehicle for nutrient distribution and waste elimination. We then transition into an in-depth study of the morphophysiological aspects of the glymphatic system, a recent discovery revolutionizing the perception of waste clearance from the brain, highlighting its lymphatic-like characteristics and remarkable operations. This paper subsequently emphasizes the glymphatic system's potential implications in Alzheimer's disease (AD), discussing the connection between inefficient glymphatic clearance and AD pathogenesis. This review also elucidates the intriguing interplay between the glymphatic system and the circadian rhythm, illustrating the optimal functioning of glymphatic clearance during sleep. Lastly, we underscore the hitherto underappreciated involvement of the glymphatic system in the tumoral microenvironment, potentially impacting tumor growth and progression. This comprehensive paper accentuates the glymphatic system's pivotal role in multiple domains, fostering an understanding of the brain's waste clearance mechanisms and offering avenues for further research into neuropathological conditions.

Keywords: glymphatic system; brain physiology; astrocytes; AQP4; Alzheimer's disease; circadian rhythm; migraine aura

Citation: Ciurea, A.V.; Mohan, A.G.; Covache-Busuioc, R.-A.; Costin, H.P.; Saceleanu, V.M. The Brain's Glymphatic System: Drawing New Perspectives in Neuroscience. *Brain Sci.* **2023**, *13*, 1005. https://doi.org/10.3390/brainsci13071005

Academic Editors: Chitra Mandyam and Konstantin V. Slavin

Received: 17 May 2023
Revised: 16 June 2023
Accepted: 26 June 2023
Published: 28 June 2023

1. Introduction

The lymphatic system encompasses an intricate assembly of cells, tissues, and organs, which collectively operate in a highly organized manner to meticulously regulate the internal fluid environment of the body. This system consists of an extensive network of endothelial vessels, which function to collect the extracellular fluid and prevent its accumulation that would otherwise result in lymphedema. The lymph is funneled through the thoracic duct before draining into the subclavian vein.

In contrast to lymphatic capillaries, lymphatic vessels, through the presence of a smooth muscular layer and valves, facilitate a unidirectional flow of lymph through the lymph nodes [1]. Within the lymph nodes, the ultrafiltrate of blood plasma gathered from the extracellular matrix undergoes constant filtration and monitoring by aggregations of T and B cells, which are vital components of the immune system.

Lymphocytes, along with ancillary cells such as monocytes, macrophages, and granulocytes, comprise the central cellular components of the lymphatic system. These cells are categorized into primary lymphoid organs (bone marrow and thymus) and secondary lymphoid organs (spleen and lymph nodes) [2].

Furthermore, the lymphatic system plays an indispensable role in the immune response owing to the extensive network of lymphatic vessels which serve to detoxify tissues by draining extracellular fluids.

Conversely, the central nervous system (CNS) is characterized by an intricate structure comprising white and gray matter. Both structures incorporate glial cells, which primarily serve a supportive function, particularly abundant in white matter. Microglia, specialized macrophages in the CNS, are instrumental in the immune response within the brain, as they facilitate the elimination of antigens and certain metabolites. Additionally, astrocytes, star-shaped cells that are the most abundant in the human brain, play critical roles in the blood–brain barrier, nutrient transport from capillaries to neurons, and reparative processes following brain injury [3].

It is noteworthy that the CNS, despite being among the most metabolically active systems in the body, accounting for approximately 20% of the body's energy expenditure, lacks a traditional lymphatic system for waste removal [4].

2. CSF—Localization, Production, Role

The CNS presents in its structure four cavities named ventricles filled with an ultrafiltered blood plasma called cerebrospinal fluid (CSF). Inside each of the four ventricles, a structure called the choroid plexus, consisting of modified ependymal cells, takes part into the secretion of the CSF that travels through the subarachnoid space, in order to end up in the sagittal sinus, by means of the arachnoid granulations [5]. Since the CSF is then secreted in the ventricular system, mostly by the 2 lateral ventricles, clear surveillance of the components of this liquid is important, and this is where the action of the macrophages, the astrocytes and the tight junctions between the choroid ependymal cells, which creates a barrier between the fenestrated capillary and the CSF system, factors in [6].

The difference between the CSF and blood plasma is that the concentration of sodium, chloride and magnesium is higher in the CSF while the concentration of potassium and calcium is higher in the blood plasma. Moreover, the CSF only contains a considerably small quantity of proteins and cells [7]. The first role of the CSF is shock absorption, preventing the brain from hitting the skull, therefore reducing the probability of an injury [8]. Secondly, the CSF can be regarded as buoyant, since it is capable of reducing the weight of the brain from 1400 to 50 g [9]. Thus, the possibility of an injury is diminished for both the CNS and the vasculature, which could be constricted by the weight. However, in recent years, a role of the CSF which is not so clear yet was discovered. It was shown that the CSF plays a role in maintaining the homeostasis of the brain's interstitial space, in order to maintain the proper functioning of the neurons [10].

Since this role is attributed in the body to the ultrafiltered plasma, representing waste removal fluid, via the lymph and later via the venous system, the brain does not allow this production due to the blood–brain barrier (BBB). For this reason, it becomes more and more clear that the CSF replaces the lymph. As a comparison between the lymphatic system and the glymphatic one, both the ultrafiltrate of plasma and the CSF can occur at the arterial bed of the microvasculature. However, the former will drain in the lymphatic system after clearing the interstitial space, while the other will end up in the paravenous space [11]. However, even if both the CSF and the lymphatic fluid drain through the lymphatic system and ganglia after that, the former can take two different paths. Firstly, it can drain in the venous system indirectly through the lymphatic network, through either the spinal nerve roots or the cribriform plate. Secondly, the CSF can end up directly in the venous sinuses using the arachnoid villi (arachnoid granulations), which help the efflux of the CSF from the subarachnoidian space [12].

3. The Morphophysiological Aspects of the Glymphatic System

It appears that the glymphatic system of the CNS plays a crucial role in maintaining the homeostasis of the brain parenchyma. Nonetheless, one important factor for the entire CNS is regarded as the perivascular space (PVS), which represents the space between the astrocytes end-foot and the smooth muscle and vascular endothelium of the brain vasculature. Further on, the CSF from the PVS will drain through the dural lymphatic vessels to the cervical lymphatic nodes [13].

Both the arteriole and the astrocytes play a key role in determining the influxes of the CSF through the brain parenchyma, since the vasodilatation or vasoconstriction of the arterial end, as well as the swelling of the astrocyte end-foot, can alter the bulk flow of the CSF to the paravenous space.

According to a study conducted by Mestre et al., (2020) [14] that reviewed the post-ischemic edema showed that due to the spreading depolarization that leads to a vasoconstriction of the arterioles in the focal cerebral infarct, the PVS will get enlarged. This action is responsible for an increased volume of CSF, which will contribute to the swelling of the tissue, due to the gradient change, therefore doubling the speed of the CSF inflow into the tissues.

However, we should also take into account the aquaporin-4 channel (Aqp-4) found with a higher prevalence in the astrocyte end-foot that makes the link between the para-arterial space and paravenous space. Thus, the CSF can pass freely through the BBB, making its way into the interstitial place [15]. Studies showed that the inactivation of the Aqp-4 channel in the brain of the mouse, leads to an altered water balance. This abnormality has consequences such as brain edema, hydrocephalus and stroke [16]. Moreover, studies have shown that the impairment of the glymphatic system may have much more repercussions than expected.

A study conducted by Aaron J. Schain et al. [17], showed that the CSD (cortical spreading depression), which is a slowly propagating wave due to an altered brain activity regarded as responsible for inducing the migraine aura, creating a severe disruption of the ionic homeostasis. It increases the concentrations of K+, neurotransmitters such as Glutamate, as well as the synthesis of inflammatory factors such as NO synthetase and COX-2 in the brain parenchyma. All of these changes in the homeostasis lead to a sudden vasodilation, decreasing the PVS and therefore reducing the glymphatic flow, as well as contributing to the pathophysiological aspects of the migraine aura.

This discovery leads us to another very important role of the glymphatic system: the protection of the brain against the β-amyloid, the underlying protein of Alzheimer's disease.

The existence of the Aqp-4 in the structure of astroglial end-feet plays an important role in linking the para-arterial space, the interstitial space and the paravenous space. Therefore, besides maintaining a firm homeostasis of the brain parenchyma, it also takes part in clearance of the metabolic wastes produced by the brain (Figure 1) [18].

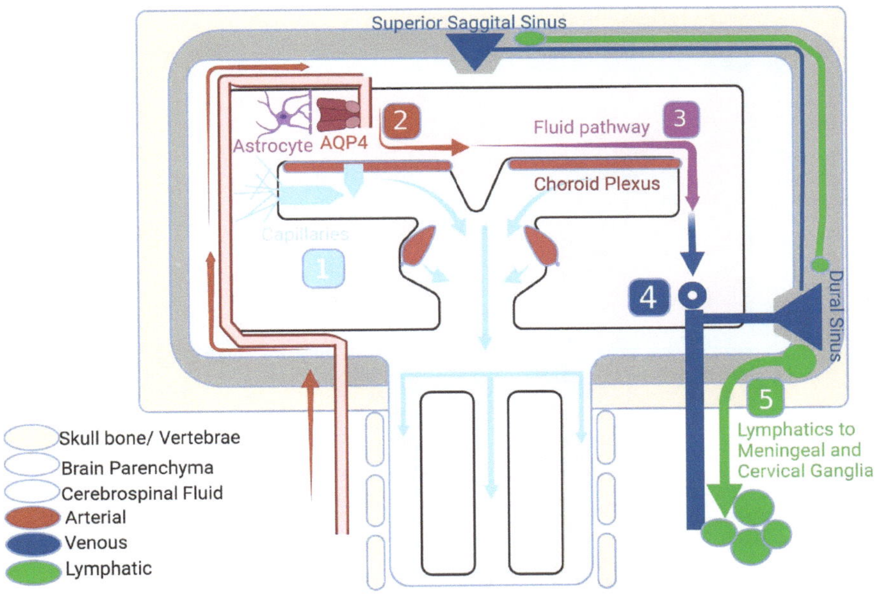

Figure 1. The pathway for fluid transport can be separated into five distinct sections: (1) the production of cerebrospinal fluid (CSF) by the choroid plexus, potentially supplemented by extrachoroidal sources such as capillary influx and metabolic water production; (2) the pulsation of arterial walls which drives CSF deep into the brain along perivascular spaces; (3) the CSF is carried into the brain parenchyma through AQP4 water channels and then dispersed throughout the neuropil; (4) interstitial fluid (ISF) and CSF mix, accumulating in the peri-venous space and subsequently draining from the brain through (5) meningeal and cervical lymphatic vessels, as well as cranial and spinal nerves.

4. The Glymphatic System in AD

Alzheimer's disease (AD), the most prevalent neurodegenerative dementia, affects over 50 million individuals globally, making it a serious public health challenge [19,20]. AD is irreversible disease marked by neurodegeneration caused by harmful buildups of extracellular amyloid plaques and intracellular neurofibrillary tangles composed of hyperphosphorylated tau protein that results in neuronal damage which eventually leads to cognitive deterioration as well as personality and behavioral changes [21].

The traditional amyloid cascade hypothesis postulates that amyloid-b accumulation is among the initial factors leading to Alzheimer's disease (AD) and that its further progression, including neurofibrillary tangle formation containing tau protein, results from an imbalance between Aβ production and clearance [22,23]. Other neurodegenerative disorders have similar proteins accumulating both inside and outside cells such as Lewy Bodies or neurites found in Parkinson's Disease; such accumulations contribute to neurodegeneration while being implicated in both pathogenesis of both AD and PD, but their exact biological linkage with their biological relationship within their respective glymphatic system remains unknown [24].

The glymphatic system may play an integral role in removing a-synuclein and thus impacting PD progression. A negative correlation has been observed between deposition of a-synuclein and expression of water channel protein AQP4 in brains of patients suffering PD, suggesting glymphatic dysfunction may contribute to protein accumulation (Figure 2) [25]. Clearance of Aβ and tau proteins into cerebrospinal fluid (CSF) serves as the basis for measurement purposes as clinical biomarkers of AD [26,27].

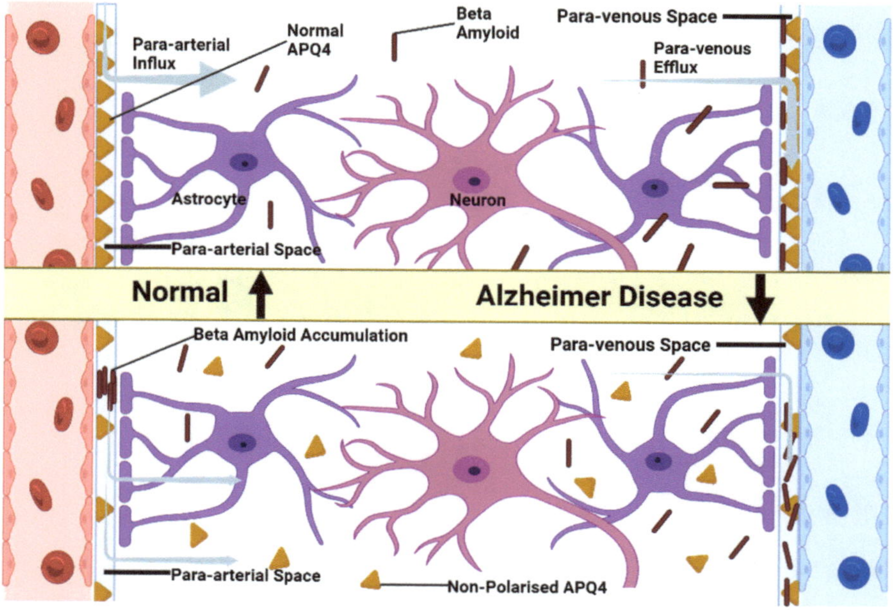

Figure 2. AQP4 aids in the elimination of waste via the glymphatic system. Under normal conditions, AQP4 is primarily located on the end-feet of astrocytes, which is referred to as polarized AQP4 expression. The cerebrospinal fluid (CSF) flows into the brain tissue through the para-arterial system, or para-arterial influx, and then exits into the veins, a process known as para-arterial efflux. However, as we age, and particularly in pathological conditions, the polarization of AQP4 diminishes, leading to an increased presence of AQP4 on parenchymal processes. This phenomenon, known as AQP4 depolarization, hampers the glymphatic system's ability to effectively clear waste, such as beta-amyloid.

Though Aβ is widely believed to travel through the blood–brain barrier (BBB), recent evidence from Nauen and Troncoso's study [28] indicates otherwise; their discovery of Aβ in human lymph nodes suggests alternative exit routes from central nervous system for Aβ and similar proteins, suggesting that it may also use the lymphatic system as part of its clearance from brain.

Aging, with its altered circadian rhythms and associated sleep deprivation, has long been recognized as a risk factor for neurodegenerative disorders such as Alzheimer's disease and Parkinson's disease. Aging also causes efficiency loss between subarachnoidal CSF exchange and brain parenchyma exchange, thus decreasing efficiency between CSF exchange and brain parenchyma exchange [29]. Evidence supporting the decline of glymphatic system function with age comes both from experimental models and patients with neurological conditions [30].

Recent work conducted by McKee and colleagues [31] has examined the relationship between circadian clock function and neurodegeneration. Their studies examined how astrocyte activation caused by deletion of core clock gene Bmal1 affects gene expression, Aβ plaque activation and deposition. Deletion disrupts clock function while inducing cell-autonomous activation in astrocytes.

Studies indicate that any impairment to the glymphatic system's ability to clear extracellular tau, a protein, may contribute to tau aggregation and neurodegeneration [32]. Furthermore, deficiency of water channel protein AQP4 could exacerbate tau accumulation further and create a vicious cycle between compromised clearance of extracellular tau and accumulation on other cells [32]. Although exact mechanisms by which impaired extracellular tau clearance exacerbates tau-related pathology remain unknown, impaired clearance among PS19 mice deficient for AQP4 could facilitate spread of pathological tau species over other cells, suggesting impaired extracellular tau clearance could promote spread over other cells resulting in pathologically spread of pathological tau species to other cells and organs [33].

Overall, both deletion of AQP4 or its pharmacological inhibition amplifies pathogenic accumulations of Aβ and tau in AD transgenic mouse models. Furthermore, recent genetic research has demonstrated that various SNPs within the AQP4 gene can influence cognitive decline following an AD diagnosis; two are linked with slower decline whereas others such as rs9951307 and rs3875089 may lead to faster cognitive deterioration post diagnosis while two others (rs3763040 and rs3763043) could promote rapid decline [34,35].

Variations in the AQP4 gene have been linked with Aβ accumulation, disease stage progression and cognitive decline, suggesting that it could serve as a biomarker to accurately predict disease burden in those within the AD spectrum. Furthermore, certain SNPs of this gene have been linked with reduced perivascular localization of AQP4 protein among AD patients [34,35].

According to a study conducted by Iliff et al. [36], the soluble β-amyloid clearance is reduced by almost 55% in case of an Aqp-4 null mouse compared to a normal wild-type mouse.

Consequently, both the perturbance of the astrocyte's vascular end foot that presents the Aqp-4 as well as a decreased glymphatic flow (which can be caused due to the CSD that produces the vasodilation, underlying the physiopathology of the migraine aura) lead to an increased risk of Alzheimer's disease development [37].

5. The Glymphatic System and the Circadian Rhythm

Recent studies demonstrate a clear correlation between sleep disruption, brain glymphatic system dysfunction and Alzheimer's disease (AD) [38–40]. Similar to lymphatic systems found elsewhere, glymphatic systems consist of para-vascular channels within the brain's blood vessels that carry cerebrospinal fluid (CSF) to capillary beds before penetrating parenchyma tissue where it mixes with interstitial fluid, collects metabolic waste and eventually returns back through paravenous space and then cervical lymphatic vessels for removal [41].

The glymphatic system plays an essential role in clearing away neurotoxic substances such as Aβ from central nervous system parenchyma [42,43]. Studies have demonstrated that over half of Aβ can be eliminated through this mechanism. Sleep can have an interesting influence on the functioning of the glymphatic system. When sleeping naturally, brain interstitial space vastly expands compared to wakefulness—possibly as a result of astroglial cell shrinkage [38,44]. An expansion of extracellular space accelerates clearance processes; for instance, mice demonstrated that Aβ clearance during sleep was twice as fast compared to wakefulness [45]. Another study demonstrated that clearance through the glymphatic system also depends on body posture—with lateral positions commonly adopted during sleeping being most efficient for clearing [46].

As Aβ clearance is impaired in both early and late stages of AD [47], it is plausible to propose that there may be a connection between impaired glymphatic system function and AD. Studies in animals and humans have documented diurnal oscillation of Aβ level in brain interstitial fluid; endogenous neuronal activity influences regional Aβ concentration [48]; decreased neuronal activity during certain sleep stages may cause this fluctuation, potentially contributing to both impaired clearance as well as disturbances in Aβ production caused by disturbed slow wave sleep patterns [49]. Thus, altered sleep quality may contribute to both its onset and progression through impaired clearance as well as disruptions of Aβ production caused by reduced endogenous neuronal activity influencing regional concentration of Aβ [49,50].

Sleep stages, specifically slow wave sleep, has been found to influence Aβ 42 levels in the cerebrospinal fluid (CSF). A study involving 36 cognitively normal and elderly subjects discovered an inverse relationship between CSF Aβ 42 levels and slow wave sleep duration, percentage of total sleep time spent sleeping on slow waves, frontal EEG leads activity while sleeping, local Aβ accumulation during low-frequency-range sleep time and reduced slow wave activity during these hours [51], suggesting there may be an association between decreased Aβ clearance or production and deficiency and slow wave sleep deficiency and reduced clearance/production/clearer production/clearance/production and slow wave sleep deficiency and slow wave sleep deficiency [52].

Nonetheless, the glymphatic system seems to operate in accordance with the circadian rhythm. The role of this rhythm is to create different neural activity peaks, in order to maintain a good balance between sleep and active phase [45].

The role of the glymphatic system is to synchronize the entire rhythmicity, since the suprachiasmatic nucleus of the hypothalamus which represents the main pacemaker is connected to the CSF. Moreover, other neurotransmitters such as VIP (Vasoactive Intestinal Peptide) and AVP (Arginine Vasopressin) are transported through the CSF. Therefore, the entire glymphatic system could be responsible for through its bulk flow to coordinate the neurons, the physiological mechanisms of the brain and in the end the entire behavior of the human [29].

6. Glymphatic System Involvement in the Tumoral Microenvironment

Studies on rodents have provided invaluable insights into the changes to glymphatic function that correlate with the development of gliomas. These animal studies observed a decreased rate of cerebrospinal fluid (CSF) efflux and restructuring of the glymphatic pathway, both indicators that could contribute to brain edema around gliomas [53,54]. Animal studies have considerably broadened our knowledge, yet it is essential that their findings can be directly applied to human glymphatic systems. A variety of noninvasive techniques have been suggested for studying this function in human bodies; structural and diffusion MRI have both been proposed as noninvasive approaches for investigating it [50,55,56].

The ALPS (Analysis Along the Perivascular Space) index, derived from diffusion tensor imaging (DTI) [50], may provide insight into diffusivity within medullary veins at the level of lateral ventricle bodies at medullary vein level and may provide an estimation of human glymphatic function. Variations in ALPS index show correlation with scores on Mini-Mental State Examination tests administered to Alzheimer's disease patients which could indicate dysfunction, while lower values have also been seen among normal pressure hydrocephalus patients which may suggest impaired functioning system due to delayed clearance of intrathecically administered gadobutrol [54,57]. Given its noninvasive nature and potential as a tool for exploring human glymphatic systems, we decided to utilize the ALPS index for exploring glymphatic function among patients with gliomas. Our research seeks to comprehend any correlations between this aspect of patient care and variables such as tumor and peritumoral brain edema volumes, tumor grades and IDH1 mutation status.

Traditional theories regarding peritumoral brain edema in gliomas suggest it occurs due to net fluid transport from intravascular compartments into brain interstitial spaces as a result of microvessel proliferation with defective inter-endothelial tight junctions [58], but this does not account for its occurrence with low grade gliomas with intact tight junctions. Research demonstrated an inverse relationship between ALPS index and peritumoral brain edema volume, suggesting brain edema associated with intra-axial tumors may also be affected by malfunction in the glymphatic system [59,60].

Recent research suggests that changes to the supporting structures of the blood–brain barrier, including astrocytes, pericytes, and microglial cells, could contribute to fluid entering the interstitium of the brain. Astrocyte coverage of brain microvessels appears to be an impediment to water movement, and AQP4 channels located on astrocytic foot processes could play an essential part in creating peritumoral brain edema. Evidence showing a correlation between peritumoral brain edema and elevated AQP4 expression levels in human gliomas astrocytes indicates that increased expression could play a key role in its pathogenesis [61,62]. As AQP4 water channels form part of the glymphatic system, we hypothesize that there may be a correlation between ALPS index values and expression of AQP4 water channels and expression levels in gliomas. Further research needs to confirm its use as an imaging marker of AQP4 expression in these tumors.

Changes in glymphatic function due to glioma growth have only been explored through limited animal studies. An investigation using an orthotopic xenograft glioma model demonstrated how tumor growth led to reduced CSF flow into extracranial spaces, leading to dysfunction and lower glymphatic flow overall. It was found that IDH1 wild-type gliomas, as measured by ALPS index scores, demonstrated significantly lower glymphatic performance than their IDH1 mutant counterparts; no human studies have reported any correlation between changes in glymphatic function and IDH1 mutation status or gene status for any human studies either way [53,63].

7. Conclusions

In conclusion, the glymphatic system of the brain is similar to the lymphatic systems of the entire body, whereas the role and mechanism of it represents critical discoveries regarding the deeper understanding of the human brain. It seems that its role is not so easy to comprehend, since it has implications in both physiological and pathological aspects of the brain, regarded as complex pathways that must be discovered.

Given AQP4's crucial role in the glymphatic system and potential implications for Alzheimer's disease, future research should include this water channel. It has been suggested as a promising therapeutic target for AD due to its effects on Aβ and tau clearance as well as neuronal function improvement, making it highly relevant in relation to both aging and neurodegeneration. AQP4, naturally upregulated with age but mislocalized in AD brains, could offer therapeutic potential by being modulated [64,65].

The blood–brain barrier (BBB) is highly selective [64], restricting drug passage from the bloodstream into extracellular fluids in the brain. Some conditions are associated with compromised BBBs that allow drug delivery; this is not the case in early-stage AD, where anti-amyloid agents such as bapineuzumab have failed clinical trials due to being unable to access amyloid plaques directly [66]; any drug designed to effectively target AQP4 and be relevant in treating AD must therefore be capable of crossing an intact BBB in order to be effective against its progression.

Even though the brain's CSF drainage function is an integrated system involving various compartments, the glymphatic system and AQP4 could serve as intervention targets in neurodegenerative diseases. Enhancing its function and efficiency could help delay or prevent protein accumulation in the brain; additionally, this mechanism could play a critical role in clearing away tau from circulation; special attention should be paid to Aβ-independent regulators of tau, such as the glymphatic system, when investigating neurodegenerative tauopathies [67].

Due to the importance of decreased molecular clearance from CSF to blood, which contributes to neurological diseases, direct measurement of CSF to blood clearance on an individual basis has become a method for personalized intrathecal drug administration in the central nervous system [68].

Moreover, we delve deeper into the intriguing relationship between the glymphatic system and migraine aura, offering new insight into its pathophysiology. This discovery might open up possibilities for research and treatments of this debilitating condition.

Furthermore, the interaction between the glymphatic system and circadian rhythms has been explored, drawing attention to their potential impact on brain health and function as well as sleep cycles and biological rhythms. Such findings could have significant ramifications for understanding and treating various neurological disorders.

Finally, it is vital to identify neuroimaging markers to detect changes in the glymphatic system, and noninvasive techniques could serve as effective tools to detect dysfunction in this area and become new potential biomarkers in AD.

Author Contributions: Conceptualization, A.V.C. and A.G.M.; methodology, V.M.S.; software, H.P.C.; validation, A.V.C., A.G.M. and R.-A.C.-B.; formal analysis, R.-A.C.-B. and H.P.C.; investigation, V.M.S.; resources, A.G.M.; data curation, H.P.C.; writing—original draft preparation, R.-A.C.-B.; writing—review and editing, R.-A.C.-B. and H.P.C.; visualization, A.V.C.; supervision, A.V.C.; project administration, V.M.S.; funding acquisition, A.G.M. All authors have read and agreed to the published version of the manuscript.

Funding: The APC was funded by the University of Oradea through an internal project. The authors would like to express their gratitude to the University of Oradea, which supported the APC.

Institutional Review Board Statement: Not applicable.

Informed Consent Statement: Not applicable.

Data Availability Statement: All data is available on PubMed.

Conflicts of Interest: The authors declare no conflict of interest.

References

1. Swartz, M.A. The physiology of the lymphatic system. *Adv. Drug Deliv. Rev.* **2001**, *50*, 3–20. [CrossRef] [PubMed]
2. Tamura, R.; Yoshida, K.; Toda, M. Current understanding of lymphatic vessels in the central nervous system. *Neurosurg. Rev.* **2019**, *43*, 1055–1064. [CrossRef] [PubMed]
3. Weber, B.; Barros, L.F. The Astrocyte: Powerhouse and Recycling Center. *Cold Spring Harb. Perspect. Biol.* **2015**, *7*, a020396. [CrossRef] [PubMed]
4. Raichle, M.E.; Gusnard, D.A. Appraising the brain's energy budget. *Proc. Natl. Acad. Sci. USA* **2002**, *99*, 10237–10239. [CrossRef]
5. Lun, M.P.; Monuki, E.S.; Lehtinen, M.K. Development and functions of the choroid plexus–cerebrospinal fluid system. *Nat. Rev. Neurosci.* **2015**, *16*, 445–457. [CrossRef]
6. Filis, A.K.; Aghayev, K.; Vrionis, F.D. Cerebrospinal Fluid and Hydrocephalus: Physiology, Diagnosis, and Treatment. *Cancer Control* **2017**, *24*, 6–8. [CrossRef]
7. Aldred, A.R.; Brack, C.M.; Schreiber, G. The cerebral expression of plasma protein genes in different species. *Comp. Biochem. Physiol. Part B Biochem. Mol. Biol.* **1995**, *111*, 1–15. [CrossRef]
8. Telano, L.N.; Baker, S. *Physiology, Cerebral Spinal Fluid*; StatPearls Publishing: Treasure Island, FL, USA, 2023. Available online: http://www.ncbi.nlm.nih.gov/books/NBK519007/ (accessed on 15 June 2023).
9. Idris, Z. Searching for the Origin through Central Nervous System: A Review and Thought which Related to Microgravity, Evolution, Big Bang Theory and Universes, Soul and Brainwaves, Greater Limbic System and Seat of the Soul. *Malays. J. Med. Sci.* **2014**, *21*, 4–11.
10. Koh, L.; Zakharov, A.; Johnston, M. Integration of the subarachnoid space and lymphatics: Is it time to embrace a new concept of cerebrospinal fluid absorption? *Cerebrospinal Fluid Res.* **2005**, *2*, 6. [CrossRef]
11. Bakker, E.N.T.P.; Naessens, D.M.P.; VanBavel, E. Paravascular spaces: Entry to or exit from the brain? *Exp. Physiol.* **2019**, *104*, 1013–1017. [CrossRef]
12. Adigun, O.O.; Al-Dhahir, M.A. *Anatomy, Head and Neck: Cerebrospinal Fluid*; StatPearls Publishing: Treasure Island, FL, USA, 2023. Available online: http://www.ncbi.nlm.nih.gov/books/NBK459286/ (accessed on 15 June 2023).
13. Thomas, J.H. Fluid dynamics of cerebrospinal fluid flow in perivascular spaces. *J. R. Soc. Interface* **2019**, *16*, 20190572. [CrossRef]
14. Mestre, H.; Du, T.; Sweeney, A.M.; Liu, G.; Samson, A.J.; Peng, W.; Mortensen, K.N.; Stæger, F.F.; Bork, P.A.R.; Bashford, L.; et al. Cerebrospinal fluid influx drives acute ischemic tissue swelling. *Science* **2020**, *367*, eaax7171. [CrossRef]
15. Engelhardt, B.; Sorokin, L. The blood–brain and the blood–cerebrospinal fluid barriers: Function and dysfunction. *Semin. Immunopathol.* **2009**, *31*, 497–511. [CrossRef]
16. Nielsen, S.; Nagelhus, E.A.; Amiry-Moghaddam, M.; Bourque, C.; Agre, P.; Ottersen, O.P. Specialized Membrane Domains for Water Transport in Glial Cells: High-Resolution Immunogold Cytochemistry of Aquaporin-4 in Rat Brain. *J. Neurosci.* **1997**, *17*, 171–180. [CrossRef]
17. Schain, A.J.; Melo-Carrillo, A.; Strassman, A.M.; Burstein, R. Cortical Spreading Depression Closes Paravascular Space and Impairs Glymphatic Flow: Implications for Migraine Headache. *J. Neurosci.* **2017**, *37*, 2904–2915. [CrossRef]
18. Mestre, H.; Mori, Y.; Nedergaard, M. The Brain's Glymphatic System: Current Controversies. *Trends Neurosci.* **2020**, *43*, 458–466. [CrossRef]
19. Partridge, L.; Deelen, J.; Slagboom, P.E. Facing up to the global challenges of ageing. *Nature* **2018**, *561*, 45–56. [CrossRef]
20. Prince, M.; Ali, G.-C.; Guerchet, M.; Prina, A.M.; Albanese, E.; Wu, Y.-T. Recent global trends in the prevalence and incidence of dementia, and survival with dementia. *Alzheimers Res. Ther.* **2016**, *8*, 23. [CrossRef]
21. Taylor, J.P.; Hardy, J.; Fischbeck, K.H. Toxic Proteins in Neurodegenerative Disease. *Science* **2002**, *296*, 1991–1995. [CrossRef]
22. Selkoe, D.J.; Hardy, J. The amyloid hypothesis of Alzheimer's disease at 25 years. *EMBO Mol. Med.* **2016**, *8*, 595–608. [CrossRef]
23. Cline, E.N.; Bicca, M.A.; Viola, K.L.; Klein, W.L. The Amyloid-β Oligomer Hypothesis: Beginning of the Third Decade. *J. Alzheimers Dis.* **2018**, *64*, S567–S610. [CrossRef] [PubMed]

24. Hardy, J.; Selkoe, D.J. The Amyloid Hypothesis of Alzheimer's Disease: Progress and Problems on the Road to Therapeutics. *Science* **2002**, *297*, 353–356. [CrossRef] [PubMed]
25. Hoshi, A.; Tsunoda, A.; Tada, M.; Nishizawa, M.; Ugawa, Y.; Kakita, A. Expression of Aquaporin 1 and Aquaporin 4 in the Temporal Neocortex of Patients with Parkinson's Disease. *Brain Pathol.* **2016**, *27*, 160–168. [CrossRef] [PubMed]
26. Carandini, T.; Arighi, A.; Sacchi, L.; Fumagalli, G.G.; Pietroboni, A.M.; Ghezzi, L.; Colombi, A.; Scarioni, M.; Fenoglio, C.; De Riz, M.A.; et al. Testing the 2018 NIA-AA research framework in a retrospective large cohort of patients with cognitive impairment: From biological biomarkers to clinical syndromes. *Alzheimers Res. Ther.* **2019**, *11*, 84. [CrossRef] [PubMed]
27. Villain, N.; Dubois, B.; Frisoni, G.B.; Rabinovici, G.D.; Sabbagh, M.N.; Cappa, S.; Bejanin, A.; Bombois, S.; Epelbaum, S.; Teichmann, M.; et al. Clinical diagnosis of Alzheimer's disease: Recommendations of the International Working Group (IWG). *Alzheimers Dement.* **2021**, *17*, e051167. [CrossRef]
28. Nauen, D.W.; Troncoso, J.C. Amyloid-beta is present in human lymph nodes and greatly enriched in those of the cervical region. *Alzheimers Dement.* **2021**, *18*, 205–210. [CrossRef]
29. Hablitz, L.M.; Plá, V.; Giannetto, M.; Vinitsky, H.S.; Stæger, F.F.; Metcalfe, T.; Nguyen, R.; Benrais, A.; Nedergaard, M. Circadian control of brain glymphatic and lymphatic fluid flow. *Nat. Commun.* **2020**, *11*, 4411. [CrossRef]
30. Fleischman, D.; Berdahl, J.P.; Zaydlarova, J.; Stinnett, S.; Fautsch, M.P.; Allingham, R.R. Cerebrospinal Fluid Pressure Decreases with Older Age. *PLoS ONE* **2012**, *7*, e52664. [CrossRef]
31. McKee, C.A.; Lee, J.; Cai, Y.; Saito, T.; Saido, T.; Musiek, E.S. Astrocytes deficient in circadian clock gene Bmal1 show enhanced activation responses to amyloid-beta pathology without changing plaque burden. *Sci. Rep.* **2022**, *12*, 1796. [CrossRef]
32. Ishida, K.; Yamada, K.; Nishiyama, R.; Hashimoto, T.; Nishida, I.; Abe, Y.; Yasui, M.; Iwatsubo, T. Glymphatic system clears extracellular tau and protects from tau aggregation and neurodegeneration. *J. Exp. Med.* **2022**, *219*, e20211275. [CrossRef]
33. Medina, M.; Avila, J. The role of extracellular Tau in the spreading of neurofibrillary pathology. *Front. Cell. Neurosci.* **2014**, *8*, 113. [CrossRef]
34. Burfeind, K.G.; Murchison, C.F.; Westaway, S.K.; Simon, M.J.; Erten-Lyons, D.; Kaye, J.A.; Quinn, J.F.; Iliff, J.J. The effects of noncoding aquaporin-4 single-nucleotide polymorphisms on cognition and functional progression of Alzheimer's disease. *Alzheimers Dementia Transl. Res. Clin. Interv.* **2017**, *3*, 348–359. [CrossRef]
35. Chandra, A.; Farrell, C.; Wilson, H.; Dervenoulas, G.; De Natale, E.R.; Politis, M. Aquaporin-4 polymorphisms predict amyloid burden and clinical outcome in the Alzheimer's disease spectrum. *Neurobiol. Aging* **2020**, *97*, 1–9. [CrossRef]
36. Iliff, J.J.; Wang, M.; Liao, Y.; Plogg, B.A.; Peng, W.; Gundersen, G.A.; Benveniste, H.; Vates, G.E.; Deane, R.; Goldman, S.A.; et al. A Paravascular Pathway Facilitates CSF Flow Through the Brain Parenchyma and the Clearance of Interstitial Solutes, Including Amyloid β. *Sci. Transl. Med.* **2012**, *4*, 147ra111. [CrossRef]
37. Weller, R.O.; Subash, M.; Preston, S.D.; Mazanti, I.; Carare, R.O. Perivascular drainage of amyloid-beta peptides from the brain and its failure in cerebral amyloid angiopathy and Alzheimer's disease. *Brain Pathol. Zurich Switz.* **2008**, *18*, 253–266. [CrossRef]
38. O'Donnell, J.; Ding, F.; Nedergaard, M. Distinct functional states of astrocytes during sleep and wakefulness: Is norepinephrine the master regulator? *Curr. Sleep Med. Rep.* **2015**, *1*, 1–8. [CrossRef]
39. Mendelsohn, A.R.; Larrick, J.W. Sleep Facilitates Clearance of Metabolites from the Brain: Glymphatic Function in Aging and Neurodegenerative Diseases. *Rejuvenation Res.* **2013**, *16*, 518–523. [CrossRef]
40. Krueger, J.M.; Frank, M.G.; Wisor, J.P.; Roy, S. Sleep function: Toward elucidating an enigma. *Sleep Med. Rev.* **2016**, *28*, 46–54. [CrossRef]
41. Ratner, V.; Zhu, L.; Kolesov, I.; Nedergaard, M.; Benveniste, H.; Tannenbaum, A. Optimal-mass-transfer-based estimation of glymphatic transport in living brain. *Proc. Spie—Int. Soc. Opt. Eng.* **2015**, *9413*, 94131J.
42. Bakker, E.N.T.P.; Bacskai, B.J.; Arbel-Ornath, M.; Aldea, R.; Bedussi, B.; Morris, A.W.J.; Weller, R.O.; Carare, R.O. Lymphatic Clearance of the Brain: Perivascular, Paravascular and Significance for Neurodegenerative Diseases. *Cell. Mol. Neurobiol.* **2016**, *36*, 181–194. [CrossRef]
43. Simon, M.J.; Iliff, J.J. Regulation of cerebrospinal fluid (CSF) flow in neurodegenerative, neurovascular and neuroinflammatory disease. *Biochim. Biophys. Acta BBA-Mol. Basis Dis.* **2015**, *1862*, 442–451. [CrossRef] [PubMed]
44. Kress, B.T.; Iliff, J.J.; Xia, M.; Wang, M.; Wei, H.S.; Zeppenfeld, D.; Xie, L.; Kang, H.; Xu, Q.; Liew, J.A.; et al. Impairment of paravascular clearance pathways in the aging brain. *Ann. Neurol.* **2014**, *76*, 845–861. [CrossRef]
45. Xie, L.; Kang, H.; Xu, Q.; Chen, M.J.; Liao, Y.; Thiyagarajan, M.; O'Donnell, J.; Christensen, D.J.; Nicholson, C.; Iliff, J.J.; et al. Sleep Drives Metabolite Clearance from the Adult Brain. *Science* **2013**, *342*, 373–377. [CrossRef] [PubMed]
46. Lee, H.; Xie, L.; Yu, M.; Kang, H.; Feng, T.; Deane, R.; Logan, J.; Nedergaard, M.; Benveniste, H. The Effect of Body Posture on Brain Glymphatic Transport. *J. Neurosci.* **2015**, *35*, 11034–11044. [CrossRef] [PubMed]
47. Tarasoff-Conway, J.M.; Carare, R.O.; Osorio, R.S.; Glodzik, L.; Butler, T.; Fieremans, E.; Axel, L.; Rusinek, H.; Nicholson, C.; Zlokovic, B.V.; et al. Clearance systems in the brain-implications for Alzheimer disease. *Nat. Rev. Neurol.* **2015**, *11*, 457–470. [CrossRef]
48. Kiviniemi, V.; Wang, X.; Korhonen, V.; Keinänen, T.; Tuovinen, T.; Autio, J.; LeVan, P.; Keilholz, S.; Zang, Y.F.; Hennig, J.; et al. Ultra-fast magnetic resonance encephalography of physiological brain activity-Glymphatic pulsation mechanisms? *J. Cereb. Blood Flow Metab.* **2016**, *36*, 1033–1045. [CrossRef]
49. Musiek, E.S. Circadian clock disruption in neurodegenerative diseases: Cause and effect? *Front. Pharmacol.* **2015**, *6*, 29. [CrossRef]

50. Taoka, T.; Masutani, Y.; Kawai, H.; Nakane, T.; Matsuoka, K.; Yasuno, F.; Kishimoto, T.; Naganawa, S. Evaluation of glymphatic system activity with the diffusion MR technique: Diffusion tensor image analysis along the perivascular space (DTI-ALPS) in Alzheimer's disease cases. *Jpn. J. Radiol.* **2017**, *35*, 172–178. [CrossRef]
51. Varga, A.W.; Wohlleber, M.E.; Giménez, S.; Romero, S.; Alonso, J.F.; Ducca, E.L.; Kam, K.; Lewis, C.; Tanzi, E.B.; Tweardy, S.; et al. Reduced Slow-Wave Sleep Is Associated with High Cerebrospinal Fluid Aβ42 Levels in Cognitively Normal Elderly. *Sleep* **2016**, *39*, 2041–2048. [CrossRef]
52. Mander, B.A.; Marks, S.M.; Vogel, J.W.; Rao, V.; Lu, B.; Saletin, J.M.; Ancoli-Israel, S.; Jagust, W.J.; Walker, M.P. β-amyloid disrupts human NREM slow waves and related hippocampus-dependent memory consolidation. *Nat. Neurosci.* **2015**, *18*, 1051–1057. [CrossRef]
53. Ma, Q.; Schlegel, F.; Bachmann, S.B.; Schneider, H.; Decker, Y.; Rudin, M.; Weller, M.; Proulx, S.T.; Detmar, M. Lymphatic outflow of cerebrospinal fluid is reduced in glioma. *Sci. Rep.* **2019**, *9*, 14815. [CrossRef]
54. Hu, X.; Deng, Q.; Ma, L.; Li, Q.; Chen, Y.; Liao, Y.; Zhou, F.; Zhang, C.; Shao, L.; Feng, J.; et al. Meningeal lymphatic vessels regulate brain tumor drainage and immunity. *Cell Res.* **2020**, *30*, 229–243. [CrossRef]
55. Ballerini, L.; Lovreglio, R.; Hernández, M.d.C.V.; Ramirez, J.; MacIntosh, B.J.; Black, S.E.; Wardlaw, J.M. Perivascular Spaces Segmentation in Brain MRI Using Optimal 3D Filtering. *Sci. Rep.* **2018**, *8*, 2132. [CrossRef]
56. Debaker, C.; Djemai, B.; Ciobanu, L.; Tsurugizawa, T.; Le Bihan, D. Diffusion MRI reveals in vivo and non-invasively changes in astrocyte function induced by an aquaporin-4 inhibitor. *PLoS ONE* **2020**, *15*, e0229702. [CrossRef]
57. Stummer, W. Mechanisms of tumor-related brain edema. *Neurosurg. Focus* **2007**, *22*, E8. [CrossRef]
58. Ringstad, G.; Vatnehol, S.A.S.; Eide, P.K. Glymphatic MRI in idiopathic normal pressure hydrocephalus. *Brain* **2017**, *140*, 2691–2705. [CrossRef]
59. Belykh, E.; Shaffer, K.V.; Lin, C.; Byvaltsev, V.A.; Preul, M.C.; Chen, L. Blood-Brain Barrier, Blood-Brain Tumor Barrier, and Fluorescence-Guided Neurosurgical Oncology: Delivering Optical Labels to Brain Tumors. *Front. Oncol.* **2020**, *10*, 739. [CrossRef]
60. Nduom, E.; Yang, C.; Merrill, M.J.; Zhuang, Z.; Lonser, R.R. Characterization of the blood-brain barrier of metastatic and primary malignant neoplasms. *J. Neurosurg.* **2013**, *119*, 427–433. [CrossRef]
61. Haj-Yasein, N.N.; Vindedal, G.F.; Eilert-Olsen, M.; Gundersen, G.A.; Skare, Ø.; Laake, P.; Klungland, A.; Thorén, A.E.; Burkhardt, J.M.; Ottersen, O.P.; et al. Glial-conditional deletion of aquaporin-4 (Aqp4) reduces blood-brain water uptake and confers barrier function on perivascular astrocyte endfeet. *Proc. Natl. Acad. Sci. USA* **2011**, *108*, 17815–17820. [CrossRef]
62. Saadoun, S.; Papadopoulos, M.C.; Davies, D.C.; Krishna, S.; Bell, B.A. Aquaporin-4 expression is increased in oedematous human brain tumours. *J. Neurol. Neurosurg. Psychiatry* **2002**, *72*, 262–265. [CrossRef]
63. Louis, D.N.; Wesseling, P.; Aldape, K.; Brat, D.J.; Capper, D.; Cree, I.A.; Eberhart, C.; Figarella-Branger, D.; Fouladi, M.; Fuller, G.N.; et al. cIMPACT-NOW update 6: New entity and diagnostic principle recommendations of the cIMPACT-Utrecht meeting on future CNS tumor classification and grading. *Brain Pathol.* **2020**, *30*, 844–856. [CrossRef] [PubMed]
64. Daneman, R.; Prat, A. The Blood–Brain Barrier. *Cold Spring Harb. Perspect. Biol.* **2015**, *7*, a020412. [CrossRef] [PubMed]
65. Zeppenfeld, D.M.; Simon, M.; Haswell, J.D.; D'abreo, D.; Murchison, C.; Quinn, J.F.; Grafe, M.R.; Woltjer, R.L.; Kaye, J.; Iliff, J.J. Association of Perivascular Localization of Aquaporin-4 With Cognition and Alzheimer Disease in Aging Brains. *JAMA Neurol.* **2017**, *74*, 91–99. [CrossRef] [PubMed]
66. Valenza, M.; Facchinetti, R.; Steardo, L.; Scuderi, C. Altered Waste Disposal System in Aging and Alzheimer's Disease: Focus on Astrocytic Aquaporin-4. *Front. Pharmacol.* **2020**, *10*, 1656. [CrossRef]
67. Buccellato, F.R.; D'Anca, M.; Serpente, M.; Arighi, A.; Galimberti, D. The Role of Glymphatic System in Alzheimer's and Parkinson's Disease Pathogenesis. *Biomedicines* **2022**, *10*, 2261. [CrossRef]
68. Hovd, M.H.; Mariussen, E.; Uggerud, H.; Lashkarivand, A.; Christensen, H.; Ringstad, G.; Eide, P.K. Population pharmacokinetic modeling of CSF to blood clearance: Prospective tracer study of 161 patients under work-up for CSF disorders. *Fluids Barriers CNS* **2022**, *19*, 55. [CrossRef]

International Journal of *Molecular Sciences*

MDPI

Review

From Homeostasis to Pathology: Decoding the Multifaceted Impact of Aquaporins in the Central Nervous System

Corneliu Toader [1,2], Calin Petru Tataru [3,4,*], Ioan-Alexandru Florian [5,*], Razvan-Adrian Covache-Busuioc [1], David-Ioan Dumitrascu [1], Luca Andrei Glavan [1], Horia Petre Costin [1], Bogdan-Gabriel Bratu [1] and Alexandru Vlad Ciurea [1,6]

1 Department of Neurosurgery, "Carol Davila" University of Medicine and Pharmacy, 020021 Bucharest, Romania; corneliu.toader@umfcd.ro (C.T.); razvan-adrian.covache-busuioc0720@stud.umfcd.ro (R.-A.C.-B.); david-ioan.dumitrascu0720@stud.umfcd.ro (D.-I.D.); luca-andrei.glavan0720@stud.umfcd.ro (L.A.G.); horia-petre.costin0720@stud.umfcd.ro (H.P.C.); bogdan.bratu@stud.umfcd.ro (B.-G.B.); prof.avciurea@gmail.com (A.V.C.)

2 Department of Vascular Neurosurgery, National Institute of Neurology and Neurovascular Diseases, 077160 Bucharest, Romania

3 Department of Opthamology, "Carol Davila" University of Medicine and Pharmacy, 020021 Bucharest, Romania

4 Central Military Emergency Hospital "Dr. Carol Davila", 010825 Bucharest, Romania

5 Department of Neurosciences, "Iuliu Hatieganu" University of Medicine and Pharmacy, 400012 Cluj-Napoca, Romania

6 Neurosurgery Department, Sanador Clinical Hospital, 010991 Bucharest, Romania

* Correspondence: calinnpetrutataru@yahoo.com (C.P.T.); florian.ioan.alexandru@gmail.com (I.-A.F.)

check for updates

Citation: Toader, C.; Tataru, C.P.; Florian, I.-A.; Covache-Busuioc, R.-A.; Dumitrascu, D.-I.; Glavan, L.A.; Costin, H.P.; Bratu, B.-G.; Ciurea, A.V. From Homeostasis to Pathology: Decoding the Multifaceted Impact of Aquaporins in the Central Nervous System. *Int. J. Mol. Sci.* **2023**, *24*, 14340. https://doi.org/10.3390/ijms241814340

Academic Editors: Pablo García-Miranda and Miriam Echevarría

Received: 2 September 2023
Revised: 15 September 2023
Accepted: 18 September 2023
Published: 20 September 2023

Abstract: Aquaporins (AQPs), integral membrane proteins facilitating selective water and solute transport across cell membranes, have been the focus of extensive research over the past few decades. Particularly noteworthy is their role in maintaining cellular homeostasis and fluid balance in neural compartments, as dysregulated AQP expression is implicated in various degenerative and acute brain pathologies. This article provides an exhaustive review on the evolutionary history, molecular classification, and physiological relevance of aquaporins, emphasizing their significance in the central nervous system (CNS). The paper journeys through the early studies of water transport to the groundbreaking discovery of Aquaporin 1, charting the molecular intricacies that make AQPs unique. It delves into AQP distribution in mammalian systems, detailing their selective permeability through permeability assays. The article provides an in-depth exploration of AQP4 and AQP1 in the brain, examining their contribution to fluid homeostasis. Furthermore, it elucidates the interplay between AQPs and the glymphatic system, a critical framework for waste clearance and fluid balance in the brain. The dysregulation of AQP-mediated processes in this system hints at a strong association with neurodegenerative disorders such as Parkinson's Disease, idiopathic normal pressure hydrocephalus, and Alzheimer's Disease. This relationship is further explored in the context of acute cerebral events such as stroke and autoimmune conditions such as neuromyelitis optica (NMO). Moreover, the article scrutinizes AQPs at the intersection of oncology and neurology, exploring their role in tumorigenesis, cell migration, invasiveness, and angiogenesis. Lastly, the article outlines emerging aquaporin-targeted therapies, offering a glimpse into future directions in combatting CNS malignancies and neurodegenerative diseases.

Keywords: aquaporins; central nervous system; fluid homeostasis; glymphatic system; neurodegenerative diseases; brain tumorigenesis; permeability assays; AQP4; AQP1; stroke; Parkinson's disease; Alzheimer's disease; neuromyelitis optica; cellular physiology; therapeutic interventions

1. Introduction

1.1. Definition and General Characteristics of Aquaporins

Aquaporins represent a ubiquitous family of integral membrane proteins that are disseminated extensively across both animal and plant kingdoms. Structurally, these proteins are characterized by a core architecture comprising six alpha-helical transmembrane domains and two additional, shorter helical elements. These components delineate cytoplasmic and extracellular vestibules, which are connected via a constricted aqueous pore (Figure 1). Notably, conserved sequence motifs, such as the asparagine-proline-alanine (NPA) motifs, are frequently observed within these shorter helical regions. Functionally, aquaporin monomers autonomously integrate into the lipid bilayer, subsequently aggregating as tetrameric complexes. Certain variants, exemplified by mammalian Aquaporin 4 (AQP4), have the capability to further organize into higher-order structures, known as orthogonal arrays of particles, within the cell membrane [1,2].

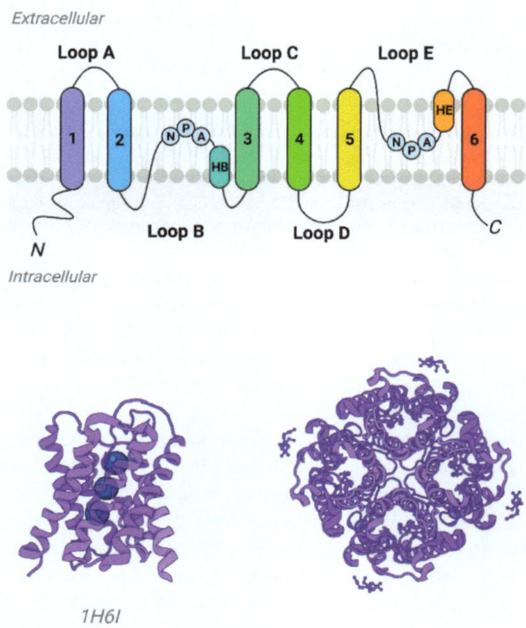

Figure 1. Aquaporin monomer membrane topography (upper panel) and crystal structure (AQP1, PDB 1j4n) showcase helices H1–H8 and water molecules (blue spheres) in the aqueous pore.

The predominant role of most aquaporins is to facilitate the translocation of water molecules across cellular membranes in reaction to osmotic gradients, which are typically established via active solute transport mechanisms. Owing to the relatively low individual throughput of water by aquaporin monomers, membranes are often saturated with a high density of these proteins—up to 10,000 per square micron—to substantially augment the membrane's overall water permeability [3]. Computational studies, particularly molecular dynamics simulations, posit that the specificity of aquaporins for water is mediated by steric constraints and electrostatic interactions within the channel's aqueous pore [4].

Additionally, a specialized subgroup of aquaporins, termed aquaglyceroporins, exhibits the capacity to transport glycerol molecules. The structural configuration of aquaglyceroporins is distinguished by a marginally larger pore diameter and a lining enriched with hydrophobic amino acid residues, in contrast to the hydrophilic nature of the pore in water-selective aquaporins. Beyond the primary functions of transporting water and, in some cases, glycerol, aquaporins are postulated to facilitate the passage of various gases (CO_2, NH_3, NO, O_2) and small solutes such as hydrogen peroxide and arsenite. Some aquaporins have even been implicated in ion transport (K^+, Cl^-), although these claims remain a subject of ongoing debate [5].

Moreover, a range of non-transport-centric functions have been ascribed to some aquaporins, encompassing roles in cellular adhesion, membrane polarization, and the modulatory regulation of protein interactions, such as those involving ion channels [5].

1.2. The Overarching Significance of Aquaporins in Cellular Physiology, with an Emphasis on Their Role in the Central Nervous System

The expression profiles of Aquaporins (AQPs), particularly AQP4 and AQP1, have garnered significant research attention within the context of the central nervous system and sensory organs, as compared to the peripheral and enteric nervous systems [6]. AQP4, the predominant water channel in the brain, spinal cord, and optic nerve, is largely expressed in the astrocyte cell plasma membrane but is distinctly localized to specialized regions such as astrocyte foot processes [7–9]. Such polarized expression patterns are hypothesized to be mediated by intracellular associations between AQP4 and α-syntrophin or through extracellular interactions with agrin. Both α-syntrophin and agrin also exhibit polarized expression in astrocyte foot processes [10].

Additionally, the formation of orthogonal arrays of particles (OAPs) by AQP4 may facilitate its polarized localization, as it requires only a single anchoring link for stabilization, as opposed to individual AQP4 tetramers [10]. In the CNS, AQP4 is predominantly localized to the subpial astrocyte processes forming the glial-limiting membrane, the perivascular astrocyte endfeet, and the basolateral membrane of ependymal and subependymal astrocyte processes [11,12]. Such strategic localization suggests that AQP4 may play a critical role in regulating water flux across the CNS-water compartment interfaces.

In specific brain regions lacking a blood–brain barrier, such as the circumventricular organs, and in hippocampal regions such as CA1 and the dentate gyrus, AQP4 manifests a more diffused expression profile [13]. This dispersed distribution may facilitate rapid water flux essential for maintaining potassium (K^+) homeostasis during neuronal activity.

AQP1, on the other hand, is prominently found in the ventricular-facing plasma membrane of choroid plexus epithelial cells, implicating its role in cerebrospinal fluid (CSF) secretion [14]. While AQP1 is ubiquitously expressed in vascular endothelial cells throughout the body, it is conspicuously absent in the cerebrovascular endothelium, except in areas such as the circumventricular organs. Intriguingly, astrocyte co-culture leads to the suppression of AQP1 mRNA expression in primary brain microvessel endothelial cells, suggesting that astrocyte-endothelial interactions may down-regulate endothelial AQP1 expression [15].

The expression of AQP9 within the brain is relatively sparse, and its precise localization remains ambiguous due to the limitations in antibody specificity [16]. However, some evidence points to its expression in specialized neural cells such as neurons in the substantia nigra, tanycytes, and a subset of astrocytes.

In the spinal cord, the expression profiles of various AQPs are notably differentiated. AQP4 is predominantly localized to perivascular astrocyte foot processes and the glial-limiting membrane, whereas AQP1 is expressed in the processes of non-myelinated neurons in the dorsal horns. AQP9 is purportedly expressed in spinal cord radial astrocytes and also potentially in the glial-limiting membrane [16].

Similar to its expression patterns in the brain and spinal cord, AQP4 is primarily found in perivascular astrocyte foot processes and the glial-limiting membrane within the optic nerve [7]. Additionally, AQP4 expression extends to astrocyte processes in CNS regions lacking a blood-brain barrier, including but not limited to the pre-laminar optic nerve head, circumventricular organs, and root entry zones in the spinal cord. This peripheral exposure of the CNS AQP4 pool suggests its potential susceptibility as an initial target for circulating

AQP4-specific antibodies, particularly in the context of neuromyelitis optica (NMO), a topic warranting further exploration [3].

2. Historical Perspective on Aquaporins

The scientific exploration of water transport mechanisms predates the molecular characterization of aquaporins, which initially focused on tissues with notable water permeability. The seminal discovery of the first aquaporin, AQP1, was made during research endeavors aimed at identifying the Rh blood group antigens. Since this initial discovery, the field has undergone substantial expansion, extending the study of aquaporins to diverse organisms. In mammals, the aquaporin family exhibits functional heterogeneity. While certain isoforms specialize exclusively in facilitating water transport, others are capable of translocating a diverse array of solutes across cellular membranes. Regulatory mechanisms also exist that modulate aquaporin function, either by influencing channel permeability or by altering subcellular localization. Despite extensive research elucidating the physiological roles of various mammalian aquaporins, significant knowledge gaps remain within this domain. Understanding the comprehensive physiological implications of this integral membrane protein family is still evolving and presents an area ripe for further scholarly inquiry [17].

2.1. *Early Studies of Water Transport*

Initial investigations into water transport mechanisms commenced with the observation that certain amphibian tissues, such as the skin and bladder—which bear functional resemblance to the mammalian kidney's collecting duct—exhibited enhanced water permeability compared to other tissues. Early work by Hans Ussing and collaborators highlighted the particular water permeability of amphibian skin [18]. Subsequent electron microscopic analyses identified structural assemblies in amphibian bladder tissues, presumed to be water channels, whose prevalence correlated with increased tissue water permeability [19].

The concept of regulated water permeability through localized water channels was initially postulated under the 'shuttle hypothesis'. In this framework, protein aggregates were detected in intracellular vesicles during periods of low water reabsorption (diuresis) and were translocated to the plasma membrane during periods of increased water reabsorption (antidiuresis) [20].

Moreover, red blood cells were recognized for their high water permeability. Investigations by A.K. Solomon and associates revealed water transport in red blood cells that exhibited low Arrhenius activation energy, implying the existence of membrane pores facilitating the water movement [21]. Further studies by Robert Macey and colleagues indicated that the water permeability of red blood cells could be chemically modulated. Specifically, water transport inhibition via $HgCl_2$ could be reversed by chemical reducing agents, suggesting the presence of a protein featuring accessible free sulfhydryl groups sensitive to mercury [22].

In the quest to ascertain the molecular identity of water channels [23], a myriad of strategies were employed by different research groups. Among these efforts were the radiolabeling studies conducted by William Harris and colleagues, who compared proteins from toad bladders exposed to antidiuretic hormones (ADH) with those unexposed, seeking to identify proteins localized to the plasma membrane in response to ADH [24]. Other approaches included expression cloning utilizing mRNA from tissues with known water transport properties and radiation inactivation techniques aiming to ascertain the molecular weight of water channels [25]. Despite these extensive efforts, the molecular identity of aquaporins remained elusive through these methodologies.

2.2. *Discovery of Aquaporin 1*

In a groundbreaking development for the understanding of water channels, Benga's group in 1985 located a water channel protein among the polypeptides in the 35–60 kDa range on the electrophoretogram of red blood cell (RBC) membrane proteins. Subsequent work by Agre and colleagues in 1988 isolated a novel 28-kDa protein from the RBC membrane, termed CHIP28 (channel-forming integral membrane protein of 28 kDa). Intriguingly, this protein also had a glycosylated component in the 35–60 kDa range, aligning with Benga's findings. It was not until 1992 that Agre's team posited that CHIP28 likely represented a functional membrane water channel. The seminal contributions of Benga's group from 1986 were largely overlooked when Peter Agre was awarded half of the 2003 Nobel Prize in Chemistry for the "discovery of water channels". This omission sparked widespread support for Benga in the scientific community [26].

The discovery of the first water channel, AQP1, was rooted in a search for Rh blood group antigens [27]. A 32-kDa protein related to one of the Rh antigens was isolated using hydroxylapatite chromatography. Interestingly, the protein did not stain well with Coomassie stain. However, upon using silver reagent, a separate 28-kDa protein was revealed. Initially thought to be a fragment of the 32-kDa protein, an antibody test disproved this hypothesis, confirming that the 28-kDa protein was unique and had properties akin to a membrane channel [28]. This protein was temporarily named CHIP28 [29]. John C. Parker, a membrane physiologist, was the first to suggest that CHIP28 might be the elusive water channel, based on its presence in red blood cells and specific renal tissues.

Further characterization led to the cloning of the cDNA encoding CHIP28, which was found to share sequence similarity with uncharacterized genes from microbes and plants [30]. Definitive evidence for CHIP28 as a water channel was obtained through its expression in Xenopus laevis oocytes, a model organism that had been previously used to study other transport proteins [31]. This marked a major milestone in the scientific understanding of membrane water channels, albeit with a backdrop of controversy regarding the acknowledgment of pioneering work.

Erich Windhager, based at Cornell Medical School, was the first to propose using oocytes for studying water transport, given their naturally low water permeability. Experiments using Xenopus laevis oocytes expressing CHIP28 presented compelling results. When placed in a dilute Modified Barth's solution with a lowered osmolarity (70 mosM as opposed to the standard 200 mosM), these oocytes swelled rapidly and burst, contrasting sharply with control oocytes, which exhibited negligible swelling [32].

The water permeability coefficient (Pf) of the CHIP28-expressing oocytes was found to be 20 times higher than that of the controls. The Arrhenius activation energy of these specialized oocytes was also lower, aligning with previous observations in native membranes. Importantly, the addition of 1 mM HgCl2 inhibited water permeability in CHIP28 oocytes, an effect reversible by a reducing agent. This was consistent with earlier studies on water-permeable tissues, reinforcing the idea that CHIP28 was instrumental in water transport [22].

To discount the possibility that CHIP28 merely activated an endogenous water channel in oocytes, purified CHIP28 was integrated into membrane proteoliposomes. The result was striking: the water permeability of these proteoliposomes increased by 50-fold, while urea and proton transport remained unaffected. It was estimated that a single CHIP28 subunit could facilitate the passage of roughly 3×10^9 water molecules per second.

As researchers discovered more homologs of CHIP28, the term "aquaporin" was introduced to denote this expanding family of water channels [33]. Accordingly, the original CHIP28 was renamed Aquaporin 1 (AQP1).

AQP1 serves as a model for understanding the aquaporin family, which spans across microorganisms, plants, and mammals. Early studies using red blood cells suggested that AQP1 exists as a homotetramer in the membrane. Each subunit of the tetramer is mostly embedded in the membrane, with the N- and C-termini of the polypeptide chain extending into the cytoplasm [29]. Negative staining electron microscopy of cells expressing AQP1 and proteoliposomes confirmed this tetrameric structure [34].

Prior work on the lens MIP protein, now identified as AQP0, had proposed that there were six domains of the protein spanning the membrane [35]. This was later confirmed for AQP1 through experiments that involved inserting an epitope tag at varying points along the polypeptide chain. The location of the epitope within the membrane was ascertained via proteolysis and antibody targeting. The experimental findings helped elucidate the architecture of the helices in the membrane, revealing them to be obversely symmetric. This led to the proposal that AQP1 features two highly conserved NPA (asparagine-proline-alanine) motifs located adjacently in the lipid bilayer, giving the protein an hourglass-like structure [36].

This model of AQP1's structure has been further validated by high-resolution studies of AQP1 and other members of the aquaporin family. The research has significantly advanced our understanding of how these proteins function at a molecular level, illuminating their role in facilitating water transport across cell membranes.

3. Distribution and Molecular Classification of Aquaporins

3.1. Overview of Various Aquaporin Isoforms Present in Mammalian Systems

In mammals, the aquaporin (AQP) family is generally made up of 12 to 15 isoforms grouped into 13 subfamilies (AQP0–12). However, humans stand out by having 18 paralogs due to tandem duplications, including four extra AQP7 pseudogenes and a second copy of AQP12. Additionally, more ancient mammalian lineages, such as Metatheria and Prototheria, have been shown to possess further classes such as AQP13–14 [37].

Animal AQPs are typically categorized into four main groups:

1. Classical or Orthodox AQPs (AQP0, 1, 2, 4, 5, 6, 14), which are mainly focused on water transport.
2. Aqua-ammoniaporins (AQP8), which are sometimes counted among the orthodox AQPs.
3. Aquaglyceroporins (AQP3, 7, 9, 10, 13), which can transport glycerol in addition to water.
4. Unorthodox AQPs (AQP11–12), also known as "superaquaporins", which have low sequence homology with other AQPs and feature unusual N-terminal NPA motifs [37–40].

Plants show even more diversity in their aquaporin families, likely due to higher genome duplication rates and their stationary lifestyles requiring more specialized water transport mechanisms. For example, the model plant species Arabidopsis thaliana has 35 isoforms, Zea mays has 38, and upland cotton (*Gossypium hirsutun*) has 71. The current record holder is oil seed rape (*Brassica napus*), with 121 full-length AQPs [41–43].

Plant AQPs are typically classified into seven subfamilies based on subcellular localization and other specific features:

1. Plasma Membrane Intrinsic Proteins (PIPs);
2. Tonoplast Intrinsic Proteins (TIPs);
3. NOD26-Like Intrinsic Proteins (NIPs);
4. Small Basic Intrinsic Proteins (SIPs);
5. Unknown Intrinsic Proteins (XIPs), which are absent in monocots and Brassicaceae;
6. Hybrid Intrinsic Proteins (HIPs) and GLpF-like intrinsic proteins (GIPs), which are found only in older plant lineages such as mosses [44–46].

The origins and evolutionary roles of these plant aquaporin families remain subjects of ongoing research. For instance, it is still debated whether NIPs are ancestral or acquired through horizontal gene transfer [37–39].

Green algae also possess a unique set of AQPs, including PIPs and GIPs, along with five other families (MIP A–E) not seen in plants [47]. A completely unique family, large intrinsic proteins (LIPs), was recently discovered in diatoms [48]. These various subfamilies can further be divided into paralog groups, enhancing the complexity and adaptability of water transport across different organisms.

The plant intrinsic protein (PIP) subfamily in plants is usually divided into PIP1 and PIP2 groups based on sequence homology [49]. These divisions are functionally significant, as the two groups have distinct capabilities in terms of membrane localization and water transport activity. In terms of nomenclature, plant major intrinsic proteins (MIPs) are usually named based on the abbreviation of their subfamily, along with two numbers denoting their phylogenetic group and order of discovery. However, this system of naming can be confusing, especially when comparing proteins across species. Two proteins with identical names from different species may not actually be orthologous [39].

Recent phylogenetic research has refined our understanding further, identifying 19 clusters of orthologous genes in flowering plants. For example, the PIP family, previously divided into two clusters, was found to be divided into three, aligning with prior classifications in Arabidopsis thaliana [38]. This has led to calls for a standardized, globally accepted nomenclature based on evolutionary relationships. Such a nomenclature could facilitate the comparison and integration of knowledge across both animal and plant MIPs.

A particular phylogenetic study highlighted the close evolutionary relationships between animal and plant MIP subfamilies when considering proteins with a high amino acid identity (>25%). The study suggested a model of vertical transfer for four ancestral subfamilies [38]. In this model, AQP1-like and PIP subfamilies are grouped together as family A; AQP8-like and TIP as family B; AQP3-like and NIP as family C; and AQP11-like and SIP as family D. This grouping is not merely theoretical but also has practical implications. For instance, it can be used to predict substrate specificities across different MIP subfamilies.

3.2. The Molecular Underpinnings of Their Selective Permeability and Functionality

Water transport assays to study Aquaporins have been conducted using a variety of model systems across species, including bacteria, yeast, and mammalian cells [50–52]. This diversity in experimental setups reflects the widespread distribution of AQPs in nature. Specifically, cellular components such as intracellular and plasma membrane vesicles, particularly from animal tissues such as the kidney and intestinal epithelia, have been essential for evaluating AQP activity [53].

One of the prominent methods for characterizing new AQP isoforms involves the use of Xenopus laevis oocytes, owing to their inherently low water permeability [32]. This heterologous expression system allows for functional studies of AQPs from various organisms. Yeast cells lacking endogenous AQPs also serve as an effective model system for this purpose. They have been used to develop a high-throughput assay to identify functionally relevant AQP mutants that can withstand freeze-thaw cycles. Additionally, AQP-transfected cell lines and Zebrafish embryos injected with AQP mRNA have been deployed for similar studies [54].

Research has not been limited to functional assays alone; purified AQPs have been reconstituted into liposomes to directly investigate their roles in water and solute transport. Moreover, the development of transgenic mice models lacking specific AQPs has enriched our understanding of these proteins [55]. Such mouse models have illuminated the multifaceted roles of AQPs in various biological functions. These range from transepithelial fluid transport and cell migration to more complex physiological processes such as brain edema, neuroexcitation, cell proliferation, skin hydration, and even metabolic activities in adipocytes.

3.3. Permeability Assays

The quantification of water or solute permeability across biological membranes serves as an indirect method to assess both the expression and functional status of Aquaporins. In general, the analytical focus for AQP activity and water permeability revolves around tracking alterations in cell or vesicle volume, which result from osmotically or pressure-driven water fluxes [56]. For solute permeability estimations, it is essential to account for both solute and water fluxes when examining volume changes, as these fluxes are instigated by their respective gradients.

Both water and solute fluxes are directly correlated to their inducing forces, represented by osmotic permeability (Pf) and solute permeability (Ps) coefficients, respectively. The kinetics of volume changes depend on the partitioning of water or solute between the aqueous pathway through the channel and diffusion across the lipid bilayer. Furthermore, analyzing permeability as a function of temperature enables the calculation of the activation energy (Ea) required for transport, a crucial metric for gauging AQP functionality. It is generally observed that water or solute fluxes via hydrophilic channel pores necessitate lower activation energy compared to fluxes traversing a hydrophobic lipid bilayer. Consequently, high permeability coupled with low Ea is indicative of AQP-mediated transport [57].

Analytical methodologies for assessing permeability commonly employ optical properties that are volume-dependent, such as light transmission, absorbance, scattering, and fluorescence [58–60]. These strategies are equally applicable to evaluating transport kinetics in both cellular models and proteoliposomes. The amalgamation of diverse biological models (cells, vesicles, and proteoliposomes) with optical detection systems offers a comprehensive toolkit for the investigation of AQP functions.

3.4. Epithelial Assays

Water permeability can be assessed in native epithelial tissues, such as those from the intestinal wall or kidney tubules, as well as in cultured epithelial cell monolayers positioned on permeable supports within Ussing chambers [61]. In these experiments, the apical and basolateral membranes of polarized cells are exposed to distinct compartments. The introduction of a membrane-impermeable solute such as sucrose or mannitol to one of these compartments instigates a transepithelial water flux. This flux is then quantified either by observing the fluid level change in a capillary tube linked to the opposite compartment or through the use of a fluorescent dye. The dye is added to the hyperosmotic compartment, and the rate of fluorescence alteration due to dye dilution serves as a measure of overall transepithelial osmotic water permeability [62].

It is crucial to note that the total transepithelial permeability is the aggregate of two distinct pathways: the cellular and the paracellular pathways. In this context, the presence of AQPs specifically influences the cellular pathway. To assess the permeability of individual membranes in these bipolar epithelial cells, isolated vesicles from either the basolateral or apical membranes can be employed. This methodology is prevalently utilized for both the identification and functional characterization of AQPs in epithelial membranes.

3.5. Osmotic Swelling Assays

The functionality of AQP-mediated water transport can be probed using X. laevis oocytes in an osmotic swelling assay [56]. Oocytes that are microinjected with AQP mRNA are exposed to hypo-osmotic conditions, and the kinetics of cellular swelling are monitored using video microscopy. To investigate solute permeability, an inwardly directed chemical gradient is established, prompting a solute influx followed by an influx of water, culminating in oocyte swelling [63]. The low intrinsic water permeability and negligible permeability for glycerol and other solutes in oocytes make this system particularly well-suited for AQP studies.

An analogous swelling assay employs erythrocytes, which express native AQPs. In humans, erythrocytes predominantly express a single aquaglyceroporin isoform, AQP3. When subjected to hyperosmotic solute gradients, such as glycerol, the ensuing influx of glycerol provokes erythrocyte swelling, eventually leading to cell hemolysis. This hemolysis can be tracked as a reduction in light absorbance at 625 nm [64]. The rate constant derived from the hemolytic process can be employed to determine glycerol permeability.

4. Functional Dynamics of Aquaporins in the Brain

4.1. Detailed Exploration of AQP4 and AQP1, Emphasizing Their Role in Maintaining Fluid Equilibrium across Neural Compartments

Aquaporin 1, localized in the choroid plexus epithelial cells, and Aquaporin 4, found in ependymal cells as well as glial limitants, are implicated in the regulation of cerebrospinal fluid homeostasis and production. The specific roles played by these individual water channels in these processes are currently under academic debate. Evidence suggests that both AQP1 and AQP4 contribute significantly to CSF production and that a concurrent mutation in both AQP1 and AQP4 genes disrupts CSF drainage and ventricular compliance. Data further underscore the role of AQP4 in extra-choroidal CSF formation, advocating for a critical and sustained balance in CSF production and absorption, facilitated by water flux between brain capillaries and interstitial fluid (ISF). Additionally, findings indicate that AQPs also participate in structural capacities related to CSF homeostasis, including the distensibility of the ventricular system [65].

Cerebrospinal fluid and interstitial fluid form integral parts of the cerebral extracellular milieu, existing in a dynamic equilibrium that bathes neurons and glial cells. This fluid balance, maintained through a tightly regulated exchange between CSF and ISF, is essential for ensuring stable brain volume as well as ionic and solute concentrations for optimal neural and glial function [66–68]. Traditionally, CSF, which is synthesized from plasma modification, has been understood to occupy the brain ventricles and subarachnoid spaces and serve primarily as a mechanical cushion for the central nervous system (CNS). It also provides a route for waste removal from neural tissues. According to the "classic view", the choroid plexus is the principal site for CSF production, with fluid circulating unidirectionally and draining via arachnoid granulations or through the nasal mucosa's lymphatic vessels [66,69].

However, recent advancements have challenged this "classic view", proposing an updated paradigm wherein: (i) extra-choroidal CSF production occurs through fluid exchange between brain capillaries and ISF; (ii) there exists noteworthy paravascular CSF/ISF flow within the brain parenchyma; and (iii) the meningeal lymphatic system plays a significant role in CSF drainage [68,70,71]. Aberrations in CSF homeostasis are strongly correlated with the pathophysiology of various CNS conditions, including hydrocephalus, cerebral edema, and ischemia. Emerging evidence further indicates that diminished CSF production or impaired drainage may contribute to the decline in brain function associated with aging or age-related neurodegenerative disorders [72,73].

Cerebral aquaporins, specifically AQP1 and AQP4, are posited to be crucial regulators of cerebrospinal fluid (CSF) and interstitial fluid (ISF) homeostasis. Given their expression profiles—AQP1 localized solely in the choroid plexus epithelial cells and AQP4 in ependymal cells as well as glial limitants—it has been simplistically theorized that AQP1 is primarily involved in CSF production while AQP4 facilitates CSF/ISF exchange and absorption [74]. Consistent with traditional theories of CSF circulation, AQP1-deficient mice exhibited reduced CSF production and intraventricular pressure, underscoring the importance of AQP1 in CSF genesis [75].

However, a study involving AQP1-null and AQP4-null mice concluded that AQP4, rather than AQP1, is instrumental in mediating water influx into the CSF, suggesting a pivotal role for AQP4 in CSF production [76]. This finding lends credence to the "Bulat–Klarica–Oreskovic hypothesis", which argues for widespread, continuous CSF formation in the brain due to water exchange between brain capillaries and ISF [70]. Moreover, the discovery of the so-called "glymphatic" system by Iliff et al., which facilitates AQP4-dependent paravascular flow between peri-arterial and peri-venous regions, has added further support to the crucial role of AQP4 in regulating CSF homeostasis [71].

Although the specific role of AQP4 within the "glymphatic" system remains a subject of ongoing discussion, there is general agreement that both AQP1 and AQP4 are indispensable for the regulation of cerebral fluid dynamics [67]. However, the exact contributions of AQP1 and AQP4 to various facets of CSF/ISF homeostasis—including CSF production and drainage, CSF/ISF interchange, and ventricular system regulation—remain to be elucidated.

4.2. *Deep Dive into the Interplay between Aquaporins, Cerebrospinal Fluid, and the Intricacies of Brain Fluid Homeostasis*

Aquaporin 1 (AQP1) and Aquaporin 4 (AQP4) serve as key water channel proteins in the central nervous system, each exhibiting unique but complementary expression patterns within cerebral tissues. Both are postulated to be central regulators of cerebral fluid homeostasis under both physiological and pathological conditions. Specifically, AQP1 is localized solely to the apical membranes of choroid plexus epithelial cells, while AQP4 is predominantly expressed in astrocytes and ependymal cells [6].

Traditionally, the choroid plexus has been viewed as the primary site for cerebrospinal fluid (CSF) production, facilitated by AQP1, with the fluid then flowing unidirectionally into the subarachnoid space for eventual reabsorption via arachnoid granulations. However, emerging evidence underscores the significance of extra-choroidal CSF formation, whereby CSF is generated through water filtration from brain parenchymal capillaries, facilitated by AQP4 [76].

In a study that performed in vivo analyses on wild-type, AQP1-null, AQP4-null, and double AQP1-AQP4-null mice to elucidate the specific contributions of AQP1 and AQP4 to critical aspects of CSF homeostasis, such as ventricular volume, CSF outflow dynamics, and ventricular compliance, Magnetic Resonance Imaging (MRI) and Intraventricular Pressure (IVP) measurements indicated that both AQP1 and AQP4 are implicated in CSF production. Interestingly, AQP1-null and AQP4-null mice exhibited comparable reductions in both ventricular volume and IVP, suggesting that choroidal AQP1 and parenchymal AQP4 contribute similarly to CSF formation. Consistent with this, double AQP-null mice displayed further reductions in ventricular volume compared to either single AQP1-null or AQP4-null mutants. Notably, despite the observed reduction in ventricular volume, double AQP-null mice maintained IVP levels comparable to those in single AQP-null mutants. These findings suggest a functional synergy between CSF production, whether facilitated by choroidal AQP1 or astroglial and ependymal AQP4, and ventricular volume to preserve intracranial pressure within homeostatic ranges [65].

The study sheds light on the role of Aquaporin 1 (AQP1) and Aquaporin 4 (AQP4) in cerebrospinal fluid (CSF) drainage dynamics. According to pressure-dependent outflow assessments, neither AQP1-null nor AQP4-null mutants exhibited significant alterations in CSF drainage, corroborating earlier findings from Verkman's group [72]. Interestingly, data from double AQP-null mutants suggest that both channels could play a role, either directly or indirectly, in regulating CSF drainage dynamics. However, further research is needed to pinpoint where and how AQP1 and AQP4 contribute to this process. Some evidence of their involvement comes from hydrocephalus models, but the exact mechanisms remain elusive [77,78]. The recently described AQP4-mediated "glymphatic system" and its interaction with meningeal lymphatics provide additional support for the role of cerebral AQPs in interstitial fluid (ISF)/CSF exchange and drainage [68,71].

Regarding ventricular compliance, data in double AQP-null mice surprisingly indicate that these channels could also influence ventricular distensibility. However, the specific mechanism behind this remains unclear and warrants further experimental exploration. Questions to address include whether this alteration is related to reduced ventricular volume, changes in the biophysical properties of cerebral parenchyma, or systemic changes that secondarily affect cerebral compliance.

In addition to AQP1 and AQP4, another channel, Aquaglyceroporin 9 (AQP9), is present in the brain. It is found in astrocytes at the glial limitans, endothelial cells in pial vessels, and specific neuronal groups [79,80]. The role of AQP9 in CSF dynamics and its impact on cerebral pathologies is a burgeoning area of research that needs further experimental and clinical investigation [81,82].

Importantly, most knowledge about the role of AQP1 and AQP4 in CSF homeostasis comes from animal models. However, growing evidence suggests that these aquaporins serve similar functions in humans and could be implicated in various pathophysiological conditions [83,84]. Additionally, a newly identified aquaporin, AQP11, has been reported to be expressed in the brain [85,86]. Future work should consider whether alterations observed in double AQP-null mice might be due to compensatory mechanisms involving other aquaporins or changes in brain water content.

5. The Glymphatic System: An Essential Framework for Brain Health

5.1. Comprehensive Breakdown of the Glymphatic System's Architecture and Its Functional Significance

The glymphatic system, a groundbreaking discovery in neuroscience, serves as a macroscopic waste clearance pathway that employs a specialized network of perivascular channels created by astroglial cells. This system is essential for the efficient removal of soluble proteins and metabolites from the central nervous system. Beyond its role in waste clearance, the glymphatic system may also assist in the distribution of vital substances such as glucose, lipids, amino acids, and neurotransmitters across the brain. Notably, the system is primarily active during sleep, suggesting that the biological necessity for sleep may stem from the brain's need to eliminate potentially harmful waste products, including β-amyloid [87].

Recent research indicates that cerebrospinal fluid (CSF) and interstitial fluid (ISF) engage in a continuous exchange, facilitated by convective influx along the periarterial space [71]. Driven by factors such as arterial pulsatility, respiratory rhythms, and pressure gradients, CSF flows from the subarachnoid space into the Virchow-Robin spaces. The perivascular space acts as a low-resistance pathway for CSF influx. The migration of CSF into the intricate brain parenchyma is enabled by Aquaporin 4 (AQP4) water channels, which are highly polarized and localized in the astrocytic endfeet that envelop the brain vasculature [71,88].

This influx of CSF into the brain parenchyma triggers convective fluxes of ISF within the tissue, leading it toward perivenous spaces adjacent to large, deep veins. Eventually, the ISF is gathered in these perivenous spaces and drains towards the cervical lymphatic system [89,90]. The system of convective fluid fluxes, characterized by its rapid exchange between CSF and ISF, has been dubbed the glymphatic system. This name reflects its functional similarities to the peripheral lymphatic system and emphasizes the critical role of glial AQP4 channels in facilitating fluid transport.

In 2012, the inner workings of the glymphatic system were detailed for the first time through in vivo studies using two-photon microscopy in mice [71]. Iliff and colleagues marked cerebrospinal fluid (CSF) with fluorescent tracers injected into the cisterna magna, which allowed them to observe that the CSF rapidly enters the brain along cortical pial arteries. This influx was followed by its movement into the Virchow-Robin spaces along penetrating arterioles. Importantly, it was found that the tracers in the CSF were not diffused randomly throughout the brain tissue. Instead, they entered via a specific periarterial pathway that is closely aligned with vascular smooth muscle cells and bounded by the perivascular astrocytic endfeet. Further ex vivo studies showed that these tracers exited the brain mainly along central deep veins and lateral-ventral caudal rhinal veins [71].

Subsequent analyses revealed that this directed movement of CSF via periarterial pathways into the brain parenchyma assists in clearing out interstitial solutes, directing them towards perivenous drainage channels. This finding has significant implications for neurodegenerative diseases, such as Alzheimer's, which are characterized by protein accumulations, including amyloid plaques and tau tangles [91,92].

To investigate the role of the glymphatic system in clearing β-amyloid, a hallmark of Alzheimer's disease, Iliff et al. injected fluorescent or radiolabeled amyloid β1–40 into the striatum of mice. Their observations confirmed that β-amyloid was rapidly eliminated from the brain via the glymphatic system's paravenous efflux pathways [71].

Moreover, imaging studies in mice lacking Aquaporin 4 (AQP4) showed significant disruptions in the glymphatic system. Specifically, there was approximately a 65% reduction in the flux of CSF through the brain tissue when compared to wild-type control mice. Additionally, the clearance rate of intrastriatally-injected radiolabeled β-amyloid was reduced by 55% [71]. These findings led to the proposition that the AQP4-mediated glymphatic pathway serves as a crucial mechanism for the removal of interstitial fluid solutes, including waste products, from the brain's parenchyma [88].

5.2. AQP4-Centric Discussion on the Glymphatic Pathway, Detailing How Aquaporin Malfunctions Might Hinder the System

Convective CSF Fluxes in Aging and Pathology

Glymphatic activity decreases sharply during aging-Recent research has shown a striking decrease in glymphatic function in older mice compared to their younger counterparts, with an estimated 80–90% reduction in activity [93]. This decline encompasses both the influx of CSF tracers into the brain and the clearance of radiolabeled β-amyloid and inulin. It has been suggested that reactive gliosis, characterized by the hypertrophy of GFAP+ astrocyte processes, may be a contributing factor to this age-related decline in glymphatic function. However, the exact mechanisms linking changes in GFAP expression to diminished glymphatic activity remain unclear [94].

In younger animals, Aquaporin 4 is localized to the astrocytic endfeet and plays a pivotal role in facilitating the exchange of cerebrospinal and interstitial fluid along periarterial influx pathways (Figure 2), as well as aiding in the clearance of interstitial solutes through perivascular drainage paths. Previous research has shown that the genetic deletion of AQP4 leads to a roughly 65% impairment in CSF-ISF exchange and a 55% reduction in the clearance of β-amyloid [95]. In aging brains, there is a partial loss of this vascular polarization of astrocytic AQP4 (Figure 2). Specifically, AQP4 is no longer solely confined to the astrocytic endfeet but is also found in the parenchymal processes of astrocytes [96].

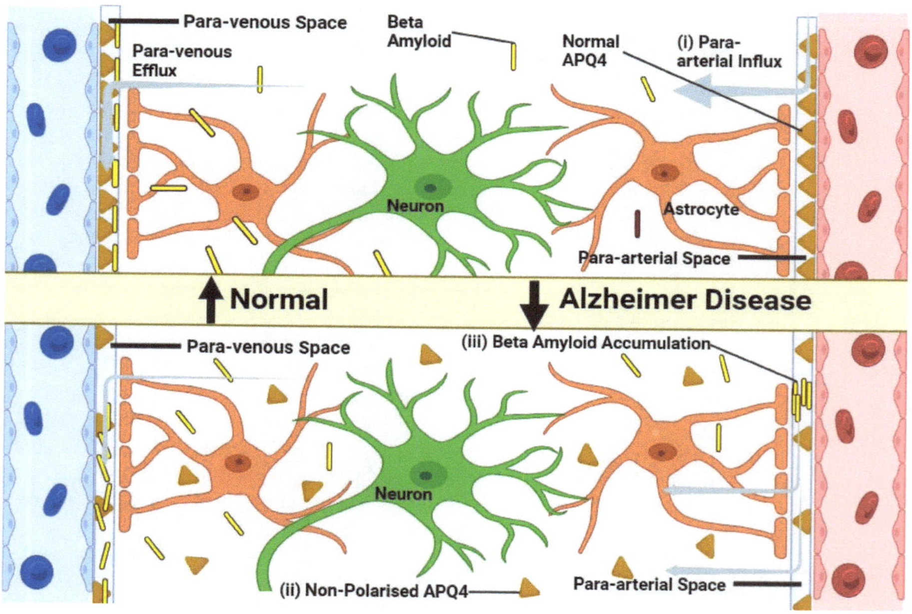

Figure 2. Model of glymphatic function in young, old and in Alzheimer's disease. (i) CSF efficiently clears brain solutes in young, healthy individuals via periarterial pathways. (ii) Aging disrupts glymphatic function, possibly due to reactive astrocytes and AQP4 de-polarization. (iii) In Alzheimer's, β-amyloid accumulates in perivascular spaces, potentially due to glymphatic impairment, further hindering waste clearance.

This loss of AQP4 polarization in older brains suggests that the age-related decline in glymphatic function could be partially attributed to dysregulation in astroglial water transport. Other factors that might contribute to the decrease in glymphatic activity with age include a 66% reduction in CSF production and a 27% decrease in CSF pressure [97].

Additionally, the arterial walls become stiffer with age, leading to a decrease in arterial pulsatility, which is one of the key drivers of glymphatic influx [98].

The observed decline in glymphatic function due to aging is especially significant given that age is the most significant risk factor for neurodegenerative diseases. A compromised glymphatic system in older individuals could potentially lead to the accumulation of misfolded and hyperphosphorylated proteins, thereby making the brain more susceptible to neurodegenerative diseases or even accelerating the progression of cognitive decline.

5.3. Correlation between Compromised Glymphatic Functionality and Neurodegenerative Disorders

Traumatic brain injuries (TBIs), commonly observed in military personnel and athletes, are associated with an elevated risk of premature dementia and Alzheimer's disease [99]. Numerous studies have indicated that both recurring traumatic incidents, as well as single events of moderate to severe head trauma, can result in ongoing neurodegeneration. The mechanism by which only certain individuals develop chronic traumatic encephalopathy after experiencing a similar degree of initial brain injury remains elusive [100]. TBIs result in the release of β-amyloid peptide and C-tau, a proteolytic derivative of MAP-tau, which is a prominent intracellular microtubule protein in axons. Notably, C-tau, due to its vast release correlating with TBI severity, serves as a brain injury biomarker [101]. A prevailing theory posits that significant surges in interstitial tau can lead to its cellular intake, initiating fibrillary aggregates. These aggregates attract more tau, fostering the formation of neurofibrillary tangles and facilitating a prion-like progression of the disease. Moreover, TBIs are connected to the development of extensive astroglial scars and the sustained activation of inherent neuroinflammation. In a study focusing on repetitive moderate TBIs, the influx of cerebrospinal fluid (CSF) into the brain was compromised in the ipsilateral hemisphere from the first day post-injury, and this reduced glymphatic functionality endured for at least 28 days post-injury. This significant decline in glymphatic activity was linked to glial scars marked by enlarged GFAP-positive processes in the ipsilateral hemispheres. Moreover, an AQP4 mislocalization from the vascular endfeet to parenchymal processes was observed, paralleling the AQP4 misplacement seen in aging processes [93]. Through intracortical injections of human tau, Iliff et al. managed to monitor the tau clearance pathway, noting that human tau congregated around large veins [97]. The residual tau in the tissue was in line with diminished glymphatic clearance, underscoring the pivotal role of CSF-mediated tau removal through glymphatic routes in mitigating subsequent neuronal damage post-TBI. Another investigation employing magnetic resonance imaging (MRI) to evaluate glymphatic function revealed that head traumas, such as subarachnoid hemorrhages, considerably weaken glymphatic function [102]. Introducing freshly isolated arterial blood into the CSF inhibited the CSF's influx pathways into the brain, excluding the cerebellum, indicating that cerebral hemorrhages can extensively suppress glymphatic functionality. In this subarachnoid hemorrhage model, the use of a tissue-type plasminogen activator, which eliminates fibrin clots, enhanced glymphatic perfusion. An embolic ischemic stroke temporarily hindered glymphatic flow shortly post-ischemia; however, functionality was naturally restored 24 h post-transient ischemia [102], suggesting that brief glymphatic function interruptions, resulting from reduced arterial pulsation or mild stroke-induced perivascular pathway blockages, can be reversible and contribute to enhanced recuperation.

6. Implications of Aquaporins in Degenerative and Acute Brain Pathologies

6.1. Profiling Each Degenerative Disease (iNPH, PD, AD) and Its Associated Aquaporin Dysregulations

6.1.1. Parkinson's Disease

Neurodegenerative diseases impact millions globally, constituting a diverse set of disorders that specifically target certain regions of the Central Nervous System (CNS). These diseases typically result in a sustained decline in cognitive or motor functions, contingent upon the distinct type of neuronal cells undergoing selective degeneration [103]. Predominantly, these pathological states are correlated with aging. Indeed, as life expectancy has risen over recent decades, the prevalence of these age-related disorders has correspondingly increased [104]. Neuronal harm and oxidative stress, which are fundamental events in the onset of these conditions, precipitate an upsurge in the production of pro-apoptotic and pro-inflammatory cytokines. Concurrently, there's a disruption in the balance of water, extracellular ions, and amino acid neurotransmitters. Given the perturbed brain homeostasis observed in many of these diseases, some researchers have postulated a potential involvement of the Aquaporin (AQP) protein family in the pathogenic mechanisms underlying these conditions [105].

6.1.2. Dopamine Regulation of AQP4 Expression

In the realm of neurodegenerative diseases, a pivotal role is played by neural stem cells. Throughout adulthood, these cells have the capability to proliferate and differentiate into either new neurons or glial cells [106]. Intriguingly, studies have highlighted that adult neural stem cells display glial-associated properties in both in vivo and in vitro environments. As a testament to this, these stem cells express GFAP, a protein considered a hallmark for fully differentiated astrocytes [107]. Notably, alterations in the count of GFAP-expressing cells have been implicated in neurodegenerative disorders, including Parkinson's disease [108].

Dopamine (DA), a central neurotransmitter, has been found to invigorate the proliferation of progenitor cells. This effect is observed not just in the striatum but also in the subventricular zone of mature brains. Building on this, a recent investigation led by Kueppers et al. postulates that DA orchestrates the proliferation of striatal astrocytes in culture through the mediation of Aquaporin-4 [109]. Their results delineate a scenario wherein DA prompts a reduction in AQP4 expression in striatal glial cells when studied in vitro.

However, the narrative surrounding AQP4's role in cellular proliferation is replete with ambiguities. Saadoun et al. observed a stable rate of proliferation in astrocytes cultured from AQP4-deficient transgenic mice. Contrasting this, Nicchia et al. identified a pronounced ~70% decline in cultured astrocyte numbers after the cells were subjected to short interference RNA (siRNA) treatments targeting AQP4 [110,111]. This underscores the necessity for in-depth in vivo lesion studies to validate these findings. Furthermore, understanding AQP4 expression in a damaged striatum is imperative, especially given observations of elevated AQP4 mRNA levels in the substantia nigra following a 6-hydroxydopamine (6-OH-DA) lesion [112].

These findings coalesce to underline the hypothesis that DA can modulate astrocyte proliferation. This strengthens the assertion that neurodegenerative diseases, where dopaminergic transmission is disrupted (such as PD), are inextricably linked with alterations in astrocyte proliferation. Conclusively, these insights open up the potential therapeutic avenue of modulating AQP4 as an intervention strategy in treating PD.

6.1.3. Mitochondrial AQP9 in PD Brains

Within the scope of neurodegenerative disorders, a potentially significant yet conjectural association has been proposed between Aquaporin-9 (AQP9) and Parkinson's disease [113]. Within the cerebral environment, AQP9—a channel facilitating the movement of water and certain solutes—is manifested in astrocytes, brain stem catecholaminergic neurons, as well as specific subgroups of midbrain dopaminergic and hypothalamic neurons. An intriguing observation is the pronounced presence of AQP9 within the mitochondrial inner membranes, hinting at its potential role in supporting neuronal metabolism. Especially compelling is the notion that aberrations in mitochondrial AQP9 within dopaminergic neurons might be linked to the heightened susceptibility of these neurons to PD [113].

Given the presumptive significance of mitochondrial AQP9, Yang et al. embarked on an exhaustive exploration to discern the putative functional repercussions of its expression [114]. Their investigations centered on determining the transport functionality of AQP9 within mitochondrial inner membranes. Through empirical assessments, they gauged the permeabilities of water and glycerol in brain mitochondria, juxtaposing these with measurements derived from tissues devoid of AQP9 expression [115]. Surprisingly, their findings indicated no discernible variances in water or glycerol permeabilities across different mitochondrial sources. These findings, in essence, cast doubts on the postulated role of aquaporins within the mitochondrial environment. Despite these insights, whether AQP9's expression and activity can be harnessed as therapeutic avenues to augment PD treatment remains an open-ended inquiry (Figure 3).

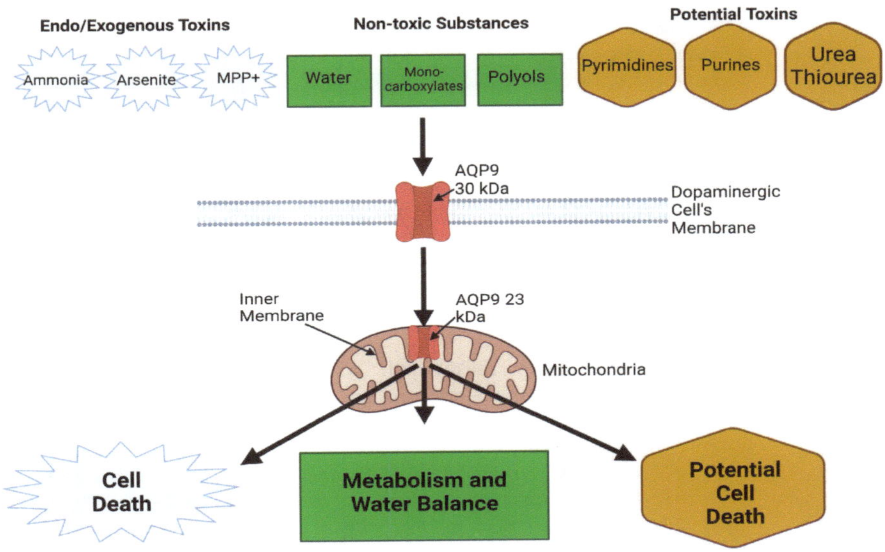

Figure 3. Suggested mechanisms for AQP9-linked dopaminergic cell death and their physiological roles.

6.1.4. Idiopathic Normal Pressure Hydrocephalus (iNPH)

Idiopathic normal pressure hydrocephalus (iNPH) is recognized as a distinct form of dementia, wherein intervention through cerebrospinal fluid diversion can yield favorable outcomes. Pioneering studies leveraging magnetic resonance imaging with CSF tracers have highlighted compromised CSF tracer clearance from specific cerebral regions, notably the entorhinal cortex, in iNPH patients. This compromised clearance mechanism, especially for waste solutes such as soluble amyloid-β, might be central to the neurodegenerative processes and cognitive decline characterizing iNPH.

The primary objective of the study at hand was to investigate potential alterations in the subcellular localization of aquaporin-4 water channels in relation to iNPH. Notably, AQP4 is known to play a pivotal role in modulating CSF flow and facilitating the glymphatic removal of brain metabolites. To accomplish this, cortical brain biopsy samples from 30 iNPH patients and 12 control subjects were scrutinized using AQP4 immunogold cytochemistry.

Utilizing electron microscopy, the study unveiled a markedly diminished presence of AQP4 water channels within the astrocytic endfoot membranes juxtaposed to cortical microvessels in iNPH patients, relative to the control group. Further analysis divulged a statistically significant association between AQP4 densities oriented perivascularly and those directed towards the parenchyma. However, the decline in AQP4 densities oriented towards the parenchyma, while observed, was not statistically significant in iNPH cases.

The findings suggest that iNPH is associated with a suppressed perivascular expression of AQP4. Such diminished AQP4 presence might adversely impact the glymphatic circulation, hindering efficient waste removal and potentially fostering neurodegeneration.

Therefore, strategies aimed at reinstating the optimal perivascular distribution of AQP4 could be postulated as innovative therapeutic avenues for addressing iNPH [116].

6.2. *Insights into Aquaporin Behavior during Acute Cerebral Events, Such as Stroke or Traumatic Injuries*

Aquaporins, water channel proteins found in the brain, have attracted substantial research attention given their potential implications in both physiological and pathological contexts [117]. Among these, AQP4 has been the focal point of numerous studies, especially in relation to various brain conditions, spanning acute afflictions such as stroke and traumatic brain injury to chronic autoimmune neurodegenerative ailments. As of now, there are no targeted therapeutic interventions specifically designed to modulate the water transport activity of these channels. However, accumulating experimental data underscores the pivotal nature of AQPs, suggesting their profound relevance in future research endeavors. For instance, curtailing water channel activity early on might ameliorate edema development in instances of brain injuries, while during the later stages of certain diseases, AQPs play a crucial role in facilitating water clearance from the brain to blood vessels [118].

A predominant characteristic of many brain diseases, ranging from stroke, traumatic brain injuries, and brain tumors to inflammation, is edema. This phenomenon denotes the accumulation of water due to imbalances in brain osmotic homeostasis. A primary ramification of edema is brain swelling, which can exacerbate secondary complications, such as compromised brain perfusion. Despite its long-standing recognition in both clinical and pre-clinical realms, the molecular and cellular intricacies underpinning edema genesis and resolution remain relatively elusive. Furthermore, current therapeutic interventions fall short of effectively curbing edema onset or progression in various brain conditions. Thus, the identification of brain AQPs provided a glimmer of optimism in the quest for novel therapeutic approaches to counteract edema. The insights gleaned over the past decade and a half regarding the potential of AQPs as therapeutic targets for edema will be delineated, followed by a brief overview of the initial phases of edema formation [117].

6.2.1. Edema Build-Up Phase: Anoxic, Ionic and Vasogenic Edema

Cerebral edema, for the past four decades, has been predominantly categorized into two principal types: cytotoxic and vasogenic [119]. Conventionally, cytotoxic edema denotes the accumulation of intracellular water without any disruption of the blood–brain barrier (BBB). In contrast, vasogenic edema emerges post-BBB disruption, instigating protein diffusion from the bloodstream into the tissue, subsequently leading to water buildup in the extracellular matrix. This longstanding bifurcation, however, is now considered an oversimplification, particularly in light of recent insights into the molecular shifts during edema onset and the evolving understanding of BBB properties. Instead, with respect to vascular brain injuries, edema's initiation phase can be more accurately segmented into three primary categories: anoxic, ionic, and vasogenic edema [120].

Anoxic edema is typified by the immediate swelling of astrocytes and neuronal dendrites ensuing moments after a deprivation of oxygen and glucose, particularly in cerebrovascular ailments. This deficiency in essential nutrients triggers significant perturbations in cellular ionic gradients due to the non-operational energy-dependent co-transporters. Consequently, this facilitates an extensive ion influx into cells, a manifestation of which is a gradual surge in the extracellular K+ levels, leading to subsequent water entry and resultant swelling initially in astrocytes and later in neuronal dendrites [121]. Rapidly following this, anoxic edema transitions into ionic edema. The depletion of vital nutrients similarly impacts the ionic gradients in endothelial cells, manifesting in changes such as transcapillary sodium flux leading to tissue edema. This endothelial distress results in an initial transient BBB leakage observed in conditions such as stroke and traumatic brain injuries (TBI) [122]. Such changes culminate in enhanced water influx through endothelial cells, leading to brain swelling, as exemplified in stroke models observed within the first 30 min of reperfusion, coupled with further intensified BBB permeability [123]. The ensuing vasogenic edema is characterized by heightened permeability to plasma proteins such as albumin, stemming from a comprehensive disruption of endothelial tight junctions, extracellular matrix degradation, and potential augmentation in transendothelial cell transport via the transcytosis mechanism [119].

However, it is pivotal to highlight that such delineation primarily pertains to brain injuries characterized by acute cerebrovascular anomalies, and might not be wholly applicable to other neurological conditions, such as brain tumors [124]. This underscores the importance of recognizing that clinical interventions focusing exclusively on osmotic challenges might not effectively treat cerebral edema. The rationale behind this is the intricate and varied molecular mechanisms steering edema development. While the precise functionalities of cerebral AQPs in this context remain a topic of active research, their strategic localization and inherent nature as water channel proteins suggest their significant implications in cerebral edema dynamics. The subsequent sections delve into the roles of AQPs in the context of this refined classification of edema formation.

6.2.2. Contribution of AQPs in Edema Formation and Resolution

Aquaporins, specifically AQP1, 4, and 9, have demonstrated alterations in their expression levels in various brain disorders, as seen in both rodent models and human specimens [125]. Of these, AQP4 has been the primary focus of many investigations, as its expression patterns appear to mirror the progression of edema in numerous neurological conditions [126]. A significant advancement in the exploration of AQP4's role in edema came with the creation of AQP4 knockout mice (AQP4−/−) by Dr. Verkman's team [127].

Interestingly, these AQP4−/− mice did not show any substantial structural or physiological deviations. However, a notable observation was the expansion of their extracellular space by roughly 20% in comparison to their wild-type (WT) counterparts [128]. The outcomes presented by Dr. Verkman's group generated a hypothesis suggesting a dual functionality for AQP4 in the context of edema: AQP4 could exacerbate edema during its formation phase, but conversely, it might play a beneficial role during the edema resolution phase. Yet, the lack of specific pharmaceutical agents capable of acutely and selectively inhibiting AQP4 made it challenging to validate this dual-role theory post-brain injury. It was only recently that this hypothesis was put to the test in vivo, utilizing a siRNA methodology.

6.2.3. AQP4 and Edema Build-Up

The role of Aquaporin-4 in cerebral edema is complex, exhibiting a spectrum of expression profiles across various pathological conditions including traumatic brain injury, ischemic stroke, and subarachnoid hemorrhage. These conditions each display unique fluctuations in AQP4 expression levels [124,126,129]. Specifically, in ischemic models involving transient middle cerebral artery occlusion, AQP4 is acutely up-regulated in the astrocyte endfeet adjoining blood vessels, reaching peak levels approximately 1 h post-stroke onset. This elevation in AQP4 expression is spatially and temporally correlated with the extent of cerebral edema and is most prominent in the peri-infarct region as well as the prospective lesion site [123,126].

This initial surge in AQP4 is implicated in the genesis of ionic cerebral edema and astrocyte swelling. Intriguingly, in more severe ischemic conditions, this up-regulation is not observed, leading to the hypothesis that under extensive tissue stress, the brain may lack the capacity for rapid AQP4 synthesis during the early reperfusion phase. Furthermore, the ischemic conditions induce a shift in the ratio of AQP4 isoforms, AQP4-m1 and AQP4-m23, suggesting a potential disruption of orthogonal arrays of particles (OAPs). The functional implications of these alterations, however, remain to be elucidated [130].

The modulation of AQP4 expression is influenced by a multitude of factors, including injury severity, model system, and age, further complicating the interpretation of its role in edema formation [129]. Genetic studies utilizing AQP4 knockout mice (AQP4−/−) have also yielded conflicting outcomes. For example, AQP4 deletion conferred protective effects in spinal cord injury models, characterized by reduced edema and lesion size in the acute phase [131]. Contrarily, another study observed improved functional outcomes in wild-type mice relative to AQP4−/− mice in long-term follow-up post-spinal cord injury [132]. Such discrepancies point to the limitations inherent in employing AQP4−/− mice as a tool to ascertain AQP4's pathophysiological roles.

In recent advancements, small interfering RNA (siRNA) targeting AQP4 (siAQP4) has been developed to transiently down-regulate AQP4 expression and has been shown to reduce water mobility [133,134]. In juvenile traumatic brain injury models, this targeted molecular intervention was associated with reduced edema and cognitive improvement at two months post-injury, suggesting siAQP4 as a potential therapeutic modality [134]. Despite these developments, the role of AQP4 in edema formation remains complex and not fully understood. Overall, findings indicate that AQP4 may have a dual function, either exacerbating or mitigating edema, depending on a variety of factors, including the specific pathological context.

6.2.4. Edema Resolution in Acute Brain Disease: Role of AQP in Water Clearance

The hypothesis that Aquaporin-4 has a dual role in cerebral edema—being deleterious during edema formation while beneficial during its resolution—has gained empirical support despite the absence of conclusive evidence. Initial evidence for this comes from experiments using AQP4 knockout mice (AQP4−/−), which revealed that intracranial pressure increased significantly when a saline solution was infused into the brain parenchyma compared to wild-type mice [8]. Furthermore, multiple studies utilizing magnetic resonance imaging (MRI) have demonstrated that elevated AQP4 expression is temporally correlated with the resolution of edema in various pathological states, including stroke, traumatic brain injury (TBI), and neuroinflammatory lesions [119,133,135].

Specifically, AQP4 expression typically escalates 48 h post-insult in models of stroke, TBI, and neuroinflammatory lesions. This augmented expression is frequently localized to the astrocyte endfeet adjacent to blood vessels, as well as the astrocyte processes and glia limitans [126,129]. Such spatial distribution of heightened AQP4 expression suggests its potential role in facilitating the clearance of edematous fluid via the subarachnoid space. For instance, in juvenile traumatic brain injury (jTBI) models, elevated levels of AQP4 in the glia limitans appear to counterbalance water accumulation at one- and three-days post-injury (as indicated by higher T2 values), leading to a normalization of both AQP4 and T2 levels by day seven [129].

In neuroinflammatory lesion models in rats, apparent diffusion coefficient (ADC) time course studies revealed a bifurcation in AQP4 expression patterns, distinguishing its minor elevation during the edema formation phase from its pronounced increase during the edema resolution phase. Importantly, peak AQP4 expression levels coincided with significantly elevated ADC values, providing further support for its role in edema resolution.

The role of other aquaporin family members, such as Aquaporin-9, in the resolution of cerebral edema warrants exploration. While AQP9 is up-regulated in reactive astrocytes seven days post-ischemic injury along the infarct border, its expression pattern does not show a strong correlation with the extent of cerebral swelling, in contrast to AQP4 [136]. Moreover, this alteration in AQP9 expression is not universally observed across various models of brain pathology. In gerbil models of ischemic stroke, AQP9 is expressed in hippocampal pyramidal neurons in regions CA1, CA2, and CA3 as early as 6 h post-stroke onset, a pattern distinct from other stroke models. Taken together, these observations suggest that AQP9 may have a limited role in edema regulation following brain injury. Notably, pyramidal neurons typically do not express this channel under physiological conditions, suggesting that its expression may be induced by metabolic stressors. Despite this, the functional implications of increased AQP9 expression in neurons and astrocytes remain unclear. A possible role for AQP9 in astrocyte energy metabolism has been suggested, indicating that enhanced AQP9 expression could facilitate glycerol utilization as an alternative metabolic substrate, thereby potentially aiding in neuronal recovery [126].

In contrast to AQP9, AQP4 appears to collaborate with other astrocytic proteins such as Connexin-43 (Cx43) and the potassium channel Kir4.1 in managing cerebral edema. Cx43 forms gap junctions between astrocytes and plays a critical role in solute and water diffusion within the astrocytic network. Intriguingly, Cx43 expression is down-regulated when AQP4 is silenced using RNA interference in primary astrocyte cultures, which could potentially impair astrocytic connectivity. Furthermore, Kir4.1 is co-localized with AQP4 in the astrocyte endfeet and has been implicated in potassium reuptake impairment in AQP4 knockout mice in epilepsy models. Notably, potassium flux has been associated with water movement during astrocyte swelling in spinal cord injury models [137–139]. Thus, it is plausible that AQP4, Cx43, and Kir4.1, all expressed in astrocytes, may function synergistically in the regulation and resolution of cerebral edema following brain injury.

6.2.5. Chronic Changes of Brain AQP: Relation with Water Homeostasis Dysfunction?

Aquaporins, traditionally implicated in water homeostasis both in physiological and pathological contexts, have been increasingly associated with a broader range of cellular functions, including cell migration and gas diffusion [140]. Specific isoforms, such as AQP1, AQP4, and AQP9, manifest elevated expression in both brain tumors and peritumoral tissues. The augmented presence of these AQPs in astrocytes within peritumoral regions may be related to edema formation, potentially due to altered tissue homeostasis and elevated metabolic rates. Additionally, these AQPs may facilitate gas diffusion, playing a role in the clearance of excess CO_2 and the diffusion of O_2. In rodent models of spinal cord injury (SCI), an upregulation of AQP1 is observed in astrocytes as well as neurons. AQP1 expression has also been documented in neuronal processes in the dorsolateral septum at various time points following juvenile traumatic brain injury (jTBI) [129]. While astrocytic expression of AQP1 may be associated with cerebrospinal fluid secretion during cyst formation, its co-localization with growth-associated protein-43 (GAP-43) in neurons suggests a role in neuroplasticity and repair after injury. Furthermore, AQP1 knockout mice exhibit reduced pain responses, highlighting its potential relevance to the cognitive pathways associated with pain sensation, although a consensus regarding AQP1's role in pain processing has yet to be established [141].

Longitudinal changes in AQP4 expression have also been noted; its expression remains elevated up to 28 days in the perilesional area following stroke and SCI [142]. AQP4 has been implicated in astrocyte migration, particularly in the formation of the glial scar, by facilitating the water influx necessary for filopodia formation and promoting cellular adhesion among astrocytes [143]. This heightened expression of AQP4 in astrocytes is likely to contribute to glial scar formation, aiding in the establishment of a new tissue barrier at the borders of post-injury cavities.

In the context of chronic neurodegenerative diseases such as Alzheimer's disease, as well as post-jTBI, a reduction in AQP4 expression has been observed around blood vessels in association with beta-amyloid deposition. This decrease in AQP4 levels could potentially disturb water homeostasis within the neurovascular unit, subsequently attenuating perivascular flow and impeding the effective clearance of toxic substances, such as beta-amyloid, from the brain tissue [71].

6.3. The Cascading Effects of Altered AQP Expression in Autoimmune Conditions, with a Focus on NMO

Aquaporins are not just pivotal in regulating water homeostasis; they also have roles in the immune system, specifically within both the innate and adaptive arms. In human blood leukocytes, AQP1 and AQP9 are expressed and show upregulation upon stimulation with lipopolysaccharide (LPS) either intravenously or in vitro [144]. AQP9 expression is also elevated in activated polymorphonuclear leukocytes in patients suffering from systemic inflammatory response syndrome (SIRS) and infective endocarditis [145].

B and T lymphocytes, key players in adaptive immunity, have been found to express AQP1, AQP3, and AQP5. Similarly, immature dendritic cells (DCs), which are integral to the innate immune system, express AQP3 and AQP5. In these immune cells, AQP expression is often correlated with their activation and proliferation. Notably, AQP9 is the most highly expressed isoform in DCs and shows further upregulation upon LPS stimulation. In a mouse model of induced colitis with AQP9 knockout (AQP9-KO), the absence of AQP9 did not offer complete protection against colitis-associated inflammation but did diminish the DC-mediated inflammatory response [146].

Human primary blood-derived macrophages and neutrophils, critical components of the innate immune system, display high levels of AQP9, which also sees upregulation at both transcript and protein levels when stimulated with LPS [147]. AQP3, another isoform, is also sensitive to LPS stimulation in monocytic THP-1 cells, a model often used to study inflammation. The inhibition or silencing of AQP3 in these cells leads to partial blockage of LPS priming and a reduction in the production of key inflammatory cytokines such as interleukin-6 (IL-6), pro-IL-1β, and tumor necrosis factor-alpha (TNF-α). This implicates a functional connection between AQP3 and Toll-like receptor 4 (TLR4) during the priming of macrophages [147].

Moreover, a separate study on THP-1 cells revealed an increase in AQP1 expression after LPS treatment, while AQP5 mRNA levels decreased [148]. Elevated AQP9 expression was also found in neutrophils in SIRS patients compared to healthy controls [149].

AQPs appear to play multifaceted roles in the immune response, particularly during inflammation. They are not only expressed in various immune cells but also undergo regulation in response to inflammatory stimuli, as summarized in Table 1. This highlights their potential significance in understanding and possibly treating inflammatory conditions.

Table 1. Regulation of immune-related AQPs during inflammation.

Gene	Species	Immune Cells	Stimuli	Regulation	References
AQP1	Human	Leucocytes	LPS	Upregulation	[144]
	Human	Monocytic THP-1 cells	LPS	Upregulation	[148]
AQP3	Human	Leucocytes	Sepsis	Downregulation	[144]
	Human	Monocytic THP-1 cells	LPS	Upregulation	[147]
AQP5	Human	Monocytic THP-1 cells	LPS	Downregulation	[148]
AQP7	Mouse	Macrophages		Unknown	[150]
AQP9	Human	Leucocytes	SIRS	Upregulation	[149]
	Mouse	Dendritic cells	LPS	Upregulation	[146]
	Human	Macrophages	*Pseudomonas aeruginosa*	Upregulation	[151]
	Human	Leucocytes	LPS	Upregulation	[152]
	Human	Monocytes	LPS	Upregulation	[147]
	Mouse	Macrophages		Unknown	[150]

Neuromyelitis optica spectrum disorders (NMOSD) are a range of inflammatory demyelinating diseases (IDDs) that primarily affect the optic nerves and spinal cord but can also extend to the brain and, in rare instances, muscles. Brain lesions in NMOSD often localize to areas with high AQP4 expression, such as the circumventricular organs (responsible for intractable nausea and vomiting) and the diencephalon (linked to sleep disorders, endocrine imbalances, and the syndrome of inappropriate antidiuresis). Up to 10% of NMOSD patients even fulfill the Barkoff criteria for multiple sclerosis when evaluated through MRI [153].

One of the hallmarks of NMOSD is the presence of autoantibodies against AQP4, known as AQP4-IgG or NMO-IgG, detected in 60–90% of NMO patients [154,155]. AQP4 is not only found on the astrocytes in the central nervous system (CNS) but also in skeletal muscle and various epithelial cells such as those in the kidney, stomach, and exocrine glands. AQP4-IgG was initially perceived as a mere marker of the disease, possibly related to astrocyte damage. However, mounting evidence now suggests that AQP4-IgG plays a pathogenic role in NMO.

When AQP4-IgG binds to AQP4 on astrocytes, this initiates a cascade of immunological responses, primarily through complement-dependent cytotoxicity. This leads to the invasion of leukocytes into the CNS, the release of cytokines, and the disruption of the blood–brain barrier. These series of events are thought to culminate in the death of oligodendrocytes (cells responsible for myelination), resulting in myelin loss and, ultimately, neuronal death. This cascade explains the neurological deficits observed in patients with NMO.

Given the rapidly evolving knowledge on the immunobiology of AQP4 autoimmunity, there is a growing need for ongoing revisions in the diagnostic criteria for NMOSD [153]. As our understanding broadens, the role of highly specific assays that can detect pathogenic AQP4-IgG targeting the extracellular domains of AQP4 becomes increasingly crucial.

The size of AQP4-IgG (autoantibodies against AQP4) presents a steric challenge when it comes to binding with AQP4, which is a tetramer consisting of four separate monomers and, by extension, four distinct water pores. This size mismatch suggests that it would be unlikely for a single AQP4 tetramer to bind with more than one AQP4-IgG molecule. Therefore, significant inhibition of water permeability by AQP4-IgG appears to be theoretically implausible.

Backing this notion, despite extensive research efforts, no small-molecule inhibitors of AQP4 have been identified to date. Several cell culture studies have concluded that AQP4-IgG does not inhibit the water permeability of AQP4 [156,157]. In experiments that used a stopped-flow light scattering method to evaluate AQP4 water permeability, neither high concentrations of serum from multiple NMO patients nor monoclonal NMO antibodies were found to significantly affect AQP4 function [158]. This approach, involving plasma membrane vesicles from AQP4-expressing cells and AQP4-reconstituted proteoliposomes, is sensitive enough to detect variations in water permeability less than 5% and is not influenced by factors such as internalization, unstirred layers, or ion/solute transport.

However, it is worth mentioning that there is one conflicting study that utilized a time-to-lysis assay on Xenopus oocytes and claimed that AQP4-IgG inhibits AQP4 water permeability. This method is considered an inaccurate surrogate for measuring osmotic water permeability [159].

Except for this one conflicting study by Hinson et al. [159], the prevailing evidence supports the idea that AQP4-IgG does not inhibit the water permeability of AQP4. Given this, the role of AQP4-IgG in the pathology of Neuromyelitis Optica Spectrum Disorders may lie elsewhere, possibly in initiating immunological cascades that lead to tissue damage rather than directly affecting water transport through AQP4.

7. Aquaporins at the Intersection of Oncology and Neurology

7.1. Unraveling the Possible Links between Aquaporin-Mediated Processes and Brain Tumorigenesis

AQP1, primarily known for its role in water transport, has additional functions that are intriguing both scientifically and medically. One of these is its capability to act as a cyclic nucleotide-gated cation channel activated mainly by cGMP and, to a lesser extent, by cAMP [160]. Research by Yu and colleagues suggests that the interaction of cGMP with an arginine-rich cytoplasmic Loop D in AQP1 leads to a conformational change that may mediate the gating of its central ion channel [161,162].

Interestingly, AQP1 overexpression has been documented in a variety of human cancers, including those of the biliary duct, bladder, brain, breast, cervix, colon, lung, nasopharynx, and prostate [163,164]. Specifically in the case of colon cancer, Moon and colleagues have shown that AQP1 is expressed in colonic adenoma and both primary and secondary colon cancer, but is absent in normal colonic mucosa [164]. This suggests that AQP1 may play a role in the early stages of colon cancer development. Additionally, the expression levels of AQP1 have been found to correlate with key clinical prognostic indicators such as histological grade, lympho-vascular invasion, and nodal involvement [163,165].

The link between AQP1 and cancer progression has garnered significant attention, leading to numerous reviews on the subject. These reviews often focus on the potential of AQP inhibitors as therapeutic agents in cancer treatment [166,167]. Given the multifaceted roles of AQP1, from water transport to ion channel gating and its association with various types of cancer, the protein appears to be a critical player in both physiology and pathophysiology, including tumorigenesis. As such, understanding its function and regulation could offer valuable insights into the development of novel therapeutic approaches for a range of diseases, including cancer (Figure 4).

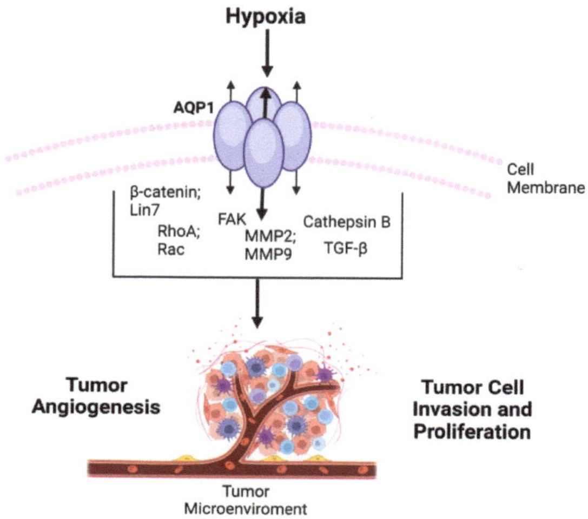

Figure 4. Overview of AQP1 in Cancer Progression-tumor development via transited molecules-emphasizing its involvement in cell migration, invasion, and angiogenesis, and its potential as a prognostic factor in various cancers.

7.2. Implications of Aquaporins in Neoplastic Cell Migration, Invasiveness, and Angiogenesis

7.2.1. AQP1-Modulated Tumor Cell Migration and Invasion

The role of AQP1 in cancer progression extends beyond its well-known function in water transport, particularly implicating it in the critical processes of tumor cell migration and invasion. Hu and Verkman found that AQP1 accelerates the migration of specific mouse melanoma and breast cancer cell lines in vitro. Notably, they observed polarized AQP1 expression at the leading edge of migrating cells. In vivo studies further revealed that AQP1 promoted cancer cell extravasation and lung metastases [168]. Jiang's work supports these findings, showing that altering AQP1 expression levels directly influences the migratory behavior of HT20 human colon cancer cells both in vitro and in vivo [169].

One proposed mechanism by which AQP1 facilitates tumor cell migration involves osmotic water flow across the plasma membrane. This flow is thought to be induced by an osmotic gradient created by actin depolymerization and active solute influx at the cell's leading edge [170]. Water influx via AQP1 then increases hydrostatic pressure, leading to local expansion of the plasma membrane and actin re-polymerization to stabilize cell membrane protrusions [6]. Chen and colleagues also provided evidence that AQP1's role extends to facilitating intracellular parasite invasion through mechanisms involving localized water influx [171].

An alternative explanation focuses on AQP1's role in changing the cell shape and volume as tumor cells navigate through confined spaces. The water flow facilitated by AQP1 helps generate the hydrostatic forces needed for this process. Actin polymerization and depolymerization as well as ionic fluxes across the membrane could support this osmotic water flow [124].

Adding a more nuanced layer to these mechanisms, Stroka and colleagues proposed an "Osmotic Engine Model", where cell migration in confined spaces is driven not just by actin and myosin interactions but also through water permeation and ion transport mediated by AQPs and Na+/H+ pumps [172].

The implications of these findings are profound, as the acquisition of migratory and invasive capabilities is a key step in cancer metastasis, which accounts for the majority of cancer-related deaths. Understanding AQP1's multifaceted role in these processes could thus offer crucial insights for the development of new therapeutic strategies targeting cancer metastasis.

The role of ion channels and transporters in cellular migration is increasingly recognized as critical. These components regulate various cellular mechanisms, such as intracellular calcium levels (Ca2+i), pH balance both inside and outside the cell, cellular membrane potential, and volume [173]. During directional cell migration, specific ion channels and transporters, including AQPs, K^+ channels, and Na^+/H^+ exchangers, tend to localize at the leading edge of cells, where they could potentially trigger osmotic water flow across the plasma membrane [174].

These ion channels and transporters are implicated in crucial stages of tumor metastasis, such as loss of cell-to-cell contacts, invasion of the surrounding stroma, and entry into and exit from blood vessels (intra- and extra-vasation) [175]. Kourghi and colleagues further deepen this narrative by showing that certain AQP1 ion channel blockers, which do not impact AQP1's water channel activity, effectively inhibited the migration of HT29 cancer cells. The degree of inhibition was directly related to the potency of the AQP1 ion channel blockage, suggesting that the ion channel properties of AQP1 alone might be sufficient for facilitating tumor cell migration in certain cases [176].

Recent research has also started to focus on the role of AQP1 in the broader tumor microenvironment. Pelagalli and colleagues found that bone marrow-derived mesenchymal stem cells (BM-MSCs)-conditioned medium increased AQP1 expression in U2OS osteosarcoma and SNU-398 hepatocellular carcinoma cells. This upregulation led to enhanced migration and invasion, which could be counteracted by the AQP1 inhibitor, tetraethylammonium chloride [177]. Given that BM-MSCs have been shown to differentiate into cancer-associated fibroblasts that further promote tumor growth and metastasis [178], AQP1 appears to play a significant role in the complex interactions between the tumor microenvironment and tumor cells.

Understanding this multifaceted role of AQP1—encompassing its ion channel properties, its function in osmotic water flow, and its involvement in the tumor microenvironment—could offer valuable insights into new therapeutic avenues for combating cancer metastasis, which is responsible for the vast majority of cancer-related deaths.

7.2.2. AQP1-Modulated Tumor Angiogenesis

Tumor angiogenesis, or the formation of new blood vessels within a tumor, serves as a lifeline for cancer cells, supplying essential nutrients and providing a pathway for metastasis [179]. Ion channels and transporters, including aquaporins such as AQP1, are increasingly being recognized as key players in this process. These proteins serve various roles, including acting as enzymes, chemical and mechanical sensors, receptors, and structural scaffolds [180]. Notably, AQP1 is preferentially upregulated in areas of astrocytomas where tumor cell infiltration occurs, hinting at its potential involvement in tumor angiogenesis [165].

A growing body of evidence further implicates AQP1 in angiogenic processes. Saadoun and colleagues found that mice lacking AQP1 and implanted with melanoma cells showed a reduced density of tumor microvessels, slower tumor growth, and increased survival [181]. This supports previous work that demonstrated diminished tumor microvascular density upon inhibiting AQP1, either through RNA interference or genetic knockouts, in various animal models [182]. Concomitant with this reduction in vascular density was a decrease in markers of angiogenesis, such as the vascular endothelial growth factor receptor 2 (VEGFR2) and the endothelial marker factor VIII [183].

On a cellular level, Saadoun and colleagues showed that endothelial cells deficient in AQP1 had impaired migration and abnormal vessel formation in vitro [181]. Moreover, silencing AQP1 in human endothelial cells (HMEC-1) led to disorganized F-actin polarization at the cell membrane's leading edge and a failure to establish a cord-like vascular network in culture. Similar findings were also noted in AQP1-silenced human melanoma cells [184]. These data suggest that AQP1 facilitates not just tumor cell migration but also that of endothelial cells, a critical feature for angiogenesis.

The link between AQP1 and endothelial cell migration may also tie into vascular permeability, an early step in angiogenesis. Clapp and Escalera proposed that AQP1 could amplify vessel permeability by enhancing cellular water transport. This, in turn, could trigger an angiogenic cascade by facilitating the extravasation of plasma proteins, which then serve as a scaffold for the migration of endothelial cells [185].

AQP1 appears to have a multifaceted role in cancer progression, impacting both tumor cells and the surrounding microenvironment. Its involvement in angiogenesis adds another layer to its importance, providing further avenues for targeted therapeutic interventions.

7.2.3. AQP1-Modulated Tumor Proliferation

While much attention has been focused on the role of AQP1 in tumor cell migration and angiogenesis, the evidence concerning its role in tumor cell proliferation is less consistent. For instance, inhibition of AQP1 activity had no impact on the proliferation of the colon cancer cell line HT29, whereas a modest 17% reduction was observed in another colon cancer cell line, HCT-116 [186]. Furthermore, overexpression of AQP1 in mouse cancer cell lines B16F10 and 4T1 did not lead to increased proliferation, although it did result in enhanced extravasation and metastases [168]. On the other hand, forced AQP1 expression was linked to heightened cell proliferation in NIH-3T3 mouse embryo fibroblasts and the rat pheochromocytoma cell line PC12 [187].

Adding another layer of complexity is the relationship between AQP1 and resistance to apoptosis. Hoque and colleagues observed that cells expressing AQP1 displayed resistance to apoptosis, possibly contributing to enhanced proliferation [187]. Specifically, AQP1 transfection in PC12 cells was associated with an altered cell cycle profile [188]—increased proportions of cells in the S and G2/M phases and elevated expression of cyclin D1 and E1, which are key proteins for cell cycle progression [189].

Cell volume dynamics are intrinsically tied to cell cycle progression and apoptosis. The cell cycle is marked by an increase in cell volume, while apoptosis involves a reduction in cell volume [190,191]. Given that AQP1-overexpressing PC12 cells exhibited greater cell size and higher intracellular complexity compared to wild-type controls, Galan-Cobo and colleagues hypothesized that these changes in cell morphology might facilitate cell cycle progression and inhibit apoptosis [189].

The role of AQP1 in tumor cell proliferation is not as clearly defined as its roles in tumor cell migration and angiogenesis. While some evidence suggests that AQP1 may contribute to enhanced proliferation and resistance to apoptosis, other studies offer conflicting results. This makes AQP1 an intriguing but complex target for potential therapeutic interventions aimed at inhibiting tumor cell proliferation. Further research is needed to fully understand the mechanisms by which AQP1 may or may not influence this aspect of tumor biology.

7.3. Assessing the Prospects of Aquaporin-Targeted Therapies in Malignancies of the CNS

While there is considerable excitement around the idea of aquaporin-targeted therapeutics, especially given promising data from animal studies, progress in this domain has been less than satisfactory [192–194]. A number of factors have contributed to this state of affairs, including conflicting reports about the inhibition of AQPs by commonly used ion transport inhibitors such as loop diuretics and antiepileptic medications. The literature is also muddled by reports of small-molecule inhibitors of AQPs that later could not be replicated in follow-up studies.

Given these inconsistencies, there is a critical need for rigorous scientific approaches to evaluating AQP function and inhibitor efficacy. For instance, comprehensive functional screens involving large collections of random, drug-like small molecules could provide valuable insights into effective AQP inhibitors. Alternatively, computational chemistry could guide smaller, more focused screens to identify potent compounds. Either way, robust validation in diverse cellular contexts will be essential to confirm the efficacy of any proposed AQP-targeting compounds.

A relatively unexplored area within this realm is the development of antibody- and peptide-based therapeutics against AQPs. There are also opportunities for research into small-molecule transcriptional regulators that can modulate AQP expression. These innovative approaches could offer alternative ways to target AQPs and their functions, potentially overcoming some of the limitations posed by existing methods and animal models.

Commercial interest in AQP-targeted therapies remains high, and the unmet need for effective treatments for various diseases provides a strong motivation for continued research in this area. Thus, the dual goals of creating viable commercial products and developing better research tools to replace or augment current transgenic animal models are likely to drive further innovations and breakthroughs in the study of AQPs.

8. Conclusions

8.1. Emerging Therapeutic Interventions and Prospects

Loss-of-function mutations in aquaporins are infrequently associated with human diseases. For example, mutations in AQP2 can lead to non-X-linked nephrogenic diabetes insipidus (NDI), a condition with an extremely low incidence (~1 in 20 million births) [195]. NDI results in severe symptoms such as polyuria and polydipsia that do not respond to antidiuretic hormones. The current treatment primarily involves water replacement and the use of thiazides to reduce urinary water loss. There is also ongoing research into pharmacological chaperone therapies and gene replacement or stem cell treatments as potential therapeutic avenues for NDI related to AQP2 mutations [196].

In contrast to AQP2, very few individuals have been identified with loss-of-function mutations in other AQPs such as AQP1, AQP3, and AQP7. For instance, individuals with AQP1 deficiency, identified through blood-group screenings, generally appear phenotypically normal but exhibit impaired urinary concentrating abilities when water-deprived—similar to AQP1-null mice [197]. Due to the rarity and variability of these conditions, there is limited information available about the roles these AQPs play in human health.

Furthermore, mutations in the major intrinsic protein (MIP, also known as AQP0) have been linked to congenital cataracts [198]. However, recent studies suggest that the primary function of MIP in the lens might be related more to cell–cell adhesion and gap-junction channel regulation rather than water transport.

Overall, the rarity of AQP deficiencies and the variability of their phenotypic expression in humans make it challenging to understand their roles fully. To date, no other disease-causing mutations in AQPs have been described in the medical literature. This highlights the need for continued research to better understand the function of AQPs in human physiology and pathology.

8.2. Final Thoughts

The study of aquaporins is an extraordinarily complex field that intersects with a wide range of scientific disciplines, from physiology and biochemistry to cell biology and biophysics. These proteins serve multifaceted roles in various tissues throughout the body, adding layers of intricacy to their understanding.

In this comprehensive review, we have embarked on an intellectual expedition to explore the many dimensions of aquaporins. We delved into their historical development, molecular classification, and the impact they have on neural health, among other areas. We have also scrutinized their involvement in a range of medical conditions, notably within the realms of oncology and neurology. The multi-faceted nature of aquaporins underscores the broad relevance and applicability they possess in the scientific and medical communities.

We earnestly aim for our work to resonate with the pressing issues and objectives of the contemporary world. In doing so, we hope to contribute meaningfully to the search for effective therapies against some of the most devastating diseases facing humanity today. Given the ubiquitous presence of aquaporins throughout the human body, these proteins offer an exciting frontier for therapeutic research, particularly in the quest to address the debilitating impacts of neurodegenerative diseases, acute cerebral events, and cancer.

As we look to the future, one key area that warrants further investigation is the role of aquaporins in conditions such as brain tumors. Given the devastating nature of these malignancies and the still-limited therapeutic options, understanding the impact and regulatory functions of aquaporins in such conditions could yield valuable insights and perhaps even revolutionary treatments.

Overall, while substantial progress has been made in understanding the roles and functions of aquaporins, significant questions remain. As such, future research should not only focus on a deeper understanding of their molecular behavior and physiological impacts but should also explore their potential as therapeutic targets. By pursuing these avenues of inquiry, we are optimistic that we can unlock new, more effective ways to address some of the most intractable health challenges of our time.

Author Contributions: Conceptualization, C.T. and A.V.C.; methodology, C.P.T.; software, R.-A.C.-B.; validation, I.-A.F., D.-I.D. and L.A.G.; formal analysis, H.P.C.; investigation, B.-G.B.; resources, I.-A.F.; data curation, R.-A.C.-B.; writing—original draft preparation, D.-I.D.; writing—review and editing, L.A.G.; visualization, H.P.C.; supervision, C.P.T.; project administration, C.T.; funding acquisition, A.V.C. All authors have read and agreed to the published version of the manuscript.

Funding: This research received no external funding.

Institutional Review Board Statement: Not applicable.

Informed Consent Statement: Not applicable.

Data Availability Statement: Not applicable.

Conflicts of Interest: The authors declare no conflict of interest.

References

1. Verkman, A.S. Aquaporins. *Curr. Biol.* **2013**, *23*, R52–R55. [CrossRef] [PubMed]
2. Yool, A.J.; Weinstein, A.M. New roles for old holes: Ion channel function in aquaporin-1. *News Physiol. Sci.* **2002**, *17*, 68–72. [CrossRef] [PubMed]
3. Papadopoulos, M.C.; Verkman, A.S. Aquaporin 4 and neuromyelitis optica. *Lancet Neurol.* **2012**, *11*, 535–544. [CrossRef] [PubMed]
4. Verkman, A.S. Aquaporins in clinical medicine. *Annu. Rev. Med.* **2012**, *63*, 303–316. [CrossRef]
5. Kruse, E.; Uehlein, N.; Kaldenhoff, R. The aquaporins. *Genome Biol.* **2006**, *7*, 206. [CrossRef]
6. Papadopoulos, M.C.; Verkman, A.S. Aquaporin water channels in the nervous system. *Nat. Rev. Neurosci.* **2013**, *14*, 265–277. [CrossRef]
7. Nagelhus, E.A.; Veruki, M.L.; Torp, R.; Haug, F.M.; Laake, J.H.; Nielsen, S.; Agre, P.; Ottersen, O.P. Aquaporin-4 water channel protein in the rat retina and optic nerve: Polarized expression in Müller cells and fibrous astrocytes. *J. Neurosci.* **1998**, *18*, 2506–2519. [CrossRef]
8. Papadopoulos, M.C.; Manley, G.T.; Krishna, S.; Verkman, A.S. Aquaporin-4 facilitates reabsorption of excess fluid in vasogenic brain edema. *FASEB J.* **2004**, *18*, 1291–1293. [CrossRef]
9. Oshio, K.; Binder, D.K.; Yang, B.; Schecter, S.; Verkman, A.S.; Manley, G.T. Expression of aquaporin water channels in mouse spinal cord. *Neuroscience* **2004**, *127*, 685–693. [CrossRef]
10. Furman, C.S.; Gorelick-Feldman, D.A.; Davidson, K.G.; Yasumura, T.; Neely, J.D.; Agre, P.; Rash, J.E. Aquaporin-4 square array assembly: Opposing actions of M1 and M23 isoforms. *Proc. Natl. Acad. Sci. USA* **2003**, *100*, 13609–13614. [CrossRef]
11. Rash, J.E.; Yasumura, T.; Hudson, C.S.; Agre, P.; Nielsen, S. Direct immunogold labeling of aquaporin-4 in square arrays of astrocyte and ependymocyte plasma membranes in rat brain and spinal cord. *Proc. Natl. Acad. Sci. USA* **1998**, *95*, 11981–11986. [CrossRef]

12. Nielsen, S.; Nagelhus, E.A.; Amiry-Moghaddam, M.; Bourque, C.; Agre, P.; Ottersen, O.P. Specialized membrane domains for water transport in glial cells: High-resolution immunogold cytochemistry of aquaporin-4 in rat brain. *J. Neurosci.* **1997**, *17*, 171–180. [CrossRef] [PubMed]
13. Hsu, M.S.; Seldin, M.; Lee, D.J.; Seifert, G.; Steinhäuser, C.; Binder, D.K. Laminar-specific and developmental expression of aquaporin-4 in the mouse hippocampus. *Neuroscience* **2011**, *178*, 21–32. [CrossRef] [PubMed]
14. Wilson, A.J.; Carati, C.J.; Gannon, B.J.; Haberberger, R.; Chataway, T.K. Aquaporin-1 in blood vessels of rat circumventricular organs. *Cell Tissue Res.* **2010**, *340*, 159–168. [CrossRef] [PubMed]
15. Dolman, D.; Drndarski, S.; Abbott, N.J.; Rattray, M. Induction of aquaporin 1 but not aquaporin 4 messenger RNA in rat primary brain microvessel endothelial cells in culture. *J. Neurochem.* **2005**, *93*, 825–833. [CrossRef] [PubMed]
16. Shields, S.D.; Mazario, J.; Skinner, K.; Basbaum, A.I. Anatomical and functional analysis of aquaporin 1, a water channel in primary afferent neurons. *Pain* **2007**, *131*, 8–20. [CrossRef] [PubMed]
17. Carbrey, J.M.; Agre, P. Discovery of the Aquaporins and Development of the Field. In *Aquaporins. Handbook of Experimental Pharmacology*; Springer: Berlin/Heidelberg, Germany, 2009; Volume 190, pp. 3–28. [CrossRef]
18. Ussing, H.H. Transport of electrolytes and water across epithelia. *Harvey Lect.* **1965**, *59*, 1–30.
19. Kachadorian, W.A.; Wade, J.B.; DiScala, V.A. Vasopressin: Induced structural change in toad bladder luminal membrane. 1975. *J. Am. Soc. Nephrol.* **2000**, *11*, 376–380.
20. Wade, J.B.; Stetson, D.L.; Lewis, S.A. ADH action: Evidence for a membrane shuttle mechanism. *Ann. N. Y. Acad. Sci.* **1981**, *372*, 106–117. [CrossRef]
21. Solomon, A.K. Characterization of biological membranes by equivalent pores. *J. Gen. Physiol.* **1968**, *51*, 335–364. [CrossRef]
22. Macey, R.I. Transport of water and urea in red blood cells. *Am. J. Physiol.* **1984**, *246 Pt 1*, C195–C203. [CrossRef]
23. van Os, C.H.; Deen, P.M.; Dempster, J.A. Aquaporins: Water selective channels in biological membranes. Molecular structure and tissue distribution. *Biochim. Biophys. Acta* **1994**, *1197*, 291–309. [CrossRef] [PubMed]
24. Harris, H.W.; Wade, J.B.; Handler, J.S. Identification of specific apical membrane polypeptides associated with the antidiuretic hormone-elicited water permeability increase in the toad urinary bladder. *Proc. Natl. Acad. Sci. USA* **1988**, *85*, 1942–1946. [CrossRef] [PubMed]
25. Benga, G.; Popescu, O.; Pop, V.I.; Holmes, R.P. p-(Chloromercuri)benzenesulfonate binding by membrane proteins and the inhibition of water transport in human erythrocytes. *Biochemistry* **1986**, *25*, 1535–1538. [CrossRef] [PubMed]
26. Benga, G. The first water channel protein (later called aquaporin 1) was first discovered in Cluj-Napoca, Romania. *Rom. J. Physiol.* **2004**, *41*, 3–20.
27. Agre, P.; Saboori, A.M.; Asimos, A.; Smith, B.L. Purification and partial characterization of the Mr 30,000 integral membrane protein associated with the erythrocyte Rh(D) antigen. *J. Biol. Chem.* **1987**, *262*, 17497–17503. [CrossRef]
28. Denker, B.M.; Smith, B.L.; Kuhajda, F.P.; Agre, P. Identification, purification, and partial characterization of a novel Mr 28,000 integral membrane protein from erythrocytes and renal tubules. *J. Biol. Chem.* **1988**, *263*, 15634–15642. [CrossRef]
29. Smith, B.L.; Agre, P. Erythrocyte Mr 28,000 transmembrane protein exists as a multisubunit oligomer similar to channel proteins. *J. Biol. Chem.* **1991**, *266*, 6407–6415. [CrossRef]
30. Preston, G.M.; Agre, P. Isolation of the cDNA for erythrocyte integral membrane protein of 28 kilodaltons: Member of an ancient channel family. *Proc. Natl. Acad. Sci. USA* **1991**, *88*, 11110–11114. [CrossRef]
31. Hediger, M.A.; Coady, M.J.; Ikeda, T.S.; Wright, E.M. Expression cloning and cDNA sequencing of the Na+/glucose co-transporter. *Nature* **1987**, *330*, 379–381. [CrossRef]
32. Preston, G.M.; Carroll, T.P.; Guggino, W.B.; Agre, P. Appearance of water channels in Xenopus oocytes expressing red cell CHIP28 protein. *Science* **1992**, *256*, 385–387. [CrossRef] [PubMed]
33. Agre, P.; Sasaki, S.; Chrispeels, M.J. Aquaporins: A family of water channel proteins. *Am. J. Physiol.* **1993**, *265 Pt 2*, F461. [CrossRef] [PubMed]
34. Verbavatz, J.M.; Brown, D.; Sabolić, I.; Valenti, G.; Ausiello, D.A.; Van Hoek, A.N.; Ma, T.; Verkman, A.S. Tetrameric assembly of CHIP28 water channels in liposomes and cell membranes: A freeze-fracture study. *J. Cell Biol.* **1993**, *123*, 605–618. [CrossRef] [PubMed]
35. Gorin, M.B.; Yancey, S.B.; Cline, J.; Revel, J.P.; Horwitz, J. The major intrinsic protein (MIP) of the bovine lens fiber membrane: Characterization and structure based on cDNA cloning. *Cell* **1984**, *39*, 49–59. [CrossRef]
36. Jung, J.S.; Preston, G.M.; Smith, B.L.; Guggino, W.B.; Agre, P. Molecular structure of the water channel through aquaporin CHIP. The hourglass model. *J. Biol. Chem.* **1994**, *269*, 14648–14654. [CrossRef]
37. Finn, R.N.; Chauvigné, F.; Hlidberg, J.B.; Cutler, C.P.; Cerdà, J. The lineage-specific evolution of aquaporin gene clusters facilitated tetrapod terrestrial adaptation. *PLoS ONE* **2014**, *9*, e113686. [CrossRef]
38. Soto, G.; Alleva, K.; Amodeo, G.; Muschietti, J.; Ayub, N.D. New insight into the evolution of aquaporins from flowering plants and vertebrates: Orthologous identification and functional transfer is possible. *Gene* **2012**, *503*, 165–176. [CrossRef]
39. Abascal, F.; Irisarri, I.; Zardoya, R. Diversity and evolution of membrane intrinsic proteins. *Biochim. Biophys. Acta* **2014**, *1840*, 1468–1481. [CrossRef]
40. Morishita, Y.; Sakube, Y.; Sasaki, S.; Ishibashi, K. Molecular mechanisms and drug development in aquaporin water channel diseases: Aquaporin superfamily (superaquaporins): Expansion of aquaporins restricted to multicellular organisms. *J. Pharmacol. Sci.* **2004**, *96*, 276–279. [CrossRef]

41. Johanson, U.; Karlsson, M.; Johansson, I.; Gustavsson, S.; Sjövall, S.; Fraysse, L.; Weig, A.R.; Kjellbom, P. The complete set of genes encoding major intrinsic proteins in Arabidopsis provides a framework for a new nomenclature for major intrinsic proteins in plants. *Plant Physiol.* **2001**, *126*, 1358–1369. [CrossRef]
42. Quigley, F.; Rosenberg, J.M.; Shachar-Hill, Y.; Bohnert, H.J. From genome to function: The Arabidopsis aquaporins. *Genome Biol.* **2002**, *3*, 1–17. [CrossRef]
43. Yuan, D.; Li, W.; Hua, Y.; King, G.J.; Xu, F.; Shi, L. Genome-Wide Identification and Characterization of the Aquaporin Gene Family and Transcriptional Responses to Boron Deficiency in Brassica napus. *Front. Plant Sci.* **2017**, *8*, 1336. [CrossRef] [PubMed]
44. Danielson, J.A.H.; Johanson, U. Unexpected complexity of the aquaporin gene family in the moss Physcomitrella patens. *BMC Plant Biol.* **2008**, *8*, 45. [CrossRef] [PubMed]
45. Gustavsson, S.; Lebrun, A.-S.; Nordén, K.; Chaumont, F.; Johanson, U. A novel plant major intrinsic protein in Physcomitrella patens most similar to bacterial glycerol channels. *Plant Physiol.* **2005**, *139*, 287–295. [CrossRef]
46. Danielson, J.A.H.; Johanson, U. Phylogeny of major intrinsic proteins. *Adv. Exp. Med. Biol.* **2010**, *679*, 19–31. [CrossRef] [PubMed]
47. Anderberg, H.I.; Danielson, J.Å.H.; Johanson, U. Algal MIPs, high diversity and conserved motifs. *BMC Evol. Biol.* **2011**, *11*, 110. [CrossRef] [PubMed]
48. Khabudaev, K.V.; Petrova, D.P.; Grachev, M.A.; Likhoshway, Y.V. A new subfamily LIP of the major intrinsic proteins. *BMC Genom.* **2014**, *15*, 173. [CrossRef]
49. Chaumont, F.; Barrieu, F.; Wojcik, E.; Chrispeels, M.J.; Jung, R. Aquaporins constitute a large and highly divergent protein family in maize. *Plant Physiol.* **2001**, *125*, 1206–1215. [CrossRef]
50. Mallo, R.C.; Ashby, M.T. AqpZ-mediated water permeability in Escherichia coli measured by stopped-flow spectroscopy. *J. Bacteriol.* **2006**, *188*, 820–822. [CrossRef]
51. Madeira, A.; Leitão, L.; Soveral, G.; Dias, P.; Prista, C.; Moura, T.; Loureiro-Dias, M.C. Effect of ethanol on fluxes of water and protons across the plasma membrane of Saccharomyces cerevisiae. *FEMS Yeast Res.* **2010**, *10*, 252–2580. [CrossRef]
52. Madeira, A.; Camps, M.; Zorzano, A.; Moura, T.F.; Soveral, G. Biophysical assessment of human aquaporin-7 as a water and glycerol channel in 3T3-L1 adipocytes. *PLoS ONE* **2013**, *8*, e83442. [CrossRef] [PubMed]
53. Mollajew, R.; Zocher, F.; Horner, A.; Wiesner, B.; Klussmann, E.; Pohl, P. Routes of epithelial water flow: Aquaporins versus cotransporters. *Biophys. J.* **2010**, *99*, 3647–3656. [CrossRef] [PubMed]
54. Ikeda, M.; Andoo, A.; Shimono, M.; Takamatsu, N.; Taki, A.; Muta, K.; Matsushita, W.; Uechi, T.; Matsuzaki, T.; Kenmochi, N.; et al. The NPC motif of aquaporin-11, unlike the NPA motif of known aquaporins, is essential for full expression of molecular function. *J. Biol. Chem.* **2011**, *286*, 3342–3350. [CrossRef] [PubMed]
55. Verkman, A.S.; Anderson, M.O.; Papadopoulos, M.C. Aquaporins: Important but elusive drug targets. *Nat. Rev. Drug Discov.* **2014**, *13*, 259–277. [CrossRef]
56. Verkman, A.S. Water permeability measurement in living cells and complex tissues. *J. Membr. Biol.* **2000**, *173*, 73–87. [CrossRef]
57. Soveral, G.; Prista, C.; Moura, T.F.; Loureiro-Dias, M.C. Yeast water channels: An overview of orthodox aquaporins. *Biol. Cell* **2010**, *103*, 35–54. [CrossRef]
58. Levin, M.H.; de la Fuente, R.; Verkman, A.S. Urearetics: A small molecule screen yields nanomolar potency inhibitors of urea transporter UT-B. *FASEB J.* **2007**, *21*, 551–563. [CrossRef]
59. Soveral, G.; Madeira, A.; Loureiro-Dias, M.C.; Moura, T.F. Membrane tension regulates water transport in yeast. *Biochim. Biophys. Acta* **2008**, *1778*, 2573–2579. [CrossRef]
60. Soveral, G.; Madeira, A.; Loureiro-Dias, M.C.; Moura, T.F. Water transport in intact yeast cells as assessed by fluorescence self-quenching. *Appl. Environ. Microbiol.* **2007**, *73*, 2341–2343. [CrossRef]
61. Clarke, L.L. A guide to Ussing chamber studies of mouse intestine. *Am. J. Physiol. Gastrointest. Liver Physiol.* **2009**, *296*, G1151–G1166. [CrossRef]
62. Levin, M.H.; Sullivan, S.; Nielson, D.; Yang, B.; Finkbeiner, W.E.; Verkman, A.S. Hypertonic saline therapy in cystic fibrosis: Evidence against the proposed mechanism involving aquaporins. *J. Biol. Chem.* **2006**, *281*, 25803–25812. [CrossRef] [PubMed]
63. Beitz, E.; Becker, D.; von Bulow, J.; Conrad, C.; Fricke, N.; Geadkaew, A.; Krenc, D.; Song, J.; Wree, D.; Wu, B. In vitro analysis and modification of aquaporin pore selectivity. In *Aquaporins. Handbook of Experimental Pharmacology*; Beitz, E., Ed.; Springer: Berlin/Heidelberg, Germany, 2009; pp. 77–92. [CrossRef]
64. Campos, E.; Moura, T.F.; Oliva, A.; Leandro, P.; Soveral, G. Lack of Aquaporin 3 in bovine erythrocyte membranes correlates with low glycerol permeation. *Biochem. Biophys. Res. Commun.* **2011**, *408*, 477–481. [CrossRef]
65. Trillo-Contreras, J.L.; Toledo-Aral, J.J.; Echevarría, M.; Villadiego, J. AQP1 and AQP4 Contribution to Cerebrospinal Fluid Homeostasis. *Cells* **2019**, *8*, 197. [CrossRef]
66. Johanson, C.E.; Duncan, J.A.; Klinge, P.M.; Brinker, T.; Stopa, E.G.; Silverberg, G.D. Multiplicity of cerebrospinal fluid functions: New challenges in health and disease. *Cerebrospinal Fluid. Res.* **2008**, *5*, 10. [CrossRef]
67. Abbott, N.J.; Pizzo, M.E.; Preston, J.E.; Janigro, D.; Thorne, R.G. The role of brain barriers in fluid movement in the CNS: Is there a 'glymphatic' system? *Acta Neuropathol.* **2018**, *135*, 387–407. [CrossRef] [PubMed]
68. Da Mesquita, S.; Fu, Z.; Kipnis, J. The Meningeal Lymphatic System: A New Player in Neurophysiology. *Neuron* **2018**, *100*, 375–388. [CrossRef] [PubMed]
69. Hladky, S.B.; Barrand, M.A. Mechanisms of fluid movement into, through and out of the brain: Evaluation of the evidence. *Fluids Barriers CNS* **2014**, *11*, 26. [CrossRef]

70. Orešković, D.; Radoš, M.; Klarica, M. Role of choroid plexus in cerebrospinal fluid hydrodynamics. *Neuroscience* **2017**, *354*, 69–87. [CrossRef]
71. Iliff, J.J.; Wang, M.; Liao, Y.; Plogg, B.A.; Peng, W.; Gundersen, G.A.; Benveniste, H.; Vates, G.E.; Deane, R.; Goldman, S.A.; et al. A paravascular pathway facilitates CSF flow through the brain parenchyma and the clearance of interstitial solutes, including amyloid β. *Sci. Transl. Med.* **2012**, *4*, 147ra111. [CrossRef]
72. Trillo-Contreras, J.L.; Ramírez-Lorca, R.; Hiraldo-González, L.; Sánchez-Gomar, I.; Galán-Cobo, A.; Suárez-Luna, N.; de Rojas-de Pedro, E.S.; Toledo-Aral, J.J.; Villadiego, J.; Echevarría, M. Combined effects of aquaporin-4 and hypoxia produce age-related hydrocephalus. *Biochim. Biophys. Acta Mol. Basis Dis.* **2018**, *1864*, 3515–3526. [CrossRef]
73. Silverberg, G.D.; Mayo, M.; Saul, T.; Rubenstein, E.; McGuire, D. Alzheimer's disease, normal-pressure hydrocephalus, and senescent changes in CSF circulatory physiology: A hypothesis. *Lancet Neurol.* **2003**, *2*, 506–511. [CrossRef] [PubMed]
74. Amiry-Moghaddam, M.; Ottersen, O.P. The molecular basis of water transport in the brain. *Nat. Rev. Neurosci.* **2003**, *4*, 991–1001. [CrossRef] [PubMed]
75. Oshio, K.; Watanabe, H.; Song, Y.; Verkman, A.S.; Manley, G.T. Reduced cerebrospinal fluid production and intracranial pressure in mice lacking choroid plexus water channel Aquaporin-1. *FASEB J.* **2005**, *19*, 76–78. [CrossRef] [PubMed]
76. Igarashi, H.; Tsujita, M.; Kwee, I.L.; Nakada, T. Water influx into cerebrospinal fluid is primarily controlled by aquaporin-4, not by aquaporin-1: 17O JJVCPE MRI study in knockout mice. *Neuroreport* **2014**, *25*, 39–43. [CrossRef]
77. Paul, L.; Madan, M.; Rammling, M.; Chigurupati, S.; Chan, S.L.; Pattisapu, J.V. Expression of aquaporin 1 and 4 in a congenital hydrocephalus rat model. *Neurosurgery* **2011**, *68*, 462–473. [CrossRef]
78. Kalani, M.Y.S.; Filippidis, A.S.; Rekate, H.L. Hydrocephalus and aquaporins: The role of aquaporin-1. *Acta Neurochir. Suppl.* **2012**, *113*, 51–54. [CrossRef]
79. Badaut, J.; Petit, J.-M.; Brunet, J.-F.; Magistretti, P.J.; Charriaut-Marlangue, C.; Regli, L. Distribution of Aquaporin 9 in the adult rat brain: Preferential expression in catecholaminergic neurons and in glial cells. *Neuroscience* **2004**, *128*, 27–38. [CrossRef]
80. Badaut, J.; Regli, L. Distribution and possible roles of aquaporin 9 in the brain. *Neuroscience* **2004**, *129*, 971–981. [CrossRef]
81. Badaut, J. Aquaglyceroporin 9 in brain pathologies. *Neuroscience* **2010**, *168*, 1047–1057. [CrossRef]
82. Liu, H.; Yang, M.; Qiu, G.P.; Zhuo, F.; Yu, W.H.; Sun, S.Q.; Xiu, Y. Aquaporin 9 in rat brain after severe traumatic brain injury. *Arq. Neuropsiquiatr.* **2012**, *70*, 214–220. [CrossRef]
83. Ringstad, G.; Vatnehol, S.A.S.; Eide, P.K. Glymphatic MRI in idiopathic normal pressure hydrocephalus. *Brain* **2017**, *140*, 2691–2705. [CrossRef] [PubMed]
84. Suzuki, Y.; Nakamura, Y.; Yamada, K.; Igarashi, H.; Kasuga, K.; Yokoyama, Y.; Ikeuchi, T.; Nishizawa, M.; Kwee, I.L.; Nakada, T. Reduced CSF Water Influx in Alzheimer's Disease Supporting the β-Amyloid Clearance Hypothesis. *PLoS ONE* **2015**, *10*, e0123708. [CrossRef] [PubMed]
85. Koike, S.; Tanaka, Y.; Matsuzaki, T.; Morishita, Y.; Ishibashi, K. Aquaporin-11 (AQP11) Expression in the Mouse Brain. *Int. J. Mol. Sci.* **2016**, *17*, 861. [CrossRef] [PubMed]
86. Xi, T.; Jin, F.; Zhu, Y.; Wang, J.; Tang, L.; Wang, Y.; Liebeskind, D.S.; Scalzo, F.; He, Z. miR-27a-3p protects against blood-brain barrier disruption and brain injury after intracerebral hemorrhage by targeting endothelial aquaporin-11. *J. Biol. Chem.* **2018**, *293*, 20041–20050. [CrossRef]
87. Jessen, N.A.; Munk, A.S.F.; Lundgaard, I.; Nedergaard, M. The Glymphatic System—A Beginner's Guide. *Neurochem. Res.* **2015**, *40*, 2583–2599. [CrossRef]
88. Iliff, J.J.; Nedergaard, M. Is there a cerebral lymphatic system? *Stroke* **2013**, *44* (Suppl. S1), S93–S95. [CrossRef]
89. Johnston, M.; Zakharov, A.; Papaiconomou, C.; Salmasi, G.; Armstrong, D. Evidence of connections between cerebrospinal fluid and nasal lymphatic vessels in humans, non-human primates and other mammalian species. *Cerebrospinal Fluid. Res.* **2004**, *1*, 2. [CrossRef]
90. Murtha, L.A.; Yang, Q.; Parsons, M.W.; Levi, C.R.; Beard, D.J.; Spratt, N.J.; McLeod, D.D. Cerebrospinal fluid is drained primarily via the spinal canal and olfactory route in young and aged spontaneously hypertensive rats. *Fluids Barriers CNS* **2014**, *11*, 12. [CrossRef]
91. Weller, R.O.; Subash, M.; Preston, S.D.; Mazanti, I.; Carare, R.O. Perivascular drainage of amyloid-beta peptides from the brain and its failure in cerebral amyloid angiopathy and Alzheimer's disease. *Brain Pathol.* **2008**, *18*, 253–266. [CrossRef]
92. Ciurea, A.V.; Mohan, A.G.; Covache-Busuioc, R.A.; Costin, H.P.; Glavan, L.A.; Corlatescu, A.D.; Saceleanu, V.M. Unraveling Molecular and Genetic Insights into Neurodegenerative Diseases: Advances in Understanding Alzheimer's, Parkinson's, and Huntington's Diseases and Amyotrophic Lateral Sclerosis. *Int. J. Mol. Sci.* **2023**, *24*, 10809. [CrossRef]
93. Kress, B.T.; Iliff, J.J.; Xia, M.; Wang, M.; Wei, H.S.; Zeppenfeld, D.; Xie, L.; Kang, H.; Xu, Q.; Liew, J.A.; et al. Impairment of paravascular clearance pathways in the aging brain. *Ann. Neurol.* **2014**, *76*, 845–861. [CrossRef] [PubMed]
94. Sabbatini, M.; Barili, P.; Bronzetti, E.; Zaccheo, D.; Amenta, F. Age-related changes of glial fibrillary acidic protein immunoreactive astrocytes in the rat cerebellar cortex. *Mech. Ageing Dev.* **1999**, *108*, 165–172. [CrossRef] [PubMed]
95. Deane, R.; Zlokovic, B.V. Role of the blood-brain barrier in the pathogenesis of Alzheimer's disease. *Curr. Alzheimer Res.* **2007**, *4*, 191–197. [CrossRef]
96. Ciurea, A.V.; Mohan, A.G.; Covache-Busuioc, R.-A.; Costin, H.P.; Saceleanu, V.M. The Brain's Glymphatic System: Drawing New Perspectives in Neuroscience. *Brain Sci.* **2023**, *13*, 1005. [CrossRef]

97. Iliff, J.J.; Wang, M.; Zeppenfeld, D.M.; Venkataraman, A.; Plog, B.A.; Liao, Y.; Deane, R.; Nedergaard, M. Cerebral arterial pulsation drives paravascular CSF-interstitial fluid exchange in the murine brain. *J. Neurosci.* **2013**, *33*, 18190–18199. [CrossRef]
98. Zieman, S.J.; Melenovsky, V.; Kass, D.A. Mechanisms, pathophysiology, and therapy of arterial stiffness. *Arterioscler. Thromb. Vasc. Biol.* **2005**, *25*, 932–943. [CrossRef]
99. Moretti, L.; Cristofori, I.; Weaver, S.M.; Chau, A.; Portelli, J.N.; Grafman, J. Cognitive decline in older adults with a history of traumatic brain injury. *Lancet Neurol.* **2012**, *11*, 1103–1112. [CrossRef] [PubMed]
100. Irimia, A.; Wang, B.; Aylward, S.R.; Prastawa, M.W.; Pace, D.F.; Gerig, G.; Hovda, D.A.; Kikinis, R.; Vespa, P.M.; Van Horn, J.D. Neuroimaging of structural pathology and connectomics in traumatic brain injury: Toward personalized outcome prediction. *Neuroimage Clin.* **2012**, *1*, 1–17. [CrossRef]
101. Gabbita, S.P.; Scheff, S.W.; Menard, R.M.; Roberts, K.; Fugaccia, I.; Zemlan, F.P. Cleaved-tau: A biomarker of neuronal damage after traumatic brain injury. *J. Neurotrauma* **2005**, *22*, 83–94. [CrossRef]
102. Gaberel, T.; Gakuba, C.; Goulay, R.; De Lizarrondo, S.M.; Hanouz, J.L.; Emery, E.; Touze, E.; Vivien, D.; Gauberti, M. Impaired glymphatic perfusion after strokes revealed by contrast-enhanced MRI: A new target for fibrinolysis? *Stroke* **2014**, *45*, 3092–3096. [CrossRef]
103. Yamada, K. Multifaceted Roles of Aquaporins in the Pathogenesis of Alzheimer's Disease. *Int. J. Mol. Sci.* **2023**, *24*, 6528. [CrossRef] [PubMed]
104. Mayeux, R. Epidemiology of neurodegeneration. *Annu. Rev. Neurosci.* **2003**, *26*, 81–104. [CrossRef]
105. Eriksen, J.L.; Wszolek, Z.; Petrucelli, L. Molecular pathogenesis of Parkinson disease. *Arch. Neurol.* **2005**, *62*, 353–357. [CrossRef] [PubMed]
106. Kernie, S.G.; Erwin, T.M.; Parada, L.F. Brain remodeling due to neuronal and astrocytic proliferation after controlled cortical injury in mice. *J. Neurosci. Res.* **2001**, *66*, 317–326. [CrossRef]
107. Eng, L.F.; Ghirnikar, R.S.; Lee, Y.L. Glial fibrillary acidic protein: GFAP-thirty-one years (1969–2000). *Neurochem. Res.* **2000**, *25*, 1439–1451. [CrossRef] [PubMed]
108. Mazurová, Y.; Rudolf, E.; Látr, I.; Osterreicher, J. Proliferation and differentiation of adult endogenous neural stem cells in response to neurodegenerative process within the striatum. *Neurodegener. Dis.* **2006**, *3*, 12–18. [CrossRef] [PubMed]
109. Kueppers, E.; Gleiser, C.; Brito, V.; Wachter, B.; Pauly, T.; Hirt, B.; Grissmer, S. AQP4 expression in striatal primary cultures is regulated by dopamine--implications for proliferation of astrocytes. *Eur. J. Neurosci.* **2008**, *28*, 2173–2182. [CrossRef]
110. Saadoun, S.; Papadopoulos, M.C.; Watanabe, H.; Yan, D.; Manley, G.T.; Verkman, A.S. Involvement of aquaporin-4 in astroglial cell migration and glial scar formation. *J. Cell Sci.* **2005**, *118 Pt 24*, 5691–5698. [CrossRef]
111. Nicchia, G.P.; Frigeri, A.; Liuzzi, G.M.; Svelto, M. Inhibition of aquaporin-4 expression in astrocytes by RNAi determines alteration in cell morphology, growth, and water transport and induces changes in ischemia-related genes. *FASEB J.* **2003**, *17*, 1508–1510. [CrossRef]
112. Vizuete, M.L.; Venero, J.L.; Vargas, C.; Ilundain, A.A.; Echevarra, M.; Machado, A.; Cano, J. Differential upregulation of aquaporin-4 mRNA expression in reactive astrocytes after brain injury: Potential role in brain edema. *Neurobiol. Dis.* **1999**, *6*, 245–258. [CrossRef]
113. Moore, D.J.; West, A.B.; Dawson, V.L.; Dawson, T.M. Molecular pathophysiology of Parkinson's disease. *Annu. Rev. Neurosci.* **2005**, *28*, 57–87. [CrossRef]
114. Yang, B.; Zhao, D.; Verkman, A.S. Evidence against functionally significant aquaporin expression in mitochondria. *J. Biol. Chem.* **2006**, *281*, 16202–16206. [CrossRef]
115. Carbrey, J.M.; Gorelick-Feldman, D.A.; Kozono, D.; Praetorius, J.; Nielsen, S.; Agre, P. Aquaglyceroporin AQP9: Solute permeation and metabolic control of expression in liver. *Proc. Natl. Acad. Sci. USA* **2003**, *100*, 2945–2950. [CrossRef]
116. Hasan-Olive, M.M.; Enger, R.; Hansson, H.-A.; Nagelhus, E.A.; Eide, P.K. Loss of perivascular aquaporin-4 in idiopathic normal pressure hydrocephalus. *GLIA* **2019**, *67*, 91–100. [CrossRef] [PubMed]
117. Badaut, J.; Fukuda, A.M.; Jullienne, A.; Petry, K.G. Aquaporin and brain diseases. *Biochim. Biophys. Acta* **2014**, *1840*, 1554–1565. [CrossRef] [PubMed]
118. Filippidis, A.S.; Carozza, R.B.; Rekate, H.L. Aquaporins in Brain Edema and Neuropathological Conditions. *Int. J. Mol. Sci.* **2016**, *18*, 55. [CrossRef]
119. Badaut, J.; Ashwal, S.; Obenaus, A. Aquaporins in cerebrovascular disease: A target for treatment of brain edema? *Cerebrovasc. Dis.* **2011**, *31*, 521–531. [CrossRef] [PubMed]
120. Berezowski, V.; Fukuda, A.M.; Cecchelli, R.; Badaut, J. Endothelial cells and astrocytes: A concerto en duo in ischemic pathophysiology. *Int. J. Cell Biol.* **2012**, *2012*, 176287. [CrossRef]
121. Dityatev, A.; Schachner, M. Extracellular matrix molecules and synaptic plasticity. *Nat. Rev. Neurosci.* **2003**, *4*, 456–468. [CrossRef]
122. Pop, V.; Sorensen, D.W.; Kamper, J.E.; Ajao, D.O.; Murphy, M.P.; Head, E.; Hartman, R.E.; Badaut, J. Early brain injury alters the blood-brain barrier phenotype in parallel with β-amyloid and cognitive changes in adulthood. *J. Cereb. Blood Flow. Metab.* **2013**, *33*, 205–214. [CrossRef]
123. Hirt, L.; Ternon, B.; Price, M.; Mastour, N.; Brunet, J.-F.; Badaut, J. Protective role of early aquaporin 4 induction against postischemic edema formation. *J. Cereb. Blood Flow. Metab.* **2009**, *29*, 423–433. [CrossRef] [PubMed]
124. Nico, B.; Ribatti, D. Role of aquaporins in cell migration and edema formation in human brain tumors. *Exp. Cell Res.* **2011**, *317*, 2391–2396. [CrossRef]

125. Badaut, J.; Brunet, J.F.; Grollimund, L.; Hamou, M.F.; Magistretti, P.J.; Villemure, J.G.; Regli, L. Aquaporin 1 and aquaporin 4 expression in human brain after subarachnoid hemorrhage and in peritumoral tissue. *Acta Neurochir. Suppl.* **2003**, *86*, 495–498. [CrossRef] [PubMed]
126. de Castro Ribeiro, M.; Hirt, L.; Bogousslavsky, J.; Regli, L.; Badaut, J. Time course of aquaporin expression after transient focal cerebral ischemia in mice. *J. Neurosci. Res.* **2006**, *83*, 1231–1240. [CrossRef]
127. Manley, G.T.; Fujimura, M.; Ma, T.; Noshita, N.; Filiz, F.; Bollen, A.W.; Chan, P.; Verkman, A. Aquaporin-4 deletion in mice reduces brain edema after acute water intoxication and ischemic stroke. *Nat. Med.* **2000**, *6*, 159–163. [CrossRef]
128. Binder, D.K.; Papadopoulos, M.C.; Haggie, P.M.; Verkman, A.S. In vivo measurement of brain extracellular space diffusion by cortical surface photobleaching. *J. Neurosci.* **2004**, *24*, 8049–8056. [CrossRef] [PubMed]
129. Fukuda, A.M.; Pop, V.; Spagnoli, D.; Ashwal, S.; Obenaus, A.; Badaut, J. Delayed increase of astrocytic aquaporin 4 after juvenile traumatic brain injury: Possible role in edema resolution? *Neuroscience* **2012**, *222*, 366–378. [CrossRef]
130. Suzuki, M.; Iwasaki, Y.; Yamamoto, T.; Konno, H.; Yoshimoto, T.; Suzuki, J. Disintegration of orthogonal arrays in perivascular astrocytic processes as an early event in acute global ischemia. *Brain Res.* **1984**, *300*, 141–145. [CrossRef]
131. Saadoun, S.; Waters, P.; Bell, B.A.; Vincent, A.; Verkman, A.S.; Papadopoulos, M.C. Intra-cerebral injection of neuromyelitis optica immunoglobulin G and human complement produces neuromyelitis optica lesions in mice. *Brain* **2010**, *133 Pt 2*, 349–361. [CrossRef]
132. Kimura, A.; Hsu, M.; Seldin, M.; Verkman, A.S.; Scharfman, H.E.; Binder, D.K. Protective role of aquaporin-4 water channels after contusion spinal cord injury. *Ann. Neurol.* **2010**, *67*, 794–801. [CrossRef]
133. Badaut, J.; Ashwal, S.; Adami, A.; Tone, B.; Recker, R.; Spagnoli, D.; Ternon, B.; Obenaus, A. Brain water mobility decreases after astrocytic aquaporin-4 inhibition using RNA interference. *J. Cereb. Blood Flow. Metab.* **2011**, *31*, 819–831. [CrossRef] [PubMed]
134. Fukuda, A.M.; Adami, A.; Pop, V.; Bellone, J.A.; Coats, J.S.; Hartman, R.E.; Ashwal, S.; Obenaus, A.; Badaut, J. Posttraumatic reduction of edema with aquaporin-4 RNA interference improves acute and chronic functional recovery. *J. Cereb. Blood Flow. Metab.* **2013**, *33*, 1621–1632. [CrossRef] [PubMed]
135. Tourdias, T.; Mori, N.; Dragonu, I.; Cassagno, N.; Boiziau, C.; Aussudre, J.; Brochet, B.; Moonen, C.; Petry, K.G.; Dousset, V. Differential aquaporin 4 expression during edema build-up and resolution phases of brain inflammation. *J. Neuroinflamm.* **2011**, *8*, 143. [CrossRef]
136. Badaut, J.; Hirt, L.; Granziera, C.; Bogousslavsky, J.; Magistretti, P.J.; Regli, L. Astrocyte-specific expression of aquaporin-9 in mouse brain is increased after transient focal cerebral ischemia. *J. Cereb. Blood Flow. Metab.* **2001**, *21*, 477–482. [CrossRef] [PubMed]
137. Nagelhus, E.A.; Mathiisen, T.M.; Ottersen, O.P. Aquaporin-4 in the central nervous system: Cellular and subcellular distribution and coexpression with KIR4.1. *Neuroscience* **2004**, *129*, 905–913. [CrossRef] [PubMed]
138. Binder, D.K.; Yao, X.; Zador, Z.; Sick, T.J.; Verkman, A.S.; Manley, G.T. Increased seizure duration and slowed potassium kinetics in mice lacking aquaporin-4 water channels. *GLIA* **2006**, *53*, 631–636. [CrossRef] [PubMed]
139. Dibaj, P.; Kaiser, M.; Hirrlinger, J.; Kirchhoff, F.; Neusch, C. Kir4.1 channels regulate swelling of astroglial processes in experimental spinal cord edema. *J. Neurochem.* **2007**, *103*, 2620–2628. [CrossRef]
140. Herrera, M.; Garvin, J.L. Aquaporins as gas channels. *Pflugers Arch.* **2011**, *462*, 623–630. [CrossRef]
141. Cauli, B.; Hamel, E. Revisiting the role of neurons in neurovascular coupling. *Front. Neuroenergetics* **2010**, *2*, 9. [CrossRef]
142. Nesic, O.; Lee, J.; Ye, Z.; Unabia, G.C.; Rafati, D.; Hulsebosch, C.E.; Perez-Polo, J.R. Acute and chronic changes in aquaporin 4 expression after spinal cord injury. *Neuroscience* **2006**, *143*, 779–792. [CrossRef]
143. Hiroaki, Y.; Tani, K.; Kamegawa, A.; Gyobu, N.; Nishikawa, K.; Suzuki, H.; Walz, T.; Sasaki, S.; Mitsuoka, K.; Kimura, K.; et al. Implications of the aquaporin-4 structure on array formation and cell adhesion. *J. Mol. Biol.* **2006**, *355*, 628–639. [CrossRef] [PubMed]
144. Vassiliou, A.G.; Maniatis, N.A.; Orfanos, S.E.; Mastora, Z.; Jahaj, E.; Paparountas, T.; Armaganidis, A.; Roussos, C.; Aidinis, V.; Kotanidou, A. Induced expression and functional effects of aquaporin-1 in human leukocytes in sepsis. *Crit. Care* **2013**, *17*, R199. [CrossRef] [PubMed]
145. Thuny, F.; Textoris, J.; Ben Amara, A.; El Filali, A.; Capo, C.; Habib, G.; Raoult, D.; Mege, J.L. The gene expression analysis of blood reveals S100A11 and AQP9 as potential biomarkers of infective endocarditis. *PLoS ONE* **2012**, *7*, e31490. [CrossRef] [PubMed]
146. De Santis, S.; Serino, G.; Fiorentino, M.R.; Galleggiante, V.; Gena, P.; Verna, G.; Liso, M.; Massaro, M.; Lan, J.; Troisi, J.; et al. Corrigendum: Aquaporin-9 Contributes to the Maturation Process and Inflammatory Cytokine Secretion of Murine Dendritic Cells. *Front. Immunol.* **2019**, *10*, 216. [CrossRef]
147. da Silva, I.V.; Cardoso, C.; Martínez-Banaclocha, H.; Casini, A.; Pelegrín, P.; Soveral, G. Aquaporin-3 is involved in NLRP3-inflammasome activation contributing to the setting of inflammatory response. *Cell Mol. Life Sci.* **2021**, *78*, 3073–3085. [CrossRef]
148. Rump, K.; Brendt, P.; Frey, U.H.; Schäfer, S.T.; Siffert, W.; Peters, J.; Adamzik, M. Aquaporin 1 and 5 expression evoked by the β2 adrenoreceptor agonist terbutaline and lipopolysaccharide in mice and in the human monocytic cell line THP-1 is differentially regulated. *Shock* **2013**, *40*, 430–436. [CrossRef]
149. Matsushima, A.; Ogura, H.; Koh, T.; Shimazu, T.; Sugimoto, H. Enhanced expression of aquaporin 9 in activated polymorphonuclear leukocytes in patients with systemic inflammatory response syndrome. *Shock* **2014**, *42*, 322–326. [CrossRef]
150. Zhu, N.; Feng, X.; He, C.; Gao, H.; Yang, L.; Ma, Q.; Guo, L.; Qiao, Y.; Yang, H.; Ma, T. Defective macrophage function in aquaporin-3 deficiency. *FASEB J.* **2011**, *25*, 4233–4239. [CrossRef]

151. Holm, A.; Karlsson, T.; Vikström, E. Pseudomonas aeruginosa lasI/rhlI quorum sensing genes promote phagocytosis and aquaporin 9 redistribution to the leading and trailing regions in macrophages. *Front. Microbiol.* **2015**, *6*, 915. [CrossRef]
152. Talwar, S.; Munson, P.J.; Barb, J.; Fiuza, C.; Cintron, A.P.; Logun, C.; Tropea, M.; Khan, S.; Reda, D.; Shelhamer, J.H.; et al. Gene expression profiles of peripheral blood leukocytes after endotoxin challenge in humans. *Physiol. Genomics* **2006**, *25*, 203–215. [CrossRef]
153. Pittock, S.J.; Lucchinetti, C.F. Neuromyelitis optica and the evolving spectrum of autoimmune aquaporin-4 channelopathies: A decade later. *Ann. N. Y. Acad. Sci.* **2016**, *1366*, 20–39. [CrossRef] [PubMed]
154. Jarius, S.; Wildemann, B. AQP4 antibodies in neuromyelitis optica: Diagnostic and pathogenetic relevance. *Nat. Rev. Neurol.* **2010**, *6*, 383–392. [CrossRef] [PubMed]
155. Lennon, V.A.; Kryzer, T.J.; Pittock, S.J.; Verkman, A.S.; Hinson, S.R. IgG marker of optic-spinal multiple sclerosis binds to the aquaporin-4 water channel. *J. Exp. Med.* **2005**, *202*, 473–477. [CrossRef]
156. Melamud, L.; Fernández, J.M.; Rivarola, V.; Di Giusto, G.; Ford, P.; Villa, A.; Capurro, C. Neuromyelitis Optica Immunoglobulin G present in sera from neuromyelitis optica patients affects aquaporin-4 expression and water permeability of the astrocyte plasma membrane. *J. Neurosci. Res.* **2012**, *90*, 1240–1248. [CrossRef] [PubMed]
157. Nicchia, G.P.; Mastrototaro, M.; Rossi, A.; Pisani, F.; Tortorella, C.; Ruggieri, M.; Lia, A.; Trojano, M.; Frigeri, A.; Svelto, M. Aquaporin-4 orthogonal arrays of particles are the target for neuromyelitis optica autoantibodies. *GLIA* **2009**, *57*, 1363–1373. [CrossRef]
158. Rossi, A.; Ratelade, J.; Papadopoulos, M.C.; Bennett, J.L.; Verkman, A.S. Consequences of NMO-IgG binding to aquaporin-4 in neuromyelitis optica. *Proc. Natl. Acad. Sci. USA* **2012**, *109*, E1511, author reply E1512. [CrossRef]
159. Hinson, S.R.; Romero, M.F.; Popescu, B.F.; Lucchinetti, C.F.; Fryer, J.P.; Wolburg, H.; Fallier-Becker, P.; Noell, S.; Lennon, V.A. Molecular outcomes of neuromyelitis optica (NMO)-IgG binding to aquaporin-4 in astrocytes. *Proc. Natl. Acad. Sci. USA* **2012**, *109*, 1245–1250. [CrossRef]
160. Boassa, D.; Yool, A.J. Single amino acids in the carboxyl terminal domain of aquaporin-1 contribute to cGMP-dependent ion channel activation. *BMC Physiol.* **2003**, *3*, 12. [CrossRef]
161. Tomita, Y.; Dorward, H.; Yool, A.J.; Smith, E.; Townsend, A.R.; Price, T.J.; Hardingham, J.E. Role of Aquaporin 1 Signalling in Cancer Development and Progression. *Int. J. Mol. Sci.* **2017**, *18*, 299. [CrossRef]
162. Yu, J.; Yool, A.J.; Schulten, K.; Tajkhorshid, E. Mechanism of gating and ion conductivity of a possible tetrameric pore in aquaporin-1. *Structure* **2006**, *14*, 1411–1423. [CrossRef]
163. Saadoun, S.; Papadopoulos, M.C.; Davies, D.C.; Bell, B.A.; Krishna, S. Increased aquaporin 1 water channel expression in human brain tumours. *Br. J. Cancer* **2002**, *87*, 621–623. [CrossRef] [PubMed]
164. Moon, C.; Soria, J.C.; Jang, S.J.; Lee, J.; Hoque, M.O.; Sibony, M.; Trink, B.; Chang, Y.S.; Sidransky, D.; Mao, L. Involvement of aquaporins in colorectal carcinogenesis. *Oncogene* **2003**, *22*, 6699–6703. [CrossRef]
165. El Hindy, N.; Bankfalvi, A.; Herring, A.; Adamzik, M.; Lambertz, N.; Zhu, Y.; Siffert, W.; Sure, U.; Sandalcioglu, I.E. Correlation of aquaporin-1 water channel protein expression with tumor angiogenesis in human astrocytoma. *Anticancer. Res.* **2013**, *33*, 609–613. [PubMed]
166. Monzani, E.; Shtil, A.A.; La Porta, C.A.M. The water channels, new druggable targets to combat cancer cell survival, invasiveness and metastasis. *Curr. Drug Targets* **2007**, *8*, 1132–1137. [CrossRef] [PubMed]
167. Verkman, A.S.; Hara-Chikuma, M.; Papadopoulos, M.C. Aquaporins--new players in cancer biology. *J. Mol. Med.* **2008**, *86*, 523–529. [CrossRef]
168. Hu, J.; Verkman, A.S. Increased migration and metastatic potential of tumor cells expressing aquaporin water channels. *FASEB J.* **2006**, *20*, 1892–1894. [CrossRef]
169. Jiang, Y. Aquaporin-1 activity of plasma membrane affects HT20 colon cancer cell migration. *IUBMB Life* **2009**, *61*, 1001–1009. [CrossRef]
170. Oster, G.F.; Perelson, A.S. The physics of cell motility. *J. Cell Sci. Suppl.* **1987**, *8*, 35–54. [CrossRef]
171. Chen, X.-M.; O'Hara, S.P.; Huang, B.Q.; Splinter, P.L.; Nelson, J.B.; LaRusso, N.F. Localized glucose and water influx facilitates Cryptosporidium parvum cellular invasion by means of modulation of host-cell membrane protrusion. *Proc. Natl. Acad. Sci. USA* **2005**, *102*, 6338–6343. [CrossRef]
172. Stroka, K.M.; Jiang, H.; Chen, S.H.; Tong, Z.; Wirtz, D.; Sun, S.X.; Konstantopoulos, K. Water permeation drives tumor cell migration in confined microenvironments. *Cell* **2014**, *157*, 611–623. [CrossRef]
173. Schwab, A.; Fabian, A.; Hanley, P.J.; Stock, C. Role of ion channels and transporters in cell migration. *Physiol. Rev.* **2012**, *92*, 1865–1913. [CrossRef] [PubMed]
174. Klein, M.; Seeger, P.; Schuricht, B.; Alper, S.L.; Schwab, A. Polarization of Na^+/H^+ and Cl^-/HCO_3^- exchangers in migrating renal epithelial cells. *J. Gen. Physiol.* **2000**, *115*, 599–608. [CrossRef] [PubMed]
175. Schwab, A.; Stock, C. Ion channels and transporters in tumour cell migration and invasion. *Philos. Trans. R. Soc. Lond. B Biol. Sci.* **2014**, *369*, 20130102. [CrossRef] [PubMed]
176. Kourghi, M.; Pei, J.V.; De Ieso, M.L.; Flynn, G.; Yool, A.J. Bumetanide Derivatives AqB007 and AqB011 Selectively Block the Aquaporin-1 Ion Channel Conductance and Slow Cancer Cell Migration. *Mol. Pharmacol.* **2016**, *89*, 133–140. [CrossRef]
177. Pelagalli, A.; Nardelli, A.; Fontanella, R.; Zannetti, A. Inhibition of AQP1 Hampers Osteosarcoma and Hepatocellular Carcinoma Progression Mediated by Bone Marrow-Derived Mesenchymal Stem Cells. *Int. J. Mol. Sci.* **2016**, *17*, 1102. [CrossRef]

178. Jung, Y.; Kim, J.K.; Shiozawa, Y.; Wang, J.; Mishra, A.; Joseph, J.; Berry, J.E.; McGee, S.; Lee, E.; Sun, H.; et al. Recruitment of mesenchymal stem cells into prostate tumours promotes metastasis. *Nat. Commun.* **2013**, *4*, 1795. [CrossRef]
179. Zetter, B.R. Angiogenesis and tumor metastasis. *Annu. Rev. Med.* **1998**, *49*, 407–424. [CrossRef]
180. Munaron, L. Systems biology of ion channels and transporters in tumor angiogenesis: An omics view. *Biochim. Biophys. Acta* **2015**, *1848 Pt B*, 2647–2656. [CrossRef]
181. Saadoun, S.; Papadopoulos, M.C.; Hara-Chikuma, M.; Verkman, A.S. Impairment of angiogenesis and cell migration by targeted aquaporin-1 gene disruption. *Nature* **2005**, *434*, 786–792. [CrossRef] [PubMed]
182. Nicchia, G.P.; Stigliano, C.; Sparaneo, A.; Rossi, A.; Frigeri, A.; Svelto, M. Inhibition of aquaporin-1 dependent angiogenesis impairs tumour growth in a mouse model of melanoma. *J. Mol. Med.* **2013**, *91*, 613–623. [CrossRef] [PubMed]
183. Esteva-Font, C.; Jin, B.-J.; Verkman, A.S. Aquaporin-1 gene deletion reduces breast tumor growth and lung metastasis in tumor-producing MMTV-PyVT mice. *FASEB J.* **2014**, *28*, 1446–1453. [CrossRef]
184. Monzani, E.; Bazzotti, R.; Perego, C.; La Porta, C.A.M. AQP1 is not only a water channel: It contributes to cell migration through Lin7/beta-catenin. *PLoS ONE* **2009**, *4*, e6167. [CrossRef]
185. Clapp, C.; de la Escalera, G.M. Aquaporin-1: A novel promoter of tumor angiogenesis. *Trends Endocrinol. Metab.* **2006**, *17*, 1–2. [CrossRef]
186. Dorward, H.S.; Du, A.; Bruhn, M.A.; Wrin, J.; Pei, J.V.; Evdokiou, A.; Price, T.J.; Yool, A.J.; Hardingham, J.E. Pharmacological blockade of aquaporin-1 water channel by AqB013 restricts migration and invasiveness of colon cancer cells and prevents endothelial tube formation in vitro. *J. Exp. Clin. Cancer Res.* **2016**, *35*, 36. [CrossRef]
187. Hoque, M.O.; Soria, J.C.; Woo, J.; Lee, T.; Lee, J.; Jang, S.J.; Upadhyay, S.; Trink, B.; Monitto, C.; Desmaze, C.; et al. Aquaporin 1 is overexpressed in lung cancer and stimulates NIH-3T3 cell proliferation and anchorage-independent growth. *Am. J. Pathol.* **2006**, *168*, 1345–1353. [CrossRef]
188. Klebe, S.; Griggs, K.; Cheng, Y.; Driml, J.; Henderson, D.W.; Reid, G. Blockade of aquaporin 1 inhibits proliferation, motility, and metastatic potential of mesothelioma in vitro but not in an in vivo model. *Dis. Markers* **2015**, *2015*, 286719. [CrossRef]
189. Galán-Cobo, A.; Ramírez-Lorca, R.; Echevarría, M. Role of aquaporins in cell proliferation: What else beyond water permeability? *Channels* **2016**, *10*, 185–201. [CrossRef]
190. Kunzelmann, K. Ion channels and cancer. *J. Membr. Biol.* **2005**, *205*, 159–173. [CrossRef]
191. Pedersen, S.F.; Hoffmann, E.K.; Novak, I. Cell volume regulation in epithelial physiology and cancer. *Front. Physiol.* **2013**, *4*, 233. [CrossRef]
192. Beitz, E.; Golldack, A.; Rothert, M.; von Bülow, J. Challenges and achievements in the therapeutic modulation of aquaporin functionality. *Pharmacol. Ther.* **2015**, *155*, 22–35. [CrossRef]
193. Frigeri, A.; Nicchia, G.P.; Svelto, M. Aquaporins as targets for drug discovery. *Curr. Pharm. Des.* **2007**, *13*, 2421–2427. [CrossRef] [PubMed]
194. Jeyaseelan, K.; Sepramaniam, S.; Armugam, A.; Wintour, E.M. Aquaporins: A promising target for drug development. *Expert. Opin. Ther. Targets* **2006**, *10*, 889–909. [CrossRef]
195. Deen, P.M.; Verdijk, M.A.; Knoers, N.V.; Wieringa, B.; Monnens, L.A.; van Os, C.H.; van Oost, B.A. Requirement of human renal water channel aquaporin-2 for vasopressin-dependent concentration of urine. *Science* **1994**, *264*, 92–95. [CrossRef] [PubMed]
196. Tamarappoo, B.K.; Verkman, A.S. Defective aquaporin-2 trafficking in nephrogenic diabetes insipidus and correction by chemical chaperones. *J. Clin. Investig.* **1998**, *101*, 2257–2267. [CrossRef]
197. King, L.S.; Choi, M.; Fernandez, P.C.; Cartron, J.P.; Agre, P. Defective urinary concentrating ability due to a complete deficiency of aquaporin-1. *N. Engl. J. Med.* **2001**, *345*, 175–179. [CrossRef] [PubMed]
198. Chepelinsky, A.B. Structural function of MIP/aquaporin 0 in the eye lens; genetic defects lead to congenital inherited cataracts. In *Handbook of Experimental Pharmacology*; Springer: Berlin/Heidelberg, Germany, 2009; pp. 265–297. [CrossRef]

cells

MDPI

Editorial

An Important Step in Neuroscience: Camillo Golgi and His Discoveries

Vicentiu Mircea Saceleanu [1,2], Razvan-Adrian Covache-Busuioc [3,*], Horia-Petre Costin [3,*], Luca-Andrei Glavan [3,*] and Alexandru Vlad Ciurea [3,4,5]

1 Neurosurgery Department, Sibiu County Emergency Hospital, 550245 Sibiu, Romania
2 Department of Neurosurgery, "Lucian Blaga" University of Medicine, 550024 Sibiu, Romania
3 Department of Neurosurgery, "Carol Davila" University of Medicine and Pharmacy, 020021 Bucharest, Romania
4 Neurosurgery Department and Scientific, Sanador Clinical Hospital, 70000 Bucharest, Romania
5 Department of Medical Sciences, Honorary Member of the Romanian Academy, 010071 Bucharest, Romania
* Correspondence: razvan_adrian69@yahoo.ro (R.-A.C.-B.); horiacostin2001@yahoo.com (H.-P.C.); glavan.luca@gmail.com (L.-A.G.)

Citation: Saceleanu, V.M.; Covache-Busuioc, R.-A.; Costin, H.-P.; Glavan, L.-A.; Ciurea, A.V. An Important Step in Neuroscience: Camillo Golgi and His Discoveries. *Cells* **2022**, *11*, 4112. https://doi.org/10.3390/cells11244112

Received: 1 December 2022
Accepted: 12 December 2022
Published: 18 December 2022

1. General Data

Camillo Golgi (Figure 1) is one of the most prestigious personalities of modern medicine. His success is due to the revolutionary research he has conducted in fields such as cell biology, histology, anatomy, neurology, neuroscience and parasitology. Thus, his prestige is due to the numerous discoveries that support modern science.

Figure 1. Picture of Camillo Golgi. Image in public domain and free from copyright issues.

He was born on 7 July 1843 in the town of Corteno in the province of Brescia, Italy. Inspired by his family, Camillo Golgi began his studies in medicine at the University of Pavia, finishing them in 1865 [1,2]. Since his father was also a medical doctor, the only purpose for Camillo Golgi was to help him; however, the general vision of Golgi concerning medicine was changed when he started his collaboration with his mentor, the doctor Cesare Lombroso (1835–1909), the founder of the "Italian school of positivist criminology". Under his guidance, Camillo Golgi wrote his dissertation based on the etiology of mental illness [3], thus obtaining his medical degree in 1868. Thanks to this collaboration, Camillo Golgi was attracted by the desire to explore the human brain, a research area of interest, which was lacking many breakthroughs at the time.

Subsequently, due to numerous antithetical perspectives between Cesare Lombroso and Camillo Golgi, the collaboration between them ceased. This is when Camillo Golgi met a true personality of science at that time, Giulio Bizzozero (1846–1901), a well-known pioneer of histology, famous for his studies in the field of microbiology (helicobacter pylori) and for describing the platelets [3]. The collaboration was very beneficial for Camillo Golgi because Giulio Bizzozero mentored him in the field of optic microscopy, influencing his further research.

In 1872, Camillo Golgi became Chief of the Hospital of the Chronically Ill in Abbiategrasso, near Milan. This is where he invented, in 1873, the "Black Reaction" [4]. Later on, in 1879, he was appointed Chief of Anatomy at the University of Siena for one year. Returning to Pavia, he was appointed Professor of General Pathology and the honorary chief of the medical hospital in Pavia [5]. Camillo Golgi held the position of Rector of the University of Pavia two times: the first time between 1893–1896, and later on 1901–1909. The year 1918 marks his retirement, his research in his laboratory continuing until 1923. Despite an existence full of glorious and impactful achievements, Camillo Golgi quietly passed away at the age of 82, on 21 January 1926, in Pavia, the city where he spent a major part of his life.

2. Neuroscience Contribution

The two influences exerted by his mentors, Giulio Bizzozero and Cesare Lombroso, shaped his research emphasizing the structure of the nervous system. Governed by the question "What are the cytoarchitectonics of the fascinating nervous system?", Camillo Golgi began research in this area.

In the 19th century, the general aspects regarding the morphology of the nervous cell were not entirely elucidated. Earlier that century, Otto Friedrich Karl Deiters (1834–1863) managed to demonstrate a certain continuity between the nerve cell and its extensions—axons and dendrites [6]. Additionally, the methods of fixation for the histological specimens at that time were inappropriate for capturing the pretentious nervous tissue under a microscope; hematoxylin or carmine did not provide satisfactory results, the reason being the extraordinarily small space between the cells of the nervous tissue.

That was the moment when, in the year 1873, Camillo Golgi made a revolutionary discovery, which totally changed the trajectory of the neurological field. Camillo Golgi was studying some specimens of nervous tissue, fixed according to a new technique. Small pieces of nervous tissue were kept in aldehyde, and later they were hardened in potassium dichromate. After this, silver nitrate reacted with pieces of the specimen that have been hardened in potassium dichromate. This coloration technique was later called the "Black Reaction"—"La Reazione Nera" in Italian, or, as it is now known in the scientific community, "Golgi's impregnation"—"Golgi Staining".

There are two advantages of using this technique: the first one is that the silver chromate will form microparticles inside the cells, while the second one is that the "Black Reaction" only impregnates just a couple of the cells inside the histological sample, in comparison with other methods. This is very useful, especially for studying the histology of the nervous system, as it allowed Golgi to easily differentiate axons from dendrites. Unlike the classical methods, the fixation by using solutions with metals determines the preferential precipitation on certain structures, through a mechanism that is not even to this day completely elucidated [7].

A year before the invention of the "Black Reaction", Joseph von Gerlach (1820–1896), a German anatomist, formulated the reticular theory. It claims that the nervous system would actually function as a syncytium due to the multitude of dendritic ramifications, therefore it would, in fact, function as a whole [8]. Glial cells, particularly the astrocytes, create an enormous network that connects vast parts of the nervous system, by modulating and regulating synapses altogether.

On the other hand, the enunciation of the reticular theory also implies the description of its rival, "The Neural Theory" (or neuron doctrine), postulated by Santiago Ramón y Cajal. It assumes that the neuron functions as a unit in itself, not as an entire system. The statement is based on studies conducted by the Spanish researcher, who observed a discontinuity between neuronal extensions, later called synaptic [8].

Cajal was amazed by the quality of Camillo Golgi's tissues impregnated using the Black Reaction method, and in fact, more than half of his sketches were made using this method. Moreover, Cajal even made certain improvements to the protocol [7]. Thus, in 1891, "The Neural Theory" was formulated by the Spanish researcher, a theory which denounces The Reticular Theory as being obsolete.

The debate between the two theories was the foundation of modern neuroscience, and for about a century, "The Neural Theory" was the central dogma of this field. For more than a century, the reticular theory was shadowed by the concept postulated by Cajal. However, it is important to state that glial cells were an underdeveloped subject, the spotlight being on the neurons and their morpho-physiological particularities. We believe that the truth is a comprehension of the best arguments presented by both theories, as numerous recent studies have revealed new information about the functioning of the nervous system. A functional syncytium created by gap junctions between the astrocytes has been identified. Astrocytes, in turn, can form gap junctions with oligodendrocytes as well. The bonds between the astrocytes are tightened at the level of the gap junction most often by the Cx43 protein [9]. This functional syncytium can be described, briefly, by an extrapolation: an astrocyte is in simultaneous contact with about 140 thousand nervous cells [10], and currently, it is considered that the number of astrocytes exceeds the number of neurons in the nervous system of mammals [11]. At the same time, linking these cells through gap junctions will decrease the electrical resistance between cells, allowing joint depolarization and creating a similar behavior for all the components of the system, functioning therefore as a singular unit [12]. The property of astrocytes to work as a syncytial unit represents a key point in synchronization of the nervous system network regarding the brain state [13].

It is interesting to note that, in 1903, Camillo Golgi published a paper entitled "Opera Omnia", a work in which numerous anatomical nerve structures were sketched. Among them are the cerebellum and the hippocampus (Figure 2), formations that present numerous glial cells, explaining this way the fierce support of the reticular theory by the Italian researcher [14].

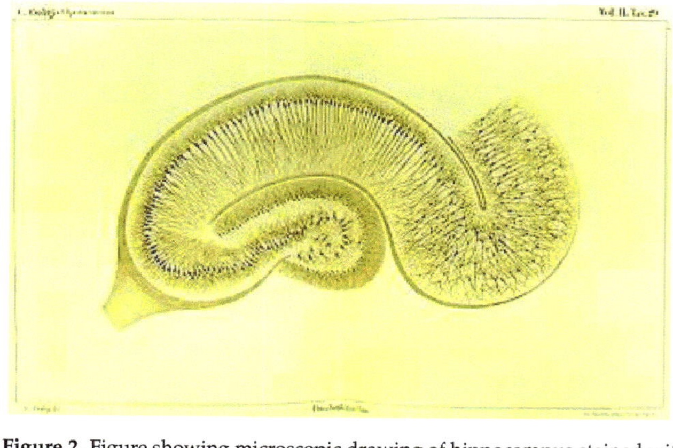

Figure 2. Figure showing microscopic drawing of hippocampus stained with Black Reactio. Histological plate prepared by Camilo Golgi. Image in public domain and free from copyright issues. Source of Image: Wikimedia Commons.

3. Important Discoveries

On the basis of "The Black Reaction", Camillo Golgi published in 1874 the work "Sulla fina anatomia del cervelletto umano" [15]. In this work, the researcher made a detailed description regarding the cytoarchitectonics of the nervous system, simultaneously with the description of a nervous cell that bears his name up to this moment: Golgi cell. The scientist noticed the presence of a developed axonal plexus, a criterion that is still used nowadays to observe a Golgi cell on a histological sample. In the same paper, Golgi outlines the hypothesis of a local interconnectivity, such as nervous cells functioning as a whole, enhanced and synchronized by the glial cells, which can be found in the entire nervous system. One of the functions of Golgi cells is that, being GABAergic and glycinergic, they inhibit neuronal circuits in the vestibulocerebellar loop [16].

In 1878, he discovered two new types of sensory receptors. First of all, he discovered the Golgi–Mazzoni corpuscles, that encapsulate the nerve ending, being fine pressure receptors that are located only in the fingertips. These structures are similar to the Vater-Pacini receptors, except for their localization. Second of all, another discovery, which took place in 1878, was called the Golgi tendon organ, located in the muscle–tendon junction and representing a muscular proprioceptor [3].

Starting in 1879, Camillo Golgi was appointed Professor of General Pathology at the hospital in Pavia, thus having the opportunity to study the in vivo evolution of malaria. By associating the clinical symptomatology developed by his patients with the evolution of the parasite in the blood, the researcher managed to elaborate a description of the cycle that takes place in the human erythrocyte regarding Plasmodium, now known as the Golgi cycle [17].

Later that century, Camillo Golgi made new discoveries regarding the kidney's histology. He noticed that in each nephron, the ascending part of the Henle loop returns back to the cortical, juxtamedullary zone. This part comes in close contact with the Malpighi corpuscle, and was later called the juxtaglomerular apparatus, important in regulating blood pressure through renin secretion.

In 1898, while analyzing the Purkinje cells in the cerebellum, Camillo Golgi observed what we now call the Golgi apparatus, through silver nitrate staining. Immediately after this, he discovered that the Golgi apparatus could also be found inside the neural cells of the spinal ganglia. This event started an avalanche of adjacent research that proved the presence of the Golgi apparatus in numerous other types of structures. Initially, it was called "internal reticular apparatus", due to the fact that, at that time, the function of the Golgi apparatus could not be stated, and it was assumed for a long time that the organelle was actually a fixation artifact. This whole situation was called "The Golgi Controversy" [18]. The dilemma was solved in 1954, when the first observation of this organelle was made through electron microscopy. Subsequently, the uneven division of the phosphatases on the trans face of the Golgi apparatus suggests its segmentation into several parts. Other studies have involved the use of the staining method used by the Italian scientist when describing the organelle and observing the specimens, this time using electron microscopy. The result was the discovery of the cis face of this organelle, which was the only part that was highlighted by this method [19]. This organelle was a central pillar of the Nobel lecture of researcher George E. Palade, in which he explained the mechanism of extracellular secretion, concerning the subject of the acinar pancreatic cell [20].

4. Recognition

Camillo Golgi's prestige has been recognized by numerous international faculties and scientific societies. They paid tribute to the scientist by choosing him as Doctor Honoris Causa of the Universities of Cambridge, Athens, Paris (Université de la Sorbonne), Geneva and Kristiania (Oslo), also being a Nobel laureate in 1906, along with Santiago Ramón y Cajal, an award obtained in recognition of their work regarding the structure of the nervous system.

Additionally, both scientists held Nobel lectures that summarized their findings: Santiago Ramón y Cajal: "The Structure and Connexions of Neurons" and Camillo Golgi: "The neuron doctrine—theory and facts".

5. Conclusions

Despite the fact that Santiago Ramón y Cajal is proclaimed as the father of modern neuroscience because he promoted "The Neuron Doctrine", we believe that researcher

Camillo Golgi deserves at least as much recognition. Numerous recent studies justify the return of the reticular theory in the spotlight of neuroscience, bringing it back to relevancy.

If Santiago Ramón y Cajal's work is a marble statue, Camillo Golgi's work was the podium of the statue, but also the marble block from which it was carved. "La Reazione Nera", through the silver impregnation of nervous structures is the testament of Golgi's genius, the gold standard for neuroscience research.

Author Contributions: Conceptualization, A.V.C. and R.-A.C.-B.; Methodology, V.M.S. and H.-P.C.; Investigation, L.-A.G. and V.M.S.; Software, R.-A.C.-B. and H.-P.C. All authors have read and agreed to the published version of the manuscript.

Funding: This research received no external funding.

Institutional Review Board Statement: Not applicable.

Informed Consent Statement: Not applicable.

Data Availability Statement: Not applicable.

Conflicts of Interest: The authors declare no conflict of interest.

References

1. Ghosh, S.K. Camillo Golgi (1843–1926): Scientist extraordinaire and pioneer figure of modern neurology. *Anat. Cell Biol.* **2020**, *53*, 385–392. Available online: https://acbjournal.org/journalDOIx.php?id=10.5115/acb.20.196 (accessed on 1 January 2020). [CrossRef] [PubMed]
2. Pannese, E. The Black Reaction. *Brain Res. Bull* **1996**, *41*, 343–349. Available online: https://www.sciencedirect.com/science/article/pii/S0361923096001773 (accessed on 1 January 2020). [CrossRef] [PubMed]
3. Mazzarello, P. Camillo Golgi's Scientific Biography. *J. Hist. Neurosci.* **1999**, *8*, 121–131. Available online: https://www.tandfonline.com/doi/abs/10.1076/jhin.8.2.121.1836 (accessed on 1 January 2020). [CrossRef] [PubMed]
4. Kruger, L. The sensory neuron and the triumph of Camillo Golgi. *Brain Res. Rev.* **2007**, *55*, 406–410. Available online: https://www.sciencedirect.com/science/article/pii/S0165017307000239 (accessed on 1 January 2020). [CrossRef] [PubMed]
5. Pannese, E. The contribution of Camillo Golgi to our understanding of the structure of the nervous system. In *Archives Italiennes de Biologie*; Pietrini, P., Ghelarducci, B., Ricciardi, E., d'Ascanio, P., Aloe, L., Kubin, L., Eds.; Pisa University Press: Pisa, Italy, 2007; pp. 111–115.
6. Glickstein, M. Golgi: A life in science. *Brain* **2012**, *135*, 301–303. [CrossRef]
7. Bentivoglio, M.; Cotrufo, T.; Ferrari, S.; Tesoriero, C.; Mariotto, S.; Bertini, G.; Berzero, A.; Mazzarello, P. The Original Histological Slides of Camillo Golgi and His Discoveries on Neuronal Structure. *Front. Neuroanat.* **2019**, *13*, 3. Available online: https://www.frontiersin.org/articles/10.3389/fnana.2019.00003 (accessed on 1 January 2020). [CrossRef] [PubMed]
8. Saceleanu, V.M.; Mohan, A.G.; Covache-Busuioc, R.A.; Costin, H.P.; Ciurea, A.V. Wilhelm von Waldeyer: Important Steps in Neural Theory, Anatomy and Citology. *Brain Sci.* **2022**, *12*, 224. Available online: https://www.mdpi.com/2076-3425/12/2/224 (accessed on 1 January 2020). [CrossRef] [PubMed]
9. Scemes, E.; Spray, D.C. The astrocytic syncytium. In *Advances in Molecular and Cell Biology*; Elsevier: Amsterdam, The Netherlands, 2003; pp. 165–179. Available online: https://www.sciencedirect.com/science/article/pii/S1569255803310070 (accessed on 1 January 2020).
10. Bushong, E.A.; Martone, M.E.; Jones, Y.Z.; Ellisman, M.H. Protoplasmic Astrocytes in CA1 Stratum Radiatum Occupy Separate Anatomical Domains. *J. Neurosci.* **2002**, *22*, 183. Available online: http://www.jneurosci.org/content/22/1/183.abstract (accessed on 1 January 2020). [CrossRef] [PubMed]
11. Khakh, B.S.; Sofroniew, M.V. Diversity of astrocyte functions and phenotypes in neural circuits. *Nat. Neurosci.* **2015**, *18*, 942–952. [CrossRef] [PubMed]
12. Kiyoshi, C.M.; Zhou, M. Astrocyte syncytium: A functional reticular system in the brain. *Neural. Regen. Res.* **2019**, *14*, 595. Available online: https://journals.lww.com/nrronline/Fulltext/2019/14040/Astrocyte_syncytium__a_functional_reticular_system.10.aspx (accessed on 1 January 2020). [PubMed]
13. Pacholko, A.G.; Wotton, C.A.; Bekar, L.K. Astrocytes—The Ultimate Effectors of Long-Range Neuromodulatory Networks? *Front. Cell Neurosci.* **2020**, *14*, 581075. Available online: https://www.frontiersin.org/articles/10.3389/fncel.2020.581075 (accessed on 1 January 2020). [CrossRef] [PubMed]
14. Jones, E.G. Golgi, Cajal and the Neuron Doctrine. *J. Hist. Neurosci.* **1999**, *8*, 170–178. Available online: https://www.tandfonline.com/doi/abs/10.1076/jhin.8.2.170.1838 (accessed on 1 January 2020). [CrossRef] [PubMed]
15. Golgi, C. Sulla fina anatomia del cervelletto umano. In *Stabilimento dei fratelli Rechiedei*; Editore Libraio Milano: Milan, Italy, 1874.
16. Galliano, E.; Mazzarello, P.; D'Angelo, E. Discovery and rediscoveries of Golgi cells. *J. Physiol.* **2010**, *588*, 3639–3655. [CrossRef] [PubMed]

17. Golgi, C. Sul ciclo evolutivo dei parassiti malarici nella febbre terzana: Diagnosi differenziale tra i parassiti endoglobulari malarici della terzana e quelli della quartana. *Arch. Sci. Med. Torino.* **1889**, *13*, 173–196.
18. Bentivoglio, M. The Discovery of the Golgi Apparatus. *J. Hist. Neurosci.* **1999**, *8*, 202–208. Available online: https://www.tandfonline.com/doi/abs/10.1076/jhin.8.2.202.1833 (accessed on 1 January 2020). [CrossRef] [PubMed]
19. Farquhar, M.G.; Palade, G.E. The Golgi apparatus: 100 years of progress and controversy. *Trends Cell Biol.* **1998**, *8*, 2–10. Available online: https://pubmed.ncbi.nlm.nih.gov/9695800 (accessed on 1 January 2020). [CrossRef] [PubMed]
20. Palade, G. Intracellular Aspects of the Process of Protein Synthesis. *Science* **1975**, *189*, 347–358. Available online: http://www.jstor.org/stable/1740547 (accessed on 1 January 2020). [CrossRef] [PubMed]

brain sciences

MDPI

Review

Unraveling the Multifaceted Role of the Golgi Apparatus: Insights into Neuronal Plasticity, Development, Neurogenesis, Alzheimer's Disease, and SARS-CoV-2 Interactions

Corneliu Toader [1,2], Lucian Eva [3,4,*], Razvan-Adrian Covache-Busuioc [1,*], Horia Petre Costin [1], Luca-Andrei Glavan [1], Antonio Daniel Corlatescu [1] and Alexandru Vlad Ciurea [1,5]

1 Department of Neurosurgery, "Carol Davila" University of Medicine and Pharmacy, 020021 Bucharest, Romania; corneliu.toader@umfcd.ro (C.T.); horia-petre.costin0720@stud.umfcd.ro (H.P.C.); luca-andrei.glavan0720@stud.umfcd.ro (L.-A.G.); antonio.corlatescu0920@stud.umfcd.ro (A.D.C.); prof.avciurea@gmail.com (A.V.C.)
2 Department of Vascular Neurosurgery, National Institute of Neurology and Neurovascular Diseases, 077160 Bucharest, Romania
3 Faculty of Medicine, "Dunarea de Jos" University of Galati, 800201 Galați, Romania
4 Emergency Clinical Hospital "Prof. dr. N. Oblu", 700309 Iasi, Romania
5 Neurosurgery Department, Sanador Clinical Hospital, 010991 Bucharest, Romania
* Correspondence: elucian73@yahoo.com (L.E.); razvan-adrian.covache-busuioc0720@stud.umfcd.ro (R.-A.C.-B.)

Abstract: This article critically evaluates the multifunctional role of the Golgi apparatus within neurological paradigms. We succinctly highlight its influence on neuronal plasticity, development, and the vital trafficking and sorting mechanisms for proteins and lipids. The discourse further navigates to its regulatory prominence in neurogenesis and its implications in Alzheimer's Disease pathogenesis. The emerging nexus between the Golgi apparatus and SARS-CoV-2 underscores its potential in viral replication processes. This consolidation accentuates the Golgi apparatus's centrality in neurobiology and its intersections with both neurodegenerative and viral pathologies. In essence, understanding the Golgi's multifaceted functions harbors profound implications for future therapeutic innovations in neurological and viral afflictions.

Keywords: Golgi apparatus; neuronal plasticity; neuronal development; neurogenesis; Alzheimer's disease; SARS-CoV-2

Citation: Toader, C.; Eva, L.; Covache-Busuioc, R.-A.; Costin, H.P.; Glavan, L.-A.; Corlatescu, A.D.; Ciurea, A.V. Unraveling the Multifaceted Role of the Golgi Apparatus: Insights into Neuronal Plasticity, Development, Neurogenesis, Alzheimer's Disease, and SARS-CoV-2 Interactions. *Brain Sci.* **2023**, *13*, 1363. https://doi.org/10.3390/brainsci13101363

Academic Editor: Deepshikha Pande Katare

Received: 1 August 2023
Revised: 16 September 2023
Accepted: 20 September 2023
Published: 23 September 2023

1. Introduction

The complexity of the nervous system's architecture is underpinned by processes of neuronal maturation and synaptic adaptability. Central to these intricate mechanisms is an oft-overlooked cellular entity in neurobiological discussions: the Golgi apparatus. Functioning as the cellular "sorting hub", the Golgi apparatus is indispensable in protein categorization and adjustments crucial to the dynamic equilibrium of neuronal structure, function, and synaptic adaptability [1]. Contemporary research has illuminated potential associations between the Golgi apparatus and the pathogenesis of Alzheimer's, particularly emphasizing the role of the Golgi matrix protein GM130 (a morphological determinant protein situated on the cis-face of the Golgi apparatus) [2–4]. Novel insights are surfacing, linking this protein to its Golgi-based counterpart and the pathological shifts characteristic of Alzheimer's, thus ushering in innovative investigative trajectories in our pursuit of understanding this devastating neurodegenerative disorder [4].

The Golgi apparatus is not only pivotal in protein categorization and alterations but also stands out as a central orchestrator of neuronal maturation. The nuanced orchestration of axonal and dendritic differentiation is facilitated by this organelle, paving the way for growth and distinctiveness within neural circuits. Moreover, its influence is unmistakably evident in synaptic plasticity, fundamental processes governing learning and memory within the nervous system [5].

Overall, this paper aims to portray the importance of the Golgi apparatus in key aspects of neurobiology, such as neurogenesis, neural development, and neural plasticity, while also elucidating the pathophysiology of the Golgi apparatus in SARS-CoV-2 infection and Alzheimer's disease.

2. Golgi Apparatus in Axonal Development

Cerebral cells, comprising neurons and glia, are characterized by intricate architectures tailored to execute designated roles. Neurons establish synapses to facilitate electrical communication, oligodendrocytes envelop axons with myelin sheaths for insulating properties, and astrocytes bridge vascular networks [6,7]. These specialized morphologies necessitate distinct routes for protein trafficking, prominently featuring the Golgi apparatus outposts (GOPs) (Figure 1). In specialized cellular contexts, Golgi outposts assume critical functions essential for the sculpting of distinct cellular morphologies and architectures. Notably, within neurons, these organelles are pivotal for the formation of dendritic branches [8]. Concurrently, in muscle cells, they are implicated in the genesis of grid-like microtubule lattices, and in oligodendrocytes, they facilitate the formation of microtubules that encircle the myelin sheath in a spiral configuration [9]. Intriguingly, notwithstanding their crucial roles, to date, Golgi outposts have been exclusively identified in vivo and within primary cultured cells, evading detection in immortalized cell lines [10].

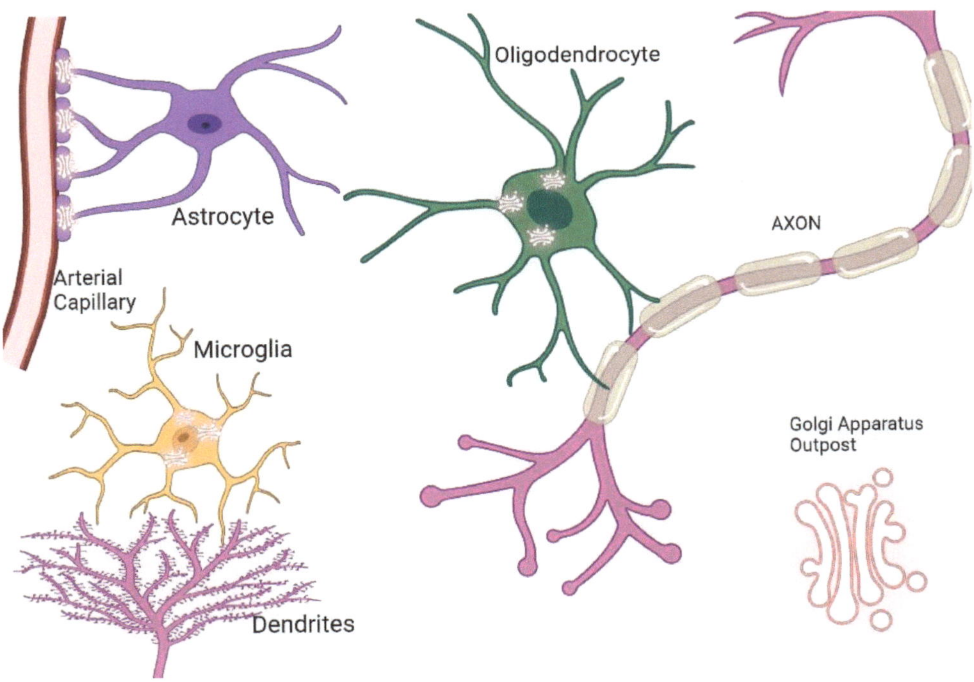

Figure 1. The distribution and function of Golgi outposts in different types of glial cells. In oligodendrocytes, these outposts are located in processes and the myelin sheath, where they nucleate new microtubules along radial and lamellar microtubules. In microglia, Golgi outposts contribute to the nucleation of new microtubules and the establishment of branched processes. In astrocytes, Golgi outposts are observed in endfeet contacting blood vessels as evidenced by electron microscopy, although their specific function is yet to be determined.

In summary, the secretory pathway initiates in the endoplasmic reticulum (ER), where nascent proteins are synthesized. Subsequently, these proteins traverse through the Golgi apparatus (GA) for further maturation, processing, and sorting. They are then dispatched to subsequent post-Golgi/trans-Golgi network (TGN) compartments, ultimately reaching their designated locations such as endosomes, lysosomes, and the plasma membrane [11]. Transmembrane proteins synthesized on the RER are incorporated into the ER membrane subsequently exiting through specialized sites termed ERES (ER exit site) with the aid of COPII vesicles [12]. Following the detachment of COPII proteins [13], these vesicles employ SNAREs (Soluble NSF Attachment Protein Receptors) to amalgamate with tubular assemblies termed ER-Golgi intermediate compartments (ERGICs). Additionally, certain proteins can retrogress from the ER via COPI vesicles initiated at the Golgi apparatus. Within the ERGIC, proteins undergo processing and subsequently navigate to the Golgi apparatus for further modifications, encompassing glycosylation and proteolysis. This journey encompasses the cis-Golgi network (CGN), medial Golgi cisternae, and the trans-Golgi network (TGN) [14].

Upon reaching the TGN, proteins adopt diverse processing trajectories contingent upon their intended function and localization. Vesicles emanating from the Golgi, laden with secreted or transmembrane proteins, can bifurcate and merge with plasma membranes, undergoing subsequent processing. This can culminate in the extracellular secretion of proteins or their integration into plasma membranes. Additionally, vesicles laden with secretory proteins can be conserved proximal to plasma membranes or endosomes, awaiting release upon specific stimuli. Moreover, proteins destined for endosomes and lysosomes undergo modification via an oligosaccharide marker termed mannose 6-phosphate, and subsequently depart the TGN enclosed within clathrin-coated vesicles [15]. Consequently, the selection of secretory routes is determined by the intended role, ultimate destination, and inherent nature of the respective cargo proteins [16].

3. Golgi Apparatus Involvement in Dendritic Formation

Neuronal maturation is characterized by the augmentation of the plasma membrane during processes such as polarization and outgrowth. The localization of membrane constituents is contingent upon the secretory pathway, a sequence encompassing a myriad of organelles such as the endoplasmic reticulum (ER), ER-Golgi intermediate compartment, cis-Golgi, medial Golgi, and trans-Golgi, among others [17]. These specialized organelles orchestrate the synthesis and delivery of novel membrane lipids and proteins; however, the understanding of their spatial arrangement, functionality, and regulatory mechanisms within neurons remains nascent.

In non-neuronal cells, the architecture of the secretory pathway organelles is notably conserved. The ER permeates the cell and includes distinct locales termed "ER exit sites". Here, the coat protein complex II-coated vesicles, laden with newly synthesized cargo, commence their journey to the Golgi complex for subsequent protein processing and sorting. Ultimately, post-Golgi carriers facilitate the conveyance to the plasma membrane [18,19].

Contemporary research underscores that dendrites, the specialized neuronal protrusions, possess inherent secretory faculties. For instance, vesicles emanating from the trans-Golgi network partake in calcium-mediated exocytosis within dendrites [20]. Notably, glycine and glutamate receptors are rapidly discernible on dendritic surfaces, potentially originating from dendritic compartments housing secretory proteins. Dendrites also exhibit the ability to integrate sugars and translate membrane protein-encoding mRNA, hinting at their Golgi-mimetic functions [21]. However, it remains an enigma whether dendritic trafficking encompasses the entire secretory continuum, from ER to Golgi, or is exclusive to the latter stages involving post-Golgi vesicles.

Dendritic structures house distinct Golgi units adept at secretory cargo transport [21]. Quantitative analyses reveal that within the hippocampal neuronal cohort studied, 70% exhibited dendritic Golgi apparatus. An intriguing observation highlighted that most neurons had Golgi units localized to a singular dendrite, while a minor proportion exhibited multiple dendrites with Golgi presence. Overall, 51% of hippocampal neurons possessed Golgi in at least one dendrite, 19.5% in two, and 29.5% were devoid of dendritic Golgi [22]. Interestingly, the presence or absence of Golgi in specific dendrites was not contingent upon the diameter of their proximal extensions.

Contrasting starkly with other membrane-bound organelles ubiquitously found within dendrites, such as ER exit sites, endosomes, and mitochondria, Golgi bodies in dendrites presented a distinct pattern. Utilizing GM130 as a marker, Golgi stacks were conspicuously identified in proximal regions of pyramidal neurons' apical dendrites but were seldom evident on basolateral extensions [23].

A limited subset of neuronal dendrites houses Golgi outposts, typically restricted to one per neuron. These outposts have garnered significant scholarly attention, especially given their potential role as non-centrosomal microtubule organizing centers (MTOCs). Outposts are capable of synthesizing microtubules at considerable distances from their nuclei, facilitating the establishment of microtubule networks. Such configurations are evident in cells with distinct structural attributes, as seen in *Drosophila* neurons, murine muscle cells, and rodent oligodendrocytes. Furthermore, Golgi outposts have been implicated in various pathologies, including muscular dystrophy and Parkinson's disease. Their distribution is especially pronounced in pyramidal neurons, where they predominantly localize to the apical dendrite [24].

The elaboration and extension of dendrites and axons potentially necessitate coordinated modifications in cytoskeletal organization and membrane transport mechanisms. Specifically, molecules such as RhoA [25] and MAP2 [26] appear to be predisposed to augment dendritic growth. Concurrently, a commensurate adaptation in membrane provisioning might be requisite to facilitate their distinctive morphologies.

Intriguingly, neurons also harbor unique secretory conduits in dendrites, which encompass both ER and Golgi outposts [16]. Yet, a conspicuous absence of Golgi outposts in axons prompts inquiries regarding the potential role of the polarized distribution of Golgi proteins in delineating the differential maturation trajectories of dendrites and axons. Contemporary research intimates a proactive involvement of secretory trafficking in the genesis of specialized dendritic compartments. Nevertheless, the extent to which these secretory pathways modulate the differential elongation patterns of dendrites and axons remains enigmatic (as illustrated in Figure 2). Furthermore, the implications of dendritic Golgi outposts in dendritic elaboration, and their putative role in sculpting the distinct morphological attributes of dendrites and axons, await elucidation [27].

Imaging analyses on cultured hippocampal neurons have elucidated the pivotal role of dendritic Golgi outposts (GOPs) in trafficking subsequent to the endoplasmic reticulum, notably via the observation of GFP-VSV-G ts045 transport [27,28]. Intriguingly, NMDA receptors seem to follow a distinct sorting pathway from AMPA receptors at the ER, predominantly associating with dendritic GOPs instead of transiting through somatic Golgi apparatuses (GAs). This intimates that mini-GAs within GOPs might serve as strategic platforms to locally dispatch synaptic receptors, potentially facilitating synaptic plasticity [29,30]. GOPs are instrumental in influencing dendritic expansion and branching dynamics [29,30], likely by orchestrating the allocation of cargo to disparate dendritic branches [28]. Investigations on *Drosophila* sensory neurons have illuminated that perturbations in the Lava-lamp adaptor, or its mutations, culminate in diminished GOPs and a consequent reduction in dendritic branches [31]. For *Drosophila* dendritic arborization neurons, it has been posited that GOPs orchestrate dendritic morphology by acting as epicenters for microtubule nucleation [30]. However, contemporary findings have emphasized that γ-tubulin actually governs dendritic microtubule nucleation, independent of GOPs, across diverse *Drosophila* neuronal classes [32].

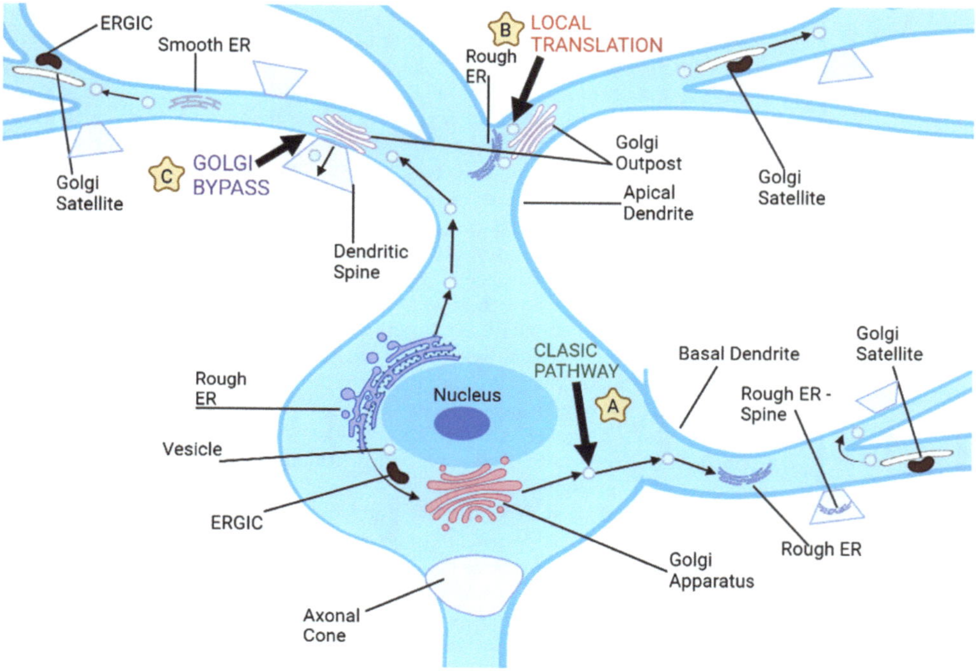

Figure 2. Neuronal dendrites provide multiple pathways for cargo transport and secretory processes, and there are three distinct routes used for dendritic cargo transportation. (**A**) The classic secretion pathway involves protein translation in neuronal cell bodies, followed by exit from endoplasmic reticulum and further processing by the ER-Golgi intermediate compartments and cell body Golgi. Vesicles then transport these proteins to dendrites or other parts of a neuron for storage and transport. (**B**) Local dendritic translation involves synthesizing proteins on dendritic rough ER before sending them off for further modification by dendritic Golgi outposts. After leaving Golgi outposts, proteins are transported via post-Golgi vesicles which may either travel along dendrites or fuse with either dendritic plasma membranes or synaptic spines. (**C**) Golgi bypass pathway allows cargo synthesized on the cell body ER to bypass its Golgi organelle and be transported directly to dendrites for modification by Golgi outposts before being packaged into post-Golgi vesicles destined for synapses or plasma membrane.

Pertaining to GOP genesis, a plausible mechanism might encompass localized de novo synthesis from the ER, reminiscent of the Golgi reconstitution observed at cellular extremities in non-neuronal entities [33]. Alternatively, GOPs could emerge post somatic Golgi fragmentation due to heightened neuronal activity, with remnants of the degraded Golgi potentially serving as blueprints for the subsequent assembly of satellite Golgi arrays [34]. Formation of GOPs might also ensue from the primary somatic GA, prompting inquiries into the participation of Golgi fission machinery components such as LIMK1 [35] and Protein Kinase D1 (PKD1) [36], and the potential involvement of their proximal regulators and distal effectors. Notably, evidence underscores the potential influence of a RhoA-Rho kinase (Rock) signaling cascade in the orchestration of polarized GOPs during neuronal morphogenesis.

The Golgi apparatus inherently harbors the capability to act as a reservoir of membrane constituents vital for localized cellular expansion [27]. Dendrites replete with Golgi elements consistently manifest greater length and intricate architecture as compared to their Golgi-devoid counterparts. Empirical assessments have ascertained that an overwhelming

86% of neurons under examination showcased Golgi elements within their most elongated dendrite [22].

In mature neurons, a hallmark characteristic resides in their intricately branched dendritic arbors, which conspicuously surpass the complexity of dendrites devoid of Golgi outposts.

Collectively, these revelations accentuate that a non-uniform Golgi distribution resonates with dendritic expansion patterns. The enhanced length and complexity of Golgi-endowed dendrites underscore the indispensable role of this organelle in sculpting dendritic architecture.

4. BARS Regulation of Golgi Trafficking

The CtBP (C-terminal-binding protein) family, comprising CtBP1 and CtBP2, is pivotal in a gamut of biological undertakings such as development, differentiation, oncogenesis, and apoptosis, chiefly functioning as transcriptional co-repressors within the nuclear compartment [37]. BARS (Brefeldin A ADP-Ribosylated Substrate) emerges as a salient mediator in membrane fission activities including macropinocytosis, fluid-phase endocytosis, COPI-coated vesicle genesis, and an array of post-Golgi carrier and ribbon events during mitosis. The intricacies of BARS in fission are well-chronicled in cellular studies, where it is posited as a key intermediary in basolateral post-Golgi carrier formation. Specifically, MDCK cells, manifesting polarized epithelial attributes, display hindered Vesicular Stomatitis Virus G (VSVG) transport upon BARS inhibition, attributed to tubular carriers laden with this payload remaining adhered to the Trans Golgi Network (TGN). This fission process orchestrated by BARS at the TGN entails a multifaceted ensemble of proteins [38,39].

Through in situ hybridization evaluations, it is apparent that CtBP family constituents are profusely expressed across the neural framework. Notably, CtBP1 and CtBP2 showcase divergent distribution spectra within mature cerebral structures, distinguished by expression magnitudes, territorial expression blueprints, subcellular targeting proficiencies, and intracellular positioning paradigms. Mouse-based investigations with deletions in one or both CtBP genes intimate the existence of both common and unique functionalities for these proteins [40,41]. Of particular significance is the discernment that concurrent excision of CtBP1 and CtBP2 in murine models precipitates retarded development of forebrain and midbrain structures, a sequela of modulated transcription factor activities in their absence [42,43].

Diminished BARS expression is correlated with marked curtailment in hippocampal neuronal growth, migration, and the multipolar-to-bipolar transitional dynamics in cortical neurons. Notably, these effects can be mitigated by co-expression of fission-inefficient mutants exhibiting reduced nuclear positioning or fission vigor. Contrarily, accentuated growth is observed upon co-expression of the said mutants, an effect which can be counteracted by the concurrent expression of aforementioned mutants [39].

The meticulous orchestration of membrane constituents along the secretory trajectory is cardinal for neuronal tasks such as augmentation, polarity inception, sustenance, synaptic plasticity, and cellular migration [44,45]. This conduit is initiated at the rough endoplasmic reticulum, subsequently traversing the Golgi machinery where membrane proteins designated for axons and dendrites are segregated into specific carriers for extracellular dispatch [46]. Upon departure from the trans-Golgi matrix, they are relayed to plasma membranes via molecular propellants, exemplified by kinesin superfamily entities and myosins. Despite this knowledge, the specialized apparatus governing their TGN exit remains inadequately delineated. Nevertheless, it is established that BARS, an integral affiliate of the CtBP assemblage, is indispensable during neuronal ontogeny [47].

Both LIMK1 and PKD1 are quintessential components of the TGN fission apparatus in polarized epithelial structures, overseeing membrane protein transference either to apical or basolateral facets [48,49]. Concurrently, these proteins, evident within neurons situated at GAs, partake in dynamin-facilitated fission of Golgi conduits, thereby engendering Golgi outposts. Ensuing research ought to delve into the mechanisms by which BARS engenders dendritic carrier fissions at TGNs and probe the putative role of BARS in the fission of Golgi conduits during the generation of Golgi outposts [50].

5. Golgi Apparatus and Synaptic Plasticity

Rare diseases, particularly genetic anomalies, exert a pronounced effect on the nervous system, culminating in manifestations such as neurodegeneration and behavioral perturbations [51]. These infrequent neurological conditions furnish a platform to elucidate hitherto uncharted cellular dynamics integral to neuronal performance. This is especially pertinent in disorders associated with ATP7A/ATP7B genes and COG complex subunit genes. The proteins expressed from these genes predominantly localize within the Golgi apparatus under standard conditions [52]. Over the course of evolution, organisms have intricately modulated copper metabolism and transport. Central to this regulatory framework are ATP7A and ATP7B proteins. These proteins, classified under P-type Cu-transporting ATPases, harness ATP hydrolysis to facilitate the transmembrane movement of copper ions. Essentially, these copper pumps either extrude surplus copper from cells or allocate it to copper-reliant enzymes. The physiological roles of Cu-transporting ATPases are multifaceted, spanning dietary copper excretion through bile, placental transport, and lactational secretion to modulating resistance against select anti-cancer therapeutics [53–56].

Mutational alterations in the ATP7A gene underpin Menkes disease, typified by systemic copper deficiency stemming from hindered intestinal copper assimilation [57]. The clinical picture of Menkes disease predominantly emerges in childhood, exhibiting a spectrum of systemic and neurologically linked symptoms. The latter include intellectual impairments and gray matter neurodegeneration [58]. Such clinical manifestations may be ascribed to perturbations in copper-reliant enzymes that traverse the Golgi apparatus or reside within mitochondria [59]. Paradoxically, cell-specific aberrations in ATP7A precipitate intracellular copper accumulation, a scenario arising despite the overarching copper deficiency evident in Menkes patients [60]. In a contrasting paradigm, mutations in the ATP7B gene underlie Wilson's disease. This disorder is marked by hepatic impediments to copper excretion, ensuing in systemic copper excess, hepatotoxicity, psychiatric manifestations, and neurodegeneration, particularly in the lenticular domain [59].

Differential expressions of ATP7 and COG complex subunits can significantly modulate synaptic morphology, mitochondrial constituents, and neurotransmission in response to stimulatory or plasticity-inducing cues. A deeper exploration is imperative to elucidate these interconnected dynamics. Mitochondria, pivotal cellular organelles, mediate synaptic operations through diverse modalities. They serve as calcium reservoirs [61], facilitate ATP synthesis [62], generate Krebs cycle intermediates influencing neurotransmission [63], and engage in glutamate metabolism [64]. Additionally, mitochondria confer metabolic adaptability by diversifying carbon inputs into the Krebs cycle, including sources such as pyruvate, glutamate, or fatty acids [64].

The mitochondria's competence in modulating synaptic calcium concentrations might shed light on certain neurotransmission phenotypes, particularly in contexts of ATP7 overexpression or disrupted copper efflux. Brief episodes of high-frequency neuronal excitation can engender synaptic plasticity via facilitation, augmentation, and potentiation, cumulatively enhancing transmission efficiency, termed synaptic enhancement [65]. Facilitation hinges on the preservation of optimal calcium concentrations post-influx, while concurrently augmenting it with basal calcium during neuronal activity and inducing exocytosis of readily releasable vesicles in response to neural excitation [66–68]. In contrast, potentiation seeks a balance between exocytosis and endocytosis [69].

Post-tetanic facilitation and augmentation typically involve a diminution in residual calcium [65]. Basal calcium concentrations appear cardinal for post-tetanic potentiation (PTP), with activity-mediated calcium elevations instigating phosphorylation-oriented signaling persisting beyond its ephemeral phase [69]. Given the involvement of mitochondria during PTP evolution, their potential role here merits consideration [70,71].

Mitochondria's indispensable role in modulating calcium concentrations at neuronal terminals is paramount for synaptic plasticity and the inception of short-term memories. They act as calcium buffers postconditioning [72]. However, in synapses manifesting ATP7 overexpression or compromised copper efflux pathways, any enhancements post-sustained tetanic or post-tetanic stimuli are negated. It is salient that upon diminishing COG complex expression, both mitochondrial content and neurotransmission phenotypes in ATP7 hyperexpressing terminals are concurrently ameliorated, substantiating the hypothesis of mitochondria-mediated mechanisms dictating these phenotypes [73,74].

6. LARGE Gene Interactions with Golgi Apparatus—Consequences and Implications

The LARGE gene, prominently expressed in the hippocampus compared to other tissues, plays a pivotal role in cognitive functions [61]. Mutations within this gene give rise to human congenital muscular dystrophy type 1D, characterized by severe cognitive impairments, atypical electroretinogram results, and nuanced structural cerebral anomalies [62,63]. LARGEmyd mice, harboring truncation mutations of the LARGE gene, exhibit neurological manifestations akin to humans bearing such mutations, including sensorineural hearing loss, retinal transmission deficits, neurodevelopmental irregularities, and attenuated long-term potentiation [LTP] [64,65]. Recent research underscores the potential of aberrant synaptic operations as the underlying cause of such cognitive impediments [64].

Intellectual disability, historically termed mental retardation, is delineated by pronounced cognitive (IQ $\leq$ 70) and adaptive behavior deficits, manifested as restricted conceptual, social, or practical skills prior to 18 years of age [75]. This disability might also correlate with other cognitive deteriorations across one's lifetime, including dementia of a neurodegenerative origin. Numerous X-linked cerebral disorders related to intellectual disability have proteins, instrumental in synaptic signaling pathways, as plausible etiological agents [76–78]. Fragile X syndrome, a predominant genetically inherited intellectual disability, is directly associated with AMPA-R dysregulation. Proteins PAK3 and OPHN1, implicated in intellectual disability, oversee synaptic AMPA-R expression and stability [74]. Mutations in its GluA3 subunit have been identified in individuals with X-linked intellectual disability [79]. However, the specifics of these perturbations in AMPA-R dynamics and their contribution to cognitive dysfunction remain largely enigmatic.

The LARGE gene is instrumental in regulating AMPA-R—a synaptic receptor crucial for synaptic plasticity and long-term potentiation, vital for memory consolidation and learning. Mice devoid of this protein manifest compromised LTP, potentially correlated with neurodevelopmental aberrations such as anomalous neuronal migratory patterns [80].

Experiments aimed at excising dystroglycan from murine brains yield outcomes reminiscent of congenital muscular dystrophies, such as lissencephaly. This implies a central role of dystroglycan in the central nervous system, suggesting that the anomalies observed in LARGEmyd mice might echo mechanisms operational in humans with similar dystrophies [81].

LARGEmyd mice present with irregular stratification within their cerebral and cerebellar structures. Atypical neurons are evident across both regions. Additionally, traces of neurons in the external granular layer, presumably attributable to migratory defects, are observed along with congregations of inappropriately located neurons within the white matter or beneath the pial surface. Such abnormalities mirror those seen in specific muscular dystrophies, including Fukuyama congenital muscular dystrophy (FCMD) and Muscle–Eye–Brain disease (MEB) [63,82,83].

Microscopic evaluations have identified dystroglycan in the hippocampus, particularly within postsynaptic assemblies crafted by mossy fibers on pyramidal neurons [84]. Prior investigations revealed that dystroglycan suppression, achieved using GFAP-Cre, results in diminished LTP, proposing its role in synaptic functionality [85]. Synaptic adaptability alterations might arise due to diminished dystroglycan in either neurons or glial cells—fundamental to the orchestration of synaptic operations [86]. The potential contribution of developmental cerebral anomalies in influencing the electrophysiological attributes of the GFAP-Cre/DG-null brain cannot be dismissed [64].

Moreover, research has pinpointed the LARGE protein's role in thwarting AMPA-R localization by impeding its transit from the Golgi apparatus to the cellular exterior. A deficiency in LARGE culminates in an AMPA-R overabundance at synapses, hindering hippocampal LTP. Experiments utilizing animal models exposed to both knockout and knockdown vectors bearing small hairpin RNA target LARGE unveil defects in associative fear memory formation. This highlights LARGE's significance in memory creation through its governance of AMPA-R movement from the Golgi to the cellular facade [80]. Collectively, these insights emphasize its cardinal role in memory genesis by modulating synaptic AMPA-R localization within the hippocampus [87].

7. Golgi Matrix Protein 130 (GM130)

7.1. Role of GM130 in the Golgi Apparatus

The Golgi Matrix Protein 130 (GM130) emerged as the inaugural matrix protein deemed indispensable for preserving the Golgi apparatus's morphology, identified in a 1995 screen for proteins affiliated with the Golgi apparatus (GA) [88]. The GOLGA2 gene is responsible for the transcription of GM130, a Golgin protein that has been the focus of extensive investigations. Acting as a Golgi architectural protein, GM130 adheres firmly to Golgi membranes, safeguarding their sophisticated configuration. Additionally, GM130 is integral to several cellular endeavors, encompassing: vesicular fusion between the Golgi and ER [89]; mitotic spindle organization and cellular bifurcation; microtubule nucleation at the Golgi; and compartmental layout in dendritic Golgi outposts [90].

7.2. Pathological Dimensions of GM130

Mutations culminating in GM130 inactivation or diminished expression have been identified in patients grappling with diverse ailments, including colorectal and breast cancers. Intriguingly, an elevated GM130 expression in gastric cancer correlates with reduced patient survival [91]. Diacylglycerol acyltransferase I (DGAT1) may serve as a therapeutic modality against prostate cancer by adjusting microtubule-organizing centers and GM130 concentrations, thereby undermining microtubule structural integrity [92,93]. The presence of Mannose N-glycans and the Golgi residency of α-mannosidase 1A could serve as prognostic markers for aggressive prostate carcinoma cells [94]. Interference in GM130 and GRASP65 binding precipitates its degradation, triggering Golgi fragmentation and acute pancreatitis in murine models [95]. Cells bereft of NAGLU exhibit heightened GM130 expression, leading to the elongation and expansion of the Golgi and anomalous lysosomal accrual; restoring GM130 levels can ameliorate these pathological hallmarks [96]. The mammalian GA is instrumental in modulating surfactant protein secretion by pulmonary epithelial cells [97].

7.3. Neurological Implications of GM130 Dysfunction

Furthermore, GM130 ablation in murine nervous systems can precipitate the gradual demise of Purkinje neurons in the cerebellum, resulting in discernible motor dysfunctions, imbalanced postures, and severe tail-suspension rotational behaviors, mirroring cerebellar ataxia. In advanced stages, paralytic manifestations with degenerative traits can be evident. Zebrafish bearing mutations that deactivate GM130 demonstrate pronounced skeletal muscle anomalies and progressive microcephaly (MCPH) [2]. Individuals harboring homozygous variants of such mutations manifest analogous clinical presentations, including myofibrillar degeneration, hypotonia, growth impediments, and degenerative features [98].

GM130 exerts a fundamental influence on nervous system ontogeny, orchestrating myriad biological operations vital for its optimal functionality. Among these are: preservation of the Golgi apparatus's structural integrity, efficient intracellular protein and lipid trafficking, mitotic activities such as mitotic segregation, and regulatory mechanisms overseeing cellular migration/polarization, along with proficient glycosylation processes [99].

The Golgi apparatus, a cornerstone of the endomembrane network, exhibits structural deviations linked to assorted neurodegenerative conditions. As an intrinsic constituent of the GA matrix, GM130 is paramount in conserving its ribbon-esque architecture [100]. An aberrant GA morphology frequently coincides with a diminished matrix protein expression. The assembly of the Golgi ribbon necessitates seamless integration of ER-to-Golgi carriers (EGCs or ERGICS) into GA layers, contingent on the uninterrupted shuttling of GM130 between the cis-Golgi and EGCs. In the absence of GM130, this process falters, leading to vesicular membrane accumulation, curtailment of smooth ER vesicles, and Golgi ribbon disintegration [101].

8. Alzheimer's Disease and GM130

8.1. Neuronal Development and the Role of GM130 in Neurodegenerative Diseases

During embryogenesis of neurons, the Golgi apparatus emerges as a pivotal non-centrosome-associated microtubule organizing center. Cells forming dendrites leverage these structures for direct cargo transport to nascent dendritic plasma membranes and localized microtubule nucleation to support dendrite elongation [23,102]. Neurodegenerative conditions, including Alzheimer's Disease (AD), Parkinson's Disease (PD), Amyotrophic Lateral Sclerosis (ALS), and Spinocerebellar Ataxia Type 2 (SCA2), are characterized by a disrupted Golgi architecture. GM130 ensures the integrity of the Golgi's structure, anchoring transport vesicles crucial for streamlined endoplasmic reticulum (ER)-to-Golgi transport [101]. Furthermore, GM130 is indispensable for precise Golgi positioning, cytoskeletal modulation, and establishment of neuronal Golgi satellites [23]. Aberrations in GM130 functionality, manifested as vesicular accumulation and Golgi disarray, correlate with neuronal impairment and cell apoptosis [103]. Investigations utilizing human induced pluripotent stem cells and neurons with compromised GM130 functionality elucidate adverse repercussions on cell polarity, motility, migration, neurogenesis, and neuritogenesis [104]. Given its multifaceted physiological implications, GM130 is instrumental in a plethora of neuropathological conditions.

8.2. Alzheimer's Disease: Pathogenesis and Golgi Disruption

Alzheimer's Disease is characterized as a progressive, age-associated cognitive deterioration paired with distinct neuropathological signatures. Initial manifestations encompass diminished memory consolidation capabilities, which evolve into broader cognitive and behavioral anomalies. Contemporary therapeutic strategies for AD encompass acetylcholinesterase inhibitors for cognitive augmentation and nonsteroidal anti-inflammatory agents to potentially decelerate disease progression and ameliorate cognitive deficits [105].

Central to AD pathogenesis is the aggregation of amyloid-β (Aβ) peptides, derivatives of amyloid precursor protein (APP) cleavage, implicated in neurodegenerative pathways in AD patients [106,107]. Aβ peptides, functioning as cell surface ligands, are pivotal for neurite extension, cellular adhesion, and synaptogenesis, among other cellular processes [108]. Fluorescent microscopic analyses of hippocampal specimens from transgenic mice models of AD, possessing the APP Sweden and PS1 deletion mutations, highlighted disrupted Golgi morphologies in contrast to the intact perinuclear formations in control mice [109–112]. Additionally, early AD progression is marked by neuronal Golgi fragmentation and spatial redistribution. Ultrastructural assessments delineate perturbed Golgi lamellae with diminished diameters and proximal vesicular congregates [113]. Such Golgi disorganization could compromise protein trafficking dynamics across Golgi membranes, adversely affecting APP processing and elevating Aβ synthesis [114]. Herein, GM130's core functionality in preserving Golgi architecture becomes evident.

Cdk5 appears to be instrumental in AD-associated Golgi disintegration. Evidenced by Cdk5 phosphorylation loci on GM130, unchecked Cdk5 activity might instigate Golgi fragmentation, acting as a substrate for Cdc2. Analogous to the early prophase, where Cdc2-mediated GM130 phosphorylation persists through metaphase to anaphase culminating in fragmentation, with subsequent dephosphorylation in telophase promoting Golgi reconstitution, Cdk5 might exert a similar modulatory influence [115].

9. Interplay between the Golgi Apparatus and SARS-CoV-2: Possible Associations with Alzheimer's Disease

The enveloped SARS-CoV-2 virus exploits the host cell's secretory pathway for its lifecycle processes, including replication, assembly, and egress. Within the host cellular environment, the virus harnesses three non-structural proteins, namely, Nsp3, 4, and 6, to reconfigure the endoplasmic reticulum, resulting in the creation of double-membrane vesicles instrumental for viral RNA replication. The ER-Golgi intermediate compartment (ERGIC) and the Golgi apparatus serve as the milieu for virion assembly, where processes such as spike protein glycosylation and furin cleavage transpire. Efficient functioning of this assembly hub is contingent on a protein stack formed by GRASP55 and GRASP65. Deficiency in these GRASP proteins culminates in Golgi fragmentation (GF), enhancing trafficking, albeit with ramifications including altered lysosomal enzyme sorting, protein glycosylation modifications, and disruptions to cellular functions such as adhesion and proliferation [116–118].

In the context of SARS-CoV-2 infections, GRASP55, integral to various stress responses, undergoes down-regulation, whereas GRASP65 remains largely stable [119]. Research underscores GRASP55's salience in viral infections, potentially attributable to its role in Golgi architecture. Corroborating this, diminished GRASP55 expression impedes the translocation of the spike protein to the cell's exterior, implying that virus-induced down-regulation of GRASP55 expedites viral trafficking [120]. Such observations align with prior work highlighting increased protein trafficking in the wake of GRASP55 deficiency [121]. Furthermore, GRASP55 expression diminishes ACE2 surface presence, insinuating its potential impact on the SARS-CoV-2 entry pathway.

Electron microscopy has elucidated marked disparities in Golgi configurations between infected and non-infected cells. The Golgi apparatus in non-infected cells exhibits a pronounced density proximal to the nucleus with extended cisternae. In contrast, SARS-CoV-2-afflicted cells display substantial Golgi disarray, with predominant vesiculation of Golgi membranes. Intriguingly, copious viral particles inhabit the expansive Golgi lumen, suggesting the virus's disruptive transit through the Golgi [122]. Moreover, in a neuropathological examination of 20 COVID-19 cases, six (three biopsies and three autopsies) exhibited white matter abnormalities on MRI, presenting microhemorrhages suggestive of small artery diseases. These cases displayed COVID-19-associated cerebral microangiopathy (CCM) with distinct perivascular changes and evidence of blood–brain barrier compromise. Despite the absence of fibrinoid necrosis and viral RNA in the brain, the SARS-CoV-2 spike protein was identified in the brain endothelial cells' Golgi apparatus, associating with the furin protease. Cultured endothelial cells resisted SARS-CoV-2 replication, and the protein's distribution differed from that in pneumocytes [123].

Alzheimer's disease (AD) presents a parallel, with its brain specimens often exhibiting Golgi fragmentation, attributed to heightened neuronal activity. Golgi anomalies are implicated in several AD-related pathologies, encompassing aberrant protein sorting and glycosylation, hindered lysosomal/autophagosomal degradation, and an escalation in Aβ peptide production. Notably, Aβ oligomer accumulation triggers GF through cyclin-dependent protein kinase 5 (CDK5) activation, which subsequently phosphorylates GRASP65, elevating Golgi growth factor. This series of events amplifies APP trafficking and, in turn, Aβ peptide production [124].

GRASP65's role is pivotal in mitigating amyloid-beta generation by promoting non-amyloidogenic a-cleavage of the amyloid precursor protein. This unveils an intricate interplay between augmented amyloidogenic cleavage of APP and tau phosphorylation, with CDK5 being the primary kinase regulating tau phosphorylation and influencing Aβ formation [125].

Conversely, GRASP55 modulates the secretion and aggregation of neurotoxic proteins via autophagy. Ordinarily, GRASP55 undergoes O-linked N-acetylglucosamine (O-GlcNAc) modification, impacting nucleocytoplasmic proteins. However, under glucose-deprivation, GRASP55's O-GlcNAcylation wanes, prompting its relocation to function as a membrane tether for autophagosome–lysosome fusion, thereby augmenting unconventional secretion of neurotoxic entities, such as tau [106,126–128].

Appreciating the ramifications of SARS-CoV-2 infection on neural cells at the molecular and cellular echelons remains paramount. With Golgi irregularities providing an investigative fulcrum, probing their association with Alzheimer's pathophysiology and clinical dementia mandates delving into the molecular underpinnings. Such endeavors can elucidate mechanisms wherein Aβ fibrils interfere with protein trafficking across neural cell types, linking observed aberrations in glycosylation and lysosomal activity to these disruptions (Figure 3).

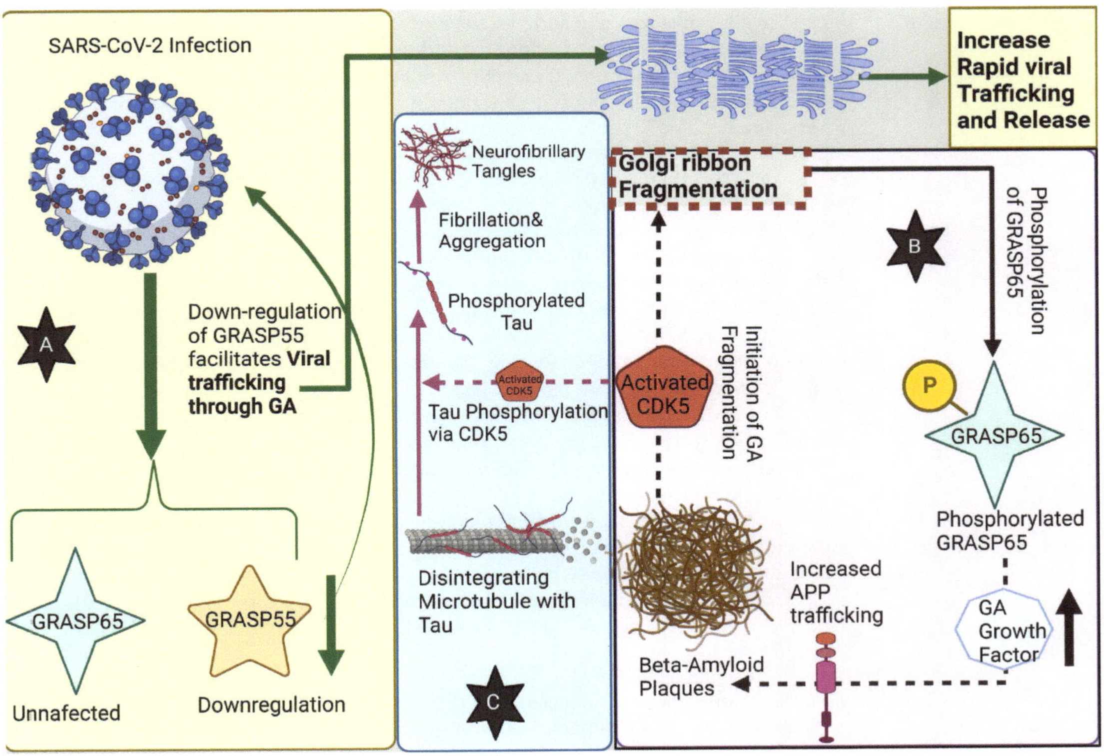

Figure 3. (**A**) SARS-CoV-2 infection down-regulates GRASP55, whereas GRASP65 remains unaffected. Down-regulation of GRASP55 facilitates viral trafficking through the Golgi apparatus (GA) leading to its fragmentation; GA fragmentation then leads to a rapid increase in viral trafficking and release. (**B**) Beta-amyloid plaques lead to GA fragmentation, through CDK5 activation, that in turn increases the phosphorylation of GRASP65, which will stimulate the GA growth factor synthesis and release, contributing to more APP (Amyloid Precursor Protein) trafficking, increasing the beta-amyloid plaque formation from Aβ oligomers. (**C**) Moreover, the activation of CDK5 contributes to Tau protein phosphorylation, thus promoting its fibrillation and aggregation into neurofibrillary tangles.

10. Conclusions

The Golgi apparatus, frequently marginalized in neurobiological discourse, has recently emerged as pivotal to a plethora of processes within the nervous system. This intricate organelle is integral to multifaceted biological phenomena encompassing neurogenesis, Alzheimer's Disease, and SARS-CoV-2 infections.

Its cardinal function in axonal and dendritic dynamics, as well as synaptic plasticity, underscores its indispensability during neuronal maturation. Such involvement epitomizes the depth and specificity mandated by these mechanisms. As we broaden our comprehension of the Golgi apparatus's functions within the nervous system, we inch closer to deciphering its elaborate mechanisms and the repercussions of neuronal malfunctions. This enriched understanding not only propels foundational neuroscience research but also illuminates prospective therapeutic interventions against formidable neurodegenerative disorders, notably Alzheimer's.

Exploring the myriad functions of the Golgi apparatus in neuronal health and pathology unveils transformative horizons in neuroscience. Grasping the intricate nexus between this pivotal organelle, neural circuitry, and the trajectory of neurological disorders could pave the way for innovative therapeutic strategies, especially in the context of Alzheimer's, which afflicts countless individuals globally. A nuanced comprehension of the interplay between the GM130 protein and the Golgi apparatus might pioneer treatments with the potential to attenuate, halt, or even reverse disease progression.

Delving into the Golgi apparatus's role in synaptic plasticity could wield profound implications for diverse cognitive dysfunctions and neuronal traumas, providing fresh insights into memory consolidation, learning, and neural rejuvenation.

Furthermore, elucidating the ramifications of Golgi apparatus fragmentation in viral pathologies could spotlight innovative therapeutic avenues for myriad infections. Beyond SARS-CoV-2, pathogens such as Zika, HIV, and influenza capitalize on the host cell's Golgi apparatus. By intensifying research endeavors in these domains, potential retroviral therapeutics could emerge, offering significant impacts on formidable infections such as HIV or novel interventions against Zika.

Prospective research trajectories ought to recognize this organelle's integral role in cellular life cycles. Crafting novel pharmaceutical agents that modify the virus–Golgi interaction, whilst preserving cellular cytoarchitecture, could represent a groundbreaking scientific advancement.

As investigative endeavors into the Golgi's myriad functionalities intensify, it becomes patently evident that this organelle transcends mere cellular infrastructure, asserting itself as a cornerstone in diverse biological and pathological paradigms. Progressing in our grasp of these dimensions demands a collective endeavor from the international scientific fraternity, paired with an unwavering dedication to unlocking the intricacies of this cellular nexus.

Author Contributions: Conceptualization, C.T. and L.E.; methodology, R.-A.C.-B.; software, L.-A.G.; validation, A.V.C., R.-A.C.-B. and H.P.C.; formal analysis, A.D.C.; investigation, R.-A.C.-B.; resources, L.-A.G.; data curation, L.-A.G.; writing—original draft preparation, H.P.C.; writing—review and editing, A.D.C.; visualization, C.T.; supervision, L.E.; project administration, A.V.C.; funding acquisition, A.V.C. All authors have read and agreed to the published version of the manuscript.

Funding: This research received no external funding.

Institutional Review Board Statement: Not applicable.

Informed Consent Statement: Not applicable.

Data Availability Statement: All data are available via online libraries such as PubMed.

Conflicts of Interest: The authors declare no conflict of interest.

References

1. Liu, J.; He, J.; Huang, Y.; Xiao, H.; Jiang, Z.; Hu, Z. The Golgi apparatus in neurorestoration. *J. Neurorestoratol.* **2019**, *7*, 116–128. [CrossRef]
2. Liu, C.; Mei, M.; Li, Q.; Roboti, P.; Pang, Q.; Ying, Z.; Gao, F.; Lowe, M.; Bao, S. Loss of the golgin GM130 causes Golgi disruption, Purkinje neuron loss, and ataxia in mice. *Proc. Natl. Acad. Sci. USA* **2017**, *114*, 346–351. [CrossRef] [PubMed]
3. Elsherbini, A.; Zhu, Z.; Quadri, Z.; Crivelli, S.M.; Ren, X.; Vekaria, H.J.; Tripathi, P.; Zhang, L.; Zhi, W.; Bieberich, E. Novel Isolation Method Reveals Sex-Specific Composition and Neurotoxicity of Small Extracellular Vesicles in a Mouse Model of Alzheimer's Disease. *Cells* **2023**, *12*, 1623. [CrossRef]
4. Haukedal, H.; Corsi, G.I.; Gadekar, V.P.; Doncheva, N.T.; Kedia, S.; De Haan, N.; Chandrasekaran, A.; Jensen, P.; Schiønning, P.; Vallin, S.; et al. Golgi fragmentation—One of the earliest organelle phenotypes in Alzheimer's disease neurons. *Front. Neurosci.* **2023**, *17*, 1120086. [CrossRef]
5. Crawford, K.; Leonenko, G.; Baker, E.; Grozeva, D.; Lan-Leung, B.; Holmans, P.; Williams, J.; O'Donovan, M.C.; Escott-Price, V.; Ivanov, D.K. Golgi apparatus, endoplasmic reticulum and mitochondrial function implicated in Alzheimer's disease through polygenic risk and RNA sequencing. *Mol. Psychiatry* **2023**, *28*, 1327–1336. [CrossRef] [PubMed]
6. Martinez-Curiel, R.; Jansson, L.; Tsupykov, O.; Avaliani, N.; Aretio-Medina, C.; Hidalgo, I.; Monni, E.; Bengzon, J.; Skibo, G.; Lindvall, O.; et al. Oligodendrocytes in human induced pluripotent stem cell-derived cortical grafts remyelinate adult rat and human cortical neurons. *Stem Cell Rep.* **2023**, *18*, 1643–1656. [CrossRef]
7. Hösli, L.; Zuend, M.; Bredell, G.; Zanker, H.S.; Porto De Oliveira, C.E.; Saab, A.S.; Weber, B. Direct vascular contact is a hallmark of cerebral astrocytes. *Cell Rep.* **2022**, *39*, 110599. [CrossRef]
8. Ori-McKenney, K.M.; Jan, L.Y.; Jan, Y.-N. Golgi Outposts Shape Dendrite Morphology by Functioning as Sites of Acentrosomal Microtubule Nucleation in Neurons. *Neuron* **2012**, *76*, 921–930. [CrossRef]
9. Oddoux, S.; Zaal, K.J.; Tate, V.; Kenea, A.; Nandkeolyar, S.A.; Reid, E.; Liu, W.; Ralston, E. Microtubules that form the stationary lattice of muscle fibers are dynamic and nucleated at Golgi elements. *J. Cell Biol.* **2013**, *203*, 205–213. [CrossRef]
10. Wang, J.; Fourriere, L.; Gleeson, P.A. Local Secretory Trafficking Pathways in Neurons and the Role of Dendritic Golgi Outposts in Different Cell Models. *Front. Mol. Neurosci.* **2020**, *13*, 597391. [CrossRef]
11. Jamieson, J.D.; Palade, G.E. Synthesis, intracellular transport, and discharge of secretory proteins in stimulated pancreatic exocrine cells. *J. Cell Biol.* **1971**, *50*, 135–158. [CrossRef]
12. Stephens, D.J.; Lin-Marq, N.; Pagano, A.; Pepperkok, R.; Paccaud, J.-P. COPI-coated ER-to-Golgi transport complexes segregate from COPII in close proximity to ER exit sites. *J. Cell Sci.* **2000**, *113*, 2177–2185. [CrossRef]
13. Wang, T.; Hanus, C.; Cui, T.; Helton, T.; Bourne, J.; Watson, D.; Harris, K.M.; Ehlers, M.D. Local Zones of Endoplasmic Reticulum Complexity Confine Cargo in Neuronal Dendrites. *Cell* **2012**, *148*, 309–321. [CrossRef]
14. Mikhaylova, M.; Bera, S.; Kobler, O.; Frischknecht, R.; Kreutz, M.R. A Dendritic Golgi Satellite between ERGIC and Retromer. *Cell Rep.* **2016**, *14*, 189–199. [CrossRef]
15. Stoorvogel, W.; Oorschot, V.; Geuze, H.J. A novel class of clathrin-coated vesicles budding from endosomes. *J. Cell Biol.* **1996**, *132*, 21–33. [CrossRef]
16. Kemal, S.; Richardson, H.S.; Dyne, E.D.; Fu, M. ER and Golgi trafficking in axons, dendrites, and glial processes. *Curr. Opin. Cell Biol.* **2022**, *78*, 102119. [CrossRef] [PubMed]
17. Teixeira, A.I.; Ilkhanizadeh, S.; Wigenius, J.A.; Duckworth, J.K.; Inganäs, O.; Hermanson, O. The promotion of neuronal maturation on soft substrates. *Biomaterials* **2009**, *30*, 4567–4572. [CrossRef] [PubMed]
18. Watson, P.; Townley, A.K.; Koka, P.; Palmer, K.J.; Stephens, D.J. Sec16 Defines Endoplasmic Reticulum Exit Sites and is Required for Secretory Cargo Export in Mammalian Cells. *Traffic* **2006**, *7*, 1678–1687. [CrossRef] [PubMed]
19. Tang, B.L.; Ong, Y.S.; Huang, B.; Wei, S.; Wong, E.T.; Qi, R.; Horstmann, H.; Hong, W. A Membrane Protein Enriched in Endoplasmic Reticulum Exit Sites Interacts with COPII. *J. Biol. Chem.* **2001**, *276*, 40008–40017. [CrossRef]
20. Mallard, F.; Antony, C.; Tenza, D.; Salamero, J.; Goud, B.; Johannes, L. Direct Pathway from Early/Recycling Endosomes to the Golgi Apparatus Revealed through the Study of Shiga Toxin B-fragment Transport. *J. Cell Biol.* **1998**, *143*, 973–990. [CrossRef]
21. Horton, A.C.; Ehlers, M.D. Dual Modes of Endoplasmic Reticulum-to-Golgi Transport in Dendrites Revealed by Live-Cell Imaging. *J. Neurosci.* **2003**, *23*, 6188–6199. [CrossRef] [PubMed]
22. Horton, A.C.; Rácz, B.; Monson, E.E.; Lin, A.L.; Weinberg, R.J.; Ehlers, M.D. Polarized Secretory Trafficking Directs Cargo for Asymmetric Dendrite Growth and Morphogenesis. *Neuron* **2005**, *48*, 757–771. [CrossRef]
23. Zhou, W.; Chang, J.; Wang, X.; Savelieff, M.G.; Zhao, Y.; Ke, S.; Ye, B. GM130 Is Required for Compartmental Organization of Dendritic Golgi Outposts. *Curr. Biol.* **2014**, *24*, 1227–1233. [CrossRef]
24. Valenzuela, A.; Meservey, L.; Nguyen, H.; Fu, M. Golgi Outposts Nucleate Microtubules in Cells with Specialized Shapes. *Trends Cell Biol.* **2020**, *30*, 792–804. [CrossRef] [PubMed]
25. Lee, T.; Winter, C.; Marticke, S.S.; Lee, A.; Luo, L. Essential Roles of Drosophila RhoA in the Regulation of Neuroblast Proliferation and Dendritic but Not Axonal Morphogenesis. *Neuron* **2000**, *25*, 307–316. [CrossRef]
26. Harada, A.; Teng, J.; Takei, Y.; Oguchi, K.; Hirokawa, N. MAP2 is required for dendrite elongation, PKA anchoring in dendrites, and proper PKA signal transduction. *J. Cell Biol.* **2002**, *158*, 541–549. [CrossRef]
27. Prigozhina, N.L.; Waterman-Storer, C.M. Protein Kinase D-Mediated Anterograde Membrane Trafficking Is Required for Fibroblast Motility. *Curr. Biol.* **2004**, *14*, 88–98. [CrossRef] [PubMed]

28. Hanus, C.; Ehlers, M.D. Secretory Outposts for the Local Processing of Membrane Cargo in Neuronal Dendrites. *Traffic* **2008**, *9*, 1437–1445. [CrossRef] [PubMed]
29. Zheng, Y.; Wildonger, J.; Ye, B.; Zhang, Y.; Kita, A.; Younger, S.H.; Zimmerman, S.; Jan, L.Y.; Jan, Y.N. Dynein is required for polarized dendritic transport and uniform microtubule orientation in axons. *Nat. Cell Biol.* **2008**, *10*, 1172–1180. [CrossRef]
30. Ye, B.; Zhang, Y.; Song, W.; Younger, S.H.; Jan, L.Y.; Jan, Y.N. Growing Dendrites and Axons Differ in Their Reliance on the Secretory Pathway. *Cell* **2007**, *130*, 717–729. [CrossRef]
31. Satoh, D.; Sato, D.; Tsuyama, T.; Saito, M.; Ohkura, H.; Rolls, M.M.; Ishikawa, F.; Uemura, T. Spatial control of branching within dendritic arbors by dynein-dependent transport of Rab5-endosomes. *Nat. Cell Biol.* **2008**, *10*, 1164–1171. [CrossRef]
32. Nguyen, M.M.; McCracken, C.J.; Milner, E.S.; Goetschius, D.J.; Weiner, A.T.; Long, M.K.; Michael, N.L.; Munro, S.; Rolls, M.M. γ-Tubulin controls neuronal microtubule polarity independently of Golgi outposts. *Mol. Biol. Cell* **2014**, *25*, 2039–2050. [CrossRef]
33. Cole, N.B.; Sciaky, N.; Marotta, A.; Song, J.; Lippincott-Schwartz, J. Golgi dispersal during microtubule disruption: Regeneration of Golgi stacks at peripheral endoplasmic reticulum exit sites. *Mol. Biol. Cell* **1996**, *7*, 631–650. [CrossRef] [PubMed]
34. Thayer, D.A.; Jan, Y.N.; Jan, L.Y. Increased neuronal activity fragments the Golgi complex. *Proc. Natl. Acad. Sci. USA* **2013**, *110*, 1482–1487. [CrossRef]
35. Salvarezza, S.B.; Deborde, S.; Schreiner, R.; Campagne, F.; Kessels, M.M.; Qualmann, B.; Caceres, A.; Kreitzer, G.; Rodriguez-Boulan, E. LIM Kinase 1 and Cofilin Regulate Actin Filament Population Required for Dynamin-dependent Apical Carrier Fission from the *Trans* -Golgi Network. *Mol. Biol. Cell* **2009**, *20*, 438–451. [CrossRef] [PubMed]
36. Bossard, C.; Bresson, D.; Polishchuk, R.S.; Malhotra, V. Dimeric PKD regulates membrane fission to form transport carriers at the TGN. *J. Cell Biol.* **2007**, *179*, 1123–1131. [CrossRef]
37. Corda, D.; Colanzi, A.; Luini, A. The multiple activities of CtBP/BARS proteins: The Golgi view. *Trends Cell Biol.* **2006**, *16*, 167–173. [CrossRef] [PubMed]
38. Bonazzi, M.; Spanò, S.; Turacchio, G.; Cericola, C.; Valente, C.; Colanzi, A.; Kweon, H.S.; Hsu, V.W.; Polishchuck, E.V.; Polishchuck, R.S.; et al. CtBP3/BARS drives membrane fission in dynamin-independent transport pathways. *Nat. Cell Biol.* **2005**, *7*, 570–580. [CrossRef]
39. Valente, C.; Turacchio, G.; Mariggiò, S.; Pagliuso, A.; Gaibisso, R.; Di Tullio, G.; Santoro, M.; Formiggini, F.; Spanò, S.; Piccini, D.; et al. A 14-3-3γ dimer-based scaffold bridges CtBP1-S/BARS to PI(4)KIIIβ to regulate post-Golgi carrier formation. *Nat. Cell Biol.* **2012**, *14*, 343–354. [CrossRef]
40. Furusawa, T.; Moribe, H.; Kondoh, H.; Higashi, Y. Identification of CtBP1 and CtBP2 as Corepressors of Zinc Finger-Homeodomain Factor δEF1. *Mol. Cell Biol.* **1999**, *19*, 8581–8590. [CrossRef]
41. Hübler, D.; Rankovic, M.; Richter, K.; Lazarevic, V.; Altrock, W.D.; Fischer, K.-D.; Gundelfinger, E.D.; Fejtova, A. Differential Spatial Expression and Subcellular Localization of CtBP Family Members in Rodent Brain. *PLoS ONE* **2012**, *7*, e39710. [CrossRef] [PubMed]
42. Hildebrand, J.D.; Soriano, P. Overlapping and Unique Roles for C-Terminal Binding Protein 1 (CtBP1) and CtBP2 during Mouse Development. *Mol. Cell Biol.* **2002**, *22*, 5296–5307. [CrossRef] [PubMed]
43. Acosta-Baena, N.; Tejada-Moreno, J.A.; Arcos-Burgos, M.; Villegas-Lanau, C.A. *CTBP1* and *CTBP2* mutations underpinning neurological disorders: A systematic review. *Neurogenetics* **2022**, *23*, 231–240. [CrossRef]
44. Horton, A.C.; Ehlers, M.D. Neuronal Polarity and Trafficking. *Neuron* **2003**, *40*, 277–295. [CrossRef]
45. Collins, M.O.; Husi, H.; Yu, L.; Brandon, J.M.; Anderson, C.N.G.; Blackstock, W.P.; Choudhary, J.S.; Grant, S.G.N. Molecular characterization and comparison of the components and multiprotein complexes in the postsynaptic proteome. *J. Neurochem.* **2006**, *97*, 16–23. [CrossRef]
46. Chen, M.; Xu, L.; Wu, Y.; Soba, P.; Hu, C. The organization and function of the Golgi apparatus in dendrite development and neurological disorders. *Genes Dis.* **2023**, *10*, 2425–2442. [CrossRef] [PubMed]
47. Hirokawa, N.; Niwa, S.; Tanaka, Y. Molecular Motors in Neurons: Transport Mechanisms and Roles in Brain Function, Development, and Disease. *Neuron* **2010**, *68*, 610–638. [CrossRef]
48. Quassollo, G.; Wojnacki, J.; Salas, D.A.; Gastaldi, L.; Marzolo, M.P.; Conde, C.; Bisbal, M.; Couve, A.; Cáceres, A. A RhoA Signaling Pathway Regulates Dendritic Golgi Outpost Formation. *Curr. Biol.* **2015**, *25*, 971–982. [CrossRef]
49. Bisbal, M.; Conde, C.; Donoso, M.; Bollati, F.; Sesma, J.; Quiroga, S.; Díaz Añel, A.; Malhotra, V.; Marzolo, M.P.; Cáceres, A. Protein Kinase D Regulates Trafficking of Dendritic Membrane Proteins in Developing Neurons. *J. Neurosci.* **2008**, *28*, 9297–9308. [CrossRef]
50. Pagliuso, A.; Valente, C.; Giordano, L.L.; Filograna, A.; Li, G.; Circolo, D.; Turacchio, G.; Marzullo, V.M.; Mandrich, L.; Zhukovsky, M.A.; et al. Golgi membrane fission requires the CtBP1-S/BARS-induced activation of lysophosphatidic acid acyltransferase δ. *Nat. Commun.* **2016**, *7*, 12148. [CrossRef]
51. Sanders, S.J.; Sahin, M.; Hostyk, J.; Thurm, A.; Jacquemont, S.; Avillach, P.; Douard, E.; Martin, C.L.; Modi, M.E.; Moreno-De-Luca, A.; et al. A framework for the investigation of rare genetic disorders in neuropsychiatry. *Nat. Med.* **2019**, *25*, 1477–1487. [CrossRef] [PubMed]
52. Kaler, S.G. ATP7A-related copper transport diseases—Emerging concepts and future trends. *Nat. Rev. Neurol.* **2011**, *7*, 15–29. [CrossRef] [PubMed]
53. Terada, K.; Schilsky, M.L.; Miura, N.; Sugiyama, T. ATP7B (WND) protein. *Int. J. Biochem. Cell Biol.* **1998**, *30*, 1063–1067. [CrossRef] [PubMed]

54. Bartee, M.Y.; Lutsenko, S. Hepatic copper-transporting ATPase ATP7B: Function and inactivation at the molecular and cellular level. *BioMetals* **2007**, *20*, 627. [CrossRef]
55. Dierick, H.A.; Ambrosini, L.; Spencer, J.; Glover, T.W.; Mercer, J.F.B. Molecular Structure of the Menkes Disease Gene (ATP7A). *Genomics* **1995**, *28*, 462–469. [CrossRef]
56. Tümer, Z. An Overview and Update of *ATP7A* Mutations Leading to Menkes Disease and Occipital Horn Syndrome. *Hum. Mutat.* **2013**, *34*, 417–429. [CrossRef] [PubMed]
57. Tümer, Z.; Møller, L.B.; Horn, N. Mutation Spectrum of ATP7A, the Gene Defective in Menkes Disease. In *Copper Transport and Its Disorders*; Leone, A., Mercer, J.F.B., Eds.; Advances in Experimental Medicine and Biology; Springer: Boston, MA, USA, 1999; Volume 448, pp. 83–95. [CrossRef]
58. Guthrie, L.M.; Soma, S.; Yuan, S.; Silva, A.; Zulkifli, M.; Snavely, T.C.; Greene, H.F.; Nunez, E.; Lynch, B.; De Ville, C.; et al. Elesclomol alleviates Menkes pathology and mortality by escorting Cu to cuproenzymes in mice. *Science* **2020**, *368*, 620–625. [CrossRef]
59. Lutsenko, S.; Barnes, N.L.; Bartee, M.Y.; Dmitriev, O.Y. Function and Regulation of Human Copper-Transporting ATPases. *Physiol. Rev.* **2007**, *87*, 1011–1046. [CrossRef]
60. Morgan, M.T.; Bourassa, D.; Harankhedkar, S.; McCallum, A.M.; Zlatic, S.A.; Calvo, J.S.; Meloni, G.; Faundez, V.; Fahrni, C.J. Ratiometric two-photon microscopy reveals attomolar copper buffering in normal and Menkes mutant cells. *Proc. Natl. Acad. Sci. USA* **2019**, *116*, 12167–12172. [CrossRef]
61. Peyrard, M.; Seroussi, E.; Sandberg-Nordqvist, A.-C.; Xie, Y.-G.; Han, F.-Y.; Fransson, I.; Collins, J.; Dunham, I.; Kost-Alimova, M.; Imreh, S.; et al. The human *LARGE* gene from 22q12.3-q13.1 is a new, distinct member of the glycosyltransferase gene family. *Proc. Natl. Acad. Sci. USA* **1999**, *96*, 598–603. [CrossRef]
62. Clarke, N.F.; Maugenre, S.; Vandebrouck, A.; Urtizberea, J.A.; Willer, T.; Peat, R.A.; Gray, F.; Bouchet, C.; Manya, H.; Vuillaumier-Barrot, S.; et al. Congenital muscular dystrophy type 1D (MDC1D) due to a large intragenic insertion/deletion, involving intron 10 of the LARGE gene. *Eur. J. Hum. Genet.* **2011**, *19*, 452–457. [CrossRef] [PubMed]
63. Longman, C. Mutations in the human *LARGE* gene cause MDC1D, a novel form of congenital muscular dystrophy with severe mental retardation and abnormal glycosylation of α-dystroglycan. *Hum. Mol. Genet.* **2003**, *12*, 2853–2861. [CrossRef] [PubMed]
64. Satz, J.S.; Ostendorf, A.P.; Hou, S.; Turner, A.; Kusano, H.; Lee, J.C.; Turk, R.; Nguyen, H.; Ross-Barta, S.E.; Westra, S.; et al. Distinct Functions of Glial and Neuronal Dystroglycan in the Developing and Adult Mouse Brain. *J. Neurosci.* **2010**, *30*, 14560–14572. [CrossRef]
65. Holzfeind, P.J. Skeletal, cardiac and tongue muscle pathology, defective retinal transmission, and neuronal migration defects in the Large^{myd} mouse defines a natural model for glycosylation-deficient muscle—Eye—Brain disorders. *Hum. Mol. Genet.* **2002**, *11*, 2673–2687. [CrossRef] [PubMed]
66. Jo, S.; Lee, K.-H.; Song, S.; Jung, Y.-K.; Park, C.-S. Identification and functional characterization of cereblon as a binding protein for large-conductance calcium-activated potassium channel in rat brain. *J. Neurochem.* **2005**, *94*, 1212–1224. [CrossRef]
67. Ou, X.-M.; Lemonde, S.; Jafar-Nejad, H.; Bown, C.D.; Goto, A.; Rogaeva, A.; Albert, P.R. Freud-1: A Neuronal Calcium-Regulated Repressor of the 5-HT1A Receptor Gene. *J. Neurosci.* **2003**, *23*, 7415–7425. [CrossRef]
68. Piton, A.; Michaud, J.L.; Peng, H.; Aradhya, S.; Gauthier, J.; Mottron, L.; Champagne, N.; Lafreniere, R.G.; Hamdan, F.F.; S2D team. Mutations in the calcium-related gene IL1RAPL1 are associated with autism. *Hum. Mol. Genet.* **2008**, *17*, 3965–3974. [CrossRef]
69. Vaillend, C.; Poirier, R.; Laroche, S. Genes, plasticity and mental retardation. *Behav. Brain Res.* **2008**, *192*, 88–105. [CrossRef] [PubMed]
70. Lee, D.; Lee, K.-H.; Ho, W.-K.; Lee, S.-H. Target Cell-Specific Involvement of Presynaptic Mitochondria in Post-Tetanic Potentiation at Hippocampal Mossy Fiber Synapses. *J. Neurosci.* **2007**, *27*, 13603–13613. [CrossRef]
71. Tang, Y.; Zucker, R.S. Mitochondrial Involvement in Post-Tetanic Potentiation of Synaptic Transmission. *Neuron* **1997**, *18*, 483–491. [CrossRef]
72. Paidi, R.; Nthenge-Ngumbau, D.; Singh, R.; Kankanala, T.; Mehta, H.; Mohanakumar, K. Mitochondrial Deficits Accompany Cognitive Decline Following Single Bilateral Intracerebroventricular Streptozotocin. *Curr. Alzheimer Res.* **2015**, *12*, 785–795. [CrossRef] [PubMed]
73. Stout, A.K.; Raphael, H.M.; Kanterewicz, B.I.; Klann, E.; Reynolds, I.J. Glutamate-induced neuron death requires mitochondrial calcium uptake. *Nat. Neurosci.* **1998**, *1*, 366–373. [CrossRef] [PubMed]
74. Boda, B.; Alberi, S.; Nikonenko, I.; Node-Langlois, R.; Jourdain, P.; Moosmayer, M.; Parisi-Jourdain, L.; Muller, D. The Mental Retardation Protein PAK3 Contributes to Synapse Formation and Plasticity in Hippocampus. *J. Neurosci.* **2004**, *24*, 10816–10825. [CrossRef]
75. Shree, A.; Shukla, P.C. Intellectual Disability: Definition, classification, causes and characteristics. *Learn. Community-Int. J. Educ. Soc. Dev.* **2016**, *7*, 9. [CrossRef]
76. Hayashi, S.; Inoue, Y.; Hattori, S.; Kaneko, M.; Shioi, G.; Miyakawa, T.; Takeichi, M. Loss of X-linked Protocadherin-19 differentially affects the behavior of heterozygous female and hemizygous male mice. *Sci. Rep.* **2017**, *7*, 5801. [CrossRef]
77. Takano, K.; Liu, D.; Tarpey, P.; Gallant, E.; Lam, A.; Witham, S.; Alexov, E.; Chaubey, A.; Stevenson, R.E.; Schwartz, C.E.; et al. An X-linked channelopathy with cardiomegaly due to a CLIC2 mutation enhancing ryanodine receptor channel activity. *Hum. Mol. Genet.* **2012**, *21*, 4497–4507. [CrossRef] [PubMed]

78. Srivastava, S.; McMillan, R.; Willis, J.; Clark, H.; Chavan, V.; Liang, C.; Zhang, H.; Hulver, M.; Mukherjee, K. X-linked intellectual disability gene CASK regulates postnatal brain growth in a non-cell autonomous manner. *Acta Neuropathol. Commun.* **2016**, *4*, 30. [CrossRef]
79. Wu, Y.; Arai, A.C.; Rumbaugh, G.; Srivastava, A.K.; Turner, G.; Hayashi, T.; Suzuki, E.; Jiang, Y.; Zhang, L.; Rodriguez, J.; et al. Mutations in ionotropic AMPA receptor 3 alter channel properties and are associated with moderate cognitive impairment in humans. *Proc. Natl. Acad. Sci. USA* **2007**, *104*, 18163–18168. [CrossRef]
80. Seo, B.A.; Cho, T.; Lee, D.Z.; Lee, J.; Lee, B.; Kim, S.-W.; Shin, H.-S.; Kang, M.-G. LARGE, an intellectual disability-associated protein, regulates AMPA-type glutamate receptor trafficking and memory. *Proc. Natl. Acad. Sci. USA* **2018**, *115*, 7111–7116. [CrossRef]
81. Kano, H.; Kobayashi, K.; Herrmann, R.; Tachikawa, M.; Manya, H.; Nishino, I.; Nonaka, I.; Straub, V.; Talim, B.; Voit, T.; et al. Deficiency of α-Dystroglycan in Muscle–Eye–Brain Disease. *Biochem. Biophys. Res. Commun.* **2002**, *291*, 1283–1286. [CrossRef]
82. Takada, K.; Nakamura, H.; Takashima, S. Cortical dysplasia in Fukuyama congenital muscular dystrophy (FCMD): A Golgi and angioarchitectonic analysis. *Acta Neuropathol.* **1988**, *76*, 170–178. [CrossRef] [PubMed]
83. Takada, K.; Rin, Y.-S.; Kasagi, S.; Sato, K.; Nakamura, H.; Tanaka, J. Long survival in Fukuyama congenital muscular dystrophy: Occurrence of neurofibrillary tangles in the nucleus basalis of meynert and locus ceruleus. *Acta Neuropathol.* **1986**, *71*, 228–232. [CrossRef] [PubMed]
84. Zaccaria, M.L.; Di Tommaso, F.; Brancaccio, A.; Paggi, P.; Petrucci, T.C. Dystroglycan distribution in adult mouse brain: A light and electron microscopy study. *Neuroscience* **2001**, *104*, 311–324. [CrossRef] [PubMed]
85. Moore, S.A.; Saito, F.; Chen, J.; Michele, D.E.; Henry, M.D.; Messing, A.; Cohn, R.D.; Ross-Barta, S.E.; Westra, S.; Williamson, R.A.; et al. Deletion of brain dystroglycan recapitulates aspects of congenital muscular dystrophy. *Nature* **2002**, *418*, 422–425. [CrossRef]
86. Haydon, P.G. Glia: Listening and talking to the synapse. *Nat. Rev. Neurosci.* **2001**, *2*, 185–193. [CrossRef]
87. Portera-Cailliau, C. Which Comes First in Fragile X Syndrome, Dendritic Spine Dysgenesis or Defects in Circuit Plasticity? *Neurosci.* **2012**, *18*, 28–44. [CrossRef]
88. Nakamura, N.; Rabouille, C.; Watson, R.; Nilsson, T.; Hui, N.; Slusarewicz, P.; Kreis, T.E.; Warren, G. Characterization of a cis-Golgi matrix protein, GM130. *J. Cell Biol.* **1995**, *131*, 1715–1726. [CrossRef]
89. Wei, J.-H.; Zhang, Z.C.; Wynn, R.M.; Seemann, J. GM130 Regulates Golgi-Derived Spindle Assembly by Activating TPX2 and Capturing Microtubules. *Cell* **2015**, *162*, 287–299. [CrossRef]
90. Rivero, S.; Cardenas, J.; Bornens, M.; Rios, R.M. Microtubule nucleation at the cis-side of the Golgi apparatus requires AKAP450 and GM130. *EMBO J.* **2009**, *28*, 1016–1028. [CrossRef]
91. Baschieri, F.; Confalonieri, S.; Bertalot, G.; Di Fiore, P.P.; Dietmaier, W.; Leist, M.; Crespo, P.; Macara, I.G.; Farhan, H. Spatial control of Cdc42 signalling by a GM130–RasGRF complex regulates polarity and tumorigenesis. *Nat. Commun.* **2014**, *5*, 4839. [CrossRef]
92. Zhao, J.; Yang, C.; Guo, S.; Wu, Y. GM130 regulates epithelial-to-mesenchymal transition and invasion of gastric cancer cells via snail. *Int. J. Clin. Exp. Pathol.* **2015**, *8*, 10784–10791.
93. Nardi, F.; Franco, O.E.; Fitchev, P.; Morales, A.; Vickman, R.E.; Hayward, S.W.; Crawford, S.E. DGAT1 Inhibitor Suppresses Prostate Tumor Growth and Migration by Regulating Intracellular Lipids and Non-Centrosomal MTOC Protein GM130. *Sci. Rep.* **2019**, *9*, 3035. [CrossRef] [PubMed]
94. Cheng, P.-W.; Davidson, S.; Bhat, G. Markers of malignant prostate cancer cells: Golgi localization of α-mannosidase 1A at GM130-GRASP65 site and appearance of high mannose N-glycans on cell surface. *Biochem. Biophys. Res. Commun.* **2020**, *527*, 406–410. [CrossRef] [PubMed]
95. Li, X.; Xia, Y.; Huang, S.; Liu, F.; Ying, Y.; Xu, Q.; Liu, X.; Jin, G.; Papasian, C.J.; Chen, J.; et al. Identification of the interaction of VP1 with GM130 which may implicate in the pathogenesis of CVB3-induced acute pancreatitis. *Sci. Rep.* **2015**, *5*, 13324. [CrossRef]
96. Roy, E.; Bruyère, J.; Flamant, P.; Bigou, S.; Ausseil, J.; Vitry, S.; Heard, J.M. GM130 gain-of-function induces cell pathology in a model of lysosomal storage disease. *Hum. Mol. Genet.* **2012**, *21*, 1481–1495. [CrossRef] [PubMed]
97. Pang, Q.; Liu, C.; Qiao, Y.; Zhao, J.; Lam, S.M.; Mei, M.; Shui, G.; Bao, S.; Li, Q. GM130 regulates pulmonary surfactant protein secretion in alveolar type II cells. *Sci. China Life Sci.* **2022**, *65*, 193–205. [CrossRef]
98. Shamseldin, H.E.; Bennett, A.H.; Alfadhel, M.; Gupta, V.; Alkuraya, F.S. GOLGA2, encoding a master regulator of golgi apparatus, is mutated in a patient with a neuromuscular disorder. *Hum. Genet.* **2016**, *135*, 245–251. [CrossRef]
99. Nakamura, N. Emerging New Roles of GM130, a cis-Golgi Matrix Protein, in Higher Order Cell Functions. *J. Pharmacol. Sci.* **2010**, *112*, 255–264. [CrossRef]
100. Mitchell, S.B.; Iwabuchi, S.; Kawano, H.; Yuen, T.M.T.; Koh, J.-Y.; Ho, K.W.D.; Harata, N.C. Structure of the Golgi apparatus is not influenced by a GAG deletion mutation in the dystonia-associated gene Tor1a. *PLoS ONE* **2018**, *13*, e0206123. [CrossRef]
101. Marra, P.; Salvatore, L.; Mironov, A.; Di Campli, A.; Di Tullio, G.; Trucco, A.; Beznoussenko, G.; Mironov, A.; De Matteis, M.A. The Biogenesis of the Golgi Ribbon: The Roles of Membrane Input from the ER and of GM130. *Mol. Biol. Cell* **2007**, *18*, 1595–1608. [CrossRef]
102. Chang, S.-H.; Hong, S.-H.; Jiang, H.-L.; Minai-Tehrani, A.; Yu, K.-N.; Lee, J.-H.; Kim, J.-E.; Shin, J.-Y.; Kang, B.; Park, S.; et al. GOLGA2/GM130, cis-Golgi matrix protein, is a novel target of anticancer gene therapy. *Mol. Ther. J. Am. Soc. Gene Ther.* **2012**, *20*, 2052–2063. [CrossRef]

103. Vitry, S.; Bruyère, J.; Hocquemiller, M.; Bigou, S.; Ausseil, J.; Colle, M.-A.; Prévost, M.-C.; Heard, J.M. Storage Vesicles in Neurons Are Related to Golgi Complex Alterations in Mucopolysaccharidosis IIIB. *Am. J. Pathol.* **2010**, *177*, 2984–2999. [CrossRef]
104. Lemonnier, T.; Blanchard, S.; Toli, D.; Roy, E.; Bigou, S.; Froissart, R.; Rouvet, I.; Vitry, S.; Heard, J.M.; Bohl, D. Modeling neuronal defects associated with a lysosomal disorder using patient-derived induced pluripotent stem cells. *Hum. Mol. Genet.* **2011**, *20*, 3653–3666. [CrossRef]
105. Soria Lopez, J.A.; González, H.M.; Léger, G.C. Alzheimer's disease. In *Handbook of Clinical Neurology*; Elsevier: Amsterdam, The Netherlands, 2019; pp. 231–255. [CrossRef]
106. Lew, J.; Winkfein, R.J.; Paudel, H.K.; Wang, J.H. Brain proline-directed protein kinase is a neurofilament kinase which displays high sequence homology to p34cdc2. *J. Biol. Chem.* **1992**, *267*, 25922–25926. [CrossRef] [PubMed]
107. Paidi, R.K.; Raha, S.; Roy, A.; Pahan, K. Muscle-building supplement β-hydroxy β-methylbutyrate binds to PPARα to improve hippocampal functions in mice. *Cell Rep.* **2023**, *42*, 112717. [CrossRef]
108. Zhang, Y.; Thompson, R.; Zhang, H.; Xu, H. APP processing in Alzheimer's disease. *Mol. Brain* **2011**, *4*, 3. [CrossRef]
109. Joshi, G.; Wang, Y. Golgi defects enhance APP amyloidogenic processing in Alzheimer's disease: Hypotheses. *BioEssays* **2015**, *37*, 240–247. [CrossRef] [PubMed]
110. Jankowsky, J.L.; Fadale, D.J.; Anderson, J.; Xu, G.M.; Gonzales, V.; Jenkins, N.A.; Copeland, N.G.; Lee, M.K.; Younkin, L.H.; Wagner, S.L.; et al. Mutant presenilins specifically elevate the levels of the 42 residue β-amyloid peptide in vivo: Evidence for augmentation of a 42-specific γ secretase. *Hum. Mol. Genet.* **2004**, *13*, 159–170. [CrossRef] [PubMed]
111. Joshi, G.; Chi, Y.; Huang, Z.; Wang, Y. Aβ-induced Golgi fragmentation in Alzheimer's disease enhances Aβ production. *Proc. Natl. Acad. Sci. USA* **2014**, *111*, E1230–E1239. [CrossRef]
112. Stieber, A.; Mourelatos, Z.; Gonatas, N.K. In Alzheimer's disease the Golgi apparatus of a population of neurons without neurofibrillary tangles is fragmented and atrophic. *Am. J. Pathol.* **1996**, *148*, 415–426.
113. Baloyannis, S.J. Golgi Apparatus and Protein Trafficking in Alzheimer's Disease. *J. Alzheimers Dis.* **2014**, *42*, S153–S162. [CrossRef]
114. Joshi, G.; Bekier, M.E.; Wang, Y. Golgi fragmentation in Alzheimer's disease. *Front. Neurosci.* **2015**, *9*, 340. [CrossRef]
115. Sun, K.-H.; De Pablo, Y.; Vincent, F.; Johnson, E.O.; Chavers, A.K.; Shah, K. Novel Genetic Tools Reveal Cdk5′s Major Role in Golgi Fragmentation in Alzheimer's Disease. *Mol. Biol. Cell* **2008**, *19*, 3052–3069. [CrossRef] [PubMed]
116. Tang, D.; Yuan, H.; Wang, Y. The Role of GRASP65 in Golgi Cisternal Stacking and Cell Cycle Progression. *Traffic* **2010**, *11*, 827–842. [CrossRef] [PubMed]
117. Lane, J.D.; Lucocq, J.; Pryde, J.; Barr, F.A.; Woodman, P.G.; Allan, V.J.; Lowe, M. Caspase-mediated cleavage of the stacking protein GRASP65 is required for Golgi fragmentation during apoptosis. *J. Cell Biol.* **2002**, *156*, 495–509. [CrossRef]
118. Cheng, J.P.X.; Betin, V.M.S.; Weir, H.; Shelmani, G.M.A.; Moss, D.K.; Lane, J.D. Caspase cleavage of the Golgi stacking factor GRASP65 is required for Fas/CD95-mediated apoptosis. *Cell Death Dis.* **2010**, *1*, e82. [CrossRef] [PubMed]
119. Ireland, S.; Ramnarayanan, S.; Fu, M.; Zhang, X.; Zhang, J.; Li, J.; Emebo, D.; Wang, Y. Cytosolic Ca^{2+} Modulates Golgi Structure Through PKCα-Mediated GRASP55 Phosphorylation. *iScience* **2020**, *23*, 100952. [CrossRef]
120. Zhang, J.; Kennedy, A.; Xing, L.; Bui, S.; Reid, W.; Joppich, J.; Ahat, E.; Rose, M.; Tang, Q.; Tai, A.W.; et al. SARS-CoV-2 triggers Golgi fragmentation via down-regulation of GRASP55 to facilitate viral trafficking. *bioRxiv* **2022**. [CrossRef]
121. Xiang, Y.; Zhang, X.; Nix, D.B.; Katoh, T.; Aoki, K.; Tiemeyer, M.; Wang, Y. Regulation of protein glycosylation and sorting by the Golgi matrix proteins GRASP55/65. *Nat. Commun.* **2013**, *4*, 1659. [CrossRef]
122. Cortese, M.; Lee, J.-Y.; Cerikan, B.; Neufeldt, C.J.; Oorschot, V.M.J.; Köhrer, S.; Hennies, J.; Schieber, N.L.; Ronchi, P.; Mizzon, G.; et al. Integrative Imaging Reveals SARS-CoV-2-Induced Reshaping of Subcellular Morphologies. *Cell Host Microbe* **2020**, *28*, 853–866.e5. [CrossRef]
123. Boluda, S.; Mokhtari, K.; Mégarbane, B.; Annane, D.; Mathon, B.; Cao, A.; Adam, C.; Androuin, A.; Bielle, F.; Brochier, G.; et al. Golgi localization of SARS-CoV-2 spike protein and interaction with furin in cerebral COVID-19 microangiopathy: A clue to the central nervous system involvement? *Free Neuropathol.* **2023**, *4*, 1. [CrossRef]
124. Pei, J.-J.; Grundke-Iqbal, I.; Iqbal, K.; Bogdanovic, N.; Winblad, B.; Cowburn, R.F. Accumulation of cyclin-dependent kinase 5 (cdk5) in neurons with early stages of Alzheimer's disease neurofibrillary degeneration. *Brain Res.* **1998**, *797*, 267–277. [CrossRef] [PubMed]
125. Wang, Y.; Gandy, S. The Golgi apparatus: Site for convergence of COVID-19 brain fog and Alzheimer's disease? *Mol. Neurodegener.* **2022**, *17*, 67. [CrossRef] [PubMed]
126. Ledesma, M.D.; Correas, I.; Avila, J.; Diaz-Nido, J. Implication of brain cdc2 and MAP2 kinases in the phosphorylation of tau protein in Alzheimer's disease. *FEBS Lett.* **1992**, *308*, 218–224. [CrossRef] [PubMed]
127. Lew, J.; Wang, J.H. Neuronal cdc2-like kinase. *Trends Biochem. Sci.* **1995**, *20*, 33–37. [CrossRef] [PubMed]
128. Lovestone, S.; Reynolds, C.H.; Latimer, D.; Davis, D.R.; Anderton, B.H.; Gallo, J.-M.; Hanger, D.; Mulot, S.; Marquardt, B.; Stabel, S.; et al. Alzheimer's disease-like phosphorylation of the microtubule-associated protein tau by glycogen synthase kinase-3 in transfected mammalian cells. *Curr. Biol.* **1994**, *4*, 1077–1086. [CrossRef]

International Journal of
Molecular Sciences

MDPI

Review

Unraveling the Intricate Link: Deciphering the Role of the Golgi Apparatus in Breast Cancer Progression

Adrian Vasile Dumitru [1,2], Evelina-Elena Stoica [3,*], Razvan-Adrian Covache-Busuioc [4,*], Bogdan-Gabriel Bratu [4] and Monica-Mihaela Cirstoiu [3,5]

1 Department of Pathology, "Carol Davila" University of Medicine and Pharmacy, 050474 Bucharest, Romania; vasile.dumitru@umfcd.ro
2 Department of Pathology, University Emergency Hospital, 050098 Bucharest, Romania
3 Department of Obstetrics and Gynaecology, University Emergency Hospital, 050098 Bucharest, Romania; monica.cirstoiu@umfcd.ro
4 Department of Neurosurgery, "Carol Davila" University of Medicine and Pharmacy, 050474 Bucharest, Romania; bogdan.bratu@stud.umfcd.ro
5 Department of Obstetrics and Gynaecology, "Carol Davila" University of Medicine and Pharmacy, 050474 Bucharest, Romania
* Correspondence: evelina.almajanu@gmail.com (E.-E.S.); razvan-adrian.covache-busuioc0720@stud.umfcd.ro (R.-A.C.-B.)

Abstract: Breast cancer represents a paramount global health challenge, warranting intensified exploration of the molecular underpinnings influencing its progression to facilitate the development of precise diagnostic instruments and customized therapeutic regimens. Historically, the Golgi apparatus has been acknowledged for its primary role in protein sorting and trafficking within cellular contexts. However, recent findings suggest a potential link between modifications in Golgi apparatus function and organization and the pathogenesis of breast cancer. This review delivers an exhaustive analysis of this correlation. Specifically, we examine the consequences of disrupted protein glycosylation, compromised protein transport, and inappropriate oncoprotein processing on breast cancer cell dynamics. Furthermore, we delve into the impacts of Golgi-mediated secretory routes on the release of pro-tumorigenic factors during the course of breast cancer evolution. Elucidating the nuanced interplay between the Golgi apparatus and breast cancer can pave the way for innovative therapeutic interventions and the discovery of biomarkers, potentially enhancing the diagnostic, prognostic, and therapeutic paradigms for afflicted patients. The advancement of such research could substantially expedite the realization of these objectives.

Keywords: golgi apparatus; breast cancer; protein glycosylation; protein trafficking; cancer progression; tumor microenvironment; biomarkers therapeutic targets

Citation: Dumitru, A.V.; Stoica, E.-E.; Covache-Busuioc, R.-A.; Bratu, B.-G.; Cirstoiu, M.-M. Unraveling the Intricate Link: Deciphering the Role of the Golgi Apparatus in Breast Cancer Progression. *Int. J. Mol. Sci.* **2023**, *24*, 14073. https://doi.org/10.3390/ijms241814073

Academic Editor: Anna Kawiak

Received: 23 August 2023
Revised: 11 September 2023
Accepted: 13 September 2023
Published: 14 September 2023

1. Introduction

Camillo Golgi is credited with the identification of the Golgi apparatus, a fundamental organelle inherent to eukaryotic cells [1]. Characterized by its intricate and dynamic nature, the Golgi apparatus is pivotal in an array of cellular activities, predominantly protein modification, segregation, conveyance, and packaging before its designated delivery to specific intracellular locales. Structurally, the organelle comprises overlapping membranous sacs, termed cisternae, which possess a unique architectural design optimized for proficient modification and packaging before routing to defined cellular destinations.

Acting as a central hub, the Golgi apparatus is instrumental in the processing and classification of diverse soluble proteins and lipids, directing them to their intended cellular destinations [2]. Given its seminal position in the secretory continuum, any perturbation in its architecture or functionality can gravely impact cellular protein and lipid equilibrium. Notably, a growing body of research has demonstrated that aberrations in the Golgi apparatus are implicated in a spectrum of conditions ranging from neurodegenerative maladies [3–7] to ischemic strokes, cardiovascular ailments, pulmonary arterial hypertension, infectious diseases, and malignancies.

Intrinsically adaptable, the Golgi apparatus possesses the capacity to rapidly recalibrate in response to evolving cellular demands and extracellular cues. This involves undergoing morphological transitions, such as reorganization, fragmentation, and integration with ancillary organelles, to aptly address variances in protein production and cellular requisites. In this context, the Golgi apparatus engages in complex interplays with other cellular structures, notably the endoplasmic reticulum (ER) and endosomes, orchestrating a sophisticated intracellular transport and communication matrix [8].

The indispensability of the Golgi apparatus in cellular operations underscores the ramifications of its dysfunction on human physiology and health. Disruptions or anomalies in its form, functionality, protein shuttling, or associated metabolic pathways have been pinpointed as etiological agents in a diverse array of pathologies, including malignancies, neurodegenerative diseases, and metabolic anomalies. Consequently, deepening our comprehension of its operational dynamics and significance is imperative for unveiling the underpinnings of these disorders and crafting targeted therapeutic modalities for their efficacious management.

2. Golgi Sorting, Protein Trafficking, and Glycosylation Abnormalities

Preserving the structural coherence of the Golgi apparatus is paramount for its optimal operation, as structural perturbations could usher in an array of pathologies. Operational aberrations of the Golgi apparatus encompass modifications in its pH equilibrium, anomalous glycosylation trajectories, and compromised membrane transport. Notably, fragmentation of the Golgi has been postulated as a precursor event in cellular apoptosis [9,10]. In scenarios of pharmacological or oxidative duress, the Golgi apparatus undergoes transformations, such as cargo saturation, ion concentration disequilibrium, and irregular luminal acidity, which collectively can induce membrane transport defects. We have coined the term "Golgi stress" to encapsulate this specific Golgi apparatus response, and two well-discussed molecular pathways are the structural preservation of Golgi apparatus by the TFE3 (transcription factor binding to IGHM enhancer 3) pathway and the proteoglycan pathway, which uptake the expression of enzymes for glycosylation [11].

Glycosylation stands as a pervasive posttranslational modification of proteins and plays a pivotal role in protein-mediated signaling. The glycans situated at glycosylation loci can span a spectrum in terms of complexity, from singular sugar chains to polymers boasting over 200 sugar units. Furthermore, glycans can be subjected to auxiliary modifications, encompassing the addition of entities like phosphate, sulfate, acetate, or phosphorylcholine for further diversification. It is noteworthy that a multitude of glycans manifest branch-like structures. An N-glycan entity can house up to six branches, each embedded with several recurrent disaccharide segments. The work by Stanley et al. (2011) offers insights into the traits and operations of Golgi glycosyltransferases (GTs), encompassing their activity spectra from their initiation at the cis-Golgi to their passage through the trans-Golgi network (TGN) [12].

The glycosylation of proteins is executed at two discrete intracellular locales, each defined by unique attributes. Proteins resident in the cytosol and nucleus undergo O-GlcNAcylation, wherein singular sugar entities termed N-acetylglucosamine (GlcNAc) directly bind to serine or threonine amino acids. This mechanism is instrumental in fine-tuning protein interactions, stability, functionality, and a gamut of cellular undertakings such as transcription, metabolism, apoptosis, and organelle genesis and transport [13,14]. In contrast, within the ER and Golgi apparatus lumen, secretory and transmembrane proteins are subjected to glycosylation by affixing specific glycosaccharides, or glycans, to particular amino acid chains. This modus operandi facilitates their functional diversification, allowing them to partake in multifarious cellular events [15].

The Golgi apparatus houses an array of glycosylation enzymes capable of either cleaving monosaccharides (glycosidases) or attaching them (GTs). Intriguingly, these enzymes can form both heteromeric and homomeric assemblies [16]. Structurally, GTs are membrane-bound proteins characterized by a brief N-terminal segment, a singular membrane domain, and a luminal domain. Due to this intricate configuration, they frequently establish enzyme complexes with other active enzymes within specific glycosylation pathways. N-glycosyltransferases within the Golgi can manifest in either homomeric or heteromeric groupings. The cyclical process of these GTs entails transitions influenced by the microenvironment, oscillating between heteromeric and homomeric states. While homomeric enzyme formations are pivotal in facilitating the folding and transportation of GTs to the Golgi apparatus, the more active heteromers are predominantly utilized for streamlined glycosylation [17,18]. Noteworthy GTs include GalNAc-T2 (N-acetylgalactosaminyltransferase-2) and GalT (β1,4-galactosyltransferase), which, due to their specificity for the Golgi apparatus, highlight that any depletion of juxtanuclear Golgi staining might be indicative of the organelle's attributes and the associated membrane proteins [19].

A dysfunctional Golgi glycosylation process has been associated with invasive behavior in various cancer types, encompassing prostate and breast malignancies [20,21]. The glycosylation process within the Golgi plays a cardinal role in numerous oncogenic molecular and cellular sequences, such as signal transduction, cellular communication, dissociation and invasion of cancer cells, cell–matrix attachment, angiogenesis, immunomodulation, and metastasis [22]. Analogous to the function of epithelial cadherin in mediating epithelial cellular cohesion, the Golgi-mediated glycosylation of N-linked glycans on epithelial cadherin might influence the epithelial-to-mesenchymal transition, thereby catalyzing the emergence of metastatic outgrowths. Such a mechanism is postulated to facilitate the migratory capacity of neoplastic cells from their inception point, be it during reparative processes post-injury or other standard physiological events, and becomes instrumental in the metastatic spread and proliferation of cancer [8,23].

The GOLPH3 complex, recognized as Golgi phosphoprotein 3, stands as a pivotal molecular entity in the realm of Golgi-facilitated oncogenesis. Its centrality in cancer can be attributed to a myriad of critical functionalities. GOLPH3 not only orchestrates Golgi glycosylation pivotal for the cancerous phenotype manifestation but also amplifies the DNA (deoxyribonucleic acid) damage response, bolstering survival amidst DNA-injurious scenarios. Additionally, it synergizes with retromer elements to enhance the mTOR (mammalian target of rapamycin) signaling upon growth factor induction and facilitates cell motility by orienting the Golgi apparatus toward the cellular forefront. Beyond GOLPH3, the Golgi spectrum hosts another consequential protein, GM130 (Golgi matrix protein 130). Integral to Golgi glycosylation and membranous protein trafficking, the downregulation of GM130 culminates in autophagy, diminished angiogenesis, and suppressed tumorigenesis [6,23–25].

Dysregulated Golgi glycosylation not only holds implications for carcinogenesis but might also propel cancer progression. Given the intertwined nature of Golgi-related operations and oncology, delving into and therapeutically targeting these processes should be foundational in cancer research endeavors.

Divergences in glycosylation can engender alterations in the conformation and function of numerous membranous proteins, with particular significance to collagen, fibronectin, integrins, and laminin at the extracellular interface. The paramount role of transmembrane integrins lies in fortifying the cytoskeleton via myriad cell–cell and cell–matrix interactions, thereby catalyzing cellular maturation and proliferation [26]. Glycosylation aberrancies might culminate in the flawed anchorage of these proteins, engendering a plethora of pathologies encompassing neurodegenerative conditions, malignancies, and cardiovascular afflictions [27–29].

The Golgi apparatus, with its cardinal role in modulating core cellular mechanisms, like adhesion and migration, stands as a keystone in the panorama of cancer evolution and metastatic dissemination. A prominent influencer in these oncogenic processes is identified as phosphatidylinositol 4-phosphate (PI4P). Hence, its role in human breast cancer can markedly sway cell–cell adhesion and migratory patterns [24].

Recent investigations underscore the paramount regulatory role of PI4P in the structural and functional intricacies of the Golgi apparatus, notably affecting glycosylation and the trafficking of proteins pivotal to cell–cell adhesion. By modulating the Golgi PI4P concentrations, the localization and activity of cardinal adhesion molecules, such as E-cadherin, are affected, thereby reshaping the intensity and dynamics of cell–cell interactions. Beyond its role in adhesion, PI4P governs activities linked with enzymes crucial for the synthesis or restructuring of glycosphingolipids, which are indispensable for cell surface interactions and signaling modalities. Furthermore, PI4P is integral in governing invasive cellular motility, a critical phenomenon in oncologic metastasis. Its regulatory role in Golgi-centric vesicular trafficking and membranous dynamism facilitates the modulation of invasive cell polarization and protrusive activities, augmenting their migratory and invasive propensities. In this orchestration, PI4P collaborates with a cohort of Golgi-associated proteins and lipid-mediated signaling pathways to modulate cytoskeletal transformations and matrix degradation, thereby facilitating the metastatic voyage of cancerous cells [21,30].

The traversal of cargoes through the Golgi apparatus is a multifaceted event and remains a focal point of discourse in the scientific literature. This review sheds light on five contemporary models postulated for assessing Golgi traffic, weighing their merits and demerits. The inaugural model posits anterograde vesicular transport amidst stable compartments of the Golgi. Conversely, the second hypothesis advocates for cisternal progression/maturation, wherein Golgi cisternae transition through sequential maturation phases. The third paradigm fuses progression/maturation with heterotypic tubular conveyance between cisternae. The penultimate model champions swift protein partitioning within a heterogenous Golgi and the terminal model envisions stable compartments as precursors for ensuing cisternal development.

A meticulous analysis reveals that no singular model can holistically encapsulate all documented phenomena across varied organisms. It might be more tenable to perceive cisternal progression/maturation as a foundational and evolutionarily conserved mechanism governing Golgi traffic. Certain cellular systems might integrate heterotypic tubular transport within Golgi cisternae. A judicious exploration of these models will illuminate the intricate facets of Golgi traffic, bestowing deeper insights into its operational mechanisms and elucidating this quintessential cellular undertaking. Grasping its foundational tenets is indispensable for decoding its influence on cellular equilibrium as well as pathological states linked to protein trafficking or excretion [31].

3. Golgi Apparatus Involvement in Breast Cancer

Breast cancer, a pressing global health challenge, leads to female mortality rates, and its prevalence is anticipated to surge in the forthcoming years. Diagnostic techniques like mammography and clinical breast inspections are pivotal for its early identification. While therapeutic modalities encompass surgical interventions, chemotherapy, and radiation treatments, each come with a set of concerns. Chemotherapy, despite its efficacy in neutralizing cancerous cells, presents a suite of adverse reactions. Radiotherapy, typically paired with surgery, may inflict enduring harm to critical organs. More promising therapeutic avenues encompass the deployment of anti-ErbB2 antibodies, exemplified by trastuzumab, especially for HER2-positive breast cancer variants. Additionally, antiestrogens and aromatase inhibitors serve to suppress the manifestation of estrogen-associated genes, proffering treatment avenues with diminished side effects [32,33].

The Rab GTPases, pivotal orchestrators of vesicular transportation, hold profound implications for the malignancy and invasiveness of cancer cells. Delving into estrogen receptor-positive breast cancer cellular frameworks reveals the instrumental role of Rab27B. Its heightened expression correlates with an augmented cellular elongation and an escalated invasiveness when interacting with collagen matrices. Such effects can be counteracted through miRNA-mediated interventions. Moreover, the amplification of Rab27B expression bears a direct relation to the surge in HSP90 alpha expression, a molecular custodian pivotal for upholding the structural integrity of MMP2 [34,35].

Rab40B's influence is palpably seen in maneuvering the trafficking pathways of metalloproteases MMP2 and MMP9 within the MDA-MB-231 breast cancer cellular context, facilitating the degradation of the external cellular matrix. Another metalloprotease, MT1-MMP, falls under the regulatory domain of Rab2A, which further fuels metastatic behaviors via its interaction with the VPS39 protein and is crucial for the amalgamation and clustering of late endosomes/lysosomes.

SiRNA screening has unmasked Rab2A's regulatory influence over the Golgi transport mechanisms of surface E-cadherin in breast cancer cells. Given the pivotal role of E-cadherin loss as an oncogenic transformation indicator, these revelations accentuate the significance of Rab GTPases in dictating vesicular transportation mechanisms. Such processes wield influence over cellular structural dynamics, invasion capacities, and external matrix degradation in these particular breast cancer cells [36,37]. Refer to Figure 1 for a visual representation.

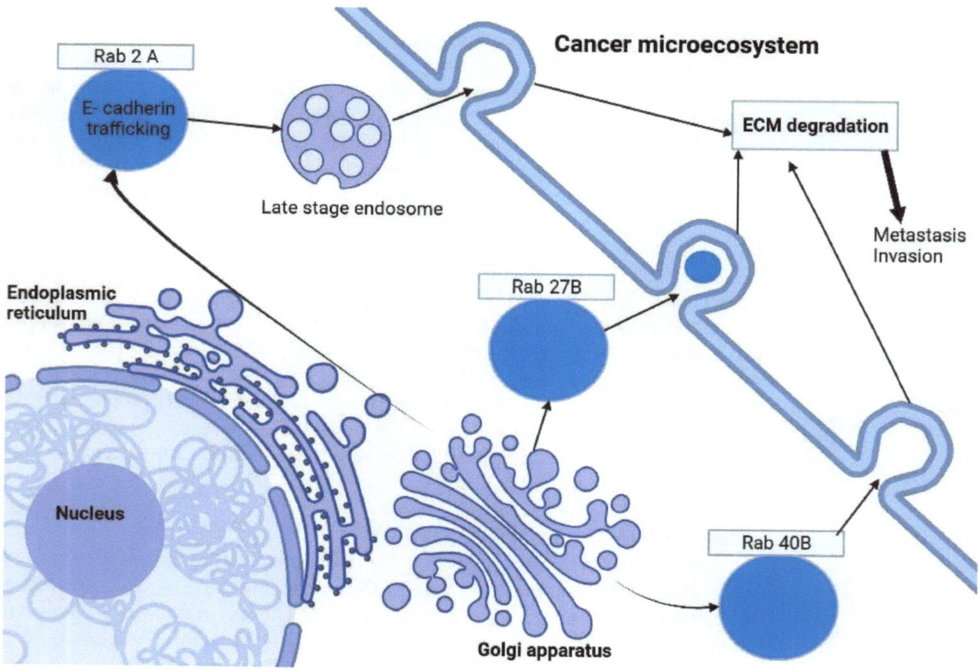

Figure 1. This diagram depicts how Rab27B, Rab2A (in a late-stage endosome form), and Rab40B from the Rab family of proteins, formed by the Golgi apparatus, are involved in breast cancer. After exocytosis, those Rab proteins play a key role in ECM (Extracellular Matrix Degradation), a representative element in the cancer microecosystem, which will further determine cell proliferation, tumoral invasion, and the formation of metastatic masses.

As highlighted in the cited study [38], GOLPH3's pronounced overexpression in breast cancer cells and tissues contrasts starkly with its presence in normal breast tissue. Escalated GOLPH3 levels correlate with advanced tumor development, metastatic spread, and a grim prognosis for breast cancer sufferers.

In the breast cancer scenario, GOLPH3 emerges as a pivotal entity, underpinning cancer cell proliferation and longevity by modulating its DNA damage response apparatus. A noteworthy interaction of GOLPH3 is with ATM (ataxia-telangiectasia mutated), a quintessential DNA damage response protein located at the Golgi. This interaction amplifies survival rates, rendering cancer cells more resilient against DNA damage-driven cellular demise. Such a mechanism equips cancer cells with heightened resistance against the genotoxic assaults unleashed by treatments like chemotherapy and radiation [39].

Furthermore, GOLPH3's tentacles extend into cancer progression by exerting regulatory control over a slew of signaling pathways, notably the PI3K-AKT-mTOR axis. GOLPH3 bolsters AKT activation, a kinase pivotal for cell proliferation and survival. This ultimately accelerates tumor growth and endows them with fortified resistance against treatments. From a therapeutic lens, targeting GOLPH3 emerges as a promising stratagem in the battle against breast cancer. A nuanced inhibition of its expression or functional prowess could prime cancer cells for increased susceptibility to DNA-damaging agents, effectively crippling tumor proliferation and metastatic spread [40].

In breast cancer patients, a surge in gene expression linked to ER-Golgi transport processes is evident, exemplified by genes like ARF4, COPB1, and USO1. These genes play an instrumental role in ferrying proteins between the ER and Golgi apparatus. To elucidate further, COPII vesicles shepherd proteins from the ER to the Golgi, whereas ARFs pilot the retrograde journey from the Golgi to the ER, which is facilitated by COPI vesicle formation [41].

Delving into the transportation dynamics of these genes reveals intriguing insights. The overexpression of ARF4, COPB1, and USO1 accelerates protein shuttling from the ER to Golgi. Introducing biotin amplifies this trafficking tempo even more, hinting at the pivotal role these ER-Golgi trafficking genes play in optimizing transportation kinetics [41].

A comprehensive meta-analysis of breast cancer cells also unravels that the expression patterns of ARF4, COPB1, and USO1 are orchestrated by the CREB3-like transcription factors. This harmonious co-expression, when disrupted, wreaks havoc on the cellular adhesive capacities, mobility, invasion potential, and the overarching metastatic traits of cancerous cells [32].

The discoveries highlight the pivotal roles that ARF4, COPB1, and USO1 undertake in breast cancer cell proliferation and invasiveness. Their significance underscores their role as key contributors to the disease's progression. Their paramount importance is further spotlighted through their integral roles in breast cancer evolution, particularly via the ER-Golgi trafficking mechanisms [42].

CREB3, a transcriptional architect, is instrumental in governing the traffic between the ER and Golgi apparatus. Its implications for breast cancer metastasis are a subject of fervent research. The quest to understand gene expression footprints steered by CREB3-mediated ER-Golgi trafficking unveils repercussions on the metastatic journey of breast cancer [43].

Evidence affirms that CREB3 activation spurs the upregulation of genes that participate in ER-Golgi trafficking, prominently featuring constituents of the COPII and COPI vesicle transportation networks. This unique genetic footprint, orchestrated by CREB3, is tethered to enhance metastatic capabilities in breast cancer cells. A slew of experimental methodologies was deployed in this research, which strove to pinpoint the direct nexus between CREB3-driven trafficking and the invasiveness inherent to breast cancer cells [44].

Moreover, the CREB3-directed trafficking signature has been painted as a harbinger of grim clinical outcomes in breast cancer patients. An upsurge in the expression of signature genes coincides with an elevated risk of metastasis and a dip in overall survival rates. This spotlight on the clinical significance accentuates its potential as a harbinger of disease prognosis in breast cancer [21].

In addition, genetic and epigenetic deviations in loci affiliated with the Golgi apparatus, such as CCDC170 (Coiled–Coil Domain Containing 170), have been entwined with susceptibilities to breast cancer. The unraveling of Golgi microtubule organization by proteins, like CCDC170, can culminate in anomalies in cell polarity and motility. These facets are quintessential for the invasiveness and metastatic prowess inherent to cancer cells [45].

4. Estrogen-Mediated Regulation of Protein Transcriptome: Impact on Vesicular Trafficking and Giant Vesicle Formation in Breast Cancer

Giant Vesicles (GVs) are vesicles, either inside or outside cells that range in size from 3 to 42 μm amd play a pivotal role in tumor proliferation. These vesicles originate mainly from ERα (estrogen receptor alpha)-negative breast cell lines and predominantly reside at the cell's edge [46].

Estrogen, pivotal in steering the transcriptome of proteins involved in vesicle movement and GV formation in breast cancer cells, regulates gene expressions essential to these processes. By modifying this transcriptome, estrogen directs the creation of vital proteins for vesicle movement and GV formation, thereby fueling growth and disease progression. As such, estrogen stands as a chief architect in this molecular realm and is linked with vesicle movement and GV formation. Estrogens, as paramount female sex hormones, are intrinsic to many physiological and pathological functions and hold a significant role in the onset of breast cancer. They operate by docking onto nuclear estrogen receptors, subsequently reshaping gene expressions. Notably, genes dictated by estrogen have been identified to sway various dimensions of cancer cell movement [47–49].

Two gene standouts, SYTL5 (Synaptotagmin-like 5) and RAB27B, regulated by 17 ss estradiol, are central to vesicle movement and exocytosis. SYTL5 functions as an intermediary molecule, collaborating with GTPases RAB27A/B. An increased presence of these genes has been documented in estrogen receptor-positive breast cancer cell lines, emphasizing their association with vesicle movement [50].

Additionally, SNX24 (Sorting Nexin 24), another estradiol-influenced gene, plays a quintessential role in endosomal categorization. GALNT4 (Polypeptide N-Acetylgalactosaminyltransferase 4) and SLC12A2 (Solute Carrier Family 12 Member 2) are foundational to the vesicle movement mechanism. Specifically, SLC12A2 is instrumental in steering the exocytosis of catecholamine from chromaffin cells, governing breast shape dynamics [51].

Wright et al. unveiled a unique vesicle species in breast cancer cells, which is known as the GV. Breast cancer cells, like MCF 7 and T47D, which robustly express the estrogen receptor alpha, rely on estradiol for their genesis. In contrast, non-ER alpha-negative cells, like MDA-MB-231/MDA-MB-468, remained untouched by estradiol in GV formation. However, in the presence of ER alpha, estradiol instigated the inception of estradiol-dependent GVs, hinting at a possible route where estradiol might spark this formation through ER alpha expression [52].

In essence, genes guided by estradiol and their interplay in vesicle movement across diverse frameworks considerably drive breast cancer cell growth and spread. This provides a compelling narrative on the multifaceted relationship between estrogen cues, gene orchestration, and vesicle movement in breast cancer. These discoveries elucidate the sophisticated dance between estrogen signaling, gene oversight, and vesicle movement, shedding light on their role in the progression of the disease [53].

5. Inhibition of Golgi-Associated Lipid Transfer Proteins (LTPs) as Potential Targets for Disease Intervention

The Golgi complex (GC) is pivotal in lipid biosynthesis and distribution. This incorporates both vesicle transport and non-vesicular pathways via Lipid Transfer Proteins (LTPs) like CERT (ceramide transfer protein), OSBP (oxysterol-binding protein), and FAPP2 (four-phosphate adaptor protein 2). Each of these proteins boasts distinct transport capabilities: CERT shuttles ceramide, OSBP transfers cholesterol, and FAPP2 moves GlcCer. All these proteins carry an N-terminal PH domain, enabling them to bind with PI4P for effective delivery within the GC. Some inhibitors targeting these processes are emerging as potential antiviral and anticancer therapeutics [54].

CERT specializes in carrying ceramide from the ER to the TGN, where it undergoes conversion to sphingomyelin [55]. Its inhibition or depletion results in ceramide accumulation, catalyzing ceramide-induced ER stress. This phenomenon primes various cancer cells, including ovarian, colorectal, and HER2-positive breast cancer cells, for enhanced vulnerability to chemotherapy. HPA-12, a CERT inhibitor, works by hampering CERT's recruitment during viral or parasitic invasions. This leads to augmented ceramide levels, rendering cancer cells more susceptible to paclitaxel-induced cell death, especially in contexts where these cells are resistant to traditional paclitaxel treatments or when there are infections or parasitic interferences [56].

OSBP1, a specialist in binding oxysterol, is central to the swap of cholesterol for PI4P between the ER and GC. OSBP, alongside ORP4L (oxysterol binding protein (OSBP)-related protein 4L), has been pinpointed as a target for various anticancer agents, including ORPphilin compounds like cephalostatin 1, OSW-1, Ritterazine B, and Schweinfurthin A, due to their profound effects on lipid metabolism. Moreover, Itraconazole, primarily an antifungal, demonstrates anticancer efficacy by targeting OSBP. Summarily, the GC's equilibrium in lipid concentrations hinges on both vesicle-based transport and the action of lipid transfer agents, like CERT and OSBP. Strategies that target these proteins are emerging as promising avenues in the development of novel anticancer and antiviral treatments [57].

6. Rho-Related BTB Domain Containing 1 (RhoBTB1) Drives Breast Cancer Growth and Metastasis through Methyltransferase-like 7B (METTL7B) Regulation

In breast cancer cells and patient samples, there is a marked downregulation of RhoBTB1 (Rho-Related BTB Domain Containing 1) expression, hinting at its probable tumor suppressor properties. Scientific inquiry reveals that a lack of RhoBTB1 leads to the disintegration of the Golgi structure, compromising its functions. This results in anomalies in protein glycosylation and associated trafficking pathways [58].

Adding another layer, METTL7B (Methyltransferase-like 7B) operates downstream of RhoBTB1. Recognized for its role in protein glycosylation, METTL7B experiences an uptick in its expression in breast cancer cells. Its heightened presence is, unfortunately, an indicator of grim outcomes for cancer patients. Through various functional studies, it is discerned that a surge in METTL7B expression counteracts the invasive tendencies of breast cancer cells and concurrently rectifies the Golgi disintegration brought on by RhoBTB1 insufficiency [59].

This interplay implies that RhoBTB1 acts as a brake on METTL7B, ensuring the structural and functional integrity of the Golgi apparatus and curtailing the invasion of breast cancer cells. Disturbances in this delicate balance seem to facilitate aggressive tendencies in breast cancer cells [60].

The narratives surrounding RhoBTB1 and RhoBTB2, and their associations with breast cancer, are subjects of intrigue, but their precise roles remain shrouded in mystery. RhoBTB3's part in breast cancer deterrence is even less understood. To delve deeper into these ambiguities, this research utilized bioinformatics tools, like Oncomine and cBioportal, and aimed to elucidate the roles and potential prognostic value of RhoBTB3 and Col1a1 in the context of breast cancer. Comprehensive methodologies including qRT-PCR analysis, immunoblotting assays, and a range of functional tests, such as invasion and proliferation assays, and flow cytometry were employed. The endgame was to pinpoint their contribution to the trajectory of breast cancer and evaluate their worth as prognostic indicators [61].

7. Possible Therapeutic Targets Regarding Breast Cancer

Breast cancer presents itself as a multifaceted ailment that is characterized by diverse subtypes and molecular variations. This heterogeneity necessitates a wide spectrum of treatment approaches targeting different pathways and molecules (Table 1). A breakdown of these targeted pathways and molecules follows.

Table 1. Pharmaceutical agents for breast cancer treatment with Golgi apparatus implications.

Type of Breast Cancer	Drug Name	Development Stage	Golgi Apparatus Mechanism
Estrogen receptor-postive	Tamoxifen Letrozole	FDA-approved FDA-approved	During post-Golgi recycling pathways, SCAMP1 determines cellular modifications against the progression of cancerous processes
HER2-positive	Trastuzumab Pertuzumab Adotrastuzumab emtansine	FDA-approved FDA-approved FDA-approved	
HER2-negative	Palbociclib Robociclib Abemaciclib	FDA-approved FDA-approved FDA-approved	Trafficking HER2 from the Golgi to the endocytic cellular pathways leads to a decreased level of HER2 expression
BRCA1/2	Olaparib Talazoparib Nitraparib	FDA-approved FDA-approved FDA-approved	Inhibition of Poly (ADP-ribose) Polymerase (PARP)
Triple-negative	Pembrolizumab Atezolizumab	FDA-approved FDA-approved	AXL (tyrosine kinase) has a polarized localization at the Golgi apparatus

In estrogen receptor-positive breast cancers, the cancer cells thrive on estrogen signals. To curtail this growth, endocrine therapies are deployed. Notable examples include selective estrogen receptor modulators, like Tamoxifen, and aromatase inhibitors, such as Letrozole, which have proven adept at inhibiting cancer cell proliferation [62].

Breast cancers characterized by excessive HER2 expression are labeled "HER2-positive". This overexpression catalyzes rapid cell growth and survival. Targeted therapies, specifically designed to obstruct HER2 signaling, are the countermeasures employed here. Drugs like trastuzumab (Herceptin), pertuzumab (Perjeta), and ado-trastuzumab emtansine (Kadcyla) are paramount in battling these types of breast cancers [63].

When it comes to both estrogen receptor-positive and HER2 breast cancers, there is an interesting dynamic at play involving SCAMP1 (Secretory Carrier-Associated Membrane Protein 1). Acting in post-Golgi recycling pathways, this protein, deemed a tumor suppressor, enhances the trafficking of MTSS1 (metastasis suppressor protein 1) to the cell's surface. Once there, MTSS1 kickstarts Rac1-GTP, promoting cell–cell adhesion, thereby forming a defensive front against cancer progression and invasion [64].

Lastly, for breast cancers carrying mutations in the BRCA1/2 genes, Poly (ADP-ribose) Polymerase (PARP) inhibitors come into play. Drugs like olaparib, talazoparib, and niraparib capitalize on DNA repair deficiencies in these cancers. By targeting cells that show homologous recombination inadequacy, they demonstrated significant clinical benefits for patients with BRCA mutation-positive breast cancer cases [65].

Breast cancer's complexity necessitates a wide array of targeted therapeutic interventions to tackle its multifarious molecular pathways. Several key molecules and pathways in breast cancer therapy include:

Cyclin-Dependent Kinase 4/6 (CDK4/6)—Drugs such as palbociclib, ribociclib, and abemaciclib, have been developed to inhibit CDK4 and CDK6 activities, which are pivotal for the progression of the cell cycle. Their introduction has yielded promising results, especially for patients with hormone receptor-positive and HER2-negative advanced breast cancer, showcasing significant improvements in progression-free survival [66].

The PI3K/AKT/mTOR Pathway—A recurrent aberration observed in breast cancer patients is a malfunction in this particular signaling pathway. As a consequence, inhibitors targeting its constituents, like PI3K inhibitors (e.g., alpelisib) and mTOR inhibitors (e.g., everolimus), have emerged as particularly potent against certain breast cancer variants [67].

Immune Checkpoint Inhibitors—These are groundbreaking agents, such as pembrolizumab and atezolizumab. They are engineered to target immune checkpoints, notably PD-1/PD-L1 (Programmed Cell Death Ligand 1), amplifying the immune system's capacity to identify and destroy cancer cells. Their efficacy has been particularly notable in cases of advanced triple-negative breast cancer [68].

Moreover, it is worth highlighting that the Golgi apparatus has sparked interest as a potential molecular target in breast cancer therapy. A growing number of therapeutic avenues, including antibody–drug conjugates, nanoparticles equipped to deliver antibody drugs, conjugate drug therapies, and advanced immunotherapies, are under scrutiny to optimize cancer patient outcomes [69]. The selection of a treatment strategy is contingent on myriad factors, including the patient's tumor subtype, its stage, and substage classification.

8. Conclusions

Breast cancer's progression and initiation intricately intertwine with the functioning of the Golgi apparatus. Alterations in its structural and operational framework have deep implications, manifesting in the acceleration of tumor growth, invasive behavior, and metastasis. A significant causative factor is the malfunctioning of Golgi-associated proteins, like matrix proteins, lipid transporters, and resident enzymes, leading to anomalies in glycosylation, protein trafficking, and signal relay—fundamental hallmarks of breast cancer.

CCDC170 stands out as a critical genetic component interlinked with the Golgi system with profound effects on breast cancer proliferation. Any perturbations within this structure might trigger cellular transformations, escalating the risk of cell invasion and potential metastasis.

Golgi-associated proteins, extending from matrix proteins to lipid transfer agents and enzymes resident within the Golgi, are complicit in fostering tumor growth, invasiveness, and metastatic spread. They modulate vital cellular functions, including adhesion, migration, and proliferation—fundamental processes underpinning the progression of breast cancer.

The intricate nexus between the Golgi apparatus and breast cancer is a treasure trove of insights. It reveals the underlying molecular intricacies and illuminates potential therapeutic interventions. By zeroing in on specific Golgi-mediated processes, such as protein glycosylation, vesicle transport, and interactions between Golgi and microtubules, we unveil novel therapeutic horizons. These might halt tumor expansion, avert metastatic spread, and enhance the prognosis for breast cancer patients.

However, the enigma of the Golgi apparatus and its relationship with breast cancer demands further exploration. By deciphering its elaborate processes and their role in cancer evolution, we are better poised to develop advanced diagnostic instruments, tailor treatments more precisely, and ensure a brighter prognosis for those battling breast cancer.

Author Contributions: Conceptualization, E.-E.S. and M.-M.C.; methodology, E.-E.S., A.V.D. and B.-G.B.; software, R.-A.C.-B.; validation, R.-A.C.-B. and E.-E.S.; formal analysis, A.V.D. and M.-M.C.; investigation, E.-E.S. and A.V.D.; resources, E.-E.S.; data curation, B.-G.B.; writing—original draft preparation, B.-G.B.; writing—review and editing, R.-A.C.-B. and B.-G.B.; visualization, E.-E.S.; supervision, M.-M.C.; project administration, M.-M.C.; funding acquisition, not applicable. All authors have read and agreed to the published version of the manuscript.

Funding: This research received no external funding.

Institutional Review Board Statement: Not applicable.

Informed Consent Statement: Not applicable.

Data Availability Statement: All data are available online in libraries such as PubMed.

Acknowledgments: Publication of this paper was supported by the "Carol Davila" University of Medicine and Pharmacy, through the institutional program Publish not Perish.

Conflicts of Interest: The authors declare no conflict of interest.

References

1. Rios, R.M.; Bornens, M. The Golgi apparatus at the cell centre. *Curr. Opin. Cell Biol.* **2002**, *15*, 60–66. [CrossRef] [PubMed]
2. Saceleanu, V.M.; Covache-Busuioc, R.-A.; Costin, H.-P.; Glavan, L.-A.; Ciurea, A.V. An Important Step in Neuroscience: Camillo Golgi and His Discoveries. *Cells* **2022**, *11*, 4112. [CrossRef] [PubMed]
3. Rendón, W.O.; Martínez-Alonso, E.; Tomás, M.; Martínez-Martínez, N.; Martínez-Menárguez, J.A. Golgi fragmentation is Rab and SNARE dependent in cellular models of Parkinson's disease. *Histochem. Cell Biol.* **2012**, *139*, 671–684. [CrossRef] [PubMed]
4. Brandstaetter, H.; Kruppa, A.J.; Buss, F. Huntingtin is required for ER-to-Golgi transport and for secretory vesicle fusion at the plasma membrane. *Dis. Model. Mech.* **2014**, *7*, 1335–1340. [CrossRef] [PubMed]

5. Yuan, D.; Liu, C.; Hu, B. Dysfunction of Membrane Trafficking Leads to Ischemia-Reperfusion Injury After Transient Cerebral Ischemia. *Transl. Stroke Res.* **2017**, *9*, 215–222. [CrossRef] [PubMed]
6. Li, T.; You, H.; Mo, X.; He, W.; Tang, X.; Jiang, Z.; Chen, S.; Chen, Y.; Zhang, J.; Hu, Z. GOLPH3 Mediated Golgi Stress Response in Modulating N2A Cell Death upon Oxygen-Glucose Deprivation and Reoxygenation Injury. *Mol. Neurobiol.* **2015**, *53*, 1377–1385. [CrossRef]
7. Ciurea, A.V.; Mohan, A.G.; Covache-Busuioc, R.A.; Costin, H.P.; Glavan, L.A.; Corlatescu, A.D.; Saceleanu, V.M. Unraveling Molecular and Genetic Insights into Neurodegenerative Diseases: Advances in Understanding Alzheimer's, Parkinson's, ALS, and Huntington's Disease. Medicine and Pharmacology. 2023. Available online: https://www.preprints.org/manuscript/202305.1229/v1 (accessed on 5 June 2023).
8. Liu, J.; Huang, Y.; Li, T.; Jiang, Z.; Zeng, L.; Hu, Z. The role of the Golgi apparatus in disease (Review). *Int. J. Mol. Med.* **2021**, *47*, 38. [CrossRef]
9. Sundaramoorthy, V.; Sultana, J.M.; Atkin, J.D. Golgi fragmentation in amyotrophic lateral sclerosis, an overview of possible triggers and consequences. *Front. Neurosci.* **2015**, *9*, 400. [CrossRef]
10. Gonatas, N.K.; Stieber, A.; Gonatas, J.O. Fragmentation of the Golgi apparatus in neurodegenerative diseases and cell death. *J. Neurol. Sci.* **2006**, *246*, 21–30. [CrossRef]
11. Jiang, Z.; Hu, Z.; Zeng, L.; Lu, W.; Zhang, H.; Li, T.; Xiao, H. The role of the Golgi apparatus in oxidative stress: Is this organelle less significant than mitochondria? *Free Radic. Biol. Med.* **2011**, *50*, 907–917. [CrossRef]
12. Stanley, P. Golgi glycosylation. *Cold Spring Harb. Perspect. Biol.* **2011**, *3*, a005199. [CrossRef]
13. Bond, M.R.; Hanover, J.A. A little sugar goes a long way: The cell biology of O-GlcNAc. *J. Cell Biol.* **2015**, *208*, 869–880. [CrossRef]
14. Torres, C.R.; Hart, G.W. Topography and polypeptide distribution of terminal N-acetylglucosamine residues on the surfaces of intact lymphocytes. Evidence for O-linked GlcNAc. *J. Biol. Chem.* **1984**, *259*, 3308–3317. [CrossRef] [PubMed]
15. Zhang, X.; Wang, Y. Glycosylation Quality Control by the Golgi Structure. *J. Mol. Biol.* **2016**, *428*, 3183–3193. [CrossRef] [PubMed]
16. Yamaguchi, N.; Fukuda, M.N. Golgi retention mechanism of beta-1,4-galactosyltransferase. Membrane-spanning domain-dependent homodimerization and association with alpha- and beta-tubulins. *J. Biol. Chem.* **1995**, *270*, 12170–12176. [CrossRef] [PubMed]
17. Schmitz, K.R.; Liu, J.; Li, S.; Setty, T.G.; Wood, C.S.; Burd, C.G.; Ferguson, K.M. Golgi Localization of Glycosyltransferases Requires a Vps74p Oligomer. *Dev. Cell* **2008**, *14*, 523–534. [CrossRef] [PubMed]
18. Hassinen, A.; Kellokumpu, S. Organizational Interplay of Golgi N-Glycosyltransferases Involves Organelle Microenvironment-Dependent Transitions between Enzyme Homo- and Heteromers. *J. Biol. Chem.* **2014**, *289*, 26937–26948. [CrossRef]
19. Storrie, B.; White, J.; Röttger, S.; Stelzer, E.H.; Suganuma, T.; Nilsson, T. Recycling of Golgi-resident Glycosyltransferases through the ER Reveals a Novel Pathway and Provides an Explanation for Nocodazole-induced Golgi Scattering. *J. Cell Biol.* **1998**, *143*, 1505–1521. [CrossRef]
20. Petrosyan, A.; Holzapfel, M.S.; Muirhead, D.E.; Cheng, P.W. Restoration of compact Golgi morphology in advanced prostate cancer enhances susceptibility to galectin-1-induced apoptosis by modifying mucin O-glycan synthesis. *Mol. Cancer Res.* **2014**, *12*, 1704–1716. [CrossRef]
21. Tokuda, E.; Itoh, T.; Hasegawa, J.; Ijuin, T.; Takeuchi, Y.; Irino, Y.; Fukumoto, M.; Takenawa, T. Phosphatidylinositol 4-Phosphate in the Golgi Apparatus Regulates Cell–Cell Adhesion and Invasive Cell Migration in Human Breast Cancer. *Cancer Res* **2014**, *74*, 3054–3066. [CrossRef]
22. Pinho, S.S.; Reis, C.A. Glycosylation in cancer: Mechanisms and clinical implications. *Nat. Rev. Cancer* **2015**, *15*, 540–555. [CrossRef]
23. Baschieri, F.; Confalonieri, S.; Bertalot, G.; Di Fiore, P.P.; Dietmaier, W.; Leist, M.; Crespo, P.; Macara, I.G.; Farhan, H. Spatial control of Cdc42 signalling by a GM130–RasGRF complex regulates polarity and tumorigenesis. *Nat. Commun.* **2014**, *5*, 4839. [CrossRef] [PubMed]
24. Rizzo, R.; Parashuraman, S.; D'Angelo, G.; Luini, A. GOLPH3 and oncogenesis: What is the molecular link? *Tissue Cell* **2017**, *49 Pt A*, 170–174. [CrossRef]
25. Xing, M.; Peterman, M.C.; Davis, R.L.; Oegema, K.; Shiau, A.K.; Field, S.J.; Chapnick, D.A.; Liu, X.; Luo, M.E.K.; Koch, S.; et al. GOLPH3 drives cell migration by promoting Golgi reorientation and directional trafficking to the leading edge. *Mol. Biol. Cell* **2016**, *27*, 3828–3840. [CrossRef] [PubMed]
26. Thomas, D.; Rathinavel, A.K.; Radhakrishnan, P. Altered glycosylation in cancer: A promising target for biomarkers and therapeutics. *Biochim. Biophys. Acta Rev. Cancer* **2021**, *1875*, 188464. [CrossRef] [PubMed]
27. Menni, C.; Gudelj, I.; Macdonald-Dunlop, E.; Mangino, M.; Zierer, J.; Bešić, E.; Joshi, P.K.; Trbojević-Akmačić, I.; Chowienczyk, P.J.; Spector, T.D.; et al. Glycosylation Profile of Immunoglobulin G Is Cross-Sectionally Associated with Cardiovascular Disease Risk Score and Subclinical Atherosclerosis in Two Independent Cohorts. *Circ. Res.* **2018**, *122*, 1555–1564. [CrossRef] [PubMed]
28. Schedin-Weiss, S.; Winblad, B.; Tjernberg, L.O. The role of protein glycosylation in Alzheimer disease. *FEBS J.* **2013**, *281*, 46–62. [CrossRef] [PubMed]
29. Stowell, S.R.; Ju, T.; Cummings, R.D. Protein glycosylation in cancer. *Annu. Rev. Pathol.* **2015**, *10*, 473–510. [CrossRef]
30. Durrant, L.G.; Noble, P.; Spendlove, I. Immunology in the clinic review series; focus on cancer: Glycolipids as targets for tumour immunotherapy. *Clin. Exp. Immunol.* **2012**, *167*, 206–215. [CrossRef]
31. Glick, B.S.; Luini, A. Models for Golgi Traffic: A Critical Assessment. *Cold Spring Harb. Perspect. Biol.* **2011**, *3*, a005215. [CrossRef]

32. Mughees, M.; Chugh, H.; Wajid, S. Vesicular trafficking–related proteins as the potential therapeutic target for breast cancer. *Protoplasma* **2020**, *257*, 345–352. [CrossRef] [PubMed]
33. Chevalier, C.; Roche, S.; Bénistant, C. Vesicular trafficking regulators are new players in breast cancer progression: Role of TOM1L1 in ERBB2-dependent invasion. *Mol. Cell. Oncol.* **2016**, *3*, e1182241. [CrossRef] [PubMed]
34. Shahabi, A.; Naghili, B.; Ansarin, K.; Zarghami, N. The relationship between microRNAs and Rab family GTPases in human cancers. *J. Cell. Physiol.* **2019**, *234*, 12341–12352. [CrossRef]
35. Li, Z.; Fang, R.; Fang, J.; He, S.; Liu, T. Functional implications of Rab27 GTPases in Cancer. *Cell Commun. Signal.* **2018**, *16*, 44. [CrossRef] [PubMed]
36. Jacob, A.; Jing, J.; Lee, J.; Schedin, P.; Gilbert, S.M.; Peden, A.A.; Junutula, J.R.; Prekeris, R. Rab40b regulates MMP2 and MMP9 trafficking during invadopodia formation and breast cancer cell invasion. *J. Cell Sci.* **2013**, *126*, 4647–4658. [CrossRef]
37. Kajiho, H.; Kajiho, Y.; Frittoli, E.; Confalonieri, S.; Bertalot, G.; Viale, G.; Di Fiore, P.P.; Oldani, A.; Garre, M.; Beznoussenko, G.V.; et al. RAB2A controls MT1-MMP endocytic and E-cadherin polarized Golgi trafficking to promote invasive breast cancer programs. *EMBO Rep.* **2016**, *17*, 1061–1080. [CrossRef]
38. Buschman, M.D.; Rahajeng, J.; Field, S.J. GOLPH3 Links the Golgi, DNA Damage, and Cancer. *Cancer Res.* **2015**, *75*, 624–627. [CrossRef]
39. Farber-Katz, S.E.; Dippold, H.C.; Buschman, M.D.; Peterman, M.C.; Xing, M.; Noakes, C.J.; Tat, J.; Ng, M.M.; Rahajeng, J.; Cowan, D.M.; et al. DNA Damage Triggers Golgi Dispersal via DNA-PK and GOLPH3. *Cell* **2014**, *156*, 413–427. [CrossRef]
40. Ciccia, A.; Elledge, S.J. The DNA Damage Response: Making It Safe to Play with Knives. *Mol. Cell* **2010**, *40*, 179–204. [CrossRef]
41. Howley, B.V.; Link, L.A.; Grelet, S.; El-Sabban, M.; Howe, P.H. A CREB3-regulated ER–Golgi trafficking signature promotes metastatic progression in breast cancer. *Oncogene* **2018**, *37*, 1308–1325. [CrossRef]
42. Nakai, W.; Kondo, Y.; Saitoh, A.; Naito, T.; Nakayama, K.; Shin, H.-W.; Tanaka, Y.; Ono, N.; Shima, T.; Tanaka, G.; et al. ARF1 and ARF4 regulate recycling endosomal morphology and retrograde transport from endosomes to the Golgi apparatus. *Mol. Biol. Cell* **2013**, *24*, 2570–2581. [CrossRef]
43. Sampieri, L.; Di Giusto, P.; Alvarez, C. CREB3 Transcription Factors: ER-Golgi Stress Transducers as Hubs for Cellular Homeostasis. *Front. Cell Dev. Biol.* **2019**, *7*, 123. [CrossRef] [PubMed]
44. Yadavalli, T.; Sharma, P.; Wu, D.; Kapoor, D.; Shukla, D. CREB3 Plays an Important Role in HPSE-Facilitated HSV-1 Release in Human Corneal Epithelial Cells. *Viruses* **2022**, *14*, 1171. [CrossRef] [PubMed]
45. Jiang, P.; Li, Y.; Poleshko, A.; Medvedeva, V.; Baulina, N.; Zhang, Y.; Zhou, Y.; Slater, C.M.; Pellegrin, T.; Wasserman, J.; et al. The Protein Encoded by the CCDC170 Breast Cancer Gene Functions to Organize the Golgi-Microtubule Network. *EBioMedicine* **2017**, *22*, 28–43. [CrossRef] [PubMed]
46. Wright, P.K.; Jones, S.B.; Ardern, N.; Ward, R.; Clarke, R.B.; Sotgia, F.; Lisanti, M.P.; Landberg, G.; Lamb, R. 17β-estradiol regulates giant vesicle formation via estrogen receptor-alpha in human breast cancer cells. *Oncotarget* **2014**, *5*, 3055–3065. [CrossRef]
47. Osborne, C.K.; Schiff, R. Estrogen-Receptor Biology: Continuing Progress and Therapeutic Implications. *J. Clin. Oncol.* **2005**, *23*, 1616–1622. [CrossRef]
48. Cunliffe, H.E.; Ringnér, M.; Bilke, S.; Walker, R.L.; Cheung, J.M.; Chen, Y.; Meltzer, P.S. The gene expression response of breast cancer to growth regulators: Patterns and correlation with tumor expression profiles. *Cancer Res.* **2003**, *63*, 7158–7166.
49. DeRoo, B.J.; Korach, K.S. Estrogen receptors and human disease. *J. Clin. Investig.* **2006**, *116*, 561–570. [CrossRef]
50. Wright, P.K.; May, F.E.; Darby, S.; Saif, R.; Lennard, T.W.; Westley, B.R. Estrogen regulates vesicle trafficking gene expression in EFF-3, EFM-19 and MCF-7 breast cancer cells. *Int. J. Clin. Exp. Pathol.* **2009**, *2*, 463–475.
51. Lacey, J.; Webster, S.J.; Heath, P.R.; Hill, C.J.; Nicholson-Goult, L.; Wagner, B.E.; Khan, A.O.; Morgan, N.V.; Makris, M.; Daly, M.E. Sorting nexin 24 is required for α-granule biogenesis and cargo delivery in megakaryocytes. *Haematologica* **2022**, *107*, 1902–1913. [CrossRef]
52. Thompson, E.W.; Torri, J.; Johnson, M.D.; Boulay, V.; Wright, A.; Lippman, M.E.; Steeg, P.S.; Clarke, R.; Brünner, N. The invasive and metastatic properties of hormone-independent but hormone-responsive variants of MCF-7 human breast cancer cells. *Clin. Exp. Metastasis* **1993**, *11*, 15–26. [CrossRef] [PubMed]
53. Safe, S. Transcriptional activation of genes by 17 beta-estradiol through estrogen receptor-Sp1 interactions. *Vitam. Horm.* **2001**, *62*, 231–252. [PubMed]
54. Peretti, D.; Dahan, N.; Shimoni, E.; Hirschberg, K.; Lev, S. Coordinated Lipid Transfer between the Endoplasmic Reticulum and the Golgi Complex Requires the VAP Proteins and Is Essential for Golgi-mediated Transport. *Mol. Biol. Cell* **2008**, *19*, 3871–3884. [CrossRef]
55. Chung, L.H.; Liu, D.; Liu, X.T.; Qi, Y. Ceramide Transfer Protein (CERT): An Overlooked Molecular Player in Cancer. *Int. J. Mol. Sci.* **2021**, *22*, 13184. [CrossRef]
56. Santos, C.; Fleury, L.; Rodriguez, F.; Markus, J.; Berkeš, D.; Daïch, A.; Ausseil, F.; Baudoin-Dehoux, C.; Ballereau, S.; Génisson, Y. The CERT antagonist HPA-12: First practical synthesis and individual binding evaluation of the four stereoisomers. *Bioorganic Med. Chem.* **2015**, *23*, 2004–2009. [CrossRef]
57. Zerbinatti, C.V.; Cordy, J.M.; Chen, C.-D.; Guillily, M.; Suon, S.; Ray, W.J.; Seabrook, G.R.; Abraham, C.R.; Wolozin, B. Oxysterol-binding protein-1 (OSBP1) modulates processing and trafficking of the amyloid precursor protein. *Mol. Neurodegener.* **2008**, *3*, 5. [CrossRef]

58. McKinnon, C.M.; Mellor, H. The tumor suppressor RhoBTB1 controls Golgi integrity and breast cancer cell invasion through METTL7B. *BMC Cancer* **2017**, *17*, 145. [CrossRef]
59. Berthold, J.; Schenková, K.; Ramos, S.; Miura, Y.; Furukawa, M.; Aspenström, P.; Rivero, F. Characterization of RhoBTB-dependent Cul3 ubiquitin ligase complexes—Evidence for an autoregulatory mechanism. *Exp. Cell Res.* **2008**, *314*, 3453–3465. [CrossRef] [PubMed]
60. Yadav, S.; Puri, S.; Linstedt, A.D. A primary role for Golgi positioning in directed secretion, cell polarity, and wound healing. *Mol. Biol. Cell* **2009**, *20*, 1728–1736. [CrossRef]
61. Kim, K.; Kim, Y.-J. RhoBTB3 Regulates Proliferation and Invasion of Breast Cancer Cells via Col1a1. *Mol. Cells* **2022**, *45*, 631–639. [CrossRef]
62. Fuentes, N.; Silveyra, P. Estrogen receptor signaling mechanisms. *Adv. Protein Chem. Struct. Biol.* **2019**, *116*, 135–170. [CrossRef]
63. Iqbal, N.; Iqbal, N. Human Epidermal Growth Factor Receptor 2 (HER2) in Cancers: Overexpression and Therapeutic Implications. *Mol. Biol. Int.* **2014**, *2014*, 852748. [CrossRef]
64. Spano, D.; Colanzi, A. Golgi Complex: A Signaling Hub in Cancer. *Cells* **2022**, *11*, 1990. [CrossRef] [PubMed]
65. Wiggans, A.J.; Cass, G.K.; Bryant, A.; Lawrie, T.A.; Morrison, J. Poly(ADP-ribose) polymerase (PARP) inhibitors for the treatment of ovarian cancer. *Cochrane Database Syst. Rev.* **2015**, *2015*, CD007929. [CrossRef] [PubMed]
66. Shah, M.; Nunes, M.R.; Stearns, V. CDK4/6 Inhibitors: Game Changers in the Management of Hormone Receptor–Positive Advanced Breast Cancer? *Oncology* **2018**, *32*, 216–222. [PubMed]
67. Peng, Y.; Wang, Y.; Zhou, C.; Mei, W.; Zeng, C. PI3K/Akt/mTOR Pathway and Its Role in Cancer Therapeutics: Are We Making Headway? *Front. Oncol.* **2022**, *12*, 819128. [CrossRef]
68. Shiravand, Y.; Khodadadi, F.; Kashani, S.M.A.; Hosseini-Fard, S.R.; Hosseini, S.; Sadeghirad, H.; Ladwa, R.; O'byrne, K.; Kulasinghe, A. Immune Checkpoint Inhibitors in Cancer Therapy. *Curr. Oncol.* **2022**, *29*, 3044–3060. [CrossRef]
69. Subhan, M.A.; Torchilin, V.P. Advances in Targeted Therapy of Breast Cancer with Antibody-Drug Conjugate. *Pharmaceutics* **2023**, *15*, 1242. [CrossRef]

brain sciences

MDPI

Viewpoint

Wilhelm von Waldeyer: Important Steps in Neural Theory, Anatomy and Citology

Vicentiu Mircea Saceleanu [1,2], Aurel George Mohan [3,4,*], Razvan Adrian Covache-Busuioc [5], Horia Petre Costin [5] and Alexandru Vlad Ciurea [6,7]

1 Neurosurgery Department, Sibiu County Emergency Hospital, 550245 Sibiu, Romania; vicentiu.saceleanu@gmail.com
2 Neurosurgery at "Lucian Blaga" University of Medicine, 550024 Sibiu, Romania
3 Department of Neurosurgery, Bihor County Emergency Clinical Hospital, 410167 Oradea, Romania
4 Neurosurgery, Faculty of Medicine, Oradea University, 410610 Oradea, Romania
5 General Medicine, "Carol Davila" University of Medicine and Pharmacy, 020021 Bucharest, Romania; razvan_adrian69@yahoo.ro (R.A.C.-B.); horiacostin2001@yahoo.com (H.P.C.)
6 Neurosurgery, "Carol Davila" University of Medicine and Pharmacy, 020021 Bucharest, Romania; prof.avciurea@gmail.com
7 Neurosurgery Department and Scientific Director at the Sanador Clinical Hospital, 70000 Bucharest, Romania
* Correspondence: mohanaurel@yahoo.com

Abstract: Heinrich Wilhelm Gottfried von Waldeyer-Harz is regarded as a significant anatomist who helped the entire medical world to discover and develop new techniques in order to improve patient treatment as well as decrease death rates. He discovered fascia propria recti in 1899, which is important in total mesorectal excision which improves cancer treatment as well as outcomes. He played an important role in developing the neuron theory which states that the nervous system consists of multiple individual cells, called neurons, which currently stands as the basis of the impulse transmission of neurons. Waldeyer was also interested in cytology, where he made a substantial contribution, being the first who adopted the name "Chromosome". Therefore, he accelerated the progress of what it is now known as Genetics. In conclusion, starting from the Fascia propria recti and continuing with great discoveries in cytology and neuron theory, Wilhelm von Waldeyer represents a key person in what we today call medicine.

Keywords: neuron theory; Waldeyer history; Waldeyer citology; Waldeyer anatomy

check for updates

Citation: Saceleanu, V.M.; Mohan, A.G.; Covache-Busuioc, R.A.; Costin, H.P.; Ciurea, A.V. Wilhelm von Waldeyer: Important Steps in Neural Theory, Anatomy and Citology. *Brain Sci.* **2022**, *12*, 224. https://doi.org/10.3390/brainsci12020224

Academic Editor: Stephen D. Meriney

Received: 7 January 2022
Accepted: 3 February 2022
Published: 6 February 2022

1. Introduction

Heinrich Wilhelm Gottfried von Waldeyer-Hartz (Figure 1) was a well-known German anatomist who is known for his efforts in giving both the medical world and humanity an acquaintance in multiple fields of study such as anatomy, embryology and pathology, which today plays a vital role in treating genetic diseases and cancer.

He was born on 6 October 1836 in Hehlen, a small village near Brunswick. He completed his studies at the Gymnasium Theodorianum in Paderborn, were he obtained a graduation diploma in 1856, which attested his eligibility of attending university courses. Further on, he attended the Universität Göttingen were he focused his studies on mathematics and natural sciences [1].

This is a place that played an important role in the future of Waldeyer, because he met the recognised anatomist Jakob Henle (1809–1885), who discovered the loop of Henle which has great importance in the kidney's physiology. Waldeyer was so impressed with Henle's work that he entered medical school in 1857. In 1861, he acquired his doctorate diploma based on his thesis entitled "De claviculae articulis e functione" [2].

Figure 1. Heinrich Wilhelm Gottfried von Waldeyer-Hartz (1836–1921).

Later on, in 1867, he became a professor of pathological anatomy in Breslau, then Waldeyer became a full professor in Strasbourg in 1872, and in 1883, he moved to Berlin where he lived for more than 33 years working at the Institute of Anatomy [3].

Before his death on 23 January 1921, his desire was for his hand, skull and brain to be preserved at the Institute of Anatomy in Berlin in order to be studied and examined. Hans Virchow was the one who dissected the hands and published an entire detailed description of the anatomy of this body part of Waldeyer (Figure 2) [4]. However, the studies based on his brain and skull were not assigned to Virchow and were not found. The entire idea behind his desire to donate his body parts to the Institute was a popular decision among well-known people from the medical world of that era due to a belief that distinguished signs could be seen in tremendously intelligent people's brains [2].

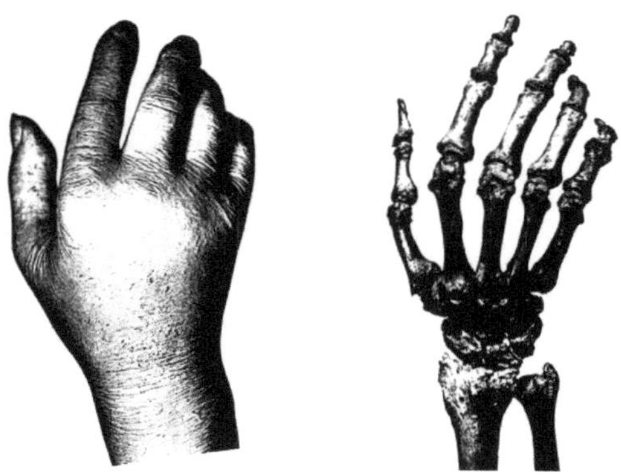

Figure 2. Waldeyer's right hand (**left**) and the skeleton of the right hand (**right**) according to Hans Virchow's description (1923) [4].

2. Waldeyer's Medical Contributions

One of Waldeyer's biggest anatomical contribution was represented by the "Fascia propria recti", which at the time of discovery in 1899, did not present such a great medical interest, but in the last century, its importance in surgical practice has grown higher and higher due to its importance in total mesorectal excision, a surgical procedure involved in rectal cancer treatment. It was described for the first time by Professor Bill Heald in 1982, at the Basingstoke District Hospital in the United Kingdom [5].

Fascia propria recti is regarded as a thin layer of connective tissue which lies between the presacral fascia and the rectal proper fascia. It is also known as the rectosacral fascia, according to its position, defining the retrorectal space in two compartments: a superior and an inferior one [6].

The great debate that appeared around the fascia propria recti is whether Waldeyer was the first one who discovered it or not. In the first edition of the anatomical book "Traité d'Anatomie Humaine", revised by P. Poirier, which was probably published in 1894, but definitely between 1892 and 1896 in Paris, Toma Ionescu described this fascia for the first time, under the name of "rectal sheath", about 5 years before the name of "fascia propria recti" was spread around the entire medical world [7].

It is not clear why Toma Ionescu was not perceived as the first anatomist who discovered it, but some probable theories suggest that it was mostly because of the big difference between Toma Ionescu and Waldeyer's age. Waldeyer was at that time with 25 years older than Toma Ionescu and was already one of the most well-known anatomists across the world, having a wider influence and a recognised reputation [7].

Nevertheless, French authors carry the entire merit of giving Toma Ionescu the credits for his discovery, considering him as the first one who claimed the name for the rectosacral fascia.

3. The Neuron Theory

The neuron theory, which is also called the neuron doctrine, represents an idea of Santiago Ramón y Cajal, which states that the nervous system consists of multiple individual cells called "neurons", which have an individual structure and function, working together in order to create a singular and refined machinery that controls the entire human body [8].

However, the path to achieve this concept was not an easy one and Wilhelm von Waldeyer played an important role in expressing the neuron theory.

The history of the neuron theory starts back in 1873, when Camillo Golgi invented a new staining method, known as "la reazione nera" ("black reaction"), later called Golgi staining technique in his honour [9].

This method was used for microscopic research, which at that time was difficult due to the lack of staining techniques. Therefore, the new method discovered by Golgi played a vital role in the discovery of the nervous system, because he could differentiate the dendrites from the axon of the neuron.

Thus, he observed an entire network of neurons in the grey matter and proposed what was called "The reticular theory". This concept proposed that the entire cerebro-spinal axis was one continuous neural network that acted as a single organ. This theory represented the main idea of how the entire nervous system works, but in reality, the truth was totally different from what Camillo Golgi proposed [10].

In 1887, Santiago Ramón y Cajal used the Golgi staining technique to study the neural network, making a discovery that would change the entire approach of the nervous system (Figure 3). He discovered that between the neurons, there is not a continuous link, but instead there is a space between them, which is now known as the synaptic cleft. This was the moment "The neuron theory" was born [11].

Golgi's concept of a continuous nervous system was therefore obsolete, and even though he never agreed with Cajal's theory, Waldeyer was a firm supporter of it. Moreover, the impact of Waldeyer's contributions is mostly represented by naming the nervous cells "neurons", which comes from the Greek word "sinew" [12].

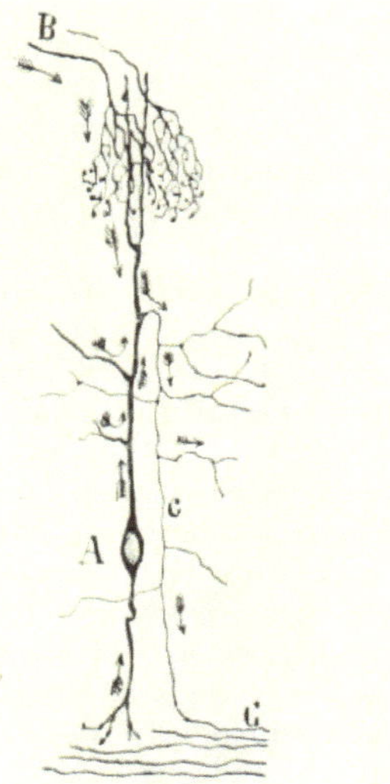

Figure 3. A graphic description of a synapse involved in vision (optic lobe), according to Cajal (1909), using the Golgi staining method. The arrows indicate the direction of the nervous impulse. (**A**) neural body, (**B**) afferent fibres. (**C**) axonal fibre [11].

Therefore, the neuron theory constituted a solid base for the following discoveries in terms of impulse transmission, as well as structural and functional particularities of the neurons, which could be later described using electronic microscopy in order to make a clear statement about the way the entire nervous system works [13].

However, we do not have to assume that Golgi's reticular theory was entirely wrong. Nowadays, studies have determined that there is an intense interconnectedness between neurons and astrocytes, and even if we could describe the nervous system as a network composed of many independent cells, it is much more important to assume that it works as a unitary and perfectly coordinated system.

4. Waldeyer's Contributions to Cytology

Cytology in the 19th century represented a controversial study subject due to the lack of information, as well as the absence of the lab techniques needed in order to analyse the structure of the cell and its mechanisms. However, different studies were conducted and step by step, the researchers of that time made rapid progress. Waldeyer, played an important role in refining the cytology.

In 1888, Waldeyer published an article intitled "Über Karyokinese und ihre Beziehungen zu den Befruchtungsvorgängen" ("About karyokinesis and its relationships with the fertilization processes") [14], which signifies what is now called a review article, since it has 210 references, used to objectify a vast amount of information in just one paper.

Among the scientists of that century who were citated in this extended review, names including Rudolf Virchow, Theodor Boveri, Oskar Hertwig, Edouard-Gerard Balbiani, Walther Flemming and many others provided both theoretical and experimental information which was used to enhance the explanation of the entire fertilization and karyokinesis process.

One of the most relevant information that can be extracted from this article, is the word "Chromosomen" (in German) [14], which was translated into English under the form we use today of "Chromosome". Before this name was introduced by Wilhelm von Waldeyer, the name "Chromatinelemente" (Chromatic elements) was proposed by Theodor Boveri and used by the entire scientific community [2].

However, there was a long way that had to be followed in order to reach the chromosome discovery. First of all, near the middle of the 19th century, Theodor Schwann (1810–1882) and Matthias Schleiden (1804–1881) were regarded as the discoverers of cell theory, suggested in 1838–1839. They strongly believed that the cells were produced de novo from a substance called "cytoblastem", which did not have a specific structure [15].

Even though this theory sounds aberrating these days, the scientific community did not have any strong experimental or theoretical information to correlate, and this lack of data led to a wrong perception of the cytokinesis.

The theory was not rejected until 1855, when Rudolf Virchow (1821–1902) stated "omnis cellula e cellula" [15], which means that every new cell is created from a pre-existing one through division (Figure 4).

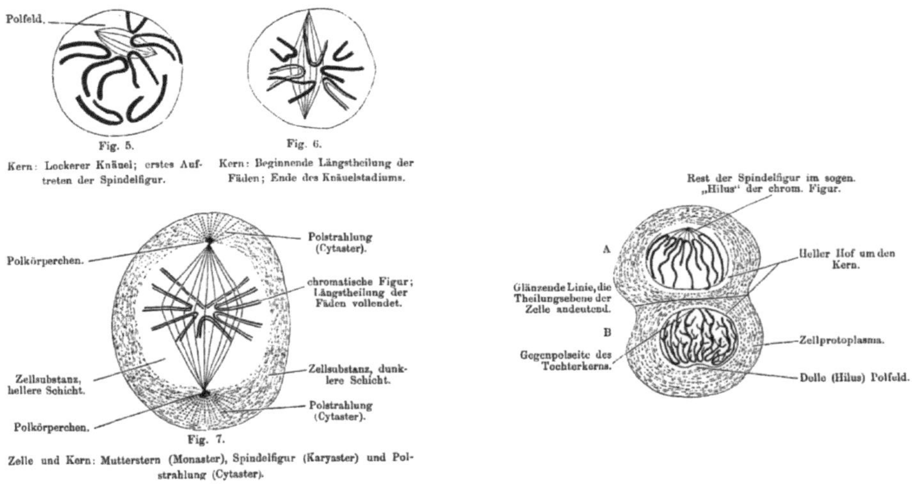

Figure 4. Wilhelm von Waldeyer's representation of cytokinesis and karyokinesis, according to the extensive review entitled "Über Karyokinese und ihre Beziehungen zu den Befruchtungsvorgängen" [14]. Figures 5–7 in ref. [14].

Nevertheless, after Virchow's statement became clear for the entire scientific community, the new debate about nucleus division became more and more thought-provoking for all researchers. It could be seen at that time that during cell division, the nucleus disappears and appears immediately along with the birth of new daughter cells.

Some of the scientists of that time considered that karyokinesis was actually a "generatio spontanea" inside the cells, while others, such as Walther Flemming, were supporters of the indirect division (amitosis) of the nucleus. Later on in 1917, Oscar Hertwig made a discovery that the role of the chromosomes is represented by their hereditary information, helping the cells to become specialised according to the indications made by the chromosomes [16].

Therefore, the 19th century represented a century of discovery that laid the foundation for future research. Starting from cell theory and going up to the main role of chromosomes in deciding the fate of the cell, many experimental advancements were made and opened a new path in understanding the basic functions of the cell.

Moreover, Waldeyer represented a key person in the development of the cytology, and in addition to familiarising the name "chromosomes" throughout the world, his exceptional microscopy methods led to important observations in fertilization and karyokinesis, being able to describe even the way polar bodies are formed during oogenesis.

Author Contributions: Conceptualization, A.V.C. and R.A.C.-B.; Methodology, V.M.S. and H.P.C.; Investigation, A.G.M. and V.M.S.; Software R.A.C.-B. and H.P.C. All authors have read and agreed to the published version of the manuscript.

Funding: This research received no external funding.

Informed Consent Statement: Not applicable.

Conflicts of Interest: The authors declare no conflict of interest.

References

1. Winkelmann, A. Wilhelm von Waldeyer-Hartz (1836–1921): An anatomist who left his mark. *Clin. Anat.* **2007**, *20*, 231–234. [CrossRef] [PubMed]
2. Scheuerlein, H.; Henschke, F.; Köckerling, F. Wilhelm von Waldeyer-Hartz—A Great Forefather: His Contributions to Anatomy with Particular Attention to "His" Fascia. *Front. Surg.* **2017**, *4*, 74. [CrossRef] [PubMed]
3. Von Waldeyer, W. Wilhelm von Waldeyer (1836–1921). *Nature* **1936**, *138*, 579. [CrossRef]
4. Virchow, H. Die Hände von Wilhelm von Waldeyer-Hartz. *Z. Anat Entwickl.* **1923**, *68*, 1–28. [CrossRef]
5. Knol, J.; Keller, D.S. Total Mesorectal Excision Technique-Past, Present, and Future. *Clin. Colon Rectal Surg.* **2020**, *33*, 134–143. [CrossRef] [PubMed]
6. Aeberhard, P.; Fasolini, F. *Total Mesorectal Excision for Cancer of the Rectum BT-Rectal Cancer*; Schlag, P.M., Ed.; Springer: Berlin/Heidelberg, Germany, 1998; pp. 66–70.
7. Vasilescu, C. Toma Ionescu sau Wilhelm von Waldeyer? Teaca rectului sau fascia propria recti? Istoria unei idei călătoare. *Chirurgia* **2010**, *105*, 305–315. [PubMed]
8. Fodstad, H. The neuron theory. *Stereotact. Funct. Neurosurg.* **2002**, *77*, 20–24. [CrossRef] [PubMed]
9. López-Muñoz, F.; Boya, J.; Alamo, C. Neuron theory, the cornerstone of neuroscience, on the centenary of the Nobel Prize award to Santiago Ramón y Cajal. *Brain Res. Bull.* **2006**, *70*, 391–405. [CrossRef] [PubMed]
10. Cimino, G. Reticular theory versus neuron theory in the work of Camillo Golgi. *Physis Riv. Internazionale Stor. Sciza* **1999**, *36*, 431–472.
11. y Cajal, S.R. *Histologie du Système Nerveux de L'homme & Des Vertébrés*; Maloine: Paris, France, 1909; Volume 1. Available online: https://www.biodiversitylibrary.org/item/103261 (accessed on 1 December 2021).
12. Mehta, A.R.; Mehta, P.R.; Anderson, S.P.; MacKinnon, B.L.H.; Compston, A. Grey matter: Etymology and the neuron(e). *Brain* **2020**, *143*, 374–379. [CrossRef] [PubMed]
13. Guillery, R.W. Relating the neuron doctrine to the cell theory. Should contemporary knowledge change our view of the neuron doctrine? *Brain Res. Rev.* **2007**, *55*, 411–421. [CrossRef] [PubMed]
14. Waldeyer, W. Ueber Karyokinese und ihre Beziehungen zu den Befruchtungsvorgängen. *Arch. Mikrosk. Anat.* **1888**, *32*, 1. [CrossRef]
15. Cremer, T.; Cremer, C. Centennial of Wilhelm Waldeyer's introduction of the term "chromosome" in 1888. *Cytogenet. Genome Res.* **1988**, *48*, 65. [CrossRef]
16. Ribatti, D. An historical note on the cell theory. *Exp. Cell Res.* **2018**, *364*, 1–4. [CrossRef] [PubMed]

neurology international

MDPI

Review

Migraine: Advances in the Pathogenesis and Treatment

Horia Pleș [1], Ioan-Alexandru Florian [2,*], Teodora-Larisa Timis [3,*], Razvan-Adrian Covache-Busuioc [4], Luca-Andrei Glavan [4], David-Ioan Dumitrascu [4], Andrei Adrian Popa [4], Andrei Bordeianu [4] and Alexandru Vlad Ciurea [4]

1 Department of Neurosurgery, Centre for Cognitive Research in Neuropsychiatric Pathology (NeuroPsy-Cog), "Victor Babeș" University of Medicine and Pharmacy, 300041 Timișoara, Romania; ples.horia@umft.ro

2 Department of Neurosciences, "Iuliu Hatieganu" University of Medicine and Pharmacy, 400012 Cluj-Napoca, Romania

3 Department of Physiology, "Iuliu Hatieganu" University of Medicine and Pharmacy, 400012 Cluj-Napoca, Romania

4 Neurosurgery Department, "Carol Davila" University of Medicine and Pharmacy, 020021 București, Romania; razvan-adrian.covache-busuioc0720@stud.umfcd.ro (R.-A.C.-B.); luca-andrei.glavan0720@stud.umfcd.ro (L.-A.G.); david-ioan.dumitrascu0720@stud.umfcd.ro (D.-I.D.); andreiadrianpopa@stud.umfcd.ro (A.A.P.); andrei.bordeianu@stud.umfcd.ro (A.B.); prof.avciurea@gmail.com (A.V.C.)

* Correspondence: florian.ioan.alexandru@gmail.com (I.-A.F.); doratimis@gmail.com (T.-L.T.)

Abstract: This article presents a comprehensive review on migraine, a prevalent neurological disorder characterized by chronic headaches, by focusing on their pathogenesis and treatment advances. By examining molecular markers and leveraging imaging techniques, the research identifies key mechanisms and triggers in migraine pathology, thereby improving our understanding of its pathophysiology. Special emphasis is given to the role of calcitonin gene-related peptide (CGRP) in migraine development. CGRP not only contributes to symptoms but also represents a promising therapeutic target, with inhibitors showing effectiveness in migraine management. The article further explores traditional medical treatments, scrutinizing the mechanisms, benefits, and limitations of commonly prescribed medications. This provides a segue into an analysis of emerging therapeutic strategies and their potential to enhance migraine management. Finally, the paper delves into neuromodulation as an innovative treatment modality. Clinical studies indicating its effectiveness in migraine management are reviewed, and the advantages and limitations of this technique are discussed. In summary, the article aims to enhance the understanding of migraine pathogenesis and present novel therapeutic possibilities that could revolutionize patient care.

Keywords: migraine pathogenesis; molecular markers; calcitonin gene-related peptide (CGRP); migraine treatment; neuromodulation

check for updates

Citation: Pleș, H.; Florian, I.-A.; Timis, T.-L.; Covache-Busuioc, R.-A.; Glavan, L.-A.; Dumitrascu, D.-I.; Popa, A.A.; Bordeianu, A.; Ciurea, A.V. Migraine: Advances in the Pathogenesis and Treatment. *Neurol. Int.* **2023**, *15*, 1052–1105. https://doi.org/10.3390/neurolint15030067

Academic Editor: Yasushi Shibata

Received: 1 August 2023
Revised: 28 August 2023
Accepted: 28 August 2023
Published: 31 August 2023

1. Introduction

An overwhelming majority of the global population, approximately 95%, have suffered from a headache at some point in their lives, with an alarming annual prevalence that suggests nearly half of all adults have experienced a headache within a given year [1]. The ramifications of this health issue extend far beyond personal discomfort, with headaches accounting for one-tenth of consultations with general practitioners [2], and a significant portion, one-third, of referrals to neurologists [3]. Moreover, acute medical admissions related to headaches are alarmingly high, constituting one in every five cases.

The World Health Organization recognizes the debilitating nature of headaches, including them among the top ten global causes of disability. Interestingly, in women, the prevalence and impact of headaches are even more pronounced, ranking among the top five causes of disability [4]. It is pertinent to note that the debilitating impact of headaches is comparable to chronic conditions, such as arthritis and diabetes, and its severity exceeds that of conditions like asthma [5,6].

Taking the United Kingdom as an illustrative example, the socioeconomic implications of migraine, a specific type of headache, are vast. An estimated 25 million workdays are lost annually due to migraine, creating an indirect economic burden of nearly GBP 2 billion per year. This figure does not even include the direct healthcare costs associated with managing headaches, such as medication expenses, consultations with general practitioners, referrals to specialists, and visits to emergency care facilities [7].

Quantifying the influence of headaches, particularly migraine, on an individual's quality of life can be challenging. Yet, it is clear from reported data that the impact is substantial. A significant percentage, approximately 75% of patients, experience functional disability during a migraine attack. Additionally, half of the sufferers require the assistance of family members or friends during an attack, causing a significant disruption to their social lives [8]. The ripple effect of headaches extends beyond the individuals suffering from them, impacting society at large and requiring serious attention for more effective management strategies [9].

Migraine, a long-term headache disorder punctuated by episodic bouts, is characterized by repeated instances of severe headaches that present with unique associated symptoms. These include photophobia, a heightened sensitivity to light, and phonophobia, an increased sensitivity to sound [10]. The classification of episodic migraine—an intermittent but recurring form of this disorder—hinges on the frequency with which a patient experiences these debilitating headaches.

In the majority of cases, patients undergo fewer than 15 episodes of headaches per month, a condition identified as episodic migraine. Conversely, there is a subset of individuals who face a more frequent occurrence of headaches—on 15 or more days each month, spanning over three months. Importantly, at least eight of these days should either meet the diagnostic criteria for migraine without the accompanying aura or show responsiveness to treatment specifically designed for migraine. The International Headache Society recognizes this latter classification as chronic migraine [11].

Chronic migraine, although less common when compared to its episodic counterpart, remains a pervasive and incapacitating issue [12]. It poses a significant burden on those afflicted with the condition, dramatically impacting their daily lives and well-being [13,14]. This persistent form of migraine continues to be a widespread challenge, necessitating ongoing research and improved therapeutic strategies to ease the strain it puts on sufferers [15].

1.1. *Brief Overview of Migraine as a Prevalent Neurological Condition*

Over the past three decades, there has been a significant upsurge in the worldwide prevalence of migraine. As highlighted by the Global Burden of Disease (GBD) 2019 study, the estimated global occurrence of migraine escalated from 721.9 million (with a 95% uncertainty interval (UI) of 624.9–833.4) in 1990 to a staggering 1.1 billion (95% UI: 0.98–1.3) in 2019. The percentual shifts in global age-standardized prevalence rate and years lived with disability (YLDs) over these nearly three decades were recorded at 1.7 (95% UI: 0.7–2.8) and 1.5 (95% UI: −4.4 to 3.3), respectively [16].

In this time frame, the sharpest escalations in the age-standardized prevalence per 100,000 individuals were recorded in East Asia with a 7.9% increase (95% UI: 4.3–12%), and in Andean Latin America with an increase of 6.7% (95% UI: 2.1– 11.9%). Conversely, the most significant decreases were seen in high-income North America [−2.2% (95% UI: −5.3 to 1.1%)] and Southeast Asia [−2.2% (95% UI: −3 to −1.4%)]. Moreover, the age-standardized YLD rate due to migraine also saw an increase from 517.6 (95% UI: 82.0–1169.1) in 1990 to 525.5 (95% UI: 78.8–1194.0) in 2019.

The incidence of migraine consistently appeared higher in females than in males across all age groups. In 2019, the global age-standardized prevalence rate for females was 17,902.5 (95% UI: 15,588.3, 20,531.7) per 100,000 populations, in comparison to 10,337.6 (95% UI: 8948.0, 12,013.0) for males [16].

Notably, the most frequent incidence of migraine, both in terms of rate and absolute number of new cases, was seen in the age bracket of 10–14 years for both genders. Over the course of 2019, the number of YLDs due to migraine began to increase from birth, reaching a peak in the 30–34 age group, after which it slowly receded for both sexes [17].

Contrary to expectations, socioeconomic status did not appear to have a direct correlation with the burden of migraine. The study did not reveal any discernible link between the socio-demographic index (SDI) and the YLD rate associated with migraine. This lack of association suggests that migraine does not discriminate based on socioeconomic status, further underlining the pervasive nature of this debilitating condition [17].

The prevalence of migraine has been reported to fluctuate between 2.6% and 21.7%, with an estimated average prevalence close to 12%. However, these figures vary significantly across different nations and even between individual studies conducted within the same country [18–21].

Notably, there appears to be a strong familial connection among individuals suffering from migraine, suggesting that genetic factors significantly contribute to the risk of developing this condition [20,22–25]. Supporting this theory, twin studies have indicated that migraine represent a complex genetic disease that involves an intricate interplay between genetic and environmental factors. Remarkably, the heritability of migraine has been estimated to be as high as 65% [26–29].

However, in spite of the robust genetic implications suggested by these studies and several large-scale genome-wide association studies (GWAS) conducted over the years, the scientific community has yet to conclusively identify specific candidate genes responsible for migraine. A recent systematic re-evaluation of 27 proposed candidate genes found none to be statistically significant [30].

Interestingly, the prevalence of migraine among neurologists is significantly higher when compared to the general population, reaching prevalence rates as high as 48.6% in some studies. This elevated prevalence is most likely attributable to enhanced self-recognition of migraine symptoms among professionals who are extensively trained and experienced in diagnosing and treating the condition. This assertion is supported by a study revealing that just over half of individuals who were diagnosed with migraine actually recognized their headache as a migraine [31].

1.2. Prevalence of Migraine in Pediatric Patients

From a meta-analysis, in which data were sourced from 40 studies encompassing a sample of 15,626 pediatric and adolescent individuals diagnosed with migraines, an 11% prevalence rate was noted, displaying considerable heterogeneity. Among these, 27 studies delineated migraine prevalence based on gender. The aggregated prevalence rate for females stood at 4%, whereas for males it was 3%. Specific data concerning MwoA (migraine without aura) and MwA (migraine with aura) were gleaned from 13 studies, which covered 3481 and 1322 subjects diagnosed with MwoA and MwA, respectively. Prevalence for MwoA was identified at 8% and for MwA at 3%, with marked heterogeneity for both. Only six studies offered data on chronic migraines, revealing a prevalence that fluctuated between 0.2% and 12% [32].

From a separate dataset of 31 studies, information was extracted involving 13,105 pediatric and adolescent subjects diagnosed with TTH (tension-type headache). This cohort exhibited a prevalence of 17%, with notable heterogeneity. Out of these studies, 23 offered a gender-based breakdown of TTH prevalence, yielding a consolidated prevalence rate of 11% for females and 9% for males. Limited data on episodic and chronic TTH were derived from 7 studies, which presented a prevalence range of 4–29% and 0.2–12.9%, respectively [32].

Another set of data, obtained from 40 studies, encompassed 76,782 pediatric and adolescent participants diagnosed with primary headaches in general. The overall prevalence was determined at 62%, with significant heterogeneity observed. Gender-based prevalence data for primary headaches, extracted from 29 studies, showed an aggregated prevalence rate of 38% for females and 27% for males [32].

1.3. Medical Treatments of Migraine in Children

Recent advancements in the pharmaceutical sector have introduced a selective 5-HT1F agonist, lasmiditan, which serves as an efficacious acute treatment for adults, demonstrating no vasoconstrictor activity. This drug is currently under investigation for its applicability in pediatric populations. Additionally, several novel calcitonin gene-related peptide (CGRP) antibodies and antagonists, which have demonstrated efficacy in both the acute treatment and prevention of migraines in adults, are now being assessed in pediatric clinical trials. In adult medical practices, there is an increasing inclination towards peripheral nerve blocks and botulinum toxin; however, the need for robust evidence supporting their efficacy in children is paramount. Furthermore, the introduction of electroceuticals—therapeutic electric devices—has broadened the treatment horizon. These devices include the external trigeminal nerve stimulator (e-TNS), non-invasive vagal nerve stimulator (nVNS), single-pulse transcranial magnetic stimulator (sTMS), and remote electrical neuromodulation device (REN). Presently, substantial evidence supporting their effectiveness in pediatric populations remains elusive; furthermore, while significant progress has been observed, it predominantly benefits the adult demographic. There is an imperative need to expedite migraine research focusing on children [33].

2. Pathogenesis of Migraine: Role of Molecular Markers in Identifying Migraine Triggers and Mechanisms

2.1. Definition and Significance of Molecular Markers

Biomarkers, in the realm of medical and biological research, are defined as quantifiable indicators of biological conditions, representing either physical manifestations or results obtained from laboratory tests that correlate with biological processes. These markers have the potential to serve critical diagnostic or prognostic functions [34]. A more explicit definition of biomarkers was proposed during a conference hosted by the US Food and Drug Administration. In this context, biomarkers are characterized as quantifiable attributes that can be objectively measured and assessed, providing insights into standard biological, pathological, or pharmacological processes [35].

This clear and precise definition paves the way for a bifurcation of biomarkers into the following two unique types: diagnostic and therapeutic. Diagnostic biomarkers serve as flags for pathological conditions and bear a close association with the risk of developing a disease and its severity. They aid in identifying the presence of a disease and gauging its stage or intensity, thus playing a crucial role in guiding clinical decision-making [36].

On the other hand, therapeutic biomarkers hold a different but equally important role. They provide information on a treatment's response, effectively serving as indicators of the efficacy or success of a therapeutic intervention. These biomarkers help clinicians tailor treatments to individual patients, allowing for personalized medicine approaches. They offer a chance to predict whether a patient is likely to respond positively to a particular treatment, making them a powerful tool in the management and treatment of diseases. By providing an early indication of the effectiveness of a therapeutic regimen, these markers can guide healthcare professionals in adjusting treatments as necessary, minimizing the trial-and-error aspect of disease management and increasing the probability of successful outcomes [15].

Biomarkers represent objective physical traits that can be harnessed to illuminate and distinguish the biological nature and mechanisms of various diseases and syndromes. Essentially, they provide snapshots of the body's physiological state and can offer valuable insights into health and disease processes. Biomarkers have an extensive range of potential manifestations, which can include but are certainly not limited to, results obtained from the examination of blood, urine, muscle, nerve, skin, or cerebrospinal fluid [37].

Additionally, biomarkers may also be identified in the form of genes or gene products. These genetic markers offer a unique insight into an individual's inherent disease susceptibility or resistance and can often illuminate potential therapeutic pathways. Likewise, biomarkers can be identified through advanced imaging techniques such as X-rays, magnetic resonance imaging (MRI), or computed tomographic (CT) scans. These imaging biomarkers can provide a visual representation of disease progression, allowing clinicians to identify anatomical or functional changes in the body over time [34].

Another fascinating domain of biomarkers lies in the realm of electrophysiological measurements, such as those generated by electrocardiograms (ECGs), electroencephalograms (EEGs), or nerve conduction studies. These types of biomarkers record the electrical activity of the heart, brain, or nerves, respectively, offering a unique insight into the physiological function of these systems.

An important issue worth mentioning is those paraclinical investigations offer a new avenue for the management of migraine but are not proven to be of high sensibility and sensitivity for daily physician's practice. Even though neuroimaging and functional analyses of the brain activity might give a broader point of view regarding therapeutic possibilities, those should not be taken into consideration as absolute clinical criteria.

Ultimately, a biomarker could be virtually any characteristic that can be detected, quantified, and expressed in terms of physical qualities. These could include diverse measures, such as height, weight, depth, voltage, luminescence, resistance, viscosity, width, length, volume, or area. Each of these measures contributes to the vast array of biomarkers that hold promise for enhancing our understanding of diseases and guiding the development of effective therapeutic interventions. The utilization of such a wide array of biomarkers allows for a comprehensive, multi-faceted approach to understanding and treating diseases, ultimately leading to more effective and personalized healthcare solutions [38].

2.2. Identification of Potential Molecular Markers Associated with Migraine

The National Institutes of Health Biomarkers Definitions Working Group, in 1998, presented a definition for biomarkers. As per their definition, a biomarker refers to "a characteristic that can be objectively measured and evaluated as an indicator of normal biological processes, pathogenic processes, or pharmacological responses to a therapeutic intervention" [35]. Biomarkers may be classified based on their functional roles, such as diagnostic, therapeutic, risk, progression, and prognostic indicators.

The 'ideal' biomarker is characterized by the following features [39]:

- High sensitivity and specificity: this ensures that the biomarker can accurately identify individuals with a specific condition, and also correctly rule out those without the condition;
- High predictive value: the biomarker should be able to accurately forecast the course of the disease, providing valuable insights for disease management;
- Analytical stability: the biomarker should remain consistent over time and across different conditions, thereby ensuring reliable results;
- Easy, cost-effective, and minimally invasive analysis: the method of assessing the biomarker should be simple, economical, and cause minimal discomfort to the patient;
- Repeatability of method: the assessment method should yield consistent results when repeated, thereby ensuring the reliability of the biomarker.

In the context of migraine, however, there are no validated biomarkers due to the absence of substance or genetic variants that are exclusively associated with this condition or the lack of comprehensive studies on potential biomarkers.

2.2.1. Markers of Inflammation and Oxidative Stress

The markers of inflammation and oxidative stress have been associated with migraine in several studies. Proinflammatory cytokines, such as interleukin-1 (IL-1) and interleukin-6 (IL-6), have been implicated in this condition [40]. It has been found that the level of IL-1α is elevated in the blood of children suffering from migraine with aura (MA) [40]. Similarly, adults with MA have been found to exhibit higher plasma levels of IL-1β during headache-free periods and early stages of attacks as compared to those suffering from migraine without aura (MO) [40,41].

The concentration of IL-6 is reported to increase during the initial two hours of a migraine attack. Additionally, the levels of IL-10 and tumor necrosis factor alpha (TNF-α) are also found to be elevated during these attacks. It is believed that other inflammatory markers associated with vascular dysfunction, such as homocysteine (Hcy) and matrix metalloproteinase-9 (MMP-9), are also elevated in the blood of individuals with migraine [15].

Elevated serum Hcy concentration has been linked to migraine with aura (MA), and some studies have noted a relationship between increased Hcy levels and higher frequency and severity of migraine; however, these findings are not supported by all research. Hyperhomocysteinemia (elevated Hcy) is hypothesized to initiate migraine with aura attacks through changes in pain threshold [42,43].

2.2.2. Markers Associated with Pain Transmission and Emotions

Biochemical research has revealed several metabolic irregularities in the synthesis of neuromodulators and neurotransmitters associated with migraine, particularly migraine without aura (MO). Alterations in the metabolic pathway of tyrosine, for example, lead to abnormal production of neurotransmitters like noradrenaline (NE) and dopamine (DA). This process results in an increase in the levels of trace amines, such as tyramine, octopamine, and synephrine. Such changes compromise mitochondrial function and elevate glutamate concentrations within the central nervous system (CNS), as can be seen in Table 1 [43].

These imbalances in the neurotransmitter and neuromodulator levels within the dopaminergic and noradrenergic synapses of pain pathways could potentially activate the trigeminovascular system (TGVS), causing the release of the calcitonin gene-related peptide (CGRP). This chain of events is believed to directly trigger migraine attacks [44,45].

CGRP plays a key role in transmitting pain signals and promoting inflammation. Its release is stimulated by the activation of TGVS and severe migraine episodes. Infusion of CGRP has been observed to provoke migraine-like attacks in patients with migraine with aura (MA). It has been reported that during inter-attack periods, the saliva and plasma levels of CGRP in migraine patients are significantly higher compared to healthy individuals [43].

Research conducted on cultured trigeminal neurons suggests that migraine treatment strategies can inhibit CGRP transcription and curtail its release, while tumor necrosis factor alpha (TNF-α) may stimulate the transcription of this peptide [15]. Another study proposes that high levels of CGRP in saliva may correlate with a significantly improved response to rizatriptan treatment, suggesting that CGRP could serve as a valuable therapeutic marker [46].

Glutamate, which could potentially activate pathways involving both TGVS and cortical spreading depression (CSD), has been found in elevated concentrations in the plasma, platelets, and cerebrospinal fluid (CSF) of migraine sufferers, including those with chronic migraine. Research suggests that a reduction in plasma glutamate levels could be a marker of a positive response to prophylactic treatment in MO patients [43].

Serotonin (5-HT) release from platelets into the plasma may be implicated in the pathophysiology of the aura phase of migraine. Izzati-Zade observed a depletion of 5-HT stored in platelets during migraine attacks; moreover, a pattern has been observed in which the plasma level of 5-HT decreases between migraine attacks and the level of the corresponding metabolite, hydroxyindoleacetic acid (5-HIAA), increases. This pattern reverses during migraine attacks [47,48]. This correlation suggests that low 5-HT levels might enable the activation of the trigeminovascular nociceptive pathway triggered by CSD, thus supporting the hypothesis that migraines are a syndrome of low serotonergic disposition.

Additionally, a significantly higher concentration of hypocretin-1, a wakefulness-promoting neuropeptide, has been detected in the CSF of patients with chronic migraine, and this has been observed to correlate with painkiller usage [49,50]. Elevated hypocretin-1 levels may be indicative of the early stages of a migraine attack. Conversely, a study involving patients with cluster headaches reported reduced hypocretin-1 levels in the CSF, suggesting that low hypocretin-1 concentrations might reflect insufficient antinociceptive activity in the hypothalamus [51].

New therapeutic targets for migraine treatment, such as CGRP receptor antagonists, anti-CGRP antibodies, 5-HT1F agonists, glutamate antagonists, and dual hypocretin-1 receptor antagonists, are currently under investigation in phase II clinical trials [52,53]. These emerging therapies reflect the continuous exploration and evolution of our understanding of migraine pathophysiology.

Table 1. Molecules with altered CSF (cerebrospinal fluid) concentrations in patients with migraine.

Molecule	Migraine Type (Chronic Migraine [CM]/Episodic Migraine [EM])	Action in Relation to Migraine
Sodium [54,55]	EM	• During a migraine, there is an increase in cerebrospinal fluid (CSF) sodium concentration, while the blood plasma sodium concentration remains unchanged. Additionally, sodium excursions may follow a temporal pattern that worsens migraine in susceptible patients
Homocysteine [56]	EM	• High levels of homocysteine are potentially linked to migraine with aura and an increased risk of cardiovascular events in patients with migraine
3,4-Dihydroxyphenylacetic acid (DOPAC) [57]	EM	• Related with dopaminergic activity • Positive correlation between the concentration of DOPAC (3,4-dihydroxyphenylacetic acid) and the intensity of migraine, whether with or without aura
Phosphatidylcholine-specific phospholipase C [58]	EM	• The process involves the hydrolysis of phosphatidylcholine, resulting in the production of important second messengers, diacylglycerol, and phosphorylcholine
Transforming growth factor-β1 [59]	EM, CM	• An anti-inflammatory cytokine
Interleukin-1 receptor antagonist [59]	EM, CM	• Proinflammatory cytokine
Monocyte chemoattractant protein-1 [59]	EM, CM	• Proinflammatory cytokine
Corticotrophin-releasing factor [60]	CM, MOH	• May be involved in activation of hypocretin/orexin system.
Orexin-A (also referred to as hypocretin-1) [60]	CM, MOH	• Involved in the maintenance and regulation of various physiological functions, including arousal, sleep, appetite, drinking behavior, central control of autonomic activity, certain endocrine responses, and pain modulation
Glial cell line-derived neurotrophic factor [61]	CM	• It may play a role in pain relief by regulating the expression of sodium channel subunits, capsaicin VR1 receptors, and substance P release • Reduced levels found in patients with migraine
Somatostatin [61]	CM	• Regulatory anti-inflammatory and antinociceptive peptide
Glutamate [62]	CM	• The primary excitatory neurotransmitter in the central nervous system. It has been linked to various migraine-related processes, including cortical spreading depression, trigeminovascular activation, and central sensitization.

Table 1. *Cont.*

Molecule	Migraine Type (Chronic Migraine [CM]/Episodic Migraine [EM])	Action in Relation to Migraine
Tumor necrosis factor-α [63]	CM	• A proinflammatory cytokine that plays a significant role in brain inflammatory and immune processes, as well as in the initiation of pain
Taurine [64]	EM, CM	• Inhibitory effect on neuronal activity and vasodilating properties
Glycine [64]	EM, CM	• Inhibitory neurotransmitter
Glutamine [64]	EM, CM	• May be involved with initiation and propagation of spreading cortical depression
Neuropeptide Y [65]	Acute migraine	• Strong vasoconstrictor

3. Other Biomarkers Associated with Increased Risk for Migraine

3.1. Genetic Markers and Migraine

Many scientific investigations have striven to identify specific genetic mutations or polymorphisms that might contribute to an increased risk of developing migraine. However, as of now, none of these findings have been implemented in standard clinical practice. One rare subtype of migraine, known as familial hemiplegic migraine (FHM), which is characterized by aura and transient hemiplegia, has a well-understood genetic basis. There are three known genes where mutations have been linked with FHM—CACNA1A (FHM1), ATP1A2 (FHM2), and SCN1A (FHM3)—and this condition is inherited in an autosomal-dominant fashion [66].

The identified mutations connected to FHM lead to alterations in calcium and sodium channel functions, which are integral components of neuronal communication and excitability. Interestingly, these genetic variants have also been associated with other neurological disorders, including ataxia and childhood epilepsy [66]. Nevertheless, these mutations have not shown a strong correlation with common forms of migraine (with or without aura) or other types of headaches.

A recent study discerned a significant genetic correlation linking migraine risk to intracranial volume ($rG = -0.11$, $P = 1 \times 10^{-3}$). This correlation was not observed in relation to any subcortical region. Notwithstanding, the study pinpointed concurrent genomic overlap between migraines and all brain structures. Gene enrichment in these mutual genomic regions indicated potential associations with neuronal signaling and vascular regulation. Furthermore, the research suggested a potential causative link between reduced overall brain volume, as well as the volume of the hippocampus and ventral diencephalon, and heightened migraine risk. Additionally, a causative correlation was proposed between heightened migraine risk and an expanded amygdala volume. Through the utilization of comprehensive genome-wide association studies, the study illuminated shared genetic pathways influencing both migraine risk and various brain structures. This suggests that variances in brain morphology in individuals with elevated migraine susceptibility could be rooted in genetics. Delving deeper into these findings offers support to the neurovascular premise of migraine origin, highlighting prospective therapeutic avenues [67].

Another study elucidated the following genes associated with familial hemiplegic migraine [68]:

- FHM1: CACNA1A—This gene undergoes missense mutations resulting in a gain of function, alongside rare large exonic deletions or deletions at the 5′ non-coding end promoter. It codes for the Alpha-1 subunit of the neuronal Cav2.1 (P/Q type) voltage-gated calcium channels, crucial for modulating neuronal excitability at the presynaptic end of glutamatergic synapses;

- FHM2: ATP1A2—Characterized by missense mutations, rare small deletions, or truncating mutations and frameshifts. It encodes the catalytic alpha-2 subunit of glial and neuronal ATP-dependent transmembrane Na+/K+ pumps, pivotal for extracellular K+ clearance and establishing a Na+ gradient, which is indispensable for glutamate reuptake;
- FHM3: SCN1A—Experiences missense mutations (gain of function) and is responsible for the Alpha-1 subunit of neuronal Nav1.1 voltage-gated sodium channels. It plays a key role in propelling action potentials of cortical neurons, predominantly in GABAergic inhibitory interneurons;
- FMH4: PRRT2—Noted for missense mutations, this gene codes for the pre-synaptic proline-rich transmembrane protein. It interacts with the synaptosomal-associated protein 25 (SNAP25), implying a potential role in merging synaptic vesicles with the plasma membrane.

Two FHM1 knock-in (KI) transgenic mouse models have been established as per references [69,70]. The KI model for the R192Q mutation, linked with pure FHM1, does not exhibit clinical anomalies. In contrast, the KI for the S218L mutation, attributed to severe FHM1, presents cerebellar ataxia, transient hemiparesis, and epilepsy. As outlined in [71], these FHM1-KI mice demonstrate heightened CaV2.1 currents and neurotransmitter release, an imbalance in cortical neurotransmission, amplified excitatory transmission in the visual cortex, and a higher vulnerability to cortical spreading depression (CSD).

Various models of FHM2-KI transgenic mice have been developed. Heterozygous transgenic mice [72] display no clinical changes but have an elevated predisposition to CSD. Mice with a partial knock-out (KO) of ATP1A2 also demonstrate a heightened vulnerability to CSD [73]. Another model with a complete KO of ATP1A2 in astrocytes manifests episodic paralysis and spontaneous CSD waves coupled with diminished EEG activity. Aberrations in brain metabolism were observed with increased levels of serine and glycine. Interestingly, a diet devoid of serine and glycine curtailed paralysis episodes in these mutants [74].

For FHM3, multiple SCN1A mutations have been documented, with the majority being missense alterations leading to enhanced function [75]. A mouse model harboring the L1649Q variant exhibited an increased susceptibility to CSD, attributed to Na+ channel inactivation defects and augmented Na+ currents, causing hyperactivity in inhibitory interneurons.

With respect to FHM4, mutations in PRRT2 have been discovered in numerous instances as referenced in [76]. A significant portion of these cases were pure FHM, while others exhibited accompanying epilepsy, cognitive impairments, or dyskinesia. PRRT2-KO mice displayed paroxysmal abnormal movements early in life, progressing to unusual audiogenic motor behaviors in adulthood and a reduced seizure threshold. Notably, both human and mouse homozygous KO-PRRT2 neurons in culture exhibited hyperactive NaV1.2 and NaV1.6 channels, inferring PRRT2's inhibitory effect on voltage-gated sodium channels as described in [77].

Researchers have also employed genome-wide association studies (GWAS) to pinpoint genes linked to an elevated susceptibility for migraine (see Table 2). In one such investigation, genetic information from 5122 individuals afflicted with migraine and 18,108 control participants was scrutinized. This scrutiny led to the identification of several specific genetic variations known as single-nucleotide polymorphisms (SNPs), which displayed significant connections to migraine. Noteworthy among these were rs2651899 (positioned on chromosome 1p36.32, close to the PRDM16 gene), rs10166942 (situated on 2q37.1, near TRPM8), and rs11172113 (positioned on 12q13.3, near LRP1). It is important to highlight that although rs2651899 and rs10166942 could be differentiated between migraine and non-migraine headaches, these three SNPs did not exhibit exclusivity for migraine with or without aura, nor were they tied to specific migraine characteristics. Nonetheless, the biological significance of these connections is substantiated by the established functions of TRPM8 in neuropathic pain and LRP1 in glutamatergic synaptic transmission [78].

Another GWAS pinpointed the following two susceptibility loci for migraine without aura: MEF2D and TGFBR2 [79]. It is important to bear in mind that the results from GWAS carried out have not overlapped so far, and larger-scale studies are necessary to confirm and expand the findings of smaller investigations and to permit the use of meta-analytical methodologies.

A migraine GWAS study from 2021 [80] identified 79 independent loci significantly correlated with migraine. This study was ethnically diverse, encompassing participants of East Asian, African American, and Hispanic/Latino origin, and consisted of 28,852 cases versus 525,717 controls.

The latest migraine GWAS from 2022 by Hautakangas et al. comprised 102,084 cases against 771,257 controls. This study unearthed 123 unique loci associated with migraines, 86 of which were newly discovered post the 2016 GWAS. Further studies even expanded independent SNPs to 167. The 2022 GWAS [81] underscored both vascular and CNS tissues/cell types. Newly detected loci encoded migraine drug targets, such as CGRP (CALCA/CALCB) and serotonin 1F receptor (HTR1F). Significantly, CGRP is the objective for CGRP antibodies, and HTR1F is targeted by ditans. Moreover, an in-depth assessment of roughly 30,000 patients from the 2022 GWAS with a precise migraine diagnosis revealed unique risk variants for specific migraine types.

The research presented thus far suggests that, aside from FHM, we are only at the preliminary stage of identifying genes significantly associated with migraine risk [82]. This observation is further illustrated by the inconsistent findings from studies investigating specific associations in migraine patients with and without aura (summarized in Table 2). For instance, one study found a significant association between a polymorphism in the gene encoding the dopamine D2 receptor (see Table 2) and migraine without aura [83]. Meanwhile, another study supported the association of DBH and SLC6A3 genes with migraine with aura [62]. Contradictorily, other investigations did not corroborate these associations in migraine patients, whether with or without aura. This variability is not unusual in genetic studies investigating diseases with a multifactorial etiology. As such, further research is needed to unravel the complex genetic underpinnings of migraine [15].

Table 2. Genetic mutations/polymorphisms associated with increased risk for migraine and relation to migraine.

Gene Product	Migraine Type/Features	Action in Relation to Migraine
Dopamine type 2 (D2) receptor [23]	Migraine with and without aura [23,84]	• Vasoconstriction • Reduces trigeminal nerve activation • Inhibits release of vasoactive neuropeptide • Interrupts pain transmission centrally
Glutathione S-transferase [85]	Migraine without aura [85]	• Increases susceptibility to environmental xenobiotic-induced migraine attacks in GSTM1 genotype
Dopamine type 4 (D4) receptor [86]	Migraine without aura [86]	• A potential genetic association exists between dopamine D4 receptor gene and migraine without aura
Tumor necrosis factor-α [87]	Migraine without aura [88]	• Proinflammatory cytokine
Methyltetrahydrofolate reductase (MTHFR) C677T allele [89]	Migraine with aura [89]	• MTHFR C677T polymorphism may increase homocysteine levels associated with migraine with aura
Dopamine β-hydroxylase gene [90]	Migraine with aura [90]	• An intracellular enzyme catalyzing the conversion of dopamine to noradrenaline; imbalance may increase susceptibility to migraine
Angiotensin-converting enzyme allele [91]	Migraine with and without aura [92–94]	• Involved in vasoconstriction and vascular remodeling

Table 2. *Cont.*

Gene Product	Migraine Type/Features	Action in Relation to Migraine
Hypocretin receptor 1 [95]	Migraine without aura [95]	• Neuropeptide generated within the clusters of nerve cells in the hypothalamus that could potentially play a role in feelings of tiredness, frequent yawning, heightened drowsiness, and strong urges for food linked to migraine.
Syntaxin 1A [96]	Migraine without aura [96]	• Involved in the control of brain chemicals, such as serotonin and gamma-aminobutyric acid (GABA).
Cytochrome P450 (CYP) 1A2 [97]	Chronic migraine [97]	• CYP1A2*1F is connected to excessive use of triptan medications, and among those who misuse these drugs, it also impacts the drug response.

3.2. Recent Genetic Findings and Migraine

In a newly published family-based association study, significant markers connected to migraine were discovered, alongside genes believed to contribute to or modify the phenotypic expression of migraine within a substantial region of chromosome 6p12.2–p21.1. This region is recognized by the locus name MIGR3. Regrettably, due to the vastness of this area of interest, it is currently not feasible to pinpoint a singular gene; however, it is anticipated that future investigations employing more refined sequencing methodologies will eventually lead to the identification of a promising candidate gene implicated in migraine [98,99].

Despite the growing body of evidence suggesting that genetic factors play a pivotal role in the development of migraine, efforts to uncover the specific genes responsible for the common forms of migraine have only yielded modest success. As scientific collaboration expands on a global scale, the chances of identifying additional genetic variants linked to migraine are likely to increase. Furthermore, the unraveling of the genetic intricacies underlying polygenic diseases could potentially shed new light on the molecular pathways implicated in the pathophysiology of migraine. By extending our understanding of the genetic aspects of migraine, we may pave the way for the development of more effective diagnostic tools and therapeutic interventions.

3.3. Inflammatory Indicators and Migraine

Interleukins, specifically IL-1 and IL-6, have been linked with the occurrence of migraine. These cytokines, characterized by their proinflammatory nature, are believed to play a role in vascular dysfunction. Studies indicate that children experiencing migraine have raised plasma levels of IL-1α compared to those who do not suffer from migraine, and these concentrations are markedly higher in individuals with migraine accompanied by aura compared to those without aura [40]. Furthermore, adults experiencing aura migraine have significantly elevated plasma levels of IL-1β during periods free from headaches and during the early onset of migraine attacks, in comparison to individuals with migraine that do not present with aura [41]. IL-6 levels also exhibit a surge in the initial two hours of a migraine attack when measured from blood samples taken from the jugular vein [50].

Other cytokines, including IL-10 and tumor necrosis factor alpha (TNF-α), have shown associations with migraine. During migraine attacks, there are elevated serum levels of IL-10 and TNF-α [100]; moreover, between attacks, TNF-α levels in plasma are higher in children who suffer from migraine compared to those who do not. The connection between TNF-α and migraine is particularly noteworthy, given the repeated association of elevated levels of this cytokine with endothelial dysfunction [63]. While some studies propose that patients with migraine may have compromised endothelial function, others contradict these findings [15].

Further inflammatory markers, which are considered to be linked to vascular dysfunction, are found to be elevated in the blood of migraine patients. Research has demonstrated that average plasma levels of C-reactive protein and homocysteine are higher in children who suffer from migraine compared to those who are not plagued by headaches [54]. Evidence also suggests that premenopausal women with migraine, especially those with aura, show signs of increased endothelial activation—a component of endothelial dysfunction—evidenced by elevated levels of von Willebrand factor, C-reactive protein, nitrate/nitrite, and tissue-type plasminogen activator antigen [55]. Markers linked to vascular repair and remodeling processes have also shown an association with migraine.

Investigations in both human subjects and animal models have proposed that matrix metalloproteinase-9 (MMP-9) might protect against the development and destabilization of plaques [61]. Moreover, patients experiencing migraine have been found to have significantly higher plasma levels of MMP-9 compared to healthy individuals and those with tension-type headaches. The average plasma MMP-9 levels were highest in subjects who had their blood samples taken between two and four days post their latest attack, implying that the elevated MMP-9 might be an indication of structural damage and subsequent remodeling associated with migraine attacks [64].

While the findings summarized here point to a relationship between various inflammatory mediators and migraine, additional research aimed at understanding the biological implications of these inflammatory mediators is necessary to confirm the validity of these potential indicators.

3.4. Contribution of Imaging Techniques in Understanding Migraine Pathology

3.4.1. Overview of Imaging Methods Used in Migraine Research

The advent of neuroimaging technologies has brought significant advancements in our understanding of migraine mechanisms and has enabled us to pinpoint secondary structural and functional impacts resulting from migraine. Imaging conducted during a migraine episode has helped the scientific community progress from a strictly vascular understanding of migraine pathophysiology, to a neurovascular theory, and currently towards a central nervous system (CNS) model.

Through the lens of neuroimaging, we have gained substantial ground towards unearthing the elusive "migraine generator"—the structure that triggers the initiation of a migraine episode. The investigative power of neuroimaging has shed light on the role of central sensitization in the pathophysiology of individual migraine attacks and in the progression of the disease. It has also enhanced our understanding of medication overuse headaches, along with the mechanisms by which abortive and prophylactic medications for migraine work [58].

One pivotal discovery has been the identification of cortical spreading depression (CSD) during migraine, with or perhaps without the accompaniment of an aura. This phenomenon is a wave of hyperactivity followed by a wave of inhibition in neuronal activities, which spreads across the cortex of the brain. It has been increasingly recognized as an important part of migraine pathophysiology [65].

Moreover, it has become evident through neuroimaging that individuals prone to migraine undergo structural and functional brain alterations in the periods between migraine attacks. These alterations appear to be correlated with both the duration of the disease and its severity, suggesting a possible link between more severe disease manifestation and persistent abnormalities between migraines. In essence, the affliction may not be limited to the episodes of the migraine attack but may present as a continuous, cyclic process with long-term impacts on brain structure and function. Neuroimaging has revolutionized the exploration and understanding of migraine, enabling us to visualize the structural and functional impacts of this condition on the brain, refine our understanding of its pathophysiology, and develop more effective therapeutic strategies. However, despite these strides, the complex nature of migraine warrants further research to fully understand the intricate interplay between genetic, environmental, and neurobiological factors in the manifestation and progression of the disease [101].

3.4.2. Findings and Insights Gained through Imaging Studies

Given the episodic and largely unpredictable nature of individual migraine attacks, conducting imaging during a spontaneous migraine has posed substantial challenges. To circumvent this issue, some researchers have induced migraine attacks in subjects by exposing them to known triggers, such as photic stimulation, physical exertion, or nitroglycerin [102]. A handful of investigators have successfully captured the onset of a migraine headache, while others have performed imaging immediately after the headache's initiation. Nevertheless, the total number of studies that have managed to image a migraine attack in progress remains relatively small.

4. Central Sensitization

The phenomenon of central sensitization in individuals prone to migraine leads to heightened pain perception during a migraine, a condition known as cutaneous allodynia, and may contribute to the progression from episodic to chronic migraine [102]. Approximately 65% of migraine sufferers develop cutaneous allodynia during individual headache episodes [103–105]. Patients with allodynia experience the skin becoming painfully sensitive to stimuli that are normally harmless, such as a light touch. Developing ways to block or reverse central sensitization could potentially alleviate migraine pain and lower the likelihood of episodic migraine evolving into chronic migraine.

One of the challenges in neuroimaging of central sensitization is differentiating between changes that arise from the increased pain sensation of cutaneous allodynia and those from structures that may specifically mediate the onset and maintenance of central sensitization. Recent advances in functional magnetic resonance imaging (fMRI) studies offer some progress in this regard. Utilizing the heat/capsaicin model of sensitization, fMRI studies have identified activation in the midbrain reticular formation region that appears specific to central sensitization [106,107]. Investigators propose that this activation occurs in the nucleus cuneiformis and the rostral superior colliculi/periaqueductal gray area.

Further fMRI studies investigated the influence of gabapentin, a medication commonly used to treat nerve pain, on brain activations following painful mechanical stimulation of normal skin compared to skin with capsaicin-induced secondary hyperalgesia [92]. Under both conditions, gabapentin reduced activations in the operculoinsular cortex. Interestingly, it was only in the presence of central sensitization that gabapentin was able to reduce activations in the brainstem and suppress stimulus-induced deactivations, suggesting that gabapentin might be more effective at reducing painful transmission when central sensitization is present. These insights set the foundation for additional investigations aimed at pinpointing the site where gabapentin acts to affect central sensitization. Unraveling this could provide valuable information for the development of future therapies aimed at inhibiting central sensitization, thereby offering a potential new avenue for migraine treatment [101].

Insights into Migraine Pathology: From Current Pathophysiological Understanding to Peripheral Interactions and Plasma Protein Extravasation

A number of laboratory studies conducted during the 1990s postulated that the pain associated with migraine might arise from a sterile, neurogenically mediated inflammation of the dura mater, the thick membrane that surrounds the brain. Evidence of neurogenic plasma extravasation, the process whereby plasma proteins pass out of small blood vessels into surrounding tissues, has been observed during the electrical stimulation of the trigeminal ganglion in rat models. Interestingly, this process of extravasation can be halted by substances such as ergot alkaloids, indomethacin, acetylsalicylic acid, and the serotonin 5HT1B/1D agonist, sumatriptan [84].

Adding to this, preclinical research has suggested that a phenomenon known as cortical spreading depression, a wave of hyperactivity followed by a wave of inhibition in the brain, could act as a potent trigger for the activation of trigeminal neurons [85]. However, this notion has been a subject of controversy and ongoing debate in the scientific community.

Notably, post-stimulation of the trigeminal ganglion, researchers have observed structural changes in the dura mater, including mast cell degranulation, a process by which mast cells release granules rich in histamine and other molecules, and modifications in postcapillary venules, including platelet aggregation [86].

While it is widely accepted that such changes—particularly the initiation of a sterile inflammatory response—would be likely to cause pain, it remains uncertain whether these alterations alone are sufficient or whether they necessitate the presence of other stimulators or promoters. One of the limitations of neurogenic dural plasma extravasation as a theory is its inability to predict whether novel therapeutic targets would be effective in either the acute or preventative treatment of migraine. Indeed, the blockade of neurogenic plasma protein extravasation (PPE) has not been proven to be a reliable indicator of antimigraine efficacy in humans. This observation is substantiated by the unsuccessful outcomes of clinical trials of several potential treatments such as substance P, neurokinin 1 receptor antagonists, specific PPE blockers, CP122,288 and 4991w93, an endothelin antagonist, a neurosteroid, and an inhibitor of the inducible form of nitric oxide synthase (iNOS) named GW274150 [87].

5. Investigations into Neuropeptides

Through the application of electrical stimulation to the trigeminal ganglion, observable increases in extracerebral blood flow and the local release of calcitonin gene-related peptide (CGRP) and substance P (SP) have been noted in both human and cat subjects. In felines, such stimulation not only enhances the cerebral blood flow but also prompts the release of vasoactive intestinal polypeptide (VIP), a potent vasodilator peptide, via the greater superficial petrosal branch of the facial nerve. Intriguingly, the VIP ergic innervation of cerebral vessels is mainly anterior as opposed to posterior, which may make these regions more susceptible to spreading depression, possibly accounting for the common posterior onset of aura symptoms. A more specific pain-inducing area, the superior sagittal sinus, when stimulated, raises the cerebral blood flow and jugular vein CGRP levels. In human studies, elevated CGRP levels have been observed during the headache phase of severe migraine, though not in less intense attacks, as well as in cluster headaches and chronic paroxysmal hemicranias, corroborating the hypothesis that the trigeminovascular system might serve a protective function in such conditions. Migraine triggered by nitric oxide (NO) donors, which mimic typical migraine, also lead to CGRP increases that can be blocked by sumatriptan, as is the case in spontaneous migraine. Significantly, certain compounds that have been proven ineffective for migraine treatment, such as the conformationally restricted analogs of sumatriptan, CP122,288, and zolmitriptan, 4991w93, were also unable to inhibit CGRP release following superior sagittal sinus stimulation in cats. The development and successful trials of specific non-peptide CGRP receptor antagonists underscore the significance of this as a novel principle in the treatment of acute migraine. Nonetheless, considering the variability, it is unlikely to serve as a reliable migraine biomarker. Also, the lack of effect of CGRP receptor antagonists on plasma protein extravasation (PPE) explains, in part, why this model has not successfully translated into human therapeutic strategies [88].

Migraine triggers in patients include cAMP-mediated mechanisms via cilostazol, even when the CGRP receptor is blocked with erenumab. Additionally, cranial artery dilation from cilostazol remains unaffected by CGRP receptor blockage. These insights imply that migraine attacks induced by cAMP do not need CGRP receptor activation, hinting at potential novel avenues for mechanism-based migraine drug development [89].

In the realm of preclinical research, there is evidence suggesting that PACAP-specific active transport systems cross the blood–brain barrier (BBB) [90]. Yet, after crossing the BBB, PACAP isoforms either degrade quickly or re-enter the bloodstream, pointing to its primary peripheral effect. In vitro data [91] highlighted PACAP38's ability to relax vascular smooth muscle cells post-abluminal application, but not after luminal application in cerebral arteries. In contrast, in vivo tests showed no significant change in regional cerebral blood flow due to PACAP38 intravenous infusion [93].

Throbbing headaches during migraines probably stem from pain signals from both intra- and extracranial (when vasodilated) vessels, especially arteries. No studies have focused on selective VIP blockage for migraine treatment yet. However, recent findings hint that prolonged vasodilation from VIP might induce migraine-like episodes, suggesting that VIP blockage could be a promising migraine treatment [94].

The receptors (AM1, AM2, or CGRP) that might mediate migraine-like reactions due to adrenomedullin remain unidentified [95]. AM22-52, the only known adrenomedullin antagonist, appears limited in its ability to antagonize adrenomedullin effects in rat cells, though it has shown potential in inhibiting CGRP effects [96]. No specific treatments to counteract adrenomedullin or its receptors exist currently. However, an in vitro research [97] indicated that the CGRP-receptor targeting antibody erenumab and the CGRP-receptor antagonist telcagepant opposed not only CGRP but also adrenomedullin signaling at the CGRP receptor.

Research on arresting the NO-cGMP cascade for drug development shows potential. A mouse study [56] illustrated that the sGC stimulator VL-102 induced both acute and prolonged hyperalgesia. This effect was blocked by the sGC inhibitor (ODQ) and by several antimigraine drugs (sumatriptan, topiramate, and propranolol).

Lastly, migraine patient studies have identified mutations in the TRESK potassium channel. TRESK works by inhibiting TREK1 and TREK2, amplifying the TG's excitability. As mentioned in [59,60], decreasing TG excitability using the TREK1/TREK2 agonist ML67-33 countered an NO donor-triggered migraine-like phenotype in mice similarly to the CGRP receptor antagonist olcegepant. Furthermore, it entirely reversed TG-induced facial allodynia in rats due to NO donors.

Our understanding of alternative targets leading to intracranial artery vasodilation is expanding. However, we have yet to develop successful therapies to tackle CGRP-independent mechanisms. For instance, an antibody designed to inhibit PACAP (a peptide part of the VIP, secretin, and glucagon superfamily) was developed as an alternative treatment but did not show efficacy in trials. Additionally, the results from attempts to block NO-induced reactions have been mixed. Using glibenclamide did not alleviate headaches caused by PACAP38 and levcromakalim. Still, TRPV1 agonists like capsaicin and civamide demonstrated some effectiveness due to their capacity to desensitize nerve endings hosting these channels. There is a pressing need for more research to craft alternative targeted migraine therapies, such as those focusing on VIP, amylin, adrenomedullin, PDE3, PDE5, calcium channels, and ASICs. In theory, targeting the most downstream elements like KATP channels, being the cascade's "final link", might yield better results; however, this could also bring about severe and unwanted side effects. The high induction rate of levcromakalim possibly being a result of this remains to be confirmed [57].

Changes in the connection between the hypothalamus and brainstem with the spinal trigeminal nuclei and the dorsal rostral pons have been observed during the premonitory phase of a migraine, lasting up to 48 h before pain begins [108]. The exact process that makes the hypothalamus 'overactive' in migraine situations, leading to the sensitization of trigeminal nociceptors, remains undefined. Moreover, the hypothalamus houses chemosensitive neurons that can recognize metabolic alterations in the brain and the body. External stimuli causing disruptions in balance and the brain's inherent biorhythm could potentially push the brain toward a migraine episode through hypothalamic activation [109].

NSAIDs are a reliable choice for managing acute migraine flare-ups, but care must be taken due to potential side effects like stomach issues and kidney problems. Beta-blockers serve as effective preventative measures against migraines, but they come with their own drawbacks, such as causing dizziness and fatigue. Moreover, they are not recommended for patients with specific health conditions like asthma, heart failure, and certain cardiac rhythm disorders. While calcium channel blockers have been considered for migraine prevention, the current evidence does not strongly support their use for this purpose. Anti-seizure medications, like topiramate and divalproex sodium, and certain antidepressants, namely, venlafaxine and amitriptyline, have been found effective in preventing migraine attacks, but users must be wary of associated side effects. In deciding on a treatment course, it is crucial to weigh the potential benefits against the risks. Open dialogue between the patient and the physician will ensure the most suitable therapeutic choice is made [110].

6. Headache Physiology: Central Connections and the Trigeminocervical Complex

6.1. Migraine Neuronal Activation and Therapeutic Implications

Utilizing Fos immunohistochemistry, researchers have been able to discern the activation of cells by detecting Fos protein expression within the trigeminocervical complex. Following the irritation of meningeal with blood, there was a significant upregulation of Fos expression within the trigeminal nucleus caudalis. Additionally, upon stimulation of the superior sagittal sinus, Fos-like immunoreactivity was observed not only in the trigeminal nucleus caudalis but also in the dorsal horn at the C1 and C2 levels in both feline and simian subjects [58]. These findings are consistent with results obtained from 2-deoxyglucose analyses in congruent experiments [65]. Similarly, the activation of the greater occipital nerve, an offshoot of the C2, amplifies the metabolic activity in the aforementioned regions. It has been documented in animal-based studies that it is feasible to directly obtain readings from trigeminal neurons receiving input from both the supratentorial trigeminal and the greater occipital nerve. A mere 5 min stimulation of the greater occipital nerve resulted in a pronounced escalation in response to supratentorial dural stimuli, with the effects lasting for more than 60 minutes [101]. Conversely, the stimulation of the dura mater of the middle meningeal artery utilizing mustard oil as a C fiber irritant augmented the responses to occipital muscle stimulation [102]. Additional data derived from the Fos technique posit that such interactions likely necessitate the activation of the NMDA subtype of glutamate receptors [103]. Taken together, these findings suggest that the cervical and ophthalmic inputs intersect at the level of the second-order neuron [104]. It is worth nothing that bilateral Fos expression was observed when a lateralized structure, specifically the middle meningeal artery, was stimulated in both feline and simian models [105]. This particular group of neurons from the superficial laminae of the trigeminal nucleus caudalis and C1/2 dorsal horns is functionally recognized as the "trigeminocervical" complex. Such insights indicate that the transmission of nociceptive information from the trigeminovascular system predominantly occurs via the most caudal cells, which provides an anatomical elucidation for the referral of migraine-associated pain to the posterior cranial region. It is imperative to highlight that pharmacological experimentation has unveiled that migraine-abating drugs, such as ergot derivatives, acetylsalicylic acid, sumatriptan, eletriptan, naratriptan, rizatriptan, zolmitriptan, and novel CGRP receptor antagonists, have the potential to modulate these second-order neurons, thereby decreasing their activity [106]. This proposes another plausible avenue for therapeutic interventions in migraine. The modus operandi of triptans is believed to engage the 5-HT1B, 5-HT1D, and 5-HT1F receptor subtypes, which correlates with the positioning of these receptors on peptidergic nociceptors.

Subsequent exploration into neuropeptides has underscored the potential of CGRP receptor antagonists for migraine treatment. Elevated levels of CGRP and SP subsequent to trigeminal ganglion stimulation, observed in both humans and felines, might serve as a protective mechanism in severe migraine, cluster headaches, and chronic paroxysmal hemicranias. Additionally, an upswing in CGRP levels was documented during a NO donor-triggered migraine episode, but this surge was mitigated by sumatriptan, further corroborating the implication of the trigeminovascular system in these conditions. Nevertheless, specific compounds, such as CP122,288 and 4991w93, which were ineffective against migraine, failed to inhibit CGRP release post-superior sagittal sinus stimulation in feline subjects. In spite of these observations, the advent and ensuing triumphant clinical trial outcomes of particular CGRP receptor antagonists for acute migraine have emphasized the significance of this therapeutic strategy [64,107].

In studies involving the trigeminocervical complex, Fos immunohistochemistry has been utilized to identify activated cells by mapping the Fos protein expression. This method revealed an increased Fos expression in the trigeminal nucleus caudalis after meningeal irritation with blood and in the trigeminal nucleus caudalis and the C1 and C2 levels in the dorsal horn of cats and monkeys after stimulation of the superior sagittal sinus [111]. Moreover, the stimulation of the greater occipital nerve, a branch of C2, resulted in increased metabolic activity in these regions [112]. This suggests that inputs from the cervical and ophthalmic regions converge at the level of the second-order neuron. Pharmacological research indicates that drugs like ergot derivatives, acetylsalicylic acid, sumatriptan, eletriptan, naratriptan, rizatriptan, zolmitriptan, and CGRP receptor antagonists may help reduce cell activity at these second-order neurons and therefore could be a potential therapeutic approach for migraine [88].

Studies into serotonin–5 HT1F receptor agonists and their relation to migraine have shown that some triptans, including naratriptan, are also potent 5 HT1F receptor agonists. This suggests that 5 HT1F activation could potentially inhibit trigeminal nucleus Fos activation and neuronal firing in response to dural stimulation without affecting cranial vascular effects. These findings further support the idea that vascular mechanisms are not necessarily required for acute migraine treatments [113].

Glutamatergic transmission in the trigeminocervical complex has also been explored as a potential target for antimigraine drugs. The family of glutamate receptors (GluRs) is particularly interesting, with studies showing that NMDA receptor channel blockers can reduce nociceptive trigeminovascular transmission in vivo. Furthermore, the AMPA/kainate receptor antagonists CNQX and 2,3-Dioxo-6-nitro-1,2,3,4-tetrahydrobenzoquinoxaline-7-sulfonamide decreased Fos protein expression after the activation of structures involved in nociceptive pathways [114]. Notably, the iGluR5 kainate receptors may play a role in trigeminovascular physiology, suggesting a potential target for future treatments. In clinical trials, the iGluR5 kainate receptor antagonist LY466195 demonstrated efficacy in acute migraine treatment, reinforcing the pursuit of glutamate targets for treatment, albeit with caution regarding potential side effects [115].

6.2. Discussion of Key Pathological Processes Involved in Migraine

6.2.1. Neurophysiology of Migraine as a Backdrop to Imaging

The utilization of neurophysiological techniques in migraine patients has yielded significant knowledge about the condition. These techniques prioritize time resolution over spatial resolution and, until the advent of MRI, and to some extent even now, provided a higher chance for repeated trials. Research across the visual, somatosensory, auditory, and nociceptive domains has consistently shown activation patterns that differ markedly from non-migraine. These findings have led to the theory that thalamocortical dysrhythmia plays a significant role in migraine pathophysiology [116,117] (see Figure 1).

An intriguing observation from these studies is the abnormal habituation in migraine patients between attacks, as exemplified by the increased intensity of auditory evoked potentials in these periods [118]. Interestingly, this abnormality seems to normalize just before a migraine attack [119]. It is worth noting that this metric appears to be serotonin-dependent and can be modulated by triptans, which are serotonin 5-HT1B/1D receptor agonists [120]. The amplification of the passive "oddball" auditory event-related potential and an interictal habituation deficit measured by the nociceptive blink reflex, further suggest that the brain of a migraineur does not habituate in the same way as a non-migraineur's brain does [121].

These observations have given rise to the idea that the brain of a person with migraine reacts more intensely, rather than simply being hyperexcitable [121].

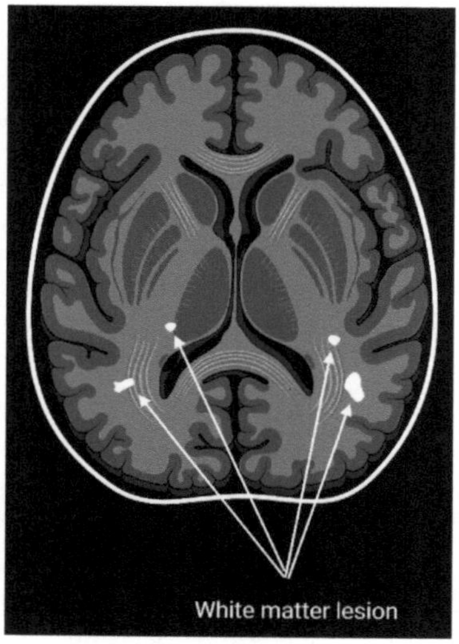

Figure 1. The two main types of lesions found in migraine include: white matter hyperintensities and silent brain infarcts.

6.2.2. Inter-Attack Imaging Studies

1. Structural Studies

Numerous research endeavors have revealed structural disparities between the brains of those who suffer from migraine and those who do not. Structural investigations, often cross-sectional, should be analyzed with consideration to the functional interactions of pain-processing areas and the trigeminal system. Voxel-based morphometry has showcased decreased grey matter in areas involved in pain processing, such as the anterior cingulate cortex, amygdala, insula, operculum, and the frontal, temporal, and precentral gyri. Interestingly, grey matter reduction in the anterior cingulate cortex has been found to be correlated with the frequency of migraine episodes [122]. On the other hand, a greater grey matter volume was seen in the bilateral caudate nuclei in high-frequency migraine sufferers compared to those with low-frequency migraine [123]. Additionally, there was an observed thickening in the somatosensory cortex, particularly in the portion responsible for mapping sensations from the head and face. This thickening was more pronounced compared to individuals who did not exhibit these changes and served as controls [124]. Research involving diffusion tensor imaging has revealed decreased fractional anisotropy in the thalamocortical pathway of individuals with migraine. Specifically, this reduction was observed in the ventral trigeminothalamic tract for those experiencing aura, and in the ventrolateral periaqueductal gray (PAG) for those without aura [124]. Other studies found only minor diffusivity changes in grey matter, while changes in white matter and brain volume were similar in both groups. Nevertheless, a more extensive investigation discovered that individuals with migraine accompanied by aura exhibited a shorter T1 relaxation time in the thalamus compared to those without aura and individuals without migraine who were in good health. In a separate study, a reduction in brain volume was detected across multiple regions when comparing migraine sufferers to control subjects. Significantly, it remains uncertain whether these alterations stem from recurrent migraine episodes or are intricately connected to the underlying mechanisms of migraine. It is worth noting that certain changes tend to revert to a more typical state during a migraine, implying a potential connection between the most recent attack and these structural modifications [125]. Taken together, these findings suggest that the structural alterations seen in migraine patients, particularly in the anterior cingulate cortex and the trigeminal somatosensory system, reflect the brain's ability to develop migraine attacks and may underlie the progression of the disorder.

2. Functional Studies

Functional studies serve to supplement structural brain imaging, focusing on the resting migrainous brain and its response to external stimuli.

(A) Metabolism and Receptor Pharmacology

Functional differences between two groups at rest can be evaluated using 18F-FDG PET to assess regional brain metabolism. Kim et al. [126] found that migraine was associated with reduced metabolism in central pain processing areas, suggesting a dysfunction of central pain processing in the interictal state. No area showed hypermetabolism.

(B) Stimulated Blood Flow Changes

Photophobia, a common non-head pain symptom in migraine, can result in light being perceived as overly bright or painful. Individuals suffering from migraine, even when not experiencing an attack, have been shown to tolerate less luminance than healthy individuals [127]. A study using H215O-PET revealed that exposure to different light intensities activated the visual cortex in migraine but not in controls. Moreover, applying trigeminal pain activated the same areas in control subjects, suggesting a facilitation of the retino-geniculate-cortical pathway of visual processing and/or a dysfunction of visual association areas causing photophobia. Other studies found differences in response to heat stimuli and painful heat stimulation in migraine compared to controls. Demarquay et al. [128] evaluated olfactory processing in migraine and found unique cortical responses associated with olfactory hypersensitivity. These findings suggest that the brains of migraine respond differently to external stimuli compared to healthy controls, possibly due to pre-existing functional abnormalities that worsen during a migraine attack, leading to a "dys-excitable" state [129].

(C) Studies focused on resting-state brain activity

Resting-state studies offer a unique perspective on brain function, especially in the context of disorders like migraine. These investigations focus on how the brain operates when it is not performing any particular task, thereby providing insights into the intrinsic communication patterns within the brain. They represent a significant departure from other neuroimaging techniques as follows: while structural brain imaging helps identify disparities in grey and white matter, and stimulus-driven functional magnetic resonance imaging (fMRI) pinpoints distinct dysfunctional areas, resting-state studies evaluate the interaction, or "cross-talk", among different brain regions [121].

One such study by Mainero et al. found that individuals suffering from migraine exhibit heightened connectivity between the periaqueductal gray (PAG) and multiple areas significant to nociceptive and somatosensory processing. These findings were further linked with the frequency of migraine episodes, signifying the pathophysiological relevance of this enhanced connectivity in the modulation of pain during migraine [130].

Another vital aspect of migraine is the presence of cutaneous allodynia, considered a reflection of central sensitization during migraine attacks. When comparing resting-state connectivity between migraine with and without cutaneous allodynia, distinctive patterns were identified. Specifically, connectivity differences were noted between the PAG/nucleus cuneiformis and various discriminative pain processing centers, such as the brainstem, thalamus, insula, cerebellum, and higher-order pain-modulating areas located in frontal and temporal regions [131]. This evidence suggests that individual symptoms during a migraine attack might be determined by abnormal communication between pain-modulating areas during interictal periods.

While all participants in the study had migraine with normal routine investigation results, the presence of ictal allodynia seems to delineate different subtypes of migraine. This implies that migraine itself could also be pathophysiologically diverse. In terms of headache phase studies, seed-based resting-state fMRI has exhibited increased connectivity between primary visual and auditory cortices and the right dorsal anterior insula, and between the dorsal pons and the bilateral anterior insulae. Interestingly, these findings did not correlate with migraine frequency, suggesting that these changes were inherent characteristics of migraine pathophysiology rather than episodic manifestations [121].

Expanding the scope, resting functional connectivity of certain brain regions, including the right middle temporal, posterior insula, middle cingulate, left ventromedial prefrontal, and bilateral amygdala regions, was found to effectively distinguish between the brains of migraine and non-migraine [132]. Despite being grounded in clinical observation, these techniques have the potential to yield insights into migraine biology, contribute to the development of NextGen treatments, and offer biomarkers of change, particularly for preventive studies.

Resting-state studies are not limited to the seed-based approach, which focuses on the connectivity of specific "seed" areas, such as the PAG or nucleus cuneiformis. Task-free resting-state studies employing independent component analysis without a priori hypothesis have also been conducted. One such study by Tessidore et al. [133] examined the default mode network in patients with migraine without aura. They found decreased connectivity in the prefrontal and temporal regions of the default mode network among these patients. The authors speculated that this could indicate a dysfunction of the default mode network, potentially tied to maladaptive responses to stress or environmental triggers, which are often characteristic of migraine.

In conclusion, being a migraineur suggests the presence of nuanced differences in brain structure and function, even outside of active migraine attacks. Notably, most areas showing such differences belong to the non-specific pain processing areas or the trigeminal system. A significant challenge moving forward is understanding how these differences predispose individuals to migraine, and identifying which structures drive the transition from the interictal phase, through the premonitory phase, to the headache phase, and eventually, the postdrome period that returns to the interictal phase.

(D) Studies focusing on mitochondrial energy metabolism

Considering the recognition of migraine as a component of mitochondrial cytopathies, exploring how this biological aspect influences the onset and development of migraine provides a promising research avenue. Although initial studies on mitochondrial DNA did not find any typical MELAS or MERRF mutations, a successfully conducted randomized controlled trial of riboflavin (vitamin B2) as a preventive treatment for migraine supports the hypothesis that metabolic dysfunction could increase susceptibility in some patients [134].

Utilizing 31P-NMR spectroscopy, Welch et al. [135] identified changes in the phosphate metabolism in patients with migraine with aura during an attack. Later studies using the same technique found similar metabolic shifts in patients with migraine without aura and even in children [136]. Furthermore, by employing 3T MRI and 31P-NMR spectroscopy, researchers were able to identify alterations in energy metabolism in the occipital cortex of patients with migraine without aura [137]. Given the observed variations in energy changes among patients, these findings could potentially explain some, but not all, of the biological mechanisms contributing to the manifestation of migraine.

3. Premonitory Phase Studies

From a clinical perspective, the premonitory phase—the transitional period between the asymptomatic interictal phase and the onset of a headache attack—is crucial for understanding what triggers migraine. An essential fMRI study in this regard examined the activation and deactivation patterns induced by the trigemino-nociceptive stimulation of the nasal mucosa as the day of the headache approached [138]. In comparison to control subjects, interictal migraine showed reduced activation of the spinal trigeminal nuclei. Interestingly, this deactivation demonstrated the following cyclic behavior throughout a migraine interval: normalization prior to the next attack and a significant reduction of deactivation during the attack. This cyclical behavior may reflect the brain's increased susceptibility to initiate the next attack, with the identification of its pacemaker critical to our understanding of the initiation of a migraine attack.

Clinically, the earliest indicators of an impending migraine attack are known as premonitory symptoms, which manifest before the onset of head pain and signal to the patient that a headache is imminent. These symptoms, likely tied to the hypothalamus [139], include concentration problems, fatigue, irritability, and depression. A recent study by Maniyar et al. [140] induced migraine attacks in eight patients who could predict the onset of a headache by a pronounced premonitory phase. During this phase, which occurs before the onset of head pain, H215O-PET showed activation of the hypothalamus, midbrain ventral tegmental area, and the PAG. This functional representation of premonitory symptoms hints at the potential role of the hypothalamus in triggering migraine. Additional data from a single patient tracked with BOLD-fMRI over a 30-day period showed increased hypothalamic responses as the attack approached, and the effects were coupled with the dorsolateral pons [141]. Additionally, the hypothalamus could play a crucial role in non-headache symptoms during the pain phase since its activation was observed in spontaneous migraine attacks using H215O-PET [142]. Interestingly, activations reported in trigeminal-autonomic cephalalgias are more posterior than those reported in migraine [121].

4. The Aura Phase

Typically, a visual aura in the context of a migraine presents itself before the onset of the headache phase, although there are instances where it coincides with the headache or even occurs without any headache at all. The manifestation of this aura often begins as a scintillating or blind spot situated in the center of the individual's field of vision [121]. Personal experiences reported by Lashley [143] indicated that this visual disturbance or scotoma progressively expanded over a period of approximately one hour, moving in a C-shaped trajectory towards one side's temporal visual field. Based on his observations, the estimated speed of this phenomenon over the visual cortex was calculated to be approximately 3 mm per minute.

A few years later, the concept of a potential underlying mechanism emerged from the work of Leão, who stimulated the cortices of rabbits electrically. He observed a depression in the electroencephalogram (EEG) readings that spread out from the stimulation site at a similar speed of 3 mm per minute. Leão postulated that this cortical phenomenon could possibly serve as the foundation for the migraine aura [144]. This theory sustained for several decades that the occurrence of a typical visual aura could be associated with this phenomenon, termed "cortical spreading depression" (CSD) [145].

The validation of the CSD occurrence in humans was conjectural until a groundbreaking study by Olesen et al. [146]. They injected Xenon-133 into the carotid artery during a human migraine aura and found an observable progressive alteration of regional cerebral blood flow (rCBF). Fast forward two decades later, patients who could self-trigger their visual aura or who were capable of reaching a medical facility during the early stages of a visual aura were included in a functional magnetic resonance imaging (fMRI) study using checkerboard stimulation. The change in blood oxygenation level-dependent (BOLD) signal in the visual cortex in response to checkerboard stimulation during the progression of a visual aura showed characteristics similar to those of the CSD observed in animal models. This included a signal spread at a velocity of roughly 3.5 mm per minute, aligning with the earlier clinical predictions and the CSD observed in rabbit cortices [147]. These findings suggest that the visual aura experienced in migraine might indeed be the result of a CSD-like event. Additionally, the study by Hadjikhani et al. [148] pinpointed the origin of this unique response to checkerboard stimulation to be located in the visual association cortex V3A.

5. The Headache Stage

Renowned as the most prominent symptom of a migraine attack, the headache stage is often the defining phase for many patients. To diagnose a migraine, however, additional symptoms need to be present. These can range from nausea, photophobia (light sensitivity), phonophobia (sound sensitivity), to sensitivity to movement [149]. The imaging patterns observed during a migraine are typically a composite of these symptoms, with some possibly reflecting individual symptoms like head pain, photophobia, or allodynia (an increased response to pain), and others hinting at underlying mechanisms that trigger the migraine.

(a) The Experience of Head Pain: The complexity of primary headache disorders extends beyond head pain that is typically triggered by harmful stimuli on the skin. However, the sensation of pain is a shared experience in harmful head pain and spontaneous migraine attacks. Therefore, the markers identified in functional brain imaging of experimental head pain should also be observable in migraine headaches. Thus, any additional regions highlighted in primary headache disorders may provide specific insights into migraine, potentially revealing symptoms beyond head pain or even mechanisms that drive migraine attacks. Functional brain imaging of harmful pain in the head is a significant focus as it could improve our understanding of functional brain imaging of migraine. May et al. [150] used H215O-PET to measure rCBF in seven healthy subjects after injecting a small amount of capsaicin into the forehead. They noticed an increase in rCBF in several brain areas during the pain state, including the bilateral insula, the anterior cingulate cortex, the cavernous sinus, and the cerebellum. Notably, there was no activation of the brainstem.

(b) Migraine Attacks: Over the past two decades, a seminal study employing positron emission tomography sought to evaluate regional cerebral blood flow utilizing 15C-labeled O2 inhalation in a cohort of nine individuals experiencing spontaneous right-lateralized migraine episodes. Relative to the non-painful interlude, the migraine episodes were concomitant with augmented rCBF in regions, such as the cingulate cortex, auditory association cortex, and the parieto-occipital juncture proximate to the visual association cortex. Furthermore, the migraine-afflicted state manifested with escalated rCBF in the midbrain, the dorsal rostral pons adjacent to the periaqueductal gray, and the raphe nuclei [128]. Contrasting the generalized pain signature derived from the capsaicin experiment, this investigation attributed diverse migraine-related symptoms to distinct cerebral domains as follows: the experience of cephalic pain was associated with the cingulate cortex, photophobia was linked to the visual association cortex, and phonophobia was ascribed to the auditory association cortex. The cessation of these symptoms was synchronous with the waning of the aforementioned signals. Yet, the elevated rCBF in the brainstem endured during the nascent non-painful phase, implying that this anatomical region might not merely be symptomatic but could also typify a dysfunction pivotal for initiating or perpetuating a migraine episode. Clinical research, coupled with fundamental studies, further bolsters the centrality of the brainstem in the pathogenesis of migraine. An illustrative point being the emergence of migraine episodes in individuals previously devoid of migraine who underwent deep brain stimulation targeting the PAG for unrelated pain conditions. Moreover, the progressive accumulation of iron in the PAG over the disease's tenure intimates the indispensable role of the PAG in migraine genesis. Corroborating this notion, a plethora of animal-centric studies have delineated how brainstem nuclei, specifically the PAG and raphe nuclei, profoundly modulate trigeminovascular pathways in laboratory-induced migraine paradigms [91,129].

Functional brain imaging employing advanced techniques of enhanced spatial and temporal resolutions have corroborated the pivotal role of the brain stem in the pathophysiology of migraine. In a comparative study, Bahra et al. [130] distinguished migraine from cluster headaches, underscoring the specificity of brain stem activation to migraine. Examining the lateralization of this activation during unilateral migraine episodes, Afridi et al. [131] ascertained that the activation was ipsilateral to the side of the headache. This suggests the possibility that unilateral migraine might be attributed to unilateral dysfunction of the brain stem. In a preceding discourse, Maniyar et al. [117] detected activation in the dorsal rostral pons, the PAG, and the hypothalamus during the preliminary premonitory phase of migraine, buttressing the hypothesis of a central "migraine mediator" located in these regions. Melding the clinical manifestations of migraine—typified by altered sensory, nociceptive, photic, acoustic, and olfactory perceptions—with functional imaging insights, it becomes evident that either the brain stem, hypothalamic structures, or a combination of both are instrumental in migraine pathophysiology. Such structures potentially pinpoint the anatomical epicenters of cerebral dysfunction engendering the multifaceted dynamics of migraine episodes.

(c) Photophobia: Denuelle et al. [119] conducted research on eight individuals suffering from migraine, evaluating them during headache episodes, post-sumatriptan alleviation, and interictal periods. Utilizing continuous light stimulation, they discerned that low luminance provocation elevated rCBF as indicated by H215O-PET scans. During the headache phase, hyperperfusion was detected in the cuneus, and post-relief, both the cuneus and lingual gyrus demonstrated this phenomenon; conversely, such changes were absent interictally. This might insinuate an augmented excitability of the visual cortex amidst migraine occurrences. Notably, even post-headache alleviation, the persistence of this hyperperfusion, unrelated to the headache's presence, hints at the structural foundation of photophobia potentially residing in primary and ancillary visual cortices. Moreover, zones responsive to minimal luminance during migraine were equivalently reactive interictally to escalated luminous intensities, further substantiating the cyclic nature underpinning migraine and their concomitant symptoms.

In studies targeting the premonitory phase, focusing on non-painful symptoms, pivotal insights have emerged, emphasizing the separation of such phenomena from pain while highlighting their integral role in migraine biology. When contrasting individuals with provoked premonitory symptoms, those exhibiting photophobia (or perhaps more aptly termed photic hypersensitivity due to the absence of pain) demonstrated activation within the extrastriate visual cortex, specifically Brodman area 18 [117]. Interictal connectivity within the visual system, manifested in the lingual gyrus, was also identified using resting-state methodologies. Furthermore, in experiments distinguishing migraine with and without nausea, those experiencing nausea showcased activation in the rostral dorsal medulla encompassing regions like the nucleus tractus solitarius, the dorsal motor nucleus of the vagus nerve, and the nucleus ambiguous; moreover, activation in the PAG was also evident. These research endeavors have enriched our comprehension of cerebral regions implicated in migraine, unequivocally indicating mechanisms transcending mere pain dependency [91].

Different visual migrainous phenomena are associated with dysfunctions in different areas of the visual association cortex. For instance, the cuneus and lingual gyrus are involved in photophobia, while V3A might be the origin of a typical visual aura. When comparing the imaging results during the migraine premonitory phase, such as hypothalamic and brain stem activation, with those of cortical activation during a typical migraine aura, it appears likely that the aura and migraine are distinct phenomena [121].

6. Blood–Brain Barrier (BBB): The integrity of the BBB in migraine

In postdrome, patients often describe fatigue, difficulty with concentration, and a need for sleep. Some patients also report a feeling of elation and well-being and a return of appetite. The symptoms are not always perceived as bothersome and are commonly overlooked by patients and doctors. Often, the patient is just relieved that the headache phase is over.

More work is needed to understand the nature and cause of the postdromal phase. It would be especially helpful to understand the brain's role in postdromal phase symptoms, whether the brain goes back to normal after a migraine attack, and if not, why not. Also, more understanding of how the brain recovers and how quickly it recovers would be very helpful. Whether this phase represents a therapeutic opportunity is unknown but should be explored [121].

6.3. Calcitonin Gene-Related Peptide (CGRP) in Migraine

6.3.1. Role of CGRP in Migraine Development and Progression

1. Introduction to CGRP and its significance in migraine

Calcitonin gene-related peptide (CGRP) is an incredibly potent neuropeptide comprised of 37 amino acids. It serves as a vasodilator and is produced within neurons located in both the peripheral and central nervous systems. This neuro-peptide binds to a complex heterodimer receptor, which is primarily composed of a class B G-protein coupled receptor, commonly referred to as CLR (calcitonin receptor-like receptor) [151].

Within the central nervous system, empirical research has identified elevated levels of CGRP in the blood and saliva of patients who experience certain headache disorders. Such disorders include migraine and cluster headaches, as well as neuralgias like trigeminal neuralgia, chronic paroxysmal hemicranias, and even rhinosinusitis. It is noteworthy that the levels of CGRP remain heightened during a migraine episode and continue to be elevated in-between these attacks for patients suffering from chronic migraine. Additionally, studies have revealed that exogenous infusions of CGRP can initiate a migraine episode [152].

CGRP's role extends to the pathogenesis of migraine, which is an intricate neurovascular disorder. This is typically characterized by a throbbing or pounding headache, which affects one side of the head. It is often accompanied by other symptoms, such as photophobia (light sensitivity), phonophobia (sound sensitivity), nausea, vomiting, and even disability. Additionally, the duration of a typical migraine episode can last between 4 and 72 h [153]. Researchers have found that CGRP triggers the release of vasoactive neuropeptides in trigeminal neurons, leading to vasodilation of the cerebral vasculature, thereby contributing to the emergence of a migraine.

The Food and Drug Administration (FDA) has sanctioned several medications specifically targeting CGRP or its receptor for the management and prevention of migraine. Prominent among these are monoclonal antibodies including erenumab, eptinezumab, galcanezumab, and fremanezumab, which zero in on the CGRP receptor. In addition, CGRP receptor antagonists, like rimegepant and ubrogepant, have been incorporated into therapeutic regimens. Two other receptor antagonists, atogepant and vazegepant, remain under clinical evaluation and anticipate FDA endorsement. Prophylactic interventions for both episodic and chronic migraine commonly incorporate erenumab, eptinezumab, galcanezumab, and fremanezumab. For addressing acute migraine manifestations, with or without the presence of an aura, rimegepant and ubrogepant are the preferred choices. Presently, the efficacy of galcanezumab and fremanezumab in precluding cluster headaches is a subject of active research [135].

Interestingly, CGRP has been shown to have cardio-protective properties in pathological conditions. For instance, research conducted on rodent models of various cardiovascular diseases has revealed this beneficial action of CGRP. Human studies also support this notion by showing that CGRP can reduce afterload and increase inotropy, which are potentially cardioprotective effects, particularly in cases of heart failure. Despite these findings, no drugs have yet been developed to harness this cardio-protective effect of CGRP on the cardiovascular system [154].

Recent research has also begun to uncover CGRP's involvement in numerous other physiological and pathological phenomena. These include peripheral nerve regeneration, Alzheimer's disease, regulation of vascular tone in mesenteric arteries, and even pregnancy. Despite these promising findings, no medications have been developed to date to leverage these potential beneficial effects of CGRP [155].

Preclinical studies have demonstrated that calcitonin gene-related peptide (CGRP) exhibits activity in both the central and peripheral nervous systems (CNS and PNS), making it a crucial element in the pathophysiology of migraine. In the periphery, CGRP acts on a number of targets, such as mast cells, blood vessels, glial cells, and trigeminal afferents located in the meninges, along with neural cell bodies and satellite glia found in the trigeminal ganglia. Within the meninges, CGRP is thought to contribute to neurogenic inflammation by stimulating mast cells to release neuron-sensitizing agents. This cascade effect can lead to enhanced vasodilation in the dura. Consequently, the modulation of neural activity within the meninges may instigate a feedback loop, ultimately leading to peripheral sensitization of nociceptors [156]. The notion of CGRP playing a peripheral role in migraine is strongly supported by the effectiveness of systemically administered CGRP-targeting monoclonal antibodies, which exhibit poor permeability to the blood–brain barrier (BBB) [157]. It is clear that peripheral sensitization is critical for CGRP's actions and likely establishes the foundation for CGRP actions in the CNS.

Within the central nervous system (CNS), the distribution of CGRP and its receptor spans various pathways postulated to play pivotal roles in migraine pathophysiology. Situated externally to the blood–brain barrier (BBB), the trigeminal ganglion extends its projections to the trigeminal nucleus caudalis (TNC). From here, second-order neurons transmit signals to the posterior thalamic area (PTA), an umbrella term denoting all nuclei within this specific thalamic region. Serving ostensibly as a hub for sensory integration, the PTA exhibits functional aberrations during migraine occurrences. Neurons in the thalamus receive inputs from both the TNC and retinal ganglion cells. Crucial rodent studies have accentuated the import of the PTA in photophobia's onset, proposing its role as a nexus for light and pain integration [139].

Situated in select nuclei of the PTA, CGRP and its receptors are postulated to be integral to this pathway—a hypothesis fortified by research showing that CGRP infusion into the PTA augments neuronal activity. Additionally, ascending pathways carrying somatosensory and nociceptive stimuli converge upon the CGRP-expressing neurons located in the subparafascicular and intralaminar nuclei. In human subjects, during migraine attacks, activation is discerned in the posterior thalamus, which also manifests altered connectivity patterns with various brain areas. Taken together, this suggests that the neuromodulatory actions of CGRP, observed in distinct neural networks, might be instrumental in rendering the PTA hyperresponsive to sensory input [140].

Further elucidating this sensory hyperreactivity, pathways potentially involve the parabrachial nucleus (PBN), colloquially termed the "general alarm" system. Functioning as an intermediary for pain and assorted sensory signals en route to the forebrain, the PBN receives direct extensions from the trigeminal nucleus, influencing the emotional facet of pain. Given the abundance of CGRP in the PBN and its extensive projections to brain regions implicated in migraine pathophysiology, modified signaling within this conduit might underlie the heightened sensory perception characteristic of migraine. CGRP's dual—peripheral and central—actions likely synergize to precipitate a migraine. Considering the disorder's multifaceted nature, it is improbable that a singular action of CGRP singularly instigates migraine [141].

2. Mechanisms by which CGRP contributes to migraine symptoms

The role of CGRP in migraine symptomatology is primarily deciphered through preclinical investigations. Recognized as a paramount instigator of migraine, CGRP can elicit an array of migraine-reminiscent symptoms in animals that parallel the effects observed when humans are infused with CGRP, encompassing pain-related symptoms. In rodents, mechanical hypersensitivity, an often-reported migraine symptom, can be instigated following CGRP administration. For instance, CGRP's dural administration in mice provoked periorbital touch hypersensitivity, whereas its intrathecal introduction led to heightened pain sensitivity in rat hindpaws and amplified mechanical allodynia in mice upon pinch [140].

For an extended period, the challenge of gauging spontaneous pain in animals persisted due to the absence of an apt assessment method. However, in 2010, Mogil and his team pioneered a method, illustrating that specific pain forms could be evaluated via facial grimace scales without necessitating an evoked response [142]. Building on this foundation, our research demonstrated that injecting CGRP peripherally in mice culminates in spontaneous, migraine-analogous pain, which sumatriptan could substantially mitigate. We employed an uninterrupted, objective appraisal of eye closure to assess the grimace induced by CGRP. Our findings also elucidated that CGRP-induced pain was not influenced by light levels, positing that pain and light aversion, another symptom induced by CGRP, operate independently [143].

Photophobia, characterized by an augmented sensitivity or discomfort in light conditions usually deemed non-painful, stands as a diagnostic hallmark of migraine. Those afflicted with migraine often perceive even subdued light as unsettling, gravitating away from luminous environments [144]. This human experience has been adeptly mirrored in a mouse paradigm where exposure to light becomes aversive post-CGRP administration, both centrally and peripherally, in conventional mice. This aversive reaction to light can be palliated by triptans, insinuating that the murine aversion to light resonates with human migraine. Intriguingly, in a specialized mouse model sensitized to CGRP—where human RAMP1 (an essential component of the CGRP receptor) is overexpressed within the nervous system—a mere 55 lux light intensity suffices to trigger light aversion post central CGRP administration. Notably, this aversive reaction to light is not an offshoot of anxiety, as evidenced by the unaffected performance of these mice in a light-agnostic anxiety assessment (the open field test) [145,146]. Furthermore, post-CGRP injection, these mice exhibit diminished mobility, but this inertia is predominantly observed in dimly lit sections of their enclosure. Such a preference for darker locales and a proclivity to rest mirrors human behavioral tendencies during migraine episodes.

6.3.2. CGRP as a Therapeutic Target

1. Overview of CGRP-targeted treatments in migraine management—Evaluation of the effectiveness of CGRP inhibitors

Over recent years, a plethora of molecules aiming to obstruct CGRP signaling pathways have been developed, with the objective of mitigating migraine symptoms. The first molecules that showed promise were CGRP receptor antagonists, known as "gepants." These substances demonstrate a high affinity for the canonical CGRP receptor, blocking the CGRP from binding and obstructing the subsequent signal transduction. Importantly, gepants do not incite direct vasoconstriction, making them potentially safer than triptans for a migraine population that statistically exhibits a higher prevalence of cardiovascular diseases [158].

Numerous clinical trials have established that both intravenous and orally administered gepants can effectively alleviate acute migraine symptoms (Table 3); however, the efficacy of gepants in preventing migraine is currently a matter of debate. Some clinical trials had to be halted due to adverse effects, while others are still in progress [159]. The development of some gepants was ceased for various reasons. For example, olcegepant demonstrated low oral bioavailability, and both telcagepant and MK-3207 were discontinued due to liver toxicity associated with frequent use [159].

In spite of these initial safety concerns, the apparent efficacy of gepants has encouraged continued efforts to devise safe molecules that block CGRP. Currently, three gepants—rimegepant, ubrogepant, and atogepant—are still under clinical development. In phase 2b clinical trials, the efficacy of rimegepant for acute migraine treatment was assessed using various endpoints, such as freedom from pain, migraine, photophobia, phonophobia, and nausea remission [76]. Medium doses of rimegepant (75, 150, and 300 mg) were found to be significantly more effective than the placebo, and unlike the previously terminated gepants, rimegepant did not exhibit any adverse effects on liver function. Interestingly, a higher dose of rimegepant (600 mg) did not yield significant benefits over the placebo, leading the researchers to hypothesize that this could be attributed to inherent variability among patients randomized to this dose group [160].

Post this investigation, a series of three phase 3 double-blind, randomized, placebo-controlled trials (NCT03235479, NCT03237845, NCT03461757) alongside a safety investigation (NCT03266588) were commenced, with outcomes yet to be unveiled [150]. In parallel, ubrogepant showcased a favorable dose-effect correlation for acute migraine treatment in a phase 2b double-blind, randomized, placebo-controlled trial [151], exhibiting negligible side effects. However, these findings are somewhat overshadowed by the heightened placebo group response and the study's restricted patient count. Two subsequent phase 3 double-blind, randomized, placebo-controlled trials (NCT02867709, NCT02828020) wrapped up in December 2017 and February 2018, respectively. The initial findings echo the results of the earlier phase 2b study. Atogepant, possessing a molecular structure distinct from its gepant counterparts, is currently under examination for migraine prevention. Initial data from a phase 2b/3 trial (NCT02848326) indicate that adults administered atogepant underwent a more pronounced decline in their monthly migraine days average compared to those receiving a placebo. There were no reported grave side effects tied to the treatment. As of the time this review was drafted, ubrogepant emerged as the inaugural gepant to secure FDA sanctioning for acute migraine intervention, encompassing cases with or devoid of an aura [140].

Table 3. Clinical trials investigating drugs that target CGRP (calcitonin gene-related peptide): both completed and currently underway.

Drug Name, Type of Molecule	Indication (Acute or Prophylactic)	Development Stage
Atogepant, CGRP antagonist	Prophylactic	FDA-approved
BI 44370 TA, CGRP antagonist	Acute	Abandoned
Eptinezumab, CGRP monoclonal antibody	Prophylactic	FDA-approved EMA-approved
Erenumab, CGRP receptor monoclonal antibody	Prophylactic	FDA-approved
Fremanezumab, CGRP monoclonal antibody	Prophylactic	FDA-approved
Galcanezumab, CGRP monoclonal antibody	Prophylactic	FDA-approved
MK-3207, CGRP antagonist	Acute	Abandoned for liver toxicity
Olcegepant, CGRP antagonist	Acute	Abandoned for lack of oral availability
Rimegepant, CGRP antagonist	Acute	FDA-approved
Telcagepant, CGRP antagonist	Acute and prophylactic	Abandoned for liver toxicity
Ubrogepant, CGRP antagonist	Acute	FDA-approved

Monoclonal antibodies aimed at CGRP (such as fremanezumab, galcanezumab, and eptinezumab) and its receptor (like erenumab) form a separate molecular category adept at obstructing CGRP signaling pathways. Three among these antibodies (fremanezumab, galcanezumab, and erenumab) recently achieved FDA endorsement for preventive migraine therapy, while a verdict on a fourth candidate, eptinezumab, is anticipated in 2020. Remarkably, about half of the patients administered these antibodies witnessed a 50% downturn in their migraine days. Notably, no discernible efficacy disparity was observed across antibodies, irrespective of whether they latch onto the receptor or isolate CGRP. These antibodies also maintain their therapeutic effectiveness for an extended period beyond a month post-application, thus qualifying as prophylactic agents administered monthly or even on a quarterly schedule to patients. This mode of application stands in stark contrast to the daily oral dosing demanded by gepants [140].

Drawing from extensive clinical trials and nearly a year's presence in the market, CGRP and its receptor antibodies seem to have a good safety profile and are generally well-tolerated. Yet, the ramifications of the prolonged CGRP blockade remain to be understood. A glimmer of optimism emerges from Amgen/Novartis's findings, which indicate that their antibody remained safe up to the three-year mark in an ongoing five-year open-label study [152]. Moreover, an examination focusing on patients diagnosed with angina did not highlight any detrimental effects of the antibody [153]. Still, this study is not without its constraints. It predominantly featured male participants in a disorder that chiefly affects females, roped in patients with stable angina pectoris instead of those with microvascular disease (who would better mirror the vulnerable population), and gauged drug implications rather prematurely before the receptor antibody could adequately attach to the receptor (see Figure 2) [154].

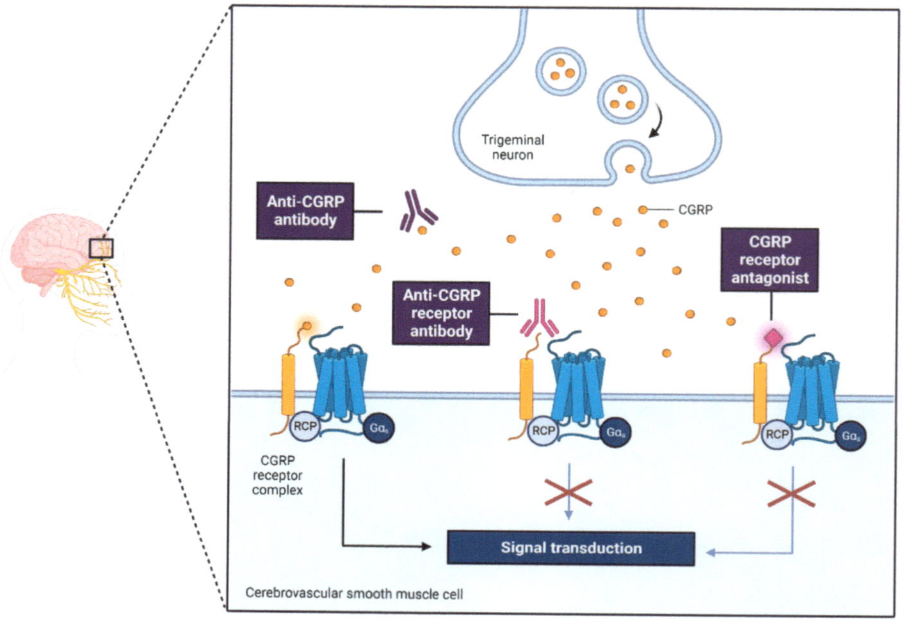

Figure 2. Calcitonin gene-related peptide-targeting drugs for migraine.

Cardiovascular risks loom large, especially considering that migraine patients are identified as high-risk candidates for stroke and cardiovascular complications. The following pressing question arises: might a CGRP blockade amplify the intensity of a stroke episode? A recent case study highlighted a patient who encountered an ischemic incident post the administration of a CGRP antagonist medication [155]. Still, it is pivotal to understand that conclusions can not be hastily drawn from a solitary patient's experience. Rigorous, long-duration studies centered on cardiovascular wellness are imperative. A prudent starting point would be animal-based research endeavors, delving into the impacts of CGRP blockade on ischemic conditions (see Figure 2).

6.4. Medical Treatment of Migraine

6.4.1. Overview of Conventional Medical Treatments for Migraine

1. Medications commonly prescribed for migraine relief—Discussion of their mechanisms of action and limitations (Figure 3)

Therapeutic Interventions for Migraine Cessation

(a) Anti-Inflammatory Agents: NSAIDs and Acetaminophen

Non-steroidal anti-inflammatory drugs (NSAIDs) serve as the primary selection for mitigating the intensity and duration of migraines and are supported by an extensive body of evidence. Various NSAIDs, such as ibuprofen, naproxen sodium, acetylsalicylic acid (ASA), diclofenac potassium, aspirin, tolfenamic acid, piroxicam, ketoprofen, and ketorolac, have all exhibited their efficacy in treating migraine through evidence gleaned from randomized controlled trials and systematic reviews. Acetaminophen, as well as a combination formula of acetaminophen, aspirin, and caffeine, has also displayed significant efficacy in the acute treatment of migraine [161].

Mechanism of Action

NSAIDs primarily act by hindering the synthesis of prostaglandins. They act to reversibly block cyclooxygenase (COX) enzymes 1 and 2. The NSAIDs that inhibit prostaglandin E2 synthesis are particularly effective in mitigating acute migraine attacks. Aspirin, for instance, serves as an irreversible inhibitor of both COX 1 and 2 enzymes.

The complete mode of action of acetaminophen is not fully understood yet; however, it is believed to exert its effects on central processes, like enhancing the serotonergic descending inhibitory pathways. Acetaminophen may also interact with opioidergic systems, eicosanoid systems, and the nitric oxide-containing pathways [162].

(b) Triptans

The U.S. Food and Drug Administration (FDA) has given the green light to a total of seven triptans specifically designed for the immediate relief of migraine episodes. This lineup includes sumatriptan, eletriptan, naratriptan, zolmitriptan, rizatriptan, frovatriptan, and almotriptan. In terms of pricing, triptans tend to be on the higher side compared to NSAIDs. As a result, they are generally selected as a treatment strategy when alternatives like NSAIDs or acetaminophen do not produce the desired outcomes, or when the intensity of the migraine demands their application [158].

Mechanism of Action

Triptans function as serotonin-receptor agonists. They possess high affinity for the 5-HT1B and 5-HT1D receptors, and they have variable affinity for the 5-HT1F receptors. The supposed mode of action involves binding to postsynaptic 5-HT1B receptors located on the smooth muscle cells of blood vessels and to presynaptic 5-HT1D receptors situated on trigeminal nerve terminals and dorsal horn neurons [163].

(c) Antiemetics

Antiemetics are frequently chosen for migraine treatment when symptoms include nausea or vomiting. These medications can be administered either alongside NSAIDs or triptans, or used as monotherapy. Metoclopramide and prochlorperazine are two commonly employed antiemetics. Metoclopramide has the most significant body of evidence supporting its efficacy in treating migraine and is less likely to cause extrapyramidal side effects compared to prochlorperazine. Other antiemetics used for migraine management include domperidone, promethazine, and chlorpromazine [161].

Mechanism of Action

Metoclopramide is a benzamide that antagonizes the D2 receptor at lower doses and the 5HT-3 receptor at higher doses, providing both antiemetic and migraine relief effects. Both prochlorperazine and chlorpromazine function as dopamine antagonists, interacting with the D2 receptor, which helps in mitigating the symptoms of migraine and controlling nausea and vomiting.

(d) Ergotamines

With the advent of triptans, the usage of ergotamines has declined as triptans have shown superior efficacy. Dihydroergotamine has shown some efficacy in treating migraine, while the effectiveness of ergotamine remains unclear. A systematic review revealed that dihydroergotamine was not as effective as triptans; however, when combined with an antiemetic, dihydroergotamine was found to be as effective as ketorolac, opiates, or valproate [164]. Dihydroergotamine might be a useful alternative when patients do not respond to other medications, including triptans.

Mechanism of Action

Ergotamines, similar to triptans, are potent agonists of 5-HT 1b/1d receptors. Their mechanism of action is thought to involve the constriction of the presumed pain-causing intracranial extracerebral blood vessels at the 5-HT1B receptors and inhibition of trigeminal neurotransmission at both peripheral and central 5-HT1D receptors; moreover, they interact with other serotonin, adrenergic, and dopamine receptors. They induce the constriction of peripheral and cranial blood vessels [165].

Interventions Aimed at Migraine Prevention

(a) Beta-Blockers

Beta-blockers, such as propranolol, timolol, bisoprolol, metoprolol, atenolol, and nadolol, have been explored for their prophylactic role in preventing migraine attacks and have shown positive results in clinical studies. However, beta-blockers exhibiting intrinsic sympathomimetic activity, like acebutolol, alprenolol, oxprenolol, and pindolol, do not appear to demonstrate efficacy for migraine prevention [166].

Mechanism of Action

The exact mechanisms underlying the preventive effect of beta-blockers on migraine are not completely understood. One prevailing theory suggests that their migraine prevention abilities may be linked to their beta-1 mediated effects, which inhibit the release of noradrenaline and activity of tyrosine hydroxylase, thereby contributing to their prophylactic action. Other potential mechanisms may involve the serotonergic blockade, inhibiting thalamic activity, and blocking the effect of nitrous oxide.

(b) Antiepileptic Drugs

Several antiepileptic drugs (AEDs) have been investigated for their efficacy in migraine prevention, with topiramate and valproate demonstrating the most substantial evidence of effectiveness [166].

Mechanism of Action

The precise mode of action of antiepileptic drugs in preventing migraine remains elusive. For topiramate, it is known to block several channels, such as voltage-dependent sodium and calcium channels. In addition, it has been shown to reduce glutamate-mediated excitatory neurotransmission, enhance the inhibition mediated by GABA-A, inhibit carbonic anhydrase activity, and decrease CGRP secretion from trigeminal neurons, all of which could potentially contribute to its preventive effects on migraine. Similarly, for valproate, a multifaceted approach is likely involved in preventing migraine. These mechanisms might encompass enhancing GABAergic inhibition, blocking excitatory ion channels, and downregulating the expression of CGRP in brain tissue. These multifactorial actions collectively could underpin the migraine preventive effects of these antiepileptic drugs.

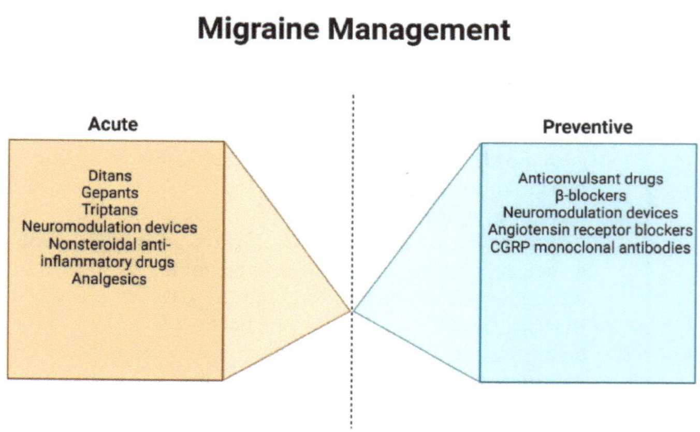

Figure 3. Migraine management.

6.4.2. Emerging Therapeutic Approaches in Migraine Management

1. Introduction to novel medications and treatment strategies and Potential benefits and challenges associated with these approaches

Goals of present research

The goals of present research in migraine treatment are manifold and multifaceted, particularly in the context of elucidating and exploring the role of peptides or their respective receptors that emerge during a migraine episode. Simultaneously, there is an unflagging pursuit of methods to obstruct the activation of the trigeminovascular system and the receptors of the neurotransmitters, which are intricately linked to the cascade of events leading to a migraine. These avenues of exploration are deemed paramount for the inception of innovative, effective, and targeted pharmacological interventions. These treatments are intended to serve both as acute therapeutic options and preventive strategies against migraine.

In light of the challenges and limitations associated with the use of triptans, a class of drugs conventionally employed in the treatment of migraine, there is a concerted effort within research and development programs to discover and develop new acute treatments. The aim here is to ensure these novel treatments are as effective as, if not more so, than triptans, whilst also being better tolerated by patients; furthermore, these treatments should ideally possess a distinctive, migraine-specific neural mechanism of action. An ideal feature of these new treatments would be the ability to avoid manipulating the vascular tone, thereby reducing potential side effects associated with vascular modulation [167].

In pursuit of these objectives, two primary biochemical pathways have been subjected to rigorous scrutiny—the calcitonin gene-related peptide (CGRP) pathway and the serotonin pathway. This intensive research has yielded two promising new classes of drugs, known as the gepants and the ditans, respectively. Both of these drug families are targeted interventions designed to alleviate the symptoms of migraine. In addition to their potential as acute treatments, the studies focusing on the CGRP pathway have also unearthed potential preventative strategies. This has culminated in the development of anti-CGRP monoclonal antibodies, which are being recognized as potential prophylactic treatments for migraine.

These promising new treatments are currently in an advanced stage of development, undergoing rigorous clinical trials. Some of these therapies are anticipated to be launched into the market in the near future, ushering in a new era of migraine management. This continued research endeavors to deliver more personalized and effective therapeutic options, minimizing the distress and impairment that migraine bring into the lives of those afflicted by them [167].

6.5. Competitive Environment

6.5.1. The Role of Calcitonin Gene-Related Peptide in Migraine

Regarded as one of the most potent vasodilators known, the calcitonin gene-related peptide (CGRP) exists in the following two forms within the human body: the alpha-CGRP, a 37-amino acid peptide that is predominantly expressed in primary sensory neurons of the dorsal root ganglia, trigeminal ganglia, and vagal ganglia; and the beta-CGRP, which is found primarily within intrinsic enteric neurons. The ubiquity of CGRP is seen in its widespread distribution across the cerebral and cerebellar cortex, thalamus, hypothalamus, inhibitory nociceptive nuclei of the brainstem, trigemino-cervical complex, and the trigeminovascular system [168].

Within the trigeminal ganglia, CGRP is found in cells that generate thinly myelinated A-delta fibers as well as unmyelinated C-fibers. Receptors for CGRP are present within the cortical and subcortical structures mentioned earlier. On trigeminal fibers, these receptors operate as autoreceptors, thereby governing CGRP release. Elevated levels of CGRP are detected during migraine attacks, although some research suggests conflicting evidence [167].

Notably, an intravenous infusion of CGRP was found to induce migraine attacks in about 60% of the patients studied. Interestingly, patients diagnosed with familial hemiplegic migraine, a rare type of migraine accompanied by aura, showed no sensitivity to CGRP. This could potentially be attributed to alterations in the levels of CGRP within their trigeminal system [169].

Experimental activation of trigeminal ganglion cells has been found to cause the release of CGRP. This release is inhibited in a dose-dependent manner by 5-HT1B/D agonists, underscoring the importance of the trigeminal system as a potential target for CGRP receptor antagonists and triptans. CGRP, in addition to its vascular effects, has emerged as a key regulator of neuronal function, significantly influencing neurotransmitter systems like the glutamatergic system [170].

Drawing from these findings, drugs that modulate CGRP activity have shown promise in the future treatment of migraine. These include CGRP receptor antagonists, which compete with the body's naturally occurring CGRP at receptor binding sites and have been shown to be effective in the treatment of acute migraine attacks. Other approaches to modulate CGRP activity, such as the development of monoclonal antibodies against CGRP and the CGRP receptor, have been introduced recently [167].

6.5.2. CGRP Receptor Antagonists (The Gepants)

CGRP receptor antagonists, referred to as gepants, are small compounds that vie with the body's endogenous CGRP for receptor binding sites. The ability of these CGRP receptor antagonists to cross the blood–brain barrier remains uncertain. Despite their promise, the development journey of new emerging CGRP antagonists has been beset by challenges. Initial antagonists, such as olcegepant (BIBN4096BS), telcagepant (MK-0974), and MK-3207, demonstrated efficacy as acute treatments for migraine. However, they were burdened by unfavorable safety profiles.

In an initial proof-of-concept study in acute migraine treatment, the intravenously administered olcegepant at a dosage of 2.5 mg significantly outperformed placebo at a 2-h response rate, marking a potential breakthrough in acute migraine treatment [171]. Following this, telcagepant, an orally administered CGRP receptor antagonist, underwent a phase II proof-of-concept study that showcased the efficacy of a 300–600 mg dose. However, when telcagepant was tried as a daily preventive migraine treatment, it led to liver enzyme derangement, leading to the discontinuation of trials [172]. MK-3207, the third oral CGRP receptor antagonist developed and tested in migraine, showed superiority to placebo above the dose of 10 mg in 2-h pain freedom, but was discontinued due to liver toxicity issues.

A promising CGRP receptor antagonist, BI 44370 TA, was utilized in a phase II study to evaluate its safety, tolerability, and efficacy in the treatment of an acute migraine attack in episodic migraine sufferers [173]; however, studies on this agent have been discontinued as well.

6.5.3. Ubrogepant (MK-1602)

Ubrogepant (MK-1602) is a novel oral CGRP receptor antagonist that is chemically distinct from both telcagepant and MK-3207. The safety and efficacy of ubrogepant at varying doses was evaluated in a Phase IIb, multicenter, randomized, double-blind, placebo-controlled trial. Results showed a positive trend across all doses of ubrogepant for the 2-h pain freedom endpoint. Importantly, there were no observed post-treatment elevations of ALT > 3 ULN hand and no other abnormal laboratory values of clinical relevance, as found with the earlier CGRP antagonists. The success of this study led to the initiation of phase 3 clinical trials. Positive preliminary efficacy and safety results of two phase III multicenter randomized, double-blind, placebo-controlled clinical trials comparing ubrogepant 50 mg and 100 mg versus placebo (Achieve 1) and ubrogepant 25 mg and 50 mg versus placebo (Achieve 2) were recently presented at the American Headache Society (AHS) conference [174]. Despite these promising findings, further data on consistency of effect and safety in patients for whom triptans are contraindicated are needed to solidify its role as a viable alternative to triptans.

6.5.4. Rimegepant (BMS-927711): Overview and Clinical Trials

Rimegepant is a pioneering and unique calcitonin gene-related peptide (CGRP) receptor antagonist that bears a distinct chemical structure from telcagepant. In the sphere of migraine treatment, rimegepant's effectiveness and safety have been evaluated in a rigorous phase II clinical trial. This trial was double-blind, randomized, placebo-controlled, and dose-ranging, involving a total of 885 participants.

The participants of the study were allocated randomly to receive one of six doses of BMS-927711 (10 mg, 25 mg, 75 mg, 150 mg, 300 mg, or 600 mg), sumatriptan (100 mg), or a placebo for the treatment of moderate to severe migraine attacks. The study was designed with the primary endpoint of achieving pain freedom two hours after the dose. Secondary endpoints were more comprehensive, including an endpoint consisting of the absence of headache pain, as well as a lack of symptoms, such as photophobia, phonophobia, and nausea, at two hours post-dose.

Along with these endpoints, the trial studied several other secondary efficacy and safety measures. Interestingly, a higher proportion of participants who received rimegepant 150 mg achieved the primary endpoint of being pain-free at two hours, amounting to 32.9%, which was significantly higher than the percentage observed for other doses of rimegepant ($p < 0.001$). For instance, the respective proportions were 31.4% for the 75 mg dose, 29.7% for the 300 mg dose, and 15.3% for the placebo group. Sumatriptan 100 mg proved superior to all doses of rimegepant, with a success rate of 35%.

With regards to the secondary efficacy endpoint of total migraine freedom, the dose of rimegepant 75 mg was the most effective, with a success rate of 28.2%. This dose was statistically superior to the placebo. However, sumatriptan 100 mg outperformed each dose of rimegepant at this secondary endpoint. The trial also reported that the proportion of patients who were headache-free for up to 24 h after dosing was higher for several doses of rimegepant and for sumatriptan compared to the placebo group.

In terms of safety, most adverse events (AEs) were mild to moderate, and none of the patients had to discontinue due to AEs. Two patients experienced increased hepatic enzymes reported as an adverse event, one in the rimegepant group and the other in the placebo group.

The results of this trial suggest that rimegepant's effectiveness is similar to sumatriptan 100 mg in treating migraine attacks, but with potentially fewer triptan-related side effects, such as paresthesia and chest discomfort. A phase III trial comparing the efficacy of rimegepant 75 mg with a placebo has recently been concluded. Moreover, an ongoing prospective multicenter open-label long-term safety study is expected to complete recruitment by late 2019. These two studies will contribute to understanding the consistency and safety of rimegepant in migraine therapy [160].

6.5.5. Atogepant (AGN-241689): The Future of Migraine Prevention

Atogepant, a small molecule with a distinct structure similar to that of ubrogepant, is the only CGRP receptor antagonist currently under investigation for migraine prevention. Its higher potency and longer half-life compared to ubrogepant make it suitable for preventive treatment.

The safety, efficacy, and tolerability of atogepant were evaluated in a phase II/III multicenter, randomized, double-blind, placebo-controlled, parallel-group study (NCT02848326). In this trial, adult patients were randomized to receive placebo, 10 mg QD, 30 mg QD, 30 mg BID, 60 mg QD, and 60 mg BID, respectively, and were treated for 12 weeks for the prevention of episodic migraine.

The primary efficacy endpoint was the change from baseline in mean monthly migraine/probable migraine headache days across the 12-week treatment period. All active treatment groups showed a statistically significant reduction from baseline in the primary efficacy parameter. In terms of safety, atogepant was well tolerated, with the most common adverse events being nausea, fatigue, constipation, nasopharyngitis, and urinary tract infection. The liver safety profile for atogepant was similar to placebo, with no indications of hepatotoxicity with daily administration over 12 weeks. The development program of this treatment is set to advance to the next stage [167].

6.5.6. Anti-CGRP and Anti-CGRP Receptor Monoclonal Antibodies

The CGRP pathway has also been targeted through the development of antibodies against CGRP and the CGRP receptor. This marks the first use of engineered antibodies in the field of migraine, and their development was initially met with some skepticism.

These monoclonal antibodies (mAbs), including galcanezumab (LY2951742), a fully humanized mAb anti-CGRP, fremanezumab (TEV-48125), a fully humanized mAb anti-CGRP, and eptinezumab, a genetically engineered humanized anti-CGRP antibody, target the ligand to prevent the binding of CGRP to its receptor. Erenumab (AMG 334) is a fully humanized mAb that targets the CGRP receptor.

These compounds have a favorable pharmacological profile that includes a long half-life, the absence of a vasoconstrictive effect or other significant hemodynamic changes [80]. Because of their high molecular weight, they do not cross the blood–brain barrier, suggesting a reduced likelihood of central nervous system-related side effects, which are commonly observed with current pharmacological prophylaxis treatments used in migraine.

Further enhancing their potential for long-term patient compliance, these mAbs can be administered either subcutaneously (sc) or intravenously (IV) at different rates ranging from once every three months to twice a month, depending on the compound.

In a series of methodologically similar randomized, double-blind, placebo-controlled Phase II and III clinical trials, the efficacy and safety of these novel treatments in the prevention of episodic and chronic migraine (CM) were explored [167].

6.5.7. Erenumab (AMG334)

Erenumab, marketed as Aimovig, is a novel migraine prophylactic belonging to the monoclonal antibody class of drugs. It operates by specifically targeting and blocking the calcitonin gene-related peptide (CGRP) receptor, a crucial element in the neurochemical pathway believed to play a role in migraine pathogenesis. The initial efficacy of erenumab was established through a phase II trial where participants suffering from episodic migraine received monthly doses of 70 mg over three months. The study outcomes exhibited a promising reduction in the number of monthly migraine days by 3.4 days compared to placebo [167].

This was further substantiated in the STRIVE study, a multicenter, phase III trial, which compared two doses of erenumab (70 mg and 140 mg) with a placebo over six months. The primary objective was to assess the change in the mean number of migraine days per month, while secondary objectives were to evaluate reductions in the severity of migraine and their impacts on physical function and daily activities. Both doses showed statistically significant reductions in the number of migraine days and severity compared to placebo. Notably, patients' disability scores, measuring the impact of migraine on daily activities, were also significantly improved [175].

A similar study, the ARISE trial, evaluated only the 70 mg dose of erenumab and confirmed its superior efficacy compared to placebo; however, unlike the STRIVE study, it did not reveal a significant improvement in the migraine disability scores. In both trials, erenumab demonstrated an excellent safety and tolerability profile with common side effects being minor, such as upper respiratory tract infection, injection site pain, and nasopharyngitis.

The long-term safety and efficacy of erenumab were further evaluated in a 5-year open-label extension of the phase II clinical trial. Here, participants with a history of inadequate response to up to two previous preventive treatments received erenumab 70 mg. This study demonstrated a reduction of 5 migraine days on average from an initial baseline of 8.8 migraine days per month. In addition, significant improvements were observed in disability and quality of life scores, indicating the long-term benefit of erenumab treatment [176].

Erenumab's performance was also assessed in chronic migraine patients in a randomized, double-blind, placebo-controlled phase II clinical trial. The participants received subcutaneous injections of either placebo, erenumab 70 mg, or erenumab 140 mg every

4 weeks for 12 weeks. Erenumab significantly reduced the number of monthly migraine days and monthly acute migraine treatments compared to placebo, thereby validating its preventive role in chronic migraine management [177]. As a result of these studies, erenumab was granted FDA approval in May 2018 for the prevention of migraine in adults.

6.5.8. Galcanezumab (LY2951742)

Like erenumab, galcanezumab is a humanized monoclonal antibody but differs in its mechanism. It inhibits the activity of CGRP by binding to the ligand itself rather than the receptor. The phase II proof-of-concept trials conducted in episodic migraine patients showed a mean reduction of 4.2 monthly migraine days with galcanezumab (150 mg) compared to a reduction of 3.0 days in the placebo group. The most commonly reported adverse events were erythema, upper respiratory tract infections, and abdominal pain [178].

A subsequent phase IIb clinical trial assessed the superiority of galcanezumab at varying doses (5, 50, 120, 300 mg) administered subcutaneously monthly for three months compared to placebo. The primary outcome was the mean change in migraine days from week 9 to 12 post-randomization. The 120 mg dosage significantly reduced migraine headache days compared with placebo [179].

Two phase III trials, EVOLVE-1 and EVOLVE-2, confirmed the efficacy of galcanezumab in reducing migraine days and improving the quality of life. Participants received monthly subcutaneous injections of galcanezumab at doses of 120 and 240 mg versus placebo for 6 months. Both studies met the primary and secondary efficacy endpoints at 6 months for both doses. The REGAIN study (NCT02614261), a 3-month double-blind study with a 9-month open-label extension for preventing migraine in chronic migraine patients, echoed the previous findings. The study met the primary endpoint at 3 months, showing that galcanezumab was significantly more effective in reducing monthly migraine days than placebo. Notably, it also showed a higher incidence of injection site reactions, erythema, and sinusitis in the galcanezumab groups [180]. These results highlight the promising role of galcanezumab in managing migraine, with long-term safety and efficacy data eagerly awaited.

6.5.9. Examination of Fremanezumab (TEV48125)

Fremanezumab, a fully humanized monoclonal antibody that targets the calcitonin gene-related peptide (CGRP), has been investigated for its potential use in the prevention of migraine. Its development was initially spurred by promising results from phase I studies, echoing the progress of other monoclonal antibodies against CGRP. To evaluate its efficacy and safety, a phase IIb study was carried out. The experimental protocol included a multicenter, randomized, double-blind, placebo-controlled design where participants were assigned to receive either 225 mg or 675 mg of subcutaneous (sc) TEV-48125 or a placebo. This was to be administered every 28 days for three months.

The primary objective of the study was to observe the mean decrease from the baseline in the number of migraine days during the third treatment cycle (weeks 9–12). In addition to this, safety parameters were also evaluated. In post-hoc analyses, investigators assessed the percentage of participants who achieved at least a 50% and 75% decrease in the number of migraine days relative to baseline. Both dosage levels of TEV48125 were found to meet the primary efficacy outcome, and no issues regarding safety or tolerability were identified. The most common adverse events associated with the treatment were mild pain or erythema at the injection site.

Following this, a phase III trial was conducted to further investigate the preventative effects of TEV-48125 (fremanezumab) in episodic migraine. This study was a randomized, double-blind, placebo-controlled, parallel-group trial that tested monthly sc fremanezumab injections of 225 mg or 675 mg following a quarterly dose regimen, against a placebo. The primary outcome was the mean change in the number of monthly migraine days per month over a 12-week period. Secondary efficacy endpoints included the proportion of patients achieving at least a 50% reduction in the mean number of monthly migraine days from baseline to week 12, along with changes in migraine-related disability scores [181].

Both the monthly and quarterly regimens of fremanezumab met the primary efficacy endpoint, showing superiority over the placebo in reducing mean migraine days. No significant difference was found between the two fremanezumab regimens. Adverse events were most commonly injection site reactions, but the proportion of participants who discontinued due to these events was small (2%). This study also shed light on the possibility of using a single dose therapy given quarterly for migraine prevention, which is significant as the results were similar to those from monthly injections. This may open up new potential for multi-injection regimens in migraine prevention [182].

TEV-48125 (fremanezumab) was also tested for chronic migraine (CM) in a multicenter, randomized, double-blind, placebo-controlled, phase IIb study. Here, TEV-48125 was administered at different doses from those used in the episodic migraine trials as follows: 675 mg in the first treatment cycle and 225 mg in the second and third treatment cycles, or 900 mg monthly for three months, against a placebo. The efficacy endpoints for this trial varied from those of the episodic migraine studies. The primary outcome was the change from baseline in the number of headache-hours during the third treatment cycle (weeks 9–12), while the secondary endpoint was the change in the number of moderate or severe headache days. Both doses showed a significant reduction in the number of headache-hours compared to placebo and a significantly greater reduction in mean number of headache days [183].

After the promising results of the phase II study, a randomized, double-blind, placebo-controlled, parallel-group trial was conducted to further confirm the efficacy of fremanezumab for CM prevention. Participants with CM were randomized to receive fremanezumab either quarterly (a single dose of 675 mg at baseline and placebo at weeks 4 and 8) or monthly (675 mg at baseline and 225 mg at weeks 4 and 8) or a placebo. The primary endpoint was the mean change from baseline in the average number of headache days, as defined by the International Headache Society (IHS). Both doses met the primary endpoint and showed a significantly greater percentage of participants obtaining at least a 50% reduction in headache days compared to placebo.

Fremanezumab was found to be associated with a higher incidence of injection-site reactions than placebo, but the severity of these reactions did not significantly differ among the trial arms. Following these studies, fremanezumab (Ajovy) received FDA approval on 14 September 2018, making it the second anti-CGRP monoclonal antibody approved for preventing migraine in adults. Notably, it became the first drug of its kind to offer both quarterly and monthly dosing options [184].

6.5.10. Eptinezumab (ALD403)

Eptinezumab (ALD403) is an innovative product of genetic engineering, specifically, a humanized antibody that is designed to target both isoforms of human CGRP. CGRP, a molecule implicated in the pathophysiology of migraine, serves as a significant focal point for novel migraine therapeutics, including eptinezumab (see Figure 4) [65].

This novel therapeutic agent has been put through a phase II proof-of-concept study to evaluate its potential in managing episodic migraine. This study was designed with the primary objective of ascertaining the safety of eptinezumab following the intravenous administration of a single 1000 mg dose. As secondary objectives, the researchers investigated efficacy outcomes and gauged the extent of disability induced by migraine at the 12-week mark following infusion. They were particularly interested in discerning changes in the frequency of migraine days from the baseline to weeks 5–8.

Participants in the trial experienced an average of 8.4–8.8 migraine days per month at baseline, and adverse events were reported by 57% of eptinezumab recipients, compared to 52% in the placebo group. The adverse events included conditions like upper respiratory tract infection, urinary tract infection, fatigue, back pain, nausea, vomiting, and arthralgia, most of which were of mild to moderate severity. Importantly, none of the serious adverse events reported were linked to the study drug.

When it came to efficacy, eptinezumab outperformed the placebo by achieving a statistically significant reduction in the mean number of migraine days from the baseline to weeks 5–8. The results also underscored a significant finding—a large proportion of participants receiving eptinezumab achieved a 50% reduction in migraine days at weeks 5–8. Interestingly, this trial observed a high placebo response rate, which might have been influenced by the intravenous administration of the drug, a departure from the delivery methods used in previous trials with anti-CGRP monoclonal antibodies.

Motivated by these promising findings, a phase III study was designed. Titled PROMISE 1, this randomized, double-blind, placebo-controlled trial sought to evaluate the efficacy and safety of various doses of eptinezumab in participants suffering from episodic migraine. The primary endpoint of this study was to identify changes in the mean frequency of migraine days over weeks 1–12, relative to a 28-day baseline period. In this study, participants were randomized to receive one of three doses of eptinezumab (300 mg, 100 mg, or 30 mg) or a placebo. The medication was administered via intravenous infusion every 12 weeks.

In the PROMISE 1 study, all doses of eptinezumab met the primary efficacy endpoint, demonstrating a significantly greater reduction in migraine days compared to the placebo group. Further, a significantly greater proportion of participants administered with eptinezumab experienced a 50% reduction in migraine days, providing more evidence for the drug's efficacy. The study did not find any significant safety issues, further supporting the suitability of eptinezumab as a therapeutic option.

Several studies with anti-CGRP monoclonal antibodies, including eptinezumab, have highlighted a remarkably rapid response to the active drug, usually noticeable within the first month following administration. Eptinezumab, in particular, demonstrated the ability to reduce migraine from the very first day post-administration and maintained similar improvement levels at 4- and 12-weeks post-infusion.

The evaluation of eptinezumab's efficacy, safety, and tolerability continued with phase II and phase III trials in patients with chronic migraine (CM). In a randomized, double-blind, placebo-controlled phase II study, different doses of eptinezumab were administered to test their effects against a placebo. Unique to this study, the primary endpoint was the percentage of patients achieving a 75% reduction in migraine days per month from baseline to week 12. Eptinezumab doses of 300 mg and 100 mg significantly outperformed the placebo group in achieving this endpoint.

These results paved the way for a phase III trial, known as PROMISE 2, to further investigate the safety and efficacy of eptinezumab for the prevention of chronic migraine. In this study, patients received either eptinezumab (300 mg or 100 mg) or a placebo, administered via infusion every 12 weeks. This study found that both doses of eptinezumab outperformed the placebo in reducing the mean number of monthly migraine days during the 12-week, double-blind treatment period. Notably, as early as day one post-infusion, a significant percentage of patients receiving eptinezumab demonstrated a reduction in migraine prevalence that was sustained through day 28.

Taken together, these studies affirm the potential of eptinezumab as a promising, efficacious, and well-tolerated therapeutic option for the management of episodic and chronic migraine. The drug's safety profile aligns well with that observed in previous studies, further bolstering its candidacy as an effective addition to the array of available migraine therapeutics [185].

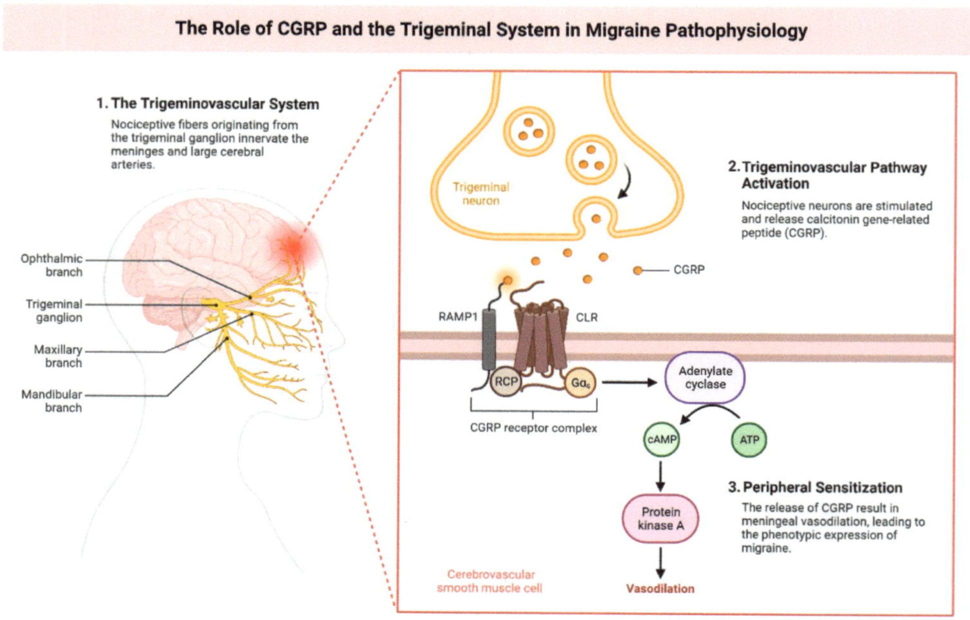

Figure 4. The role of CGRP and the trigeminal system in migraine pathophysiology.

6.6. Neuromodulation in Migraine

6.6.1. Definition and Concept of Neuromodulation in Migraine Therapy

1. Explanation of neuromodulation as a treatment modality

Neuromodulation is an innovative biomedical technique that manipulates or modulates the activities of the central or peripheral nervous system through the use of various stimuli such as electrical, magnetic, or chemical agents [186]. The technique hinges on the targeted delivery of these stimuli to specific neural sites within the body, with the express aim of altering nerve activity. Fundamentally, neuromodulation is non-destructive, reversible, and adaptable, highlighting its potential as a safe and effective approach for therapeutic interventions.

Recently, there has been an increased recognition of neuromodulation as a potentially superior treatment for migraine, compared to traditional medication therapies. The efficacy and safety of this technology are driving a paradigm shift in how both clinicians and patients approach the management of migraine. Instead of relying solely on pharmacological interventions, there is a growing interest in non-invasive neuromodulation therapies.

Among these therapies, non-invasive vagus nerve stimulation (nVNS) and single-pulse transcranial magnetic stimulation have gained notable attention due to their demonstrated effectiveness and safety profiles. These methods represent significant advancements in neuromodulation and are changing the landscape of migraine management [187].

In the modern healthcare context, neuromodulation technology holds substantial promise, particularly for vulnerable patient populations. For instance, expectant mothers, who need to avoid certain medications due to potential harm to the fetus, and patients who struggle with tolerating medications or find them ineffective, may greatly benefit from non-invasive neuromodulation therapies. The potential benefits extend not just to the realm of improved health outcomes, but also to the sphere of healthcare economics.

In specific circumstances, non-pharmacological neuromodulation techniques may prove to be a cost-effective alternative to traditional treatment modalities [188]. By reducing dependency on medications, these techniques can help circumvent the long-term costs associated with drug therapy, such as costs related to side effects and long-term use. Furthermore, as these therapies improve in effectiveness and efficiency, they may help reduce the indirect costs of migraine, such as lost productivity and reduced quality of life.

In conclusion, neuromodulation represents a pioneering field in neuroscience that has the potential to revolutionize the treatment of migraine and other neurological disorders. Through the targeted use of electrical, magnetic, or chemical stimuli, this technique can modulate nerve activity in a way that is non-destructive, reversible, and adaptable. As research in this area continues to unfold, the healthcare industry will likely see an increasing shift toward these non-invasive, cost-effective, and patient-friendly treatment options.

2. Overview of different types of neuromodulation techniques

Neuromodulation operates through the application of electrical or magnetic pulses to interact with or stimulate the central or peripheral pain pathways. This technique essentially targets the pain mechanisms in the human body with the aim of reducing the intensity of experienced pain. The application of electrical or magnetic stimuli can alter central neurotransmitters when they engage with pain circuits [189]. These modifications, in the context of treating acute migraine attacks, may potentially halt the processes that lead to the onset of an attack. For preventative purposes, these neuromodulatory changes aim to lessen the central sensitization that culminates in chronic headaches.

Traditional neuromodulation techniques engage the neurological system either centrally or peripherally through the skin, employing a changing magnetic field or an electric current to manipulate the mechanisms associated with headache-related pain. Both modes of delivery demonstrate rapid effectiveness, making them suitable for addressing acute symptoms; moreover, sustained application of these modalities might confer long-term preventative benefits [190]. Figure 5 offers a visualization of various neuromodulation techniques alongside their respective targets or sites of action.

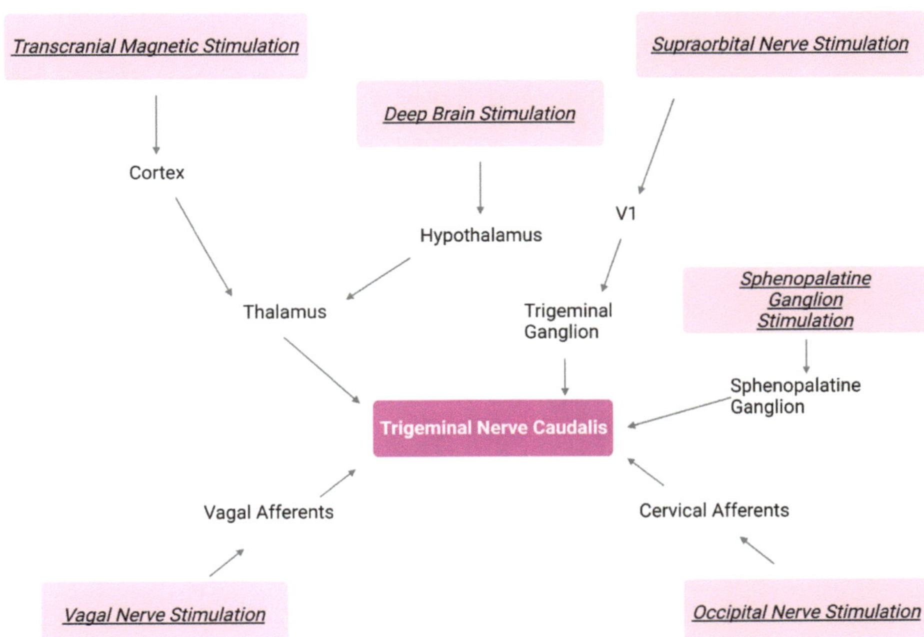

Figure 5. Several neuromodulation techniques and their respective site of action or targets.

Neuromodulation is implemented via a device that modifies brain cell activity utilizing electrical or magnetic stimulations. These devices exhibit diversity in their modes of operation. Some are designed to halt attacks, while others are utilized for preventative purposes. The commonality among them lies in the principle of altering the activity within nerve pathways. These devices, alternatively referred to as stimulators, can be categorized based on their operational parameters. Neuromodulation devices may employ magnetic, electrical, or temperature-changing stimuli. They may be invasive or non-invasive, and their design may range from portable, easy-to-use devices to those that necessitate surgical placement [187].

In the realm of neuromodulation, different techniques and devices offer a range of options to tailor treatments according to the patient's specific needs. As our understanding of the mechanisms behind migraine deepens, the applications of neuromodulation will continue to evolve, potentially offering more effective and personalized treatments for patients suffering from migraine and other neurological conditions.

3. Neuromodulation Modalities

With the emergence of neuromodulation and the recognition of its potential for preventative treatment in chronic pain conditions, such as migraine and cluster headaches, there has been significant development in the field of non-invasive neuromodulation techniques. These advancements offer alternative treatments that pose minimal risk to the patient while maximizing the potential for pain management and relief. Not only have several of these techniques successfully passed through clinical trial phases, but they have also found their way into the marketplace where they are actively being utilized for patient treatment.

Simultaneously, numerous other neuromodulation techniques are currently in various stages of clinical trials, showing the ongoing growth and exploration in this sector of medicine. These techniques, aimed at managing acute migraine attacks and chronic pain, are progressing towards market utilization, continually broadening the treatment options available for these conditions [191].

Figure 6 offers a visual representation of the different non-invasive neuromodulation techniques employed in the management of acute migraine episodes. This visual aid helps provide a better understanding of the variety and extent of non-invasive techniques available, each with its distinct advantages and specific use-cases, contributing to the expansive repertoire of neuromodulation methods.

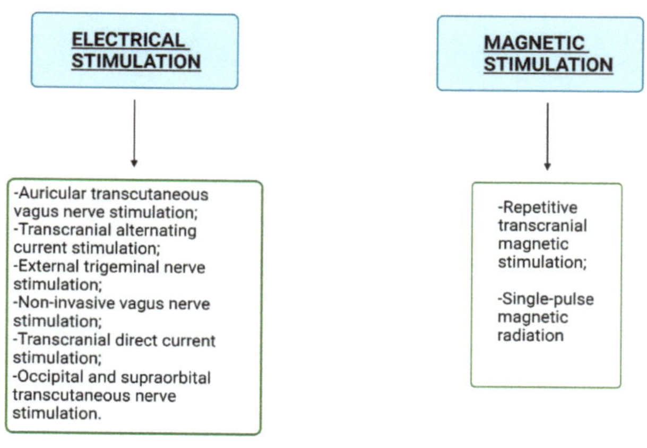

Figure 6. Different non-invasive neuromodulations in the management of acute attacks of migraine.

The evolution and diversity in neuromodulation techniques symbolize the adaptability and progressive nature of contemporary medicine. Through continuous research and development, these techniques are becoming increasingly refined, potentially providing better, more targeted treatment for patients suffering from acute migraine and other chronic pain conditions. It signifies a remarkable shift in our approach to pain management, with techniques that prioritize patient safety and comfort while not compromising on effectiveness [187].

6.6.2. Effectiveness of Neuromodulation in Managing Migraine: Review of Clinical Studies on the Use of Neuromodulation for Migraine—Assessment of the Benefits and Limitations of Neuromodulation

1. Vagus Nerve Stimulation

The gammaCore Sapphire™, crafted by electroCore, Inc. (Basking Ridge, NJ, USA), is a handheld non-invasive vagus nerve stimulator (nVNS) designed for transcutaneous application to the vagus nerve on either side of the neck. The FDA first granted clearance to this device in April 2017, initially for the treatment of acute pain associated with episodic cluster headaches in adults. Now, it has expanded FDA approval, covering both the acute and preventive treatment of migraine-related pain for individuals aged 12 or older.

Four major nVNS trials have been conducted concerning migraine, consisting of two studies for episodic migraine (EM), one tailored for chronic migraine (CM), and another one incorporating CM. The EVENT study, which was the first nVNS trial for CM, primarily sought to assess safety and tolerability in a pilot feasibility format. This randomized double-blind sham-controlled research focused on CM prevention. Though the efficacy endpoints (a change in monthly headache days after 2 months; −1.4 vs. −0.2, $p = 0.59$) did not show statistical significance, there were hints of potential benefits with prolonged use. After an 8-month treatment period, 15 subjects witnessed a mean change of −7.9 (95% CI −11.9 to −3.8, $p < 0.01$) [186].

Another noteworthy trial was the PREMIUM II study, a randomized sham-controlled double-blind study that included both CM and EM subjects. Unfortunately, it was cut short due to the COVID-19 pandemic. With a total of 300 participants enrolled and a modified intention-to-treat (mITT) subgroup of 113 subjects analyzed, the PREMIUM II study discovered a non-significant reduction in monthly migraine days (verum vs. sham: −3.1 vs. −2.3 headache days, $p = 0.233$) in the mITT cohort. However, 44.9% of verum-treated individuals experienced at least a 50% decrease in migraine days, in contrast to 26.8% in the sham group ($p = 0.048$) [187].

Clinical trials have demonstrated that the gammaCore Sapphire™ is safe and generally well-tolerated, with no significant treatment-linked adverse events. Some common side effects reported in the CM trials were facial pain, gastrointestinal symptoms, and upper respiratory tract infections. While the preliminary data indicate that nVNS may hold promise for treating CM, further exploration and evidence are needed to solidify these findings.

2. Remote Electronic Neuromodulation

Nerivio® (Theranica Bio-Electronics Ltd., Montclair, NJ, USA) is a remote electronic neuromodulation (REN) device that has been FDA-approved for the acute treatment of migraine (both EM and CM) in patients aged 12 years and older. The device comprises an armband that emits electronic stimuli controlled by a smartphone app. It is recommended that patients begin use within 60 min of the onset of a migraine or aura. The stimulation lasts for 45 min and can be adjusted for intensity by the individual user via the app, which also features a migraine diary for logging headaches and usage sessions.

Following two randomized double-blind sham-controlled trials that led to the initial FDA clearance of Nerivio for use in EM in 2019, two subsequent open-label observational studies (TCH-005, TCH-006) contributed to the FDA clearance for CM in 2020. In the TCH-005 study, 42 subjects were enrolled, with 210 evaluable treatments carried out by 38 participants. The TCH-006 study evaluated a total of 493 evaluable treatments from 91 participants out of the 126 enrolled [192].

The device-related adverse events were generally related to topical peripheral sensations such as warmth, itching, arm pain, redness, and numbness.

3. Electrical Stimulation of the Trigeminal Nerve

The CEFALY DUAL device, a next-generation evolution of the original electrical trigeminal nerve stimulation (eTNS) unit designed by CEFALY-Technology (Seraing, Belgium), received the stamp of approval from the FDA for the acute as well as preventive management of migraine in adults. As of late 2020, this device is available over-the-counter. CEFALY DUAL offers two customizable settings—an ACUTE program, which utilizes a 100 Hz frequency over a 60 min session for instant relief, and a PREVENTIVE program, which operates at a lower frequency of 60 Hz for 20 min daily to thwart the onset of migraine [193,194].

The therapeutic potential of the CEFALY device for both acute and preventive treatment of migraine is supported by two randomized controlled trials, with one trial specifically mentioning CM [8]. The ACME study, a randomized double-blind sham-controlled trial involving patients with migraine (likely including but not specifying CM), revealed that using CEFALY during the onset of a headache provided more significant pain reduction compared to the sham group (-3.46 ± 2.32 vs. -1.78 ± 1.89; $p < 0.001$). The study likely comprised mostly patients with episodic migraine (EM), as those who had used Botox in the previous four months were excluded [193,194].

Three open-label observation studies, two involving CEFALY and one using supraorbital transcutaneous electrical nerve stimulation (TENS), have specifically targeted CM. In these studies, the CEFALY device was found to decrease monthly migraine days and acute medication consumption. The adverse events were minimal and reversible, with the most common being paresthesia (2.03%), changes in arousal (mostly fatigue, sometimes insomnia, 0.82%), headache (0.52%), and skin allergy to the electrode (0.09%) [13]. These findings suggest that eTNS could be beneficial for CM treatment [192].

4. Single-Pulse Transcranial Magnetic Stimulation

eNeura Inc.'s sTMS mini™, hailing from Baltimore, MD, USA, is an innovative device that administers single-pulse stimulation to the rear of the head. This stimulation method is believed to modify the excitability of the cerebral cortex by halting cortical spreading depolarization waves and curbing thalamocortical signaling [190]. Initially conceived for episodic migraine patients exhibiting aura, the device has secured FDA clearance for both the acute and prophylactic treatment of migraine in individuals aged 12 years and up. The preventive regimen necessitates a twice-daily treatment involving 4 pulses (a pair of consecutive pulses, a pause of 15 min, followed by another 2 pulses). For acute treatment, the procedure entails three successive pulses when a migraine begins, with an option to administer more pulses at quarter-hour intervals if required [191].

The FDA endorsement of this device came on the heels of a randomized, double-blind, sham-controlled trial. Following this, an open-label observational investigation, dubbed the ESPOUSE study, was conducted. This study saw participation from 13 (amounting to 10%) chronic migraine sufferers who used the device daily for both preventive measures (the aforementioned 4 pulses administered twice daily) and acute relief (3 pulses with the possibility of repetition every 15 min for two subsequent sessions). Although the specific effects of daily sTMS application on chronic migraine were not delineated in detail, the research indicated that sTMS effectively slashed monthly headache days, reduced the need for acute medication, and brought down the Headache Impact Test-6 (HIT-6) score. The treatment regimen involving sTMS was generally well-received by participants. The predominant side effects cataloged were sensations of lightheadedness, tingling, and the occurrence of tinnitus [191].

5. Combined Occipital and Trigeminal Nerve Stimulation

The Relivion® is a user-operated stimulation device that secured FDA clearance in the early months of 2021. It is adept at delivering electrical surges to six branches spanning both the occipital and trigeminal nerves. The device comes equipped with pre-set settings, allowing for six treatment cycles (providing uninterrupted stimulation for a 48-h window) as and when needed, primarily for delivering swift relief from migraine episodes. The green light for its market entry was largely due to findings from a widespread study dubbed the RIME study. This investigation involved 131 episodic migraine sufferers who underwent stimulation for a one-hour period [187].

A look back at previous studies revealed that the concurrent stimulation of both occipital and trigeminal nerves, executed via an implanted mechanism, yielded positive results both in the immediate aftermath and over extended periods. Specifically, 75% (or 4 out of 16) of the subjects with stubborn CM observed short-term benefits, while half of the participants (8 out of 16) reported sustained relief [192]. These initial findings hint at the potential of Relivion® as a viable therapeutic avenue for CM. However, a more comprehensive dataset is required to substantiate these preliminary yet promising outcomes.

7. Investigational Devices

7.1. Transcranial Direct Current Stimulation (tDCS)

Transcranial direct current stimulation (tDCS) employs a gentle current, typically ranging from 1 to 2 milliamperes, delivered through sponge electrodes arranged in a variety of montages. This technology has been investigated as a potential treatment for headache disorders. Despite the exact mechanism of action remaining unclear, it is believed that tDCS may influence network-level neural information processing without directly affecting neural spiking or membrane potential. This process could potentially modify brain connectivity, thereby enhancing placebo effects while reducing nocebo effects. However, it is important to note that many of the tDCS trials for migraine were pilot studies of low to moderate quality. These trials have employed varying stimulation duration, current ampere, polarity, montage, and the number of sessions, thus, a universally optimized stimulation protocol has not yet been established [192].

In recent years, spanning half a decade, four rigorously designed, sham-controlled trials have honed in on transcranial direct current stimulation (tDCS) as a possible treatment route for chronic migraine (CM). These studies, distinct in design, likely adhered to a single-blind methodology. Complementing these, two studies adopting an open-label approach were undertaken. Noteworthy findings were presented by Andrade and team, pinpointing that anodal stimulation, targeted either at the left primary motor area (M1) or the dorsolateral prefrontal cortex (DLPFC), as opposed to sham procedures, significantly diminished HIT-6 scores and alleviated pain intensity [193]. In a contrasting study spearheaded by Dalla Volta and colleagues, it was discovered that cathodal stimulation, specifically targeting the coolest forehead point, surpassed sham stimulation in terms of reducing monthly headache occurrences, the frequency of attacks, and their duration. Adding to this narrative, two niche open-label research initiatives underscored a notable reduction in headache episodes observed 30 days post anodal stimulation [194]. In stark contrast, findings documented by Cerrahoglu Sirin and team indicated no discernible difference in the monthly headache frequencies a month subsequent to either anodal or sham stimulation [195]. Further amplifying the narrative of inconclusive results, a comprehensive investigation by Grazzi et al. failed to delineate any significant differences when comparing anodal tDCS, cathodal tDCS, and sham stimulation at the 6 and 12-month mark. This was specifically observed in CM patients who were in the throes of abrupt medication withdrawal due to overuse [196].

Conclusively, the true efficacy of tDCS in offering a preventive strategy against CM, especially when gauged several months post-intervention, remains shrouded in ambiguity. For future tDCS trials to shed more clarity and establish therapeutic consistency, there is an imperative need to adopt a standardized protocol, meticulously outlining the polarity, montage, number of sessions, repetition intervals, and the definitive endpoints.

7.2. Repetitive Transcranial Magnetic Stimulation (rTMS)

Repetitive transcranial magnetic stimulation (rTMS) devices deliver a sequence of rapid pulses. These pulses are engineered to generate a minute, focused cortical electric current targeting specific brain areas like the M1 and DLPFC. Intriguingly, these areas are pivotal in modulating motor-thalamus-brainstem as well as prefrontal-thalamic-cingulate signaling pathways. As a general rule of thumb, high-frequency stimulations, ranging from 5 to 20 Hz, are believed to amplify cortical excitability, interact with diverse neurotransmitter and opioidergic networks, and mold neuronal plasticity [197,198]. The therapeutic potential of rTMS has been officially recognized by the FDA for the management of severe depression and obsessive-compulsive disorders. Even with a solid safety track record, the efficacy of rTMS in tackling pain disorders, post-traumatic headaches, and primary headache syndromes is currently under the microscope. Recent comprehensive reviews propose that high frequency rTMS, specifically targeting the motor cortex, might emerge as a promising migraine management strategy. However, the need of the hour is robust, high-quality randomized controlled trials (RCTs) with unified protocols to substantiate this therapeutic proposition [199,200].

In the recent half-decade, both open-label and randomized controlled trials have delved deep into the potential of rTMS for CM management [187]. Though certain open-label initiatives incorporated CM, they steered clear of presenting efficacy data and, hence, will not be the focus here [201]. An exploratory study helmed by Rapinesi et al. broadcasted a marked dip in migraine episodes, dependency on rescue medications, pain intensity, and scores measuring depression. This was observed after the combination of deep TMS, targeting the left DLPFC with a specific protocol, and conventional treatments [202]. A comparative study by Shehata et al., juxtaposing rTMS with onabotulinumtoxin A, revealed that rTMS, designed with a specific frequency and targeting the left M1, echoed the therapeutic efficacy of onabotulinumtoxin-A in managing CM; however, the rTMS effects waned noticeably after a span of eight weeks [203]. A research endeavor by Kalita et al., contrasting left M1 rTMS over a three-month period in CM and chronic tension-type headache patients, underscored a significant reduction in headache episodes post rTMS treatment within the group; however, when pitting the two groups against each other, the results were not statistically significant [204]. Granato et al., in their research with CM patients also grappling with medication overuse headache (MOH), could not identify any clear advantages of rTMS over its sham counterpart, especially when metrics like monthly headache days, symptomatic drug dependency, and Migraine Disability Assessment (MIDAS) scores were evaluated after a 120-day span [205]. An intriguing facet of this study was the sham stimulator's capacity to mimic the vibratory sensation of actual stimulation. Whether this mimicry carries any real therapeutic value remains an enigma.

7.3. Occipital Nerve Stimulation (ONS)

Occipital nerve stimulation (ONS) is a form of treatment that involves the use of an implantable device with electrodes positioned near the occipital nerves and a pulse generator in the chest. This type of treatment has been researched for many years for conditions like occipital neuralgia and refractory chronic migraine (CM). While the precise mechanism is not fully understood, it is believed that ONS may counterbalance trigeminally mediated central sensitization or restore the lost conditional pain modulation in these patients. A number of multi-center randomized sham-controlled trials, published over five years ago, showed improvements in areas such as headache frequency, intensity, and disability [192].

Within the last five years, open-label studies have been published on the use of ONS for chronic migraine [195–197]. However, the specific parameters of the stimulation and the study endpoints were quite varied between studies. Miller et al. analyzed a group of 53 intractable CM patients and discovered an 8.51-day reduction in monthly moderate-to-severe headache days after bilateral ONS electrodes implantation [195]. Similarly, Garcia-Ortega et al. studied 37 refractory CM patients and reported significant pain reduction [196]. However, it is worth noting that up to 20% of patients reported adverse events such as infection, lead migration, and stimulation-related symptoms one year after treatment [197].

Despite the potential side effects, ONS seems to be a promising device for chronic migraine; however, no ONS device has been cleared by the FDA for use in treating migraine yet.

7.4. Spinal Cord Stimulation (SCS)

Spinal cord stimulation (SCS) is another method that has been utilized to manage intractable headache. It involves placing SCS electrodes into the high-cervical epidural space (C2/3) with a pulse generator implanted subcutaneously. This technique has shown promising results, with patients experiencing a significant reduction in mean pain intensity and a decrease in the median number of migraine days. Nevertheless, issues such as infections and lead dislocations have been reported. Despite these concerns, the effectiveness of SCS for chronic migraine is still under investigation [198].

Another interesting approach being explored is the exposure to green light. This might help modulate nociception and anxiety. While non-green light stimuli seemed to exacerbate pain intensity during a migraine attack, exposure to green light was seen to reduce pain intensity in around 20% of the patients. In a small crossover study, daily green light exposure resulted in a significant reduction in the number of headaches in CM patients. Due to its potential effectiveness and safety, this approach warrants further investigation [192].

8. Clinical Perspective

There are presently almost half a dozen devices cleared by the U.S. Food and Drug Administration (FDA) specifically for treating migraine. Interestingly, the clearance for several of these devices has also been extended to cover adolescent patients aged 12 or older. However, it is crucial to note that not all of these FDA-cleared devices have been thoroughly investigated for use in chronic migraine (CM). The only randomized sham-controlled trial conducted for CM was undertaken utilizing the gammaCore device, and unfortunately, it failed to meet its primary efficacy endpoint.

Various open-label observational studies have employed devices, such as CEFALY or Nerivio, to assess pain reduction in patients with CM. However, these studies lack blinding and are subject to potential selection and reporting biases. Furthermore, certain studies have included both episodic migraine (EM) and CM patients, yet failed to provide a detailed breakdown of the number of CM cases or any response data specifically related to CM.

In addition, the absence of established trial guidelines that address the unique aspects and challenges associated with neuromodulation device trials for migraine has led to substantial variation in study endpoints, types of control, and the populations analyzed (intention-to-treat vs. per-protocol). As a result, it can be difficult to compare the outcomes of different studies. In an effort to address these issues, the International Headache Society has published recommendations for evaluating neuromodulation devices in both the acute and preventive treatment of migraine [199].

As the application of these devices becomes more common, we anticipate the emergence of larger, higher-quality studies that adhere to established clinical trial guidelines. These studies are essential to fully establish the benefits of these devices in the treatment of CM. Furthermore, the conduct of high-quality trials should also ideally encourage insurance companies to broaden their coverage to include more neuromodulation devices.

9. Discussion

Migraine remains an intriguing neurological disorder with complex mechanisms and pathophysiology that are still being researched extensively. Migraines are believed to be hereditary conditions characterized by increased responsiveness of cortical and subcortical networks; however, the triggers and factors contributing to its onset and progression remain unknown.

Migraine's multidimensionality can be seen through the following various phases: premonitory phase, headache pain, postdromal phase, and sometimes aura phase. Each of these involves interactions among hypothalamus nuclei, cortical regions, and trigeminovascular pathways, resulting in characteristic symptoms experienced during an attack.

Despite these challenges, migraine research has made significant advances, providing light on potential therapeutic targets and validating CGRP as an effective target for both acute and preventative treatments—giving hope of the better management of migraine episodes as well as improved quality of life for affected individuals.

However, much work remains to be performed in understanding its intricate processes and developing effective treatment and prevention measures. More awareness, additional funding for research, collaboration among scientists, healthcare providers, and patients, as well as increased collaboration, will all play an integral role in increasing understanding and alleviating migraine's burden.

As we discover more about migraine, it is crucial that we provide empathy, support, and personalized care to those living with this neurological challenge. By harnessing the collective knowledge and dedication of scientific communities worldwide, we can make significant advances in migraine research that ultimately benefit millions living with this neurological disorder [200].

Author Contributions: Conceptualization, H.P. and A.V.C.; methodology, T.-L.T.; soft-ware, I.-A.F.; validation, R.-A.C.-B., L.-A.G. and H.P.; formal analysis, D.-I.D.; investigation, A.A.P.; resources, A.B.; data curation, L.-A.G.; writing—original draft preparation, R.-A.C.-B. and A.V.C.; writing—review and editing, R.-A.C.-B., L.-A.G. and A.V.C.; visualization, A.B.; supervision, D.-I.D.; project administration, A.V.C.; funding acquisition, A.V.C. All authors have read and agreed to the published version of the manuscript.

Funding: This research received no external funding.

Institutional Review Board Statement: Not applicable.

Informed Consent Statement: No requirement for ethical approval.

Data Availability Statement: All Data is available on PubMed.

Conflicts of Interest: The authors declare no conflict of interest.

References

1. Stovner, L.; Hagen, K.; Jensen, R.; Katsarava, Z.; Lipton, R.; Scher, A.; Steiner, T.; Zwart, J.-A. The Global Burden of Headache: A Documentation of Headache Prevalence and Disability Worldwide. *Cephalalgia* **2007**, *27*, 193–210. [CrossRef] [PubMed]
2. Bateman, D. The future of neurology services in the UK. *Pract. Neurol.* **2011**, *11*, 134–135. [CrossRef] [PubMed]
3. Sender, J.; Bradford, S.; Watson, D.; Lipscombe, S.; Rees, T.; Manley, R.; Dowson, A. Setting up a Specialist Headache Clinic in Primary Care: General Practitioners with a Special Interest (GPwSI) in Headache. *Headache Care* **2004**, *1*, 165–171. [CrossRef]
4. WHO. *The World Health Report 2001: Mental Health: New Understanding, New Hope*; WHO: Geneva, Switzerland, 2001.
5. Solomon, G.D.; Price, K.L. Burden of Migraine: A Review of its Socioeconomic Impact. *Pharmacoeconomics* **1997**, *11*, 1–10. [CrossRef] [PubMed]
6. Terwindt, G.M.; Ferrari, M.D.; Tijhuis, M.; Groenen, S.M.A.; Picavet, H.S.J.; Launer, L.J. The impact of migraine on quality of life in the general population: The GEM study. *Neurology* **2000**, *55*, 624–629. [CrossRef] [PubMed]
7. Steiner, T.; Scher, A.; Stewart, W.; Kolodner, K.; Liberman, J.; Lipton, R. The Prevalence and Disability Burden of Adult Migraine in England and their Relationships to Age, Gender and Ethnicity. *Cephalalgia* **2003**, *23*, 519–527. [CrossRef]
8. Clarke, C.E.; MacMillan, L.; Sondhi, S.; Wells, N.E. Economic and social impact of migraine. *QJM Mon. J. Assoc. Physicians* **1996**, *89*, 77–84. [CrossRef]
9. Ahmed, F. Headache disorders: Differentiating and managing the common subtypes. *Br. J. Pain* **2012**, *6*, 124–132. [CrossRef]
10. Haut, S.R.; Bigal, E.M.; Lipton, R.B. Chronic disorders with episodic manifestations: Focus on epilepsy and migraine. *Lancet Neurol.* **2006**, *5*, 148–157. [CrossRef]
11. Olesen, J. Preface to the Second Edition. *Cephalalgia* **2004**, *24*, 9–10. [CrossRef]
12. Stewart, W.F.; Wood, G.C.; Manack, A.; Varon, S.F.; Buse, D.C.; Lipton, R.B. Employment and Work Impact of Chronic Migraine and Episodic Migraine. *J. Occup. Environ. Med.* **2010**, *52*, 8–14. [CrossRef] [PubMed]
13. Natoli, J.; Manack, A.; Dean, B.; Butler, Q.; Turkel, C.; Stovner, L.; Lipton, R. Global prevalence of chronic migraine: A systematic review. *Cephalalgia* **2009**, *30*, 599–609. [CrossRef]
14. Blumenfeld, A.; Varon, S.; Wilcox, T.; Buse, D.; Kawata, A.; Manack, A.; Goadsby, P.; Lipton, R. Disability, HRQoL and resource use among chronic and episodic migraineurs: Results from the International Burden of Migraine Study (IBMS). *Cephalalgia* **2010**, *31*, 301–315. [CrossRef] [PubMed]

15. Durham, P.; Papapetropoulos, S. Biomarkers Associated with Migraine and Their Potential Role in Migraine Management. *Headache J. Head Face Pain* **2013**, *53*, 1262–1277. [CrossRef] [PubMed]
16. Safiri, S.; Pourfathi, H.; Eagan, A.; Mansournia, M.A.; Khodayari, M.T.; Sullman, M.J.M.; Kaufman, J.; Collins, G.; Dai, H.; Bragazzi, N.L.; et al. Global, regional, and national burden of migraine in 204 countries and territories, 1990 to 2019. *Pain* **2021**, *163*, e293–e309. [CrossRef] [PubMed]
17. Amiri, P.; Kazeminasab, S.; Nejadghaderi, S.A.; Mohammadinasab, R.; Pourfathi, H.; Araj-Khodaei, M.; Sullman, M.J.M.; Kolahi, A.-A.; Safiri, S. Migraine: A Review on Its History, Global Epidemiology, Risk Factors, and Comorbidities. *Front. Neurol.* **2022**, *12*, 800605. [CrossRef]
18. Burch, R.C.; Loder, S.; Loder, E.; Smitherman, T.A. The Prevalence and Burden of Migraine and Severe Headache in the United States: Updated Statistics from Government Health Surveillance Studies. *Headache J. Head Face Pain* **2015**, *55*, 21–34. [CrossRef]
19. Lipton, R.B.; Stewart, W.F.; Diamond, S.; Diamond, M.L.; Reed, M. Prevalence and Burden of Migraine in the United States: Data from the American Migraine Study II. *Headache J. Head Face Pain* **2001**, *41*, 646–657. [CrossRef]
20. Merikangas, K.R. Contributions of Epidemiology to Our Understanding of Migraine. *Headache J. Head Face Pain* **2013**, *53*, 230–246. [CrossRef]
21. Wang, S.-J. Epidemiology of migraine and other types of headache in Asia. *Curr. Neurol. Neurosci. Rep.* **2003**, *3*, 104–108. [CrossRef]
22. Dzoljic, E.; Vlajinac, H.; Sipetic, S.; Marinkovic, J.; Grbatinic, I.; Kostic, V. A survey of female students with migraine: What is the influence of family history and lifestyle. *Int. J. Neurosci.* **2013**, *124*, 82–87. [CrossRef] [PubMed]
23. Hernandez-Latorre, M.; Roig, M. Natural History of Migraine in Childhood. *Cephalalgia* **2000**, *20*, 573–579. [CrossRef] [PubMed]
24. Russell, M.B.; Olesen, J. Increased familial risk and evidence of genetic factor in migraine. *BMJ* **1995**, *311*, 541–544. [CrossRef] [PubMed]
25. Stewart, W.F.; Staffa, J.; Lipton, R.B.; Ottman, R. Familial risk of migraine: A population-based study. *Ann. Neurol.* **1997**, *41*, 166–172. [CrossRef] [PubMed]
26. Gervil, M.; Ulrich, V.; Kaprio, J.; Olesen, J.; Russell, M.B. The relative role of genetic and environmental factors in migraine without aura. *Neurology* **1999**, *53*, 995. [CrossRef]
27. Honkasalo, M.-L.; Kaprio, J.; Winter, T.; Heikkila, K.; Sillanpaa, M.; Koskenvuo, M. Migraine and Concomitant Symptoms Among 8167 Adult Twin Pairs. *Headache J. Head Face Pain* **1995**, *35*, 70–78. [CrossRef]
28. Mulder, E.J.; van Baal, C.; Gaist, D.; Kallela, M.; Kaprio, J.; Svensson, D.A.; Nyholt, D.R.; Martin, N.G.; MacGregor, A.J.; Cherkas, L.F.; et al. Genetic and Environmental Influences on Migraine: A Twin Study Across Six Countries. *Twin Res.* **2003**, *6*, 422–431. [CrossRef]
29. Svensson, D.A.; Larsson, B.; Waldenlind, E.; Pedersen, N.L.; Msc, D.A.S. Shared Rearing Environment in Migraine: Results from Twins Reared Apart and Twins Reared Together. *Headache J. Head Face Pain* **2003**, *43*, 235–244. [CrossRef]
30. De Vries, B.; Anttila, V.; Freilinger, T.; Wessman, M.; Kaunisto, A.M.; Kallela, M.; Artto, V.; Vijfhuizen, L.S.; Göbel, H.; Dichgans, M.; et al. Systematic re-evaluation of genes from candidate gene association studies in migraine using a large genome-wide association data set. *Cephalalgia* **2015**, *36*, 604–614. [CrossRef]
31. Yeh, W.Z.; Blizzard, L.; Taylor, B.V. What is the actual prevalence of migraine. *Brain Behav.* **2018**, *8*, e00950. [CrossRef]
32. Onofri, A.; Pensato, U.; Rosignoli, C.; Wells-Gatnik, W.; Stanyer, E.; Ornello, R.; Chen, H.Z.; De Santis, F.; Torrente, A.; Mikulenka, P.; et al. Primary headache epidemiology in children and adolescents: A systematic review and meta-analysis. *J. Headache Pain* **2023**, *24*, 8. [CrossRef] [PubMed]
33. Loh, N.R.; Whitehouse, W.P.; Howells, R. What is new in migraine management in children and young people. *Arch. Dis. Child.* **2022**, *107*, 1067–1072. [CrossRef] [PubMed]
34. Loder, E.; Rizzoli, P. Biomarkers in Migraine: Their Promise, Problems, and Practical Applications. *Headache J. Head Face Pain* **2006**, *46*, 1046–1058. [CrossRef] [PubMed]
35. Biomarkers Definitions Working Group; Atkinson, A.J., Jr.; Colburn, W.A.; DeGruttola, V.G.; DeMets, D.L.; Downing, G.J.; Hoth, D.F.; Oates, J.A.; Peck, C.C.; Spilker, B.A.; et al. Biomarkers and surrogate endpoints: Preferred definitions and conceptual framework. *Clin. Pharmacol. Ther.* **2001**, *69*, 89–95. [CrossRef]
36. Phillips, K.A.; Van Bebber, S.; Issa, A.M. Diagnostics and biomarker development: Priming the pipeline. *Nat. Rev. Drug Discov.* **2006**, *5*, 463–469. [CrossRef]
37. Lesko, L.; Atkinson, A. Use of Biomarkers and Surrogate Endpoints in Drug Development and Regulatory Decision Making: Criteria, Validation, Strategies. *Annu. Rev. Pharmacol. Toxicol.* **2001**, *41*, 347–366. [CrossRef]
38. Packer, C.S. Biochemical markers and physiological parameters as indices for identifying patients at risk of developing pre-eclampsia. *J. Hypertens.* **2005**, *23*, 45–46. [CrossRef]
39. Mayeux, R. Biomarkers: Potential uses and limitations. *NeuroRx* **2004**, *1*, 182–188. [CrossRef]
40. Boćkowski, L.; Sobaniec, W.; Żelazowska-Rutkowska, B. Proinflammatory Plasma Cytokines in Children with Migraine. *Pediatr. Neurol.* **2009**, *41*, 17–21. [CrossRef]
41. Kaciński, M.; Gergont, A.; Kubik, A.; Steczkowska-Klucznik, M. Proinflammatory cytokines in children with migraine with or without aura. *Przeglad Lek.* **2005**, *62*, 1276–1280.
42. Kara, I.; Sazci, A.; Ergul, E.; Kaya, G.; Kilic, G. Association of the C677T and A1298C polymorphisms in the 5,10 methylenetetrahydrofolate reductase gene in patients with migraine risk. *Mol. Brain Res.* **2003**, *111*, 84–90. [CrossRef] [PubMed]

43. Kowalska, M.; Prendecki, M.; Kozubski, W.; Lianeri, M.; Dorszewska, J. Molecular factors in migraine. *Oncotarget* **2016**, *7*, 50708–50718. [CrossRef] [PubMed]
44. D'Andrea, G.; D'Arrigo, A.; Carbonare, M.D.; Leon, A. Pathogenesis of Migraine: Role of Neuromodulators. *Headache J. Head Face Pain* **2012**, *52*, 1155–1163. [CrossRef] [PubMed]
45. D'andrea, G.; D'amico, D.; Bussone, G.; Bolner, A.; Aguggia, M.; Saracco, M.G.; Galloni, E.; De Riva, V.; Colavito, D.; Leon, A.; et al. The role of tyrosine metabolism in the pathogenesis of chronic migraine. *Cephalalgia* **2013**, *33*, 932–937. [CrossRef] [PubMed]
46. Cady, R.K.; Vause, C.V.; Ho, T.W.; Bigal, M.E.; Durham, P.L. Elevated Saliva Calcitonin Gene-Related Peptide Levels During Acute Migraine Predict Therapeutic Response to Rizatriptan. *Headache J. Head Face Pain* **2009**, *49*, 1258–1266. [CrossRef]
47. Sicuteri, F.; Testi, A.; Anselmi, B. Biochemical Investigations in Headache: Increase in the Hydroxyindoleacetic Acid Excretion During Migraine Attacks. *Int. Arch. Allergy Immunol.* **1961**, *19*, 55–58. [CrossRef]
48. Ferrari, M.D.; Odink, J.; Tapparelli, C.; Van Kempen, G.; Pennings, E.J.; Bruyn, G.W. Serotonin metabolism in migraine. *Neurology* **1989**, *39*, 1239. [CrossRef]
49. Tsujino, N.; Sakurai, T. Orexin/Hypocretin: A Neuropeptide at the Interface of Sleep, Energy Homeostasis, and Reward System. *Pharmacol. Rev.* **2009**, *61*, 162–176. [CrossRef]
50. Sarchielli, P.; Rainero, I.; Coppola, F.; Rossi, C.; Mancini, M.; Pinessi, L.; Calabresi, P. Involvement of Corticotrophin-Releasing Factor and Orexin-A in Chronic Migraine and Medication-Overuse Headache: Findings from Cerebrospinal Fluid. *Cephalalgia* **2008**, *28*, 714–722. [CrossRef]
51. Barloese, M.; Jennum, P.; Lund, N.; Knudsen, S.; Gammeltoft, S.; Jensen, R. Reduced CSF hypocretin-1 levels are associated with cluster headache. *Cephalalgia* **2014**, *35*, 869–876. [CrossRef]
52. Tso, A.R.; Goadsby, P.J. New Targets for Migraine Therapy. *Curr. Treat. Options Neurol.* **2014**, *16*, 318. [CrossRef]
53. Diener, H.-C.; Charles, A.; Goadsby, P.J.; Holle, D. New therapeutic approaches for the prevention and treatment of migraine. *Lancet Neurol.* **2015**, *14*, 1010–1022. [CrossRef] [PubMed]
54. Nelson, K.B.; Richardson, A.K.; He, J.; Lateef, T.M.; Khoromi, S.; Merikangas, K.R. Headache and Biomarkers Predictive of Vascular Disease in a Representative Sample of US Children. *Arch. Pediatr. Adolesc. Med.* **2010**, *164*, 358–362. [CrossRef] [PubMed]
55. Tietjen, G.E.; Herial, N.A.; White, L.; Utley, C.; Kosmyna, J.M.; Khuder, S.A. Migraine and Biomarkers of Endothelial Activation in Young Women. *Stroke* **2009**, *40*, 2977–2982. [CrossRef] [PubMed]
56. Ben Aissa, M.; Tipton, A.F.; Bertels, Z.; Gandhi, R.; Moye, L.S.; Novack, M.; Bennett, B.M.; Wang, Y.; Litosh, V.; Lee, S.H.; et al. Soluble guanylyl cyclase is a critical regulator of migraine-associated pain. *Cephalalgia* **2017**, *38*, 1471–1484. [CrossRef]
57. Al-Hassany, L.; Boucherie, D.M.; Creeney, H.; van Drie, R.W.A.; Farham, F.; Favaretto, S.; Gollion, C.; Grangeon, L.; Lyons, H.; Marschollek, K.; et al. Future targets for migraine treatment beyond CGRP. *J. Headache Pain* **2023**, *24*, 76. [CrossRef]
58. Matharu, M.S.; Bartsch, T.; Ward, N.; Frackowiak, R.S.J.; Weiner, R.; Goadsby, P.J. Central neuromodulation in chronic migraine patients with suboccipital stimulators: A PET study. *Brain* **2004**, *127*, 220–230. [CrossRef]
59. Royal, P.; Andres-Bilbe, A.; Prado, P.Á.; Verkest, C.; Wdziekonski, B.; Schaub, S.; Baron, A.; Lesage, F.; Gasull, X.; Levitz, J.; et al. Migraine-Associated TRESK Mutations Increase Neuronal Excitability through Alternative Translation Initiation and Inhibition of TREK. *Neuron* **2019**, *101*, 232–245.e6. [CrossRef]
60. Prado, P.; Landra-Willm, A.; Verkest, C.; Ribera, A.; Chassot, A.-A.; Baron, A.; Sandoz, G. TREK channel activation suppresses migraine pain phenotype. *iScience* **2021**, *24*, 102961. [CrossRef]
61. Lim, C.S.; Shalhoub, J.; Gohel, M.S.; Shepherd, A.C.; Davies, A.H. Matrix Metalloproteinases in Vascular Disease-A Potential Therapeutic Target. *Curr. Vasc. Pharmacol.* **2010**, *8*, 75–85. [CrossRef]
62. Todt, U.; Netzer, C.; Toliat, M.; Heinze, A.; Goebel, I.; Nürnberg, P.; Göbel, H.; Freudenberg, J.; Kubisch, C. New genetic evidence for involvement of the dopamine system in migraine with aura. *Hum. Genet.* **2009**, *125*, 265–279. [CrossRef]
63. Naya, M.; Tsukamoto, T.; Morita, K.; Katoh, C.; Furumoto, T.; Fujii, S.; Tamaki, N.; Tsutsui, H. Plasma Interleukin-6 and Tumor Necrosis Factor-.ALPHA. Can Predict Coronary Endothelial Dysfunction in Hypertensive Patients. *Hypertens. Res.* **2007**, *30*, 541–548. [CrossRef] [PubMed]
64. Imamura, K.; Takeshima, T.; Fusayasu, E.; Nakashima, K. Increased Plasma Matrix Metalloproteinase-9 Levels in Migraineurs. *Headache J. Head Face Pain* **2007**, *48*, 135–139. [CrossRef] [PubMed]
65. Olesen, J.; Friberg, L.; Iversen, H.K.; Lassen, N.A.; Andersen, A.R.; Karle, A.; Olsen, T.S. Timing and topography of cerebral blood flow, aura, and headache during migraine attacks. *Ann. Neurol.* **1990**, *28*, 791–798. [CrossRef] [PubMed]
66. Thomsen, L.L.; Kirchmann, M.; Bjornsson, A.; Stefansson, H.; Jensen, R.M.; Fasquel, A.C.; Petursson, H.; Stefansson, M.; Frigge, M.L.; Kong, A.; et al. The genetic spectrum of a population-based sample of familial hemiplegic migraine. *Brain* **2006**, *130*, 346–356. [CrossRef] [PubMed]
67. Mitchell, B.L.; Diaz-Torres, S.; Bivol, S.; Cuellar-Partida, G.; Gormley, P.; Anttila, V.; Winsvold, B.S.; Palta, P.; Esko, T.; Pers, T.H.; et al. Elucidating the relationship between migraine risk and brain structure using genetic data. *Brain* **2022**, *145*, 3214–3224. [CrossRef]
68. Grangeon, L.; Lange, K.S.; Waliszewska-Prosół, M.; Onan, D.; Marschollek, K.; Wiels, W.; Mikulenka, P.; Farham, F.; Gollion, C.; Ducros, A.; et al. Genetics of migraine: Where are we now. *J. Headache Pain* **2023**, *24*, 12. [CrossRef]
69. van den Maagdenberg, A.M.; Pietrobon, D.; Pizzorusso, T.; Kaja, S.; Broos, L.A.; Cesetti, T.; van de Ven, R.C.; Tottene, A.; van der Kaa, J.; Plomp, J.J.; et al. A Cacna1a Knockin Migraine Mouse Model with Increased Susceptibility to Cortical Spreading Depression. *Neuron* **2004**, *41*, 701–710. [CrossRef]

70. Maagdenberg, A.M.J.M.v.D.; Pizzorusso, T.; Kaja, S.; Terpolilli, N.; Shapovalova, M.; Hoebeek, F.E.; Barrett, C.F.; Gherardini, L.; van de Ven, R.C.G.; Todorov, B.; et al. High cortical spreading depression susceptibility and migraine-associated symptoms in $Ca_V2.1$ S218L mice. *Ann. Neurol.* **2010**, *67*, 85–98. [CrossRef]
71. Meneghetti, N.; Cerri, C.; Vannini, E.; Tantillo, E.; Tottene, A.; Pietrobon, D.; Caleo, M.; Mazzoni, A. Synaptic alterations in visual cortex reshape contrast-dependent gamma oscillations and inhibition-excitation ratio in a genetic mouse model of migraine. *J. Headache Pain* **2022**, *23*, 125. [CrossRef]
72. Leo, L.; Gherardini, L.; Barone, V.; De Fusco, M.; Pietrobon, D.; Pizzorusso, T.; Casari, G. Increased Susceptibility to Cortical Spreading Depression in the Mouse Model of Familial Hemiplegic Migraine Type 2. *PLoS Genet.* **2011**, *7*, e1002129. [CrossRef] [PubMed]
73. Unekawa, M.; Ikeda, K.; Tomita, Y.; Kawakami, K.; Suzuki, N. Enhanced susceptibility to cortical spreading depression in two types of Na^+,K^+-ATPase $\alpha2$ subunit-deficient mice as a model of familial hemiplegic migraine 2. *Cephalalgia* **2017**, *38*, 1515–1524. [CrossRef] [PubMed]
74. Smith, S.E.; Chen, X.; Brier, L.M.; Bumstead, J.R.; Rensing, N.R.; Ringel, A.E.; Shin, H.; Oldenborg, A.; Crowley, J.R.; Bice, A.R.; et al. Astrocyte deletion of $\alpha2$-Na/K ATPase triggers episodic motor paralysis in mice via a metabolic pathway. *Nat. Commun.* **2020**, *11*, 6164. [CrossRef] [PubMed]
75. Bertelli, S.; Barbieri, R.; Pusch, M.; Gavazzo, P. Gain of function of sporadic/familial hemiplegic migraine-causing SCN1A mutations: Use of an optimized cDNA. *Cephalalgia* **2018**, *39*, 477–488. [CrossRef]
76. Suzuki-Muromoto, S.; Kosaki, R.; Kosaki, K.; Kubota, M. Familial hemiplegic migraine with a PRRT2 mutation: Phenotypic variations and carbamazepine efficacy. *Brain Dev.* **2020**, *42*, 293–297. [CrossRef]
77. Michetti, C.; Castroflorio, E.; Marchionni, I.; Forte, N.; Sterlini, B.; Binda, F.; Fruscione, F.; Baldelli, P.; Valtorta, F.; Zara, F.; et al. The PRRT2 knockout mouse recapitulates the neurological diseases associated with PRRT2 mutations. *Neurobiol. Dis.* **2016**, *99*, 66–83. [CrossRef]
78. Chasman, D.I.; Schürks, M.; Anttila, V.; De Vries, B.; Schminke, U.; Launer, L.J.; Terwindt, G.M.; van den Maagdenberg, A.M.J.M.; Fendrich, K.; Völzke, H.; et al. Genome-wide association study reveals three susceptibility loci for common migraine in the general population. *Nat. Genet.* **2011**, *43*, 695–698. [CrossRef]
79. Freilinger, T.; International Headache Genetics Consortium; Anttila, V.; de Vries, B.; Malik, R.; Kallela, M.; Terwindt, G.M.; Pozo-Rosich, P.; Winsvold, B.; Nyholt, D.R.; et al. Genome-wide association analysis identifies susceptibility loci for migraine without aura. *Nat. Genet.* **2012**, *44*, 777–782. [CrossRef]
80. Choquet, H.; Yin, J.; Jacobson, A.S.; Horton, B.H.; Hoffmann, T.J.; Jorgenson, E.; Avins, A.L.; Pressman, A.R. New and sex-specific migraine susceptibility loci identified from a multiethnic genome-wide meta-analysis. *Commun. Biol.* **2021**, *4*, 864. [CrossRef]
81. Hautakangas, H.; Winsvold, B.S.; Ruotsalainen, S.E.; Bjornsdottir, G.; Harder, A.V.E.; Kogelman, L.J.A.; Thomas, L.F.; Noordam, R.; Benner, C.; Gormley, P.; et al. Genome-wide analysis of 102,084 migraine cases identifies 123 risk loci and subtype-specific risk alleles. *Nat. Genet.* **2022**, *54*, 152–160. [CrossRef]
82. Cutrer, F.M.; Black, D.F. Imaging Findings of Migraine. *Headache J. Head Face Pain* **2006**, *46*, 1095–1107. [CrossRef] [PubMed]
83. Del Zompo, M.; Cherchi, A.; Palmas, M.A.; Ponti, M.; Bocchetta, A.; Gessa, G.L.; Piccardi, M.P. Association between dopamine receptor genes and migraine without aura in a Sardinian sample. *Neurology* **1998**, *51*, 781–786. [CrossRef] [PubMed]
84. Moskowitz, M.A.; Cutrer, F.M. Sumatriptan: A Receptor-Targeted Treatment for Migraine. *Annu. Rev. Med.* **1993**, *44*, 145–154. [CrossRef]
85. Bolay, H.; Reuter, U.; Dunn, A.K.; Huang, Z.; Boas, D.A.; Moskowitz, M.A. Intrinsic brain activity triggers trigeminal meningeal afferents in a migraine model. *Nat. Med.* **2002**, *8*, 136–142. [CrossRef] [PubMed]
86. Dimtriadou, V.; Buzzi, M.; Moskowitz, M.; Theoharides, T. Trigeminal sensory fiber stimulation induces morphological changes reflecting secretion in rat dura mater mast cells. *Neuroscience* **1991**, *44*, 97–112. [CrossRef]
87. Peroutka, S.J. Neurogenic inflammation and migraine: Implications for the therapeutics. *Mol. Interv.* **2005**, *5*, 304–311. [CrossRef]
88. Goadsby, P.J. Pathophysiology of migraine. *Ann. Indian Acad. Neurol.* **2012**, *15*, 15. [CrossRef]
89. Do, T.P.; Deligianni, C.; Amirguliyev, S.; Snellman, J.; Lopez, C.L.; Al-Karagholi, M.A.-M.; Guo, S.; Ashina, M. Second messenger signaling bypasses CGRP receptor blockade to provoke migraine attacks in humans. *Brain* **2023**, awad261. [CrossRef]
90. Amin, F.M.; Schytz, H.W. Transport of the pituitary adenylate cyclase-activating polypeptide across the blood-brain barrier: Implications for migraine. *J. Headache Pain* **2018**, *19*, 35. [CrossRef]
91. Grände, G.; Nilsson, E.; Edvinsson, L. Comparison of responses to vasoactive drugs in human and rat cerebral arteries using myography and pressurized cerebral artery method. *Cephalalgia* **2012**, *33*, 152–159. [CrossRef]
92. Iannetti, G.D.; Zambreanu, L.; Wise, R.G.; Buchanan, T.J.; Huggins, J.P.; Smart, T.S.; Vennart, W.; Tracey, I. Pharmacological modulation of pain-related brain activity during normal and central sensitization states in humans. *Proc. Natl. Acad. Sci. USA* **2005**, *102*, 18195–18200. [CrossRef] [PubMed]
93. Birk, S.; Sitarz, J.T.; Petersen, K.A.; Oturai, P.S.; Kruuse, C.; Fahrenkrug, J.; Olesen, J. The effect of intravenous PACAP38 on cerebral hemodynamics in healthy volunteers. *Regul. Pept.* **2007**, *140*, 185–191. [CrossRef] [PubMed]
94. Olesen, J.; Burstein, R.; Ashina, M.; Tfelt-Hansen, P. Origin of pain in migraine: Evidence for peripheral sensitisation. *Lancet Neurol.* **2009**, *8*, 679–690. [CrossRef] [PubMed]

95. Ghanizada, H.; Al-Karagholi, M.A.-M.; Arngrim, N.; Mørch-Rasmussen, M.; Walker, C.S.; Hay, D.L.; Ashina, M. Effect of Adrenomedullin on Migraine-like Attacks in Patients 2ith Migraine: A Randomized Crossover Study. *Neurology* **2021**, *96*, e2488–e2499. [CrossRef]
96. Hinson, J.P.; Kapas, S.; Smith, D.M. Adrenomedullin, a Multifunctional Regulatory Peptide. *Endocr. Rev.* **2000**, *21*, 138–167. [CrossRef] [PubMed]
97. Bhakta, M.; Vuong, T.; Taura, T.; Wilson, D.S.; Stratton, J.R.; Mackenzie, K.D. Migraine therapeutics differentially modulate the CGRP pathway. *Cephalalgia* **2021**, *41*, 499–514. [CrossRef]
98. Carlsson, A.; Forsgren, L.; Nylander, P.-O.; Hellman, U.; Forsman-Semb, K.; Holmgren, G.; Holmberg, D.; Holmberg, M. Identification of a susceptibility locus for migraine with and without aura on 6p12.2-p21.1. *Neurology* **2002**, *59*, 1804–1807. [CrossRef]
99. Oterino, A.; Toriello, M.; Valle, N.; Castillo, J.; Alonso-Arranz, A.; Bravo, Y.; Ruiz-Alegria, C.; Quintela, E.; Pascual, J. The Relationship Between Homocysteine and Genes of Folate-Related Enzymes in Migraine Patients. *Headache J. Head Face Pain* **2010**, *50*, 99–168. [CrossRef]
100. Perini, F.; D'Andrea, G.; Galloni, E.; Pignatelli, F.; Billo, G.; Alba, S.; Bussone, G.; Toso, V. Plasma Cytokine Levels in Migraineurs and Controls. *Headache J. Head Face Pain* **2005**, *45*, 926–931. [CrossRef]
101. Schwedt, T.J.; Dodick, D.W. Advanced neuroimaging of migraine. *Lancet Neurol.* **2009**, *8*, 560–568. [CrossRef]
102. Burstein, R.; Yamamura, H.; Malick, A.; Strassman, A.M.; Goadsby, P.J.; Holland, P.R.; Martins-Oliveira, M.; Hoffmann, J.; Schankin, C.; Akerman, S.; et al. Chemical Stimulation of the Intracranial Dura Induces Enhanced Responses to Facial Stimulation in Brain Stem Trigeminal Neurons. *J. Neurophysiol.* **1998**, *79*, 964–982. [CrossRef]
103. Burstein, R.; Yarnitsky, D.; Goor-Aryeh, I.; Ransil, B.J.; Bajwa, Z.H. An association between migraine and cutaneous allodynia. *Ann. Neurol.* **2000**, *47*, 614–624. [CrossRef] [PubMed]
104. Jakubowski, M.; Silberstein, S.; Ashkenazi, A.; Burstein, R. Can allodynic migraine patients be identified interictally using a questionnaire. *Neurology* **2005**, *65*, 1419–1422. [CrossRef] [PubMed]
105. Bigal, M.E.; Ashina, S.; Burstein, R.; Reed, M.L.; Buse, D.; Serrano, D.; Lipton, R.B.; On behalf of the AMPP Group. Prevalence and characteristics of allodynia in headache sufferers: A population study. *Neurology* **2008**, *70*, 1525–1533. [CrossRef] [PubMed]
106. Zambreanu, L.; Wise, R.G.; Brooks, J.C.; Iannetti, G.D.; Tracey, I. A role for the brainstem in central sensitisation in humans. Evidence from functional magnetic resonance imaging. *Pain* **2005**, *114*, 397–407. [CrossRef]
107. Lee, M.C.; Zambreanu, L.; Menon, D.K.; Tracey, I. Identifying Brain Activity Specifically Related to the Maintenance and Perceptual Consequence of Central Sensitization in Humans. *J. Neurosci.* **2008**, *28*, 11642–11649. [CrossRef] [PubMed]
108. Schulte, L.H.; Mehnert, J.; May, A. Longitudinal Neuroimaging over 30 Days: Temporal Characteristics of Migraine. *Ann. Neurol.* **2020**, *87*, 646–651. [CrossRef]
109. Gross, E.C.; Lisicki, M.; Fischer, D.; Sándor, P.S.; Schoenen, J. The metabolic face of migraine—From pathophysiology to treatment. *Nat. Rev. Neurol.* **2019**, *15*, 627–643. [CrossRef]
110. Ingram, E.E.; Bocklud, B.E.; Corley, S.C.; Granier, M.A.; Neuchat, E.E.; Ahmadzadeh, S.; Shekoohi, S.; Kaye, A.D. Non-CGRP Antagonist/Non-Triptan Options for Migraine Disease Treatment: Clinical Considerations. *Curr. Pain Headache Rep.* **2023**, 1–6. [CrossRef]
111. Goadsby, P.J.; Hoskin, K.L. The distribution of trigeminovascular afferents in the nonhuman primate brain Macaca nemestrina: A c-fos immunocytochemical study. *J. Anat.* **1997**, *190*, 367–375. [CrossRef]
112. Bartsch, T.; Goadsby, P.J. Stimulation of the greater occipital nerve induces increased central excitability of dural afferent input. *Brain* **2002**, *125*, 1496–1509. [CrossRef] [PubMed]
113. Shepheard, S.; Edvinsson, L.; Cumberbatch, M.; Williamson, D.; Mason, G.; Webb, J.; Boyce, S.; Hill, R.; Hargreaves, R. Possible Antimigraine Mechanisms of Action of the $5HT_{1F}$ Receptor Agonist LY334370. *Cephalalgia* **1999**, *19*, 851–858. [CrossRef]
114. Andreou, A.P.; Goadsby, P.J. Therapeutic potential of novel glutamate receptor antagonists in migraine. *Expert Opin. Investig. Drugs* **2009**, *18*, 789–803. [CrossRef] [PubMed]
115. Weiss, B.; Alt, A.; Ogden, A.M.; Gates, M.; Dieckman, D.K.; Clemens-Smith, A.; Ho, K.H.; Jarvie, K.; Rizkalla, G.; Wright, R.A.; et al. Pharmacological Characterization of the Competitive GLU_{K5} Receptor Antagonist Decahydroisoquinoline LY466195 in Vitro and in Vivo. *J. Pharmacol. Exp. Ther.* **2006**, *318*, 772–781. [CrossRef] [PubMed]
116. Coppola, G.; Vandenheede, M.; Di Clemente, L.; Ambrosini, A.; Fumal, A.; De Pasqua, V.; Schoenen, J. Somatosensory evoked high-frequency oscillations reflecting thalamo-cortical activity are decreased in migraine patients between attacks. *Brain* **2004**, *128*, 98–103. [CrossRef]
117. de Tommaso, M.; Ambrosini, A.; Brighina, F.; Coppola, G.; Perrotta, A.; Pierelli, F.; Sandrini, G.; Valeriani, M.; Marinazzo, D.; Stramaglia, S.; et al. Altered processing of sensory stimuli in patients with migraine. *Nat. Rev. Neurol.* **2014**, *10*, 144–155. [CrossRef]
118. Wang, W.; Timsit-Berthier, M.; Schoenen, J. Intensity dependence of auditory evoked potentials is pronounced in migraine: An indication of cortical potentiation and low serotonergic neurotransmission. *Neurology* **1996**, *46*, 1404. [CrossRef]
119. Judit, Á.; Sándor, P.; Schoenen, J. Habituation of Visual and Intensity Dependence of Auditory Evoked Cortical Potentials Tends to Normalize Just Before and During the Migraine Attack. *Cephalalgia* **2000**, *20*, 714–719. [CrossRef]

120. Proietti-Cecchini, A.; Áfra, J.; Schoenen, J. Intensity Dependence of the Cortical Auditory Evoked Potentials as A Surrogate Marker of Central Nervous System Serotonin Transmission in Man: Demonstration of A Central Effect for the $5Ht_{1B/1D}$ Agonist Zolmitriptan (311C90, Zomig®). *Cephalalgia* **1997**, *17*, 849–854. [CrossRef]
121. Goadsby, P.J.; Holland, P.R.; Martins-Oliveira, M.; Hoffmann, J.; Schankin, C.; Akerman, S. Pathophysiology of Migraine: A Disorder of Sensory Processing. *Physiol. Rev.* **2017**, *97*, 553–622. [CrossRef]
122. Valfrè, W.; Rainero, I.; Bergui, M.; Pinessi, L. Voxel-Based Morphometry Reveals Gray Matter Abnormalities in Migraine. *Headache J. Head Face Pain* **2007**, *48*, 109–117. [CrossRef]
123. Maleki, N.; Becerra, L.; Nutile, L.; Pendse, G.; Brawn, J.; Bigal, M.; Burstein, R.; Borsook, D. Migraine Attacks the Basal Ganglia. *Mol. Pain* **2011**, *7*, 71. [CrossRef] [PubMed]
124. DaSilva, A.F.; Granziera, C.; Tuch, D.S.; Snyder, J.; Vincent, M.; Hadjikhani, N. Interictal alterations of the trigeminal somatosensory pathway and periaqueductal gray matter in migraine. *NeuroReport* **2007**, *18*, 301–305. [CrossRef] [PubMed]
125. Coppola, G.; Tinelli, E.; Lepre, C.; Iacovelli, E.; Di Lorenzo, C.; Di Lorenzo, G.; Serrao, M.; Pauri, F.; Fiermonte, G.; Bianco, F.; et al. Dynamic changes in thalamic microstructure of migraine without aura patients: A diffusion tensor magnetic resonance imaging study. *Eur. J. Neurol.* **2013**, *21*, 287-e13. [CrossRef] [PubMed]
126. Kim, J.; Kim, S.; Suh, S.-I.; Koh, S.-B.; Park, K.-W.; Oh, K. Interictal Metabolic Changes in Episodic Migraine: A Voxel-Based FDG-PET Study. *Cephalalgia* **2009**, *30*, 53–61. [CrossRef] [PubMed]
127. Vanagaite, J.; Pareja, J.; St⊘Ren, O.; White, L.; Sanc, T.; Stovner, L. Light-Induced Discomfort and Pain in Migraine. *Cephalalgia* **1997**, *17*, 733–741. [CrossRef] [PubMed]
128. Demarquay, G.; Royet, J.; Mick, G.; Ryvlin, P. Olfactory Hypersensitivity in Migraineurs: A $H_2{}^{15}O$-PET Study. *Cephalalgia* **2008**, *28*, 1069–1080. [CrossRef]
129. Stankewitz, A.; May, A. The phenomenon of changes in cortical excitability in migraine is not migraine-specific–A unifying thesis. *Pain* **2009**, *145*, 14–17. [CrossRef]
130. Mainero, C.; Bs, J.B.; Hadjikhani, N. Altered functional magnetic resonance imaging resting-state connectivity in periaqueductal gray networks in migraine. *Ann. Neurol.* **2011**, *70*, 838–845. [CrossRef]
131. Schwedt, T.J.; Berisha, V.; Chong, C.D. Temporal Lobe Cortical Thickness Correlations Differentiate the Migraine Brain from the Healthy Brain. *PLoS ONE* **2015**, *10*, e0116687. [CrossRef]
132. Chong, C.D.; Gaw, N.; Fu, Y.; Li, J.; Wu, T.; Schwedt, T.J. Migraine classification using magnetic resonance imaging resting-state functional connectivity data. *Cephalalgia* **2016**, *37*, 828–844. [CrossRef] [PubMed]
133. Tessitore, A.; Russo, A.; Giordano, A.; Conte, F.; Corbo, D.; De Stefano, M.; Cirillo, S.; Cirillo, M.; Esposito, F.; Tedeschi, G. Disrupted default mode network connectivity in migraine without aura. *J. Headache Pain* **2013**, *14*, 89. [CrossRef] [PubMed]
134. Schoenen, J.; Jacquy, J.; Lenaerts, M. Effectiveness of high-dose riboflavin in migraine prophylaxis A randomized controlled trial. *Neurology* **1998**, *50*, 466–470. [CrossRef] [PubMed]
135. Welch, K.; Levine, S.R.; D'Andrea, G.; Schultz, L.R.; Helpern, J.A. Preliminary observations on brain energy metabolism in migraine studied by in vivo phosphorus 31 NMR spectroscopy. *Neurology* **1989**, *39*, 538. [CrossRef] [PubMed]
136. Montagna, P.; Cortelli, P.; Monari, L.; Pierangeli, G.; Parchi, P.; Lodi, R.; Iotti, S.; Frassineti, C.; Zaniol, P.; Lugaresi, E.; et al. ^{31}P-Magnetic resonance spectroscopy in migraine without aura. *Neurology* **1994**, *44*, 666. [CrossRef]
137. Reyngoudt, H.; Paemeleire, K.; Descamps, B.; De Deene, Y.; Achten, E. 31P-MRS demonstrates a reduction in high-energy phosphates in the occipital lobe of migraine without aura patients. *Cephalalgia Int. J. Headache* **2011**, *31*, 1243–1253. [CrossRef]
138. Stankewitz, A.; Aderjan, D.; Eippert, F.; May, A. Trigeminal Nociceptive Transmission in Migraineurs Predicts Migraine Attacks. *J. Neurosci.* **2011**, *31*, 1937–1943. [CrossRef]
139. Krowicki, Z.K.; Kapusta, D.R. Microinjection of Glycine into the Hypothalamic Paraventricular Nucleus Produces Diuresis, Natriuresis, and Inhibition of Central Sympathetic Outflow. *J. Pharmacol. Exp. Ther.* **2011**, *337*, 247–255. [CrossRef]
140. Maniyar, F.H.; Sprenger, T.; Monteith, T.; Schankin, C.; Goadsby, P.J. Brain activations in the premonitory phase of nitroglycerin-triggered migraine attacks. *Brain* **2013**, *137*, 232–241. [CrossRef]
141. Schulte, L.H.; May, A. The migraine generator revisited: Continuous scanning of the migraine cycle over 30 days and three spontaneous attacks. *Brain* **2016**, *139*, 1987–1993. [CrossRef]
142. Denuelle, M.; Fabre, N.; Payoux, P.; Chollet, F.; Geraud, G. Hypothalamic Activation in Spontaneous Migraine Attacks. *Headache: J. Head Face Pain* **2007**, *47*, 1418–1426. [CrossRef]
143. Lashley, K.S. Patterns of cerebral integration indicated by the scotomas of migraine. *Arch. Neurol. Psychiatry* **1941**, *46*, 331. [CrossRef]
144. Leo, A.A.P.; Morison, R.S. Propagation of spreading cortical depression. *J. Neurophysiol.* **1945**, *8*, 33–45. [CrossRef]
145. Lauritzen, M.; Dreier, J.P.; Fabricius, M.; Hartings, A.J.; Graf, R.; Strong, A.J. Clinical Relevance of Cortical Spreading Depression in Neurological Disorders: Migraine, Malignant Stroke, Subarachnoid and Intracranial Hemorrhage, and Traumatic Brain Injury. *J. Cereb. Blood Flow Metab.* **2010**, *31*, 17–35. [CrossRef] [PubMed]
146. Olesen, J.; Larsen, B.; Lauritzen, M. Focal hyperemia followed by spreading oligemia and impaired activation of rcbf in classic migraine. *Ann. Neurol.* **1981**, *9*, 344–352. [CrossRef] [PubMed]
147. Leao, A.A.P.; de Baaij, J.H.F.; Hoenderop, J.G.J.; Bindels, R.J.M.; Funke, F.; Kron, M.; Dutschmann, M.; Müller, M.; Goadsby, P.J.; Mayevsky, A.; et al. Spreading depression of activity in the cerebral cortex. *J. Neurophysiol.* **1944**, *7*, 359–390. [CrossRef]

148. Hadjikhani, N.; del Rio, M.S.; Wu, O.; Schwartz, D.; Bakker, D.; Fischl, B.; Kwong, K.K.; Cutrer, F.M.; Rosen, B.R.; Tootell, R.B.H.; et al. Mechanisms of migraine aura revealed by functional MRI in human visual cortex. *Proc. Natl. Acad. Sci. USA* **2001**, *98*, 4687–4692. [CrossRef]
149. Headache Classification Committee of the International Headache Society (IHS). The International Classification of Headache Disorders, 3rd edition (beta version). *Cephalalgia* **2013**, *33*, 629–808. [CrossRef]
150. May, A.; Kaube, H.; Büchel, C.; Eichten, C.; Rijntjes, M.; Jüptner, M.; Weiller, C.; Diener, C.H. Experimental cranial pain elicited by capsaicin: A PET study. *Pain* **1998**, *74*, 61–66. [CrossRef]
151. Russell, F.A.; King, R.; Smillie, S.-J.; Kodji, X.; Brain, S.D.; Pressly, J.D.; Soni, H.; Jiang, S.; Wei, J.; Liu, R.; et al. Calcitonin Gene-Related Peptide: Physiology and Pathophysiology. *Physiol. Rev.* **2014**, *94*, 1099–1142. [CrossRef]
152. Stoker, K.; Baker, D.E. Erenumab-aooe. *Hosp. Pharm.* **2018**, *53*, 363–368. [CrossRef] [PubMed]
153. MacGregor, E.A. Migraine. *Ann. Intern. Med.* **2017**, *166*, ITC49–ITC64. [CrossRef] [PubMed]
154. Roy, A.; Adinoff, B.; DeJong, J.; Linnoila, M. Cerebrospinal fluid variables among alcoholics lack seasonal variation. *Acta Psychiatr. Scand.* **1991**, *84*, 579–582. [CrossRef] [PubMed]
155. Rashid, A.; Manghi, A. Calcitonin Gene-Related Peptide Receptor. In *StatPearls*; StatPearls Publishing: Treasure Island, FL, USA, 2023. Available online: http://www.ncbi.nlm.nih.gov/books/NBK560648/ (accessed on 24 July 2023).
156. Messlinger, K.; Russo, A.F. Current understanding of trigeminal ganglion structure and function in headache. *Cephalalgia* **2018**, *39*, 1661–1674. [CrossRef]
157. Edvinsson, J.C.A.; Warfvinge, K.; Krause, D.N.; Blixt, F.W.; Sheykhzade, M.; Edvinsson, L.; Haanes, K.A. C-fibers may modulate adjacent Aδ-fibers through axon-axon CGRP signaling at nodes of Ranvier in the trigeminal system. *J. Headache Pain* **2019**, *20*, 105. [CrossRef]
158. Sacco, S.; Kurth, T. Migraine and the Risk for Stroke and Cardiovascular Disease. *Curr. Cardiol. Rep.* **2014**, *16*, 524. [CrossRef]
159. Negro, A.; Martelletti, P. Gepants for the treatment of migraine. *Expert Opin. Investig. Drugs* **2019**, *28*, 555–567. [CrossRef]
160. Marcus, R.; Goadsby, P.J.; Dodick, D.; Stock, D.; Manos, G.; Fischer, T.Z. BMS-927711 for the acute treatment of migraine: A double-blind, randomized, placebo controlled, dose-ranging trial. *Cephalalgia* **2013**, *34*, 114–125. [CrossRef]
161. Becker, W.J. Acute Migraine Treatment in Adults. *Headache J. Head Face Pain* **2015**, *55*, 778–793. [CrossRef]
162. Smith, H.S. Potential analgesic mechanisms of acetaminophen. *Pain Physician* **2009**, *12*, 269–280. [CrossRef]
163. Taylor, F.R.; Kaniecki, R.G. Symptomatic Treatment of Migraine: When to Use NSAIDs, Triptans, or Opiates. *Curr. Treat. Options Neurol.* **2010**, *13*, 15–27. [CrossRef] [PubMed]
164. Colman, I.; Brown, M.D.; Innes, G.D.; Grafstein, E.; Roberts, T.E.; Rowe, B.H. Parenteral Dihydroergotamine for Acute Migraine Headache: A Systematic Review of the Literature. *Ann. Emerg. Med.* **2005**, *45*, 393–401. [CrossRef] [PubMed]
165. Silberstein, S.D.; McCrory, D.C.; Silberstein, F.S.D. Ergotamine and Dihydroergotamine: History, Pharmacology, and Efficacy. *Headache J. Head Face Pain* **2003**, *43*, 144–166. [CrossRef] [PubMed]
166. Sprenger, T.; Viana, M.; Tassorelli, C. Current Prophylactic Medications for Migraine and Their Potential Mechanisms of Action. *Neurotherapeutics* **2018**, *15*, 313–323. [CrossRef]
167. Lambru, G.; Andreou, A.P.; Guglielmetti, M.; Martelletti, P. Emerging drugs for migraine treatment: An update. *Expert Opin. Emerg. Drugs* **2018**, *23*, 301–318. [CrossRef]
168. Assas, M.B. Anti-migraine agents from an immunological point of view. *J. Transl. Med.* **2021**, *19*, 23. [CrossRef]
169. Mathew, R.; Andreou, A.P.; Chami, L.; Bergerot, A.; van den Maagdenberg, A.M.; Ferrari, M.D.; Goadsby, P.J. Immunohistochemical characterization of calcitonin gene-related peptide in the trigeminal system of the familial hemiplegic migraine 1 knock-in mouse. *Cephalalgia* **2011**, *31*, 1368–1380. [CrossRef]
170. Biella, G.; Panara, C.; Pecile, A.; Sotgiu, M. Facilitatory role of calcitonin gene-related peptide (CGRP) on excitation induced by substance P (SP) and noxious stimuli in rat spinal dorsal horn neurons. An iontophoretic study in vivo. *Brain Res.* **1991**, *559*, 352–356. [CrossRef]
171. Olesen, J.; Diener, H.-C.; Husstedt, I.W.; Goadsby, P.J.; Hall, D.; Meier, U.; Pollentier, S.; Lesko, L.M. Calcitonin Gene–Related Peptide Receptor Antagonist BIBN 4096 BS for the Acute Treatment of Migraine. *N. Engl. J. Med.* **2004**, *350*, 1104–1110. [CrossRef]
172. Ho, T.W.; Connor, K.M.; Zhang, Y.; Pearlman, E.; Koppenhaver, J.; Fan, X.; Lines, C.; Edvinsson, L.; Goadsby, P.J.; Michelson, D. Randomized controlled trial of the CGRP receptor antagonist telcagepant for migraine prevention. *Neurology* **2014**, *83*, 958–966. [CrossRef]
173. Diener, H.-C.; Barbanti, P.; Dahlöf, C.; Reuter, U.; Habeck, J.; Podhorna, J. BI 44370 TA, an oral CGRP antagonist for the treatment of acute migraine attacks: Results from a phase II study. *Cephalalgia* **2010**, *31*, 573–584. [CrossRef] [PubMed]
174. Voss, T.; Lipton, R.B.; Dodick, D.W.; Dupre, N.; Ge, J.Y.; Bachman, R.; Assaid, C.; Aurora, S.K.; Michelson, D. A phase IIb randomized, double-blind, placebo-controlled trial of ubrogepant for the acute treatment of migraine. *Cephalalgia* **2016**, *36*, 887–898. [CrossRef] [PubMed]
175. Goadsby, P.J.; Reuter, U.; Hallström, Y.; Broessner, G.; Bonner, J.H.; Zhang, F.; Sapra, S.; Picard, H.; Mikol, D.D.; Lenz, R.A. A Controlled Trial of Erenumab for Episodic Migraine. *N. Engl. J. Med.* **2017**, *377*, 2123–2132. [CrossRef] [PubMed]
176. Ashina, M.; Dodick, D.; Goadsby, P.J.; Reuter, U.; Silberstein, S.; Zhang, F.; Gage, J.R.; Cheng, S.; Mikol, D.D.; Lenz, R.A. Erenumab (AMG 334) in episodic migraine: Interim analysis of an ongoing open-label study. *Neurology* **2017**, *89*, 1237–1243. [CrossRef]

177. Tepper, S.; Ashina, M.; Reuter, U.; Brandes, J.L.; Doležil, D.; Silberstein, S.; Winner, P.; Leonardi, D.; Mikol, D.; Lenz, R. Safety and efficacy of erenumab for preventive treatment of chronic migraine: A randomised, double-blind, placebo-controlled phase 2 trial. *Lancet Neurol.* **2017**, *16*, 425–434. [CrossRef]
178. Dodick, D.W.; Goadsby, P.J.; Spierings, E.L.H.; Scherer, J.C.; Sweeney, S.P.; Grayzel, D.S. Safety and efficacy of LY2951742, a monoclonal antibody to calcitonin gene-related peptide, for the prevention of migraine: A phase 2, randomised, double-blind, placebo-controlled study. *Lancet Neurol.* **2014**, *13*, 885–892. [CrossRef]
179. Skljarevski, V.; Oakes, T.M.; Zhang, Q.; Ferguson, M.B.; Martinez, J.; Camporeale, A.; Johnson, K.W.; Shan, Q.; Carter, J.; Schacht, A.; et al. Effect of Different Doses of Galcanezumab vs Placebo for Episodic Migraine Prevention: A Randomized Clinical Trial. *JAMA Neurol.* **2018**, *75*, 187–193. [CrossRef]
180. Skljarevski, V.; Matharu, M.; Millen, B.A.; Ossipov, M.H.; Kim, B.-K.; Yang, J.Y. Efficacy and safety of galcanezumab for the prevention of episodic migraine: Results of the EVOLVE-2 Phase 3 randomized controlled clinical trial. *Cephalalgia* **2018**, *38*, 1442–1454. [CrossRef]
181. Bigal, M.E.; Dodick, D.W.; Rapoport, A.M.; Silberstein, S.D.; Ma, Y.; Yang, R.; Loupe, P.S.; Burstein, R.; Newman, L.C.; Lipton, R.B. Safety, tolerability, and efficacy of TEV-48125 for preventive treatment of high-frequency episodic migraine: A multicentre, randomised, double-blind, placebo-controlled, phase 2b study. *Lancet Neurol.* **2015**, *14*, 1081–1090. [CrossRef]
182. Dodick, D.W.; Silberstein, S.D.; Bigal, M.E.; Yeung, P.P.; Goadsby, P.J.; Blankenbiller, T.; Grozinski-Wolff, M.; Yang, R.; Ma, Y.; Aycardi, E. Effect of Fremanezumab Compared with Placebo for Prevention of Episodic Migraine. *JAMA* **2018**, *319*, 1999–2008. [CrossRef]
183. Bigal, M.E.; Dodick, D.W.; Krymchantowski, A.V.; VanderPluym, J.H.; Tepper, S.J.; Aycardi, E.; Loupe, P.S.; Ma, Y.; Goadsby, P.J. TEV-48125 for the preventive treatment of chronic migraine: Efficacy at early time points. *Neurology* **2016**, *87*, 41–48. [CrossRef]
184. Silberstein, S.D.; Dodick, D.W.; Bigal, M.E.; Yeung, P.P.; Goadsby, P.J.; Blankenbiller, T.; Grozinski-Wolff, M.; Yang, R.; Ma, Y.; Aycardi, E. Fremanezumab for the Preventive Treatment of Chronic Migraine. *N. Engl. J. Med.* **2017**, *377*, 2113–2122. [CrossRef] [PubMed]
185. Dodick, D.W.; Goadsby, P.J.; Silberstein, S.D.; Lipton, R.B.; Olesen, J.; Ashina, M.; Wilks, K.; Kudrow, D.; Kroll, R.; Kohrman, B.; et al. Safety and efficacy of ALD403, an antibody to calcitonin gene-related peptide, for the prevention of frequent episodic migraine: A randomised, double-blind, placebo-controlled, exploratory phase 2 trial. *Lancet Neurol.* **2014**, *13*, 1100–1107. [CrossRef] [PubMed]
186. Hasselmo, M.E. Neuromodulation: Acetylcholine and memory consolidation. *Trends Cogn. Sci.* **1999**, *3*, 351–359. [CrossRef]
187. Tiwari, V.; Agrawal, S. Migraine and Neuromodulation: A Literature Review. *Cureus* **2022**, *14*, e31223. [CrossRef]
188. Barloese, M.; Lambru, G. Methodological Difficulties in Clinical Trials Assessing Neuromodulation Devices in the Headache Field. In *Neuromodulation in Headache and Facial Pain Management*; Lambru, G., Lanteri-Minet, M., Eds.; Springer International Publishing: Cham, Switzerland, 2020; pp. 227–239. [CrossRef]
189. Oshinsky, M.L.; Murphy, A.L.; Hekierski, H.; Cooper, M.; Simon, B.J. Noninvasive vagus nerve stimulation as treatment for trigeminal allodynia. *Pain* **2014**, *155*, 1037–1042. [CrossRef]
190. Puledda, F.; Shields, K. Non-Pharmacological Approaches for Migraine. *Neurotherapeutics* **2018**, *15*, 336–345. [CrossRef]
191. Reuter, U.; McClure, C.; Liebler, E.; Pozo-Rosich, P. Non-invasive neuromodulation for migraine and cluster headache: A systematic review of clinical trials. *J. Neurol. Neurosurg. Psychiatry* **2019**, *90*, 796–804. [CrossRef]
192. Yuan, H.; Chuang, T.-Y. Update of Neuromodulation in Chronic Migraine. *Curr. Pain Headache Rep.* **2021**, *25*, 71. [CrossRef]
193. Schoenen, J.; Vandersmissen, B.; Jeangette, S.; Herroelen, L.; Vandenheede, M.; Gérard, P.; Magis, D. Migraine prevention with a supraorbital transcutaneous stimulator: A randomized controlled trial. *Neurology* **2013**, *80*, 697–704. [CrossRef]
194. E Chou, D.; Yugrakh, M.S.; Winegarner, D.; Rowe, V.; Kuruvilla, D.; Schoenen, J. Acute migraine therapy with external trigeminal neurostimulation (ACME): A randomized controlled trial. *Cephalalgia* **2018**, *39*, 3–14. [CrossRef]
195. Miller, S.; Watkins, L.; Matharu, M. Long-term outcomes of occipital nerve stimulation for chronic migraine: A cohort of 53 patients. *J. Headache Pain* **2016**, *17*, 68. [CrossRef]
196. Garcia-Ortega Rn, R.; Rn, T.E.; Moir, L.; Aziz, T.Z.; Green, A.L.; Fitzgerald, J.J. Burst Occipital Nerve Stimulation for Chronic Migraine and Chronic Cluster Headache. *Neuromodul. Technol. Neural Interface* **2018**, *22*, 638–644. [CrossRef]
197. Ashkan, K.; Sokratous, G.; Göbel, H.; Mehta, V.; Gendolla, A.; Dowson, A.; Wodehouse, T.; Heinze, A.; Gaul, C. Peripheral nerve stimulation registry for intractable migraine headache (RELIEF): A real-life perspective on the utility of occipital nerve stimulation for chronic migraine. *Acta Neurochir.* **2020**, *162*, 3201–3211. [CrossRef]
198. De Agostino, R.; Federspiel, B.; Cesnulis, E.; Sandor, P.S. High-Cervical Spinal Cord Stimulation for Medically Intractable Chronic Migraine. *Neuromodul. Technol. Neural Interface* **2015**, *18*, 289–296. [CrossRef]
199. Tassorelli, C.; Diener, H.-C.; Silberstein, S.D.; Dodick, D.W.; Goadsby, P.J.; Jensen, R.H.; Magis, D.; Pozo-Rosich, P.; Yuan, H.; Martinelli, D.; et al. Guidelines of the International Headache Society for clinical trials with neuromodulation devices for the treatment of migraine. *Cephalalgia* **2021**, *41*, 1135–1151. [CrossRef]
200. Dodick, D.W. A Phase-by-Phase Review of Migraine Pathophysiology. *Headache J. Head Face Pain* **2018**, *58*, 4–16. [CrossRef]
201. Blom, R.M.; Figee, M.; Vulink, N.; Denys, D. Update on repetitive transcranial magnetic stimulation in obsessive-compulsive disorder: Different targets. *Curr. Psychiatry Rep.* **2011**, *13*, 289–294. [CrossRef]
202. Lan, L.; Zhang, X.; Li, X.; Rong, X.; Peng, Y. The efficacy of transcranial magnetic stimulation on migraine: A meta-analysis of randomized controlled trails. *J. Headache Pain.* **2017**, *18*, 1–7. [CrossRef] [PubMed]

203. Shehata, H.S.; Esmail, E.H.; Abdelalim, A.; El-Jaafary, S.; Elmazny, A.; Sabbah, A.; Shalaby, N.M. Repetitive transcranial magnetic stimulation versus botulinum toxin injection in chronic migraine prophylaxis: A pilot randomized trial. *J. Pain Res.* **2016**, *9*, 771–777. [CrossRef]
204. Kalita, J.; Laskar, S.; Bhoi, S.K.; Misra, U.K. Efficacy of single versus three sessions of high rate repetitive transcranial magnetic stimulation in chronic migraine and tension-type headache. *J. Neurol.* **2016**, *263*, 2238–2246. [CrossRef] [PubMed]
205. Granato, A.; Fantini, J.; Monti, F.; Furlanis, G.; Musho Ilbeh, S.; Semenic, M.; Manganotti, P. Dramatic placebo effect of high frequency repetitive TMS in treatment of chronic migraine and medication overuse headache. *J. Clin. Neurosci.* **2019**, *60*, 96–100. [CrossRef] [PubMed]

biomedicines

MDPI

Review

A Comprehensive Review on Neuroimmunology: Insights from Multiple Sclerosis to Future Therapeutic Developments

Lucian Eva [1], Horia Pleș [2,*], Razvan-Adrian Covache-Busuioc [3,*], Luca Andrei Glavan [3], Bogdan-Gabriel Bratu [3], Andrei Bordeianu [3], David-Ioan Dumitrascu [3], Antonio Daniel Corlatescu [3] and Alexandru Vlad Ciurea [3]

[1] Clinical Emergency Hospital "Prof. Dr. Nicolae Oblu", 700309 Iasi, Romania; elucian73@yahoo.com
[2] Department of Neurosurgery, Centre for Cognitive Research in Neuropsychiatric Pathology (NeuroPsy-Cog), "Victor Babeș" University of Medicine and Pharmacy, 300041 Timisoara, Romania
[3] Department of Neurosurgery, "Carol Davila" University of Medicine and Pharmacy, 020021 București, Romania; luca-andrei.glavan0720@stud.umfcd.ro (L.A.G.); bogdan.bratu@stud.umfcd.ro (B.-G.B.); andrei.bordeianu@stud.umfcd.ro (A.B.); david-ioan.dumitrascu0720@stud.umfcd.ro (D.-I.D.); antonio.corlatescu0920@stud.umfcd.ro (A.D.C.); prof.avciurea@gmail.com (A.V.C.)
* Correspondence: ples.horia@umft.ro (H.P.); razvan-adrian.covache-busuioc0720@stud.umfcd.ro (R.-A.C.-B.)

Abstract: This review delves into neuroimmunology, focusing on its relevance to multiple sclerosis (MS) and potential treatment advancements. Neuroimmunology explores the intricate relationship between the immune system and the central nervous system (CNS). Understanding these mechanisms is vital for grasping the pathophysiology of diseases like MS and for devising innovative treatments. This review introduces foundational neuroimmunology concepts, emphasizing the role of immune cells, cytokines, and blood–brain barrier in CNS stability. It highlights how their dysregulation can contribute to MS and discusses genetic and environmental factors influencing MS susceptibility. Cutting-edge research methods, from omics techniques to advanced imaging, have revolutionized our understanding of MS, offering valuable diagnostic and prognostic tools. This review also touches on the intriguing gut–brain axis, examining how gut microbiota impacts neuroimmunological processes and its potential therapeutic implications. Current MS treatments, from immunomodulatory drugs to disease-modifying therapies, are discussed alongside promising experimental approaches. The potential of personalized medicine, cell-based treatments, and gene therapy in MS management is also explored. In conclusion, this review underscores neuroimmunology's significance in MS research, suggesting that a deeper understanding could pave the way for more tailored and effective treatments for MS and similar conditions. Continued research and collaboration in neuroimmunology are essential for enhancing patient outcomes.

Keywords: multiple sclerosis; neuroimmunology; therapeutic development

Citation: Eva, L.; Pleș, H.; Covache-Busuioc, R.-A.; Glavan, L.A.; Bratu, B.-G.; Bordeianu, A.; Dumitrascu, D.-I.; Corlatescu, A.D.; Ciurea, A.V. A Comprehensive Review on Neuroimmunology: Insights from Multiple Sclerosis to Future Therapeutic Developments. *Biomedicines* **2023**, *11*, 2489. https://doi.org/10.3390/biomedicines11092489

Academic Editor: Sasa Cucnik

Received: 9 August 2023
Revised: 2 September 2023
Accepted: 6 September 2023
Published: 8 September 2023

1. Introduction

1.1. Background

Neuroimmunology is an interdisciplinary field that brings together knowledge from biology, immunology, chemistry, neurology, pathology, psychiatry, and virology to examine the intricate interrelations between the central nervous system (CNS) and immune system (IS), its interactions during various developmental stages, as well as maintaining homeostasis or responding to injuries. Neuroimmunologists primarily seek to develop strategies to treat or even prevent neuroimmunological diseases by understanding these complex interactions in depth [1].

Tradition holds that the immune system and brain operate independently, separated by the blood–brain barrier (BBB). Yet, over recent decades, these long-held beliefs have been vigorously disproved, and evidence now exists to demonstrate otherwise. Not only does the nervous system receive communication from immune cells directly, but brain signals actively regulate immune functions—resulting in inflammation occurring elsewhere outside the central nervous system [2].

An intriguing fact about neuroimmunology is that it did not receive its introduction on PubMed until 1982—coinciding with both its inaugural congress in Stresa, Italy, as well as the launch of the *Journal of Neuroimmunology* that same year. Neuroimmunology research has traditionally centered around multiple sclerosis (MS). It is essential to acknowledge, however, that immune responses can also be seen in many other conditions like Guillain–Barré syndrome (GBS), white matter diseases, psychiatric disorders, infections, trauma, and neurodegenerative diseases that are traditionally considered more "cell autonomous".

1.2. Purpose and Scope of the Review

Over time, neuroimmunology has grown into an interdisciplinary field focused on understanding the complex relationship between nervous and immune systems. Thanks to extensive research in this field, significant progress has been made in uncovering mechanisms associated with various neurological conditions—multiple sclerosis (MS) being one such disorder that sheds light on this complicated interrelationship [3].

This review seeks to assess the current level of understanding in neuroimmunology, specifically regarding multiple sclerosis research. By compiling recent discoveries, clinical data, and advances in this area of science, this comprehensive overview aims to provide an in-depth knowledge of both MS's pathophysiological mechanisms and their significance relative to other neurological conditions.

1.3. Methodology of Review Selection

For this comprehensive review on neuroimmunology, with particular reference to insights gained from multiple sclerosis for future therapeutic advancement, a methodical and rigorous approach was taken in selecting and assessing literature sources. A specific methodology for selecting and assessing literature sources has been devised:

(a) Search Strategy: An extensive literature search was performed using multiple electronic databases such as PubMed, Scopus, Web of Science, and Google Scholar. Our search strategy comprised pertinent keywords and phrases related to neuroimmunology, multiple sclerosis, neuroinflammation, immune system dysfunction, central nervous system therapies, as well as therapeutic interventions with no restrictions placed on publication dates, allowing a broad representation.

(b) Inclusion and Exclusion Criteria:

To ensure the quality and relevance of this review, specific inclusion and exclusion criteria were used during the selection process. Articles meeting certain criteria were selected for inclusion: studies or research articles focused on neuroimmunology; publications providing insight into pathophysiology, etiology, clinical aspects, and immunology implications related to multiple sclerosis as they apply to neuroimmunology; clinical trials/experimental/observational research investigating therapeutic developments/interventions regarding MS/related neuroinflammatory conditions; and publications available either in English or with accessible English translations to ensure comprehension.

Subsequently, certain articles were disqualified: non-peer-reviewed materials such as conference abstracts or unpublished manuscripts were removed; studies with limited relevance to neuroimmunology of multiple sclerosis or not meeting its primary objectives were disqualified, as were articles focusing solely on non-neurological immune-related disorders or general immunology without direct ties to neuroimmunology, which were specifically excluded to ensure reliability in our findings. This process ensured our results would stand up against scrutiny.

2. The Basics of Neuroimmunology

2.1. Definition and Overview

Neuroimmunology is an interdisciplinary field devoted to understanding the complex relationships and interactions between the nervous system (including the brain and spinal cord) and immune system (including antibodies and their targets) in maintaining equilibrium in our bodies and responding to infections, injuries, or diseases that affect these vital systems. Neuroimmunology offers an examination of this topic that spans numerous fields [4].

Multiple sclerosis (MS) is one of the most prevalent disabling neurological afflictions among young adults, typically appearing between 20 and 40 years of age.

MS is an autoimmune condition in which immune system cells, normally responsible for protecting against viruses, bacteria, and abnormal cells in the body, attack myelin in the central nervous system (brain, optic nerves, and spinal cord). Myelin acts as a protective substance by creating sheaths (myelin sheaths) around nerve fibers (axons).

MS is a chronic condition that varies considerably among its victims, from mild cases with limited disability to progressive decline leading to greater disability over time. Most commonly seen are intermittent symptoms surfacing followed by periods of relative quiescence or dormancy and then either partial or full recovery—MS is rarely fatal. Individuals diagnosed tend to have life expectancies comparable to the general population [5].

2.2. Key Players in Neuroimmunology: Cells and Molecules

As part of their shared response to environmental challenges, the nervous and immune systems have formed an interdependent communication mechanism between themselves in response to environmental challenges. Neurons exhibit various receptors found on immune cells, such as Toll-like receptors (TLRs) and inflammatory cytokine receptors found on immune cells; this allows immune cells to influence and regulate neuronal activity—for instance, using IL-1β to sensitize sensory neurons during inflammation while managing pain levels [6].

Immune cells are capable of sensing signals from neurons by expressing receptors for neurotransmitters and neuropeptides produced by neurons; for example, innate lymphoid cells express such receptors for neuropeptides as Calcitonin Gene-Related Peptide (CGRP) and Neuromedin U (NMU). This mutual sensing between immune cells and neurons has proved highly advantageous, decreasing costs related to dealing with certain insults and helping coordinate complex host responses more efficiently.

Furthermore, the microbiome—or collection of microorganisms that live inside your body—plays a key role in both neuronal activation and immune development. Immune cells and neurons both interact directly or indirectly with microbes present in the environment, making their composition key for shaping neuronal programming and maturation. In turn, this affects various aspects of intestinal physiology, including visceral pain management, gut motility regulation, and related functions. This interaction between the nervous system, immune system, and microbiome plays an intricate, evolutionary, beneficial process that plays its part in controlling host responses overall [7–9].

2.3. Neuroimmune Communication Pathways

Recent research has demonstrated that the peripheral immune system and nervous system can communicate effectively by using similar molecular signaling cues.

Veiga-Fernandes and Pachnis' work presents the idea of an "enteric neuroimmune cell unit," an internal sensory organ responsible for protecting and maintaining intestinal integrity and function. This enteric neuroimmune system plays an essential role in providing both innate defenses and memory responses against certain pathogens. During gestation, extracellular signals coordinate the development of enteric neuroprogenitor and hematopoietic cells, resulting in this complex network. Postnatal development of this system takes place through colonization with commensal microbes that colonize the gut, leading to mutual signals between neuronal–glial cells and tissue-resident immune cells that stimulate the maturation of an enteric neuroimmune system [10].

It is believed that gut microbiomes have an influence over peripheral immune cells and central nervous system (CNS)-residing cells, as well as having an influence over brain development and disease progression. They stress the role of commensal microbes in producing short-chain fatty acids and aryl hydrocarbon ligands that may alter glial cell development and function within the CNS. Neuroendocrine mediators produced through the hypothalamic–pituitary–adrenal axis can influence intestinal permeability, immune-cell activation, and the composition of gut microbiomes. Dysbiosis of microbiota has been observed in neurological and psychiatric conditions, suggesting it could contribute to their causes; however, more research needs to be conducted in order to uncover its exact nature and causal role.

Prinz and Priller examine how peripheral immune cells may enter the CNS under pathological circumstances. A healthy CNS does not normally contain blood-borne immune cells, as immune surveillance is provided by tissue-resident microglia, meningeal macrophages, and perivascular macrophages, which produce their own immunological mediators. In conditions like autoimmune diseases, infections, or injuries, where blood–brain barrier permeability changes, activated adaptive immune cells are allowed into CNS via fenestrated capillaries, contributing towards disease progression [11].

Engelhardt and colleagues investigate the CNS's immunological advantages towards peripheral immune cells under normal circumstances, distinguishing between lymphatic drainage of cerebrospinal fluid that bathes meninges and cerebral ventricles from its lack of drainage in interstitial fluid bathing CNS parenchymal tissues, such as interstitial drainage. They provide a detailed anatomical account of both human and rodent brain barriers limiting access to CNS parenchyma, which prevent lymphatic drainage of cerebrospinal fluid from meninges/ventricles, while discussing both trafficking of cells/solutes through these barrier sites via their perivascular areas allowing lymphatic drainage or non-drainage channels [12].

Further research explores neurological impairments caused by acute infections caused by neurotropic pathogens. Acute infections trigger the release of pro-inflammatory cytokines by astrocytes, microglia, and leukocytes into the CNS from cells like astrocytes and microglia, which release these inflammatory agents into circulation, leading to potential effects on blood–brain barrier integrity as well as symptoms like fatigue, hypersomnia, cognitive difficulties, or chronic inflammation that persists beyond antimicrobial treatments such as treating pathogens like memory deficits, depression or mood disorders—raising questions regarding the involvement of immunological factors in different neurological diseases [13].

2.4. The Role of the Immune System in Maintaining Neuronal Health

The immune system plays a pivotal role in protecting and maintaining neuronal health in multiple ways. One mechanism by which it affects neuronal activity is through signaling molecules known as cytokines that control immune responses and may have an effect on neuronal activity—proinflammatory cytokines such as IL-1, IL-6, and TNFα are believed to trigger fever responses during infections by increasing body temperature to counter infection symptoms [14].

Immune cells become activated upon contact with infectious agents, producing proinflammatory cytokines, which then induce the production and release of prostaglandins in the brain—specifically PGE_2. PGE_2 plays an essential role here, acting on thermosensitive neurons in the hypothalamus to induce fever; specifically, its effect is known to activate thermosensitive neurons within this area and induce fever by inducing thermosensitivity responses from thermosensitive neurons. This then triggers further heat production via thermosensitive neurons within the liver, followed by a sustained phase by brain astrocyte production of IL-6, which further stimulates PGE_2 production; all this action provides a sustained phase fever response [15].

Cytokines play a pivotal role in activating the hypothalamic–pituitary–adrenal (HPA) axis. When immune cells release cytokines in response to pathogens, PGE_2 is produced within brain vasculature and interacts with catecholaminergic neurons before projecting onto corticotropin-releasing hormone-containing neurons of the hypothalamus; these send projections onto corticotropin-releasing hormone (CRH), leading to elevated levels of ACTH and corticosterone. Additionally, cytokines directly influence the pituitary gland to increase the release of ACTH. This complex process demonstrates just how powerfully these proteins exert diverse effects upon this part of our immune response as well as stress regulation mechanisms [16].

Longer exposure to cytokines may result in glucocorticoid resistance, making the body less responsive to their effects and diminishing their effects on homeostasis and stress responses. Resistance predominantly manifests itself at the level of the hippocampus and impairs regulation of the HPA axis, consequently compromising one's ability to effectively respond to stressors and maintain homeostasis [17].

Communication between the immune and nervous systems, enabled by cytokines, plays an integral part in protecting neuronal health and orchestrating physiological responses during infections or inflammation events. A thorough understanding of these complex pathways is vital to unlocking their underlying mechanisms as well as their profound implications for overall health and disease development; such knowledge could pave the way for novel therapeutic interventions aimed at maintaining harmony between these vital systems for increased human wellbeing [14].

3. The Neuroimmunology of Multiple Sclerosis

3.1. Pathophysiology of MS: Neuroimmunological Perspective

Multiple sclerosis (MS) is a devastating autoimmune condition affecting the central nervous system (CNS), comprising both the brain and spinal cord. MS is caused by immune attacks targeting myelin sheaths encasing nerve fibers axons in the CNS. These attacks cause inflammation, demyelination (depleted myelin levels), and damage to nerve fibers encased within it, resulting in inflammation, demyelination (depletion of myelin), and damage [18].

MS is a debilitating disease, targeting not only white matter axons covered in myelin but also nerve cell bodies in the gray matter of the brain and optic nerves to transmit information between the eye and brain. Over time, MS can lead to cortical atrophy, which shrinks the outermost layer of the cerebral cortex.

Individuals living with MS often experience various symptoms based on the intensity, location, and extent of inflammation, as well as the plaques' distribution across specific body areas, including the brain stem, cerebellum (associated with balance and coordination), spinal cord, optic nerves, and white matter surrounding brain ventricles (fluid-filled cavities). The variations in plaque distribution contribute to various degrees of symptoms being experienced by MS patients [5].

Research indicates a potential link between disrupted energy metabolism in the CNS of MS patients. Higher blood pyruvate levels during fasting and after meals in MS patients experiencing relapses had been found, as well as a potential flaw in pyruvate metabolism in MS. Other researchers have also noted elevated fasting pyruvate levels in MS. However, some studies showed normal fasting lactate levels or elevated levels in only a few MS patients. An unusual increase in blood pyruvate after glucose consumption has been reported. These findings point towards potential irregularities in pyruvate metabolism in MS patients. Moreover, the rise in Krebs cycle acids, such as alpha-ketoglutarate when fasting and citrate after glucose consumption, in MS patients reinforces the idea of disrupted pyruvate metabolism being linked to MS progression. Increased pyruvate and α-ketoglutarate levels in MS were also observed. Elevated enzyme activity, including enolase, pyruvate kinase, lactate dehydrogenase (Ldh), and aldolase in the CSF of those with disseminated sclerosis, suggests they could be markers of active demyelination.

3.2. The Role of the Immune System in MS

3.2.1. The Blood–Brain Barrier

One striking characteristic of MS lesions is the disruption of the blood–brain barrier (BBB), making its understanding a key aspect of understanding its pathogenesis [19]. The BBB serves as both a functional and anatomical separation between blood flow in the central nervous system (CNS) and neurons located outside. It consists of the vascular wall, CNS astrocytes covered with glia limitans covering them, and the perivascular space between [20]. Furthermore, its functioning provides many essential functions essential for proper brain function, such as the regulation of proper ionic concentrations.

While the term "barrier" might connote something static, the blood–brain barrier (BBB) is actually dynamic, providing processes like immunological surveillance. Lesions occur when leukocytes migrate into the brain, causing inflammation. This process comprises two steps, starting with initial migration across postcapillary venules into the perivascular space, followed by passage through the glia limitans to brain parenchyma [20]. The perivascular space provides monocytes with an area for normal immunosurveillance while also serving as lymphatic drainage. Furthermore, animal models show two additional routes through which leukocytes enter the CNS. Understanding these pathways provides insight into the workings of the blood–cerebrospinal fluid (CSF) barrier (BBB), including its function in immune cell infiltration during MS (Figure 1).

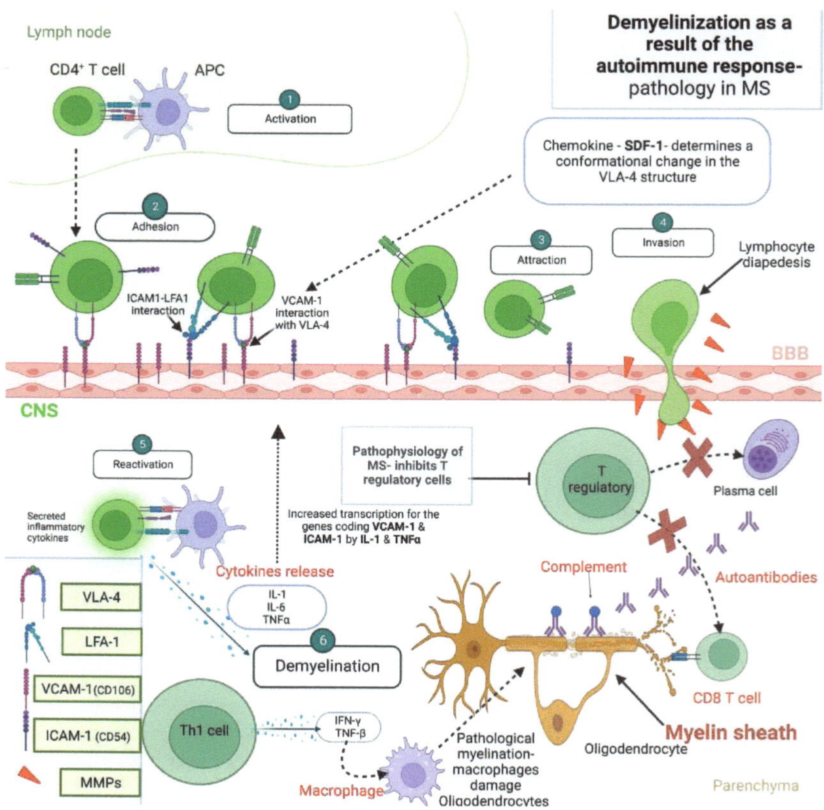

Figure 1. The immunopathogenic model of multiple sclerosis (MS) posits that a confluence of genetic and environmental variables, including viral infections, bacterial exposure, and superantigens, synergistically contribute to the heightened activation of myelin-reactive T cells within the circulatory

system of individuals afflicted with MS. This phenomenon is further exacerbated by the elevated expression of endothelial adhesion molecules, namely intercellular adhesion molecule-1 (ICAM-1) and vascular cell adhesion molecule-1 (VCAM-1), which facilitate the translocation of T cells into the central nervous system (CNS). The motility of these T cells is augmented by chemokines and the synthesis of matrix metalloproteinases, which enzymatically degrade extracellular matrix proteins to smooth the migration pathway. This collectively culminates in the T cells' successful transgression of the blood–brain barrier (BBB). Upon entry into the CNS, T cells become activated via interactions with antigen-presenting cells, initiating the release of pro-inflammatory and cytotoxic mediators. These substances inflict tissue damage and compromise the integrity of the myelin sheath through several mechanisms, including cytokine-mediated injury, antibody-mediated phagocytosis of myelin antigens by macrophages, complement system activation, and direct cellular damage mediated by CD4+ and CD8+ T cells. Regulatory changes at the transcriptional level, instigated by cytokines such as tumor necrosis factor-alpha (TNF-α) and interleukin-1 (IL-1), lead to the upregulation of VCAM-1 in endothelial cells. Notably, the integrin very late antigen-4 (VLA-4) does not bind to its designated ligands, including VCAM-1 and fibronectin, unless activated by chemotactic agents or additional stimuli, often synthesized by endothelial cells or other cellular entities at the site of inflammation. One such activating chemokine is stromal cell-derived factor-1 (SDF-1). ICAM-1, a type of intercellular adhesion molecule, is constitutively present in nominal concentrations on leukocytes and endothelial cells. In response to cytokine stimulation, however, these concentrations markedly escalate. ICAM-1 serves as a ligand for lymphocyte function-associated antigen-1 (LFA-1), an integrin present in leukocytes. LFA-1 plays a pivotal role in leukocyte emigration, facilitating their egress from the bloodstream and entry into tissues. Moreover, LFA-1 partakes in the processes of cytotoxic T-cell-mediated and antibody-mediated cellular destruction, enacted by granulocytes and monocytes. The mechanism governing T-lymphocyte infiltration into the CNS is explicable via the elevated expression of adhesion molecules on the endothelial cells constituting the BBB in MS cases. This assertion is further substantiated by the observation of these molecules on inflammatory cells such as macrophages and lymphocytes in MS lesions. It is a widely accepted notion that cytokines released by Th1 cells, including interferon-gamma (IFN-γ) and TNF-beta (TNF-β), are capable of macrophage activation, subsequently causing damage to oligodendrocytes and leading to pathologic alterations in myelination.

3.2.2. The Role of T Cells

T cells play an essential role in the pathogenesis of multiple sclerosis (MS), as evidenced by EAE animal models and HLA class II genes that strongly associate with MS. EAE, which mimics MS in animals, can be caused by immunizing rodents with myelin peptides, leading to an immune response targeting myelin in the central nervous system (CNS). Initial investigations suggested that auto-reactive CD4+ T cells producing interferon-gamma (IFN-γ) were the main contributors to inflammation in MS lesions; however, more recent investigations have thrown other cell types, particularly T helper 17 (Th17) cells, into focus as primary aggressors of MS lesions. These Th17 cells are responsible for producing pro-inflammatory cytokines such as IL-17 and IL-6, while their activity is governed by IL-23. These findings demonstrate the complex immune response involved in MS pathogenesis as well as Th17 cells' participation in driving inflammation processes [21].

CD4+ Th17 cells play an essential part in the development of EAE. MS lesions exhibit both CD4+ and CD8+ cells, which express IL-17. Th17 cells use the CCR6 chemokine receptor to travel from the choroid plexus into the cerebrospinal fluid (CSF) and perivascular spaces, inducing inflammation and damaging neurons and glial cells along their path. Th17 cells play a pivotal role in EAE and MS pathogenesis by secreting GM-CSF, exacerbating an already intense inflammatory response. Furthermore, they break through the blood–brain barrier (BBB) with their release of IL-17 and IL-22, breaking it down further by drawing in immune cells into the CNS through disruption of BBB. These findings demonstrate their crucial contribution to both diseases' pathologies, contributing to inflammation and damage seen within MS lesions [22].

Chemokines play an essential role in recruiting immune cells across the blood–brain barrier (BBB) during an episode of inflammation. MS lesions involve specific endothelial-wall-expressed chemokines interacting with receptors present on T cells to enable their extravasation into the CNS. Our research focused on investigating the expression of different chemokine receptors on T-cell clones derived from both blood and cerebrospinal fluid of an MS patient treated with glatiramer acetate (GA). CCR_4, CCR_5, CCR_6, and $CXCR_3$ emerged as particularly pertinent receptors to investigate. Their activation coincided with T-cell migration patterns that corresponded with specific receptor expression patterns, suggesting these receptors might play an essential role in migration and infiltration into the CNS during MS inflammation [23].

Based on this study's findings, reduced glatiramer acetate (GA) efficacy may be attributable to inflammation and early activation of GA-reactive cells. A better understanding of T-cell migration into the central nervous system (CNS) holds great promise in potentially increasing GA's effectiveness and ultimately bettering treatment outcomes for MS patients. By exploring mechanisms governing T-cell infiltration into CNS, we may open doors for more targeted and efficient therapies that benefit individuals living with MS [24].

3.2.3. The Role of B Cells

Immunoglobulin G1, present in most diagnosed multiple sclerosis (MS) patients' cerebrospinal fluid (CSF), may point to B cells as potential agents in its pathogenesis. Myelin-reactive antibodies have been detected, yet their exact significance remains uncertain, similar to antigens' significance. Immunoglobulin and complement are present in lesions associated with MS, suggesting their possible pathogenicity. Additionally, in patients at later stages of MS, B lymphoid follicles, T cells, and antigen-presenting cells have been detected in the meninges, suggesting potential contributions by B cells through antigen presentation, cell interactions, or immunoglobulin production by plasma cells. However, it must be kept in mind that B-cell activity in MS could be a result of an autoimmune reaction and not be the sole trigger of multiple sclerosis. More research needs to be conducted in order to fully comprehend how B cells play into multiple sclerosis pathogenesis [25].

Experimental animal models like EAE have well-documented antigens responsible for disease, such as myelin proteins. Unfortunately, for humans with multiple sclerosis (MS), specific triggers that trigger this illness remain undecided despite numerous research efforts examining various candidate antigens, but none have emerged as definitive culprits [26].

Due to similarities between EAE and MS, myelin or myelin-derived peptides were initially considered promising candidates as potential causal agents. However, responses to antigens have proven nonspecific, suggesting the involvement of multiple antigens or epitopes that spread once the disease begins. One recently proposed antigen is ab-crystallin, an antibody-binding protein not naturally found in human myelin but observed in early active MS lesions. Patients diagnosed with MS possess antibodies against ab-crystallin in their cerebrospinal fluid (CSF). Studies on mice demonstrated that knocking out its gene caused more intense EAE with elevated cytokine levels, suggesting its protective role has been compromised by an autoimmune response that disrupted it [27].

In an animal model of MS, antibodies targeting neurofascin cause axonal damage, disrupt neuronal conduction, and worsen the disease. Neurofascin is a cell adhesion molecule found on the surface of neurons. It plays a critical role in the formation and maintenance of nodes of Ranvier, which are essential for the rapid conduction of nerve impulses. Any disruption to these nodes can impair nerve signaling, leading to neurological symptoms. For an autoantibody in the bloodstream to damage the CNS, it must first cross the blood–brain barrier (BBB) and then bind to its target within the CNS. In MS, both conditions are met. The disease increases the BBB's permeability, allowing antibodies to enter the CNS. Moreover, these antibodies can recognize and bind to the native structure of NF186, a type of neurofascin. These antibodies can bind to NF186 in vivo, potentially intensifying axonal damage in MS patients, especially those with high antibody levels.

While 20–30% of MS patients have high levels of these antibodies, they are also found in some healthy individuals. This is not surprising, as similar patterns are seen with other autoimmune responses in MS. In conclusion, we have identified neurofascin as a potential target in some MS patients. Antibodies against neurofascin can cause axonal damage in CNS inflammatory diseases, suggesting potential new therapeutic avenues.

Neurofascin, another candidate antigen expressed on neuronal axons and linked with MS, may play a part in contributing to axonal damage. The presence of antibodies against neurofascin in MS patients suggests such damage may contribute to MS pathogenesis and treatment interventions. Even so, further research must continue in this field to understand the antigenic targets for MS disease pathogenesis and potential therapeutic interventions more deeply [28].

3.2.4. The Role of NK Cells

NK cells are large granular lymphocytes with the capacity to spontaneously eliminate target cells without prior sensitization, as well as serving a regulatory function by secreting cytokines and engaging in cell-to-cell interactions. Functionally, natural killer (NK) cells play an essential role in initiating immune responses against viral infections and slowing tumor growth. Their actions are tightly regulated through activating and inhibitory receptors, which guide their responses to stimuli. NK cells play an essential role in our immune system's defense against infections and cancerous growth by identifying stress-induced ligands on target cells using specific receptors like natural cytotoxicity receptors and C-type lectin receptor NKG2D. Through their unique capabilities, they serve as a frontline defense against infections or cancerous growth [29].

NK cells' role in multiple sclerosis (MS) remains an area of intensive investigation, with both positive and negative correlations observed in studies. Studies with an experimental animal model of MS (EAE) indicate possible protective benefits from their depletion in terms of more severe EAE symptoms and increased CNS pathology due to CD4+ T cells being killed directly by them; however, conflicting results have also been reported showing both beneficial and detrimental impacts from their involvement with EAE [26].

Functional activity of NK cells tends to be lower among MS patients compared with healthy individuals, leading to decreased numbers in their blood. Lower numbers have been linked to higher risks of relapse and lesion formation for MS patients. While its exact role in MS pathogenesis remains complex and multifaceted, evidence indicates NK cells could exert a major impact on disease outcomes. Further investigation will help shed more light on this intricate relationship between immune response mechanisms and NK cells, potentially providing insights into new therapeutic strategies and approaches for managing MS [30].

Researchers have suggested that interactions between NK cells and dendritic cells (DCs) could play an integral role in understanding their impact on MS (Figure 2). Studies indicate that GA (glatiramer acetate), an effective MS treatment medication, can alter these communications between NK cells and DCs to decrease autoantigenicity presentation to autoreactive T cells and thus lessen inflammation [31].

Studies conducted with mice with experimental autoimmune encephalomyelitis (EAE), an animal model of MS treated with GA, support the concept that GA enhances natural killer cell (NK cell) cytotoxicity. Furthermore, in MS patients undergoing GA treatment, their NK cells demonstrated significantly enhanced antitumor and DC killing abilities when compared to their pre-treatment levels, suggesting interactions between NK cells and DCs as potentially being key pathways underlying its effects in MS management and meriting further exploration to better understand how this drug acts when treating this disease [32].

Multiple sclerosis (MS) involves many complex interactions between NK cells and MS, which may be both positive and negative in nature. To fully grasp their impact, further research must be conducted, particularly to understand the specific effects of various subsets of NK cells in relation to MS development and response to treatments; such deeper knowledge will reveal insights into potential therapeutic strategies or interventions that may help manage MS more effectively.

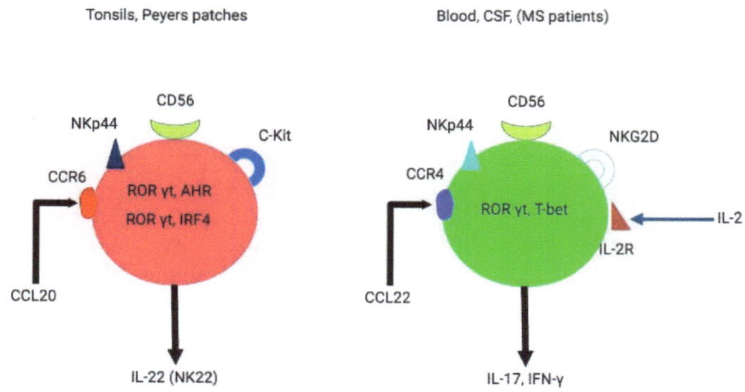

Figure 2. Comparison between NK22 cells and NK17/NK1 cells. NK22 cells are primarily located in tissues such as tonsils and Peyer's patches, while NK17/NK1 cells originate from human blood when activated in vitro with IL-2 and are notably present in the CSF of MS patients even in the absence of activation. NK22 cells predominantly express the NK cytotoxicity receptor NKp44. NK17/NK1 cells express receptors NKp30, NKG2D, NKp44, NKp46, and CD158, although the expression of NKp44, NKp46, and CD158 is less pronounced compared to other NK cell cytotoxicity receptors. NK22 cells express CCR6, and both secrete and respond to CCL20/MIP-3α. NK17/NK1 cells, on the other hand, express CCR4 and both secrete and respond to CCL22/MDC. NK22 cells express transcription factors RORγ, AHR, RORα, and IRF4. NK17/NK1 cells express RORγ and T-bet, leading them to secrete both IL-17 and IFN-γ.

4. Immune Mechanisms in Neurological Disorders

4.1. Immune-Related Neurological Disorders: MS and Beyond

Autoimmune disorders such as multiple sclerosis (MS) involve the immune system directly attacking body cells. In MS, immune cells target and attack the myelin sheath surrounding axons to cause demyelination and disability. Historically, this disease was classified as either relapsing-remitting or progressive; now, however, these classifications overlap, creating a spectrum of disease manifestations [33].

Relapsing symptoms in MS are associated with acute inflammation, often due to the infiltration of immune cells from the periphery into the central nervous system (CNS). Progressive symptoms, however, tend to be caused by neuroinflammation. Immunotherapy has proven moderately successful at treating MS relapsing symptoms; however, due to targeting peripheral immune system components only, it thus fails to address progressive symptoms. Immunotherapy's success at treating relapsing symptoms underscores how important immune interactions between central and peripheral systems are in MS onset and progression, opening potential avenues for improving therapeutic targeting for progressive MS cases [34].

Although much about MS's etiology remains unknown, experimental autoimmune encephalomyelitis (EAE) models have become an invaluable way of studying this illness and uncovering potential mechanisms. A typical one involves CD4+ T cells migrating into the CNS. These infiltrations may result from disruptions in either the blood–brain barrier

(BBB) or blood–cerebrospinal fluid barrier (BCSFB) and may include T cells that recognize host myelin as antigen during infiltration or become activated later. These reactive T cells release proinflammatory cytokines that trigger microglia activation and attract peripheral macrophages that contribute to myelin destruction. Microglia become activated when they release proinflammatory cytokines, exacerbating neuroinflammation further and disrupting the BBB, leading to greater infiltration of peripheral immune cells into the central nervous system. Neuroinflammation caused by microglia and astrocytes persists throughout disease progression as neurons undergo demyelination and degeneration, which makes understanding these complex mechanisms essential to progressing knowledge about MS and developing more effective treatment approaches [35].

Microglia were once thought to be solely pro-inflammatory cells; however, recent discoveries show they also play an anti-infiltrative function by restricting macrophage infiltration into the central nervous system (CNS). Studies have also uncovered activation phenotypes resembling "disease-associated microglia", suggesting similarities with other neurodegenerative conditions. Furthermore, their interaction is essential in multiple sclerosis development and progression and presents an attractive target for therapeutic interventions [36].

Neutrophils have emerged as key players in the severity of multiple sclerosis (MS), as their count correlates directly to disease activity. Studies using MS models have also demonstrated that depleting neutrophils leads to reduced disease severity, suggesting their potentially detrimental role in an autoimmune disorder like MS. $CXCR_2$ signaling might play a part in mediating neutrophil-induced damage and its interactions with microglia, providing a promising therapeutic target for managing MS [37].

So, MS research highlights the critical role played by immune cell interactions in the development of multiple sclerosis (MS). Recognizing and targeting these interactions may provide effective therapies to impede or reverse MS progression and lead to better patient outcomes.

4.2. The Role of Neuroimmunity in the Pathogenesis of Neurological Disorders

Dysregulated immune responses in the central nervous system (CNS) play an essential part in inducing neuroinflammation and contributing to various neurological conditions, such as Alzheimer's, Parkinson's, multiple sclerosis, and stroke. Microglia cells secrete inflammatory proteins that cause neuronal damage while worsening symptoms, and understanding neuroimmunity offers valuable insights into potential therapeutic targets for these complex neurological illnesses [35].

5. Clinical Neuroimmunology Research

5.1. Recent Advancements in Clinical Research of Neuroimmunological Disorders

Recent advances in neuroimmunology research have made substantial strides, providing unprecedented insights into the underlying mechanisms, diagnosis, and treatment of neurological conditions involving the immune system. Advancements in technology, including sophisticated imaging techniques such as MRI and PET scans, as well as high-throughput genomics and proteomics analyses, have enabled more in-depth investigation of the relationship between nervous and immune systems. PET is a potent functional imaging technique that can explore both healthy and affected brains. It offers a non-invasive way to measure specific biological markers, enhancing our comprehension of intricate central nervous system conditions like multiple sclerosis (MS). While MRI remains pivotal in tracking MS's clinical progression, PET has traditionally played a supplementary role. Nevertheless, recent advancements in PET imaging present opportunities to examine the MS brain in ways MRI cannot. PET can delve into the root causes of neuroinflammation, neuronal issues, demyelination, and remyelination in MS. Additionally, PET's ability to quantitatively assess molecular targets might be beneficial in future drug development clinical trials. However, the widespread adoption of PET is constrained by the significant expenses associated with cyclotrons and radiochemical labs. Discovering novel biomarkers has proven immensely helpful in early disease diagnosis and monitoring disease progression. Immunotherapies that target specific immune components have demonstrated promising results in clinical trials, revolutionizing neuroimmunological disorders management. These advances hold great promise in improving patient outcomes while furthering our understanding of these complex conditions [1].

5.2. Methodologies Used in Clinical Neuroimmunology Research

Clinical neuroimmunology research is an interdisciplinary field that draws upon clinical, immunological, and neurobiological methods to gain a comprehensive understanding of neuroimmunological disorders. This field conducts large-scale epidemiological studies to study their prevalence and risk factors associated with them, and advanced imaging technologies like MRI allow researchers to visualize brain and spinal cord lesions as well as track disease progression over time [38].

Immunological assays, such as flow cytometry and cytokine profiling, play an integral part in characterizing immune cell populations and their activation status in relation to neurological disorders under investigation. Genetic studies are performed in order to identify susceptibility genes as well as their influence in contributing to their development and progression [39].

Clinical trials play an integral part in assessing the safety and efficacy of various immunomodulatory and neuroprotective therapies used to treat neuroimmunological disorders in patients. Employing different research methodologies together contributes to expanding our knowledge about these complex conditions while developing interventions designed to enhance patient outcomes [40].

5.3. Challenges and Opportunities in Current Research

As clinical neuroimmunology research advances, however, it faces several difficulties. One key hurdle stems from the diverse nature of neurological conditions and associated immune responses. Due to this diversity, it becomes challenging to create universal therapeutic approaches that work for all patients [41].

Another barrier lies within the blood–brain barrier, which hinders the effective delivery of immunotherapies to the central nervous system (CNS). Finding appropriate strategies to overcome this barrier will improve treatment results and thus increase treatment efficacy. As part of clinical research, evaluating the long-term safety and efficacy of immunomodulatory treatments remains a paramount focus. A comprehensive evaluation is necessary to safeguard patient wellbeing as well as identify long-term benefits from therapies like these. Lacking validated biomarkers for disease prognosis and treatment response poses another difficulty: without reliable indicators, it becomes hard to tailor treatments specifically to individual patient needs and delay personalized medicine approaches. Ethical considerations in clinical research require careful consideration, from obtaining informed consent from patients to addressing data privacy concerns. Maintaining a balance between scientific advancement and patient welfare/privacy rights must always be prioritized for responsible research conduct [42].

5.4. Case Studies: Application of Clinical Research in Managing MS and Other Neurological Disorders

Clinical neuroimmunology research has had a substantial effect on the management of multiple sclerosis (MS) and other neurological conditions. Disease-modifying therapies, including interferon-beta, glatiramer acetate, monoclonal antibodies targeting B cells or T cells, and stem cell transplantation or gene therapies, have proven highly successful at reducing relapse rates and slowing progression. Early diagnosis with advanced techniques like MRI scans or cerebrospinal fluid analysis allows timely treatment initiation, leading to better long-term results, while stem cell transplantation or gene therapies show promise at slowing disease progression or even stimulating repair within MS patients [43].

Neuroimmunological conditions such as neuromyelitis optica spectrum disorder (NMOSD) and autoimmune encephalitis have seen great strides made through clinical research, with specific autoantibodies having been identified, providing targeted therapies. Rituximab, a B-cell-depleting agent, has been effective at managing NMOSD attacks while decreasing disability progression and recurrences, and corticosteroids, intravenous immunoglobulins, and plasmapheresis therapies have all proved beneficial when treating patients suffering from autoimmune encephalitis patients [44,45].

Neuroimmunology research holds the promise of further improving the management of complex neurological conditions, leading to improved patient outcomes and an enhanced quality of life.

6. Neuroimmunological Aspects of Health and Diseases

6.1. The Impact of Neuroimmunology on General Health

The Global Burden of Diseases (GBD) includes various non-communicable conditions like cardiovascular diseases, cancer, and chronic obstructive pulmonary disease. All three share risk factors like smoking and diet as well as pathophysiological causes like oxidative stress, inflammation, and excessive sympathetic activity [46]. This article proposes an innovative solution to predict, understand, prevent, and potentially treat these illnesses by employing neuroimmunology with vagal neuro-modulation as its foundation (Figure 3).

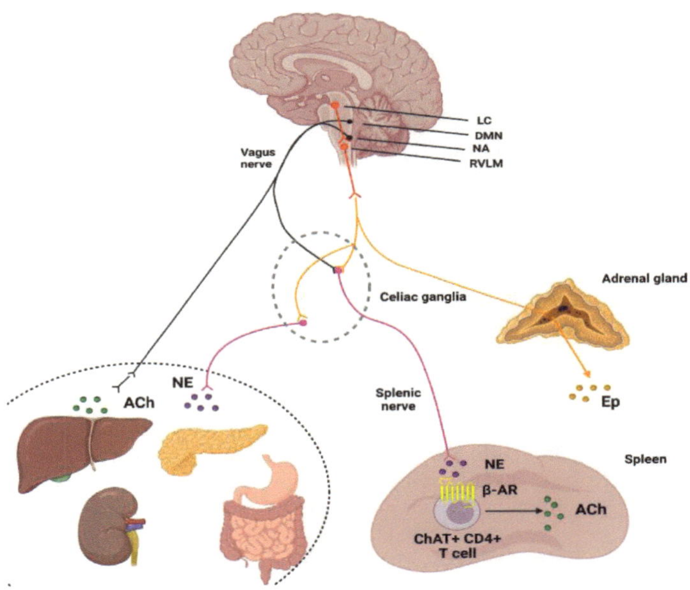

Figure 3. Autonomic neurons involved in regulating immune responses. Neurons of the efferent vagus nerve have their origins in the dorsal motor nucleus of the vagus (DMN) and the nucleus ambiguus (NA) located within the brainstem's medulla oblongata. These initial cholinergic neurons extend to various visceral organs in the thoracic and abdominal regions, encompassing the lungs, heart, liver, gastrointestinal system, kidneys, and pancreas. They engage with postganglionic vagal neurons either close to or inside the organs they innervate, with these neurons primarily releasing acetylcholine. The preganglionic neurons of the vagus nerve also end at the celiac ganglia and the superior mesenteric ganglion, which is where the splenic nerve begins.

Vagal nerve activity plays an integral part in controlling frontal brain activity, which in turn influences unhealthy lifestyle behaviors. The aim of this article is to introduce this paradigm to medicine and public health practitioners while emphasizing its significance for disease prevention and management. Epidemiological evidence shows that increased vagal activity, as measured by greater heart rate variability (HRV), independently predicts decreased GBD risk as well as better outcomes in those already living with GBD conditions [47].

Neuroimmunology is the study of how the nervous and immune systems interact, playing key roles in neural development, homeostasis, plasticity, and behavior modification [1]. While its potential benefits for human health and the treatment of neurological and psychiatric disorders are evident, translating research findings to clinical applications remains challenging due to knowledge gaps, optimal intervention timing issues, and a lack of tools that enable visualization or modulation.

As a way of meeting these challenges and pushing neuroimmunology forward, we propose ten research questions that, if explored in depth, could yield tangible advances over the short to medium term. These questions cover various aspects of neuroimmune interactions and their implication for health, disease, and various stages of life. As key themes in answering each question, we emphasize four cross-cutting themes critical to effective investigation: (i) understanding the two-way interactions of neuroimmune interactions; (ii) considering biological context (healthy state, disease state, and lifespan aspects); (iii) employing appropriate tools and technologies for better understanding underlying mechanisms; and (iv) translating research findings into practical clinical applications [48].

Though these questions do not encompass every knowledge gap in neuroimmunology, they address areas with immediate and broad impacts. Our goal in outlining research priorities is to foster collaboration among current and future research teams, working cross-disciplinarily towards furthering neuroimmunology for significant advancements in human health and disease treatment.

6.2. The Role of Neuroimmunology in Non-Neurological Diseases

The brain has long been considered immunologically unique due to the limited immune reaction seen here compared to peripheral tissues. Early studies demonstrated that transplanted foreign tissues survive longer after transplanting them into the brain than elsewhere, which suggests some degree of immune privilege [49,50]. One theory behind this phenomenon attributed this to the blood–brain barrier (BBB), which was thought to prevent immune cells and molecules from entering from the bloodstream into the brain via lymphatic vessels in its vicinity—further supporting T-cell immunity's surveillance by T cells.

Recent evidence is increasingly challenging this assumption and suggests that T cells do visit the brain. Although the BBB provides protection from passive protein entry but not from active cell entry, studies suggest drainage of cerebrospinal fluid and intracerebrally injected antigens to deep cervical lymph nodes is used as a connection between brain immune systems and peripheral ones [51,52].

Becher et al. (2000) illustrate the immunoreactivity of the brain as being in an active state. Thus, attention has shifted towards creating and maintaining an immunosuppressive microenvironment through factors like TGFβ, prostaglandins, and neurotrophins that are released by astrocytes and neurons releasing TGFβ, prostaglandins, and neurotrophins which serve to suppress immune responses while mitigating excess inflammation within the brain [53].

B cells play an integral part in immune responses, both positively and negatively regulating them. B cells have been associated with neuroimmunological diseases like myasthenia gravis (MG), multiple sclerosis (MS), Guillain–Barré syndrome (GBS), and Lambert–Eaton myasthenic syndrome (LEMS), with B cells playing a part in their pathogenesis by producing antigen-specific antibodies which stimulate T-cell activation and cytokine production while contributing to optimal T-cell activation and production by T cells [54].

Though research on B cells and autoantibodies in neuroimmunological diseases remains relatively scarce, ongoing investigations in this field continue to shed light on their pathogenic roles and gradual involvement. Understanding how B cells contribute to disease development and progression is key for advancements in this field, with promising clinical applications and greater insight into underlying causes for neuroimmunological conditions expected as B-cell research expands further [55]. Eventually, it is hoped that further insight from research on neuroimmunological conditions can lead to improved treatments or management methods.

6.3. Neuroimmunological Perspectives in Chronic Diseases

An experiment was carried out in order to determine the presence of T cells in the brain parenchyma of individuals with Alzheimer's disease (AD), other degenerative dementias, and control subjects using semi-quantitative analysis of immunohistochemically stained tissue sections. T cells were present in all cases examined; notably, more T cells could be observed in AD cases than in any other cases studied [56].

T cells found in AD patients' brains appeared activated but not fully differentiated, without evidence of antigen-triggered clonal expansion. Instead, local inflammation may explain this pattern of T-cell accumulation and activation within AD brain tissue. Our findings shed light on their possible involvement in AD pathology as well as suggest a possible role inflammation plays in driving T-cell activity within this setting; further investigation should elucidate their exact mechanisms and implications in Alzheimer's disease [57].

Autoimmunity is a significant contributor to diseases affecting both the central and peripheral nervous systems. The causes and clinical manifestations of these conditions vary considerably. Among the key causes is dysregulated complement activation at specific sites—this has been implicated as one cause in various neurological conditions like Guillain–Barré syndrome and neuromyelitis optica, where autoantibodies trigger activation of the complement system and damage self-tissues. Understanding this interaction is integral for understanding the pathogenesis behind these neurological diseases [58].

Neuroimmunological diseases and their treatments pose challenges to the immune system, elevating the risk of infections and severe illnesses. Consequently, vaccinations play a crucial role in the clinical management of these conditions. However, the diverse array of immunotherapies utilized in treating neuroimmunological diseases, especially multiple sclerosis and neuromyelitis optica spectrum disorders, can also impact immune responses to vaccines [59].

Polyspecific intrathecal immune response (PSIIR), commonly referred to as the "MRZ Reaction" (M = measles, R = rubella, and Z = zoster; optionally herpes simplex virus, HSV), refers to intrathecal immunoglobulin synthesis against two or more unrelated viruses found in cerebrospinal fluid [60]. While widely recognized as an indicator for multiple sclerosis (MS), PSIIR, an autoimmune–inflammatory neurological disease typically affecting early adulthood with widespread damage throughout its central nervous system (CNS), remains incompletely understood across other CAINDs that show positive results.

6.4. The Influence of Environmental and Lifestyle Factors on Neuroimmune Health

Behavioral neuroscience has been profoundly shaped by the understanding that brain development occurs throughout adolescence and adulthood. Studies over time have uncovered extensive research that indicates that adolescence is characterized by temporary differences, leading to greater risks, rewards, and susceptibility for affective disorders (as discussed here and further below). Importantly, behaviors and mental health disorders do not arise solely from neuronal activity in the brain itself. Instead, communication occurs between peripheral factors like immunity. Neuronal and immune interactions play a pivotal role in regulating cognitive and behavioral functioning as well as any dysfunctions throughout life, so for an accurate understanding of adolescent brain development, we propose conducting in-depth developmental investigations of both peripheral and central immune mechanisms [61].

Due to the devastating impacts of substance use disorders on health and wellbeing, extensive research has been carried out to uncover factors contributing to their progression from initial drug use to pathological drug use. Pinpointing these risk factors is essential in creating strategies for mitigating risks and decreasing prevalence rates among the population. We will review emerging research that explores interactions among peripheral immune system elements, gut microbiome composition, and central nervous system functions in driving pathological drug use [62].

Multiple sclerosis (MS) is a chronic, progressive, inflammatory, and degenerative condition that impacts millions of people worldwide, mostly impacting the central nervous system. The interplay among glial, neural, and immune cells plays an integral part in driving MS pathology, providing opportunities for therapeutic interventions [63]. We explore contributing risk factors behind MS development, current treatments available to modify it, and potential emerging technologies to fill clinical gaps, as well as identify novel avenues of therapy targets in this review article.

MS displays gender disparities in both immune reactions and neurodegeneration, influencing disease vulnerability and its progression. Interestingly, while women seem more prone to MS due to immune system differences, these differences do not align with those in the CNS. This raises the following query: if women have a higher disease incidence and stronger peripheral immune responses, why do they not experience quicker disability progression? In fact, men seem to progress faster. We theorize that gender-specific factors might have distinct impacts on the immune system compared to the CNS, leading to differences in disease susceptibility versus the rate of disability progression.

7. Developments in the Immunotherapy of Neurological Disorders

7.1. Current Approaches to Immunotherapy in Neurological Disorders

Imaging plays an integral role in diagnosing brain metastases, with MRI being the preferred imaging technique due to its greater sensitivity compared to CT. It effectively reveals the size, number, and distribution of central nervous system (CNS) lesions, such as solid or pseudospherical lesions. Brain metastases typically appear at grey–white junctions, with cerebral hemispheres being most prevalent (80%), followed by the cerebellum (15%) and brainstem (5%) [64]. Certain primary malignancies like melanomas, choriocarcinomas, germ cell tumors, thyroid cancers, or renal cell carcinomas are more susceptible than others to hemorrhage-related brain metastases.

Postoperative MRI can help distinguish between residual tumor material and blood byproducts after surgical removal of CNS lesions, according to data from malignant glioma studies. Changes in tumor diameter after stereotactic radiosurgery (SRS) can vary widely: approximately one-third of lesions experience transient increases in volume that usually stabilize or return to their original size over time [65].

Immunotherapy holds promise for treating neurological diseases, similar to its successes in treating cancer and autoimmune conditions. By harnessing immune responses, along with their capacity for tissue repair support, immunotherapy may hold great promise as an approach [66].

Multiple sclerosis (MS) was once treated by treating symptoms alone; however, thanks to recent advancements in our understanding of MS's underlying mechanisms, treatment options are expanding quickly in an attempt to slow or stop its progression. Immunological therapies have proven particularly successful at this, offering hope of better disease control before severe neurological disabilities manifest themselves. Ongoing research capitalizes on the advances made in immunoregulation understanding by offering more targeted and selective immunological treatments targeting myelin antigens while decreasing side effects [67].

7.2. Advances in MS Immunotherapy

Multiple sclerosis (MS) remains unknown with certainty, although evidence points towards a role for the immune system in its progression. Genetic and environmental risk factors both play a part in susceptibility [68]. Researchers have extensively researched experimental autoimmune encephalomyelitis as an animal model for MS that displays

T-cell-mediated inflammation within the central nervous system that often results in demyelination or nerve fiber damage.

Multiple sclerosis (MS) immunotherapy can lead to numerous side effects, which could impact treatment decisions based on an analysis of risks and benefits. One such adverse reaction is progressive multifocal leukoencephalopathy (PML), most frequently seen after taking natalizumab. Factors such as prior immunosuppressive therapy, presence of John Cunningham virus (JCV) antibodies, and treatment duration exceeding two years increase the risk of natalizumab-associated PML. Its incidence among MS patients treated with all three risk factors is estimated at 13 cases for every 1000 cases treated with this medicine. PML cases have also been associated with treatments of fingolimod and dimethylfumarate, leading to significant considerations as to whether early intervention with potent anti-inflammatory drugs such as fingolimod and dimethylfumarate is the optimal approach to prevent long-term disability. There has been considerable debate regarding which sequence and combination therapy would best address long-term disability, and ethical considerations related to placebo-controlled trials for relapsing MS have further complicated matters where withholding treatment is seen as unethical [69].

Current understandings of multiple sclerosis (MS) as an autoimmune disease have been informed by research on MS lesions and animal models, especially experimental autoimmune encephalomyelitis (EAE) [70]. Such studies not only provided key insight into potential mechanisms but also into fundamental immunological processes.

This article is divided into two sections. The first addresses essential immunological concepts and offers an imagined scenario for the immunopathogenesis of multiple sclerosis (MS). The second provides a critical review of various biotech-based immunotherapies with particular attention paid to their immunological principles, clinical evidence, and potential challenges.

One promising approach involves coupling intact proteins or multiple myelin-derived peptides to single cells like splenocytes or erythrocytes to enable simultaneous targeting of various T-cell specificities. This concept is especially significant for MS antigen-specific immunotherapy as tolerance of multiple T-cell epitopes is believed to be essential in effective disease treatment due to epitope spreading [71].

Altered peptide ligands (APLs) are modified versions of antigenic peptides that can switch their response from being an agonist to antagonist by disrupting specific hydrogen bond interactions with T-cell receptors. This has been shown to reduce Th1 responses significantly. When conjugated to reduced mannan, this cyclic APL induces significant Th1 reduction and moderate Th2 responses [72].

Another study using APLs of linear and cyclic MBP83-99 analogs, MBP83-99(A91, A96), conjugated to reduced mannan, showed an ability to redirect immune responses away from Th1 towards Th2 [73]. Thus, using reduced mannan as an immunotherapy strategy offers potential immunotherapeutic solutions for MS patients.

7.3. *Challenges in Developing Effective Immunotherapies*

Recent clinical research in multiple sclerosis (MS) has provided valuable insights into treating clinically isolated syndromes and secondary progression, as well as the effectiveness of different immunomodulatory therapies—findings with significant implications for optimizing the care of MS patients [74].

At present, treatment of autoimmune diseases typically includes corticosteroids and immunosuppressive therapies. More refined approaches have also been incorporated, such as interferon β (IFN-β) and glatiramer acetate (GA), and both medications have been employed successfully in managing MS [75].

However, despite progress made in MS research, significant challenges still exist. Progressive MS remains untreatable, and efforts to repair injured axons and protect neurons are limited [76]. Recent advances have laid the foundation for developing targeted immunotherapies based on genetic and genomic discoveries, hopefully expanding patient therapeutic options.

MS is characterized by an adaptive immune response involving regulatory T cells (Tregs), with functional deficiencies being seen within this network of Tregs being linked with disease activity and progression. Thus, inducing Tregs as part of MS immunotherapy should be one key goal [77].

Since 1993, when interferons became available as treatments for MS, significant progress has been made in immunotherapy for MS. More specific immunomodulatory drugs have become available, showing increased efficacy tailored to each person with MS.

Targeting immune elements involved in the immunologic cascade is one approach to treating MS, immunoactive drugs that focus on either the innate or adaptive immune response have been developed. We discuss humoral-targeted immunotherapies for MS, such as Rituximab, Ocrelizumab, and Ofatumumab, that show promise as B-cell-depleting agents—some agents, such as Atacicept, were discontinued due to increased inflammation activity compared with the placebo [78].

7.4. Future Directions in Immunotherapeutic Strategies for Neurological Disorders

Precision oncology and immunotherapy hold great promise in developing more effective, well-tolerated therapies against highly aggressive forms of glioblastoma cancer. This review showcases recent advancements in treatment strategies as well as possible future directions of these approaches to precision oncology and immunotherapy treatments for glioblastoma treatment [79].

As we seek to gain a better understanding of the extracellular mechanism of MS, equal emphasis should be given to preclinical research targeting the intracellular deposition of MS and its potential role in relieving neuronal dysfunction [80]. When conducting ongoing and future clinical investigations assessing immunotherapies for Parkinson's disease, careful patient selection criteria, outcome criteria evaluation, and thorough pharmacodynamic assessments must all be implemented so as to accurately evaluate the efficacy of immunotherapies.

Glioma, a leading primary intracranial malignancy, poses many difficulties due to limited treatment options and poor survival rates. Immunotherapy's proven success with other cancers has opened the door for similar therapies that seek to activate patients' immune systems to target and eliminate gliomas. Furthermore, this review highlights novel concepts and advanced technologies that hold promise in designing more effective immunotherapies, providing potential blueprints for more efficient glioma treatments [81].

Current treatments for advanced melanoma involve immunotherapy using anti-PD1 antibodies or targeted therapy with BRAF and MEK inhibitors, with immunomodulatory agents like LAG_3, TIM_3, OX_{40}, CD_{137}, IDO, and GITR being explored as potential therapeutic solutions. Research is being conducted into fully exploiting available treatments while simultaneously creating new drugs; however, an optimal first-line treatment remains uncertain [82].

Psychotropic medications have been employed to address psychiatric symptoms associated with NMDAR encephalitis; however, their effectiveness may be limited. There has been little investigation of their possible use as adjunct therapies to enhance glutamate and GABA neurotransmission. In this regard, it should be remembered that schizophrenia differs slightly from anti-NMDAR encephalitis, where there is an irreversible loss of surface NMDARs [83].

Future investigations should focus on understanding the significance of microglial subpopulation depletion across various disease stages. A comprehensive understanding is crucial before considering microglia depletion as a potential clinical intervention to resolve neuroinflammation and promote recovery, and genetic tools like CX3CR1CreER mice or targeted knockdown strategies can offer valuable insight. They may even aid in developing tailored therapeutic approaches [84].

8. Emerging Fields in Neuroimmunology

8.1. Neuroimmunology in Aging and Neurodegenerative Disorders

Chronic and acute stress can have profound impacts on health and functional wellbeing, with neuroimmunological dysregulation and inflammation serving as major contributing factors in numerous disease states [85]. Psychosocial strain and negative emotions have been associated with increased levels of pro-inflammatory biomarkers; immunosenescence—an aging process marked by declining immune function that leads to higher morbidity rates and susceptibility for fatal illnesses—particularly impacts neurodegenerative conditions like Parkinson's disease and diabetes.

Recent studies on brain health and its effect on other body tissues remain incompletely understood; however, recent evidence points towards immune cells playing an integral part in brain aging. Studies of both aging mice and humans show an increase in type I interferon (IFN) response in the choroid plexus (CP), an area responsible for cognitive function and neurogenesis, and neutralizing age-related type I IFN responses partially restores cognitive function as well as hippocampal neurogenesis [86].

As individuals age, their brain's immune cells, known as microglia, may become hyperreactive and display age-related pro-inflammatory biases. This phenomenon may be affected by both intrinsic factors (e.g., increased priming) and environmental ones, such as amplified danger signals, cytokines, or altered glymphatic function affecting microglia cells [87].

Aging and traumatic brain injuries (TBIs) may increase the priming of microglia, which leads to presymptomatic neurodegenerative diseases like Alzheimer's. Risk increases with age and may also be affected by previous head trauma [88,89].

Aging has an adverse impact on the blood–brain barrier (BBB), rendering it more permeable and permitting increased immune activation, expression of stress-induced and inflammatory genes, and infiltration of CNS immune cells into the central nervous system (CNS). Such age-related changes in immunity could contribute to the ineffective clearance of toxic protein aggregates associated with neurodegenerative disorders [90].

8.2. Pediatric Neuroimmunology

Recent advancements in pediatric neuroimmunology have seen considerable advances. The detection of conformationally correct myelin oligodendrocyte glycoprotein (MOG) antibodies using cell-based assays has led to an increasing identification of children with various monophasic and relapsing phenotypes; however, debate remains as to the severity and outcome in this rapidly expanding spectrum of MOG-related demyelination [91].

Research efforts have primarily centered around studying Rituximab's effects in pediatric neuromyelitis optica (NMO)/NMO spectrum disorders (NMOSD) patients and exploring its relationship to B-cell repopulation and relapses. Rituximab has proven effective at preventing relapses; however, the risk increases with B-cell repopulation—redosing before B-cell repopulation may help further decrease its likelihood [92].

Another area of investigation includes exploring how serum neurofilament light chain (sNfL) could serve as an accurate biomarker of disease activity and treatment response in pediatric patients with multiple sclerosis (MS). Studies suggest sNfL could serve as an accurate biomarker for monitoring disease activity and response among young MS patients, potentially helping predict severity levels as well as guide decision-making regarding treatment decisions in young MS cases [93].

Studies conducted among pediatric neuroimmunology populations demonstrated significant variance in 2-m and 4-m ETDRS chart scores, even after accounting for factors like optic neuritis (ON), vision correction, gender, and age, as revealed by the GEE model [94].

8.3. Neuroimmunoendocrinology: The Interactions between the Nervous, Immune, and Endocrine Systems

Bidirectional feedback communication among stress response, immune system, and endocrine system is integral for adaption to stressful stimuli, maintenance of homeostasis, and overall survival. Advancements have been made in understanding the molecular, cellular, and systemic physiological mechanisms underlying this communication—especially regarding immunity. Numerous neuroendocrine mediators like cortisol, estrogen, testosterone, DHEA catecholamines, corticotropin-releasing hormone, and adenosine have been identified as having immunomodulatory activities [95].

The nervous and immune systems share intimate connections, with close interactions regulating systemic homeostasis via the production and secretion of regulatory peptides (e.g., peptide hormones, cytokines, chemokines, and integrins). These molecules play an integral part in maintaining tissue homeostasis. Similarly, their production in both brain areas as well as central organs of immunity/endocrinology systems is comparable [96].

Maintaining homeostasis requires three regulating systems: nervous, endocrine, and immune. Interactions among these three have long been recognized, giving rise to neuroendocrinology. More recently, however, their combination is emerging as an exciting and rapidly expanding area of research [97].

These communication pathways encompass an array of interactions among cells, tissues, and organs with implications for mitochondrial functioning. Disruptions to interactions between mitochondria and neuroendocrine–immune systems contribute to Alzheimer's and Parkinson's diseases as pathologies [98].

Recent understandings of Parkinson's disease have included immuno-inflammation and oxidative and nitrosative stress as key features, which are also seen in depression, somatization, and peripheral inflammation. We present evidence supporting their relevance in such conditions such as depression and somatization that are typically thought of as "comorbidities," proposing instead that depression and somatization experienced throughout a lifetime, during the prodromal phase and concurrent with Parkinson's, may actually play a significant role in its origins and progression rather than simply being "psychiatric" symptoms [99].

Interactions between neuroendocrine and immune systems have long been a focus of intense study and represent a highly promising area of investigation. Ample data are available that provide new insights into their bidirectional signal exchanges [100].

The hypothalamic–pituitary–adrenal (HPA) axis plays a key role in our body's response to stress. Corticotropin-releasing factor (CRH), adrenocorticotropin (ACTH), and glucocorticoids play an integral part in stress response mechanisms, modulation of proinflammatory cytokines production, regulation of peripheral immune response regulation and neuroimmunoendocrine interactions regulating peripheral immunity response regulation; targeting this system could provide therapeutic approaches both human and experimental forms of Chagas disease [101].

8.4. The Role of Microbiota in Neuroimmunology

Recent interest in gut microbiota's role in human health has renewed attention on its relationship to the brain–gut axis, yet its exploration dates back over three centuries [102]. We explore its historical development here with special reference to microbiota–neuroimmune communication, specifically highlighting that gut microbiota plays an integral part in pathogenicity for NMOSD (neuromyelitis optica spectrum disorder), suggesting microbiota interventions, such as diet changes, probiotics, antibiotics, or even fecal bacterial transplantation, could serve as treatment options for NMOSD patients [103].

Parkinson's disease (PD) has long been linked with gut microbiota dysbiosis. Studies have demonstrated how imbalanced microbiota contributes to exacerbating symptoms. Germ-free (GF) animal models offer invaluable insight into this relationship between microbiota and various neurobiological and neurodevelopmental disorders—autism, obsessive compulsive disorder (OCD), depression, and anxiety, among others—and their microbiomes. GF rodents exhibit distinct behaviors when compared with conventional rodents that often exhibit anxiety- or depressive-like characteristics [103].

Th_{17} cells, an integral component of the immune response, can be stimulated by commensal microbiota present in the gut. Signaling between the nervous system and gastrointestinal tract appears to be impaired in conditions like irritable bowel syndrome (IBS), often described as a disorder of the gut–brain axis with multiple communication systems involved like microbiota–host crosstalk and neuroimmune interactions. Emerging evidence shows IBS patients possess different gut microbiomes than healthy individuals and, more specifically, dysbiotic colonic microbiota or mycobiota are found more frequently in those suffering from hypersensitivity [104].

Indeed, gut microbiota plays an indispensable role in regulating physiological and metabolic pathways. We will explore significant and current connections between gut microbiota and other systems in the body in subsequent sections. Any shifts in metabolic or physiological functions caused by changes to gut microbiota can create system-wide imbalances. Studies on interactions between host–microbiota interactions involving intestinal epithelium/immune system interactions have been extensively researched. Moreover, there has been increasing understanding regarding interactions between gut microbiota/neuroimmune system interactions, thus shedding light on their importance to overall health/wellbeing [105].

9. Conclusions

9.1. Summary of Key Findings

Multiple sclerosis (MS) research in neuroimmunology has brought forth significant advances, deepening our knowledge and improving treatment approaches for this condition:

(a) Immune Dysregulation: Multiple sclerosis is characterized by autoreactive T cells attacking the central nervous system, leading to inflammation and nerve fiber demyelination. Understanding their activation and regulation is crucial to discovering potential therapeutic targets for treatment;

(b) Cells Play an Important Role in Multiple Sclerosis Pathogenesis: B cells have emerged as key contributors to MS pathogenesis. By producing autoantibodies and modulating T-cell responses, these B cells play an integral part in tissue damage and neuroinflammation. Depletion therapies targeting B cells have shown promise as a possible treatment option for MS;

(c) Genetic Susceptibility: Genome-wide association studies have linked various genetic variations with MS susceptibility, enabling personalized therapy approaches and insights into genetic factors contributing to its cause;

(d) Environmental Triggers: Research has shed light on how environmental triggers like smoking, vitamin D deficiency, and viral infections interact to increase MS risk and severity. Researchers have also investigated the interaction between genetic predisposition and environmental influences in driving this process forward;

(e) Neuroprotective Strategies: Because MS leads to neurodegeneration, researchers are investigating neuroprotective measures that preserve nerve function and facilitate repair processes. Researching molecules involved with remyelination and axonal support could reveal therapeutic targets;

(f) Immunomodulatory Medications: Immune-system-targeted disease-modifying therapies have become a cornerstone of MS treatment, demonstrating significant slowing in disease progression and enhanced patient outcomes through early use.

(g) Studies have shed light on how gut microbiota influences immune responses and possibly contributes to MS. Understanding this relationship could lead to novel treatment approaches aimed at altering gut bacteria composition and functioning;

(h) Neuroimmunological studies have provided valuable insights into multiple sclerosis by shedding light on its gut–brain axis, genetic predispositions, environmental influences, B-cell involvement, immune dysregulation, and neuroprotective treatments as key elements. These discoveries hold great promise for developing more targeted and effective treatment strategies and ultimately improving the quality of life among those living with this complex neurological disorder.

9.2. Implications for Future Research

Future studies in neuroimmunology hold great promise for furthering our understanding of multiple sclerosis (MS) and developing more effective therapies. Key areas of investigation may include the following:

(a) Uncovering Underlying Mechanisms: Exploring the interrelations between MS's immune and neurological systems can provide insight into specific physiological and molecular processes driving disease development while pinpointing key immune cell types, cytokines, and chemokines associated with inflammation and demyelination, which will allow targeted therapies;
(b) Neuroprotective Strategies: Although immune dysregulation is currently at the core of MS treatments, future research should prioritize neuroprotective approaches that preserve and restore damaged nerve cells. Deliberate identification of chemicals or pathways that promote neuronal survival and remyelination is key for slowing disability progression;
(c) Microbiome and Gut–Brain Axis: Understanding the role of gut microbiomes and gut–brain axis in MS pathophysiology can lead to innovative treatments that control immune responses and decrease disease activity, including interventions that regulate gut microbiomes;
(d) Precision Medicine: Advancements in molecular profiling, biomarker discovery, and tailored treatment strategies hold great potential to optimize medication selection treatment efficacy and minimize side effects in MS patients;
(e) Immunological Tolerance and Immunomodulation: Generating novel immunological tolerance mechanisms against myelin antigens could revolutionize MS therapy. Investigating innovative immunomodulatory strategies such as antigen-specific therapy or immune cell modulation could slow disease progression;
(f) Innovative Imaging Technologies: State-of-the-art imaging techniques like optical coherence tomography and advanced MRI methods offer more precise and early diagnostic indicators, monitor disease progression, and assess treatment efficacy in real time;
(g) Drug Repurposing: Investigating the therapeutic potential of existing neuroprotective or immunomodulatory medications can hasten the discovery of new MS treatments.

Future neuroimmunology research for MS should prioritize advanced imaging techniques, uncovering disease mechanisms, examining genetic and environmental influences, devising neuroprotective and precision medicine strategies, and exploring gut–brain axis/microbiome interactions while considering drug repurposing strategies—these efforts will allow for the creation of personalized, efficient therapies to enhance the quality of life for MS patients while lessening the burden of this debilitating condition.

9.3. Final Thoughts

An Exhaustive Analysis of Neuroimmunology: Lessons learned from multiple sclerosis and upcoming therapeutic advancements provides an essential resource on the complex interplay between immune and neurological systems, with multiple sclerosis (MS) as its focal point. It offers contemporary insight into its pathophysiology, clinical manifestations, and available treatment options, providing readers with a thorough overview.

This review offers invaluable insight into the neuroimmunological mechanisms underlying MS. It may address how immune cells, cytokines, and signaling molecules play a role in central nervous system inflammation responses and demyelination. Furthermore, genetic and environmental factors may play a part in contributing to its onset and its heterogeneous presentations in different patients' cases.

One of the hallmarks of this comprehensive review is its detailed examination of current MS treatments. This likely involves discussing immunomodulatory drugs, symptom-relieving medication, and disease-modifying therapies. Furthermore, it may provide insight into emerging therapies or research methodologies that hold the promise of more precise medicines in the future.

As this review highlights, its primary message may be the importance of ongoing research and collaboration among neuroscientists, immunologists, and clinicians to fully comprehend neuroimmunology's application in MS treatment. As our understanding of MS deepens, the potential exists for more tailored therapeutic advancements that could potentially enhance the quality of life for those living with MS.

Overall, this in-depth analysis serves as an indispensable resource for academics, medical professionals, and anyone wanting deeper insights into neuroimmunology and its application in treating multiple sclerosis today and in the future.

Author Contributions: Conceptualization, H.P. and A.V.C.; methodology, L.E.; soft-ware, H.P. and A.V.C.; validation, R.-A.C.-B., L.A.G. and H.P.; formal analysis, D.-I.D.; investigation, A.D.C. and D.-I.D.; resources, A.B.; data curation, L.A.G., B.-G.B. and R.-A.C.-B.; writing—original draft preparation, R.-A.C.-B. and A.V.C.; writing—review and editing, R.-A.C.-B., L.A.G. and A.V.C.; visualization, B.-G.B.; supervision, L.E.; project administration, A.V.C.; funding acquisition, A.V.C. All authors have read and agreed to the published version of the manuscript.

Funding: This research received no external funding.

Data Availability Statement: All Data is available on PubMed.

Conflicts of Interest: The authors declare no conflict of interest.

References

1. Nutma, E.; Willison, H.; Martino, G.; Amor, S. Neuroimmunology—The past, present and future. *Clin. Exp. Immunol.* **2019**, *197*, 278–293. [CrossRef] [PubMed]
2. Kim, S.; Kim, H.; Yim, Y.S.; Ha, S.; Atarashi, K.; Tan, T.G.; Longman, R.S.; Honda, K.; Littman, D.R.; Choi, G.B.; et al. Maternal gut bacteria promote neurodevelopmental abnormalities in mouse offspring. *Nature* **2017**, *549*, 528–532. [CrossRef] [PubMed]
3. Pachner, A.R. The Neuroimmunology of Multiple Sclerosis: Fictions and Facts. *Front. Neurol.* **2021**, *12*, 796378. [CrossRef] [PubMed]
4. Caldwell, L.J.; Subramaniam, S.; MacKenzie, G.; Shah, D.K. Maximising the potential of neuroimmunology. *Brain Behav. Immun.* **2020**, *87*, 189–192. [CrossRef]
5. Multiple Sclerosis. National Institute of Neurological Disorders and Stroke. Available online: https://www.ninds.nih.gov/health-information/disorders/multiple-sclerosis (accessed on 24 July 2023).
6. Kraus, A.; Buckley, K.M.; Salinas, I. Sensing the world and its dangers: An evolutionary perspective in neuroimmunology. *eLife* **2021**, *10*, e66706. [CrossRef]
7. Aguilera-Lizarraga, J.; Florens, M.V.; Viola, M.F.; Jain, P.; Decraecker, L.; Appeltans, I.; Cuende-Estevez, M.; Fabre, N.; Van Beek, K.; Perna, E.; et al. Local immune response to food antigens drives meal-induced abdominal pain. *Nature* **2021**, *590*, 151–156. [CrossRef]
8. Binshtok, A.M.; Wang, H.; Zimmermann, K.; Amaya, F.; Vardeh, D.; Shi, L.; Brenner, G.J.; Ji, R.-R.; Bean, B.P.; Woolf, C.J.; et al. Nociceptors Are Interleukin-1β Sensors. *J. Neurosci.* **2008**, *28*, 14062–14073. [CrossRef]
9. Nagashima, H.; Mahlakõiv, T.; Shih, H.Y.; Davis, F.P.; Meylan, F.; Huang, Y.; Harrison, O.J.; Yao, C.; Mikami, Y.; Urban, J.F., Jr.; et al. Neuropeptide CGRP Limits Group 2 Innate Lymphoid Cell Responses and Constrains Type 2 Inflammation. *Immunity* **2019**, *51*, 682–695.e6. [CrossRef]
10. Veiga-Fernandes, H.; Pachnis, V. Neuroimmune regulation during intestinal development and homeostasis. *Nat. Immunol.* **2017**, *18*, 116–122. [CrossRef]
11. Prinz, M.; Priller, J. The role of peripheral immune cells in the CNS in steady state and disease. *Nat. Neurosci.* **2017**, *20*, 136–144. [CrossRef]
12. Engelhardt, B.; Vajkoczy, P.; Weller, R.O. The movers and shapers in immune privilege of the CNS. *Nat. Immunol.* **2017**, *18*, 123–131. [CrossRef] [PubMed]
13. Neuroimmune communication. *Nat. Immunol.* **2017**, *18*, 115. [CrossRef] [PubMed]
14. Dantzer, R. Neuroimmune Interactions: From the Brain to the Immune System and Vice Versa. *Physiol. Rev.* **2018**, *98*, 477–504. [CrossRef] [PubMed]
15. Besedovsky, H.; Sorkin, E.; Keller, M.; Müller, J. Changes in blood hormone levels during the immune response. *Proc. Soc. Exp. Biol. Med.* **1975**, *150*, 466–470. [CrossRef]
16. Besedovsky, H.; del Rey, A.; Sorkin, E.; Dinarello, C.A. Immunoregulatory feedback between interleukin-1 and glucocorticoid hormones. *Science* **1986**, *233*, 652–654. [CrossRef]

17. Pace, T.W.W.; Hu, F.; Miller, A.H. Cytokine-Effects on Glucocorticoid Receptor Function: Relevance to Glucocorticoid Resistance and the Pathophysiology and Treatment of Major Depression. *Brain Behav. Immun.* **2007**, *21*, 9–19. [CrossRef]
18. Dobson, R.; Giovannoni, G. Multiple sclerosis—A review. *Eur. J. Neurol.* **2019**, *26*, 27–40. [CrossRef]
19. Engelhardt, B. The blood-central nervous system barriers actively control immune cell entry into the central nervous system. *Curr. Pharm. Des.* **2008**, *14*, 1555–1565. [CrossRef]
20. Bechmann, I.; Galea, I.; Perry, V.H. What is the blood-brain barrier (not)? *Trends Immunol.* **2007**, *28*, 5–11. [CrossRef]
21. Dolei, A.; Garson, J.A.; Arru, G.; Clerici, M.; Germi, R.; Marche, P.N.; Perron, H. Multiple sclerosis-associated retrovirus and related human endogenous retrovirus-W in patients with multiple sclerosis. *J. Neuroimmunol.* **2014**, *266*, 87–88. [CrossRef]
22. Tzartos, J.S.; Friese, M.A.; Craner, M.J.; Palace, J.; Newcombe, J.; Esiri, M.M.; Fugger, L. Interleukin-17 production in central nervous system-infiltrating T cells and glial cells is associated with active disease in multiple sclerosis. *Am. J. Pathol.* **2008**, *172*, 146–155. [CrossRef] [PubMed]
23. Vollmer, T.; Panitch, H.; Bar-Or, A.; Dunn, J.; Freedman, M.S.; Gazda, S.; Campagnolo, D.; Deutsch, F.; Arnold, D. Glatiramer acetate after induction therapy with mitoxantrone in relapsing multiple sclerosis. *Mult. Scler.* **2008**, *14*, 663–670. [CrossRef] [PubMed]
24. Høglund, R.A.; Maghazachi, A.A. Multiple sclerosis and the role of immune cells. *World J. Exp. Med.* **2014**, *4*, 27–37. [CrossRef]
25. Glatigny, S.; Bettelli, E. Experimental Autoimmune Encephalomyelitis (EAE) as Animal Models of Multiple Sclerosis (MS). *Cold Spring Harb. Perspect. Med.* **2018**, *8*, a028977. [CrossRef]
26. Procaccini, C.; De Rosa, V.; Pucino, V.; Formisano, L.; Matarese, G. Animal models of Multiple Sclerosis. *Eur. J. Pharmacol.* **2015**, *759*, 182–191. [CrossRef]
27. van Noort, J.M.; Verbeek, R.; Meilof, J.F.; Polman, C.H.; Amor, S. Autoantibodies against alpha B-crystallin, a candidate autoantigen in multiple sclerosis, are part of a normal human immune repertoire. *Mult. Scler.* **2006**, *12*, 287–293. [CrossRef] [PubMed]
28. Mathey, E.K.; Derfuss, T.; Storch, M.K.; Williams, K.R.; Hales, K.; Woolley, D.R.; Al-Hayani, A.; Davies, S.N.; Rasband, M.N.; Olsson, T.; et al. Neurofascin as a novel target for autoantibody-mediated axonal injury. *J. Exp. Med.* **2007**, *204*, 2363–2372. [CrossRef]
29. Öberg, L.; Johansson, S.; Michaëlsson, J.; Tomasello, E.; Vivier, E.; Kärre, K.; Höglund, P. Loss or mismatch of MHC class I is sufficient to trigger NK cell-mediated rejection of resting lymphocytes in vivo—Role of KARAP/DAP12-dependent and -independent pathways. *Eur. J. Immunol.* **2004**, *34*, 1646–1653. [CrossRef]
30. Cella, M.; Fuchs, A.; Vermi, W.; Facchetti, F.; Otero, K.; Lennerz, J.K.M.; Doherty, J.M.; Mills, J.C.; Colonna, M. A human natural killer cell subset provides an innate source of IL-22 for mucosal immunity. *Nature* **2009**, *457*, 722–725. [CrossRef]
31. Al-Ani, M.; Elemam, N.M.; Hundt, J.E.; Maghazachi, A.A. Drugs for Multiple Sclerosis Activate Natural Killer Cells: Do They Protect Against COVID-19 Infection? *Infect. Drug Resist.* **2020**, *13*, 3243–3254. [CrossRef]
32. Maghazachi, A.A. On The Role of Natural Killer Cells in Neurodegenerative Diseases. *Toxins* **2013**, *5*, 363–375. [CrossRef] [PubMed]
33. Zuroff, L.R.; Benjamins, J.A.; Bar-Or, A.; Lisak, R.P. Inflammatory mechanisms underlying cortical injury in progressive multiple sclerosis. *Neuroimmunol. Neuroinflamm.* **2021**, *8*, 111. [CrossRef]
34. Baecher-Allan, C.; Kaskow, B.J.; Weiner, H.L. Multiple Sclerosis: Mechanisms and Immunotherapy. *Neuron* **2018**, *97*, 742–768. [CrossRef] [PubMed]
35. Passaro, A.P.; Lebos, A.L.; Yao, Y.; Stice, S.L. Immune Response in Neurological Pathology: Emerging Role of Central and Peripheral Immune Crosstalk. *Front. Immunol.* **2021**, *12*, 676621. [CrossRef]
36. Geloso, M.C.; Corvino, V.; Marchese, E.; Serrano, A.; Michetti, F.; D'Ambrosi, N. The Dual Role of Microglia in ALS: Mechanisms and Therapeutic Approaches. *Front. Aging Neurosci.* **2017**, *9*, 242. [CrossRef]
37. Khaw, Y.M.; Cunningham, C.; Tierney, A.; Sivaguru, M.; Inoue, M. Neutrophil-selective deletion of Cxcr2 protects against CNS neurodegeneration in a mouse model of multiple sclerosis. *J. Neuroinflamm.* **2020**, *17*, 49. [CrossRef]
38. Hemond, C.C.; Bakshi, R. Magnetic Resonance Imaging in Multiple Sclerosis. *Cold Spring Harb. Perspect. Med.* **2018**, *8*, a028969. [CrossRef]
39. Qiu, J.G.; Mei, X.L.; Chen, Z.S.; Shi, Z. Cytokine detection by flow cytometry. *Methods Mol. Biol.* **2014**, *1172*, 235–242. [CrossRef]
40. Mahajan, K.R.; Ontaneda, D. The Role of Advanced Magnetic Resonance Imaging Techniques in Multiple Sclerosis Clinical Trials. *Neurotherapeutics* **2017**, *14*, 905–923. [CrossRef]
41. Misra, M.K.; Damotte, V.; Hollenbach, J.A. The immunogenetics of neurological disease. *Immunology* **2018**, *153*, 399–414. [CrossRef]
42. Minen, M.T.; Law, E.F.; Harriott, A.; Seng, E.K.; Hranilovich, J.; Szperka, C.L.; Wells, R.E. Challenges to successful research careers in neurology. *Neurology* **2020**, *95*, 349–359. [CrossRef] [PubMed]
43. Piehl, F. Current and emerging disease-modulatory therapies and treatment targets for multiple sclerosis. *J. Intern. Med.* **2021**, *289*, 771–791. [CrossRef]
44. Hanif, N.; Anwer, F. Rituximab. In *StatPearls*; StatPearls Publishing: St. Petersburg, FL, USA, 2023. Available online: http://www.ncbi.nlm.nih.gov/books/NBK564374/ (accessed on 25 July 2023).
45. Neuromyelitis Optica. National Institute of Neurological Disorders and Stroke. Available online: https://www.ninds.nih.gov/health-information/disorders/neuromyelitis-optica (accessed on 25 July 2023).

46. Benziger, C.P.; Roth, G.A.; Moran, A.E. The Global Burden of Disease Study and the Preventable Burden of NCD. *Glob. Heart* **2016**, *11*, 393–397. [CrossRef] [PubMed]
47. Gidron, Y.; Deschepper, R.; De Couck, M.; Thayer, J.F.; Velkeniers, B. The Vagus Nerve Can Predict and Possibly Modulate Non-Communicable Chronic Diseases: Introducing a Neuroimmunological Paradigm to Public Health. *J. Clin. Med.* **2018**, *7*, 371. [CrossRef] [PubMed]
48. MacKenzie, G.; Subramaniam, S.; Caldwell, L.J.; Fitzgerald, D.; A Harrison, N.; Hong, S.; Irani, S.R.; Khandaker, G.M.; Liston, A.; E Miron, V.; et al. Research priorities for neuroimmunology: Identifying the key research questions to be addressed by 2030. *Wellcome Open Res.* **2021**, *6*, 194. [CrossRef] [PubMed]
49. Medawar, P.B. Immunity to homologous grafted skin; the fate of skin homografts transplanted to the brain, to subcutaneous tissue, and to the anterior chamber of the eye. *Br. J. Exp. Pathol.* **1948**, *29*, 58–69. [PubMed]
50. Barker, C.F.; Billingham, R.E. Immunologically privileged sites. *Adv. Immunol.* **1977**, *25*, 1–54.
51. Cserr, H.F.; Knopf, P.M. Cervical lymphatics, the blood-brain barrier and the immunoreactivity of the brain: A new view. *Immunol. Today* **1992**, *13*, 507–512. [CrossRef]
52. Yamamoto, S.; Kubotsu, K.; Kida, M.; Kondo, K.; Matsuura, S.; Uchiyama, S.; Yonekawa, O.; Kanno, T. Automated homogeneous liposome-based assay system for total complement activity. *Clin. Chem.* **1995**, *41*, 586–590. [CrossRef]
53. Becher, B.; Prat, A.; Antel, J.P. Brain-immune connection: Immuno-regulatory properties of CNS-resident cells. *Glia* **2000**, *29*, 293–304. [CrossRef]
54. Correale, J.; Mix, E.; Olsson, T.; Kostulas, V.; Fredrikson, S.; Hojeberg, B.; Link, H. CD5+ B-cells and CD4-8-T-cells in neuroimmunological diseases. *J. Neuroimmunol.* **1991**, *32*, 123–132. [CrossRef]
55. Karim, M.; Wang, Y.F. The Study Progress of B Cells and Neuroimmunological Diseases. *J. Neurol. Neurophysiol.* **2016**, *7*, 369. [CrossRef]
56. Gate, D.; Saligrama, N.; Leventhal, O.; Yang, A.C.; Unger, M.S.; Middeldorp, J.; Chen, K.; Lehallier, B.; Channappa, D.; De Los Santos, M.B.; et al. Clonally expanded CD8 T cells patrol the cerebrospinal fluid in Alzheimer's disease. *Nature* **2020**, *577*, 399–404. [CrossRef]
57. Togo, T.; Akiyama, H.; Iseki, E.; Kondo, H.; Ikeda, K.; Kato, M.; Oda, T.; Tsuchiya, K.; Kosaka, K. Occurrence of T cells in the brain of Alzheimer's disease and other neurological diseases. *J. Neuroimmunol.* **2002**, *124*, 83–92. [CrossRef] [PubMed]
58. Pilch, K.S.; Spaeth, P.J.; Yuki, N.; Wakerley, B.R. Therapeutic complement inhibition: A promising approach for treatment of neuroimmunological diseases. *Expert. Rev. Neurother.* **2017**, *17*, 579–591. [CrossRef]
59. Winkelmann, A.; Loebermann, M.; Barnett, M.; Hartung, H.P.; Zettl, U.K. Vaccination and immunotherapies in neuroimmunological diseases. *Nat. Rev. Neurol.* **2022**, *18*, 289–306. [CrossRef]
60. Brauchle, F.; Rapp, D.; Senel, M.; Huss, A.; Dreyhaupt, J.; Klose, V.; Süße, M.; Stürner, K.H.; Leypoldt, F.; Tumani, H.; et al. Clinical associations and characteristics of the polyspecific intrathecal immune response in elderly patients with non-multiple sclerosis chronic autoimmune-inflammatory neurological diseases—A retrospective cross-sectional study. *Front. Neurol.* **2023**, *14*, 1193015. Available online: https://www.frontiersin.org/articles/10.3389/fneur.2023.1193015 (accessed on 25 July 2023). [CrossRef] [PubMed]
61. Brenhouse, H.C.; Schwarz, J.M. Immunoadolescence: Neuroimmune development and adolescent behavior. *Neurosci. Biobehav. Rev.* **2016**, *70*, 288–299. [CrossRef]
62. Lucerne, K.E.; Osman, A.; Meckel, K.R.; Kiraly, D.D. Contributions of neuroimmune and gut-brain signaling to vulnerability of developing substance use disorders. *Neuropharmacology* **2021**, *192*, 108598. [CrossRef] [PubMed]
63. Charabati, M.; Wheeler, M.A.; Weiner, H.L.; Quintana, F.J. Multiple sclerosis: Neuroimmune crosstalk and therapeutic targeting. *Cell* **2023**, *186*, 1309–1327. [CrossRef]
64. Svokos, K.A.; Salhia, B.; Toms, S.A. Molecular Biology of Brain Metastasis. *Int. J. Mol. Sci.* **2014**, *15*, 9519–9530. [CrossRef]
65. Suh, J.H.; Kotecha, R.; Chao, S.T.; Ahluwalia, M.S.; Sahgal, A.; Chang, E.L. Current approaches to the management of brain metastases. *Nat. Rev. Clin. Oncol.* **2020**, *17*, 279–299. [CrossRef]
66. Villoslada, P.; Moreno, B.; Melero, I.; Pablos, J.L.; Martino, G.; Uccelli, A.; Montalban, X.; Avila, J.; Rivest, S.; Acarin, L.; et al. Immunotherapy for neurological diseases. *Clin. Immunol.* **2008**, *128*, 294–305. [CrossRef] [PubMed]
67. Vollmer, T.L. Multiple Sclerosis: New Approaches to Immunotherapy. *Neuroscientist* **1996**, *2*, 127–136. [CrossRef]
68. Hemmer, B.; Nessler, S.; Zhou, D.; Kieseier, B.; Hartung, H.P. Immunopathogenesis and immunotherapy of multiple sclerosis. *Nat. Rev. Neurol.* **2006**, *2*, 201–211. [CrossRef]
69. Kammona, O.; Kiparissides, C. Recent Advances in Antigen-Specific Immunotherapies for the Treatment of Multiple Sclerosis. *Brain Sci.* **2020**, *10*, 333. [CrossRef] [PubMed]
70. Hohlfeld, R.; Wekerle, H. Autoimmune concepts of multiple sclerosis as a basis for selective immunotherapy: From pipe dreams to (therapeutic) pipelines. *Proc. Natl. Acad. Sci. USA* **2004**, *101*, 14599–14606. [CrossRef]
71. Brain Sciences | Free Full-Text | Novel Approaches in the Immunotherapy of Multiple Sclerosis: Cyclization of Myelin Epitope Peptides and Conjugation with Mannan. Available online: https://www.mdpi.com/2076-3425/11/12/1583 (accessed on 25 July 2023).
72. Lulu, S.; Waubant, E. Humoral-Targeted Immunotherapies in Multiple Sclerosis. *Neurotherapeutics* **2013**, *10*, 34–43. [CrossRef] [PubMed]
73. Karussis, D. Immunotherapy of Multiple Sclerosis. *BioDrugs* **2013**, *27*, 113–148. [CrossRef]

74. Weiner, H.L. The challenge of multiple sclerosis: How do we cure a chronic heterogeneous disease? *Ann. Neurol.* **2009**, *65*, 239–248. [CrossRef]
75. Hauser, S.L.; Chan, J.R.; Oksenberg, J.R. Multiple sclerosis: Prospects and promise. *Ann. Neurol.* **2013**, *74*, 317–327. [CrossRef] [PubMed]
76. Martin, R.; Stürzebecher, C.S.; McFarland, H.F. Immunotherapy of multiple sclerosis: Where are we? Where should we go? *Nat. Immunol.* **2001**, *2*, 785–788. [CrossRef] [PubMed]
77. Kitz, A.; Singer, E.; Hafler, D. Regulatory T Cells: From Discovery to Autoimmunity. *Cold Spring Harb. Perspect. Med.* **2018**, *8*, a029041. [CrossRef] [PubMed]
78. Ovchinnikov, A.; Findling, O. An overview of pivotal trials and real-world evidence for CD20-depleting therapy in multiple sclerosis. *Wien. Med. Wochenschr.* **2022**, *172*, 359–364. [CrossRef] [PubMed]
79. Miyauchi, J.T.; Tsirka, S.E. Advances in immunotherapeutic research for glioma therapy. *J. Neurol.* **2018**, *265*, 741–756. [CrossRef] [PubMed]
80. Chatterjee, D.; Kordower, J.H. Immunotherapy in Parkinson's disease: Current status and future directions. *Neurobiol. Dis.* **2019**, *132*, 104587. [CrossRef] [PubMed]
81. Wang, H.; Xu, T.; Huang, Q.; Jin, W.; Chen, J. Immunotherapy for Malignant Glioma: Current Status and Future Directions. *Trends Pharmacol. Sci.* **2020**, *41*, 123–138. [CrossRef]
82. Kozłowski, M.; Borzyszkowska, D.; Cymbaluk-Płoska, A. The Role of TIM-3 and LAG-3 in the Microenvironment and Immunotherapy of Ovarian Cancer. *Biomedicines* **2022**, *10*, 2826. [CrossRef]
83. Kuppuswamy, P.S.; Takala, C.R.; Sola, C.L. Management of psychiatric symptoms in anti-NMDAR encephalitis: A case series, literature review and future directions. *Gen. General. Hosp. Psychiatry* **2014**, *36*, 388–391. [CrossRef]
84. Shi, J.; Hua, L.; Harmer, D.; Li, P.; Ren, G. Cre Driver Mice Targeting Macrophages. *Methods Mol. Biol.* **2018**, *1784*, 263–275. [CrossRef]
85. McAllister, A.K.; Patterson, P.H. Neuroimmunology in Brain Development and Disease. *Dev. Neurobiol.* **2012**, *72*, 1269–1271. [CrossRef] [PubMed]
86. Park, H.S.; Park, M.J.; Kwon, M.S. Central Nervous System-Peripheral Immune System Dialogue in Neurological Disorders: Possible Application of Neuroimmunology in Urology. *Int. Neurourol. J.* **2016**, *20* (Suppl. S1), S8–S14. [CrossRef]
87. Norden, D.M.; Muccigrosso, M.M.; Godbout, J.P. Microglial priming and enhanced reactivity to secondary insult in aging, and traumatic CNS injury, and neurodegenerative disease. *Neuropharmacology* **2015**, *96*, 29–41. [CrossRef] [PubMed]
88. Archer, T.; Fredriksson, A.; Schütz, E.; Kostrzewa, R.M. Influence of Physical Exercise on Neuroimmunological Functioning and Health: Aging and Stress. *Neurotox. Res.* **2011**, *20*, 69–83. [CrossRef]
89. Fonken, L.K.; Gaudet, A.D. Neuroimmunology of healthy brain aging. *Curr. Opin. Neurobiol.* **2022**, *77*, 102649. [CrossRef] [PubMed]
90. Lucin, K.M.; Wyss-Coray, T. Immune activation in brain aging and neurodegeneration: Too much or too little? *Neuron* **2009**, *64*, 110–122. [CrossRef]
91. Tenembaum, S.N. Pediatric demyelinating disease and anti-MOG antibody. *Clin. Exp. Neuroimmunol.* **2021**, *12*, 7–21. [CrossRef]
92. Nosadini, M.; Alper, G.; Riney, C.J.; Benson, L.A.; Mohammad, S.S.; Ramanathan, S.; Nolan, M.; Appleton, R.; Leventer, R.J.; Deiva, K.; et al. Rituximab monitoring and redosing in pediatric neuromyelitis optica spectrum disorder. *Neurol.-Neuroimmunol. Neuroinflamm.* **2016**, *3*, e188. [CrossRef]
93. Reinert, M.-C.; Benkert, P.; Wuerfel, J.; Michalak, Z.; Ruberte, E.; Barro, C.; Huppke, P.; Stark, W.; Kropshofer, H.; Tomic, D.; et al. Serum neurofilament light chain is a useful biomarker in pediatric multiple sclerosis. *Neurol.-Neuroimmunol. Neuroinflamm.* **2020**, *7*, e749. [CrossRef]
94. Sguigna, P.V.; McCreary, M.C.; Conger, D.L.; Graves, J.S.; Benson, L.A.; Waldman, A.T.; Greenberg, B.M.; on behalf of the PERCEPTION Collaboration. Utilization of Visual Acuity Retroilluminated Charts for the Assessment of Afferent Visual System Dysfunction in a Pediatric Neuroimmunology Population. *J. Neuro-Ophthalmol.* **2021**, *41*, 19. [CrossRef]
95. Polyakova, V.O.; Kvetnoy, I.M.; Anderson, G.; Rosati, J.; Mazzoccoli, G.; Linkova, N.S. Reciprocal Interactions of Mitochondria and the Neuroimmunoendocrine System in Neurodegenerative Disorders: An Important Role for Melatonin Regulation. *Front. Physiol.* **2018**, *9*, 199. Available online: https://www.frontiersin.org/articles/10.3389/fphys.2018.00199 (accessed on 25 July 2023). [CrossRef] [PubMed]
96. Akmaev, I.G. Current concepts of the interactions of regulating systems: Nervous, endocrine and immune. *Usp. Fiziol. Nauk.* **1996**, *27*, 3–20.
97. Kvetnoy, I.M. Neuroimmunoendocrinology: Where is the field for study? *Neuroendocrinol. Lett.* **2002**, *23*, 119–120.
98. Wilder, R.L. Neuroimmunoendocrinology of the Rheumatic Diseases. *Ann. N. Y. Acad. Sci.* **2002**, *966*, 13–19. [CrossRef] [PubMed]
99. Pérez-Aguilar, M.; Rondón, R. Neuroimmunoendocrine System in Health and Disease. *EC Microbiol.* **2018**, *14*, 2–6.
100. Anderson, G.; Maes, M. Neurodegeneration in Parkinson's Disease: Interactions of Oxidative Stress, Tryptophan Catabolites and Depression with Mitochondria and Sirtuins. *Mol. Neurobiol.* **2014**, *49*, 771–783. [CrossRef]
101. Akmayev, I.G. Neuroimmunoendocrinology: Evidence and hypothesis. *Probl. Endocrinol.* **1997**, *43*, 3–9. [CrossRef]
102. Nguyen, M.; Palm, N.W. Gut instincts in neuroimmunity from the eighteenth to twenty-first centuries. *Semin. Immunopathol.* **2022**, *44*, 569–579. [CrossRef]

103. Cui, C.; Ruan, Y.; Qiu, W. Potential role of the gut microbiota in neuromyelitis optica spectrum disorder: Implication for intervention. *J. Clin. Neurosci.* **2020**, *82*, 193–199. [CrossRef]
104. van Thiel, I.A.M.; de Jonge, W.J.; Chiu, I.M.; van den Wijngaard, R.M. Microbiota-neuroimmune cross talk in stress-induced visceral hypersensitivity of the bowel. *Am. J. Physiol.-Gastrointest. Liver Physiol.* **2020**, *318*, G1034–G1041. [CrossRef]
105. Maranduba, C.M.d.C.; De Castro, S.B.R.; de Souza, G.T.; Rossato, C.; da Guia, F.C.; Valente, M.A.S.; Rettore, J.V.P.; Maranduba, C.P.; de Souza, C.M.; Carmo, A.M.R.D.; et al. Intestinal Microbiota as Modulators of the Immune System and Neuroimmune System: Impact on the Host Health and Homeostasis. *J. Immunol. Res.* **2015**, *2015*, e931574. [CrossRef] [PubMed]

biomolecules

MDPI

Review

Cannabinoids in Medicine: A Multifaceted Exploration of Types, Therapeutic Applications, and Emerging Opportunities in Neurodegenerative Diseases and Cancer Therapy

Victor Voicu [1,2], Felix-Mircea Brehar [3,4,*], Corneliu Toader [4,5,*], Razvan-Adrian Covache-Busuioc [4], Antonio Daniel Corlatescu [4], Andrei Bordeianu [4], Horia Petre Costin [4], Bogdan-Gabriel Bratu [4], Luca-Andrei Glavan [4] and Alexandru Vlad Ciurea [4,6]

1 Pharmacology, Toxicology and Clinical Psychopharmacology, "Carol Davila" University of Medicine and Pharmacy in Bucharest, 020021 Bucharest, Romania; victor.voicu@yahoo.com
2 Medical Section within the Romanian Academy, 010071 Bucharest, Romania
3 Neurosurgery Department, Emergency Clinical Hospital Bagdasar-Arseni, 041915 Bucharest, Romania
4 Department of Neurosurgery, "Carol Davila" University of Medicine and Pharmacy, 020021 Bucharest, Romania; razvan-adrian.covache-busuioc0720@stud.umfcd.ro (R.-A.C.-B.); antonio.corlatescu0920@stud.umfcd.ro (A.D.C.); andrei.bordeianu@stud.umfcd.ro (A.B.); horia-petre.costin0720@stud.umfcd.ro (H.P.C.); bogdan.bratu@stud.umfcd.ro (B.-G.B.); luca-andrei.glavan0720@stud.umfcd.ro (L.-A.G.); prof.avciurea@gmail.com (A.V.C.)
5 Department of Vascular Neurosurgery, National Institute of Neurology and Neurovascular Diseases, 077160 Bucharest, Romania
6 Neurosurgery Department, Sanador Clinical Hospital, 010991 Bucharest, Romania
* Correspondence: felixbrehar@yahoo.com (F.-M.B.); corneliu.toader@umfcd.ro (C.T.)

Citation: Voicu, V.; Brehar, F.-M.; Toader, C.; Covache-Busuioc, R.-A.; Corlatescu, A.D.; Bordeianu, A.; Costin, H.P.; Bratu, B.-G.; Glavan, L.-A.; Ciurea, A.V. Cannabinoids in Medicine: A Multifaceted Exploration of Types, Therapeutic Applications, and Emerging Opportunities in Neurodegenerative Diseases and Cancer Therapy. *Biomolecules* **2023**, *13*, 1388. https://doi.org/10.3390/biom13091388

Academic Editor: Balapal S. Basavarajappa

Received: 14 August 2023
Revised: 8 September 2023
Accepted: 12 September 2023
Published: 14 September 2023

Abstract: In this review article, we embark on a thorough exploration of cannabinoids, compounds that have garnered considerable attention for their potential therapeutic applications. Initially, this article delves into the fundamental background of cannabinoids, emphasizing the role of endogenous cannabinoids in the human body and outlining their significance in studying neurodegenerative diseases and cancer. Building on this foundation, this article categorizes cannabinoids into three main types: phytocannabinoids (plant-derived cannabinoids), endocannabinoids (naturally occurring in the body), and synthetic cannabinoids (laboratory-produced cannabinoids). The intricate mechanisms through which these compounds interact with cannabinoid receptors and signaling pathways are elucidated. A comprehensive overview of cannabinoid pharmacology follows, highlighting their absorption, distribution, metabolism, and excretion, as well as their pharmacokinetic and pharmacodynamic properties. Special emphasis is placed on the role of cannabinoids in neurodegenerative diseases, showcasing their potential benefits in conditions such as Alzheimer's disease, Parkinson's disease, Huntington's disease, and multiple sclerosis. The potential antitumor properties of cannabinoids are also investigated, exploring their potential therapeutic applications in cancer treatment and the mechanisms underlying their anticancer effects. Clinical aspects are thoroughly discussed, from the viability of cannabinoids as therapeutic agents to current clinical trials, safety considerations, and the adverse effects observed. This review culminates in a discussion of promising future research avenues and the broader implications for cannabinoid-based therapies, concluding with a reflection on the immense potential of cannabinoids in modern medicine.

Keywords: cannabinoids; neurodegenerative diseases; phytocannabinoids; antitumor properties; endocannabinoids; cannabinoid receptors

1. Introduction

1.1. Background on Cannabinoids

The burgeoning field of research surrounding Cannabis sativa has identified Δ9-tetrahydrocannabinol (THC) as one of its key active compounds. This compound, along with a plethora of related molecules, both phytochemical and synthetic, exhibits an affinity for specific neuronal binding sites, contributing to an altered mood and perceptions. These binding locations are particularly abundant in neural structures such as the substantia nigra, hippocampus, and cerebellum. One of the clinical manifestations associated with excessive cannabis consumption is cannabinoid hyperemesis syndrome (CHS), a condition characterized by recurrent episodes of severe vomiting. This syndrome is exacerbated by the consumption of high doses of cannabinoids and has imposed a growing burden on healthcare systems, particularly in the United States. Despite the increasing prevalence of CHS, the medical community has yet to reach a consensus on optimal treatment strategies [1,2].

Beyond their well-documented medicinal applications, cannabinoids have been the subject of intensive investigations aimed at exploring their therapeutic potential for a variety of medical conditions, including pain, addiction, obesity, and inflammation, among others. Recent discoveries have expanded our understanding of the pharmacology of cannabinoids by revealing the existence of non-CB1 and non-CB2 orphan G-protein-coupled receptors such as GPR18, GPR55, and GPR119. These receptors operate in conjunction with the established CB1 and CB2 receptors but have unique characteristics, including allosteric binding and biased signaling, which could lead to distinct functional outcomes. A particularly intriguing line of inquiry has revealed the presence of CB1 receptors within the mitochondria of striated and cardiac muscles, implicating them in the modulation of intramitochondrial signaling and respiratory processes [3–5].

1.2. Importance of Endogenous Cannabinoids

The endocannabinoid system (ECS) represents a complex neuromodulatory network that is of paramount significance to the central nervous system (CNS), synaptic plasticity, and adaptive responses to both endogenous and environmental stimuli. Comprising cannabinoid receptors (CBRs), endogenously synthesized cannabinoids (endocannabinoids), and enzymatic pathways for their synthesis and degradation, the ECS serves as a critical mechanism for neuromodulation. While CB1 cannabinoid receptors are the most commonly implicated receptors in these interactions, other receptors such as CB2, transient receptor potential channels, and peroxisome proliferator-activated receptors also participate [5,6].

Two endogenous cannabinoids that have attracted significant academic attention are 2-arachidonoyl glycerol (2-AG) and arachidonoyl ethanolamide (anandamide). Despite sharing molecular similarities, these cannabinoids diverge in their synthetic and degradative enzymatic pathways, leading to distinct physiological and pathophysiological roles [7,8].

The societal ubiquity of cannabis consumption has fueled a wealth of research into the physiological and pathophysiological functions of endocannabinoids. Marijuana's prevalence as a widely consumed substance in Western societies contributed to the discovery of the ECS and elucidated its involvement in a plethora of physiological processes. This intricate system comprises G-protein-coupled CBRs that are activated by lipid mediators, commonly referred to as endocannabinoids (eCBs). These eCBs are not only synthesized from cannabis but also encompass a variety of biochemical constituents including precursors, enzymes, and transporters. Research has revealed an extensive distribution of components of the ECS throughout various bodily regions and organs, underscoring its fundamental role in physiology and the potential for targeted interventions for a range of human ailments [9–12].

Historically, Cannabis sativa (marijuana) has been employed to stimulate appetite, but rigorous scientific scrutiny of its molecular mechanisms gained momentum following the identification of THC in the late 1960s. Although marred by societal disapproval due to misuse, empirical evidence has highlighted the therapeutic potential of marijuana and its derivatives. Specifically, they have been found to enhance appetite for sweet foods. The elucidation of distinct CBRs and their endogenous ligands has provided a robust physiological framework for understanding the myriad biological effects mediated by marijuana and other cannabinoids [13,14].

Recent advancements in the field have illuminated the existence of a plethora of naturally occurring compounds that serve as binding partners to CBRs. eCBs bear a functional resemblance to endorphins and have been detected in a range of mammalian species, including humans. Notably, eCBs have been identified in a diverse array of tissues such as the CNS, peripheral nerves, and reproductive and immune organs like the uterus, leukocytes, spleen, and testicles. Anandamide, one of the earliest-discovered eCBs, is present in notably high concentrations in uterine tissue [15]. This suggests a pivotal role in reproductive processes, a notion corroborated via extensive investigations. Empirical studies have revealed anandamide's crucial involvement in orchestrating implantation processes. A diminished enzymatic activity responsible for the breakdown of anandamide has been frequently associated with early pregnancy loss [16,17]. The growing body of scientific literature on eCBs has notably concentrated on the study of anandamide, reaffirming its significance in both physiological and pathophysiological contexts.

1.3. *Overview of Neurodegenerative Diseases and Cancer*

THC demonstrates interactions with CBRs, chiefly CB1 and CB2, which are naturally activated by eCBs. The compound plays a multifaceted role in various physiological and pathological domains, including the modulation of the release of neurotransmitters, the regulation of pain perception, and the functioning of the cardiovascular, digestive, and hepatic systems. Nonetheless, THC's psychotropic effects, which are mainly mediated via the activation of CB1 receptors in the brain, have considerably restricted its clinical applicability. Contrastingly, the cannabis plant is replete with cannabinoids that exhibit minimal to no psychotropic activity, many of which have demonstrated therapeutic potential surpassing that of THC. Among these, cannabidiol (CBD) has gained prominence for its prospective utility in treating conditions such as inflammation, diabetes, cancer, affective disorders, and neurodegenerative diseases. Another cannabinoid, D9-tetrahydrocannabivarin (THCV), shows promise in addressing issues like epilepsy and obesity [18,19].

Neurological disorders, inclusive of neurodegenerative diseases and traumatic brain injuries, pose considerable challenges to healthcare due to their impact on cognitive, motor, and behavioral functions. While diverse therapeutic strategies have been explored, none have provided definitive results. However, accumulated evidence indicates that cannabinoids may offer a novel pathway for treatment. Research elucidates the in vivo potential of both natural and synthetic cannabinoids in ameliorating cognitive decline and motor impairments. Animal models have demonstrated the efficacy of cannabinoids in enhancing neurobehavioral function, improving working memory, and reducing neurological deficits through mechanisms such as modulating inflammation, mitigating edema, and preserving the neuronal structure [20,21].

CBD's antioxidative properties have revealed its potential in combating neurodegenerative and cardiovascular disorders. Moreover, animal studies have showcased CBD's anticancer properties. The co-administration of THC with radiation therapy has also been observed to induce higher rates of autophagy and apoptosis in cancer cells. The National Cancer Institute acknowledges the therapeutic potential of Cannabis sativa, particularly THC and CBD, in alleviating various symptoms associated with cancer, including pain, appetite loss, nausea, and anxiety. CBD's complex pharmacological profile allows it to act as an adaptogen and modulator, interacting intricately with the receptor proteins CB1 and CB2, among other sites [22].

CBD is increasingly being recognized for its potential as an immunomodulatory entity. Empirical studies substantiate its efficacy in engendering immunosuppression against non-infectious inflammatory conditions, such as inflammatory bowel disease, rheumatoid arthritis, and neurodegenerative disorders. Moreover, CBD has exhibited immunoprotective qualities against viral infections, including COVID-19. Its interactions with an array of cellular targets and signaling pathways have been found to induce specific anti-cancer responses, which is in alignment with its principal role in ECS-mediated homeostasis [23].

Case reports in the medical literature affirm the therapeutic viability of cannabinoids derived from Cannabis sativa. However, the clinical adoption of these compounds is often hindered by the psychotropic side effects that are predominantly attributable to THC. Advancements in the understanding of the ECS, including the discovery of new receptors, ligands, and mediators, have facilitated the exploration of novel therapeutic avenues that could mitigate the adverse psychotropic effects associated with certain plant constituents. Such scientific innovations have catalyzed the development of FDA-approved medications that are revolutionizing contemporary medical treatment modalities. For instance, Nabiximols, an FDA-sanctioned amalgam of THC and non-psychoactive CBD, has demonstrated utility in alleviating the pain and spasticity related to multiple sclerosis [24,25]. Additionally, DRO and Nabilone have gained FDA approval for their effectiveness in countering chemotherapy-induced nausea and vomiting in cancer patients. Notably, DRO has also secured regulatory endorsement for its role in managing anorexia among AIDS patients [26–29].

1.4. Significance of Studying Cannabinoid Effects

In a comprehensive meta-analysis incorporating 211 studies, the binding affinities of cannabinoid receptor ligands at human (Hs) and rat (Rn) CB1 and CB2 receptors were examined. Methodologies in line with the Cochrane procedures guided this nonclinical investigation. Meta-regression techniques were utilized to identify data variances due to methodological factors. The Ki values for THC exhibited discernible differences between HsCB1 and RnCB1. The Kd values for CP55,940 and WIN55,212-2 also displayed significant discrepancies between HsCB1 and RnCB1, as well as between HsCB1 and HsCB2. Moreover, SR141716A exhibited affinity to both sets of CBRs [30].

Another exhaustive analysis considered 91 publications consisting of 104 individual studies with 9958 participants. These studies ranged from randomized controlled trials (RCTs) to observational research and covered a variety of pain-related conditions including neuropathic pain, fibromyalgia, and chronic non-cancer pain (CNCP). Pooled event rates (PERs) revealed that cannabinoids were statistically more effective than a placebo for achieving a 30% reduction in pain, but no significant difference was found for a 50% reduction in pain [31].

A discernible uptick has been observed in the usage of synthetic cannabinoid products among adolescents. A study concentrated on the self-reported psychoactive and somatic ramifications of synthetic cannabinoid use among adolescents. Notably, all participants indicated experiencing euphoria and memory alterations. A significant majority, 82%, also reported negative mood shifts. The concurrent use of marijuana and alcohol was noted by 91% of the subjects. Intriguingly, a robust correlation was observed between the frequency of synthetic cannabinoid use and the number of other drugs consumed ($r = 0.896$, $p < 0.05$). Consequently, the study concluded that adolescent users of synthetic cannabinoids report substantial psychoactive effects [32].

Previous research has shown that the stimulation of CB1 receptors affects both motility and food intake in rodent models and also has implications for human gastrointestinal (GI) function; however, specific effects on human GI transit times and sensations of fullness remain undetermined. To shed light on this, a double-blind, randomized study involving 30 healthy volunteers assessed the effects of DRO versus a placebo through a series of diagnostic tests, including the Ensure® Satiation test and scintigraphic transit testing [33]. In summary, the study posits that the ECS within the human gastrointestinal tract can be modulated by the non-selective cannabinoid receptor (CBR) agonist DRO to decelerate gastric emptying. The study further advocates for subsequent investigations involving both selective and non-selective cannabinoid antagonists to substantiate these initial findings. Owing to preliminary evidence indicating gender-specific variations in gastric emptying and fasting gastric volume in response to the acute administration of DRO, it is recommended that future research on cannabinoids incorporate gender stratification to rigorously assess their impact [33].

2. Cannabinoids: Types and Mechanisms of Action

2.1. Phytocannabinoids

The historical and medicinal relevance of Cannabis sativa is rich, encompassing both therapeutic and recreational applications. With over 120 C21 terpenophenolic compounds known as phytocannabinoids, Cannabis sativa is a prime source of bioactive natural compounds. THC, discovered in 1971, remains dominant among these, and its discovery led to the identification of the ECS, which comprises CB1 and CB2 receptors. Despite its psychotropic effects limiting its medical utility, THC, along with other phytocannabinoids, holds promise for treating conditions like pain, anxiety, and cachexia. Contemporary research is exploring the biosynthesis of phytocannabinoids in various species including Cannabis, Rhododendron, and Radula, as well as the potential for engineering cannabinoids with enhanced properties via synthetic biology strategies [34,35].

Similar bioactive constituents, namely phytocannabinoids, are also present in hashish and marijuana, both of which are derivatives of Cannabis sativa L. Traditional pharmacology focused primarily on these compounds interacting with CB1 and CB2 receptors. However, newer insights suggest a more complex interaction profile involving multiple targets. The molecular pharmacology of key phytocannabinoids, particularly THC and CBD, is a focal point in understanding their diverse range of actions [36,37].

Intricacies in the composition of phytocannabinoids involve a range of pathways and variations in side-chain composition as well as the degree of isoprenyl residue oligomerization. The complexity of these compounds extends to their varying origins, which include not just higher plants but also liverworts and fungi. Factors like heat, light, and atmospheric oxygen can induce non-enzymatic alterations in these compounds, affecting key constituents like CBG, CBD, THC, and CBC. Not confined to CBRs, these bioactive molecules engage with a variety of targets, such as thermo-TRPs and transcription factors like PPARs, suggesting their potential as an investigational class of drugs with actions beyond the ECS [38–40].

The limitations of conventional antiepileptic medications, characterized by suboptimal efficacy and adverse side effects, render the exploration of alternative therapies imperative. Phytocannabinoids, notably THC and CBD, offer a promising avenue in this context as they have exhibited anticonvulsant properties with comparatively fewer adverse effects in both preclinical and initial human studies. With the growing global acceptance of cannabis-derived products as medical interventions, an understanding of their neurochemical mechanisms of action is essential. THC functions as a partial agonist at the cannabinoid 1 and 2 receptors (CB1/2), leading to typical outcomes such as euphoria and relaxation. However, it may also induce dysphoria, anxiety, and manifestations of psychosis in certain cases [41].

CBD and its propyl analog, cannabidivarin (CBDV), have been the focus of increasing scientific inquiry due to their wide spectrum of therapeutic attributes, including anti-inflammatory, anti-nausea, anti-tumor, anti-convulsant, anxiolytic, and neuroprotective qualities. Despite the plethora of molecular targets with which phytocannabinoids interact across various body systems, a comprehensive understanding of their mechanisms of action remains elusive. The nematode C. elegans serves as a vital model organism in this context, sharing approximately 60% of the genes associated with human pathologies and exhibiting remarkable neural circuitry and G-protein-coupled receptor (GPCR) signaling similarities to mammals [42].

With respect to major depressive disorder, compelling evidence suggests that the activation of CB1 receptors may function as a protective mechanism in humans, either directly or indirectly. This proposition is further supported by the negative mood effects, including depression and suicidal thoughts, observed in obese patients treated with CB1 antagonists. Moreover, the silencing of CB1 receptors in specific neural circuits has been shown to elevate susceptibility to stress, potentially triggering a cascade of stress-related disorders, including depression [43,44].

Despite the long-standing recognition of the therapeutic potential of phytocannabinoids and their antioxidative capabilities, the operational mechanisms remain underexplored. Recent investigations have employed density functional theory (DFT) calculations to scrutinize the radical scavenging abilities of CBD and cannabidiolic acid (CBDA). These studies highlight the effectiveness of these compounds in neutralizing hydroperoxyl radicals in polar physiological environments, albeit with diminished efficacy in lipid-rich media. This focus has also extended to investigating the antiradical properties of eight key compounds from all the primary families of phytocannabinoids, including cannabinol (CBN), THC, cannabichromene (CBC), cannabicyclol (CBL), cannabielsoin (CBE), CBD, cannabifuran (CBF), and cannabigerol (CBG) [45].

The intricate interactions between the CBRs CB1 and CB2 have revealed a fascinating phenomenon: the formation of complex molecular assemblies known as CB1/2RHet complexes. These complexes are noteworthy for their potential to modulate CB1R-mediated effects. Investigations have been extended into the impact of various cannabinoid compounds on the formation of these receptor complexes [46]. A model utilizing HEK-293T cells was employed for this purpose; these cells were subjected to transfection with a consistent quantity of CB1R-RLuc cDNA while varying amounts of CB2R-GFP2 cDNA were also introduced. The resulting data were plotted onto a saturable bioluminescence resonance energy transfer (BRET) curve. Key observations from this experiment include the BRETmax value, determined to be 214 ± 15, and the BRET50 values, identified as 48 ± 9, both of which suggest targeted and specific interactions between CB1R and CB2R [47].

2.2. Endocannabinoids

The isolation of a cannabinoid receptor ligand, arachidonylethanolamide or anandamide, from porcine brain tissue has provided significant insights into its physiological effects. Anandamide effectively interferes with radiolabeled cannabinoid ligand binding to synaptosomal membranes from rat brains. Moreover, the compound exhibits dose-dependent inhibitory actions in electrically stimulated mouse vas deferens [48]. Further investigations reveal that anandamide acts as a cannabinoid agonist, effectively inhibiting the forskolin-induced activation of adenylate cyclase in N18TG2 cells, similar to the effects of HU-210, while (+)-HU-211 demonstrates minimal impact [1]. The multifaceted actions of anandamide encompass the blockage of voltage-gated calcium channels, the activation of inwardly rectifying potassium currents, G-protein binding, and the induction of multiple cellular signaling pathways, among other effects [49,50].

2-Arachidonoylglycerol (2-AG), a molecular variant of monoacylglycerol, bears a structural resemblance to anandamide. It is unique in attaching arachidonic acid at the second position of its glycerol framework. Studies indicate that 2-AG binds to CBRs on synaptosomal membranes derived from rat brain tissue, albeit with reduced potency compared to anandamide [51]. This raises the possibility that arachidonic-acid-containing monoacylglycerols could function as endogenous ligands for CBRs in specific neural contexts. Unlike conventional neurotransmitters, which act through vesicular secretion from synaptic terminals, anandamide and 2-AG may be synthesized on demand via the stimulation-triggered cleavage of separate phospholipid precursors located within neuronal membranes [52].

Research indicates the presence of cannabinergic modulation within the basal ganglia, as evidenced by the effects of the CB1 receptor antagonist SR141716 [53]. This antagonist led to increased locomotion in mice and induced stereotypies in rats. The activation of CBRs has also been found to significantly reduce electrically-induced dopamine release in rat striatal slices and to potentiate the symptoms of neuroleptic-induced catalepsy [54]. Importantly, blocking the CBRs effectively removed the inhibitory regulation mediated by endogenously released anandamide, thereby amplifying quinpirole-induced motor activation [55]. The absence of any observable effects of SR141716A when administered alone, at doses similar to those used to augment quinpirole-induced motor activation, suggests that anandamide's behavioral effects may be dependent on D2 receptor stimulation, potentially countering dopamine-D2-facilitated psychomotor activity [56].

Anandamide has been shown to activate transient receptor potential (TRP) channels, specifically TRPV1, under certain conditions. While both CBRs and TRP channels seem to contribute to its effects, their individual roles appear to be variable [57]. Anandamide is also known to activate alpha and gamma peroxisome proliferator-activated receptors (PPARs), which have a substantial impact on gene transcription. The inhibition of their degradation via FAAH increases the levels of anandamide as well as other N-acylamides that modulate PPARα receptors [58,59].

Both THC and anandamide are classified as low efficacy agonists. Under specific conditions, such as low receptor density or limited post-receptor effectors, these compounds may function as antagonists by negating the CB1 receptor signaling initiated by 2-AG [60].

The ECS is a complex network that includes endogenous cannabinoids (eCBs) like N-arachidonoylethanolamine (anandamide or AEA) and 2-arachidonoylglycerol (2-AG), biosynthetic enzymes such as NAPE-specific phospholipase D and Diacylglycerol lipase- α, and degradative enzymes like fatty acid amide hydrolase (FAAH) and monoacylglycerol lipase (MAGL) [61–64]. The receptors for these substances, termed cannabinoid receptors (CBRs), also form integral components of this system. Notably, these eCBs interact not just with the primary CBR subtypes (CB1R and CB2R) but also with various other receptors, including transient receptor potential vanilloid type 1 (TRPV1) cation channels, GTP-binding protein-coupled receptor GPR55, abnormal-CBD receptor, and peroxisome proliferator-activated receptors (PPARs) [65,66].

eCBs serve as critical regulators of synaptic transmission through various physiological feedback mechanisms designed to counteract either the overexcitation or inhibition of synapses. These mechanisms include retrograde signaling, which leads to the depolarization-induced suppression of inhibition (DSI) at GABAergic synapses and the depolarization-induced suppression of excitation (DSE) at glutamatergic synapses. The presynaptic location of CB1R allows eCBs to influence other neurotransmitters, such as opioid peptides, acetylcholine, and 5-hydroxytryptamine (5-HT), even if CB1Rs may not be expressed in nigrostriatal dopaminergic neurons [67,68]. Nonetheless, the functionality of these neurons can be profoundly affected either by the activation or blockade of ECS components present in nearby neuronal subpopulations like GABAergic, glutamatergic, and opioidergic neurons that interconnect with dopaminergic neurons. Moreover, dopaminergic neurons can produce extracellular peptide-binding proteins, enhancing retrograde signaling at both excitatory and inhibitory synapses [69,70].

Additional theories suggest that eCBs modulate dopamine (DA) transmission through interactions with TRPV1 receptors and the formation of heteromers with metabotropic receptors such as dopamine D1 and D2 receptors. The presence of CB2R also implies a direct role of eCBs in modulating dopamine transmission, thus expanding the scope of their physiological and potentially therapeutic roles [71,72].

2.3. *Synthetic Cannabinoids*

Phytocannabinoids, predominantly comprising THC, originate from plant sources such as cannabis. On the other hand, synthetic cannabinoids present in products like Spice include various compounds such as naphthoylindoles, benzoylindones, and phenylacetylindoles [73]. Notably, the composition of synthetic cannabinoids can vary substantially across different Spice products and even within the same batch or package (European Monitoring Centre for Drugs and Drug Addiction 2009). Beyond synthetic cannabinoids, Spice formulations may include a range of other substances such as additives, preservatives, fatty acids, amides, esters, and additional psychoactive compounds like the benzodiazepine phenazepam and an active metabolite of tramadol. Some formulations have even been found to contain Salvia divinorum, Kratom, or cannabis, although the impact of these additional substances on the overall effects of Spice remains unclear [74].

Neurologically, Spice has been associated with a diverse array of symptoms that include tremors, ataxia, nystagmus, fasciculations, and hypertonicity, as well as hyperflexion and hyperextension. Cognitive impairments affecting attention, concentration, and memory have also been reported, along with a compromised ability to operate machinery. Palpitations frequently accompany feelings of panic, complicating efforts to discern whether these symptoms stem from underlying anxiety. Even after the acute phase of palpitations subsides, residual irregularities may persist. Additional observed symptoms include xerostomia (commonly known as "cotton mouth"), reddened conjunctiva, changes in pupil size leading to either constriction (miosis) or dilation (mydriasis), heightened sensitivity to light, and persistent coughing and inflammation or injury to the lungs [73].

2.4. Cannabinoid Receptors and Signaling Pathways

CB-1R receptors are ubiquitously distributed throughout the nervous system, with pronounced concentrations in regions such as the hippocampus, association cortex, cerebellum, and basal ganglia, among others [75,76]. In contrast, CB-2R receptors are principally localized in gastrointestinal and lymphatic tissues, as well as specific CNS locations such as the dorsal nucleus of the vagus nerve, spinal trigeminal nuclei and nucleus ambiguous [76,77]. Both CB-1R and CB-2R function as G-protein-coupled receptors, modulating the release of neurotransmitters like glutamate, dopamine, and acetylcholine through the inhibition of adenyl cyclase activity via G0/Gi proteins. Additional neurotransmitter pathways, including serotonergic, GABAergic, and NMDA (N-methyl-D-aspartate), are indirectly modulated. A noteworthy observation pertains to the TRPV1 receptor, which has been identified in the basal ganglia through advanced imaging techniques [78]. The ECS employs feedback mechanisms to regulate synaptic transmission, affecting cell development, differentiation, and apoptosis via the MAPK/ERK pathway [79].

Recent studies have classified CBRs in both rats and humans as members of the seven-transmembrane GTP-binding protein-coupled receptor family. Investigations demonstrated that exogenous cannabinoids could suppress forskolin or secretin-induced adenylate cyclase activity and inhibit the opening of the N-type calcium channel, processes that are rendered ineffectual when pretreated with pertussis toxin (PTX), thereby implicating the G1/G0 GTP-binding proteins in these signaling pathways [80].

Furthermore, the CB1 receptor has been identified to contain 472 amino acids in humans and 473 in rats [81]. A second cannabinoid receptor, CB2, was also successfully isolated and was found to have a similar architecture consisting of 360 amino acids [82]. Despite sharing only 44% overall similarity, both receptors share 68% resemblance in their transmembrane domains and are coupled to Gi/Go proteins. Various compounds, including those with antagonist or inverse-agonist properties such as SR141716A for CB1 and SR144528 for CB2, have been developed to interact with these receptors [83]. It is posited that CB1 is fundamentally implicated in the regulation of cognition, memory, and motor activities [84].

3. The Pharmacology of Cannabinoids

3.1. Absorption, Distribution, Metabolism, and Excretion (ADME)

In the realm of drug discovery, a comprehensive selection of target and ligand molecules from cyano-bacterial species was carried out based on their biological and pharmacological attributes. These selections were further refined through homology modeling, molecular docking, and molecular dynamics (MD) simulations. A highlight was the utilization of an in silico tool, Maestro v10.2's Quikprop, for the assessment of their absorption, distribution, metabolism, and excretion (ADME) properties. This computational approach adhered to well-established guidelines such as the Rule of Five and considered both physicochemical parameters and toxicology measures. The insights gained facilitated the accurate prediction of pharmacokinetic properties, corroborating the vital role of in silico methods in drug discovery processes [85].

In evaluating the pharmacodynamic, pharmacokinetic, and toxicity profiles of the selected cannabinoids, ADME/TOPKAT prediction proved to be highly instructive. The compounds exhibited varied levels of human intestinal absorption, blood–brain barrier penetration, and solubility. Furthermore, the compounds also displayed varying degrees of plasma protein binding and hepatotoxicity. These analyses collectively contributed to a nuanced understanding of the complex pharmacokinetic and pharmacodynamic profiles of the studied cannabinoid compounds [86].

The integral role of ADME studies in both drug discovery and development is well-documented. These studies not only predict human pharmacokinetic properties but also help establish correlations with pharmacodynamic assessments in commonly employed animal models for nonclinical investigations. The employment of specialized methodologies, such as [14C]-S-777469, has proven to be invaluable for the in-depth analysis of specific agonists like S-777469, offering nuanced insights into their behavior within human systems and broadening the understanding of their pharmacological profiles [87].

In the context of pharmacology, it has been observed that serious illness substantially alters all facets of drug disposition, including absorption, distribution, metabolism, and excretion (ADME). Such alterations manifest in multiple dimensions, from disrupted oral absorption and bioavailability to shifts in drug distribution patterns and metabolic pathways. These modifications suggest that pharmacokinetic models based on healthy subjects may not be wholly applicable to those with illnesses, thus necessitating more nuanced approaches for this demographic [88].

In studies focusing on CBRs, the compounds were stringently assessed for their binding affinities to human and mouse CB2 receptors (CB2Rs), as well as their selectivity towards human CB1 receptors (CB1Rs). Among the tested compounds, "Compound 2f" stood out for its strong affinity and selectivity for CB2R. This compound was further subjected to advanced metabolic pathway analyses, including incubation with human and rat liver microsomes. Additional in vivo tests were conducted to evaluate the metabolic stability of [18F]2f, thereby adding to the compound's potential candidacy for PET tracer applications [89].

Regarding the neuroprotective potential of cannabinoids, extensive analyses were carried out on three specific receptors (CB1, CB2, and CB3) and selected phytocannabinoids, including THC and CBD, as well as endogenous cannabinoids like anandamide (AEA) and 2-arachidonoylglycerol (2-AG). These studies shed light on the intricate interplay between cannabinoids and their molecular targets. Furthermore, the ADME profiles of these compounds indicate favorable drug-like characteristics, thereby supporting their potential applicability in the treatment of neurodegenerative or other neurological conditions [90].

The foundational premise for the detection of drugs in sweat is anchored in the pharmacokinetic understanding of a drug's absorption, distribution, metabolism, and excretion (ADME) cycle. During this cycle, a fraction of the drug is anticipated to be excreted through sweat. The presence of lipophilic compounds in the bloodstream is modulated by factors such as their pKa (acid dissociation constant) values and the pH of the fluids they enter. Employing a modified Henderson–Hasselbach equation, which incorporates both pKa and pH, allows for the theoretical determination of the fluid/plasma concentration ratio (F/P ratio). Passive diffusion, which is governed by concentration gradients, typically enables drugs to permeate sweat, with only the unbound fractions making this transition. Additionally, the more acidic pH of sweat compared to blood creates a tendency for basic drugs to accumulate within its layers [91].

As for the pharmacokinetics of inhaled cannabinoids, inhalation and intravenous administration yield similar profiles. Post inhalation, peak plasma concentrations of THC and CBD are rapidly achieved, typically within 3–10 min. The bioavailability of inhaled THC is estimated to range between 10 and 35%, a value influenced by various factors including inhalation patterns, breath-holding durations, and the specifications of the inhalation device used. CBD, when inhaled, exhibits an average systemic bioavailability of approximately 31%, presenting a plasma concentration–time profile analogous to that of THC [92–94].

3.2. Pharmacokinetics and Pharmacodynamics

The administration of natural cannabis products and cannabinoids predominantly occurs via inhalation or oral consumption, whereas methods like rectal administration, sublingual intake, transdermal application, eye drops, or aerosols are generally considered to have limited practical utility. The pharmacokinetics of THC in particular are strongly influenced by the mode of administration. For instance, inhalation leads to a swift increase in plasma concentration and psychotropic effects manifest within seconds to minutes, reaching a peak within 15–30 min and dissipating over a span of two to three hours. Contrastingly, oral administration yields delayed psychotropic effects that appear 30 to 90 min post consumption, peaking between 2 and 3 h and lasting 4–12 h, contingent on the dosage and specific effects [95].

The pharmacokinetics of novel psychoactive substances (NPSs) serve as a crucial framework for understanding organismal responses to drug administration. While forensic casework offers data on cannabinoid concentrations in human users, it offers restricted insights into the pharmacokinetics of individual samples. Preclinical research utilizing laboratory animals offers more comprehensive data concerning the biological effects of synthetic cannabinoids, although it often surfaces after these substances have exited market circulation. In light of these considerations, the pharmacokinetics and pharmacodynamics of 5F-MDMB-PICA, an FDA-approved synthetic cannabinoid that is popular in the USA, have been characterized. A validated analytical methodology has been established that is capable of quantifying 5F-MDMB-PICA and its primary metabolites in rat plasma. Previous studies have identified 12 and 22 metabolites of 5F-MDMB-PICA in vitro [96].

Although CBD is often presumed to mitigate some of the undesirable side effects of THC, such as its anxiety-inducing properties, controlled clinical investigations have yielded inconsistent conclusions. Some studies indicate that CBD can attenuate specific acute effects of THC, while others suggest that CBD may augment THC's pharmacodynamics, resulting in more profound drug effects. Yet other findings imply that CBD might not alter either the pharmacodynamics or pharmacokinetics of THC [97].

The distribution of cannabinoids within bodily tissues is significantly influenced by their lipophilic nature. THC, for instance, exhibits a substantial distribution volume (ranging from 5.7–10 L/kg) that is attributable to its lipophilic properties. Similarly, CBD's distribution volume is notable, allowing for its efficient penetration into the brain, adipose tissue, and various organs. The chronic consumption of cannabinoids tends to result in gradual tissue accumulation, further amplifying their distribution volume. The metabolism of cannabinoids is primarily hepatic, although extra-hepatic metabolism also occurs in other organs such as the brain, intestines, and lungs. Cytochrome P450 (CYP 450) enzymes, which are predominantly found in liver tissue, play a vital role in the metabolic breakdown of THC into its main components through decarboxylation, epoxidation, and oxidation processes, leading to D11-hydroxy-THC (D11-OH-THC) and D11-carboxy-THC (D11-COOH-THC). Tissues expressing CYP 450 enzymes also contribute to the extra-hepatic metabolism of THC [98,99].

3.3. Factors Influencing Cannabinoid Effects

The question of marijuana serving as a gateway drug has been the subject of extensive inquiry. A particular computational model replicates observed phenomena frequently cited to substantiate the gateway effect. However, the model does not indicate a direct causal relationship between marijuana use and the initiation of using hard drugs. Another facet of the argument focuses on the relative risk associated with the user's age at the time of the initiation of marijuana use. This variant of relative risk associates the initiation of hard drug use with user characteristics like age rather than solely marijuana use and thus fails to provide compelling evidence for the existence of a gateway effect [100].

Legal frameworks and public perception play pivotal roles in the variations in the availability of synthetic cannabis plant material. Originating primarily in China, these new psychoactive substances (NPSs) are influenced by law enforcement efforts, media coverage, and legislative changes in both their country of origin and in destination countries, affecting their global availability [101].

Maternal marijuana usage during gestation has been scrutinized for its potential impact on the neurobehavioral development of offspring. Animal models reveal enduring negative consequences associated with cannabis exposure during gestation and lactation, particularly with the rise of cannabis use among adolescents. The long-term administration of cannabinoid agonists in the periadolescent phase in animals has been correlated with enduring behavioral changes and increased susceptibility to conditions like psychosis or other neuropsychiatric disorders [102].

Research involving heavy adolescent users of cannabis suggests prolonged deficits in learning and working memory which endure up to six weeks post cessation. These findings are particularly concerning given the ongoing process of neuromaturation during adolescence. Rodent models corroborate this, showing more pronounced memory impairments in animals exposed to cannabinoids during adolescence as opposed to later exposure. Moreover, adult humans who initiated cannabis use in their adolescent years experience greater cognitive dysfunction compared to those who initiated it later. This adds credence to the hypothesis that adolescents might be more susceptible to the neurocognitive disruptions associated with chronic and heavy marijuana usage, although the role of preexisting risk factors remains an area for further investigation [103].

Scholarly investigations into the potential impact of permissive state medical marijuana laws (MMLs) on recreational cannabis use have yielded inconclusive results. One underexplored avenue is the effect of MMLs on the average potency of consumed marijuana. It is theorized that heightened potency could indirectly influence individual consumption patterns as less material would be needed to achieve intoxication, potentially reducing overall usage. This line of inquiry, while theoretically compelling, has yet to gain substantial attention in academic circles [104]. Cannabinoids, initially synthesized in acidic forms such as THC, CBD, CBN, CBG, CBC, and CBND, have showcased considerable therapeutic promise. Their documented benefits range from alleviating nausea in chemotherapy patients and enhancing appetite in HIV-positive individuals to reducing spasticity in adults with multiple sclerosis. Additional potential applications include antitumor effects and the management of conditions like glaucoma, epilepsy, and schizophrenia [105]. An understanding of both the pharmacokinetic and pharmacodynamic properties of cannabinoids is essential to fully appreciating their biological impacts. While pharmacodynamic studies have confirmed anti-inflammatory, antiviral, and anticancer properties, pharmacokinetic attributes can vary considerably among individuals. Numerous factors, including prior consumption habits, pharmacogenetics, body size, health status, diet, and microbiome composition, along with dosage and the route of administration, influence the cannabinoids' pharmacokinetic profiles. Empirical research employing subjective self-reports, cognitive task assessments, and neurophysiological evaluations like electroencephalography (EEG) and event-related potentials (ERPs) has elucidated some effects of THC consumption. Compared to placebo conditions, THC-infused cigarettes were associated with expected shifts in mood, behavior, and brain activity. These alterations included diminished task performance and attenuated EEG power and ERP components linked to attentional processes during memory-intensive tasks. Importantly, these effects largely lacked dose dependence. Furthermore, variations in the concentrations of other cannabinoids like CBC and CBD did not significantly influence these outcomes, underscoring the primary bioactive role of THC and its metabolites and affirming the utility of EEG/ERP as biomarkers of its impact [106].

4. Cannabinoids and Neurodegenerative Diseases

4.1. Alzheimer's Disease

In the context of Alzheimer's disease (AD), in which existing therapies offer limited efficacy, research is gradually turning toward the endogenous cannabinoid system as a promising therapeutic target. This system comprises CB1 and CB2 receptors, intrinsic ligands, and enzymes for synthesizing and degrading eCBs. Experimental models of Alzheimer's have demonstrated the potential of activating CB1 and CB2 receptors with non-psychoactive agonists to produce favorable outcomes. These include attenuating the deleterious effects of beta-amyloid peptides and tau phosphorylation while promoting brain repair mechanisms. Although much of this evidence is derived from animal models simulating the pathology of AD, preliminary clinical data supports the role of cannabinoids in ameliorating the behavioral symptoms associated with Alzheimer's disease, particularly when using THC analogs such as nabilone or DRO. Notably, adverse effects like euphoria, somnolence, and fatigue were generally manageable and did not necessitate the cessation of treatment [107,108].

Moreover, cannabinoids possess neuroprotective properties that could be crucial in treating AD. For instance, they can diminish tau phosphorylation and mitigate the negative impact of beta-amyloid-induced oxidative stress while promoting neurotrophin expression and neurogenesis. THC itself shows the capacity to inhibit acetylcholinesterase activity, potentially slowing the progression of the disease [109,110]. CBRs on microglial cells also present a unique intervention point for mitigating AD-associated neuroinflammation without inducing psychoactive effects [111,112]. Epidemiological data further corroborate ECS's role in AD, especially as nonsteroidal anti-inflammatory drugs (NSAIDs) have been observed to reduce risk. The findings suggest that eCBs may offer a protective mechanism against beta-amyloid-induced damage. Indeed, recent studies indicate that inhibiting endocannabinoid uptake could reverse beta-amyloid-induced neurotoxicity and cognitive impairment, emphasizing the therapeutic potential of augmenting endocannabinoid levels in the brain [113–115].

Despite skepticism around the psychoactive properties of cannabis, accumulating evidence suggests that THC, CBD, and synthetic analogs hold therapeutic potential in ameliorating memory impairment associated with AD. Specifically, these substances have displayed consistent efficacy in rodent and human studies, validating the CB1 receptor as a candidate for therapeutic targeting. Furthermore, the anti-inflammatory capabilities of cannabis and THC align with the need to mitigate neuroinflammation in neurodegenerative diseases like AD. To harness the therapeutic utility of THC effectively, it is crucial to discern its medical attributes from its recreational effects, as supported by preclinical and clinical findings [116].

Additional research has illuminated the chronic administration of THC and CBD as promising in mitigating memory impairments during the advanced stages of the pathology of AD, as demonstrated in APP/PS1 mice. Interestingly, these compounds failed to exert similar benefits during the early stages of the disease, leaving Ab deposition and gliosis unaltered. Therapeutic outcomes in aged APP/PS1 mice correlated with improved synaptic function, which was characterized by specific changes in metabotropic glutamate receptor 2/3 and GABA-A Ra1 levels. Both CB1 receptor agonists and THC have been shown to induce the release of brain-derived neurotrophic factor (BDNF), implying that this mechanism could be central to THC's neuroprotective properties. Given BDNF's role in regulating synaptic plasticity, its upregulation could offer a therapeutic pathway to restore synaptic function in AD patients. However, it should be noted that any therapeutic endeavor involving cannabinoids must consider the long preclinical phase of AD and the necessity for early diagnosis for therapies to be effective [117–120].

The therapeutic capacity of THC in targeting intraneuronal amyloid-beta (Ab) has been found to be promising, although its psychoactive attributes pose social challenges. A recent study sought to examine a selection of non-psychoactive cannabinoids for their neuroprotective qualities. Utilizing a well-regarded Alzheimer's disease drug discovery platform, these compounds underwent thorough assessments via assays that evaluate toxicities pertinent to the aging brain, such as proteotoxicity, trophic support loss, energy depletion, and oxidative stress. The study also scrutinized the cannabinoids' effects on microglial inflammation. Preliminary results indicate that many of these cannabinoids manifest significant neuroprotective properties across a range of assessments, suggesting that they are viable candidates for clinical applications in treating neurodegenerative disorders [121].

The potential therapeutic roles of cannabinoids in addressing late-onset Alzheimer's disease (LOAD) and other prevalent conditions among the elderly have increasingly captured scholarly interest. A host of in vitro and in vivo investigations corroborate the capacity of cannabinoids to mitigate oxidative stress, neuroinflammation, and the formation of hallmark LOAD markers like amyloid plaques and neurofibrillary tangles. Additionally, population-based studies suggest that cannabinoids may ameliorate symptoms commonly associated with dementia, such as behavioral disturbances. This comprehensive review elaborates on the burgeoning body of evidence suggesting cannabinoids' potential utility in treating LOAD while also offering critical insights into their efficacy, safety, and pharmacokinetics when administered as treatment in dementia-afflicted populations [122].

4.2. Parkinson's Disease

The burgeoning interest in cannabinoid treatment for alleviating Parkinsonian symptoms, such as dyskinesia and tremors, has been significantly propelled by media coverage and anecdotal evidence disseminated via online platforms. Carroll et al. executed a rigorous randomized, double-blind, placebo-controlled crossover trial utilizing a standardized whole-plant extract with a specific THC concentration and a THC:CBD ratio of approximately 2:1, with the dosage tailored to individual body weight. Despite the double-blind design, a majority (71%) of the 17 participating Parkinson's disease patients were able to correctly identify their treatment arm. The study's primary findings indicated that the oral cannabis extract was well-tolerated but yielded no notable changes in Parkinsonian symptoms. Key outcome measures, such as the Unified Parkinson's Disease Rating Scale (UPDRS) and Rush Dyskinesia Rating Scale, along with secondary outcomes like pain scores and sleep quality assessments, failed to evince any significant treatment effects on levodopa-induced dyskinesia (LID) [123].

In the realm of neuroprotection for Parkinson's disease (PD), cannabinoids exhibit considerable promise, particularly in mitigating factors like excitotoxicity, calcium influx, glial activation, and oxidative damage—all of which are implicated in the progressive degeneration of nigral neurons [124,125]. Although preclinical evidence is robust in supporting the neuroprotective potential of cannabinoids, clinical investigations remain markedly limited. Despite the pressing necessity for innovative therapeutic approaches that extend beyond dopaminergic replacement therapy and the lack of existing effective neuroprotective strategies, the clinical exploration of cannabinoid-based treatments has been constrained. This dearth of clinical studies persists despite the compelling preclinical data, underscoring the urgent need for further research to bridge the gap between preclinical promise and clinical applicability in the treatment of PD [126].

The ECS has been highlighted as a key player in the functioning of the basal ganglia, which is critically implicated in movement disorders such as Parkinson's disease (PD). While preclinical studies suggest that the modulation of cannabinoid (CB) signaling may alleviate motor symptoms—including levodopa-induced dyskinesias (LIDs)—the clinical translation of these findings remains insufficiently explored [127,128]. LIDs often manifest as a result of repetitive dopamine receptor stimulation, leading to a heightened sensitivity in CB receptor-related striatal signaling. Despite advancements in understanding the molecular interplay between cannabinoids (CBs) and dopamine (DA), their applicability in treating PD and LIDs is still an area that demands further investigation [123,129,130].

A limited number of clinical studies exist, such as a Class III randomized, double-blind, placebo-controlled crossover trial exploring the effect of nabilone, a CB1 and CB2 agonist.

The trial demonstrated a reduction in both Rush Dyskinesia Disability Scale scores and the total LID time, suggesting the drug's potential therapeutic efficacy. In the context of PD, levodopa-induced dyskinesia (LID) has been shown to involve hyperactivity in the lateral segment of the globus pallidus (GPl). The activation of CBRs in this region can modulate the reuptake of GABA and potentially enhance neurotransmission, which could theoretically ameliorate the symptoms of dyskinesia. Supporting this hypothesis, a controlled clinical study—a randomized, double-blind, placebo-controlled crossover trial involving seven PD patients—provided empirical evidence that the cannabinoid receptor agonist nabilone significantly reduced occurrences of LID [66]. Despite the insights gained from preclinical research, definitive clinical evidence regarding the therapeutic efficacy of cannabinoid therapies for PD and associated LIDs is still sparse, warranting further in-depth clinical evaluations [131].

In a study employing a rat model with 6-hydroxydopamine-induced lesions, the post-lesion-onset administration of THC led to a resurgence of neuronal injury two weeks after the cessation of the cannabinoid treatment. This finding prompts questions concerning the nature of THC's protective effects against 6-hydroxydopamine toxicity—whether they are inherently neuroprotective and sustained post treatment or merely transient upregulatory responses. Concurrent investigations probed alterations in the efficacy of CB1 receptors within the caudate putamen and substantia nigra two weeks post toxin administration. Emerging research substantiates that the prolonged hyperactivation of these receptors parallels observations in other Parkinson's disease models. Importantly, the chronic administration of THC appeared to significantly attenuate dopaminergic neuronal injury in hemiparkinsonian rats, corroborating previous evidence of the neuroprotective properties of cannabinoids—whether plant-derived, synthetic, or endogenous—across various in vivo and in vitro models of neuronal damage [132].

The collective body of evidence suggests a role for CB2 receptor activation in mitigating inflammation and neuronal degeneration within a 6-hydroxydopamine (6-OHDA) model of Parkinson's disease (PD). An investigation into the impacts of cannabinoids on neuronal survival post 6-OHDA exposure highlighted microglia-mediated effects. CB receptor agonists such as 8-THC and 9-THC have been demonstrated to suppress the release of proinflammatory cytokines, including tumor necrosis factor-alpha (TNF-α) and interleukins, from human monocytes [133]. Additional compounds, such as WIN-55,212-2, CBD, and JWH-133 (a selective CB2 receptor agonist), have been reported to counteract ATP-induced increases in intracellular calcium concentrations in N13 microglial cell lines. The effects of JWH-133 and WIN-55,212-2 were completely nullified by the CB2 antagonist SR 144528, underscoring their CB2-receptor dependency. Interestingly, such antagonistic effects were absent in CBD-treated cells, indicating the existence of CB2-independent mechanisms that potentially contribute to the observed neuroprotective effects [134].

4.3. *Huntington's Disease*

Since the 1980s, there has been a remarkable surge in the field of cannabinoid pharmacology, culminating in the development of innovative cannabinoid-based pharmaceuticals to address an array of medical conditions. Notable medications that emerged during this period include Cesamet (nabilone) and Marinol, which received approval for the treatment of chemotherapy-induced nausea and vomiting in oncology patients and anorexia–cachexia in the context of AIDS therapy, respectively. The most recent addition to this burgeoning domain is Sativex, an oromucosal spray developed by GW Pharmaceuticals Plc, comprising equimolar concentrations of THC and CBD for optimal efficacy. Beyond MS, the multifaceted pharmacological profile of Sativex®—including its demonstrated analgesic, antitumoral, anti-inflammatory, and neuroprotective properties in preclinical settings—has catalyzed ongoing research into its applicability for treating additional neurological disorders [135,136].

One neurodegenerative condition that has garnered particular attention in this context is Huntington's disease (HD). HD is an autosomal-dominant disorder characterized by the presence of excessive CAG repeats in one allele, leading to polyglutamine (polyQ) expansion in the huntingtin protein. The disease predominantly affects striatal and cortical neurons and may manifest clinically as chorea or dementia. In an investigation into the role of CB2 receptors in excitotoxicity-induced striatal neurodegeneration, the anti-inflammatory compound minocycline was administered to CB2 receptor-deficient mice. The study found a significant reduction in excitotoxicity-induced seizures and enhanced motor coordination, and balance, as indicated by performance on a RotaRod test. Moreover, minocycline alleviated glial activation and decreased the loss of medium-sized spiny neurons in these CB2-receptor knockout mice. These results underscore the significance of CB2 receptors in mediating microglia-driven neuroinflammatory processes and suggest that the efficacy of HU-308, a CB2 receptor agonist, may rely substantially on the modulation of microglial activation [137].

In the realm of Huntington's disease (HD), a particular focus has been placed on the potential use of cannabinoids in treating dystonia, a frequent motor symptom. One study revealed that a cohort of early-onset HD patients experienced a notable amelioration of the symptoms of dystonia upon the initiation of cannabinoid treatment, affirming the therapeutic promise of cannabinoids in mitigating motor dysfunctions, especially dystonia, in early-onset HD cases [138].

An additional study sought to elucidate the neuroprotective capabilities of cannabinoids within the context of neurodegenerative disorders, using an experimental model that mimicked the mitochondrial complex II deficiency frequently observed in HD. Here, the effects of THC, a nonselective cannabinoid receptor agonist, and SR141716, a specific CB1 receptor antagonist, were examined during malonate-induced striatal toxicity. Contrary to expectations, both THC and SR141716 exacerbated malonate-induced lesions. These findings indicate the complexity of manipulating the ECS for neuroprotection in HD and suggest that targeting highly selective CB1 receptor agonists may be necessary to effectively mitigate neurodegeneration [132].

In a longitudinal study that employed a R6/1 mouse model to investigate the effects of various cannabinoid treatments on Huntington's disease (HD), several key observations were made over a 20-week period. Although changes in female body weight were statistically insignificant and consequently omitted from the report, marked disparities were noted in male cohorts. Specifically, wild-type mice exhibited a consistent pattern of weight gain, whereas in R6/1 mice, weight gain reached a plateau after the twelfth week. Various eight-week treatment regimens with HU210, THC, or URB597, initiated at the 12-week mark, failed to mitigate behavioral deficits in the R6/1 mice. However, molecular assays indicated that URB597 effectively preserved CB1 receptors in the striatum, while HU210 resulted in an aggregation of ubiquitin-positive protein [139] (Table 1).

Table 1. A summary of molecular changes occurring in R6/1 and WT mice at 20 wk of age following chronic cannabinoid drug treatment compared to vehicle treatment.

	Brain Region	**Drug**	**R6/1**	**WT**
Aggregate number	Striatum	HU210	Increased	N/A
CB1 ligand binding	Striatum	URB597	Increased	None
	Hippocampus	THC	None	Decreased
CB1 mRNA	Striatum	HU210 URB597	None	Decreased
$GABA_A$ ligand binding	Globus pallidus	URB597	None	Increased
5HT2A ligand binding	Striatum	HU210 URB597	None	Decreased
	Hippocampus	URB597	None	Decreased
	Motor cortex	URB597	None	Decreased

The study challenges the extant literature on the subject by revealing no significant downregulation of CB1 receptors due to chronic drug therapy. This discrepancy may be attributable to variations in treatment durations and the specific brain regions analyzed. Moreover, the study underscores the unexpected increase in seizure events in the R6/1 mice following HU210 treatment, thereby raising questions regarding the safety and appropriateness of utilizing highly potent cannabinoid agonists in HD therapy (Figure 1).

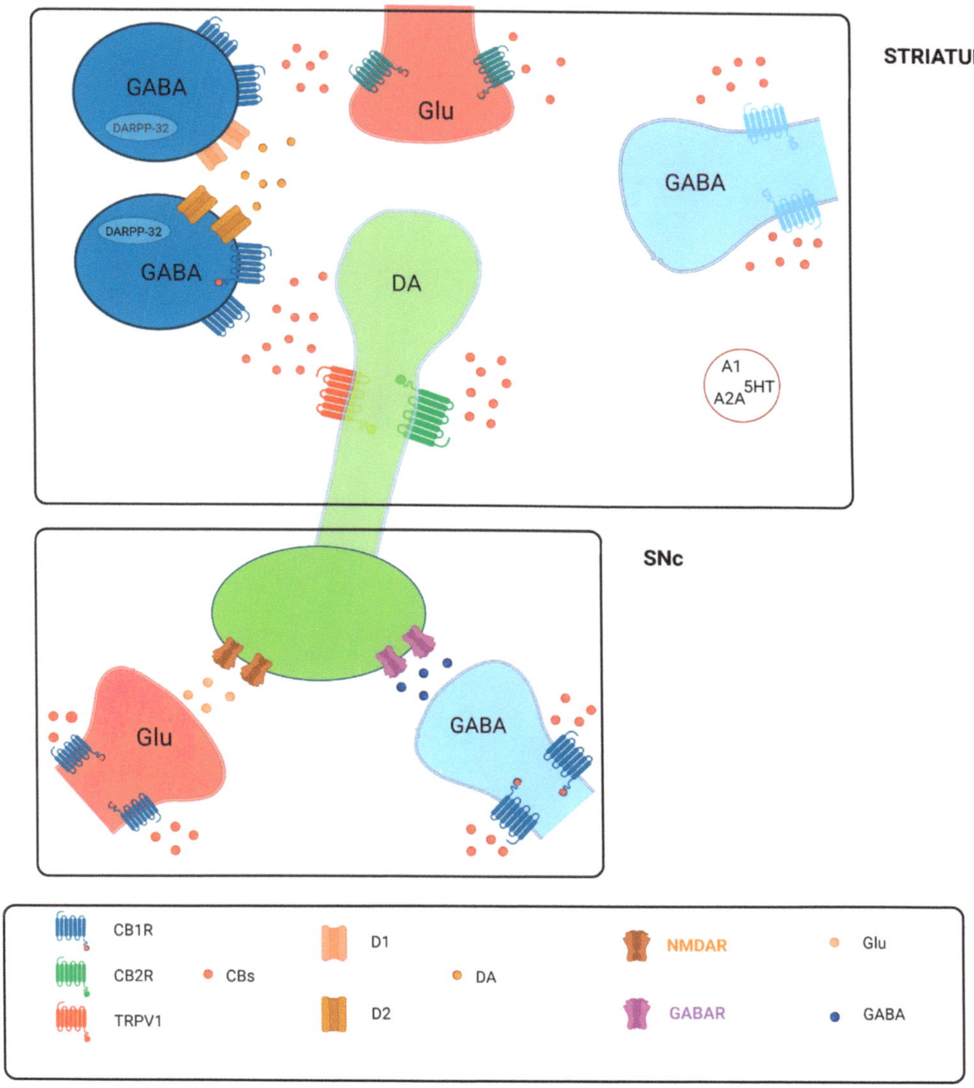

Figure 1. The distribution of the CB1R and CB2R in the striatum of the rat.

Collectively, these observations offer nuanced insights into the intricate relationship between cannabinoids and HD, thereby accentuating the need for further research. The study highlights the necessity to clarify the role of CB1 receptors in disease progression and assess the viability of targeted therapeutic interventions [139] (Table 2).

Table 2. Cannabinoids and neurodegenerative diseases.

	CB_1 Receptor	CB_2 Receptor	Endocannabinoid Levels	Endocannabinoid Synthesis	Endocannabinoid Degradation
Alzheimer's disease	CB_1 receptor expression initially increased, followed by a decline during disease progression [140]. CB_1 receptor was functionally impaired [141].	CB_2 receptor increased in the entorhinal cortex and parahippocamus [142].	Decreased AEA levels in the midfrontal and temporal cortex [143].	DGLalfa and DGLbeta levels were increased in AD patients (Braak stage IV) [144].	Increased FAAH levels [145]. Increased MGL levels in AD patients (Braak stage IV) [144].
Parkinson's disease	CB_1 receptor expression decreased in the substantia nigra. CB_1 receptor expression increased in dopaminergic projecting areas [146]		AEA levels increased in cerebrospinal fluid. A sevenfold increase in 2AG levels in the globus pallidus [147].	-	Decreased levels of anandamide membrane transporter and FAAH [145].
Huntington's disease	CB_1 receptor expression decreased in the caudate nucleus, putamen, and globus pallidus [148].	CB_2 receptor expression increased in striatal microglia [137].	AEA and 2AG levels decreased in the striatum. AEA levels increased and 2AG levels decreased in the cortex [145].	NAPE-PLD and DGL levels decreased in the striatum [145].	FAAH levels increased and MGL levels decreased in the cortex [145].

4.4. Multiple Sclerosis

The potential therapeutic efficacy of cannabinoids in treating the symptoms associated with multiple sclerosis (MS) and spinal cord injury has gained some empirical support, primarily from a limited set of eight clinical trials and a solitary case study focused on spinal cord injury. Among these, five studies concentrated on the administration of oral THC, offering preliminary evidence of its beneficial effects for symptom relief in both MS and spinal cord injury. Although the current corpus of data does not definitively establish cannabis or particular cannabinoids as effective treatments for muscle spasticity, spasms, or pain in these conditions, it provides a crucial foundation for future research endeavors [149].

Indeed, given the inadequacies of existing treatment modalities for MS, a growing number of patients are exploring alternative therapeutic approaches, including cannabis extracts. There is increasing empirical validation for anecdotal claims of symptom alleviation through the use of cannabinoids, especially concerning muscle stiffness, spasms, neuropathic pain, sleep disturbances, and bladder issues. However, investigations targeting symptoms such as tremors and nystagmus have yet to yield favorable outcomes. In terms of safety profiles, cannabinoids generally appear to be well tolerated, with no major safety concerns reported during the testing phases. Moreover, improved tolerability has been observed with extended and gradual dosing regimens [150]. Recent advances in research methodologies and trial designs are actively being deployed to overcome existing limitations. Furthermore, burgeoning evidence suggests that cannabinoids may exert a modulatory influence on the core physiological processes pertinent to MS, including anti-inflammatory mechanisms, remyelination support, and neuroprotective functions. Consequently, ongoing clinical trials are investigating whether cannabinoids could not only provide symptomatic relief but also mitigate the progression of disability in MS patients, aligning with emerging insights in this research domain [150].

Emerging evidence suggests that the antispasticity effect observed in CREAE mice following treatment with AM374, an irreversible fatty acid amide hydrolase inhibitor, may be mediated through CB1/CB2 receptors. Studies have shown that the compound's antispasticity impact is somewhat diminished with combined pre-treatment using SR141716A and SR144528, similar to findings involving R-(+)-WIN55212. However, the individual administration of SR141716A or SR144528 has been reported to exacerbate spasticity in CREAE mice. Despite these limitations, AM374 still holds promise for alleviating spasticity by stimulating endogenous CBRs. Additionally, it is noteworthy that multiple sclerosis patients using an oromucosal spray or oral DRO cannabis extracts can self-adjust dosages to minimize side effects without compromising therapeutic benefits. Moreover, there is evidence that the frequency of adverse events diminishes over the course of treatment without leading to a rapid adaptation to the medicinal effects of cannabinoids [151,152].

The utility of cannabinoids as immunosuppressive agents in chronic inflammatory conditions is further substantiated by research employing Theiler's murine encephalomyelitis virus (TMEV), an animal model that mimics human multiple sclerosis. The administration of synthetic cannabinoids like WIN 55,212-2, ACEA, and JWH-015 during the established phase of TMEV infection yielded marked and sustained improvement in neurological deficits. Specifically, treatment with WIN 55,212-2, a non-selective CB1/CB2 agonist, led to a significant enhancement in RotaRod performance in TMEV-infected mice both immediately after treatment and 25 days post treatment. Similar outcomes were noted with ACEA, a selective CB1 agonist. Although the CB2 selective agonist JWH-015 did not achieve full restoration of motor function, significant improvement was still observed. Importantly, these functional gains persisted for at least 25 days after the cessation of treatment, underscoring the potential long-term benefits of cannabinoid therapy [153].

4.5. Mechanisms of Cannabinoid Action in Neurodegeneration

In the realm of neurodegenerative diseases such as Alzheimer's, Parkinson's, and Huntington's (AD-PD-HD), neuroinflammation has been identified as a pivotal element contributing to neuronal degeneration. Various studies have demonstrated the efficacy of cannabinoids in ameliorating this inflammatory burden. For example, JWH015, a selective CB2 receptor agonist, has been shown to counteract the upregulation of CD40 in interferon-gamma-treated mouse microglial cells via interfering with the JAK/STAT pathway. This action further inhibits the production of proinflammatory cytokines while simultaneously promoting Ab phagocytosis. Additionally, compounds like CBD and the synthetic cannabinoids WIN 55212-2 and JWH-133 have been implicated in attenuating ATP-induced increases in intracellular Ca^{2+}, a critical factor in microglial activation and the onset of inflammatory responses [111,134].

Acute neurodegeneration, such as the neurodegeneration resulting from cerebral ischemia due to stroke, trauma, or cardiac arrest, calls for immediate intervention. The existing research corroborates the neuroprotective properties of cannabinoids (CBs) in such scenarios, particularly in ameliorating secondary damage following the initial injury. Intriguingly, endogenous cannabinoids exhibit increased levels of production following brain trauma, suggesting a potential role in mitigating secondary injuries. For instance, levels of anandamide (AEA), an endocannabinoid, are elevated following controlled blood flow disruptions, a change that is attributed to the decreased expression and activity of the FAAH enzyme. Moreover, studies involving models of middle cerebral artery occlusion (MCAO) as well as clinical studies on stroke patients have corroborated the trend of elevated anandamide levels. Interestingly, levels of 2-arachidonoylglycerol (2-AG), another endocannabinoid, are noted to increase following physical traumas like concussive head injuries or seizures. However, its levels were found to decrease in mice subjected to MCAO-induced ischemia [83,154–156].

In the recent literature, cannabinoids have emerged as potent medicinal agents, particularly in the realms of appetite stimulation and antiemetic treatment for conditions such as cancer and AIDS. THC and CBD, the principal active compounds within the cannabinoid class, interact with G-protein-coupled receptors—CB1 receptors predominantly located in the CNS and CB2 receptors mainly found in immune cells. The body's endogenous cannabinoids (ECBs), notably anandamide, virodhamine, and 2-arachidonoylglycerol (2-AG), bind to these receptors, facilitating various physiological responses such as cognitive function and pain perception. Elevated ECB levels have been observed in several neurodegenerative diseases, suggesting their potentially neuroprotective roles. Intriguingly, cannabinoids can function as agonists, antagonists, or inverse agonists when binding to CBRs, thereby modulating their neuroprotective effects. For instance, CBD has been shown to mitigate the toxicity caused by beta-amyloid peptides, resulting in reduced levels of reactive oxygen species, lipid peroxidation, and pro-apoptotic proteins. Additionally, CBD has demonstrated an ability to downregulate the production of proinflammatory cytokines and specific secretase enzymes. These attributes are promising for the use of cannabinoids in the treatment of neurodegenerative conditions like Parkinson's and Alzheimer's diseases [157] (Figure 2).

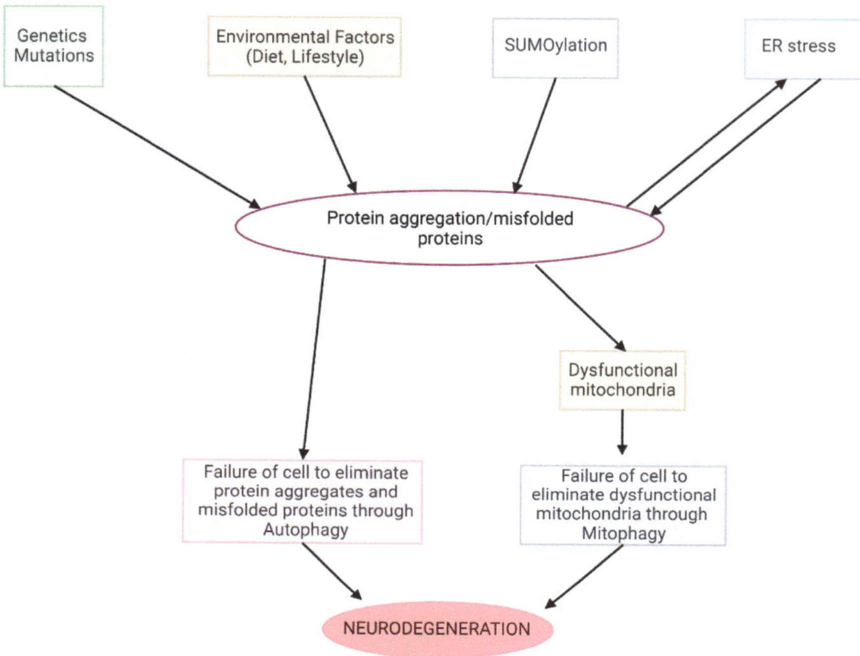

Figure 2. The mechanism of the protein aggression and degeneration that lead to neurodegenerative phenomena.

Over the past 15 years, scholarly attention has concentrated on elucidating the neuroprotective potential of agents that target the ECS, encompassing cannabinoid agonists, endocannabinoid degradation inhibitors, and allosteric modulators. These compounds have demonstrated a capacity to neutralize a variety of neurotoxic elements, including excitotoxicity, oxidative stress, and inflammation, thereby supporting neuronal health and longevity. Given the complexity of neurodegenerative disorders—which are often characterized by the simultaneous occurrence of multiple deleterious stimuli—an efficacious neuroprotective strategy necessitates a multifaceted approach to counteract these cytotoxic agents. In this context, cannabinoids distinguish themselves via their versatile neuroprotective properties. Unlike other compound classes under investigation for neuroprotective potential—such as antioxidants, N-methyl-D-aspartate (NMDA) receptor antagonists, and calcium channel blockers—cannabinoids offer a more comprehensive range of protective attributes. As a result, they have emerged as viable candidates for therapeutic interventions in conditions like stroke and traumatic brain injuries (TBI). It is imperative to note that most existing studies were conducted using animal models and often involve the administration of cannabinoids prior to the introduction of potentially cytotoxic factors. The relevance of such administration sequences to human pathology requires cautious interpretation. Among the various cannabinoids showing promise in preclinical models are Dexanabinol (HU-211), a synthetic compound with structural similarities to traditional cannabinoids but lacking cannabinoid receptor affinity; nonselective synthetic agonists like HU-210, WIN 55,212-2, TAK-937, and BAY 38-7271; and phytocannabinoids, including THC and

CBD. The endogenous counterparts, such as 2-arachidonoylglycerol (2-AG) and anandamide, also hold significant therapeutic potential. These compounds have frequently been observed to confer a range of neuroprotective outcomes, including improved neurological function, diminished infarct sizes, and reduced edema and inflammation, alongside the modulation of immunomodulatory responses [158].

Cannabinoids demonstrate impressive antioxidative abilities by engaging the CB1 and CB2 CBRs to counteract free radical damage and modulate the production of ROS, as well modulating antioxidative defense mechanisms. The activation of CB1 receptors =initiates complex signaling pathways that play an essential role in supporting antioxidative responses and cellular survival, including the phosphoinositide 3-kinase (PI3K)/Akt, mitogen-activated protein kinase (MAPK), and Nrf2 pathways [159–162]. Furthermore, the activation of CB1 receptors governs key aspects of glutamatergic signaling such as activating the N-methyl-D-aspartate (NMDA)-receptor-activated regulation of calcium influx and the orchestration of Ca^{2+}-dependent signaling cascades. CB2 receptors' neuroprotective abilities derive from their capacity to reduce microglial activation and the release of pro-oxidative and proinflammatory agents. CB2 activation therefore plays a crucial role in mitigating neuroinflammation's potentially damaging effects while also creating an environment that is conducive to cell integrity and sustained wellbeing. By harmonizing all of these intricate mechanisms, cannabinoids exert a multifaceted neuroprotective influence that promotes balance within an ecosystem conducive to sustained well-being [163] (Figure 3).

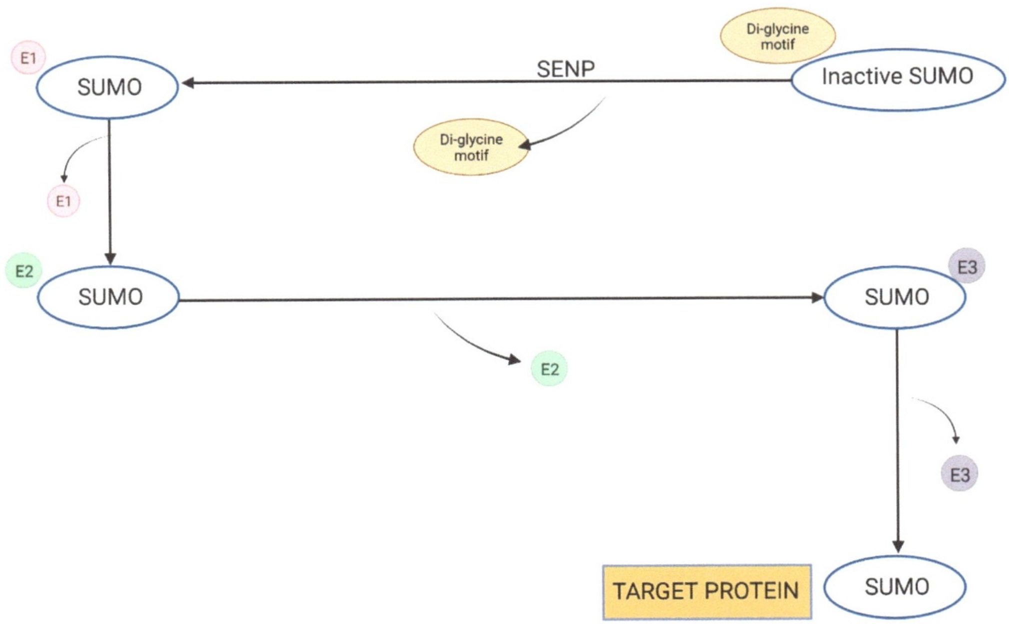

Figure 3. The enzymatic steps leading from inactive SUMO to activated SUMO.

5. Cannabinoids and Cancer

5.1. Antitumor Effects of Cannabinoids

The investigation into cannabinoids as potential anticancer agents has expanded in recent years, although it remains relatively nascent. Limited to a small number of human studies, including one phase I/II clinical trial and three experimental studies, the body of evidence does reveal some promise. One of these studies distinguished itself through a rigorous methodological approach, aligning closely with the evaluation criteria outlined by the Cochrane Collaboration Manual. This meticulousness facilitated a more reliable interpretation of its experimental methodologies and outcomes, accentuating the need for further high-quality research to substantiate the antitumor effects of cannabinoids. Beyond merely serving as palliative agents in cancer treatment, cannabinoids hold potential as primary or adjunctive antineoplastic agents. However, the need for an expansive array of well-designed clinical trials remains critical for validating the antitumor efficacy of cannabinoids in oncological settings [164].

Recent advancements have particularly spotlighted the antiproliferative attributes of CBD, a nonpsychoactive cannabinoid. In a focused in vitro study examining the effects of CBD on U87 and U373 human glioma cell lines, a significant reduction in the mitochondrial oxidative metabolism was observed, along with a decrease in cell viability. The antiproliferative impact was noted within 24 h of exposure to CBD and was partially attenuated by specific agents like SR144528 and α-tocopherol. Intriguingly, other cannabinoid antagonists failed to reverse CBD's effects. For the first time, the study linked CBD's antiproliferative activity with the induction of apoptosis, which was confirmed via a cytofluorimetric analysis and single-strand DNA staining. Furthermore, in vivo studies on nude mice implanted with U87 human glioma cells demonstrated significant tumor reductions following the subcutaneous administration of CBD, reinforcing its potential role as an antineoplastic agent. These findings contribute substantially to our understanding of CBD's antitumor properties, both in vitro and in vivo, advocating for its further exploration as a potential antineoplastic agent [165].

The therapeutic potential of Cannabis sativa, particularly its bioactive components like cannabinoids and terpenes, has garnered substantial attention in contemporary research. In a study involving female C57BL/6 mice treated with azoxymethane (AOM) and dextran sulfate sodium (DSS), THC exhibited both anti-inflammatory and antitumoral properties [166]. THC administration led to marked reductions in the severity of inflammation and tumor formation, as evidenced via the hematoxylin and eosin staining of the colonic tissue. Additionally, THC was found to mitigate the production of interleukin-22, a cytokine implicated in inflammation-driven colon cancer, by intraepithelial cells. Both cannabinoids and terpenes such as β-caryophyllene, limonene, and myrcene have demonstrated promise in inducing apoptosis, inhibiting cell proliferation, and suppressing angiogenesis in colorectal cancer (CRC). Of significance is the synergistic interaction between cannabinoids and terpenes, which may amplify therapeutic efficacy in treating CRC [167].

In a separate investigation focused on elucidating the antitumoral mechanisms of cannabinoid compounds, particularly those that are high in CBD, three extracts of Cannabis sativa were evaluated. The study centered on their effects on cell mortality, cytochrome C oxidase activity, and lipid composition in SH-SY5Y neuroblastoma cells. The results indicated that these extracts induce cell mortality by inhibiting the activity of cytochrome C oxidase. Importantly, this cytotoxicity was comparable to the cytotoxicity induced by known cannabinoid agonists like WIN55,212-2. While this effect could be partially attenuated by the selective CB1 receptor antagonist AM281 and antioxidants like α-tocopherol, it underscores the critical role of oxidative stress in mediating the antitumoral properties of cannabinoids. Furthermore, the extracts with high CBD contents revealed diverse antitumoral effects against human neuroblastoma cells which appeared to operate via multiple mechanisms, not only by affecting cannabinoid receptor activity but also by disrupting mitochondrial electron transport and increasing oxidative stress. Interestingly, the study suggested that whole-plant extracts may offer superior antitumoral effects compared to isolated cannabinoids. However, the study did not account for the potential mitigating impact of antioxidants, such as α-tocopherol. This omission is noteworthy since α-tocopherol, a well-known antioxidant commonly used to alleviate adverse reactions in chemotherapy, could potentially diminish the antitumoral efficacy of cannabinoid-based treatments [168].

5.2. Cannabinoids in Cancer Therapy

The expression levels of cannabinoid receptors (CB-Rs), particularly CB1-R and CB2-R, in breast cancer tissues have been illuminated through microarray technology analysis. The findings indicate that while CB1-R immunoreactivity was observed in 28% of carcinoma samples, a staggering 72% displayed CB2-R immunoreactivity. This is in stark contrast to non-transformed mammary tissues, which showed negligible immunoreactivity for both CB1-R and CB2-R. The association between elevated CB2-R expression and increased tumor aggressiveness is noteworthy. For instance, tumors devoid of estrogen and/or progesterone receptors, which generally have a poorer prognosis, frequently exhibit elevated levels of CB2-R. This trend is also seen in particularly challenging triple-negative tumors, which are characterized by their lack of both steroid hormone receptors and HER2/neu receptors. These tumors often display high CB2-R levels which correlate with poor differentiation, an increased likelihood of early local recurrence, and distant metastasis. The therapeutic landscape for breast cancer could potentially be revolutionized by targeting CB-Rs, particularly CB2-R and CB1-R. This avenue may offer effective treatment options for patients who experience recurrence post anti-HER2-targeted therapies. Beyond CB1-R and CB2-R, other CB-Rs like GPR55 also merit attention. The elevated expression of GPR55 has been observed in metastatic MDA-MB-231 cells, and its proliferative effects are thought to be linked to extracellular signal-regulated kinase (ERK) activation and the subsequent expression of the c-FOS proto-oncogene. Furthermore, cannabinoids (CBs) present potential therapeutic agents for challenging HER2-expressing breast tumors. Combining CBs with targeted therapies like lapatinib, a tyrosine kinase inhibitor, may potentiate antitumoral effects and enhance synergy with conventional chemotherapy agents such as cisplatin. Empirical studies have corroborated the synergistic effect between CBs and other oncologic agents including cisplatin [169–171].

From a translational standpoint, the synergistic potential of cannabinoids with existing chemotherapy treatments should not be overlooked. Preclinical studies have demonstrated that CBD and THC in particular can enhance the effectiveness of conventional chemotherapies. Although the scientific literature has yet to present data on the possible synergies between FAAH or MAGL inhibitors and classical chemotherapy or immunotherapies, cannabinoids have already been successfully employed in a clinical setting to mitigate the side effects associated with chemotherapy, such as nausea, vomiting, and pain. Recent work has also indicated the utility of MAGL inhibitors like MJN110 in reversing chemotherapy-induced neuropathy. Consequently, future research endeavors should prioritize combination studies with traditional chemotherapy agents to evaluate potential synergistic effects on tumor growth inhibition and metastasis reduction while simultaneously assessing the ability to alleviate chemotherapy-induced side effects [172,173].

In parallel, CBD has garnered an increasing amount of research interest for its analgesic properties in neurologically mediated conditions. One notable pharmacological formulation, Nabiximols (Sativex), which is a composite of CBD and THC, has gained regulatory approval in specific jurisdictions for mitigating spasticity associated with multiple sclerosis and as an adjunct in cancer-related pain management. CBD's interaction profile is broad, encompassing not just the canonical CB1R and CB2R but also other receptors like TRPVs, 5-HT1A, GPR55, and PPARg. In the realm of oncology, CBD has exhibited anticancer properties through various mechanisms, including the induction of apoptosis and the inhibition of cell migration and metastasis across diverse cancer types [174].

Adding to the complexity of the cannabinoid landscape are compounds like cannabigerol (CBG), O-1602, and URB-602, which have shown promising anti-neoplastic effects in experimental models, notably in decreasing tumor volume and averting the formation of aberrant crypt foci (ACF) [175].

The ECS has emerged as a focal point of medical research owing to its regulatory role in an array of physiological and pathological processes, encompassing pain modulation and memory formation. Deviations in the activity of the ECS have been identified across a gamut of medical conditions, ranging from oncological to neurodegenerative disorders such as Parkinson's disease, Huntington's chorea, and multiple sclerosis (MS). Consequently, pharmacological interventions aiming to modulate the activity of the ECS have gained considerable momentum, often employing plant-derived or synthetic cannabinoids as active agents. Such pharmacological strategies have yielded tangible benefits in clinical contexts such as AIDS-related cachexia and MS-associated spasticity, among other palliative care applications. Prominent examples of these pharmaceutical agents include Sativex, a standard plant extract formulation of nabiximols, and synthetic compounds like Nabilone (Cesamet) and DRO (Marinol). While preliminary evidence suggests a potential utility of oral cannabinoids in ameliorating chemotherapy-induced nausea and vomiting (CINV), further empirical investigations are requisite to substantiate and consolidate this therapeutic application [176].

5.3. Potential Mechanisms of Cannabinoid-Mediated Anticancer Effects

Cannabinoids' neuroprotective and antioxidant effects are produced via several complex mechanisms, the primary one of which is their effect on mitochondrial function. CB receptors typically reside on cell membranes. However, 30% of neuronal mitochondria contain CB1 receptors on their outer membranes, evidence that cannabinoids play an integral role in energy balance through the modulation of the mitochondrial electron transport chain (mETC), thus impacting learning processes as well as other physiological processes. The activation of the mitochondrial CB1 receptor pathway involves multiple components, including the Gai protein, soluble-adenylyl cyclase (sAC), and protein kinase A (PKA) [177,178]. Studies have also demonstrated that cannabinoids influence OXPHOS via non-receptor mechanisms, as supported by previous research [179]. Another study provides further evidence of a correlation between the inhibition of cytochrome C oxidase activity in SH-SY5Y cell lines and the concentration of THC in Cannabis sativa extracts and their ability to modulate the metabolism as well as the cannabinoids' involvement in mitochondria-related toxicity and oxidative stress [168]. Cannabinoids' production of ROS has been shown to cause changes to cell membranes, including the peroxidation of lipids that affect normal and cancer cells alike [180].

Limonene, a cyclic monoterpene found in citrus fruit peel oils, has been demonstrated to exert notable anticancer properties both in vitro and in vivo across various types of cancer. It can reduce tumor growth while simultaneously inducing apoptosis through multiple pathways. Limonene displayed significant cytotoxicity against T24 human bladder cancer cells by inducing G2/M-phase cell cycle arrest, decreasing migration and invasion, increasing apoptosis rates, and upregulating Bax/caspase-3 expression levels while attenuating Bcl-2 [181]. Limonene produced changes in gene regulation related to apoptosis, signal transduction, inflammation, and DNA repair within HepG2 cells. D-limonene demonstrated similar results in colon cancer cells, where it inhibited cell viability by inducing apoptosis through intrinsic pathway activation and suppressing PI3K/Akt activity [182]. For gastric cancer cells, however, the activation of the mitochondria-mediated intrinsic pathway was evidenced. Notably, the combination of limonene and berberine yielded amazing anticancer effects that surpassed their individual potencies [183]. Neuroblastoma cells were observed to exhibit autophagy through lipidated Light chain 3 (LC3) independent of the generation of ROS or ERK activation and in conjunction with decreased levels of p62 protein [184]. Lung cancer cell lines also displayed autophagy; D-limonene showed promising results at curtailing tumor growth in murine models [185]. D-limonene caused cell apoptosis through two distinct mechanisms in murine T-cell lymphoma cells: at lower concentrations, it caused the production of H_2O_2 and activated the ERK pathway, while at higher concentrations, it inhibited protein farnesylation and the production of O_2 [186]. Niosomes containing 20uM D-limonene showed significant cytotoxicity against HepG2 cell lines as well as other cell lines; when combined with docetaxel, the effect was further amplified through an escalation in the production of ROS and an increase in the expression of apoptotic proteins, suggesting the involvement of the mitochondrial apoptosis pathway [187].

In an exploration of the mechanistic underpinnings of CBD and CBG treatments, one study focused on their impact on the expression of genes that are pertinent to cannabinoid activity and the pathobiology of mesothelioma. Notably absent from this analysis was CNR2 as its expression was not observed across any mesothelioma cell lines. The treatments with both CBD and CBG led to the substantial upregulation of key genes associated with cannabinoid activity and the pathology of mesothelioma across the three mesothelioma cell lines examined [188]. Specifically, noteworthy upregulations were observed for cannabinoid CB1 receptor (CNR1), G-protein-coupled receptor 55 (GPR55), and 5-HT1a receptor (5HTR1A) when compared to vehicle-treated controls, with an approximately 50-fold increase in the expression of GPR55. Interestingly, the CBD treatment had a variable impact on the mRNA expression of transient receptor potential vanilloid type 1 (TRPV1). While it influenced the expression of TRPV1 across most mesothelioma cell lines, an exception was noted in the case of H2452 cells [188]. The expression of TRPV2 or peroxisome proliferator-activated receptor gamma (PPARG) demonstrated cell-line-dependent variability. Moreover, CBD treatment led to a reduction in the endogenous CXCR4 agonist C-X-C motif chemokine 12 (CXCL12), while CBG's effects on CXCL12 expression varied across different cell lines. To delve deeper into the mechanistic landscape, gene pathway analyses were conducted. Both CBD and CBG were observed to similarly influence cell cycle regulation pathways. Intriguingly, the Gaq/PLC signaling pathways may have been disrupted through the upregulation of GPR55 receptors by cannabinoids, affecting calcium homeostasis [189]. Furthermore, CBG appeared to stimulate the nuclear factor of activated T cells (NFAT) signaling pathways, a group of transcription factors that could also be activated via GPR55 receptors. Among the salient findings was the consistent activation of nuclear factor kappa-light-chain-enhancer of activated B cells (NF-kB) by both CBD and CBG. NF-kB is implicated in multiple inflammatory pathways, including the release of various cytokines and chemokines (such as CXCL12-CXCR4), along with cell cycle regulators, anti-apoptotic factors, and adhesion molecules. Overall, these preliminary results underscore the impact of CBD and CBG on human mesothelioma cell lines, indicating a clear avenue for further investigation [190].

In a subsequent investigation, a research team explored the antitumoral potential of WIN 55,212-2 against pediatric osteosarcoma. The study delineated that WIN 55,212-2 induced cell cycle arrest and prompted the upregulation of crucial markers of endoplasmic reticulum stress such as GRP78, CHOP, and TRB3, followed by autophagy [191–193]. These findings align with previously reported mechanisms in adult cancers. For example, Fisher et al. conducted a study on the impact of both THC and CBD on pediatric neuroblastoma cells, revealing significant reductions in cell viability. Additionally, CBD was observed to inhibit xenograft growth in vivo. Although the precise mechanisms underlying CBD's antitumoral effects remain to be elucidated, the compound induced apoptosis in neuroblastoma cells both in vitro and in vivo, effectively causing cell death directly and indirectly [194].

Collectively, these preclinical findings demonstrate the prospective anticancer efficacy of cannabinoids against a range of pediatric cancers, albeit via multiple mechanisms. It is important to recognize that pediatric cancers are heterogeneous, originating from various cell types and tissues and often driven by specific mutations. One limitation of these studies is their reliance on long-term cultured cell lines that may not fully represent the complexity of human cancers. Furthermore, a dearth of studies have corroborated their findings through animal models or orthotopically xenografted models, consequently failing to replicate the authentic tissue contexts of these malignancies [195]. Moreover, no clinical trials have assessed the potential antitumoral effects of cannabinoids in the treatment of pediatric cancer. Anecdotal evidence exists, however, suggesting potential benefits. For instance, Foroughi et al. reported two cases of female patients experiencing spontaneous regressions of low-grade glioma coinciding with cannabis inhalation. While a retrospective study linked the expression of CB1R to tumor regression, suggesting a plausible mechanism, Foroughi et al. did not investigate CB1R expression in their reported cases [196,197].

A growing body of literature supports the notion that the therapeutic benefits of cannabis are not merely attributable to individual constituents but are the result of synergistic interactions among various compounds within the plant. For instance, evidence has emerged that a holistic botanical preparation of cannabis exhibits greater potency in both in vitro and in vivo models for breast cancer treatment compared to isolated 9-THC [198]. Furthermore, in vivo experiments yielded compelling results, showing additive effects when cannabis terpenes such as α-humulene and β-pinene were combined with WIN55,212-2 in mouse models [199]. These findings suggest that the inclusion of terpenes enhances the activity of isolated cannabinoids, likely through a synergistic mechanism. Parallel lines of inquiry have delved into the molecular mechanisms underlying potential pharmacological interventions against prostate cancer. Various prostate cancer cell lines, including PC3, DU145, and LNCaP, have demonstrated reduced migration, implicating cannabinoids as promising agents in combating cancer cell motility. Specifically, the CB1 agonist WIN-55,212 was observed to decrease the activity of RhoA GTPase, a critical regulator of cell migration [200]. This led to a subsequent disruption in actin/myosin microfilament formation and a reduction in cell migration. Reinforcing RhoA protein activity resulted in an increase in microfilament formation and cell spreading, whereas the exogenous CB1 agonist anandamide mimicked the reduction by disrupting actin/myosin microfilaments. In this context, Roberto et al. reported significant, dose-dependent decreases in the migration and invasion capabilities of PC3 and DU145 cells when treated with the synthetic cannabinoid WIN-55,212 [201].

6. Clinical Applications and Challenges

6.1. Cannabinoids as Therapeutic Agents

Emerging combined cancer therapies have generated widespread interest in the use of cannabis botanicals for treating glioblastoma (GB). Their polypharmaceutical nature offers distinct advantages over current therapies. They may complement standard-of-care treatments more effectively by fully harnessing their anticancer properties. Moreover, they have off-target effects that are less toxic than those of traditional chemotherapeutics. The use of cannabinoids has already proven successful as palliative care in many GB patients. Cannabinoids as anticancer agents can be seen in numerous academic publications, demonstrating tumor-specific cytotoxic and cytostatic effects in experimental models as well as clinical studies, including those on GB patients [202]. Furthermore, GB stands out in that its aggressive infiltration within the brain parenchyma limits metastatic spread outside it; this paradoxical behavior underscores both its complexity and resistance to treatment. Cancer stem cells (CSCs) play an essential role in the resistance of GB to therapy, with active DNA repair mechanisms and efficient xenobiotic export systems being key determinants [203]. Therapy-resistant CSCs may lie dormant in protective niches before becoming aggressive cells that trigger tumor regrowth elsewhere in the brain. The findings of Lah et al. show that all three cannabinoids induced a significant apoptotic rate of approximately 30% among GB cancer stem cells at their respective IC50 concentrations, suggesting the significant inhibition of cell viability mechanisms as cytotoxins, with the significant inhibition of cell viability mechanisms being a key mechanism of their cytotoxicity and the inhibition of viability mechanisms. Notable among CBG's signaling effects is the activation of caspase-3/7, which is further amplified with CBD and temozolomide (TMZ) [204].

Experiments using a spinal nerve ligation neuropathic pain model revealed that administering CB1 receptor-selective agonist ACEA led to decreased mechanically evoked responses in spinal neurons. This effect could only be prevented with CB1 antagonists such as AT1077. When applied systemically and locally in rats suffering chemotherapeutic-agent-induced neuropathic pain, both the systemic and local administration of ACEA demonstrated attenuated mechanical allodynia without inducing psychoactive side effects at the administered doses [205].

Another compound, the CB1/CB2 dual agonist CRA13, demonstrated powerful anti-hyperalgesic properties in an animal model of neuropathic pain. Both the oral administration and local injection of CRA13 were effective at reversing mechanical hyperalgesia caused by established mechanical hyperalgesia. Importantly, its anti-hyperalgesic action occurred via peripheral CB1 receptors, as evidenced by its responsiveness to CB1 antagonist [206]. Preclinical studies showed AZD1940, an orally active mixed CB1/CB2 receptor agonist, to have an analgesic effect in both inflammatory and neuropathic pain models without leading to the development of tolerance or a high level of brain uptake at effective antinociceptive doses in rats or primates [207]. Its analgesic action was CB1-receptor-dependent, acting peripherally without leading to the formation of tolerance; the brain uptake at effective antinociceptive doses was low. However, clinical studies of AZD1940 yielded mixed results. While preclinical trials demonstrated promising effects against capsaicin-induced pain and hyperalgesia in human trials, its analgesic properties failed to reduce post-operative dental extraction pain in healthy subjects, and mild-to-moderate gastrointestinal and CNS side effects were reported in clinical studies of this compound [207–209]. While certain compounds showed promising analgesic and peripheral-site-of-action effects in animal studies, their translation into human clinical trials yielded variable outcomes and side effects. Furthermore, harnessing cannabinoids as pain management solutions is both possible and difficult due to how intricately connected CBRs and pain pathways are.

Cannabis has long been used to treat various conditions, and one such ancient use was to manage epilepsy. Epilepsy, a chronic neurological condition affecting millions worldwide, is characterized by recurrent seizures which are often coupled with cognitive impairments and mood disturbances. Efforts at management typically revolve around modulating neuronal ion channels as well as GABA/glutamate receptors, yet approximately one-third of epileptic patients remain resistant to the current treatments available to them [210]. Cannabinoids may help alleviate epileptic seizures due to the presence of CB1 receptors in key brain areas involved in partial seizure initiation, such as the hippocampus and amygdala [211]. The studies conducted to date have highlighted the significant anticonvulsive properties of various cannabinoids, especially CBD, and more recently, CBDV/D9-THCV [212]. Unfortunately, however, its exact antiepileptic mechanisms remain unknown due to its relatively low affinity for CB1 and CB2 receptors. CBD may exert its effects through several pathways, including interactions with the equilibrative nucleoside transporter and GPR55, TPRV-1, and 5-HT1A receptors, as well as the a3 and a1 glycine receptors [213,214]. Another possible antiepileptic mechanism of CBD could involve its interactions with mitochondrial Na^{2+}/Ca^{2+} exchanger [215].

CBD's therapeutic potential goes well beyond epilepsy, with applications across a broad range of both nonpsychiatric and psychiatric disorders including anxiety, depression, bipolar-disorder psychosis, and sleep disturbances, and a significant amount of research into its pharmacological effects across various biological systems has been undertaken to understand its mechanisms of action as medicine. Animal models suggest that CBD may produce anxiolytic-like effects by activating post-synaptic 5-HT1A receptors located in key brain regions associated with defensive responses, such as the dorsal periaqueductal grey dorsal periaqueductal grey bed nucleus of the stria terminalis and the medial prefrontal cortex [216].

Due to opioids' shortcomings and risks, cannabis' potential as an effective pain remedy has garnered increasing consideration. Legalizing cannabis has proven its safety potential by leading to a decrease in opioid overdose deaths. In general, healthcare providers have shown an encouraging outlook towards the therapeutic uses of cannabis for patients, actively helping to facilitate access to medical cannabis. Patient perspectives support this sentiment, with many believing in cannabis' effectiveness as an analgesic and considering it an alternative medication to opioids. Some patients perceive cannabis to be both safe and effective for treating multiple medical conditions, prompting their cannabis use. Sometimes patients combine cannabis use with prescription drugs; the impact on overall well-being remains uncertain in such instances [217].

Secondary metabolites derived from both cannabinoids and non-cannabinoids have vast therapeutic potential across a wide range of conditions, from cancers, diabetes, cardiovascular issues, neurodegenerative disorders, inflammatory diseases, and viral infections to neurodegeneration disorders and viral infections. Unlike THC—one of the best-studied cannabinoids—most of these phytochemicals lack psychotropic effects, enabling them to provide therapeutic benefits without creating the psychoactive responses associated with THC [218].

6.2. Clinical Trials and Evidence-Based Medicine

Until now, limited and disparate research findings have illuminated the effects of cannabinoids on mental health during the prepubertal stages. Studies have demonstrated that providing CBD during peri-pubertal periods may reduce the behavioral abnormalities seen in animal models of schizophrenia. Preclinical evidence also shows that exposure to both THC and stress during peri-adolescence could result in impaired fear extinction in adulthood for mice, though this was not evident among animals who only received either THC or stress alone [219]. Therefore, further clinical investigations must be conducted in order to ascertain whether concurrent exposure to cannabis and stress during teenage years might contribute to long-term anxiety disorders or pathological fear in adulthood. Studies have demonstrated that frequent cannabis users may suffer neurocognitive deficits, including reduced psychomotor speed and working memory, but these effects can be effectively and affordably improved through aerobic fitness activities—showing the potential of physical fitness to ameliorate the cognitive deficits associated with cannabis consumption in adolescents [220–222].

Multiple sclerosis (MS) spasticity therapy aims to increase functional capacity, facilitate rehabilitation, prevent contractures, and alleviate discomfort among individuals diagnosed with MS. Cannabinoids stand out as notable interventions within neurological disorders, particularly MS. Cannabinoids have proven to be particularly successful at managing MS-related spasticity in several recently conducted studies, demonstrating its benefits among complementary medicine approaches like pharmaceutical cannabinoids. Nabiximols has emerged from multiple rigorous randomized controlled clinical trials against placebos to gain approval as an effective medication for alleviating spasticity-related symptoms. Notably, one recent enriched-design methodology study demonstrated that adding Nabiximols provided more effective relief from MS spasticity than simply adjusting an anti-spasticity medication regimen. Another investigation known as SAVANT explored the use of oromucosal Nabiximols as adjunctive therapy against moderate-to-severe spasticity symptoms [223–227].

Studies exploring means of reducing opioid doses for managing chronic pain may not always produce reliable findings due to the incomplete reporting of dose adjustments and analgesic outcomes. Notable recent analyses have not yielded evidence that cannabis exerts any opioid-sparing properties. Preclinical evidence points toward cannabinoids' effectiveness for treating inflammatory bowel diseases; preclinical findings demonstrate CBD's protective impact against intestinal inflammation. Yet more rigorous clinical trials must still be conducted on a larger scale to ascertain whether cannabinoids or their derivatives offer any advantages when treating individuals afflicted with IBDs [228,229].

In a landmark interventional pilot clinical trial, the first of its kind to document anti-inflammatory effects following cannabinoid administration in humans, a focus was placed on individuals living with HIV (PWH) and undergoing antiretroviral therapy (ART). The study successfully completed treatments involving eight participants who were administered oral cannabinoids. The results displayed significant reductions in surrogate markers linked to gut mucosal damage, systemic inflammation, immune cell activation, fatigue, and cellular senescence. These initial outcomes advocate for extended investigations through larger clinical trials, aiming to determine the feasibility of using cannabinoid capsules to mitigate chronic inflammation in PWH on ART [230]. Within the scope of this pilot trial, safety and tolerability were the principal concerns. Additionally, the study probed the impact of oral cannabinoids on the integrity of the gut mucosal barrier. This was achieved by monitoring the dynamics of REG-3a and I-FABP throughout the treatment period. REG-3a is instrumental in modulating interactions between humans and gut microbiota, whereas I-FABP is released upon the death of enterocytes. Elevated plasma levels of these markers were previously noted among pregnant women on ART; however, these levels were observed to decline following a 12-week regimen of oral cannabinoid therapy. This corroborates the results of earlier studies elucidating the beneficial impacts of CBD and palmitoylethanolamide on gut mucosal permeability, which are attributed to CBD's activation of CB1R in the ECS [231]. Although substantial anti-inflammatory benefits were evident in the trial, the precise mechanisms underlying these effects remain unclear. An additional layer of complexity is introduced by the dual nature of THC as both a partial agonist and antagonist for CB1R. Furthermore, THC can interact with other endocannabinoid receptors to exert anti-inflammatory effects [230,232].

6.3. Safety Considerations and Adverse Effects

Notable precautions regarding cannabis and THC revolve around their neuropsychiatric side effects, which often determine their maximum tolerated dose and can result in discontinuation. Clinical trials conducted using DRO have demonstrated its potential to exacerbate conditions like mania, depression, and schizophrenia. Recent meta-analyses conducted on cannabinoids demonstrated an almost threefold increased likelihood of experiencing adverse mental or nervous system effects compared to comparator groups, although individual symptoms such as anxiety or depression did not show statistically significant variations [233]. The FDA recommends pre-screening patients before initiating THC therapy as well as medical cannabis treatments; pre-screening should extend across both treatments. Furthermore, THC has the potential for the development of a dependency among individuals with histories of substance use disorders involving nicotine, alcohol, opioids, or illicit drugs [234]. While recreational users may seek certain effects specifically from THC use, clinical trial participants have reported adverse side effects like disorientation, dissociation, euphoria, and hallucinations, which can pose particular dangers to medically vulnerable populations such as older adults [235]. Recent prescribing information for newly approved products has highlighted the elevated risk of suicidal behavior and ideation associated with psychoactive medications, such as antihypertensives, antidepressants, and opioids [236]. As with CBD recommendations—though with more emphasis being given to THC-containing products—individuals experiencing depression or those who are taking medications which share similar risk should exercise extreme caution when considering medical cannabis products as alternatives. Antipsychotic and antidepressant users should prioritize selections with reduced potential for drug—drug interactions (DDIs) in keeping with the principle of prudent medication management.

Cannabinoids are widely perceived as harmless substances by the general population, and any long-term health implications are overlooked. Comparing cannabinoid users and non-users within the broader population highlights their potentially negative impact on cognitive function. There is an increasing amount of evidence linking acute cannabinoid use to deficits in neurocognitive decision making across areas like processing speed, sustained attention span, verbal fluency, and executive functioning. Over time, chronic cannabinoid consumption among teenagers and young adults has shown adverse impacts across various cognitive domains like learning memory, attention, executive function, and psychomotor speed [237,238].

Cannabinoids' growing prevalence, both recreationally and clinically, has increased the possibility of the co-administration of cannabinoids with selective serotonin reuptake inhibitors (SSRIs), potentially leading to adverse outcomes. A comprehensive analysis was performed on adverse event reports submitted through the FAERS of the U.S. Food and Drug Administration's Adverse Event Reporting System; the results demonstrated significant instances in which cannabis or its derivatives caused adverse events that demonstrated an interaction risk between this substance and other medications. Notably, adverse events reported have shown an upward trend over time due to the increased availability of marijuana-derived products, both prescription and over-the-counter (OTC) THC/CBD for medical and recreational use. Although direct clinical interactions remain an evolving area of research, one case report has hinted at an association between cannabis hyperemesis syndrome and the concurrent use of an SSRI medication [239]. By analyzing a vast dataset encompassing nearly 15 million patient reports from the FDA, specific cohorts were formed in order to analyze the frequencies of side effects. This included medications (sertraline, escitalopram, and citalopram), cannabinoids (THC, CBD, and other cannabinoids) and combinations which were metabolized via CYPC219 [240]. The frequencies of sertraline side effects served as baseline for comparison against the cannabinoids/combinations cohorts; 23 side effects each, with an occurrence rate exceeding 5% on sertraline or escitalopram labels, were selected for benchmarking purposes. Established pharmacovigilance metrics, relative risks, and safety signals were employed to identify potential associations between drug or combination side effects and strict statistical methodologies, such as Benjamini–Hochberg–Yekutieli tests, to ensure accurate statistical significance with a threshold set at 0.05 while also compensating for the false discovery rate (FDR) [241]. As is evident by an increase in adverse events associated with the co-administration of cannabinoids and SSRIs, careful consideration must be paid when researching potential interactions between them [240].

7. Future Perspectives and Conclusions

7.1. Promising Avenues for Future Research

THC interacts with the ECS in an interesting fashion by acting as both a partial agonist of CB1 and CB2 receptors and an agonist for GPR55 receptors, while CBD acts as an antagonist or negative allosteric modulator of these same receptors, leading to its ability to modulate THC's psychotropic effects when co-administered. While CBD displays weak binding to CB1 and CB2 receptors at therapeutic doses, its influence on GPR55 curtails intracellular calcium release, thus potentially mitigating the neuronal hyperactivity associated with conditions like epilepsy. The administration of CBD has been linked to elevated serum levels of anandamide (AEA), potentially contributing to its therapeutic effects in schizophrenia patients [242]. Unfortunately, however, its exact mechanism for elevating AEA levels remains incompletely understood and requires further study. Some evidence suggests that both THC and CBD preferentially bind fatty-acid-binding proteins that are essential for the intracellular transport of anandamide and its subsequent degradation via FAAH within cells, contrary to rodent studies which demonstrated an inhibition of FAAH activity. Such results highlight the limitations of animal models as research tools in cannabinoid research [243]. Cannabinoids such as THC and CBD exhibit potency anti-inflammatory actions through COX-2 inhibition, leading to a reduced production of pro-inflammatory prostaglandins [244]. This may indirectly raise endocannabinoid levels, thereby contributing to their antiepileptic properties. CBD's influence on CYP isoenzymes in the brain may further modulate the production of specific eicosanoids like EETs, EET-EAs, and HETE-EAs which may exert indirect influence upon receptors via the regulation of downstream eicosanoid pathways; its influence on CYP isoenzymes, in addition to its inhibition of their production, show its intricate interaction with various molecular pathways [245]. CBD's ability to decrease 5-LOX activity and its related metabolites in human tumor cells raises the possibility that CBD could have anti-seizure properties [246]. This possibility is particularly intriguing given the possible link between the targeted inhibitors of Cys-LT synthesis and reduced seizure risk, although further validation will likely be required before this can be established [247].

Regarding anxiety modulation, due to their widespread presence in key regions of the brain associated with emotional responses such as the prefrontal cortex, amygdala, and hippocampus, CB1Rs have attracted increased scrutiny as potential sources for reducing anxiety. CB1R has been linked with controlling behavioral responses associated with altered emotional states [248,249]. Studies have demonstrated that cannabinoid agonists exhibit a two-pronged approach to controlling anxiety and stress, with lower doses showing attenuated effects, while higher ones could induce anxiety-inducing responses. CB1R-mediated responses involve diverse molecular mechanisms, including contradictory roles played by different areas in modulating anxiety, the activation of specific CB1R populations on GABAergic or glutamatergic neurons, and the potential engagement of non-CB1R-related pathways [250–252]. Environmental factors, including stress-induced changes in GABA responses to CB1R agonists, also play a part. CB1Rs are found not only on GABAergic and glutamatergic neurons but are also present in raphe nuclei to influence serotonergic nerve cell function. Studies involving mice lacking CB1R-deficient serotonergic neurons have revealed increased anxiety levels and diminished socialization skills. Neuroinflammatory processes have long been implicated in anxiety and depression. Both CB1R and CB2R possess the power to alleviate neuroinflammation, offering multiple avenues for anxiolytic effects. Notably, CB2R polymorphisms in Japanese individuals with depression have been linked to altered behavior. The antisense oligonucleotide targeting of CB2R mRNA expression in mice led to reduced anxiety-like behaviors. Studies involving CB2R-overexpressing mice and spontaneously anxious mice have indicated that CB2R can play an essential role in modulating anxiety-like behaviors through its interactions with GABA receptors. The deletion of CB2R from dopaminergic neurons located in the ventral tegmental area (VTA) of mice was demonstrated to significantly affect anxiety, depression, and psychomotor behavior. Given the adverse neurological reactions associated with CB1R antagonism by rimonabant, other therapeutic avenues involving CB2R modulation may offer promising solutions [253–257].

7.2. Implications for Cannabinoid-Based Therapies

Immunity plays an integral role in protecting against foreign agents and pathogens, so for over four decades, researchers have investigated how cannabinoids impact immune responses against pathogens. Notably, in 1977, a groundbreaking study by Bradley et al. demonstrated how THC combined with lipopolysaccharide (LPS) caused increased toxicity while amplifying the lethality of heat-killed bacteria [258]. Subsequent investigations by this same group explored the effects of THC and cannabis extracts on host resistance to Listeria monocytogenes and herpes simplex virus, ultimately showing decreased levels of pathogen resistance in subsequent investigations [259,260]. Subsequent studies revealed the roles of the ECS in initiating immune responses against pathogens, with specific CB2 genotypes correlating with susceptibility to certain viral illnesses. Conversely, in vitro studies demonstrated the microbicidal activity of cannabinoids against various bacteria and fungi, as well as some instances of the regulation of viral pathogenesis by cannabinoids [261]; moreover, the oral administration of cannabis increased survival in murine models of malaria with enhanced host immunity, while increased levels were detected within their lungs and intestines [262]. This was further evidenced by increased host immunity, which was observed with elevated endocannabinoid levels which increased survival when orally administered to infected animals, indicating its protective nature. Cannabinoid-based treatments for infectious diseases are determined by two key elements: their anti-inflammatory and pathogen-targeting capabilities [263,264].

Another intriguing area is vaccination and how Cannabis/CBD treatments impact vaccine-related immunity. Dotsey et al. explored this connection using a transient CB2 blockade on immune reactions of young and aged mice undergoing vaccination; intensified antigen-specific immune reactions were noted following immunization [265]. However, a prospective study of humoral/cellular immune responses during hepatitis B vaccination among habitual marijuana smokers did not reveal significant alterations to the development of systemic immunity [266].

Inflammatory bowel disease (IBD) represents a category of autoimmune gastrointestinal disorders that includes ulcerative colitis (UC) and Crohn's disease (CD). Evidence from various studies has substantiated the anti-inflammatory potential of cannabinoids in mitigating colitis in murine models. Interestingly, these effects were found to be reversed when CBRs were either blocked or deficient. In the realm of preclinical research, phytocannabinoids have been employed in models of gastrointestinal inflammation, and their efficacy has further been examined in clinical trials involving IBD patients [267]. A study utilizing a DSS-induced murine model of colitis revealed a particular sensitivity to cannabinoid-based interventions. Intriguingly, cannabis extract treatments outperformed their pure-cannabinoid counterparts, possibly due to the synergistic interactions and distinct anti-inflammatory attributes conferred by other phytochemicals present in the whole plant [267]. Macrophages, which are aberrantly regulated in IBD, play a pivotal role in the pathogenesis of the disease. These cells are notably abundant in the inflamed mucosa of IBD patients and display an altered phenotype and function compared to normal conditions, such as the elevated expression of co-stimulatory molecules and the production of the inflammatory cytokines IL-12 and IL-23. The research indicated that cannabinoid-based treatments were effective in inhibiting the infiltration of macrophages into the colons of DSS-induced mice, with the severity of the disease directly correlating with the average number of infiltrating macrophages. Moreover, different cannabinoid treatments had varying effects on cytokine levels in murine models. CBD was observed to reduce interferon-beta and interleukin-6 levels, whereas THC primarily diminished interleukin-6 levels. CBDE exerted an impact on TNF-alpha and IL-6, while THCE significantly curtailed the levels of all three investigated cytokines. Importantly, all the cannabinoid treatments examined influenced IL-6 and TNF-alpha, key cytokines associated with IBD. CBDE emerged as particularly effective in downregulating TNF-alpha levels, which is noteworthy considering the frequent clinical usage of anti-TNF agents due to the cytokine's central role in disease pathogenesis in both human and murine models [267,268].

In a comprehensive analysis aimed at evaluating the impact of cannabinoids on lymphocyte function, in vitro methodologies were employed. The study particularly focused on cannabis extracts enriched in CBD Botanical Drug Substance (CBD BDS) or THC Botanical Drug Substance (THC BDS), with concentrations ranging from 20% to 30% of each. The inclusion of these cannabis extracts alongside pure cannabinoids served a dual purpose. Firstly, it acknowledged the prevalence of cannabis-based medicines over isolated cannabinoids in patient care. Secondly, it allowed for an exploration of the putative advantages conferred by the entourage effect. Activated lymphocyte proliferation was the primary outcome measure under investigation. Anti-CD3 antibodies were employed to activate mouse splenocytes (C57BL/6 or BALB/c), which were subsequently exposed to various concentrations of pure cannabinoids or their botanical drug substance counterparts. An FACS analysis was then utilized to assess cell proliferation. Interestingly, it was revealed that the pure cannabinoids were more potent in inhibiting lymphocyte activation compared to cannabis extracts. Among the pure cannabinoids, CBD was found to be more effective at curtailing proliferation than THC irrespective of the form in which it was administered [269]. Similar outcomes were also observed in human peripheral blood mononuclear cells (PBMCs). Upon the activation of anti-CD3, an increase in the CD8 cell percentage was noted, an effect that could be efficaciously mitigated by CBD, CBD BDS, and, to a lesser extent, THC BDS treatments [269].

7.3. Concluding Remarks on the Potential of Cannabinoids

Cannabis is an impressively complex plant, boasting more than 100 cannabinoids in addition to various terpenes and flavonoids. Our understanding of its effects is further complicated by cannabinoids' demonstrated activity across numerous receptors. This characteristic bestows cannabinoids—and by extension, cannabis itself—with the label of being promiscuous drugs; though often seen as disadvantageous, the promiscuity of a drug actually provides distinct advantages, the most important of which is the ability to engage various pathways within an illness with just a single therapeutic agent. Medical cannabis has seen a rapid expansion in recent years as more patients turn to using it as a solution for various ailments. With more patients turning to this botanical remedy for treatment purposes, a growing demand exists among the scientific and medical communities to investigate how cannabis orchestrates its effects within the body; this goes beyond simply understanding potential merits and risks. Optimal routes of administration vary according to each condition and must also be investigated carefully. As the use of medical cannabis continues to expand, research initiatives become ever more necessary. These endeavors must not only dissect how the components of cannabis interact within the body but also establish safe routes of application for various medical scenarios. Furthermore, an understanding of all of its complexities must be obtained before effective use can occur within clinical environments [270].

Cannabinoids and eCBs have quickly become one of the most exciting areas of biomedical and chemical research, witnessing over 1000 publications annually with an expected upward trajectory. The investigation into cannabinoid delivery systems has also witnessed tremendous activity, with companies filing patents relating to localized or transdermal administration. Innovative formulation strategies provide an effective avenue for producing swift systemic effects with sustained long-term effects, as evidenced by the potential synergy between intranasal cannabinoid sprays and patches for fast absorption and an immediate systemic impact. Furthermore, compelling findings include using terpenes from cannabis sources (CBD and THC) as penetration enhancers to increase the efficacy of therapeutic constituents, further underscoring quality control's important role in shaping the composition, dosage, and safety profiles of cannabis-derived components. Amid today's evolving therapeutic paradigms are possibilities for innovative therapeutic paradigms combining established cannabinoids in new applications with engineered cannabinoid derivatives. Nanotechnology presents particularly encouraging avenues, with SEDDS representing one promising route towards realizing the clinical use of Cannabis, involving both oral and pulmonary routes of administration [271]. The integration of carbon nanotubes is still in its infancy but holds promise as an efficient delivery system; however, thorough scrutiny is required to optimize cost-effectiveness and the long-term safety of nano-delivery systems before adopting them into mainstream applications. As the surge in interest in cannabinoids coincides with advances in pharmacology, pharmaceuticals, and technology, an enabling environment is set up for the creation of innovative therapeutic strategies that leverage established cannabinoids and their synthetic derivatives. Science meets innovation while quality and safety issues come together to shape cannabinoid-based interventions for therapy in the near future.

Author Contributions: Conceptualization, V.V. and A.V.C.; methodology, F.-M.B.; software, B.-G.B.; validation, C.T., R.-A.C.-B. and A.D.C.; formal analysis, H.P.C.; investigation, A.B.; resources, H.P.C.; data curation, R.-A.C.-B.; writing—original draft preparation, B.-G.B.; writing—review and editing, L.-A.G.; visualization, A.D.C.; supervision, A.V.C.; project administration, V.V.; funding acquisition, C.T. All authors have read and agreed to the published version of the manuscript.

Funding: This research received no external funding.

Conflicts of Interest: The authors declare no conflict of interest.

Abbreviations

The following abbreviations are used in this manuscript:

CHS	Cannabinoid Hyperemesis Syndrome
THC	Δ9-tetrahydrocannabinol THC
ECS	The Endocannabinoid System
CNS	Central nervous system
eCBs	Endocannabinoids
CBD	Cannabidiol
THCV	D9-tetrahydrocannabivarin
ROS	Reactive Oxygen Species
CBRs	Cannabinoid Receptors
DRO	Dronabinol
CBG	Cannbigerol
CBC	Cannabichromene
CBDV	Cannabidivarin

GPCR	G Protein-Coupled Receptor
CBDA	Cannabidiolic Acid
CBN	Cannabinol
CBL	Cannabicyclol
CBE	Cannabielsoin
CBF	Cannabifuran
TRP	Transient Receptor Potential
PPAR	Peroxisome Proliferator-Activated Receptors
TRPV1	Transient Receptor Potential Vanilloid Type 1
DSI	Depolarization-Induced Suppression of Inhibition
DSE	Depolarization-Induced Suppression of Excitation
5-HT	5-hydroxytryptamine
CYP450	Cytochrome P450
D11-OH-THC	D11-hydroxy-THC
D11-COOH-THC	D11-carboxy-THC
CBND	Cannabinodiol
EEG	Electroencephalography
ERP	Event Related Potentials
AD	Alzheimer's Disease
Ab	Amyloid Beta
PSEN1	Presenilin-1
PSEN2	Presenilin-2
BDNF	Brain-Derived Neurotrophic Factor
LOAD	Late-Onset Alzheimer's Disease
PD	Parkinson's disease
LIDs	Levodopa-Induced Dyskinesias
GPl	Globus Pallidus
HD	Huntington's Disease
MS	Multiple Sclerosis
CREA	Chronic Relapsing Experimental Allergic Encephalomyelitis
AEA	Anandamide
FAAH	Fatty Acid Amide Hydrolase
MCAO	Middle Cerebral Artery Occlusion
NMDA	N-methyl-D-aspartate
PEA	Palmitoylethanolamide
OEA	Oleoylethanolamide
AraS	N-arachidonoyl-L-serine
MAPK	Mitogen-Activated Protein Kinase
CRC	Colorectal Cancer
MAGL	Monoacylglycerol Lipase
LC3	Light Chain 3
NFAT	Nuclear Factor of Activated T Cells
LNCaP	Lymph Node Carcinoma of the Prostate
GB	Glioblastoma
TMZ	Temozolomide
CSCs	Cancer Stem Cells
PWH	People Living with HIV
ART	Antiretroviral Therapy
DDIs	Drug–Drug Interactions
SSRIs	Selective Serotonin Reuptake Inhibitors
OTC	Over the Counter
IBD	Inflammatory Bowel Disease
UC	Ulcerative Colitis
CD	Crohn's Disease
BMT	Bone Marrow Transplant
EAE	Experimental Autoimmune Encephalomyelitis
MM	Medical Marijuana

References

1. Vogel, Z.; Barg, J.; Levy, R.; Saya, D.; Heldman, E.; Mechoulam, R. Anandamide, a Brain Endogenous Compound, Interacts Specifically with Cannabinoid Receptors and Inhibits Adenylate Cyclase. *J. Neurochem.* **1993**, *61*, 352–355. [CrossRef] [PubMed]
2. Hopkins, C.Y.; Gilchrist, B.L. A Case of Cannabinoid Hyperemesis Syndrome Caused by Synthetic Cannabinoids. *J. Emerg. Med.* **2013**, *45*, 544–546. [CrossRef] [PubMed]
3. Sawzdargo, M.; Nguyen, T.; Lee, D.K.; Lynch, K.R.; Cheng, R.; Heng, H.H.Q.; George, S.R.; O'Dowd, B.F. Identification and cloning of three novel human G protein-coupled receptor genes GPR52, ΨGPR53 and GPR55: GPR55 is extensively expressed in human brain. *Mol. Brain Res.* **1999**, *64*, 193–198. [CrossRef] [PubMed]
4. Overton, H.A.; Babbs, A.J.; Doel, S.M.; Fyfe, M.C.T.; Gardner, L.S.; Griffin, G.; Jackson, H.C.; Procter, M.J.; Rasamison, C.M.; Tang-Christensen, M.; et al. Deorphanization of a G protein-coupled receptor for oleoylethanolamide and its use in the discovery of small-molecule hypophagic agents. *Cell Metab.* **2006**, *3*, 167–175. [CrossRef] [PubMed]
5. Hansen, K.B.; Rosenkilde, M.M.; Knop, F.K.; Wellner, N.; Diep, T.A.; Rehfeld, J.F.; Andersen, U.B.; Holst, J.J.; Hansen, H.S. 2-Oleoyl Glycerol Is a GPR119 Agonist and Signals GLP-1 Release in Humans. *J. Clin. Endocrinol. Metab.* **2011**, *96*, E1409–E1417. [CrossRef]
6. Luk, T.; Jin, W.; Zvonok, A.; Lu, D.; Lin, X.-Z.; Chavkin, C.; Makriyannis, A.; Mackie, K. Identification of a potent and highly efficacious, yet slowly desensitizing CB1 cannabinoid receptor agonist: An efficacious, slowly desensitizing CB1 agonist. *Br. J. Pharmacol.* **2004**, *142*, 495–500. [CrossRef]
7. Gonsiorek, W.; Lunn, C.; Fan, X.; Narula, S.; Lundell, D.; Hipkin, R.W. Endocannabinoid 2-arachidonyl glycerol is a full agonist through human type 2 cannabinoid receptor: Antagonism by anandamide. *Mol. Pharmacol.* **2000**, *57*, 1045–1050.
8. Mackie, K.; Devane, W.A.; Hille, B. Anandamide, an endogenous cannabinoid, inhibits calcium currents as a partial agonist in N18 neuroblastoma cells. *Mol. Pharmacol.* **1993**, *44*, 498–503.
9. Hanuš, L.; Abu-Lafi, S.; Fride, E.; Breuer, A.; Vogel, Z.; Shalev, D.E.; Kustanovich, I.; Mechoulam, R. 2-Arachidonyl glyceryl ether, an endogenous agonist of the cannabinoid CB_1 receptor. *Proc. Natl. Acad. Sci. USA* **2001**, *98*, 3662–3665. [CrossRef]
10. Porter, A.C.; Sauer, J.-M.; Knierman, M.D.; Becker, G.W.; Berna, M.J.; Bao, J.; Nomikos, G.G.; Carter, P.; Bymaster, F.P.; Leese, A.B.; et al. Characterization of a Novel Endocannabinoid, Virodhamine, with Antagonist Activity at the CB_1 Receptor. *J. Pharmacol. Exp. Ther.* **2002**, *301*, 1020–1024. [CrossRef]
11. Huang, S.M.; Bisogno, T.; Trevisani, M.; Al-Hayani, A.; De Petrocellis, L.; Fezza, F.; Tognetto, M.; Petros, T.J.; Krey, J.F.; Chu, C.J.; et al. An endogenous capsaicin-like substance with high potency at recombinant and native vanilloid VR1 receptors. *Proc. Natl. Acad. Sci. USA* **2002**, *99*, 8400–8405. [CrossRef] [PubMed]
12. Huang, S.M.; Bisogno, T.; Petros, T.J.; Chang, S.Y.; Zavitsanos, P.A.; Zipkin, R.E.; Sivakumar, R.; Coop, A.; Maeda, D.Y.; De Petrocellis, L.; et al. Identification of a New Class of Molecules, the Arachidonyl Amino Acids, and Characterization of One Member That Inhibits Pain. *J. Biol. Chem.* **2001**, *276*, 42639–42644. [CrossRef] [PubMed]
13. Michael Rossi, A.; Kuehnle, J.C.; Mendelson, J.H. Marihuana and mood in human volunteers. *Pharmacol. Biochem. Behav.* **1978**, *8*, 447–453. [CrossRef] [PubMed]
14. Abel, E.L. Cannabis: Effects on hunger and thirst. *Behav. Biol.* **1975**, *15*, 255–281. [CrossRef]
15. Schmid, P.C.; Paria, B.C.; Krebsbach, R.J.; Schmid, H.H.O.; Dey, S.K. Changes in anandamide levels in mouse uterus are associated with uterine receptivity for embryo implantation. *Proc. Natl. Acad. Sci. USA* **1997**, *94*, 4188–4192. [CrossRef]
16. Paria, B.C.; Huet-Hudson, Y.M.; Dey, S.K. Blastocyst's state of activity determines the "window" of implantation in the receptive mouse uterus. *Proc. Natl. Acad. Sci. USA* **1993**, *90*, 10159–10162. [CrossRef]
17. Paria, B.C.; Deutsch, D.D.; Dey, S.K. The uterus is a potential site for anandamide synthesis and hydrolysis: Differential profiles of anandamide synthase and hydrolase activities in the mouse uterus during the periimplantation period. *Mol. Reprod. Dev.* **1996**, *45*, 183–192. [CrossRef]
18. Kunos, G.; Osei-Hyiaman, D.; Bátkai, S.; Sharkey, K.A.; Makriyannis, A. Should peripheral CB_1 cannabinoid receptors be selectively targeted for therapeutic gain? *Trends Pharmacol. Sci.* **2009**, *30*, 1–7. [CrossRef]
19. Scutt, A.; Williamson, E.M. Cannabinoids Stimulate Fibroblastic Colony Formation by Bone Marrow Cells Indirectly via CB2 Receptors. *Calcif. Tissue Int.* **2007**, *80*, 50–59. [CrossRef]
20. Mori, M.A.; Meyer, E.; Soares, L.M.; Milani, H.; Guimarães, F.S.; De Oliveira, R.M.W. Cannabidiol reduces neuroinflammation and promotes neuroplasticity and functional recovery after brain ischemia. *Prog. Neuro-Psychopharmacol. Biol. Psychiatry* **2017**, *75*, 94–105. [CrossRef]
21. Pazos, M.R.; Cinquina, V.; Gómez, A.; Layunta, R.; Santos, M.; Fernández-Ruiz, J.; Martínez-Orgado, J. Cannabidiol administration after hypoxia–ischemia to newborn rats reduces long-term brain injury and restores neurobehavioral function. *Neuropharmacology* **2012**, *63*, 776–783. [CrossRef] [PubMed]
22. Hartsel, J.A.; Eades, J.; Hickory, B.; Makriyannis, A. Cannabis sativa and Hemp. In *Nutraceuticals*; Elsevier: Amsterdam, The Netherlands, 2016; pp. 735–754, ISBN 978-0-12-802147-7. [CrossRef]
23. Aziz, A.; Nguyen, L.C.; Oumeslakht, L.; Bensussan, A.; Ben Mkaddem, S. Cannabinoids as Immune System Modulators: Cannabidiol Potential Therapeutic Approaches and Limitations. *Cannabis Cannabinoid Res.* **2023**, *8*, 254–269. [CrossRef] [PubMed]
24. Coghe, G.; Pau, M.; Corona, F.; Frau, J.; Lorefice, L.; Fenu, G.; Spinicci, G.; Mamusa, E.; Musu, L.; Massole, S.; et al. Walking improvements with nabiximols in patients with multiple sclerosis. *J. Neurol.* **2015**, *262*, 2472–2477. [CrossRef] [PubMed]

25. Flachenecker, P.; Henze, T.; Zettl, U.K. Nabiximols (THC/CBD Oromucosal Spray, Sativex®) in Clinical Practice—Results of a Multicenter, Non-Interventional Study (MOVE 2) in Patients with Multiple Sclerosis Spasticity. *Eur. Neurol.* **2014**, *71*, 271–279. [CrossRef] [PubMed]
26. Gonzalez-Rosales, F.; Walsh, D. Intractable nausea and vomiting due to gastrointestinal mucosal metastases relieved by tetrahydrocannabinol (Dronabinol). *J. Pain Symptom Manag.* **1997**, *14*, 311–314. [CrossRef]
27. Hernandez, S.L.; Sheyner, I.; Stover, K.T.; Stewart, J.T. Dronabinol Treatment of Refractory Nausea and Vomiting Related to Peritoneal Carcinomatosis. *Am. J. Hosp. Palliat. Care* **2015**, *32*, 5–7. [CrossRef]
28. Beal, J.E.; Olson, R.; Laubenstein, L.; Morales, J.O.; Bellman, P.; Yangco, B.; Lefkowitz, L.; Plasse, T.F.; Shepard, K.V. Dronabinol as a treatment for anorexia associated with weight loss in patients with AIDS. *J. Pain Symptom Manag.* **1995**, *10*, 89–97. [CrossRef]
29. DeJesus, E.; Rodwick, B.M.; Bowers, D.; Cohen, C.J.; Pearce, D. Use of Dronabinol Improves Appetite and Reverses Weight Loss in HIV/AIDS-Infected Patients. *J. Int. Assoc. Physicians AIDS Care* **2007**, *6*, 95–100. [CrossRef]
30. McPartland, J.M.; Glass, M.; Pertwee, R.G. Meta-analysis of cannabinoid ligand binding affinity and receptor distribution: Interspecies differences: Ligand affinity and receptor distribution of CB1. *Br. J. Pharmacol.* **2007**, *152*, 583–593. [CrossRef]
31. Stockings, E.; Campbell, G.; Hall, W.D.; Nielsen, S.; Zagic, D.; Rahman, R.; Murnion, B.; Farrell, M.; Weier, M.; Degenhardt, L. Cannabis and cannabinoids for the treatment of people with chronic noncancer pain conditions: A systematic review and meta-analysis of controlled and observational studies. *Pain* **2018**, *159*, 1932–1954. [CrossRef]
32. Castellanos, D.; Singh, S.; Thornton, G.; Avila, M.; Moreno, A. Synthetic Cannabinoid Use: A Case Series of Adolescents. *J. Adolesc. Health* **2011**, *49*, 347–349. [CrossRef]
33. Esfandyari, T.; Camilleri, M.; Ferber, I.; Burton, D.; Baxter, K.; Zinsmeister, A.R. Effect of a cannabinoid agonist on gastrointestinal transit and postprandial satiation in healthy human subjects: A randomized, placebo-controlled study. *Neurogastroenterol. Motil.* **2006**, *18*, 831–838. [CrossRef] [PubMed]
34. Toyota, M.; Shimamura, T.; Ishii, H.; Renner, M.; Braggins, J.; Asakawa, Y. New Bibenzyl Cannabinoid from the New Zealand Liverwort Radula marginata. *Chem. Pharm. Bull.* **2002**, *50*, 1390–1392. [CrossRef] [PubMed]
35. Iwata, N.; Kitanaka, S. New Cannabinoid-Like Chromane and Chromene Derivatives from Rhododendron anthopogonoides. *Chem. Pharm. Bull.* **2011**, *59*, 1409–1412. [CrossRef] [PubMed]
36. Morales, P.; Hurst, D.P.; Reggio, P.H. Molecular Targets of the Phytocannabinoids: A Complex Picture. In *Phytocannabinoids*; Kinghorn, A.D., Falk, H., Gibbons, S., Kobayashi, J., Eds.; Progress in the Chemistry of Organic Natural Products; Springer International Publishing: Cham, Switzerland, 2017; Volume 103, pp. 103–131, ISBN 978-3-319-45539-6. [CrossRef]
37. Howlett, A.C. International Union of Pharmacology. XXVII. Classification of Cannabinoid Receptors. *Pharmacol. Rev.* **2002**, *54*, 161–202. [CrossRef]
38. Kogan, N.M.; Rabinowitz, R.; Levi, P.; Gibson, D.; Sandor, P.; Schlesinger, M.; Mechoulam, R. Synthesis and Antitumor Activity of Quinonoid Derivatives of Cannabinoids. *J. Med. Chem.* **2004**, *47*, 3800–3806. [CrossRef]
39. Granja, A.G.; Carrillo-Salinas, F.; Pagani, A.; Gómez-Cañas, M.; Negri, R.; Navarrete, C.; Mecha, M.; Mestre, L.; Fiebich, B.L.; Cantarero, I.; et al. A Cannabigerol Quinone Alleviates Neuroinflammation in a Chronic Model of Multiple Sclerosis. *J. Neuroimmune Pharmacol.* **2012**, *7*, 1002–1016. [CrossRef]
40. Carrillo-Salinas, F.J.; Navarrete, C.; Mecha, M.; Feliú, A.; Collado, J.A.; Cantarero, I.; Bellido, M.L.; Muñoz, E.; Guaza, C. A Cannabigerol Derivative Suppresses Immune Responses and Protects Mice from Experimental Autoimmune Encephalomyelitis. *PLoS ONE* **2014**, *9*, e94733. [CrossRef]
41. Dos Santos, R.G.; Hallak, J.E.C.; Crippa, J.A.S. Neuropharmacological Effects of the Main Phytocannabinoids: A Narrative Review. In *Cannabinoids and Neuropsychiatric Disorders*; Murillo-Rodriguez, E., Pandi-Perumal, S.R., Monti, J.M., Eds.; Advances in Experimental Medicine and Biology; Springer International Publishing: Cham, Switzerland, 2021; Volume 1264, pp. 29–45, ISBN 978-3-030-57368-3. [CrossRef]
42. Shrader, S.H.; Tong, Y.-G.; Duff, M.B.; Freedman, J.H.; Song, Z.-H. Involvement of dopamine receptor in the actions of non-psychoactive phytocannabinoids. *Biochem. Biophys. Res. Commun.* **2020**, *533*, 1366–1370. [CrossRef]
43. Smaga, I.; Bystrowska, B.; Gawlinski, D.; Przegalinski, E.; Filip, M. The Endocannabinoid/Endovanilloid System and Depression. *Curr. Neuropharmacol.* **2014**, *12*, 462–474. [CrossRef]
44. Shen, C.-J.; Zheng, D.; Li, K.-X.; Yang, J.-M.; Pan, H.-Q.; Yu, X.-D.; Fu, J.-Y.; Zhu, Y.; Sun, Q.-X.; Tang, M.-Y.; et al. Cannabinoid CB1 receptors in the amygdalar cholecystokinin glutamatergic afferents to nucleus accumbens modulate depressive-like behavior. *Nat. Med.* **2019**, *25*, 337–349. [CrossRef] [PubMed]
45. Boulebd, H. Insights on the antiradical capacity and mechanism of phytocannabinoids: H-abstraction and electron transfer processes in physiological media and the influence of the acid-base equilibrium. *Phytochemistry* **2023**, *208*, 113608. [CrossRef] [PubMed]
46. Hua, T.; Vemuri, K.; Pu, M.; Qu, L.; Han, G.W.; Wu, Y.; Zhao, S.; Shui, W.; Li, S.; Korde, A.; et al. Crystal Structure of the Human Cannabinoid Receptor CB1. *Cell* **2016**, *167*, 750–762.e14. [CrossRef] [PubMed]
47. Navarro, G.; Varani, K.; Lillo, A.; Vincenzi, F.; Rivas-Santisteban, R.; Raïch, I.; Reyes-Resina, I.; Ferreiro-Vera, C.; Borea, P.A.; Sánchez De Medina, V.; et al. Pharmacological data of cannabidiol- and cannabigerol-type phytocannabinoids acting on cannabinoid CB1, CB2 and CB1/CB2 heteromer receptors. *Pharmacol. Res.* **2020**, *159*, 104940. [CrossRef]

48. Devane, W.A.; Hanuš, L.; Breuer, A.; Pertwee, R.G.; Stevenson, L.A.; Griffin, G.; Gibson, D.; Mandelbaum, A.; Etinger, A.; Mechoulam, R. Isolation and Structure of a Brain Constituent That Binds to the Cannabinoid Receptor. *Science* **1992**, *258*, 1946–1949. [CrossRef]
49. Wang, H.; Matsumoto, H.; Guo, Y.; Paria, B.C.; Roberts, R.L.; Dey, S.K. Differential G protein-coupled cannabinoid receptor signaling by anandamide directs blastocyst activation for implantation. *Proc. Natl. Acad. Sci. USA* **2003**, *100*, 14914–14919. [CrossRef]
50. Maccarrone, M.; Rossi, S.; Bari, M.; De Chiara, V.; Fezza, F.; Musella, A.; Gasperi, V.; Prosperetti, C.; Bernardi, G.; Finazzi-Agrò, A.; et al. Anandamide inhibits metabolism and physiological actions of 2-arachidonoylglycerol in the striatum. *Nat. Neurosci.* **2008**, *11*, 152–159. [CrossRef]
51. Piomelli, D.; Beltramo, M.; Giuffrida, A.; Stella, N. Endogenous Cannabinoid Signaling. *Neurobiol. Dis.* **1998**, *5*, 462–473. [CrossRef]
52. Liu, J.; Wang, L.; Harvey-White, J.; Osei-Hyiaman, D.; Razdan, R.; Gong, Q.; Chan, A.C.; Zhou, Z.; Huang, B.X.; Kim, H.-Y.; et al. A biosynthetic pathway for anandamide. *Proc. Natl. Acad. Sci. USA* **2006**, *103*, 13345–13350. [CrossRef]
53. González, S.; Mena, M.A.; Lastres-Becker, I.; Serrano, A.; De Yébenes, J.G.; Ramos, J.A.; Fernández-Ruiz, J. Cannabinoid CB1 receptors in the basal ganglia and motor response to activation or blockade of these receptors in parkin-null mice. *Brain Res.* **2005**, *1046*, 195–206. [CrossRef]
54. Bidaut-Russell, M.; Hewlett, A.C. Cannabinoid Receptor-Regulated Cyclic AMP Accumulation in the Rat Striatum. *J. Neurochem.* **1991**, *57*, 1769–1773. [CrossRef]
55. Fan, S.-F.; Yazulla, S. Inhibitory interaction of cannabinoid CB1 receptor and dopamine D2 receptor agonists on voltage-gated currents of goldfish cones. *Vis. Neurosci.* **2004**, *21*, 69–77. [CrossRef]
56. Navarro, M.; Hernández, E.; Muñoz, R.M.; Del Arco, I.; Villanúa, M.A.; Carrera, M.R.A.; Rodríguez De Fonseca, F. Acute administration of the CB1 cannabinoid receptor antagonist SR 141716A induces anxiety-like responses in the rat. *NeuroReport* **1997**, *8*, 491–496. [CrossRef]
57. Muller, C.; Lynch, D.L.; Hurst, D.P.; Reggio, P.H. A Closer Look at Anandamide Interaction With TRPV1. *Front. Mol. Biosci.* **2020**, *7*, 144. [CrossRef]
58. Russo, R.; LoVerme, J.; La Rana, G.; D'Agostino, G.; Sasso, O.; Calignano, A.; Piomelli, D. Synergistic antinociception by the cannabinoid receptor agonist anandamide and the PPAR-α receptor agonist GW7647. *Eur. J. Pharmacol.* **2007**, *566*, 117–119. [CrossRef]
59. Bouaboula, M.; Hilairet, S.; Marchand, J.; Fajas, L.; Fur, G.L.; Casellas, P. Anandamide induced PPARγ transcriptional activation and 3T3-L1 preadipocyte differentiation. *Eur. J. Pharmacol.* **2005**, *517*, 174–181. [CrossRef]
60. Falenski, K.W.; Thorpe, A.J.; Schlosburg, J.E.; Cravatt, B.F.; Abdullah, R.A.; Smith, T.H.; Selley, D.E.; Lichtman, A.H.; Sim-Selley, L.J. FAAH$^{-/-}$ Mice Display Differential Tolerance, Dependence, and Cannabinoid Receptor Adaptation after Δ9-Tetrahydrocannabinol and Anandamide Administration. *Neuropsychopharmacology* **2010**, *35*, 1775–1787. [CrossRef]
61. Xiang, W.; Shi, R.; Kang, X.; Zhang, X.; Chen, P.; Zhang, L.; Hou, A.; Wang, R.; Zhao, Y.; Zhao, K.; et al. Monoacylglycerol lipase regulates cannabinoid receptor 2-dependent macrophage activation and cancer progression. *Nat. Commun.* **2018**, *9*, 2574. [CrossRef]
62. Ueda, N.; Tsuboi, K.; Uyama, T.; Ohnishi, T. Biosynthesis and degradation of the endocannabinoid 2-arachidonoylglycerol. *BioFactors* **2011**, *37*, 1–7. [CrossRef]
63. Palese, F.; Pontis, S.; Realini, N.; Piomelli, D. NAPE-specific phospholipase D regulates LRRK2 association with neuronal membranes. In *Advances in Pharmacology*; Elsevier: Amsterdam, The Netherlands, 2021; Volume 90, pp. 217–238, ISBN 978-0-12-822516-5.
64. Hillard, C.J.; Edgemond, W.S.; Jarrahian, A.; Campbell, W.B. Accumulation of N-Arachidonoylethanolamine (Anandamide) into Cerebellar Granule Cells Occurs via Facilitated Diffusion. *J. Neurochem.* **2002**, *69*, 631–638. [CrossRef]
65. Pertwee, R.G. GPR55: A new member of the cannabinoid receptor clan?: Commentary. *Br. J. Pharmacol.* **2007**, *152*, 984–986. [CrossRef]
66. Sharir, H.; Abood, M.E. Pharmacological characterization of GPR55, a putative cannabinoid receptor. *Pharmacol. Ther.* **2010**, *126*, 301–313. [CrossRef] [PubMed]
67. Ferreira, S.G.; Teixeira, F.M.; Garção, P.; Agostinho, P.; Ledent, C.; Cortes, L.; Mackie, K.; Köfalvi, A. Presynaptic CB1 cannabinoid receptors control frontocortical serotonin and glutamate release—Species differences. *Neurochem. Int.* **2012**, *61*, 219–226. [CrossRef] [PubMed]
68. Heifets, B.D.; Chevaleyre, V.; Castillo, P.E. Interneuron activity controls endocannabinoid-mediated presynaptic plasticity through calcineurin. *Proc. Natl. Acad. Sci. USA* **2008**, *105*, 10250–10255. [CrossRef]
69. Chung, Y.C.; Bok, E.; Huh, S.H.; Park, J.-Y.; Yoon, S.-H.; Kim, S.R.; Kim, Y.-S.; Maeng, S.; Hyun Park, S.; Jin, B.K. Cannabinoid Receptor Type 1 Protects Nigrostriatal Dopaminergic Neurons against MPTP Neurotoxicity by Inhibiting Microglial Activation. *J. Immunol.* **2011**, *187*, 6508–6517. [CrossRef]
70. Fernández-Ruiz, J. The endocannabinoid system as a target for the treatment of motor dysfunction: Endocannabinoids and motor disorders. *Br. J. Pharmacol.* **2009**, *156*, 1029–1040. [CrossRef]
71. Kannampalli, P.; Sengupta, J.N. Role of principal ionotropic and metabotropic receptors in visceral pain. *J. Neurogastroenterol. Motil.* **2015**, *21*, 147–158. [CrossRef] [PubMed]

72. García, M.C.; Cinquina, V.; Palomo-Garo, C.; Rábano, A.; Fernández-Ruiz, J. Identification of CB_2 receptors in human nigral neurons that degenerate in Parkinson's disease. *Neurosci. Lett.* **2015**, *587*, 1–4. [CrossRef]
73. Spaderna, M.; Addy, P.H.; D'Souza, D.C. Spicing things up: Synthetic cannabinoids. *Psychopharmacology* **2013**, *228*, 525–540. [CrossRef]
74. Ogata, J.; Uchiyama, N.; Kikura-Hanajiri, R.; Goda, Y. DNA sequence analyses of blended herbal products including synthetic cannabinoids as designer drugs. *Forensic Sci. Int.* **2013**, *227*, 33–41. [CrossRef]
75. Navarrete, F.; García-Gutiérrez, M.S.; Aracil-Fernández, A.; Lanciego, J.L.; Manzanares, J. Cannabinoid CB_1 and CB_2 Receptors, and Monoacylglycerol Lipase Gene Expression Alterations in the Basal Ganglia of Patients with Parkinson's Disease. *Neurotherapeutics* **2018**, *15*, 459–469. [CrossRef]
76. Sierra, S.; Luquin, N.; Rico, A.J.; Gómez-Bautista, V.; Roda, E.; Dopeso-Reyes, I.G.; Vázquez, A.; Martínez-Pinilla, E.; Labandeira-García, J.L.; Franco, R.; et al. Detection of cannabinoid receptors CB_1 and CB_2 within basal ganglia output neurons in macaques: Changes following experimental parkinsonism. *Brain Struct. Funct.* **2015**, *220*, 2721–2738. [CrossRef]
77. Duncan, M.; Mouihate, A.; Mackie, K.; Keenan, C.M.; Buckley, N.E.; Davison, J.S.; Patel, K.D.; Pittman, Q.J.; Sharkey, K.A. Cannabinoid CB_2 receptors in the enteric nervous system modulate gastrointestinal contractility in lipopolysaccharide-treated rats. *Am. J. Physiol.-Gastrointest. Liver Physiol.* **2008**, *295*, G78–G87. [CrossRef] [PubMed]
78. Morgese, M.G.; Cassano, T.; Cuomo, V.; Giuffrida, A. Anti-dyskinetic effects of cannabinoids in a rat model of Parkinson's disease: Role of CB_1 and TRPV1 receptors. *Exp. Neurol.* **2007**, *208*, 110–119. [CrossRef]
79. Valjent, E.; Pagès, C.; Rogard, M.; Besson, M.-J.; Maldonado, R.; Caboche, J. Δ9-tetrahydrocannabinol-induced MAPK/ERK and Elk-1 activation *in vivo* depends on dopaminergic transmission: THC and ERK activation *in vivo*. *Eur. J. Neurosci.* **2001**, *14*, 342–352. [CrossRef] [PubMed]
80. Mackie, K.; Hille, B. Cannabinoids inhibit N-type calcium channels in neuroblastoma-glioma cells. *Proc. Natl. Acad. Sci. USA* **1992**, *89*, 3825–3829. [CrossRef]
81. Leterrier, C.; Bonnard, D.; Carrel, D.; Rossier, J.; Lenkei, Z. Constitutive Endocytic Cycle of the CB_1 Cannabinoid Receptor. *J. Biol. Chem.* **2004**, *279*, 36013–36021. [CrossRef]
82. Griffin, G.; Tao, Q.; Abood, M.E. Cloning and pharmacological characterization of the rat $CB_{(2)}$ cannabinoid receptor. *J. Pharmacol. Exp. Ther.* **2000**, *292*, 886–894.
83. Sugiura, T.; Waku, K. Cannabinoid Receptors and Their Endogenous Ligands. *J. Biochem.* **2002**, *132*, 7–12. [CrossRef] [PubMed]
84. Herkenham, M.; Lynn, A.B.; Little, M.D.; Johnson, M.R.; Melvin, L.S.; De Costa, B.R.; Rice, K.C. Cannabinoid receptor localization in brain. *Proc. Natl. Acad. Sci. USA* **1990**, *87*, 1932–1936. [CrossRef] [PubMed]
85. Vijayakumar, S.; Manogar, P.; Prabhu, S.; Pugazhenthi, M.; Praseetha, P.K. A pharmacoinformatic approach on Cannabinoid receptor 2 (CB2) and different small molecules: Homology modelling, molecular docking, MD simulations, drug designing and ADME analysis. *Comput. Biol. Chem.* **2019**, *78*, 95–107. [CrossRef]
86. Altyar, A.E.; Youssef, F.S.; Kurdi, M.M.; Bifari, R.J.; Ashour, M.L. The Role of *Cannabis sativa* L. as a Source of Cannabinoids against Coronavirus 2 (SARS-CoV-2): An In Silico Study to Evaluate Their Activities and ADMET Properties. *Molecules* **2022**, *27*, 2797. [CrossRef]
87. Sekiguchi, K.; Kanazu, T.; Takeuchi, M.; Hasegawa, H.; Yamaguchi, Y. Non-clinical evaluation of the metabolism, pharmacokinetics and excretion of S-777469, a new cannabinoid receptor 2 selective agonist. *Xenobiotica* **2014**, *44*, 48–58. [CrossRef] [PubMed]
88. Cvan Trobec, K.; Kerec Kos, M.; Trontelj, J.; Grabnar, I.; Tschirner, A.; Palus, S.; Anker, S.D.; Springer, J.; Lainscak, M. Influence of cancer cachexia on drug liver metabolism and renal elimination in rats: Drug liver metabolism and renal elimination in cancer cachexia. *J. Cachexia Sarcopenia Muscle* **2015**, *6*, 45–52. [CrossRef] [PubMed]
89. Caillé, F.; Cacheux, F.; Peyronneau, M.-A.; Jego, B.; Jaumain, E.; Pottier, G.; Ullmer, C.; Grether, U.; Winkeler, A.; Dollé, F.; et al. From Structure–Activity Relationships on Thiazole Derivatives to the *In Vivo* Evaluation of a New Radiotracer for Cannabinoid Subtype 2 PET Imaging. *Mol. Pharm.* **2017**, *14*, 4064–4078. [CrossRef] [PubMed]
90. Basha, F.H.; Waseem, M.; Srinivasan, H. Promising Action of Cannabinoids on ER Stress-Mediated Neurodegeneration: An In Silico Investigation. *J. Env. Pathol. Toxicol. Oncol.* **2022**, *41*, 39–54. [CrossRef]
91. De La Torre, R.; Pichini, S. Usefulness of Sweat Testing for the Detection of Cannabis Smoke. *Clin. Chem.* **2004**, *50*, 1961–1962. [CrossRef] [PubMed]
92. Solowij, N.; Broyd, S.J.; Van Hell, H.H.; Hazekamp, A. A protocol for the delivery of cannabidiol (CBD) and combined CBD and Δ9-tetrahydrocannabinol (THC) by vaporisation. *BMC Pharmacol. Toxicol.* **2014**, *15*, 58. [CrossRef]
93. Martin, J.H.; Schneider, J.; Lucas, C.J.; Galettis, P. Exogenous Cannabinoid Efficacy: Merely a Pharmacokinetic Interaction? *Clin. Pharmacokinet.* **2018**, *57*, 539–545. [CrossRef]
94. Ohlsson, A.; Lindgren, J.-E.; Andersson, S.; Agurell, S.; Gillespie, H.; Hollister, L.E. Single-dose kinetics of deuterium-labelled cannabidiol in man after smoking and intravenous administration. *Biol. Mass. Spectrom.* **1986**, *13*, 77–83. [CrossRef]
95. Grotenhermen, F. Pharmacokinetics and Pharmacodynamics of Cannabinoids: *Clin. Pharmacokinet.* **2003**, *42*, 327–360. [CrossRef]
96. Krotulski, A.J.; Garibay, N.; Walther, D.; Walton, S.E.; Mohr, A.L.A.; Logan, B.K.; Baumann, M.H. Pharmacokinetics and pharmacodynamics of the synthetic cannabinoid, 5F-MDMB-PICA, in male rats. *Neuropharmacology* **2021**, *199*, 108800. [CrossRef] [PubMed]

97. Zamarripa, C.A.; Spindle, T.R.; Surujunarain, R.; Weerts, E.M.; Bansal, S.; Unadkat, J.D.; Paine, M.F.; Vandrey, R. Assessment of Orally Administered Δ9-Tetrahydrocannabinol When Coadministered with Cannabidiol on Δ9-Tetrahydrocannabinol Pharmacokinetics and Pharmacodynamics in Healthy Adults: A Randomized Clinical Trial. *JAMA Netw. Open* **2023**, *6*, e2254752. [CrossRef]
98. Lucas, C.J.; Galettis, P.; Schneider, J. The pharmacokinetics and the pharmacodynamics of cannabinoids. *Br. J. Clin. Pharmacol.* **2018**, *84*, 2477–2482. [CrossRef] [PubMed]
99. Diao, X.; Scheidweiler, K.B.; Wohlfarth, A.; Zhu, M.; Pang, S.; Huestis, M.A. Strategies to distinguish new synthetic cannabinoid FUBIMINA (BIM-2201) intake from its isomer THJ-2201: Metabolism of FUBIMINA in human hepatocytes. *Forensic Toxicol.* **2016**, *34*, 256–267. [CrossRef]
100. Morral, A.R.; McCaffrey, D.F.; Paddock, S.M. Reassessing the marijuana gateway effect: Marijuana gateway effect. *Addiction* **2002**, *97*, 1493–1504. [CrossRef] [PubMed]
101. Stansfield, C.R.; Somerville, R.F.; Hassan, V.R.; Kolbe, E.; Partington, H.K.; Walsh, K.A.J.; Johnson, C.S. Effects of external influences on synthetic cannabinoid trends in New Zealand, 2014 to 2020. *Forensic Sci. Int.* **2020**, *316*, 110485. [CrossRef]
102. Day, N.L.; Goldschmidt, L.; Day, R.; Larkby, C.; Richardson, G.A. Prenatal marijuana exposure, age of marijuana initiation, and the development of psychotic symptoms in young adults. *Psychol. Med.* **2015**, *45*, 1779–1787. [CrossRef]
103. Tapert, S.; Schweinsburg, A.; Brown, S. The Influence of Marijuana Use on Neurocognitive Functioning in Adolescents. *Curr. Drug Abus. Rev.* **2008**, *1*, 99–111. [CrossRef]
104. Sevigny, E.L.; Pacula, R.L.; Heaton, P. The effects of medical marijuana laws on potency. *Int. J. Drug Policy* **2014**, *25*, 308–319. [CrossRef]
105. Agarwal, C.; Máthé, K.; Hofmann, T.; Csóka, L. Ultrasound-Assisted Extraction of Cannabinoids from *Cannabis sativa* L. Optimized by Response Surface Methodology. *J. Food Sci.* **2018**, *83*, 700–710. [CrossRef] [PubMed]
106. Ilan, A.B.; Gevins, A.; Coleman, M.; ElSohly, M.A.; De Wit, H. Neurophysiological and subjective profile of marijuana with varying concentrations of cannabinoids: *Behav. Pharmacol.* **2005**, *16*, 487–496. [CrossRef] [PubMed]
107. Volicer, L.; Stelly, M.; Morris, J.; McLaughlin, J.; Volicer, B.J. Effects of dronabinol on anorexia and disturbed behavior in patients with Alzheimer's disease. *Int. J. Geriatr. Psychiatry* **1997**, *12*, 913–919. [CrossRef]
108. Walther, S.; Mahlberg, R.; Eichmann, U.; Kunz, D. Delta-9-tetrahydrocannabinol for nighttime agitation in severe dementia. *Psychopharmacology* **2006**, *185*, 524–528. [CrossRef] [PubMed]
109. Inestrosa, N.C.; Alvarez, A.; Pérez, C.A.; Moreno, R.D.; Vicente, M.; Linker, C.; Casanueva, O.I.; Soto, C.; Garrido, J. Acetylcholinesterase Accelerates Assembly of Amyloid-β-Peptides into Alzheimer's Fibrils: Possible Role of the Peripheral Site of the Enzyme. *Neuron* **1996**, *16*, 881–891. [CrossRef]
110. Eubanks, L.M.; Rogers, C.J.; Beuscher; Koob, G.F.; Olson, A.J.; Dickerson, T.J.; Janda, K.D. A Molecular Link between the Active Component of Marijuana and Alzheimer's Disease Pathology. *Mol. Pharm.* **2006**, *3*, 773–777. [CrossRef]
111. Ehrhart, J.; Obregon, D.; Mori, T.; Hou, H.; Sun, N.; Bai, Y.; Klein, T.; Fernandez, F.; Tan, J.; Shytle, R.D. Stimulation of cannabinoid receptor 2 (CB_2) suppresses microglial activation. *J. Neuroinflamm.* **2005**, *2*, 29. [CrossRef]
112. Facchinetti, F.; Del Giudice, E.; Furegato, S.; Passarotto, M.; Leon, A. Cannabinoids ablate release of TNF? in rat microglial cells stimulated with lypopolysaccharide. *Glia* **2003**, *41*, 161–168. [CrossRef]
113. Stella, N.; Schweitzer, P.; Piomelli, D. A second endogenous cannabinoid that modulates long-term potentiation. *Nature* **1997**, *388*, 773–778. [CrossRef]
114. Marsicano, G.; Goodenough, S.; Monory, K.; Hermann, H.; Eder, M.; Cannich, A.; Azad, S.C.; Cascio, M.G.; Gutiérrez, S.O.; Van Der Stelt, M.; et al. CB1 Cannabinoid Receptors and On-Demand Defense Against Excitotoxicity. *Science* **2003**, *302*, 84–88. [CrossRef]
115. Van Der Stelt, M.; Mazzola, C.; Esposito, G.; Matias, I.; Petrosino, S.; Filippis, D.D.; Micale, V.; Steardo, L.; Drago, F.; Iuvone, T.; et al. Endocannabinoids and β-amyloid-induced neurotoxicity in vivo: Effect of pharmacological elevation of endocannabinoid levels. *Cell. Mol. Life Sci.* **2006**, *63*, 1410–1424. [CrossRef] [PubMed]
116. Koppel, J.; Vingtdeux, V.; Marambaud, P.; d'Abramo, C.; Jimenez, H.; Stauber, M.; Friedman, R.; Davies, P. CB_2 Receptor Deficiency Increases Amyloid Pathology and Alters Tau Processing in a Transgenic Mouse Model of Alzheimer's Disease. *Mol. Med.* **2014**, *20*, 29–36. [CrossRef] [PubMed]
117. Aso, E.; Sánchez-Pla, A.; Vegas-Lozano, E.; Maldonado, R.; Ferrer, I. Cannabis-Based Medicine Reduces Multiple Pathological Processes in AβPP/PS1 Mice. *J. Alzheimer's Dis.* **2014**, *43*, 977–991. [CrossRef] [PubMed]
118. Aso, E.; Andrés-Benito, P.; Ferrer, I. Delineating the Efficacy of a Cannabis-Based Medicine at Advanced Stages of Dementia in a Murine Model. *J. Alzheimer's Dis.* **2016**, *54*, 903–912. [CrossRef] [PubMed]
119. Marsicano, G.; Lafenêtre, P. Roles of the Endocannabinoid System in Learning and Memory. In *Behavioral Neurobiology of the Endocannabinoid System*; Kendall, D., Alexander, S., Eds.; Current Topics in Behavioral Neurosciences; Springer: Berlin/Heidelberg, Germany, 2009; Volume 1, pp. 201–230, ISBN 978-3-540-88954-0. [CrossRef]
120. Butovsky, E.; Juknat, A.; Goncharov, I.; Elbaz, J.; Eilam, R.; Zangen, A.; Vogel, Z. In vivo up-regulation of brain-derived neurotrophic factor in specific brain areas by chronic exposure to Delta9-tetrahydrocannabinol. *J. Neurochem.* **2005**, *93*, 802–811. [CrossRef] [PubMed]
121. Schubert, D.; Kepchia, D.; Liang, Z.; Dargusch, R.; Goldberg, J.; Maher, P. Efficacy of Cannabinoids in a Pre-Clinical Drug-Screening Platform for Alzheimer's Disease. *Mol. Neurobiol.* **2019**, *56*, 7719–7730. [CrossRef]

122. Ahmed, A.; Van Der Marck, M.; Van Den Elsen, G.; Olde Rikkert, M. Cannabinoids in late-onset Alzheimer's disease. *Clin. Pharmacol. Ther.* **2015**, *97*, 597–606. [CrossRef]
123. Carroll, C.B.; Bain, P.G.; Teare, L.; Liu, X.; Joint, C.; Wroath, C.; Parkin, S.G.; Fox, P.; Wright, D.; Hobart, J.; et al. Cannabis for dyskinesia in Parkinson disease: A randomized double-blind crossover study. *Neurology* **2004**, *63*, 1245–1250. [CrossRef]
124. Kim, S.H.; Won, S.J.; Mao, X.O.; Jin, K.; Greenberg, D.A. Molecular Mechanisms of Cannabinoid Protection from Neuronal Excitotoxicity. *Mol. Pharmacol.* **2006**, *69*, 691–696. [CrossRef]
125. Pérez-Olives, C.; Rivas-Santisteban, R.; Lillo, J.; Navarro, G.; Franco, R. Recent Advances in the Potential of Cannabinoids for Neuroprotection in Alzheimer's, Parkinson's, and Huntington's Diseases. In *Cannabinoids and Neuropsychiatric Disorders*; Murillo-Rodriguez, E., Pandi-Perumal, S.R., Monti, J.M., Eds.; Advances in Experimental Medicine and Biology; Springer International Publishing: Cham, Switzerland, 2021; Volume 1264, pp. 81–92, ISBN 978-3-030-57368-3. [CrossRef]
126. Garcia-Arencibia, M.; Garcia, C.; Fernandez-Ruiz, J. Cannabinoids and Parkinsons Disease. *CNS Neurol. Disord.-Drug Targets* **2009**, *8*, 432–439. [CrossRef]
127. Brotchie, J.M. Nondopaminergic mechanisms in levodopa-induced dyskinesia. *Mov. Disord.* **2005**, *20*, 919–931. [CrossRef] [PubMed]
128. Ferrer, B.; Asbrock, N.; Kathuria, S.; Piomelli, D.; Giuffrida, A. Effects of levodopa on endocannabinoid levels in rat basal ganglia: Implications for the treatment of levodopa-induced dyskinesias. *Eur. J. Neurosci.* **2003**, *18*, 1607–1614. [CrossRef] [PubMed]
129. Mesnage, V.; Houeto, J.L.; Bonnet, A.M.; Clavier, I.; Arnulf, I.; Cattelin, F.; Le Fur, G.; Damier, P.; Welter, M.L.; Agid, Y. Neurokinin B, Neurotensin, and Cannabinoid Receptor Antagonists and Parkinson Disease: *Clin. Neuropharmacol.* **2004**, *27*, 108–110. [CrossRef] [PubMed]
130. Chagas, M.H.N.; Zuardi, A.W.; Tumas, V.; Pena-Pereira, M.A.; Sobreira, E.T.; Bergamaschi, M.M.; Dos Santos, A.C.; Teixeira, A.L.; Hallak, J.E.; Crippa, J.A.S. Effects of cannabidiol in the treatment of patients with Parkinson's disease: An exploratory double-blind trial. *J. Psychopharmacol.* **2014**, *28*, 1088–1098. [CrossRef] [PubMed]
131. Sieradzan, K.A.; Fox, S.H.; Hill, M.; Dick, J.P.R.; Crossman, A.R.; Brotchie, J.M. Cannabinoids reduce levodopa-induced dyskinesia in Parkinson's disease: A pilot study. *Neurology* **2001**, *57*, 2108–2111. [CrossRef]
132. Lastres-Becker, I.; Molina-Holgado, F.; Ramos, J.A.; Mechoulam, R.; Fernández-Ruiz, J. Cannabinoids provide neuroprotection against 6-hydroxydopamine toxicity in vivo and in vitro: Relevance to Parkinson's disease. *Neurobiol. Dis.* **2005**, *19*, 96–107. [CrossRef]
133. Klegeris, A.; Bissonnette, C.J.; McGeer, P.L. Reduction of human monocytic cell neurotoxicity and cytokine secretion by ligands of the cannabinoid-type CB_2 receptor: Antineurotoxic action of cannabinoids. *Br. J. Pharmacol.* **2003**, *139*, 775–786. [CrossRef]
134. Martín-Moreno, A.M.; Reigada, D.; Ramírez, B.G.; Mechoulam, R.; Innamorato, N.; Cuadrado, A.; De Ceballos, M.L. Cannabidiol and Other Cannabinoids Reduce Microglial Activation In Vitro and In Vivo: Relevance to Alzheimer's Disease. *Mol. Pharmacol.* **2011**, *79*, 964–973. [CrossRef]
135. López-Sendón Moreno, J.L.; García Caldentey, J.; Trigo Cubillo, P.; Ruiz Romero, C.; García Ribas, G.; Alonso Arias, M.A.A.; García De Yébenes, M.J.; Tolón, R.M.; Galve-Roperh, I.; Sagredo, O.; et al. A double-blind, randomized, cross-over, placebo-controlled, pilot trial with Sativex in Huntington's disease. *J. Neurol.* **2016**, *263*, 1390–1400. [CrossRef]
136. Russo, E.; Guy, G.W. A tale of two cannabinoids: The therapeutic rationale for combining tetrahydrocannabinol and cannabidiol. *Med. Hypotheses* **2006**, *66*, 234–246. [CrossRef]
137. Palazuelos, J.; Aguado, T.; Pazos, M.R.; Julien, B.; Carrasco, C.; Resel, E.; Sagredo, O.; Benito, C.; Romero, J.; Azcoitia, I.; et al. Microglial CB2 cannabinoid receptors are neuroprotective in Huntington's disease excitotoxicity. *Brain* **2009**, *132*, 3152–3164. [CrossRef] [PubMed]
138. Saft, C.; Von Hein, S.M.; Lücke, T.; Thiels, C.; Peball, M.; Djamshidian, A.; Heim, B.; Seppi, K. Cannabinoids for Treatment of Dystonia in Huntington's Disease. *J. Huntington's Dis.* **2018**, *7*, 167–173. [CrossRef] [PubMed]
139. Dowie, M.J.; Howard, M.L.; Nicholson, L.F.B.; Faull, R.L.M.; Hannan, A.J.; Glass, M. Behavioural and molecular consequences of chronic cannabinoid treatment in Huntington's disease transgenic mice. *Neuroscience* **2010**, *170*, 324–336. [CrossRef]
140. Farkas, S.; Nagy, K.; Palkovits, M.; Kovács, G.G.; Jia, Z.; Donohue, S.; Pike, V.; Halldin, C.; Máthé, D.; Harkany, T.; et al. [125I]SD-7015 reveals fine modalities of CB1 cannabinoid receptor density in the prefrontal cortex during progression of Alzheimer's disease. *Neurochem. Int.* **2012**, *60*, 286–291. [CrossRef] [PubMed]
141. Ramírez, B.G.; Blázquez, C.; Del Pulgar, T.G.; Guzmán, M.; De Ceballos, M.L. Prevention of Alzheimer's Disease Pathology by Cannabinoids: Neuroprotection Mediated by Blockade of Microglial Activation. *J. Neurosci.* **2005**, *25*, 1904–1913. [CrossRef] [PubMed]
142. Solas, M.; Francis, P.T.; Franco, R.; Ramirez, M.J. CB_2 receptor and amyloid pathology in frontal cortex of Alzheimer's disease patients. *Neurobiol. Aging* **2013**, *34*, 805–808. [CrossRef] [PubMed]
143. Jung, K.-M.; Sepers, M.; Henstridge, C.M.; Lassalle, O.; Neuhofer, D.; Martin, H.; Ginger, M.; Frick, A.; DiPatrizio, N.V.; Mackie, K.; et al. Uncoupling of the endocannabinoid signalling complex in a mouse model of fragile X syndrome. *Nat. Commun.* **2012**, *3*, 1080. [CrossRef]
144. Mulder, J.; Zilberter, M.; Pasquaré, S.J.; Alpár, A.; Schulte, G.; Ferreira, S.G.; Köfalvi, A.; Martín-Moreno, A.M.; Keimpema, E.; Tanila, H.; et al. Molecular reorganization of endocannabinoid signalling in Alzheimer's disease. *Brain* **2011**, *134*, 1041–1060. [CrossRef]
145. Bisogno, T. Endogenous Cannabinoids: Structure and Metabolism. *J. Neuroendocr.* **2008**, *20*, 1–9. [CrossRef]

146. Van Laere, K.; Casteels, C.; Lunskens, S.; Goffin, K.; Grachev, I.D.; Bormans, G.; Vandenberghe, W. Regional changes in type 1 cannabinoid receptor availability in Parkinson's disease in vivo. *Neurobiol. Aging* **2012**, *33*, 620.e1–620.e8. [CrossRef]
147. Pisani, V.; Madeo, G.; Tassone, A.; Sciamanna, G.; Maccarrone, M.; Stanzione, P.; Pisani, A. Homeostatic changes of the endocannabinoid system in Parkinson's disease: eCB Changes in PD. *Mov. Disord.* **2011**, *26*, 216–222. [CrossRef] [PubMed]
148. Glass, M.; Dragunow, M.; Faull, R.L.M. The pattern of neurodegeneration in Huntington's disease: A comparative study of cannabinoid, dopamine, adenosine and GABAA receptor alterations in the human basal ganglia in Huntington's disease. *Neuroscience* **2000**, *97*, 505–519. [CrossRef] [PubMed]
149. Mechoulam, R. The Pharmacohistory of Cannabis Sativa. In *Cannabinoids as Therapeutic Agents*, 1st ed.; Mechoulam, R., Ed.; Chapman and Hall/CRC: Boca Raton, FL, USA, 2019; pp. 1–20, ISBN 978-0-429-26066-7.
150. Zajicek, J.P.; Apostu, V.I. Role of Cannabinoids in Multiple Sclerosis. *CNS Drugs* **2011**, *25*, 187–201. [CrossRef] [PubMed]
151. Pertwee, R.G. Pharmacology of cannabinoid CB1 and CB2 receptors. *Pharmacol. Ther.* **1997**, *74*, 129–180. [CrossRef] [PubMed]
152. Zajicek, J.; Ball, S.; Wright, D.; Vickery, J.; Nunn, A.; Miller, D.; Cano, M.G.; McManus, D.; Mallik, S.; Hobart, J. Effect of dronabinol on progression in progressive multiple sclerosis (CUPID): A randomised, placebo-controlled trial. *Lancet Neurol.* **2013**, *12*, 857–865. [CrossRef]
153. Arévalo-Martín, Á.; Vela, J.M.; Molina-Holgado, E.; Borrell, J.; Guaza, C. Therapeutic Action of Cannabinoids in a Murine Model of Multiple Sclerosis. *J. Neurosci.* **2003**, *23*, 2511–2516. [CrossRef]
154. Panikashvili, D.; Simeonidou, C.; Ben-Shabat, S.; Hanuš, L.; Breuer, A.; Mechoulam, R.; Shohami, E. An endogenous cannabinoid (2-AG) is neuroprotective after brain injury. *Nature* **2001**, *413*, 527–531. [CrossRef]
155. Franklin, A.; Parmentier-Batteur, S.; Walter, L.; Greenberg, D.A.; Stella, N. Palmitoylethanolamide Increases after Focal Cerebral Ischemia and Potentiates Microglial Cell Motility. *J. Neurosci.* **2003**, *23*, 7767–7775. [CrossRef]
156. Muthian, S.; Rademacher, D.J.; Roelke, C.T.; Gross, G.J.; Hillard, C.J. Anandamide content is increased and CB_1 cannabinoid receptor blockade is protective during transient, focal cerebral ischemia. *Neuroscience* **2004**, *129*, 743–750. [CrossRef]
157. Hetz, C.; Saxena, S. ER stress and the unfolded protein response in neurodegeneration. *Nat. Rev. Neurol.* **2017**, *13*, 477–491. [CrossRef]
158. Fernández-Ruiz, J.; Moro, M.A.; Martínez-Orgado, J. Cannabinoids in Neurodegenerative Disorders and Stroke/Brain Trauma: From Preclinical Models to Clinical Applications. *Neurotherapeutics* **2015**, *12*, 793–806. [CrossRef] [PubMed]
159. Kang, K.A.; Wang, Z.H.; Zhang, R.; Piao, M.J.; Kim, K.C.; Kang, S.S.; Kim, Y.W.; Lee, J.; Park, D.; Hyun, J.W. Myricetin Protects Cells against Oxidative Stress-Induced Apoptosis via Regulation of PI3K/Akt and MAPK Signaling Pathways. *Int. J. Mol. Sci.* **2010**, *11*, 4348–4360. [CrossRef] [PubMed]
160. Jazvinšćak Jembrek, M.; Radovanović, V.; Vlainić, J.; Vuković, L.; Hanžić, N. Neuroprotective effect of zolpidem against glutamate-induced toxicity is mediated via the PI3K/Akt pathway and inhibited by PK11195. *Toxicology* **2018**, *406–407*, 58–69. [CrossRef] [PubMed]
161. Jazvinšćak Jembrek, M.; Vlainić, J.; Čadež, V.; Šegota, S. Atomic force microscopy reveals new biophysical markers for monitoring subcellular changes in oxidative injury: Neuroprotective effects of quercetin at the nanoscale. *PLoS ONE* **2018**, *13*, e0200119. [CrossRef]
162. Zakharova, I.O.; Sokolova, T.V.; Bayunova, L.V.; Zorina, I.I.; Rychkova, M.P.; Shpakov, A.O.; Avrova, N.F. The Protective Effect of Insulin on Rat Cortical Neurons in Oxidative Stress and Its Dependence on the Modulation of Akt, GSK-3beta, ERK1/2, and AMPK Activities. *Int. J. Mol. Sci.* **2019**, *20*, 3702. [CrossRef]
163. Sánchez-Blázquez, P.; Rodríguez-Muñoz, M.; Vicente-Sánchez, A.; Garzón, J. Cannabinoid Receptors Couple to NMDA Receptors to Reduce the Production of NO and the Mobilization of Zinc Induced by Glutamate. *Antioxid. Redox Signal.* **2013**, *19*, 1766–1782. [CrossRef]
164. Rocha, F.C.M.; Dos Santos Júnior, J.G.; Stefano, S.C.; Da Silveira, D.X. Systematic review of the literature on clinical and experimental trials on the antitumor effects of cannabinoids in gliomas. *J. Neurooncol.* **2014**, *116*, 11–24. [CrossRef]
165. Massi, P.; Vaccani, A.; Ceruti, S.; Colombo, A.; Abbracchio, M.P.; Parolaro, D. Antitumor Effects of Cannabidiol, a Nonpsychoactive Cannabinoid, on Human Glioma Cell Lines. *J. Pharmacol. Exp. Ther.* **2004**, *308*, 838–845. [CrossRef]
166. Becker, W.; Alrafas, H.R.; Wilson, K.; Miranda, K.; Culpepper, C.; Chatzistamou, I.; Cai, G.; Nagarkatti, M.; Nagarkatti, P.S. Activation of Cannabinoid Receptor 2 Prevents Colitis-Associated Colon Cancer through Myeloid Cell De-activation Upstream of IL-22 Production. *iScience* **2020**, *23*, 101504. [CrossRef]
167. Silva-Reis, R.; Silva, A.M.S.; Oliveira, P.A.; Cardoso, S.M. Antitumor Effects of *Cannabis sativa* Bioactive Compounds on Colorectal Carcinogenesis. *Biomolecules* **2023**, *13*, 764. [CrossRef]
168. Sánchez-Sánchez, L.; García, J.; Fernández, R.; Noskova, E.; Egiguren-Ortiz, J.; Gulak, M.; Ochoa, E.; Laso, A.; Oiarbide, M.; Santos, J.I.; et al. Characterization of the Antitumor Potential of Extracts of *Cannabis sativa* Strains with High CBD Content in Human Neuroblastoma. *Int. J. Mol. Sci.* **2023**, *24*, 3837. [CrossRef]
169. Ligresti, A.; Moriello, A.S.; Starowicz, K.; Matias, I.; Pisanti, S.; De Petrocellis, L.; Laezza, C.; Portella, G.; Bifulco, M.; Di Marzo, V. Antitumor Activity of Plant Cannabinoids with Emphasis on the Effect of Cannabidiol on Human Breast Carcinoma. *J. Pharmacol. Exp. Ther.* **2006**, *318*, 1375–1387. [CrossRef] [PubMed]
170. Sultan, A.S.; Marie, M.A.; Sheweita, S.A. Novel mechanism of cannabidiol-induced apoptosis in breast cancer cell lines. *Breast* **2018**, *41*, 34–41. [CrossRef] [PubMed]

171. Leyva-Illades, D.; DeMorrow, S. Orphan G protein receptor GPR55 as an emerging target in cancer therapy and management. *Cancer Manag. Res.* **2013**, *5*, 147–155. [CrossRef] [PubMed]
172. Niphakis, M.J.; Cognetta, A.B.; Chang, J.W.; Buczynski, M.W.; Parsons, L.H.; Byrne, F.; Burston, J.J.; Chapman, V.; Cravatt, B.F. Evaluation of NHS Carbamates as a Potent and Selective Class of Endocannabinoid Hydrolase Inhibitors. *ACS Chem. Neurosci.* **2013**, *4*, 1322–1332. [CrossRef] [PubMed]
173. Curry, Z.A.; Wilkerson, J.L.; Bagdas, D.; Kyte, S.L.; Patel, N.; Donvito, G.; Mustafa, M.A.; Poklis, J.L.; Niphakis, M.J.; Hsu, K.-L.; et al. Monoacylglycerol Lipase Inhibitors Reverse Paclitaxel-Induced Nociceptive Behavior and Proinflammatory Markers in a Mouse Model of Chemotherapy-Induced Neuropathy. *J. Pharmacol. Exp. Ther.* **2018**, *366*, 169–183. [CrossRef]
174. Seltzer, E.S.; Watters, A.K.; MacKenzie, D.; Granat, L.M.; Zhang, D. Cannabidiol (CBD) as a Promising Anti-Cancer Drug. *Cancers* **2020**, *12*, 3203. [CrossRef]
175. Pagano, E.; Borrelli, F.; Orlando, P.; Romano, B.; Monti, M.; Morbidelli, L.; Aviello, G.; Imperatore, R.; Capasso, R.; Piscitelli, F.; et al. Pharmacological inhibition of MAGL attenuates experimental colon carcinogenesis. *Pharmacol. Res.* **2017**, *119*, 227–236. [CrossRef]
176. Mücke, M.; Weier, M.; Carter, C.; Copeland, J.; Degenhardt, L.; Cuhls, H.; Radbruch, L.; Häuser, W.; Conrad, R. Systematic review and meta-analysis of cannabinoids in palliative medicine: Cannabinoids in palliative medicine. *J. Cachexia Sarcopenia Muscle* **2018**, *9*, 220–234. [CrossRef]
177. Noskova, E.; Fernández, R.; García, J.; Ochoa, E.; Domínguez-Fernández, C.; Adell, A.; Laso, A.; Andrés, M.F.; González-Coloma, A.; Astigarraga, E.; et al. Screening System of *Cannabis sativa* Extracts Based on Their Mitochondrial Safety Profile Using Cytochrome c Oxidase Activity as a Biomarker. *Int. J. Mol. Sci.* **2023**, *24*, 1315. [CrossRef]
178. Harkany, T.; Horvath, T.L. (S)Pot on Mitochondria: Cannabinoids Disrupt Cellular Respiration to Limit Neuronal Activity. *Cell Metab.* **2017**, *25*, 8–10. [CrossRef] [PubMed]
179. Fišar, Z.; Singh, N.; Hroudová, J. Cannabinoid-induced changes in respiration of brain mitochondria. *Toxicol. Lett.* **2014**, *231*, 62–71. [CrossRef] [PubMed]
180. Han, K.H.; Lim, S.; Ryu, J.; Lee, C.-W.; Kim, Y.; Kang, J.-H.; Kang, S.-S.; Ahn, Y.K.; Park, C.-S.; Kim, J.J. CB_1 and CB_2 cannabinoid receptors differentially regulate the production of reactive oxygen species by macrophages. *Cardiovasc. Res.* **2009**, *84*, 378–386. [CrossRef] [PubMed]
181. Ye, Z.; Liang, Z.; Mi, Q.; Guo, Y. Limonene terpenoid obstructs human bladder cancer cell (T24 cell line) growth by inducing cellular apoptosis, caspase activation, G2/M phase cell cycle arrest and stops cancer metastasis. *J. BUON Off. J. Balk. Union Oncol.* **2020**, *25*, 280–285.
182. Hafidh, R.R.; Hussein, S.Z.; MalAllah, M.Q.; Abdulamir, A.S.; Abu Bakar, F. A High-throughput Quantitative Expression Analysis of Cancer-related Genes in Human HepG2 Cells in Response to Limonene, a Potential Anticancer Agent. *Curr. Cancer Drug Targets* **2018**, *18*, 807–815. [CrossRef]
183. Zhang, X.-Z.; Wang, L.; Liu, D.-W.; Tang, G.-Y.; Zhang, H.-Y. Synergistic Inhibitory Effect of Berberine and *d*-Limonene on Human Gastric Carcinoma Cell Line MGC803. *J. Med. Food* **2014**, *17*, 955–962. [CrossRef]
184. Russo, R.; Cassiano, M.G.V.; Ciociaro, A.; Adornetto, A.; Varano, G.P.; Chiappini, C.; Berliocchi, L.; Tassorelli, C.; Bagetta, G.; Corasaniti, M.T. Role of D-Limonene in Autophagy Induced by Bergamot Essential Oil in SH-SY5Y Neuroblastoma Cells. *PLoS ONE* **2014**, *9*, e113682. [CrossRef]
185. Yu, X.; Lin, H.; Wang, Y.; Lv, W.; Zhang, S.; Qian, Y.; Deng, X.; Feng, N.; Yu, H.; Qian, B. D-limonene exhibits antitumor activity by inducing autophagy and apoptosis in lung cancer. *OncoTargets Ther.* **2018**, *11*, 1833–1847. [CrossRef]
186. Del Toro-Arreola, S.; Flores-Torales, E.; Torres-Lozano, C.; Del Toro-Arreola, A.; Tostado-Pelayo, K.; Guadalupe Ramirez-Dueñas, M.; Daneri-Navarro, A. Effect of d-limonene on immune response in BALB/c mice with lymphoma. *Int. Immunopharmacol.* **2005**, *5*, 829–838. [CrossRef]
187. Manassero, C.A.; Girotti, J.R.; Mijailovsky, S.; García De Bravo, M.; Polo, M. In vitro comparative analysis of antiproliferative activity of essential oil from mandarin peel and its principal component limonene. *Nat. Prod. Res.* **2013**, *27*, 1475–1478. [CrossRef]
188. Colvin, E.K.; Hudson, A.L.; Anderson, L.L.; Kumar, R.P.; McGregor, I.S.; Howell, V.M.; Arnold, J.C. An Examination of the Anti-Cancer Properties of Plant Cannabinoids in Preclinical Models of Mesothelioma. *Cancers* **2022**, *14*, 3813. [CrossRef]
189. Lauckner, J.E.; Jensen, J.B.; Chen, H.-Y.; Lu, H.-C.; Hille, B.; Mackie, K. GPR55 is a cannabinoid receptor that increases intracellular calcium and inhibits M current. *Proc. Natl. Acad. Sci. USA* **2008**, *105*, 2699–2704. [CrossRef] [PubMed]
190. Kurosawa, M.; Arakaki, R.; Yamada, A.; Tsunematsu, T.; Kudo, Y.; Sprent, J.; Ishimaru, N. NF-κB2 Controls the Migratory Activity of Memory T Cells by Regulating Expression of CXCR4 in a Mouse Model of Sjögren's Syndrome. *Arthritis Rheumatol.* **2017**, *69*, 2193–2202. [CrossRef] [PubMed]
191. Notaro, A.; Sabella, S.; Pellerito, O.; Di Fiore, R.; De Blasio, A.; Vento, R.; Calvaruso, G.; Giuliano, M. Involvement of PAR-4 in Cannabinoid-Dependent Sensitization of Osteosarcoma Cells to TRAIL-Induced Apoptosis. *Int. J. Biol. Sci.* **2014**, *10*, 466–478. [CrossRef] [PubMed]
192. Hernández-Tiedra, S.; Fabriàs, G.; Dávila, D.; Salanueva, Í.J.; Casas, J.; Montes, L.R.; Antón, Z.; García-Taboada, E.; Salazar-Roa, M.; Lorente, M.; et al. Dihydroceramide accumulation mediates cytotoxic autophagy of cancer cells via autolysosome destabilization. *Autophagy* **2016**, *12*, 2213–2229. [CrossRef] [PubMed]

193. Salazar, M.; Carracedo, A.; Salanueva, Í.J.; Hernández-Tiedra, S.; Lorente, M.; Egia, A.; Vázquez, P.; Blázquez, C.; Torres, S.; García, S.; et al. Cannabinoid action induces autophagy-mediated cell death through stimulation of ER stress in human glioma cells. *J. Clin. Investig.* **2009**, *119*, 1359–1372. [CrossRef]
194. Fisher, T.; Golan, H.; Schiby, G.; PriChen, S.; Smoum, R.; Moshe, I.; Peshes-Yaloz, N.; Castiel, A.; Waldman, D.; Gallily, R.; et al. In Vitro and In Vivo Efficacy of Non-Psychoactive Cannabidiol in Neuroblastoma. *Curr. Oncol.* **2016**, *23*, 15–22. [CrossRef]
195. Borrell, B. How accurate are cancer cell lines? *Nature* **2010**, *463*, 858. [CrossRef]
196. Sredni, S.T.; Huang, C.-C.; Suzuki, M.; Pundy, T.; Chou, P.; Tomita, T. Spontaneous involution of pediatric low-grade gliomas: High expression of cannabinoid receptor 1 (CNR1) at the time of diagnosis may indicate involvement of the endocannabinoid system. *Childs Nerv. Syst.* **2016**, *32*, 2061–2067. [CrossRef]
197. Foroughi, M.; Hendson, G.; Sargent, M.A.; Steinbok, P. Spontaneous regression of septum pellucidum/forniceal pilocytic astrocytomas—Possible role of Cannabis inhalation. *Childs Nerv. Syst.* **2011**, *27*, 671–679. [CrossRef]
198. Blasco-Benito, S.; Seijo-Vila, M.; Caro-Villalobos, M.; Tundidor, I.; Andradas, C.; García-Taboada, E.; Wade, J.; Smith, S.; Guzmán, M.; Pérez-Gómez, E.; et al. Appraising the "entourage effect": Antitumor action of a pure cannabinoid versus a botanical drug preparation in preclinical models of breast cancer. *Biochem. Pharmacol.* **2018**, *157*, 285–293. [CrossRef]
199. LaVigne, J.; Hecksel, R.; Streicher, J.M. In Defense of the "Entourage Effect": Terpenes Found in *Cannabis sativa* Activate the Cannabinoid Receptor 1 *In Vivo*. *FASEB J.* **2020**, *34*, 1. [CrossRef]
200. Nithipatikom, K.; Gomez-Granados, A.D.; Tang, A.T.; Pfeiffer, A.W.; Williams, C.L.; Campbell, W.B. Cannabinoid Receptor Type 1 (CB_1) Activation Inhibits Small GTPase RhoA Activity and Regulates Motility of Prostate Carcinoma Cells. *Endocrinology* **2012**, *153*, 29–41. [CrossRef]
201. Roberto, D.; Klotz, L.H.; Venkateswaran, V. Cannabinoid WIN 55,212-2 induces cell cycle arrest and apoptosis, and inhibits proliferation, migration, invasion, and tumor growth in prostate cancer in a cannabinoid-receptor 2 dependent manner. *Prostate* **2019**, *79*, 151–159. [CrossRef]
202. Allister, S.D.M.; Chan, C.; Taft, R.J.; Luu, T.; Abood, M.E.; Moore, D.H.; Aldape, K.; Yount, G. Cannabinoids selectively inhibit proliferation and induce death of cultured human glioblastoma multiforme cells. *J. Neurooncol.* **2005**, *74*, 31–40. [CrossRef] [PubMed]
203. Doherty, G.J.; De Paula, B.H.R. Cannabinoids in glioblastoma multiforme—Hype or hope? *Br. J. Cancer* **2021**, *124*, 1341–1343. [CrossRef] [PubMed]
204. Lah, T.T.; Novak, M.; Pena Almidon, M.A.; Marinelli, O.; Žvar Baškovič, B.; Majc, B.; Mlinar, M.; Bošnjak, R.; Breznik, B.; Zomer, R.; et al. Cannabigerol Is a Potential Therapeutic Agent in a Novel Combined Therapy for Glioblastoma. *Cells* **2021**, *10*, 340. [CrossRef] [PubMed]
205. Stachtari, C.C.; Thomareis, O.N.; Tsaousi, G.G.; Karakoulas, K.A.; Chatzimanoli, F.I.; Chatzopoulos, S.A.; Vasilakos, D.G. Interaction of a Cannabinoid-2 Agonist with Tramadol on Nociceptive Thresholds and Immune Responses in a Rat Model of Incisional Pain. *Am. J. Ther.* **2016**, *23*, e1484–e1492. [CrossRef] [PubMed]
206. Gardin, A.; Kucher, K.; Kiese, B.; Appel-Dingemanse, S. Cannabinoid Receptor Agonist 13, a Novel Cannabinoid Agonist: First in Human Pharmacokinetics and Safety. *Drug Metab. Dispos.* **2009**, *37*, 827–833. [CrossRef] [PubMed]
207. Kalliomäki, J.; Annas, P.; Huizar, K.; Clarke, C.; Zettergren, A.; Karlsten, R.; Segerdahl, M. Evaluation of the analgesic efficacy and psychoactive effects of AZD1940, a novel peripherally acting cannabinoid agonist, in human capsaicin-induced pain and hyperalgesia. *Clin. Exp. Pharmacol. Physiol.* **2013**, *40*, 212–218. [CrossRef]
208. Banister, S.D.; Krishna Kumar, K.; Kumar, V.; Kobilka, B.K.; Malhotra, S.V. Selective modulation of the cannabinoid type 1 (CB_1) receptor as an emerging platform for the treatment of neuropathic pain. *Med. Chem. Commun.* **2019**, *10*, 647–659. [CrossRef] [PubMed]
209. Kalliomäki, J.; Segerdahl, M.; Webster, L.; Reimfelt, A.; Huizar, K.; Annas, P.; Karlsten, R.; Quiding, H. Evaluation of the analgesic efficacy of AZD1940, a novel cannabinoid agonist, on post-operative pain after lower third molar surgical removal. *Scand. J. Pain* **2013**, *4*, 17–22. [CrossRef] [PubMed]
210. Bradford, H.F. Glutamate, GABA and epilepsy. *Prog. Neurobiol.* **1995**, *47*, 477–511. [CrossRef] [PubMed]
211. Karler, R.; Turkanis, S.A. The Cannabinoids as Potential Antiepileptics. *J. Clin. Pharmacol.* **1981**, *21*, 437S–448S. [CrossRef]
212. Iannotti, F.A.; Hill, C.L.; Leo, A.; Alhusaini, A.; Soubrane, C.; Mazzarella, E.; Russo, E.; Whalley, B.J.; Di Marzo, V.; Stephens, G.J. Nonpsychotropic Plant Cannabinoids, Cannabidivarin (CBDV) and Cannabidiol (CBD), Activate and Desensitize Transient Receptor Potential Vanilloid 1 (TRPV1) Channels in Vitro: Potential for the Treatment of Neuronal Hyperexcitability. *ACS Chem. Neurosci.* **2014**, *5*, 1131–1141. [CrossRef]
213. Kollipara, R.; Langille, E.; Tobin, C.; French, C.R. CBD can be combined with additional cannabinoids for optimal seizure reduction and requires GPR55 for its anticonvulsant effects. *bioRxiv* **2023**. [CrossRef]
214. Goerl, B.; Watkins, S.; Metcalf, C.; Smith, M.; Beenhakker, M. Cannabidiolic acid exhibits entourage-like improvements of anticonvulsant activity in an acute rat model of seizures. *Epilepsy Res.* **2021**, *169*, 106525. [CrossRef]
215. Ryan, D.; Drysdale, A.J.; Lafourcade, C.; Pertwee, R.G.; Platt, B. Cannabidiol Targets Mitochondria to Regulate Intracellular Ca^{2+} Levels. *J. Neurosci.* **2009**, *29*, 2053–2063. [CrossRef]
216. Chaves, Y.C.; Genaro, K.; Crippa, J.A.; Da Cunha, J.M.; Zanoveli, J.M. Cannabidiol induces antidepressant and anxiolytic-like effects in experimental type-1 diabetic animals by multiple sites of action. *Metab. Brain Dis.* **2021**, *36*, 639–652. [CrossRef]

217. Lucas, P.; Walsh, Z.; Crosby, K.; Callaway, R.; Belle-Isle, L.; Kay, R.; Capler, R.; Holtzman, S. Substituting cannabis for prescription drugs, alcohol and other substances among medical cannabis patients: The impact of contextual factors: Cannabis substitution. *Drug Alcohol. Rev.* **2016**, *35*, 326–333. [CrossRef]
218. Kostanda, E.; Khatib, S. Biotic stress caused by Tetranychus urticae mites elevates the quantity of secondary metabolites, cannabinoids and terpenes, in *Cannabis sativa* L. *Ind. Crops Prod.* **2022**, *176*, 114331. [CrossRef]
219. Loss, C.M.; Teodoro, L.; Rodrigues, G.D.; Moreira, L.R.; Peres, F.F.; Zuardi, A.W.; Crippa, J.A.; Hallak, J.E.C.; Abílio, V.C. Is Cannabidiol During Neurodevelopment a Promising Therapy for Schizophrenia and Autism Spectrum Disorders? *Front. Pharmacol.* **2021**, *11*, 635763. [CrossRef] [PubMed]
220. Karoly, H.C.; Milburn, M.A.; Brooks-Russell, A.; Brown, M.; Streufert, J.; Bryan, A.D.; Lovrich, N.P.; DeJong, W.; Cinnamon Bidwell, L. Effects of High-Potency Cannabis on Psychomotor Performance in Frequent Cannabis Users. *Cannabis Cannabinoid Res.* **2022**, *7*, 107–115. [CrossRef]
221. Wallace, A.L.; Wade, N.E.; Lisdahl, K.M. Impact of 2 Weeks of Monitored Abstinence on Cognition in Adolescent and Young Adult Cannabis Users. *J. Int. Neuropsychol. Soc.* **2020**, *26*, 776–784. [CrossRef] [PubMed]
222. Gonzalez, R.; Schuster, R.M.; Mermelstein, R.J.; Vassileva, J.; Martin, E.M.; Diviak, K.R. Performance of young adult cannabis users on neurocognitive measures of impulsive behavior and their relationship to symptoms of cannabis use disorders. *J. Clin. Exp. Neuropsychol.* **2012**, *34*, 962–976. [CrossRef]
223. Collin, C.; Ehler, E.; Waberzinek, G.; Alsindi, Z.; Davies, P.; Powell, K.; Notcutt, W.; O'Leary, C.; Ratcliffe, S.; Nováková, I.; et al. A double-blind, randomized, placebo-controlled, parallel-group study of Sativex, in subjects with symptoms of spasticity due to multiple sclerosis. *Neurol. Res.* **2010**, *32*, 451–459. [CrossRef] [PubMed]
224. Markovà, J.; Essner, U.; Akmaz, B.; Marinelli, M.; Trompke, C.; Lentschat, A.; Vila, C. Sativex ® as add-on therapy vs. further optimized first-line ANTispastics (SAVANT) in resistant multiple sclerosis spasticity: A double-blind, placebo-controlled randomised clinical trial. *Int. J. Neurosci.* **2019**, *129*, 119–128. [CrossRef]
225. Meuth, S.G.; Henze, T.; Essner, U.; Trompke, C.; Vila Silván, C. Tetrahydrocannabinol and cannabidiol oromucosal spray in resistant multiple sclerosis spasticity: Consistency of response across subgroups from the SAVANT randomized clinical trial. *Int. J. Neurosci.* **2020**, *130*, 1199–1205. [CrossRef]
226. Wade, D.T.; Collin, C.; Stott, C.; Duncombe, P. Meta-analysis of the efficacy and safety of Sativex (nabiximols), on spasticity in people with multiple sclerosis. *Mult. Scler.* **2010**, *16*, 707–714. [CrossRef]
227. Serpell, M.G.; Notcutt, W.; Collin, C. Sativex long-term use: An open-label trial in patients with spasticity due to multiple sclerosis. *J. Neurol.* **2013**, *260*, 285–295. [CrossRef]
228. Ambrose, T.; Simmons, A. Cannabis, Cannabinoids, and the Endocannabinoid System—Is there Therapeutic Potential for Inflammatory Bowel Disease? *J. Crohn's Colitis* **2019**, *13*, 525–535. [CrossRef] [PubMed]
229. Kienzl, M.; Storr, M.; Schicho, R. Cannabinoids and Opioids in the Treatment of Inflammatory Bowel Diseases. *Clin. Transl. Gastroenterol.* **2020**, *11*, e00120. [CrossRef] [PubMed]
230. Mboumba Bouassa, R.-S.; Comeau, E.; Alexandrova, Y.; Pagliuzza, A.; Yero, A.; Samarani, S.; Needham, J.; Singer, J.; Lee, T.; Bobeuf, F.; et al. Effects of Oral Cannabinoids on Systemic Inflammation and Viral Reservoir Markers in People with HIV on Antiretroviral Therapy: Results of the CTN PT028 Pilot Clinical Trial. *Cells* **2023**, *12*, 1811. [CrossRef] [PubMed]
231. Couch, D.G.; Cook, H.; Ortori, C.; Barrett, D.; Lund, J.N.; O'Sullivan, S.E. Palmitoylethanolamide and Cannabidiol Prevent Inflammation-induced Hyperpermeability of the Human Gut In Vitro and In Vivo—A Randomized, Placebo-controlled, Double-blind Controlled Trial. *Inflamm. Bowel Dis.* **2019**, *25*, 1006–1018. [CrossRef]
232. Mboumba Bouassa, R.-S.; Needham, J.; Nohynek, D.; Singer, J.; Lee, T.; Bobeuf, F.; Samarani, S.; Del Balso, L.; Paisible, N.; Vertzagias, C.; et al. Safety and Tolerability of Oral Cannabinoids in People Living with HIV on Long-Term ART: A Randomized, Open-Label, Interventional Pilot Clinical Trial (CTNPT 028). *Biomedicines* **2022**, *10*, 3168. [CrossRef]
233. Whiting, P.F.; Wolff, R.F.; Deshpande, S.; Di Nisio, M.; Duffy, S.; Hernandez, A.V.; Keurentjes, J.C.; Lang, S.; Misso, K.; Ryder, S.; et al. Cannabinoids for Medical Use: A Systematic Review and Meta-analysis. *JAMA* **2015**, *313*, 2456. [CrossRef]
234. Stuyt, E. The Problem with the Current High Potency THC Marijuana from the Perspective of an Addiction Psychiatrist. *Mo. Med.* **2018**, *115*, 482–486.
235. Tramer, M.R. Cannabinoids for control of chemotherapy induced nausea and vomiting: Quantitative systematic. *BMJ* **2001**, *323*, 16. [CrossRef]
236. Gobbi, G.; Atkin, T.; Zytynski, T.; Wang, S.; Askari, S.; Boruff, J.; Ware, M.; Marmorstein, N.; Cipriani, A.; Dendukuri, N.; et al. Association of Cannabis Use in Adolescence and Risk of Depression, Anxiety, and Suicidality in Young Adulthood: A Systematic Review and Meta-analysis. *JAMA Psychiatry* **2019**, *76*, 426. [CrossRef]
237. Sarne, Y. Beneficial and deleterious effects of cannabinoids in the brain: The case of ultra-low dose THC. *Am. J. Drug Alcohol. Abus.* **2019**, *45*, 551–562. [CrossRef]
238. Emrich, H.M.; Leweke, F.M.; Schneider, U. Towards a Cannabinoid Hypothesis of Schizophrenia: Cognitive Impairments Due to Dysregulation of the Endogenous Cannabinoid System. *Pharmacol. Biochem. Behav.* **1997**, *56*, 803–807. [CrossRef] [PubMed]
239. Iacopetti, C.L.; Packer, C.D. Cannabinoid Hyperemesis Syndrome: A Case Report and Review of Pathophysiology. *Clin. Med. Res.* **2014**, *12*, 65–67. [CrossRef] [PubMed]
240. Vaughn, S.; Strawn, J.; Poweleit, E.; Sarangdhar, M.; Ramsey, L. The Impact of Marijuana on Antidepressant Treatment in Adolescents: Clinical and Pharmacologic Considerations. *J. Pers. Med.* **2021**, *11*, 615. [CrossRef]

241. Benjamini, Y.; Yekutieli, D. The control of the false discovery rate in multiple testing under dependency. *Ann. Statist.* **2001**, *29*, 1165–1188. [CrossRef]
242. Deutsch, D.G. A Personal Retrospective: Elevating Anandamide (AEA) by Targeting Fatty Acid Amide Hydrolase (FAAH) and the Fatty Acid Binding Proteins (FABPs). *Front. Pharmacol.* **2016**, *7*, 370. [CrossRef]
243. Elmes, M.W.; Kaczocha, M.; Berger, W.T.; Leung, K.; Ralph, B.P.; Wang, L.; Sweeney, J.M.; Miyauchi, J.T.; Tsirka, S.E.; Ojima, I.; et al. Fatty Acid-binding Proteins (FABPs) Are Intracellular Carriers for Δ9-Tetrahydrocannabinol (THC) and Cannabidiol (CBD). *J. Biol. Chem.* **2015**, *290*, 8711–8721. [CrossRef] [PubMed]
244. Takeda, S.; Misawa, K.; Yamamoto, I.; Watanabe, K. Cannabidiolic Acid as a Selective Cyclooxygenase-2 Inhibitory Component in Cannabis. *Drug Metab. Dispos.* **2008**, *36*, 1917–1921. [CrossRef]
245. Devinsky, O.; Patel, A.D.; Thiele, E.A.; Wong, M.H.; Appleton, R.; Harden, C.L.; Greenwood, S.; Morrison, G.; Sommerville, K.; On behalf of the GWPCARE1 Part A Study Group. Randomized, dose-ranging safety trial of cannabidiol in Dravet syndrome. *Neurology* **2018**, *90*, e1204–e1211. [CrossRef]
246. Massi, P.; Valenti, M.; Vaccani, A.; Gasperi, V.; Perletti, G.; Marras, E.; Fezza, F.; Maccarrone, M.; Parolaro, D. 5-Lipoxygenase and anandamide hydrolase (FAAH) mediate the antitumor activity of cannabidiol, a non-psychoactive cannabinoid. *J. Neurochem.* **2008**, *104*, 1091–1100. [CrossRef]
247. Ghosh, A.; Chen, F.; Thakur, A.; Hong, H. Cysteinyl Leukotrienes and Their Receptors: Emerging Therapeutic Targets in Central Nervous System Disorders. *CNS Neurosci. Ther.* **2016**, *22*, 943–951. [CrossRef]
248. Rossi, S.; De Chiara, V.; Musella, A.; Sacchetti, L.; Cantarella, C.; Castelli, M.; Cavasinni, F.; Motta, C.; Studer, V.; Bernardi, G.; et al. Preservation of Striatal Cannabinoid CB_1 Receptor Function Correlates with the Antianxiety Effects of Fatty Acid Amide Hydrolase Inhibition. *Mol. Pharmacol.* **2010**, *78*, 260–268. [CrossRef]
249. De Chiara, V.; Angelucci, F.; Rossi, S.; Musella, A.; Cavasinni, F.; Cantarella, C.; Mataluni, G.; Sacchetti, L.; Napolitano, F.; Castelli, M.; et al. Brain-Derived Neurotrophic Factor Controls Cannabinoid CB_1 Receptor Function in the Striatum. *J. Neurosci.* **2010**, *30*, 8127–8137. [CrossRef]
250. Rubino, T.; Guidali, C.; Vigano, D.; Realini, N.; Valenti, M.; Massi, P.; Parolaro, D. CB1 receptor stimulation in specific brain areas differently modulate anxiety-related behaviour. *Neuropharmacology* **2008**, *54*, 151–160. [CrossRef] [PubMed]
251. Rey, A.A.; Purrio, M.; Viveros, M.-P.; Lutz, B. Biphasic Effects of Cannabinoids in Anxiety Responses: CB1 and GABAB Receptors in the Balance of GABAergic and Glutamatergic Neurotransmission. *Neuropsychopharmacology* **2012**, *37*, 2624–2634. [CrossRef] [PubMed]
252. Haller, J.; Bakos, N.; Szirmay, M.; Ledent, C.; Freund, T.F. The effects of genetic and pharmacological blockade of the CB_1 cannabinoid receptor on anxiety: Anxiety and cannabinoids. *Eur. J. Neurosci.* **2002**, *16*, 1395–1398. [CrossRef]
253. Su, F.; Yi, H.; Xu, L.; Zhang, Z. Fluoxetine and S-citalopram inhibit M1 activation and promote M2 activation of microglia in vitro. *Neuroscience* **2015**, *294*, 60–68. [CrossRef]
254. Onaivi, E.S.; Ishiguro, H.; Gong, J.-P.; Patel, S.; Meozzi, P.A.; Myers, L.; Perchuk, A.; Mora, Z.; Tagliaferro, P.A.; Gardner, E.; et al. Functional Expression of Brain Neuronal CB2 Cannabinoid Receptors Are Involved in the Effects of Drugs of Abuse and in Depression. *Ann. N. Y. Acad. Sci.* **2008**, *1139*, 434–449. [CrossRef]
255. García-Gutiérrez, M.S.; Manzanares, J. Overexpression of CB2 cannabinoid receptors decreased vulnerability to anxiety and impaired anxiolytic action of alprazolam in mice. *J. Psychopharmacol.* **2011**, *25*, 111–120. [CrossRef]
256. García-Gutiérrez, M.S.; García-Bueno, B.; Zoppi, S.; Leza, J.C.; Manzanares, J. Chronic blockade of cannabinoid CB2 receptors induces anxiolytic-like actions associated with alterations in GABAA receptors: Chronic blockade of CB2 receptors is anxiolytic. *Br. J. Pharmacol.* **2012**, *165*, 951–964. [CrossRef]
257. Liu, Q.-R.; Canseco-Alba, A.; Zhang, H.-Y.; Tagliaferro, P.; Chung, M.; Dennis, E.; Sanabria, B.; Schanz, N.; Escosteguy-Neto, J.C.; Ishiguro, H.; et al. Cannabinoid type 2 receptors in dopamine neurons inhibits psychomotor behaviors, alters anxiety, depression and alcohol preference. *Sci. Rep.* **2017**, *7*, 17410. [CrossRef] [PubMed]
258. Bradley, S.G.; Munson, A.E.; Dewey, W.L.; Harris, L.S. Enhanced susceptibility of mice to combinations of delta 9-tetrahydrocannabinol and live or killed gram-negative bacteria. *Infect. Immun.* **1977**, *17*, 325–329. [CrossRef] [PubMed]
259. Morahan, P.S.; Klykken, P.C.; Smith, S.H.; Harris, L.S.; Munson, A.E. Effects of cannabinoids on host resistance to Listeria monocytogenes and herpes simplex virus. *Infect. Immun.* **1979**, *23*, 670–674. [CrossRef] [PubMed]
260. Bradley, S.G. Drugs of Abuse and Infections. In *Drugs of Abuse, Immunity, and Immunodeficiency*; Friedman, H., Specter, S., Klein, T.W., Eds.; Advances in Experimental Medicine and Biology; Springer: Boston, MA, USA, 1991; Volume 288, pp. 119–125, ISBN 978-1-4684-5927-2. [CrossRef]
261. Kraft, B.; Wintersberger, W.; Kress, H.G. Cannabinoid receptor-independent suppression of the superoxide generation of human neutrophils (PMN) by CP55 940, but not by anandamide. *Life Sci.* **2004**, *75*, 969–977. [CrossRef] [PubMed]
262. Akinola, O.; Ogbeche, E.O.; Olumoh-Abdul, H.A.; Alli-Oluwafuyi, A.O.; Oyewole, A.L.; Amin, A.; AbdulMajeed, W.I.; Olajide, O.J.; Nafiu, A.B.; Njan, A.A.; et al. Oral Ingestion of *Cannabis sativa*: Risks, Benefits, and Effects on Malaria-Infected Hosts. *Cannabis Cannabinoid Res.* **2018**, *3*, 219–227. [CrossRef] [PubMed]
263. Bachmann, M.F.; Oxenius, A. Interleukin 2: From immunostimulation to immunoregulation and back again. *EMBO Rep.* **2007**, *8*, 1142–1148. [CrossRef]
264. Ito, T.; Connett, J.M.; Kunkel, S.L.; Matsukawa, A. The linkage of innate and adaptive immune response during granulomatous development. *Front. Immun.* **2013**, *4*, 10. [CrossRef]

265. Dotsey, E.; Ushach, I.; Pone, E.; Nakajima, R.; Jasinskas, A.; Argueta, D.A.; Dillon, A.; DiPatrizio, N.; Davies, H.; Zlotnik, A.; et al. Transient Cannabinoid Receptor 2 Blockade during Immunization Heightens Intensity and Breadth of Antigen-specific Antibody Responses in Young and Aged mice. *Sci. Rep.* **2017**, *7*, 42584. [CrossRef]
266. Sohal, A.; Chaudhry, H.; Dhaliwal, A.; Sharma, R.; Singla, P.; Sodhi, S.; Gupta, G.; Dukovic, D.; Kowdley, K. S1364 Effect of Cannabis on Mortality and Resource Utilisation in Patients with Acute on Chronic Liver Failure: Nationwide Analysis. *Am. J. Gastroenterol.* **2022**, *117*, e979–e980. [CrossRef]
267. Yekhtin, Z.; Khuja, I.; Meiri, D.; Or, R.; Almogi-Hazan, O. Differential Effects of D9 Tetrahydrocannabinol (THC)- and Cannabidiol (CBD)-Based Cannabinoid Treatments on Macrophage Immune Function In Vitro and on Gastrointestinal Inflammation in a Murine Model. *Biomedicines* **2022**, *10*, 1793. [CrossRef]
268. Aswad, M.; Hamza, H.; Pechkovsky, A.; Zikrach, A.; Popov, T.; Zohar, Y.; Shahar, E.; Louria-Hayon, I. High-CBD Extract (CBD-X) Downregulates Cytokine Storm Systemically and Locally in Inflamed Lungs. *Front. Immunol.* **2022**, *13*, 875546. [CrossRef]
269. Khuja, I.; Yekhtin, Z.; Or, R.; Almogi-Hazan, O. Cannabinoids Reduce Inflammation but Inhibit Lymphocyte Recovery in Murine Models of Bone Marrow Transplantation. *Int. J. Mol. Sci.* **2019**, *20*, 668. [CrossRef] [PubMed]
270. Manoharan, R.; Kemper, J.; Young, J. Exploring the medical cannabis prescribing behaviours of New Zealand physicians. *Drug Alcohol. Rev.* **2022**, *41*, 1355–1366. [CrossRef] [PubMed]
271. Knaub, K.; Sartorius, T.; Dharsono, T.; Wacker, R.; Wilhelm, M.; Schön, C. A Novel Self-Emulsifying Drug Delivery System (SEDDS) Based on VESIsorb®Formulation Technology Improving the Oral Bioavailability of Cannabidiol in Healthy Subjects. *Molecules* **2019**, *24*, 2967. [CrossRef] [PubMed]

brain sciences

Review

Cognitive Crescendo: How Music Shapes the Brain's Structure and Function

Corneliu Toader [1,2], Calin Petru Tataru [3,4,*], Ioan-Alexandru Florian [5,*], Razvan-Adrian Covache-Busuioc [1,*], Bogdan-Gabriel Bratu [1], Luca Andrei Glavan [1], Andrei Bordeianu [1], David-Ioan Dumitrascu [1] and Alexandru Vlad Ciurea [1,6]

1 Department of Neurosurgery, "Carol Davila" University of Medicine and Pharmacy, 020021 Bucharest, Romania; corneliu.toader@umfcd.ro (C.T.); bogdan.bratu@stud.umfcd.ro (B.-G.B.); luca-andrei.glavan0720@stud.umfcd.ro (L.A.G.); andrei.bordeianu@stud.umfcd.ro (A.B.); david-ioan.dumitrascu0720@stud.umfcd.ro (D.-I.D.); prof.avciurea@gmail.com (A.V.C.)
2 Department of Vascular Neurosurgery, National Institute of Neurology and Neurovascular Diseases, 077160 Bucharest, Romania
3 Department of Opthamology, "Carol Davila" University of Medicine and Pharmacy, 020021 Bucharest, Romania
4 Central Military Emergency Hospital "Dr. Carol Davila", 010825 Bucharest, Romania
5 Department of Neurosciences, "Iuliu Hatieganu" University of Medicine and Pharmacy, 400012 Cluj-Napoca, Romania
6 Neurosurgery Department, Sanador Clinical Hospital, 010991 Bucharest, Romania
* Correspondence: calinpetrutataru@yahoo.com (C.P.T.); florian.ioan.alexandru@gmail.com (I.-A.F.); razvan-adrian.covache-busuioc0720@stud.umfcd.ro (R.-A.C.-B.)

Abstract: Music is a complex phenomenon with multiple brain areas and neural connections being implicated. Centuries ago, music was discovered as an efficient modality for psychological status enrichment and even for the treatment of multiple pathologies. Modern research investigations give a new avenue for music perception and the understanding of the underlying neurological mechanisms, using neuroimaging, especially magnetic resonance imaging. Multiple brain areas were depicted in the last decades as being of high value for music processing, and further analyses in the neuropsychology field uncover the implications in emotional and cognitive activities. Music listening improves cognitive functions such as memory, attention span, and behavioral augmentation. In rehabilitation, music-based therapies have a high rate of success for the treatment of depression and anxiety and even in neurological disorders such as regaining the body integrity after a stroke episode. Our review focused on the neurological and psychological implications of music, as well as presenting the significant clinical relevance of therapies using music.

Keywords: pitch perception; cognitive enhancement; memory encoding; limbic system; rehabilitation; music-based therapies

Citation: Toader, C.; Tataru, C.P.; Florian, I.-A.; Covache-Busuioc, R.-A.; Bratu, B.-G.; Glavan, L.A.; Bordeianu, A.; Dumitrascu, D.-I.; Ciurea, A.V. Cognitive Crescendo: How Music Shapes the Brain's Structure and Function. *Brain Sci.* **2023**, *13*, 1390. https://doi.org/10.3390/brainsci13101390

Academic Editors: Raffaella Nori, Massimiliano Palmiero and Laura Piccardi

Received: 25 August 2023
Revised: 25 September 2023
Accepted: 27 September 2023
Published: 29 September 2023

1. Introduction

The inherent complexity of music renders it a multifaceted subject that eludes simple definitions. While many describe it as an ordered arrangement of sounds, musical elements such as harmony or the bass line require intricate understanding and considerable effort to master. In this research, our focus is on the neurological and psychological benefits of music listening, especially the potential usage of musical therapies, and how the brain might respond during varied activities set within a musical context [1].

Music is a universal phenomenon that utilizes a myriad of brain resources. Engaging with music is among the most cognitively demanding tasks a human can undergo, and it is identified across all cultures; therefore, it underscores its fundamental human nature [2]. The proclivity to create and appreciate music is ubiquitous among humans, permeating daily life across diverse societies [3]. This inherent connection to musical expression is deeply intertwined with human identity and experience. Molnar-Szakacs further emphasizes music's unique capacity to evoke memories, stimulate emotions, and enrich social interactions [3]. Historical examples underscore the therapeutic potential of music. For instance, Johann Sebastian Bach's Goldberg Variations (BWV 988) was purportedly composed to alleviate a count's insomnia, underscoring music's therapeutic potential [4–6]. The profound emotional impact of music, whether it be the melancholy evoked by a nocturne from F. Chopin or the elation induced by W. A. Mozart, has inspired ongoing research into its relationship with emotions and psychological disorders [7]. Fundamental to understanding music are the concepts of pitch perception, rhythm perception, and tonality perception.

1.1. Pitch Perception

Predominantly processed in the auditory cortex, pitch perception pertains to the brain's handling of sound information. The auditory cortex features a tonotopic map wherein specific regions are sensitive to distinct frequencies. Human auditory perception ranges from 20 to 20,000 Hz, with distinct pitches resonating at precise locations on the basilar membrane. Yost et al. expound that understanding pitch necessitates a grasp of the biomechanical mechanisms and neurological shifts in sound as well as the diverse ways pitch can be conceptualized and potentially quantified [8]. Often, pitch is defined as the attribute of sound that sequences it from low to high levels. Musically, pitch aids in recognizing melodies and discerning intervals, with quantification methods ranging from equal-temperament tuning scales to the perceptive mel scale [9].

For instance, a standard 1000 Hz tone delivered at a 40 dB sound pressure level corresponds to 100 mels on the mel scale. It is important to note that variations in perceived pitch proportionately influence mel values. Much of pitch perception research delves into complex sounds, with the pitch of basic tones like sinusoids determined by frequency. Intricacies in encoding high-frequency and low-frequency tonal signals differentiate them, and while amplitude modulation is absent in simple tonal sounds, temporal mechanisms might play a role in low-frequency pitch perception [10].

In summary, understanding sound transformations, coupled with a range of definitions and measurement techniques, is imperative for accurate pitch perception. This encompasses melody recognition capacity, interval discernment, and frequency perception, with various mechanisms, both spectral and temporal, influencing pitch perception [11].

1.2. Rhythm Perception

Beat perception engages specific brain regions associated with motor planning and timing, notably the basal ganglia and the supplementary motor area. Interestingly, even passive listening to music can activate these neural domains [12]. The ability to discern a steady pulse underlying a rhythmic stimulus defines beat perception. This inherent pulse, which rhythmically structures the music, is an elemental consistency that the human cognitive apparatus innately detects. By accentuating beats in specific patterns, we can synchronize our movements (e.g., dancing or foot tapping) and regulate our temporal perception, culminating in the creation of meter. Rhythmic perception necessitates a combination of interval-based (absolute) timing and beat-based (relative) timing. While interval-based timing is observed in both humans and various animal species, beat-based timing might be unique to humans [13,14].

Motor theories centered on timing are primarily focused on beat-based timing. Active motor engagement seems to actively mold our perception of beats. For instance, the negative mean asynchrony effect, where one's taps often precede the actual beat, underscores the pivotal role of anticipation in beat-based timing. Humans establish rhythmic timing anticipations and maintain a versatile perception of the intrinsic rhythmic architecture, even when confronted with alterations in tempo. Notably, rhythm perception is not merely passive; it is influenced by an individual's active cognitive processing and volitional control, underpinned by metric interpretation [15]. Moreover, the very act of motor engagement shapes the perception of beats, manifests bodily movements, enhances temporal perception, and influences interpretations of ambiguous rhythms. Both overt motor actions and their covert counterparts play a role in refining perceptual sharpness. Even in scenarios devoid of visible motion, there is accumulating evidence that motor engagement modulates the perception of beat and meter. Contemporary research posits that the motor system not only influences beat perception but can also augment synchronicity with music [13]. Faster movements can also modulate the perceived pace of music segments [16].

To encapsulate, beat perception involves recognizing a steady pulse amidst rhythmic stimuli, a process that is dynamically shaped by motor activity, conscious modulation, adaptive tempo perception, and anticipatory mechanisms. Remarkably, even in scenarios devoid of overt motion, our sense of rhythm and meter remains intricately linked with the motor system [17,18].

1.3. Tonality Perception

The comprehension of key and harmony in music engages distinct neural domains, including the auditory, prefrontal, and parietal cortices. Scientific investigations are currently delving deeper into understanding the brain's intricacies in processing musical harmony. The notion of harmony primarily stems from the amalgamation of sounds in Western tonal music. Within this musical paradigm, pitches are hierarchically arranged based on their congruence within a specific tonal context. Scales utilized in Western tonal compositions emanate from this pitch hierarchy. While the behavioral science community acknowledges the hierarchical essence of pitch organization, the neural substrates underpinning it remain a realm of exploration [19].

In a distinct study centered on J. S. Bach's compositions, researchers probed the psychological relevance of musicians' conception of tonality. Here, musically trained listeners were tasked with singing the first scale that resonated with them post hearing snippets from Bach's Preludes in The Well-Tempered Clavier. The selected tonic (starting note) and mode (major/minor) were then juxtaposed against Bach's original specifications. The data revealed that listeners could often discern the designated tonic and mode merely from the initial quartet of notes. However, as the piece progressed, there was a marked tendency to gravitate toward tonalities divergent from the original key, notably within the initial eight bars. By the concluding quartet of bars, the original tonic was often reaffirmed. Such findings not only spotlight the cognitive intricacies of tonality perception but also align with the postulations of music theorists regarding tonal discernment by listeners [20].

Tonality serves as the linchpin in music, underpinning the creation and comprehension of musical constructs such as melodies. A contemporary dynamic theory on musical tonality posits a nonlinear response of auditory neuron networks to musical stimuli. This tonal cognition, the intrinsic interconnections perceived amidst tones, arises from the robust and harmonious associations among brain frequencies, a phenomenon attributable to nonlinear resonance [21,22].

2. Materials and Methods

We conducted a comprehensive search on PubMed database for the most relevant articles regarding music studies, musicology mechanisms, and music-based therapies. For the search formula, we used the following terms: "pitch perception", "rhythm perception", "tonality perception", "memory encoding", "limbic system", "neuroplasticity", "motor coordination", "evoked memories", "rehabilitation", and "music-based therapies". Initially, PubMed database showed 341 studies. Furthermore, each title of those articles was reviewed to include minimally one of the searching terms. Those studies that did not respect the inclusion criteria or were focused on other subjects besides musicology were excluded. After the analyses, only 132 studies were included in our study.

In this comprehensive review segment, we delve into existing studies, results, and theoretical postulations regarding the neurological implications of music and its therapeutic applications. The aim is to furnish a meticulous analysis of the current state of knowledge within this field, accentuating pivotal research endeavors, methodologies, and discoveries.

Subdivisions within this section are delineated based on thematic content, research domains, or specific dimensions of the topic.

3. Results

3.1. Emotion and Reward Mechanisms in Musical Perception

Music possesses the unique capability to induce profound emotional responses, often intertwined with personal memories of significance. The neuroscientific underpinnings of this phenomenon suggest that music's emotive power is rooted in the activation of the brain's reward system. Notable neural regions involved include the nucleus accumbens and the ventromedial prefrontal cortex, elucidating the intrinsically rewarding and emotionally charged nature of musical experiences.

3.1.1. The Interplay of Music with the Limbic System

Central to our emotional resonance with music is the limbic system, an intricate assembly of neural circuits and pathways. Key components of this system, such as the amygdala—responsible for emotional processing—and the hippocampus—integral to memory consolidation—become activated during musical exposure (Table 1). Such neural activities account for the evocative power of music to invoke vivid emotional and mnemonic experiences. The consequential effects can be observed when an individual is emotionally transported to a distinct temporal or spatial context upon hearing a particular musical piece or when a gamut of emotions is experienced in response to auditory stimuli [23].

Table 1. Brain areas activated during music listening. Auditory cortices from temporal lobe and limbic system areas are the most frequently implicated brain regions in music processing, as well as other eloquent areas depicted in the table.

	Region (Brodmann Area)
Temporal cortex	
Right	Primary auditory cortex (41) Secondary auditory cortex (22 and 42) Superior temporal sulcus (21 or 22) Temporal pole (22 or 38) Middle temporal gyrus (21)
Left	Primary auditory cortex (41) Superior temporal sulcus (21 or 22)
Limbic areas	
Right	Anterior insula Hippocampus
Left	Retrosplenial cortex (29 or 30) Anterior cingulate cortex (32) Anterior insula Subcallosal cingulate gyrus (11 or 25)
Others brain areas	
	Lingual gyrus (18 and 19) Inferior parietal lobule (39)

During passive listening to unfamiliar yet positively perceived music, there was a spontaneous activation in both the limbic and paralimbic regions. Consistent with prior research on passive auditory experiences, primary and secondary auditory cortices displayed activations, corroborating findings from studies that analyzed listening to either monophonic or harmonized auditory sequences [24]. Furthermore, there were observed activations in the temporal pole, subcallosal cingulate gyrus, affective segment of the anterior cingulate cortex, retrosplenial cortex, hippocampus, anterior insula, and nucleus accumbens. It is plausible that these observed neuroanatomical patterns are a result of the intricate musical nature of the stimuli, which were highly favored by the participants. There is a prevailing theory suggesting that the left hemisphere predominantly facilitates positive emotions. This is in line with our findings that indicate a predominance of limbic and paralimbic activations on the left side, potentially mirroring the participants' positive aesthetic reactions. The acquired functional neuroanatomical insights augment existing literature on music–emotion interplay, especially those employing high-temporal-resolution methodologies such as electroencephalography and magnetoencephalography [23].

Contrasting minor with major melodies showed multiple activation sites (Figure 1) with the right parahippocampal gyrus (RPHG) being an eloquent brain area (Figure 2). Another discernible activation, when subjected to cluster-level correction, spanned both the left and right ventral anterior cingulate cortex (VACC) (BA 24) and extended into the left medial frontal gyrus (LMFG) within the medial prefrontal cortex (BA 10) (Figure 3). Remarkably, the inverse contrast (major over minor) did not yield significant activations. In a peak-voxel analysis, the response to the chromatic scale was intermediary when juxtaposed against the major and minor mode melodies for three of the aforementioned regions. These differential responses between the chromatic scale and melodies were not statistically significant, with an exception. Within the LMFG, the chromatic scale evoked the most prominent (least negative) response, trailed by the minor and subsequently the major mode. Notably, the contrast between the chromatic scale and the major mode was statistically significant in this context [25].

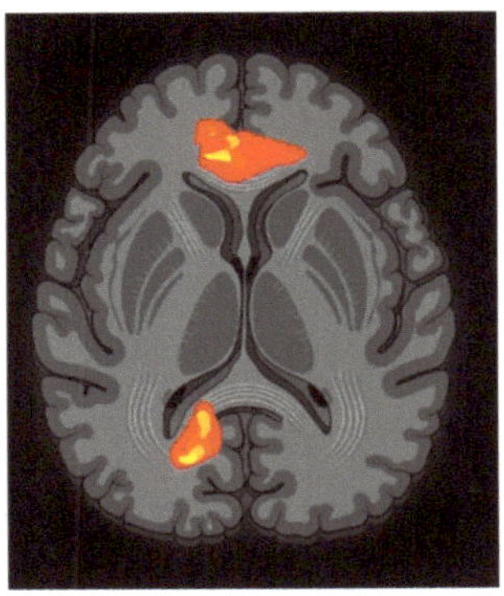

Figure 1. Activation pattern during music listening task. The transversal MRI sequence shows the overall cerebral activation pattern. The lower part of the image will be further explained in Figure 2, while the upper part will be specifically described in Figure 3.

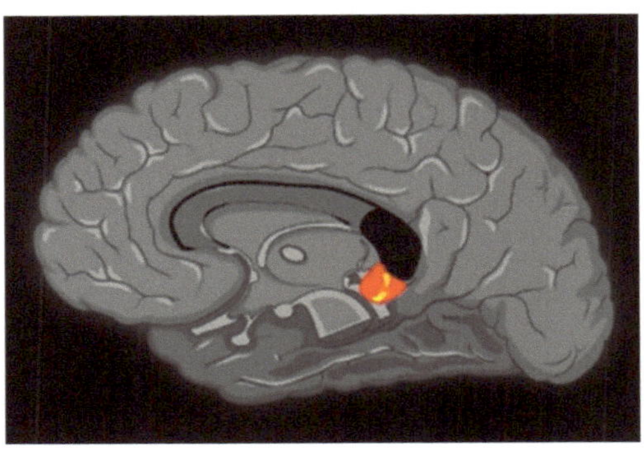

Figure 2. A sagittal MRI sequence is shown, which depicts significant neural activity in the right parahippocampal gyrus.

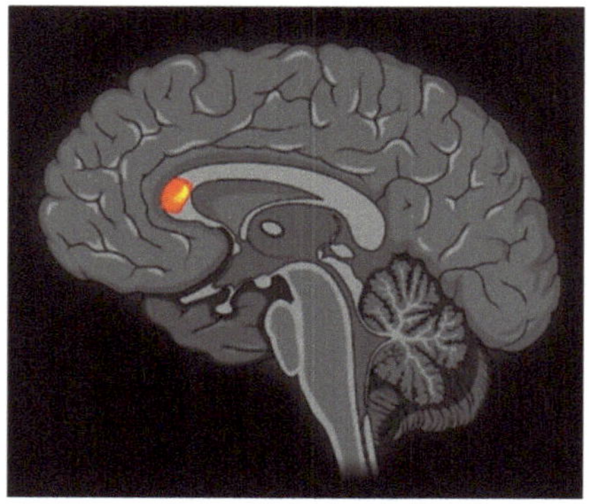

Figure 3. A sagittal MRI sequence is shown, which shows significant neural activity in the right anterior cingulate cortex (BA 24), left anterior cingulate cortex (BA 24), and left medial frontal gyrus (BA 10).

VACC activation is generally associated with affective processing, while its dorsal counterpart is linked with cognitive functions [26]. Moreover, the existing literature indicates that the VACC displays heightened sensitivity to emotional content characterized by negativity or sadness [27]. The observed engagement of the VACC might be consistent with the perception of minor mode melodies as possessing a sadder tonality in comparison to major melodies. Notably, prior neuroimaging research on mode-based contrasts has not reported VACC activation in contrasts between minor and major modes [28].

The detected involvement of the left medial frontal gyrus (LMFG) may be attributed to its robust neural connectivity with the anterior cingulate cortex and other limbic systems. Such a connectivity profile underscores the proposed function of the medial prefrontal cortex as an integrative nexus for emotional input from these associated regions [29].

Research encompassing neuropsychology, neurophysiology, and health science domains suggests that patients in a low-awareness state exhibit both anatomical and behavioral divergences in response to auditory stimuli. These differences underline the auditory channel's pivotal role in evaluating such patients. More specifically, the distinct auditory responses between individuals in a vegetative state (VS) and those in a minimally conscious state (MCS) when exposed to emotionally significant auditory stimuli imply that interventions incorporating personally resonant auditory content could lead to discernible outcomes, thus aiding diagnosis. However, diagnostic endeavors are often confounded by non-intentional emotional, or "limbic", reactions observed in VS patients [30].

Multiple studies have documented elevated neural activity in MCS patients when exposed to emotionally significant auditory cues, suggesting these individuals possess the capability for discriminatory auditory responses. For instance, Boly et al. observed that stimuli like distress calls or a patient's own name elicited more extensive neural activations compared to irrelevant noises [31]. In addition, cognitive-evoked potentials in response to an individual's own name differed from those induced by other names, reinforcing the clinical premise and observational data that personally significant stimuli are more likely to induce pronounced behavioral alterations [32].

3.1.2. Music Seems to Encourage Enhanced Connectivity between the Auditory and Emotional Regions of the Brain

Listening to music engages not only the auditory cortex, responsible for sound processing but also several emotional centers within the brain. For instance, a musical composition perceived as melancholic might enhance the connectivity between the auditory cortex and the hippocampus, a region integral to memory and emotional processing. This interconnection can trigger the recollection of somber memories or evoke feelings of sadness.

Activations were prominently observed bilaterally in the anterior sections of the middle and superior temporal gyri. Prior research has identified the anterior temporal lobe's involvement in comprehension at the sentence level, distinct from the temporal assimilation of significant auditory cues. Notably, this region's activation is rather selective for sentence-level stimuli. It does not exhibit pronounced responses to unstructured meaningful auditory cues like word lists or random sequences of environmental noises. Nevertheless, it does react to both coherent sentences and nonsensical pseudoword sentences. The study's authors noted the unresolved question of whether this region also becomes active during musical engagements [33].

Positron emission tomography (PET) studies focused on auditory imagery for music have documented the active involvement of the supplementary motor areas (SMAs) during image generation. This indicates the SMA's potential role in an internalized "singing" process during auditory imagery tasks [34,35]. However, these studies did not explicitly associate SMA activity with the rhythmic elements of music. Notably, research involving patients with SMA lesions unequivocally demonstrates their difficulties in replicating rhythms [36]. The observed diminishing correlation of SMA activity with rhythmical performance following each alteration in the degree of temporal deviations from the reference interval ratio (DRIR) mirrors the decline in SMA activity as a motor task is reiterated. This parallel highlights the analogous motor-related neural activations during both motor activities and musical perception [37].

3.2. Motor Systems

Engaging in musical activities necessitates intricate motor tasks that demand precise timing and coordination. The cerebellum, an integral part of the brain dedicated to timing and motor coordination, demonstrates heightened activity among musicians. Other motor-related regions, such as the premotor cortex and the basal ganglia, play pivotal roles in both producing and perceiving music. Comprehensive motor systems, spanning from fine motor skills to broad motor coordination, are crucial for regulating the physical actions inherent in playing a musical instrument or singing [38].

Fine Motor Control: Precision in playing musical instruments necessitates exceptional motor control, specifically in muscles such as the fingers and hands.

Finger Dexterity: Musicians cultivate nuanced finger motions, granting them the capability to adeptly handle their instrument's keys, strings, or frets. This proficiency enables diverse pitch generation and the execution of intricate melodies or chords. Notably, pianists, aspiring to master compositions like Liszt's "Transcendental Studies", S. 139, or Beethoven's Piano Sonata No. 21 "Waldstein", Op. 53, commonly practice upward of 6 h daily [39].

Hand Coordination: Instruments such as pianos or guitars necessitate meticulous coordination between hands. A harmonious interplay is required where one hand typically manages the melody or leads, while the counterpart offers harmonic or rhythmic accompaniment [40].

Embouchure Control: Wind instrument performers, encompassing flutists and trumpeters, are reliant on meticulous muscle control of their mouth and lips for tone production and airflow modulation [41].

Gross Motor Coordination: Distinct from precision-centered fine motor control, gross motor coordination emphasizes the integration of larger muscle group activities.

Body Movement: Many musicians incorporate physical gestures to accentuate rhythm or enhance their presentations, such as rhythmically swaying or foot tapping [42].

Posture and Breathing: Vocalists and wind instrument practitioners stress the importance of appropriate posture and breath management. Optimal posture underpins efficient breathing, ensuring voice projection and breath modulation [43].

Sensorimotor Integration: An intimate synergy between motor coordination and sensory feedback is paramount for musical endeavors.

Visual Feedback: Musicians harness visual indicators like music notations or the synchronized actions of co-performers to facilitate timing coordination and group harmonization [44].

Tactile Feedback: Musicians depend on tactile sensations and muscle memory, underpinning finger positioning and pressure modulation on their instruments [45].

Auditory Feedback: By closely monitoring their auditory output, musicians can fine-tune pitch, pace, and tonal quality. This auditory feedback loop enables real-time adjustments, promoting accuracy [46].

In summation, the intricate interplay of fine motor skills, gross motor coordination, and sensorimotor integration embodies the complexity of musical performance. Through relentless training and practice, musicians refine their motor capabilities, striving for both mastery and evocative expression.

3.2.1. Music and Rhythm Processing

Music, at its core, engages our motor systems predominantly through the element of rhythm. The basal ganglia and the supplementary motor area (SMA) stand out as pivotal neural regions governing rhythm processing. Specifically, the basal ganglia take center stage in organizing movements, determining timing and sequencing, and forecasting forthcoming rhythmic beats [47,48].

To delve deeper into rhythm cognition within music, one should familiarize oneself with the PRISM framework. This framework elucidates three central mechanisms: precise auditory processing, synchronization of brain oscillations to rhythmic stimuli, and the interplay between sensory perception and motor action known as sensorimotor coupling. Collectively, these mechanisms facilitate rhythm processing in both musical and speech domains [48,49].

Accurate Auditory Processing: This entails discerning minute time deviations and provides the bedrock for rhythm perception, enabling the detection of intricate temporal patterns.

Brain Oscillation Synchronization: This mechanism concerns the brain's ability to anticipate ensuing events and conform to hierarchical rhythm structures. It ensures the alignment of rhythmic components, contributing to the holistic rhythm experience [47].

Sensorimotor Coupling: This establishes a link between perception and execution, implicating the motor system in tasks like timing, prediction, and integrating auditory cues with motor actions.

The PRISM framework offers an innovative lens through which rhythm processing in music and speech is perceived. By illuminating shared neural mechanisms between music and speech, this model enriches our understanding of rhythm processing, thereby opening avenues for further research, particularly in the arena of speech and language impediments [48].

However, beyond the neurocognitive realm, rhythm perception and production are intertwined with cultural nuances. While cognitive and physiological components might offer a universal rhythm perception baseline, cultural experiences undeniably play a significant role. Infants, for instance, exhibit an inclination toward rhythmic patterns emblematic of their culture's music, suggesting cultural influences even at infancy [50]. Cultural aspects also influence language rhythm perception, with speech patterns often mirroring a given culture's musical rhythms. A comparison between Western and East African music presents a stark difference in rhythm complexities and significance, highlighting the cultural diversity in rhythm processing [51].

Furthermore, cultural disparities might not only dictate how rhythm is perceived but also the range of rhythmic frequencies one aligns with. For instance, African music's inherent metrical ambiguity might afford listeners the flexibility to engage with multiple rhythmic levels, diverging from the more rigid Western musical counterparts [52].

In summary, rhythm's multifaceted nature intertwines neural processing with cultural nuances. Cultural exposure and familiarity undoubtedly mold our rhythmic preferences and processing capabilities, underscoring the intricate relationship binding music, language, and societal constructs [14].

3.2.2. Music and Motor Coordination

Playing a musical instrument, especially the piano, is a testament to the intricate dance of our motor systems. Brain scans of musicians highlight heightened activity in regions like the motor cortex and cerebellum, both critical for motion and coordination. Notably, the cerebellum emerges as the linchpin for fine motor control and timing, skills indispensable to instrumentalists. Music is not just an art; it reshapes the brain. Lifelong musical tutelage can cause an enlarged motor cortex and cerebellum, imprinting physical markers of musical expertise [53].

One striking feature of our motor system is its redundancy. With a plethora of joints and muscles at our disposal, multiple movement combinations can yield the same outcome. Renowned pianists like Martha Argerich and Dinu Lipatti exemplify this by leveraging redundancy to achieve specific acoustic effects, each using unique motor configurations [54]. This fluidity arises from neuroplasticity, where the neuromuscular system continually reshapes itself, enhancing the finesse of advanced motor activities. By juxtaposing skilled versus novice pianists, researchers probe into the interplay of neuroplasticity, motor redundancy, and the nuanced organization of piano-playing movements. While gauging the long-term impact of training remains challenging, such comparisons offer valuable glimpses into the artistry of motor skills [55].

The redundancy in pianists' motor systems is multilayered. They can achieve the same note with various force and movement patterns at the fingertip, navigate multiple joint combinations to produce identical fingertip movements, and leverage various forces to generate the same joint rotation [56]. Amidst this intricate web, muscular torque stands out. It is birthed from the balance of forces exerted by opposing muscles around a joint. Given the motor system's richness, pianists have countless ways to strike a single note. Masters of the craft excel in navigating this maze by optimizing energy use, achieving physiological efficiency. Their prowess is evident in their enhanced coordination, minimal muscle discomfort, and adeptness at offsetting mechanical interactions [50].

How pianists employ joint rotations and balance various forces exemplifies the interplay of kinematic and kinetic configurations. Elite pianists adopt strategies like optimized postures and sequential joint movements, optimizing movement and conserving muscle energy. By harnessing gravity, they also conserve energy when pressing keys, further showcasing motor redundancy. A consistent finding in studies contrasting expert versus novice pianists is the former's unique upper limb motion organization, honed through rigorous practice. Such an organization is attuned to physiological efficiency, minimizing energy costs for known tasks. It is no surprise then that seasoned pianists, even in demanding performances, manage to retain their performance quality, all while fending off muscle fatigue [57].

In a fascinating dive into the world of jazz improvisation, Setzler M and colleagues explore how mutual coupling influences the coordination dynamics of professional jazz performers. The study revolves around understanding the interplay of rhythmic and tonal patterns as musicians exchange and spontaneously produce musical elements. With expert pianists from the vibrant New York City jazz circuit as participants, the study juxtaposes a unique one-way scenario, where a pianist improvises to a pre-recorded duet, against two dynamic duo conditions: a coupled setting where both pianists are improvising in real-time. While the one-way setup showcases unilateral coordination, in the duo scenario, the pianists adjust to each other's rhythms and tones. The catch? The improvisations are uninhibited by any predefined song structure, key, or tempo [58].

The study dives deep into the data, examining parameters like tonal consonance (how harmoniously musical combinations sound) and onset density (the extent of rhythmic activity). The findings are illuminating: when pianists are connected and responding to each other in the duo setup, they consistently exhibit enhanced coordinated behavior. They create more harmonious tonal structures and display heightened rhythmic synchronization, compared to the unilateral one-way condition. Notably, these observations align with both the pianists' personal experiences and the auditory preferences of lay listeners [58].

But why does this matter? The implications of this research are manifold. Firstly, it propels the domain of collaborative action studies and music technology. By understanding the nuances of how mutual coupling impacts musical coordination, we gain insights into complex, unrestrained coordination typical of stellar artistic performances. Such insights go beyond controlled lab environments. Moreover, the findings can shape the future of interactive music systems, potentially revolutionizing how ensemble performances are evaluated in musical training. The study's roster boasts 28 seasoned pianists, all with robust backgrounds in jazz improvisation, along with a diverse listening panel comprising both jazz maestros and undergraduate psychology students. In essence, this research provides a valuable lens into the intricate dance of coordination during musical improvisation, shedding light on how it elevates the quality of the resultant melodies [59].

3.2.3. Music and Rehabilitation

Music's healing touch has progressively found its way into motor rehabilitation, offering a rhythmic respite to those grappling with motor skill challenges. Music-based therapeutic interventions, for instance, have emerged as powerful tools for stroke patients, helping them regain lost motor functions. The rhythmic predictability embedded within music seems to have a harmonious effect on patients with Parkinson's disease, addressing their movement-related issues like gait and timing disruptions. This rhythmic auditory stimulation (RAS), as is known, offers an external rhythmic pulse that works wonders in steadying and regulating motor timing. This incorporation of music in treating age-related neurological ailments is backed by numerous studies [6].

The global surge in age-related neurological disorders, propelled by an aging population, has escalated the economic burdens associated mainly with non-acute treatments. This has ignited the quest for cost-efficient rehabilitative methods to complement traditional approaches like physiotherapy. While there is a limit to how much adult brain neurogenesis can contribute to healing, functional restoration does not share this limitation. Shifting from targeted training of impaired functions, some modern methods are championing a holistic rise in brain activity through sensory and cognitive stimulations [4].

Research has illuminated how musical pursuits like playing an instrument can reshape the brain. Even mere listening to music has been observed to bolster neuronal connections in certain brain areas, such as the auditory and visual cortices. Music's therapeutic touch extends to post-operative recovery as well, alleviating pain and anxiety and reducing the dependence on painkillers [60]. Certified music therapists employ both active and receptive music-based therapies, encompassing musical expressions ranging from singing to playing instruments [61]. While initial studies revolved around music's impact on acquired brain injuries, comprehensive investigations into its effect on major neurological diseases are still unfolding [6].

The review delves into music-based therapies' impact on ailments like stroke, dementia, Parkinson's, epilepsy, and multiple sclerosis, gauging the therapies' efficacy through randomized controlled trials. The "effect size" metric offers insights into the degree of improvement observed [5].

Further, the study zooms in on the potential of dance and RAS in rehabilitating individuals with cerebral palsy (CP) [62]. Preliminary evidence champions the benefits of dance and RAS in enhancing physical functionalities, especially areas like balance, walking, and cardiorespiratory fitness in CP patients. Despite the extensive categories in the International Classification of Functioning, Disability and Health (ICF), there remain research voids, especially in areas concerning participation and environmental factors [63]. Bridging these gaps, the review synthesizes quantitative rehabilitation findings within the ICF framework, pinpointing further research avenues. It concludes by celebrating dance and RAS's potential in enhancing not just physical processes but also emotional expression, social interactions, and overall well-being [64].

3.2.4. Entrainment

Entrainment, a captivating phenomenon where we unconsciously synchronize our movements to an external rhythm, emerges as an inherent human response when engaged with music. This almost involuntary response—be it foot tapping or dancing—is not just about moving to the beat. It is an intricate interplay of various brain regions responsible for auditory processing, motor functions, and even prediction [65].

Music, a rich tapestry of sensory, cognitive, and emotional experiences, is not just about the melody or rhythm. When we engage with music, it evokes a spectrum of emotions—from joy and sorrow to more nuanced feelings like wonder or nostalgia. These complex emotions do not necessarily fit into conventional neuroscientific emotion categories, leaving a vast realm still largely unexplored [66]. The authors delve deep into these intricate emotions, suggesting they are possibly birthed from the confluence of multiple brain areas, including those responsible for attention, motor functions, and memory, intertwined with emotional and motivational pathways. Such an understanding holds profound implications, especially in therapeutic realms, potentially aiding conditions marred by attention, motor, or affective disruptions [67,68].

"New Music" presents another dimension to our musical discourse. Unlike its classical counterpart, defining "New Music" is like capturing lightning in a bottle—it is ever-evolving, challenging norms, and shunning traditional tonality and rhythms. The listener, when immersed in the world of New Music, must recalibrate their cognitive tools to truly appreciate this avant-garde genre. While it is a mosaic of styles, some dominant shades include the second Vienna School, electronic synthesis, microtonal music, and more [9,69]. Branching further, genres like Ambient Music and Postclassical Minimal emerge, each with its unique essence.

Recognizing the need for a deeper dive into "New Music" and its neurological interplay, a dedicated research topic was launched, casting a wide net from embodied cognition to technological impacts and even neuroimaging techniques like EEG and fMRI [9]. The selected studies ventured into diverse terrains—from tempo perceptions, the philosophy of sound objects, and networked music performances to the nuances of atonal music, especially with pioneers like Arnold Schönberg at its helm [70]. Truly appreciating New Music mandates unconventional cognitive frameworks, from embodiment to heightened attention to recurring or absent elements. Functional brain imaging, though still in its nascent stages, promises insights into our cerebral engagement with these novel musical narratives. While the current discourse sheds light on New Music's mysteries, a harmonious symphony of extensive and collaborative research is imperative for a deeper understanding [71].

3.3. Memory

Music and memory share an intimate bond. Often, a song can trigger a cascade of vivid memories, while melodies and lyrics, even from years past, can be effortlessly recalled. Such connections correlate with activations in areas like the hippocampus, pivotal in memory storage and retrieval.

3.3.1. Memory Encoding with Music

Harnessing a song's melody and rhythm can be a powerful mnemonic device. Information set to a catchy tune tends to stick, an approach adopted in education to teach topics ranging from languages to science.

Smith et al. (1985) posited a compelling idea—using music as a backdrop during the encoding of words can be a catalyst in context-dependent memory during retrieval. This effectively boosts the recall of the encoded words [72]. Extending this thought, there is mounting evidence that suggests music's potency in facilitating episodic encoding of events [73]. Across various studies, employing musical stimuli like background tunes or sung texts consistently showed improvements in verbal memory for both standard [74] and clinical groups [75,76]. However, while these studies underscore music's ability to enhance the recall of encoded items, most have not delved into the musical context during the retrieval process. Among those that did, outcomes have been mixed [77].

Using fNIRS studies, it has been found that musical backdrops during verbal material encoding can bolster both item and source memory, linked to the modulation of prefrontal cortex activity [78]. However, some limitations exist, primarily since these studies only compared musical contexts to silence, leaving unanswered questions regarding the impact of non-musical auditory stimulations on memory.

Contrary to the majority, El Haj et al. (2014) proposed that musical backgrounds might impede source memory performance across age groups, adding more layers to the ongoing debate [77].

Ferreri et al. (2015) shed more light on the subject, indicating that specifically a musical backdrop (and not just any sound) can enhance verbal encoding. The ongoing discussion shifts to which specific elements of music augment memory. Past research has indicated that factors like perceptual characteristics, the emotional undertone, and interpretive variations in musical stimuli play pivotal roles in boosting memory and learning [79]. Adding depth to this understanding, it is noted that emotional inputs modulate musical memory akin to their influence in other domains [80].

A salient aspect of the music–memory nexus is the role of rewarding stimuli in cognitive tasks. Music stands tall as one of the most rewarding stimuli, and recent insights suggest its potential in augmenting cognitive performance [81].

3.3.2. Evoking Memories

Music has an uncanny ability to immerse us back into past moments, often reviving the very emotions we felt during those times. This phenomenon arises because music not only captures the essence of our emotional state when memories form but also acts as a potent cue to rekindle them. Thus, a mere tune or lyric can instantaneously propel us to a distinct time or place, evoking associated feelings.

This intricate bond between music and memory has led to the term "musical memory". This refers to the unique connection between certain songs and personal experiences, elucidating why particular melodies can instantaneously remind us of specific past events or people.

Music-evoked autobiographical memories (MEAMs) are often charged with intense emotions—be it joy, excitement, or nostalgia [82]. For instance, a study by Janata et al. (2007) found that popular music-triggered MEAMs were profoundly emotional. They noted that when participants resonated deeply with a song, they were more inclined to associate it with a personal memory [83]. Neuroimaging research supports the emotionally charged nature of MEAMs and illuminates music's ability to evoke memories of varying specificity [84]. Such revelations underscore music's prominence as a memory catalyst.

Research also explores music's role in memory recall among Alzheimer's patients. For instance, Foster and Valentine (2001) noted that Alzheimer's patients retrieved more personal memories post music exposure compared to when exposed to white noise or silence [85]. Similarly, a study by Irish et al. (2006) found that Alzheimer's patients exhibited enhanced episodic memory recall when exposed to Vivaldi's Spring from the Four Seasons [86]. However, since music in these studies played in the background and not as a direct memory cue, the results showcase music's influence on memory recall but do not differentiate music-evoked memories from those elicited by other stimuli.

When pitting memories triggered by music against those by faces, the former emerged as more vivid. However, the total number of internal details remained consistent across both. The primary distinction was in external details—face-induced memories contained more such details, often rich in semantic information about the pictured individual [87]. Interestingly, gender dynamics were evident in memory retrieval; women consistently described more vivid autobiographical memories than men, regardless of the cue. Several studies have hypothesized that this could be attributed to gender-specific encoding styles, with women registering memories more intricately. Additionally, Piefke et al. (2005) proposed that men and women employ distinct cognitive strategies during memory retrieval [88]. Another variable impacting the vividness of autobiographical memories is age. Typically, older adults recall memories that are less specific and contain fewer episodic details compared to their younger counterparts [89].

3.3.3. Neurological Basis

Our understanding of music's influence on the brain is intricate, involving numerous regions that process auditory information, emotions, and memories.

During music perception, the auditory cortex plays a central role, processing the sound. Simultaneously, areas associated with emotional responses, like the amygdala, and memory, such as the hippocampus, become activated. The medial prefrontal cortex is particularly interesting; it springs into action when we hear familiar tunes. It is also significant to note that this region is one of the last to degenerate in Alzheimer's disease, hinting at its role in the robust link between music and autobiographical memories.

Delving into the neural mechanics of music performance, Langheim et al. (2002) discovered activations in various brain areas, including the supplementary motor and premotor regions, right superior parietal lobule, right inferior frontal gyrus, bilateral midfrontal gyri, and the bilateral lateral cerebellum, during imagined musical performance. Notably, they did not observe activation in primary sensorimotor areas and auditory cortices [90]. The specific activation of the right inferior frontal gyrus is believed to be tied to music production. This idea is reinforced by other studies highlighting the involvement of this area in selective attention, working memory, and motor synchronization with auditory cues [91,92].

In another exploration, Nirkko et al. (2001) demonstrated that playing a musical sequence on a violin led to activation in several brain regions. Notably, they highlighted the involvement of bilateral fronto-opercular regions, suggesting their role in timed motor sequences present in both music and language production [93].

Another crucial region, the superior temporal gyrus, processes complex patterns formed by individual musical notes [94]. Platel et al. (1997) observed that the activation of specific parts of this gyrus, coupled with the left inferior frontal gyrus, indicates semantic access to melodic elements [95].

Popescu et al. (2004) noted early activations around primary and secondary auditory cortices, as well as in posterior parietal areas post stimulus onset [96]. These regions are critical for language and music processing. Furthermore, activations in the supramarginal and postcentral gyri have been associated with processing the basic attributes of sound [97]. Meanwhile, music listening's impact on the precuneus has been documented in several studies, emphasizing its role in sound processing [98].

In sum, our brain's reaction to music, whether in perception or production, is a symphony of neural activations across multiple regions, underpinning our rich emotional and cognitive experiences with melodies.

3.3.4. Music in Therapeutics

Music, renowned for its potent link with memory, has been harnessed therapeutically in numerous medical scenarios. In conditions like Alzheimer's disease and dementia, where individuals frequently grapple with short-term memory loss, familiar tunes can rekindle past memories and experiences. This often enhances mood and bolsters social interactions. In the realm of stroke rehabilitation, music therapy has been instrumental in assisting patients in regaining verbal memory (Figure 4).

The potential of music-based interventions in the neurological realm is immense, particularly in mending motor or cognitive functions. However, the design of these interventions often targets a specific pathology group. Among these, the evidence is most compelling for bolstering motor skills in stroke patients. It is imperative to approach these findings with caution; there is a risk of attributing improvements solely to the interventions, overlooking the role of natural recovery. Some studies, for instance, that employed bimanual piano training or gait training to musical cues, may not have utilized the most accurate measures to gauge improvements in coordination, dexterity, or balance. Nevertheless, the adaptability of music-based interventions in a clinical setting is noteworthy. They can be tailored to the individual, offering both a progression in therapy and personalization in treatment choice [99].

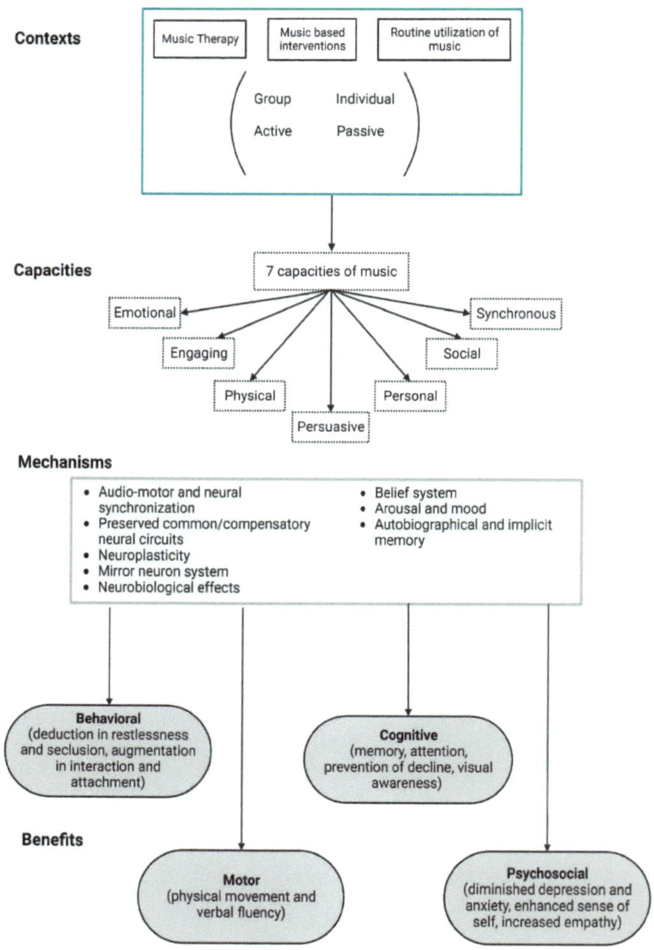

Figure 4. This image depicts examples of the main possibilities of clinical therapies using music. The given context is music therapies and daily music listening in various situations, such as in groups or individually and active or passive listening. Music offers multiple cognitive advantages and might be perceived in multiple ways which are described as "capacities". Underlying mechanisms of music processing were aforementioned in this study, audio-motor functions and neuroplasticity being of high interest. Multiple behavioral-cognitive benefits, as well as motricity and psychological status, are highly improved [61,104].

One salient area where music-based interventions shine is in addressing cognitive–motor interference, a common challenge in many neurological ailments [100]. Parallel executive deficits can sometimes hinder the effective rehabilitation of cognitive or motor shortcomings [101].

Here, music-based approaches emerge as dual-task training, transcending mere motor or cognitive training. Consider, for instance, an intervention utilizing a musical instrument. Here, the act of producing music, which involves moving parts of the body like the fingers (motor system), dovetails with the cognitive system, which processes new musical data, such as rhythm or pitch. This is especially pertinent, as significant cognitive–motor disruptions arising from such dual tasks are prevalent in many neurological conditions and can escalate the risk of falls [102,103].

In essence, the therapeutic power of music, weaving together cognitive and motor systems, can be a beacon of hope in the multifaceted landscape of neurological rehabilitation.

3.4. *Language Processing*

Music and language are intricately linked, both weaving together structured sequences of sound and resonating in similar domains of the brain's left hemisphere. Evidence suggesting that musical training can bolster language skills lends weight to the idea that the neural mechanisms underpinning both might overlap.

3.4.1. Musical Training's Influence on Language Skills

It has been established that musical education can enrich language abilities. This is likely due to the shared demands both disciplines place on discerning differences in pitch, timing, and tone. Musicians often exhibit heightened skills in phonetic discrimination (the capacity to differentiate between speech sounds), enhanced verbal memory, and superior reading capabilities. Furthermore, rhythmic competencies, which are sharpened through music, correlate with improved reading and linguistic prowess.

The nexus between musical training and intelligence has been a subject of rigorous debate in recent times. Music training and its duration have been consistently linked with higher intelligence across age groups, from children to adults [105,106]. This relationship is evident in various studies, where notable differences in non-verbal reasoning skills emerge between those with and without musical training [107,108]. Moreover, a correlation exists between non-verbal intelligence and musical aptitude. However, a potential confounder is the fact that children who receive music lessons often hail from affluent backgrounds, which could potentially skew the interpretation of these findings, especially in studies relying on correlation [109,110].

Recent data reveal that consistent participation in music playschools augments the development of phoneme processing abilities and vocabulary in children aged 5–6. In contrast, dance lessons did not exhibit a comparable impact. The disparities in children's development crystallized over our two-year monitoring period. Interestingly, children exposed to both musical and dance education did not display a noticeable edge in vocabulary development. One theory is that these children had relatively high scores at the onset, leading to a potential ceiling effect as the study progressed. However, by the study's conclusion, children who only attended music playschool and initially exhibited lower scores gravitated toward the higher-scoring group. This suggests that music-centric activities might especially benefit children who initially lag in linguistic tasks, at least within the observed age bracket of 5–6 years [111].

In essence, the confluence of music and language is undeniable, and the enriching impact of musical training on linguistic skills is evident, shedding light on the profound interconnectedness of these domains.

3.4.2. Music and Speech Prosody

Music's ties to the prosody of speech are compelling. Prosody, encompassing pitch, rhythm, and volume, is pivotal for embedding emotion and context in speech. Notably, these facets are fundamental to music, underlining a profound link between the musical and the expressive elements of language.

However, the waters are murkier when exploring pitch. While studies on pitch perception distinctly delineate between global and local processing, research on human voice recognition often treats pitch as a unified acoustic/perceptual element. The role of pitch in identifying talkers remains an enigma. Idiosyncratic prosodic alterations, especially the dynamics of the F0 contour, prove useful for distinguishing speakers [112]. However, absolute pitch height is another identifier, rooted in the individual's unique laryngeal structure. For instance, by adjusting the pitch of synthetic speech, one can shift listeners' perception of the number of dialogue participants [113]. A focus on individual differences can illuminate pitch perception's role in talker identification, potentially disentangling the different ways pitch is processed in this context. Earlier studies indicate that global and local pitch processing can be separated, especially when linked to other linguistic proficiencies like reading [114]. Connecting differences in global versus local pitch perception with listeners' variability in talker identification can provide a clearer understanding of pitch's role in this process.

Long-term musical training is reputed to enhance pitch discernment [115]. This perceptual edge extends beyond musical pitch, touching the linguistic realm. A compelling connection between music and language is evident in studies examining lexical tone processing. For instance, musical training or aptitude can predict non-tonal language speakers' prowess in identifying lexical tones [116] and in mimicking them, as well as their competency in learning them [117].

These findings, which underscore improved talker identification via experience in music and language, carry significant ramifications for understanding auditory perception's adaptability. The evidence suggests that long-term engagement with music or consistent lexical tone use can augment listeners' pitch sensitivity. This challenges the rigid compartmentalization of cognitive systems dedicated to music, language, and talker identification, pointing toward a more fluid, interconnected cognitive landscape [118]. In essence, music and language are not just standalone entities but intertwined realms, each enriching the other.

3.4.3. Therapeutic Applications

The bond between music and language not only provides insight into cognitive function but has also been harnessed for therapeutic means. A salient example is melodic intonation therapy (Table 2), a method designed to assist aphasia patients (those who have lost language abilities typically due to brain damage, often resulting from a stroke) in regaining their speech. By engaging a patient's preserved musical processing abilities, the therapy facilitates language recovery.

Table 2. Optimal profile for a patient with high responsiveness to melodic intonation therapy.

Significantly restricted speech ability, or nonfluent speaking
Left-hemisphere stroke, usually unilateral
Possibility to reproduce words while singing well-recognized songs
Moderate integrity of auditory function
Continuously failed attempts to speak
Motivated patients with a great psychological stability
Difficult capacity of repetition

However, the therapeutic application of this connection has seen a myriad of interpretations. Initial accounts [119] present deviations from the original protocol, indicating the use of three pitches instead of the initially outlined two. Anecdotal evidence further showcases this diversity: therapists, based on observational data from across the U.S., each bring their own flair to the technique. Variations range from employing two pitches with specific intervals and crafting unique melodies for phrases incorporating multiple pitches to using the piano as an accompaniment or even tapping a sequence of notes on a patient's arm as words or phrases are sung. Such diverse interpretations, while possibly tapping into right hemisphere regions pivotal for speech, might deter therapists lacking a musical foundation from adopting the therapy, given its intricate nature [120].

Additionally, the act of tapping the left hand could activate the right hemisphere's sensorimotor network, responsible for both hand and mouth movements [121]. This action might bolster sound–motor mapping—an essential facet of meaningful vocal exchanges. Moreover, akin to the consistent beat of a metronome, tapping could offer a rhythmic guide, ensuring regular pacing and ongoing cues for the production of syllables [122]. In essence, the nuanced interplay between music and language has profound therapeutic potential, albeit varied in its execution.

4. Discussion

Music training has been identified as a catalyst for neurological transformation, exemplifying the phenomenon of neuroplasticity. Notably, individuals with a background in music often exhibit more pronounced auditory and motor regions compared to their non-musical counterparts. Such changes have far-reaching implications, encompassing areas like memory enhancement and heightened attention [123].

4.1. *Anatomical Adaptations*

Long-term involvement in musical endeavors can result in discernible anatomical shifts within the brain. Such transformations mirror the refined skills inherent to musicians, encompassing areas like auditory discernment, sound-associated emotional interpretation, and intricate motor control. Musicians, for instance, typically possess a more substantial corpus callosum—the neural bridge uniting the brain's two halves. This could arise from the necessity of synchronized hand movements or the amalgamation of sensory–motor data. Moreover, regions governing motor functions, like the precentral gyrus, often exhibit greater development in musicians. Likewise, areas pivotal for auditory functions, such as the superior temporal gyrus, are frequently more evolved [124].

4.2. *Operational Modifications*

Beyond anatomical alterations, enduring musical training can usher in functional adaptations. When undertaking specific tasks, musicians often demonstrate unique brain activation patterns, emphasizing the brain's adaptability in response to persistent training. For instance, the auditory cortex in musicians may exhibit heightened activity during music perception, signifying their adeptness in deconstructing musical elements [125].

4.3. *Neurochemical Interactions and Neuronal Growth*

Musical interactions influence more than just the brain's physical contours; they also modulate its internal chemistry. Engaging with musical elements can spur dopamine release, linked with pleasure sensations, serotonin, regulating mood, and oxytocin, associated with social trust and bonding [126]. Furthermore, music might bolster neurogenesis, or the genesis of novel neurons. Preliminary animal research suggests that music exposure can amplify hippocampal neurogenesis, a core component in learning and memory. While promising, particularly concerning conditions like Alzheimer's, further studies are imperative for a comprehensive grasp [127].

4.4. *Cognitive Enhancement through Music*

Music-induced neuroplasticity can elevate cognitive prowess, transcending just musical abilities. Musical children often outpace their non-musical peers in areas like reading, linguistics, and mathematical proficiency. Additionally, their attention span, memory, and executive functionality are frequently more advanced. Such augmentations are theorized to emerge from the transfer effect, where proficiency in one domain (e.g., music) amplifies skills in another (e.g., math). In essence, the cognitive tools sharpened by musical immersion—such as pattern detection and motor coordination—might be applicable across diverse domains [128].

4.5. Therapeutic Application of Music

The neurological adaptability influenced by music has been harnessed therapeutically, especially in neurorehabilitation post traumatic events like strokes [129]. Music-centric therapies can instigate restorative neuroplasticity. An illustration is music-supported therapy, wherein patients rehabilitate motor functions by playing musical instruments. Playing instruments mandates recurrent, meticulous movements, essential for reinstating motor command. Furthermore, the intrinsic reward of music amplifies patient motivation [130]. Another intervention, melodic intonation therapy (MIT), targets non-fluent aphasia patients, aiding their speech recovery. This method capitalizes on the brain's adaptive potential, utilizing unharmed singing capacities to reinvigorate linguistic prowess [131].

An important point of view for an efficient rehabilitation process is using comprehensive approaches, especially in those patients who suffered a myocardial infarction or an ischemic stroke. In a recent study [132] focused on the effect of implementing robot-assisted physiotherapy technology for heart infarction treatment, great results were obtained in ADLs (activities of daily living) and motor functions. Moreover, in ischemic stroke scenarios, multidisciplinary combined healthcare management provides a better outcome, and by utilizing therapeutic modalities and behavioral-cognitive tests, assessing psychomotor status, and implementing robotic-based therapies, significant results are obtained [133]. Therefore, all the available therapeutical possibilities have to be used according to the patient's status for a decrease in morbidity and mortality, as well as the patient's ability improvement and reintegration into society. In this context, the capacity of music to reconfigure our brains, sharpening various abilities, provides an outstanding avenue as a therapeutic tool in healthcare situations.

5. Conclusions

The compendium of research synthesized in this review, titled "Cognitive Crescendo: How Music Shapes the Brain's Structure and Function", serves as a seminal contribution to the burgeoning interdisciplinary field at the intersection of musicology, cognitive neuroscience, and clinical psychology. By dissecting a range of subtopics—from rudimentary perceptual features such as pitch, rhythm, and tonality to complex interactions involving emotion, memory, and motor systems—the review offers a comprehensive, integrative framework for understanding how music orchestrates a vast array of neurocognitive processes.

One of the salient contributions of this review is its focus on the bidirectional interactions between music and the limbic system, which has elucidated the underlying neurobiological mechanisms by which music modulates emotional states. The evidence for enhanced connectivity between auditory and emotional regions of the brain brings a new layer of complexity to our understanding of affective regulation and provides fertile ground for future investigation into targeted music-based therapeutic interventions. Regarding motor systems and coordination, the review casts a spotlight on the neural entrainment mechanisms that facilitate synchrony between external rhythmic stimuli and internal neural oscillators. These findings are particularly germane for envisaging music-based rehabilitation paradigms, and the integration of rhythmic elements could revolutionize existing therapeutic approaches.

Furthermore, the review explicates the linguistic dividends of musical training, providing compelling empirical support for shared neural resources between musical and language processing. The implications here are not merely academic but could inform educational curricula that seek to leverage musical training for enhanced linguistic and cognitive skills in children and adults alike. As a corollary to the wide-ranging topics covered, this review also outlines a number of prospective avenues for research. For instance, the operational modifications and neurochemical interactions triggered by chronic exposure to music demand longitudinal studies to ascertain the sustainability of these neural changes. There is also a discernible gap in the literature concerning how these cognitive enhancements translate to real-world skills and well-being, an area ripe for further empirical inquiry.

Another promising avenue for exploration pertains to the therapeutic applications of music. While the existing literature, as summarized in this review, posits a strong case for music as a potent therapeutic tool, the exact protocols, durations, and modalities through which optimal therapeutic outcomes can be achieved remain to be standardized.

In summation, this review serves as both an analytical repository and a conceptual springboard, illuminating the multifaceted ways in which music interacts with the human cognitive apparatus. Its contributions are manifold, offering academic, clinical, and pedagogical insights that advance our understanding of the potent neurocognitive effects of musical engagement. By highlighting nascent areas warranting further exploration, this review not only synthesizes current knowledge but also catalyzes future interdisciplinary research aimed at decoding the myriad ways music intricately shapes our brains and our lives.

Author Contributions: Conceptualization, C.T. and C.P.T.; methodology, C.P.T. and I.-A.F.; software, R.-A.C.-B.; validation, B.-G.B.; formal analysis, L.A.G.; investigation, C.T. and I.-A.F.; resources, A.B.; data curation, D.-I.D.; writing—original draft preparation, R.-A.C.-B. and B.-G.B.; writing—review and editing, B.-G.B. and L.A.G.; visualization, C.T.; supervision, I.-A.F.; project administration, A.V.C.; funding acquisition, A.V.C. All authors have read and agreed to the published version of the manuscript.

Funding: This research received no external funding.

Institutional Review Board Statement: Not applicable.

Informed Consent Statement: Not applicable.

Data Availability Statement: All data are available online in libraries such as PubMed.

Conflicts of Interest: The authors declare no conflict of interest.

References

1. Vuust, P.; Heggli, O.A.; Friston, K.J.; Kringelbach, M.L. Music in the brain. *Nat. Rev. Neurosci.* **2022**, *23*, 287–305. [CrossRef] [PubMed]
2. Zatorre, R.J.; Chen, J.L.; Penhune, V.B. When the brain plays music: Auditory–motor interactions in music perception and production. *Nat. Rev. Neurosci.* **2007**, *8*, 547–558. [CrossRef] [PubMed]
3. Molnar-Szakacs, I.; Overy, K. Music and mirror neurons: From motion to emotion. *Soc. Cogn. Affect. Neurosci.* **2006**, *1*, 235–241. [CrossRef] [PubMed]
4. Leonardi, S.; Cacciola, A.; De Luca, R.; Aragona, B.; Andronaco, V.; Milardi, D.; Bramanti, P.; Calabrò, R.S. The role of music therapy in rehabilitation: Improving aphasia and beyond. *Int. J. Neurosci.* **2018**, *128*, 90–99. [CrossRef]
5. Tomassini, V.; Matthews, P.M.; Thompson, A.J.; Fuglø, D.; Geurts, J.J.; Johansen-Berg, H.; Jones, D.K.; Rocca, M.A.; Wise, R.G.; Barkhof, F.; et al. Neuroplasticity and functional recovery in multiple sclerosis. *Nat. Rev. Neurol.* **2012**, *8*, 635–646. [CrossRef]
6. Sihvonen, A.J.; Särkämö, T.; Leo, V.; Tervaniemi, M.; Altenmüller, E.; Soinila, S. Music-based interventions in neurological rehabilitation. *Lancet Neurol.* **2017**, *16*, 648–660. [CrossRef]
7. Mula, M.; Trimble, M.R. Music and madness: Neuropsychiatric aspects of music. *Clin. Med.* **2009**, *9*, 83–86. [CrossRef]
8. Yost, W.A. Pitch perception. *Atten. Percept. Psychophys.* **2009**, *71*, 1701–1715. [CrossRef]
9. Leek, M.R.; Summers, V. Pitch strength and pitch dominance of iterated rippled noises in hearing-impaired listeners. *J. Acoust. Soc. Am.* **2001**, *109*, 2944–2954. [CrossRef]
10. Arnaud, L.; Gracco, V.; Ménard, L. Enhanced perception of pitch changes in speech and music in early blind adults. *Neuropsychologia* **2018**, *117*, 261–270. [CrossRef]
11. Lewis, M.S.; Gallun, F.J.; Gordon, J.; Lilly, D.J.; Crandell, C. A Pilot Investigation Regarding Speech-Recognition Performance in Noise for Adults with Hearing Loss in the FM + HA Listening Condition. *TVR* **2010**, *110*, 31–54. [CrossRef]
12. Nozaradan, S.; Schwartze, M.; Obermeier, C.; Kotz, S.A. Specific contributions of basal ganglia and cerebellum to the neural tracking of rhythm. *Cortex* **2017**, *95*, 156–168. [CrossRef] [PubMed]
13. Ross, J.M.; Iversen, J.R.; Balasubramaniam, R. Motor simulation theories of musical beat perception. *Neurocase* **2016**, *22*, 558–565. [CrossRef]
14. Cameron, D.J.; Bentley, J.; Grahn, J.A. Cross-cultural influences on rhythm processing: Reproduction, discrimination, and beat tapping. *Front. Psychol.* **2015**, *6*, 366. [CrossRef] [PubMed]

15. Demarin, V.; Bedeković, M.R.; Puretić, M.B.; Pašić, M.B. Arts, Brain and Cognition. *Psychiatr. Danub.* **2016**, *28*, 343–348.
16. Grahn, J.A. The Role of the Basal Ganglia in Beat Perception: Neuroimaging and Neuropsychological Investigations. *Ann. N. Y. Acad. Sci.* **2009**, *1169*, 35–45. [CrossRef]
17. Grahn, J.A.; Brett, M. Rhythm and Beat Perception in Motor Areas of the Brain. *J. Cogn. Neurosci.* **2007**, *19*, 893–906. [CrossRef]
18. Grahn, J.A.; Henry, M.J.; McAuley, J.D. FMRI investigation of cross-modal interactions in beat perception: Audition primes vision, but not vice versa. *NeuroImage* **2011**, *54*, 1231–1243. [CrossRef]
19. Sauvé, S.A.; Cho, A.; Zendel, B.R. Mapping Tonal Hierarchy in the Brain. *Neuroscience* **2021**, *465*, 187–202. [CrossRef]
20. Cohen, A.J. Tonality and perception: Musical scales primed by excerpts from the Well-Tempered Clavier of J. S. Bach. *Psychol. Res.* **1991**, *53*, 305–314. [CrossRef]
21. Large, E.W. Tonality and Nonlinear Resonance. *Ann. N. Y. Acad. Sci.* **2005**, *1060*, 53–56. [CrossRef]
22. Large, E.W.; Almonte, F.V. Neurodynamics, tonality, and the auditory brainstem response: Neurodynamics, tonality, and the ABR. *Ann. N. Y. Acad. Sci.* **2012**, *1252*, E1–E7. [CrossRef]
23. Brown, S.; Martinez, M.J.; Parsons, L.M. Passive music listening spontaneously engages limbic and paralimbic systems. *NeuroReport* **2004**, *15*, 2033–2037. [CrossRef]
24. Zatorre, R.; Evans, A.; Meyer, E. Neural mechanisms underlying melodic perception and memory for pitch. *J. Neurosci.* **1994**, *14*, 1908–1919. [CrossRef]
25. Green, A.C.; Bærentsen, K.B.; Stødkilde-Jørgensen, H.; Wallentin, M.; Roepstorff, A.; Vuust, P. Music in minor activates limbic structures: A relationship with dissonance? *NeuroReport* **2008**, *19*, 711–715. [CrossRef]
26. Bush, G.; Luu, P.; Posner, M.I. Cognitive and emotional influences in anterior cingulate cortex. *Trends Cogn. Sci.* **2000**, *4*, 215–222. [CrossRef]
27. Allman, J.M.; Hakeem, A.; Erwin, J.M.; Nimchinsky, E.; Hof, P. The anterior cingulate cortex. The evolution of an interface between emotion and cognition. *Ann. N. Y. Acad. Sci.* **2001**, *935*, 107–117. [CrossRef]
28. Pallesen, K.J. Emotion Processing of Major, Minor, and Dissonant Chords: A Functional Magnetic Resonance Imaging Study. *Ann. N. Y. Acad. Sci.* **2005**, *1060*, 450–453. [CrossRef]
29. Krawczyk, D.C. Contributions of the prefrontal cortex to the neural basis of human decision making. *Neurosci. Biobehav. Rev.* **2002**, *26*, 631–664. [CrossRef]
30. Magee, W.L. Music as a diagnostic tool in low awareness states: Considering limbic responses. *Brain Inj.* **2007**, *21*, 593–599. [CrossRef]
31. Boly, M.; Faymonville, M.-E.; Peigneux, P.; Lambermont, B.; Damas, F.; Luxen, A.; Lamy, M.; Moonen, G.; Maquet, P.; Laureys, S. Cerebral processing of auditory and noxious stimuli in severely brain injured patients: Differences between VS and MCS. *Neuropsychol. Rehabil.* **2005**, *15*, 283–289. [CrossRef]
32. Wilson, S.L.; Powell, G.E.; Brock, D.; Thwaites, H. Vegetative state and responses to sensory stimulation: An analysis of 24 cases. *Brain Inj.* **1996**, *10*, 807–818. [CrossRef]
33. Humphries, C.; Willard, K.; Buchsbaum, B.; Hickok, G. Role of anterior temporal cortex in auditory sentence comprehension: An fMRI study. *Neuroreport* **2001**, *12*, 1749–1752. [CrossRef]
34. Zatorre, R.J.; Halpern, A.R.; Perry, D.W.; Meyer, E.; Evans, A.C. Hearing in the Mind's Ear: A PET Investigation of Musical Imagery and Perception. *J. Cogn. Neurosci.* **1996**, *8*, 29–46. [CrossRef]
35. Halpern, A.R. When That Tune Runs Through Your Head: A PET Investigation of Auditory Imagery for Familiar Melodies. *Cereb. Cortex* **1999**, *9*, 697–704. [CrossRef]
36. Halsband, U.; Ito, N.; Tanji, J.; Freund, H.-J. The role of premotor cortex and the supplementary motor area in the temporal control of movement in man. *Brain* **1993**, *116*, 243–266. [CrossRef]
37. Dammers, J.; Ioannides, A.A. Neuromagnetic Localization of CMV Generators Using Incomplete and Full-Head Biomagnetometer. *NeuroImage* **2000**, *11*, 167–178. [CrossRef]
38. Janka, Z. Musica et medicina. *Orv. Hetil.* **2019**, *160*, 403–418. [CrossRef]
39. Stegemöller, E.L.; Ferguson, T.D.; Zaman, A.; Hibbing, P.; Izbicki, P.; Krigolson, O.E. Finger tapping to different styles of music and changes in cortical oscillations. *Brain Behav.* **2021**, *11*, e2324. [CrossRef]
40. Stegemöller, E.L.; Izbicki, P.; Hibbing, P. The influence of moving with music on motor cortical activity. *Neurosci. Lett.* **2018**, *683*, 27–32. [CrossRef]
41. O'Kelly, J.; James, L.; Palaniappan, R.; Taborin, J.; Fachner, J.; Magee, W.L. Neurophysiological and Behavioral Responses to Music Therapy in Vegetative and Minimally Conscious States. *Front. Hum. Neurosci.* **2013**, *7*, 884. [CrossRef]
42. Washburn, A.; Wright, M.J.; Chafe, C.; Fujioka, T. Temporal Coordination in Piano Duet Networked Music Performance (NMP): Interactions Between Acoustic Transmission Latency and Musical Role Asymmetries. *Front. Psychol.* **2021**, *12*, 707090. [CrossRef]
43. Platte, S.L.; Gollhofer, A.; Gehring, D.; Willimann, J.; Schuldt-Jensen, M.; Lauber, B. The Effect of Different Preparatory Conducting Gestures on Breathing Behavior and Voice Quality of Choral Singers. *J. Voice* **2022**, S0892-1997(22)00155-2. [CrossRef]
44. Schultz, B.G.; Palmer, C. The roles of musical expertise and sensory feedback in beat keeping and joint action. *Psychol. Res.* **2019**, *83*, 419–431. [CrossRef]
45. Pfordresher, P.Q. Musical training and the role of auditory feedback during performance: Pfordresher. *Ann. N. Y. Acad. Sci.* **2012**, *1252*, 171–178. [CrossRef]

46. Martin, S.; Mikutta, C.; Leonard, M.K.; Hungate, D.; Koelsch, S.; Shamma, S.; Chang, E.F.; Millán, J.D.R.; Knight, R.T.; Pasley, B.N. Neural Encoding of Auditory Features during Music Perception and Imagery. *Cereb. Cortex* **2018**, *28*, 4222–4233. [CrossRef]
47. Link, E.; Baumann, E.; Klimmt, C. Explaining Online Information Seeking Behaviors in People with Different Health Statuses: German Representative Cross-sectional Survey. *J. Med. Internet Res.* **2021**, *23*, e25963. [CrossRef]
48. Fiveash, A.; Bedoin, N.; Gordon, R.L.; Tillmann, B. Processing rhythm in speech and music: Shared mechanisms and implications for developmental speech and language disorders. *Neuropsychology* **2021**, *35*, 771–791. [CrossRef]
49. Ladányi, E.; Persici, V.; Fiveash, A.; Tillmann, B.; Gordon, R.L. Is atypical rhythm a risk factor for developmental speech and language disorders? *WIREs Cogn. Sci.* **2020**, *11*, e1528. [CrossRef]
50. Furuya, S.; Altenmüller, E. Acquisition and reacquisition of motor coordination in musicians: Virtuosity and disorder of musicians. *Ann. N. Y. Acad. Sci.* **2015**, *1337*, 118–124. [CrossRef]
51. Bunketorp-Käll, L.; Pekna, M.; Pekny, M.; Blomstrand, C.; Nilsson, M. Effects of horse-riding therapy and rhythm and music-based therapy on functional mobility in late phase after stroke. *NRE* **2019**, *45*, 483–492. [CrossRef]
52. Temperley, D. Meter and Grouping in African Music: A View from Music Theory. *Ethnomusicology* **2000**, *44*, 65. [CrossRef]
53. Galińska, E. Music therapy in neurological rehabilitation settings. *Psychiatr. Pol.* **2015**, *49*, 835–846. [CrossRef] [PubMed]
54. Jamshidzad, M.; Maghsoudipour, M.; Zakerian, S.A.; Bakhshi, E.; Coh, P. Impact of music type on motor coordination task performance among introverted and extroverted students. *Int. J. Occup. Saf. Ergon.* **2020**, *26*, 444–449. [CrossRef] [PubMed]
55. Münte, T.F.; Altenmüller, E.; Jäncke, L. The musician's brain as a model of neuroplasticity. *Nat. Rev. Neurosci.* **2002**, *3*, 473–478. [CrossRef]
56. Bharathi, G.; Jayaramayya, K.; Balasubramanian, V.; Vellingiri, B. The potential role of rhythmic entrainment and music therapy intervention for individuals with autism spectrum disorders. *J. Exerc. Rehabil.* **2019**, *15*, 180–186. [CrossRef]
57. Winges, S.A.; Furuya, S.; Faber, N.J.; Flanders, M. Patterns of muscle activity for digital coarticulation. *J. Neurophysiol.* **2013**, *110*, 230–242. [CrossRef]
58. Setzler, M.; Goldstone, R. Coordination and Consonance Between Interacting, Improvising Musicians. *Open Mind* **2020**, *4*, 88–101. [CrossRef]
59. Demos, A.P.; Carter, D.J.; Wanderley, M.M.; Palmer, C. The Unresponsive Partner: Roles of Social Status, Auditory Feedback, and Animacy in Coordination of Joint Music Performance. *Front. Psychol.* **2017**, *8*, 149. [CrossRef]
60. Särkämö, T. Cognitive, emotional, and neural benefits of musical leisure activities in aging and neurological rehabilitation: A critical review. *Ann. Phys. Rehabil. Med.* **2018**, *61*, 414–418. [CrossRef]
61. Brancatisano, O.; Baird, A.; Thompson, W.F. Why is music therapeutic for neurological disorders? The Therapeutic Music Capacities Model. *Neurosci. Biobehav. Rev.* **2020**, *112*, 600–615. [CrossRef] [PubMed]
62. López-Ortiz, C.; Gaebler-Spira, D.J.; Mckeeman, S.N.; Mcnish, R.N.; Green, D. Dance and rehabilitation in cerebral palsy: A systematic search and review. *Dev. Med. Child Neurol.* **2019**, *61*, 393–398. [CrossRef] [PubMed]
63. Särkämö, T.; Tervaniemi, M.; Laitinen, S.; Numminen, A.; Kurki, M.; Johnson, J.K.; Rantanen, P. Cognitive, Emotional, and Social Benefits of Regular Musical Activities in Early Dementia: Randomized Controlled Study. *Gerontologist* **2014**, *54*, 634–650. [CrossRef] [PubMed]
64. Klang Ibragimova, N.; Pless, M.; Adolfsson, M.; Granlund, M.; Björck-Åkesson, E. Using content analysis to link texts on assessment and intervention to the International Classification of Functioning, Disability and Health-version for Children and Youth (ICF-CY). *J. Rehabil. Med.* **2011**, *43*, 728–733. [CrossRef] [PubMed]
65. Koelsch, S.; Jäncke, L. Music and the heart. *Eur. Heart J.* **2015**, *36*, 3043–3049. [CrossRef]
66. Vuilleumier, P.; Trost, W. Music and emotions: From enchantment to entrainment: Musical emotions and entrainment. *Ann. N. Y. Acad. Sci.* **2015**, *1337*, 212–222. [CrossRef]
67. Thaut, M.H. The discovery of human auditory–motor entrainment and its role in the development of neurologic music therapy. In *Progress in Brain Research*; Elsevier: Amsterdam, The Netherlands, 2015; Volume 217, pp. 253–266, ISBN 978-0-444-63551-8.
68. Thaut, M.H.; McIntosh, G.C.; Hoemberg, V. Neurobiological foundations of neurologic music therapy: Rhythmic entrainment and the motor system. *Front. Psychol.* **2015**, *5*, 1185. [CrossRef]
69. Moens, B.; Leman, M. Alignment strategies for the entrainment of music and movement rhythms: Alignment of music and movement rhythms. *Ann. N. Y. Acad. Sci.* **2015**, *1337*, 86–93. [CrossRef]
70. Wollman, I.; Arias, P.; Aucouturier, J.-J.; Morillon, B. Neural entrainment to music is sensitive to melodic spectral complexity. *J. Neurophysiol.* **2020**, *123*, 1063–1071. [CrossRef]
71. Lundy, T.; James, C.; Reybrouck, M. Editorial: Neuroscience and new music: Assessing behavioral and cerebral aspects of its perception, cognition, induction, and entrainment. *Front. Psychol.* **2022**, *13*, 979570. [CrossRef]
72. Smith, S.M. Background Music and Context-Dependent Memory. *Am. J. Psychol.* **1985**, *98*, 591. [CrossRef]
73. Balch, W.R.; Bowman, K.; Mohler, L.A. Music-dependent memory in immediate and delayed word recall. *Mem. Cogn.* **1992**, *20*, 21–28. [CrossRef] [PubMed]
74. Kang, H.J.; Williamson, V.J. Background music can aid second language learning. *Psychol. Music.* **2014**, *42*, 728–747. [CrossRef]
75. Racette, A. Making non-fluent aphasics speak: Sing along! *Brain* **2006**, *129*, 2571–2584. [CrossRef] [PubMed]
76. Moussard, A.; Bigand, E.; Belleville, S.; Peretz, I. Music as an Aid to Learn New Verbal Information in Alzheimer's Disease. *Music. Percept.* **2012**, *29*, 521–531. [CrossRef]

77. El Haj, M.; Omigie, D.; Clément, S. Music causes deterioration of source memory: Evidence from normal ageing. *Q. J. Exp. Psychol.* **2014**, *67*, 2381–2391. [CrossRef]
78. Ferreri, L.; Bigand, E.; Perrey, S.; Muthalib, M.; Bard, P.; Bugaiska, A. Less Effort, Better Results: How Does Music Act on Prefrontal Cortex in Older Adults during Verbal Encoding? An fNIRS Study. *Front. Hum. Neurosci.* **2014**, *8*, 301. [CrossRef]
79. Ferreri, L.; Bigand, E.; Bugaiska, A. The positive effect of music on source memory. *Musicae Sci.* **2015**, *19*, 402–411. [CrossRef]
80. Eschrich, S.; Münte, T.F.; Altenmüller, E.O. Unforgettable film music: The role of emotion in episodic long-term memory for music. *BMC Neurosci.* **2008**, *9*, 48. [CrossRef]
81. Adcock, R.A.; Thangavel, A.; Whitfield-Gabrieli, S.; Knutson, B.; Gabrieli, J.D.E. Reward-Motivated Learning: Mesolimbic Activation Precedes Memory Formation. *Neuron* **2006**, *50*, 507–517. [CrossRef]
82. Barrett, F.S.; Grimm, K.J.; Robins, R.W.; Wildschut, T.; Sedikides, C.; Janata, P. Music-evoked nostalgia: Affect, memory, and personality. *Emotion* **2010**, *10*, 390–403. [CrossRef] [PubMed]
83. Janata, P.; Tomic, S.T.; Rakowski, S.K. Characterisation of music-evoked autobiographical memories. *Memory* **2007**, *15*, 845–860. [CrossRef] [PubMed]
84. Ford, J.H.; Addis, D.R.; Giovanello, K.S. Differential neural activity during search of specific and general autobiographical memories elicited by musical cues. *Neuropsychologia* **2011**, *49*, 2514–2526. [CrossRef] [PubMed]
85. Foster, N.A.; Valentine, E.R. The Effect of Auditory Stimulation on Autobiographical Recall in Dementia. *Exp. Aging Res.* **2001**, *27*, 215–228. [CrossRef] [PubMed]
86. Irish, M.; Cunningham, C.J.; Walsh, J.B.; Coakley, D.; Lawlor, B.A.; Robertson, I.H.; Coen, R.F. Investigating the Enhancing Effect of Music on Autobiographical Memory in Mild Alzheimer's Disease. *Dement. Geriatr. Cogn. Disord.* **2006**, *22*, 108–120. [CrossRef] [PubMed]
87. Belfi, A.M.; Karlan, B.; Tranel, D. Music evokes vivid autobiographical memories. *Memory* **2016**, *24*, 979–989. [CrossRef]
88. Piefke, M.; Weiss, P.H.; Markowitsch, H.J.; Fink, G.R. Gender differences in the functional neuroanatomy of emotional episodic autobiographical memory. *Hum. Brain Mapp.* **2005**, *24*, 313–324. [CrossRef]
89. Levine, B.; Svoboda, E.; Hay, J.F.; Winocur, G.; Moscovitch, M. Aging and autobiographical memory: Dissociating episodic from semantic retrieval. *Psychol. Aging* **2002**, *17*, 677–689. [CrossRef]
90. Langheim, F. Cortical Systems Associated with Covert Music Rehearsal. *NeuroImage* **2002**, *16*, 901–908. [CrossRef]
91. Rao, S.M.; Harrington, D.L.; Haaland, K.Y.; Bobholz, J.A.; Cox, R.W.; Binder, J.R. Distributed Neural Systems Underlying the Timing of Movements. *J. Neurosci.* **1997**, *17*, 5528–5535. [CrossRef]
92. Carter, C.S.; Mintun, M.; Cohen, J.D. Interference and Facilitation Effects during Selective Attention: An H215O PET Study of Stroop Task Performance. *NeuroImage* **1995**, *2*, 264–272. [CrossRef] [PubMed]
93. Nirkko, A.C.; Baader, A.P.; Loevblad, K.-O.; Milani, P.; Wiesendanger, M. Cortical representation of music production in violin players: Behavioral assessment and functional imaging of finger sequencing, bimanual coordination and music specific brain activation. *NeuroImage* **2000**, *11*, S106. [CrossRef]
94. Griffiths, T.D. The Neural Processing of Complex Sounds. *Ann. N. Y. Acad. Sci.* **2001**, *930*, 133–142. [CrossRef] [PubMed]
95. Platel, H. The structural components of music perception. A functional anatomical study. *Brain* **1997**, *120*, 229–243. [CrossRef]
96. Popescu, M.; Otsuka, A.; Ioannides, A.A. Dynamics of brain activity in motor and frontal cortical areas during music listening: A magnetoencephalographic study. *NeuroImage* **2004**, *21*, 1622–1638. [CrossRef]
97. Hughes, H.C.; Darcey, T.M.; Barkan, H.I.; Williamson, P.D.; Roberts, D.W.; Aslin, C.H. Responses of Human Auditory Association Cortex to the Omission of an Expected Acoustic Event. *NeuroImage* **2001**, *13*, 1073–1089. [CrossRef]
98. Nakamura, S.; Sadato, N.; Oohashi, T.; Nishina, E.; Fuwamoto, Y.; Yonekura, Y. Analysis of music–brain interaction with simultaneous measurement of regional cerebral blood flow and electroencephalogram beta rhythm in human subjects. *Neurosci. Lett.* **1999**, *275*, 222–226. [CrossRef]
99. Moumdjian, L.; Sarkamo, T.; Leone, C.; Leman, M.; Feys, P. Effectiveness of music-based interventions on motricity or cognitive functioning in neurological populations: A systematic review. *Eur. J. Phys. Rehabil. Med.* **2017**, *53*, 466–482. [CrossRef]
100. Wang, X.P.; Wang, X.H.; Bao, Y.C.; Zhou, J.N. Apolipoprotein E genotypes in Chinese patients with Wilson's disease. *QJM* **2003**, *96*, 541–542. [CrossRef]
101. Skidmore, E.R.; Whyte, E.M.; Holm, M.B.; Becker, J.T.; Butters, M.A.; Dew, M.A.; Munin, M.C.; Lenze, E.J. Cognitive and Affective Predictors of Rehabilitation Participation After Stroke. *Arch. Phys. Med. Rehabil.* **2010**, *91*, 203–207. [CrossRef]
102. Kim, G.Y.; Han, M.R.; Lee, H.G. Effect of Dual-task Rehabilitative Training on Cognitive and Motor Function of Stroke Patients. *J. Phys. Ther. Sci* **2014**, *26*, 1–6. [CrossRef]
103. Sosnoff, J.J.; Sung, J. Reducing falls and improving mobility in multiple sclerosis. *Expert Rev. Neurother.* **2015**, *15*, 655–666. [CrossRef] [PubMed]
104. Thompson, W.F.; Schlaug, G. The healing power of music. *Sci. Am. Mind* **2015**, *26*, 32–41. [CrossRef]
105. Schellenberg, E.G. Examining the association between music lessons and intelligence: Music lessons and intelligence. *Br. J. Psychol.* **2011**, *102*, 283–302. [CrossRef] [PubMed]
106. Schellenberg, E.G.; Mankarious, M. Music training and emotion comprehension in childhood. *Emotion* **2012**, *12*, 887–891. [CrossRef] [PubMed]
107. Bergman Nutley, S.; Darki, F.; Klingberg, T. Music practice is associated with development of working memory during childhood and adolescence. *Front. Hum. Neurosci.* **2014**, *7*, 926. [CrossRef]

108. Forgeard, M.; Winner, E.; Norton, A.; Schlaug, G. Practicing a Musical Instrument in Childhood is Associated with Enhanced Verbal Ability and Nonverbal Reasoning. *PLoS ONE* **2008**, *3*, e3566. [CrossRef]
109. Swaminathan, S.; Schellenberg, E.G.; Khalil, S. Revisiting the association between music lessons and intelligence: Training effects or music aptitude? *Intelligence* **2017**, *62*, 119–124. [CrossRef]
110. Corrigall, K.A.; Schellenberg, E.G.; Misura, N.M. Music Training, Cognition, and Personality. *Front. Psychol.* **2013**, *4*, 222. [CrossRef]
111. Linnavalli, T.; Putkinen, V.; Lipsanen, J.; Huotilainen, M.; Tervaniemi, M. Music playschool enhances children's linguistic skills. *Sci. Rep.* **2018**, *8*, 8767. [CrossRef]
112. Mary, L.; Yegnanarayana, B. Extraction and representation of prosodic features for language and speaker recognition. *Speech Commun.* **2008**, *50*, 782–796. [CrossRef]
113. Magnuson, J.S.; Nusbaum, H.C. Acoustic differences, listener expectations, and the perceptual accommodation of talker variability. *J. Exp. Psychol. Hum. Percept. Perform.* **2007**, *33*, 391–409. [CrossRef]
114. Foxton, J.M.; Talcott, J.B.; Witton, C.; Brace, H.; McIntyre, F.; Griffiths, T.D. Reading skills are related to global, but not local, acoustic pattern perception. *Nat. Neurosci.* **2003**, *6*, 343–344. [CrossRef] [PubMed]
115. Bidelman, G.M.; Hutka, S.; Moreno, S. Tone Language Speakers and Musicians Share Enhanced Perceptual and Cognitive Abilities for Musical Pitch: Evidence for Bidirectionality between the Domains of Language and Music. *PLoS ONE* **2013**, *8*, e60676. [CrossRef] [PubMed]
116. Delogu, F.; Lampis, G.; Belardinelli, M.O. From melody to lexical tone: Musical ability enhances specific aspects of foreign language perception. *Eur. J. Cogn. Psychol.* **2010**, *22*, 46–61. [CrossRef]
117. Wong, P.C.M.; Perrachione, T.K. Learning pitch patterns in lexical identification by native English-speaking adults. *Appl. Psycholinguist.* **2007**, *28*, 565–585. [CrossRef]
118. Xie, X.; Myers, E. The impact of musical training and tone language experience on talker identification. *J. Acoust. Soc. Am.* **2015**, *137*, 419–432. [CrossRef]
119. Sparks, R.W.; Holland, A.L. Method: Melodic Intonation Therapy for Aphasia. *J. Speech Hear. Disord.* **1976**, *41*, 287–297. [CrossRef] [PubMed]
120. Norton, A.; Zipse, L.; Marchina, S.; Schlaug, G. Melodic Intonation Therapy: Shared Insights on How It Is Done and Why It Might Help. *Ann. N. Y. Acad. Sci.* **2009**, *1169*, 431–436. [CrossRef]
121. Gentilucci, M.; Volta, R.D. Spoken Language and arm Gestures are Controlled by the same Motor Control System. *Q. J. Exp. Psychol.* **2008**, *61*, 944–957. [CrossRef] [PubMed]
122. Lahav, A.; Saltzman, E.; Schlaug, G. Action Representation of Sound: Audiomotor Recognition Network While Listening to Newly Acquired Actions. *J. Neurosci.* **2007**, *27*, 308–314. [CrossRef]
123. Gaser, C.; Schlaug, G. Brain Structures Differ between Musicians and Non-Musicians. *J. Neurosci.* **2003**, *23*, 9240–9245. [CrossRef] [PubMed]
124. Schlaug, G. Musicians and music making as a model for the study of brain plasticity. In *Progress in Brain Research*; Elsevier: Amsterdam, The Netherlands, 2015; Volume 217, pp. 37–55, ISBN 978-0-444-63551-8.
125. Sorati, M.; Behne, D.M. Audiovisual Modulation in Music Perception for Musicians and Non-musicians. *Front. Psychol.* **2020**, *11*, 1094. [CrossRef] [PubMed]
126. Harvey, A.R. Links Between the Neurobiology of Oxytocin and Human Musicality. *Front. Hum. Neurosci.* **2020**, *14*, 350. [CrossRef] [PubMed]
127. Matziorinis, A.M.; Koelsch, S. The promise of music therapy for Alzheimer's disease: A review. *Ann. N. Y. Acad. Sci.* **2022**, *1516*, 11–17. [CrossRef]
128. Degé, F. Music Lessons and Cognitive Abilities in Children: How Far Transfer Could Be Possible. *Front. Psychol.* **2021**, *11*, 557807. [CrossRef]
129. Huang, W.-H.; Dou, Z.-L.; Jin, H.-M.; Cui, Y.; Li, X.; Zeng, Q. The Effectiveness of Music Therapy on Hand Function in Patients With Stroke: A Systematic Review of Randomized Controlled Trials. *Front. Neurol.* **2021**, *12*, 641023. [CrossRef]
130. Ripollés, P.; Rojo, N.; Grau-Sánchez, J.; Amengual, J.L.; Càmara, E.; Marco-Pallarés, J.; Juncadella, M.; Vaquero, L.; Rubio, F.; Duarte, E.; et al. Music supported therapy promotes motor plasticity in individuals with chronic stroke. *Brain Imaging Behav.* **2016**, *10*, 1289–1307. [CrossRef]
131. Van Der Meulen, I.; Van De Sandt-Koenderman, M.W.M.E.; Heijenbrok, M.H.; Visch-Brink, E.; Ribbers, G.M. Melodic Intonation Therapy in Chronic Aphasia: Evidence from a Pilot Randomized Controlled Trial. *Front. Hum. Neurosci.* **2016**, *10*, 533. [CrossRef]
132. Bartík, P.; Vostrý, M.; Hudáková, Z.; Šagát, P.; Lesňáková, A.; Dukát, A. The Effect of Early Applied Robot-Assisted Physiotherapy on Functional Independence Measure Score in Post-Myocardial Infarction Patients. *Healthcare* **2022**, *10*, 937. [CrossRef]
133. Vostrý, M.; Zilcher, L. Combined Therapy for Patients after Ischemic Stroke from the Point View of Comprehensive Rehabilitation. *J. Educ. Cult. Soc.* **2020**, *11*, 119–125. [CrossRef]

Journal of
Clinical Medicine

MDPI

Review

Neuropsychiatric and Neuropsychological Aspects of Alcohol-Related Cognitive Disorders: An In-Depth Review of Wernicke's Encephalopathy and Korsakoff's Syndrome

Lucian Eva [1,2], Felix-Mircea Brehar [3,4,*], Ioan-Alexandru Florian [5,*], Razvan-Adrian Covache-Busuioc [4], Horia Petre Costin [4], David-Ioan Dumitrascu [4], Bogdan-Gabriel Bratu [4], Luca-Andrei Glavan [4] and Alexandru Vlad Ciurea [4,6]

1 Faculty of Medicine, Dunarea de Jos University, 800010 Galati, Romania; elucian73@yahoo.com
2 Department of Neurosurgery, Clinical Emergency Hospital "Prof. Dr. Nicolae Oblu", 700309 Iasi, Romania
3 Department of Neurosurgery, Clinical Emergency Hospital "Bagdasar-Arseni", 041915 Bucharest, Romania
4 Department of Neurosurgery, "Carol Davila" University of Medicine and Pharmacy, 020021 Bucharest, Romania; razvan-adrian.covache-busuioc0720@stud.umfcd.ro (R.-A.C.-B.); horia-petre.costin0720@stud.umfcd.ro (H.P.C.); david-ioan.dumitrascu0720@stud.umfcd.ro (D.-I.D.); bogdan.bratu@stud.umfcd.ro (B.-G.B.); luca-andrei.glavan0720@stud.umfcd.ro (L.-A.G.); prof.avciurea@gmail.com (A.V.C.)
5 Department of Neurosciences, "Iuliu Hatieganu" University of Medicine and Pharmacy, 400012 Cluj-Napoca, Romania
6 Neurosurgery Department, Sanador Clinical Hospital, 010991 Bucharest, Romania
* Correspondence: felixbrehar@yahoo.com (F.-M.B.); florian.ioan.alexandru@gmail.com (I.-A.F.)

Abstract: Alcohol-related cognitive disorders have long been an area of study, yet they continue to pose challenges in the diagnosis, treatment, and understanding of underlying neuropsychiatric mechanisms. The present article offers a comprehensive review of Wernicke's Encephalopathy and Korsakoff's Syndrome, two conditions often seen on a continuum of alcohol-related brain damage. Drawing on current medical literature, neuroimaging studies, and clinical case reports, we explore the neuropsychiatric and neuropsychological profiles, symptomatology, and differential diagnoses of these disorders. We delve into the biochemical pathways implicated in the development of WE and KS, notably thiamine deficiency and its impact on neurotransmitter systems and neural networks. The article also addresses the challenges in early diagnosis, often complicated by non-specific symptoms and co-occurring psychiatric conditions. Furthermore, we review the current state of treatment protocols, including pharmacological and non-pharmacological interventions. Finally, the article highlights gaps in current knowledge and suggests directions for future research to improve diagnosis, treatment, and patient outcomes. Understanding the nuanced interplay between the neuropsychiatric and neuropsychological aspects of WE and KS is crucial for both clinicians and researchers alike, in order to provide effective treatment and to advance our understanding of these complex conditions.

Keywords: Wernicke's Encephalopathy; Korsakoff's Syndrome; alcohol-related cognitive disorders; neuropsychiatric symptoms; neuroimaging research

Citation: Eva, L.; Brehar, F.-M.; Florian, I.-A.; Covache-Busuioc, R.-A.; Costin, H.P.; Dumitrascu, D.-I.; Bratu, B.-G.; Glavan, L.-A.; Ciurea, A.V. Neuropsychiatric and Neuropsychological Aspects of Alcohol-Related Cognitive Disorders: An In-Depth Review of Wernicke's Encephalopathy and Korsakoff's Syndrome. *J. Clin. Med.* **2023**, *12*, 6101. https://doi.org/10.3390/jcm12186101

Academic Editor: Roy P.C. Kessels

Received: 14 August 2023
Revised: 16 September 2023
Accepted: 18 September 2023
Published: 21 September 2023

1. Introduction

1.1. Wernicke's Encephalopathy (WE) and Korsakoff's Syndrome (KS): Brief History and Linkage

Carl Wernicke first described WE, now named in his honor. This condition featured symptoms like lethargy, ophthalmoplegia, ataxia and cognitive impairment. At roughly the same time, but independently from Wernicke's, work Sergei Korsakoff presented his doctoral thesis entitled "Alcoholic Paralysis", discussing a particular form of memory loss found among chronic alcoholism cases known as circumscribed amnesia.

Over fifty years later, it took an unusually long time for researchers to uncover a shared cause between these two disorders: thiamine deficiency was identified as being at the core of both conditions, contributing to their symptoms in both WE and KS and becoming recognized as their link.

Now we understand more fully the temporal relationship between WE and KS, with two distinct phases:

1. WE typically presents with acute/phase lesions to areas such as the periventricular nuclei, thalami and components of Papez's circuit (including mammillary bodies).
2. Korsakoff's Syndrome progresses into its chronic phase with lesions developing into more permanent bilateral damage that causes global amnesia.

Noting this correlation, the combined condition is often referred to as Wernicke-Korsakoff Syndrome (WKS). Amnesia played an instrumental role in connecting WKS to neuropsychology; further exploration into memory processes as well as identification of distinct neural substrates responsible for different aspects of memory function ensued from this linkage [1].

KS often develops in those who have experienced WE but failed to receive appropriate thiamine replacement therapy promptly and appropriately. The most recognizable feature of KS is an amnesia which may be profound; combined with additional cognitive and behavioral impairments that often appear with more serious cases, its effects can significantly impact an individual's daily functioning and quality of life [2]. Research endeavors examining the scope, arrangement and essence of episodic memory impairments within KS have significantly advanced our understanding of memory as an abstract concept. Furthermore, these studies have shed light on memory being not an all-inclusive function, and have demonstrated its multidimensional nature. Furthermore, examination of KS has highlighted diencephalic structures' vital importance to memory processes—leading researchers into further exploring distinct brain structures or neural pathways responsible for individual memory processes to create more intricate mnemonic components [3].

1.2. Connection to Alcohol-Related Cognitive Disorders

KS and WE are two distinct neurological conditions; however, they frequently share similarities due to a deficiency of thiamine (vitamin B1) which often stems from chronic alcohol misuse. Current diagnostic frameworks for cognitive impairment related to alcohol consumption can be divided into two primary syndromes: Wernicke-Korsakoff syndrome (WKS) and alcohol/related dementia (ARD) [4].

WE can be identified clinically by three distinct symptoms, including changes to mental status, dysfunction in oculomotor capabilities and issues with cerebellar function. Note, however, that not all patients exhibit all three symptoms at once. According to Western studies, up to 90% of WE cases can be linked to alcohol misuse [5]. Left undiagnosed or untreated, WE can progress into KS, which occurs in 56–84% of those with an alcoholism history and possibly less frequently among cases unrelated to alcohol consumption. The main cognitive manifestation of KS is intense amnesia, characterized by difficulties forming new memories (anterograde) and recalling old ones (retrograde), although it can also produce other cognitive and behavioral symptoms [6]. Alcohol-related dementia (ARD), as an identifiable clinical entity, has generated much discussion, due to the uncertainty regarding its causes, lack of diagnostic criteria and difficulties assessing affected populations. Due to this lack of clarity regarding ARD classification as a distinct clinical entity, some scholars prefer the term 'alcohol-related brain damage' which encompasses neurocognitive impairments caused by chronic alcohol consumption such as Wernicke-Korsakoff Syndrome and ARD [4].

Uncertainty remains as to whether alcohol-related dementia (ARD) results directly from neurotoxic effects of alcohol consumption (neurotoxicity hypothesis), alternative pathologies like thiamine deficiency or from multiple interlinked factors. Oslin et al. have developed preliminary diagnostic criteria for ARD. Their five-year period with men drinking an average of 35 standard drinks weekly (30 for women) over this five-year period should be enough to induce this disorder [7].

Diagnosing alcohol-related dementia (ARD) can be complex, due to other related conditions, including head trauma, substance abuse, co-occurring psychiatric disorders, and elevated cardiovascular risk factors. Although neuropsychological characteristics of ARD have received limited study, observations indicate a more comprehensive cognitive decline compared to WKS [4].

This comprehensive review covers many facets of alcohol-related cognitive disorders. These aspects include diagnostic criteria, the cognitive aftermath and the neuropathology underlying these disorders, and concurrent physical health conditions and trends in epidemiology, as well as available treatment approaches, the impact on quality of life for affected individuals, mental capacity considerations, and the role neuroimaging plays in both diagnosis and management.

This comprehensive article's overall aim is to serve as an indispensable reference for researchers, healthcare practitioners, and clinicians. Its objective is to illuminate various dimensions of WE and Korsakoff's Syndrome so as to advance early-detection practices, enhance patient outcomes, and raise awareness regarding these debilitating conditions.

2. Assessment and Diagnostic Work-Up

1. Criteria for Diagnosing WE: Diagnosing WE can be challenging, given the various ways it manifests clinically; however, the latest diagnostic criteria suggest at least two of four signs should be present—diet inadequacy, abnormalities in oculomotor function abnormalities, deficits in cerebellar function and altered mental state or mild impairment of memory are indicators. While not all patients will necessarily exhibit all these classic signs, an elevated sense of suspicion is necessary in order to accurately confirm diagnosis [8].

Caine et al. [8] recently developed operational criteria for the clinical diagnosis of WE called the Caine criteria (Table 1). While these criteria could greatly facilitate diagnosis, histopathological evidence still forms the cornerstone of WE diagnosis [9].

2. Criteria for Diagnosing Korsakoff's Syndrome (KS): Recognizing Korsakoff's Syndrome, often the result of chronic thiamine deficiency or WE, requires the identification of certain markers. These include both anterograde (inability to form new memories) and retrograde amnesia, with difficulty recalling past memories; anterograde amnesia is often accompanied by confabulations (fabricated or altered memories used to fill memory gaps), which cannot be attributed to any other medical cause; furthermore patients might exhibit executive dysfunction or have limited awareness regarding memory impairments affecting them.

At present, there is a lack of consensus on a universal definition or diagnostic criteria for Korsakoff's Syndrome (KS) [10]. According to the *Diagnostic and Statistical Manual of Mental Disorders, Fifth Edition (DSM-5)*, KS is characterized as "an alcohol-induced major neurocognitive disorder with amnestic confabulatory features" [11]. This current classification is problematic, since it erroneously presumes that the etiology of KS is exclusively alcohol-related, thereby limiting its diagnostic scope since it fails to establish a direct relationship with Wernicke's Encephalopathy (WE). *The International Classification of Diseases, Tenth Edition (ICD-10)*, offers a separate categorization for alcoholic and non-alcoholic forms of KS, but this further complicates the diagnostic landscape. Non-alcoholic forms are categorized under organic mental disorders, whereas those induced by alcohol abuse fall under the F10.26 code category, thereby rendering the application of ICD-10 criteria more cumbersome [12].

One objective of the present research was to formulate an exhaustive definition and outline potential diagnostic criteria for KS [2].

Table 1. The operational criteria for the diagnosis of WE, as elaborated by Caine et al [8].

Symptom or Sign	As Evidenced by One or More of the Following
Nutritional deficiencies	– Undernutrition
	– A record of significantly impaired nutritional intake
	– An unusual thiamine status
Oculomotor impairments	– Nystagmus
	– Gaze palsy
	– Ophthalmoplegia
Cerebellar impairment	– Abnormalities of past pointing
	– Instability or ataxia
	– Dysdiadokokinesia
	– Impaired heel–shin testing
Either an altered mental state	– Disorientation in two of three fields
	– Confused
	– An abnormal digit span
	– Comatose
or	or
Mild memory dysfunction	– Failure to remember two or more words in the four-item memory test
	– Impairment on more elaborate neuropsychological tests of memory function

Notes: When two out of these four criteria apply, the clinical diagnosis of WE is made. Abbreviation: WE, Wernicke's encephalopathy.

2.1. Neuroimaging

Early neuroradiological studies of WE from the 1970s relied on computed tomography (CT) to identify third ventricle enlargement, but struggled with limitations in detecting edema or localized damage [13,14]. The advent of magnetic resonance imaging (MRI) revolutionized this field, offering higher sensitivity to tissue-water content [15]. MRI was more effective in diagnosing WE, identifying abnormalities in 7 of 15 studied WE patients compared to just 2 of 15 through CT. It also highlighted further affected regions like the periaqueductal gray matter and mammillary bodies [16]. MRI's sensitivity thus underscores its importance in detecting subtle neuropathological changes, especially in alcohol abusers without overt symptoms of WE.

Initial MR techniques for enhancing the visibility of edematous lesions in WE focused on T2-weighted late-echo sequences. Bilateral hyperintensity in areas like mammillary bodies and thalamic nuclei is a key neuroradiological feature of acute WE, observed in both alcoholic and nonalcoholic cases [17,18]. These MRI findings are consistent with post-mortem reports [19,20]. Neuropathology affecting both brain hemispheres could exacerbate clinical symptoms. Although less common in WE, a specific MR study noted excess fluid in the central pons of patients with Wernicke-Korsakoff Syndrome (WKS), correlating with markers like macrocytic anemia and cognitive fluency in non-WKS alcoholic patients [21].

The FLAIR (MR fluid-attenuated inversion recovery) sequence marked a significant technological leap from traditional T2 methods by incorporating additional T1 contrast mechanisms. FLAIR effectively suppresses signals from cerebrospinal fluid (CSF), enhancing the visibility of edematous tissue regions [22]. One case study involving hyperemesis gravidarum documented high signal intensity in the mammillary bodies and hypothalamus, which subsided following thiamine treatment [23]. FLAIR imaging of nonalcoholic WE cases showed hyperintense signals in specific brain regions. Follow-up assessments indicated recovery in cases without prior cortical damage, while no improvement was observed in those with existing damage [17].

Magnetic resonance diffusion-weighted imaging (DWI) has shown high accuracy in detecting brain pathology related to WE. Interestingly, what would intuitively appear as high diffusivity levels actually show up as hyperintense on DWI due to a phenomenon known as the T2 shine-through effect [24]. This counterintuitive brightness represents low, rather than high, diffusivity levels.

Unlu et al. (2006) found that WE often features abnormalities in periventricular and thalamic tissues [25]. A case study showed that a distinct DWI bright signal originated from abnormally low diffusivity in the cerebellum, which improved with thiamine treatment but left lingering motor impairment [26]. Acute WE studies consistently reveal increased DWI signal intensity and decreased diffusivity as captured via apparent diffusion coefficient (ADC) images [26]. Although DWI signals can be affected by the T2 shine-through effect, combining DWI with ADC imaging allows for a more comprehensive characterization of WE lesions as they evolve from high- to low-diffusivity states.

2.2. Biomarkers

Alcohol-related health complications are widespread, yet frequently go undetected. Precise screening for alcohol misuse hinges on tests that can both sensitively and specifically spot those engaging in dangerous drinking habits and track their behavior. By coupling laboratory indicators like AST, ALT, MCV, and GGT with self-reported data, it becomes feasible to identify and monitor those engaging in harmful drinking patterns [27].

While much research on alcohol dependence has primarily centered on men, studies tailored to the unique experiences of women underscore their heightened susceptibility. This vulnerability stems from differences in physiology, metabolism, hormonal interactions, and additional concerns during pregnancy [28].

It is worth spotlighting several key alcohol biomarkers:

1. Blood alcohol concentration (BAC): Determined from samples of blood, breath, saliva, or urine, BAC can confirm alcohol intoxication. Notably, men and women often display significant variances in their BAC levels.
2. Gamma-glutamyltransferase (GGT): A spike in GGT signals potential early-stage liver damage from alcohol. It is a useful marker for alcohol-triggered harm, although it might not be as effective in discerning dangerous drinking habits in younger individuals and women.
3. Transaminases AST and ALT: A rise in ALT points to liver damage. The AST/ALT ratio serves as a tool for identifying alcohol-related liver diseases, though its accuracy can waver across different age segments.
4. Mean corpuscular volume (MCV): a heightened MCV can be an indicator of consistent excessive drinking, particularly in women.
5. Carbohydrate-deficient transferrin (CDT): this marker is adept at indicating heavy alcohol consumption but may not be as effective for women, especially those expecting.
6. Fatty acid ethyl esters (FAEEs): they efficiently differentiate between casual and excessive drinkers and can pinpoint alcohol consumption during pregnancy.
7. Ethyl glucuronide (EtG): this marker excels at detecting excessive drinking instances, even if other traditional indicators might miss them.
8. Phosphatidylethanol (PEth): a highly specific and sensitive sign of prolonged alcohol consumption, PEth can be detected in the blood for up to two weeks post drinking.
9. Thiamine and its esters: Chronic alcohol users often have a thiamine deficiency. By directly measuring thiamine and its esters in red blood cells, this deficiency can be gauged.

Leveraging these biomarkers can offer invaluable data on alcohol misuse, subsequently facilitating diagnosis and treatment, especially for women and during their pregnancies.

Even though many of these markers are used to determine the health of the liver in general, they should be taken into consideration when trying to diagnose WE or KS.

Markers like CDT, EtG, FAEEs and Peth should also be kept in mind when trying to diagnose these two diseases, being of great value in interpreting the etiology of the liver damage, and thus detecting alcoholism.

To summarize, pinpointing conditions like WE and KS involves a comprehensive evaluation of clinical symptoms, cognitive tests, and neuroimaging to detect the brain anomalies tied to thiamine deficiency. The relentless pursuit of dependable biomarkers and innovative tools is poised to significantly boost early detection and management strategies for these intricate neurological conditions [29].

3. Neuropsychiatric and Cognitive Sequelae

As previously highlighted, WE and KS are typically perceived as being interrelated, since both originate from a deficiency in thiamine. Consequently, medical professionals should simultaneously assess both conditions. While WE is conventionally associated with a combination of cognitive disturbances, ophthalmoplegia, and ataxia, it is worth noting that less than a fifth of patients will manifest all these symptoms [30].

WE patients frequently exhibit impairments in neuropsychological functioning that become increasingly evident as their physical condition improves. Neuropsychological functioning encompasses various cognitive abilities ranging from fundamental processes like attention and concentration to more intricate ones like memory, executive functioning, and reasoning; all these higher-order cognitive processes play a key role in overall quality of life for an individual, although unfortunately clinicians sometimes overlook this assessment [31].

When delving deeper into KS, in contrast to Wernicke's encephalopathy, specific cognitive deficiencies surface. These encompass anterograde amnesia, which severely hampers the ability to assimilate new information, and retrograde amnesia, as well as executive function disorders leading to reduced self-restraint and challenges in areas like judgment, strategizing, and problem resolution [32]. A key manifestation of anterograde amnesia is confabulation, where individuals unconsciously fill memory gaps with fabricated details [33].

Assessing memory function—an essential factor for maintaining quality of life (QOL)—was conducted via PGI-BBD evaluation once the physical condition had improved. Korsakoff memory impairments were evident immediately upon assessment, correlating with initial assessments; however, cases where assessments could take place post improvement proved challenging, due to symptoms overlapping globally with memory dysfunction and memory deficits; pinpointing exactly when cognitive confusion recedes, paving the way for memory deficits, was often difficult and required in-depth discussion amongst participants.

Memory dysfunction was so debilitating in these initial cases that these individuals experienced difficulty maintaining an acceptable quality of life (QOL), accompanied by noticeable occupational impairments. Early administration of thiamine led to improvements in ataxia and ophthalmoplegia; however, confusion and neurological dysfunction persisted for an extended period, according to neuropsychological assessments [34], even when confusion eventually resolved itself in later phases; deficits in memory function, as well as the capacity for learning new material, persisted for an extended timeframe.

In KS, while anterograde amnesia stands out as a major symptom, the patient's remote memory is also impacted. This leads to retrograde amnesia that affects both general knowledge—such as facts from the news—and personal autobiographical events [35]. For those with KS, their memory loss can span numerous years or even multiple decades, though memories from early life, like childhood, tend to remain intact [36]. This pattern of losing more recent memories while retaining distant ones in KS and other conditions has been termed a "temporal gradient". Théodule Ribot first documented this observation in 1881, leading to its designation as Ribot's law. While many studies on KS identify a pronounced temporal gradient, some suggest an even memory impairment across all past timelines [37]. These differing findings may arise from methodological challenges or the type of memory being examined: general vs. personal [38]. Autobiographical episodic memories entail personal experiences set in specific places and times, like recalling the exact moment of meeting a significant other on a balmy evening at a poolside bar. Fewer studies focus on these autobiographical memories, due to their complex nature compared to general memories. A recent investigation by Rensen et al. [38] confirmed that the temporal gradient also influences episodic autobiographical memory, supporting earlier research findings.

WE and Korsakoff's Syndrome can have severely disabling consequences. Impaired memory and executive function make even basic daily tasks daunting for patients. This impairment often results in challenges in sustaining relationships and jobs. The hindered ability to recall new events or acquire new skills drastically affects their independence and life quality. Emotional health deteriorates as patients confront their cognitive restrictions and the loneliness stemming from social disconnect [2].

In essence, both WE and Korsakoff's Syndrome arise primarily from a thiamine deficiency, frequently linked to prolonged alcohol consumption. These conditions present with psychiatric symptoms and profound cognitive deficits, and significantly hinder daily activities and overall life satisfaction. Prompt identification, thiamine replenishment, and holistic rehabilitation are vital to counteract these conditions' effects.

4. Neuropathology of Wernicke's Encephalopathy and Korsakoff's Syndrome

Thiamine is prevalent in various parts of the body, notably in skeletal muscles, liver, heart, kidney, and brain. The balance of thiamine in the body is maintained through adequate dietary intake, absorption in the intestines, reuptake in the kidneys, and storage and release in the liver, when necessary. The brain has a safety buffer of thiamine, and noticeable neuropsychiatric symptoms only emerge when thiamine levels drop below 20% of the usual amount [39].

The primary active form of thiamine in the central nervous system is thiamine pyrophosphate (TPP). It acts as a critical component or cofactor for three significant enzymes in glucose metabolism: transketolase, pyruvate dehydrogenase, and α-ketoglutarate dehydrogenase. These enzymes participate in crucial metabolic processes like the pentose phosphate pathway, glycolysis, and the citric acid cycle. These processes produce vital molecules for functions in neurons and glial cells, such as nucleic acids, neurotransmitters, myelin, and energy-rich compounds like ATP [40]. Moreover, because of its involvement in cellular energy processes, thiamine aids in defending cells against oxidative stress. In its triphosphorylated form (TTP), thiamine is essential for nerve membrane functionality (Figure 1).

Thiamine Deficiency in Relation to Alcohol

Chronic alcohol consumption can lead to thiamine deficiency due to several reasons:

1. Alcoholics often opt for diets high in carbohydrates but low in vitamins [41].
2. Without proper supplementation, thiamine reserves are exhausted within 2–3 weeks.
3. Acute alcohol consumption can hamper thiamine absorption from the gut.
4. Alcohol's impact on renal epithelial cells results in greater thiamine loss through the kidneys.
5. Chronic alcoholic liver disease can slash the liver's thiamine storage capability by up to 73%.
6. Alcohol can reduce the enzymatic activity of thiamine pyrophosphokinase (TPK), which in turn lowers the amount of available TPP.
7. Reduced TPK activity further impedes the facilitated diffusion of thiamine into cells.
8. Alcohol can decrease the absorption of thiamine produced by colonic bacteria.
9. Many alcoholics have hypomagnesemia, a deficiency of magnesium, which is a vital cofactor in thiamine utilization.

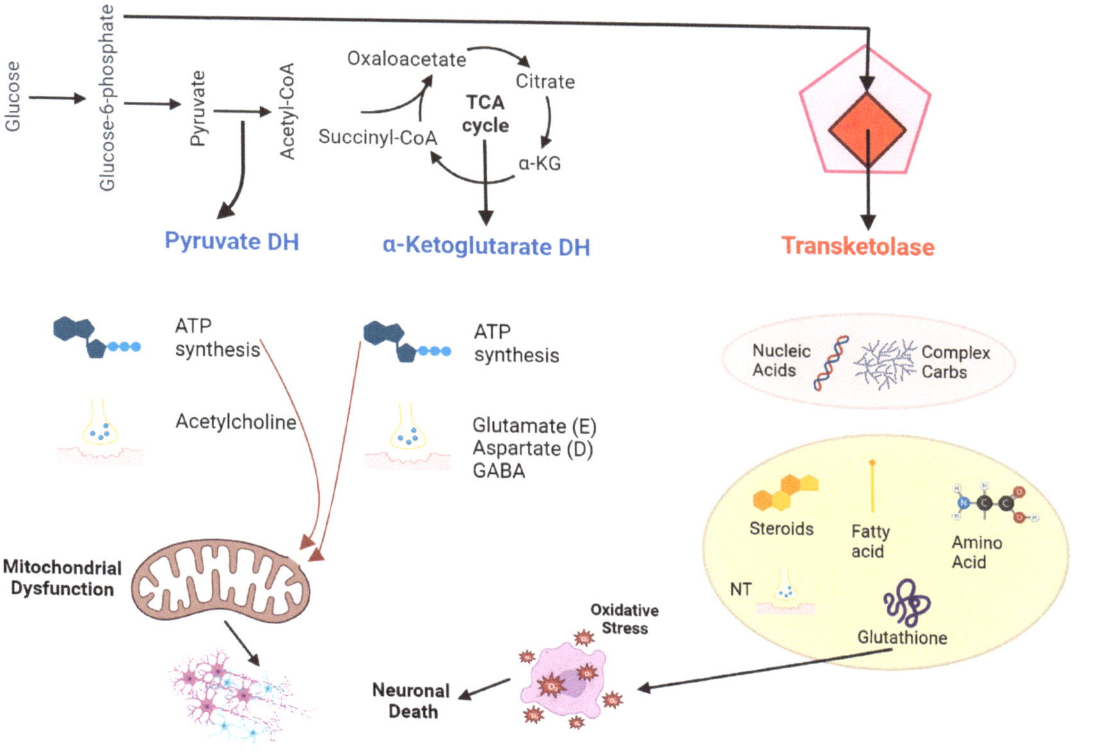

Figure 1. Pathophysiology of thiamine deficiency in Wernicke-Korsakoff syndrome.

In essence, alcohol can induce a state of thiamine deficiency through both direct and indirect mechanisms [42].

In situations where Wernicke–Korsakoff Syndrome (WKS) arises without alcohol as a factor, thiamine deficiency can be attributed to one of these four reasons: limited supply, hindered usage, increased consumption, or augmented loss of thiamine. Limited supply often happens during times of starvation, malnourishment, poor absorption, excessive losses, or vomiting. Hindered usage takes place when the body struggles to process thiamine due to diminished enzyme activity or coenzyme deficiencies. Thiamine consumption increases when there is a spike in the body's glucose metabolism, such as in instances of heightened metabolic rate, enhanced carbohydrate metabolism, or situations with swift cell regeneration. Interestingly, even when on a diet considered sufficient, thiamine deficiency can still occur under these conditions. Specifically, in the context of cancers, particularly the types associated with rapid cellular growth like leukemia and lymphoma, there is a surge in thiamine consumption [43]. Cancer can often lead to a reduced thiamine intake, either due to the disease's direct effects, loss of appetite, chemotherapy side effects, or vomiting (see Table 2) [44]. If patients are started on total parenteral nutrition (TPN) without the necessary thiamine supplementation, they can end up being deficient. Lastly, augmented thiamine loss is observed in hemodialysis, since thiamine is expelled into the dialysate [45].

Table 2. Clinical Examples of Medical Comorbidities Associated with Thiamine Deficiency, Grouped by Mechanism.

Mechanism of Thiamine Deficiency	Etiology	Clinical Examples
Accelerated usage	Hypermetabolic state	Systemic illness Infection/sepsis Thyrotoxicosis Pregnancy
	Rapid cell turnover/high cell density	Hematological malignancy Fast-growing tumor
	Excess glucose metabolism	Seizures, following rapid influx of glucose Alcohol withdrawal
Impaired utilization	Decreased enzyme activity	Malignancy Co-factor deficiency Chemotherapy-induced
Increased losses	Iatrogenic	Hemodialysis
Decreased availability	Malabsorption	Bariatric surgery Crohn's disease
	Malnutrition	Anorexia nervosa Hunger strike Fad dieting Bariatric surgery Homebound elderly TPN (if not supplemented)
	Starvation	Anorexia nervosa Hunger strike GI obstruction Chemotherapy Systemic illness
	Vomiting	Hyperemesis gravidarum Chemotherapy-induced Status post abdominal surgery GI obstruction Pancreatitis Bariatric surgery

For an in-depth exploration of the physiological impacts of thiamine deficiency, refer to the work by Sechi and Serra. When the brain experiences a lack of thiamine, it can result in cell-damaging swelling and an increase in astrocyte volume within just 4 days. Between the 7th and 10th day, a drop in transketolase activity brings about dysfunction in endothelial cells, leading to the production of nitrous oxide and the spilling of intracellular glutamate into surrounding spaces. This results in disrupted osmotic balances and the generation of free radicals, further causing swelling and increased permeability in the blood-brain Barrier (BBB). By the time 14 days of thiamine deficiency have passed, the damage to neuronal DNA and increased lactic acid lead to permanent structural harm and neuronal cell death [46]. During the process of alcohol withdrawal, the heightened sensitivity of the NMDA-receptor can compound the neurotoxic effects, releasing even more glutamate. Some brain regions are more susceptible to these damaging effects than others. For instance, in one study, the mammillary bodies displayed signs of damage in every examined case. However, the reasons for the particular vulnerability of areas like the mammillary bodies, the region around the aqueduct, and the tectum are not entirely clear, as factors like the embryonic origin of cells, blood flow patterns, and tissue characteristics do not seem to provide a full explanation. As scientists delve deeper into the molecular mechanisms behind cell damage resulting from thiamine deficiency, the reasons behind the selective harm to specific brain regions, including the mammillary bodies, remain a mystery [47].

5. Somatic Comorbidity and Epidemiology

If WE is suspected, immediate treatment is crucial. On average, symptoms of this condition, such as lethargy, confusion, and difficulty walking that may lead to a fall, manifest around 3 to 4 days prior to a subsequent hospitalization [48]. However, within a hospital setting, WE can be triggered by refeeding syndrome in just 2 to 3 days [49]. This is because most patients with thiamine deficiency may have neglected proper nutrition for months, and in some cases, might not have consumed any food for days or even weeks.

For such patients, it is essential to reintroduce calories slowly and under the guidance of a dietitian. Continuous monitoring is necessary in the initial days of hospitalization, and includes regular checks of blood glucose levels and electrolytes. Refeeding syndrome is typified by imbalances in water and electrolytes, especially low levels of phosphorus, potassium, and magnesium, alongside glucose intolerance, signs of thiamine deficiency, and excessive fluid retention [50]. Administering glucose or reintroducing carbohydrates without supplementing thiamine can pose a risk of triggering Wernicke encephalopathy, particularly in individuals with already low thiamine levels.

Criteria that indicate a high likelihood of developing refeeding syndrome can be found in the recommendations set forth by the American Society for Parenteral and Enteral Nutrition (ASPEN). If no abnormal laboratory results are observed for elements like phosphate, calcium, potassium, magnesium, and glucose, monitoring in relation to refeeding syndrome can cease after 72 h.

Various other risk factors can also play a significant role in leading to thiamine deficiency. These include infections, esophageal narrowing (Barrett's esophagus), colitis, and, notably, the renal loss of thiamine experienced in conditions like diabetes mellitus or nephropathy [51]. Loss of appetite and vomiting can both result from and exacerbate thiamine deficiency [52].

Examination of Epidemiological Data and Demographics

There is a limited amount of epidemiological information regarding cognitive disorders tied to alcohol. To truly grasp the impact of diseases like Wernicke-Korsakoff syndrome (WKS) and alcohol-related dementia (ARD), it is essential to have updated data on their incidence and mortality rates within populations [53].

The current knowledge about the frequency and occurrence of WE and alcohol-related dementia (ARD) has significant gaps. Most of our insights into these conditions are based on research from the 1970s and 1990s, which might now be considered outdated. Autopsy-based data from the past indicated that WE impacted between 0.4% and 2.8% of the overall population, and an alarming 12.5% to 59% of those with alcoholism or deaths related to alcohol consumption [54]. However, modern research is not without its own challenges in methodology and reach. For KS, data from the Netherlands estimated a prevalence of 3–4.8 per 10,000 people, based on healthcare records. Meanwhile, data from Glasgow suggested an annual KS rate of 12.5 to 81.25 per million inhabitants between 1990 and 1995.

The frequency of ARD shows significant variance across studies. Some assessments derive their estimates from associating dementia rates with alcohol use patterns or by examining particular groups. Among various research studies, ARD prevalence varied wildly, from 0.119% in general hospital stays to a staggering 25.6% in elderly patients with alcoholism being treated in substance abuse clinics [55]. Broader population-focused research indicated prevalence rates as low as 0.0066% in those aged 30–64 and as high as 0.7% in US Medicare beneficiaries aged 68 or older.

The future outcomes and prolonged mortality trends of both WKS and ARD remain underexplored. A previous study examining 245 WKS patients reported an immediate mortality rate of 17%, with subsequent deaths primarily attributed to infections, liver diseases, and cancers [56]. Contemporary studies highlight an acute mortality rate of 5.3% to 10% for those diagnosed with WE. Hospital patients diagnosed with alcohol-linked WE or KS exhibited a median survival rate of 8 years and a death rate of 7.4 per 100 person-years. The predominant causes of death were bacterial infections and cancers [55].

6. Pharmacological and Nonpharmacological Treatments

Alcohol impacts the gut in a way that unpredictably hinders thiamine absorption. To ensure an adequate concentration of thiamine in the bloodstream, it should be given through injection or infusion. The reasoning for elevated blood levels of thiamine links back to how it is transported to the central nervous system (CNS). As noted, thiamine's secondary method of reaching the CNS, through simple diffusion, does not meet the expected quantity. Theoretically, with increased serum levels, thiamine can penetrate the CNS more effectively [57].

The exact definition of a "high dose" of thiamine remains undefined. Back in 1950, Victor and Adams opted for 100 mg of thiamine to combat WE, using their best judgment of what might be considered a high dose [58]. Some experts advise treating alcoholic WE patients with 500 mg or more, although this suggestion is built on sparse evidence. The sole published randomized trial evaluating thiamine for treating WKS contrasted five thiamine doses (ranging from 5 to 200 mg daily) over two days for patients undergoing alcohol withdrawal treatment [59]. Those on the highest dose showcased superior performance in delayed alternation, a task pinpointing WKS-related cognitive issues. Yet the findings did not exhibit a consistent dose–response relationship. An updated 2008 Cochrane review concluded that evidence is too scant to give clear guidance on thiamine's dosage, frequency, delivery method, or duration for treating or preventing WKS in alcoholics.

Both the Royal College of Physicians (RCP) and the European Federation of Neurological Societies have released guidelines on preventing WE in alcoholics (refer to Table 3). Despite the distinctions between them, both emphasize the need for vigilance when it comes to WE and advocate for robust treatment. This stance arises from the diagnostic challenges in a clinical context and the favorable risk–benefit profile of using high thiamine doses [60].

The safety profile of intravenous (IV) thiamine was examined in a study where 1070 doses of 100 mg thiamine hydrochloride were rapidly given to 989 patients [61]. Minor reactions, including transient arm burning lasting from a few seconds to minutes occurred in about 1.02% of cases. One major reaction was observed, which was widespread itching without other symptoms, occurring at a rate of 0.093%. Although there have been documented cases of severe allergic reactions to IV thiamine, the likelihood is extremely low, with a recorded risk of one reaction per five million thiamine vials in the UK [42]. A survey from emergency department doctors estimated that out of around 300,000 patients receiving thiamine through injection, none experienced severe allergic reactions. In contrast, there is a 1–10% risk of allergic reactions to penicillin, a 2–3% risk of a reaction to contrast agents, and a 1–18% risk of reaction to streptokinase. Hence, thiamine is typically regarded as safe for injectable use [62].

For alcoholics, replenishing magnesium is crucial, as their diminished magnesium levels can render thiamine treatment ineffective.

The appropriate timing for administering thiamine in relation to carbohydrate has been a topic of discussion. Given thiamine's role in glucose metabolism, a spike in metabolic rate can lead to a relative thiamine deficiency, potentially triggering WKS. Several case studies and animal experiments have reported such outcomes [63]. However, a recent review of WKS emergency management posits that a single dose of glucose will not induce this condition, suggesting that in emergencies glucose administration should not be postponed because of thiamine [58].

Table 3. Comparison of Guidelines for Diagnosis and Treatment of Suspected WE.

	Who to Treat	Route	Frequency	Duration
European Federation of Neurological Societies (EFNS)	People with a combination of any two of the following: (1) inadequate nutrition (2) oculomotor abnormalities (3) cerebellar dysfunction (4) mild memory problem	IVpreferred	TID	Until the symptoms are resolved.
Royal College of Physicians (RCP)	Signs of alcohol abuse with one of the following: (1) acute confusion (2) reduced level of consciousness (3) ataxia (4) ophthalmoplegia (5) memory problems (6) hypothermia with hypotension	IV	TID	Three days. If a positive response is observed, continue with a daily dose of 250 mg administered intravenously or intramuscularly for 5 days, or until clinical improvement stops.

6.1. *Effectiveness and Limitations of Nonpharmacological Interventions*

Emerging evidence is challenging the long-standing view that Korsakoff's Syndrome (KS) is an unmodifiable condition resistant to further cognitive recovery. This paradigm shift is bolstered by a growing body of research advocating the efficacy of memory rehabilitation in KS patients [32]. Among various interventions, compensatory memory-enhancement techniques, which encompass traditional aids like agendas and memory cards, as well as digital tools such as smartphones and smartwatches, appear to be particularly promising. A meta-analysis of six studies highlights the fact that these techniques are most effective when (1) the therapeutic objectives are clearly defined, (2) adequate time is committed to individualized patient instruction, and (3) these aids are cohesively integrated into comprehensive learning methodologies.

In this context, errorless learning emerges as a theoretically ideal approach that aligns well with the unique cognitive strengths and vulnerabilities of KS patients. The core principle of this strategy lies in minimizing the opportunity for errors during the acquisition phase of learning. By eliminating the margin for guesswork, errorless learning fortifies the patient's procedural memory system against the incorporation of maladaptive or erroneous strategies. This is particularly crucial for KS patients whose episodic memory impairments would otherwise be incapable of correcting such errors [64]. Although empirical studies directly comparing errorless learning with traditional trial-and-error methods are sparse and yield mixed results, the technique holds promise, especially within specific learning contexts. Importantly, the utility of errorless learning extends beyond the domain of procedural memory, potentially offering advantages in the realm of semantic memory as well. Given that KS patients often suffer from episodic memory deficits, the errorless learning approach compensates by reinforcing and leveraging more robust procedural and semantic memory systems, thereby providing a more holistic strategy for cognitive rehabilitation.

6.2. *Emerging Therapies and Future Directions*

6.2.1. Thiamine in the Treatment and Prevention of Wernicke-Korsakoff Syndrome for Alcohol Abusers

While WKS is conventionally treated with thiamine once diagnosed, the efficacy of this treatment, especially regarding cognitive symptoms, remains uncertain. Current guidelines concerning the dosage and treatment duration with thiamine are largely based on best guesses. We aimed to find randomized controlled trials that either contrasted thiamine with placebos or other treatments, or assessed varying thiamine treatments. From our search, two studies met the prerequisites for inclusion. However, one did not provide any analyzable data, and the other's analysis was hampered by design flaws and inadequate data presentation. Consequently, there is an absence of solid evidence from these trials to guide doctors on the optimal thiamine treatment parameters to prevent or address WKS stemming from alcohol misuse [65].

6.2.2. How the Intervention Might Work

This analysis delves into WKS resulting from alcohol abuse, which is the predominant cause in developed nations. Between 30% and 80% of those abusing alcohol exhibit either clinical or biochemical indications of a thiamine deficit. Notably, alcohol seems to amplify the thiamine quantity essential for effective treatment compared to cases where the deficiency stems primarily from nutritional reasons [66]. Even with data pointing towards a significant undetected neuropathological impact due to thiamine shortage, comprehensive studies evaluating thiamine's therapeutic potency in alcohol abusers remain scant. Rapid improvements in ataxia and eye movement anomalies with thiamine administration have been noted, yet its influence on memory remains ambiguous. One comprehensive look into this area revealed the absence of systematic, placebo-controlled studies exploring the usage of injectable B-complex vitamins (thiamine-inclusive) in alcohol abuse scenarios. Hence, before suggesting varied treatment strategies for those at risk of WE and those currently afflicted by it, more research is imperative [67].

7. Quality of Life and Mental Capacity

The term quality of life (QoL) is frequently associated with an individual's overall well-being. Despite the apparent increase in patients with KS, there is a significant gap in our understanding of their QoL. Presently, most insights are derived from dementia care. This study aims to juxtapose various QoL facets in KS patients against those with dementia from identical care settings, in a bid to deepen our grasp on the socio-emotional dimensions of KS [68].

Measurements: The QUALIDEM scale was employed to evaluate Quality of Life (QoL). To contrast the QoL of patients with KS against those with dementia, multivariate linear regression analysis was utilized, considering the variables "age", "gender", and "nursing home".

Results: Out of the 147 participants, 72 (48.9%) had a KS diagnosis. KS patients exhibited a higher overall QoL. On the QUALIDEM subscales, KS patients outperformed dementia patients in areas such as "Restless tense behavior", "Social relations", and "Having something to do". There was a noticeable inclination for KS patients to score higher on "Positive affect" and lower on the "Feeling at home" subscale.

Conclusions: KS presents distinct QoL disparities when compared to dementia. Those with KS tend to experience richer social connections and heightened positive emotions than their dementia-afflicted counterparts. While dementia patients exhibit more restless tendencies, KS patients generally feel a lesser sense of belonging in nursing homes. The findings underscore the necessity for tailored nursing homes and care programs to meet the unique requirements of both patient groups.

Individuals diagnosed with KS often exhibit stronger social connections and a higher frequency of positive emotions compared to those with dementia. On the other hand, dementia patients tend to demonstrate increased restless behaviors in contrast to their

KS counterparts. The variances between the two patient groups might stem from their distinct cognitive impairments. Specifically, KS is characterized by deep-seated amnesia and often executive dysfunction, necessitating structured and well-organized daily routines. Consequently, we advocate for tailored nursing homes and care plans designed to address the unique requirements of KS patients [68].

8. Patient Care, Nursing, and Mental Capacity

8.1. Assessment of the Patient

Evaluations should rule out other potential reasons for the patient's symptoms. This could require drug tests, checks for anemia, leukemia, and variations in blood sugar levels. It is essential to conduct blood tests, including y-glutamyl transpeptidase, serum B1, pyruvate level, and an assessment of thiamine status by measuring erythrocyte transketolase activity, as these can assist in the diagnosis.

Furthermore, EEG is needed only when trying to monitor patients with seizures, while CT or MRI scans can identify brain alterations [69].

8.2. Management of the Patient

Whenever patients exhibit the described signs and symptoms, a potential diagnosis of Wernicke's/Korsakoff's should be taken into consideration, regardless of their age. These indicators might not always be clear-cut. If other health issues can be dismissed, a history of alcohol consumption should be obtained, if not already documented [70].

If WE or KS is confirmed, the primary treatment involves a high-dose thiamine injection, ideally intravenous, with a maximum limit of 300 mg for intense deficiencies. This replenishes the body's essential vitamin B1 levels, potentially averting brain injury or even death. After this intervention, the patient's condition should be re-evaluated. While the risk of an allergic response to high thiamine doses is minimal, nursing staff should be vigilant and ready to address anaphylactic reactions [71].

Daily electrolyte levels should be assessed and tracked for a duration of 7–10 days. Monitoring both food and fluid consumption and excretion is crucial. Due to the difficulties in metabolizing and absorbing vital nutrients caused by extended alcohol use, a treatment plan should incorporate parenteral nutrition [72].

It is essential to avoid triggering the "refeeding" syndrome. This occurs when patients suddenly consume high-protein calories beyond their standard capacity, potentially leading to cardiac issues [50].

Administering glucose before thiamine is discouraged, given the potential imbalances in electrolytes and fluids and a possible surge in insulin levels due to the abrupt nutritional influx.

9. Neuroimaging Research and Future Directions

Advanced Imaging Techniques

Previous studies employing magnetic resonance imaging (MRI) to investigate alcoholism have predominantly focused on cerebral structures, often sidelining the cerebellum. However, advancements in neuroimaging techniques, particularly the utilization of voxel-based morphometry (VBM) [62], have expanded the scope of investigation to include a comprehensive analysis of multiple brain regions, such as the diencephalon, midbrain, and cerebellum. The efficacy of VBM has been substantiated in the examination of various neurological and psychiatric conditions, including schizophrenia [73], aging, and Alzheimer's disease [74].

In the realm of alcoholism research, VBM has been employed to elucidate structural brain differences between alcohol-dependent and non-dependent controls [75]. One study conducted a comprehensive analysis of gray and white matter in a cohort of 22 alcohol-dependent individuals, contrasting their neuroanatomy with age- and gender-matched controls. The findings revealed marked reductions in gray matter volume in regions such as the thalamus, posterior hippocampus, and frontal cortical areas among the alcohol-dependent subjects. Concurrently, the study observed white matter atrophy in the pons and cerebellum. A subsequent investigation compared gray matter volumes in 45 abstinent alcoholics and 50 controls, identifying significant reductions in the lateral prefrontal cortex, medial frontal cortex, and posterior cingulate gyrus in the alcoholic group [75].

Such morphological alterations in gray matter have also been documented in other substance-dependent populations, such as cocaine, cannabis [76], methamphetamine, and heroin users [77]. However, research on polysubstance abusers remains limited [78]. Noteworthy studies in this niche have highlighted reduced gray matter volume in the bilateral prefrontal lobes, as identified by Liu et al. [77], and in the medial orbital frontal cortex, as observed by Tanabe et al. [78].

Prior research conducted at the NIH Clinical Center focused on the forebrain volumes of alcohol-dependent patients, categorizing them into monosubstance and poly-substance abusers [79]. This research revealed minimal differences in overall gray and white matter volumes between the two subcategories of alcoholics. Nonetheless, the study had limitations in its narrow focus on forebrain structures, excluding potential changes in the diencephalon or midbrain—regions previously implicated in alcoholism effects [69].

A more recent investigation employed VBM to conduct a nuanced examination of regional brain volumes between alcohol-dependent subjects and non-dependent controls. The study postulated that alcohol-dependent individuals would manifest reduced gray matter volume in specific brain regions, such as the frontal lobes and cerebellar cortex. The investigation also aimed to identify whether volume differences existed within subregions between alcoholics who were polysubstance abusers and those who were exclusively alcohol-dependent. Given the absence of global gray matter volume differences in prior studies [79], the study endeavored to unearth regional volume discrepancies.

The use of VBM is particularly advantageous for differentiating Wernicke-Korsakoff Syndrome (WKS) from general alcohol-related cognitive disorders. VBM's capacity for detailed regional analysis enables a more nuanced understanding of the specific structural brain changes attributed to WKS. This level of specificity provides crucial insights into the distinct pathophysiology of WKS, thereby facilitating more targeted therapeutic interventions compared to broader alcohol-induced cognitive impairments.

10. Conclusions

Nearly a century and a half has elapsed since Carl Wernicke first identified the symptoms indicative of persistent thiamine deficiency. Despite this lengthy period, the condition often remains overlooked, underdiagnosed, and inadequately treated. The challenge of accurately diagnosing the condition arises from its symptoms, which closely mimic those of other conditions like acute alcohol intoxication. This diagnostic ambiguity makes rapid treatment paramount, often taking precedence over an exhaustive diagnosis. Intravenous thiamine replenishment remains the gold standard for quick treatment, although healthcare providers have traditionally been hesitant to administer it. For instance, a retrospective analysis found that merely one-fifth of head injury patients at risk of thiamine deficiency were actually treated with thiamine, and of these, the majority were given low doses orally for a limited duration.

This hesitancy is compounded by misconceptions surrounding the rarity of the condition and the notion that a diagnosis requires the presence of a classic symptom triad. Swift and decisive treatment with intravenous thiamine is crucial to prevent irreversible brain damage due to biochemical and metabolic imbalances. An estimated 80% of those experiencing WE, the acute and treatable phase, may evolve into Korsakoff's syndrome, a chronic condition marked by persistent memory impairments and potential confabulation. In the most severe cases, the syndrome can be fatal, accounting for approximately 20% of such cases.

Though the optimal thiamine administration regimen remains to be definitively established, identifying at-risk individuals and debunking common myths about Wernicke-Korsakoff syndrome can play a significant role in reducing both morbidity and mortality in this vulnerable population. Future research should focus on fine-tuning thiamine administration protocols to improve further outcomes.

Author Contributions: Conceptualization, L.E. and A.V.C.; methodology, F.-M.B.; software, I.-A.F.; validation, R.-A.C.-B., H.P.C. and D.-I.D.; formal analysis, B.-G.B.; investigation, L.E.; resources, F.-M.B.; data curation, I.-A.F.; writing—original draft preparation, R.-A.C.-B.; writing—review and editing, L.-A.G.; visualization, H.P.C.; supervision, D.-I.D.; project administration, B.-G.B.; funding acquisition, A.V.C. All authors have read and agreed to the published version of the manuscript.

Funding: This research received no external funding.

Institutional Review Board Statement: Not applicable.

Informed Consent Statement: Not applicable.

Data Availability Statement: Not applicable.

Conflicts of Interest: The authors declare no conflict of interest.

References

1. Sullivan, E.V.; Fama, R. Wernicke's encephalopathy and Korsakoff's syndrome revisited. *Neuropsychol. Rev.* **2012**, *22*, 69–71. [CrossRef] [PubMed]
2. Arts, N.J.; Walvoort, S.J.; Kessels, R.P. Korsakoff's syndrome: A critical review. *Neuropsychiatr. Dis. Treat.* **2017**, *13*, 2875–2890. [CrossRef] [PubMed]
3. Fama, R.; Pitel, A.-L.; Sullivan, E.V. Anterograde Episodic Memory in Korsakoff Syndrome. *Neuropsychol. Rev.* **2012**, *22*, 93–104. [CrossRef]
4. Ridley, N.J.; Draper, B.; Withall, A. Alcohol-related dementia: An update of the evidence. *Alzheimer's Res. Ther.* **2013**, *5*, 3–8. [CrossRef]
5. Isenberg-Grzeda, E.; Kutner, H.E.; Nicolson, S.E. Wernicke-Korsakoff-Syndrome: Under-Recognized and Under-Treated. *Psychosomatics* **2012**, *53*, 507–516. [CrossRef]
6. Chandrakumar, A.; Bhardwaj, A.; 'T Jong, G.W. Review of thiamine deficiency disorders: Wernicke encephalopathy and Korsakoff psychosis. *J. Basic Clin. Physiol. Pharmacol.* **2019**, *30*, 153–162. [CrossRef]
7. Oslin, D.; Atkinson, R.M.; Smith, D.M.; Hendrie, H. Alcohol related dementia: Proposed clinical criteria. *Int. J. Geriatr. Psychiatry* **1998**, *13*, 203–212. [CrossRef]
8. Caine, D.; Halliday, G.M.; Kril, J.J.; Harper, C.G. Operational criteria for the classification of chronic alcoholics: Identification of Wernicke's encephalopathy. *J. Neurol. Neurosurg. Psychiatry* **1997**, *62*, 51–60. [CrossRef]
9. Harper, C.; Dixon, G.; Sheedy, D.; Garrick, T. Neuropathological alterations in alcoholic brains. Studies arising from the New South Wales Tissue Resource Centre. *Prog. Neuro-Psychopharmacol. Biol. Psychiatry* **2003**, *27*, 951–961. [CrossRef]
10. Gerridzen, I.J.; Brink, W.G.M.D.; Depla, M.F.; Verschuur, E.M.L.; Veenhuizen, R.B.; van der Wouden, J.C.; Hertogh, C.M.P.M.; Joling, K.J. Prevalence and severity of behavioural symptoms in patients with Korsakoff syndrome and other alcohol-related cognitive disorders: A systematic review. *Int. J. Geriatr. Psychiatry* **2017**, *32*, 256–273. [CrossRef]
11. American Psychiatric Association (Ed.) *Diagnostic and Statistical Manual of Mental Disorders: DSM-5-TR, Fifth Edition, Text Revision*; American Psychiatric Association Publishing: Washington, DC, USA, 2022.
12. World Health Organization. *ICD-10: International Statistical Classification of Diseases and Related Health Problems, 10th Revision, 2010 Edition*; World Health Organization: Geneva, Switzerland, 2011.
13. Shimamura, A.; Jernigan, T.; Squire, L. Korsakoff's syndrome: Radiological (CT) findings and neuropsychological correlates. *J. Neurosci.* **1988**, *8*, 4400–4410. [CrossRef] [PubMed]
14. Gallucci, M.; Bozzao, A.; Splendiani, A.; Masciocchi, C.; Passariello, R. Wernicke encephalopathy: MR findings in five patients. *Am. J. Neuroradiol.* **1990**, *11*, 887–892. [CrossRef] [PubMed]
15. Antunez, E.; Estruch, R.; Cardenal, C.; Nicolas, J.M.; Fernandez-Sola, J.; Urbano-Marquez, A. Usefulness of CT and MR imaging in the diagnosis of acute Wernicke's encephalopathy. *Am. J. Roentgenol.* **1998**, *171*, 1131–1137. [CrossRef]
16. Sullivan, E.V.; Pfefferbaum, A. Neurocircuitry in alcoholism: A substrate of disruption and repair. *Psychopharmacol.* **2005**, *180*, 583–594. [CrossRef] [PubMed]
17. Zhong, C.; Jin, L.; Fei, G. MR Imaging of Nonalcoholic Wernicke Encephalopathy: A Follow-Up Study. *Am. J. Neuroradiol.* **2005**, *26*, 2301–2305. [PubMed]
18. Schroth, G.; Wichmann, W.; Valavanis, A. Blood-brain-barrier disruption in acute Wernicke encephalopathy: MR findings. *J. Comput. Assist. Tomogr.* **1991**, *15*, 1059–1061. [CrossRef]
19. Bae, S.-J.; Lee, H.K.; Lee, J.-H.; Choi, C.G.; Suh, D.C. Wernicke's Encephalopathy: Atypical Manifestation at MR Imaging. *Am. J. Neuroradiol.* **2001**, *22*, 1480–1482.
20. Baker, K.; Harding, A.; Halliday, G.; Kril, J.; Harper, C. Neuronal loss in functional zones of the cerebellum of chronic alcoholics with and without Wernicke's encephalopathy. *Neuroscience* **1999**, *91*, 429–438. [CrossRef]

21. Sullivan, E.V.; Pfefferbaum, A. Magnetic resonance relaxometry reveals central pontine abnormalities in clinically asymptomatic alcoholic men. *Alcohol. Clin. Exp. Res.* **2001**, *25*, 1206–1212. [CrossRef]
22. Ashikaga, R.; Araki, Y.; Ono, Y.; Itami, N.; Akamatsu, M.; Kinoshita, K.; Ishida, O. FLAIR appearance of Wernicke encephalopathy. *Radiat. Med.* **1997**, *15*, 251–253.
23. Maeda, M.; Tsuchida, C.; Handa, Y.; Ishii, Y. Fluid attenuated inversion recovery (FLAIR) imaging in acute Wernicke encephalopathy. *Radiat. Med.* **1995**, *13*, 311–313. [PubMed]
24. Gillard, J.H.; Waldman, A.D.; Barker, P.B. *Clinical MR Neuroimaging: Diffusion, Perfusion and Spectroscopy*; Cambridge University Press: Cambridge, UK, 2004.
25. Unlu, E.; Cakir, B.; Asil, T. MRI findings of Wernicke encephalopathy revisited due to hunger strike. *Eur. J. Radiol.* **2006**, *57*, 43–53. [CrossRef] [PubMed]
26. Lapergue, B.; Klein, I.; Olivot, J.; Amarenco, P. Diffusion weighted imaging of cerebellar lesions in Wernicke's encephalo-pathy. *J. Neuroradiol.* **2006**, *33*, 126–128. [CrossRef]
27. Aalto, M.; Seppä, K. Use of laboratory markers and the audit questionnaire by primary care physicians to detect alcohol abuse by patients. *Alcohol Alcohol.* **2005**, *40*, 520–523. [CrossRef]
28. Mancinelli, R.; Guiducci, M.S. Women and alcohol: Biological vulnerability. *Ann. Ist. Super. Sanita* **2004**, *40*, 19–23.
29. Wijnia, J.W.; Oudman, E. Biomarkers of delirium as a clue to diagnosis and pathogenesis of Wernicke–Korsakoff syndrome. *Eur. J. Neurol.* **2013**, *20*, 1531–1538. [CrossRef]
30. Sharp, C.S.; Wilson, M.P.; Nordstrom, K. Psychiatric Emergencies for Clinicians: Emergency Department Management of Wernicke-Korsakoff Syndrome. *J. Emerg. Med.* **2016**, *51*, 401–404. [CrossRef]
31. Swain, S.P.; Behura, S.S. Neuropsychological functioning in Wernicke's encephalopathy. *Ind. Psychiatry J.* **2015**, *24*, 99–103. [CrossRef]
32. Oudman, E.; Nijboer, T.C.W.; Postma, A.; Wijnia, J.W.; Van der Stigchel, S. Procedural Learning and Memory Rehabilitation in Korsakoff's Syndrome–a Review of the Literature. *Neuropsychol. Rev.* **2015**, *25*, 134–148. [CrossRef]
33. Johnson, J.M.; Fox, V. Beyond Thiamine: Treatment for Cognitive Impairment in Korsakoff's Syndrome. *Psychosomatics* **2018**, *59*, 311–317. [CrossRef]
34. Thomson, A.D.; Marshall, E.J. The Treatment of Patients at Risk of Developing Wernicke's Encephalopathy in the Community. *Alcohol Alcohol.* **2006**, *41*, 159–167. [CrossRef]
35. Kopelman, M. Retrograde amnesia in patients with diencephalic, temporal lobe or frontal lesions. *Neuropsychologia* **1999**, *37*, 939–958. [CrossRef]
36. Moscovitch, M.; Nadel, L.; Winocur, G.; Gilboa, A.; Rosenbaum, R.S. The cognitive neuroscience of remote episodic, semantic and spatial memory. *Curr. Opin. Neurobiol.* **2006**, *16*, 179–190. [CrossRef] [PubMed]
37. Race, E.; Verfaellie, M. Remote Memory Function and Dysfunction in Korsakoff's Syndrome. *Neuropsychol. Rev.* **2012**, *22*, 105–116. [CrossRef] [PubMed]
38. Rensen, Y.C.M.; Kessels, R.P.C.; Migo, E.M.; Wester, A.J.; Eling, P.A.T.M.; Kopelman, M.D. Personal semantic and episodic autobiographical memories in Korsakoff syndrome: A comparison of interview methods. *J. Clin. Exp. Neuropsychol.* **2017**, *39*, 534–546. [CrossRef]
39. Spector, R.; Johanson, C.E. REVIEW: Vitamin transport and homeostasis in mammalian brain: Focus on Vitamins B and E. *J. Neurochem.* **2007**, *103*, 425–438. [CrossRef] [PubMed]
40. Martin, P.R.; Singleton, C.K.; Hiller-Sturmhöfel, S. The Role of Thiamine Deficiency in Alcoholic Brain Disease. *Alcohol Res. Health J. Natl. Inst. Alcohol Abus. Alcohol.* **2003**, *27*, 134–142.
41. Agabio, R. Thiamine Administration in Alcohol-Dependent Patients. *Alcohol Alcohol.* **2005**, *40*, 155–156. [CrossRef]
42. Thomson, A.D.; Cook, C.C.H.; Touquet, R.; Henry, J.A. The Royal College of Physicians Report on Alcohol: Guidelines for Managing Wernicke's Encephalopathy in the Accident and Emergency Department. *Alcohol Alcohol.* **2002**, *37*, 513–521. [CrossRef]
43. Yae, S.; Okuno, S.; Onishi, H.; Kawanishi, C. Development of Wernicke encephalopathy in a terminally ill cancer patient consuming an adequate diet: A case report and review of the literature. *Palliat. Support. Care* **2005**, *3*, 333–335. [CrossRef]
44. Kuo, S.-H.; Debnam, J.M.; Fuller, G.N.; de Groot, J. Wernicke's Encephalopathy: An Underrecognized and Reversible Cause of Confusional State in Cancer Patients. *Oncology* **2008**, *76*, 10–18. [CrossRef] [PubMed]
45. Ueda, K.; Takada, D.; Mii, A.; Tsuzuku, Y.; Saito, S.K.; Kaneko, T.; Utsumi, K.; Iino, Y.; Katayama, Y. Severe thiamine deficiency resulted in Wernicke's encephalopathy in a chronic dialysis patient. *Clin. Exp. Nephrol.* **2006**, *10*, 290–293. [CrossRef] [PubMed]
46. Sechi, G.; Serra, A. Wernicke's encephalopathy: New clinical settings and recent advances in diagnosis and management. *Lancet Neurol.* **2007**, *6*, 442–455. [CrossRef] [PubMed]
47. Thomson, A.D.; Marshall, E.J. The Natural History and Pathophysiology of Wernicke's Encephalopathy and Korsakoff's Psychosis. *Alcohol Alcohol.* **2006**, *41*, 151–158. [CrossRef]
48. Wijnia, J.W.; Oudman, E.; van Gool, W.A.; Wierdsma, A.I.; Bresser, E.L.; Bakker, J.; van de Wiel, A.; Mulder, C.L. Severe Infections are Common in Thiamine Deficiency and May be Related to Cognitive Outcomes: A Cohort Study of 68 Patients with Wernicke-Korsakoff Syndrome. *Psychosomatics* **2016**, *57*, 624–633. [CrossRef]
49. Hershkowitz, E.; Reshef, A.; Munich, O.; Yosefi, B.; Markel, A. Thiamine Deficiency in Self-Induced Refeeding Syndrome, an Undetected and Potentially Lethal Condition. *Case Rep. Med.* **2014**, *2014*, 605707. [CrossRef]

50. Mehanna, H.M.; Moledina, J.; Travis, J. Refeeding syndrome: What it is, and how to prevent and treat it. *BMJ* **2008**, *336*, 1495–1498. [CrossRef]
51. Larkin, J.R.; Zhang, F.; Godfrey, L.; Molostvov, G.; Zehnder, D.; Rabbani, N.; Thornalley, P.J. Glucose-Induced Down Regulation of Thiamine Transporters in the Kidney Proximal Tubular Epithelium Produces Thiamine Insufficiency in Diabetes. *PLoS ONE* **2012**, *7*, e53175. [CrossRef]
52. Oudman, E.; Wijnia, J.W.; Oey, M.; van Dam, M.; Painter, R.C.; Postma, A. Wernicke's encephalopathy in hyperemesis gravidarum: A systematic review. *Eur. J. Obstet. Gynecol. Reprod. Biol.* **2019**, *236*, 84–93. [CrossRef]
53. Palm, A.; Vataja, R.; Talaslahti, T.; Ginters, M.; Kautiainen, H.; Elonheimo, H.; Suvisaari, J.; Lindberg, N.; Koponen, H. Incidence and mortality of alcohol-related dementia and Wernicke-Korsakoff syndrome: A nationwide register study. *Int. J. Geriatr. Psychiatry* **2022**, *37*, 1–9. [CrossRef]
54. Naidoo, D.P.; Bramdev, A.; Cooper, K. Wernicke's encephalopathy and alcohol-related disease. *Postgrad. Med. J.* **1991**, *67*, 978–981. [CrossRef] [PubMed]
55. Cheng, C.; Huang, C.-L.; Tsai, C.-J.; Chou, P.-H.; Lin, C.-C.; Chang, C.-K. Alcohol-Related Dementia: A Systemic Review of Epidemiological Studies. *Psychosomatics* **2017**, *58*, 331–342. [CrossRef] [PubMed]
56. Goodman, R.A.; Lochner, K.A.; Thambisetty, M.; Wingo, T.S.; Posner, S.F.; Ling, S.M. Prevalence of dementia subtypes in United States Medicare fee-for-service beneficiaries, 2011–2013. *Alzheimer's Dement.* **2017**, *13*, 28–37. [CrossRef] [PubMed]
57. Greenwood, J.; Love, E.R.; Pratt, O.E. Kinetics of thiamine transport across the blood-brain barrier in the rat. *Perspect. Surg.* **1982**, *327*, 95–103. [CrossRef] [PubMed]
58. Donnino, M.W.; Vega, J.; Miller, J.; Walsh, M. Myths and Misconceptions of Wernicke's Encephalopathy: What Every Emergency Physician Should Know. *Ann. Emerg. Med.* **2007**, *50*, 715–721. [CrossRef]
59. Ambrose, M.L.; Bowden, S.C.; Whelan, G. Thiamin Treatment and Working Memory Function of Alcohol-Dependent People: Preliminary Findings. *Alcohol. Clin. Exp. Res.* **2001**, *25*, 112–116. [CrossRef] [PubMed]
60. Day, E.; Bentham, P.; Callaghan, R.; Kuruvilla, T.; George, S. Thiamine for Wernicke-Korsakoff Syndrome in people at risk from alcohol abuse. In *Cochrane Database of Systematic Reviews*; The Cochrane Collaboration, Ed.; John Wiley & Sons, Ltd.: Chichester, UK, 2004. [CrossRef]
61. Wrenn, K.D.; Murphy, F.; Slovis, C.M. A toxicity study of parenteral thiamine hydrochloride. *Ann. Emerg. Med.* **1989**, *18*, 867–870. [CrossRef] [PubMed]
62. Galvin, R.; Bråthen, G.; Ivashynka, A.; Hillbom, M.; Tanasescu, R.; Leone, M.A. EFNS guidelines for diagnosis, therapy and prevention of Wernicke encephalopathy. *Eur. J. Neurol.* **2010**, *17*, 1408–1418. [CrossRef]
63. Schabelman, E.; Kuo, D. Glucose before Thiamine for Wernicke Encephalopathy: A Literature Review. *J. Emerg. Med.* **2012**, *42*, 488–494. [CrossRef]
64. Clare, L.; Jones, R.S.P. Errorless Learning in the Rehabilitation of Memory Impairment: A Critical Review. *Neuropsychol. Rev.* **2008**, *18*, 1–23. [CrossRef]
65. Day, E.; Bentham, P.W.; Callaghan, R.; Kuruvilla, T.; George, S. Thiamine for prevention and treatment of Wernicke-Korsakoff Syndrome in people who abuse alcohol. *Cochrane Database Syst. Rev.* **2013**, *2013*, CD004033. [CrossRef] [PubMed]
66. Thomson, A.D.; Cook, C.C.H.; Guerrini, I.; Sheedy, D.; Harper, C.; Marshall, E.J. Wernicke's encephalopathy: 'Plus ca change, plus c'est la meme chose'. *Alcohol Alcohol.* **2008**, *43*, 180–186. [CrossRef] [PubMed]
67. Cook, C.C.H. Prevention and treatment of Wernicke-Korsakoff syndrome. *Alcohol Alcohol.* **2000**, *35* (Suppl. S1), 19–20. [CrossRef] [PubMed]
68. Oudman, E.; Zwart, E. Quality of Life of Patients with Korsakoff's Syndrome and Patients with Dementia: A Cross-Sectional Study. *J. Am. Med. Dir. Assoc.* **2012**, *13*, 778–781. [CrossRef]
69. Zuccoli, G.; Gallucci, M.; Capellades, J.; Regnicolo, L.; Tumiati, B.; Giadás, T.C.; Bottari, W.; Mandrioli, J.; Bertolini, M. Wernicke Encephalopathy: MR Findings at Clinical Presentation in Twenty-Six Alcoholic and Nonalcoholic Patients. *Am. J. Neuroradiol.* **2007**, *28*, 1328–1331. [CrossRef]
70. Smith, I.D.; Hillman, A. Management of alcohol Korsakoff syndrome. *Adv. Psychiatr. Treat.* **1999**, *5*, 271–278. [CrossRef]
71. Jacobson, J.L.; Jacobson, A. (Eds.) *Psychiatric Secrets*; Secret Series; Hanley & Belfus: Philadelphia, PA, USA; Mosby: St. Louis, MO, USA, 1996.
72. Hamilton, H. *Total Parenteral Nutrition: A Practical Guide for Nurses*; Churchill Livingstone: Edinburgh, UK, 2000.
73. Salgado-Pineda, P.; Fakra, E.; Delaveau, P.; McKenna, P.; Pomarol-Clotet, E.; Blin, O. Correlated structural and functional brain abnormalities in the default mode network in schizophrenia patients. *Schizophr. Res.* **2011**, *125*, 101–109. [CrossRef]
74. Derflinger, S.; Sorg, C.; Gaser, C.; Myers, N.; Arsic, M.; Kurz, A.; Zimmer, C.; Wohlschläger, A.; Mühlau, M. Grey-Matter Atrophy in Alzheimer's Disease is Asymmetric but not Lateralized. *J. Alzheimer's Dis.* **2011**, *25*, 347–357. [CrossRef]
75. Rando, K.; Hong, K.-I.; Bhagwagar, Z.; Li, C.-S.R.; Bergquist, K.; Guarnaccia, J.; Sinha, R. Association of Frontal and Posterior Cortical Gray Matter Volume with Time to Alcohol Relapse: A Prospective Study. *Am. J. Psychiatry* **2011**, *168*, 183–192. [CrossRef]
76. Cousijn, J.; Wiers, R.W.; Ridderinkhof, K.R.; van den Brink, W.; Veltman, D.J.; Goudriaan, A.E. Grey matter alterations associated with cannabis use: Results of a VBM study in heavy cannabis users and healthy controls. *NeuroImage* **2012**, *59*, 3845–3851. [CrossRef]
77. Liu, H.; Hao, Y.; Kaneko, Y.; Ouyang, X.; Zhang, Y.; Xu, L.; Xue, Z.; Liu, Z. Frontal and cingulate gray matter volume reduction in heroin dependence: Optimized voxel-based morphometry. *Psychiatry Clin. Neurosci.* **2009**, *63*, 563–568. [CrossRef] [PubMed]

78. Tanabe, J.; Tregellas, J.R.; Dalwani, M.; Thompson, L.; Owens, E.; Crowley, T.; Banich, M. Medial Orbitofrontal Cortex Gray Matter Is Reduced in Abstinent Substance-Dependent Individuals. *Biol. Psychiatry* **2009**, *65*, 160–164. [CrossRef] [PubMed]
79. Bjork, J.M., Grant, S.J., Hommer, D.W. Cross-Sectional Volumetric Analysis of Brain Atrophy in Alcohol Dependence: Effects of Drinking History and Comorbid Substance Use Disorder. *Am. J. Psychiatry* **2003**, *160*, 2038–2045. [CrossRef] [PubMed]

brain sciences

MDPI

Opinion

Perspectives and Implications of Coanda Effect in Aneurysms

Vicentiu-Mircea Saceleanu [1,2], Razvan-Adrian Covache-Busuioc [3,*], Luca Andrei Glavan [3], Antonio-Daniel Corlatescu [3] and Alexandru Vlad Ciurea [3,4,5]

1 Neurosurgery Department, Sibiu County Emergency Hospital, 550245 Sibiu, Romania; vicentiu.saceleanu@gmail.com
2 Department of Neurosurgery, "Lucian Blaga" University of Medicine, 550024 Sibiu, Romania
3 Neurosurgery Department, "Carol Davila" University of Medicine and Pharmacy, 020021 Bucharest, Romania; glavan.luca@gmail.com (L.-A.G.); antonio.corlatescu@gmail.com (A.-D.C.); prof.avciurea@gmail.com (A.V.C.)
4 Neurosurgery Department, Sanador Clinical Hospital, 70000 Bucharest, Romania
5 Romanian Academy, Calea Victoriei 125, 010071 Bucharest, Romania
* Correspondence: razvan-adrian.covache-busuioc0720@stud.umfcd.ro

Abstract: It is yet unknown how the formation of an aneurysm inside the human body occurs. Thus, understanding and analyzing the Coanda effect will result in a better overview of the overall fluid mechanics that develop inside such a structure, leading not only to better treatment plans, but also to diminished postoperative risks. This paper presents how the fluid behaves in this situation, and takes into consideration how this physical phenomenon influences the hemodynamics inside numerous anatomical regions, located in the central nervous system, where aneurysms usually develop. Analyzing the three main areas in which cerebral aneurysms form, the Coanda effect can potentially lead to the rupture of the aneurysm by changing the blood flow trajectory; this should be taken into consideration when choosing a treatment plan, especially in postoperative care. In addition, there are other factors that can influence the evolution of an aneurysm, such as its shape, size, localization and the patient's health condition. Understanding and analyzing the Coanda effect will result in a better overview of the overall fluid mechanics that develop inside such a structure, leading not only to better treatment plans, but also to diminished postoperative risks.

Keywords: Coanda effect; aneurysm; hemodynamics; Willis Polygon; pathology

Citation: Saceleanu, V.-M.; Covache-Busuioc, R.-A.; Glavan, L.-A.; Corlatescu, A.-D.; Ciurea, A.V. Perspectives and Implications of Coanda Effect in Aneurysms. *Brain Sci.* **2023**, *13*, 966. https://doi.org/10.3390/brainsci13060966

Academic Editors: Andrew P Carlson and Alfonso Lagares

Received: 24 May 2023
Revised: 6 June 2023
Accepted: 16 June 2023
Published: 19 June 2023

1. General Data

1.1. Historical Facts

Henri Coanda was born in Bucharest on the 7th of June 1886. His early education consisted mostly of military training due to the fact that his father was a general in the Romanian Army. At just 19 years of age, he constructed his first prototype of a rocket-powered airplane. Later that decade, Henri Coanda created the "Coanda 1910", an airplane without a propeller, following his graduation from " Institut Supérieur de l'Aéronautique et de l'Espace".

Through the years, he invented other aircrafts, including "Coanda 1911" and "Bristol-Coanda", both with propellers, and "Coanda 1916", which he invented while working at "Saint-Chamond" in Saint Denis.

Coanda's inventions not only had a great impact on further discoveries in the aeronautical domain, which were used in the First World War, but also in fields such as medicine, architecture, transport and construction, among others.

In 1930 [1], he obtained a French patent for the observation and description of the "Coanda Effect", a physical phenomenon that, besides medicine, has numerous appliances in fields such as aeronautics, fluids, and meteorology. Henri Coanda died on the 25th of November 1972.

1.2. The Physics of the Coanda Effect

The Coanda effect is a physical phenomenon that describes the tendency of a fluid or gas to attach to a surface after passing through an asymmetrical narrowing, exerting force on neighboring fluid or gas. We believe that the Coanda effect is overlooked when taking into consideration the formation and development of numerous types of aneurysms, and also the postsurgical events that can occur after clipping an aneurysm. This physical phenomenon is of great importance, both when the blood passes through a narrowing of the lumen, as well as when the blood travels through an arterial passage that is physiological or pathologically enlarged in comparison to the rest of the artery.

Recognizing the effect of the Coanda effect in aneurysms is vitally important for several reasons. First, it provides insights into its genesis. When blood flow is attached to irregular surfaces within vessel walls due to this effect, the shear stress levels may be altered, leading to remodeling that contributes to aneurysm formation. Furthermore, understanding these effects can assist with identifying individuals at a greater risk of aneurysm development and suggest preventative measures that should be implemented accordingly.

From a medical standpoint, a narrowing inside the body develops, for example, when a cholesterol plaque is created. In addition, being a protuberance (convex structure), it will divert the blood flow on the opposite side of the artery, creating a pressure gradient between the blood jet and the rest of the lumen. When this phenomenon develops, the blood flow will become adherent to the wall, creating further pressure on the afferent wall while also recruiting adjacent liquid particles into the stream. It is important to mention that the Coanda effect occurs for Reynold's numbers as small as 50, meaning that it is present both in laminar and turbulent flows. Also, the Coanda effect is viable for both high and extremely low pressure [2]. It is important to note that if the narrowing of the vessel is developed close to a junction of an artery, the blood flow will tend to become asymmetrical, thus creating a flux gradient between the two ramifications. The effect on the boundary wall makes the blood tend to stick to the wall and create a pressure gradient, with the jet adhering more to a ramification; this results in the possibility of some blood joining the ramification where the pressure is higher, creating a retrograde flow (Figure 1).

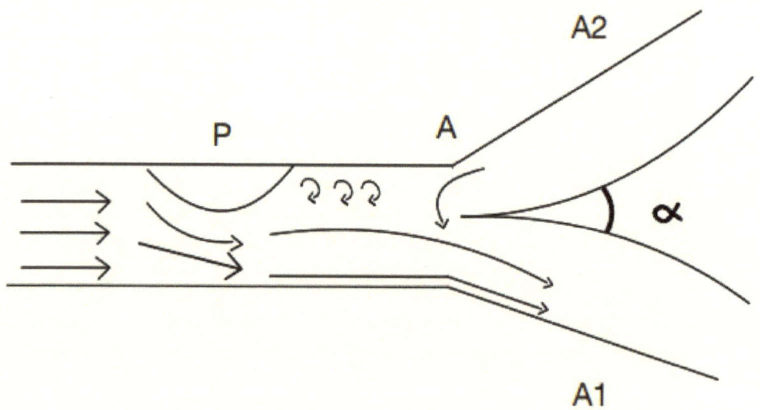

Figure 1. P—Protuberance, A—point of junction, A1—artery 1, A2—artery 2.

According to [2], the protuberance will create an asymmetry inside the blood vessel, resulting in the jet being diverted to the opposite wall; thus, the jet will create, according to the Coanda effect, a pressure gradient that will cause a retrograde flow from the A2 artery to the A1 artery, a phenomenon that can lead to dangerous ischemic damage.

However, when the surface of the blood vessel is concave the phenomenon will become opposite to the one previously described. The blood jet will tend to attach to the concave surface due to the low pressure inside the pit, thus creating small-scale swirls that exert pressure on the arterial wall. This event will divert the blood flow on the wall with the concavity, and, according to the Coanda effect, will promote neighboring liquid to join the newly created jet. This will, once again, result in an asymmetry of the blood flow if this event happens before a junction (Figure 2).

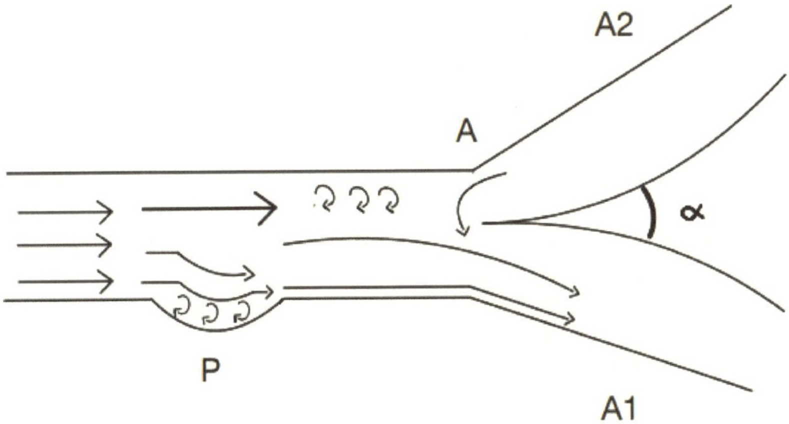

Figure 2. P—Concavity, A—point of junction, A1—artery 1, A2—artery 2.

2. Typical Aneurysms

An aneurysm is a vascular structure that implies a weakening of a blood vessel (with all its layers) that can be localized along the route of both arteries and veins. It takes form as a dilatation of the blood vessel and can be classified based on various factors such as location (abdominal, thoracic, cerebral or peripheral), shape (fusiform, saccular), size (small, large, giant) and cause (congenital, infectious, gained, tumorous, dissecting, traumatic). A special classification of aneurysms is based on their association with the surrounding vessels; this has been proven to be useful in treating cerebral aneurysms [3].

3. Typical Form of Cranial Aneurysms

Regarding cranial aneurysms, based on their form, they can be either fusiform or saccular.

Fusiform—This type of intracranial aneurysm is represented by cylindrical dilatations formed at a relatively short distance on the vessel and is usually associated with atherosclerosis and a high blood pressure [4].

Coanda effect in fusiform aneurysm—In this case, the Coanda effect appears on the blood vessel wall due to the presence of concavities. This physical phenomenon will promote the stream to behave in a certain way: a part of the jet will be bound to one wall, another part will be bound to the other side, and a part of the jet will flow through the center of the blood vessel (in the case of a bilateral fusiform aneurysm). The volume of the stream that will be bound to one side or another can be influenced by a number of factors, such variations in the pressure, a more specific level of contraction force in the walls, the number of particles involved, the temperature of the environment, the flow rate, the viscosity of the blood, and also the size and shape of the aneurysm [5]. Inside these types of concavities, micro swirls will be created, applying continuous pressure upon the vessel's walls. In time, this will lead to an overall increase in the size of the aneurysm.

Another important factor is how the aneurysm is clipped intraoperatively; if the aneurysm is clipped too tight or the placement is not perfect, the Coanda effect will produce a deviation in the jet and an asymmetrical blood flow. This effect causes the blood to follow the curved walls of the vessel, contributing to an increase in the pressure inside the blood vessel [2]; as a result, it will progressively expand, potentially leading to a potential rupture. The vascular complications are caused not only by an increase in the pressure, but also by the fact that the Coanda effect causes the blood to stagnate in certain parts of the aneurysm, resulting in the formation of thrombi. In addition to the progression of the aneurysm, taking the Coanda effect into consideration can change the treatment approach, especially when using endovascular techniques. In this case, the catheter is guided by this physical phenomenon on the wall and can be used to direct the blood flow away from the aneurysm. In this way, the risk of rupture is lowered and the normal jet stream is restored [6].

Saccular—This type of intracranial aneurysm is represented by small sac-like dilatations in an artery that are usually associated with the weakening of the blood vessel wall caused by an increase in the hemodynamic pressure, as well as the distension of the artery [7]. It can also be associated with tumorous structures or with infections such as tuberculosis [8].

Coanda Effect in Saccular Aneurysms

Our understanding of saccular aneurysms develops from convex structures inside the blood vessels. As mentioned before, convex structures appear inside the body when, for example, a cholesterol plaque forms inside an artery. The blood will interact with the convex structure and stick to the opposite wall of the artery, according to the Coanda effect. Subsequently, the resistance of the afferent wall decreases and, in time, will lead to the formation of a concavity. As previously stated, due to the Coanda effect, a concavity will cause the blood flow to adhere to it, a vicious cycle that will, in time, lead to the formation of a saccular aneurysm.

It Is mandatory to perform surgery on an aneurysm under conditions of hypotension; thus, after clipping an aneurysm, if the clip is placed too tightly and too close to the base of the aneurysm, post-operatively we may identify a protrusion inside the vessel, generating other life-threatening complications.

4. Special Types of Brain Aneurysm and How They Are Affected by the Coanda Effect

Serpentine (Giant)—This type of aneurysm is a giant aneurysm that is formed by an evolving saccular or fusiform type. It has a complex form and a sinuous vascular channel. From an angiographic perspective, the features presented by a serpentine aneurysm are as follows: a diameter greater than 25 mm; an undulating intra-aneurysmal vascular channel; and a common association with partial thrombosis [9]. The Coanda effect appears when the blood jet flows through the aneurysm and can cause the stream to bind to the walls of the blood vessel. As a result, areas of stagnating blood will appear, thus increasing the risk of blood clots forming within the aneurysm. The micro swirls that form within the environment of the blood vessel will also contribute to increasing the pressure of the jet stream and to its deviation on the opposite wall from the one on which the aneurysm is situated. There are a number of techniques that address the Coanda effect in serpentine aneurysms, such as surgical clipping or using stents to redirect the blood flow and minimize the risk of blood clots appearing; these methods are very relevant considering the increased risk of stroke compared to other types of aneurysms.

5. Coanda Effect Inside the Main Aneurysmal Areas of the Willis Polygon

1. Anterior communicating artery

Aneurysms appear commonly at the junction between the anterior communicating artery, the pericallosal artery and the anterior cerebral artery. Recent studies [10] have shown that for these types of aneurysms, the most common approach is clipping, because endovascular treatment is often too inefficient when performed in this segment. The parallel clipping of these types of aneurysms will lead to the formation of a protuberance in the walls of the anterior cerebral artery; this protuberance will divert the majority of the flow either through the pericallosal artery, through the anterior communicating artery, or into the opposite pericallosal artery. Therefore, the Coanda effect can lead to serious ischemic damage in postoperative care (Figure 3).

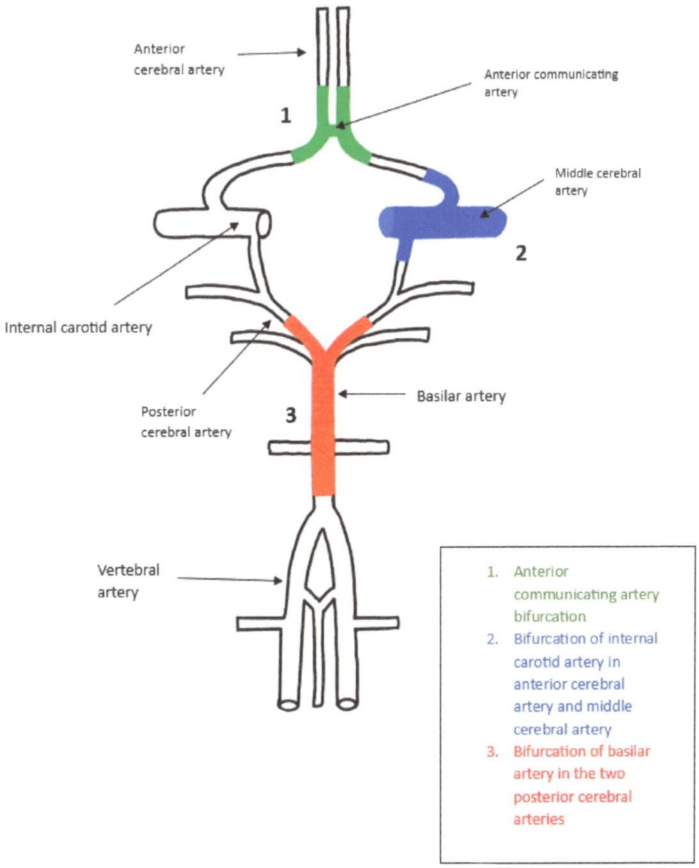

Figure 3. Coanda Effect Inside the Main Aneurysmal Areas of the Willis Polygon.

2. Bifurcation of Internal Carotid Artery in Anterior Cerebral Artery and Middle Cerebral Artery

ICA aneurysm is a type of cerebral aneurysm that develops in the internal carotid artery; this is one of the main arteries that supplies blood to the brain and is mostly saccular, considering the shape. The junction in this region is formed by the internal carotid artery, anterior cerebral artery and the middle cerebral artery, thus forming a T-shape with an alpha angle of approximately 180 degrees. The constriction of the wall appears on the internal carotid artery, but it has to be at a very high level in order for the Coanda effect to appear; however, if it appears, the blood flow will adhere to the opposite wall of the artery and will be lead to the anterior cerebral artery [2]. The main risk is represented by the progression of the aneurysm, which can lead to rupture as well as the risk of blood clots forming inside the aneurysm; therefore, the localization of the aneurysm at this bifurcation is relevant and can influence the form treatment. Lower aneurysms can be clipped using any technique, while higher ones need to be clipped at the right angle; otherwise, the Coanda effect can appear, leading to other treatment plans needing to be addressed in order to minimize the risk that arises with the occurrence of this phenomenon, as well as to a deviation in the blood flow. Other factors that can influence the manifestation of fluid mechanics disturbances, as well as the Coanda effect, are the size, location, and shape of the aneurysm, as well as the patient's age, overall health state, and other medical conditions (Figure 3).

3. Bifurcation of Basilar artery in the two Posterior Cerebral arteries

A basilar artery aneurysm is a type of cerebral aneurysm that appears in the basilar artery, which is a large blood vessel inside the brain that supplies the brainstem and the cerebellum. These aneurysms are extremely rare and are mostly saccular. The junction in this region is Y-shaped, which makes it a textbook case of the Coanda effect, which causes the blood flow to deviate to the opposite wall of the aneurysm into the posterior cerebral artery. In this case, the Coanda effect plays a most significant role in determining the blood stream's trajectory change, as well as the progression of the aneurysm. Treatment options include clipping, flow diversion and endovascular coiling, reducing the main risk of rupture in this area. In some cases, for example, after clipping, a convection can appear below the clipping area, which can determine the occurrence of another aneurysm; in this instance, multiple treatment plans should be applied. The treatment plan also can depend on other factors, such as the shape or size of the aneurysm, the patient's health condition and the risk of rupture (Figure 3).

6. Discussion

It is of great importance, when looking at the human body as a whole, to take into consideration all the physical phenomena that occur naturally inside it. The tendency of the fluid to attach to a part of the artery can lead to a deviation in the blood flow, resulting in serious ischemic damage in the absence of thrombosis [11]. This effect can explain both how aneurysms are created, and also how they develop. Using actual methods, such as Doppler echocardiography, computing tomography and magnetic resonance imaging, we can detect the Coanda effect and use different treatment plans in order to minimize the risks that can occur.

Understanding how the Coanda effect influences hemodynamics might be an invaluable asset when performing surgical aneurysm treatments and postoperative interventions. First and foremost, understanding this phenomenon allows medical practitioners to approach aneurysm clipping procedures with greater caution and precision. Taking into account the intricate dynamics of blood flow during such procedures, surgeons can make more informed decisions during clipping processes, thus ensuring adequate aneurysm treatment while mitigating the risks associated with complications.

Further study of the Coanda effect in aneurysms is essential to our comprehension of their complex hemodynamics. Although its influence on blood flow within aneurysms has long been recognized, more needs to be explored and understood regarding this phenomenon's role in aneurysm formation, progression, and rupture.

Incorporating knowledge of the Coanda effect into clinical practice guidelines will allow healthcare providers to benefit from consistent protocols and recommendations for managing aneurysms. Such guidelines may outline considerations to account for its impact, enabling physicians to tailor treatment strategies according to this knowledge, thus furthering consistency, uniformity, and improving treatment outcomes overall.

Overall, an extensive examination of the Coanda effect in aneurysms could significantly advance our knowledge. By deciphering how it affects aneurysm hemodynamics and discovering its underlying mechanisms, a detailed investigation could pave the way for improved diagnostic, therapeutic and preventive strategies that will ultimately benefit those suffering from these complex vascular conditions.

Author Contributions: Conceptualization, A.V.C. and R.-A.C.-B.; Methodology, V.-M.S. and L.-A.G.; Validation: A.V.C. and V.-M.S.; Formal analysis, A.-D.C., R.-A.C.-B. and A.V.C.; Investigation, V.-M.S., L.-A.G. and A.-D.C.; writing—original draft preparation, A.V.C. and R.-A.C.-B.; writing—review and editing, V.-M.S. and L.-A.G.; visualization, A.-D.C. and L.-A.G.; supervision, A.V.C. and V.-M.S.; project administration, R.-A.C.-B. and A.V.C.; funding acquisition, V.-M.S. and A.V.C. All authors have read and agreed to the published version of the manuscript.

Funding: This research received no external funding.

Institutional Review Board Statement: Not applicable.

Informed Consent Statement: No requirement for ethical approval.

Data Availability Statement: All Data is available on PubMed.

Conflicts of Interest: The authors declare no conflict of interest.

References

1. Coandă, H. Procédé de Propulsion dans un Fluide. Brevet d'invention France. No. 762.688, 23 November 1932.
2. Robinson, J.L.; Roberts, A. Operative treatment of aneurysms and Coanda effect: A working hypothesis. *J. Neurol. Neurosurg. Psychiatry* **1972**, *35*, 804–809. Available online: http://jnnp.bmj.com/content/35/6/804.abstract (accessed on 10 February 2023). [CrossRef] [PubMed]
3. Pritz, M.B. Cerebral Aneurysm Classification Based on Angioarchitecture. *J. Stroke Cerebrovasc. Dis.* **2011**, *20*, 162–167. [CrossRef] [PubMed]
4. Echiverri, H.C.; Rubino, F.A.; Gupta, S.R.; Gujrati, M. Fusiform aneurysm of the vertebrobasilar arterial system. *Stroke* **1989**, *20*, 1741–1747. [CrossRef] [PubMed]
5. Camuffo, D. Microclimate and Atmospheric Variables. In *Microclimate for Cultural Heritage*; Elsevier: Amsterdam, The Netherlands, 2019; pp. 3–14.
6. Barletta, E.A.; Ricci, R.L.; Silva, R.D.G.; Gaspar, R.H.M.L.; Araújo, J.F.M.; Neves, M.W.F.; de Aquino, J.L.B.; Belsuzarri, T.A.B. Fusiform aneurysms: A review from its pathogenesis to treatment options. *Surg. Neurol. Int.* **2018**, *9*, 189. [CrossRef] [PubMed]
7. Frösen, J.; Tulamo, R.; Paetau, A.; Laaksamo, E.; Korja, M.; Laakso, A.; Niemelä, M.; Hernesniemi, J. Saccular intracranial aneurysm: Pathology and mechanisms. *Acta Neuropathol.* **2012**, *123*, 773–786. [CrossRef] [PubMed]
8. Foutrakis, G.N.; Yonas, H.; Sclabassi, R.J. Saccular Aneurysm Formation in Curved and Bifurcating Arteries. *Am. J. Neuroradiol.* **1999**, *20*, 1309–1317. Available online: http://www.ajnr.org/content/20/7/1309.abstract (accessed on 30 January 2023). [PubMed]
9. Christiano, L.D.; Gupta, G.; Prestigiacomo, C.J.; Gandhi, C.D. Giant serpentine aneurysms. *Neurosurg. Focus* **2009**, *26*, E5. [CrossRef] [PubMed]
10. Chen, J.; Li, M.; Zhu, X.; Chen, Y.; Zhang, C.; Shi, W.; Chen, Q.; Wang, Y. Anterior Communicating Artery Aneurysms: Anatomical Considerations and Microsurgical Strategies. *Front. Neurol.* **2020**, *11*, 1020. [CrossRef] [PubMed]
11. Guntheroth, W.; Miyaki-Hull, C. Coanda effect on ductal flow in the pulmonary artery. *J. Ultrasound Med.* **1999**, *18*, 203–206. Available online: https://onlinelibrary.wiley.com/doi/abs/10.7863/jum.1999.18.3.203 (accessed on 14 February 2023). [CrossRef] [PubMed]

brain sciences

MDPI

Review

Intracranial Aneurysms and Genetics: An Extensive Overview of Genomic Variations, Underlying Molecular Dynamics, Inflammatory Indicators, and Forward-Looking Insights

Corneliu Toader [1,2], Lucian Eva [3,4,*], Bogdan-Gabriel Bratu [1,*], Razvan-Adrian Covache-Busuioc [1], Horia Petre Costin [1], David-Ioan Dumitrascu [1], Luca-Andrei Glavan [1], Antonio Daniel Corlatescu [1] and Alexandru Vlad Ciurea [1,5]

[1] Department of Neurosurgery, "Carol Davila" University of Medicine and Pharmacy, 020021 Bucharest, Romania; corneliu.toader@umfcd.ro (C.T.); razvan-adrian.covache-busuioc0720@stud.umfcd.ro (R.-A.C.-B.); horia-petre.costin0720@stud.umfcd.ro (H.P.C.); david-ioan.dumitrascu0720@stud.umfcd.ro (D.-I.D.); luca-andrei.glavan0720@stud.umfcd.ro (L.-A.G.); antonio.corlatescu0920@stud.umfcd.ro (A.D.C.); prof.avciurea@gmail.com (A.V.C.)

[2] Department of Vascular Neurosurgery, National Institute of Neurology and Neurovascular Diseases, 077160 Bucharest, Romania

[3] Department of Neurosurgery, Dunarea de Jos University, 800010 Galati, Romania

[4] Department of Neurosurgery, Clinical Emergency Hospital "Prof. Dr. Nicolae Oblu", 700309 Iasi, Romania

[5] Neurosurgery Department, Sanador Clinical Hospital, 010991 Bucharest, Romania

* Correspondence: elucian73@yahoo.com (L.E.); bogdan.bratu@stud.umfcd.ro (B.-G.B.)

Abstract: This review initiates by outlining the clinical relevance of IA, underlining the pressing need to comprehend its foundational elements. We delve into the assorted risk factors tied to IA, spotlighting both environmental and genetic influences. Additionally, we illuminate distinct genetic syndromes linked to a pronounced prevalence of intracranial aneurysms, underscoring the pivotal nature of genetics in this ailment's susceptibility. A detailed scrutiny of genome-wide association studies allows us to identify key genomic changes and locations associated with IA risk. We further detail the molecular and physiopathological dynamics instrumental in IA's evolution and escalation, with a focus on inflammation's role in affecting the vascular landscape. Wrapping up, we offer a glimpse into upcoming research directions and the promising horizons of personalized therapeutic strategies in IA intervention, emphasizing the central role of genetic insights. This thorough review solidifies genetics' cardinal role in IA, positioning it as a cornerstone resource for professionals in the realms of neurology and genomics.

Keywords: intracranial aneurysm; genetic syndromes; genome-wide association studies; molecular mechanisms; inflammatory status; genomic modifications

check for updates

Citation: Toader, C.; Eva, L.; Bratu, B.-G.; Covache-Busuioc, R.-A.; Costin, H.P.; Dumitrascu, D.-I.; Glavan, L.-A.; Corlatescu, A.D.; Ciurea, A.V. Intracranial Aneurysms and Genetics: An Extensive Overview of Genomic Variations, Underlying Molecular Dynamics, Inflammatory Indicators, and Forward-Looking Insights. *Brain Sci.* **2023**, *13*, 1454. https://doi.org/10.3390/brainsci13101454

Academic Editor: Teng Jiang

Received: 15 August 2023
Revised: 22 September 2023
Accepted: 10 October 2023
Published: 12 October 2023

1. Introduction

About 80% of all non-traumatic subarachnoid hemorrhages (SAHs) are caused by intracranial aneurysms (IAs), commonly referred to as saccular or berry aneurysms. Between 2% and 5% of people have intracranial aneurysms. Of these, 0.7% to 1.9% will experience a rupture leading to subarachnoid hemorrhage [1,2].

An SAH is the predominant initial sign of cerebral aneurysms in both adults and kids. In children, the incidence of an SAH varies between 1.9% and 4.6%. The growing detection of SAHs in children can probably be attributed to better diagnostic tools and heightened clinical vigilance [3].

However, the occurrence of an IA seems to increase with age [4]. For those over 30 years old, the likelihood of having an IA ranges from 3.6% to 6.5% [5]. Women are more predisposed to aneurysms than men, with a 3:1 ratio in cases of unruptured IAs. About 70–75% of IAs appear individually, whereas 25–30% occur as multiple aneurysms [6,7].

Among adults, unruptured IAs have a prevalence rate of about 2% to 6%. They generally do not show symptoms and are usually discovered during MRI or CT scans for other neurological reasons or when screening high-risk individuals [8].

The frequency of SAHs due to IAs varies globally. Finland and Japan have the highest rates, between 22.5 and 32 per 100,000 people annually. Globally, this rate stands at 9.1 per 100,000 individuals every year [9]. An SAH strikes early in life and is fatal, contributing to more than 25% of years of life lost for stroke sufferers below 65 years of age [10].

The majority of individuals are unaware that they have an aneurysm until it bursts. Some might experience precursor symptoms like pain around the eye, nerve paralysis in the face, or headaches and neck pain from a minor aneurysm leak, known as a "sentinel bleed" [11].

2. Risk Factors Associated with IAs

Habits and health conditions such as active smoking, a high blood pressure, and excessive alcohol consumption are recognized as separate risk factors that can contribute to the development of an aneurysmal subarachnoid hemorrhage [12]. Additional risk factors include an advancing age and being female, since the yearly occurrence rate for men was determined to be 4.5 per 100,000 with a confidence interval (CI) of 3.1 to 5.8, while for women, it was 7.1 per 100,000 with a CI of 5.4 to 8.7. The risk for women, when compared to men, had a relative value of 1.6 with a CI of 1.1 to 2.3 [13]. Moreover, having a family history of intracranial aneurysms, and the consumption of drugs that stimulate the sympathetic nervous system, commonly known as sympathomimetic drugs, are other risk factors [14]. Genetic conditions can further heighten the risk of an aneurysmal SAH. For instance, those with autosomal dominant polycystic kidney disease or vascular Ehlers-Danlos syndrome (often referred to as type IV EDS) are more susceptible.

Recent research, like the PHASES study, indicates that one's geographical origin, such as being of Finnish or Japanese descent, can be a strong determinant for aneurysmal rupture. This potentially suggests that genetic predispositions, associated with certain regions or ethnicities, can play a significant role in rupture risk [15]. The 1999 study by the MARS group provided clarity on the importance of screening close relatives of a patient diagnosed with an IA. Their findings underscored the preventive value of screening: for every 149 first-degree relatives screened using Magnetic Resonance Angiography (MRA), one SAH could be averted. To prevent a single SAH-related death, 298 patients would need to be screened [16]. A family history of IA stands out as the most telling risk marker for the condition. To illustrate, family members from a family with at least two patients diagnosed with IA have about a four-fold increased risk of having an IA, as shown by MRA screening, compared to the general public [4].

Although several inheritable conditions, including autosomal dominant polycystic kidney disease, neurofibromatosis type I, Marfan syndrome, multiple endocrine neoplasia type I, pseudoxanthoma elasticum, hereditary hemorrhagic telangiectasia, and Ehlers-Danlos syndrome types II and IV, have links to the formation of IAs, they represent less than 1% of all IAs in the general population. As such, these genetic syndromes cannot account for the majority of familial clusters of IA cases. However, when family histories are thoroughly researched, familial occurrences of IA seem to be more frequent than initially thought. By studying extended family trees over multiple generations based on affected individuals, we might quickly identify a significant number of familial cases [17].

2.1. Genetic Syndromes with Increased Intracranial Aneurysm Incidence

Autosomal Dominant Polycystic Kidney Disease (ADPKD): Through comprehensive literary reviews, it has been established that individuals with ADPKD have a roughly 11% prevalence rate of intracranial aneurysms [18]. ADPKD's genetic origin lies in the loss-of-function of either the polycystin 1 (PKD1) or polycystin 2 (PKD2) genes. This primarily results in the development of cysts in the kidneys, which can impair renal function and eventually progress to kidney failure [19]. Beyond the kidneys, ADPKD can impact other organs, particularly the liver. A common comorbidity among ADPKD patients is hypertension, which is a known risk factor not just for IA, but for other cardiovascular diseases as well [20]. The life expectancy for those with ADPKD is notably reduced, with PKD1 and PKD2 patients in Europe averaging lifespans of 53 and 69 years, respectively [21].

Microcephalic Osteodysplastic Primordial Dwarfism Type II (MOPDII): This is the most prevalent variant of microcephalic primordial dwarfism. It presents with an exceptionally short stature, microcephaly, and unique facial attributes. The features that distinguish it from other primordial dwarfism types, which might require medical attention, encompass irregular dental structures, a fragile bone skeletal dysplasia leading to hip deformities or scoliosis, tendencies towards insulin resistance or diabetes, chronic kidney ailments, heart defects, and a broad spectrum of vascular diseases. This wide-ranging vascular ailment category encompasses neurovascular conditions like moyamoya vasculopathy and intracranial aneurysms (potentially causing strokes), early-onset coronary artery diseases (potentially resulting in premature heart attacks), and kidney-related vascular issues. The frequent occurrence of hypertension is complicated due to multiple potential origins tied to the array of associated conditions [22].

Observationally, from birth onwards, 25 out of 47 individuals were identified with intracranial aneurysms (30 of them had multiple aneurysms), 22 showed signs of moyamoya vasculopathy, 17 exhibited both conditions, and 17 had neither. Roughly 50% of those diagnosed with either moyamoya vasculopathy or aneurysms eventually experienced a stroke, be it ischemic or hemorrhagic. The risk of an aneurysm remained fairly consistent throughout childhood, while the risk associated with moyamoya vasculopathy was accentuated before the age of five [23].

Type IV Ehlers–Danlos Syndrome (Vascular Subtype): This autosomal dominant condition, affecting between 1 in 50,000 to 200,000 individuals, is a connective tissue disorder. It is chiefly marked by pronounced vascular fragility, heightening the risk of a hemorrhage and premature death [24]. The underlying cause of vascular EDS is the presence of mutations or variants in the Col3a1 gene. This gene is responsible for producing a protein that plays a crucial role in reinforcing connective tissue throughout the body [25].

Microcephalic/Majewski's Osteodysplastic Primordial Dwarfism, Type II (MOPD2): This exceedingly rare autosomal recessive disorder is characterized by a notably short stature, alongside skeletal abnormalities, especially a significantly small head size in proportion to the body [26]. Given the scant occurrence of MOPD2, establishing a precise prevalence of IA in affected individuals remains challenging. However, current evidence indicates that as many as half of those with MOPD2 might develop an IA [27]. The genetic basis for MOPD2 is traced back to mutations in the Pericentrin 1 (PCNT) gene. This gene encodes for a protein that is central to chromosomal segregation. Additionally, a loss-of-function in PCNT disrupts cilia formation in epithelial cells and impacts PDK2 positioning at the basal bodies [28]. The intricate interaction between PCNT and PDK2 and the potential implications of PCNT insufficiency in an IA becomes even more evident when considering the presence of rare PCNT variants in multiple IA-affected families. Some family members also presented with kidney cysts [29].

However, in these families, it is crucial to understand that other genetic variants, apart from PCNT, may exist. These additional variants might play significant roles in disease manifestation and progression.

Loeys–Dietz Syndrome (LDS): LDS shares similarities with type IV Ehlers–Danlos, being an autosomal dominant connective tissue disease. Caused by mutations in the genes of the Tgf-β pathway (mainly TGFBR1, TGFBR2, SMAD3, and TGFB2), patients exhibit severe vascular anomalies. Arterial aneurysm dissections and bleeding events, which often appear early in life, are frequent complications. A significant portion, roughly one-third, of LDS-related deaths, result from cerebrovascular hemorrhages. The incidence of an IA among those with LDS ranges from 10% to 28% [30].

Marfan Syndrome is among the prevalent hereditary disorders impacting connective tissue. It is an autosomal dominant condition, manifesting in approximately 1 out of every 3000 to 5000 individuals. The underlying genetic flaw is in the FBN1 gene on chromosome 15, which is responsible for producing fibrillin, a vital connective tissue protein. The heightened risk associated with the aorta in males has been recognized for a while, yet its underlying reason remains unclear. In this context, the influence of gender was evident across all molecular categories. Patients with PTC displayed a trend towards earlier surgeries, and the lifetime aortic risk for males was double compared to females who possessed in-frame pathogenic variations [31]. Though some studies suggest a link between Marfan's syndrome and IAs, autopsies and family analyses with multiple members having both an IA and Marfan's syndrome do not consistently support this association [32,33].

Neurofibromatosis type 1: NF1 is a condition marked by tumors in the skin and nervous system. The two primary forms, types 1 and 2, are both inherited in an autosomal dominant manner. Type 1, often referred to as von Recklinghausen disease, is typified by features like neurofibromas, cafe-au-lait spots, freckling, and optic gliomas. In contrast, type 2 is distinguished by the presence of bilateral vestibular schwannomas and meningiomas [34].

In a 2005 research study by Schievink et al. [35], 39 patients with neurofibromatosis type 1 were examined, with an average age of 30.4 years, ranging from 3 to 64 years. Incidental intracranial aneurysms were found in 2 (5%) out of the 39 patients through MRI scans. When focusing only on the 22 patients who had an MRI, the detection rate rose to 9%. This rate was notably higher ($p < 0.005$) than the 0% detection rate in the control group of 526 individuals, therefore concluding the existence of a correlation between NF1 as a genetic risk factor for IA development.

CDKN2BAS: The 9p21.3 locus first gained attention during a 2007 genome-wide association study (GWAS) related to cardiovascular disease. In this study, it was highlighted that the specific genetic variations associated with cardiovascular diseases also had a connection with intracranial aneurysms (IAs). Interestingly, the 9p21.3 region itself does not house any known protein-coding genes. However, in its vicinity—just a few kilobases away—are protein-coding genes, including CDKN2A and CDKN2B [36]. It is worth mentioning that the 9p21.3 gene cluster (CDKN2A-CDKN2B) has emerged as a frequently identified region in GWASs for several prevalent conditions, such as heart diseases, type 2 diabetes, brain tumors (gliomas), and skin cancers (basal cell carcinomas) [37].

One significant discovery came from a study that pinpointed the CDKN2BAS SNP (rs6475606) as a contributing factor for IA vulnerability, particularly in the US population. This was a monumental finding, as understanding genetic predispositions can be the key to predicting and potentially preventing the onset of conditions like IAs [38].

Subsequent GWAS research involving both Asian and Caucasian populations has further underscored this genetic link. Numerous single-nucleotide polymorphisms (SNPs) of CDKN2BAS have been identified as being associated with intracranial aneurysms. This consistent finding across diverse populations suggests a robust genetic connection and underscores the importance of this genetic region in understanding, and potentially addressing, IAs [39].

Other Syndromes: Several conditions, such as multiple endocrine neoplasia type I [40], pseudoxanthoma elasticum [41], and hereditary hemorrhagic telangiectasia (HHT) [42], are frequently highlighted in relation to intracranial berry aneurysm. While there are individual patient case studies documenting the existence of intracranial aneurysms in these conditions, the evidence is not as comprehensive as it is for other syndromes that were previously discussed [43].

2.2. Genetic Implications

A predominant strategy for discerning genetic risk factors related to intracranial aneurysms (IAs) entails genotyping polymorphisms within sporadic cases and controls. The genetic markers frequently employed in these endeavors are termed single nucleotide polymorphisms (SNPs). Genetic association investigations can be segmented into two main categories: Candidate Gene Association Studies (CGASs) and genome-wide association studies (GWASs). A CGAS involves the examination of select common polymorphisms in genes, specifically chosen based on existing biological evidence suggesting their relevance to IA development. Additionally, candidate genes for IAs have been sourced from genetic studies focusing on connective tissue disorders like vascular Ehlers–Danlos syndrome; known genetic diseases where an IA is a manifestation, such as polycystic kidney disease; and from gene expression studies [44].

A plethora of recognized mendelian disorders is known to elevate the risk of IA development. Over recent decades, there has been a recalibration in our understanding of genetic causation. Previously, the paradigm centered on single gene mutations that are directly correlated with distinct clinical phenotypes. However, the contemporary perspective emphasizes the intricate interactions between singular gene mendelian phenotypes and allelic variants located elsewhere in the genome, termed transacting regulatory elements. Consequently, our understanding of genetic inheritance in diseases has evolved from a binary approach to a more nuanced, spectrum-oriented view [45].

The proclivity towards IAs, stemming from these genetic anomalies, likely emanates from structural deficiencies in the brain's vasculature due to the malfunctioning of specific genes. In illuminating this, a recent meta-analysis concerning the ACE insertion/deletion (I/D) polymorphism (rs4646994) demonstrated a linkage between the I allele and increased IA susceptibility. Specifically, individuals possessing ACE I/I and I/D genetic configurations displayed a heightened IA risk. The exact mechanisms by which the ACE I allele influences the development of an IA remain nebulous [46].

Enriching the discourse, Yasuno and colleagues amalgamated more datasets, thereby analyzing a larger sample consisting of 5891 cases and 14,181 controls. Their findings revealed a total of five IA-specific loci. Among these, the SOX17 (8q11.23; rs9298506) and CDKN2BAS1 (9p21; rs1333040) loci were previously identified, and their association with IAs was fortified in this research. Although the SNP rs10958409 (8q11.23; rs10958409) did not manifest an association, the study unveiled three novel IA associations with CNNM2 (10q24.32; rs12413409), STARD13 (13q13.1; rs9315204), and RBBP8 (18q11.2; rs11661542). It is pivotal to acknowledge that certain SNPs, despite achieving genome-wide significance like rs700651 and rs11661542, did not exhibit replicability in subsequent studies and thus were not deemed significant in the encompassing meta-analyses [47] (Figure 1 and Table 1).

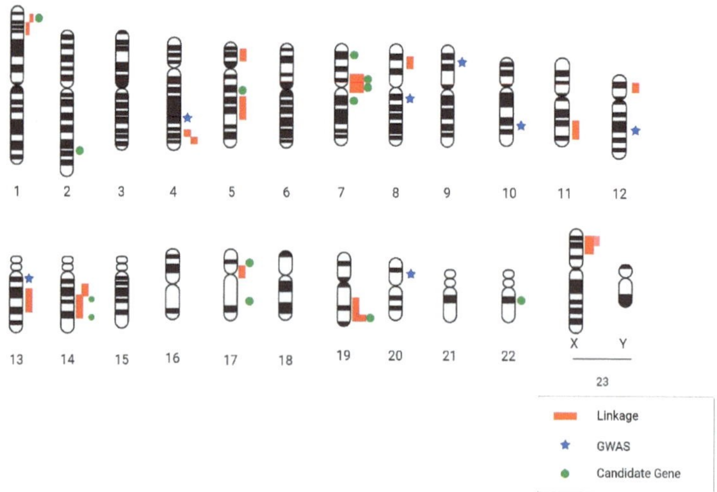

Figure 1. A genetic blueprint highlighting locations related to intracranial aneurysms is presented. Next to the chromosome diagrams, dark red lines point out regions identified through DNA linkage research. Light red lines suggest regions supported by family links to two specific spots. Blue star symbols mark the places where SNPs were identified in wide-ranging genetic studies, while green circular markers show where SNPs were pinpointed in targeted gene association studies.

Table 1. Linkage analyses of DNA related to intracranial aneurysms.

Chromosomal Region	Study Design	LOD Score	Genetic Marker	Phenotype IDs and OMIM Locus
1p36.21-p36.13	Non-parametric	3.18	D1S2826-D1S234	609122; ANIB3
1p34.21-p36.13	Family, AD	4.2		609122; ANIB3
4q32.2	Non-parametric	2.5	Rs1458149	
4q32.3	Parametric	2.6		
5p15.2-p14.3	Family, AD	3.57	D5S1954	610213; ANIB4
5q22-q31	Affected sib pair	2.24	D5S471-D5S2010	
7q11	AR	2.34	D7S2421	105800; ANIB1
7q11	Affected sib pair	3.22	D7S2415-D7S657	105800; ANIB1
8p22	Family, AD	3.61	D8S552	614252; ANIB11
11q24-q25	Family	4.3	rs618176-rs1940033	612161; ANIB7
12p12.3	Parametric	3.1		
13q14.12-q21.1	Family	4.56	rs7983420-rs17054625	
14q23-q31	Family	3.0	rs235991-rs2373098	612162; ANIB8
14q22	Affected sib pair	2.31	D14S258-D14S74	
17cen	Non-parametric	3.0	D17S921-D17S1800	
19q13	Non-parametric	2.15	D19S198-D19S596	608542; ANIB2
19q13	Affected sib pair with covariates	5.70	D19S178-D19S545	608542; ANIB2
19q13.3	Affected only, parametric	4.10	D19S198-D19S902	608542; ANIB2
Xp22	Non-parametric	2.16	DXS987-DXS7593	300870; ANIB5
Xp22	Affected sib pair	2.08	DXS987	300870; ANIB5
Xp22.32-p22.2	Non-parametric	4.54	DXS6807-DXS1224	300870; ANIB5

AD = autosomal dominant; AR = autosomal recessive.

There exist certain genetic syndromes that are transmitted through an autosomal recessive manner or stem from spontaneous genetic alterations. These are known to be associated with intracranial aneurysms, albeit at a lesser frequency compared to more prevalent syndromes. An exemplar of such conditions is pseudoxanthoma elasticum. This disorder is linked to mutations found within the ABCC6 gene situated on chromosome 16p13 [48]. One prevailing theory proposes that IAs in individuals with this syndrome originate due to irregularities in the elastic lamina of the intracranial blood vessels, with the intracranial internal carotid artery being a primary site [41]. Whole exome sequencing (WES) has spotlighted the potential significance of the PCNT gene in relation to cerebrovascular diseases. The protein encoded by this gene plays a pivotal role in orchestrating microtubule nucleation and ensuring the orderly progression of the cell cycle. Importantly, mutations where both copies of the gene are dysfunctional lead to a condition known as microcephalic osteodysplastic primordial dwarfism type II. Astonishingly, in over half of these cases, it is surmised that individuals bearing specific mutations in the PCNT gene are susceptible to IAs. Intriguingly, these genetic variations might also amplify the risk of a single individual manifesting multiple IAs [27].

While these genetic anomalies offer invaluable insights into the genesis of aneurysms, comprehending the specific risks tied to aneurysm rupture in these distinct syndromes necessitates further exploration. Beyond merely identifying the genetic markers related to the onset or presence of IAs, there exists a trove of data ripe for analysis to pinpoint genetic locations that might dictate the propensity for aneurysmal rupture and the subsequent clinical aftermath of a subarachnoid hemorrhage (SAH). The post-rupture genetic landscape of the disease becomes murkier, as there could be other genetic factors influencing various aspects of SAH care. One such aspect is vasospasm—a contraction of blood vessels—which is widely acknowledged as a key contributor to grievous outcomes due to delayed cerebral ischemia (DCI) [49]. Presently, there is a noticeable scarcity of comprehensive studies elucidating outcomes post aneurysm rupture. The limited studies available in the scientific literature often lack the rigorous sample sizes required for drawing incontrovertible conclusions. Most of these outcome-centric studies place their emphasis on the ever-fluctuating gene expression profiles. This dynamic nature contrasts with static genetic anomalies, which could potentially offer a constant measure of risk, hence posing challenges in consistently assessing the genetic basis of the outcomes [50].

2.3. Genome-Wide Association Studies

Genome-wide linkage analyses are sophisticated methods utilized to pinpoint the specific chromosomal location, referred to as loci, that houses the gene that is responsible for certain diseases. To accomplish this, these analyses probe specific genetic markers, such as single nucleotide polymorphism (SNP) and Variable Number Tandem Repeat (VNTR) [51]. Thanks to the versatility of the linkage analysis method, researchers have managed to unveil genes that drive the onset of complex disorders. These range from metabolic conditions like diabetes and obesity to cardiovascular issues such as hypertension. The breakthroughs provided by these analyses have deepened our understanding of the genetic underpinnings of such multifaceted ailments [52,53]. Over time, numerous studies have leveraged linkage analysis to unearth insights about IAs. The allure of non-parametric methodologies is particularly potent when it comes to IAs. This is primarily because the penetrance—referring to the likelihood that an individual with a mutation will display symptoms—may not be absolute. In simpler terms, not every individual harboring the mutation will be afflicted. Summarily, non-parametric linkage studies have unveiled several loci that may play roles in the formation and potential rupture of IAs. However, only a select few, namely 1p34.3-p36.13, 7q11, 19q13.3, and Xp22, have seen consistent results across various populations [54–56].

Current research furnishes the most compelling evidence concerning regions on chromosomal arms 7q and 19q. Numerous independent studies have reported linkage hits on these arms, signaling their potential significance in IAs. Alg et al., in a comprehensive meta-analysis involving a staggering 116,000 participants, spotlighted 19 SNPs that exhibited a significant correlation with IAs under at least one genetic model. GWASs emerged as particularly illuminating, uncovering 12 robustly associated SNPs. Some of the most telling associations are related to chromosomes 9, 8, and 4. Meanwhile, the CGAS method identified eight significant SNPs, with SERPINA3 and two collagen-related variants showcasing the strongest links. Notably, 9 out of the 19 SNPs exhibited significant statistical variance, mandating random-effects and sensitivity evaluations for those reported in over two publications [57].

A recent genome-wide association study, centered on both familial and sporadic IAs in a Caucasian population, identified six SNPs in the 9p21.3 region as being significantly associated with IA. Of these, one (rs6475606) achieved a stellar level of statistical significance for its association with both types of IAs. Additionally, the study reaffirmed the association of the rs1333040 SNP with IAs, bolstering the belief in its relevance [58].

The endothelin receptor type A (often abbreviated as EDNRA) gene encodes a receptor that is activated by endothelins, a group of proteins instrumental in regulating both the constriction and dilation of blood vessels, particularly following any hemodynamic disturbances. Specifically, Endothelin-1, which is the primary variant present in vascular smooth muscle cells, is responsible for triggering EDNRA. This endothelin signaling pathway becomes particularly active at sites of vascular injuries, leading to heightened cell proliferation—a critical step in the body's reparative process [59]. There is a potential that a diminished activity, or downregulation, of EDNRA signaling might be a precursor to compromised vascular repair mechanisms. When this repair process does not function optimally, it could make the vascular system more susceptible to the formation of aneurysms following injuries or disturbances. Adding credence to this theory is the functional analysis of the EDNRA gene variants. Specifically, the rs6841581 risk allele has been shown to have reduced transcriptional activity. This essentially means that the gene's ability to produce its associated protein might be diminished, which could potentially contribute to the aforementioned vulnerabilities in the vascular repair system [60].

Continuing research using genome-wide association studies recently highlighted another intriguing discovery. The SNP denoted as rs6841581A.G, located on chromosome 4q31.23, which codes for the EDNRA gene, has caught researchers' attention. This particular SNP displayed a significant association with intracranial aneurysms. What is particularly noteworthy is that this association was not restricted to just one ethnic group or population; it emerged as a consistent pattern across Dutch, Finnish, and Japanese populations. This widespread correlation accentuates the potential global relevance and importance of the EDNRA gene in understanding and potentially addressing IAs [61] (Table 2).

Table 2. Key research findings on intracranial aneurysm genetics: loci, candidate genes, and associated functions.

Type of Genetic Analysis	Loci	Candidate Genes	Functions
GWAS	13q13.1	STARD 13	Movement of endothelial cells
GWAS	18q11.2	RBBP8	Cell cycle
GWAS	2q	PLCL1, BOLL	Similarity to phospholipase C, positioned after VEGFR2 in the signaling pathway
GWAS	4q31.23	EDNRA	Vasoconstriction
GWAS	8q21.3	SOX17	Endothelial sprouting
GWAS	9p21.3	CDKN2A/B	Smooth muscle proliferation
GWAS	10q24.32	CNNM2	Epithelial absorption of Mg^{2+}
GWL	1p34.3-p36.13Xp22	PERLECAN	Promote endothelial cell growth and renewal, maintain the endothelial barrier function, and inhibit smooth muscle cell proliferation
GWL	7q11, 14q22, 5q22	ELASTIN	Elasticity of the parietal vessel
GWL	11q24, 14q23		
GWL	19q13, Xp22		

Abbreviations: GWL = genome wide linkage; GWAS = genome wide association studies.

3. Molecular and Physiopathological Mechanisms Implicated in IA Formation and Progression

Human IA samples serve as a treasure trove for researchers, offering them a unique glimpse into the intricate molecular mechanisms underpinning an IA's onset and rupture. At the heart of IA pathology is the degradation of the internal elastic lamina (IEL), which sits as a protective barrier separating the intima from the media layers of blood vessels. Alongside this, various other signs, like an uneven vessel inner surface, thickening due to myointimal growth, the chaotic nature of the muscular media, reduced cell presence, and heightened inflammatory cell infiltration, paint a complex picture of an IA's internal environment [62].

As a critical component in arteries, the IEL comprises elastic fibers that grant flexibility to these vessels. In a healthy state, the IEL presents as consistent and intact in intracranial arteries. However, when confronted with an IA, the IEL tends to become fragmented or torn, or may even vanish completely, which is especially evident at the aneurysm's base. This degradation is a significant red flag pointing towards an IA's pathology [63].

One of the more striking aspects of IAs, whether from human patients or those mimicked in animal models, is the formation of pronounced outward bulges and deep inward crevices in the intimal lining. These anomalies become evident through the intricate imaging of transmission electron microscopy [64].

The degradation of the IEL acts as a trigger. It allows for the migration of smooth muscle cells from the media to the intima layer. Here, they multiply, leading to myointimal hyperplasia, resulting in a pronounced thickening of the intima layer, which is a recurring characteristic in IA samples [65].

IAs mark a shift in the medial layer's cellular configuration. Here, smooth muscle cells transition or "switch" from their primary contractile state to a more synthetic one. This new state is marked by heightened inflammatory and remodeling tendencies. In the context of an IA, there is a heightened influx of inflammatory cells. Both animal models and human samples consistently reveal the presence of cells like macrophages, T cells, B cells, and neutrophils. Significantly, macrophages, apart from their standard functions, release matrix-degrading enzymes such as MMP2 and MMP9. These enzymes, alongside various cytokines, serve to attract a greater volume of inflammatory cells to the site [64].

Microarray-based mRNA profiling stands out as a holistic tool, allowing for the clear demarcation of molecular markers in both healthy and diseased states. While arteries have their distinct mRNA expression profiles, the same type of artery from varied individuals tends to have more similarities than different arteries within one individual. This highlights the necessity for meticulous control tissue selection when conducting expression studies [66].

To truly discern the molecular markers specific to an intracranial artery, researchers utilized RNA samples from both IA-affected and unaffected arteries, applying two microarray platforms. Intriguingly, almost half of the genes on these platforms were expressed in these arteries. Further, DNA linkage studies spotlighted around 800 diverse genes present in intracranial arteries. Utilizing advanced bioinformatics tools like GO and KEGG, researchers found a significant clustering around biological pathways, including the likes of Notch and MAPK signaling channels [67].

Genome-wide studies focusing on both mRNA and miRNA expressions offer a comprehensive, unbiased pathway to dive into an IA's pathophysiology. Tools like GO and KEGG not only facilitate a more structured understanding of gene expression, but also throw light on the intricate web of interconnected pathways, providing a holistic view of the underlying pathology defining an IA.

4. Adherens Junction, MAPK, and Notch Signaling, Which Are Functionally Relevant to the Pathogenesis of an IA

Adherens Junction: When delving into the intricate world of intracranial aneurysms, the most pivotal biological pathway that emerges is the adherens junction pathway. These adherens junctions, far from being static structures, are vibrant, multifaceted assemblies comprising both adhesive and signaling molecules. Their primary roles are to oversee and maintain the inhibition of endothelial cell growth in response to contact. Additionally, they play pivotal roles in regulating vascular permeability, ensuring a controlled environment against inflammatory cells and solutes. Another feather in their cap is their responsibility in steering the formation of new blood vessels during angiogenesis. Zooming in on vascular endothelial cells, adherens junctions wear the badge of honor in preserving the vessel wall's sanctity. They also play crucial roles in remodeling the endothelium during both physiological and pathological events. An interesting tangent to this is how proteins within these junctions can channel signals to β-catenin, which can then make its journey to the nucleus. Once there, it synergizes with transcription factors to govern the expression of genes that are instrumental in the endothelial cell's life cycle, influencing their growth, differentiation, and programmed cell death [68]. What catches one's eye is the CTNND2 gene. Located on chromosome 5p and aligned with the IA locus, it encodes for catenin delta-2. This protein could potentially be the puppet master in regulating adhesion molecules within the brain's confines. Drawing a parallel, another gene, CTNNA1, which encodes catenin alpha-1, is also seen as a prominent positional IA candidate. This particular gene might be tasked with ensuring that the integrity of the intracranial arteries remains uncompromised [69].

In certain IA patients, a rather unique phenomenon can be observed: the presence of red blood cells not just in the intima but also the media. One could theorize that such dissections could stem from underlying defects in the cohesion and integrity between endothelial cells, involving the trifecta of adherens, tight, and gap junctions. Offering further insight, mouse models have demonstrated that the absence of SOX17 in endothelial cells paves the way for IA formation. Furthermore, it disrupts VE-cadherin, a pivotal element in adherens junctions. This gives rise to the notion that proper cell-to-cell adhesion might be a protective barrier against an IA's formation. Diving deeper, it is believed that SOX17, acting as a transcription factor, might have a diverse set of downstream targets, with VE-cadherin and its regulators being among them [70].

Another intriguing gene that emerges in the context of IAs is THSD1. This gene holds the reigns to the assembly of adherens junctions, as evidenced by reduced or misplaced VE-cadherin in endothelial cells devoid of THSD1. As researchers peer into the future, a tantalizing prospect is uncovering whether THSD1 also operates under the influence of SOX17. This could potentially unveil an additional layer of governance over VE-cadherin's activity within endothelial cells [71].

Taking a deeper dive into THSD1 reveals it as a single-span transmembrane protein that intricately intertwines with FAK signaling and cellular adhesion. Proteomic studies originally spotlighted Thsd1 as a potential partner of talin based on mass spectrometry analyses of talin immunoprecipitation [72].

Current databases detailing protein–protein interactions have not yet painted a complete picture concerning Thsd1 interactions. A significant reason behind this gap might be the predominant presence of THSD1 in endothelial cells, thereby limiting the breadth of available data. But given the monumental role that talin plays in integrin-mediated focal adhesion, it is plausible that THSD1 could influence endothelial focal adhesion and FAK signaling. To further bolster this hypothesis, experiments on human umbilical vein endothelial cells have pointed towards THSD1′s regulatory role over focal adhesion stability. When THSD1 was silenced using RNA interference techniques, there was a noticeable dip in the number of focal adhesions, paralleled by a reduced capability of cells to adhere to collagen. Interestingly, subsequent rescue experiments threw light on how most THSD1 variants exhibited a diminished ability in both focal adhesion and cell attachment [62] (Figure 2).

MAPK Signaling: The MAPK (Mitogen-Activated Protein Kinase) pathway is a central hub of biological signaling, with a vast interconnected network. This intricate pathway splits into three main branches: the classical MAPK pathway, the c-Jun NH2-terminal kinase (JNK) and p38 MAPK pathway, and the ERK5 pathway. All of these play crucial roles in managing a cell's life cycle, covering everything from its proliferation and differentiation to inflammation and programmed cell death. When it comes to the cardiovascular system, MAPKs are present in various vascular cell types such as cardiomyocytes, vascular endothelial cells, and vascular smooth muscle cells (VSMCs). They are entrusted with the pivotal task of controlling cardiovascular signal transduction pathways. These kinases act as translators, converting signals from cytokines, growth factors, and environmental stressors (like ischemia and shear stress) into cellular responses. In-depth studies of the cardiovascular system have underscored the significance of these MAPK cascades, which govern various vital functions, including regulating vascular endothelial cell permeability, producing cytokines, modulating vasomotor function, and mediating the effects of reperfusion injuries [73]. In the context of intracranial aneurysms, 18 identified candidate genes have been found that are intricately linked with MAPK signaling. Some of the key players include MAPK7, TGFB3, and CDC42. Interestingly, genes encoding heat shock proteins, such as HSPA9B and HSPB1, also make their presence felt in IA. Further widening the MAPK network in intracranial arteries, other significant genes include EGFR, TGFBR1, and IL1B. Within this intricate web, HSP5A emerges as a potential candidate gene for IAs, with a major role in negatively regulating JNK. Similarly, DUSP10 encodes a dual-specificity MAPK phosphatase, functioning primarily in both innate and adaptive immune responses by inhibiting JNK and thus influencing the transcription factor activator protein-1 (AP1) [67].

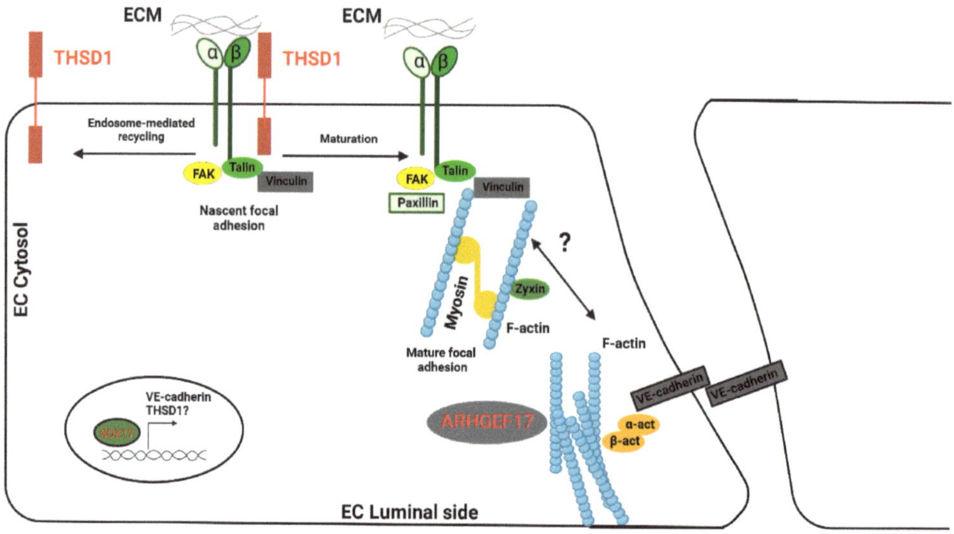

Figure 2. Diagram showing IA genes in vascular endothelial cells. Three key IA genes, THSD1, SOX17, and ARHGEF17, are emphasized in red. THSD1 interacts directly with the integrin complex via talin at emerging focal adhesions. As these focal adhesions mature, THSD1 relocates to a new emerging focal adhesion site through an endosome-based recycling mechanism. Without THSD1, there are impairments in focal adhesion, which is vital for the actin cytoskeleton and for influencing various subsequent pathways. SOX17 acts as a transcriptional regulator, affecting VE-cadherin expression. A decrease in VE-cadherin weakens cell-to-cell binding and raises permeability. ARHGEF17, a guanidine exchange factor, may play a role in the restructuring of the actin cytoskeleton.

Notch Signaling: The Notch signaling pathway wears multiple hats. It is instrumental in the development of cardiovascular structures within mammals, facilitating the formation of arteries and veins by acting upon endothelial and smooth muscle cells. This pathway's significance does not wane in adult vascular systems. Mutations in the NOTCH3 gene have been associated with cerebral autosomal dominant arteriopathy with subcortical infarcts and leukoencephalopathy (CADASIL). This condition stands out as the predominant hereditary stroke disorder [74]. Underlying CADASIL is the progressive degeneration of VSMCs within arteries. Mutations within the NOTCH3 gene lead to the anomalous accumulation of the Notch3 receptor at the VSMCs' cytoplasmic membrane within intracranial vessels [75]. CADASIL patients exhibit disruptions in the typical anchoring of VSMCs to the extracellular matrix and neighboring cells. Coupled with alterations to the cytoskeleton, this might set off the chain reaction, leading to VSMC degeneration [76]. VSMC loss brings about abnormalities in the endothelial cell tight and gap junctions. Such adhesion pathways are interwoven with others, and are responsible for maintaining cell junctions, involving not just adherens junctions but also cell adhesion molecules, actin cytoskeleton regulation, and the previously mentioned MAPK and Ca2 signaling pathways [77].

An intracranial aneurysm exhibits marked differences in the collagen gene expression levels, with a spike in transcription for TIMP-3. These observations hint at extensive extracellular matrix (ECM) remodeling within the aneurysm, aligning with the findings of elevated levels of matrix metalloproteases (MMPs)-2 and -9 in aneurysmal tissues [78].

Intracranial aneurysms showcase an overexpression of the factors believed to be active participants in collagen metabolism, including big-h3 and CTGF. The exact role of big-h3 remains an enigma, though it is thought to potentially bridge collagen interactions within the ECM [79]. Though big-h3's exact function remains a topic of ongoing research, it is postulated that it might serve as a bridge, facilitating or stabilizing the interactions between collagen and other integral ECM structures [80].

Another molecule of interest in this context is SPARC (osteonectin), a counter adhesive glycoprotein found across various tissues, from vascular endothelium and smooth muscles to fibroblasts. Beyond inhibiting endothelial cell adhesion and proliferation, elevated SPARC protein and mRNA levels have been documented in renal vascular injury scenarios [81]. The intricate dance of ECM remodeling may be choreographed, in part, by specific interactions between SPARC and type I collagen, highlighting the delicate balance and interplay within the vascular system [82]. Intriguingly, within the realm of acidic proteins, another significant member is hevin. Located primarily in the high endothelial venules of lymphoid tissues, hevin possesses antiadhesive properties. What is captivating is its striking similarity to SPARC. Drawing a parallel with SPARC, hevin's expression is noticeable within intracranial aneurysms, yet it remains conspicuously absent in STA [83].

Intracranial aneurysms reveal an overexpression of several factors that align with extracellular matrix (ECM) remodeling. However, the presence of some of these factors is less anticipated. A case in point is OSF-2, a transcription factor that is typically associated with cells from the osteoblast lineage, hinting at its primary role in bone-related functions [84]. While OSF-2 has been implicated in the activation process of collagenase-3 (MMP-13) during bone development, its association with arterial conditions remains largely uncharted territory. As of now, no conclusive evidence ties it to arterial pathologies [85]. Venturing deeper into the cellular world, cathepsins emerge as notable entities. Specifically, Cathepsin B, a lysosomal cysteine protease, plays a pivotal role in intracellular protein breakdown. Its involvement has been noted in conditions like cancer and chronic inflammatory diseases, particularly those affecting airways and joints. On the other hand, Cathepsin D, an aspartyl endoproteinase, boasts a ubiquitous presence in lysosomes [86]. While cathepsins are primarily recognized for their protein-degrading prowess in lysosomes and phagosomes, their influence extends beyond this. They play a cardinal role in activating biologically active protein precursors within the prelysosomal compartments of specialized cells [87]. The association of cathepsins with tumor progression, especially in the context of invasion and metastasis, is hard to overlook. Their ability to dissolve the ECM grants them a pivotal role in such processes. Due to this, cathepsins have found utility as prognostic markers, offering insights into the potential metastatic tendencies of tumors [88]. Although cathepsins exhibit characteristics that are synonymous with tissue remodeling in vascular structures, their roles in arterial disorders have not been thoroughly explored. In the past, various studies have pointed towards their potential involvement in the pathogenesis of abdominal aortic aneurysms, prompting a closer examination of their significance in arterial conditions [89]. However, recent studies show that in the progression of abdominal aortic aneurysm (AAA) formation, there is a marked elevation in the expression and enzymatic activity of cathepsins, as substantiated by several studies [90,91]. Through the application of immunohistochemical techniques and a Western blot analysis, it was elucidated that cathepsin L is absent or negligibly expressed in healthy vascular tissues. However, in contrast, AAA tissues exhibited robustly positive staining for cathepsin L. Further, a quantitative assessment of the mRNA levels revealed that cathepsin L expression in AAA lesions was augmented by 22% relative to that in normal aortic tissues. Moreover, when comparing protein levels in AAA lesions to those in aorta tissues acquired from patients diagnosed with an artery occlusion, cathepsin H exhibited a substantial increase, registering at 330% higher in AAA lesions. This significant disparity underscores the potential role and importance of cathepsins in the pathogenesis and progression of an AAA [92].

Inflammation Contribution

A defining characteristic of an intracranial aneurysm is the pervasive infiltration of inflammatory cells, which significantly contributes to its pathophysiology [64]. While various inflammatory cells find their way into the IA wall, macrophages stand out as the predominant infiltrators. Interestingly, both unruptured and ruptured IAs experience macrophage infiltration. However, ruptured IAs present a considerably higher degree of infiltration, highlighting a close link between vascular inflammation and the likelihood of IA rupture [93]. There is a pronounced correlation between the degree of leukocyte infiltration and degenerative changes in the IA wall. Such alterations include the notable loss of medial smooth muscle cells and the degradation of the extracellular matrix [94]. Studies utilizing animal models, such as mice and rats, have shed light on macrophages' roles in IAs. Notably, the early stages of experimentally induced IAs in these displayed animals can stymie the degenerative changes observed in the IA wall. Moreover, mast cell degranulation has been linked to the enhanced expression and activation of matrix metalloproteinases (MMPs)-2 and -9, and inducible nitric oxide synthase (iNOS) in primary cultured SMCs. Given mast cells' established roles in allergic inflammation, they seem to play integral roles in an IA's inflammatory response [95]. An apparent link exists between the humoral immune reaction, driven by inflammatory cells, and an IA's progression. Notably, antibodies like IgM and IgG predominantly line the luminal side of human IAs [96]. Tulamo et al.'s studies on human IA walls emphasized complement activation, especially in ruptured cases. They posited that this activation is correlated with an IA rupture and wall deterioration. As complement activation also triggers the release of pro-inflammatory cytokines, it might act as an ignition point for inflammatory cascades within the IA wall [97].

Endothelial cells are pivotal for maintaining vascular equilibrium. Significant morphological alterations in ECs within the IA wall have been documented, with noticeable gap formations at EC junctions. Furthermore, proteins that are integral for tight junctions, like occludin and zona occludens-1 (ZO-1), have been shown to be reduced in early stages of experimentally induced rat IAs [98]. Beyond mere morphological alterations, the functional impairment of the endothelium has garnered attention in an IA's pathogenesis. Endothelial dysfunction is manifested through the activation of proinflammatory genes. For instance, certain chemokines and cell adhesion molecules, expressed in ECs within the IA wall, facilitate macrophage infiltration. Monocyte chemotactic protein-1 (MCP-1) stands out for its pivotal role in directing monocyte/macrophage traffic to areas impacted by various vascular conditions [99]. Early-stage rat IA formation showcases an upregulation in MCP-1 gene expression. Experiments with MCP-1-deficient mice revealed a marked decline in IA formation. Furthermore, treatments employing a dominant negative mutant of MCP-1 effectively hindered IA progression in rats [100]. To anchor monocytes/macrophages, vascular cell adhesion molecule-1 (VCAM-1) plays a significant role, fostering the firm adhesion of monocytes to ECs. Investigations into human IA samples have reported VCAM-1 expression, with gene expression profiling further underscoring its heightened presence in IAs [101]. The exact role that VCAM-1 plays in the development of an intracranial aneurysm is still under investigation. However, VCAM-1 is emerging as a pivotal molecule bridging endothelial dysfunction with the accumulation of macrophages. The transcriptional factor family, nuclear factor-kappa B (NF-κB), is integral to this process. It oversees the expression of numerous genes, including VCAM-1 and MCP-1, in response to inflammatory stimuli [102]. The signaling connection between prostaglandin E2 and prostaglandin E receptor 2 serves as a conduit between hemodynamic stress and IA genesis, primarily through the activation of NF-κB. PGE2 itself is birthed from arachidonic acid, thanks to the sequential actions of cyclooxygenase and PGE synthase. Of the many variants of these two enzymes, COX-2 and microsomal PGES1 (mPGES1)—which are usually upregulated in numerous inflammatory diseases—also exhibited increased expression in the walls of both rat and human IAs. This expression pattern coincided with that of EP2 [103]. Studies have shown that both an EP2 deficiency and COX-2 inhibition led to a significant curtailing of NF-κB activation and IA development. Intriguingly, COX-2 expression in the IA wall was suppressed by NF-κB inhibition, hinting at a positive feedback loop between COX-2 and NF-κB via EP2. Given that excessive shear stress can stimulate COX-2 and EP2 in cultured endothelial cells, it is plausible that this feedback mechanism, initiated by such stress, perpetuates chronic inflammation in the IA wall [104]. Research spearheaded by

Jayaraman and colleagues has uncovered an elevation in TNF-α expression in human IA samples using both reverse transcription polymerase chain reaction (RT-PCR) and Western blotting methodologies [105]. The augmented presence of TNF-α was further corroborated in experimentally induced rat IAs. This increase was parallel to the heightened activity of the TNF-α converting enzyme (TACE), an enzyme that is instrumental in releasing TNF-α [106]. In the context of atherosclerosis, TLR-4 has been known to stimulate NF-κB activation in arterial walls. This expression was noted in the endothelial cell layers of both human and mouse IAs [107]. The concurrent expression patterns of TLR4 and NF-κB activation provide a solid ground for the hypothesis that the TNF-α/TLR4/NF-κB pathway could be instrumental in IA genesis. This, along with other observations, underscores the significance of TNF-α in IA pathogenesis [108]. Moreover, a study conducted by Wei et al. (2020) [109] showed that there is a connection between coronary artery ectasia (CAE) and inflammation, as well as between coronary artery aneurysm (CAA) and atherosclerosis, which are being intricately studied and are of interest in cardiovascular research. A notable characteristic of endothelial dysfunction is the diminished expression of eNOS. Rat IAs demonstrated a decline in eNOS expression when juxtaposed with the contralateral cerebral arterial wall devoid of IAs [110] (Figure 3).

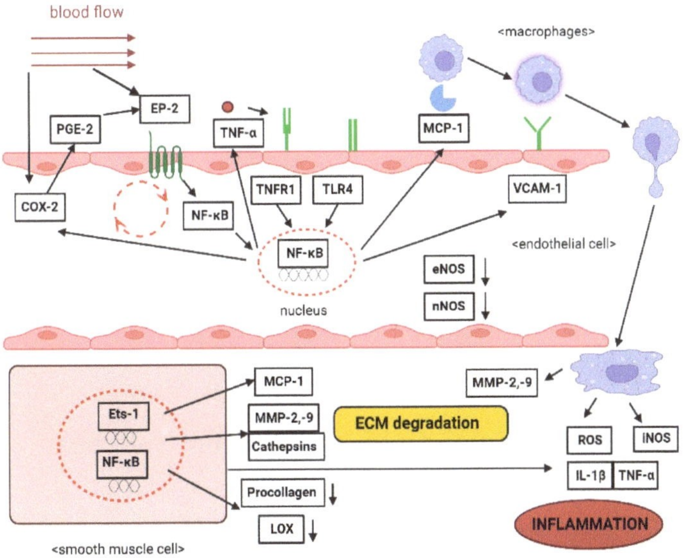

Figure 3. Inflammation pathways in the formation of intracranial aneurysms. COX-2 stands for cyclooxygenase-2; MCP-1 is known as monocyte chemotactic protein-1; PGE-2 refers to prostaglandin E2; MMP-2, -9 represent metalloproteinase types 2 and 9; EP-2 denotes prostaglandin receptor 2; ROS stands for reactive oxygen species; TNF-α is short for tumor necrosis factor-alpha; iNOS indicates inducible nitric oxide synthase; IL-1β refers to interleukin 1 beta; VCAM-1 represents vascular cell adhesion molecule-1; LOX is known as lysyl oxidase; NF-κB stands for nuclear factor kappa B; ECM is an abbreviation for extracellular matrix; TNFR1 is short for tumor necrosis factor receptor 1; TLR4 denotes Toll-like receptor 4; eNOS indicates endothelial nitric oxide synthase; nNOS stands for neuronal nitric oxide synthase.

IL-6, a pivotal proinflammatory cytokine, has been of interest in IA studies. Some researchers have shown an association between the IL-6 promoter polymorphism—572G>C—and IAs in Caucasians [111]. Moreover, the IL-6-572GG genotype was linked to a heightened IA risk in Chinese populations [112]. Both IL-12A and IL-12B seem to be implicated in IA susceptibility, either independently or jointly. Mounting evidence accentuates the importance of inflammatory processes in the onset and progression of IAs. Recent research has drawn connections between polymorphisms in the TP53 gene, inflammation, and inflammatory diseases [113]. Research has illuminated a significant association between IA and the rs6841581 polymorphism on chromosome 4q31.23, located just ahead of the endothelin receptor type A. Additionally, there is augmented evidence connecting IA to two other chromosomal regions: 12q22 and 20p12.1. These discoveries suggest that targeting the endothelin pathway could be pivotal for both the prevention and treatment of IAs [114].

A plethora of gene candidates implicated in the development of intracranial aneurysms are intertwined with the inflammatory cascade. This includes matrix metalloproteinases, transforming growth factor-beta (TGFβ) proteins, and endothelial nitric oxide synthase. Their presence underlines the functional significance of inflammation in the development of aneurysms. One primary mechanism involves the inflammatory response that becomes activated due to endothelial dysfunction and the degradation of the structural integrity of the cell wall [64]. There is a proposed model that outlines the process of aneurysm development initiated by inflammation. In this model, when inflammatory mediators are recruited to the cell wall, they disturb the internal elastic lamina, setting the stage for aneurysm formation. As the inflammation persists, the cellular wall undergoes more damage, leading to cell death and structural decay, and increasing the propensity for rupture. This model further gains strength when considering the role of nuclear factor-kappa B. Accumulating research underscores NF-κB's importance in the genesis of cerebral aneurysms, particularly through its facilitation of the inflammatory process that recruits and activates macrophages [115,116]. The role of the immune response in the development and rupture risk of IAs is not just theoretical; empirical evidence also supports this claim. Advancements in RNA sequencing technologies have unearthed the involvement of certain pathways and components in IA pathogenesis. Specifically, studies spotlighted the lysosomal pathway and various immunoglobulins as key elements in the context of IAs. These findings strengthen the hypothesis that the immune system's response is intricately woven into the tapestry of IA development and the subsequent risk of rupture [117].

Moreover, a key component that can contribute to the formation of an aneurysm through an inflammatory pathway is represented by the Renin–Angiotensin–Aldosterone System (RAAS).

A research study conducted by Cassis et al. (2009) [118] underscores a notable observation: ANG II-induced abdominal aortic aneurysms (AAAs) manifest through pathways that are not contingent upon blood pressure fluctuations. Furthermore, this investigation corroborates the notion that the exacerbation of atherosclerosis, when ANG II is introduced, transpires irrespective of the hypertensive effects attributed to ANG II.

It is scientifically fascinating to acknowledge that the infusion of ANG II can trigger two differentiated vascular malfunctions, both of which manifest without direct correlation to increased blood pressure metrics. Delving into the intricacies of ANG II's broader implications, it is plausible to postulate that its primary impact might be anchored in inflammatory processes inherent to these variegated vascular dysfunctions.

In a broader therapeutic context, these findings illuminate a promising avenue. Medications that target and inhibit the renin–angiotensin system might present a dual advantage. According to Zhong et al. (2022) [119], for hypertensive patients with intracranial aneurysms, the administration of RAAS inhibitors notably reduced the risk of rupture, irrespective of their effect on blood pressure control.

5. Conclusions and Future Perspectives

The results from comprehensive genomic expression studies related to intracranial aneurysms have laid the groundwork for subsequent investigations. There is evidence suggesting that microRNAs play crucial roles in regulating vascular remodeling. This knowledge could be instrumental in pinpointing targets for potential treatments that might stabilize or decelerate the progression of an IA. It is also essential to identify the transcription factors that govern the transcriptional regulation of gene expression. The concentration of messenger RNA and microRNA in the blood samples of individuals with an IA might offer diagnostic solutions, acting as biomarkers. Such data could assist in discerning markers indicative of an impending rupture. Moreover, employing techniques like laser-capture microdissection to segregate distinct cell groups found in the IA wall would facilitate a comprehensive understanding of their roles in the pathobiology of IAs.

Intracranial aneurysms are perceived as multifaceted conditions typified by their onset, progression, and potential rupture. They are influenced by a myriad of genetic and external risk determinants that may affect one or multiple stages of the disease. While determining the impact of all recognized determinants on the likelihood of an individual developing an IA, or on the progression and potential rupture of an existing IA, it is imperative to also account for the interplay among these factors. This includes considering external risk determinants like tobacco usage. Undertaking such extensive analyses necessitates large cohorts with exhaustive data, demanding the collaboration of multiple research centers. The overarching objective of these molecular and genetic explorations is to harness the derived knowledge to construct advanced risk evaluation algorithms. These algorithms could anticipate an individual's susceptibility to develop an IA, or monitor its potential progression or rupture, thereby proving to be invaluable in a clinical setting.

Another vital objective is to innovate therapeutic approaches based on insights into the mechanisms underlying the disease, with the aim to either preclude the onset or impede the progression of an IA. There is also a potential genetic commonality across diverse aneurysm types, as the same genetic variant on the 9p21 locus is associated with an increased risk for conditions such as myocardial infarction, abdominal aortic aneurysms (AAAs), and IAs. This suggests a predisposition towards a dysfunctional response to vascular injuries.

In recent times, the FIA consortium has propelled the study of the genetics of hereditary forms of IAs into the realm of whole-exome sequencing. Extensive familial histories, inclusive of American, Australian, and New Zealander populations affected by IAs, were amassed to scrutinize the entire coding sequences from fifty individuals, irrespective of whether they had an IA. Several pertinent genetic mutations were pinpointed, and there are ongoing efforts to further evaluate the affected genes [120].

There exists a comprehensive collection of studies in the scientific literature that associates genetic mutations and variations with the development and progression of intracranial aneurysms. The quality of these studies varies. It is likely that certain inherited genetic variations, with differing degrees of expressivity, contribute to the risk of IAs. The intricacies and full spectrum of interactions among multiple genes in relation to IAs still need to be explored in depth. Several genes have been identified as potential key players and deserve continued scrutiny. The rationale behind considering immune-altering agents for treatment might be based on the genetic underpinnings that drive IAs. Future research should focus on unbiased, comprehensive genomic analyses to uncover the roles of novel genes in the development and rupture of aneurysms. Such studies could offer crucial insights into potential biomarkers for this condition. As research progresses, there is a pressing need to create predictive models that integrate well-established genetic alterations. Given the advanced tools at our disposal in this era after the mapping of the human genome, comprehensive genomic studies involving large groups of patients diagnosed with IAs are essential to deepen our grasp on the genetic intricacies of this multifaceted disease.

In spite of significant strides in intracranial aneurysm research, the precise mechanisms through which blood flow induces inflammation remain elusive. It is unequivocal that changes in localized blood flow play pivotal roles in endothelial malfunction and adverse remodeling of the IA wall. The introduction of computational flow dynamics (CFD) in the early 2000s in IA research marked a significant advancement, offering a robust method for assessing flow characteristics within IAs and the associated mechanical properties of the aneurysm wall. Looking ahead, merging experimental approaches from molecular biology with engineering techniques holds promise to revolutionize our understanding of how flow dynamics influence biological alterations in the aneurysm wall. The specific mechanisms leading to IA rupture are still not entirely understood. Comparing ruptured and unruptured IAs has yielded insights into certain histopathological characteristics of those that rupture. In most unruptured IAs, inflammation and deterioration in the blood vessel wall undergo a natural healing process. Persistent inflammatory processes within the aneurysm wall, maintained by reinforcing feedback mechanisms, could provide insights into the molecular foundations of IA ruptures. In the coming years, imaging techniques such as MRI that can depict active inflammation within the aneurysm wall might become a pivotal diagnostic tool in identifying high-risk IAs that are more susceptible to rupture.

Moreover, screening should also be taken into account, since it plays an important role in preventing the rupturing of the aneurysm. Present guidelines do not recommend routine screening for intracranial aneurysms in the general public. However, they advise using CTA/MRA to screen for intracranial aneurysms in people who have at least two family members with intracranial aneurysms or subarachnoid hemorrhages. Additionally, screening is considered reasonable for patients with conditions like autosomal dominant polycystic kidney disease, fibromuscular dysplasia (FMD), aortic aneurysms, coarctation of the aorta, and microcephalic osteodysplastic primordial dwarfism, especially if they have a family history of these conditions [121].

Author Contributions: Conceptualization, C.T. and B.-G.B.; methodology, R.-A.C.-B.; software, R.-A.C.-B.; validation, H.P.C.; formal analysis, L.-A.G.; investigation, B.-G.B.; resources, D.-I.D.; data curation, D.-I.D.; writing—original draft preparation, R.-A.C.-B.; writing—review and editing, B.-G.B.; visualization, A.D.C.; supervision, L.E.; project administration, A.V.C.; funding acquisition, A.V.C. All authors have read and agreed to the published version of the manuscript.

Funding: This research received no external funding.

Institutional Review Board Statement: Not applicable.

Informed Consent Statement: Not applicable.

Data Availability Statement: All data is available online on libraries such as PubMed.

Conflicts of Interest: The authors declare no conflict of interest.

References

1. Brown, R. Unruptured Intracranial Aneurysms. *Semin. Neurol.* **2010**, *30*, 537–544. [CrossRef] [PubMed]
2. Rinkel, G.J.E.; Djibuti, M.; Algra, A.; Van Gijn, J. Prevalence and Risk of Rupture of Intracranial Aneurysms: A Systematic Review. *Stroke* **1998**, *29*, 251–256. [CrossRef] [PubMed]
3. Levy, M.L.; Levy, D.M.; Manna, B. Pediatric Cerebral Aneurysm. In *StatPearls*; StatPearls Publishing: Treasure Island, FL, USA, 2023. Available online: http://www.ncbi.nlm.nih.gov/books/NBK537085/ (accessed on 10 August 2023).
4. Menghini, V.V.; Brown, R.D.; Sicks, J.D.; O'Fallon, W.M.; Wiebers, D.O. Incidence and prevalence of intracranial aneurysms and hemorrhage in Olmsted County, Minnesota, 1965 to 1995. *Neurology* **1998**, *51*, 405–411. [CrossRef]
5. Ujiie, H.; Sato, K.; Onda, H.; Oikawa, A.; Kagawa, M.; Takakura, K.; Kobayashi, N. Clinical analysis of incidentally discovered unruptured aneurysms. *Stroke* **1993**, *24*, 1850–1856. [CrossRef] [PubMed]
6. The International Study of Unruptured Intracranial Aneurysms Investigators. Unruptured Intracranial Aneurysms—Risk of Rupture and Risks of Surgical Intervention. *N. Engl. J. Med.* **1998**, *339*, 1725–1733. [CrossRef] [PubMed]
7. Schackert, H.K.; Schackert, G.; Krex, D. Genesis of Cerebral Aneurysms—An Update. *Acta Neurochir.* **2001**, *143*, 429–449. [CrossRef] [PubMed]
8. Nakagawa, T.; Hashi, K. The incidence and treatment of asymptomatic, unruptured cerebral aneurysms. *J. Neurosurg.* **1994**, *80*, 217–223. [CrossRef]
9. De Rooij, N.K.; Linn, F.H.H.; Van Der Plas, J.A.; Algra, A.; Rinkel, G.J.E. Incidence of subarachnoid haemorrhage: A systematic review with emphasis on region, age, gender and time trends. *J. Neurol. Neurosurg. Psychiatry* **2007**, *78*, 1365–1372. [CrossRef]
10. Johnston, S.C.; Selvin, S.; Gress, D.R. The burden, trends, and demographics of mortality from subarachnoid hemorrhage. *Neurology* **1998**, *50*, 1413–1418. [CrossRef]
11. Grobelny, T.J. Brain Aneurysms: Epidemiology, Treatment Options, and Milestones of Endovascular Treatment Evolution. *Dis. Mon.* **2011**, *57*, 647–655. [CrossRef]
12. Kubota, M.; Yamaura, A.; Ono, J. Prevalence of risk factors for aneurysmal subarachnoid haemorrhage: Results of a Japanese multicentre case control study for stroke. *Br. J. Neurosurg.* **2001**, *15*, 474–478. [CrossRef] [PubMed]
13. Clarke, M. Systematic review of reviews of risk factors for intracranial aneurysms. *Neuroradiology* **2008**, *50*, 653–664. [CrossRef] [PubMed]

14. Kernan, W.N.; Viscoli, C.M.; Brass, L.M.; Broderick, J.P.; Brott, T.; Feldmann, E.; Morgenstern, L.B.; Wilterdink, J.L.; Horwitz, R.I. Phenylpropanolamine and the Risk of Hemorrhagic Stroke. *N. Engl. J. Med.* **2000**, *343*, 1826–1832. [CrossRef] [PubMed]
15. Greving, J.P.; Wermer, M.J.H.; Brown, R.D.; Morita, A.; Juvela, S.; Yonekura, M.; Ishibashi, T.; Torner, J.C.; Nakayama, T.; Rinkel, G.J.E.; et al. Development of the PHASES score for prediction of risk of rupture of intracranial aneurysms: A pooled analysis of six prospective cohort studies. *Lancet Neurol.* **2014**, *13*, 59–66. [CrossRef] [PubMed]
16. Raaymakers, T.W.M. Aneurysms in relatives of patients with subarachnoid hemorrhage: Frequency and risk factors. *Neurology* **1999**, *53*, 982. [CrossRef] [PubMed]
17. Nahed, B.V.; Bydon, M.; Ozturk, A.K.; Bilguvar, K.; Bayrakli, F.; Gunel, M. Genetics of intracranial aneurysms. *Neurosurgery* **2007**, *60*, 213–226. [CrossRef] [PubMed]
18. Cagnazzo, F.; Gambacciani, C.; Morganti, R.; Perrini, P. Intracranial aneurysms in patients with autosomal dominant polycystic kidney disease: Prevalence, risk of rupture, and management. A systematic review. *Acta Neurochir.* **2017**, *159*, 811–821. [CrossRef] [PubMed]
19. Pirson, Y. Extrarenal Manifestations of Autosomal Dominant Polycystic Kidney Disease. *Adv. Chronic Kidney Dis.* **2010**, *17*, 173–180. [CrossRef]
20. Vlak, M.H.; Algra, A.; Brandenburg, R.; Rinkel, G.J. Prevalence of unruptured intracranial aneurysms, with emphasis on sex, age, comorbidity, country, and time period: A systematic review and meta-analysis. *Lancet Neurol.* **2011**, *10*, 626–636. [CrossRef]
21. Harburger, D.S.; Calderwood, D.A. Integrin signalling at a glance. *J. Cell Sci.* **2009**, *122*, 159–163. [CrossRef]
22. Duker, A.; Jackson, A.; Bober, M.B. Microcephalic Osteodysplastic Primordial Dwarfism Type II. In *GeneReviews®*; Adam, M.P., Mirzaa, G.M., Pagon, R.A., Wallace, S.E., Bean, L.J., Gripp, K.W., Amemiya, A., Eds.; University of Washington: Seattle, WA, USA, 1993. Available online: http://www.ncbi.nlm.nih.gov/books/NBK575926/ (accessed on 10 August 2023).
23. Duker, A.L.; Kinderman, D.; Jordan, C.; Niiler, T.; Baker-Smith, C.M.; Thompson, L.; Parry, D.A.; Carroll, R.S.; Bober, M.B. Microcephalic osteodysplastic primordial dwarfism type II is associated with global vascular disease. *Orphanet J. Rare Dis.* **2021**, *16*, 231. [CrossRef] [PubMed]
24. De Paepe, A.; Malfait, F. The Ehlers-Danlos syndrome, a disorder with many faces. *Clin. Genet.* **2012**, *82*, 1–11. [CrossRef] [PubMed]
25. Superti-Furga, A.; Gugler, E.; Gitzelmann, R.; Steinmann, B. Ehlers-Danlos syndrome type IV: A multi-exon deletion in one of the two COL3A1 alleles affecting structure, stability, and processing of type III procollagen. *J. Biol. Chem.* **1988**, *263*, 6226–6232. [CrossRef] [PubMed]
26. Bober, M.B.; Khan, N.; Kaplan, J.; Lewis, K.; Feinstein, J.A.; Scott, C.I.; Steinberg, G.K. Majewski Osteodysplastic Primordial Dwarfism Type II (MOPD II): Expanding the vascular phenotype. *Am. J. Med. Genet. A* **2010**, *152A*, 960–965. [CrossRef] [PubMed]
27. Li, F.-F.; Wang, X.-D.; Zhu, M.-W.; Lou, Z.-H.; Zhang, Q.; Zhu, C.-Y.; Feng, H.-L.; Lin, Z.-G.; Liu, S.-L. Identification of two novel critical mutations in PCNT gene resulting in microcephalic osteodysplastic primordial dwarfism type II associated with multiple intracranial aneurysms. *Metab. Brain Dis.* **2015**, *30*, 1387–1394. [CrossRef]
28. Jurczyk, A.; Gromley, A.; Redick, S.; Agustin, J.S.; Witman, G.; Pazour, G.J.; Peters, D.J.M.; Doxsey, S. Pericentrin forms a complex with intraflagellar transport proteins and polycystin-2 and is required for primary cilia assembly. *J. Cell Biol.* **2004**, *166*, 637–643. [CrossRef] [PubMed]
29. Lorenzo-Betancor, O.; Blackburn, P.R.; Edwards, E.; Vázquez-do-Campo, R.; Klee, E.W.; Labbé, C.; Hodges, K.; Glover, P.; Sigafoos, A.N.; Soto, A.I.; et al. *PCNT* point mutations and familial intracranial aneurysms. *Neurology* **2018**, *91*, e2170–e2181. [CrossRef] [PubMed]
30. Kim, S.T.; Brinjikji, W.; Kallmes, D.F. Prevalence of Intracranial Aneurysms in Patients with Connective Tissue Diseases: A Retrospective Study. *Am. J. Neuroradiol.* **2016**, *37*, 1422–1426. [CrossRef]
31. Arnaud, P.; Milleron, O.; Hanna, N.; Ropers, J.; Ould Ouali, N.; Affoune, A.; Langeois, M.; Eliahou, L.; Arnoult, F.; Renard, P.; et al. Clinical relevance of genotype–phenotype correlations beyond vascular events in a cohort study of 1500 Marfan syndrome patients with FBN1 pathogenic variants. *Genet. Med.* **2021**, *23*, 1296–1304. [CrossRef]
32. Conway, J.E.; Hutchins, G.M.; Tamargo, R.J. Marfan Syndrome Is Not Associated With Intracranial Aneurysms. *Stroke* **1999**, *30*, 1632–1636. [CrossRef]
33. Van Den Berg, J.S.P.; Limburg, M.; Hennekam, R.C.M. Is Marfan Syndrome Associated With Symptomatic Intracranial Aneurysms? *Stroke* **1996**, *27*, 10–12. [CrossRef] [PubMed]
34. Le, C.; Bedocs, P.M. Neurofibromatosis. In *StatPearls*; StatPearls Publishing: Treasure Island, FL, USA, 2023. Available online: http://www.ncbi.nlm.nih.gov/books/NBK459329/ (accessed on 10 August 2023).
35. Schievink, W.I.; Riedinger, M.; Maya, M.M. Frequency of incidental intracranial aneurysms in neurofibromatosis type 1. *Am. J. Med. Genet. Part A* **2005**, *134A*, 45–48. [CrossRef] [PubMed]
36. Broadbent, H.M.; Peden, J.F.; Lorkowski, S.; Goel, A.; Ongen, H.; Green, F.; Clarke, R.; Collins, R.; Franzosi, M.G.; Tognoni, G.; et al. Susceptibility to coronary artery disease and diabetes is encoded by distinct, tightly linked SNPs in the ANRIL locus on chromosome 9p. *Hum. Mol. Genet.* **2008**, *17*, 806–814. [CrossRef] [PubMed]
37. Zhou, S.; Dion, P.A.; Rouleau, G.A. Genetics of Intracranial Aneurysms. *Stroke* **2018**, *49*, 780–787. [CrossRef] [PubMed]

38. Salvi, E.; Kutalik, Z.; Glorioso, N.; Benaglio, P.; Frau, F.; Kuznetsova, T.; Arima, H.; Hoggart, C.; Tichet, J.; Nikitin, Y.P.; et al. Genomewide Association Study Using a High-Density Single Nucleotide Polymorphism Array and Case-Control Design Identifies a Novel Essential Hypertension Susceptibility Locus in the Promoter Region of Endothelial NO Synthase. *Hypertension* **2012**, *59*, 248–255. [CrossRef] [PubMed]
39. Yu, T.; Jiang, H.; Fan, Y.; Xu, Y.; Wang, N. The association of CDKN2BAS gene polymorphisms and intracranial aneurysm: A meta-analysis. *Medicine* **2020**, *99*, e23209. [CrossRef] [PubMed]
40. Adachi, K.; Kudo, M.; Chen, M.N.; Nakazawa, S.; Wakabayashi, I. Cerebral Aneurysm Associated with Multiple Endocrine Neoplasia, Type 1—Case Report. *Neurol. Med. Chir.* **1993**, *33*, 309–311. [CrossRef]
41. Bock, A.; Schwegler, G. Intracerebral haemorrhage as first manifestation of Pseudoxanthoma elasticum. *Clin. Neurol. Neurosurg.* **2008**, *110*, 262–264. [CrossRef]
42. Willemse, R.B.; Mager, J.J.; Westermann, C.J.J.; Overtoom, T.T.C.; Mauser, H.; Wolbers, J.G. Bleeding risk of cerebral vascular malformations in hereditary hemorrhagic telangiectasia. *J. Neurosurg.* **2000**, *92*, 779–784. [CrossRef]
43. Hitchcock, E.; Gibson, W.T. A Review of the Genetics of Intracranial Berry Aneurysms and Implications for Genetic Counseling. *J. Genet. Couns.* **2017**, *26*, 21–31. [CrossRef]
44. Tromp, G.; Weinsheimer, S.; Ronkainen, A.; Kuivaniemi, H. Molecular basis and genetic predisposition to intracranial aneurysm. *Ann. Med.* **2014**, *46*, 597–606. [CrossRef] [PubMed]
45. Nadeau, J.H. Modifier genes in mice and humans. *Nat. Rev. Genet.* **2001**, *2*, 165–174. [CrossRef] [PubMed]
46. You, C.; Chen, Z.; Ma, J.; Cen, Y.; Liu, Y. The angiotensin converting enzyme insertion/deletion polymorphism and intracranial aneurysm: A meta-analysis of case-control studies. *Neurol. India* **2013**, *61*, 293. [CrossRef] [PubMed]
47. Yasuno, K.; Bilguvar, K.; Bijlenga, P.; Low, S.-K.; Krischek, B.; Auburger, G.; Simon, M.; Krex, D.; Arlier, Z.; Nayak, N.; et al. Genome-wide association study of intracranial aneurysm identifies three new risk loci. *Nat. Genet.* **2010**, *42*, 420–425. [CrossRef] [PubMed]
48. Moitra, K.; Garcia, S.; Jaldin, M.; Etoundi, C.; Cooper, D.; Roland, A.; Dixon, P.; Reyes, S.; Turan, S.; Terry, S.; et al. ABCC6 and Pseudoxanthoma Elasticum: The Face of a Rare Disease from Genetics to Advocacy. *Int. J. Mol. Sci.* **2017**, *18*, 1488. [CrossRef] [PubMed]
49. Samuel, N.; Radovanovic, I. Genetic basis of intracranial aneurysm formation and rupture: Clinical implications in the postgenomic era. *Neurosurg. Focus* **2019**, *47*, E10. [CrossRef] [PubMed]
50. Vergouwen, M.D.I.; Frijns, C.J.M.; Roos, Y.B.W.E.M.; Rinkel, G.J.E.; Baas, F.; Vermeulen, M. Plasminogen Activator Inhibitor-1 4G Allele in the 4G/5G Promoter Polymorphism Increases the Occurrence of Cerebral Ischemia After Aneurysmal Subarachnoid Hemorrhage. *Stroke* **2004**, *35*, 1280–1283. [CrossRef] [PubMed]
51. Xiang, C.; Liu, S.; Fan, Y.; Wang, X.; Jia, Y.; Li, L.; Cong, S.; Han, F. Single nucleotide polymorphisms, variable number tandem repeats and allele influence on serotonergic enzyme modulators for aggressive and suicidal behaviors: A review. *Pharmacol. Biochem. Behav.* **2019**, *180*, 74–82. [CrossRef]
52. Concannon, P.; Erlich, H.A.; Julier, C.; Morahan, G.; Nerup, J.; Pociot, F.; Todd, J.A.; Rich, S.S. Type 1 Diabetes: Evidence for Susceptibility Loci from Four Genome-Wide Linkage Scans in 1,435 Multiplex Families. *Diabetes* **2005**, *54*, 2995–3001. [CrossRef]
53. Wilson, F.H.; Disse-Nicodème, S.; Choate, K.A.; Ishikawa, K.; Nelson-Williams, C.; Desitter, I.; Gunel, M.; Milford, D.V.; Lipkin, G.W.; Achard, J.-M.; et al. Human Hypertension Caused by Mutations in WNK Kinases. *Science* **2001**, *293*, 1107–1112. [CrossRef]
54. Nahed, B.V.; Seker, A.; Guclu, B.; Ozturk, A.K.; Finberg, K.; Hawkins, A.A.; DiLuna, M.L.; State, M.; Lifton, R.P.; Gunel, M. Mapping a Mendelian Form of Intracranial Aneurysm to 1p34.3-p36.13. *Am. J. Hum. Genet.* **2005**, *76*, 172–179. [CrossRef] [PubMed]
55. Yamada, S.; Utsunomiya, M.; Inoue, K.; Nozaki, K.; Inoue, S.; Takenaka, K.; Hashimoto, N.; Koizumi, A. Genome-Wide Scan for Japanese Familial Intracranial Aneurysms: Linkage to Several Chromosomal Regions. *Circulation* **2004**, *110*, 3727–3733. [CrossRef] [PubMed]
56. Onda, H.; Kasuya, H.; Yoneyama, T.; Takakura, K.; Hori, T.; Takeda, J.; Nakajima, T.; Inoue, I. Genomewide-Linkage and Haplotype-Association Studies Map Intracranial Aneurysm to Chromosome 7q11. *Am. J. Hum. Genet.* **2001**, *69*, 804–819. [CrossRef] [PubMed]
57. Alg, V.S.; Sofat, R.; Houlden, H.; Werring, D.J. Genetic risk factors for intracranial aneurysms: A meta-analysis in more than 116,000 individuals. *Neurology* **2013**, *80*, 2154–2165. [CrossRef] [PubMed]
58. Foroud, T.; Koller, D.L.; Lai, D.; Sauerbeck, L.; Anderson, C.; Ko, N.; Deka, R.; Mosley, T.H.; Fornage, M.; Woo, D.; et al. Genome-Wide Association Study of Intracranial Aneurysms Confirms Role of Anril and SOX17 in Disease Risk. *Stroke* **2012**, *43*, 2846–2852. [CrossRef] [PubMed]
59. Wang, X.; Douglas, S.A.; Louden, C.; Vickery-Clark, L.M.; Feuerstein, G.Z.; Ohlstein, E.H. Expression of Endothelin-1, Endothelin-3, Endothelin-Converting Enzyme-1, and Endothelin-A and Endothelin-B Receptor mRNA After Angioplasty-Induced Neointimal Formation in the Rat. *Circ. Res.* **1996**, *78*, 322–328. [CrossRef] [PubMed]
60. Low, S.-K.; Takahashi, A.; Cha, P.-C.; Zembutsu, H.; Kamatani, N.; Kubo, M.; Nakamura, Y. Genome-wide association study for intracranial aneurysm in the Japanese population identifies three candidate susceptible loci and a functional genetic variant at EDNRA. *Hum. Mol. Genet.* **2012**, *21*, 2102–2110. [CrossRef] [PubMed]

61. Yasuno, K.; Bakırcıoğlu, M.; Low, S.-K.; Bilgüvar, K.; Gaál, E.; Ruigrok, Y.M.; Niemelä, M.; Hata, A.; Bijlenga, P.; Kasuya, H.; et al. Common variant near the endothelin receptor type A (*EDNRA*) gene is associated with intracranial aneurysm risk. *Proc. Natl. Acad. Sci. USA* **2011**, *108*, 19707–19712. [CrossRef]
62. Keedy, A. An overview of intracranial aneurysms. *McGill J. Med. MJM Int. Forum Adv. Med. Sci. Stud.* **2006**, *9*, 141–146. [CrossRef]
63. Krings, T.; Mandell, D.M.; Kiehl, T.-R.; Geibprasert, S.; Tymianski, M.; Alvarez, H.; terBrugge, K.G.; Hans, F.-J. Intracranial aneurysms: From vessel wall pathology to therapeutic approach. *Nat. Rev. Neurol.* **2011**, *7*, 547–559. [CrossRef]
64. Chalouhi, N.; Ali, M.S.; Jabbour, P.M.; Tjoumakaris, S.I.; Gonzalez, L.F.; Rosenwasser, R.H.; Koch, W.J.; Dumont, A.S. Biology of Intracranial Aneurysms: Role of Inflammation. *J. Cereb. Blood Flow Metab.* **2012**, *32*, 1659–1676. [CrossRef] [PubMed]
65. Fennell, V.S.; Kalani, M.Y.S.; Atwal, G.; Martirosyan, N.L.; Spetzler, R.F. Biology of Saccular Cerebral Aneurysms: A Review of Current Understanding and Future Directions. *Front. Surg.* **2016**, *3*, 43. [CrossRef] [PubMed]
66. Tromp, G.; Kuivaniemi, H. Developments in Genomics to Improve Understanding, Diagnosis and Management of Aneurysms and Peripheral Artery Disease. *Eur. J. Vasc. Endovasc. Surg.* **2009**, *38*, 676–682. [CrossRef] [PubMed]
67. Weinsheimer, S.; Lenk, G.M.; Van Der Voet, M.; Land, S.; Ronkainen, A.; Alafuzoff, I.; Kuivaniemi, H.; Tromp, G. Integration of expression profiles and genetic mapping data to identify candidate genes in intracranial aneurysm. *Physiol. Genom.* **2007**, *32*, 45–57. [CrossRef] [PubMed]
68. Bazzoni, G.; Dejana, E. Endothelial Cell-to-Cell Junctions: Molecular Organization and Role in Vascular Homeostasis. *Physiol. Rev.* **2004**, *84*, 869–901. [CrossRef]
69. Verlaan, D.J. A new locus for autosomal dominant intracranial aneurysm, ANIB4, maps to chromosome 5p15.2-14.3. *J. Med. Genet.* **2006**, *43*, e31. [CrossRef] [PubMed]
70. Nakajima-Takagi, Y.; Osawa, M.; Oshima, M.; Takagi, H.; Miyagi, S.; Endoh, M.; Endo, T.A.; Takayama, N.; Eto, K.; Toyoda, T.; et al. Role of SOX17 in hematopoietic development from human embryonic stem cells. *Blood* **2013**, *121*, 447–458. [CrossRef] [PubMed]
71. Haasdijk, R.A.; Den Dekker, W.K.; Cheng, C.; Tempel, D.; Szulcek, R.; Bos, F.L.; Hermkens, D.M.A.; Chrifi, I.; Brandt, M.M.; Van Dijk, C.; et al. *THSD1* preserves vascular integrity and protects against intraplaque haemorrhaging in ApoE$^{-/-}$ mice. *Cardiovasc. Res.* **2016**, *110*, 129–139. [CrossRef]
72. De Hoog, C.L.; Foster, L.J.; Mann, M. RNA and RNA Binding Proteins Participate in Early Stages of Cell Spreading through Spreading Initiation Centers. *Cell* **2004**, *117*, 649–662. [CrossRef]
73. Khan, T.A.; Bianchi, C.; Ruel, M.; Voisine, P.; Sellke, F.W. Mitogen-activated protein kinase pathways and cardiac surgery. *J. Thorac. Cardiovasc. Surg.* **2004**, *127*, 806–811. [CrossRef]
74. Joutel, A.; Andreux, F.; Gaulis, S.; Domenga, V.; Cecillon, M.; Battail, N.; Piga, N.; Chapon, F.; Godfrain, C.; Tournier-Lasserve, E. The ectodomain of the Notch3 receptor accumulates within the cerebrovasculature of CADASIL patients. *J. Clin. Investig.* **2000**, *105*, 597–605. [CrossRef] [PubMed]
75. Bianchi, S.; Dotti, M.T.; Federico, A. Physiology and pathology of notch signalling system. *J. Cell. Physiol.* **2006**, *207*, 300–308. [CrossRef] [PubMed]
76. Ruchoux, M.M.; Domenga, V.; Brulin, P.; Maciazek, J.; Limol, S.; Tournier-Lasserve, E.; Joutel, A. Transgenic mice expressing mutant Notch3 develop vascular alterations characteristic of cerebral autosomal dominant arteriopathy with subcortical infarcts and leukoencephalopathy. *Am. J. Pathol.* **2003**, *162*, 329–342. [CrossRef] [PubMed]
77. Brulin, P.; Godfraind, C.; Leteurtre, E.; Ruchoux, M.-M. Morphometric analysis of ultrastructural vascular changes in CADASIL: Analysis of 50 skin biopsy specimens and pathogenic implications. *Acta Neuropathol.* **2002**, *104*, 241–248. [CrossRef] [PubMed]
78. Kim, S.C.; Singh, M.; Huang, J.; Prestigiacomo, C.J.; Winfree, C.J.; Solomon, R.A.; Connolly, E.S.C. Matrix Metalloproteinase-9 in Cerebral Aneurysms. *Neurosurgery* **1997**, *41*, 642–647. [CrossRef] [PubMed]
79. Peters, D.G.; Kassam, A.B.; Feingold, E.; Heidrich-O'Hare, E.; Yonas, H.; Ferrell, R.E.; Brufsky, A. Molecular Anatomy of an Intracranial Aneurysm: Coordinated Expression of Genes Involved in Wound Healing and Tissue Remodeling. *Stroke* **2001**, *32*, 1036–1042. [CrossRef] [PubMed]
80. Gibson, M.A.; Kumaratilake, J.S.; Cleary, E.G. Immunohistochemical and Ultrastructural Localization of MP78/70 (βig-h3) in Extracellular Matrix of Developing and Mature Bovine Tissues. *J. Histochem. Cytochem.* **1997**, *45*, 1683–1696. [CrossRef]
81. Pichler, R.H.; Hugo, C.; Shankland, S.J.; Reed, M.J.; Bassuk, J.A.; Andoh, T.F.; Lombardi, D.M.; Schwartz, S.M.; Bennett, W.M.; Alpers, C.E.; et al. SPARC is expressed in renal interstitial fibrosis and in renal vascular injury. *Kidney Int.* **1996**, *50*, 1978–1989. [CrossRef]
82. Iruela-Arispe, M.L.; Vernon, R.B.; Wu, H.; Jaenisch, R.; Sage, E.H. Type I collagen-deficient Mov-13 mice do not retain SPARC in the extracellular matrix: Implications for fibroblast function. *Dev. Dyn.* **1996**, *207*, 171–183. [CrossRef]
83. Girard, J.-P.; Springer, T.A. Cloning from purified high endothelial venule cells of hevin, a close relative of the antiadhesive extracellular matrix protein SPARC. *Immunity* **1995**, *2*, 113–123. [CrossRef]
84. Ducy, P.; Zhang, R.; Geoffroy, V.; Ridall, A.L.; Karsenty, G. Osf2/Cbfa1: A Transcriptional Activator of Osteoblast Differentiation. *Cell* **1997**, *89*, 747–754. [CrossRef] [PubMed]
85. Jiménez, M.J.; Balbín, M.; López, J.M.; Alvarez, J.; Komori, T.; López-Otín, C. Collagenase 3 is a target of Cbfa1, a transcription factor of the runt gene family involved in bone formation. *Mol. Cell. Biol.* **1999**, *19*, 4431–4442. [CrossRef] [PubMed]
86. Barrett, A.J. Human cathepsin B1. Purification and some properties of the enzyme. *Biochem. J.* **1973**, *131*, 809–822. [CrossRef] [PubMed]

87. Diment, S.; Martin, K.J.; Stahl, P.D. Cleavage of parathyroid hormone in macrophage endosomes illustrates a novel pathway for intracellular processing of proteins. *J. Biol. Chem.* **1989**, *264*, 13403–13406. [CrossRef] [PubMed]
88. Barrett, A.J. Cathepsin D: The Lysosomal Aspartic Proteinase. In *Novartis Foundation Symposia*; Evered, D., Whelan, J., Eds.; Wiley: Hoboken, NJ, USA, 1980; Volume 75, pp. 37–50. [CrossRef]
89. Gacko, M.; Chyczewski, L. Activity and localization of cathepsin B, D and G in aortic aneurysm. *Int. Surg.* **1997**, *82*, 398–402. [PubMed]
90. Abisi, S.; Burnand, K.G.; Waltham, M.; Humphries, J.; Taylor, P.R.; Smith, A. Cysteine protease activity in the wall of abdominal aortic aneurysms. *J. Vasc. Surg.* **2007**, *46*, 1260–1266. [CrossRef] [PubMed]
91. Lohoefer, F.; Reeps, C.; Lipp, C.; Rudelius, M.; Zimmermann, A.; Ockert, S.; Eckstein, H.-H.; Pelisek, J. Histopathological analysis of cellular localization of cathepsins in abdominal aortic aneurysm wall. *Int. J. Exp. Pathol.* **2012**, *93*, 252–258. [CrossRef] [PubMed]
92. Wang, Y.; Tang, C.; Qin, Y. Cathepsins: A new culprit behind abdominal aortic aneurysm. *Regen. Med. Res.* **2013**, *1*, 5. [CrossRef]
93. Frösen, J.; Piippo, A.; Paetau, A.; Kangasniemi, M.; Niemelä, M.; Hernesniemi, J.; Jääskeläinen, J. Remodeling of Saccular Cerebral Artery Aneurysm Wall Is Associated With Rupture: Histological Analysis of 24 Unruptured and 42 Ruptured Cases. *Stroke* **2004**, *35*, 2287–2293. [CrossRef]
94. Kataoka, K.; Taneda, M.; Asai, T.; Kinoshita, A.; Ito, M.; Kuroda, R. Structural Fragility and Inflammatory Response of Ruptured Cerebral Aneurysms: A Comparative Study Between Ruptured and Unruptured Cerebral Aneurysms. *Stroke* **1999**, *30*, 1396–1401. [CrossRef]
95. Ishibashi, R.; Aoki, T.; Nishimura, M.; Hashimoto, N.; Miyamoto, S. Contribution of Mast Cells to Cerebral Aneurysm Formation. *Curr. Neurovasc. Res.* **2010**, *7*, 113–124. [CrossRef] [PubMed]
96. Tulamo, R.; Frösen, J.; Junnikkala, S.; Paetau, A.; Pitkäniemi, J.; Kangasniemi, M.; Niemelä, M.; Jääskeläinen, J.; Jokitalo, E.; Karatas, A.; et al. Complement activation associates with saccular cerebral artery aneurysm wall degeneration and rupture. *Neurosurgery* **2006**, *59*, 1069–1077. [CrossRef] [PubMed]
97. Tulamo, R.; Frösen, J.; Junnikkala, S.; Paetau, A.; Kangasniemi, M.; Peláez, J.; Hernesniemi, J.; Niemelä, M.; Meri, S. Complement system becomes activated by the classical pathway in intracranial aneurysm walls. *Lab. Investig.* **2010**, *90*, 168–179. [CrossRef] [PubMed]
98. Tada, Y.; Yagi, K.; Kitazato, K.T.; Tamura, T.; Kinouchi, T.; Shimada, K.; Matsushita, N.; Nakajima, N.; Satomi, J.; Kageji, T.; et al. Reduction of endothelial tight junction proteins is related to cerebral aneurysm formation in rats. *J. Hypertens.* **2010**, *28*, 1883–1891. [CrossRef] [PubMed]
99. Boring, L.; Gosling, J.; Cleary, M.; Charo, I.F. Decreased lesion formation in CCR2$^{-/-}$ mice reveals a role for chemokines in the initiation of atherosclerosis. *Nature* **1998**, *394*, 894–897. [CrossRef] [PubMed]
100. Aoki, T.; Kataoka, H.; Ishibashi, R.; Nozaki, K.; Egashira, K.; Hashimoto, N. Impact of Monocyte Chemoattractant Protein-1 Deficiency on Cerebral Aneurysm Formation. *Stroke* **2009**, *40*, 942–951. [CrossRef]
101. Kadirvel, R.; Ding, Y.-H.; Dai, D.; Lewis, D.A.; Raghavakaimal, S.; Cloft, H.J.; Kallmes, D.F. Gene Expression Profiling of Experimental Saccular Aneurysms Using Deoxyribonucleic Acid Microarrays: Fig 1. *Am. J. Neuroradiol.* **2008**, *29*, 1566–1569. [CrossRef]
102. Collins, T.; Read, M.A.; Neish, A.S.; Whitley, M.Z.; Thanos, D.; Maniatis, T. Transcriptional regulation of endothelial cell adhesion molecules: NF-kappa B and cytokine-inducible enhancers. *FASEB J. Off. Publ. Fed. Am. Soc. Exp. Biol.* **1995**, *9*, 899–909. [CrossRef]
103. Tilley, S.L.; Coffman, T.M.; Koller, B.H. Mixed messages: Modulation of inflammation and immune responses by prostaglandins and thromboxanes. *J. Clin. Investig.* **2001**, *108*, 15–23. [CrossRef]
104. Hasan, D.; Hashimoto, T.; Kung, D.; Macdonald, R.L.; Winn, H.R.; Heistad, D. Upregulation of Cyclooxygenase-2 (COX-2) and Microsomal Prostaglandin E_2 Synthase-1 (mPGES-1) in Wall of Ruptured Human Cerebral Aneurysms: Preliminary Results. *Stroke* **2012**, *43*, 1964–1967. [CrossRef]
105. Jayaraman, T.; Berenstein, V.; Li, X.; Mayer, J.; Silane, M.; Shin, Y.S.; Niimi, Y.; Kılıç, T.; Gunel, M.; Berenstein, A. Tumor Necrosis Factor α is a Key Modulator of Inflammation in Cerebral Aneurysms. *Neurosurgery* **2005**, *57*, 558–564. [CrossRef] [PubMed]
106. Aoki, T.; Fukuda, M.; Nishimura, M.; Nozaki, K.; Narumiya, S. Critical role of TNF-alpha-TNFR1 signaling in intracranial aneurysm formation. *Acta Neuropathol. Commun.* **2014**, *2*, 34. [CrossRef] [PubMed]
107. Björkbacka, H. Multiple roles of Toll-like receptor signaling in atherosclerosis. *Curr. Opin. Lipidol.* **2006**, *17*, 527–533. [CrossRef] [PubMed]
108. Aoki, T.; Nishimura, M.; Ishibashi, R.; Kataoka, H.; Takagi, Y.; Hashimoto, N. Toll-like receptor 4 expression during cerebral aneurysm formation: Laboratory investigation. *J. Neurosurg.* **2010**, *113*, 851–858. [CrossRef] [PubMed]
109. Wei, W.; Wang, X.; Huang, Z.; Li, X.; Luo, Y. Difference in inflammation, atherosclerosis, and platelet activation between coronary artery aneurysm and coronary artery ectasia. *J. Thorac. Dis.* **2020**, *12*, 5811–5821. [CrossRef] [PubMed]
110. Aoki, T.; Nishimura, M.; Kataoka, H.; Ishibashi, R.; Nozaki, K.; Miyamoto, S. Complementary inhibition of cerebral aneurysm formation by eNOS and nNOS. *Lab. Investig.* **2011**, *91*, 619–626. [CrossRef] [PubMed]
111. Zhang, B.; Dhillon, S.; Geary, I.; Howell, W.M.; Iannotti, F.; Day, I.N.M.; Ye, S. Polymorphisms in Matrix Metalloproteinase-1, -3, -9, and -12 Genes in Relation to Subarachnoid Hemorrhage. *Stroke* **2001**, *32*, 2198–2202. [CrossRef] [PubMed]
112. Liu, Y.; Sun, J.; Wu, C.; Cao, X.; He, M.; You, C. The Interleukin-6-572G/C Gene Polymorphism and the Risk of Intracranial Aneurysms in a Chinese Population. *Genet. Test. Mol. Biomark.* **2012**, *16*, 822–826. [CrossRef]

113. Li, L.-J.; Pan, X.-M.; Sima, X.; Li, Z.-H.; Zhang, L.-S.; Sun, H.; Zhu, Y.; Liang, W.-B.; Gao, L.-B.; Zhang, L. Interactions of interleukin-12A and interleukin-12B polymorphisms on the risk of intracranial aneurysm. *Mol. Biol. Rep.* **2012**, *39*, 11217–11223. [CrossRef]
114. Mohan, D.; Munteanu, V.; Coman, T.; Ciurea, A.V. Genetic factors involves in intracranial aneurysms–actualities. *J. Med. Life* **2015**, *8*, 336–341.
115. Aoki, T.; Kataoka, H.; Shimamura, M.; Nakagami, H.; Wakayama, K.; Moriwaki, T.; Ishibashi, R.; Nozaki, K.; Morishita, R.; Hashimoto, N. NF-κB Is a Key Mediator of Cerebral Aneurysm Formation. *Circulation* **2007**, *116*, 2830–2840. [CrossRef] [PubMed]
116. Pawlowska, E.; Szczepanska, J.; Wisniewski, K.; Tokarz, P.; Jaskólski, D.; Blasiak, J. NF-κB-Mediated Inflammation in the Pathogenesis of Intracranial Aneurysm and Subarachnoid Hemorrhage. Does Autophagy Play a Role? *Int. J. Mol. Sci.* **2018**, *19*, 1245. [CrossRef] [PubMed]
117. Kleinloog, R.; Verweij, B.H.; Van Der Vlies, P.; Deelen, P.; Swertz, M.A.; De Muynck, L.; Van Damme, P.; Giuliani, F.; Regli, L.; Van Der Zwan, A.; et al. RNA Sequencing Analysis of Intracranial Aneurysm Walls Reveals Involvement of Lysosomes and Immunoglobulins in Rupture. *Stroke* **2016**, *47*, 1286–1293. [CrossRef] [PubMed]
118. Cassis, L.A.; Gupte, M.; Thayer, S.; Zhang, X.; Charnigo, R.; Howatt, D.A.; Rateri, D.L.; Daugherty, A. ANG II infusion promotes abdominal aortic aneurysms independent of increased blood pressure in hypercholesterolemic mice. *Am. J. Physiol.-Heart Circ. Physiol.* **2009**, *296*, H1660–H1665. [CrossRef] [PubMed]
119. Zhong, P.; Lu, Z.; Li, Z.; Li, T.; Lan, Q.; Liu, J.; Wang, Z.; Chen, S.; Huang, Q. Effect of Renin-Angiotensin-Aldosterone System Inhibitors on the Rupture Risk Among Hypertensive Patients With Intracranial Aneurysms. *Hypertension* **2022**, *79*, 1475–1486. [CrossRef] [PubMed]
120. Foroud, T.; for the FIA Study Investigators. Whole Exome Sequencing of Intracranial Aneurysm. *Stroke* **2013**, *44*, S26–S28. [CrossRef] [PubMed]
121. Toth, G.; Cerejo, R. Intracranial aneurysms: Review of current science and management. *Vasc. Med.* **2018**, *23*, 276–288. [CrossRef]

brain sciences

MDPI

Editorial

The Surgical Odyssey: Romania's Contributions to Pituitary Gland Procedures

Corneliu Toader [1,2,*], Andrei-Adrian Popa [2,*], Razvan-Adrian Covache-Busuioc [2], Bogdan-Gabriel Bratu [2] and Alexandru Vlad Ciurea [2,3,4]

1 Department of Neurosurgery, National Institute of Neurology and Neurovascular Diseases, 077160 Bucharest, Romania

2 Department of Neurosurgery "Carol Davila", University of Medicine and Pharmacy, 020021 Bucharest, Romania; razvan-adrian.covache-busuioc0720@stud.umfcd.ro (R.-A.C.-B.); bogdan.bratu@stud.umfcd.ro (B.-G.B.); prof.avciurea@gmail.com (A.V.C.)

3 Department of Neurosurgery, Sanador Clinical Hospital, 010991 Bucharest, Romania

4 Medical Science Section, Romanian Academy, 060021 Bucharest, Romania

* Correspondence: corneliu.toader@umfcd.ro (C.T.); andreiadrianpopa@stud.umfcd.ro (A.-A.P.)

Abstract: The pituitary gland, a puzzling medical subject up until the 20th century, had its early pathologies first documented in the 19th century by Pierre Marie and Hutchinson, where the gland's meaningful study was hindered by its hard-to-reach location. This paper revisits the pioneering work of Romanian doctors such as Gheorghe Marinescu, Nicolae Paulescu, and Grigore T. Popa in surgical techniques targeting the pituitary gland. Marinescu's 1892 experiment, albeit unsuccessful, laid the groundwork for future research in this area. Before Paulescu, surgical attempts could be classified into three types: oral, cranial, and sphenopalatine fossa approaches—all of which were notably dangerous and often resulted in fatal bleeding. Paulescu was the first to successfully and safely perform a complete in vivo hypophysectomy, opting for an innovative subtemporal method. He also conducted extensive research over four years to identify the gland's essential functions. Later, a 1938 study by Popa and Harris demonstrated a temporal approach to the hypothalamo-hypophysial region in a rabbit. These groundbreaking contributions significantly influenced the trajectory of pituitary gland surgery.

Keywords: pituitary gland; Romanian physicians; surgical approach; hypophysectomy; Nicolae Paulescu; Gheorghe Marinescu; Grigore T. Popa; medical history; in vivo; hemorrhage; subtemporal approach; oral approach; cranial approach; sphenopalatine fossa

check for updates

Citation: Toader, C.; Popa, A.-A.; Covache-Busuioc, R.-A.; Bratu, B.-G.; Ciurea, A.V. The Surgical Odyssey: Romania's Contributions to Pituitary Gland Procedures. *Brain Sci.* **2023**, *13*, 1431. https://doi.org/10.3390/brainsci13101431

Received: 2 September 2023
Accepted: 20 September 2023
Published: 8 October 2023

1. Early Milestones in Pituitary Gland History

The initial investigations into the role of the pituitary gland in human physiology were predominantly symptomatic in nature, focusing on the clinical manifestations resulting from the gland's dysfunction. In the late 19th century, a pioneering revelation was made by French physician Pierre Marie, who identified acromegaly as a distinct pathological condition [1]. While Marie initially failed to discern the explicit relationship between acromegaly and pituitary abnormalities, it was in the year 1890 that a seminal publication was released, co-authored by Marie and Romanian neurologist Gheorghe Marinescu (1863–1938) (Figure 1).

This body of work was instrumental in delineating the anatomopathological features of acromegaly, specifically emphasizing its correlation with perturbations in pituitary morphology, including glandular hyperplasia and interstitial sclerosis. Although Marie and Marinescu initially postulated that acromegaly was the result of pituitary insufficiency rather than hyperfunction, the crux of their contributions lay in establishing the fundamental link between the pituitary gland and acromegaly, thereby catalyzing subsequent research trajectories within endocrinology and neurobiology [1,2].

Figure 1. Gheorghe Marinescu (1863–1938).

The scholarly pursuits of Gheorghe Marinescu did not culminate with this collaborative endeavor; he continued to expand upon his initial findings. In 1892, Marinescu published another pivotal paper detailing an experimental hypophysectomy performed on a feline subject. Utilizing a transpalatal approach, he employed thermal cauterization to excise the pituitary gland. While the experiment was compromised by subsequent infection, Marinescu concluded that his data corroborated the feasibility of pituitary gland ablation through an oral route in felines, sustaining survival for a period of several weeks post-operation. The methodological shortcomings notwithstanding, Marinescu's experiment provided invaluable impetus for future investigations within the realm of pituitary research.

Later citations of Marinescu's work, particularly by researchers such as Paulescu, underscore its impact on the field. Paulescu recognized Marinescu as the inaugural investigator to physically interact with, and experimentally manipulate, the pituitary gland. He further noted that Marinescu's experimental endeavors lent credence to the hypothesis—initially proposed in conjunction with Pierre Marie—that pituitary gland alterations could potentially be implicated in the pathogenesis of acromegaly [3]. Thus, the collaborative and independent contributions of Pierre Marie and Gheorghe Marinescu served as cornerstones in the foundational understanding of pituitary gland function and its role in systemic endocrine disorders, notably acromegaly.

The scientific inquiry into the pituitary gland presented substantial methodological challenges, owing in part to the anatomical complexity and inaccessibility of its location within the cranial cavity. Early attempts to approach this endocrine organ surgically were fraught with difficulty, primarily because the surgical techniques available at the time risked damaging adjacent neurological structures. One of the earliest recorded endeavors to surgically access a pituitary adenoma was undertaken by Sir Victor Horsley in 1889. Despite his pioneering efforts, the procedure was ultimately deemed unsuccessful due to excessive force applied during the retraction of the frontal and temporal lobes, which led to undesirable collateral damage [4].

In 1903, a momentous leap in pituitary research was achieved by Romanian physiologist Nicolae Paulescu (1869–1931) (Figure 2). By employing a methodological approach that was meticulous enough to avoid the complications that had plagued earlier attempts, Paulescu successfully performed a total pituitary ablation on a canine subject. Subsequently, in his 1907 publication, he conclusively demonstrated the irreplaceable role of the pituitary gland in sustaining life [5].

Figure 2. Nicolae Paulescu (1869–1931).

The progression of surgical methodologies seeking to access the pituitary gland continued to evolve during the early 20th century. In 1907, Hermann Schloffer (1868–1937) introduced the first trans-sphenoidal approach to address pituitary growths, thereby offering an alternative that mitigated some of the risks associated with intracranial procedures [6]. This method was further refined by Harvey Cushing (1869–1939), an American neurosurgeon who became a seminal figure in pituitary research. Cushing initially perfected the trans-sphenoidal approach through a sublabial route, chronicling his findings in a series of publications released in 1912, 1914, and 1922. However, in 1929, he transitioned to a different surgical approach, the transfrontal method, adopting this subfrontal route as his exclusive technique for pituitary adenomectomy [7].

The history of pituitary gland research represents a chronology of methodological refinement, evolving from initial efforts to understand its physiopathological role, through to the implementation of surgical interventions, ranging from subtemporal approaches to trans-sphenoidal partial lesionectomies, and finally, to subfrontal adenomectomies. This trajectory underscores the incremental advancements that have been made in both the conceptual understanding of the pituitary gland and the surgical techniques employed to study and treat it.

2. Pituitary Gland Surgery: The Enigma and the Skill

The genesis of Nicolae Paulescu's (1869–1931) scholarly interest in the pituitary gland remains an unchronicled facet of his career. Nonetheless, this curiosity led him to devise an idiosyncratic surgical method to access this endocrine organ: the transtemporal approach. Paulescu's journey in developing effective surgical approaches for pituitary gland research can be seen as a series of methodological explorations, each with its own distinct advantages and shortcomings.

Initially, Paulescu, in collaboration with Dr. Paul Reyener, experimented with an oral approach to studying the pituitary gland. Conducted in Paris during the years 1897 and 1898, within the physiology laboratories of the Sorbonne, these experiments utilized canine subjects. Regrettably, the outcomes were suboptimal, with all animal subjects succumbing to either hemorrhage or suppurative meningeal infection. Given the catastrophic results, the investigators ultimately decided to discontinue this line of surgical approach.

In addition to the oral approach, other surgical routes were also carefully scrutinized and ultimately deemed untenable for varying reasons. For instance, a cranial approach via the upper part of the skull was dismissed by Paulescu, due to its propensity to induce extensive cerebral lesions. The sphenopalatine approach, although contemplated, was similarly ruled out, primarily because it did not afford an adequate visualization of the gland and carried an elevated risk of neurological damage.

Consequently, Paulescu opted for an innovative surgical route, the subtemporal approach, which emerged as a more viable option given the limitations of alternative techniques. After a series of preliminary attempts that commenced in Paris during the years 1897 and 1898, Paulescu successfully executed a complication-free total pituitary ablation on a canine subject on 11 March 1903. The operation was facilitated by the surgical expertise of Dr. Balacescu.

Paulescu's contributions to pituitary gland research can be viewed as a meticulously constructed iterative process. Beginning with an exploration of various surgical approaches, each with distinct challenges and limitations, he ultimately devised the subtemporal approach, which demonstrated both the feasibility and a reduced risk of complications. This methodological refinement represents a significant milestone in the annals of endocrinology and neurosurgery, setting the stage for further advancements in our understanding of pituitary physiology and pathology.

Nicolae Paulescu elucidated his pioneering surgical methodology for accessing the pituitary gland in a seminal work entitled *L'Hypophyse de Cerveau,* published in 1908 (Figure 3). The operative procedure was delineated in nine elaborate steps, providing a comprehensive framework for pituitary ablation [5].

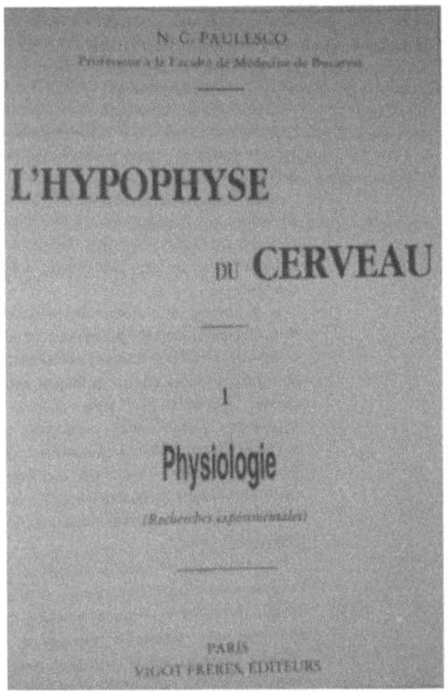

Figure 3. Paulescu's work: *L'Hypophyse du Cerveau*—1908.

The nine steps of Paulescu's operative procedure for accessing the pituitary gland are as follows:

1. Cutaneous Incision and Muscle Dissection: The procedure begins with a midline incision on the skin, commencing slightly below the eyebrows and extending posteriorly to a few finger-widths behind the occipital protuberance. The platysma muscle,

which adheres to the deep face of the skin, is longitudinally sectioned and disinserted. Sterilized compresses are employed to secure the skin edges.

2. Temporal Muscle Incision: On the left temporal region, a semicircular incision parallel to the superior insertion of the temporal muscle is made, extending from the external orbital apophysis to 2–3 cm below and lateral to the occipital protuberance. Subsequent dissection reveals the underlying temporal bone, and this is mirrored on the right side with slight variations.
3. Zygomatic Arch Manipulation: The zygomatic arch on the right side is cut at both ends and maneuvered outward and downward. This affords greater surgical accessibility to the temporal region.
4. Cranial Trepanation: A trephine is employed to create holes in the parietal bones on either side, through which further bone sections are made. Specialized forceps are used to minimize diploic hemorrhage.
5. Dura Mater Incision: A longitudinal incision is made on the dura mater, parallel to the previous bone incisions, creating a dural flap.
6. Temporal Lobe Elevation: A specialized instrument is inserted beneath the right temporal lobe to elevate it gently. The pituitary gland, recognizable by its characteristic red-yellow hue, becomes visible once the lobe is sufficiently elevated.
7. Pituitary Gland Extraction: Employing a specialized curette, the pituitary gland is carefully detached from its posterior anchoring and then fully excised from the base of the skull.
8. Dural and Muscular Closure: After pituitary removal and hemostasis, the dural flap is repositioned (Paulescu initially used thin catgut sutures for this, but later abandoned them, deeming them unnecessary). The zygomatic arch and temporal muscle are then sutured back into place.
9. Final Closure and Dressing: The incisions are sutured and sterilizing compresses are applied to the surgical site. Special attention is given to protect the eyes and ears of the animal with sterilizing cotton wool, and a compressive dressing is applied to the head, allowing unimpeded respiration and deglutition.

This elaborate surgical schema, developed by Paulescu, marked a significant advancement in the field of neuroendocrinology, providing a sophisticated and structured approach for pituitary gland access and removal. The methodological rigor of Paulescu's operative procedure laid the groundwork for subsequent innovations in pituitary surgery and significantly enhanced our understanding of endocrine physiology and pathology.

3. Pioneering Anesthesia in Pituitary Gland Surgery

In the early 20th century, Nicolae Paulescu developed an innovative anesthesia protocol to facilitate his groundbreaking pituitary surgeries. He initially induced anesthesia using ether and then maintained the narcotic state with periodic administrations of chloroform. Paulescu found that this dual-agent approach mitigated complications and appeared to reduce the tendency for hemorrhages as compared to using ether alone. This practice was thoroughly documented in 1908, subsequent to its initial employment in 1907 [5].

Paulescu's surgical technique was via a transtemporal approach, wherein the temporal lobe was carefully moved to provide access to the pituitary gland. His seminal research yielded compelling evidence for the critical role of the pituitary in maintaining life. Paulescu found that all dogs subjected to total pituitary ablation died within 24 hours post-surgery. However, some animals survived considerably longer after partial hypophysectomy—two lived for five months, and another for a year. These observations led him to conclude that the pituitary gland is indeed indispensable for life.

In the following years, Harvey Cushing and other leading neurosurgeons, including Crowe and Homans in 1910 and Reford in 1909, replicated Paulescu's pioneering work. They utilized Paulescu's surgical techniques to perform an extensive series of hypophysectomies on dogs [8]. The results corroborated Paulescu's conclusions: the pituitary gland was crucial for survival.

The validation of Paulescu's work by Cushing and his team not only lent credibility to Paulescu's groundbreaking contributions, but also amplified the global understanding of the critical role of the pituitary gland in physiological regulation. Collectively, these early surgical explorations laid an important foundation for the modern field of neuroendocrinology, demonstrating the pituitary's indispensable role in sustaining life and paving the way for subsequent advancements in both surgical techniques and endocrine therapies.

4. The Legacy of a Lost Approach: A Timely Reassessment

Although Nicolae Paulescu's subtemporal approach to the pituitary gland marked a milestone as the first in vivo approach with long-term survival and without complications, it did not become the primary surgical technique for addressing pituitary pathology. Instead, Harvey Cushing and Oskar Hirsch pioneered the trans-sphenoidal approach in 1909, which later gained wider acceptance [7].

Cushing acknowledged Paulescu's contributions in his own reports, particularly emphasizing Paulescu's discovery that retaining a fragment of the pituitary gland is essential for survival [8]. The significance of Paulescu's work was also recognized by Sir Sharpey Shafer in his 1926 treatise, "The Endocrine Organs". According to Shafer, Paulescu was the first to definitively show that the complete removal of the pituitary gland results in fatal outcomes. This landmark finding has been verified across various classes of vertebrates and was subsequently confirmed by Cushing and other researchers, including Bield, B.A. Houssay, Asoli and Lagnani, Blair Bell, and Dotti.

Further cementing Paulescu's legacy, Evelyn Anderson and Webb Haymaker highlighted his contributions in their 1974 treatise, "Progress in Brain Research," particularly in the chapter on "Hypothalamic and Pituitary Research". According to them, Paulescu was among the early trailblazers in the field and was notably the first to develop a superior operative technique for pituitaryectomy in dogs [9].

Overall, although Paulescu's subtemporal approach may not have been adopted as the primary surgical technique for treating pituitary disorders, his groundbreaking work established critical principles concerning the gland's physiological importance and introduced a pioneering surgical method. His contributions have been acknowledged by subsequent generations of researchers and clinicians, affirming his role in the historical development of neuroendocrinology and pituitary surgery.

5. Temporal Approach—Revisiting Traditional Views

Another significant step forward in the evolution of experimental research with regard to the pituitary gland took place in 1938, with the work of Grigore T. Popa and Geoffrey Wingfield Harris. Popa had previously gained attention for his landmark publication in *The Lancet* in 1930, which was co-authored with Una Fielding, where he described the hypothalamo-hypophyseal portal system for the first time. This discovery set the stage for his later work with Harris, aiming to improve access to the hypothalamo-hypophyseal region for further study [10].

Moreover, their 1938 publication, *A Technique for Operations on the Hypothalamo-Hypophysial Region of the Rabbit* [10], introduced a four-stage temporal approach on rabbits to better expose the region of interest:

First Stage: The authors initiated the process with the surgical removal of a part of the zygomatic arch. The skin and superficial fascia were incised, and the periosteum over the zygomatic arch was carefully reflected. The arch was then removed, with caution taken not to damage large vessels in the masseter muscle. If hemorrhage occurred, it could be controlled by applying pressure.

Second Stage: The procedure continued with the removal of the upper ramus of the mandible. The overlying masseter muscle was cut, and the periosteum was removed. The mandibular articular surface was separated from the mandible through gentle pressure, followed by a bone cut.

Third Stage: The next stage involved exposing and removing part of the squamous temporal bone. The pterygomaxillary fossa was cleared, and the skull was accessed in the temporo-sphenoidal region. Once reached, the dura mater was incised and reflected to expose the underlying structures.

Fourth Stage: The final stage involved retracting the temporal lobe of the brain to expose the hypothalamo-hypophyseal region. Upon retraction, the tentorium cerebelli became visible, and the hypothalamus could be identified above it. In this exposed region, the small internal carotid artery was visible, and the hypophyseal stalk was located postero-medially to this artery.

The research by Popa and Harris further advanced the field's understanding of the hypothalamo-hypophyseal system and introduced a new surgical technique that facilitated the exploration of this critical region. Their work stands as a notable contribution to the lineage of research that aims to better understand the physiological and anatomical complexities of the pituitary gland and its surrounding structures.

6. Conclusions

Initially regarded as a relatively unimportant anatomical structure, the pituitary gland's significance has become increasingly evident, thanks to the seminal contributions of pioneering researchers and surgeons. The evolving understanding of this essential gland is a story of intellectual pursuit, marked by breakthroughs and advances that have fundamentally shifted our understanding of its role in physiology.

The journey to unravel the complexities of the pituitary—also known as the hypophysis—began with early trailblazers like Pierre Marie and Gheorghe Marinescu. They laid the initial groundwork by investigating the correlation between acromegaly, a disorder characterized by the abnormal growth of bones and tissues, and the pituitary gland. Their work hinted at the critical functions that this once-overlooked organ might serve, setting the stage for further investigations.

Subsequently, Nicolae Constantin Paulescu emerged as a key figure who dispelled the notion of the pituitary as an inconsequential organ. Through rigorous experimentation, Paulescu conclusively demonstrated that the pituitary is indispensable for life itself. His surgical techniques for pituitaryectomy in dogs paved the way for more in-depth studies, effectively shifting the perception of the gland from that of a mere anatomical curiosity to an organ of vital importance.

The story, however, does not end with Paulescu. Surgeons and researchers like Victor Horsley and Hermann Schloffer carried the baton forward, each contributing valuable insights and methodologies. However, it was perhaps Harvey Cushing who most dramatically advanced the field by pioneering new surgical approaches to pituitary disorders. His work, in collaboration with others on his team, not only validated earlier findings pertaining to the pituitary's essential functions, but also opened a new chapter in medical history—the era of surgical interventions focused on the pituitary gland.

Thus, what began as an exploration of a seemingly inconsequential organ eventually evolved into a rich, multidisciplinary field of study. Today, the pituitary is recognized not merely as a gland, but as a cornerstone of the endocrine system, influencing a range of physiological processes, from growth and metabolism to the functioning of other glands. This sea of change in pituitary gland knowledge is a testament to the generations of scholars and practitioners who dedicated themselves to unraveling the mysteries of the pituitary gland. Their collective work has not only deepened our biological understanding, but also introduced surgical and therapeutic options that continue to improve and save lives to this day.

Author Contributions: Conceptualization, C.T., A.-A.P.; Writing—original draft, R.-A.C.-B.; Validation, A.V.C.; Methodolody, B.-G.B.; Investigation, A.-A.P., R.-A.C.-B.; Writing—review and editing, C.T., B.-G.B.; Supervision, A.V.C. All authors have read and agreed to the published version of the manuscript.

Funding: This research received no external funding.

Conflicts of Interest: The authors declare no conflict of interest.

References

1. Marie, P. Sur deux cas d'acromegalie: Hipertrophie singuliere non congenitale des extremites superieures, inferieures et cephaliques. *Rev. Med.* **1886**, *6*, 297–333.
2. Marie, P.; Marinesco, G. Sur l'anatomie pathologique de l'acromegalie. *Arch. Exp. Anat. Path.* **1891**, *3*, 539–565.
3. Marinesco, G. De la destruction de la glande pituitaire chez le chat. *CR Seances Soc. Biol. Fil.* **1892**, *44*, 509–510.
4. Landolt, A.M. History of pituitary surgery. In *A History of Neurosurgery in Its Scientific and Professional Contexts*; Greenblatt, S.H., Forcht Dagi, T., Epstein, M.H., Eds.; American Association of Neurological Surgeons: Park Ridge, IL, USA, 1997; pp. 373–400.
5. Paulesco, N.C. *L'hypophyse Du Cerveau. I. Physiologie*; Vigot Freères: Paris, France, 1908.
6. Schloffer H: Erfolgreiche Operation eines Hypophysentumors auf nasalem Wege. *Wien Klin. Wchnschr.* **1907**, *20*, 621–624.
7. Liu, J.K.; Das, K.; Weiss, M.H.; Laws, E.R., Jr.; Couldwell, W.T. The history and evolution of transsphenoidal surgery. *J. Neurosurg.* **2001**, *95*, 1083–1096. [CrossRef] [PubMed]
8. Cushing, H., III. Partial Hypophysectomy for Acromegaly: With Remarks on the Function of the Hypophysis. *Ann. Surg.* **1909**, *50*, 1002–1017. [CrossRef]
9. Anderson, E.; Haymaker, W. *Breakthroughs in Hypothalamic and Pituitary Research, Progress in Brain Research*; Elsevier: Amsterdam, The Netherlands, 1974; Volume 41, pp. 1–60.
10. Harris, G.W. A Technique for Operations on the Hypothalamo-Hypophysial Region of the Rabbit. *J. Anat.* **1938**, *72*, 226–233.

brain sciences

Review

Decoding Chiari Malformation and Syringomyelia: From Epidemiology and Genetics to Advanced Diagnosis and Management Strategies

Corneliu Toader [1,2], Horia Ples [3,4,*], Razvan-Adrian Covache-Busuioc [1], Horia Petre Costin [1], Bogdan-Gabriel Bratu [1], David-Ioan Dumitrascu [1], Luca Andrei Glavan [1] and Alexandru Vlad Ciurea [1,5]

1 Department of Neurosurgery, "Carol Davila" University of Medicine and Pharmacy, 020021 Bucharest, Romania; corneliu.toader@umfcd.ro (C.T.); razvan-adrian.covache-busuioc0720@stud.umfcd.ro (R.-A.C.-B.); horia-petre.costin0720@stud.umfcd.ro (H.P.C.); bogdan.bratu@stud.umfcd.ro (B.-G.B.); david-ioan.dumitrascu0720@stud.umfcd.ro (D.-I.D.); luca-andrei.glavan0720@stud.umfcd.ro (L.A.G.); prof.avciurea@gmail.com (A.V.C.)

2 Department of Vascular Neurosurgery, National Institute of Neurology and Neurovascular Diseases, 077160 Bucharest, Romania

3 Department of Neurosurgery, Centre for Cognitive Research in Neuropsychiatric Pathology 6 (NeuroPsy-Cog), 300736 Timișoara, Romania

4 Department of Neurosurgery, "Victor Babeș" University of Medicine and Pharmacy, 300041 Timișoara, Romania

5 Neurosurgery Department, Sanador Clinical Hospital, 010991 Bucharest, Romania

* Correspondence: ples.horia@umft.ro

Abstract: Chiari Malformation and Syringomyelia are neurosurgical entities that have been the subject of extensive research and clinical interest. Globally prevalent, these disorders vary demographically and have witnessed evolving temporal trends. Chiari Malformation impacts the normal cerebrospinal fluid flow, consequently affecting overall health. Key observations from canine studies offer pivotal insights into the pathogenesis of Syringomyelia and its extrapolation to human manifestations. Genetics plays a pivotal role; contemporary knowledge identifies specific genes, illuminating avenues for future exploration. Clinically, these disorders present distinct phenotypes. Diagnostically, while traditional methods have stood the test of time, innovative neurophysiological techniques are revolutionizing early detection and management. Neuroradiology, a cornerstone in diagnosis, follows defined criteria. Advanced imaging techniques are amplifying diagnostic precision. In therapeutic realms, surgery remains primary. For Chiari 1 Malformation, surgical outcomes vary based on the presence of Syringomyelia. Isolated Syringomyelia demands a unique surgical approach, the effectiveness of which is continually being optimized. Post-operative long-term prognosis and quality of life measures are crucial in assessing intervention success. In conclusion, this review amalgamates existing knowledge, paving the way for future research and enhanced clinical strategies in the management of Chiari Malformation and Syringomyelia.

Keywords: Chiari 1; Chiari 2; cerebral malformation; syringomyelia; neurosurgery; syrinx pathologies; genetic implications

Citation: Toader, C.; Ples, H.; Covache-Busuioc, R.-A.; Costin, H.P.; Bratu, B.-G.; Dumitrascu, D.-I.; Glavan, L.A.; Ciurea, A.V. Decoding Chiari Malformation and Syringomyelia: From Epidemiology and Genetics to Advanced Diagnosis and Management Strategies. *Brain Sci.* **2023**, *13*, 1658. https://doi.org/10.3390/brainsci13121658

Academic Editors: Giovanni Grasso and Woon-Man Kung

Received: 6 November 2023
Accepted: 24 November 2023
Published: 30 November 2023

1. Introduction

1.1. Brief Overview of Chiari Malformation, Syringomyelia and Related Disorders

Chiari Malformation is a structural defect in the architecture of the skull and brain, particularly involving the position of the cerebellum. The cerebellum, which is a vital part of the brain responsible for regulating balance, muscle coordination, and some cognitive functions, descends abnormally below the skull's base in individuals with this condition.

Instead of resting entirely within the skull, a portion of the cerebellum protrudes into the upper segment of the spinal canal [1].

This protrusion can result in multiple issues. Primarily, it has the potential to exert pressure on the brainstem, a critical area of the brain responsible for basic life functions like breathing and heart rate. Additionally, this malformation can hinder the normal flow of cerebrospinal fluid (CSF). CSF is a clear, protective fluid that circulates around the brain and spinal cord, cushioning them from injury and helping to maintain the brain's chemical balance. When the flow of CSF is disrupted, it can cause a range of neurological symptoms that vary in intensity and type [2].

To understand and manage the condition better, medical professionals typically classify Chiari Malformation into distinct types. These classifications are based on the malformation's severity and the extent to which the cerebellum herniates or protrudes into the spinal canal. Each type has specific characteristics and implications for diagnosis and treatment [3].

Syringomyelia refers to a cavity filled with fluid that forms within the spinal cord tissue or the central canal. Over time, various theories have been proposed to shed light on the development of syringomyelia, especially when it results from blockages in the spinal subarachnoid space. Despite a century of dedicated experimental and clinical studies, the exact pathophysiological underpinnings of syringomyelia remain elusive [4].

A central topic of debate is the origin of the fluid within the syrinx and the mechanisms driving its formation. Dominant theories suggest that this cavity, or syrinx, is filled with cerebrospinal fluid (CSF). It is believed that a surge in pressure within the subarachnoid space pushes the CSF into the syrinx. Yet, this notion is counterintuitive because increasing external pressure on a cavity would typically compress it, not fill it. Moreover, it is puzzling how CSF could infiltrate the syrinx when the pressure within the syrinx is either higher than or equal to the surrounding CSF pressure [5].

Recent innovative research has started to shift the perspective on this medical enigma. Instead of focusing on external factors, contemporary studies have begun to consider internal spinal cord dynamics. This fresh viewpoint suggests that the cavity forms due to heightened pulse pressure within the spinal cord tissue. Additionally, the fluid inside the cavity may not be CSF as traditionally believed but could be extracellular fluid. The so-called "intramedullary pulse pressure theory" offers a novel explanation that seems to account for various aspects of syringomyelia, irrespective of what causes the related lesion in the subarachnoid space [6].

The various merits and limitations of past theories on syringomyelia have been meticulously dissected in thorough reviews by experts like Klekamp and Levine.

Tethered cord syndrome is a condition where the spinal cord becomes abnormally anchored to surrounding tissues, limiting its typical movement. This anomaly can give rise to complications, including Chiari Malformation and Syringomyelia. The syndrome is characterized by motor and sensory dysfunctions resulting from the undue tension exerted on the spinal cord due to this abnormal tethering. Traditionally, it is linked with the presence of a low-situated conus medullaris. The primary mode of treatment is surgical intervention, with outcomes varying from patient to patient. Even though tethered spinal cord syndrome is seldom diagnosed in emergency settings, emergency physicians must be vigilant and consider this condition in patients exhibiting symptoms reminiscent of cauda equina syndrome [7].

The aim of this comprehensive review is to put into perspective the classifications of Chiari malformation and related disorders, the therapy management of these pathologies and their genetic etiology. Therefore, this paper's main goal is to decipher the past, present, and future perspectives of Chiari malformation.

1.2. Materials and Methods

The search process was conducted using specific databases such as PubMed and Web of Science. Search terms used were "MRI imaging in Chiari malformations", "Syringomyelia",

"Syrinx Pathologies", and "Chiari malformations type I, II, III, and IV", among others. The inclusion criteria specified articles written in English that referred to diagnostic imaging, surgical interventions, and outcomes of Chiari malformations. The initial search resulted in 369 articles, and after removing all duplicates, they were reviewed regarding their relevance, selecting 121 of them meeting our criteria.

1.2.1. Tonsillar Configuration

The extent of tonsillar descent does not necessarily align with symptom severity, as a significant portion (around 30%) of patients with pronounced tonsillar descent may not display any symptoms [8]. Instead, the shape of the tonsils has been considered more indicative [9,10]. Pronounced compression results in peg-like tonsils that could further impede CSF flow. Such peg-shaped tonsils are more prevalent in patients with a herniation greater than 5 mm (85%), in contrast to those with rounded or intermediary forms [11].

1.2.2. Magnetic Resonance Imaging (MRI)—Craniocervical Junction—Dynamic Evaluation

Dynamic Flow Studies

Dynamic studies play a role in evaluating CM-I symptoms and anticipating surgical outcomes. Healthy individuals exhibit CSF flow at the craniocervical junction that alternates in a cranial and caudal direction, mirroring cardiac and respiratory-induced variations in intracranial blood volume [12]. The spinal arachnoid space serves as a cushion, moderating intracranial pressure spikes [13]. Due to enhanced intracranial compliance, children exhibit quicker caudal velocities [14]. Phase contrast cine MRI has pinpointed notable variances in CSF velocity in symptomatic CM-I patients. Specifically, a heightened peak velocity at the FM and diminished overall volume motion were observed alongside flow jets displaying regions dominated by flow in a single direction or exhibiting concurrent bidirectional flow [2,15,16]. In symptomatic CM-I patients, unusual pulsatile actions of the cerebellar tonsils were noted, with post-surgical improvements in tonsillar pulsation amplitude and arachnoid space reduction [17,18].

CSF dynamic investigations aim to differentiate between symptomatic and asymptomatic CM-I patients, but research outcomes have been mixed [19,20]. However, in predicting surgical enhancements, CSF velocity patterns in CM-I patients might be pivotal [2,21–23]. Interestingly, patients who showcased normal preoperative hindbrain CSF flow had an almost five-fold increased likelihood of post-operative symptom resurgence, regardless of their tonsillar herniation extent or syringomyelia presence. In contrast, full CSF flow blockage prior to surgery correlated with sustained symptom relief [22]. Currently, dynamic studies assess craniocervical junction obstructions and can be indicative of a patient's aptness for surgery. While not universally applicable, identifying obstructed flow can help forecast favorable surgical outcomes in borderline scenarios or during follow-up evaluations in cases of symptom reemergence.

1.2.3. Magnetic Resonance Imaging (MRI)—Spinal Evaluation

Syringomyelia

A significant proportion of individuals with symptomatic CM-I, as many as 50%, might develop a syrinx [24]. The preferred imaging method for visualizing syrinxes is sagittal MR T2-weighted scans of the entire spinal cord, supplemented by axial T2 views. When a syrinx appears without an accompanying CM-I, contrast-enhancing sequences become essential to investigate any linked tumors, although they are less relevant when CM-I is evident [25]. Syrinx development is more frequent in patients who have pronounced tonsillar herniation and CSF flow obstruction, predominantly appearing between the C4 and C6 spinal levels [26,27]. A terminal syrinx, situated in the tail end of the spinal cord, often correlates with conditions like a tethered cord or spinal dysraphism. Recognizing a syrinx via pre-surgery imaging, even if it does not manifest symptoms, is instrumental for surgical planning and post-surgical success evaluation.

1.3. Tethered Cord Syndrome

Tethered Cord Syndrome (TCS) is present in about 14% of CM-I patients [28]. While the phrase "tethered cord" denotes a fixed segment of the spinal cord, TCS specifically alludes to the lumbar-level anchoring of the spinal cord [29]. It is typically diagnosed when the conus medullaris is positioned below L2. However, there are other diagnostic criteria, such as a thick or fatty filum, spina bifida oculta, terminal syringomyelia, lower thoracic scoliosis, and a dorsal arrangement of the filum in prone or upright MRI scans [22]. Lumbar MRI aids in identifying the conus medullaris level, the thickness of the filum terminale, and any related dysraphic conditions. A CT scan for intricate bone anomalies and electrophysiological evaluations for urological complications might be deemed necessary in certain instances. Although surgically addressing a radiologically apparent tethered cord in patients with symptomatic CM-I and/or terminal syrinx is a recognized treatment approach, some experts suggest intervening on a regular filum when treating CM-I patients, although this method is debated [30,31].

2. Epidemiology of Chiari Malformation, Syringomyelia and Related Disorders

2.1. Global Prevalence and Distribution: Demographics Affected

Chiari I malformation (CM) often coexists with a spinal cord syrinx. However, gauging the actual prevalence of CM and syrinx is challenging due to the absence of a flawless diagnostic tool applicable to the entire population. Instead, medical professionals often depend on approximations, primarily sourced from extensive retrospective studies examining brain and spinal imagery [32].

In imaging-based prevalence studies, a CM diagnosis is typically based on the cerebellar tonsil protruding 5 mm or more beneath the foramen magnum. Estimates suggest CM affects anywhere from 0.24% to 3.6% of individuals. This variance can be attributed to differences in the diagnostic sensitivity for CM and the diverse populations studied [33].

Age plays a significant role in CM's prevalence. Research indicates that CM is more commonly found in children. Notably, MRI scans reveal that the position of the cerebellar tonsil varies with age, descending during early life and rising again in adulthood [9].

Imaging studies offer the most accurate prevalence estimates for CM, especially when considering the many asymptomatic individuals who meet the imaging diagnostic criteria. Interestingly, females are more likely to have CM based on imaging results. They also tend to exhibit a lower position of the cerebellar tonsil across all ages when compared to males. On the other hand, factors like obesity or an increased body mass index do not seem to correlate with CM's prevalence in imagery [8].

In summary, CM prevalence differs across age and gender demographics. Truly estimating its occurrence within the general population remains a complex task. Ongoing research aims to gain a clearer understanding of how age impacts tonsil positioning and the related conditions to better grasp CM's epidemiology.

2.2. Temporal Trends

In studying Chiari malformation (CM) and associated disorders, temporal trends refer to the analysis of how the prevalence, incidence, or specific features of these conditions have transformed over various periods or generations.

For instance, a study conducted by Luzzi et al. [34], spanning from January 2015 to December 2019, highlighted several findings. Among these, the surgery was found to effectively alleviate headaches and mitigate symptoms like dysesthetic pain, weakness, and dissociated sensory loss within a six-month period post-operation. Nevertheless, there was a limited improvement noted in atrophy and spasticity post-surgery. These findings can be referenced in Table 1.

Table 1. The study reported on the preoperative symptoms and postoperative outcomes of patients who underwent surgery.

Preoperative Symptoms	Postoperative Outcomes	Atrophy and Spasticity
Dissociated sensory loss, headache, lower cranial nerve dysfunction, and weakness.	The headache immediately disappeared after surgery, indicating a successful resolution of this symptom.	Atrophy and spasticity were largely unaffected by surgery, suggesting that the treatment may not have a significant impact on these symptoms.
The involvement of C2–C5 metameres	All treated patients experienced a full recovery within 6 months after surgery.	

A substantial amount of research has been directed towards understanding the connections between syringomyelia and other diseases. However, there is a noticeable deficit in exhaustive and unbiased accounts of the research progress of syringomyelia. The present study endeavored to perform a bibliometric analysis to bridge this gap, charting the research trajectory of syringomyelia and identifying emergent themes over the past two decades.

Between January 2003 and August 2022, an impressive 9556 authors from 66 nations contributed to a total of 1902 research papers on syringomyelia, published across 518 scholarly journals. The majority of these contributions originate from the United States, China, the United Kingdom, and Japan, with the United States taking the lead. Both Nanjing University and the University of Washington stand out as the most prolific contributors. Among individual researchers, Dr. Claire Rusbridge boasts the highest publication count, and Miholat leads in co-citations. The Journal of Neurosurgery is prominent in the most co-cited articles, primarily in the domains of neurology, surgery, and biology. Notably, terms like syringomyelia, Chiari-I malformation, children, surgical treatment, and spinal cord emerged as high-frequency keywords.

Over the past twenty years, there has been a consistent upward trend in the publication of articles focusing on syringomyelia. Current research gravitates toward understanding the age at which the disease manifests, evaluating potential therapeutic strategies, the efficacy of surgical interventions, recurrence prevention, and pain delay. The therapeutic surgical approaches to the ailment and exploration into advanced treatment modalities are at the forefront of contemporary research. Key areas of interest also encompass the link between trauma and inherent factors, practical applications, post-operative recurrence, and potential complications. Insights from these areas could pave the way for groundbreaking therapeutic solutions for syringomyelia in the future [35].

3. Pathophysiology of Chiari: Hydrodynamics of Cerebro-Spinal Fluid Flow

3.1. An Overview of the Normal Cerebro-Spinal Fluid Flow

The cerebrospinal fluid (CSF) is a clear liquid derived primarily from blood plasma and is found within the brain's ventricles and the subarachnoid spaces of both the skull and spine. This fluid plays several crucial roles, including delivering nutrients to the brain, facilitating waste removal, and offering a protective buffer for the brain. In adults, the total volume of CSF is approximately 150 mL, of which around 125 mL is located in the subarachnoid spaces and 25 mL in the ventricles. The main source of CSF production is the choroid plexus, although there are other less understood contributors. In adults, the amount of CSF produced varies between individuals but typically falls within 400 to 600 mL daily [36]. This continuous production ensures the CSF is replaced four to five times daily in a typical young adult. As people age or in certain neurodegenerative conditions, a decline in CSF circulation might lead to a buildup of metabolites. The precise composition of CSF is meticulously maintained, and any deviations in its constituents can serve as significant indicators for diagnostic evaluations [37].

3.2. Changes Observed in Chiari Malformation

Chiari malformation Type I (CMI) is a complex anomaly of the craniospinal system. Traditionally, it is identified through radiology as a downward displacement or herniation of the cerebellar tonsils (CTH) by more than 3–5 mm beneath the foramen magnum (FM) into the spinal subarachnoid space (SSS). However, in-depth retrospective studies have indicated that the depth of CTH does not always align with the severity of CMI symptoms. Interestingly, patients with pronounced CTH can sometimes exhibit only minor neurological symptoms and vice versa [38].

From a fluid dynamics perspective, CTH result in a narrowing at the craniovertebral junction (CVJ). This narrowing impedes the rhythmic flow of cerebrospinal fluid (CSF) between the brain and spinal subarachnoid spaces. While the CSF flow is approximately 1 cc per heartbeat, the restriction at the CVJ can amplify CSF pressure gradients, leading to significant neurological issues [39].

Currently, MRI techniques are under development to gauge CSF pressure gradients without invasive methods. However, these techniques are not yet mainstream and require further validation [40]. Theoretically, changes in CSF pressure gradients connect to factors like resistance to CSF motion, CSF flow rates, neural tissue movement, and the overall adaptability of the craniospinal system. It is believed that evaluating these biomechanical factors might offer insights into the conditions at the CVJ in CMI patients [41].

Many patients diagnosed with syringomyelia accompanying Chiari I malformation demonstrate a two-phase systolic-diastolic CSF flow pattern, as captured in cine phase-contrast MRI. Posterior fossa decompression (PFD) can rapidly decrease these flow rates within the syrinx and at the FM. These flow rates can potentially predict positive outcomes, especially in terms of swift recovery from symptoms like headaches, pain, weakness, cranial nerve issues, and specific sensory losses [34].

Chiari II malformation (CM-II), also referred to as Arnold-Chiari malformation, represents a congenital anomaly that is frequently identified and characterized by a constellation of neuroanatomical abnormalities. These include a beaked appearance of the midbrain and caudal displacement of the cerebellar tonsils and vermis, in conjunction with spinal myelomeningocele. There is a prevalent misconception that posits CM-II as a mere exacerbation of Chiari I malformation (CM-I); however, these entities are distinct, albeit with some similar radiological presentations. A notable correlation exists between myelomeningocele and CM-II, with the latter condition often co-occurring with hydrocephalus [42].

A spectrum of additional pathological features is associated with CM-II, such as cerebellar dysplasia, caudal elongation of the pons and medulla oblongata, and caudal migration of the fourth ventricle into the cervical spinal canal. Magnetic Resonance Imaging (MRI) remains the cornerstone of diagnostic evaluation, offering a detailed assessment of the patient's neuroanatomy [43].

Therapeutic approaches for CM-II are predominantly surgical, aiming to address structural anomalies and ameliorate associated symptoms. The prognosis for individuals with CM-II is variable and is contingent upon the severity of the anatomical malformations and the clinical manifestations exhibited by the patient.

Among the spectrum of Chiari malformations, Type III is acknowledged as the least common variant. Magnetic Resonance Imaging (MRI) serves as a non-invasive diagnostic modality for Chiari Type III malformation. During the prenatal phase, the implementation of MRI, specifically employing a Single Shot Fast Spin Echo (SSFSE) sequence, can provide crucial insights following sonographic indications of this malformation [44]. This imaging technique empowers obstetricians to anticipate potential delivery complications, strategize the method of delivery, and engage neurosurgical collaborators. The symptomatic manifestations of Chiari Type III are diverse, encompassing hypotonia, hyperreflexia, seizures, developmental delays, central apnea, dysphagia, and dystonia, with symptom severity not necessarily corresponding to the extent of hindbrain or cervical cord herniation. The prognosis for individuals with Chiari Type III malformation is not universally unfavorable but hinges on multiple factors, including herniation locale, encephalocele constituents, sac coverage, associated anomalies, and the age at surgical intervention. The timing of neurosurgical measures is contingent upon the child's stability and factors like encephalocele size, neurological symptom progression, sac coverage integrity, and the risk of infection associated with compromised skin coverage. Prognostic assessment is informed through the constellation of neurological deficits at birth, such as respiratory distress, hypotonia, and dysphagia [45]. MRI plays a pivotal role in gauging the extent of herniation, which may inform symptom severity predictions, particularly given that critical medulla oblongata impairment often results in spontaneous breathing challenges. Therefore, a prudent approach is advised in the classification of Chiari Type III malformations, recognizing that the presence of a fluid-filled sac in the nuchal region does not unequivocally constitute this condition. Variability in presentation warrants meticulous scrutiny for accurate identification, which may alter prognostic expectations and therapeutic decisions.

In his work of 1895, Hans Chiari expanded the nosology of congenital hindbrain anomalies with the introduction of a fourth category, which he designated as "Chiari IV malformation". This addition to the pre-existing triad of malformations—Chiari I, II, and III—was predicated on the pathological findings observed in two patients. Distinct from the herniation characteristic of the posterior cranial fossa contents into the spinal canal, which is a hallmark of the other Chiari malformations, the Chiari IV subtype is defined by cerebellar hypoplasia in the absence of such herniation [46].

The delineation of Chiari IV malformation with Chiari elucidated a separate clinical entity, thereby refining the understanding and classification of cerebellar developmental anomalies.

The Chiari zero malformation (CM0), an infrequent subclass within the Chiari malformation spectrum, is characterized by the absence of hindbrain herniation—a defining feature of other Chiari malformations [47]. Initially, CM0's hallmark was the presence of syringomyelia, which was observed to resolve following posterior fossa decompression. However, contemporary findings have led to a revision in the diagnostic criteria, with the presence of syringomyelia no longer deemed a prerequisite for diagnosis. Although uncommon, there is also an established association between CM0 and syringobulbia [48,49].

This reevaluation of the diagnostic framework for CM0 reflects an evolving understanding of the condition and underscores the variability of its presentation. The acknowledgment of CM0 as a distinct clinical entity despite the absence of cerebellar herniation represents an advanced comprehension of the Chiari malformation spectrum.

3.3. Implications of These Changes on Overall Health

Any enlargement within the central nervous system can elevate venous pressure. This is because veins, being compressible, can experience decreased blood flow when there is a rise in CNS volume. This can potentially lead to an incremental increase in the cerebrospinal fluid (CSF) volume. Any condition that limits the space available for venous volume can trigger venous insufficiency. Healthy CSF circulation aids in optimizing venous drainage by regulating pressure within the central nervous system, facilitating its movement between the head and the spine. Conversely, any obstruction to this flow can spike localized pressures, hampering venous drainage [50].

Chiari malformations are tied to herniation of the hindbrain, often attributed to a disparity where spinal pressures are lower than those in the cranium. This leads to symptoms related to the hindbrain, often stemming from compression of the cerebellum and brainstem. When spinal damage arises from a Chiari malformation, the core issue is typically an underdeveloped posterior fossa, leading to heightened spinal pressures. This restricted posterior fossa space obstructs the CSF's movement from the spine to the brain as blood flows into the central nervous system during motion. As a result, periodic spikes in spinal pressure, especially during movements, can harm the spinal cord. It is believed that this underdevelopment of the posterior fossa, which begins in the fetal stage, can lead to syringomyelia post-birth and, subsequently, spinal cord damage in conditions like spina bifida. There is also a theory that hydrocephalus might be a byproduct of this posterior fossa underdevelopment. Here, pressure increases due to obstructed CSF flow from the brain to the spine, and in cases like anencephaly, this can lead to brain injury [38].

The prevailing understanding of dysraphism is that it results from diminished central nervous system pressure and the harmful effects of amniotic fluid on the CNS. The perspective presented here leans toward viewing spina bifida as a progressive fetal hydrocephalus manifestation. It is suggested that inadequacies in mesodermal growth can influence both the closure of the neural tube and the pressure within the CNS, culminating in dysraphism [51].

4. Pathogenesis of Syringomyelia: Lessons from Observations in Dogs

4.1. Summarized Key Findings from Canine Studies

Chiari-like malformation (CM) and syringomyelia (SM) frequently occur in small toy breed dogs, especially in the Cavalier King Charles Spaniel (CKCS), often leading to severe clinical symptoms [52].

Various terms such as caudal occipital malformation syndrome (COMS), occipital hypoplasia, Chiari malformation, and hindbrain herniation have been used to describe CM in veterinary literature. However, the Chiari-like Malformation and Syringomyelia Working Group has recently reached a consensus to refer to this condition as Chiari-like malformation (or CM) when discussing canines. Notably, this malformation is observed as an inherent trait in the CKCS breed, with a staggering occurrence rate of 100% [53].

CM/SM is characterized by a series of structural deformities in the skull and occipital bone, causing compression and, in certain instances, a posterior shift of the cerebellum (CM), as well as the presence of fluid inside the spinal cord tissue, known as SM.

Given the widespread nature of CM and SM within the CKCS breed, multiple studies have been conducted to analyze breed-specific skull shapes and dimensions to shed light on their role in causing SM [54]. A detailed breakdown of recent morphometric analyses, their findings, and potential implications on SM's onset can be found in Table 2.

Table 2. Significance of Published Morphometric Studies in Understanding the Etiology of Syringomyelia. The table presents the essential traits that amplify the likelihood of CM/SM development, with a focus on dog breeds in these studies.

Fundamental Characteristic	Mechanism	Supportive Findings	Citations
The early fusion of the spheno-occipital synchondrosis leads to brachycephalicism and miniaturization.	Easly fusion leads to reduced skull length, triggering compensatory elongation of other calvarial bones.	• A smaller skull width:length ratio guards against SM. • Dogs with a less prominent, caudally distributed cranium shape have protective attributes. • CM dogs show a shorter distance between the FM and the pons. • CKCS dogs with CM have a shorter spheno-occipital junction to atlas length and a reduced spheno-occipital angle.	[55,56]
	Overcrowding of the whole brain leads to the displacement of the cerebellum and brainstem towards the caudal region.	• Dogs with CM display shorter cerebral:cranial length than control brachycephalic dogs. • Greater cerebellar herniation links to decreased cerebral:cranial length. • CM-affected dogs show reduced FM-to-pons distance. • CM in CKCS dogs leads to rostral forebrain flattening and a distinct combo of shortened basicranium with heightened cranial height.	[57,58]
	The cause of overcrowding is linked to a smaller caudal cranial fossa.	• CKCS with SM have smaller CCF volume than mesaticephalic dogs. • Reduced caudal CCF volume in CM/SM CKCS compared to CM alone.	[27,57,59]

Table 2. *Cont.*

Fundamental Characteristic	Mechanism	Supportive Findings	Citations
CM leads to secondary effects that raise uncertainty about their impact on the development of syringomyelia.	CM leads to the herniation of the cerebellum and brainstem.	• CKCS dogs with CM/SM show stronger cerebellar pulsations during systole, potentially affecting CSF flow further; • Higher medullary kinking index links to SM presence and severity.	[60,61]
	Occipital hypoplasia undergoes gradual development.	• As the foramen magnum (FM) gets bigger, cerebellar herniation increases. • Over time, both the FM height and cerebellar herniation length consistently increase.	[60]
	A shortened skull base can decrease the size of the jugular foramen and increase ICP.	• CKCS with CM/SM have a smaller JF volume than CM alone. Moreover, venous congestion may influence CSF pulse pressures and result in SM.	[26,62]
Clinical signs related to CM/SM are impacted by issues in the craniocervical junction.	Simultaneous CJA influence both symptoms and SM progression	• CKCS shows a greater occurrence of AOO compared to other small toy breeds. • Atlantoaxial bands are linked to increased SM severity and noticeable clinical symptoms.	[63,64]
Brain parenchyma size results in overcrowding.	Overcrowding occurs as a consequence of an enlarged cerebellum	• CKCSs have larger caudal fossa parenchyma than other small breed dogs with similar CCF sizes; • CKCS with SM have a larger CCF parenchyma size but similar CCF size than those without; • The cerebellum is the larger portion of the CCF parenchyma.	[58]

AOO, atlanto-occipital overlapping; CKCS, Cavalier King Charles Spaniel; CCF, caudal cranial fossa; CM, Chiari-like malformation; CSF, cerebrospinal fluid; CJA, craniocervical junction abnormalities; SM, Syringomyelia; FM, foramen magnum; ICP, intracranial pressure; JF, jugular foramen.

4.2. Extrapolation to Human Pathogenesis

CM in canines presents a naturally occurring counterpart to CMI in humans, offering valuable insights into the human SM research sphere.

Caudal Cranial Fossa (CCF) Anatomy

The CCF is the internal cranial space housing the cerebellum, pons, and medulla oblongata. Inside, it is bordered by the tentorium cerebelli on the top and front, and its base stretches from the petrosal crests and dorsum sellae to the foramen magnum. Externally, it is framed by the triangular occipital bones: the supraoccipital, basioccipital, and paired exoccipitals. The posterior fossa's volume is pertinent to CMI, as it is notably reduced in children with both CMI and SM [65].

Researchers have employed a 3D volumetric approach to contrast the craniocerebral volumes of CKCS with those of other small breeds and Labrador retrievers. Interestingly, while CKCS had a CCF volume comparable to smaller breeds, the tissue volumes within this CCF were akin to those in Labradors. This hints at a potential volumetric congestion of the CCF. Subsequent discoveries affirmed that this increased congestion correlated with the existence and severity of SM [66].

This leads to the hypothesis that CKCS's brains, irrespective of their short-skulled (brachycephalic) nature, might be disproportionately large for their cranial encasement. A study by Shaw et al. [67] revealed that the CKCS possessed a larger cerebellum in comparison to both smaller breeds and Labradors. The cerebellum's volume also showed an association with SM. These structural anomalies might contribute to aberrant cerebellar functionality, with a recent analysis indicating CKCS exhibits varied gait patterns indicative of ataxia [68].

Two primary hypotheses have been put forth to explain this volumetric disparity. The first suggests that premature cranial suture closure (craniosynostosis) disrupts regular skull growth trajectories. This theory aligns with observations in human CMI patients, where there is an underdevelopment of cranial base bones [69]. In CKCS, a short basioccipital bone was associated with SM [70]. Studies have indicated that CKCS tend to exhibit notably earlier skull base growth plate (synchondroses) sealing when compared to both short-skulled (brachycephalic) and medium-skulled (mesaticephalic) breeds [71].

The second theory postulates a communication breakdown between the distinct cartilage-forming mesodermal precursors responsible for the occipital bones and the sealing neural tube, resulting in a limited volume for the CCF tissue. In human embryos, the CCF's growth mirrors that of the cranial fossae, seemingly independent of cerebellar development. Given that cerebellar expansion is a late-stage event in fetal development and continues post-birth in species like dogs and cats, it is plausible that the CCF might not sufficiently accommodate CKCS's relatively large cerebellum. This congestion is especially pronounced in the CCF's posterior region, potentially altering CSF dynamics [72].

5. The Role of Genetics in Chiari Malformation and Syringomyelia

5.1. Current Understanding of the Genetic Basis

While CM was once believed to be an isolated disease, research involving families suggests that genetics play a role in its development. The presence of the disease in multiple family members points to a possible genetic link, and pinpointing the exact genes or mutations responsible could enhance diagnostic and therapeutic approaches [73].

The term "familial aggregation" denotes the appearance of a disease or condition within several family members, suggesting its incidence is higher than what might be expected in the general population. With respect to CM, studies indicate that immediate family members (like parents, siblings, and children) of those afflicted have a significantly increased risk of developing the condition, underscoring the potential role of genetic predisposition [74].

The nuances and severity of CM can differ even among relatives, hinting at the potential influence of multiple genetic variations. In certain families, CM might be passed down as an autosomal dominant trait, where a single mutated gene from one parent is enough to manifest the disease. However, in other scenarios, the development of CM could arise from a combination of several genetic elements and environmental factors, leading to a more intricate inheritance pattern [75].

5.2. Identified Genes and Their Impact

The study by AvŞar T et al. [74] examines two families with members diagnosed with CMI. In the first family, surgical intervention for CMI was performed on two female siblings, while in the second family, both the mother and the second son underwent surgery for the same condition. The predominant clinical symptoms among these individuals were occipital headaches that intensified during straining or post-coughing, difficulty swallowing, extremity numbness, and an impaired ability to distinguish between hot and cold sensations, particularly in the legs.

Members of both families who did not exhibit symptoms were given cranial MRI scans, and the diagnosis of CMI was excluded for them. Every patient across both families was treated using a standard surgical method known as posterior fossa decompression combined with an extensive duraplasty.

The data from the microarray was analyzed using two primary methodologies. Firstly, single nucleotide variations (SNVs) from both affected and unaffected family members were juxtaposed. The mutations discovered in the symptomatic members of both families are detailed in a table. Secondly, a different table (Table 3) showcases chromosomal differences across all members, both affected and unaffected, from the two families.

Table 3. List of common single nucleotide variations in both families.

Chr. Location	Gene	Biological Process/ Gene Ontology[a]	Variant Class[c]	Enhanced Expression[d]
9q34.11	USP20	Endocytosis, Ubl conjugation pathway	Intronic	Low tissue specificity
5q31.1	TRPC7	Calcium transport	Intronic	Adrenal gland, brain, intestine, kidney, pituitary gland, testis
9q33.2	TRAF1	Apoptosis	Intronic	Low tissue specificity
5q31.3	SLC4A9	Anion transmembrane transporter activity	Missense	Kidney, heart
9q33.2	PHF19	Chromatin regulator	Intronic	Low tissue specificity
9q33.3	OLFML2A	Protein homodimerization activity	Missense	Low tissue specificity
5q31.3	NR3C1	Apoptosis, cell cycle, transcription regulation	Intronic	Low Tissue Specificity
13q33.3	MYO16	Motor activity, actin binding	Intronic	Brain
9q33.3	MVB12B	Protein transport	Intronic	Brain
9q34.11	LOC101929331	N/A	Intronic	N/A
5q31.3	LOC101926941	N/A	Intronic	N/A
5q31.1	LOC100996485	N/A	Intronic	N/A
17q21.33	LOC100288866	N/A	Intronic	Low tissue specificity
5q23.1	LINC00992	N/A	Intronic	Pancreas, colon
3p24.1	LINC00693	N/A	Intronic	Brain
7q22.3	LHFPL3–AS2	N/A	Intronic	Kidney
7q22.2	LHFPL3	N/A	Intronic	Brain
5q32	HTR4	G protein-coupled receptor activity	Intronic	Brain, heart muscle, intestine, pituitary gland
5q31.1	FSTL4	Calcium ion binding, metal ion binding	Intronic	Brain
5q31.3	FGF1	Angiogenesis, differentiation	Intronic	Brain, heart muscle, kidney
13q33.3	FAM155A	Calcium ion import across plasma membrane	Intronic	Brain, pituitary gland
13q.34	COL4A2	Basal membrane formation	Intronic	Placenta
13q.34	COL4A1	Basal membrane formation	5′UTR	Placenta
9q32	COL27A1	Extracellular matrix structural constituent	Intronic	Brain, uterine, cervix
9q33.2	CNTRL	Cell cycle, cell division	Intronic	Low tissue specificity
9q33.1	BRINP1	Inhibits cell proliferation with negative regulation of the G1/S transition	Intronic	Brain
9q33.1	ASTN2	Protein transport	Intronic	Low tissue specificity
5q31.3	ARHGAP26	GTPase activity	Intronic	Low tissue specificity
5q32	ADRB2	G protein-coupled receptor activity	Intronic	Blood

Single nucleotide variations (SNVs) were assessed in both families, and shared variations were cataloged in a table. The majority of these variations were found within introns.

Still, two missense variations and a single 5′UTR variation were identified. None of the variations highlighted in the table have been previously marked as clinically significant in the ClinVar database [76].

5.3. Potential Avenues for Future Research

Investigating the genetic underpinnings of Chiari malformation and syringomyelia is an evolving field of research. Potential research directions include:

1. Genome-wide Association Studies (GWAS):

Initiate expansive GWAS to pinpoint genetic variants correlated with the onset of Chiari malformation and syringomyelia. This could reveal specific genes or pathways implicated in these conditions [77].

2. Gene Expression Analysis:

Assess the gene expression patterns in individuals diagnosed with Chiari malformation and syringomyelia in comparison to their healthy counterparts. This could shed light on molecular pathways that are disrupted and potential treatment targets [78].

3. Functional Experiments:

Undertake functional experiments to discern the biological impact of genetic variants linked to Chiari malformation and syringomyelia. This could encompass experiments using cell cultures or organoids as disease models [79].

4. Exploring Gene Therapy:

Investigate the feasibility of gene therapy as a therapeutic approach for Chiari malformation and syringomyelia. Cutting-edge gene-editing tools like CRISPR/Cas9 present novel avenues for rectifying genetic mutations that cause diseases [80].

By exploring these research paths, we can enhance our comprehension of the genetic factors underlying Chiari malformation and syringomyelia. This could pave the way for refined diagnostic techniques, specialized treatments, and, ultimately, improved patient care and prognosis.

6. Clinical Phenotypes in Chiari and Syringomeylia

Presentation and Clinical Features in Chiari Malformation

Clinical manifestations of CMI are varied. Symptoms can encompass head, neck, and back discomfort, pain in the shoulders (cape pain), limb pain that's not linked to nerve roots, weakness, tingling sensations, balance disturbances, double vision, ringing in the ears, hearing impairment, fainting, slurred speech, difficulty swallowing, urinary issues, and disrupted sleep [81]. Observable clinical signs can include issues with cranial nerves (like nystagmus, swallowing difficulties, and sleep apnea), compression of the brainstem (resulting in fainting, hearing loss, and heart rate abnormalities), cerebellar indications (like coordination problems), and spinal cord complications (such as heightened reflexes and spasticity) [81].

The most commonly observed symptom in CMI is a distinct headache in the back of the head or upper neck region, characterized as sharp or pulsating and intensified by actions like coughing, Valsalva maneuvers, changes in posture, or physical activity [82]. This type of headache is classified by the International Headache Society as 7.7, specifically attributed to Chiari malformation type I (Q0.70).

There have been studies aiming to distinguish the various headache types in CMI and their causal relationships. Pascual et al. studied 50 CMI patients and found that 52% experienced headaches [83]. Based on IHS criteria, these headaches varied in type. While 14 patients had the typical CMI-associated headache, others experienced migraines, tension-related headaches, or even trigeminal neuralgia. Interestingly, the severity of pain was related to the extent of the cerebellar tonsil herniation but not to the deformity of the occipital bone. Toldo et al. identified that among 45 young patients with CMI, the primary symptom was typically a headache, with the classic CMI type being most prevalent. If a headache coexisted with three other CMI clinical signs, it was a strong indicator of pronounced tonsillar displacement [84].

Wu et al. undertook a retrospective evaluation of 49 children with CMI, all under 14 years old [85]. The most frequent symptoms were headaches, neck pain, and coordination issues. However, only three experienced the classic CMI headache. The study found no significant correlation between the severity of symptoms and the degree of tonsillar herniation or MRI CSF flow abnormalities. This sample, however, was relatively small.

Pujol et al. proposed that rather than the size of the cerebellar tonsils, the extent of their movement could be the determinant of cough-induced headaches [17]. Patients experiencing such headaches showed greater tonsillar motion than those who did not. The conclusion was that the intensity of tonsillar movement and the reduction of the arachnoid space were linked to this specific symptom but not to the occurrence of syringomyelia. Such findings prompt further exploration into CMI's origins.

Furthermore, Wu et al. [85] suggested that clinical manifestations might arise when scarring and adhesive formations occur in the arachnoidal layers at the foramen magnum, potentially due to continuous contact of the cerebellar tonsils with the bone. This might amplify hindbrain compression, producing signs and symptoms and even initiating the development of syringomyelia. This is among the various hypotheses proposed to explain the genesis of CMI.

This study aimed to evaluate the clinical presentations, imaging results, treatment outcomes, and importance of post-traumatic syringomyelia (PTS).

The study group was composed of nine males aged between 30 and 68, with an average age of 51.2 years. When injured, their average age was 27.7 years, with ages ranging from 20 to 45 years. Injury causes included motor vehicle crashes for four participants, falls for another four, and a spinal injury for the remaining one. Among them, seven had spinal fractures, one had a spinal dislocation, and another had an SCI. The period between the initial trauma and the emergence of new symptoms varied widely, from 3 to 44 years, with an average of 21.9 years. The most common new symptom was motor weakness, observed in five patients. This was followed by sensory issues and pain in four patients and urinary dysfunction in one patient [86].

7. Diagnostic Investigations: Old and New Neurophysiological Methods

7.1. From Traditional Diagnostic Methods to Newer Neurophysiological Techniques—Comparison and Evaluation of Effectiveness

7.1.1. Introduction

For diagnosing CM-I, neuroradiology plays a pivotal role, providing insight into the related anatomical structures and fluid dynamics. Techniques include magnetic resonance imaging (MRI) with dynamic and upright perspectives, alongside myelography and computed tomography (CT) [87].

The growing reliance on neuroimaging has led to an increase in the number of individuals termed "victims of contemporary imaging technology" [88]. While over 1% of the population receives a CM-I diagnosis, the majority of these cases are incidental discoveries that do not necessitate intervention [9,89]. Receiving a Chiari malformation diagnosis can stir anxiety among patients and might be attributed as the cause of a wide range of symptoms. However, many of these symptoms do not see improvement even after surgical intervention. The extent of tonsillar herniation is not always a reliable indicator of its clinical importance or an indicator of functional impairment. The foundational step in any assessment is a thorough clinical history gathering, followed by a comprehensive physical and neurological evaluation (Figure 1).

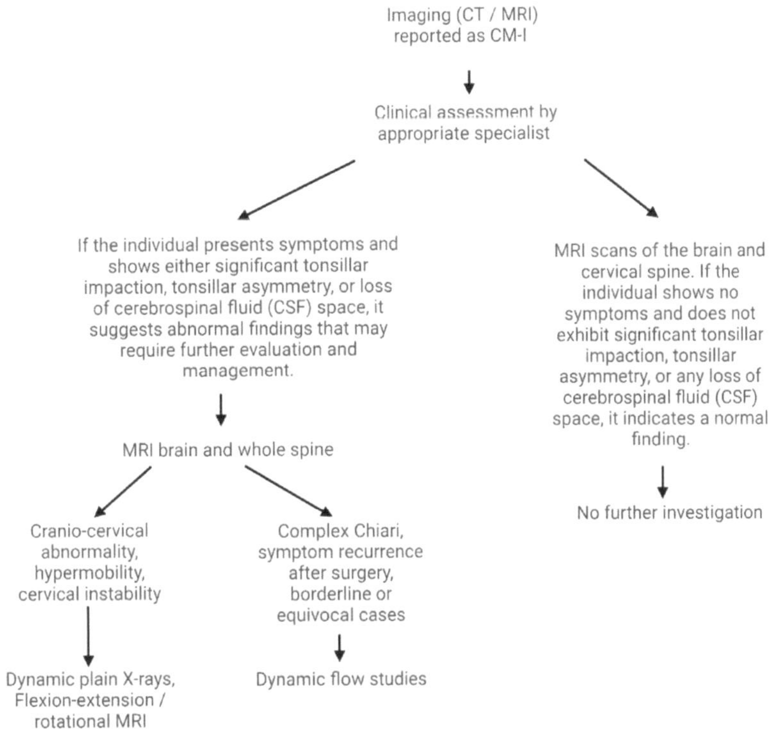

Figure 1. Diagnostic flowchart of Chiari malformation.

7.1.2. Computed Tomography (CT)

Despite the proliferation and accessibility of advanced neuroimaging techniques, many patients initially receive their diagnosis via a CT scan. Even though the majority of patients eventually undergo MRI, CT scanning remains indispensable for evaluating bone structures, especially in those with innate or developed bony irregularities at the craniocervical junction. Such scans can also assist with dynamic investigations (refer to the section on dynamic mobility studies). Various anomalies, including basilar invagination, platybasia, Klippel-Feil, atlanto-occipital assimilation, and other intricate abnormalities, might be identified [8]. Furthermore, CT can be employed in CT myelography to examine potential hidden spinal CSF leaks. Although MR myelography might offer greater precision, there are existing apprehensions regarding the use of intrathecal gadolinium [90].

7.1.3. Magnetic Resonance Imaging (MRI)—Brain

Tonsillar Herniation

For the primary assessment of CM-I, MRI scans of the brain and cervical spine are the preferred imaging methods. While a CM-I diagnosis is typically made on the T1 or T2 sagittal midline MRI view, evaluating the vertical distance between the tip of the herniating tonsil and the foramen magnum (McRae's line), it is crucial to recognize that tonsils are three-dimensional and can vary in size, shape, and extension (Figure 2). Therefore, coronal views can offer additional critical data for diagnosis and surgical planning [10].

7.2. Newer Neurophysiological Techniques

Several grading systems based on tonsillar descent and patient age have emerged. Aboulezz et al. (1985) [91] proposed that tonsillar positions be deemed normal up to 3 mm, borderline between 3 and 5 mm, and abnormal beyond 5 mm. Age-dependent thresholds were subsequently suggested: 6 mm for up to 10 years, 5 mm for ages 10–30, 4 mm for ages 30–70, and 3 mm for those over 70 [11,92]. This upward shift of the cerebellar tonsils with age may be more linked to the overall reduction in brain volume over time than to CM-I's inherent characteristics.

A specific millimeter value for tonsillar descent can be misleading [93], except when it indicates a growing intracranial pressure or persistent spinal subarachnoid hypotension. Terms like Chiari 0 and Chiari 0.5 have been introduced, referring to patients with 0 mm or <5 mm of tonsillar descent, respectively, who still exhibit related symptoms and benefit from surgery. Comprehensive radiological imaging should primarily focus on thoroughly examining the 3D anatomy and CSF dynamics at the craniocervical junction.

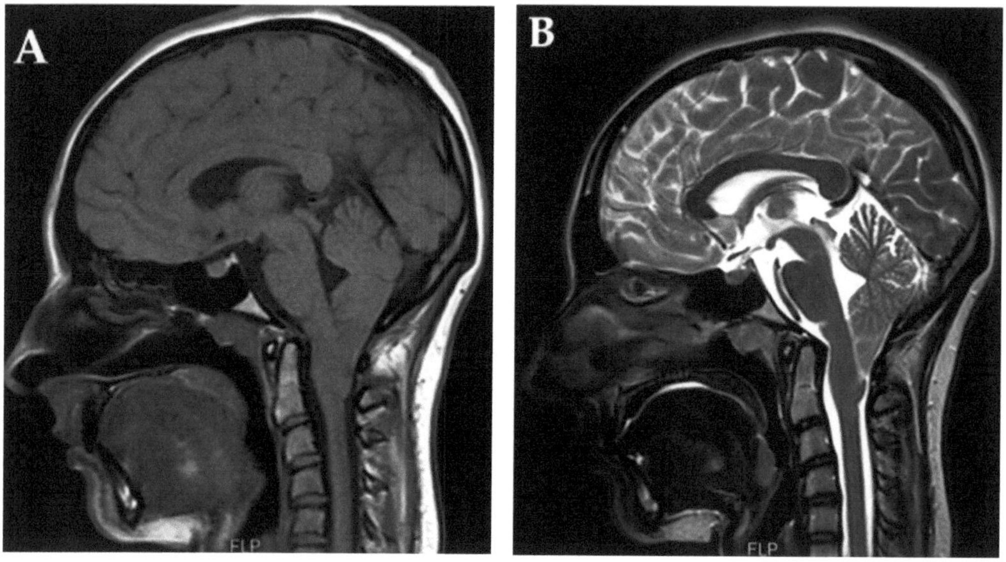

Figure 2. MRI in sagittal T1-weighted sequence (**A**) and T2-weighted sequence (**B**) are shown. Both sequences illustrate a moderate hypoplasia with the characteristic of Chiari malformation peg-like appearance of the cerebellar tonsils; moreover, important downward herniation below the McRae line is depicted (Personal case of Assoc. Prof. Horia Ples MD. PhD.).

7.2.1. The Size of Posterior Cranial Fossa (PCF)

The most commonly believed origin of a CM-I is an abnormality in the paraxial mesoderm, which results in a developmental disparity between neural and bony components [94]. The size of the PCF can be evaluated using distinct linear markers. In CM-I patients, the lengths of elements like the clivus, supraocciput, and exocciput tend to be shorter than in their healthy counterparts, though normal values can vary substantially [94–96]. A category defined as classical CM-I has been proposed for those with occipital bone hypoplasia and a smaller PCF volume without other causal factors [97]. Another research group developed a prediction model for CM-I symptomatology, independent of tonsillar herniation level, with an accuracy of 93% sensitivity and 92% specificity, based on measurements like the osseous PCF area and clival length [98].

Patients with symptoms resembling Chiari but without significant cerebellar tonsillar herniation exhibited similar morphological findings [38,99]. Nevertheless, certain studies found no link between the size of the PCF and clinical symptoms [100]. Given these inconsistent results, the present consensus is that linear measurements of the PCF do not add much value to the radiological assessment.

Newer neurophysiological techniques.

Explorations into volumetric analysis show promise. The standard volume of the posterior fossa in healthy individuals stands at about 190 mL [97]. A volume ratio, which is the brain volume relative to the cranial volume in the PCF, was examined and found to be significantly larger in CM-I patients compared to healthy ones [95]. Moreover, a smaller PCF to supratentorial volume ratio was associated with better post-surgery outcomes, and this was also true for the craniectomy extent and the PCF volume increase [101]. Predictions about the necessary craniectomy extent and optimal PCF volume increase could be drawn from pre-surgery MRI data [102]. Even with these potential benefits in diagnosis and prediction, PCF volume assessments are seldom employed in clinics due to the intricate volume calculation process, though this could change as automatic segmentation technologies advance.

7.2.2. Hydrocephalus

Chiari proposed that CMs might stem from prolonged hydrocephalus. However, only about 7–11% of CM-I patients present with hydrocephalus or idiopathic intracranial hypertension (IIH) [94,103]. The causal dynamics could differ among patients. For instance, hydrocephalus might result from the obstruction of the foramen of Magendie and the associated CM-I impeding the IVth ventricle outflow, or hydrocephalus or IIH might cause tonsils to herniate downwards, leading to a CM-I [104,105] (Figure 3). In initial patient evaluations, symptoms and indicators of hydrocephalus are examined. If detected, the primary treatment target becomes the hydrocephalus over the CM-I [106]. Tools like MRI scans, MR or CT venography, intracranial pressure monitors, and venous pressure measurements can guide treatment decisions.

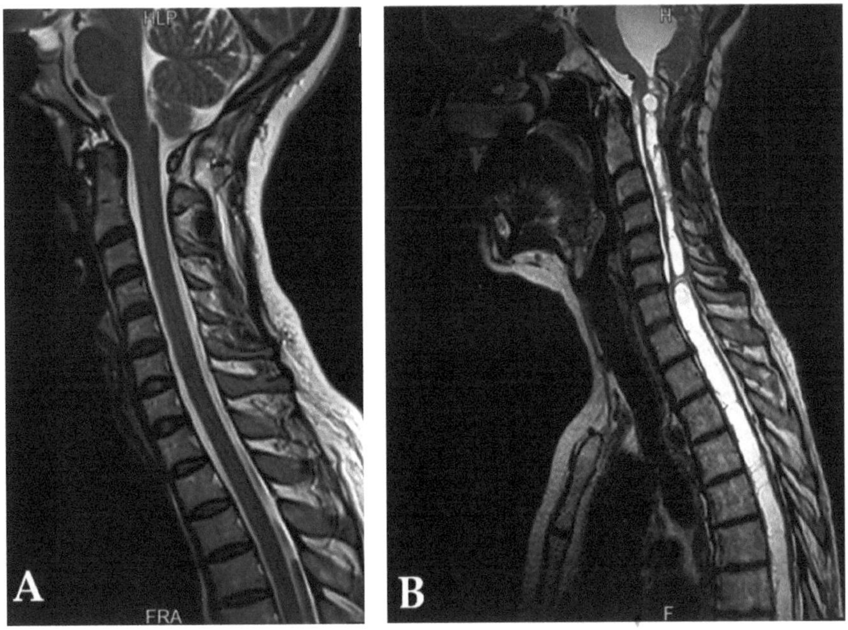

Figure 3. MRI in sagittal T2-weighted sequence (**A**) and T1-Gadolinium sequence (**B**) are shown. Cerebellar tonsillar ectopia is revealed and a significant syrinx cavities filled with cerebrospinal fluid is depicted (Personal case of Assoc. Prof. Horia Ples MD. PhD.).

8. Surgery in Chiari 1 Malformation with and without Syringomyelia

8.1. Indications for Surgery

The availability of multiple surgical treatments for CM-I underscores that no single procedure is ideal for every patient. Numerous clinical series indicate that a combination of suboccipital craniectomy, the removal of C1's posterior arch, and augmentative duraplasty is the foundational surgical approach. This is frequently applied to most Chiari I patients and is commonly adopted by many neurosurgeons. The combination of suboccipital posterior fossa decompression and atlas laminectomy is seen as the standard surgical strategy for the majority of symptomatic CM-I patients [107,108]. Clinical series have reported between 95 and 97% improvements in their patients' preoperative symptoms [107,108]. Nonetheless, positive results have been documented from alternative procedures, such as modified osseous posterior fossa decompression [109,110]. Sindou et al.'s [109,111] recommendation for a comprehensive suboccipital craniectomy and broader foramen magnum opening has not shown superiority over the conventional suboccipital craniectomy. Moreover, the extended craniectomy might introduce additional risks, including potential vascular injuries, longer surgery durations, and a heightened likelihood of postoperative CSF leaks.

In the same vein, extensive arachnoidal adhesion lysis has not conclusively shown benefits over duraplasty alone. However, in reoperations, using arachnoid dissection can be beneficial in addressing adhesions and restoring CSF flow. An essential aspect to note is that detecting arachnoidal veils or posterior fossa compartmentalization on preoperative MRI suggests the need for more intensive arachnoidal dissection to rectify compromised CSF flow [112]. But in initial operations, especially in children, arachnoidal adhesions are infrequent. Furthermore, the dissection might provoke additional postoperative arachnoidal adhesions. The matter of cerebellar tonsil reduction or removal remains debated. The paucity of comparative studies hampers the evaluation of this technique's efficacy. In situations where cerebellar tonsils are significantly affected, their removal might aid CSF circulation. But, surgeons must be wary of nearby posterior inferior cerebellar arteries during this procedure. An additional consideration is that tonsil removal might lead to postoperative complications like nausea and vomiting. Alden et al.'s [110] and Valentini et al.'s [113] findings incorporated tonsil removal in their procedures, which aligned with outcomes from series that avoided tonsil resection. Another technique involves obstructing CSF flow via the obex in CM-I management [114]. Its efficacy in improving patient outcomes remains uncertain.

8.2. Current Surgical Techniques and Their Outcomes

Reports indicate that foramen magnum decompression (FMD) for syringomyelia related to Chiari I malformation leads to the shrinkage of the syrinx cavity and relief from neurological symptoms [115,116].

Pain throughout the body, including the limbs, is a primary concern for syringomyelia patients. The underlying causes remain somewhat elusive, and only a few studies have thoroughly examined the relationship between syrinx location and body pain in this malformation [117,118].

Surgical interventions for syringomyelia linked to Chiari I malformation aim to diminish the syrinx size via enhancing cerebrospinal fluid flow. Employing procedures like FMD and syringosubarachnoid shunts, there has been progress in achieving this objective. Nevertheless, the clinical symptoms do not always align with the reduction in the syrinx size, posing challenges in treatment [117]. Pain stands out as a primary symptom of syringomyelia. With the prevalent use of MRI, early detection has improved. However, some patients continue to experience persistent pain. The exact mechanisms causing pain related to syringomyelia remain a subject of inquiry. Currently, predicting post-surgical improvements is challenging, and treatment outcomes can be unpredictable. The type of pain associated with syringomyelia is known as deafferentation pain and is believed to be linked with spontaneous pain, hyperesthesia, allodynia, or dysesthesia. Disruptions in the pain pathway anywhere from the spinal cord's dorsal horn to the cerebral cortex might trigger this pain. In syringomyelia cases, the dorsal horn's involvement has been highlighted.

Nakamura et al. observed that the syrinx's shape at the spinal cord level corresponded with the pain site's dermatome. They found that post-surgery pain is often persistent in the deviated type, where the MRI shows the syrinx on the spinal cord's posterolateral side. Here, the syrinx's location aligns with the spinal cord's dorsal horn gray matter, indicating a potentially irreversible alteration in the dorsal horn [118].

Milhorat et al. [117] pointed out that when there is persistent pain, the syrinx often reaches the spinal cord's dorsal horn, where there is a heightened concentration of substance P in the Rex I–III layers of the same section. They proposed that the spinal dorsal horn plays a role in syrinx development and pain onset.

They also found that among the enlarged-type syrinxes, those that are either resolved post-surgery or transition to a central type likely originate from the expanded central canal of the spinal cord. It was noted that nerve fibers in the afferent pathway, specifically those traversing the anterior gray commissure (lateral spinothalamic tract), tend to experience functional recovery, leading to relief from deafferentation pain. This is attributed to the syrinx size reduction and the subsequent decompression. It is theorized that a deviated-type syrinx might either extend from the enlarged central canal to the dorsal horn or originate directly at the dorsal horn, independent of the central canal. In both scenarios, if the dorsal horn neurons sustain irreversible damage, the deafferentation pain remains unrelieved, regardless of the post-surgery reduction of the syrinx size.

Literature is limited concerning the relationship between the syrinx's shape, pain, and the surgical outcomes of syringomyelia linked to Chiari I malformation. There is also a notable scarcity of studies using a quantitative VAS (Visual Analog Scale) to assess pain before and after surgery.

In the study in question, it was demonstrated that in cases with a deviated-type syrinx, the dermatome level in the upper limb, corresponding to the pain location, aligned with the spinal cord level where the syrinx deviated. The pain intensity was further quantified using the VAS score. The findings revealed that individuals with a pre- and post-surgery deviated-type syrinx on MRI usually experienced more severe pain compared to those with other syrinx types. This suggests that the pain's intensity, pre- and post-surgery, is heightened when the syrinx deviates towards the spinal dorsal horn, as visualized on an MRI. Patients suffering from syrinx-related pain, especially the deviation type, were diagnosed early and received prompt surgical intervention. The pain associated with a syrinx seems to hinder daily activities significantly. Exploring the pain related to syringomyelia appears to be a crucial area of future research [117].

A succinct synthesis of a retrospective analysis was conducted by Goel et al. [119–121] on a cohort of 388 patients diagnosed with Chiari formation (CF), with a focus on the application of atlantoaxial fixation. It encapsulates the postoperative clinical enhancements, corroborated using the radiological diminution of syrinx dimensions, and underscores a paradigm shift in the etiopathogenetic understanding of CF, positing atlantoaxial instability as a pivotal factor (Table 4).

Table 4. Surgical approach and clinical outcomes of Chiari type 1 malformation.

Category	Details
Patient Cohort	388 patients with Chiari formation
Surgical Approach	Atlantoaxial fixation
Clinical Outcomes	99.4% of patients showed immediate postoperative and sustained improvement
Radiological Outcomes	Reduction in syrinx size in 65 out of 221 patients in the immediate post-operative phase; significant syrinx size reduction in 95 out of 110 cases on delayed post-operative scans

Table 4. *Cont.*

Category	Details
Pathogenesis Perspective	Proposed atlantoaxial instability as a nodal point of pathogenesis for Chiari 1 formation
Treatment Goals	Achieve firm atlantoaxial fixation resulting in segmental arthrodesis; no foramen magnum decompression or syrinx manipulation
Surgical Technique	Lateral mass plate and screw fixation; avoidance of metal spacers post-2013 in favor of bone grafts for realignment and arthrodesis
Postoperative Management	Hard cervical collar for 3 months to facilitate bone fusion
Complications	Vertebral artery injury in a few cases; technical difficulties due to complex craniovertebral junction anatomy
Improvement Indicators	Immediate postoperative improvements in clinical symptoms such as voice volume, breathing, pain relief, and motor function; progressive improvement over time
Long-term Observations	Reversal of spinal deformities and recovery from major presenting symptoms in the im-mediate postoperative period
Clinical Assessment	Utilized Goel clinical grading scale, JOA score, VAS, and patient self-assessment; reviewed by independent neurosurgeons
Radiological Assessment	Postoperative CT and MRI to evaluate syrinx size reduction and tonsillar herniation re-gression
Considerations for Pediatric Patients	Symptoms and alterations in pediatric cases likely depend on the onset and degree of at-lantoaxial instability

8.3. Conclusions

The morphology of the cavity and related pain of syringomyelia linked to Chiari I malformation were prospectively analyzed using the VAS score, both before and after surgery. The findings indicate that pre- and post-surgery pain tends to be more pronounced when the syrinx shifts towards the spinal dorsal horn, as observed on an MRI [55].

9. Surgical Strategies in Isolated Syringomyelia

9.1. Indications for Surgery

A variety of surgical interventions have been proposed to treat post-traumatic syringomyelia. However, some techniques, like omental grafting, are now rarely performed [56]. The two primary methods employed are direct drainage of the syrinx cavity and the reconstruction of the spinal subarachnoid pathways. The syrinx can be drained into spinal subarachnoid channels or the pleural or peritoneal spaces. However, direct drainage often encounters issues, such as blockage of the tubes, prompting follow-up surgeries. These procedures typically involve a myelotomy, which can result in the loss of dorsal column function in individuals who have retained some neurological function in their lower limbs. Moreover, syringomyelia cavities can be compartmentalized, complicating or even preventing the proper placement of a drainage catheter. Lastly, even if the syrinx is effectively drained, new cavities can form adjacent to the original one unless the root cause is addressed.

Reconstructing the subarachnoid channels seeks to restore the flow of cerebrospinal fluid, addressing the root cause of syrinx formation. If successful, this technique can lead to the complete or near-complete and lasting collapse of the syrinx. This surgical procedure also allows for the simultaneous placement of a drainage tube in the syrinx, if desired. In fact, some successful outcomes from primary drain insertions might actually stem from the surgical exposure itself, like the laminectomy and the restoration of spinal CSF pathways [57].

9.2. Overview of Different Surgical Strategies/Effectiveness and Outcomes

Diverse perspectives surround the causes and surgical treatments of syringomyelia and Chiari malformations. A debated aspect is the use of surgical adjuncts for treating symptomatic Chiari malformation patients. While cervicomedullary decompression is the conventional surgical method, the benefits of dural patch grafting, intradural dissection, and fourth ventricular shunting are still under scrutiny by several experts [58]. The approach and significance of preventative surgery in patients without symptoms is another contentious area. Due to advancements in MR imaging, patients with Chiari I malformations are being diagnosed at younger ages and often present with milder or no neurological indications [8]. Consequently, neurosurgeons increasingly find themselves assessing patients who exhibit tonsillar herniation or syringomyelia but are asymptomatic.

9.2.1. Study Profile

A vast majority, 78% of participants, dedicated over half of their professional practice to pediatric neurosurgery, with 55% dedicating over three-quarters. A minor 5% allocated less than a quarter to pediatric neurosurgery.

On average, each participant annually evaluated ten patients with confirmed syringomyelia and operated on seven of them.

Cumulatively, the respondents had performed 4049 procedures for syringomyelia, averaging 56 operations per respondent throughout their career [59].

9.2.2. Monitoring Asymptomatic Patients

Additionally, 63% of the surveyed believed that patients with asymptomatic syringomyelia seldom develop symptoms.

The majority suggested semi-annual neurological exams (84%) and MRI scans (75%).

A small fraction (16%) proposed cine-mode MRI for initial or subsequent assessments.

Some recommended biannual neurometrics (5%) or initial somatosensory evoked potentials (4%).

Fewer participants advised physical restrictions for asymptomatic patients with syringomyelia (31%) or Chiari malformation (36%), with avoidance of contact sports being the most common. This figure rose to 42% when Chiari malformation co-existed with syringomyelia [60,61].

9.2.3. Criteria for Surgery

A mere 9% always advocated for preventative surgery in syringomyelia or Chiari malformation cases.

At least 83% wouldn't operate on asymptomatic individuals unless they manifested related symptoms.

Nonetheless, 61% favored surgical intervention if an MRI revealed syrinx growth without clinical progression [62].

9.2.4. Treatment Preferences

For symptomatic patients with just syringomyelia and no Chiari malformation signs, 71% preferred shunting. Amongst them, the syringosubarachnoid shunt was the top choice (72%). Multiple shunt types were chosen by some respondents.

While all participants endorsed surgery for symptomatic Chiari malformation patients, the ideal procedure remained contested.

25% supported bone decompression without dural manipulation for Chiari I malformations, while a majority opted for a suboccipital decompression with dural patch grafting. Intradural dissection recommendations marginally increased for cases with concurrent syringomyelia.

For Chiari II malformations, tonsillar manipulations were less favored.

Most participants (75%) who utilized intradural dissection excluded a fourth ventricular stent from their procedure. An overwhelming 95% advocated for a dural patch graft in at least one suggested surgical approach, with pericranial and bovine grafts being the most popular [109].

9.2.5. Surgical Outcomes

According to the respondents, syringomyelia patients usually displayed improved syrinx appearances post-surgery. In contrast, Chiari malformation patients mainly experienced pain relief [63].

Multiple theories surround Chiari malformations' onset and cause. A significant majority, 78%, believed Chiari I and II malformations had distinct genetic or pathogenic origins.

10. Outcome Measures in Chiari and Syringomyelia Long-Term Follow-Up

10.1. Parameters for Assessing Outcomes

Post-surgical evaluation for Chiari malformation type 1 (CM1) is challenging due to the absence of a consistent and dependable rating system. CM1 symptoms can vary widely from patient to patient. Some pronounced indicators, such as sudden falls, swallowing difficulties leading to aspiration, breathlessness, and the occurrence of a syrinx, strongly suggest the need for posterior fossa decompression (PFD). However, it is ambiguous whether other signs and symptoms justify surgical intervention. The potential benefits of such interventions are typically gauged based on past results, which do not always provide a clear picture. In response, Aliaga et al. introduced a straightforward and measurable outcome evaluation tool named the Chicago Chiari Outcome Scale (CCOS), which was applied retrospectively to 146 patients [120]. Each aspect is rated on a scale from 1 to 4, with different levels of severity or improvement associated with each rating, as follows:

Pain: The level of pain experienced by individuals with Chiari malformation. The scale ranges from 1 to 4, with higher values indicating worse pain:

1—Worse.

2—Unchanged and refractory to medication.

3—Significantly improved or efficient medication.

4—Resolved.

Non-pain Functionality: This column evaluates the individual's ability to perform daily activities and attend events, with higher values indicating more significant limitations:

1—Inability to attend.

2—Moderate impairment (<50% attendance).

3—Mild impairment (>50% attendance).

4—Fully functional.

Complications: The analysis of the presence and control of complications related to the disease is revealed in this column. Higher values indicate more persistent or serious complications:

1—Persistent complication, poorly controlled.

2—Persistent complication, well controlled.

3—Transient complication.

4—Uncomplicated course.

Total Score: The last column displays the cumulative score achieved by adding up the scores from the previous three columns. It provides an overview of the overall outcome for individuals with Chiari malformation:

4—Incapacitated outcome.

8—Impaired outcome.

12—Functional outcome.

16—Excellent outcome.

Earlier studies typically categorized outcomes as "improved," "unchanged," or "worse," or a similar categorization. Yet, the CCOS offers a more comprehensive view, capturing the intricacies of each outcome. It recognizes that not all symptoms of a patient presenting multiple Chiari signs might be alleviated post-surgery.

10.2. Long-Term Prognosis Post-Surgical Interventions

The CCOS employs four post-surgery categories, each with a 4-point scale, culminating in a total potential score. When juxtaposed with traditional broad-brush evaluations of "improved", "unchanged", and "worse", we discerned a dependable correlation between these general outcome assessments and the individual scores across the four categories. Those deemed improved primarily received scores of 4 and 3. The unchanged group mainly earned 3s and 2s, whereas the worsened group predominantly scored 2s and 1s. Moreover, we identified a consistent correlation between the traditional I/U/W outcome classification and the aggregated scores of patients. Predominantly, patients who showed improvement post-PFD registered cumulative scores ranging from 13 to 16. A threshold score of 13 conveniently demarcates the "improved" designation. Out of 28 patients who scored 13, 27 were categorized as improved and one as unchanged. From this group, 21 (all under the improved bracket) had one category scoring a perfect 4, with the rest at 3, suggesting significant improvement in at least three categories and a stellar result in one. The score distribution for the other seven was two 4's, one 3, and one 2, suggesting that while there was enhancement in some areas, one category remained static. Nevertheless, this score set still comfortably qualifies under the "improved" bracket. Remarkably, no patients recorded a score pattern of three 4's coupled with a 1, which would total 13. Such a distribution was improbable as it would imply a sharp decline in one domain, even when the other areas showed complete resolution [108].

While the CCOS scores generally aligned with the I/U/W evaluations, there were some deviations. Some patients received scores that did not quite match their I/U/W evaluations. For instance, patients classified as "improved" in the I/U/W system but who only secured scores of 10, 11, or 12 on the CCOS typically had more severe conditions prior to PFD. As a result, their final CCOS scores were relatively low, even though they showed an "improved" outcome compared to their starting point on the I/U/W scale.

Conversely, those labeled "unchanged" on the I/U/W scale yet who had scores of 7 or 8 on the CCOS typically presented with pronounced symptoms before PFD. Despite their stagnant status, their preoperative conditions did not allow for a score above 8. This highlights the CCOS's tendency to emphasize the absolute outcome over relative improvement compared to pre-surgical conditions.

This CCOS characteristic becomes valuable when assessing a patient's absolute progress against the severity of their condition before the surgery—something the I/U/W evaluation cannot offer. An outlier was an "unchanged" patient in the I/U/W system who achieved a score of 13 on the CCOS. This patient, having minor pain and very few non-pain symptoms before PFD, saw a major decrease in non-pain symptoms post-surgery. However, their pain symptoms returned in full. Compared to other patients classified as "unchanged" under I/U/W, this individual had a relatively milder condition before PFD, allowing them to amass a higher CCOS score than most of their counterparts.

The significance of the CCOS becomes evident here: a patient could achieve a high score with or without PFD. This brings up a crucial question about the necessity of the PFD procedure in such cases [64].

11. Conclusions

In this detailed exposition, we undertake an extensive analysis of Chiari malformation and syringomyelia, ranging from their epidemiology and genetic underpinnings to sophisticated diagnostic and therapeutic modalities. Through methodical examination, we have deepened our comprehension of the pathophysiology, clinical manifestations, diagnostic methodologies, neuroimaging techniques, surgical treatments, and resultant outcomes pertinent to these intricate neurological entities.

We examined the multifaceted interaction of genetic predispositions, environmental determinants, and developmental aberrations that contribute to the onset of Chiari malformation and syringomyelia. Grasping this multifactorial etiology enhances our ability to discern potential at-risk populations, thereby facilitating prophylactic interventions and tailored therapeutic modalities.

Furthermore, we detailed the varied clinical presentations of these disorders, ranging from nuanced neurological indications to severe symptomatology. We underscored the significance of cutting-edge diagnostic instruments, especially neuroimaging modalities, in deriving prompt and precise diagnoses. This diagnostic acuity empowers clinicians to formulate and administer patient-specific therapeutic regimens.

We elaborated on surgical therapies, elucidating the manifold surgical techniques and their ensuing outcomes. By dissecting the intricacies of procedures, from posterior fossa decompression to syrinx drainage modalities, we have illuminated the technical nuances and post-operative prognosis.

Throughout our discourse, we advocated for the establishment of uniform outcome metrics to gauge therapeutic effectiveness and longitudinal patient well-being. Instituting these robust evaluative benchmarks enables more judicious clinical decision-making, optimizing patient-centric care, and refining therapeutic paradigms.

In summation, our meticulous exploration offers a roadmap for prospective research avenues and clinical stewardship in the domain of Chiari malformation and syringomyelia. The knowledge lacunae identified herald avenues for advanced research, encompassing targeted genetic explorations, innovative therapeutic strategies, and advancements in neuroimaging modalities. By championing a holistic and collaborative methodology, we aim to further elucidate these neurological anomalies and ameliorate the prognosis for those afflicted.

Ultimately, this exposition stands as an invaluable compendium for academicians, practitioners, and healthcare stakeholders, offering an exhaustive insight into Chiari malformation and syringomyelia. By leveraging this collective wisdom and fostering innovation, we edge closer to demystifying these disorders, providing solace and hope to those grappling with Chiari malformation and syringomyelia.

Funding: This research received no external funding.

Institutional Review Board Statement: The study was conducted in accordance with the Declaration of Helsinki, and approved by the Ethics Committee of Centre for Cognitive Research in Neuropsychiatric Pathology 6 (NeuroPsy-Cog), Timișoara, Romania.

Informed Consent Statement: Informed consent was obtained from all subjects involved in the study.

Data Availability Statement: All data is available online on libraries such as PubMed.

Conflicts of Interest: The authors declare no conflict of interest.

References

1. Bejjani, G.K. Definition of the adult Chiari malformation: A brief historical overview. *Neurosurg. Focus* **2001**, *11*, 1–8. [CrossRef] [PubMed]
2. Armonda, R.A.; Citrin, C.M.; Foley, K.T.; Ellenbogen, R.G. Quantitative cine-mode magnetic resonance imaging of Chiari I malformations: An analysis of cerebrospinal fluid dynamics. *Neurosurgery* **1994**, *35*, 214–223, discussion 223–224. [CrossRef] [PubMed]
3. Appleby, A.; Foster, J.B.; Hankinson, J.; Hudgson, P. The diagnosis and management of the Chiari anomalies in adult life. *Brain J. Neurol.* **1968**, *91*, 131–140. [CrossRef] [PubMed]
4. Klekamp, J. The pathophysiology of syringomyelia—Historical overview and current concept. *Acta Neurochir. (Wien)* **2002**, *144*, 649–664. [CrossRef]
5. Levine, D.N. The pathogenesis of syringomyelia associated with lesions at the foramen magnum: A critical review of existing theories and proposal of a new hypothesis. *J. Neurol. Sci.* **2004**, *220*, 3–21. [CrossRef]
6. Greitz, D. Unraveling the riddle of syringomyelia. *Neurosurg. Rev.* **2006**, *29*, 251–264. [CrossRef]
7. Catmull, S.; Ashurst, J. Tethered Cord Syndrome. *Clin. Pract. Cases Emerg. Med.* **2019**, *3*, 297–298. [CrossRef]
8. Elster, A.D.; Chen, M.Y. Chiari I malformations: Clinical and radiologic reappraisal. *Radiology* **1992**, *183*, 347–353. [CrossRef]
9. Meadows, J.; Kraut, M.; Guarnieri, M.; Haroun, R.I.; Carson, B.S. Asymptomatic Chiari Type I malformations identified on magnetic resonance imaging. *J. Neurosurg.* **2000**, *92*, 920–926. [CrossRef]

10. Spinos, E.; Laster, D.W.; Moody, D.M.; Ball, M.R.; Witcofski, R.L.; Kelly, D.L. MR evaluation of Chiari I malformations at 0.15 T. *Am. J. Neuroradiol.* **1985**, *6*, 203–208. [CrossRef]
11. Smith, B.W.; Strahle, J.; Bapuraj, J.R.; Muraszko, K.M.; Garton, H.J.; Maher, C.O. Distribution of cerebellar tonsil position: Implications for understanding Chiari malformation. *J. Neurosurg.* **2013**, *119*, 812–819. [CrossRef] [PubMed]
12. Wilson, M.H. Monro-Kellie 2.0: The dynamic vascular and venous pathophysiological components of intracranial pressure. *J. Cereb. Blood Flow Metab.* **2016**, *36*, 1338–1350. [CrossRef] [PubMed]
13. Whedon, J.M.; Glassey, D. Cerebrospinal fluid stasis and its clinical significance. *Altern. Ther. Health Med.* **2009**, *15*, 54–60. [PubMed]
14. Ohara, S.; Nagai, H.; Matsumoto, T.; Banno, T. MR imaging of CSF pulsatory flow and its relation to intracranial pressure. *J. Neurosurg.* **1988**, *69*, 675–682. [CrossRef] [PubMed]
15. Haughton, V.M.; Korosec, F.R.; Medow, J.E.; Dolar, M.T.; Iskandar, B.J. Peak systolic and diastolic CSF velocity in the foramen magnum in adult patients with Chiari I malformations and in normal control participants. *AJNR Am. J. Neuroradiol.* **2003**, *24*, 169–176. [PubMed]
16. Quigley, M.F.; Iskandar, B.; Quigley, M.A.; Nicosia, M.; Haughton, V. Cerebrospinal Fluid Flow in Foramen Magnum: Temporal and Spatial Patterns at MR Imaging in Volunteers and in Patients with Chiari I Malformation. *Radiology* **2004**, *232*, 229–236. [CrossRef]
17. Pujol, J.; Roig, C.; Capdevila, A.; Pou, A.; Marti-Vilalta, J.L.; Kulisevsky, J.; Escartin, A.; Zannoli, G. Motion of the cerebellar tonsils in Chiari type I malformation studied by cine phase-contrast MRI. *Neurology* **1995**, *45*, 1746–1753. [CrossRef]
18. Wolpert, S.M.; Bhadelia, R.A.; Bogdan, A.R.; Cohen, A.R. Chiari I malformations: Assessment with phase-contrast velocity MR. *AJNR Am. J. Neuroradiol.* **1994**, *15*, 1299–1308.
19. Hofkes, S.K.; Iskandar, B.J.; Turski, P.A.; Gentry, L.R.; McCue, J.B.; Haughton, V.M. Differentiation between Symptomatic Chiari I Malformation and Asymptomatic Tonsilar Ectopia by Using Cerebrospinal Fluid Flow Imaging: Initial Estimate of Imaging Accuracy. *Radiology* **2007**, *245*, 532–540. [CrossRef]
20. Krueger, K.D.; Haughton, V.M.; Hetzel, S. Peak CSF Velocities in Patients with Symptomatic and Asymptomatic Chiari I Malformation. *Am. J. Neuroradiol.* **2010**, *31*, 1837–1841. [CrossRef]
21. Bhadelia, R.A.; Bogdan, A.R.; Wolpert, S.M.; Lev, S.; Appignani, B.A.; Heilman, C.B. Cerebrospinal fluid flow waveforms: Analysis in patients with Chiari I malformation by means of gated phase-contrast MR imaging velocity measurements. *Radiology* **1995**, *196*, 195–202. [CrossRef] [PubMed]
22. McGirt, M.J.; Nimjee, S.M.; Fuchs, H.E.; George, T.M. Relationship of cine phase-contrast magnetic resonance imaging with outcome after decompression for Chiari I malformations. *Neurosurgery* **2006**, *59*, 140–146, discussion 140–146. [CrossRef] [PubMed]
23. Ventureyra, E.C.G.; Aziz, H.A.; Vassilyadi, M. The role of cine flow MRI in children with Chiari I malformation. *Childs Nerv. Syst.* **2003**, *19*, 109–113. [CrossRef] [PubMed]
24. Holly, L.T.; Batzdorf, U. Chiari malformation and syringomyelia. *J. Neurosurg. Spine* **2019**, *31*, 619–628. [CrossRef] [PubMed]
25. Timpone, V.M.; Patel, S.H. MRI of a Syrinx: Is Contrast Material Always Necessary? *Am. J. Roentgenol.* **2015**, *204*, 1082–1085. [CrossRef]
26. Strahle, J.; Muraszko, K.M.; Kapurch, J.; Bapuraj, J.R.; Garton, H.J.L.; Maher, C.O. Chiari malformation Type I and syrinx in children undergoing magnetic resonance imaging. *J. Neurosurg. Pediatr.* **2011**, *8*, 205–213. [CrossRef]
27. Strahle, J.; Muraszko, K.M.; Garton, H.J.L.; Smith, B.W.; Starr, J.; Kapurch, J.R.; Maher, C.O. Syrinx location and size according to etiology: Identification of Chiari-associated syrinx. *J. Neurosurg. Pediatr.* **2015**, *16*, 21–29. [CrossRef]
28. Milhorat, T.H.; Bolognese, P.A.; Nishikawa, M.; Francomano, C.A.; McDonnell, N.B.; Roonprapunt, C.; Kula, R.W. Association of Chiari malformation type I and tethered cord syndrome: Preliminary results of sectioning filum terminale. *Surg. Neurol.* **2009**, *72*, 20–35. [CrossRef]
29. Hertzler, D.A.; DePowell, J.J.; Stevenson, C.B.; Mangano, F.T. Tethered cord syndrome: A review of the literature from embryology to adult presentation. *Neurosurg. Focus* **2010**, *29*, E1. [CrossRef]
30. Seki, T.; Hida, K.; Yano, S.; Sasamori, T.; Hamauch, S.; Koyanagi, I.; Houkin, K. Surgical Outcome of Children and Adolescents with Tethered Cord Syndrome. *Asian Spine J.* **2016**, *10*, 940–944. [CrossRef]
31. Tandon, V.; Garg, K.; Kumar, R.; Mahapatra, A.; Sharma, B. Management of adult tethered cord syndrome: Our experience and review of literature. *Neurol. India* **2014**, *62*, 137. [CrossRef]
32. Kahn, E.N.; Muraszko, K.M.; Maher, C.O. Prevalence of Chiari I Malformation and Syringomyelia. *Neurosurg. Clin. N. Am.* **2015**, *26*, 501–507. [CrossRef] [PubMed]
33. Aitken, L.A.; Lindan, C.E.; Sidney, S.; Gupta, N.; Barkovich, A.J.; Sorel, M.; Wu, Y.W. Chiari Type I Malformation in a Pediatric Population. *Pediatr. Neurol.* **2009**, *40*, 449–454. [CrossRef] [PubMed]
34. Luzzi, S.; Giotta Lucifero, A.; Elsawaf, Y.; Elbabaa, S.K.; Del Maestro, M.; Savioli, G.; Galzio, R.; Gragnaniello, C. Pulsatile cerebrospinal fluid dynamics in Chiari I malformation syringomyelia: Predictive value in posterior fossa decompression and insights into the syringogenesis. *J. Craniovertebral Junction Spine* **2021**, *12*, 15–25. [CrossRef]
35. Wu, J.; Ji, S.; Niu, P.; Zhang, B.; Shao, D.; Li, Y.; Xie, S.; Jiang, Z. Knowledge mapping of syringomyelia from 2003 to 2022: A bibliometric analysis. *J. Clin. Neurosci.* **2023**, *110*, 63–70. [CrossRef] [PubMed]

36. Sakka, L.; Coll, G.; Chazal, J. Anatomy and physiology of cerebrospinal fluid. *Eur. Ann. Otorhinolaryngol. Head Neck Dis.* **2011**, *128*, 309–316. [CrossRef]
37. Spector, R.; Robert Snodgrass, S.; Johanson, C.E. A balanced view of the cerebrospinal fluid composition and functions: Focus on adult humans. *Exp. Neurol.* **2015**, *273*, 57–68. [CrossRef]
38. Sekula, R.F.; Jannetta, P.J.; Casey, K.F.; Marchan, E.M.; Sekula, L.K.; McCrady, C.S. Dimensions of the posterior fossa in patients symptomatic for Chiari I malformation but without cerebellar tonsillar descent. *Cerebrospinal Fluid Res.* **2005**, *2*, 11. [CrossRef]
39. Matsumae, M.; Hirayama, A.; Atsumi, H.; Yatsushiro, S.; Kuroda, K. Velocity and pressure gradients of cerebrospinal fluid assessed with magnetic resonance imaging: Clinical article. *J. Neurosurg.* **2014**, *120*, 218–227. [CrossRef]
40. Yiallourou, T.I.; Kröger, J.R.; Stergiopulos, N.; Maintz, D.; Martin, B.A.; Bunck, A.C. Comparison of 4D Phase-Contrast MRI Flow Measurements to Computational Fluid Dynamics Simulations of Cerebrospinal Fluid Motion in the Cervical Spine. *PLoS ONE* **2012**, *7*, e52284. [CrossRef]
41. Shaffer, N.; Martin, B.A.; Rocque, B.; Madura, C.; Wieben, O.; Iskandar, B.J.; Dombrowski, S.; Luciano, M.; Oshinski, J.N.; Loth, F. Cerebrospinal fluid flow impedance is elevated in Type I Chiari malformation. *J. Biomech. Eng.* **2014**, *136*, 021012. [CrossRef] [PubMed]
42. Sarnat, H.B. Disorders of segmentation of the neural tube: Chiari malformations. *Handb. Clin. Neurol.* **2008**, *87*, 89–103. [CrossRef] [PubMed]
43. Stevenson, K.L. Chiari Type II malformation: Past, present, and future. *Neurosurg. Focus* **2004**, *16*, 1–7. [CrossRef]
44. Lee, R.; Tai, K.S.; Cheng, P.W.; Lui, W.M.; Chan, F.L. Chiari III malformation: Antenatal MRI diagnosis. *Clin. Radiol.* **2002**, *57*, 759–761. [CrossRef] [PubMed]
45. Caldarelli, M.; Rea, G.; Cincu, R.; Di Rocco, C. Chiari type III malformation. *Childs Nerv. Syst. ChNS Off. J. Int. Soc. Pediatr. Neurosurg.* **2002**, *18*, 207–210. [CrossRef]
46. Tubbs, R.S.; Demerdash, A.; Vahedi, P.; Griessenauer, C.J.; Oakes, W.J. Chiari IV malformation: Correcting an over one century long historical error. *Childs Nerv. Syst.* **2016**, *32*, 1175–1179. [CrossRef] [PubMed]
47. McDougall, C.M.; Alarfaj, A.K.; Jack, A.S.; Souster, J.; Broad, R.W. Klippel-Feil syndrome in association with Chiari zero malformation in the setting of acute traumatic spinal cord injury. *Interdiscip. Neurosurg.* **2017**, *9*, 1–3. [CrossRef]
48. Bogdanov, E.I.; Faizutdinova, A.T.; Heiss, J.D. Posterior cranial fossa and cervical spine morphometric abnormalities in symptomatic Chiari type 0 and Chiari type 1 malformation patients with and without syringomyelia. *Acta Neurochir. (Wien)* **2021**, *163*, 3051–3064. [CrossRef]
49. Sandu, R.B.; Pantiru, M.I.; Cosman, M.; Poeata, I. Syringobulbia and syringomyelia in a case with Chiari 0 malformation successfully treated by posterior fossa reconstruction: Case presentation and literature review. *Romanian Neurosurg.* **2019**, *33*, 239–242. [CrossRef]
50. Czosnyka, M.; Czosnyka, Z.; Momjian, S.; Pickard, J.D. Cerebrospinal fluid dynamics. *Physiol. Meas.* **2004**, *25*, R51–R76. [CrossRef]
51. Williams, H. A unifying hypothesis for hydrocephalus, Chiari malformation, syringomyelia, anencephaly and spina bifida. *Cerebrospinal Fluid Res.* **2008**, *5*, 7. [CrossRef] [PubMed]
52. Parker, J.E.; Knowler, S.P.; Rusbridge, C.; Noorman, E.; Jeffery, N.D. Prevalence of asymptomatic syringomyelia in Cavalier King Charles spaniels. *Vet. Rec.* **2011**, *168*, 667. [CrossRef] [PubMed]
53. Wijnrocx, K.; Van Bruggen, L.W.L.; Eggelmeijer, W.; Noorman, E.; Jacques, A.; Buys, N.; Janssens, S.; Mandigers, P.J.J. Twelve years of chiari-like malformation and syringomyelia scanning in Cavalier King Charles Spaniels in the Netherlands: Towards a more precise phenotype. *PLoS ONE* **2017**, *12*, e0184893. [CrossRef] [PubMed]
54. Hechler, A.C.; Moore, S.A. Understanding and Treating Chiari-like Malformation and Syringomyelia in Dogs. *Top. Companion Anim. Med.* **2018**, *33*, 1–11. [CrossRef]
55. Ono, A.; Numasawa, T.; Wada, K.; Yokoyama, T.; Takeuchi, K.; Suetsuna, F.; Ueyama, K.; Toh, S. Surgical outcomes of foramen magnum decompression for syringomyelia associated with Chiari I malformation: Relation between the location of the syrinx and body pain. *J. Orthop. Sci.* **2010**, *15*, 299–304. [CrossRef]
56. Sgouros, S.; Williams, B. Management and outcome of posttraumatic syringomyelia. *J. Neurosurg.* **1996**, *85*, 197–205. [CrossRef]
57. Ushewokunze, S.O.S.; Gan, Y.C.; Phillips, K.; Thacker, K.; Flint, G. Surgical treatment of post-traumatic syringomyelia. *Spinal Cord* **2010**, *48*, 710–713. [CrossRef]
58. Krieger, M.D.; McComb, J.G.; Levy, M.L. Toward a simpler surgical management of Chiari I malformation in a pediatric population. *Pediatr. Neurosurg.* **1999**, *30*, 113–121. [CrossRef]
59. Akbari, S.H.A.; Rizvi, A.A.; CreveCoeur, T.S.; Han, R.H.; Greenberg, J.K.; Torner, J.; Brockmeyer, D.L.; Wellons, J.C.; Leonard, J.R.; Mangano, F.T.; et al. Socioeconomic and demographic factors in the diagnosis and treatment of Chiari malformation type I and syringomyelia. *J. Neurosurg. Pediatr.* **2022**, *29*, 288–297. [CrossRef]
60. Sakushima, K.; Tsuboi, S.; Yabe, I.; Hida, K.; Terae, S.; Uehara, R.; Nakano, I.; Sasaki, H. Nationwide survey on the epidemiology of syringomyelia in Japan. *J. Neurol. Sci.* **2012**, *313*, 147–152. [CrossRef]
61. Ciaramitaro, P.; Migliaretti, G.; Ferraris, M.; Garnero, A.; Morana, G.; Carucci, P.; Stura, I.; Massaro, F.; Garbossa, D. Syringomyelia Associated with Chiari 1 Malformation in Adults: Positive Outcome Predictors after Posterior Fossa Decompression with Duraplasty. *J. Clin. Med.* **2023**, *12*, 3019. [CrossRef]
62. Siasios, J.; Kapsalaki, E.Z.; Fountas, K.N. Surgical management of patients with Chiari I malformation. *Int. J. Pediatr.* **2012**, *2012*, 640127. [CrossRef]

63. Haroun, R.I.; Guarnieri, M.; Meadow, J.J.; Kraut, M.; Carson, B.S. Current Opinions for the Treatment of Syringomyelia and Chiari Malformations: Survey of the Pediatric Section of the American Association of Neurological Surgeons. *Pediatr. Neurosurg.* **2000**, *33*, 311–317. [CrossRef]
64. Attenello, F.J.; McGirt, M.J.; Gathinji, M.; Datoo, G.; Atiba, A.; Weingart, J.; Carson, B.; Jallo, G.I. Outcome of Chiari-associated syringomyelia after hindbrain decompression in children: Analysis of 49 consecutive cases. *Neurosurgery* **2008**, *62*, 1307–1313, discussion 1313. [CrossRef]
65. Sgouros, S.; Kountouri, M.; Natarajan, K. Posterior fossa volume in children with Chiari malformation Type I. *J. Neurosurg. Pediatr.* **2006**, *105*, 101–106. [CrossRef]
66. Driver, C.J.; Rusbridge, C.; Cross, H.R.; McGonnell, I.; Volk, H.A. Relationship of brain parenchyma within the caudal cranial fossa and ventricle size to syringomyelia in cavalier King Charles spaniels. *J. Small Anim. Pract.* **2010**, *51*, 382–386. [CrossRef]
67. Shaw, T.A.; McGonnell, I.M.; Driver, C.J.; Rusbridge, C.; Volk, H.A. Increase in Cerebellar Volume in Cavalier King Charles Spaniels with Chiari-like Malformation and Its Role in the Development of Syringomyelia. *PLoS ONE* **2012**, *7*, e33660. [CrossRef]
68. Olsen, E.; Suiter, E.J.; Pfau, T.; McGonnell, I.M.; Matiasek, K.; Giejda, A.; Volk, H.A. Cavalier King Charles Spaniels with Chiari-like malformation and Syringomyelia have increased variability of spatio-temporal gait characteristics. *BMC Vet. Res.* **2017**, *13*, 159. [CrossRef]
69. Cinalli, G.; Spennato, P.; Sainte-Rose, C.; Arnaud, E.; Aliberti, F.; Brunelle, F.; Cianciulli, E.; Renier, D. Chiari malformation in craniosynostosis. *Childs Nerv. Syst.* **2005**, *21*, 889–901. [CrossRef]
70. Carrera, I.; Dennis, R.; Mellor, D.J.; Penderis, J.; Sullivan, M. Use of magnetic resonance imaging for morphometric analysis of the caudal cranial fossa in Cavalier King Charles Spaniels. *Am. J. Vet. Res.* **2009**, *70*, 340–345. [CrossRef] [PubMed]
71. McBratney-Owen, B.; Iseki, S.; Bamforth, S.D.; Olsen, B.R.; Morriss-Kay, G.M. Development and tissue origins of the mammalian cranial base. *Dev. Biol.* **2008**, *322*, 121–132. [CrossRef]
72. Driver, C.J.; Volk, H.A.; Rusbridge, C.; Van Ham, L.M. An update on the pathogenesis of syringomyelia secondary to Chiari-like malformations in dogs. *Vet. J.* **2013**, *198*, 551–559. [CrossRef]
73. Nagy, L.; Mobley, J.; Ray, C. Familial Aggregation of Chiari Malformation: Presentation, Pedigree, and Review of the Literature. *Turk. Neurosurg.* **2016**, *26*, 315–320. [CrossRef]
74. AvŞar, T.; ÇaliŞ, Ş.; Yilmaz, B.; Demİrcİ OtluoĞlu, G.; Holyavkİn, C.; KiliÇ, T. Genome-wide identification of Chiari malformation type I associated candidate genes and chromosomal variations. *Turk. J. Biol. Turk Biyol. Derg.* **2020**, *44*, 449–456. [CrossRef]
75. Merello, E.; Tattini, L.; Magi, A.; Accogli, A.; Piatelli, G.; Pavanello, M.; Tortora, D.; Cama, A.; Kibar, Z.; Capra, V.; et al. Exome sequencing of two Italian pedigrees with non-isolated Chiari malformation type I reveals candidate genes for cranio-facial development. *Eur. J. Hum. Genet. EJHG* **2017**, *25*, 952–959. [CrossRef]
76. Landrum, M.J.; Lee, J.M.; Benson, M.; Brown, G.R.; Chao, C.; Chitipiralla, S.; Gu, B.; Hart, J.; Hoffman, D.; Jang, W.; et al. ClinVar: Improving access to variant interpretations and supporting evidence. *Nucleic Acids Res.* **2018**, *46*, D1062–D1067. [CrossRef]
77. Ancot, F.; Lemay, P.; Knowler, S.P.; Kennedy, K.; Griffiths, S.; Cherubini, G.B.; Sykes, J.; Mandigers, P.J.J.; Rouleau, G.A.; Rusbridge, C.; et al. A genome-wide association study identifies candidate loci associated to syringomyelia secondary to Chiari-like malformation in Cavalier King Charles Spaniels. *BMC Genet.* **2018**, *19*, 16. [CrossRef]
78. Urbizu, A.; Toma, C.; Poca, M.A.; Sahuquillo, J.; Cuenca-León, E.; Cormand, B.; Macaya, A. Chiari Malformation Type I: A Case-Control Association Study of 58 Developmental Genes. *PLoS ONE* **2013**, *8*, e57241. [CrossRef]
79. Urbizu, A.; Garrett, M.E.; Soldano, K.; Drechsel, O.; Loth, D.; Marcé-Grau, A.; Mestres I Soler, O.; Poca, M.A.; Ossowski, S.; Macaya, A.; et al. Rare functional genetic variants in COL7A1, COL6A5, COL1A2 and COL5A2 frequently occur in Chiari Malformation Type 1. *PLoS ONE* **2021**, *16*, e0251289. [CrossRef]
80. Xiaoshuai, L.; Qiushi, W.; Rui, W. Advantages of CRISPR-Cas9 combined organoid model in the study of congenital nervous system malformations. *Front. Bioeng. Biotechnol.* **2022**, *10*, 932936. [CrossRef]
81. McVige, J.W.; Leonardo, J. Imaging of Chiari type I malformation and syringohydromyelia. *Neurol. Clin.* **2014**, *32*, 95–126. [CrossRef]
82. Stephen, D.; Silberstein, R.B.; Lipton, D.W.; Dodick, S.D.; Lipton, R.B.; Dodick, D.; Wolff, H.G. (Eds.) *Wolff's Headache and Other Head Pain*, 8th ed.; Oxford University Press: Oxford, UK; New York, NY, USA, 2008; ISBN 978-0-19-532656-7.
83. Pascual, J.; Oterino, A.; Berciano, J. Headache in type I Chiari malformation. *Neurology* **1992**, *42*, 1519. [CrossRef]
84. Toldo, I.; Tangari, M.; Mardari, R.; Perissinotto, E.; Sartori, S.; Gatta, M.; Calderone, M.; Battistella, P.A. Headache in Children With C hiari I Malformation. *Headache J. Head Face Pain* **2014**, *54*, 899–908. [CrossRef]
85. Wu, Y.W.; Chin, C.T.; Chan, K.M.; Barkovich, A.J.; Ferriero, D.M. Pediatric Chiari I malformations: Do clinical and radiologic features correlate? *Neurology* **1999**, *53*, 1271–1276. [CrossRef]
86. Kim, H.G.; Oh, H.S.; Kim, T.W.; Park, K.H. Clinical Features of Post-Traumatic Syringomyelia. *Korean J. Neurotrauma* **2014**, *10*, 66–69. [CrossRef]
87. Tam, S.K.P.; Chia, J.; Brodbelt, A.; Foroughi, M. Assessment of patients with a Chiari malformation type I. *Brain Spine* **2022**, *2*, 100850. [CrossRef]
88. Hayward, R. VOMIT (victims of modern imaging technology)—An acronym for our times. *Bmj* **2003**, *326*, 1273. [CrossRef]
89. Langridge, B.; Phillips, E.; Choi, D. Chiari Malformation Type 1: A Systematic Review of Natural History and Conservative Management. *World Neurosurg.* **2017**, *104*, 213–219. [CrossRef]

90. Chazen, J.L.; Talbott, J.F.; Lantos, J.E.; Dillon, W.P. MR myelography for identification of spinal CSF leak in spontaneous intracranial hypotension. *AJNR Am. J. Neuroradiol.* **2014**, *35*, 2007–2012. [CrossRef]
91. Aboulezz, A.O.; Sartor, K.; Geyer, C.A.; Gado, M.H. Position of cerebellar tonsils in the normal population and in patients with Chiari malformation: A quantitative approach with MR imaging. *J. Comput. Assist. Tomogr.* **1985**, *9*, 1033–1036. [CrossRef]
92. Mikulis, D.J.; Diaz, O.; Egglin, T.K.; Sanchez, R. Variance of the position of the cerebellar tonsils with age: Preliminary report. *Radiology* **1992**, *183*, 725–728. [CrossRef]
93. Jussila, M.-P.; Nissilä, J.; Vakkuri, M.; Olsén, P.; Niinimäki, J.; Leinonen, V.; Serlo, W.; Salokorpi, N.; Suo-Palosaari, M. Preoperative measurements on MRI in Chiari 1 patients fail to predict outcome after decompressive surgery. *Acta Neurochir. (Wien)* **2021**, *163*, 2005–2014. [CrossRef]
94. Milhorat, T.H.; Chou, M.W.; Trinidad, E.M.; Kula, R.W.; Mandell, M.; Wolpert, C.; Speer, M.C. Chiari I malformation redefined: Clinical and radiographic findings for 364 symptomatic patients. *Neurosurgery* **1999**, *44*, 1005–1017. [CrossRef]
95. Nishikawa, M.; Sakamoto, H.; Hakuba, A.; Nakanishi, N.; Inoue, Y. Pathogenesis of Chiari malformation: A morphometric study of the posterior cranial fossa. *J. Neurosurg.* **1997**, *86*, 40–47. [CrossRef]
96. Karagöz, F.; Izgi, N.; Sencer, S.K. Morphometric measurements of the cranium in patients with Chiari type I malformation and comparison with the normal population. *Acta Neurochir. (Wien)* **2002**, *144*, 165–171. [CrossRef]
97. Milhorat, T.H.; Nishikawa, M.; Kula, R.W.; Dlugacz, Y.D. Mechanisms of cerebellar tonsil herniation in patients with Chiari malformations as guide to clinical management. *Acta Neurochir. (Wien)* **2010**, *152*, 1117–1127. [CrossRef]
98. Urbizu, A.; Poca, M.-A.; Vidal, X.; Rovira, A.; Sahuquillo, J.; Macaya, A. MRI-based Morphometric Analysis of Posterior Cranial Fossa in the Diagnosis of Chiari Malformation Type I. *J. Neuroimaging* **2014**, *24*, 250–256. [CrossRef]
99. Dufton, J.A.; Habeeb, S.Y.; Heran, M.K.; Mikulis, D.J.; Islam, O. Posterior fossa measurements in patients with and without Chiari I malformation. *Can. J. Neurol. Sci.* **2011**, *38*, 452–455. [CrossRef]
100. Stovner, L.J.; Bergan, U.; Nilsen, G.; Sjaastad, O. Posterior cranial fossa dimensions in the Chiari I malformation: Relation to pathogenesis and clinical presentation. *Neuroradiology* **1993**, *35*, 113–118. [CrossRef]
101. Badie, B.; Mendoza, D.; Batzdorf, U. Posterior fossa volume and response to suboccipital decompression in patients with Chiari I malformation. *Neurosurgery* **1995**, *37*, 214–218. [CrossRef]
102. Noudel, R.; Gomis, P.; Sotoares, G.; Bazin, A.; Pierot, L.; Pruvo, J.-P.; Bordet, R.; Roche, P.-H. Posterior fossa volume increase after surgery for Chiari malformation Type I: A quantitative assessment using magnetic resonance imaging and correlations with the treatment response: Clinical article. *J. Neurosurg.* **2011**, *115*, 647–658. [CrossRef]
103. Tubbs, R.S.; McGirt, M.J.; Oakes, W.J. Surgical experience in 130 pediatric patients with Chiari I malformations. *J. Neurosurg.* **2003**, *99*, 291–296. [CrossRef]
104. Orakdogen, M.; Emon, S.T.; Erdogan, B.; Somay, H. Fourth Ventriculostomy in Occlusion of the Foramen of Magendie Associated with Chiari Malformation and Syringomyelia. *NMC Case Rep. J.* **2015**, *2*, 72–75. [CrossRef]
105. Decq, P.; Guérinel, C.L.; Sol, J.-C.; Brugières, P.; Djindjian, M.; Nguyen, J.-P. Chiari I malformation: A rare cause of noncommunicating hydrocephalus treated by third ventriculostomy. *J. Neurosurg.* **2001**, *95*, 783–790. [CrossRef]
106. Hayhurst, C.; Osman-Farah, J.; Das, K.; Mallucci, C. Initial management of hydrocephalus associated with Chiari malformation Type I-syringomyelia complex via endoscopic third ventriculostomy: An outcome analysis. *J. Neurosurg.* **2008**, *108*, 1211–1214. [CrossRef]
107. Hankinson, T.; Tubbs, R.S.; Wellons, J.C. Duraplasty or not? An evidence-based review of the pediatric Chiari I malformation. *Childs Nerv. Syst. ChNS Off. J. Int. Soc. Pediatr. Neurosurg.* **2011**, *27*, 35–40. [CrossRef]
108. Nash, J.; Cheng, J.S.; Meyer, G.A.; Remler, B.F. Chiari type I malformation: Overview of diagnosis and treatment. *WMJ Off. Publ. State Med. Soc. Wis.* **2002**, *101*, 35–40.
109. Sindou, M.; Gimbert, E. Decompression for Chiari type I-malformation (with or without syringomyelia) by extreme lateral foramen magnum opening and expansile duraplasty with arachnoid preservation: Comparison with other technical modalities (Literature review). *Adv. Tech. Stand. Neurosurg.* **2009**, *34*, 85–110. [CrossRef]
110. Alden, T.D.; Ojemann, J.G.; Park, T.S. Surgical treatment of Chiari I malformation: Indications and approaches. *Neurosurg. Focus* **2001**, *11*, E2. [CrossRef]
111. Sindou, M.; Chávez-Machuca, J.; Hashish, H. Cranio-cervical decompression for Chiari type I-malformation, adding extreme lateral foramen magnum opening and expansile duroplasty with arachnoid preservation. Technique and long-term functional results in 44 consecutive adult cases—Comparison with literature data. *Acta Neurochir. (Wien)* **2002**, *144*, 1005–1019. [CrossRef]
112. Tubbs, R.S.; Lyerly, M.J.; Loukas, M.; Shoja, M.M.; Oakes, W.J. The pediatric Chiari I malformation: A review. *Childs Nerv. Syst. ChNS Off. J. Int. Soc. Pediatr. Neurosurg.* **2007**, *23*, 1239–1250. [CrossRef] [PubMed]
113. Valentini, L.; Visintini, S.; Saletti, V.; Chiapparini, L.; Estienne, M.; Solero, C.L. Treatment for Chiari 1 malformation (CIM): Analysis of a pediatric surgical series. *Neurol. Sci. Off. J. Ital. Neurol. Soc. Ital. Soc. Clin. Neurophysiol.* **2011**, *32* (Suppl. S3), S321–S324. [CrossRef] [PubMed]
114. Hoffman, H.J.; Neill, J.; Crone, K.R.; Hendrick, E.B.; Humphreys, R.P. Hydrosyringomyelia and its management in childhood. *Neurosurgery* **1987**, *21*, 347–351. [CrossRef] [PubMed]
115. Munshi, I.; Frim, D.; Stine-Reyes, R.; Weir, B.K.; Hekmatpanah, J.; Brown, F. Effects of posterior fossa decompression with and without duraplasty on Chiari malformation-associated hydromyelia. *Neurosurgery* **2000**, *46*, 1384–1389, discussion 1389–1390. [CrossRef] [PubMed]

116. Ono, A.; Suetsuna, F.; Ueyama, K.; Yokoyama, T.; Aburakawa, S.; Takeuchi, K.; Numasawa, T.; Wada, K.; Toh, S. Cervical spinal motion before and after surgery in patients with Chiari malformation type I associated with syringomyelia. *J. Neurosurg. Spine* **2007**, *7*, 473–477. [CrossRef]
117. Milhorat, T.H.; Kotzen, R.M.; Mu, H.T.; Capocelli, A.L.; Milhorat, R.H. Dysesthetic pain in patients with syringomyelia. *Neurosurgery* **1996**, *38*, 940–946, discussion 946–947. [CrossRef]
118. Nakamura, M.; Chiba, K.; Nishizawa, T.; Maruiwa, H.; Matsumoto, M.; Toyama, Y. Retrospective study of surgery-related outcomes in patients with syringomyelia associated with Chiari I malformation: Clinical significance of changes in the size and localization of syrinx on pain relief. *J. Neurosurg.* **2004**, *100*, 241–244. [CrossRef]
119. Goel, A. Basilar invagination, syringomyelia and Chiari formation and their relationship with atlantoaxial instability. *Neurol. India* **2018**, *66*, 940. [CrossRef]
120. Goel, A.; Kaswa, A.; Shah, A. Atlantoaxial Fixation for Treatment of Chiari Formation and Syringomyelia with No Craniovertebral Bone Anomaly: Report of an Experience with 57 Cases. *Acta Neurochir. Suppl.* **2019**, *125*, 101–110. [CrossRef]
121. Goel, A.; Jadhav, D.; Shah, A.; Rai, S.; Dandpat, S.; Vutha, R.; Dhar, A.; Prasad, A. Chiari 1 Formation Redefined-Clinical and Radiographic Observations in 388 Surgically Treated Patients. *World Neurosurg.* **2020**, *141*, e921–e934. [CrossRef]

brain sciences

MDPI

Review

Frontiers of Cranial Base Surgery: Integrating Technique, Technology, and Teamwork for the Future of Neurosurgery

Corneliu Toader [1,2], Lucian Eva [3,4,*], Catalina-Ioana Tataru [5,6,*], Razvan-Adrian Covache-Busuioc [1], Bogdan-Gabriel Bratu [1,*], David-Ioan Dumitrascu [1], Horia Petre Costin [1], Luca-Andrei Glavan [1] and Alexandru Vlad Ciurea [1,7]

1 Department of Neurosurgery, "Carol Davila" University of Medicine and Pharmacy, 020021 Bucharest, Romania; corneliu.toader@umfcd.ro (C.T.); razvan-adrian.covache-busuioc0720@stud.umfcd.ro (R.-A.C.-B.); david-ioan.dumitrascu0720@stud.umfcd.ro (D.-I.D.); horia-petre.costin0720@stud.umfcd.ro (H.P.C.); luca-andrei.glavan0720@stud.umfcd.ro (L.-A.G.); prof.avciurea@gmail.com (A.V.C.)
2 Department of Vascular Neurosurgery, National Institute of Neurology and Neurovascular Diseases, 077160 Bucharest, Romania
3 Department of Neurosurgery, Dunarea de Jos University, 800010 Galati, Romania
4 Department of Neurosurgery, Clinical Emergency Hospital "Prof. Dr. Nicolae Oblu", 700309 Iasi, Romania
5 Department of Ophthalmology, "Carol Davila" University of Medicine and Pharmacy, 020021 Bucharest, Romania
6 Clinical Hospital of Ophthalmological Emergencies, 010464 Bucharest, Romania
7 Neurosurgery Department, Sanador Clinical Hospital, 010991 Bucharest, Romania
* Correspondence: elucian73@yahoo.com (L.E.); catalina-ioana.tataru@umfcd.ro (C.-I.T.); bogdan.bratu@stud.umfcd.ro (B.-G.B.)

Abstract: The landscape of cranial base surgery has undergone monumental transformations over the past several decades. This article serves as a comprehensive survey, detailing both the historical and current techniques and technologies that have propelled this field into an era of unprecedented capabilities and sophistication. In the prologue, we traverse the historical evolution from rudimentary interventions to the state-of-the-art neurosurgical methodologies that define today's practice. Subsequent sections delve into the anatomical complexities of the anterior, middle, and posterior cranial fossa, shedding light on the intricacies that dictate surgical approaches. In a section dedicated to advanced techniques and modalities, we explore cutting-edge evolutions in minimally invasive procedures, pituitary surgery, and cranial base reconstruction. Here, we highlight the seamless integration of endocrinology, biomaterial science, and engineering into neurosurgical craftsmanship. The article emphasizes the paradigm shift towards "Functionally" Guided Surgery facilitated by intraoperative neuromonitoring. We explore its historical origins, current technologies, and its invaluable role in tailoring surgical interventions across diverse pathologies. Additionally, the digital era's contributions to cranial base surgery are examined. This includes breakthroughs in endoscopic technology, robotics, augmented reality, and the potential of machine learning and AI-assisted diagnostic and surgical planning. The discussion extends to radiosurgery and radiotherapy, focusing on the harmonization of precision and efficacy through advanced modalities such as Gamma Knife and CyberKnife. The article also evaluates newer protocols that optimize tumor control while preserving neural structures. In acknowledging the holistic nature of cranial base surgery, we advocate for an interdisciplinary approach. The ecosystem of this surgical field is presented as an amalgamation of various medical disciplines, including neurology, radiology, oncology, and rehabilitation, and is further enriched by insights from patient narratives and quality-of-life metrics. The epilogue contemplates future challenges and opportunities, pinpointing potential breakthroughs in stem cell research, regenerative medicine, and genomic tailoring. Ultimately, the article reaffirms the ethos of continuous learning, global collaboration, and patient-first principles, projecting an optimistic trajectory for the field of cranial base surgery in the coming decade.

Citation: Toader, C.; Eva, L.; Tataru, C.-I.; Covache-Busuioc, R.-A.; Bratu, B.-G.; Dumitrascu, D.-I.; Costin, H.P.; Glavan, L.-A.; Ciurea, A.V. Frontiers of Cranial Base Surgery: Integrating Technique, Technology, and Teamwork for the Future of Neurosurgery. *Brain Sci.* **2023**, *13*, 1495. https://doi.org/10.3390/brainsci13101495

Academic Editors: Nicola Montano and Joao Paulo Almeida

Received: 3 September 2023
Revised: 10 October 2023
Accepted: 18 October 2023
Published: 23 October 2023

Keywords: cranial base surgery; minimally invasive techniques; intraoperative neuromonitoring; advanced imaging; robotics in neurosurgery; radiosurgery; gamma knife; cyberknife; interdisciplinary collaboration; functional guidance; patient-centric care; endocrinology; biomaterial science; machine learning; future of neurosurgery

1. Prologue: An Overview

1.1. Study Design and Methodology

The endeavor for this research was based on a comprehensively point of view regarding the evolution of multimodel treatment of skull base pathologies, current updates, and emerging computational systems. Firstly, we delved into microscopic and endoscopic assisted approaches, as well as intraoperative electrophysiological monitoring and adjuvant therapies, comparing the postoperative outcome, and long-term prognosis in different therapies. We performed an analytical exploration on Web of Science and PubMed datasets, using the following terms: "Skull Base Surgery", "Minimally Invasive Techniques", "Endoscopic Approach", "Intraoperative Neuromonitoring", "Cranioplasty", "Biomaterials", "Radiosurgery"," Stereotactic surgery", "Gamma Knife", "CyberKnife", "Patient-Centric Care", "Machine Learning", "Deep Learning", and "Neurosurgery Future". Our review emphasized the historical timeline of skull base treatment, showcasing the milestone discoveries.

1.2. Synthesizing the Historical Timeline: From Rudimentary Techniques to the Forefront of Neurosurgical Intervention

In the period antedating the ubiquitous presence of antibiotics, pioneering neurosurgeons such as Schloffer, Cushing, and Hirsch stood at the forefront of advancements in cranial surgery. Their endeavors were characterized by the exploration and development of avant-garde techniques to gain access to intracranial structures. The transnasal approach, specifically targeting the pituitary fossa, is a notable example of such pioneering work. The relatively low mortality rate of this approach, standing at an impressive 5%, was, however, marred by the omnipresent threat of meningitis as the chief cause of death. Despite its efficacy, the transnasal approach was gradually eclipsed as Cushing and his contemporaries gravitated towards the transcranial route over the ensuing one to two decades. This transition, driven by the imperative to circumvent the inherent infectious risks of the former technique, exemplifies the dynamic nature of neurosurgical techniques that evolve in tandem with emerging medical challenges and exigencies [1].

The genesis of Anterior Skull Base (ASB) surgery as a distinct field is anchored in the innovations of the 1940s. Dandy's instrumental contributions are emblematic of this era, particularly his surgical strategy via the anterior cranial fossa for the excision of orbital tumors and his subsequent expansion of the resection to incorporate the ethmoidal regions. In a parallel trajectory, Ray and McLean championed a novel combined transorbital and transcranial method for addressing retinoblastomas. Adding to the burgeoning body of work in this domain, in 1954 a comprehensive transcranial-transfacial approach was described tailored for managing malignancies located in the paranasal sinuses and their immediate anatomical vicinities. This surgical blueprint, having gained considerable traction and endorsement during 1960', underscored the emergence of ASB surgery as a specialized niche within the broader realms of neurosurgery and head and neck surgical disciplines. Moreover, during an exposition on craniofacial resection, meticulously crafted for the treatment of ethmoid carcinoma and inclusive of the cribriform plate resection, augmented the repertoire of ASB surgical techniques. In subsequent discourse and practice, the anterior craniofacial resection (ACFR) acquired a reputation as the quintessential intervention for ASB tumors, especially those with origins in the paranasal sinuses and encroachments into the skull base [2].

Pertaining to the lateral avenues to the skull base, the early 20th century was defined by a renaissance of surgical approaches, with the innovations and contributions of Harvey

Cushing at its epicenter. Cushing's pioneering methodologies were highlighted by the inception of the extended bilateral suboccipital technique, crafted specifically for the resection of tumors. Anchored in his conviction, he asserted that the surgical intervention should be judiciously limited to the tumor's nucleus to safeguard the functional integrity of the cranial nerves and to obviate undue perturbations to the intricate vascular networks of the brainstem. Furthermore, Cushing, with a perspicacious insight, underscored the quintessence of adopting precision-driven and nuanced surgical modalities, judicious modulation of cerebrospinal fluid dynamics, and a minimization of cerebellar manipulation to create an optimal surgical milieu. Such foundational tenets, propounded by Cushing, have been instrumental in sculpting the landscape of present-day surgical praxis and remain deeply ingrained in the ethos of contemporary skull base surgery [3].

A subsequent landmark in cranial surgical evolution was heralded by the advent of microvascular surgery in the 1960s. Pioneered by Jacobsen and Suarez, and later honed by Nakagama and his associates, this innovative paradigm ushered in a new chapter in cranial reconstructive strategies. Central to this was the concept of free flap transplantation, which, over the decades, has burgeoned into a sine qua non for cranial base reconstructions following exenteration procedures. While the utility of free grafts and regional flaps is undeniable in the reconstitution and morphological restoration of these multifaceted cranial zones, the supremacy of free flap transplantation lies in its unparalleled capability to effectuate a definitive demarcation between the intracranial enclave and the upper alimentary and respiratory tracts [4].

Spanning the annals of medical history, surgical interventions with the skull base as the focal point have invariably found a nexus in the disciplines of neurosurgery and otolaryngology. The surgical gamut in this context is extensive, encapsulating procedures such as resection of paragangliomas with an epicenter in the skull base and consequent endocranial extension, transnasal hypophysectomies, and diverse methodologies including cranio-facial and transfrontal techniques tailored for afflictions and neoplasms of the anterior cranial vault. Moreover, the choice between fronto-temporal and suboccipital trajectories has been predicated upon nuanced anatomical and pathological determinants. A pivotal inflection point in this narrative was reached in the 1980s, marking the apotheosis of microscopic neurosurgery as a globally acknowledged expertise. This monumental transition can be ascribed, in no small measure, to the erudite scholarship and pioneering endeavors of luminaries such as Malis and Yasargil. Their seminal expositions on the underlying dogmas and technical matrices of skull base surgery instigated a doctrinal metamorphosis, unequivocally enshrining microscopic neurosurgery as a cornerstone in the integrated therapeutic strategies for skull base pathologies [5].

2. Comprehensive Analysis of the Anterior, Middle, and Posterior Cranial Fossa

The anterior skull base constitutes a convex anatomical structure intricately composed of frontal, ethmoid, and sphenoid bones. Functioning as a partition, this thin osseous layer serves to segregate the intracranial contents from the sinonasal and orbital anatomical features. In terms of its specific anatomical constituents, the frontal bone forms the posterior wall of the frontal sinus and the roof of the orbit. The ethmoid bone, on the other hand, contributes to the architecture of the ethmoid sinus roof and the cribriform plate. Lastly, the planum sphenoidale and the anterior clinoid processes of the sphenoid bone establish the posterior component of the ASB. Vascular structures, specifically the posterior and anterior ethmoid arteries, demonstrate unique trajectories; the former typically courses almost directly from lateral to medial within the bone, whereas the latter exhibits more variability and may course obliquely from posterolateral to anteromedial. It is imperative to meticulously identify these vascular landmarks to avoid the potentially catastrophic complication of retrobulbar hemorrhage. Moreover, the cribriform plate features small bony channels that facilitate the passage of olfactory filae, accompanied by dural invaginations. This anatomical characteristic renders the region susceptible to both iatrogenic and spontaneous cerebrospinal fluid (CSF) leaks. The ethmoid roof, often being exceedingly thin, is also a frequent site for iatrogenic CSF leaks [6].

In a different anatomical context, dissection of the middle cerebral fossa (MCF) floor unveils a collection of crucial anatomical structures, which include the arcuate eminence, the greater superficial petrosal nerve (GSPN), the middle meningeal artery (MMA) and its corresponding foramen spinosum, the gasserian ganglion of the trigeminal nerve, the superior petrosal sinus (SPS), and the petrous internal carotid artery. To facilitate surgical dissection, three distinct anatomical landmarks within the MCF floor can be identified: Glasscock's triangle, Kawase's triangle, and Trautmann's triangle. Glasscock's triangle is demarcated by the foramen spinosum, V3 (the mandibular branch of the trigeminal nerve), and the groove for the GSPN, and is most notably associated with the location of the petrous internal carotid artery [7,8]. Kawase's triangle, conversely, defines the region for bone removal medial to the internal carotid artery and is bordered by the gasserian ganglion, cochlea, GSPN, and carotid artery [9]. Trautmann's triangle, located posterior to the internal auditory canal (IAC), is demarcated by the semicircular canals, the jugular bulb, and the adjacent posterior fossa dura in the vicinity of the sigmoid sinus and serves as a guide for posterior petrosectomy [10].

3. The Surgical Vanguard: Advanced Techniques and Modalities

3.1. Evolution and Optimization of Minimally Invasive Surgical Avenues

In the evolving landscape of neurosurgery, the initial adoption of endoscopy was notably slow-paced, a phenomenon that can be largely attributed to the concurrent rise and standardization of the surgical microscope within neurosurgical procedures. This process of standardization was not only effective but also influential enough to relegate the development of endoscopic methods to a lower priority within the scientific community [11].

This technological bifurcation manifested itself prominently in the surgical strategies targeting conditions related to the anterior skull base. Over the course of medical history, a panoply of surgical approaches—ranging from transcranial to transfacial methodologies—has been utilized, either as standalone techniques or in composite form. These procedures are often labeled as "aggressive" due to their invasiveness and complexity. They are frequently employed in oncological cases, particularly in patients whose overall health is already compromised. Interestingly, these aggressive methods have also been considered for the treatment of relatively benign conditions, such as cerebrospinal fluid fistula lesions [12].

However, the advent of Functional Endoscopic Sinus Surgery (FESS) marked a paradigmatic shift in the surgical management of anterior skull base pathologies. What commenced as an endoscopic technique primarily for diagnostic purposes eventually metamorphosed into an array of specialized surgical methodologies. Advances in endoscopic technology catalyzed the emergence of endoscopic endonasal approaches, providing surgeons with a more nuanced spectrum of options. These endoscopic techniques have been successfully adapted for a range of surgical applications, extending from transsphenoidal pituitary interventions to more elaborate endoscopic excisions involving the skull base [13,14].

Therefore, endoscopic methodologies have redefined the surgical portfolio, introducing minimally invasive options and thereby engendering a revolution in the treatment strategies for conditions involving the anterior skull base. The advancements in endoscopic technology offer a compelling alternative to traditional approaches, holding significant promise for both improving patient outcomes and broadening the scope of treatable conditions.

3.2. Innovations in Pituitary Surgery: A Confluence of Endocrinology and Neurosurgical Finesse

Pituitary adenomas, neuroendocrine tumors arising from the anterior pituitary gland, are traditionally classified as either functional or nonfunctional based on their endocrine secretory profiles. Among the functional adenomas, prolactinomas are most prevalent. For these tumors, dopamine-agonist pharmacotherapy serves as the cornerstone of treatment.

Surgical intervention is generally considered a secondary option, typically reserved for those patients who exhibit pharmacological resistance despite dose escalation or who cannot tolerate medication-induced side effects. In contrast, adenomas that secrete adrenocorticotropic hormone (ACTH), leading to Cushing's disease, or growth hormone, resulting in acromegaly, are predominantly managed with transsphenoidal surgical resection. The rates of biochemical remission postsurgery are substantially influenced by factors such as tumor size and the degree of invasiveness [15].

Postoperative endocrinological outcomes (Figure 1) following transsphenoidal surgery for non-functioning pituitary macroadenomas (NFPAs) present an intricate landscape. Interestingly, the incidence of at least one new hormonal deficiency postoperatively is lower compared to the rate of recovery for at least one preexisting hormonal axis. Among the hormone secretory reserves, ACTH appears to be the most susceptible to postoperative deficit, while the thyroid-stimulating hormone (TSH) secretory reserve is relatively resilient. Gender differences also come into play, with men exhibiting a higher likelihood of recovery from preexisting central hormonal deficiencies subsequent to surgical intervention. Furthermore, the presence of hyperprolactinemia emerges as the most potent predictor for the restoration of pituitary function. These postoperative outcomes underscore the viability of surgical intervention for hypopituitarism resulting from NFPAs, owing to the promising rates of functional recovery and the relatively modest risk of inducing new hormonal deficiencies [16].

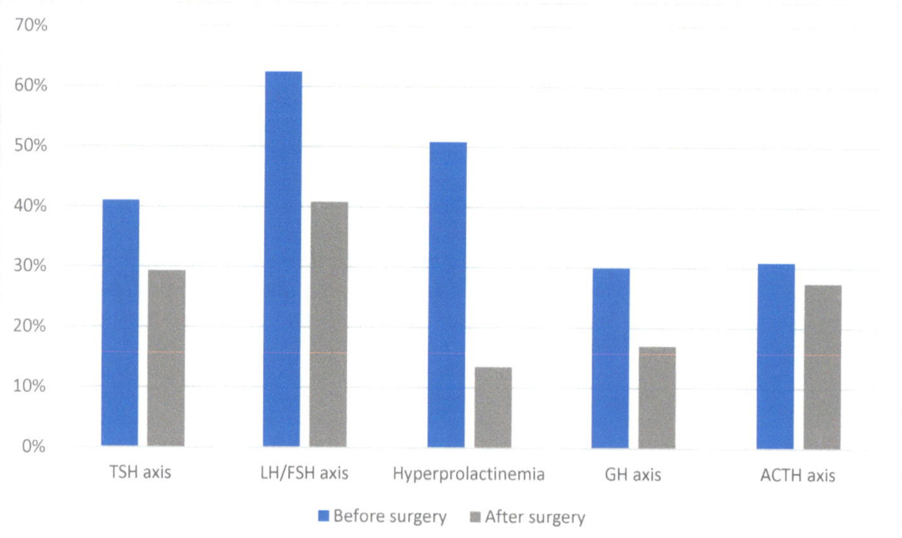

Figure 1. Endocrinological preoperative status compared to postoperative outcome, significant improvement of all pituitary hormones levels is shown.

The therapeutic approach to pituitary adenomas is guided by the hormonal activity of the tumor, among other factors. For prolactin-secreting tumors, dopamine-agonist therapy is the primary treatment choice, while adrenocorticotropic hormone and growth hormone-secreting adenomas are often managed surgically via a transsphenoidal approach. Surgical options also hold promise for non-functional pituitary adenomas, particularly given the favorable postoperative hormonal recovery rates and low incidence of new hormonal deficiencies. This tailored approach, influenced by tumor functionality and patient-specific variables, represents the current state-of-the-art in pituitary adenoma management.

Recent meta-analyses have shed further light on the efficacy of endoscopic surgical techniques for pituitary adenomas. One such meta-analysis by Tabaee and colleagues [17] reported a gross tumor resection rate of 78% in a cohort of 821 patients. The hormonal remission rates for various adenoma types following endoscopic resection were either comparable to or surpassed those achieved through microsurgical techniques. A subsequent meta-analysis by Doward [18] included additional studies and reaffirmed these results, suggesting higher hormonal remission rates for functional microadenomas treated with endoscopic techniques (84%) compared to microsurgery (77%). This differential was even more significant for macroadenomas, with endoscopic techniques achieving a 70% remission rate versus 45% for microsurgical approaches.

The enhanced panoramic visualization and illumination provided by pure endoscopic techniques offer several advantages over traditional microsurgical approaches. These include the ability to operate in areas previously deemed challenging to access, such as the cavernous sinus, suprasellar region, planum sphenoidale, olfactory groove, and retroclival lesions [19]. These endoscopic advances not only increase the surgical field but also minimize nasal mucosal trauma, thereby reducing the need for nasal packing and postoperative discomfort unless significant cerebrospinal fluid leaks or mucosal bleeding occur [20].

Based on the existing literature, it is posited that the methodology employed by Bircher in accessing the cavernous sinus was pioneering in nature. Executed in 1892, this avant-garde approach was undertaken in response to a clinical presentation of thrombophlebitis in a female patient [21]. Moreover, an analytical endeavor aimed to understand the microsurgical and endoscopic anatomy of the cavernous sinus (CS), sixteen cadaveric craniums underwent detailed anatomical dissections. Of these, six were designated for transcranial evaluations, wherein three had their supratentorial cerebral contents excised for enhanced access to the CS and its contiguous structures [22]. Conversely, the remaining three retained their brains in situ during the study. Simultaneously, another subset of six specimens underwent endoscopic examination of the CS. Interestingly, four specimens were subjected to dual analyses—both transcranial and endoscopic—for parallel observations. Post-dissection assessments revealed that, while the CS and its pertinent anatomical structures were meticulously delineated in all craniums—ten transcranially and ten endoscopically—insights into their interrelationships were also garnered. It was discerned that, though cadaveric models afford uncomplicated microscopic and endoscopic visualizations of the CS, live surgical interventions within the CS present considerable challenges necessitating advanced surgical dexterity. Such cadaveric explorations underscore the feasibility of tailored approaches, be it transcranial microsurgery, endonasal endoscopy, or a symbiotic combination, contingent upon the specific pathological presentation and its spatial attributes [22].

In summary, endoscopic techniques in the surgical management of pituitary adenomas offer several advantages, including higher rates of tumor resection and hormonal remission, especially for challenging adenomas such as macroadenomas. The utility of these techniques, coupled with reduced postoperative morbidity, suggests a shifting paradigm in the surgical management of pituitary adenomas, offering both patients and clinicians more effective treatment options.

3.3. *Cranial Base Reconstruction: A Mosaic of Biomaterial Science, Engineering, and Surgical Craftsmanship*

The primary objective of skull base reconstruction is to create a durable, watertight barrier between the intradural contents and the external environment. This priority stems from the severe complications that can arise from persistent cerebrospinal fluid fistula, including meningitis and pneumocephalus, both of which can increase mortality over time [23]. Secondary goals include the closure of dead space, functional restoration, and aesthetic improvement. Various methods have evolved for reconstructing post-craniectomy defects using both synthetic and natural materials, aimed at both mitigating postoperative complications and enhancing cosmetic outcomes [22,23].

While open skull base surgery has traditionally been the standard of care for ablative margin control and definitive reconstruction, endoscopic skull base surgery has witnessed significant growth in popularity and is swiftly becoming the new standard in many centers. However, open approaches remain indispensable for certain complex conditions such as specific malignant tumors, larger composite defects, significant craniofacial trauma, osteoradionecrosis, and failed prior endoscopic reconstruction [24–26].

The choice of reconstructive approach often depends on the specific anatomical considerations. Anterior defects with minor dural damage or an intact bony ledge might do well with simpler reconstruction methods such as multilayered acellular alloplastic materials and free grafts, benefitting from the weight of the anterior intracranial contents to help seal underlay grafts or flaps [27]. Conversely, large posterior defects involving extensive bone and dural damage are usually more challenging to manage. These often necessitate the use of robust vascularized tissue and meticulous postoperative CSF pressure management, sometimes requiring permanent or temporary CSF diversion [28].

Additionally, the extent of the defect—including volume, loss of bony buttresses, and the presence of high-flow CSF leaks—must be thoroughly evaluated to tailor the surgical approach. In summary, both open and endoscopic techniques have distinct advantages and limitations, and the choice between them is guided by the specific needs of the case, taking into account factors such as the size and location of the defect, the risk of complications, and aesthetic considerations.

Various materials, both autologous and synthetic, have been employed in skull base reconstruction, serving different functional and anatomical needs. Autogenous grafts such as nasal mucoperichondrium and mucoperiosteum, tensor fascia lata, temporoparietal fascia, calvarial bone, and abdominal adipose tissue have all been documented for use in reconstruction [29]. On the synthetic front, noncellular materials such as DuraGen, AlloDerm, DuraSeal, and hydroxyapatite cements have been utilized [28,29].

For anterior skull base reconstructions, vascularized locoregional flaps have emerged as the go-to option. Specifically, the nasal septal flap, based on the posterior septal artery, has greatly advanced endoscopic skull base surgery. Its reliability, versatility, and low morbidity have made it the first-line choice in endoscopic reconstruction, significantly lowering the rates of CSF leaks [30,31].

Soft tissue donor sites for free flaps in skull base reconstruction have also diversified. The rectus abdominus is a well-established choice, known for its reliable deep inferior epigastric pedicle, as well as offering both large skin area and muscle bulk suitable for filling dead space [31,32]. Similar characteristics are found in the latissimus dorsi flap. Recently, the anterolateral thigh has become increasingly popular for its low morbidity, versatility, and long reliable vascular pedicle. Options such as the vastus lateralis muscle or vascularized tensor fascia lata can also be utilized either alone or in combination, adding to the arsenal of reconstructive choices [33–35].

The evolution of materials and techniques in skull base reconstruction has significantly broadened the surgical toolkit, offering a range of autogenous and synthetic materials for varying needs. These developments have contributed to more effective, reliable, and low-morbidity reconstructive options, enhancing both functional and cosmetic outcomes.

4. The Renaissance of "Functionally" Guided Surgery: Intraoperative Neuromonitoring

4.1. Historical Overview and Technological Breakthroughs

The history of Intraoperative Neuromonitoring (IONM) traces back to 1898, when Dr. Fedor Krause in Berlin used monopolar faradic stimulation during an acoustic nerve neurectomy. Krause's work marked the earliest instance of visual observation of nerve activity during surgery. The technique gained significant momentum in the 1960s, when it was adapted for thyroid surgeries by Flisberg and Lindholm, and also for parotid and ear surgeries through facial nerve stimulators developed by Parsons and Hilger [36].

In contemporary medical practice, IONM has become a staple in various surgical disciplines, especially those involving close proximity to critical nerves. Predominant among these are thyroid surgeries—where the vagus and recurrent laryngeal nerves are monitored—parotidectomy, and surgeries of the posterior cranial fossa where facial nerve monitoring is crucial. During these operations, the surgeon employs a stimulator probe to accurately identify and differentiate the nerve from surrounding tissues. When the probe is placed onto the nerve, a circuit is closed that triggers either visual or auditory cues each time the nerve is contacted. This type of monitoring is often referred to as intermittent IONM (iIONM) [37].

IONM has evolved from its humble beginnings to become an integral part of modern surgery. It provides surgeons with real-time feedback, enhancing surgical precision and thereby potentially reducing post-operative complications. Its applications have been widely adopted in surgeries that risk nerve injury, making it a standard practice in many medical institutions.

4.2. Mechanisms, Modalities, and the Paradigm Shift towards Real-Time Functional Feedback

Brainstem Auditory Evoked Potentials (BAEPs) are bioelectric neural activities triggered by the stimulation of the vestibulocochlear nerve [38]. These potentials are particularly challenging to distinguish from the background electrical activity of the brain due to their relatively small amplitude [37,38]. To separate the BAEPs from this "noise", thousands of samples of the electric stimulus are gathered and averaged, allowing for a clearer identification of the auditory evoked potential.

In BAEP recordings, data are collected from multiple points along the vestibular nerve pathway as it moves from the peripheral to the central nervous system [39,40]. The peaks in these recordings are categorized as Waves I through V, which correspond to different anatomical locations—from the peripheral cochlear nerve to the inferior colliculus [41,42].

During BAEP monitoring, electrodes are placed on the scalp and earlobes. An auditory stimulator then emits acoustic clicks to the ear being operated on, delivered through an earphone-transducer setup. The electrical pulse rate for these clicks is typically set between 20 to 50 per second. Before the operation starts, the stimulus intensity, usually measured in decibels, is adjusted to a level where the patient can hear the clicks. The final stimulus is then set at an intensity a few decibels higher than this initial threshold. To ensure focused monitoring, white noise is applied to the contralateral ear at a lower intensity to mask its response [41,42].

BAEPs are a critical tool for monitoring neural activity related to auditory functions during surgical procedures. They allow for real-time tracking of auditory pathway integrity, which is particularly useful in surgeries where the auditory nerve might be at risk. This method is complex and requires precise setup and interpretation, but its importance in safeguarding auditory function during surgical procedures is well recognized [43,44].

The monitoring of somatosensory spinal pathways, specifically the dorsal column-medial lemniscus, relies on subcortical and cortical responses to continuous electrical stimulation of peripheral nerves such as the tibial, peroneal, ulnar, or median nerve. This method of Intraoperative Neuromonitoring is commonly used and easy to implement, having no contraindications. It can be particularly useful for monitoring the posterior spine approach in spinal deformity surgeries, boasting a sensitivity range of 25–92% and a specificity of 96–100% [45]. However, it does have limitations, such as a time lag (1–20 min) in data interpretation due to signal averaging, making it possible for an injury to go undetected until it becomes irreversible. It is also less effective for monitoring patients with pre-existing neurologic deficits or in situations involving isolated motor pathway or nerve root injuries, which are better detected by Motor Evoked Potentials (MEPs) or Electromyography (EMG) [46].

MEPs are particularly sensitive for monitoring motor pathways in the anterior or central regions of the spinal cord and nerve roots. They serve as highly reliable indicators of corticospinal tract injuries and have proven especially useful for detecting spinal cord ischemia during spinal deformity correction [45]. This form of monitoring involves real-time, intermittent stimulation of the motor cortex and subsequent recording at muscles, preferably those rich in corticospinal tract innervations such as distal limb muscles. Transcranial stimulation can be either magnetic or, more commonly in surgical settings, electric (Transcranial Electric Motor Evoked Potentials or TceMEP). The electromyography signals, also known as Compound Motor Action Potentials (CMAP), are typically acquired through needle electrodes inserted bilaterally into the upper limbs. These serve as controls to differentiate systemic, anesthesia, and positioning-related changes [47].

Both somatosensory and motor evoked potentials offer valuable insights into neural integrity during spinal surgeries, albeit with distinct advantages and limitations. While somatosensory monitoring is generally easier to implement and can provide information about both sensory and motor pathways, MEPs offer real-time, direct monitoring of motor pathways, making them invaluable in surgeries where motor function is at high risk.

4.3. Neuromonitoring in Diverse Pathologies: Customized Approaches for Tailored Surgical Interventions

Direct stimulation of the facial nerve during posterior cranial fossa surgery has been explored by Amano, who used a ball-type electrode to stimulate the root exit zone of the facial nerve. This method was shown to be potentially useful for assessing the state of the facial nerve during surgery. By examining variables such as amplitude preservation ratio and the last maximal amplitude, the method could predict the likelihood of facial nerve palsy postoperatively according to the House–Brackmann (HB) grade [48].

Multipulse Transcranial Electric Stimulation (TES) provides another approach to continuous monitoring of the facial nerve. A cup electrode placed on the skull sends out clusters of electrical pulses that stimulate the corticobulbar pathway, allowing real-time monitoring of facial nerve function through facial nerve muscle motor evoked potentials (FNMEP). This method has been found to accurately predict the postoperative state of the facial nerve, with patients maintaining at least 50% of the baseline amplitude generally experiencing no more than mild deterioration in facial nerve function postoperatively [49].

In contrast to active continuous Intraoperative Neuromonitoring (acIONM), there are methods described as passive continuous IONM (pcIONM) that do not involve direct stimulation but rather analyze natural discharge patterns that occur during the surgical procedure. Free-running electromyography (EMG) is one such method used to monitor the facial nerve during neurosurgery. In this technique, patterns such as spikes, bursts, and trains in the EMG signal are analyzed to provide insights into nerve function. Prass and Lüders described different types of EMG signal patterns such as spikes, bursts, and trains, which they observed during posterior fossa surgeries on 30 patients [50].

Multiple methods exist for intraoperatively monitoring the facial nerve during posterior cranial fossa surgery. Each has its unique advantages and disadvantages. Direct stimulation offers the ability to assess the facial nerve's function at specific times during the procedure, while continuous methods such as TES allow for ongoing, real-time monitoring. Passive methods such as free-running EMG offer a non-intrusive way to monitor the nerve by analyzing its natural activity during surgery.

Despite advances in Intraoperative Neuromonitoring, the retention rates for vestibulocochlear nerve function are not as favorable as those for the facial nerve. This discrepancy could arise from the intrinsic challenges of preserving auditory function, especially when dealing with large tumors and those that have extensive infiltration into the cerebellopontine angle [49–51].

In the context of spine surgery, the choice of monitoring modality depends on the approach used and the specific risks involved. For posterior approaches, somatosensory evoked potentials (SSEPs) may be sufficient. However, for anterior approaches, transcranial motor evoked potentials (MEPs) are typically recommended due to the risk of anterior spinal artery syndrome. Where nerve root or spinal cord deficits are a major concern, additional modalities such as spontaneous and triggered electromyography may be valuable. Multi-modal IONM is highly recommended for procedures such as spine deformity surgery or those involving intradural tumors [52,53]. Anesthesia should be adjusted to allow for the best possible IONM recordings, with specific anesthetic agents contraindicated for certain types of monitoring [54].

IONM is not just limited to spine or cranial surgeries. It is also used in a variety of other surgical fields such as vascular and cardiothoracic. Its utility extends to preventing perioperative peripheral nerve injury (PPNI), which could occur due to excess mechanical pressure and torsion on the limbs and neck during surgery [55].

The use of IONM is crucial for optimizing outcomes in various types of surgery. While it has shown great promise, there is still room for improvement, particularly in monitoring the vestibulocochlear nerve. The choice of monitoring technique should be tailored to the specific surgical approach and the associated risks, and multi-modal IONM is often recommended for complex cases.

5. The Digital Surgeon: Technological Synergies in Cranial Base Surgery

5.1. Endoscopy in the New Era: Advanced Imaging, Robotic Assistance, and Augmented Reality Overlays

Advancements in skull base surgery are increasingly leveraging the capabilities of virtual reality (VR) and augmented reality (AR). For instance, color-coded stereotactic VR models can be custom-tailored for individual surgical cases, providing a simulated operating field for surgeons and trainees [56]. These models offer invaluable opportunities for surgical education and preoperative simulations. Furthermore, VR technology can be integrated into real-time operative settings by overlaying 3D images onto microscopic or endoscopic views, thus enhancing spatial navigation capabilities for the surgeon [57].

AR technology appears to offer particular benefits to less experienced medical professionals. These systems serve not just as educational tools but also as potential substitutes for existing neural navigation technology. AR can offer both contextual information about underlying structures and direct patient perspectives, potentially revolutionizing conventional neural navigation systems [58].

Beyond surgery, AR also has applications in non-surgical and clinical management at the skull base. For example, it is used for ablating damaged nasal tissue and offers guidance on basic surgical plans and navigational protocols [59]. In cranio-maxillofacial procedures, AR plays a significant role in reconstructing cheekbones and offering data on the underlying structure, albeit without the capability for real-time modifications [60]. Many AR applications superimpose precollected, immersive data onto real endoscopic camera images. However, fields that lie outside the endoscopic view remain hidden to the medical team, necessitating further adaptations to fully realize the technology's potential.

Moreover, the application of Augmented Reality in clinical settings, particularly in the management of base-of-the-skull pathologies, has been gaining significant attention in the medical community, as evidenced by multiple academic conferences exploring its potential [59,60]. A specific clinical model has been proposed, offering an extended observational perspective of the area under examination [61]. In this model, endoscopic images are displayed centrally, while the projection external to the endoscopic field of view is rendered virtually, utilizing pre-existing computerized tomography data. Such an integrated AR framework suggests that, following technological advancements and methodological refinements, AR applications may become increasingly prevalent across a broader spectrum of clinical scenarios necessitating heightened alertness and precision [62].

When it comes to the design of an ideal AR device for clinical applications, certain rigorous criteria must be met to ensure its functional efficacy and safety. The system should feature a focus marker and device alignment capabilities that are intuitive and minimally intrusive, particularly for the medical professional using it. Calibration adjustments should be undertaken before the initiation of the clinical procedure to minimize undue burden or cognitive load on the healthcare provider [63].

Furthermore, conventional imaging techniques that focus solely on two-dimensional visual data may suffer from limitations in perceived depth, thereby potentially compromising the practitioner's situational awareness and decision making accuracy. To mitigate such limitations, it is advisable to incorporate depth cues to enhance the perceptual veracity of the rendered images [64]. Additionally, in applications where virtual 3D objects are superimposed onto endoscopic images, it becomes imperative to maintain parallax when the viewing position changes in order to preserve spatial relationships and depth perception.

In terms of data presentation, meticulous attention must be devoted to the structural layout of the AR interface. Inadequate design considerations can obscure critical information or induce visual discomfort, thereby diminishing the user experience and potentially compromising clinical outcomes. Therefore, it is essential to engage in an iterative design process, incorporating user feedback and empirical data, to optimize the AR interface and data presentation for the specialized needs of clinical practice.

5.2. *Data-Driven Neurosurgery: Machine Learning, AI-Assisted Diagnosis, and Surgical Planning*

The application of Radiomics in oncological diagnostics has emerged as a transformative approach in recent years, particularly in the preoperative assessment of various neoplastic conditions including prostate cancer, lung cancer, and an array of brain tumors such as gliomas, meningiomas, and brain metastases [63–65]. Traditional diagnostic methodologies that rely predominantly on qualitative assessments made by radiologists based on "visible" features, Radiomics facilitates the quantitative extraction of high-dimensional features as parametric data from radiographic images [66,67].

The incorporation of machine learning algorithms further enhances the analytical capabilities of Radiomics, offering unprecedented insights into the pathophysiological characteristics of lesions that are otherwise challenging to discern through conventional visual inspection [68,69]. Several studies have demonstrated the utility of Radiomics-based machine learning in the differential diagnosis of various brain tumors, thus indicating its prospective application in clinical decision making [70].

In the feature selection domain, Least Absolute Shrinkage and Selection Operator (LASSO) has been noted for its effectiveness in handling high-dimensional Radiomics data, particularly when the sample sizes are relatively limited [71,72]. LASSO distinguishes itself by its ability to avoid overfitting, making it an optimal choice for robust feature selection in Radiomics analyses.

Additionally, Linear Discriminant Analysis (LDA) serves as another valuable machine learning classification algorithm tailored for Radiomics applications. LDA seeks to identify and delineate boundaries around clusters belonging to distinct classes and projects these statistical entities into a lower-dimensional space to maximize class discriminatory power. Notably, it has been reported to retain substantial class discrimination information while reducing dimensionality [73–75].

Radiomics has extended its utility beyond diagnostic applications to prognostic evaluations, as exemplified in its role in both the diagnosis and treatment control rate prediction for chordoma [76]. Chordoma, a disease notorious for its refractory nature necessitating multiple surgical interventions and radiotherapeutic treatments, poses unique challenges for sustained disease control. In this context, Radiomic models built on features describing both the morphological shape and the genomic heterogeneity of the tumor have demonstrated superior performance in predicting the effectiveness of radiotherapy for tumor control. Such predictive capabilities underscore the potential benefits of Radiomics in enabling more targeted, efficient treatment regimens for diseases such as chordoma, thereby potentially reducing the need for repetitive, invasive procedures.

In another application, Radiomics-based machine learning algorithms have been shown to assist significantly in the preoperative differential diagnosis between germinoma and choroid plexus papilloma [77]. These two types of primary intracranial tumors often present with overlapping clinical manifestations and radiological features, yet they require markedly different treatment modalities. In addressing this diagnostic conundrum, high-performance prediction models have been developed using sophisticated feature selection methodologies and classifiers. These models suggest that Radiomics can offer a non-invasive diagnostic strategy with substantial reliability.

Notably, the application of Radiomics and machine learning in these scenarios holds the promise of revolutionizing the approach to image-based diagnosis and personalized clinical decision making. By leveraging advanced computational techniques to analyze complex, high-dimensional radiographic data, Radiomics provides a more nuanced understanding of tumor characteristics and treatment responses. This computational approach thereby opens avenues for more accurate, timely, and individualized therapeutic strategies, significantly enhancing the quality of patient care in oncological settings.

In the realm of skull base neurosurgery, machine learning (ML) methods, including neural network models (NNs) (Figure 2), have been rigorously applied to a comprehensive, multi-center, prospective database to predict the occurrence of Cerebrospinal Fluid Rhinorrhoea (CSFR) following endonasal surgical procedures [78]. The predictive capabilities of NNs surpass those of traditional statistical models and other ML techniques in accurately forecasting CSFR events. Notably, NNs have also revealed intricate relationships between specific risk factors and surgical repair techniques that influence CSFR, relationships that remained elusive when examined through conventional statistical approaches. As these predictive models continue to evolve through the integration of more extensive and granular datasets, refined NN architectures, and external validation processes, they hold the promise of significantly impacting future surgical decision making. Such next-generation models may provide invaluable support for more personalized patient counseling and tailored treatment plans.

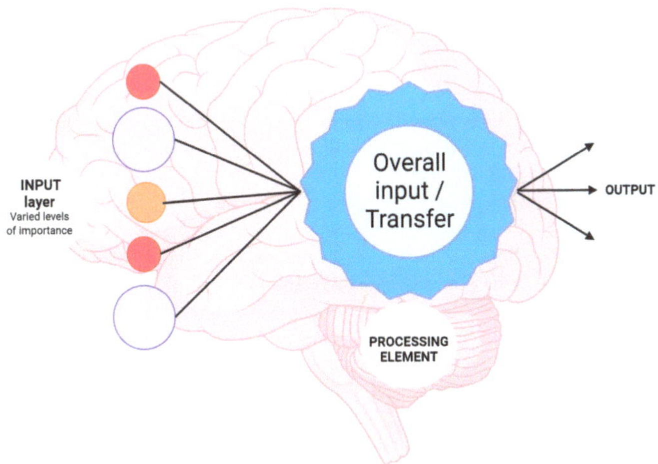

Figure 2. Mechanisms of neural network processing are shown. Input layer refers to heterogenous data which will be analyzed by the neural network incorporated algorithms. Further, output information is obtained, offering new avenues for biomedical fields.

Regarding automated image segmentation in surgical navigation applications, although there is a high correlation between the automated segmentation and the anatomical landmarks in question, the Dice Coefficient (DC)—a measure commonly used to assess the performance of the segmentation task—was not deemed to be particularly high [79]. Various factors contribute to this finding, including the complexity of anatomical pathways, the absence of clearly delineated contours in certain regions, and inherent variations arising from manual segmentation. These limitations cast doubt on the utility of the DC as a standalone metric for objectively evaluating the performance of this specific task. However, the low average Hausdorff Distance (HD) on the testing dataset better encapsulates the high accuracy of the automated segmentation, bolstering its credibility for applications such as surgical navigation.

In summary, the application of machine learning, and particularly neural networks, appears to be a game-changer in predicting complex clinical outcomes such as CSFR following skull base neurosurgery. Meanwhile, automated image segmentation remains a challenging task, warranting a more nuanced approach to performance assessment than merely relying on singular statistical measures such as the Dice Coefficient. These advancements signify not only the growing impact of computational methods in medicine but also the necessity for ongoing refinement and validation to ensure these techniques meet the highest standards of clinical efficacy and safety.

6. Radiosurgery and Radiotherapy: Harmonizing Precision and Efficacy

6.1. An In-Depth Exploration of Radiosurgical Modalities: Gamma Knife, CyberKnife, and Beyond

Stereotactic radiosurgery has emerged as a pivotal treatment modality for various types of lateral skull base lesions, with perhaps its most significant impact being on the management of glomus jugulare tumors [80]. Many medical centers have adopted this approach as the first-line treatment for growing symptomatic tumors due to its lower morbidity compared to traditional surgical interventions, coupled with comparable rates of disease control.

Additionally, the efficacy of radiosurgery in treating skull base meningiomas has been well documented, with long-term follow-up data indicating impressive tumor growth control rates ranging from 92 to 98% [81]. While determining the optimal radiation dosage is critical, findings suggest that doses greater than 12 Gy to the tumor margin are essential for effective control. Suboptimal doses, specifically less than 12 Gy, have been associated with a significant tumor growth rate during one-year follow-up periods [82]. It is generally recommended that the minimum effective radiation dose for skull base meningiomas should be between 13 and 17 Gy, although the suitability of lower doses remains a topic of ongoing debate [83].

Importantly, some nuances exist within the treatment paradigm. For instance, Lee et al. noted that previously resected tumors might pose challenges in accurate radiological delineation due to postoperative changes such as meningeal enhancements or fat signals, which could be mistaken for tumor tissue [83]. Moreover, Zachenhofer and colleagues posited that tumor regrowth often occurs outside the targeted radiosurgical volume, a phenomenon possibly attributable to microscopic remnants of tumor cells within the adjacent dura mater that are not included in the radiosurgical target [84].

Stereotactic radiosurgery offers a promising avenue for managing various types of skull base tumors, including glomus jugulare and meningiomas, with both short-term and long-term efficacy. However, it is crucial to consider factors such as optimal radiation dosage and potential challenges related to the radiological delineation of previously resected tumors. These complexities underscore the necessity for personalized treatment strategies and underscore the importance of ongoing research to fine tune radiosurgical approaches for maximum clinical benefit.

The management of benign meningiomas using radiosurgery must be approached with caution given the potential for malignant transformation. up to 2% of benign meningiomas could transform into malignant forms, and others have found that 28.5% of recurrent benign meningiomas were actually atypical or anaplastic [83,84]. These statistics underline the importance of long-term monitoring and potential re-evaluation of treatment plans [85,86].

Radiosurgical interventions also present challenges related to cranial nerve sensitivity, most notably the optic and trigeminal nerves. The optic nerve is particularly vulnerable to radiation, requiring careful dose planning. Leber et al. suggested that doses below 10 Gy could be safely administered to the optic nerve without complications, but doses between

10–15 Gy carry a 26.5% risk of optic neuropathy [87]. Thus, tumors close to or compressing the optic apparatus are less amenable to radiosurgical treatment, as delivering an effective dose could jeopardize optic nerve function.

Various guidelines have been proposed for dosing the optic apparatus, with Morita et al. allowing for short segments to receive between 12–16 Gy [88], and Stafford et al. reporting no optic neuropathy with a 12 Gy dose [89]. Therefore, radiosurgery might be more appropriately indicated for tumors situated at least 5 mm away from the chiasm and optic nerve.

The trigeminal nerve also shows significant sensitivity to radiation, with various studies reporting the development of trigeminal neuropathy post-treatment. For example, Lee et al. found that 4% of their patients developed this condition, with a portion experiencing permanent deficits [83]. Moreover, Chang et al. reported that although 86% of patients initially experienced pain relief, about half suffered pain recurrence during the follow-up period [90]. Radiation doses exceeding 19 Gy were found to be associated with a high incidence of trigeminal neuropathy [88].

The utility and efficacy of radiosurgery for skull base meningiomas appear to be influenced by several factors including tumor size, dose, and fractionation. Single-session radiosurgery has been reported to yield a five-year actuarial tumor control rate of 88.6% for large skull base meningiomas (>8 cm^3). However, tumor control rates tend to decrease with increasing tumor volume, specifically tumoral volumes $\geq$ 14 cm^3 [91]. With a median dose of 10 Gy (ranging between 8–10 Gy), the five-year and ten-year tumor growth control rates were 78% and 70%, respectively. Notably, only 6% of patients experienced permanent radiation injury with an 84-month follow-up [92].

The CyberKnife® platform offers a technological advancement in frameless robotic radiosurgery, enabling high precision and conformal intracranial tumor targeting. It allows for easy fractionation of treatment, thereby minimizing toxicity, especially when adjacent organs-at-risk (OAR) have low radiation tolerance [93]. However, the outcomes for larger, malignant tumors remain less predictable, with both tumor size and type affecting the treatment outcome [94].

In the case of smaller, radiosensitive tumors such as vestibular schwannomas and meningiomas, radiosurgery has been largely effective with minimal acute toxicity. However, areas for improvement include symptom management and late morbidity. The presence of larger tumors, less optimal dose/fractionation, and other risk factors such as previous cranial radiotherapy can lead to increased treatment-related toxicity [95].

Given these findings, the focus for smaller radiosensitive tumors should be on optimizing dose prescription and fractionation schedules. Careful planning that includes vigilance over multiple dose indices for susceptible OAR may help minimize late toxicity and optimize functional preservation. For tumors of other pathological types, which tend to be larger and/or more radioresistant, initial efforts should aim at increasing local control while minimizing toxicity through optimized dose and fractionation scheduling.

In summary, while radiosurgery has shown promising outcomes for skull base meningiomas and other cranial tumors, there are challenges that need to be addressed. Tumor size, type, and proximity to critical structures such as OAR can impact the efficacy and safety of treatment. Consequently, individualized treatment plans, leveraging advanced technologies such as CyberKnife® and ongoing research, will be key to improving outcomes.

6.2. Radiotherapy Advancements: Modulating Doses, Fractions, and Protocols for Optimal Tumor Control and Preservation of Neural Structures

Fractionated stereotactic radiotherapy (FSRT) has emerged as another viable option for the treatment of large skull base meningiomas. Studies have shown that FSRT can offer five-year tumor growth control rates ranging between 93–96%. In terms of toxicity, late clinical toxicity has been reported to be relatively low, falling in the range of 1.6–5.5%. The treatment generally involves delivering radiation doses of 50–56.8 Gy for tumor volumes averaging between 35.4–52.5 cm^3. The mean duration of follow-up in these studies was between 35–42 months [96].

In a more recent study that compared single-session gamma knife surgery (GKS) with fractionated GKS (FGKS) for meningiomas having a volume greater than 10 cm^3, FGKS appeared to show a marginally higher overall five-year tumor control rate (92.9% for FGKS vs. 88.1% for single-session GKS). However, it is important to note that the difference in the control rates between the two groups was not statistically significant ($p = 0.389$). The mean tumor volume for the single-session GKS group was 15.2 cm^3, while for the FGKS group, it was 21 cm^3. The FGKS group also included 16 skull base meningiomas [97].

Fractionated radiation therapy, which involves daily treatments usually spanning several weeks, is a commonly employed strategy for treating certain types of tumors, including WHO grade I meningiomas that are located close to sensitive areas such as the optic chiasm or optic nerves [98]. This approach is backed by evidence showing that external beam radiation therapy (EBRT) can deliver effective doses that control the tumor while preserving visual function [99] (Table 1).

Table 1. Pertinent studies on the use of additional radiotherapy in managing WHO grade II and III meningiomas.

Studies	Treatments	Histology	Results	Reference
Aghi et al., 2009	Surgery (TR), surgery + EBRT	WHO II	Relapse rates at 5 years were 41% and it was reduced to 0% with the inclusion of EBRT.	[100]
Attia et al., 2012	SRS	WHO II	TS exceeds 50% after 5 years; A 5-year LRR of 44%	[101]
Goyal et al., 2000	Surgery, surgery + FRT	WHO II	TR = 5-year TS was 87% PTR = 5 year TS was 100%	[102]
Huffman et al., 2005	GKRS	WHO II	40% relapse between 18 and 36 months.	[103]
Dziuk et al.,1998	Primarily surgery + FRT	WHO III	The 5-year TS is 57%.	[104]
Goldsmith et al., 1994	Surgery + FRT	WHO III	58%	[105]
Rosenberg and Prayson et al., 2009	Surgery + FRT; Surgery + SRS.	WHO III	5-year TS = 47%	[106]
Mattozo et al., 2007	SRS + EBRT	WHO I–III	Grade II: 3-year RFS rate of 83% Grade III: 3-year RFS rate of 0%	[107]
Adeberg et al., 2012	EBRT, surgery + EBRT	WHO II–III	Grade II tumors: 5-year TS rate of 81% and a 5-year RFS rate of 50%. Grade III tumors: 5-year TS of 53% and a RFS rate of 13%.	[108]
Hug et al., 2000	EBRT + surgery	WHO II–III	Grade II: 5-year TS rate of 38% Grade III: 5-year TS rate of 52%	[109]

Table 1. *Cont.*

Studies	Treatments	Histology	Results	Reference
Milosevic et al., 1996	Mainly surgery + FRT	WHO II–III	5-year TS rate of 28%	[110]
Pasquier et al., 2008	Surgery + EBRT, Surgery only	WHO II–III	TR: 5-year TS = 46% RT: 5-year TS = 0%	[111]
Sughrue et al., 2010	Surgery + FRT	WHO II–III	61%, 40% after 10 years	[112,113]
Yang et al., 2008 (33 atypical), (41 anaplastic)	Surgery, Surgery + EBRT	WHO II–III	Grade II: TS of 11.9 yrs and a RFS of 11.5 yrs (cases of atypical meningiomas) Grade III: TS of 3.3 yrs and RFS of 2.7 yrs	[114]
Boskos et al., 2009	EBRT Protons and photons	WHO II–III	5-year TS = 65% 5-year LRR = 61%	[115]

(TS = total survival, RFS = recurrence-free survival, EBRT = external beam radiation therapy, LRR = local recurrence rate, PTR = partial resection (STR), TR = total resection (GTR), FRT = Fractionated radiotherapy (RT), SRS = stereotactic therapy (radiosurgery), GKR = Gamma Knife radiosurgery).

In cases where meningiomas affect the optic nerve sheath, EBRT is the treatment of choice. Many patients have reported vision improvement following this treatment. Remarkably, no other treatment modalities, including surgical interventions, have been shown to improve vision to the same extent as radiation therapy (RT) for this specific patient group. Therefore, surgical decompression is typically reserved for patients with intracranial extensions and rapidly deteriorating conditions [116].

When it comes to cavernous sinus and petroclival meningiomas, radiation therapy is often the preferred treatment option, either as a primary treatment or as an adjunct to subtotal resection. These locations are associated with a high risk of surgical morbidity if extensive resection is attempted. A recent literature review indicated that stereotactic radiosurgery (SRS) alone resulted in a relatively low recurrence risk of about 3%. In contrast, more invasive procedures such as subtotal resection (STR) and gross total resection (GTR) had recurrence risks of around 11%. Moreover, cranial nerve deficits were more commonly reported among patients who underwent surgical resection [112].

Although chordomas are generally slow-growing tumors, aggressive upfront management has shown significant benefits in long-term survival. A retrospective study conducted in France demonstrated that patients who received RT immediately following surgery had a 10-year survival rate of 65%, whereas none of the patients who only received RT at the time of recurrence survived up to 10 years [117]. In the largest series on chordomas treated with RT, conducted at Harvard University, patients were treated with 60 to 79.2 Cobalt-Gray-Equivalent (CGE), and the local control (LC) rates at 10 years were found to be 44% [118]. A recent review that aggregated data from over 400 patients found that 5-year LC rates were close to 70% and overall survival (OS) was more than 80% [119].

Soft tissue sarcomas of the skull base are usually approached with maximal surgical excision, followed by post-operative radiation therapy. Recurrence risk is higher in these cases compared to soft tissue sarcomas of the extremities, mainly because obtaining clean surgical margins is often challenging. Various radiation therapy techniques are utilized, including external beam radiation therapy (EBRT), stereotactic radiosurgery (SRS), intraoperative RT, and brachytherapy [120].

Modern advancements in EBRT include technologies such as three-dimensional conformal RT (3D-CRT) and intensity-modulated radiation therapy (IMRT). Three-dimensional conformal RT typically delivers radiation from multiple angles in a coplanar fashion, akin to the spokes on a wheel. IMRT, on the other hand, allows the intensity of radiation beams to vary at different positions. This has significantly improved the ability to treat tumors located near sensitive structures, thereby advancing the field of radiation oncology [120].

Advances in radiation therapy and aggressive upfront management strategies have shown promising results in the treatment of chordomas and soft tissue sarcomas of the skull base. These findings underscore the need for individualized, multidisciplinary treatment approaches to optimize long-term outcomes.

7. Holistic Approaches: Interdisciplinary Collaborations and Patient-Centric Care

7.1. The Ecosystem of Cranial Base Surgery: Integrating Neurology, Radiology, Oncology, and Rehabilitation

The complexities involved in the surgery of skull base meningiomas (SBMs) increasingly point to the need for a multimodal treatment approach, integrating both radiosurgery and radiation therapy. This combination aims to maximize both functional outcomes and tumor control. Advances in technology, genomics, and Radiomics are poised to greatly enhance our understanding of tumor biology. This, in turn, allows for the tailoring of treatment plans in line with the tenets of precision medicine [121].

Beside multiple implicated medical specialties, neurosurgeons need to undergo a continuous high standard training for skull base pathology. Achieving surgical proficiency is paramount for educators within the domain of skull base surgery. Diligent effort, coupled with consistent and immediate evaluative feedback, constitutes a cornerstone of successful skill acquisition. Establishing an environment rooted in patient-centric values that fosters scholastic excellence augments the efficacy of a training program. Moreover, the usage of 3D printed models of the skull are currently used as a training possibility even for during the residency program. In the case of skull base pathologies, neurosurgeons can exercise the surgical approaches, especially various types of craniotomies on those synthetic-based models. For optimal knowledge assimilation, it is imperative that both the mentor and mentee engage proactively and with deliberate intent [122–124].

Given the rapidly evolving landscape of SBM treatment—fueled by technological and scientific innovations—a specialized multidisciplinary approach has become essential for optimal patient care. This has led to the conceptualization of "Centers of Excellence", institutions specifically geared towards SBM management. These centers are not only technologically advanced but also guarantee an adequate workload for healthcare providers, ensuring they remain at the forefront of the field.

Moreover, integration of a diverse array of medical and allied health disciplines has the potential to substantially augment the quality of healthcare delivery, particularly in the context of Skull Base multidisciplinary teams. In such specialized tertiary referral centers, the amalgamation of expertise from various subspecialties not only fosters a holistic approach to patient care but also enhances the precision and efficacy of diagnosis, treatment planning, and execution.

Within these multidisciplinary frameworks, palliative care physicians contribute to symptom management and quality-of-life improvement, offering critical perspectives on end-of-life care when required. Neurosurgical anesthetists bring a refined understanding of perioperative management, particularly vital in the intricate surgeries associated with skull base anomalies. Chronic pain specialists offer insights into long-term analgesic strategies, thus contributing to sustained patient comfort and improved functionality post-surgery.

Similarly, clinical psychologists can play a pivotal role in assessing and addressing the psychological comorbidities often accompanying chronic or severe medical conditions. They provide cognitive-behavioral interventions and other psychological supports to enhance patients' coping mechanisms. Audiological scientists and hearing and/or balance therapists contribute expertise on auditory and vestibular systems, which are frequently involved in skull base pathologies. Their input can be invaluable in both the diagnostic and rehabilitative phases of care.

Additionally, maxillofacial prosthetists offer specialized interventions that focus on reconstructive options, including facial prosthetics, which can be instrumental in post-operative rehabilitation. Speech and language therapists address communication and swallowing challenges that might arise due to anatomical changes or neurological impairments associated with skull base disorders. Dietitians further enrich the multidisciplinary team by offering tailored nutritional plans, thereby optimizing patients' metabolic states for improved outcomes in both surgical and nonsurgical interventions.

This expansive collaborative approach is further fortified by interdepartmental interactions with neurosurgeons, neuroradiologists, and neuropathologists. Neurosurgeons offer specialized surgical interventions, while neuroradiologists provide crucial imaging expertise, enhancing the specificity and sensitivity of diagnostic processes. Neuropathologists contribute by offering detailed tissue diagnoses, which are vital for optimal treatment planning.

Given the complex, multifaceted nature of the conditions encountered in skull base pathology, and the necessity for sophisticated diagnostic and therapeutic modalities, the persistence of skull base multidisciplinary teams as a feature of tertiary referral centers seems not only likely but also clinically imperative. This convergence of specialized skills in a collaborative environment serves to enhance patient outcomes, facilitating a more nuanced and comprehensive standard of care [125].

7.2. Patient Narratives and Quality of Life Metrics Post-Surgery

Patients ultimately want their surgical team to cure, control, or, ideally, facilitate the prevention of disease. They favor minimally invasive approaches. When possible, they want illnesses to be treated by medicines only. If further intervention is necessary, they prefer minimally invasive surgery or radiosurgery without tissue damage. When more extensive surgery cannot be avoided, they prefer it to be without undue risk. Patients rightly place a premium on minimizing morbidity, which means no damage to the surrounding brain, cranial nerves, or blood vessels and no cosmetic deformity. Regardless of the approach, they want to minimize time away from work and family and to be treated at a reasonable cost [126].

8. Conclusions—Epilogue: Gazing into the Future Horizon

8.1. Challenges, Opportunities, and the Trajectory of Cranial Base Surgery in the Coming Decade

The adoption of 3D printing technologies is on the rise across various sectors, including neurosurgery. Current applications in this field encompass the fabrication of cranioplasty implants, educational models for tumors and aneurysms, as well as surgical planning aids [127]. Further innovation comes from the Northwestern University School of Engineering, where researchers have developed 3D-printed, patient-specific bioresorbable intravascular stents. Notably, a proof-of-concept for a 3D-printed bionic ear has been developed, featuring advanced auditory sensing capabilities for radiofrequency signals [128,129].

In the realm of cranial surgery, various robotic technologies are making headway. The NeuroArm, developed at the University of Calgary, is a remote-controlled surgical robot designed for use in an MRI suite [130]. Meanwhile, the Shinshu University NeuRobot—a joint effort involving multiple research institutions—consists of a master–slave micromanipulator system equipped with a rigid endoscope and three robotic arms, designed for minimally invasive procedures. This system has already been successfully employed in surgeries and shows potential for remote telesurgical applications, albeit with a minuscule 1 ms delay [131].

Currently, a team at the University of Washington is developing an Artificially Intelligent Neurosurgical Robotic Assistant. This autonomous robot aims to replicate the functions of a microneurosurgical assistant, such as gentle tissue manipulation and precise suction within the surgical field.

The development of an Artificially Intelligent Neurosurgical Robotic Assistant aims to create an intuitive system that can act according to the surgeon's needs, either through innate understanding or voice commands. One of the major challenges lies in understanding the nuanced interaction between the surgeon and their assistant during surgery. To this end, the team has employed convolutional neural networks to analyze the surgeon's voice and tool movements captured under a microscope [132]. Python speech application program interfaces are also used for more detailed analysis.

Instrument identification and tracking in the surgical field are performed at the pixel level, offering insights into the surgeon's intended direction of movement [133]. An integral part of the project involves adapting natural language parsing to recognize specific medical terms, making the interface between the robotic assistant and the surgeon more intuitive and efficient.

In a novel application, the team recently showcased the fusion of semi-autonomous robotic therapy with a specialized biomarker known as "tumor paint", derived from a component of scorpion toxin [134]. This biomarker specifically labels brain tumors. In studies led by Hu et al., a robotic system scanned a simulated tumor margin for spots marked as positive for tumor cells. Following the surgeon's approval, the robot then executed an automated ablation pathway to remove these areas [135].

In summary, the convergence of robotics, machine learning, and specialized biomarkers is pushing the boundaries of what is possible in neurosurgical interventions. The fusion of these technologies could revolutionize how surgical procedures are planned and executed, with the promise of more precise and potentially less invasive treatments (Table 2).

Table 2. Future Advances in Various Fields in Skull Base Surgery.

Virtual Raman Microscopy and Spectroscopy for Quick Diagnosis in the Operating Room
Stem cell-based therapies for brain and cranial nerve damage from trauma, tumors, and medical procedures
Semi-autonomous robots for use in the operating room
Regenerative medicine combined with 3D printing for creating blood vessels, bone, and facial tissues
Quick molecular and genetic assessment of tumors
New training procedures for surgeons
Nanoengineering for diagnostic and therapeutic applications
Mobile imaging in the operating room and in the intensive care unit
CRISPR CAS-9 based genetic techniques to eliminate hereditary syndromes
Anti-cancer antibodies, CAR-T-cell therapy, and immune checkpoint blockade against the neoplasms
AI applications for powered disease diagnosis in hospital and outpatient care
Advanced imaging techniques (MRI and ultrasound)
Additive manufacturing (3D printing and rapid prototyping)

8.2. Potential Breakthroughs: Stem Cell Research, Regenerative Medicine, and Genomic Tailoring

Stem cell recovery techniques are poised to play a transformative role in the treatment of surgically induced and other neurological deficits within the next two decades. These advancements could especially benefit patients requiring surgery for conditions such as vestibular schwannoma, promising better recovery of cranial nerve 7 and 8 function even for those with large or giant tumors. This promising approach could also extend to iatrogenic neurological deficits that may arise after surgeries on the brain or brainstem for tumor or vascular operations. Furthermore, understanding how stem cells interact with tumors may pave the way for the prevention and potential cure of some skull base malignant neoplasms [126].

In a notable development, a 12-month phase II, randomized controlled trial conducted in the US and Japanese centers showed that SB623 stem cells were particularly effective for patients with traumatic brain injury. These cells were implanted around the injury site, leading to significant improvements in motor function as measured by the Fugl-Meyer motor scale. The primary endpoint was reached, with an average improvement of 8.3 points as opposed to an improvement of 2.3 in the control group at 24 weeks ($p = 0.040$). This promising result has led SanBio Co., Ltd. (Tokyo, Japan), to plan further studies in phase III clinical trials, as per a personal communication from Steinberg GK in 2020 and a press release from SanBio Co., Ltd., in 2019 [136].

Like any groundbreaking medical advancement, the journey of integrating stem cells into clinical practice will require significant financial investment and time. It is also important to brace for setbacks and challenges along the way, much like the development pathways for new drugs or vaccines. Nevertheless, the prospects are exciting and could herald a new era in the treatment of neurological conditions and deficits.

8.3. Reiterating the Ethos of Continuous Learning, Global Collaboration, and Patient-First Principles

The future of skull base surgery and neurosurgery will undeniably be influenced by rapid technological advancements. While surgeons will need to be agile in integrating these new technologies into their practice, the core tenets that define a great surgeon—knowledge, innovation, technical skill, judgment, and compassion—will stand the test of time. Active engagement with emerging technologies is not just an option but a necessity, as it allows surgeons to have a direct hand in shaping the future of their field [126].

Innovation will be a linchpin in the evolution of medical practice, both now and in the future. These innovations might be subtle, influencing the minutiae of day-to-day work, or they could be groundbreaking, transforming clinical surgery, basic neurosciences, or various aspects of healthcare delivery. They might also aim at improving workflow and efficiency, revamping outpatient and hospital infrastructure, elevating patient satisfaction, or enhancing quality metrics [126].

Young surgeons carry the mantle of responsibility to not only excel in their craft but also to contribute to its progression. They must constantly aspire to leave their field better than how they found it, pushing the boundaries of what is possible and effective in medical care. Furthermore, surgeons should not shy away from roles in hospital and healthcare administration. Such involvement provides them with the opportunity to guide transformative changes, ensuring that innovation and quality improvement are not just theoretical ideals but real-world practices that enhance patient care and outcomes.

Author Contributions: Conceptualization, C.T. and C.-I.T.; methodology, C.T., R.-A.C.-B. and L.-A.G.; software, R.-A.C.-B. and H.P.C.; validation, B.-G.B.; formal analysis, H.P.C.; investigation, C.-I.T.; resources, D.-I.D.; data curation, B.-G.B. and L.-A.G.; writing—original draft preparation, R.-A.C.-B.; writing—review and editing, C.-I.T.; visualization, L.E.; supervision, L.E.; project administration, A.V.C.; funding acquisition, A.V.C. All authors have read and agreed to the published version of the manuscript.

Funding: This research received no external funding.

Institutional Review Board Statement: Not applicable.

Informed Consent Statement: Not applicable.

Data Availability Statement: All data is available online on libraries such as PubMed.

Conflicts of Interest: The authors declare no conflict of interest.

Abbreviations

ASB	Anterior Skull Base
ACFR	Anterior craniofacial resection
CSF	Cerebrospinal fluid
GSPN	Greater superficial petrosal nerve
MCF	Middle cerebral fossa
IAC	Internal auditory canal
FESS	Functional Endoscopic Sinus Surgery
ACTH	Adrenocorticotropic hormone
NFPAs	Non-functioning pituitary macroadenomas
TSH	Thyroid-stimulating hormone

IONM	Intraoperative Neuromonitoring
iIONM	Intermittent IONM
BAEPs	Brainstem Auditory Evoked Potentials
MEPs	Motor Evoked Potentials
EMG	Electromyography
TceMEP	Transcranial Electric Motor Evoked Potentials
CMAP	Compound Motor Action Potentials
HB grade	House–Brackmann
TES	Transcranial Electric Stimulation
FNMEP	Facial nerve muscle motor evoked potentials
acIONM	Intraoperative Neuromonitoring
pcIONM	Passive continuous IONM
EMG	Electromyography
SSEPs	Somatosensory evoked potentials
MEPs	Motor evoked potentials
PPNI	Perioperative peripheral nerve injury
VR	Virtual reality
AR	Augmented reality
LASSO	Least Absolute Shrinkage and Selection Operator
LDA	Linear Discriminant Analysis
ML	Machine learning
CSFR	Cerebrospinal Fluid Rhinorrhoea
NNs	Neural network models
DC	Dice Coefficient
HD	Hausdorff Distance
OAR	Organs-at-risk
FSRT	Fractionated stereotactic radiotherapy
GKS	Gamma knife surgery
FGKS	Fractionated GKS
EBRT	External beam radiation therapy
TS	Total survival
RFS	Recurrence-free survival
EBRT	External beam radiation therapy
LRR	Local recurrence rate
PTR	Partial resection
TR	Total resection
FRT	Fractionated radiotherapy
SRS	Stereotactic therapy (radiosurgery)
GKR	Gamma Knife radiosurgery
RT	Radiation therapy
SRS	Stereotactic radiosurgery
STR	Subtotal resection
GTR	Gross total resection
CGE	Cobalt-Gray-Equivalent
LC	Local control
OS	Overall survival
EBRT	External beam radiation therapy
SRS	Stereotactic radiosurgery
3D-CRT	Three-dimensional conformal RT
IMRT	Intensity-modulated radiation therapy
SBMs	Skull base meningiomas

References

1. Donald, P.J. History of Skull Base Surgery. *Skull Base* **1991**, *1*, 1–3. [CrossRef]
2. Husain, Q.; Patel, S.K.; Soni, R.S.; Patel, A.A.; Liu, J.K.; Eloy, J.A. Celebrating the golden anniversary of anterior skull base surgery: Reflections on the Past 50 Years and Its Historical Evolution. *Laryngoscope* **2013**, *123*, 64–72. [CrossRef] [PubMed]
3. McRackan, T.R.; Brackmann, D.E. Historical Perspective on Evolution in Management of Lateral Skull Base Tumors. *Otolaryngol. Clin. N. Am.* **2015**, *48*, 397–405. [CrossRef] [PubMed]

4. Nakayama, K.; Tamiya, T.; Yamamoto, K.; Akimoto, S. A simple new apparatus for small vessel anastomosisi (free autograft of the sigmoid included). *Surgery* **1962**, *52*, 918–931.
5. Mazzoni, A.; Krengli, M. Historical development of the treatment of skull base tumours. *Rep. Pract. Oncol. Radiother.* **2016**, *21*, 319–324. [CrossRef]
6. Patel, C.R.; Fernandez-Miranda, J.C.; Wang, W.-H.; Wang, E.W. Skull Base Anatomy. *Otolaryngol. Clin. N. Am.* **2016**, *49*, 9–20. [CrossRef]
7. Hitselberger, W.E.; Horn, K.L.; Hankinson, H.; Brackmann, D.E.; House, W.F. The Middle Fossa Transpetrous Approach for Petroclival Meningiomas. *Skull Base* **1993**, *3*, 130–135. [CrossRef]
8. Glasscock, M.E. Middle Fossa Approach to the Temporal Bone: An Otologic Frontier. *Arch. Otolaryngol. Head Neck Surg.* **1969**, *90*, 15–27. [CrossRef]
9. Kawase, T.; Toya, S.; Shiobara, R.; Mine, T. Transpetrosal approach for aneurysms of the lower basilar artery. *J. Neurosurg.* **1985**, *63*, 857–861. [CrossRef]
10. Shibao, S.; Borghei-Razavi, H.; Orii, M.; Yoshida, K. Anterior Transpetrosal Approach Combined with Partial Posterior Petrosectomy for Petroclival Meningiomas with Posterior Extension. *World Neurosurg.* **2015**, *84*, 574–579. [CrossRef]
11. Abbott, R. History of neuroendoscopy. *Neurosurg. Clin. N. Am.* **2004**, *15*, 1–7. [CrossRef] [PubMed]
12. McCutcheon, I.E.; Blacklock, J.B.; Weber, R.S.; DeMonte, F.; Moser, R.P.; Byers, M.; Goepfert, H. Anterior Transcranial (Craniofacial) Resection of Tumors of the Paranasal Sinuses: Surgical Technique and Results. *Neurosurgery* **1996**, *38*, 471–480. [CrossRef] [PubMed]
13. Abuzayed, B.; Tanriover, N.; Gazioglu, N.; Eraslan, B.S.; Akar, Z. Endoscopic Endonasal Approach to the Orbital Apex and Medial Orbital Wall: Anatomic Study and Clinical Applications. *J. Craniofac. Surg.* **2009**, *20*, 1594–1600. [CrossRef] [PubMed]
14. Carrau, R.L.; Jho, H.-D.; Ko, Y. Transnasal-Transsphenoidal Endoscopic Surgery of the Pituitary Gland. *Laryngoscope* **1996**, *106*, 914–918. [CrossRef]
15. Messerer, M.; De Battista, J.C.; Raverot, G.; Kassis, S.; Dubourg, J.; Lapras, V.; Trouillas, J.; Perrin, G.; Jouanneau, E. Evidence of improved surgical outcome following endoscopy for nonfunctioning pituitary adenoma removal: Personal experience and review of the literature. *Neurosurg. Focus* **2011**, *30*, E11. [CrossRef]
16. Mavromati, M.; Mavrakanas, T.; Jornayvaz, F.R.; Schaller, K.; Fitsiori, A.; Vargas, M.I.; Lobrinus, J.A.; Merkler, D.; Egervari, K.; Philippe, J.; et al. The impact of transsphenoidal surgery on pituitary function in patients with non-functioning macroadenomas. *Endocrine* **2023**, *81*, 340–348. [CrossRef]
17. Tabaee, A.; Anand, V.K.; Barrón, Y.; Hiltzik, D.H.; Brown, S.M.; Kacker, A.; Mazumdar, M.; Schwartz, T.H. Endoscopic pituitary surgery: A systematic review and meta-analysis: Clinical article. *J. Neurosurg.* **2009**, *111*, 545–554. [CrossRef]
18. Dorward, N.L. Endocrine outcomes in Endoscopic Pituitary Surgery: A Literature Review. *Acta Neurochir.* **2010**, *152*, 1275–1279. [CrossRef]
19. Ceylan, S.; Koc, K.; Anik, I. Endoscopic endonasal transsphenoidal approach for pituitary adenomas invading the cavernous sinus: Clinical article. *J. Neurosurg.* **2010**, *112*, 99–107. [CrossRef]
20. Zada, G.; Kelly, D.F.; Cohan, P.; Wang, C.; Swerdloff, R. Endonasal transsphenoidal approach to treat pituitary adenomas and other sellar lesions: An assessment of efficacy, safety, and patient impressions of the surgery. *J. Neurosurg.* **2003**, *98*, 350–358. [CrossRef]
21. Griessenauer, C.J.; Mortazavi, M.M.; Loukas, M.; Shoja, M.M.; Watanabe, K.; Tubbs, R.S. Heinrich Bircher (1850–1923) and the first description of a surgical approach to the cavernous sinus. *Childs Nerv. Syst.* **2013**, *29*, 1923–1925. [CrossRef] [PubMed]
22. Chowdhury, F.; Haque, M.; Kawsar, K.; Ara, S.; Mohammod, Q.; Sarker, M.; Goel, A. Transcranial Microsurgical and Endoscopic Endonasal Cavernous Sinus (CS) Anatomy: A Cadaveric Study. *J. Neurol. Surg. Part Cent. Eur. Neurosurg.* **2012**, *73*, 296–306. [CrossRef] [PubMed]
23. Weber, S.M.; Kim, J.H.; Wax, M.K. Role of free tissue transfer in skull base reconstruction. *Otolaryngol. Neck Surg.* **2007**, *136*, 914–919. [CrossRef] [PubMed]
24. Cabanela, M.E.; Coventry, M.B.; MacCarty, C.S.; Miller, W.E. The fate of patients with methyl methacrylate cranioplasty. *J. Bone Jt. Surg. Am.* **1972**, *54*, 278–281. [CrossRef] [PubMed]
25. Bakamjian, V.Y.; Leonard, A.G. Bone Dust Cranioplasty: Case Report. *Plast. Reconstr. Surg.* **1977**, *60*, 784–788. [CrossRef]
26. Kwon, D.; Iloreta, A.; Miles, B.; Inman, J. Open Anterior Skull Base Reconstruction: A Contemporary Review. *Semin. Plast. Surg.* **2017**, *31*, 189–196. [CrossRef]
27. Germani, R.M.; Vivero, R.; Herzallah, I.R.; Casiano, R.R. Endoscopic Reconstruction of Large Anterior Skull Base Defects using Acellular Dermal Allograft. *Am. J. Rhinol.* **2007**, *21*, 615–618. [CrossRef]
28. Sheikh, B.Y. Simple and safe method of cranial reconstruction after posterior fossa craniectomy. *Surg. Neurol.* **2006**, *65*, 63–66. [CrossRef]
29. Ishii, Y.; Tahara, S.; Teramoto, A.; Morita, A. Endoscopic Endonasal Skull Base Surgery: Advantages, Limitations, and Our Techniques to Overcome Cerebrospinal Fluid Leakage: Technical Note. *Neurol. Med. Chir.* **2014**, *54*, 983–990. [CrossRef]
30. Cappabianca, P.; Esposito, F.; Cavallo, L.M.; Messina, A.; Solari, D.; Di Somma, L.G.M.; De Divitiis, E. Use of equine collagen foil as dura mater substitute in endoscopic endonasal transsphenoidal surgery. *Surg. Neurol.* **2006**, *65*, 144–148. [CrossRef]

31. Verret, D.J.; Ducic, Y.; Oxford, L.; Smith, J. Hydroxyapatite Cement in Craniofacial Reconstruction. *Otolaryngol. Neck Surg.* **2005**, *133*, 897–899. [CrossRef] [PubMed]
32. Zanation, A.M.; Carrau, R.L.; Snyderman, C.H.; Germanwala, A.V.; Gardner, P.A.; Prevedello, D.M.; Kassam, A.B. Nasoseptal Flap Reconstruction of High Flow Intraoperative Cerebral Spinal Fluid Leaks during Endoscopic Skull Base Surgery. *Am. J. Rhinol. Allergy* **2009**, *23*, 518–521. [CrossRef] [PubMed]
33. Neligan, P.C.; Mulholland, S.; Irish, J.; Gullane, P.J.; Boyd, J.B.; Gentili, F.; Brown, D.; Freeman, J. Flap Selection in Cranial Base Reconstruction. *Plast. Reconstr. Surg.* **1996**, *98*, 1159–1166. [CrossRef]
34. Low, T.-H.H.; Lindsay, A.; Clark, J.; Chai, F.; Lewis, R. Reconstruction of maxillary defect with musculo-adipose rectus free flap: Musculo-Adipose Rectus Free Flap. *Microsurgery* **2017**, *37*, 137–141. [CrossRef]
35. Hanasono, M.M.; Sacks, J.M.; Goel, N.; Ayad, M.; Skoracki, R.J. The anterolateral thigh free flap for skull base reconstruction. *Otolaryngol. Neck Surg.* **2009**, *140*, 855–860. [CrossRef] [PubMed]
36. Flisberg, K.; Lindholm, T. Electrical stimulation of the human recurrent laryngeal nerve during thyroid operation. *Acta Otolaryngol.* **1970**, *69*, 63–67. [CrossRef] [PubMed]
37. Stankovic, P.; Wittlinger, J.; Georgiew, R.; Dominas, N.; Hoch, S.; Wilhelm, T. Continuous intraoperative neuromonitoring (cIONM) in head and neck surgery—A review. *HNO* **2020**, *68*, 86–92. [CrossRef]
38. Legatt, A.D. Mechanisms of Intraoperative Brainstem Auditory Evoked Potential Changes. *J. Clin. Neurophysiol.* **2002**, *19*, 396–408. [CrossRef]
39. Colletti, V.; Fiorino, F.G.; Mocella, S.; Policante, Z. ECochG, CNAP and ABR Monitoring during Vestibular Schwannoma Surgery. *Int. J. Audiol.* **1998**, *37*, 27–37. [CrossRef]
40. Kartush, J.M.; LaRouere, M.J.; Graham, M.D.; Bouchard, K.R.; Audet, B.V. Intraoperative Cranial Nerve Monitoring During Posterior Skull Base Surgery. *Skull Base* **1991**, *1*, 85–92. [CrossRef]
41. Petrova, L.D. Brainstem auditory evoked potentials. *Am. J. Electroneurodiagnostic Technol.* **2009**, *49*, 317–332. [CrossRef] [PubMed]
42. Simon, M.V. Neurophysiologic Intraoperative Monitoring of the Vestibulocochlear Nerve. *J. Clin. Neurophysiol.* **2011**, *28*, 566–581. [CrossRef] [PubMed]
43. Lüders, H. Surgical Monitoring with Auditory Evoked Potentials. *J. Clin. Neurophysiol.* **1988**, *5*, 261–286. [CrossRef] [PubMed]
44. Schramm, J.; Mokrusch, T.; Fahlbusch, R.; Hochstetter, A. Detailed Analysis of Intraoperative Changes Monitoring Brain Stem Acoustic Evoked Potentials. *Neurosurgery* **1988**, *22*, 694–702. [CrossRef] [PubMed]
45. Lall, R.R.; Lall, R.R.; Hauptman, J.S.; Munoz, C.; Cybulski, G.R.; Koski, T.; Ganju, A.; Fessler, R.G.; Smith, Z.A. Intraoperative neurophysiological monitoring in spine surgery: Indications, efficacy, and role of the preoperative checklist. *Neurosurg. Focus* **2012**, *33*, E10. [CrossRef] [PubMed]
46. Park, J.-H. Intraoperative neurophysiological monitoring in spinal surgery. *World J. Clin. Cases* **2015**, *3*, 765. [CrossRef] [PubMed]
47. Biscevic, M.; Sehic, A.; Biscevic, S.; Gavrankapetanovic, I.; Smrke, B.; Vukomanovic, D.; Krupic, F. Kyphosis—A risk factor for positioning brachial plexopathy during spinal surgeries. *Acta Orthop. Traumatol. Turc.* **2019**, *53*, 199–202. [CrossRef]
48. Amano, M.; Kohno, M.; Nagata, O.; Taniguchi, M.; Sora, S.; Sato, H. Intraoperative continuous monitoring of evoked facial nerve electromyograms in acoustic neuroma surgery. *Acta Neurochir.* **2011**, *153*, 1059–1067. [CrossRef]
49. Dong, C.C.J.; MacDonald, D.B.; Akagami, R.; Westerberg, B.; AlKhani, A.; Kanaan, I.; Hassounah, M. Intraoperative facial motor evoked potential monitoring with transcranial electrical stimulation during skull base surgery. *Clin. Neurophysiol.* **2005**, *116*, 588–596. [CrossRef]
50. Prass, R.L.; Lüders, H. Acoustic (loudspeaker) facial electromyographic monitoring: Part 1. Evoked electromyographic activity during acoustic neuroma resection. *Neurosurgery* **1986**, *19*, 392–400. [CrossRef]
51. Danner, C.; Mastrodimos, B.; Cueva, R.A. A Comparison of Direct Eighth Nerve Monitoring and Auditory Brainstem Response in Hearing Preservation Surgery for Vestibular Schwannoma. *Otol. Neurotol.* **2004**, *25*, 826–832. [CrossRef] [PubMed]
52. Piccirillo, E.; Hiraumi, H.; Hamada, M.; Russo, A.; De Stefano, A.; Sanna, M. Intraoperative Cochlear Nerve Monitoring in Vestibular Schwannoma Surgery—Does It Really Affect Hearing Outcome? *Audiol. Neurotol.* **2008**, *13*, 58–64. [CrossRef] [PubMed]
53. Briggs, R.J.S.; Luxford, W.M.; Atkins, J.S.; Hitselberger, W.E. Translabyrinthine Removal of Large Acoustic Neuromas. *Neurosurgery* **1994**, *34*, 785–791. [CrossRef] [PubMed]
54. Sloan, T.B.; Heyer, E.J. Anesthesia for Intraoperative Neurophysiologic Monitoring of the Spinal Cord. *J. Clin. Neurophysiol.* **2002**, *19*, 430–443. [CrossRef]
55. Biscevic, M.; Sehic, A.; Krupic, F. Intraoperative neuromonitoring in spine deformity surgery: Modalities, advantages, limitations, medicolegal issues—Surgeons' views. *EFORT Open Rev.* **2020**, *5*, 9–16. [CrossRef]
56. Alaraj, A.; Lemole, M.; Finkle, J.; Yudkowsky, R.; Wallace, A.; Luciano, C.; Banerjee, P.; Rizzi, S.; Charbel, F. Virtual reality training in neurosurgery: Review of current status and future applications. *Surg. Neurol. Int.* **2011**, *2*, 52. [CrossRef]
57. Rosahl, S.; Gharabaghi, A.; Hubbe, U.; Shahidi, R.; Samii, M. Virtual Reality Augmentation in Skull Base Surgery. *Skull Base* **2006**, *16*, 059–066. [CrossRef]

58. Liu, W.P.; Azizian, M.; Sorger, J.; Taylor, R.H.; Reilly, B.K.; Cleary, K.; Preciado, D. Cadaveric Feasibility Study of da Vinci Si–Assisted Cochlear Implant With Augmented Visual Navigation for Otologic Surgery. *JAMA Otolaryngol. Neck Surg.* **2014**, *140*, 208. [CrossRef]
59. Citardi, M.J.; Agbetoba, A.; Bigcas, J.-L.; Luong, A. Augmented reality for endoscopic sinus surgery with surgical navigation: A cadaver study: Augmented reality for endoscopic sinus surgery. *Int. Forum Allergy Rhinol.* **2016**, *6*, 523–528. [CrossRef]
60. Marmulla, R.; Hoppe, H.; Mühling, J.; Hassfeld, S. New Augmented Reality Concepts for Craniofacial Surgical Procedures. *Plast. Reconstr. Surg.* **2005**, *115*, 1124–1128. [CrossRef]
61. Kawamata, T.; Iseki, H.; Shibasaki, T.; Hori, T. Endoscopic Augmented Reality Navigation System for Endonasal Transsphenoidal Surgery to Treat Pituitary Tumors: Technical Note. *Neurosurgery* **2002**, *50*, 1393–1397. [CrossRef] [PubMed]
62. Caversaccio, M.; Langlotz, F.; Nolte, L.-P.; Häusler, R. Impact of a self-developed planning and self-constructed navigation system on skull base surgery: 10 years experience. *Acta Otolaryngol.* **2007**, *127*, 403–407. [CrossRef] [PubMed]
63. Bong, J.H.; Song, H.; Oh, Y.; Park, N.; Kim, H.; Park, S. Endoscopic navigation system with extended field of view using augmented reality technology. *Int. J. Med. Robot.* **2018**, *14*, e1886. [CrossRef] [PubMed]
64. Kalaiarasan, K.; Prathap, L.; Ayyadurai, M.; Subhashini, P.; Tamilselvi, T.; Avudaiappan, T.; Infant Raj, I.; Alemayehu Mamo, S.; Mezni, A. Clinical Application of Augmented Reality in Computerized Skull Base Surgery. *Evid. Based Complement. Alternat. Med.* **2022**, *2022*, 1335820. [CrossRef]
65. Grøvik, E.; Yi, D.; Iv, M.; Tong, E.; Rubin, D.; Zaharchuk, G. Deep learning enables automatic detection and segmentation of brain metastases on multisequence MRI. *J. Magn. Reson. Imaging* **2020**, *51*, 175–182. [CrossRef]
66. Laukamp, K.R.; Thiele, F.; Shakirin, G.; Zopfs, D.; Faymonville, A.; Timmer, M.; Maintz, D.; Perkuhn, M.; Borggrefe, J. Fully automated detection and segmentation of meningiomas using deep learning on routine multiparametric MRI. *Eur. Radiol.* **2019**, *29*, 124–132. [CrossRef]
67. Lu, C.-F.; Hsu, F.-T.; Hsieh, K.L.-C.; Kao, Y.-C.J.; Cheng, S.-J.; Hsu, J.B.-K.; Tsai, P.-H.; Chen, R.-J.; Huang, C.-C.; Yen, Y.; et al. Machine Learning–Based Radiomics for Molecular Subtyping of Gliomas. *Clin. Cancer Res.* **2018**, *24*, 4429–4436. [CrossRef]
68. Varghese, B.A.; Cen, S.Y.; Hwang, D.H.; Duddalwar, V.A. Texture Analysis of Imaging: What Radiologists Need to Know. *Am. J. Roentgenol.* **2019**, *212*, 520–528. [CrossRef]
69. Gillies, R.J.; Kinahan, P.E.; Hricak, H. Radiomics: Images Are More than Pictures, They Are Data. *Radiology* **2016**, *278*, 563–577. [CrossRef]
70. Bi, W.L.; Hosny, A.; Schabath, M.B.; Giger, M.L.; Birkbak, N.J.; Mehrtash, A.; Allison, T.; Arnaout, O.; Abbosh, C.; Dunn, I.F.; et al. Artificial intelligence in cancer imaging: Clinical challenges and applications. *CA. Cancer J. Clin.* **2019**, *69*, 127–157. [CrossRef]
71. Kickingereder, P.; Burth, S.; Wick, A.; Götz, M.; Eidel, O.; Schlemmer, H.-P.; Maier-Hein, K.H.; Wick, W.; Bendszus, M.; Radbruch, A.; et al. Radiomic Profiling of Glioblastoma: Identifying an Imaging Predictor of Patient Survival with Improved Performance over Established Clinical and Radiologic Risk Models. *Radiology* **2016**, *280*, 880–889. [CrossRef]
72. Kniep, H.C.; Madesta, F.; Schneider, T.; Hanning, U.; Schönfeld, M.H.; Schön, G.; Fiehler, J.; Gauer, T.; Werner, R.; Gellissen, S. Radiomics of Brain MRI: Utility in Prediction of Metastatic Tumor Type. *Radiology* **2019**, *290*, 479–487. [CrossRef] [PubMed]
73. Zhang, B.; Tian, J.; Dong, D.; Gu, D.; Dong, Y.; Zhang, L.; Lian, Z.; Liu, J.; Luo, X.; Pei, S.; et al. Radiomics Features of Multiparametric MRI as Novel Prognostic Factors in Advanced Nasopharyngeal Carcinoma. *Clin. Cancer Res.* **2017**, *23*, 4259–4269. [CrossRef] [PubMed]
74. Wu, S.; Zheng, J.; Li, Y.; Yu, H.; Shi, S.; Xie, W.; Liu, H.; Su, Y.; Huang, J.; Lin, T. A Radiomics Nomogram for the Preoperative Prediction of Lymph Node Metastasis in Bladder Cancer. *Clin. Cancer Res.* **2017**, *23*, 6904–6911. [CrossRef] [PubMed]
75. Ortega-Martorell, S.; Olier, I.; Julià-Sapé, M.; Arús, C. SpectraClassifier 1.0: A user friendly, automated MRS-based classifier-development system. *BMC Bioinform.* **2010**, *11*, 106. [CrossRef] [PubMed]
76. Buizza, G.; Paganelli, C.; D'Ippolito, E.; Fontana, G.; Molinelli, S.; Preda, L.; Riva, G.; Iannalfi, A.; Valvo, F.; Orlandi, E.; et al. Radiomics and Dosiomics for Predicting Local Control after Carbon-Ion Radiotherapy in Skull-Base Chordoma. *Cancers* **2021**, *13*, 339. [CrossRef] [PubMed]
77. Chen, B.; Chen, C.; Zhang, Y.; Huang, Z.; Wang, H.; Li, R.; Xu, J. Differentiation between Germinoma and Craniopharyngioma Using Radiomics-Based Machine Learning. *J. Pers. Med.* **2022**, *12*, 45. [CrossRef]
78. CRANIAL Consortium Machine learning driven prediction of cerebrospinal fluid rhinorrhoea following endonasal skull base surgery: A multicentre prospective observational study. *Front. Oncol.* **2023**, *13*, 1046519. [CrossRef]
79. Neves, C.A.; Tran, E.D.; Blevins, N.H.; Hwang, P.H. Deep learning automated segmentation of middle skull-base structures for enhanced navigation. *Int. Forum Allergy Rhinol.* **2021**, *11*, 1694–1697. [CrossRef]
80. Guss, Z.D.; Batra, S.; Limb, C.J.; Li, G.; Sughrue, M.E.; Redmond, K.; Rigamonti, D.; Parsa, A.T.; Chang, S.; Kleinberg, L.; et al. Radiosurgery of Glomus Jugulare Tumors: A Meta-Analysis. *Int. J. Radiat. Oncol.* **2011**, *81*, e497–e502. [CrossRef]
81. Eustacchio, S.; Trummer, M.; Fuchs, I.; Schröttner, O.; Sutter, B.; Pendl, G. Preservation of Cranial Nerve Function Following Gamma Knife Radiosurgery for Benign Skull Base Meningiomas: Experience in 121 Patients with Follow-up of 5 to 9.8 Years. In *Advances in Epilepsy Surgery and Radiosurgery*; Sutter, B., Schröttner, O., Eds.; Springer: Vienna, Austria, 2002; pp. 71–76, ISBN 978-3-7091-7296-4.
82. Shin, M.; Kurita, H.; Sasaki, T.; Kawamoto, S.; Tago, M.; Kawahara, N.; Morita, A.; Ueki, K.; Kirino, T. Analysis of treatment outcome after stereotactic radiosurgery for cavernous sinus meningiomas. *J. Neurosurg.* **2001**, *95*, 435–439. [CrossRef] [PubMed]

83. Lee, J.Y.K.; Niranjan, A.; McInerney, J.; Kondziolka, D.; Flickinger, J.C.; Lunsford, L.D. Stereotactic radiosurgery providing long-term tumor control of cavernous sinus meningiomas. *J. Neurosurg.* **2002**, *97*, 65–72. [CrossRef] [PubMed]
84. Zachenhofer, I.; Wolfsberger, S.; Aichholzer, M.; Bertalanffy, A.; Roessler, K.; Kitz, K.; Knosp, E. Gamma-Knife Radiosurgery for Cranial Base Meningiomas: Experience of Tumor Control, Clinical Course, and Morbidity in a Follow-Up of More than 8 Years. *Neurosurgery* **2006**, *58*, 28–36. [CrossRef]
85. Jääskeläinen, J.; Haltia, M.; Servo, A. Atypical and anaplastic meningiomas: Radiology, surgery, radiotherapy, and outcome. *Surg. Neurol.* **1986**, *25*, 233–242. [CrossRef]
86. Jääskeläinen, J.; Haltia, M.; Laasonen, E.; Wahlström, T.; Valtonen, S. The growth rate of intracranial meningiomas and its relation to histology. An analysis of 43 patients. *Surg. Neurol.* **1985**, *24*, 165–172. [CrossRef]
87. Leber, K.A.; Berglöff, J.; Pendl, G. Dose—Response tolerance of the visual pathways and cranial nerves of the cavernous sinus to stereotactic radiosurgery. *J. Neurosurg.* **1998**, *88*, 43–50. [CrossRef]
88. Morita, A.; Coffey, R.J.; Foote, R.L.; Schiff, D.; Gorman, D. Risk of injury to cranial nerves after gamma knife radiosurgery for skull base meningiomas: Experience in 88 patients. *J. Neurosurg.* **1999**, *90*, 42–49. [CrossRef]
89. Stafford, S.L.; Pollock, B.E.; Leavitt, J.A.; Foote, R.L.; Brown, P.D.; Link, M.J.; Gorman, D.A.; Schomberg, P.J. A study on the radiation tolerance of the optic nerves and chiasm after stereotactic radiosurgery. *Int. J. Radiat. Oncol.* **2003**, *55*, 1177–1181. [CrossRef]
90. Chang, J.H.; Chang, J.W.; Choi, J.Y.; Park, Y.G.; Chung, S.S. Complications after gamma knife radiosurgery for benign meningiomas. *J. Neurol. Neurosurg. Psychiatry* **2003**, *74*, 226–230. [CrossRef]
91. Starke, R.M.; Przybylowski, C.J.; Sugoto, M.; Fezeu, F.; Awad, A.J.; Ding, D.; Nguyen, J.H.; Sheehan, J.P. Gamma Knife radiosurgery of large skull base meningiomas. *J. Neurosurg.* **2015**, *122*, 363–372. [CrossRef]
92. Iwai, Y.; Yamanaka, K.; Shimohonji, W.; Ishibashi, K. Staged Gamma Knife Radiosurgery for Large Skull Base Meningiomas. *Cureus* **2019**, *11*, e6001. [CrossRef] [PubMed]
93. Adler, J.R., Jr.; Chang, S.D.; Murphy, M.J.; Doty, J.; Geis, P.; Hancock, S.L. The Cyberknife: A Frameless Robotic System for Radiosurgery. *Stereotact. Funct. Neurosurg.* **1997**, *69*, 124–128. [CrossRef] [PubMed]
94. Gwak, H.-S.; Yoo, H.-J.; Youn, S.-M.; Chang, U.; Lee, D.H.; Yoo, S.-Y.; Rhee, C.H. Hypofractionated Stereotactic Radiation Therapy for Skull Base and Upper Cervical Chordoma and Chondrosarcoma: Preliminary Results. *Stereotact. Funct. Neurosurg.* **2005**, *83*, 233–243. [CrossRef] [PubMed]
95. Wilson, H.P.; Price, P.M.; Ashkan, K.; Edwards, A.; Green, M.M.; Cross, T.; Beaney, R.P.; Davies, R.; Sibtain, A.; Plowman, N.P.; et al. CyberKnife Radiosurgery of Skull-base Tumors: A UK Center Experience. *Cureus* **2018**, *10*, e2380. [CrossRef]
96. Debus, J.; Wuendrich, M.; Pirzkall, A.; Hoess, A.; Schlegel, W.; Zuna, I.; Engenhart-Cabillic, R.; Wannenmacher, M. High Efficacy of Fractionated Stereotactic Radiotherapy of Large Base-of-Skull Meningiomas: Long-Term Results. *J. Clin. Oncol.* **2001**, *19*, 3547–3553. [CrossRef]
97. Han, M.-S.; Jang, W.-Y.; Moon, K.-S.; Lim, S.-H.; Kim, I.-Y.; Jung, T.-Y.; Jung, S. Is Fractionated Gamma Knife Radiosurgery a Safe and Effective Treatment Approach for Large-Volume (>10 cm^3) Intracranial Meningiomas? *World Neurosurg.* **2017**, *99*, 477–483. [CrossRef]
98. Uy, N.W.; Woo, S.Y.; Teh, B.S.; Mai, W.-Y.; Carpenter, L.S.; Chiu, J.K.; Lu, H.H.; Gildenberg, P.; Trask, T.; Grant, W.H.; et al. Intensity-modulated radiation therapy (IMRT) for meningioma. *Int. J. Radiat. Oncol.* **2002**, *53*, 1265–1270. [CrossRef]
99. Miralbell, R.; Linggood, R.M.; De La Monte, S.; Convery, K.; Munzenrider, J.E.; Mirimanoff, R.O. The role of radiotherapy in the treatment of subtotally resected benign meningiomas. *J. Neurooncol.* **1992**, *13*, 157–164. [CrossRef]
100. Aghi, M.K.; Carter, B.S.; Cosgrove, G.R.; Ojemann, R.G.; Amin-Hanjani, S.; Martuza, R.L.; Curry, W.T.; Barker, F.G. Long-term recurrence rates of atypical meningiomas after gross total resection with or without postoperative adjuvant radiation. *Neurosurgery* **2009**, *64*, 56–60. [CrossRef]
101. Attia, A.; Chan, M.D.; Mott, R.T.; Russell, G.B.; Seif, D.; Daniel Bourland, J.; Deguzman, A.F.; Ellis, T.L.; McMullen, K.P.; Munley, M.T.; et al. Patterns of failure after treatment of atypical meningioma with gamma knife radiosurgery. *J. Neurooncol.* **2012**, *108*, 179–185. [CrossRef]
102. Goyal, L.K.; Suh, J.H.; Mohan, D.S.; Prayson, R.A.; Lee, J.; Barnett, G.H. Local control and overall survival in atypical meningioma: A retrospective study. *Int. J. Radiat. Oncol.* **2000**, *46*, 57–61. [CrossRef] [PubMed]
103. Huffmann, B.C.; Reinacher, P.C.; Gilsbach, J.M. Gamma knife surgery for atypical meningiomas. *J. Neurosurg.* **2005**, *102*, 283–286. [CrossRef] [PubMed]
104. Dziuk, T.W.; Woo, S.; Butler, E.B.; Thornby, J.; Grossman, R.; Dennis, W.S.; Lu, H.; Carpenter, L.S.; Chiu, J.K. Malignant meningioma: An indication for initial aggressive surgery and adjuvant radiotherapy. *J. Neurooncol.* **1998**, *37*, 177–188. [CrossRef]
105. Goldsmith, B.J.; Wara, W.M.; Wilson, C.B.; Larson, D.A. Postoperative irradiation for subtotally resected meningiomas: A retrospective analysis of 140 patients treated from 1967 to 1990. *J. Neurosurg.* **1994**, *80*, 195–201. [CrossRef] [PubMed]
106. Rosenberg, L.A.; Prayson, R.A.; Lee, J.; Reddy, C.; Chao, S.T.; Barnett, G.H.; Vogelbaum, M.A.; Suh, J.H. Long-Term Experience With World Health Organization Grade III (Malignant) Meningiomas at a Single Institution. *Int. J. Radiat. Oncol.* **2009**, *74*, 427–432. [CrossRef] [PubMed]
107. Mattozo, C.A.; De Salles, A.A.F.; Klement, I.A.; Gorgulho, A.; McArthur, D.; Ford, J.M.; Agazaryan, N.; Kelly, D.F.; Selch, M.T. Stereotactic radiation treatment for recurrent nonbenign meningiomas. *J. Neurosurg.* **2007**, *106*, 846–854. [CrossRef]

108. Adeberg, S.; Hartmann, C.; Welzel, T.; Rieken, S.; Habermehl, D.; Von Deimling, A.; Debus, J.; Combs, S.E. Long-Term Outcome After Radiotherapy in Patients With Atypical and Malignant Meningiomas—Clinical Results in 85 Patients Treated in a Single Institution Leading to Optimized Guidelines for Early Radiation Therapy. *Int. J. Radiat. Oncol.* **2012**, *83*, 859–864. [CrossRef]
109. Hug, E.B.; DeVries, A.; Thornton, A.F.; Munzenrider, J.E.; Pardo, F.S.; Hedley-Whyte, E.T.; Bussiere, M.R.; Ojemann, R. Management of Atypical and Malignant Meningiomas: Role of High-dose, 3D-conformal Radiation Therapy. *J. Neurooncol.* **2000**, *48*, 151–160. [CrossRef]
110. Milosevic, M.F.; Frost, P.J.; Laperriere, N.J.; Wong, C.S.; Simpson, W.J. Radiotherapy for atypical or malignant intracranial meningioma. *Int. J. Radiat. Oncol.* **1996**, *34*, 817–822. [CrossRef]
111. Pasquier, D.; Bijmolt, S.; Veninga, T.; Rezvoy, N.; Villa, S.; Krengli, M.; Weber, D.C.; Baumert, B.G.; Canyilmaz, E.; Yalman, D.; et al. Atypical and Malignant Meningioma: Outcome and Prognostic Factors in 119 Irradiated Patients. A Multicenter, Retrospective Study of the Rare Cancer Network. *Int. J. Radiat. Oncol.* **2008**, *71*, 1388–1393. [CrossRef]
112. Sughrue, M.E.; Rutkowski, M.J.; Aranda, D.; Barani, I.J.; McDermott, M.W.; Parsa, A.T. Factors affecting outcome following treatment of patients with cavernous sinus meningiomas: Clinical article. *J. Neurosurg.* **2010**, *113*, 1087–1092. [CrossRef] [PubMed]
113. Sughrue, M.E.; Sanai, N.; Shangari, G.; Parsa, A.T.; Berger, M.S.; McDermott, M.W. Outcome and survival following primary and repeat surgery for World Health Organization Grade III meningiomas: Clinical article. *J. Neurosurg.* **2010**, *113*, 202–209. [CrossRef] [PubMed]
114. Yang, S.-Y.; Park, C.-K.; Park, S.-H.; Kim, D.G.; Chung, Y.S.; Jung, H.-W. Atypical and anaplastic meningiomas: Prognostic implications of clinicopathological features. *J. Neurol. Neurosurg. Psychiatry* **2008**, *79*, 574–580. [CrossRef] [PubMed]
115. Boskos, C.; Feuvret, L.; Noel, G.; Habrand, J.-L.; Pommier, P.; Alapetite, C.; Mammar, H.; Ferrand, R.; Boisserie, G.; Mazeron, J.-J. Combined Proton and Photon Conformal Radiotherapy for Intracranial Atypical and Malignant Meningioma. *Int. J. Radiat. Oncol.* **2009**, *75*, 399–406. [CrossRef] [PubMed]
116. Roser, F.; Nakamura, M.; Martini-Thomas, R.; Samii, M.; Tatagiba, M. The role of surgery in meningiomas involving the optic nerve sheath. *Clin. Neurol. Neurosurg.* **2006**, *108*, 470–476. [CrossRef]
117. Carpentier, A.; Polivka, M.; Blanquet, A.; Lot, G.; George, B. Suboccipital and cervical chordomas: The value of aggressive treatment at first presentation of the disease. *J. Neurosurg.* **2002**, *97*, 1070–1077. [CrossRef]
118. Terahara, A.; Niemierko, A.; Goitein, M.; Finkelstein, D.; Hug, E.; Liebsch, N.; O'Farrell, D.; Lyons, S.; Munzenrider, J. Analysis of the relationship between tumor dose inhomogeneity and local control in patients with skull base chordoma. *Int. J. Radiat. Oncol.* **1999**, *45*, 351–358. [CrossRef]
119. Amichetti, M.; Cianchetti, M.; Amelio, D.; Enrici, R.M.; Minniti, G. Proton therapy in chordoma of the base of the skull: A systematic review. *Neurosurg. Rev.* **2009**, *32*, 403–416. [CrossRef]
120. Johnson, J.; Barani, I.J. Radiotherapy for Malignant Tumors of the Skull Base. *Neurosurg. Clin. N. Am.* **2013**, *24*, 125–135. [CrossRef]
121. Ius, T.; Tel, A.; Minniti, G.; Somma, T.; Solari, D.; Longhi, M.; De Bonis, P.; Scerrati, A.; Caccese, M.; Barresi, V.; et al. Advances in Multidisciplinary Management of Skull Base Meningiomas. *Cancers* **2021**, *13*, 2664. [CrossRef]
122. Alicandri-Ciufelli, M.; Marchioni, D.; Pavesi, G.; Canzano, F.; Feletti, A.; Presutti, L. Acquisition of surgical skills for endoscopic ear and lateral skull base surgery: A staged training programme. *Acta Otorhinolaryngol. Ital.* **2018**, *38*, 151–159. [CrossRef] [PubMed]
123. James, J.; Irace, A.L.; Gudis, D.A.; Overdevest, J.B. Simulation training in endoscopic skull base surgery: A scoping review. *World J. Otorhinolaryngol.—Head Neck Surg.* **2022**, *8*, 73–81. [CrossRef] [PubMed]
124. Haynes, D.S.; Roser, F.; Brackmann, D.E.; Van Loveren, H.R. Skull Base Training and Mentorship. *Otol. Neurotol.* **2020**, *41*, e1350–e1353. [CrossRef] [PubMed]
125. Rainsbury, J.W.; Ginn, E.; De, R.; Ahmed, S.K.; Irving, R.M. The skull base multidisciplinary team approach: Our experience over the first year in three hundred and seventeen patients. *Clin. Otolaryngol.* **2012**, *37*, 470–474. [CrossRef] [PubMed]
126. Sekhar, L.N.; Juric-Sekhar, G.; Qazi, Z.; Patel, A.; McGrath, L.B.; Pridgeon, J.; Kalavakonda, N.; Hannaford, B. The Future of Skull Base Surgery: A View Through Tinted Glasses. *World Neurosurg.* **2020**, *142*, 29–42. [CrossRef]
127. Ploch, C.C.; Mansi, C.S.S.A.; Jayamohan, J.; Kuhl, E. Using 3D Printing to Create Personalized Brain Models for Neurosurgical Training and Preoperative Planning. *World Neurosurg.* **2016**, *90*, 668–674. [CrossRef]
128. Mannoor, M.S.; Jiang, Z.; James, T.; Kong, Y.L.; Malatesta, K.A.; Soboyejo, W.O.; Verma, N.; Gracias, D.H.; McAlpine, M.C. 3D Printed Bionic Ears. *Nano Lett.* **2013**, *13*, 2634–2639. [CrossRef]
129. Van Lith, R.; Baker, E.; Ware, H.; Yang, J.; Farsheed, A.C.; Sun, C.; Ameer, G. 3D-Printing Strong High-Resolution Antioxidant Bioresorbable Vascular Stents. *Adv. Mater. Technol.* **2016**, *1*, 1600138. [CrossRef]
130. Sutherland, G.R.; Lama, S.; Gan, L.S.; Wolfsberger, S.; Zareinia, K. Merging machines with microsurgery: Clinical experience with neuroArm: Clinical article. *J. Neurosurg.* **2013**, *118*, 521–529. [CrossRef]
131. Goto, T.; Miyahara, T.; Toyoda, K.; Okamoto, J.; Kakizawa, Y.; Koyama, J.-I.; Fujie, M.G.; Hongo, K. Telesurgery of Microscopic Micromanipulator System "NeuRobot" in Neurosurgery: Interhospital Preliminary Study. *J. Brain Dis.* **2009**, *1*, 45–53. [CrossRef]
132. Allan, M.; Shvets, A.; Kurmann, T.; Zhang, Z.; Duggal, R.; Su, Y.-H.; Rieke, N.; Laina, I.; Kalavakonda, N.; Bodenstedt, S.; et al. 2017 Robotic Instrument Segmentation Challenge. *arXiv* **2019**, arXiv:1902.06426. [CrossRef]
133. Russell, B.C.; Torralba, A.; Murphy, K.P.; Freeman, W.T. LabelMe: A Database and Web-Based Tool for Image Annotation. *Int. J. Comput. Vis.* **2008**, *77*, 157–173. [CrossRef]

134. Veiseh, M.; Gabikian, P.; Bahrami, S.-B.; Veiseh, O.; Zhang, M.; Hackman, R.C.; Ravanpay, A.C.; Stroud, M.R.; Kusuma, Y.; Hansen, S.J.; et al. Tumor Paint: A Chlorotoxin:Cy5.5 Bioconjugate for Intraoperative Visualization of Cancer Foci. *Cancer Res.* **2007**, *67*, 6882–6888. [CrossRef]
135. Hu, D.; Gong, Y.; Hannaford, B.; Seibel, E.J. Path planning for semi-automated simulated robotic neurosurgery. In Proceedings of the 2015 IEEE/RSJ International Conference on Intelligent Robots and Systems (IROS), Hamburg, Germany, 28 September–2 October 2015; pp. 2639–2645.
136. Kawabori, M.; Weintraub, A.H.; Imai, H.; Zinkevych, I.; McAllister, P.; Steinberg, G.K.; Frishberg, B.M.; Yasuhara, T.; Chen, J.W.; Cramer, S.C.; et al. Cell Therapy for Chronic TBI: Interim Analysis of the Randomized Controlled STEMTRA Trial. *Neurology* **2021**, *96*, e1202–e1214. [CrossRef]

biomedicines

Review

Integrative Approaches in Acute Ischemic Stroke: From Symptom Recognition to Future Innovations

Vicentiu Mircea Saceleanu [1,2], Corneliu Toader [3,4,*], Horia Ples [5,6,*], Razvan-Adrian Covache-Busuioc [3], Horia Petre Costin [3], Bogdan-Gabriel Bratu [3], David-Ioan Dumitrascu [3], Andrei Bordeianu [3], Antonio Daniel Corlatescu [3] and Alexandru Vlad Ciurea [3,7]

1 Neurosurgery Department, Sibiu County Emergency Hospital, 550245 Sibiu, Romania; vicentiu.saceleanu@gmail.com
2 Neurosurgery Department, "Lucian Blaga" University of Medicine, 550024 Sibiu, Romania
3 Department of Neurosurgery, "Carol Davila" University of Medicine and Pharmacy, 020021 Bucharest, Romania; razvan-adrian.covache-busuioc0720@stud.umfcd.ro (R.-A.C.-B.); horia-petre.costin0720@stud.umfcd.ro (H.P.C.); bogdan.bratu@stud.umfcd.ro (B.-G.B.); david-ioan.dumitrascu0720@stud.umfcd.ro (D.-I.D.); andrei.bordeianu@stud.umfcd.ro (A.B.); antonio.corlatescu0920@stud.umfcd.ro (A.D.C.); prof.avciurea@gmail.com (A.V.C.)
4 Department of Vascular Neurosurgery, National Institute of Neurology and Neurovascular Diseases, 020022 Bucharest, Romania
5 Centre for Cognitive Research in Neuropsychiatric Pathology (NeuroPsy-Cog), "Victor Babes" University of Medicine and Pharmacy, 300736 Timisoara, Romania
6 Department of Neurosurgery, "Victor Babes" University of Medicine and Pharmacy, 300041 Timisoara, Romania
7 Neurosurgery Department, Sanador Clinical Hospital, 010991 Bucharest, Romania
* Correspondence: corneliu.toader@umfcd.ro (C.T.); ples.horia@umft.ro (H.P.)

Abstract: Among the high prevalence of cerebrovascular diseases nowadays, acute ischemic stroke stands out, representing a significant worldwide health issue with important socio-economic implications. Prompt diagnosis and intervention are important milestones for the management of this multifaceted pathology, making understanding the various stroke-onset symptoms crucial. A key role in acute ischemic stroke management is emphasizing the essential role of a multi-disciplinary team, therefore, increasing the efficiency of recognition and treatment. Neuroimaging and neuroradiology have evolved dramatically over the years, with multiple approaches that provide a higher understanding of the morphological aspects as well as timely recognition of cerebral artery occlusions for effective therapy planning. Regarding the treatment matter, the pharmacological approach, particularly fibrinolytic therapy, has its merits and challenges. Endovascular thrombectomy, a game-changer in stroke management, has witnessed significant advances, with technologies like stent retrievers and aspiration catheters playing pivotal roles. For select patients, combining pharmacological and endovascular strategies offers evidence-backed benefits. The aim of our comprehensive study on acute ischemic stroke is to efficiently compare the current therapies, recognize novel possibilities from the literature, and describe the state of the art in the interdisciplinary approach to acute ischemic stroke. As we aspire for holistic patient management, the emphasis is not just on medical intervention but also on physical therapy, mental health, and community engagement. The future holds promising innovations, with artificial intelligence poised to reshape stroke diagnostics and treatments. Bridging the gap between groundbreaking research and clinical practice remains a challenge, urging continuous collaboration and research.

Keywords: acute ischemic stroke; cerebrovascular disease; stroke onset symptoms; neuroimaging; neuroradiology; endovascular thrombectomy; neurovascular treatment; revascularization; fibrinolytic therapy

Citation: Saceleanu, V.M.; Toader, C.; Ples, H.; Covache-Busuioc, R.-A.; Costin, H.P.; Bratu, B.-G.; Dumitrascu, D.-I.; Bordeianu, A.; Corlatescu, A.D.; Ciurea, A.V. Integrative Approaches in Acute Ischemic Stroke: From Symptom Recognition to Future Innovations. *Biomedicines* **2023**, *11*, 2617. https://doi.org/10.3390/biomedicines11102617

Academic Editor: Kuen-Jer Tsai

Received: 23 August 2023
Revised: 21 September 2023
Accepted: 21 September 2023
Published: 23 September 2023

1. Introduction

Cerebrovascular diseases encompass a range of conditions resulting from pathological changes in the cerebral blood vessels that lead to brain dysfunction. These dysfunctions can arise from issues such as vascular occlusion, stenosis, rupture, malformation, wall damage, or permeability alterations. Globally, these diseases are a major contributor to mortality and disability. In fact, the economic impact of cerebrovascular disease surpasses that of many other vascular conditions. Without timely preventive actions, healthcare costs for addressing these diseases could see a threefold increase [1]. Stroke, as the primary clinical manifestation of cerebrovascular disease, poses a significant strain on global healthcare resources [2]. Data from *The Global Burden of Disease Study* reveals that stroke is responsible for 11.6% of worldwide deaths, ranking as the second leading cause of death and the third primary cause of disability [3]. Despite a notable 25% reduction in global stroke deaths recently, there was an alarming rise in the total number of stroke cases, their prevalence, mortality, and disability-adjusted life years between 1990 and 2019 [4].

Stroke represents the most frequent outcome of cerebrovascular disease and arises when an artery feeding the brain becomes blocked or ruptures. This chapter initially delves into the influence of epidemiological research on our understanding of stroke. It encompasses insights into its clinical presentation, diagnosis, and treatment, as well as the associated burden, risk factors, and repercussions. We will also explore transient ischemic attack (TIA), a condition as dangerous as stroke, and its impact on stroke risk [5].

While strokes are widely recognized, cerebrovascular diseases include a broader spectrum of conditions where the brain's blood vessels are the primary focus [6]. Examples include subarachnoid hemorrhages, arterial dissections, CADASIL, arteriovenous malformations, venous sinus thrombosis, moyamoya disease, and vasculitis. This chapter's latter section will delve into these diseases, emphasizing their epidemiological aspects [7].

Not all cerebrovascular anomalies produce evident symptoms immediately. Conditions such as silent brain infarcts, white-matter lesions, and microbleeds exemplify this subclinical cerebrovascular disease category [8]. Epidemiological imaging studies aim to gauge the prevalence of these subclinical conditions in the general populace, along with their risk factors and consequences [7].

Globally, stroke stands as the second primary cause of both disability and death. Low- and middle-income countries bear the brunt of this affliction. In 2016 alone, the world saw 13.7 million new stroke cases, with approximately 87% being ischemic strokes. Within these, by a modest estimate, between 10% and 20% were due to LVO. When considering treatment, less than 5% of acute ischemic stroke patients globally underwent IVT within the appropriate therapeutic time frame, and under 100,000 MTs were carried out in that year. These figures underscore the significant disparity between the number of eligible patients and the relatively scant utilization of these advanced treatments worldwide. However, several global efforts are in motion to research interventions that could potentially enhance the care systems and bridge this disparity [9,10].

From the 1980s onward, the occurrence of ischemic stroke among young adults has been on an upward trend. This uptick correlates with a rise in vascular risk factors and substance abuse in younger individuals. Unlike their older counterparts, young adults have a broader spectrum of risk factors, which include age-specific elements like pregnancy/puerperium and the use of oral contraceptives. Lifestyle choices, such as sedentary behavior, excessive alcohol intake, and smoking, also play a role [11]. In addition to these, over 150 distinct causes of early-onset ischemic stroke are recognized, inclusive of rare monogenic disorders. Recently, there have been strides in the diagnosis and treatment of stroke in young adults. This includes the molecular identification of monogenic vasculitis due to an adenosine deaminase 2 deficiency and the transcatheter closure of the patent foramen ovale for secondary prevention. Compared to their peers of similar age and gender, these patients face a mortality rate that is four times higher, predominantly due to cardiovascular causes. Furthermore, up to 15% of these patients risk a recurrent stroke within a decade [12] (pp. 2007–2018). Specific subgroups, especially those with atherosclerosis, high-risk cardioembolic sources, and small vessel disease, are more vulnerable to adverse outcomes such as survival and recurrent vascular incidents. These young stroke survivors are also susceptible to other long-term complications like epilepsy, pain, cognitive challenges, and depression [13].

Globally, stroke ranks as the second and third foremost cause of death and disability, respectively. In the global context, 68% of all strokes are ischemic, and 32% are hemorrhagic. However, in the USA, the statistics slightly vary, with 87% of strokes being ischemic, 10% hemorrhagic, and roughly 3% categorized as subarachnoid hemorrhage [14]. While comprehensive stroke data for India remains limited, western data offers some extrapolation points. Banerjee et al.'s 2001 study highlighted that India had a crude stroke prevalence rate of 147/100,000 and an annual incidence rate of 36/100,000. Notably, women exhibited higher age-adjusted prevalence (564/100,000 for women versus 196/100,000 for men) and incidence rates (204/100,000 for women compared to 36/100,000 for men). In diverse studies, the overall stroke prevalence in India fluctuated between 147–922/100,000 [15].

On a global level, stroke has far-reaching socio-economic and health implications.

In the realm of clinical practice, susceptibility-weighted imaging (SWI) emerges as an invaluable imaging method, particularly in identifying intracerebral hemorrhage, intra-arterial thrombus, microbleed, and the hemorrhagic transformation of acute stroke. Prior research reveals that the conspicuous cortical vessels detected via SWI (PCV–SWI)—which pinpoint the cortical pia mater—mirror the scope of hypoperfusion and share a relationship with leptomeningeal collaterals in AIS. There have been efforts to ascertain the merit of PCV–SWI in gauging the state of collaterals and projecting outcomes in AIS. However, the broader clinical utility of PCV–SWI for assessing leptomeningeal collaterals prior to employing recanalization therapy is yet to reach consensus. These investigations often differ in design, patient traits, imaging techniques for estimating collateral, and often involve limited sample sizes [16]. A pivotal query remains: Can PCV–SWI serve as an alternative to mCTA for evaluating leptomeningeal collaterals and predicting patient outcomes post-recanalization therapy? [17].

Regarding the Podlaskie Voivodeship region in Poland, comprehensive examinations analyzing toxic metals in AIS patients appear scant. Our current findings underline the critical public health ramifications, spotlighting a notable trace element imbalance in acute ischemic stroke patients [18]. Of notable significance from our study was the discovery that even at moderately low exposure levels, elevated blood Cd concentrations substantially influence ischemic stroke onset. Noteworthy differences in molar ratios between AIS patients and controls were also observed, with smokers demonstrating a higher propensity for LAA etiology. Despite this, Pb blood concentrations did not exhibit significant variations between the two groups. The growing nexus between environmental contaminants and stroke has piqued medical interest [19]. Research is increasingly showcasing heavy metal exposure as a potential stroke risk factor, with emerging studies linking elevated blood-Cd levels to higher stroke occurrences. A recent systematic review and meta-analysis even suggest chronic exposure to metals like Pb, Cd, and Cu might elevate stroke risk [18].

Cerebral collaterals undeniably influence the sustenance of penumbral tissue following an acute ischemic stroke. While recent research accentuates the role of collaterals in determining whether acute ischemic stroke patients are fit for reperfusion therapy, a comprehensive grasp on their predictive importance remains elusive [20]. Our literature review concentrated on methodologies for collateral assessment and their significance in acute ischemic stroke patients undergoing reperfusion therapy. Both the therapeutic and prognostic values of collaterals in acute ischemic stroke, in the context of both intravenous thrombolysis and endovascular therapy, are encapsulated. Prospects for future studies and potential drug interventions targeting collateral enhancement are also explored. Collaterals could crucially help identify acute ischemic stroke patients poised to gain from endovascular treatment in an extended timeframe. A challenging situation in acute ischemic stroke is the collateral circles, which are not determined by clinical factors or demographic influences but are highly associated with preliminary cerebrovascular events [21]. A renewed focus on understanding the role of collaterals in acute ischemia is essential, with clinical research needed to demarcate its significance in patient selection and prognosis for acute stroke [22].

A multitude of factors play into the issue of prehospital delay. Among them, socioeconomic status (SES) has been identified by some researchers as a potential catalyst for this delay [23,24]. A handful of studies have illuminated a palpable link between individual SES and prehospital delay [25–27]. The relationship between community- or neighborhood-level SES and the same delay in AIS patients has also been probed, but the findings have not always aligned [28,29]. One plausible explanation for these disparities may lie in urbanization. Urbanization is more than just the physical migration to cities; it encompasses shifts in economic, social, and cultural facets of society. Given the variance in SES distribution across urban and rural zones, a nuanced, stratified analysis anchored on urbanization status is essential. When the broader picture is considered, area-level SES does not seem to correlate with prehospital delay in AIS patients post-covariate adjustments. However, living in socioeconomically deprived urban settings could potentially amplify the delay—a pattern not mirrored in rural locales. Urban SES, therefore, might be a roadblock to curbing prehospital delay in AIS patients. Tying it all together, the utilization of emergency medical services (EMS) stands out as a vital component in mitigating prehospital delays, irrespective of urbanization or SES gradients. For a more comprehensive approach, EMS staff should be vigilant about potential biases or stereotypes they might harbor towards low-SES patients [30].

Diving deeper into the topic, the pressing need for timely diagnoses and swift interventions cannot be overstated. This meta-analysis uncovered a substantial link between stroke etiology and the status of cerebral collaterals prior to intervention in AIS patients set for reperfusion therapy (RT). It was observed that large artery atherosclerosis (LAA) correlated with enhanced rates of beneficial pre-intervention collateral, while cardioembolic (CE) strokes were associated more with deficient pre-intervention collateral. The collateral status of AIS stands as a significant determinant in influencing post-RT outcomes [22,31]. While past meta-analyses endeavored to spotlight collateral status as an outcome predictor for endovascular stroke treatment, this particular analysis is pioneering in its effort to juxtapose collateral status with stroke etiology [32,33].

Delving into the genesis of cerebral collaterals, several environmental factors come into play, with the presence of atherosclerotic blockages that hamper cerebral blood flow standing out. Such obstructions tweak vessel dynamics, escalating shear pressure and activating specific cellular pathways that pave the way for collateral creation and vascular restructuring [34]. This mechanism aligns with findings by Rebello et al., which showcased AIS patients plagued by cervical atherosclerotic issues as having a favorable pre-intervention collateral status in contrast to those afflicted by embolic strokes due to atrial fibrillation [35]. Hassler et al. further bolster this theory by pinpointing how a pre-existing atherosclerotic blockage in the carotid artery positively correlates with improved collateral health [36]. This observation resonates with this meta-analysis, where LAA was distinctly linked with pre-intervention collateral health in AIS patients [37].

Zeroing in on patient treatment, it was noted that individuals subjected to dual causal treatments showcased more severe stroke symptoms than their counterparts, as indicated by the elevated national institutes of health stroke scale (NIHSS) scores both at the time of admission and 24 h post-diagnosis. These patients underwent thrombolytic treatments, complemented by an imperative thrombectomy. Presently, stent retrievers dominate as the cornerstone tool for thrombectomies, engineered for optimum adhesion and penetration within thromboembolic matter in arteries [38]. But it is important to note that thrombectomy, despite being a mechanical procedure, has cellular repercussions. This might play into triggering an augmented inflammatory response, evident in heightened neutrophil counts, resulting in an elevated neutrophil-to-lymphocyte ratio (NLR) among these patients [39].

In the realm of understanding the pathological intricacies of ischemic strokes, the process of diagnosis predominantly hinges on imaging techniques, primarily the use of head computer tomography followed by angiography [40]. The aftermath of an acute ischemic stroke, as current knowledge indicates, can profoundly alter multiple facets of a patient's health and overall well-being. Hence, a multidisciplinary approach is vital for evaluating the effectiveness of treatments and overseeing patient recovery. Interestingly, while advanced imaging tools offer a robust diagnosis mechanism, they often falter when detecting early-stage or mild ischemic strokes. Given the time-sensitive nature of stroke management—where every passing moment can be decisive for the patient—there is an evident demand for swift, non-invasive molecular tests that can accurately pinpoint the occurrence of an ischemic stroke. Over the years, an array of biochemical biomarkers has emerged as potential diagnostic tools. Nevertheless, many of these biomarkers have been sidelined due to challenges such as specificity concerns or fluctuations in protein biomarker dynamics, particularly post-thrombolytic therapy interventions [41,42]. Still, considering the multifaceted nature of acute ischemic stroke pathophysiology—which encompasses elements like thromboembolism, inflammation, oxidative stress, and metabolic alterations—it is plausible that pertinent molecular biomarkers could be gleaned from these varied components [43–45].

Ischemic strokes predominantly arise from events such as atherogenesis or thrombogenesis, which precipitate arterial blockages. These obstructions can drastically curtail or entirely halt the blood flow to specific brain regions, culminating in tissue degradation and cellular death. While the precise role of oxidative stress in instigating the ischemic process is yet to be unequivocally established, the prevailing consensus posits that an accumulation of reactive oxygen species—borne from an ischemia-induced oxidative imbalance—fuels oxidative stress, thereby driving associated cellular and molecular harm. In this context, gauging oxidative stress biomarkers could not only shed light on the underpinnings of ischemic strokes but also present diagnostic and prognostic tools with immense potential. Furthermore, these biomarkers might illuminate novel avenues for crafting antioxidant-centric therapeutic strategies [46]. Nonetheless, it is imperative to understand that, despite the promise they hold, existing diagnostic measures for ischemic strokes are not without flaws. The path to fully harnessing the diagnostic and prognostic prowess of oxidative stress biomarkers is still strewn with hurdles, primarily due to their non-specific nature. Concurrently, while a slew of research underscores the prospective merits of modulating oxidative homeostasis in ischemic stroke intervention and prevention, the dearth of comprehensive clinical studies—and the incongruous outcomes from the ones that do exist—cloud the establishment of concrete conclusions [47].

2. Onset and Recognition

A deep dive into the symptoms of stroke onset reveals variations influenced by the specific cerebral regions impacted. It is imperative to clarify that our position is not to advocate for pharmacotherapy to supersede the well-established speech and language therapy (SLT) in aphasia treatment [48]. Over the years, a multitude of studies have endorsed the significance of SLT, while research scrutinizing the effectiveness of pharmacotherapy has been somewhat limited in comparison [49]. The primary objective of this review is to curate a comprehensive analytical framework to discern the efficacy of cholinergic agents in aphasia. This is achieved by melding insights from neuroanatomy, neurophysiology, and neuropsychology. We are also keen to draw connections between cholinergic networks and the extensive brain regions tied to not just language but other cognitive facets that are closely aligned with linguistic abilities, such as components of working memory. Exploring the neural underpinnings of pharmacological enhancements in congruence with the neurobiology of language can potentially galvanize research on aphasia-targeted drugs. Such endeavors, if successful, could eventually revolutionize treatment avenues for individuals grappling with acquired language disorders. Investigations into both human and other primates' brains reveal that cholinergic input extends to all cortical areas, primarily emanating from the nucleus basalis of Meynert in the basal forebrain [50]. The brain boasts eight primary cholinergic cell groups that channel projections to various regions. Each of these groups, from Ch1 to Ch8, is linked to specific nuclei. Intriguingly, the Ch1–Ch4 groups stand out as the lone neurons in the adult human central nervous system that consistently manifest significant quantities of the NGF receptor [51].

Acute stroke represents a major health challenge, not just in the United States but across the globe. The US, for instance, reports an alarming 750,000-plus stroke cases annually, leading to around 140,000 fatalities [52,53] (pp. 2000–2015). The standard protocol for patients who present with acute stroke symptoms at emergency departments in the US typically involves immediate computed tomography (CT) scanning, often prioritized over an in-depth clinical evaluation. Given its prominence in early stroke detection, imaging assumes a pivotal role. Beyond mere identification, imaging can illuminate patterns of infarction that might hint at potential causes, thereby shaping immediate interventions and refining secondary preventive measures to avert future strokes [54].

A pivotal metric in this discussion is the prehospital delay (PHD), quantified in minutes from the onset of stroke symptoms to a patient's hospital admission. The inception of a stroke is marked by the first instance in which a neurological deficiency is discerned, either by the patient or an observer. Hospital arrival time is typically noted from the earliest timestamp documented in the emergency department's electronic medical record. For analytical precision, PHD is bifurcated into two subintervals: the decision delay (DD) encapsulates the time elapsed between symptom onset and the initial outreach for medical intervention. The subsequent phase, or the transport delay (TD), spans the time from seeking help to the eventual hospital admission [55].

The evaluation spanned five key areas: socio-demographic attributes, clinical specifics, patients' behavioral reactions to symptoms, cognitive reactions to the onset of symptoms, and the circumstances surrounding the stroke event. During the admission phase, the severity of the stroke was gauged using the "National Institute of Health Stroke Scale" (NIHSS). Based on this scale, strokes were categorized as either mild to moderate (a score of $\leq$16) or severe (a score of >16). In instances where a score was not immediately available, a retrospective examination of the patient's documented neurological evaluations upon hospital entry was undertaken to deduce the score [56]. Prior assessments of disability employed the "modified Rankin scale", differentiating patients as either independently functional ($\leq$2) or reliant on assistance (>2). To discern coping mechanisms, the COPE-28 questionnaire in its situational variant was employed. Self-assessed severity perception and anxiety levels were gauged via a five-tier Likert scale. A patient's prior awareness of strokes was gauged based on their familiarity with at least two symptomatic signs and two potential risk factors. To assess their responses, three potential answers were provided, only one of which was accurate. The onset date of the stroke was segmented into workdays (Monday through Friday) and weekends (comprising Saturdays, Sundays, and recognized holidays). Similarly, the onset time was divided into morning (06:00–14:00), afternoon (14:00–22:00), and nighttime (22:00–06:00) slots. Geographical categorizations were made using terms like 'rural' and 'urban', with the former referring to areas outside the city's confines where the medical facility was situated and the latter describing regions within said boundary. Modes of transportation were demarcated as either ambulances or other forms [57].

Posterior circulation strokes (PCS) account for nearly 20% of the total ischemic stroke incidents [58]. However, among patients who undergo reperfusion therapy (RT), PCS figures are relatively diminutive, ranging from 5 to 19% [59]. This discrepancy stems from the unique anatomical and hemodynamic characteristics of the posterior circulation. Features such as reduced flow speeds, variance in vessel diameter, and even disparities in clot formations contribute to distinct stroke origins and progressions in comparison to their anterior circulation counterparts [60,61]. It is worth noting that the standard FAST stroke recognition tool, which comprises face asymmetry, arm weakness, and speech irregularities, does not consistently detect PCS. In fact, about 40% of PCS cases slip through this tool's radar, whereas the BEFAST tool, which integrates balance and eye symptoms, boasts enhanced sensitivity [62,63]. PCS frequently presents with atypical symptoms like nausea and seizures, contrasting with anterior circulation strokes (ACS) [64]. This discrepancy poses challenges, as PCS sufferers often face delayed hospital admissions and diminished thrombolysis application rates. Moreover, PCS patients tend to encounter protracted management processes and experience tardier RT applications compared to ACS patients [65,66]. A noteworthy point is that the NIHSS does not encompass all PCS symptoms, occasionally leading to hesitations in applying RT to PCS. The nuanced presentations and typically lower NIHSS scores for PCS make the task of balancing RT's effectiveness against its potential risks a complex endeavor. Although PCS's chances of symptomatic intracerebral hemorrhaging are inferior to ACS, the perennial challenge of weighing RT risks against its benefits remains, especially when confronted with low NIHSS readings [67]. In such scenarios, decision-making often leans on nuanced, frequently subjective indicators that might not be explicitly addressed in standardized guidelines [68].

Studies have showcased diverse symptom presentations by analyzing different case studies. Within the reviewed literature concerning immersive virtual reality (IVR), no severe adverse reactions were identified. However, some reports from the broader IVR landscape have highlighted minor symptoms like dizziness, feelings of nausea, eye discomfort, and a sense of disorientation. As Tsirlin et al. elucidated in their comprehensive review [69], when designing virtual reality (VR) tools, ergonomic design is paramount. This is especially relevant when catering to post-stroke patients, who often grapple with specific challenges like restricted mobility. An intriguing observation from our review was that five studies highlighted the usage of VR training while the participants were seated—be it in a wheelchair or a regular chair [17]. Such details underscore the constraints, like limited postural stability and movement restrictions, that patients face post-stroke, suggesting the prudence of implementing IVR within supervised clinical settings. Even though the potential exists for IVR to be introduced into home settings, the prevailing evidence underscores its utility in regulated environments. Any shift towards domestic usage mandates rigorous safety evaluations [70].

When examining the literature concerning exertional heat stroke (EHS) survival, a significant portion is derived from case reports and series. Given the ethical constraints of inducing EHS in humans for study purposes, researchers predominantly rely on these case studies for insights. In our pursuit of a structured assessment, our team turned to the Joanna Briggs Institute (JBI) and its critical appraisal tools, which were tailored for case reports and series [71–73]. These tools, devised by the JBI, offer quality benchmarks that enhance the rigor of systematic reviews, evaluating parameters like diagnostic clarity, treatment interventions, and documented adverse events. In the context of this review, these tools proved invaluable in cherry-picking case studies and series with meticulous documentation to ascertain potential medical complications tied to EHS therapeutic interventions [73]. Two blind reviewers independently evaluated all cases, and a third reviewer synthesized the scores, mediating discrepancies when needed. Case reports were assessed out of a total of eight points, while case series had a maximum of 10 points. Our criteria for inclusion demanded that case reports achieve a score of 6/8 and case series attain 8/10. This culminated in a quality threshold of 75–80%, deemed adequate for our analysis by the reviewing team [74].

Our systematic review scoured literature focusing on acute stroke patients, particularly emphasizing the correlation between stroke locations and types vis-à-vis delirium status. Our search criteria encompassed publications from January 2010 through June 2021, spanning multiple languages. We excluded case studies with fewer than 20 patients, case-control studies, and randomized controlled trials. Our search extended across various databases, including MEDLINE, EMBASE, PsycINFO, CINAHL, and Alois. We utilized either bivariate random effects models or network meta-analysis for determining pooled relative risks, while the methodological integrity was scrutinized across eight defined parameters [75]. Our endeavor culminated in the inclusion of 31 patient samples, representing a total of 8329 patients. We unearthed intriguing patterns—delirium was notably prevalent in patients with supratentorial as opposed to infratentorial lesions, anterior circulation compared to posterior, and cortical versus subcortical lesions [76]. Interestingly, the side of the stroke (right vs. left) did not exhibit any correlation with delirium. Delirium was also discernibly higher in patients suffering from hemorrhagic strokes relative to ischemic ones and those with pre-existing qualitative atrophy. This review has shed light on the intricate links between various brain regions, stroke types, and the onset of delirium. However, one must exercise caution in drawing conclusions due to the variability across studies and the sometimes vague descriptions of lesions. Nevertheless, these findings offer a pivotal foundation for predicting delirium risks in acute stroke scenarios and can pave the way for further studies probing into the neural circuits and pathological underpinnings contributing to delirium's pathophysiology [75].

A multidisciplinary approach is becoming an increasingly integral part of healthcare, especially in areas like stroke care. Organized stroke care typically involves a dedicated team that is either stationed in a stroke ward, functions as a mobile team, or operates within a broader rehabilitation service [77]. This team is composed of various specialists, each bringing unique expertise to cater to the intricate needs of stroke patients. The effectiveness of this approach is substantiated by 23 trials that have demonstrated that, compared to other care methods, multidisciplinary stroke unit care significantly reduces mortality rates and dependency rates in patients during a median follow-up of one year [78].

The foundation of the observations in this research mainly rests on Cochrane and other systematic reviews post-2000, accompanied by other relevant quantitative and qualitative studies aimed at enhancing post-stroke recovery. One salient feature of stroke recovery is its complexity and individual variability, underscoring the critical role of healthcare professionals in collaborating efficiently. This synergy ensures that a gamut of collective knowledge and specialized skills is accessible for the betterment of stroke survivors [79].

There is a distinction between multidisciplinary and interdisciplinary functions, which is essential to understanding their contribution to stroke care. The patient journey post-stroke can be seen in three primary stages: the immediate response and emergency department phase, the inpatient care phase, and the post-discharge phase, which includes long-term support. At each of these junctures, multidisciplinary teams (MDTs) play distinct roles, shaping the trajectory of the patient's recovery [79].

Policy framers and clinical guideline developers have consistently linked MDT practices to enhancements in the quality of stroke care [80]. The national stroke strategy for countries like England, Wales, and Northern Ireland earmarks a significant portion of its quality indicators towards MDT's roles in ensuring effective service delivery and fostering improved patient outcomes. This faith in MDTs emanates partly from rigorous reviews of inpatient stroke care trials. Such reviews have unambiguously revealed the advantages of organized care in stroke units by MDTs, not just in reduced mortality but also in fostering patients' independence, as evidenced by a higher number of patients living at home a year post-stroke [78].

Despite the clear advantages, the exact dynamics of how MDTs contribute to these improved outcomes have not been firmly determined. While the ideas behind multidisciplinary and interdisciplinary services in stroke care are widely accepted [80], it is imperative to distinguish between them and understand their unique impacts on post-stroke recovery.

When classifying interventions, the primary distinguishing factors are the method of delivery and the setting. For analytical purposes, outcomes from studies with similar methodologies were grouped. In cases where the same study population was reported in separate journals, they were treated as a singular entity for mortality assessment [81–83]. To address the diverse range of home-visit interventions, they were segmented based on the facilitator, either being led by a multidisciplinary team or other healthcare providers such as physiotherapists, nurses, or occupational therapists [84,85]. Further granularity in the analysis was achieved by categorizing interventions based on their nature and follow-up duration. A prominent metric used for assessment is the Barthel index (BI), a widely recognized tool to evaluate daily life activities (ADL). Different evaluation methods were discerned in the meta-analysis, offering a comprehensive view of post-intervention outcomes [86].

The economic implications associated with post-acute care (PAC) programs for stroke patients undergoing rehabilitation remain underexplored in the academic literature. This pioneering study delves into the economic burden shouldered by stroke patients engaged in PAC rehabilitation and gauges the effectiveness of multidisciplinary PAC programs in terms of both cost and patient functional outcomes [87]. Out of the 910 patients with stroke observed from March 2014 to October 2018, they were divided into two cohorts: those receiving PAC (from two medical centers) and those not involved in PAC (from three regional hospitals and a district hospital). This allocation was achieved through propensity score matching, maintaining a 1:1 ratio. To decipher the study's economic aspects, a cost-illness framework was implemented, targeting specific cost categories. Remarkably, the direct medical costs for patients under the PAC system, which used a per-diem-based costing approach, were considerably lower than those in the non-PAC system, which utilized a fee-for-service approach ($p < 0.001$). The yearly economic load for stroke patients undergoing PAC rehabilitation stands around USD 354.3 million (based on 2019's conversion rate of NT USD 30.5 to USD 1) [88]. Furthermore, functional improvement was notably more significant in the PAC cohort compared to the non-PAC cohort. This difference was both pre- and post-a year-long rehabilitation regime ($p < 0.001$). The emphasis on early rehabilitation following a stroke is evident, as it fosters health restoration, boosts confidence, and enhances the self-care abilities of patients. Evidently, PAC rehabilitation curtails the transitional period to the rehabilitation ward, showcasing its efficiency in cost-saving and functional improvement for stroke patients [89].

A structured, multidisciplinary team intervention immediately following an acute stroke can substantially reduce functional impairments, stave off complications, and subsequently lessen extended hospital stays. Data from the get with the guidelines–stroke (GWTG-Stroke) program reveals that of the 616,982 adults diagnosed with stroke, a staggering 90% were evaluated for potential rehabilitation during the acute phase [90]. Yet, a striking disparity emerges when we scrutinize inpatient stroke rehabilitation practices in Taiwan. Here, only 34.0% utilized rehabilitation services, which, when broken down, were 33.0% for physical therapy, 19.6% for occupational therapy, and a meager 5.3% for speech therapy. This is in stark contrast to the figures from countries like the United States, Canada, the UK, and Austria, where the utilization rates hover between 59% and 75% for physical therapy, 16% and 39% for occupational therapy, and 10% and 23% for speech therapy [91]. Another challenge in Taiwan is the delivery of inpatient rehabilitation, often restricted to bedside programs in certain local hospitals devoid of dedicated rehabilitation facilities. Additionally, a skills gap is apparent, with some local hospital therapists possessing limited experience in stroke rehabilitation. A potential remedy could be refining the payment system to encourage more skilled rehabilitation providers. The findings from our study reinforce the notion that initiating intensive stroke rehabilitation early leads to outcomes that are both cost-effective and efficient. The synergy of a multidisciplinary team is pivotal in realizing these outcomes [89].

The patient's journey in managing strokes underscores the vital role of time. Indeed, when confronted with a stroke, every second matters; swift action can profoundly influence the effectiveness of interventions and significantly shape a patient's ultimate health outcomes [92]. Regrettably, delayed hospital presentations are common, often sidelining potential treatments that could be beneficial if administered promptly. A critical impediment to swift hospital arrivals is the prevalent gap in awareness: both patients and bystanders often lack an understanding of the early warning signs and risk factors associated with stroke. This gap can manifest in a myriad of ways—from patients dismissing their symptoms, holding onto the hope that they will simply dissipate with time, to outright denial of the potential severity of their condition [93,94].

What is more concerning is the discrepancy in health awareness: stroke, despite its alarming prevalence and profound implications, lags behind in public consciousness. This lack of awareness is starkly evident even when juxtaposed with other grave conditions like acute coronary syndrome (ACS), cancer, or AIDS. Worryingly, even among those who have weathered a stroke episode, a significant portion remain inadequately informed about the disease [95].

To bridge this awareness gap, several public health initiatives and campaigns have been launched. Slogans such as "Stroke Chain of Survival", "Time is Brain", and "Face, Arms, Speech, Time (FAST)" aim to hammer home the urgency of stroke and the importance of early medical intervention [96–98]. The underlying hypothesis of these campaigns is straightforward: arming the public with knowledge about stroke's warning signs can expedite the decision to summon emergency medical services, a decision that could prove life-saving [99].

Furthermore, cognizance of risk factors is not solely about timely interventions. Comprehensive awareness can also galvanize primary and secondary preventative measures. By acquainting individuals with the risks, there is an opportunity to inspire preventative lifestyle changes, which could significantly curtail future cerebrovascular issues. However, the current reality is sobering: a substantial segment of individuals at elevated risk for strokes remains in the dark about their perilous position [100]. This underscores the urgent need for continued and intensified educational outreach to the broader population.

Awareness about stroke among the general populace is pivotal. Its presence or absence can spell the difference between timely intervention for an individual suffering an acute stroke and potential delayed treatment or mismanagement [101]. This study delves into this awareness in the Silesian voivodeship, the most densely populated region of Poland. The goal was not only to gauge the extent of general knowledge about stroke but also to discern "adequate knowledge of stroke", which is a comprehensive understanding that encompasses risk factors, symptoms, and the necessary actions to take when confronted with an acute stroke. This holistic understanding is imperative for efficacious stroke management. Furthermore, pinpointing the factors that influence this adequate knowledge can be instrumental in tailoring educational strategies [102].

A custom survey has been employed, all pertinent to stroke. Beyond just querying their understanding of individual stroke aspects, we looked into their "adequate knowledge of stroke", which encapsulates their understanding of risk factors, the recognition of symptoms, and the protocol to follow during an acute stroke event.

Out of those surveyed, 834 individuals (73.5%) could accurately define what a stroke is. An encouraging 92.8% recognized a stroke as a medical emergency, with 97.5% acknowledging the need for medical intervention [103]. On the flip side, a concerning 42.4% could not pinpoint a specific stroke symptom, and only a mere 38.6% could enumerate at least two or more risk factors. This culminated in just 36.3% of the respondents having what we classify as adequate knowledge of stroke. Upon analysis, factors like the length of education, a personal connection to someone who suffered a stroke, gender, and place of residence emerged as key determinants of this comprehensive stroke knowledge [104].

1. The populace of southern Poland displays a level of stroke awareness that can be deemed inadequate, especially when considering the urgency and timeliness needed for effective stroke management.
2. Personal experiences, particularly having a friend or relative who has suffered a stroke, stand out as the most influential factor in having adequate stroke knowledge [105].

The role of emergency medical services (EMS) cannot be overstated in the realm of stroke management. By mitigating pre-hospital delays and ensuring swift in-hospital assessments, they hold the potential to be game-changers. Critical steps like recognizing stroke symptoms at the dispatch center, ensuring on-the-spot stroke diagnosis and triage, expediting patient transfer to equipped facilities, and notifying hospitals in advance can dramatically reduce pre-hospital lags. These steps not only amplify the chances of administering rtPA but are imperative for timely acute stroke care in our country [106].

One major limitation is its confinement to a singular stroke center, which poses questions about the wider applicability of our findings on a national scale. While the results are telling, the reality in other regions might be even more grim, given the paucity of stroke awareness drives, the lack of specific pre-hospital protocols for stroke, and the expansive patient base covered by many county hospitals [107]. Additionally, it was earmarked the stroke onset as the time a patient was last observed without symptoms, neglecting the moment they were first spotted with them. Moreover, it did not delve into the actions of bystanders, which could significantly impact the decision to summon EMS. Addressing these lacunae will be pivotal in subsequent studies, especially if we aim to truly discern the barriers preventing stroke patients in Romania from accessing IVT [108].

3. Neuroimaging and Neuroradiology: A Deep Dive

3.1. The Evolution of Neuroimaging in Stroke Diagnosis and Its Historical Context

The domain of neuroimaging has witnessed considerable evolution over time, playing a transformative role in the diagnosis and management of stroke, specifically in relation to RSSI in patients displaying lacunar syndromes. A stroke, simply put, is a form of injury to the brain due to interrupted blood flow, and early detection is crucial for timely and effective intervention.

To diagnose an RSSI, one needs to pinpoint lesions through neuroimaging like CT or MRI scans, which are consistent with a minor ischemic stroke in specific regions of the brain. These regions include the area served by particular deep perforating arteries such as lenticulostriate and thalamoperforating, among others. This would involve the subcortical white matter areas (like the centrum semiovale and corona radiate, to name a few) or deep gray structures, including the basal ganglia and nuclei located in the brainstem [109].

Historically, CT scans emerged as the pioneering neuroimaging technique that could discern small focal hypoattenuations, synonymous with lacunar strokes. Nevertheless, during the nascent hours following the symptom onset, CT scans found it challenging to highlight these tiny subcortical infarcts, often confusing them with pre-existing lesions in patients grappling with SVD [109].

The advent of MRI revolutionized stroke diagnostics. It facilitated nuanced morphological and topographical characterization of RSSI, offering a more detailed snapshot of the brain's structures [110]. The implementation of diffusion-weighted imaging (DWI) within MRI proved to be a game-changer. It enabled the identification of recent ischemic changes by displaying hyperintensities mere minutes after the onset of a stroke, which could remain visible for roughly 3 to 5 weeks. Meanwhile, older lesions could be discerned through other structural sequences in the MRI [111,112]. However, MRI, despite its precision, is not foolproof. Factors like the magnetic field strength, motion artifact correction, and the sequencing of image acquisition can sometimes lead to small lesions going undetected [113]. Hence, even if DWI does not highlight hyperintense lesions, one should not hastily dismiss the possibility of a lacunar stroke, especially if the patient's symptoms suggest otherwise.

Additionally, an RSSI evident on an MRI might actually be the aftermath of a larger deficit in blood flow, possibly affecting a larger territory than initially thought, as some perfusion studies have shown [114]. On occasion, perfusion deficits linked to a single perforating artery might be reversible. It is essential to remember that stroke is a complex and dynamic process, influenced by a plethora of factors like metabolic demand, time of ischemia, and collateral blood supply, to name a few. Hence, while imaging provides invaluable insights, it may not capture the entire narrative, necessitating a comprehensive clinical assessment and additional tests for a holistic understanding [115].

In a study involving 312 stroke patients who underwent CT scans, 37 displayed clinical signs of lacunar syndrome. Of these, 18 exhibited lacunar-sized infarcts on their scans, 13 had unremarkable scans, and 6 surprisingly revealed large infarcts. Intriguingly, of these six patients, five manifested pure motor hemiplegia, and one exhibited a pure sensory stroke. Both clinical evaluations and angiography unveiled potential treatable sources of emboli in both lacunar-sized and large infarcts [116]. This leads to two pivotal conclusions:

1. A clinical lacunar syndrome does not always correlate with the size of the infarct—it can sometimes be linked to a larger infarct.

2. Identifying a lacunar infarct through a CT scan does not negate the need for further angiographic studies, especially if there is a likelihood of detecting an embolic source [109].

3.2. An In-Depth Look into Neuroimaging Modalities and Their Role in Post-Stroke Recovery Prediction

Neuroimaging serves as an indispensable tool in the realm of stroke diagnosis and management. Beyond its foundational role in distinguishing ischemic strokes from hemorrhagic strokes in the acute phase, neuroimaging has been emerging as a pivotal component in decision-making for cutting-edge treatments, such as late-window thrombectomy. In this review, our focus extends beyond immediate diagnostic applications, shedding light on the potential of neuroimaging techniques in forecasting post-stroke recovery.

Integrating recovery predictions with quantifiable measurements permits the identification and development of biomarkers, which can be monumental in the treatment and management of stroke patients. As defined by the FDA-NIH biomarker working group, a biomarker is a "specifically identified metric serving as an indicator of natural biological activities, pathological processes, or responses to an intervention or treatment" [117]. This definition underscores the importance of unifying terminology across scientific disciplines. Over time, as our understanding of stroke has deepened, the term 'biomarker' has evolved. It has transitioned from being primarily a diagnostic tool to being intricately linked with therapeutic mechanisms. Presently, biomarkers encompass a broad spectrum of factors, ranging from genetic markers and molecular indicators to clinical scales and, crucially, neuroimaging and neurophysiological indicators.

Stroke recovery biomarkers sourced from neuroimaging encapsulate both structural and functional dimensions [118]. For structural evaluation, parameters such as the size of the infarct, the degree of cortical or white matter damage, the integrity of white matter, and the percentage of injury to the corticospinal tract are of paramount importance. On the other hand, functional evaluations hinge on aspects such as activation patterns within ipsilesional (same side of the brain as the lesion) and contralesional (opposite side to the lesion) regions, the balance between the hemispheres, connectivity during resting states, synchronization and desynchronization during specific tasks, and measures of cortical excitability, facilitation, and inhibition [119].

Following this, we delve into a succinct discussion on specialized techniques tailored for analyzing stroke recovery. This exploration is particularly pertinent when considering the intricate processes of angiogenesis (formation of new blood vessels) and neuroplasticity (the brain's ability to reorganize and adapt). The nuances of these techniques, offering insights into their capabilities and unique characteristics, are consolidated in Table 1 for ease of reference [120]. This table serves as a valuable resource, offering readers a comprehensive overview of the breadth and depth of neuroimaging modalities available in contemporary stroke research and recovery prediction.

3.3. Post-Stroke Angiogenesis and the Expanding Horizons of Advanced Neuroimaging

Angiogenesis following a stroke is an intricate, multi-phased procedure. It begins with gene transcription and the release of proangiogenic factors, leading to a cascade of events including the proliferation of endothelial cells and the sprouting of new vascular structures, culminating in the formation of microvessels [121]. Today's imaging methodologies can investigate an array of both structural and functional characteristics within tissues [122]. Breakthroughs in magnetic resonance imaging (MRI) have ushered in techniques capable of assessing tissue blood flow and deducing various metrics related to the vascular network, such as microvascular cerebral blood volume (CBV) and the density of these microvessels [123].

Highlighting these advancements, an experimental study conducted by Yanev et al. utilized steady-state contrast-enhanced (ssCE-) MRI with an extended blood pool circulation time to delineate vascular changes within ischemic lesions and associated regions, spanning from the subacute to the chronic stages post-cerebral stroke [124,125]. Their findings elucidated dynamic vascular regeneration in areas surrounding the lesion and ongoing neovascularization in areas linked to but not directly impacted by ischemia [126–128]. Such vascular activities could play pivotal roles in the repair and restructuring of non-neuronal tissues, influencing post-stroke recovery dynamics. The nascent stages of angiogenesis can be detected via MRI techniques as disruptions in the blood–brain barrier [129,130]. This disruption, or permeability, correlates with the proliferation of endothelial cells and the initiation of vascular sprouting. To detail the integrity of the blood–brain barrier, dynamic contrast-enhanced MRI (DCE-MRI) leveraging gadolinium chelates can be harnessed, especially when alterations in MRI signals arise due to contrast seepage into surrounding tissues [131–133].

In the broader realm of stroke management, cutting-edge neuroimaging stands as a vital asset, aiding clinicians in bypassing the time restrictions and expanding the application scope of intravenous thrombolysis (IVT) [134–136]. To understand the potential effects of employing advanced neuroimaging (AN), specifically CT/MR perfusion, on the outcomes of acute ischemic stroke (AIS) patients undergoing IVT, irrespective of the elapsed time since symptom manifestation [136,137]. Through a retrospective lens, we analyzed AIS patients who underwent IVT as a sole therapeutic intervention over a span of six years. Our focus was on discerning if there were notable differences between patients who had undergone advanced neuroimaging prior to IVT (AN+) versus those who had not (AN−). Key outcome metrics ranged from clinical safety indicators, such as intracranial hemorrhage and 3-month mortality, to efficacy measures like door-to-needle time, discharge neurological status (NIHSS-score), and 3-month functional status gauged by the modified Rankin Scale (mRS) [138]. Interestingly, while the utilization of IVT monotherapy saw an uptick in the AN+ cohort, the key metrics across both groups remained comparable, suggesting the AN+ approach does not compromise the efficacy or safety of IVT treatment.

To summarize, our exploratory study underpins the value of integrating advanced neuroimaging into the acute stroke treatment pathway for AIS patients. Not only does it augment the administration rate of IVT, but it also maintains the treatment's efficacy and safety profile, offering promising avenues for enhanced patient care [139,140].

3.4. *Advancements in Stroke Treatment and the Role of Neuroimaging*

In recent years, the landscape of stroke treatment has witnessed transformative progress. A significant stride forward was marked by the DAWN and DEFUSE 3 trials in 2018. These groundbreaking studies unveiled the effectiveness of mechanical thrombectomy beyond the conventional 6-h timeframe, extending the treatment window up to 24 h post-onset of acute stroke symptoms in patients with large vessel occlusions (LVO) [141]. This paradigm shift was rooted in judicious patient selection, emphasizing a mismatch between the infarcted core and the surrounding at-risk, yet salvageable, ischemic penumbra as depicted in perfusion images. Essentially, reperfusion treatments aim to rescue the endangered penumbra and forestall the expansion of the infarct core [142].

Highlighting the trials, the DAWN (DWI or CTP assessment with clinical mismatch in the triage of wake-up and late-presenting strokes undergoing neurointervention with trevo) trial stands out as a multi-center, randomized controlled investigation. It focused on patients presenting 6 to 24 h after the emergence of stroke symptoms and exhibiting a proximal LVO. Enrollment was based on detecting a mismatch between the identified ischemic core via DWI or CT perfusion and the degree of neurological impairment (manifested as an NIHSS score of 10 or above). The median interval between symptom onset and intervention was found to be 12.5 h. Notably, the outcomes illustrated a pronounced improvement in patients undergoing mechanical thrombectomy compared to conventional treatments: 49% of these patients exhibited minimal to no disability, a stark contrast to the 13% in the standard therapy cohort [143].

In specialized stroke centers, comprehensive clinical evaluations coupled with advanced neuroimaging techniques are routinely employed, paving the way for predictive assessments of patient trajectories. A wealth of literature delves into the interplay of clinical and neuroimaging measures, with a special emphasis on proprioception during the subacute post-stroke phase [144,145]. Notably, clinical indicators, including attentional capacities and daily functioning metrics like the behavioral inattention test (BIT) and the functional independence measure (FIM), have displayed strong correlations with proprioceptive assessments [146,147]. Neuroimaging dimensions, such as lesion volume and precise regional damage, have further enriched our understanding, linking larger lesions with deteriorated post-stroke proprioceptive outcomes [148,149]. Cutting-edge tools like voxel-based lesion-symptom mapping (VLSM) have allowed researchers to discern the statistical interrelations between affected brain areas and post-stroke proprioceptive capacities [150,151]. However, while motor recovery has been extensively studied to identify early recovery predictors, research focused on forecasting long-term proprioceptive recovery remains relatively sparse [118].

Against this backdrop, our study set out to bridge this knowledge gap, seeking to ascertain the predictive prowess of combined clinical, neuroimaging, and robotic evaluations in forecasting long-term proprioceptive outcomes. The study's central objective was to gauge the efficacy of these measures, gathered within the initial fortnight post-stroke, in forecasting proprioceptive deficits at the six-month mark [152]. Drawing from prior correlations between clinical metrics, neuroimaging results, and proprioceptive evaluations, we hypothesized that standalone clinical or neuroimaging models would offer satisfactory predictive accuracy for six-month proprioceptive outcomes [153]. Expanding on this, we further postulated that robotic metrics, either in isolation or synergized with clinical and neuroimaging data, would enhance predictive accuracy and bolster the area under the receiver-operator characteristic (ROC) curve (AUC) [154].

3.5. *Understanding Cerebral Artery Occlusions and the Evolution of Stroke Treatments*

In the realm of stroke care, a deeper understanding of cerebral artery occlusions and their clinical implications is pivotal. The year 2015 marked a significant breakthrough in this domain. Groundbreaking trials, namely MR CLEAN, ESCAPE, SWIFT PRIME, REVASCAT, and EXTEND IA, conclusively demonstrated the superiority of endovascular thrombectomy over standard medical management in treating anterior circulation large vessel occlusion strokes [155]. The potency of endovascular thrombectomy as a treatment modality is evident, with a patient 'number needed to treat' ranging from a mere 3 to 10. A set of criteria, including occlusion location (focusing on proximal anterior occlusions such as the internal carotid or middle cerebral artery), time since the onset of the stroke (ideally within an early window of 6–12 h), and an acceptable level of infarct burden (reflected by an Alberta Stroke Program Early CT Score [ASPECTS] of $\geq$6 or an infarct volume of less than 50 mL), became the basis of patient selection for these trials [156].

Subsequent trials in 2017, notably DAWN and DEFUSE-3, pushed the boundaries even further by successfully expanding the treatment window up to 24 h for a certain subset of patients. This paradigm shift has been embraced by societal and national thrombectomy guidelines, granting a Class 1A recommendation for the carefully selected patient cohort. However, the journey is far from over. Currently, randomized controlled trials are underway to study thrombectomy's applicability in stroke subpopulations previously considered ineligible. These trials are fueled by promising insights from an aggregated analysis of early trials (by the HERMES collaboration) and budding retrospective data. The focal points of these trials include patients with large vessel occlusion strokes exhibiting mild deficits (with a national institutes of health stroke scale score less than 6) or those with a substantial infarct burden (a core volume exceeding 70 mL) [157].

On a global scale, stroke stands as the second primary cause of mortality. Given the limited therapeutic arsenal against ischemic stroke, there is an urgent need to innovate and expand treatment options. In recent years, metformin, primarily known for its anti-inflammatory properties, has been spotlighted for its potential neuroprotective capabilities against ischemic damage caused by the stop and restart of blood flow (ischemia/reperfusion) [158,159]. This study ventured to delve deeper into metformin's efficacy, specifically in the context of permanent middle cerebral artery occlusion (pMCAO) without any subsequent reperfusion in rat models. Neurological aftermaths post-pMCAO were gauged using the Longa scale, a reliable metric for body movement evaluation. Additionally, the extent of brain damage and swelling were ascertained through the 2,3,5-triphenyltetrazolium chloride staining technique.

Diving into the findings, metformin administration led to marked neurological improvement and a decrease in infarct size, especially 120 h post-pMCAO. While metformin prevented neuronal loss in the ischemic cortex, its effect was not as pronounced in the striatum 48 h after pMCAO. An encouraging observation was the substantial decline in the number of total and activated microglia 48 h post-stroke upon metformin treatment. This anti-inflammatory action of metformin corresponded with a surge in interleukin 10 (IL-10) production 48 h following pMCAO. Cumulatively, this study furnishes compelling evidence for metformin's anti-inflammatory and neuroprotective roles in a pMCAO setting [160].

3.6. Harnessing Neuroradiology for Therapy Planning: Delving into Techniques and Implications

Neuroradiology stands as a pillar for therapeutic planning, especially in the intricate landscape of neurological disorders. The path to decision-making through neuroradiology is a layered one, often blending advanced technology with human acumen.

Brain magnetic resonance imaging (MRI), for instance, is instrumental in prognosticating the clinical trajectory of patients with acute ischemic stroke (AIS) [161]. In recent years, there has been a technological windfall with deep learning (DL) techniques successfully employing brain MRI images and certain biomarkers for forecasting unfavorable outcomes in AIS patients. However, an intriguing dimension that has hitherto remained unexplored is the potential of using natural language processing (NLP)-oriented machine learning (ML) algorithms. The key distinction here is the source of data: free-text reports of AIS patients derived from brain MRI scans [161].

To chart this unexplored terrain, a study was conducted focusing exclusively on English MRI reports obtained during the admission phase for AIS patients. Defining poor outcomes as a modified Rankin scale score ranging between 3 and 6, data acquisition was meticulously overseen by a team of trained healthcare professionals. The emphasis was placed on the first MRI report obtained during hospitalization. Structuring the study, the collected text dataset was systematically segmented into training and testing batches, following a 70:30 proportion.

The data underwent three levels of vectorization: word, sentence, and document levels. The nuanced "bag-of-words" model found its application at the word level, which disregarded word sequence but tallied text token repetitions. On the other hand, the "sent2vec" methodology, which took the sequence of words into account, was employed at the sentence level. Meanwhile, word embedding was applied at the document level. Alongside traditional ML algorithms, DL paradigms like the convolutional neural network (CNN), long short-term memory, and multilayer perceptron were adopted. This ensemble was evaluated against 5-fold cross-validation and grid search techniques. A consistent performance metric, the area under the receiver operating characteristic (AUROC) curve, was employed.

Analyses from 1840 AIS subjects revealed a stark reality—a hefty 35.1% grappled with poor outcomes three months post-stroke onset. The random forest emerged as the top classifier at the word-level approach with an AUROC of 0.782. However, on a broader spectrum, the document-level approach eclipsed the other two. The multi-CNN algorithm set the gold standard in classification with an AUROC of 0.805, closely trailed by the CNN algorithm at 0.799. The crux of these findings lies in the supremacy of DL algorithms, particularly in NLP-based predictions, where multi-CNN and CNN outperform other neural networks in forecasting adverse outcomes [162]. This asserts the pivotal role NLP-fueled DL can play, marking its ascendancy as a digital beacon for unstructured healthcare data predictions [163].

Transitioning from MRI to the realm of computed tomography (CT), the trio of CT, CT angiography (CTA), and CT perfusion (CTP) reign supreme in emergency departments when there is a hint of cerebrovascular compromise [164]. The pressure-cooker environment of emergency settings demands precise and rapid image interpretation. Notably, the clinical tableau of an acute stroke can be mimicked by myriad conditions. Hence, the rapid discernment of true strokes from their imitators is paramount for clinicians. With a vast array of conditions masquerading as acute strokes, the onus falls on imaging to discriminate. While some of these conditions reveal themselves quite clearly, others can pose diagnostic challenges. Pictorial representations derived from CTP serve as vital clues [165]. A series of acute stroke instances and their look-alikes were presented, emphasizing the indispensable nature of these imaging "pictograms" for radiologists. This visual encyclopedia should bolster radiologists' diagnostic prowess, ensuring they remain conversant with the nuances of diverse imaging techniques, reaping their advantages while steering clear of potential pitfalls [166].

3.7. Pioneering Neuroimaging Optimization: Charting the Path to Precision Diagnostics

The vast realm of ischemic stroke, which encompasses over 80% of all stroke occurrences, stands as a formidable adversary in the global health arena, frequently leading to mortality and long-term disabilities [167]. Administering recombinant tissue plasminogen activator (rt-PA) intravenously is an accredited countermeasure for acute ischemic strokes caused by larger arteries, provided it is employed within a 4.5-h window from the onset. Moreover, mechanical thrombectomy can serve as an intervention for large artery occlusions up to 24 h post-onset [167]. Yet, the real-world challenge lies in optimizing diagnostic processes for acute treatments. Factors such as minimizing the onset-to-needle time duration, ensuring rapid access to angioCT images, and facilitating timely magnetic resonance imaging (MRI) become bottlenecks in many healthcare frameworks. Such constraints often lead to a disparity between real-world prognosis rates and those recorded in randomized controlled trials. Notwithstanding the efficacy of applied endovascular techniques, there remains a lack of comprehensive understanding regarding certain cellular mechanisms post-reperfusion [168]. Furthermore, research areas like the changes in mitochondrial morphology and function related to reperfusion and ischemia-induced neuronal death remain relatively uncharted. A future vision in stroke research mandates an in-depth exploration of the evolving landscape of imaging techniques. It is crucial to comprehend the intricate relationship between the ever-refining imaging methodologies and factors like clot structure variability, vascular permeability, and the diverse manifestations of ischemic reperfusion damages, especially in the penumbra. Insights into these domains hold the key to devising targeted interventions that confer lasting health benefits [169].

3.8. Radiologists and Neurologists: Crafting a Symbiotic Diagnostic Journey

Seamless collaborations between radiologists and neurologists stand as a beacon of hope for patients. In our study, we meticulously screened patients using CT/CTA or MRI before initiating any intervention [170]. The focus group comprised acute ischemic stroke patients, primarily attributed to large vessel occlusions (LVO) and, more specifically, those at the M2 level. LVO categorizations included occlusions of the internal carotid artery (ICA), middle cerebral artery (MCA) at the M1 segment, intracranial vertebral artery (VA), and basilar artery (BA) as identified in CTA [171]. Our inclusion criteria encompassed: patients above 18 years of age, those with a national institutes of health stroke scale (NIHSS) score of 6 or higher (or presenting isolated aphasia), individuals who demonstrated prior functional independence using the modified ranking scale (mRS $\leq$ 2), and patients who sought medical attention within 6 h of stroke onset. Interestingly, our analysis also welcomed "wake-up" stroke patients who presented between 6 h and 24 h post-onset but exhibited a discernible mismatch between ischemic core and penumbra as per MRI readings. For all qualifying patients, intravenous thrombolysis (IVT) using rt-PA was the first line of intervention. If IVT was contraindicated, the subsequent step was mechanical thrombectomy post-CT and CTA evaluation. Crucially, intervention decisions were a collective resolution made by a cohesive team of neurologists, radiologists, and interventionists. The exclusions were patients displaying pronounced massive strokes on scans, especially encompassing more than a third of the MCA's territory. Ethical considerations were paramount, with informed consent being a requisite for all study participants [170,171].

3.9. Leveraging Deep Learning in Neuroimaging: A Paradigm Shift

Neuroimaging, especially brain CT scans, forms the bedrock of cerebral evaluations. Yet, the intricate task of interpreting emergent brain CT findings demands a high degree of expertise and can be labor-intensive for even adept neuroradiologists. Herein lies the prowess of deep learning, especially convolutional neural networks (CNN), which have been revolutionizing the medical imaging landscape [172]. Our study proposed the utilization of CNN-centric deep learning paradigms to efficaciously categorize strokes based on unenhanced brain CT imagery into normal, hemorrhage, infarction, and other diverse categories. The models under our analytical radar were CNN-2, VGG-16, and ResNet-50. These were not naive models; they had undergone prior training via transfer learning, adapting to various data magnitudes, mini-batch dimensions, and optimization algorithms. Their efficacy was put to the test with brain CT images. The findings were enlightening: when juxtaposed against other research outcomes, our models, especially CNN-2 and ResNet-50, showcased superior performance. Notably, while ResNet-50 clinched an impressive accuracy score of 0.9872, it took a tad longer to render outcomes in comparison to its counterparts. In essence, with the right hyperparameter fine-tuning, these deep learning models can be pivotal in clinical scenarios, aiding neurologists and radiologists in determining potential hemorrhagic strokes, infarctions, or other neurological manifestations [173].

Functional neuroimaging has significantly furthered our grasp of neural processes involved in post-stroke recovery and enhancements derived from brain stimulation. The variability observed among individuals in terms of recovery and response to treatment can be associated with imaging markers, notably connectivity. Cutting-edge methods in fMRI data analysis, like dynamic functional connectivity, enable exploration of stroke-induced changes in temporal network dynamics and their relationship to motor deficits. Nevertheless, we are yet to achieve a tailored approach that accounts for unique network pathology to accurately rectify specific network node dysconnectivity. Preliminary efforts, such as employing multivariate machine learning approaches to forecast motor deficits or outcomes based on initial post-stroke fMRI data, have been undertaken. Still, the validity of relying solely on a single MRI network marker for individualized predictions to ensure diagnostic precision remains a topic of discussion [172].

3.10. Comprehensive Cerebral Imaging: Collaborative Diagnostics in Emergency Care

Upon their admission to the emergency department, every patient underwent a cerebral computed tomography (CT) scan, which was conducted either with or without the administration of a contrast agent. This immediate imaging step was paramount to ensuring a swift diagnostic process [174]. The precise nature, severity, and anatomical location of the stroke were then meticulously diagnosed. This pivotal task was a collaborative effort between two medical experts: the radiologist who executed the brain imaging and the neurologist who conducted the clinical evaluation of the patient. This dual expertise ensured a comprehensive understanding of the patient's neurological condition, fostering more informed clinical decisions [175].

For a definitive diagnosis of acute ischemic stroke (AIS), the medical team adhered to the World Health Organization's long-standing definition of a stroke. Introduced back in 1970 and still deemed relevant in modern clinical practice, this definition characterizes a stroke as a "sudden manifestation of clinical symptoms pointing towards a focal (or occasionally global) disruption of cerebral functions. These symptoms persist for a duration exceeding 24 h or may even culminate in the patient's death. Notably, the only discernible causative factor for these symptoms should be of vascular origin, unless there are specific interventions like surgery or medication that might interrupt this course" [176]. This clear-cut definition, along with the combined insights from imaging and clinical examinations, solidifies the diagnostic accuracy, ensuring patients receive the most appropriate and timely care.

4. Treatment Paradigms

4.1. Pharmacological Approach

4.1.1. Exploring Intravenous Thrombolytic Agents: An In-depth Analysis of Mechanisms, Advantages, and Potential Hazards

1. Alteplase and Potential Alternatives in Stroke Treatment. Alteplase currently stands as the sole drug greenlighted by the FDA for the thrombolysis of acute ischemic stroke (AIS). However, the research horizon is dotted with other thrombolytic agents that might potentially rival or replace alteplase in the future. This comprehensive study dives deep into the potency and safety of an array of such agents—urokinase, ateplase, tenecteplase, and reteplase. Through sophisticated computational simulations entwining both pharmacokinetics and pharmacodynamics, paired with a meticulous local fibrinolysis model, we benchmarked the drugs against multiple metrics: clot dissolution timeframe, resistance to plasminogen activator inhibitor (PAI), potential risk of intracranial hemorrhage (ICH), and the latency from drug introduction to clot dissolution [177].

Our insights highlighted that urokinase boasted the most rapid clot dissolution. However, it also carried an elevated risk of ICH, linked to its propensity for excessive fibrinogen depletion in the plasma. Tenecteplase and alteplase showcased akin efficiencies in clot dissolution, but the former exhibited a reduced ICH risk and demonstrated superior resilience against PAI-1. Reteplase, interestingly, showed the most sluggish rate of fibrinolysis but left fibrinogen concentrations in systemic plasma untouched. To comprehend these intricacies further, we utilized a 1D mathematical model from prior research, offering predictions on the therapeutic results of these drugs based on their inherent properties and modes of action. The model reinforced clinical observations, advocating the possible supremacy of tenecteplase over alteplase while underscoring reteplase's limited efficacy despite its diminished ICH risk [178–180]. As AIS clinical studies inherently bear unpredictable hazards, computational models like ours could be pivotal in fine-tuning dosage regimens for multi-drug therapies or newly conceptualized drugs, given a clear understanding of their PKPD mechanisms and kinetic reactions [181].

2. Intracerebral Hemorrhage's Impact on Mortality Rates. Taking a retrospective stance, this study sifted through data from South Korea's national health insurance service database, spanning 2005–2018. The focal cohort consisted of hyperacute ischemic stroke patients who had undergone intravenous thrombolysis. A stark comparison was drawn between ICH-afflicted patients and those who avoided ICH. An alarming revelation was that within the 12-month window post-treatment, the mortality rate in the ICH cohort was more than double that of their counterparts (42.8% vs. 17.5%). This suggests that ICH post-thrombolysis can drastically heighten the risk of mortality in hyperacute ischemic stroke patients, amplifying it nearly threefold [182,183].
3. The Promise of Plasmin Nanoformulations in Ischemic Stroke Treatment. While thrombolytic therapy remains the gold standard for treating ischemic strokes, its current mainstream agent, the tissue plasminogen activator (tPA), often encounters obstacles due to an associated hemorrhage risk. Plasmin, a direct fibrinolytic agent, offers a safer hemostatic profile. However, its therapeutic potential diminishes when introduced intravenously due to rapid inactivation by anti-plasmin. To navigate this, nanoformulations have emerged as viable tools to enhance drug stability. This study unveils a groundbreaking nanoformulation for plasmin, demonstrating increased stability and heightened therapeutic efficacy, potentially redefining ischemic stroke treatment [184].

4. Decoding Blood–Brain Barrier Deficits Post-Stroke. Utilizing advanced two-photon microscopy, Knowland et al. have painted a vivid picture of changes in tight junctions (TJs) post-stroke in transgenic mouse models. Observations indicated that the blood–brain barrier (BBB) started leaking as early as 6 h post-ischemia, even though profound structural defects in TJs only became evident after 48 h. The increase in endothelial caveolae and transcytosis rate post-ischemia suggests a sequential deterioration of barrier mechanisms, highlighting the multifaceted impacts on the BBB following a stroke [185].

Top of Form

Under normal physiological conditions, the blood–brain barrier (BBB) stands as a sentinel at the frontier of the central nervous system (CNS), exercising selective permeability to maintain a stable environment. A couple of primary features define this selective permeability. First, endothelial cells (ECs) in the CNS are notable for their sparse vesicular population and an almost nonexistent rate of transcytosis. Second, there exists a sophisticated system of tight junctions (TJs) that bridge these ECs, creating an almost impenetrable seal [186].

A protein of considerable significance in this scenario is the major facilitator superfamily domain containing 2a (Mfsd2a). This protein has the unique distinction of being predominantly expressed in blood vessels that harbor the BBB. Intriguingly, its expression and activity are orchestrated by surrounding pericytes. Mfsd2a does not merely exist as a passive structural element; it plays a proactive role in crafting a specific lipid milieu. This environment is crucial as it actively inhibits the formation of caveolae vesicles within CNS ECs. By doing so, it curtails transcytosis, bolstering the integrity of the BBB [187].

However, during acute ischemic stroke (IS), this meticulously balanced environment faces upheaval. The aftermath of ischemia, as early as 6 h post-event, witnesses a surge in the number of endothelial caveolae. This spike is not merely quantitative; it ushers in an escalated rate of caveolae-mediated transcytosis. Macromolecules, which under regular circumstances would be barred entry, like albumin, now find passage via this heightened caveolae-mediated transcytosis. Meanwhile, the trusted sentinels of the CNS, the tight junctions, remain largely uncompromised initially. It is only after the 48-h mark post-ischemia that these TJs betray glaring structural anomalies [188]. These cellular-level dynamics do not occur in isolation. They are accompanied by collateral shifts in the surrounding CNS architecture. The surge in CNS endothelial vesicles coincides with discernible alterations in the pericytes' basement membrane coverage. This vascular restructuring is mirrored by astrocytic responses. The usually slender astrocytic end-feet swell, displaying signs of stress. This edema is further echoed in their mitochondrial structures, which too show signs of swelling and possible dysfunction [189]. In sum, acute ischemic stroke initiates a cascade of events that disrupt the once-steadfast BBB, compromising its selective permeability and potentially influencing disease progression and recovery (Figure 1).

4.1.2. Fibrinolytic Therapy in Acute Ischemic Stroke: A Comprehensive Analysis Understanding the Fibrinolytic Approach and its Clinical Applications

Thrombolytic therapy, aimed at breaking up clots that obstruct blood flow, is critical in the management of conditions such as acute ischemic stroke (AIS) and acute myocardial infarction (AMI). However, this treatment is not without its challenges. A prominent concern is the risk of harmful hemorrhagic complications, prompting clinicians to establish a specific time window within which this therapy can be administered with optimal safety [186]. Efforts are currently underway, through various basic and clinical studies, to innovate next-generation thrombolytic drugs that might potentially broaden this time window.

In AMI-focused clinical trials, a wealth of pharmacokinetic and pharmacodynamic (PKPD) data, like temporal plasma concentrations of tissue plasminogen activator (tPA) and other fibrinolytic proteins (e.g., fibrinogen, plasminogen, and α2-antiplasmin), have been harvested. Such data are invaluable for deepening our understanding of fibrinolysis mechanisms and discerning the impact of dosage regimens on both therapeutic efficacy and potential toxicity [187–189]. In contrast, AIS clinical trials have predominantly concentrated on clinical and neurological outcomes, such as those measured by the modified Rankin scale and NIHSS. This disparity in focus has posed challenges. Given the inherent pathological distinctions between AMI and AIS, there is an observed variance in response to tPA dosage. Consequently, it becomes challenging to transpose the PKPD insights gleaned from AMI trials directly onto AIS patients [190,191]. Notably, due to the inherent risks associated with AIS, such as a heightened propensity for intracranial hemorrhage (ICH) linked to acute brain infarction, lower tPA dosages are recommended for AIS compared to AMI. This dosage recommendation is primarily anchored in clinical outcomes rather than PKPD studies, amplifying the call for more PKPD research in AIS thrombolytic therapy. A practical solution to this dilemma might be the deployment of mathematical models, especially when real-time clinical PKPD studies on stroke patients are impractical given the time-sensitive nature of the treatment. For instance, we have pioneered a computationally adept simulation platform for AIS thrombolysis. This model, accounting for systemic drug effects and the 1D progression of clot dissolution, can predict outcomes like lysis completion time and plasma FBG levels, indicative of ICH risk, without cumbersome computational demands. This model's versatility is evident as it can emulate various therapeutic scenarios and can be tweaked to mirror different fibrinolytic drugs besides alteplase [192]. Furthermore, it can be adapted to diverse AIS conditions by adjusting local PD model parameters and clot features anchored in patient-specific data. Future ambitions for this model include expanding its scope to scrutinize the influences of diverse therapeutic and physiological factors on recanalization rates and to craft bespoke dosing regimens through model-driven optimization [193].

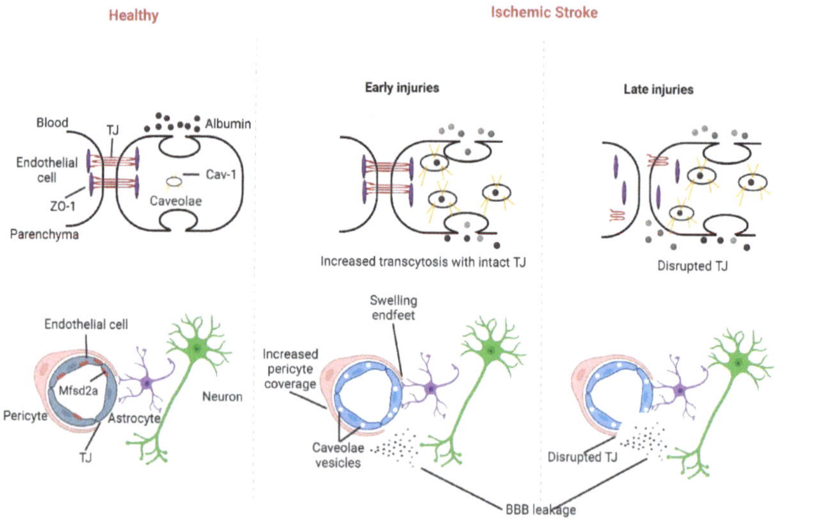

Figure 1. Illustrative Overview of Blood–Brain Barrier Dynamics in the Wake of Acute Ischemic Stroke.

At the root of ischemic strokes is an artery blockage in the brain, primarily stemming from a thrombotic or embolic event. This impediment triggers a cascade of cellular death pathways, referred to as the ischemic cascade [194]. The optimal countermeasure is swift recanalization of the occluded artery, achievable either through enzymatic fibrinolytic treatment using recombinant tissue plasminogen activators (rtPAs) to dismantle the clot, mechanical thrombectomy to physically excise the clot, or a fusion of both modalities [195]. The prognosis of ischemic stroke patients is intimately tied to rapid recanalization success; interestingly, a deficit of recanalization results in poor outcomes in both patients with acute ischemic stroke as well as in those with atrial fibrillation [196]. However, while rtPA-mediated recanalization has been documented to enhance functional outcomes, it is not without pitfalls. Shortcomings include a modest recanalization success rate (sub-30%) and the potential hemorrhagic and neurotoxic ramifications of rtPA, especially in cases where recanalization is ineffective [197–199]. Additionally, the therapeutic window for rtPA is constrained, typically to 4.5 h post-stroke onset, after which its utility diminishes and risks escalate. Thus, tools to predict patient responses to rtPA could revolutionize ischemic stroke management [200].

Complementary Medications Enhancing Mainline Therapy

In the aftermath of rtPA infusion, the body often grapples with inadequate PAI-1 levels, impairing its capacity to regulate tPA activity and, consequently, dissolve fibrin clots. Intriguingly, while rtPA can trigger platelet aggregation and bind platelet fibrinogen without a corresponding surge in thromboxanes, adjunctive antiplatelet therapy using aspirin might counteract thrombosis post-rtPA therapy. Some research insights have highlighted that rtPA activates specific platelets that deter aggregation [201]. There's a consensus, however, that such treatments are pivotal in thwarting extensive platelet activation subsequent to tPA therapy, and contemporary guidelines overwhelmingly support rtPA thrombolysis as a primary intervention [202].

Turning our attention to aphasia therapy, while speech and language therapy (SLT) alongside pharmacological interventions remain the cornerstone, augmentative therapeutic strategies are continuously sought to optimize the brain's recuperative potential, especially in stroke's chronic stages [203]. Among these, non-invasive brain stimulation (NIBS) methods, notably transcranial magnetic stimulations (TMSs) and transcranial direct current stimulations (tDCSs), hold promise. Multiple studies have documented these techniques' efficacy in enhancing linguistic functions in aphasic patients and in fostering neuroplasticity [204] Table 1.

Table 1. Non-invasive brain stimulation in post-stroke aphasia (PSA).

Methods	Major Characteristics	Advantages	Limitations	References
rTMS (repetitive transcranial magnetic stimulation)	- Generates a magnetic pulse that influences various cerebral cortex regions. - Regularly applied pulses of a specific frequency within about 15 min. - Low-frequency stimulation (<1 Hz) inhibits cortical excitability while high-frequency stimulation (>1 Hz) amplifies it.	Balances excitability across hemispheres and realigns the linguistic network.	Requires the formulation of an optimal treatment protocol; must consider individual variability.	[205–207]
tDCS (transcranial direct current stimulation)	- Non-invasive and secure method for brain stimulation. - Polarizes neuronal cell membranes, modulating cortical excitation levels. - Induced cortical changes are determined by the electrode pole.	Aids in normalizing brain activity, fostering self-recovery.	Further extensive clinical trials needed for a broader PSA group.	[208]

Neuroplasticity is the brain's inherent ability to adapt and change throughout an individual's life. Stroke stands as a predominant contributor to long-term disability, posing not just medical but also significant economic challenges globally. While a considerable amount of research in the last decade has spotlighted neuroprotection during the acute phase of a stroke, there remains a noticeable gap in studies focusing on its chronic phase [209]. The brain's capacity to reconfigure and restore the structure and functionality of its neurovascular networks is pivotal for recovery post-stroke. Employing a combination of adjuvant therapies alongside specific drugs might amplify the reparative processes, reinstating compromised brain functions [210]. Presently, various medications and rehabilitative strategies hold promise in promoting brain repair and enhancing clinical outcomes, even if initiated years after the occurrence of a stroke. Some pharmaceutical agents like citicoline, fluoxetine, niacin, and levodopa are either in active clinical use or undergoing trials [211,212]. Moreover, emerging research is delving into the potential of cell therapies, with our focus primarily on studies introducing cells in the stroke's early stages. Following this, we explore pharmaceutical interventions. Herein, we've zeroed in on cognitive, behavioral, and physical rehabilitation techniques as well as supplementary interventions aimed at neuroprotection, like non-invasive brain stimulation and the application of extremely low-frequency electromagnetic fields. Contemporary rehabilitation strategies portray a paradigm shift towards physical interventions, considering a therapeutic window extending up to six months post-stroke. However, prior research alludes to the possibility of an even more extended window for stroke recovery [213].

- Diving Deeper into Successful Pharmacological Interventions:

The identification of the ischemic penumbra in animal models and the proven efficacy of reperfusion therapies in humans once ignited optimism surrounding neuroprotection in acute stroke scenarios. However, the subsequent failures of numerous phase II and III trials led to the introduction of the STAIR recommendations, guiding both pre-clinical and clinical research [214]. Upon retrospection, it appears that the choice of agents earmarked for clinical development might not have been judicious. Despite fulfilling many STAIR criteria, the neuroprotective agent NXY-059 demonstrated no tangible benefit in a pivotal phase III study. Contemporary neuroprotective agents have yet to meet many STAIR recommendations. A myriad of unaddressed issues remain, ranging from the adoption of more representative animal models, ensuring drug distribution in human subjects, refining the evaluation metrics for neurological impairment and disability, to physiological optimization during human proof-of-concept studies. Enhancing the quality and quantity of clinical centers specializing in acute stroke research, leveraging surrogate imaging markers, and employing adaptive dose designs during phase II trials might elevate the success rates for identifying efficient neuroprotectives. Though the realm of neuroprotection in acute strokes remains fraught with challenges, its effectiveness remains a topic of debate [215]. In the face of the staggering burden that stroke presents and the limited applicability of reperfusion—currently benefiting a mere 10% of patients—there's an undeniable call for further proof-of-concept studies concerning neuroprotection, underpinned by meticulous reviews of pre-clinical data and robust phase II trial designs [216].

Inclusive evaluations of all research articles and case series, based on their evidence level, indicated that only CPSP, excluding other central pain types, was considered. Amitriptyline and lamotrigine stood out as the sole orally administered drugs exhibiting effectiveness in CPSP management in placebo-controlled studies. Though intravenous drugs like lidocaine, propofol, and ketamine demonstrated short-term CPSP control efficacy, their potential side effects and application modes render them unsuitable for prolonged treatment. The newly introduced antiepileptic drug, gabapentin, displayed promising results in managing CPSP in select patients. It is noteworthy that amitriptyline, lamotrigine, and gabapentin collectively present a more encouraging efficacy and safety profile in contrast to traditional antiepileptic medications like carbamazepine and phenytoin, which lacked any placebo-controlled evidence advocating their efficacy. The pressing need of the hour is to conduct clinical trials that can optimize the pharmacological management of CPSP [217].

Our review encapsulated 10 distinct trials. Regrettably, most of these trials were bereft of sufficient details, making methodological quality assessments challenging. The studied drugs included piracetam, bifemalane, piribedil, bromocriptine, idebenone, and Dextran 40. Preliminary findings hint that patients undergoing treatment with piracetam demonstrated a higher likelihood of language improvement upon trial completion. Notably, there were no significant differences in the incidence of adverse effects, including death, between those administered piracetam and those on a placebo. Nevertheless, the disparity in death rates between both groups raises potential concerns regarding the elevated risk associated with piracetam intake. It remains elusive whether drug treatments offer more pronounced efficacy compared to speech and language therapy or if one drug trumps another in terms of effectiveness [218].

We embarked on a systematic review, adhering to the Cochrane methodology, that focused on randomized placebo-controlled trials examining the efficacy of antidepressants in treating or preventing depressive disorders and "abnormal mood" following a stroke. Alongside mood implications, we also analyzed the potential effects of these medications on physical and other outcomes. The available data were sourced from 7 treatment trials encompassing 615 participants and 9 prevention trials that included 479 participants. Owing to notable inconsistencies in trial design, the quality of studies, and reporting methods across the different research works, we opted against combining all the outcome data. Our findings from the treatment trials revealed that while antidepressants did mitigate mood symptoms, their impact on facilitating the remission of clinical depressive disorders was nebulous at best. Moreover, the studies did not provide concrete evidence supporting the idea that antidepressants can either ward off depression or catalyze recovery post-stroke. The current body of randomized evidence is insufficient to endorse the regular use of antidepressants for averting depression or accelerating recovery after a stroke. While these drugs might uplift mood in post-stroke patients with depression, the tangible clinical significance of such minor improvements, especially in patients without major depressive disorders, remains to be elucidated. This underscores an urgent need to further investigate the precise role of antidepressants in managing stroke aftermaths [219].

Stroke stands out as a primary contributor to severe prolonged disability among adults and ranks as the world's second-leading cause of mortality [3]. The immediate aftermath of a stroke has historically witnessed an emphasis on rapid reperfusion and neuroprotection methodologies. By harnessing diverse mechanisms, pharmacological interventions have the potential to lessen disabilities in a significant proportion of survivors of acute strokes [220]. The inherent ability of the brain to adapt post-stroke, courtesy of its plasticity mechanisms, can be influenced by drugs [221]. The research landscape is rich with explorations into a myriad of therapeutic interventions, encompassing small molecules, growth factors, and monoclonal antibodies [222]. A noteworthy development was the discovery that the SSRI fluoxetine could ameliorate motor deficits in patients grappling with ischemic stroke-induced hemiplegia; intriguingly, this appeared to be independent of the patient's depressive state [223]. Given these findings, it is imperative to bolster pioneering research endeavors. This will help usher in groundbreaking pharmacological treatments that prioritize neurological recuperation after a stroke, as opposed to merely focusing on acute de-occlusion and neuroprotection. Our manuscript stems from the collaborative insights of 14 seasoned scientists, and its objectives are threefold: (1) to shed light on pivotal facets of human brain plasticity post-stroke and viable pharmacological targets for rehabilitation; (2) to foster discourse on the most fitting characteristics of clinical trials assessing post-stroke recovery drugs; and (3) to set forth recommendations for upcoming clinical trials.

Several pharmacological solutions have been conceptualized for stroke treatment. They operate either by halting molecular pathways that lead to neuronal death or by bolstering neuronal survival and rejuvenation. Bar rtPA, a majority of these therapeutic agents have not made the leap from the laboratory to successful clinical trials. However, recent strides in understanding stroke's pathophysiological nuances have reignited interest in crafting neuroprotective agents tailored for stroke intervention. This renewed interest has uncovered innovative molecular targets poised to offer tangible clinical advantages in neuroprotection and neurorestoration. Our review delves into the latest monumental advancements in stroke pharmacology, with an emphasis on ischemic stroke. We encapsulate emerging therapeutic mechanisms and cast a spotlight on the most recent clinical trial outcomes [204].

Stroke's impact on health is undeniable, and formulating an effective treatment remains a daunting task. Bar rtPA, the number of clinically approved pharmacological solutions remains meager. Yet, the past decade has witnessed remarkable breakthroughs in unraveling molecular signaling pathways and gene expression blueprints underlying stroke's pathophysiology. These revelations have paved the way for fresh therapeutic targets. Considerable energy has been channeled into conceiving innovative pharmacological agents and bolstering their transition from the lab to the clinic. Agents such as NMDAR antagonists, calcium channel blockers, anti-inflammatory compounds, neurotropic factors, monoclonal antibodies, and even non-coding RNAs have been rigorously probed in experimental setups. A handful, including minocycline and edaravone, have showcased their ability to elevate neurological outcomes in acute ischemic stroke patients. These successes have infused renewed vigor into translational studies centered on neuroprotectants. With an ever-growing arsenal of neuroprotective agents emerging from preclinical stages, there has been a noticeable uptick in clinical trials evaluating pharmacological strategies for acute stroke. A slew of these trials are currently in progress, signaling hope and momentum for the field [224] Figure 2.

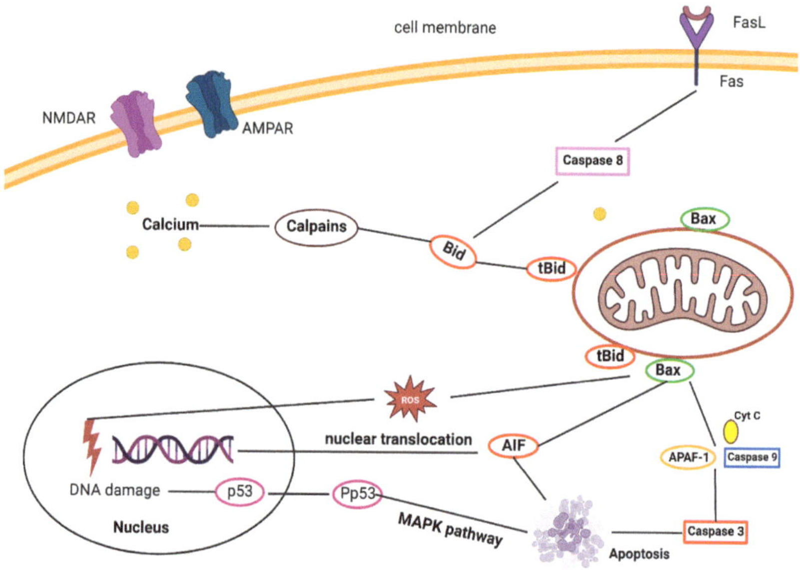

Figure 2. The intricate processes leading to ischemic neuronal cell death involve various molecular pathways and interactions. When NMDA receptors (NMDARs) and AMPA receptors (AMPARs) are stimulated, there is an elevation in the levels of calcium within the cell's cytoplasm. This heightened calcium concentration triggers the activation of enzymes known as calpains and induces dysfunction within the mitochondria. Simultaneously, the binding of Fas ligands (FasL) to their counterparts, the Fas death receptors, sets off the activation of a protein called caspase 8. These activated calpains, together with caspase 8, collaborate to cleave a protein named Bid, transforming it into its truncated version, tBid. Once formed, tBid associates with another protein, Bax, on the mitochondrial membrane. This interaction is critical as it results in the creation of pores in the membrane, leading to the expulsion of several vital molecules: cytochrome c (Cyt c), apoptosis-inducing factor (AIF), and the harmful reactive oxygen species (ROS). After their release, both ROS and AIF relocate to the cell's nucleus. Here, they play a pivotal role in damaging the DNA and initiating specific neuronal cell death pathways. A prime example of such a pathway is the phosphorylation of the protein p53, which, when phosphorylated (Pp53), activates the MAPK signaling route, pushing the cell towards apoptosis. Additionally, the expelled cytochrome c from the mitochondria plays a role outside its traditional function. In the cell's cytoplasm, cytochrome c collaborates with the apoptotic protein activating factor-1 (APAF-1) and procaspase 9 to assemble into a complex known as the apoptosome. This structure is integral to the internal apoptotic pathway, further cementing the neuron's fate towards programmed cell death.

4.2. Endovascular Thrombectomy

4.2.1. Evolution of Endovascular Strategies over Time

Endovascular treatment (EVT) for acute ischemic stroke (AIS) has experienced significant advancements, but one of the persistent challenges remains futile recanalization. After a cerebral infarction, an inflammatory response is triggered, which plays an essential role in dictating the outcome of the recanalization procedure. This study was centered on the correlation between the systemic inflammatory response index (SIRI) and the phenomenon of futile recanalization in AIS patients [225]. By retrospectively examining patients with anterior circulation proximal arterial occlusion who underwent EVT, it was determined that factors such as age, the severity of the stroke at admission, and a higher SIRI upon admission are linked with poor functional outcomes three months post-procedure [225].

Further developments in AIS treatment techniques shed light on indications for reperfusion therapy. Beyond the already established protocol for intravenous thrombolysis using rtPA within 4.5 h and mechanical thrombectomy for large artery occlusion within 6 h, recent trials have expanded these timelines. For instance, some patients can now receive reperfusion treatment up to 24 h from the onset of their symptoms, provided they fit certain criteria and undergo necessary brain imaging [226].

4.2.2. Mechanisms of Mechanical Thrombectomy

Mechanical thrombectomy has revolutionized the treatment of AIS resulting from large vessel occlusions. Over the past half-decade, numerous trials have validated its efficacy and safety for such patients, with the treatment window extending up to 24 h after the stroke onset [227]. Some landmark studies from 2015, such as MR CLEAN and ESCAPE, underscored the superiority of mechanical thrombectomy over traditional medical management in cases of anterior circulation large vessel occlusion strokes [155]. Later in 2017, studies like DAWN and DEFUSE-3 expanded the time frame for thrombectomy up to 24 h for certain patient groups [227]. Now, societal and national guidelines have incorporated these findings, recommending mechanical thrombectomy for a particular patient subset. Additionally, ongoing trials are investigating the technique's potential benefits for other stroke subpopulations, further establishing the method's relevance and potential [157].

Recent trials have spotlighted the combination of stent retriever thrombectomy with intravenous thrombolysis in AIS. It is now recommended for patients with large vessel occlusions when intravenous thrombolysis is not feasible, emphasizing the importance of rapid intervention post-symptom onset [228,229]. To ensure equitable access for all eligible patients to endovascular therapy, a call has been made for restructuring stroke care systems [230,231].

4.2.3. Spotlight on Equipment: Delving into Stent Retrievers, Aspiration Catheters, and Their Innovations

A substantial number of ischemic strokes, approximately 10%, are categorized as large hemispheric infarcts, which can lead to severe disabilities. With the intent of minimizing the disease's adverse effects, research has concentrated on multiple fronts, from pinpointing the predictors of malignant edema to improving surgical techniques. Decompressive surgery has emerged as a pivotal intervention, with studies delving into the nuances of managing large hemispheric ischemic strokes [232]. Neuroinflammatory responses are contingent upon the context, temporal progression, and nature of the neurological disturbance. Following ischemic stroke, brain injury manifests through necrosis and apoptosis, which subsequently instigates an inflammatory response mediated by the release of reactive oxygen species (ROS), chemokines, and cytokines. This inflammatory cascade originates in the microvasculature and encompasses various cell types, including innate immune cells like microglia and adaptive immune cells such as lymphocytes, leading to neuronal demise. Proinflammatory cytokines are implicated in numerous cerebral processes, directly impacting endothelial cells, neurons, and glial cells. Due to this intricate and sequential pathway, these cytokines might augment cellular movement or induce further damage. Post-ischemic stroke, microglia undergo a transition to the M1 phenotype, an inflammatory-promoting state that produces interleukin-1β (IL-1β), a proinflammatory cytokine with neurotoxic attributes. IL-1β can engage with the vascular endothelium, escalating leukocyte adherence and contributing to edema formation. Conversely, anti-inflammatory molecules like interleukin-10 (IL-10) and interleukin-4 (IL-4) act to negate the effects of proinflammatory cytokines. Under physiological circumstances, a nuanced equilibrium exists between proinflammatory and anti-inflammatory cytokines. However, in the context of stroke, a decline in anti-inflammatory IL-10 has been observed, correlating with improved patient outcomes, underscoring the intricate interplay between these inflammatory mediators during ischemia's initial phases [233]. Furthermore, the function of microglia and monocytes/macrophages during cerebral ischemia is determined by their M1/M2 polarization states. The presence of specific cytokines in the surrounding environment, such as IFN-γ for M1 and TGF-β and IL-10 for M2, dictates their polarization. A predominance of the M1 phenotype is associated with exacerbated ischemic damage, activation of hypoxia-inducible factor-1 (HIF-1), and heightened anaerobic glycolysis. A shift to the M1 phenotype in microglia, coupled with an upsurge in IL-23 production, fosters the attraction and activation of γδ T cells. These represent a distinct subset of innate T cells, potentially exacerbating the detrimental consequences of acute ischemic stroke. Astrocytes, under physiological conditions, actively uptake surplus extracellular glutamate, converting it to glutamine for neuronal utilization. However, following brain injury, astrocytic damage may compromise this glutamate uptake capacity. The precise impact of ischemia on astrocytic glutamate uptake remains ambiguous, but there are indications that the expression of the glutamate transporter, EAAT2, becomes compromised during ischemic events. Two pivotal oxidative enzymes, xanthine oxidase and NADPH oxidase, are central to the synthesis of the superoxide anion, a critical radical post-stroke. In clinical observations, an independent positive correlation has been established between serum MMP-9 levels and both initial stroke severity and subsequent clinical recovery. Moreover, the permeability of the blood–brain barrier (BBB) mediated by MMP in ischemic stroke can be attenuated by cyclooxygenase 2 inhibitors. Thus, targeting Cox-2 activity or prostaglandin E2 can serve as a strategy to thwart MMP-induced BBB disruptions [233].

A growing body of evidence underscores the pivotal role of inflammation in stroke pathogenesis, marking it as a promising therapeutic target. Nevertheless, multiple studies suggest that inflammatory cells play a dual role—both beneficial and harmful—indicating that inhibiting a specific inflammatory pathway at an inappropriate time could potentially exacerbate the disease process. A more nuanced understanding of the temporal dynamics of stroke pathophysiology could inform more precise therapeutic interventions. Furthermore, integrating preclinical research using stroke models and concurrent relevant clinical conditions, such as type 2 diabetes, preceding infections, and atherosclerosis, may bridge the gap between experimental findings and clinical applications, potentially informing future successful stroke treatments.

Investigations into rodent models for PSD were categorized based on certain criteria. These models were segmented into three categories: surgical techniques, specific structures related to cognitive function, and comorbid conditions [234]. Each category emphasizes different aspects, such as the modeling technique in "surgical technique" or stroke models combined with other conditions in the "comorbid condition" category [235].

4.2.4. The Benefits and Implications of Surgical Revascularization

Surgical revascularization stands out as a primary method to enhance cerebral hemodynamics, even though this particular facet has yet to be extensively scrutinized in randomized clinical trials [236]. The primary aim of revascularization is to mitigate the threat of ischemic incidents by augmenting cerebral blood flow (CBF) and reestablishing the cerebral vascular reserve (CVR). This is done through the creation of collateral pathways, stemming from observations that moyamoya arteriopathy (MMA) impacts the internal carotid arteries (ICAs) and their primary branches but leaves the external carotid arteries (ECAs) unscathed [236].

The Japan adult Moyamoya (JAM) trial stands as the inaugural prospective randomized controlled study specifically centered on the surgical treatment of MMA. Its findings highlighted the protective role of surgical revascularization against recurrent bleeding, primarily attributed to the post-operative diminishment of moyamoya vessels and the reduced hemodynamic stress in delicate collaterals [237]. The primary surgical intervention indicators encompass recurrent cerebral ischemic symptoms, cerebral hemodynamic deterioration (characterized by diminished regional CBF, vascular response, and perfusion reserve), and hemorrhaging resulting from the rupture of posterior collateral vessels [238–240].

For a considerable period, the surgical approach for hemorrhagic MMA was debated. However, the JAM trial has unveiled the protective effects of revascularization against subsequent bleeding, especially in patients who underwent intracranial hemorrhaging in the preceding year [239]. Delving deeper into the findings of the JAM trial, Miyamoto et al. accentuated the impact of bypass surgery on curbing the bleeding risk based on the hemorrhage site. The study illustrated a pronounced advantage of surgery for those with posterior hemorrhages in comparison to anterior hemorrhages [239,240].

In pediatric MMA cases, added vigilance is required, given the disease's aggressive nature in younger patients. Consequently, revascularization surgeries are recommended for the majority of pediatric patients and should be executed without delay [237].

4.2.5. Procedure Walkthrough: A Comprehensive Overview from Patient Selection to Post-Operative Care

A significant study comprising 333,117 patients revealed that only a minute fraction, specifically 286 (0.09%), suffered a stroke post-procedure. Since the difference between total hip arthroplasty (THA) and total knee arthroplasty (TKA) was not significant in predicting stroke occurrence, both procedures were jointly analyzed. The findings indicated that a majority, 65% of the strokes, materialized before the patients were discharged. A more detailed breakdown revealed that a quarter of these strokes took place by the first postoperative day, half by the second, and three-quarters by the ninth postoperative day. Factors like age, elevated American Society of Anesthesiologists (ASA) score, and smoking habits emerged as independent risk factors for postoperative strokes. For instance, patients in the age bracket of 60–69 had an odds ratio (OR) of 4.2, 70–79 had an OR of 8.1, and those aged 80 and above had an OR of 16.1. Similarly, patients with an ASA score equal to or greater than 3 had an OR of 1.7, while smokers had an OR of 1.6 [241].

Such findings can serve as a vital foundation for medical professionals to counsel patients about potential risks and make informed decisions to optimize patient care, ensuring the best possible outcomes after procedures like THA/TKA.

4.2.6. Early Post-Operative Strokes: An Analysis of Incidence, Risk Factors, and Outcomes

In a retrospective case-control study conducted at a university-based tertiary care hospital, researchers delved into the occurrence, risk factors, and outcomes of early post-operative strokes. Specifically, they focused on cases where strokes transpired within 24 h post-surgery. By scrutinizing medical records from 2015 to 2021, they aimed to identify trends and potential preventative measures [242].

The methodology included comparing early post-operative stroke cases with age-matched controls at a 1:3 ratio. The data covering patient demographics, events during surgery, and post-operative results were meticulously analyzed. To pinpoint risk factors that could lead to post-operative strokes, a multiple logistic regression analysis was employed.

Findings indicated that early post-operative strokes, those occurring within 24 h, had an incidence rate of 0.015% (43 out of 284,105 cases). The multivariable analysis highlighted several risk factors:

- An American Society of Anesthesiologists (ASA) physical status of $\geq$3 was associated with a higher risk (adjusted OR: 3.12) [242].

- Surgeries lasting more than 120 min also elevated the risk (adjusted OR: 10.69) [242].
- Experiencing intra-operative hypotension and the usage of inotropes/vasopressors during surgery also increased the risk (adjusted OR: 2.80) [242].

When contrasted with control groups, stroke patients exhibited increased rates of both planned and unplanned ICU admissions, longer hospital stays, more frequent use of ventilators, and a higher mortality rate. Despite the relatively low occurrence of early post-operative strokes (0.015%), these incidents lead to deteriorated clinical outcomes and a surge in mortality rates. It is imperative that medical professionals recognize and address potential risk factors and strategize for optimization to diminish the incidence of these strokes [242].

4.2.7. The Transformative Effects of Timely Endovascular Intervention: Key Insights from Case Studies

The American Heart Association and the European Stroke Organization have jointly endorsed endovascular thrombectomy (EVT) in specific cases. Their recommendations include using EVT, along with best medical practices, for adults experiencing anterior circulation stroke. The criteria encompass pre-stroke modified Rankin scale (mRS) scores below 2, occlusions in the internal carotid artery or middle cerebral artery, patient age over 18, and ASPECTS and NIHSS scores exceeding 6 [243,244].

Notwithstanding, there exists a debate concerning the use of EVT in patients presenting with NIHSS scores below 6. Those presenting after the 6-h window but meeting the criteria set by the DAWN and DIFFUSE 3 trials are also considered suitable candidates for EVT. The ongoing TESLA clinical trial is keen on enrolling patients with low ASPECT scores to further investigate the efficacy and safety of the procedure in this subgroup [245].

In a comprehensive review encompassing seven trials with 980 participants, distinctions were drawn between intravenous thrombolytic treatment and endovascular thrombectomy in cases of the anterior intracranial circulation's large vessel occlusion. Notably, all these trials employed advanced imaging techniques for patient selection. The findings reflected that intravenous thrombolytic treatment resulted in a 66% good functional outcome at a 90-day follow-up, while for endovascular thrombectomy, this figure stood at 46%. However, there was a potential increased risk of symptomatic intracranial hemorrhage associated with thrombolytic treatment [246].

In conclusion, for selected patients with acute ischemic wake-up strokes, both therapeutic interventions could significantly improve functional outcomes without amplifying the risk of mortality. Yet, it remains crucial to continue researching to determine the best patient selection criteria [246].

5. Combined Treatment Modalities

Over the past decade, our understanding and approach to stroke management have evolved significantly. This transformation has been characterized by a broader understanding of what constitutes a stroke and notable advancements in its prevention and treatment techniques. One significant breakthrough in the treatment of acute ischaemic stroke is the synergy achieved by pairing endovascular thrombectomy for large artery occlusion with intravenous alteplase, which further enhances the likelihood of patients attaining functional independence by an additional 20%. Moreover, the protective role of aspirin in thwarting early recurrent ischaemic strokes is more potent than initially believed. The panorama of preventative strategies against recurrent strokes has expanded, now offering direct oral anticoagulants as an alternative to the traditional warfarin for atrial fibrillation patients. In addition, carotid stenting has emerged as a viable option against endarterectomy for patients displaying symptomatic carotid stenosis. Delving into acute intracerebral haemorrhage, ongoing research trials are working diligently to measure the efficacy of various interventions, from acute blood pressure regulation and haemostatic therapies to minimally invasive surgical methods, anti-inflammatory treatments, and neuroprotective techniques. A rising frontier is the use of pharmacological and stem-cell treatments to promote brain regeneration and fortify rehabilitation processes, facilitating optimal functional recovery. Studies on embryonic stem cells, mesenchymal cells, and induced pluripotent stem cells have explored their capabilities in tissue regeneration, sustenance, cell migration and proliferation, neural circuitry reconstruction, and physical and behavioral restoration. Innovations in stem cell and genomic technologies have ushered in regenerative treatments designed to reconstruct neural pathways and mend neurons affected by ischemic events. Notwithstanding the descending trend in stroke-related mortalities, the global toll of stroke remains a concern. Addressing this requires an expansive preventive approach that not only encompasses individuals across the risk spectrum but also dovetails with prevention regimes for other illnesses bearing shared risk factors [247].

For over four decades, efforts to engineer treatments that shield neurons and other brain cells from the cellular and molecular ramifications of cerebral ischaemia during acute ischaemic stroke (AIS) largely proved unfruitful. Yet, with the dawn of intravenous thrombolysis coupled with endovascular thrombectomy, the medical community has ushered in a revolutionary phase of AIS treatment. Within this realm, reperfusion therapies have emerged as potent strategies [158]. Given this new backdrop, it becomes imperative to re-evaluate cytoprotective treatments as supplementary aids to reperfusion therapy. It is essential that our energies be redirected towards designing novel drugs that holistically address multiple facets of the ischaemic cascade. Drugs developed in the past should also be revisited, especially if they showcased substantial cytoprotective effects in lab settings and were deemed safe in early human trials [248]. Multiple avenues for coupling cytoprotection with reperfusion are conceivable. This review delves deep into potential targets for cytoprotective treatment, underscoring crucial considerations for future drug formulations. A spotlight is also cast on the recent ESCAPE-NA1 trial on nerinetide, marking some of the most encouraging results yet. Furthermore, we inspect innovative clinical trials that aim to discern if cytoprotective drugs can either decelerate infarct growth preceding reperfusion or mitigate the aftereffects of reperfusion, such as haemorrhagic transformation [249].

Intracranial atherosclerosis (ICAS), characterized by its progressive nature, leads to gradual stenosis and cerebral hypoperfusion, contributing significantly to both initial and recurrent stroke incidents globally. This ailment's emergence can be attributed to a myriad of factors. Enhanced angiographic imaging methodologies now enable more accurate ICAS diagnoses and facilitate the tailoring of appropriate therapeutic courses. Notably, neither intensive medication regimens nor endovascular interventions have managed to fully eliminate the recurrence of strokes in ICAS-affected individuals. The medical landscape is now observing the emergence of non-pharmacological therapies like remote ischemic conditioning and hypothermia [250]. A holistic therapy approach combining medication, endovascular procedures, and/or non-pharmacological treatments might be the key to the future of ICAS management. This piece provides an in-depth overview of ICAS, covering its epidemiology, underlying mechanisms, risk determinants, biomarkers, diagnostic imaging, and overall management [251].

The past few years have witnessed significant advancements in acute stroke therapy. With the introduction of thrombolytic treatments, advanced endovascular procedures, and specialized stroke care units, there has been a marked improvement in survival rates and prognostic outcomes for stroke victims. However, it is crucial to note that a significant portion of patients, especially those without access to these advanced interventions, continue to experience high stroke mortality, compounded with residual morbidity. For many, the aftermath of a stroke translates into severe motor and cognitive impairments, culminating in a loss of autonomy in daily life tasks [252]. Against this backdrop, recent research endeavors have been channeled to mitigate the cerebral damage caused by acute ischemia, termed 'neuroprotection.' Concurrently, there is a focus on amplifying recovery processes—encompassing plasticity, neuroregeneration, and complementing rehabilitation—to enhance recovery probabilities and facilitate a return to normal functionalities, known as 'neurorepair.' Citicoline, in particular, has exhibited therapeutic prowess at various junctures of the ischemic cascade during acute ischemic strokes and has showcased efficiency across diverse animal-based stroke models [253]. Chronic administration of citicoline has been proven to be both safe and efficacious, curtailing post-stroke cognitive deterioration and bolstering patients' functional recuperation. Sustained citicoline treatment at optimal doses has been remarkably well-received, propelling intrinsic neurogenesis and neurorepair processes, which synergize with physical therapy and rehabilitation [254].

5.1. Global Prevalence and Management of Stroke

Stroke stands as the second primary cause of death and a principal driver of disability worldwide, witnessing a surge in incidence, particularly in developing nations [255]. A substantial portion of strokes are ischaemic in nature, arising from arterial blockages. Treatment priorities lie in immediate reperfusion using intravenous thrombolysis and endovascular thrombectomy, tools that effectively reduce disability but are immensely sensitive to time. To harness the full potential of these reperfusion therapies, it is imperative to overhaul care systems to minimize treatment delays. When administered within 4.5 h post-stroke onset, intravenous thrombolysis demonstrates a significant decrease in disability rates. Furthermore, thrombolysis has been found to aid patients showing salvageable brain tissue through perfusion imaging even up to 9 h post-stroke onset, including those who wake up exhibiting stroke symptoms. Endovascular thrombectomy has been effective in curtailing disability for patients experiencing large vessel occlusion within a 6-h window post-onset, and for those pinpointed through perfusion imaging, even 24 h after stroke onset. Prevention strategies for ischaemic stroke encompass many facets akin to cardiovascular risk management seen in other medical sectors, such as monitoring blood pressure, managing cholesterol levels, and deploying antithrombotic drugs. Additionally, certain preventive methods are tailored according to the stroke's origin, like utilizing anticoagulation for atrial fibrillation or undergoing carotid endarterectomy for pronounced symptomatic carotid artery stenosis [256].

5.2. The Promise of Combined Therapies

Our objective was to evaluate the joint effects of kinesio taping (KT) and modified constraint-induced movement therapy (mCIMT) on upper limb functionality and spasticity in stroke-induced hemiplegic patients. The study, randomized and controlled, was conducted in a hospital setting on stroke patients with hemiplegia spanning 3–12 months post-stroke. A total of 35 patients participated, divided into three cohorts: sham KT with mCIMT, KT alone, or a combination of KT and mCIMT. All these treatments acted as supplementary therapies, administered alongside standard rehabilitation. The outcomes revealed significant enhancements in various scales and tests, proving that KT is beneficial for stroke patients in terms of reducing spasticity and boosting upper limb function. Importantly, coupling KT with mCIMT delivers augmented advantages in motor skills with sustained effects [257].

However, the clinical aftermath following EVT in patients suffering from proximal anterior circulation AIS can be less than ideal, even with successful recanalization. Achieving recanalization does not guarantee clinical recuperation, as reperfusion might not occur, and even when it does, it can induce damage in the brain [258]. A study pinpointed the SIRI as a determinant of unproductive endovascular reperfusion. Diving deeper into the factors and mechanisms dictating unfruitful reperfusion could pave the way for novel therapeutic avenues and strategies, potentially optimizing outcomes for acute ischemic stroke patients [225].

Moreover, frailty tends to escalate mortality and morbidity rates in patients with large vessel occlusions. With only about 20% of moderately frail patients achieving desirable outcomes, the significance of recognizing such patients and risk stratification becomes clear. The HFRS offers a means to assess these patients and provide valuable insights to their families regarding potential outcomes [259].

5.3. Advancements in Stroke Management

The realm of stroke management has undergone rapid evolution, solidifying its significance within the neurology discipline. Over time, there has been a refined understanding and classification of distinct stroke types, enhancing stratification and subsequently refining management strategies [260]. Therapeutically, the landscape has dramatically transformed. No longer solely reliant on IV TPA, today's stroke management has embraced more intricate and state-of-the-art techniques. A notable stride in this domain is EVT, which, through direct clot visualization and prompt clot retrieval methods, has remarkably enhanced patient outcomes. Moreover, preventive measures against stroke recurrence have seen marked advancements, with robust evidence now supporting the use of alternative antiplatelet therapies and oral anticoagulants, further refining patient care [261].

At the heart of top-tier stroke care lies the principle of multidisciplinary involvement, starting with prehospital care and extending through post-stroke rehabilitation. This article delves into the assessment and the latest progress in managing AIS. Beginning in the prehospital setting, moving into the emergency department, and subsequently transitioning into post-acute hospital care and rehabilitation, this article captures the full spectrum of AIS management. Furthermore, it emphasizes areas in stroke care that presently show practice gaps, necessitating more comprehensive studies. A crucial note is that, in spite of the tremendous progress in stroke management, post-stroke disability remains a prevalent issue. There is an urgent need for studies focusing on refining prehospital systems for the swift recognition and transfer of stroke patients to suitable centers, ensuring prompt treatment with thrombolytics, and delineating the potential of EVT in handling posterior circulation and distal vessel blockages [262].

5.4. Complications, Implications, and Potential Interventions

The acute damage dealt to the brain post-stroke can spur elevated levels of catecholamine and cortisol. These heightened levels, unfortunately, can trigger the apoptosis of peripheral lymphocytes, leading to functional deactivation [263]. Given their pivotal role in cellular and humoral responses, lymphocyte loss weakens defense mechanisms against pathogens, heightening susceptibility to infections—a prevalent post-stroke complication that can exacerbate the clinical trajectory [264,265]. Notably, specific lymphocyte subgroups preserve immune homeostasis, functioning as neuroprotective agents by dampening proinflammatory mediators, modulating microglial activity, restraining autoreactive cells, and fostering neurogenesis and reparative processes within ischemic regions [266,267].

Our findings augment existing evidence on plasma biomarkers predicting stroke outcomes. It is vital to recognize the limitations of solely relying on single cellular line assessments, as they might not encompass the multifaceted nature of immune responses and may be susceptible to biases stemming from conditions such as overhydration, dehydration, and blood specimen handling [268,269]. A noteworthy indicator is the SIRI, which not only correlates directly with the risk of unfruitful recanalization but also complements existing clinical predictors like age and baseline stroke severity. It encapsulates the equilibrium between innate and adaptive immunity, with elevated values hinting at heightened innate immune activity and diminished adaptive immunity. In the acute stroke phase, SIRI can symbolize the immediate inflammatory response to brain damage, shedding light on the likelihood of secondary impairments and susceptibility to ensuing complications [225].

Lastly, it is imperative to mention ischemic brainstem strokes, which account for 10% of all ischemic cerebral strokes. The onset of hemorrhagic complications spells a particularly grim prognosis. Symptoms span vertigo, cranial nerve manifestations, and both crossed and uncrossed corticospinal tract findings. In the decision-making process for brainstem stroke management, avant-garde neuroimaging techniques have emerged as indispensable. They hold potential for pinpointing patients who might benefit from thrombolysis. However, given the paucity of substantial data guiding many current recommendations, there is a palpable need for further research dedicated to acute brainstem stroke treatment [270].

5.5. Immune Dynamics Post-Stroke

The aftermath of a stroke reveals that the brain's damage is not just a simple mechanical disruption but rather intricately interlinked with immune processes. When the brain gets assaulted during a stroke, it spirals into an inflammatory state, eliciting a chain reaction involving the sympathetic nervous system and the hypothalamic-pituitary-adrenal axis. Consequently, this leads to the production of immune-suppressing compounds such as catecholamines and glucocorticoids [271].

Newer research has shed light on the brain-immunity axis, a sophisticated pathway responsible for inducing immune suppression following a stroke. The common belief is that this suppression is the brain's proactive measure to curtail additional cerebral damage. But this defensive stance comes at a cost: it leaves the patient more vulnerable to infections. Conditions like pneumonia and UTIs become frequent adversaries for stroke patients, often leading to aggravated health outcomes and a heightened mortality rate [272].

Given these findings, the medical fraternity has pivoted its focus toward minimizing the risk of post-stroke infections. The idea of preventive antibiotic treatment initially garnered interest; however, a deeper dive into clinical studies indicated that such prophylactic measures did not necessarily bring down pneumonia rates or enhance clinical results. One significant concern stems from discrepancies in pneumonia diagnosis standards across healthcare facilities. With the looming shadow of antibiotic-resistant strains, the community is nudged to explore other avenues. Immunomodulation therapies, which bolster the immune system's ability to tackle infections, appear promising. Yet, one must tread cautiously to ensure that these interventions do not inadvertently intensify cerebral damage. As we move forward, it is imperative for research to delve deeper into the precise dynamics of immune suppression post-stroke, potentially revealing novel therapeutic avenues to counteract associated infections [273,274].

6. Envisioning Comprehensive Patient Care

1. *Merging Physical Rehabilitation with Acute Stroke Care*:

Post-stroke therapies bear a dual focus: while neuroprotection aims to minimize immediate stroke-induced damage, repair therapies target cerebral restoration. The therapeutic landscape is rife with potential innovations. For instance, "mirror therapy", seamlessly blending with conventional physical therapy approaches, emerges as an easily applicable and effective technique [275]. On a similar note, neuromuscular electrical stimulation has been lauded for its ability to rejuvenate neuromuscular function and kickstart cerebral adaptability [276].

When it comes to augmenting motor function, combining transcranial magnetic stimulation with standard physical and occupational therapies has proven fruitful. The enhancement arises from a heightened stimulation of the motor cortex regions on the side opposite the hemiplegia [277]. Of notable mention is the ongoing NEST-3 trial, a rigorous exploration of the NeuroThera® Laser System's efficacy in treating acute ischemic stroke patients within the crucial 24-h window post-ictus [277]. Additionally, the horizon of stroke therapy seems bright, with robotic interventions demonstrating immense potential.

2. *Pharmacological Enhancements Post-Stroke*:

Pharmacotherapy holds promise in the post-stroke recovery phase. Specifically, a range of anti-depressants, including serotonin uptake inhibitors (SSRIs) and noradrenergic inhibitors, have shown potential for amplifying motor recovery in ischemic stroke victims [278,279]. However, the inner workings of how SSRIs achieve this remain elusive and warrant further exploration [254] Table 2.

Table 2. Stages of motor recovery.

Criteria/Stage	Cognitive Stage	Associative Stage	Automatic Stage	Citations
Primary Focus	In this stage of motor learning, the therapist helps the patient learn a piece of work.	Therapist assists the patient in task performance.	Patient is skilled and can perform tasks.	[280]
Decision Making	Decision making is based on "What to do?"	Decision making is based on "How to do a task?"	Decision making is based on "How to succeed?"	[281]
Task Execution	Learner constructs a motor program.	Patient performs and corrects errors; self-evaluation is promoted.	Complex and challenging tasks are performed to gain retention.	[254,280]
Self-evaluation and Feedback	Examine the task's demands and his ability to complete it.	Continuity proven when error becomes consistent.	Select appropriate feedback.	[280,281]
Task Perception and Memory Recall	Identify the elements and recall the memory.	Emphasize the proprioception "feel of movement".	Organize practice, self-evaluation and correction, gain retention.	[254,282]
Practice and Problem-solving	The patient then begins practicing the task, identifying and resolving problems.	Assist the learner with self-evaluation and decision-making skills.	Focus on the competitive aspect of the skills.	[280–282]

6.1. Objective Evaluation of Non-Pharmacological Therapeutic Interventions in Acute Ischemic Stroke

In assessing the efficiency of contemporary non-invasive, non-pharmacological therapeutic and rehabilitative interventions for acute ischemic stroke, it is crucial to employ objective, quantifiable measures. Integrating these metrics can make a significant difference in clinical decision-making. A myriad of tools exist for these evaluations:

1. Clinical-functional evaluation tools: These instruments, like the Glasgow coma scale (GCS), Glasgow outcome score scale (GOS), modified Rankin scale, and the national institutes of health stroke scale (NIHSS), provide key insights into a patient's neuro-functional state [283].
2. Neuroimaging and Neurophysiological Examinations: Technologies such as structural and functional magnetic resonance imaging, positron emission tomography (PET), single-photon emission computed tomography (SPECT), and repetitive/transcranial magnetic stimulation (r/TMS) give in-depth visuals and data about the brain's condition. Additionally, considering a patient's general health and neuro-functional state, tools like functional near-infrared spectroscopy (f/NIR) and diverse immuno-(cyto)/histochemical assays could also be invaluable. Similarly, pairing transcranial magnetic stimulation (TMS) with high-density EEG offers a non-invasive method to perturb and measure, providing insights into both local neuronal conditions and signal dispersion within functional networks. Notably, even in patients who appeared clinically identical (e.g., lacking residual arm functionality or lacking peripheral motor-evoked potential from standard TMS), TMS-EEG unveiled varied response patterns that correlated with subsequent recovery. This underscores the profound potential of TMS-EEG as a novel indicator of the motor network's functional reserve [284].

6.2. Cardiac Connections to Stroke

Ischemic strokes often coexist with cardiac abnormalities, like myocardial ischemia. This connection is evident in the ECG manifestations, such as depressed ST-segments, prolonged QT-interval, and alterations in T and U waves. A stroke can perturb the autonomic function, resulting in heightened sympathetic activity. Thus, variations in ECG heart rate variability serve as clinical markers, illuminating alterations in the autonomic nervous system (ANS) post-stroke [285]. Non-invasive assessment tools like electrocardiography (ECG), electroencephalography (EEG), and electromyography (EMG) offer predictive insights into stroke prognosis [286,287]. The evolution of technology has enabled machine learning to analyze vast ECG datasets, providing preliminary stroke prognosis and tracking post-stroke recovery using cardiac activity profiles [288].

6.3. *Rehabilitation after Stroke: An Ongoing Journey*

The systematic practice of daily activities has been identified as a potent strategy to boost mobility and activities of daily living (ADLs) in stroke survivors [289]. A crucial aspect of post-stroke care is the rehabilitation process. This proactive journey commences soon after hospitalization, transitioning through structured rehabilitation regimens and continuing even after the patient reintegrates into the community. Typically, a stroke patient starts low-intensity rehabilitation as early as 72 h after the event, often in specialized units or ICUs. Along with the patient, caregivers receive guidance on the path of recovery—its duration, the care trajectory, foreseeable limitations, and more—all navigated with the assistance of the healthcare team.

A study from 2002 investigated various rehabilitation strategies for acute stroke patients. Participants were divided into three intervention groups: standard care, functional task practice, and strength training. The results emphasized that task specificity and stroke severity are key considerations for post-stroke rehabilitation of the upper extremity. Significantly improved functional outcomes were achieved with targeted upper extremity treatment spanning four to six weeks. While both functional task practice and resistance strength training offered immediate advantages, the former showcased long-term benefits [290].

For acute stroke rehabilitation, interventions like early mobilization, positioning, functional mobility training, ADLs training, range of movements (ROMs), splinting, and bed mobility are vital. Mirror therapy stands out, positively influencing motor deficits, emotional wellness, visuospatial neglect, and post-stroke discomfort. As rehabilitation progresses to the sub-acute phase, the focus shifts to enhancing mobility, stamina, strength, and equilibrium. Past research has shown aerobic exercises to be beneficial even beyond the sub-acute stage [291]. Additionally, early overground bodyweight-support training has demonstrated advantages in the sub-acute phase [292]. As patients transition to the chronic stage, task-specific therapies offer persistent improvements in diverse motor deficits [293].

6.4. *Psychological and Cognitive Repercussions of Stroke*

Post-stroke cognitive impairment and dementia (PSCID) is an increasingly recognized aftermath of stroke events globally, often leading to significant morbidity and mortality. This impairment can be a consequence of various forms of stroke: ischemic, intracerebral hemorrhage, or subarachnoid hemorrhage. Even if a patient has underlying neurodegenerative pathologies—a common scenario in the elderly—cognitive decline after a stroke is typically attributed to vascular causes. Moreover, cerebral small vessel disease manifestations such as covert brain infarcts, white matter disruptions, microbleeds, and cortical microinfarcts are frequently observed in these patients and contribute further to cognitive decline.

Several factors escalate the risk of PSCID. Among them are age, lower educational attainment, socioeconomic challenges, pre-stroke cognitive or functional decline, lifelong vascular risk exposure, and prior stroke history. Additionally, the specifics of the stroke event, such as its severity, lesion volume and location, and recurrence, play a role in determining PSCID risk.

A comprehensive understanding of how acute stroke events and pre-existing brain pathologies interplay is crucial. It holds the promise of refining personalized predictions, preventive measures, interventions, and rehabilitation strategies tailored to individual needs. While the field has made progress, it still grapples with standardizing definitions, diagnostic timing, neurocognitive assessment methods, and follow-up durations post-stroke. Nevertheless, emerging insights from pathophysiology studies, advances in neuroimaging, and biomarker discoveries offer hope for clinical innovations and prospective trial designs. A pivotal strategy for optimal brain health hinges on preventing both strokes and the resultant PSCID [294].

Stroke, predominantly observed in the elderly, has profound implications for cognitive function. The repercussions, manifesting as post-stroke cognitive impairment (PSCI), not only challenge the affected individual but also place significant burdens on caregivers and society at large. The gravity of PSCI necessitates efficient management, particularly preventive approaches that address modifiable risk factors. In shedding light on PSCI, we explore its varied definitions and inconsistent prevalence findings across studies, delve into established and potential predictors, and discuss prevention and treatment avenues currently in use or under trial. The goal is to direct future research towards bridging existing knowledge gaps [295].

Ischemic strokes bear a global burden that is nearly four times greater than hemorrhagic strokes. Recent data indicates that between 25–30% of ischemic stroke survivors experience immediate or subsequent vascular cognitive impairment (VCI) or vascular dementia (VaD). Dementia, post-stroke, can cover a spectrum of cognitive disorders. Instances of cognitive dysfunction present before the actual stroke fall under pre-stroke dementia. This condition could result from vascular alterations or gradually advancing neurodegenerative processes.

Several risk factors for post-stroke cognitive impairment have been identified: aging, genetic predispositions, educational background, vascular-related comorbidities, prior transient ischemic attacks, recurrent strokes, and associated depression. Neuroimaging has revealed that silent brain infarcts, alterations in white matter, lacunar infarcts, and atrophy in the medial temporal lobe are indicators of post-stroke dementia. Historically, the neuropathological understanding of post-stroke dementia was limited. Now, it is largely attributed to VaD, with a blend of multiple underlying factors. Key contributors include microinfarction, changes in the microvascular system linked to blood–brain barrier damage, localized neuronal atrophy, and a minimal presence of co-existing neurodegenerative pathology. To mitigate the cognitive challenges post-stroke, understanding the underlying mechanisms is vital. Effective strategies for alleviation and prevention hinge upon rigorous control of vascular disease risk factors [296].

6.5. Education and Intervention: Addressing the Rising Incidence of Stroke

In recent decades, there has been a concerning rise in the incidence of strokes, notably ischemic strokes, among young adults. This surge parallels increasing rates of traditional risk factors such as hypertension, diabetes, and tobacco use, along with the consumption of illicit substances. Young adults present a unique challenge as they exhibit a higher proportion of intracerebral and subarachnoid hemorrhage when compared to older individuals. Moreover, the root causes of ischemic strokes in young adults differ significantly. One-third of these strokes have undetermined etiologies, often due to insufficient investigation into the multitude of potential causes. It is alarming to note that young individuals with premature atherosclerotic cardiovascular disease receive inadequate secondary prevention care, marked by lower utilization of antiplatelets and statin therapy. Factors such as atypical stroke symptoms, varied causes, reluctance to use statins, and clinical inertia on the part of the provider pose barriers to timely diagnosis and treatment. To mitigate the long-term implications of stroke in this demographic, immediate recognition, rigorous risk modification, and an emphasis on both primary and secondary prevention therapies are paramount [297].

Early risk factor modification holds significant promise for preempting strokes. Elevated awareness and intervention, especially regarding hypertension, can drastically diminish stroke-related morbidity and mortality. New criteria have expanded the pool of individuals classified as hypertensive, thereby widening the reach of lifestyle modifications and medical interventions. Direct oral anticoagulants, which have simplified the treatment of patients with atrial fibrillation, are now recommended as primary therapy for those with added stroke risks. In addition, it was observed that the risk factors for acute ischemic stroke are the same as those for atrial fibrillation; therefore, the underlying mechanisms of both pathologies are triggered by the same causes. The bedrock of preserving brain health through one's lifespan is the rigorous primary prevention of stroke, emphasizing a wholesome lifestyle and routine screening for predisposing factors [298].

In China, strokes account for the highest number of disability-adjusted life years lost among all diseases, with over 2 million new cases diagnosed annually. This number is projected to grow, propelled by an aging populace, persistent high-risk factors like hypertension, and subpar management. While there is enhanced access to general healthcare, specialized stroke care remains disparate, with rural areas being disproportionately disadvantaged. Hospital outcomes are better today due to the availability of reperfusion therapies and supportive care. However, there is a noticeable shortfall in adherence to secondary preventive measures and sustained care. Even though global standards of care, such as thrombolysis and dedicated stroke units, are acknowledged in China, the uptake is slow due to concerns about bleeding risks and logistical hurdles. Herbal treatments and neuroprotective drugs, although not strongly evidence-backed, are prevalent in the Chinese context. The prolific use of advanced neuroimaging has also inadvertently led to the overdiagnosis and overtreatment of 'silent strokes.' China's focus must shift towards ensuring equal access to specialized stroke care, augmenting evidence-based practices, and fostering translational research to ameliorate outcomes for stroke patients [299].

Globally, despite leaps in early diagnosis and aggressive vascular risk factor treatments, stroke retains its grim position as a principal cause of death and long-term disability. It is essential to recognize the disparity in stroke risks, occurrences, and treatments. Stroke's multifaceted nature, influenced by a myriad of additive risk factors, necessitates a tailored approach to both primary and secondary prevention, with the former targeting risk mitigation and the latter emphasizing treatments tailored to the initial stroke or transient ischemic attack's cause [300].

Addressing the escalating global disability attributed to strokes and the concurrent rise in stroke care costs necessitates a research pivot towards efficacious stroke prevention measures. Strategies span from reducing disease emergence risks (primordial prevention) and forestalling disease onset (primary prevention) to averting disease recurrence (secondary prevention) [301]. A comprehensive approach requires worldwide collaboration involving global strategies, campaigns, and effectiveness measurement of global preventive initiatives, especially in low- to middle-income countries. Findings underscore the importance of tobacco control, nutritional adequacy, and fostering healthier urban environments for primordial prevention. In contrast, primary prevention can leverage polypill approaches, mobile health technologies, and dietary interventions like salt reduction. Effective intersectoral collaboration, steered by robust government policies and campaigns, can seamlessly implement secondary prevention strategies. Tools like the WHO's non-communicable disease programs, spread across both affluent and developing nations, further underscore the promise of such collaborations [302].

6.6. Community Engagement and Support: Building Resilience among Stroke Survivors

A comprehensive review of current literature was undertaken by searching leading databases, including MEDLINE, Cochrane Central, and Web of Science. Utilizing, the extracted continuous data from randomized controlled trials (RCTs) was meticulously analyzed to discern any potential variance in outcomes [303]. Out of 15 studies involving 1339 patients that were reviewed, a mere 12 were deemed suitable for the pooled analysis. These examinations compared the results between groups undergoing telerehabilitation and those without. Intriguingly, outcomes measured through the Barthel index, Berg balance scale, Fugl–Meyer upper extremity, and the mobility subscale of the stroke impact scale presented no significant differences between the groups [304]. Furthermore, when assessing the health-related quality of life of stroke survivors, the strain on caregivers, and the overall satisfaction with the treatment provided, both groups appeared equally matched [305]. It is worth noting that telerehabilitation presents a viable alternative, especially for patients in remote or underserved locations. Nevertheless, future studies with larger sample sizes are encouraged to further delve into the intricacies of the health-related quality of life and economic implications of enhanced telerehabilitation networks [306].

Shifting our focus to home-based rehabilitation (HBR), an innovative study aimed to harness the capabilities of machine learning (ML) to track and recognize rehabilitation exercises using a smartwatch and smartphone app. This study showcased an ingenious system that integrated a commercial smartwatch with specially designed apps, using a convolutional neural network to detect home exercises [307]. An evaluation period from March 2018 to February 2019 saw two groups of stroke survivors being assessed. Remarkably, the system achieved almost near-perfect accuracy, particularly when using both accelerometer and gyroscope data. Furthermore, the study found that the HBR group exhibited significant improvements in specific clinical tests compared to the control group. The promising results suggest that such a system could be a game-changer for cost-effective, home-based care for stroke survivors.

On a global scale, strokes remain a formidable challenge, affecting over 80 million individuals and exerting substantial economic and societal costs [308]. Despite commendable strides in acute care, there remains a palpable lack of focus on post-acute stroke care. Recent evaluations have highlighted the existing gaps and challenges in post-acute care, urging a paradigm shift to encompass this critical area of stroke management [309]. Three bold recommendations were put forth: first, the establishment of criteria for rehabilitation readiness within Comprehensive Stroke Centers. Second, an expansion of the American Heart Association/American Stroke Association's guidelines to cover rehabilitation readiness and a 90-day outcome measurement. Lastly, a targeted public health campaign to amplify the message of secondary prevention and recovery. These suggestions underscore the pressing need for an all-encompassing, holistic approach to stroke care, valuing not just acute interventions but also long-term recovery and support.

7. Future Horizons in Stroke Management

7.1. On the Brink of Discovery: The Future of Diagnostic Techniques and Treatments

In the aftermath of an acute ischemic stroke (AIS), the body's inflammatory reactions can be a double-edged sword, having both potential advantages and disadvantages. Recent evidence sheds light on the inflammatory pathways and mediators that are currently being explored as potential therapeutic targets [310]. A comprehensive search was conducted on platforms like PubMed and MEDLINE until August 2016, using keywords that range from ischemic stroke to autoimmunity. The results highlighted the surge of cytokines, chemokines, and damage-associated molecular patterns (DAMPs) in AIS, which can exacerbate tissue damage during both the immediate response and the repair phase. Despite the plethora of biomarkers examined, none fully captured the intricate nature of the systemic immune response. Even though reperfusion therapies have shown promise in AIS recovery, truly effective long-term anti-inflammatory interventions remain elusive. The horizon does, however, look promising with advances in therapies such as monoclonal antibodies and cell-based treatments [310].

The burdens of IS (ischemic stroke) remain pressing, even with clinical advancements made over the years. In the context of inflammation following an IS, there is a distinction between central and peripheral responses based on origin—either the brain or peripheral tissue. Macrophages, particularly the differing responses of M1 and M2 types, have been spotlighted in recent studies post-IS. Interestingly, the spleen has been identified as a potential hotspot for inflammatory cells and cytokines after IS [311]. Beyond the commonly recognized cytokines and chemokines, new players like OPG, OPN, and autoantibodies are emerging as crucial mediators in the inflammatory environment post-IS [312]. As the medical world embraces newer treatments like monoclonal antibodies and fingolimod, it is clear that there is still much ground to cover. Some treatments have shown promise in initial clinical trial phases, while others, like cell-based therapies, could even extend their benefits beyond stroke to other neurological conditions [310].

Robotics in neurointervention is a rapidly advancing frontier. Whether it is diagnostic angiography or robot-assisted aneurysm coiling, success has been observed [313]. The rapid evolution of internet connectivity, particularly with the advent of 5G, will further catalyze the deployment of robotic systems. Such systems, while presently being operated from within the catheterization labs, have also shown promise in remote operations. A groundbreaking trial saw a telerobotic-assisted percutaneous coronary intervention being performed with the operator situated 20 miles away [314,315].

A significant stride forward in stroke prediction is the development of an ML (machine learning) pipeline, framing the prediction as a binary classification challenge. By employing a range of ML models, the study managed to achieve impressive results, particularly a low false-negative rate of 18.6% [316,317]. A significant achievement was the identification of pivotal risk factors contributing to the prediction of stroke occurrence. This study used a robust methodology, combining data resampling to address data imbalance issues with SHAP-based explainability analysis. By enhancing the understanding of stroke mechanisms through cutting-edge ML methods, this research paves the way for the development of potent diagnostic tools that can provide clinicians with a more reliable and non-invasive prediction capability [317].

7.2. Advancements in Stroke Prediction: The Role of Machine Learning and Biomarkers

Identifying reliable biomarkers and parameters that predict functional outcomes, as measured by mRS upon discharge, is crucial. To achieve this, a comprehensive ML pipeline was set in motion. In the foundational stage, data underwent normalization, wherein feature standardization occurred by eliminating the mean and scaling to unit variance. This ensured that subsequent feature selection and learning phases had a consistent reference point [318]. The Boruta library, which builds upon the principles of a random forest classifier, was utilized to select and rank features based on their importance.

The study incorporated five renowned classifiers: random forest (RF), XGBoost, multilayer perceptron (MLP), support vector machines (SVMs), and logistic regression (LR). Their inclusion was driven by a need to gauge their suitability for the task at hand. Hyperparameter tuning was integral to preventing overfitting and optimizing the efficacy of the classifiers. RF and XGBoost, both ensemble learning algorithms, were chosen owing to their commendable speed and performance [319]. In contrast, MLP, which mirrors the intricate design of human neural networks, can adeptly handle multifaceted data. Additionally, the study leveraged SVMs, which excel in high-dimensional data scenarios, and the LR model, recognized for its prowess in binary classification tasks [320].

The implementation process consisted of two distinct steps. The initial step utilized random forest regression to fill in data gaps before classification. Following this, an automated hyperparameter optimization (AutoHPO) rooted in deep neural network (DNN) methodologies was applied to predict stroke in a dataset where occurrences were unevenly distributed. Of the 43,400 medical records examined, 783 instances of stroke were identified. Notably, the prediction method reported a false negative rate of just 19.1%, marking a significant 51.5% improvement over conventional techniques. Additionally, false positive rate, accuracy, and sensitivity stood at 33.1%, 71.6%, and 67.4%, respectively [316].

7.3. The Revolutionary Impact of AI and Machine Learning in Stroke Management

In the realm of stroke diagnostics and prognostics, tools enhancing diagnosis speed and precision can be life-changing. This is where artificial intelligence and machine learning (AI/ML) promise transformative advancements. Healthcare adoption of AI/ML, currently growing at a rate of 40% annually, has the potential to yield a staggering USD 150 billion in savings by 2026 [321]. Recognizing AI/ML's revolutionary capabilities, the U.S. Food and Drug Administration (FDA) has inaugurated fresh protocols to evaluate AI/ML-empowered health tools in terms of safety and effectiveness. Various clinical avenues, ranging from liver fibrosis detection to lung cancer classification, have started harnessing the power of AI/ML algorithms [322–324].

Of significance, the FDA has green-lighted 22 AI/ML technologies designed specifically for stroke diagnosis and rehabilitation. While previous literature reviews have predominantly centered on AI/ML algorithms curated for research, there is an absence of comprehensive studies on the real-world clinical efficacy of FDA-endorsed devices tailored for stroke diagnosis and management. This review seeks to consolidate the most pertinent and current insights related to these technologies, offering an exhaustive overview of their distinctive features and contributions to refining clinical practices and outcomes. In essence, a total of 22 innovative AI/ML technologies have received FDA approval, aiming to support clinicians in diagnosing or managing strokes, or ICH [325–327].

7.4. The Technological Revolution in Stroke Diagnostics and Rehabilitation

The diagnosis and treatment of stroke have experienced unprecedented advancements with the introduction of 20 groundbreaking technologies. By prioritizing and triaging crucial scans, these innovations not only mitigate the necessity for labor-intensive tasks like segmentation but also expedite the intervention process, thereby offering patients quicker access to life-saving treatments [328]. Testament to their efficacy and acceptance, technologies like RAPID and Viz.ai have been adopted by 1800 and 900 hospitals, respectively.

Interestingly, many of these technologies seamlessly integrate into broader technological frameworks to enhance care coordination. For instance, solutions offered by companies such as Viz.ai and RapidAI are housed within an interconnected ecosystem that encompasses mobile notifications, remote CT/MRI viewing capacities, and ensures HIPAA-compliant peer communication. This holistic approach not only streamlines care coordination but also amplifies the benefits of AI/ML, creating a harmonious synergy that has substantially improved patient care outcomes. The trajectory suggests that future technologies should lean towards such integrated clinical systems. Historical data shows that technological integration and workflow simplification have notably improved outcomes, even in non-stroke medical emergencies.

Moving beyond diagnosis, AI/ML is also redefining post-stroke rehabilitation. The two current devices focused on rehabilitation harness AI/ML to innovate therapeutic techniques. These have ushered in newfound hope for patients, enabling substantial recovery after severe neurological damage. As these devices mature, they promise even greater safety and efficiency, equipping healthcare professionals with advanced tools to ensure superior patient recovery. Yet, the journey forward requires rigorous head-to-head comparisons using identical clinical datasets to guide clinicians in their choice of technology [329].

In the domain of medical imaging, the present algorithm leverages binary masks of ICA and CCA images for model training. This process mandates meticulous mask preparation, which, if overlooked, can lead to model misrecognition. However, this task is painstakingly slow and requires seasoned sonographers [330]. With the data influx, generating masks for thousands of images becomes impractical. Consequently, supervised learning faces challenges in massive data analyses [331], suggesting that unsupervised models might soon become the standard, especially in contexts where binary masks or labeled data are not mandatory. Medical image processing has vast untapped potential for such unsupervised deep learning models [332].

7.5. Deep Dive into EEG Data for Stroke Prediction

Our research ventured into deciphering stroke predictions via ML models, analyzing EEG data from both stroke-afflicted and healthy individuals during active states like walking, working, and reading. Distinct EEG band power is indicative of brain functionality and, in ischemic stroke contexts, correlates with neural damage in the brain's lesioned region. Employing cutting-edge ML models like adaptive gradient boosting, XGBoost, and

LightGBM, our study aimed to categorize active-state stroke patients. Although XGBoost and LightGBM demonstrated commendable performance metrics, the adaptive gradient boosting model surpassed them across all parameters. Its superior precision and recall metrics underscore its proficiency in distinguishing stroke patients from non-patients with heightened accuracy [286].

Further, our research delved into explainable ML methodologies, focusing on the automated diagnosis of ischemic stroke victims and healthy individuals using neural markers from active states [333]. Notably, both the Eli5 and LIME interpretable models assign paramount importance to delta and theta waves during stroke patient identification. By making the diagnostic process transparent, the results from our exploration of explainable AI have the potential to revolutionize stroke treatment and rehabilitation, in turn facilitating physicians in elucidating their diagnostic rationale [334].

7.6. The Evolution and Potential Future of Cerebrovascular Disease Management

The journey of cerebrovascular disease management has been marked by considerable advancements. Yet stroke remains a predominant cause of mortality and debilitation on a global scale. Over the past decade, the stride towards conquering this ailment has been particularly notable, especially with the inception of novel orally active anticoagulant drugs [335].

A review of these modern milestones reveals an intricate interplay of anticoagulants and antiplatelet agents in tailoring the approach to distinct ischemic stroke subtypes. These agents have been a beacon of hope in both the prevention and treatment spheres. The current clinical guidelines have a tripartite recommendation system: (a) During the early stages of an ischemic stroke, aspirin—an antiplatelet agent—is favored. (b) For recurrent prevention of noncardioembolic ischemic strokes, primarily attributed to large artery atherosclerosis, antiplatelet therapies like aspirin, clopidogrel, and dipyridamole are advocated. (c) In cases of cardioembolic ischemic strokes, particularly those linked to atrial fibrillation, anticoagulants, encompassing warfarin and the new-age NOACs, are suggested [336].

Phase III clinical trials have spotlighted NOACs like dabigatran (150 mg BID) and apixaban (5 mg BID) as being superior to the traditional warfarin in staving off stroke and systemic embolism. Rivaroxaban (20 mg QD), meanwhile, showcased equivalence in efficacy. A parallel between the bleeding tendencies of dabigatran, rivaroxaban, and warfarin was observed. However, apixaban's reduced major bleeding incidents make it stand out. As the utilization of these avant-garde anticoagulants widens, in tandem with diverse clinical investigations, it is anticipated that their efficacy in decreasing stroke instances and the complications tied to warfarin will be further emphasized.

Pivoting to the management of acute ischemic stroke, we see a domain in flux, primarily driven by the impressive outcomes of mechanical thrombectomy. Large vessel occlusion (LVO), accounting for approximately 38% of acute ischemic strokes, once brought profound distress to patients and their kin in the pre-intervention era [337]. The advent of mechanical thrombectomy heralded a revolutionary shift in LVO management, backed by a series of randomized controlled trials across nations that underscored its immense benefits. This article aims to present an exhaustive overview of LVO handling, the techniques and apparatuses employed, and potential future avenues in stroke treatment.

Historically, the mid-1990s saw the discovery of the intravenous tissue plasminogen activator (tPA), offering a modicum of benefit [338]. But its constraints—a narrow therapeutic window, minimal efficacy in vessel recanalization, and favorable outcomes—limited its universal adoption. The goal has always been to achieve results paralleling the cardiac medical domain. This aspiration became more tangible in 2015, when mechanical thrombectomy for acute ischemic strokes was rejuvenated, following the successful conclusions of five crucial RCTs [339]. These trials, spanning multiple countries, unanimously echoed the profound advantages of mechanical thrombectomy, charting a promising trajectory for cerebrovascular disease management.

7.7. Bridging Research and Clinical Practices in Stroke Rehabilitation: A Comprehensive Exploration

Stroke's prevalence and associated disability rate warrant an urgent need for effective rehabilitation methods. Rehabilitation serves as an invaluable tool for curbing the disabilities that arise from strokes. Central to crafting impactful rehabilitation strategies is the art of systematic assessment. Here, we delve into prevalent methods employed both in research and clinical settings, which range from specialized stroke rehabilitation scales to sophisticated biomedical technologies.

These technologies, from surface electromyography (sEMG) and motion analysis systems to transcranial magnetic stimulation (TMS) and magnetic resonance imaging (MRI), offer diverse avenues for understanding and assessing the stroke's impact and the progress of rehabilitation. Furthermore, innovative combinations of these techniques are charting new frontiers in assessment strategies. On the horizon, we observe pioneering experimental techniques such as the integration of artificial intelligence (AI) with optical correlation tomography (OCT) that are poised to redefine stroke rehabilitation assessment [340].

Yet, while traditional assessment scales are a staple in clinical scenarios, they often grapple with challenges related to consistency, stability, and objectivity. Biomedical techniques, despite their promise of objective data, are often stymied by the absence of comprehensive clinical studies and cohesive usage guidelines [341]. Looking forward, two trajectories are emerging in assessment development. First, there is a push towards harnessing novel technologies to deepen our understanding of post-stroke brain recovery. Second, there is momentum behind leveraging AI algorithms to replace older, manual methods, aiming for enhanced objectivity, precision, and standardization. But these pioneering technologies, still in nascent stages, need considerable research and validation before mainstream clinical adoption [342].

Shifting focus to animal models, we understand that perfectly replicating human stroke in animals is an elusive goal. The complexity and heterogeneity of ischemic stroke in humans render this task intricate. The middle cerebral artery occlusion (MCAo) model stands out due to its ability to closely mirror human ischemic strokes and its consistency in producing reproducible infarcts. While the MCAo model dominates research studies, alternative models using thromboembolic clots also hold merit, especially when evaluating thrombolytic agents. However, the current state of preclinical stroke research suffers from a low rate of translation into successful clinical outcomes. Factors such as the model choice and discrepancies between therapeutic windows in animal models versus humans can be attributive. Moreover, the prevalent use of young animals without comorbidities for these studies creates a disconnect from human stroke scenarios where elderly individuals with multiple cerebrovascular risk factors are most affected. Thus, the need for optimizing animal models to closely represent human conditions is imperative [343].

On the genetic front, strokes do not merely arise from environmental factors; they also have a hereditary component. The ever-evolving field of genetics has pinpointed numerous single-gene disorders linked to strokes and identified common variants across roughly 35 genetic loci correlated with stroke risk. Such findings have been instrumental in unveiling new stroke mechanisms and shared genetic influences across vascular conditions. The power of genetics has further illuminated causal relationships with risk factors and is pivotal in prioritizing clinical trial targets. An upcoming challenge lies in harnessing genome-wide polygenic scores to identify at-risk individuals even before any vascular risk factors manifest. Moving forward, understanding the nuances of rare genetic variants, appreciating ancestral genetic differences, integrating genetics into precision medicine, amalgamating omics data, and translating these genetic discoveries into transformative therapies will be critical [344].

8. Conclusions

8.1. Interconnected Aspects of Stroke Management: A Comprehensive Overview

The multifaceted nature of stroke management highlights the urgent need for timely, cohesive, and informed medical intervention. Acute ischemic stroke, by its very nature, places a premium on time. Ensuring optimal outcomes necessitates a seamless collaboration spanning the gamut of medical professionals, from emergency care providers to neurologists and interventional neurologists. Rapid and accurate diagnosis, expedient patient stabilization, and judicious imaging choices are cornerstones of effective care. Any delay or misstep in these processes may lead to diminished functional outcomes and even permanent paralysis [15]. As medical advancements continue to evolve, specialized stroke response teams are emerging, akin to cardiology teams that address acute myocardial infarction. However, challenges such as public awareness and hospital preparedness loom large, necessitating dedicated efforts to surmount them.

Amidst this backdrop, the past decades have seen transformative innovations heralding a renewed era in vascular neurology. This has expanded the treatment bracket and subsequently improved patient outcomes. Yet troubling trends are emerging, particularly in the US, where, despite advancements, stroke mortality rates in several states appear stagnant or even regressing. A proliferation of stroke risk factors such as diabetes, hypertension, and hyperlipidemia underscores this concerning trend [345]. To pivot this trajectory, there's a growing consensus on the pivotal role of patient education and prevention in mitigating the incidence of debilitating or fatal strokes.

Technological advancements, notably the integration of AI/ML in clinical tools, are propelling stroke care into a futuristic paradigm. As of now, the FDA has sanctioned 22 distinct AI/ML-empowered technologies tailored to aid clinicians in stroke diagnosis or management. A majority of these tools significantly expedite diagnosis, streamlining processes, and hastening life-saving interventions [329]. The widespread adoption of tools like RAPID and Viz.ai in thousands of hospitals signifies their effectiveness. Technologies emerging from unified entities, like Viz.ai or RapidAI, have introduced holistic systems that encompass mobile alerts, secure communication avenues, and remote CT/MRI viewing. By merging care coordination with AI/ML's prowess, these technologies amplify their impact. However, the onus remains on future technological solutions to be encompassed within integrated clinical platforms. Parallelly, AI-driven post-stroke rehabilitation devices are heralding innovative therapeutic methodologies, broadening recovery prospects for patients grappling with severe neurological damage. Comparative assessments of these tools will be pivotal for informed clinical decisions in the future [329].

Delving deeper, recent studies illuminate an intriguing association between gut dysbiosis and the underpinnings and outcomes of stroke. A wealth of clinical and preclinical investigations over the last decade have posited a link between gut imbalance and several stroke risk factors, including obesity, hypertension, and cardiovascular issues, among others [346]. Emerging insights suggest that post-stroke outcomes might be detrimentally affected by gut dysbiosis. Mechanisms precipitating this encompass a gamut of biological processes, including immune-driven systemic inflammation. These cascading effects compound poor outcomes post-stroke, emphasizing the potential of the gut microbiome as a promising therapeutic target.

In sum, the multi-pronged aspects of stroke management underscore the intricacies involved and the necessity for an integrated approach. From timely interventions and technological innovations to deeper insights into the gut–brain connection, the roadmap to effective stroke care demands a holistic view, ever-evolving research, and global collaborative efforts.

8.2. Unveiling the Path Forward: Stroke Research, Cellular Mechanisms, and Clinical Advances

The global health landscape recognizes stroke as the second predominant cause of both mortality and long-term disability. Alongside its clinical implications, the economic ramifications of stroke are vast. Over the past quarter century, the realm of stroke research has made substantial strides, encompassing advancements in animal experimental models, therapeutic agents, clinical trials, and rehabilitation strategies. Yet, a significant lacuna in our understanding of effective stroke treatments remains. Notwithstanding our enhanced comprehension of stroke pathophysiology, the chasm between research discoveries and their clinical applications poses a formidable challenge. The bulk of investigations have centered on re-establishing cerebral blood flow and attenuating post-ischemic neuronal deficits. But the essence lies in ensuring consistent, reliable data, mapping underlying therapeutic mechanisms, and enhancing the translational relevance of pre-clinical findings before venturing into the clinical sphere. Recent years have witnessed significant breakthroughs in stroke research. Enhancements in the choice of animal models, imaging modalities, and methodological advancements have unveiled numerous potential drug targets and therapeutic strategies. Nonetheless, the full translational promise of stroke research remains to be thoroughly explored [347].

A focal point of cerebral health is the microglia, sentinel cells intrinsic to preserving brain homeostasis. Their dual roles in both exacerbating and mitigating ischemic stroke damage, contingent upon the stroke's temporal phase, draw significant attention. Novel insights underscore the microglia's plastic nature and its adaptability to diverse functional phenotypes within the compromised brain. This adaptability extends from exacerbating inflammatory responses to fostering tissue regeneration. Therapeutic endeavors thus need to delicately balance these dual roles, emphasizing the necessity of comprehending the intricate regulatory networks governing microglial activation. Potential interventions, such as modulating the shift from M2 to M1 phenotypes in the microglia post-acute phase, offer promising therapeutic avenues. Yet, it is paramount to decipher the mechanisms steering microglia polarization within compromised tissues and avoid broad-spectrum immunosuppression that indiscriminately inactivates microglial subtypes. In essence, therapeutics targeting microglia should not only prioritize acute-phase neuroprotection but also accentuate post-acute regenerative strategies [348].

The intricacies of brain maturation serve as a cornerstone for cognitive development, underscored by neuroplasticity's manifold manifestations. While adult neurogenesis frequently occupies the research limelight, other neural alterations—including neurogenesis sans division and morphological and neurophysiological shifts—offer fertile grounds for therapeutic interventions, enabling sustained cerebral health across an individual's lifespan. Recent research efforts denote the attainability of such objectives, albeit their cellular substrates exhibit significant complexity. Emphasizing the value of comparative approaches, it is imperative to synergize findings across varied species to enrich our understanding of human neural plasticity. The nuances of comparative neuroscience are thus pivotal in deciphering fundamental biological processes in the human cerebrum [349].

The incorporation of point-of-care testing (POCT) in predicting stroke prognosis is an evolving facet of contemporary clinical practice. Although prevalent POCTs cater to general emergency department patients, bespoke stroke POCT devices are in their nascent stages. As the diagnosis and management of stroke are intrinsically time-sensitive, integrating POCTs can substantially elevate patient care during critical phases. The present-day scientific milieu is poised to craft tailored stroke POCT devices, leveraging cutting-edge immunoassays to discern stroke-specific biomarkers. With the burgeoning landscape of biomarker research, harnessing contemporary analytical instruments is indispensable for integrating POCTs seamlessly into stroke care paradigms [350].

In sum, the path ahead in stroke research, management, and clinical applications mandates continuous exploration, collaboration, and a steadfast commitment to patient-centric care. The intersections of cellular mechanisms, innovative technologies, and clinical practice beckon a holistic approach, fostering advancements in stroke therapy and prognosis Figure 3.

1. Immediate Response:
    - Upon the onset of stroke symptoms, the emergency medical services (EMS) dispatch is contacted (t = 0).
2. Initial Assessment by EMS:
    - The EMS team initiates a 'case entry protocol' to assess the patient's medical condition based on specific parameters.

3. Patient Transportation:
    - The patient is transported to the nearest hospital via ambulance, a process that ideally takes 15 min or less (t $\leq$ 15 min).
4. Traditional Stroke Identification Protocol:
    - Best Case: If the EMS has already diagnosed the patient with a stroke, they are directly sent to a specialized stroke unit.
    - Alternate Case: If not pre-diagnosed by EMS, the patient is directed to the Emergency Department (ED), where they await assessment alongside patients with various medical emergencies (t $\leq$ 30 min).
    - A neurologist then evaluates the suspected stroke patient (t $\leq$ 60 min). The most prevalent neurological assessment tool is the NIHSS, followed by a CT/MRI scan for an initial diagnosis (t $\leq$ 90 min).
    - The classification of the stroke subtype is then determined, with the process taking up to 120 min (t $\leq$ 120 min). Classification methods include the Trial of Org 10172 in Acute Stroke Treatment (TOAST), National Institute of Neurological Disorders and Stroke (NINDS), or Oxford Community Stroke Project (OCSP) schemes.
    - The administration of the tissue plasminogen activator (tPA) typically occurs within 3 h from the onset of stroke symptoms (t $\leq$ 180 min).
5. POCT-Enhanced Stroke Identification Protocol:
    - The utilization of pre-hospital POCT devices can expedite the timeline from the onset of stroke symptoms to the examination by a neurologist, reducing the wait time to 20 min or less (t $\leq$ 20 min).
    - Leveraging in-hospital POCT tools can decrease the time required for a CT/MRI scan (t $\leq$ 45 min).
    - Most critically, the window for administering tPA can be significantly shortened to between 60 and 90 min post-symptom onset (60 min $\leq$ t $\leq$ 90 min).

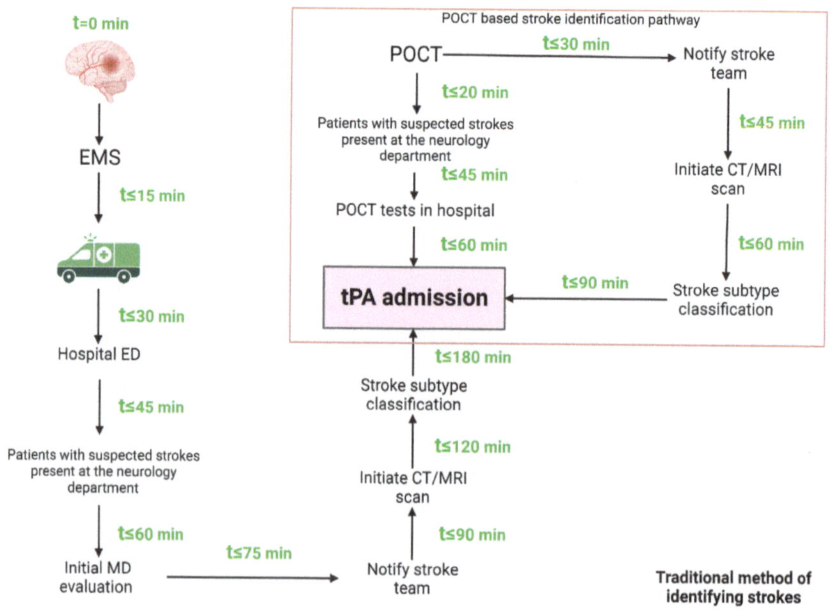

Figure 3. Comparison of Traditional Stroke Prognostic Management vs. Enhanced Approach Using Point-of-Care-Tests (POCTs).

Advancing our understanding of the cognitive dynamics following a transient ischemic attack (TIA) remains a compelling scientific imperative. Delving deeper into the cognitive trajectories post-TIA is critical, especially in discerning how the determinants of delayed-onset dementia (which emerges months or even years after a TIA) diverge from those influencing early-onset or possibly transient cognitive impairments appearing soon after the incident. A nebulous area also exists around the underlying processes that link vascular risk factors, such as diabetes or atrial fibrillation, with the ensuing risk of dementia. Additionally, the potential malleability of this risk based on intensified control of these risk factors or stroke preventive measures remains a question [351].

For comprehensive insights, rigorous research designs are essential. This involves longitudinal investigations paired with superior brain imaging techniques and cognitive evaluations. Parallelly, meticulous oversight of secondary preventive approaches will be invaluable. A notable challenge in extending these findings lies in their generalizability. For instance, the OXVASC study's outcomes predominantly focus on white study cohorts, potentially limiting the extrapolation of results to diverse ethnicities and healthcare frameworks. The nuanced task of selecting appropriate "controls" in research for a credible comparison with TIA-affected individuals also beckons attention. Historically, research studies have relied on community members who willingly participated in their experiments as control groups. Yet, this practice begs the question of the representativeness of such volunteers, particularly concerning their vascular and TIA risks or historical health profiles, and the potential predisposition towards cognitive decline [351].

This selection could inadvertently lead to a bias, wherein control groups inherently consist of individuals who might have heightened anxieties about their cognitive health risks. A viable countermeasure to this predicament might involve establishing expansive, longitudinal cohorts. Such cohorts could aim to mirror the broader demographic makeup as observed in studies like *The Canadian Longitudinal Study of Aging* or *The 1946 National Survey of Health and Development cohort from the UK*. Alternatively, some researchers might gravitate towards curating specific groups based on their projected cognitive decline risks. A case in point is the UK's PREVENT cohort, which is strategically enrolling participants spanning high, medium, and low susceptibility brackets for late-onset Alzheimer's disease. Striking a harmonious balance remains a challenge: researchers must weigh the merits of a large, encompassing population-based research design against a more targeted population that allows for a deeper dive into individual assessments and phenotypic elucidations [351].

In essence, as we endeavor to elucidate the complexities surrounding TIA and its cognitive ramifications, it is paramount that research methodologies evolve with an eye on inclusivity, precision, and generalizability to truly benefit a global audience.

Author Contributions: Conceptualization A.V.C.; methodology, H.P.; software, B.-G.B.; validation, C.T. and A.V.C.; formal analysis, H.P.C.; investigation, R.-A.C.-B.; resources, D.-I.D.; data curation, H.P.C.; writing—original draft preparation, R.-A.C.-B. and A.B.; writing—review and editing, V.M.S. and A.D.C.; visualization, H.P.; supervision, C.T.; project administration, A.V.C.; funding acquisition, V.M.S. All authors have read and agreed to the published version of the manuscript.

Funding: This research received no external funding.

Institutional Review Board Statement: Not applicable.

Informed Consent Statement: No requirement for ethical approval.

Data Availability Statement: All data are available on PubMed.

Conflicts of Interest: The authors declare no conflict of interest.

References

1. Heidenreich, P.A.; Trogdon, J.G.; Khavjou, O.A.; Butler, J.; Dracup, K.; Ezekowitz, M.D.; Finkelstein, E.A.; Hong, Y.; Johnston, S.C.; Khera, A.; et al. Forecasting the Future of Cardiovascular Disease in the United States. *Circulation* **2011**, *123*, 933–944. [CrossRef] [PubMed]

2. O'Donnell, M.J.; Chin, S.L.; Rangarajan, S.; Xavier, D.; Liu, L.; Zhang, H.; Rao-Melacini, P.; Zhang, X.; Pais, P.; Agapay, S.; et al. Global and regional effects of potentially modifiable risk factors associated with acute stroke in 32 countries (INTERSTROKE): A case-control study. *Lancet* **2016**, *388*, 761–775. [CrossRef] [PubMed]
3. Feigin, V.L.; Stark, B.A.; Johnson, C.O.; Roth, G.A.; Bisignano, C.; Abady, G.G.; Abbasifard, M.; Abbasi-Kangevari, M.; Abd-Allah, F.; Abedi, V.; et al. Global, regional, and national burden of stroke and its risk factors, 1990–2019: A systematic analysis for the Global Burden of Disease Study 2019. *Lancet Neurol.* **2021**, *20*, 795–820. [CrossRef] [PubMed]
4. Yan, F.; Yan, S.; Wang, J.; Cui, Y.; Chen, F.; Fang, F.; Cui, W. Association between triglyceride glucose index and risk of cerebrovascular disease: Systematic review and meta-analysis. *Cardiovasc. Diabetol.* **2022**, *21*, 226. [CrossRef] [PubMed]
5. Goldstein, L.B. Introduction for Focused Updates in Cerebrovascular Disease. *Stroke* **2020**, *51*, 708–710. [CrossRef]
6. Haberman, S.; Capildeo, R.; Rose, F.C. The Changing Mortality of Cerebrovascular Disease. *QJM* **1978**, *47*, 71–88. [CrossRef]
7. Portegies, M.; Koudstaal, P.; Ikram, M. Cerebrovascular disease. In *Handbook of Clinical Neurology*; Elsevier: Amsterdam, The Netherlands, 2016; Volume 138, pp. 239–261. [CrossRef]
8. de Havenon, A.; Meyer, C.; McNally, J.S.; Alexander, M.; Chung, L. Subclinical Cerebrovascular Disease: Epidemiology and Treatment. *Curr. Atheroscler. Rep.* **2019**, *21*, 39. [CrossRef]
9. Saini, V.; Guada, L.; Yavagal, D.R. Global Epidemiology of Stroke and Access to Acute Ischemic Stroke Interventions. *Neurology* **2021**, *97*, S6–S16. [CrossRef]
10. Ovbiagele, B.; Nguyen-Huynh, M.N. Stroke Epidemiology: Advancing Our Understanding of Disease Mechanism and Therapy. *Neurotherapeutics* **2011**, *8*, 319–329. [CrossRef]
11. Ohene-Frempong, K.; Weiner, S.J.; Sleeper, L.A.; Miller, S.T.; Embury, S.; Moohr, J.W.; Wethers, D.L.; Pegelow, C.H.; Gill, F.M. Cerebrovascular accidents in sickle cell disease: Rates and risk factors. *Blood* **1998**, *91*, 288–294.
12. Xu, J.; Zhang, X.; Jin, A.; Pan, Y.; Li, Z.; Meng, X.; Wang, Y. Trends and Risk Factors Associated With Stroke Recurrence in China, 2007–2018. *JAMA Netw. Open* **2022**, *5*, e2216341. [CrossRef] [PubMed]
13. Putaala, J. Ischemic Stroke in Young Adults. *Contin. Lifelong Learn. Neurol.* **2020**, *26*, 386–414. [CrossRef] [PubMed]
14. Salvadori, E.; Papi, G.; Insalata, G.; Rinnoci, V.; Donnini, I.; Martini, M.; Falsini, C.; Hakiki, B.; Romoli, A.; Barbato, C.; et al. Comparison between Ischemic and Hemorrhagic Strokes in Functional Outcome at Discharge from an Intensive Rehabilitation Hospital. *Diagnostics* **2020**, *11*, 38. [CrossRef] [PubMed]
15. Chugh, C. Acute Ischemic Stroke: Management Approach. *Indian J. Crit. Care Med.* **2019**, *23*, 140–146. [CrossRef]
16. Cao, R.; Qi, P.; Jiang, Y.; Hu, S.; Ye, G.; Zhu, Y.; Li, L.; You, Z.; Chen, J. Preliminary Application of a Quantitative Collateral Assessment Method in Acute Ischemic Stroke Patients With Endovascular Treatments: A Single-Center Study. *Front. Neurol.* **2021**, *12*, 714313. [CrossRef]
17. Oh, M.; Lee, M. Clinical Implications of Prominent Cortical Vessels on Susceptibility-Weighted Imaging in Acute Ischemic Stroke Patients Treated with Recanalization Therapy. *Brain Sci.* **2022**, *12*, 184. [CrossRef]
18. Mirończuk, A.; Kapica-Topczewska, K.; Socha, K.; Soroczyńska, J.; Jamiołkowski, J.; Chorąży, M.; Czarnowska, A.; Mitrosz, A.; Kułakowska, A.; Kochanowicz, J. Disturbed Ratios between Essential and Toxic Trace Elements as Potential Biomarkers of Acute Ischemic Stroke. *Nutrients* **2023**, *15*, 1434. [CrossRef]
19. Chen, C.; Xun, P.; Tsinovoi, C.; McClure, L.A.; Brockman, J.; MacDonald, L.; Cushman, M.; Cai, J.; Kamendulis, L.; Mackey, J.; et al. Urinary cadmium concentration and the risk of ischemic stroke. *Neurology* **2018**, *91*, e382–e391. [CrossRef]
20. Dias, F.A.; Castro-Afonso, L.H.; Zotin, M.C.Z.; Alessio-Alves, F.F.; Filho, R.K.D.V.M.; Camilo, M.R.; Nakiri, G.S.; Abud, D.G.; Pontes-Neto, O.M. Collateral Scores and Outcomes after Endovascular Treatment for Basilar Artery Occlusion. *Cerebrovasc. Dis.* **2019**, *47*, 285–290. [CrossRef]
21. Etherton, M.R.; Barreto, A.D.; Schwamm, L.H.; Wu, O. Neuroimaging Paradigms to Identify Patients for Reperfusion Therapy in Stroke of Unknown Onset. *Front. Neurol.* **2018**, *9*, 327. [CrossRef]
22. Ravindran, A.V.; Killingsworth, M.C.; Bhaskar, S. Cerebral collaterals in acute ischaemia: Implications for acute ischaemic stroke patients receiving reperfusion therapy. *Eur. J. Neurosci.* **2020**, *53*, 1238–1261. [CrossRef]
23. Macleod, M.R.; Lewis, S.C.; Dennis, M.S. Effect of deprivation on time to hospital in acute stroke. *J. Neurol. Neurosurg. Psychiatry* **2003**, *74*, 545–546. [CrossRef]
24. Cox, A.M.; McKevitt, C.; Rudd, A.G.; Da Wolfe, C. Socioeconomic status and stroke. *Lancet Neurol.* **2006**, *5*, 181–188. [CrossRef]
25. Abilleira, S.; Lucente, G.; Ribera, A.; Permanyer-Miralda, G.; Gallofré, M. Patient-related features associated with a delay in seeking care after stroke. *Eur. J. Neurol.* **2010**, *18*, 850–856. [CrossRef]
26. Eriksson, M.; Glader, E.-L.; Norrving, B.; Stegmayr, B.; Asplund, K. Acute stroke alert activation, emergency service use, and reperfusion therapy in Sweden. *Brain Behav.* **2017**, *7*, e00654. [CrossRef] [PubMed]
27. Iosif, C.; Papathanasiou, M.; Staboulis, E.; Gouliamos, A. Social factors influencing hospital arrival time in acute ischemic stroke patients. *Neuroradiology* **2011**, *54*, 361–367. [CrossRef] [PubMed]
28. Niklasson, A.; Herlitz, J.; Jood, K. Socioeconomic disparities in prehospital stroke care. *Scand. J. Trauma Resusc. Emerg. Med.* **2019**, *27*, 53. [CrossRef]
29. Kleindorfer, D.O.; Lindsell, C.J.; Broderick, J.P.; Flaherty, M.L.; Woo, D.; Ewing, I.; Schmit, P.; Moomaw, C.; Alwell, K.; Pancioli, A.; et al. Community socioeconomic status and prehospital times in acute stroke and transient ischemic attack: Do poorer patients have longer delays from 911 call to the emergency department? *Stroke* **2006**, *37*, 1508–1513. [CrossRef] [PubMed]

30. Park, H.A.; Lee, H.A.; Park, J.O. Association between Area-Level Socioeconomic Deprivation and Prehospital Delay in Acute Ischemic Stroke Patients: An Ecological Study. *Int. J. Environ. Res. Public Health* **2020**, *17*, 7392. [CrossRef] [PubMed]
31. Bhaskar, S.; Bivard, A.; Stanwell, P.; Parsons, M.; Attia, J.R.; Nilsson, M.; Levi, C. Baseline collateral status and infarct topography in post-ischaemic perilesional hyperperfusion: An arterial spin labelling study. *J. Cereb. Blood Flow Metab.* **2016**, *37*, 1148–1162. [CrossRef]
32. Leng, X.; Fang, H.; Leung, T.W.; Mao, C.; Xu, Y.; Miao, Z.; Liu, L.; Wong, K.L.; Liebeskind, D.S. Impact of Collateral Status on Successful Revascularization in Endovascular Treatment: A Systematic Review and Meta-Analysis. *Cerebrovasc. Dis.* **2015**, *41*, 27–34. [CrossRef] [PubMed]
33. Qian, J.; Fan, L.; Zhang, W.; Wang, J.; Qiu, J.; Wang, Y. A meta-analysis of collateral status and outcomes of mechanical thrombectomy. *Acta Neurol. Scand.* **2020**, *142*, 191–199. [CrossRef] [PubMed]
34. Helisch, A.; Schaper, W. Arteriogenesis The Development and Growth of Collateral Arteries. *Microcirculation* **2003**, *10*, 83–97. [CrossRef]
35. Rebello, L.C.; Bouslama, M.; Haussen, D.C.; Grossberg, J.A.; Dehkharghani, S.; Anderson, A.; Belagaje, S.R.; Bianchi, N.A.; Grigoryan, M.; Frankel, M.R.; et al. Stroke etiology and collaterals: Atheroembolic strokes have greater collateral recruitment than cardioembolic strokes. *Eur. J. Neurol.* **2017**, *24*, 762–767. [CrossRef] [PubMed]
36. Hassler, E.; Kneihsl, M.; Deutschmann, H.; Hinteregger, N.; Magyar, M.; Wießpeiner, U.; Haidegger, M.; Fandler-Höfler, S.; Eppinger, S.; Niederkorn, K.; et al. Relationship between stroke etiology and collateral status in anterior circulation large vessel occlusion. *J. Neurol.* **2020**, *267*, 3362–3370. [CrossRef]
37. Sinha, A.; Stanwell, P.; Beran, R.G.; Calic, Z.; Killingsworth, M.C.; Bhaskar, S.M.M. Stroke Aetiology and Collateral Status in Acute Ischemic Stroke Patients Receiving Reperfusion Therapy—A Meta-Analysis. *Neurol. Int.* **2021**, *13*, 608–621. [CrossRef]
38. Söderqvist, K.; Kaijser, M.; Söderman, M.; Holmin, S.; Wahlgren, N.; Andersson, T. Mechanical thrombectomy in acute ischemic stroke—Experience from 6 years of practice. *Neuroradiology* **2014**, *56*, 477–486. [CrossRef]
39. Świtońska, M.; Słomka, A.; Korbal, P.; Piekuś-Słomka, N.; Sinkiewicz, W.; Sokal, P.; Żekanowska, E. Association of Neutrophil-to-Lymphocyte Ratio and Lymphocyte-to-Monocyte Ratio with Treatment Modalities of Acute Ischaemic Stroke: A Pilot Study. *Medicina* **2019**, *55*, 342. [CrossRef]
40. Musuka, T.D.; Wilton, S.B.; Traboulsi, M.; Hill, M.D. Diagnosis and management of acute ischemic stroke: Speed is critical. *Can. Med. Assoc. J.* **2015**, *187*, 887–893. [CrossRef]
41. Martínez-Sánchez, P.; Gutiérrez-Fernández, M.; Fuentes, B.; Masjuán, J.; Cases, M.A.d.L.; Novillo-López, M.E.; Díez-Tejedor, E. Biochemical and inflammatory biomarkers in ischemic stroke: Translational study between humans and two experimental rat models. *J. Transl. Med.* **2014**, *12*, 220. [CrossRef]
42. Hill, M.D.; Jackowski, G.; Bayer, N.; Lawrence, M.; Jaeschke, R. Biochemical markers in acute ischemic stroke. *Can. Med. Assoc. J.* **2000**, *162*, 1139–1140.
43. Teppo, K.; Jaakkola, J.; Biancari, F.; Halminen, O.; Putaala, J.; Mustonen, P.; Haukka, J.; Linna, M.; Kinnunen, J.; Tiili, P.; et al. Mental health conditions and risk of first-ever ischaemic stroke and death in patients with incident atrial fibrillation: A nationwide cohort study. *Eur. J. Clin. Investig.* **2022**, *52*, e13801. [CrossRef] [PubMed]
44. Li, W.; Xiao, W.-M.; Chen, Y.-K.; Qu, J.-F.; Liu, Y.-L.; Fang, X.-W.; Weng, H.-Y.; Luo, G.-P. Anxiety in Patients With Acute Ischemic Stroke: Risk Factors and Effects on Functional Status. *Front. Psychiatry* **2019**, *10*, 257. [CrossRef] [PubMed]
45. Hoyer, C.; Schmidt, H.L.; Kranaster, L.; Alonso, A. Impact of psychiatric comorbidity on the severity, short-term functional outcome, and psychiatric complications after acute stroke. *Neuropsychiatr. Dis. Treat.* **2019**, *15*, 1823–1831. [CrossRef] [PubMed]
46. Hwang, J.-H.; Lim, S.-B. Antioxidant and Anti-inflammatory Activities of Broccoli Florets in LPS-stimulated RAW 264.7 Cells. *Prev. Nutr. Food Sci.* **2014**, *19*, 89–97. [CrossRef]
47. Kamal, F.Z.; Lefter, R.; Jaber, H.; Balmus, I.-M.; Ciobica, A.; Iordache, A.-C. The Role of Potential Oxidative Biomarkers in the Prognosis of Acute Ischemic Stroke and the Exploration of Antioxidants as Possible Preventive and Treatment Options. *Int. J. Mol. Sci.* **2023**, *24*, 6389. [CrossRef]
48. Pedersen, P.M.; Jørgensen, H.S.; Nakayama, H.; Raaschou, H.O.; Olsen, T.S. Aphasia in acute stroke: Incidence, determinants, and recovery. *Ann. Neurol.* **1995**, *38*, 659–666. [CrossRef]
49. Ellis, C.; Urban, S. Age and aphasia: A review of presence, type, recovery and clinical outcomes. *Top. Stroke Rehabil.* **2016**, *23*, 430–439. [CrossRef]
50. Butler, R.A.; Ralph, M.A.L.; Woollams, A.M. Capturing multidimensionality in stroke aphasia: Mapping principal behavioural components to neural structures. *Brain* **2014**, *137*, 3248–3266. [CrossRef]
51. Katsari, M.; Angelopoulou, G.; Laskaris, N.; Potagas, C.; Kasselimis, D. Common Neuroanatomical Substrate of Cholinergic Pathways and Language-Related Brain Regions as an Explanatory Framework for Evaluating the Efficacy of Cholinergic Pharmacotherapy in Post-Stroke Aphasia: A Review. *Brain Sci.* **2022**, *12*, 1273. [CrossRef]
52. Mozaffarian, D.; Benjamin, E.J.; Go, A.S.; Arnett, D.K.; Blaha, M.J.; Cushman, M.; de Ferranti, S.; Després, J.-P.; Fullerton, H.J.; Howard, V.J.; et al. Heart disease and stroke statistics—2015 update: A report from the American Heart Association. *Circulation* **2015**, *131*, e29–e322. [CrossRef] [PubMed]
53. Yang, Q.; Tong, X.; Schieb, L.; Vaughan, A.; Gillespie, C.; Wiltz, J.L.; King, S.C.; Odom, E.; Merritt, R.; Hong, Y.; et al. Vital Signs: Recent Trends in Stroke Death Rates—United States, 2000–2015. *MMWR Morb. Mortal. Wkly. Rep.* **2017**, *66*, 933–939. [CrossRef] [PubMed]

54. Knight-Greenfield, A.; Nario, J.J.Q.; Gupta, A. Causes of Acute Stroke. *Radiol. Clin. N. Am.* **2019**, *57*, 1093–1108. [CrossRef] [PubMed]
55. Thrift, A.G.; Thayabaranathan, T.; Howard, G.; Howard, V.J.; Rothwell, P.M.; Feigin, V.L.; Norrving, B.; Donnan, G.A.; Cadilhac, D.A. Global stroke statistics. *Int. J. Stroke* **2016**, *12*, 13–32. [CrossRef] [PubMed]
56. Montaner, J.; Alvarez-Sabín, J. NIH stroke scale and its adaptation to Spanish. *Neurologia* **2006**, *21*, 192–202.
57. Soto-Cámara, R.; González-Santos, J.; González-Bernal, J.; Martín-Santidrian, A.; Cubo, E.; Trejo-Gabriel-Galán, J.M. Factors Associated with Shortening of Prehospital Delay among Patients with Acute Ischemic Stroke. *J. Clin. Med.* **2019**, *8*, 1712. [CrossRef]
58. Tao, W.-D.; Liu, M.; Fisher, M.; Wang, D.-R.; Li, J.; Furie, K.L.; Hao, Z.-L.; Lin, S.; Zhang, C.-F.; Zeng, Q.-T.; et al. Posterior Versus Anterior Circulation Infarction. *Stroke* **2012**, *43*, 2060–2065. [CrossRef]
59. Keselman, B.; Gdovinová, Z.; Jatuzis, D.; Melo, T.P.E.; Vilionskis, A.; Cavallo, R.; Frol, S.; Jurak, L.; Koyuncu, B.; Nunes, A.P.; et al. Safety and Outcomes of Intravenous Thrombolysis in Posterior Versus Anterior Circulation Stroke. *Stroke* **2020**, *51*, 876–882. [CrossRef]
60. Lindsberg, P.J.; Pekkola, J.; Strbian, D.; Sairanen, T.; Mattle, H.P.; Schroth, G. Time window for recanalization in basilar artery occlusion: Speculative synthesis. *Neurology* **2015**, *85*, 1806–1815. [CrossRef]
61. Berndt, M.; Poppert, H.; Steiger, K.; Pelisek, J.; Oberdieck, P.; Maegerlein, C.; Zimmer, C.; Wunderlich, S.; Friedrich, B.; Boeckh-Behrens, T.; et al. Thrombus Histology of Basilar Artery Occlusions. *Clin. Neuroradiol.* **2020**, *31*, 753–761. [CrossRef]
62. Aroor, S.; Singh, R.; Goldstein, L.B. BE-FAST (Balance, Eyes, Face, Arm, Speech, Time). *Stroke* **2017**, *48*, 479–481. [CrossRef] [PubMed]
63. Chen, X.; Zhao, X.; Xu, F.; Guo, M.; Yang, Y.; Zhong, L.; Weng, X.; Liu, X. A Systematic Review and Meta-Analysis Comparing FAST and BEFAST in Acute Stroke Patients. *Front. Neurol.* **2022**, *12*, 765069. [CrossRef] [PubMed]
64. Puustjärvi, V.; Strbian, D.; Tiainen, M.; Curtze, S.; Tatlisumak, T.; Sairanen, T. Recognition of posterior circulation stroke. *Acta Neurol. Scand.* **2014**, *131*, 389–393. [CrossRef] [PubMed]
65. Halúsková, S.; Herzig, R.; Krajíčková, D.; Hamza, A.; Krajina, A.; Chovanec, V.; Lojík, M.; Raupach, J.; Renc, O.; Šimůnek, L.; et al. Acute Management Should Be Optimized in Patients with Less Specific Stroke Symptoms: Findings from a Retrospective Observational Study. *J. Clin. Med.* **2021**, *10*, 1143. [CrossRef] [PubMed]
66. Sommer, P.; Seyfang, L.; Posekany, A.; Ferrari, J.; Lang, W.; Fertl, E.; Serles, W.; Töll, T.; Kiechl, S.; Greisenegger, S. Prehospital and intra-hospital time delays in posterior circulation stroke: Results from the Austrian Stroke Unit Registry. *J. Neurol.* **2016**, *264*, 131–138. [CrossRef] [PubMed]
67. Tong, X.; Liao, X.; Pan, Y.; Cao, Y.; Wang, C.; Liu, L.; Zheng, H.; Zhao, X.; Wang, C.; Wang, Y.; et al. Intravenous thrombolysis is more safe and effective for posterior circulation stroke. *Medicine* **2016**, *95*, e3848. [CrossRef]
68. Ekkert, A.; Milmantienė, D.; Jokimaitytė, U.; Jatužis, D. Posterior Circulation Stroke Patients Receive Less Reperfusion Therapy Because of Late Arrival and Relative Contraindications: A Retrospective Study. *J. Clin. Med.* **2023**, *12*, 5181. [CrossRef]
69. Tsirlin, I.; Dupierrix, E.; Chokron, S.; Coquillart, S.; Ohlmann, T.; Kim, K.; Kim, J.; Ku, J.; Kim, D.Y.; Chang, W.H.; et al. Uses of Virtual Reality for Diagnosis, Rehabilitation and Study of Unilateral Spatial Neglect: Review and Analysis. *CyberPsychology Behav.* **2009**, *12*, 175–181. [CrossRef]
70. Cinnera, A.M.; Bisirri, A.; Chioccia, I.; Leone, E.; Ciancarelli, I.; Iosa, M.; Morone, G.; Verna, V. Exploring the Potential of Immersive Virtual Reality in the Treatment of Unilateral Spatial Neglect Due to Stroke: A Comprehensive Systematic Review. *Brain Sci.* **2022**, *12*, 1589. [CrossRef]
71. Munn, Z.; Moola, S.; Riitano, D.; Lisy, K. The development of a critical appraisal tool for use in systematic reviews addressing questions of prevalence. *Int. J. Health Policy Manag.* **2014**, *3*, 123–128. [CrossRef]
72. Porritt, K.; Gomersall, J.; Lockwood, C. JBI's Systematic Reviews: Study selection. *AJN Am. J. Nurs.* **2014**, *114*, 47–52. [CrossRef] [PubMed]
73. Munn, Z.; Tufanaru, C.; Aromataris, E. JBI's Systematic Reviews: Data extraction. *AJN Am. J. Nurs.* **2014**, *114*, 49–54. [CrossRef] [PubMed]
74. Filep, E.M.; Murata, Y.; Endres, B.D.; Kim, G.; Stearns, R.L.; Casa, D.J. Exertional Heat Stroke, Modality Cooling Rate, and Survival Outcomes: A Systematic Review. *Medicina* **2020**, *56*, 589. [CrossRef]
75. Rhee, J.Y.; Colman, M.A.; Mendu, M.; Shah, S.J.; Fox, M.D.; Rost, N.S.; Kimchi, E.Y. Associations Between Stroke Localization and Delirium: A Systematic Review and Meta-Analysis. *J. Stroke Cerebrovasc. Dis.* **2022**, *31*, 106270. [CrossRef]
76. Abawi, M.; Nijhoff, F.; Agostoni, P.; de Vries, R.; Slooter, A.J.; Emmelot-Vonk, M.H.; Voskuil, M.; Leiner, T.; Doevendans, P.A.; Stella, P.R. Effect of New Cerebral Ischemic Lesions on the Delirium Occurrence After Transcatheter Aortic Valve Replacement. *J. Am. Coll. Cardiol.* **2016**, *68*, 1489–1490. [CrossRef] [PubMed]
77. Strand, T.; Asplund, K.; Eriksson, S.; Hägg, E.; Lithner, F.; Wester, P.O. A non-intensive stroke unit reduces functional disability and the need for long-term hospitalization. *Stroke* **1985**, *16*, 29–34. [CrossRef] [PubMed]
78. Collaboration Stroke Unit Trialists; Langhorne, P. Organised inpatient (stroke unit) care for stroke. *Cochrane Database Syst. Rev.* **2001**, *3*, CD000197. [CrossRef]
79. Clarke, D.; Forster, A. Improving post-stroke recovery: The role of the multidisciplinary health care team. *J. Multidiscip. Health* **2015**, *8*, 433–442. [CrossRef]
80. Clarke, D.J. The role of multidisciplinary team care in stroke rehabilitation. *Prog. Neurol. Psychiatry* **2013**, *17*, 5–8. [CrossRef]

81. Rudd, A.G.; Wade, D.T.; Irwin, P. The National Clinical Guidelines for Stroke. *J. R Coll. Physicians Lond.* **2000**, *34*, 131–133.
82. Askim, T.; Rohweder, G.; Lydersen, S.; Indredavik, B. Evaluation of an extended stroke unit service with early supported discharge for patients living in a rural community. A randomized controlled trial. *Clin. Rehabil.* **2004**, *18*, 238–248. [CrossRef] [PubMed]
83. Fjærtoft, H.; Rohweder, G.; Indredavik, B. Stroke Unit Care Combined With Early Supported Discharge Improves 5-Year Outcome: A randomized controlled trial. *Stroke* **2011**, *42*, 1707–1711. [CrossRef]
84. Sit, J.W.; Chair, S.Y.; Choi, K.C.; Chan, C.W.; Lee, D.T.; Chan, A.W.; Cheung, J.L.; Tang, S.W.; Chan, P.S.; Taylor-Piliae, R.E. Do empowered stroke patients perform better at self-management and functional recovery after a stroke? A randomized controlled trial. *Clin. Interv. Aging* **2016**, *11*, 1441–1450. [CrossRef] [PubMed]
85. Wong, F.K.Y.; Yeung, S.M. Effects of a 4-week transitional care programme for discharged stroke survivors in Hong Kong: A randomised controlled trial. *Health Soc. Care Community* **2014**, *23*, 619–631. [CrossRef] [PubMed]
86. Wang, Y.; Yang, F.; Shi, H.; Yang, C.; Hu, H. What Type of Transitional Care Effectively Reduced Mortality and Improved ADL of Stroke Patients? A Meta-Analysis. *Int. J. Environ. Res. Public Health* **2017**, *14*, 510. [CrossRef]
87. Tseng, M.-C.; Lin, H.-J. Readmission after hospitalization for stroke in Taiwan: Results from a national sample. *J. Neurol. Sci.* **2009**, *284*, 52–55. [CrossRef]
88. Miyai, I.; Sonoda, S.; Nagai, S.; Takayama, Y.; Inoue, Y.; Kakehi, A.; Kurihara, M.; Ishikawa, M. Results of New Policies for Inpatient Rehabilitation Coverage in Japan. *Neurorehabilit. Neural Repair* **2011**, *25*, 540–547. [CrossRef]
89. Chiu, C.-C.; Wang, J.-J.; Hung, C.-M.; Lin, H.-F.; Hsien, H.-H.; Hung, K.-W.; Chiu, H.-C.; Yeh, S.-C.; Shi, H.-Y. Impact of Multidisciplinary Stroke Post-Acute Care on Cost and Functional Status: A Prospective Study Based on Propensity Score Matching. *Brain Sci.* **2021**, *11*, 161. [CrossRef]
90. Bettger, J.A.P.; Kaltenbach, L.; Reeves, M.J.; Smith, E.E.; Fonarow, G.C.; Schwamm, L.H.; Peterson, E.D. Assessing Stroke Patients for Rehabilitation During the Acute Hospitalization: Findings From the Get With The Guidelines–Stroke Program. *Arch. Phys. Med. Rehabil.* **2013**, *94*, 38–45. [CrossRef]
91. McKevitt, C.; Coshall, C.; Tilling, K.; Wolfe, C. Are There Inequalities in the Provision of Stroke Care? *Stroke* **2005**, *36*, 315–320. [CrossRef]
92. Evenson, K.R.; Foraker, R.E.; Morris, D.L.; Rosamond, W.D. A Comprehensive Review of Prehospital and In-Hospital Delay Times in Acute Stroke Care. *Int. J. Stroke* **2009**, *4*, 187–199. [CrossRef] [PubMed]
93. Mosley, I.; Nicol, M.; Donnan, G.; Thrift, A.G.; Dewey, H.M. What is Stroke Symptom Knowledge? *Int. J. Stroke* **2013**, *9*, 48–52. [CrossRef] [PubMed]
94. Jiang, B.; Ru, X.; Sun, H.; Liu, H.; Sun, D.; Liu, Y.; Huang, J.; He, L.; Wang, W. Pre-hospital delay and its associated factors in first-ever stroke registered in communities from three cities in China. *Sci. Rep.* **2016**, *6*, 29795. [CrossRef] [PubMed]
95. Ellis, C.; Barley, J.; Grubaugh, A. Poststroke Knowledge and Symptom Awareness: A Global Issue for Secondary Stroke Prevention. *Cerebrovasc. Dis.* **2013**, *35*, 572–581. [CrossRef]
96. Phipps, M.S.; Cronin, C.A. Management of acute ischemic stroke. *BMJ* **2020**, *368*, l6983. [CrossRef]
97. Saver, J.L.; Smith, E.E.; Fonarow, G.C.; Reeves, M.J.; Zhao, X.; Olson, D.M.; Schwamm, L.H.; Committee, G.W.-S.S.; Investigators. The "golden hour" and acute brain ischemia: Presenting features and lytic therapy in >30,000 patients arriving within 60 minutes of stroke onset. *Stroke* **2010**, *41*, 1431–1439. [CrossRef]
98. Harbison, J.; Hossain, O.; Jenkinson, D.; Davis, J.; Louw, S.J.; Ford, G.A. Diagnostic Accuracy of Stroke Referrals From Primary Care, Emergency Room Physicians, and Ambulance Staff Using the Face Arm Speech Test. *Stroke* **2003**, *34*, 71–76. [CrossRef]
99. Stroebele, N.; Müller-Riemenschneider, F.; Nolte, C.H.; Müller-Nordhorn, J.; Bockelbrink, A.; Willich, S.N. Knowledge of Risk Factors, and Warning Signs of Stroke: A Systematic Review from a Gender Perspective. *Int. J. Stroke* **2011**, *6*, 60–66. [CrossRef]
100. Soto-Cámara, R.; González-Bernal, J.J.; González-Santos, J.; Aguilar-Parra, J.M.; Trigueros, R.; López-Liria, R. Knowledge on Signs and Risk Factors in Stroke Patients. *J. Clin. Med.* **2020**, *9*, 2557. [CrossRef]
101. Dearborn, J.L.; McCullough, L.D. Perception of Risk and Knowledge of Risk Factors in Women at High Risk for Stroke. *Stroke* **2009**, *40*, 1181–1186. [CrossRef]
102. Lundelin, K.; Graciani, A.; García-Puig, J.; Guallar-Castillón, P.; Taboada, J.M.; Rodríguez-Artalejo, F.; Banegas, J.R. Knowledge of Stroke Warning Symptoms and Intended Action in Response to Stroke in Spain: A Nationwide Population-Based Study. *Cerebrovasc. Dis.* **2012**, *34*, 161–168. [CrossRef] [PubMed]
103. Kobayashi, A.; Czlonkowska, A.; Ford, G.A.; Fonseca, A.C.; Luijckx, G.J.; Korv, J.; de la Ossa, N.P.; Price, C.; Russell, D.; Tsiskaridze, A.; et al. European Academy of Neurology and European Stroke Organization consensus statement and practical guidance for pre-hospital management of stroke. *Eur. J. Neurol.* **2017**, *25*, 425–433. [CrossRef] [PubMed]
104. Neau, J.-P.; Ingrand, P.; Godeneche, G. Awareness within the French population concerning stroke signs, symptoms, and risk factors. *Clin. Neurol. Neurosurg.* **2009**, *111*, 659–664. [CrossRef] [PubMed]
105. Krzystanek, E.; Krzak-Kubica, A.; Świat, M.; Galus, W.; Gawryluk, J. Adequate Knowledge of Stroke Symptoms, Risk Factors, and Necessary Actions in the General Population of Southern Poland. *Brain Sci.* **2020**, *10*, 1009. [CrossRef]
106. Paul, C.L.; Ryan, A.; Rose, S.; Attia, J.R.; Kerr, E.; Koller, C.; Levi, C.R. How can we improve stroke thrombolysis rates? A review of health system factors and approaches associated with thrombolysis administration rates in acute stroke care. *Implement. Sci.* **2015**, *11*, 1–12. [CrossRef]
107. Fassbender, K.; Walter, S.; Grunwald, I.Q.; Merzou, F.; Mathur, S.; Lesmeister, M.; Liu, Y.; Bertsch, T.; Grotta, J.C. Prehospital stroke management in the thrombectomy era. *Lancet Neurol.* **2020**, *19*, 601–610. [CrossRef]

108. Terecoasă, E.O.; Radu, R.A.; Negrilă, A.; Enache, I.; Cășaru, B.; Tiu, C. Pre-Hospital Delay in Acute Ischemic Stroke Care: Current Findings and Future Perspectives in a Tertiary Stroke Center from Romania—A Cross-Sectional Study. *Medicina* **2022**, *58*, 1003. [CrossRef]
109. Nelson, R.F.; Pullicino, P.; Kendall, B.E.; Marshall, J. Computed tomography in patients presenting with lacunar syndromes. *Stroke* **1980**, *11*, 256–261. [CrossRef]
110. Rothrock, J.F.; Lyden, P.D.; Hesselink, J.R.; Brown, J.J.; Healy, M.E. Brain magnetic resonance imaging in the evaluation of lacunar stroke. *Stroke* **1987**, *18*, 781–786. [CrossRef]
111. Noguchi, K.; Nagayoshi, T.; Watanabe, N.; Kanazawa, T.; Toyoshima, S.; Morijiri, M.; Shojaku, H.; Shimizu, M.; Seto, H. Diffusion-weighted echo-planar MRI of lacunar infarcts. *Neuroradiology* **1998**, *40*, 448–451. [CrossRef]
112. Schonewille, W.J.; Tuhrim, S.; Singer, M.B.; Atlas, S.W. Diffusion-Weighted MRI in Acute Lacunar Syndromes. A clinical-radiological correlation study. *Stroke* **1999**, *30*, 2066–2069. [CrossRef] [PubMed]
113. Ay, H.; Oliveira-Filho, J.; Buonanno, F.S.; Ezzeddine, M.; Schaefer, P.W.; Rordorf, G.; Schwamm, L.H.; Gonzalez, R.G.; Koroshetz, W.J. Diffusion-weighted imaging identifies a subset of lacunar infarction associated with embolic source. *Stroke* **1999**, *30*, 2644–2650. [CrossRef] [PubMed]
114. Tan, M.Y.Q.; Singhal, S.; Ma, H.; Chandra, R.V.; Cheong, J.; Clissold, B.B.; Ly, J.; Srikanth, V.; Phan, T.G. Examining Subcortical Infarcts in the Era of Acute Multimodality CT Imaging. *Front. Neurol.* **2016**, *7*, 220. [CrossRef] [PubMed]
115. Rudilosso, S.; Rodríguez-Vázquez, A.; Urra, X.; Arboix, A. The Potential Impact of Neuroimaging and Translational Research on the Clinical Management of Lacunar Stroke. *Int. J. Mol. Sci.* **2022**, *23*, 1497. [CrossRef] [PubMed]
116. Arboix, A.; Massons, J.; García-Eroles, L.; Targa, C.; Comes, E.; Parra, O. Clinical predictors of lacunar syndrome not due to lacunar infarction. *BMC Neurol.* **2010**, *10*, 31. [CrossRef]
117. Bang, O.Y. Advances in biomarker for stroke patients: From marker to regulator. *Precis. Futur. Med.* **2017**, *1*, 32–42. [CrossRef]
118. Boyd, L.A.; Hayward, K.S.; Ward, N.S.; Stinear, C.M.; Rosso, C.; Fisher, R.J.; Carter, A.R.; Leff, A.P.; Copland, D.A.; Carey, L.M.; et al. Biomarkers of stroke recovery: Consensus-based core recommendations from the Stroke Recovery and Rehabilitation Roundtable. *Int. J. Stroke* **2017**, *12*, 480–493. [CrossRef]
119. Jones, T.A.; Kleim, J.A.; Greenough, W.T. Synaptogenesis and dendritic growth in the cortex opposite unilateral sensorimotor cortex damage in adult rats: A quantitative electron microscopic examination. *Brain Res.* **1996**, *733*, 142–148. [CrossRef]
120. Włodarczyk, L.; Cichon, N.; Saluk-Bijak, J.; Bijak, M.; Majos, A.; Miller, E. Neuroimaging Techniques as Potential Tools for Assessment of Angiogenesis and Neuroplasticity Processes after Stroke and Their Clinical Implications for Rehabilitation and Stroke Recovery Prognosis. *J. Clin. Med.* **2022**, *11*, 2473. [CrossRef]
121. Puig, J.; Pedraza, S.; Blasco, G.; Daunis-I-Estadella, J.; Prats, A.; Prados, F.; Boada, I.; Castellanos, M.; Sánchez-González, J.; Remollo, S.; et al. Wallerian Degeneration in the Corticospinal Tract Evaluated by Diffusion Tensor Imaging Correlates with Motor Deficit 30 Days after Middle Cerebral Artery Ischemic Stroke. *Am. J. Neuroradiol.* **2010**, *31*, 1324–1330. [CrossRef]
122. Lin, D.J.; Cloutier, A.M.; Erler, K.S.; Cassidy, J.M.; Snider, S.B.; Ranford, J.; Parlman, K.; Giatsidis, F.; Burke, J.F.; Schwamm, L.H.; et al. Corticospinal Tract Injury Estimated From Acute Stroke Imaging Predicts Upper Extremity Motor Recovery After Stroke. *Stroke* **2019**, *50*, 3569–3577. [CrossRef] [PubMed]
123. Kang, M.; Jin, S.; Lee, D.; Cho, H. MRI Visualization of Whole Brain Macro- and Microvascular Remodeling in a Rat Model of Ischemic Stroke: A Pilot Study. *Sci. Rep.* **2020**, *10*, 4989. [CrossRef] [PubMed]
124. Yanev, P.; Seevinck, P.R.; Rudrapatna, U.S.; Bouts, M.J.; van der Toorn, A.; Gertz, K.; Kronenberg, G.; Endres, M.; van Tilborg, G.A.; Dijkhuizen, R.M. Magnetic resonance imaging of local and remote vascular remodelling after experimental stroke. *J. Cereb. Blood Flow Metab.* **2016**, *37*, 2768–2779. [CrossRef] [PubMed]
125. Yang, Y.; Torbey, M.T. Angiogenesis and Blood-Brain Barrier Permeability in Vascular Remodeling after Stroke. *Curr. Neuropharmacol.* **2020**, *18*, 1250–1265. [CrossRef] [PubMed]
126. Almeida, S.R.M.; Vicentini, J.; Bonilha, L.; De Campos, B.M.; Casseb, R.F.; Min, L.L. Brain Connectivity and Functional Recovery in Patients With Ischemic Stroke. *J. Neuroimaging* **2016**, *27*, 65–70. [CrossRef]
127. Puig, J.; Blasco, G.; Alberich-Bayarri, A.; Schlaug, G.; Deco, G.; Biarnes, C.; Navas-Martí, M.; Rivero, M.; Gich, J.; Figueras, J.; et al. Resting-State Functional Connectivity Magnetic Resonance Imaging and Outcome After Acute Stroke. *Stroke* **2018**, *49*, 2353–2360. [CrossRef]
128. Blicher, J.U.; Near, J.; Næss-Schmidt, E.; Stagg, C.J.; Johansen-Berg, H.; Nielsen, J.F.; Østergaard, L.; Ho, Y.-C.L. GABA Levels Are Decreased After Stroke and GABA Changes During Rehabilitation Correlate With Motor Improvement. *Neurorehabilit. Neural Repair* **2014**, *29*, 278–286. [CrossRef]
129. Mazibuko, N.; Tuura, R.O.; Sztriha, L.; O'daly, O.; Barker, G.J.; Williams, S.C.R.; O'Sullivan, M.; Kalra, L. Subacute Changes in *N*-Acetylaspartate (NAA) Following Ischemic Stroke: A Serial MR Spectroscopy Pilot Study. *Diagnostics* **2020**, *10*, 482. [CrossRef]
130. Laaksonen, K.; Helle, L.; Parkkonen, L.; Kirveskari, E.; Mäkelä, J.P.; Mustanoja, S.; Tatlisumak, T.; Kaste, M.; Forss, N. Alterations in Spontaneous Brain Oscillations during Stroke Recovery. *PLoS ONE* **2013**, *8*, e61146. [CrossRef]
131. Zappasodi, F.; Pasqualetti, P.; Rossini, P.M.; Tecchio, F. Acute Phase Neuronal Activity for the Prognosis of Stroke Recovery. *Neural Plast.* **2019**, *2019*, e1971875. [CrossRef]
132. Stinear, C.M.; Byblow, W.D.; Ackerley, S.J.; Barber, P.A.; Smith, M.-C. Predicting Recovery Potential for Individual Stroke Patients Increases Rehabilitation Efficiency. *Stroke* **2017**, *48*, 1011–1019. [CrossRef]

133. Jo, J.Y.; Lee, A.; Kim, M.S.; Park, E.; Chang, W.H.; Shin, Y.-I.; Kim, Y.-H. Prediction of Motor Recovery Using Quantitative Parameters of Motor Evoked Potential in Patients With Stroke. *Ann. Rehabil. Med.* **2016**, *40*, 806–815. [CrossRef] [PubMed]
134. Pellicciari, M.C.; Bonnì, S.; Ponzo, V.; Cinnera, A.M.; Mancini, M.; Casula, E.P.; Sallustio, F.; Paolucci, S.; Caltagirone, C.; Koch, G. Dynamic reorganization of TMS-evoked activity in subcortical stroke patients. *NeuroImage* **2018**, *175*, 365–378. [CrossRef]
135. Zhu, H.; Zhang, Y.; Zhong, Y.; Ye, Y.; Hu, X.; Gu, L.; Xiong, X. Inflammation-Mediated Angiogenesis in Ischemic Stroke. *Front. Cell Neurosci.* **2021**, *15*, 652647. [CrossRef] [PubMed]
136. Callewaert, B.; Jones, E.A.V.; Himmelreich, U.; Gsell, W. Non-Invasive Evaluation of Cerebral Microvasculature Using Pre-Clinical MRI: Principles, Advantages and Limitations. *Diagnostics* **2021**, *11*, 926. [CrossRef]
137. Berge, E.; Whiteley, W.; Audebert, H.; De Marchis, G.M.; Fonseca, A.C.; Padiglioni, C.; de la Ossa, N.P.; Strbian, D.; Tsivgoulis, G.; Turc, G. European Stroke Organisation (ESO) guidelines on intravenous thrombolysis for acute ischaemic stroke. *Eur. Stroke J.* **2021**, *6*, I-LXII. [CrossRef] [PubMed]
138. Eissa, A.; Krass, I.; Levi, C.; Sturm, J.; Ibrahim, R.; Bajorek, B. Understanding the reasons behind the low utilisation of thrombolysis in stroke. *Australas. Med. J.* **2013**, *6*, 152–167. [CrossRef] [PubMed]
139. Psychogios, K.; Safouris, A.; Kargiotis, O.; Magoufis, G.; Andrikopoulou, A.; Papageorgiou, E.; Chondrogianni, M.; Papadimitropoulos, G.; Polyzogopoulou, E.; Spiliopoulos, S.; et al. Advanced Neuroimaging Preceding Intravenous Thrombolysis in Acute Ischemic Stroke Patients Is Safe and Effective. *J. Clin. Med.* **2021**, *10*, 2819. [CrossRef]
140. Tsivgoulis, G.; Katsanos, A.H.; Schellinger, P.D.; Köhrmann, M.; Caso, V.; Palaiodimou, L.; Magoufis, G.; Arthur, A.; Fischer, U.; Alexandrov, A.V. Advanced Neuroimaging in Stroke Patient Selection for Mechanical Thrombectomy. *Stroke* **2018**, *49*, 3067–3070. [CrossRef]
141. Nogueira, R.G.; Jadhav, A.P.; Haussen, D.C.; Bonafe, A.; Budzik, R.F.; Bhuva, P.; Yavagal, D.R.; Ribo, M.; Cognard, C.; Hanel, R.A.; et al. Thrombectomy 6 to 24 Hours after Stroke with a Mismatch between Deficit and Infarct. *N. Engl. J. Med.* **2018**, *378*, 11–21. [CrossRef]
142. Albers, G.W.; Marks, M.P.; Kemp, S.; Christensen, S.; Tsai, J.P.; Ortega-Gutierrez, S.; McTaggart, R.A.; Torbey, M.T.; Kim-Tenser, M.; Leslie-Mazwi, T.; et al. Thrombectomy for Stroke at 6 to 16 Hours with Selection by Perfusion Imaging. *N. Engl. J. Med.* **2018**, *378*, 708–718. [CrossRef]
143. Imran, R.; Mohamed, G.A.; Nahab, F. Acute Reperfusion Therapies for Acute Ischemic Stroke. *J. Clin. Med.* **2021**, *10*, 3677. [CrossRef] [PubMed]
144. Dukelow, S.P.; Herter, T.M.; Bagg, S.D.; Scott, S.H. The independence of deficits in position sense and visually guided reaching following stroke. *J. Neuroeng. Rehabil.* **2012**, *9*, 72. [CrossRef] [PubMed]
145. Otaka, E.; Otaka, Y.; Kasuga, S.; Nishimoto, A.; Yamazaki, K.; Kawakami, M.; Ushiba, J.; Liu, M. Reliability of the thumb localizing test and its validity against quantitative measures with a robotic device in patients with hemiparetic stroke. *PLoS ONE* **2020**, *15*, e0236437. [CrossRef] [PubMed]
146. Fisher, G.; de Oliveira, C.Q.; Verhagen, A.; Gandevia, S.; Kennedy, D. Proprioceptive impairment in unilateral neglect after stroke: A systematic review. *SAGE Open Med.* **2020**, *8*, 2050312120951073. [CrossRef]
147. Chilvers, M.J.; Hawe, R.L.; Scott, S.H.; Dukelow, S.P. Investigating the neuroanatomy underlying proprioception using a stroke model. *J. Neurol. Sci.* **2021**, *430*, 120029. [CrossRef] [PubMed]
148. Hawe, R.L.; Findlater, S.E.; Kenzie, J.M.; Hill, M.D.; Scott, S.H.; Dukelow, S.P. Differential Impact of Acute Lesions Versus White Matter Hyperintensities on Stroke Recovery. *J. Am. Heart Assoc.* **2018**, *7*, e009360. [CrossRef]
149. Kenzie, J.M.; Semrau, J.A.; Findlater, S.E.; Yu, A.Y.; Desai, J.A.; Herter, T.M.; Hill, M.D.; Scott, S.H.; Dukelow, S.P. Localization of Impaired Kinesthetic Processing Post-stroke. *Front. Hum. Neurosci.* **2016**, *10*, 505. [CrossRef]
150. Findlater, S.E.; Desai, J.A.; Semrau, J.A.; Kenzie, J.M.; Rorden, C.; Herter, T.M.; Scott, S.H.; Dukelow, S.P. Central perception of position sense involves a distributed neural network—Evidence from lesion-behavior analyses. *Cortex* **2016**, *79*, 42–56. [CrossRef]
151. Meyer, S.; Kessner, S.S.; Cheng, B.; Bönstrup, M.; Schulz, R.; Hummel, F.C.; De Bruyn, N.; Peeters, A.; Van Pesch, V.; Duprez, T.; et al. Voxel-based lesion-symptom mapping of stroke lesions underlying somatosensory deficits. *NeuroImage Clin.* **2016**, *10*, 257–266. [CrossRef]
152. Feys, H.; De Weerdt, W.; Nuyens, G.; Van De Winckel, A.; Selz, B.; Kiekens, C. Predicting motor recovery of the upper limb after stroke rehabilitation: Value of a clinical examination. *Physiother. Res. Int.* **2000**, *5*, 1–18. [CrossRef]
153. Chen, C.-L.; Tang, F.-T.; Chen, H.-C.; Chung, C.-Y.; Wong, M.-K. Brain lesion size and location: Effects on motor recovery and functional outcome in stroke patients. *Arch. Phys. Med. Rehabil.* **2000**, *81*, 447–452. [CrossRef] [PubMed]
154. Chilvers, M.J.; Rajashekar, D.; Low, T.A.; Scott, S.H.; Dukelow, S.P. Clinical, Neuroimaging and Robotic Measures Predict Long-Term Proprioceptive Impairments following Stroke. *Brain Sci.* **2023**, *13*, 953. [CrossRef] [PubMed]
155. Carvalho, A.; Cunha, A.; Rodrigues, M.; Figueiredo, S.; Paredes, L.; Gregório, T.; Morais, H.; Pinheiro, J.; Cruz, V.T.; Roriz, J.M.; et al. Mechanical Thrombectomy in Acute Ischemic Stroke: Initial Single-Center Experience and Comparison with Randomized Controlled Trials. *J. Stroke Cerebrovasc. Dis.* **2017**, *26*, 589–594. [CrossRef]
156. Raha, O.; Hall, C.; Malik, A.; D'Anna, L.; Lobotesis, K.; Kwan, J.; Banerjee, S. Advances in mechanical thrombectomy for acute ischaemic stroke. *BMJ Med.* **2023**, *2*, e000407. [CrossRef]
157. Jadhav, A.P.; Desai, S.M.; Jovin, T.G. Indications for Mechanical Thrombectomy for Acute Ischemic Stroke: Current Guidelines and Beyond. *Neurology* **2021**, *97*, S126–S136. [CrossRef]

158. Moskowitz, M.A.; Lo, E.H.; Iadecola, C. The Science of Stroke: Mechanisms in Search of Treatments. *Neuron* **2010**, *67*, 181–198. [CrossRef] [PubMed]
159. Yoo, A.J.; Pulli, B.; Gonzalez, R.G. Imaging-based treatment selection for intravenous and intra-arterial stroke therapies: A comprehensive review. *Expert Rev. Cardiovasc. Ther.* **2011**, *9*, 857–876. [CrossRef]
160. Zemgulyte, G.; Tanaka, S.; Hide, I.; Sakai, N.; Pampuscenko, K.; Borutaite, V.; Rastenyte, D. Evaluation of the Effectiveness of Post-Stroke Metformin Treatment Using Permanent Middle Cerebral Artery Occlusion in Rats. *Pharmaceuticals* **2021**, *14*, 312. [CrossRef]
161. Elkins, J.S.; Friedmanbc, C.; Boden-Albala, B.; Sacco, R.L.; Hripcsakc, G. Coding Neuroradiology Reports for the Northern Manhattan Stroke Study: A Comparison of Natural Language Processing and Manual Review. *Comput. Biomed. Res.* **2000**, *33*, 1–10. [CrossRef]
162. Rivas, R.; Montazeri, N.; Le, N.X.; Hristidis, V. Automatic Classification of Online Doctor Reviews: Evaluation of Text Classifier Algorithms. *J. Med. Internet Res.* **2018**, *20*, e11141. [CrossRef] [PubMed]
163. Heo, T.S.; Kim, Y.S.; Choi, J.M.; Jeong, Y.S.; Seo, S.Y.; Lee, J.H.; Jeon, J.P.; Kim, C. Prediction of Stroke Outcome Using Natural Language Processing-Based Machine Learning of Radiology Report of Brain MRI. *J. Pers. Med.* **2020**, *10*, 286. [CrossRef]
164. Vagal, A.; Wintermark, M.; Nael, K.; Bivard, A.; Parsons, M.; Grossman, A.W.; Khatri, P. Automated CT perfusion imaging for acute ischemic stroke: Pearls and pitfalls for real-world use. *Neurology* **2019**, *93*, 888–898. [CrossRef] [PubMed]
165. Lee, S.; Mlynash, M.; Christensen, S.; Jiang, B.; Wintermark, M.; Sträter, R.; Broocks, G.; Grams, A.; Nikoubashman, O.; Morotti, A.; et al. Hyperacute Perfusion Imaging Before Pediatric Thrombectomy: Analysis of the Save ChildS Study. *Neurology* **2023**, *100*, e1148–e1158. [CrossRef] [PubMed]
166. Haggenmüller, B.; Kreiser, K.; Sollmann, N.; Huber, M.; Vogele, D.; Schmidt, S.A.; Beer, M.; Schmitz, B.; Ozpeynirci, Y.; Rosskopf, J.; et al. Pictorial Review on Imaging Findings in Cerebral CTP in Patients with Acute Stroke and Its Mimics: A Primer for General Radiologists. *Diagnostics* **2023**, *13*, 447. [CrossRef]
167. Psychogios, K.; Tsivgoulis, G. Intravenous thrombolysis for acute ischemic stroke: Why not? *Curr. Opin. Neurol.* **2021**, *35*, 10–17. [CrossRef] [PubMed]
168. Beyeler, M.; Weber, L.; Kurmann, C.C.; Piechowiak, E.I.; Mosimann, P.J.; Zibold, F.; Meinel, T.R.; Branca, M.; Goeldlin, M.; Pilgram-Pastor, S.M.; et al. Association of reperfusion success and emboli in new territories with long term mortality after mechanical thrombectomy. *J. NeuroInt. Surg.* **2021**, *14*, 326–332. [CrossRef]
169. Popiela, T.J.; Krzyściak, W.; Pilato, F.; Ligęzka, A.; Bystrowska, B.; Bukowska-Strakova, K.; Brzegowy, P.; Muthusamy, K.; Kozicz, T. The Assessment of Endovascular Therapies in Ischemic Stroke: Management, Problems and Future Approaches. *J. Clin. Med.* **2022**, *11*, 1864. [CrossRef]
170. Pawłowski, K.; Dziadkiewicz, A.; Podlasek, A.; Klaudel, J.; Mączkowiak, A.; Szołkiewicz, M. Thrombectomy-Capable Stroke Centre—A Key to Acute Stroke Care System Improvement? Retrospective Analysis of Safety and Efficacy of Endovascular Treatment in Cardiac Cathlab. *Int. J. Environ. Res. Public Health* **2023**, *20*, 2232. [CrossRef]
171. Sato, H.; Mitsui, N.; Fujikawa, S.; Kinoshita, M.; Hori, K.; Uebayashi, M.; Kimura, T. Critical evaluation of the modified Rankin Scale for assessment of the efficacy of mechanical thrombectomy: A retrospective comparison between the modified Rankin Scale and functional independence measure. *Interv. Neuroradiol.* **2023**. [CrossRef]
172. Yeo, M.; Tahayori, B.; Kok, H.K.; Maingard, J.; Kutaiba, N.; Russell, J.; Thijs, V.; Jhamb, A.; Chandra, R.V.; Brooks, M.; et al. Evaluation of techniques to improve a deep learning algorithm for the automatic detection of intracranial haemorrhage on CT head imaging. *Eur. Radiol. Exp.* **2023**, *7*, 17. [CrossRef] [PubMed]
173. Chen, Y.-T.; Chen, Y.-L.; Chen, Y.-Y.; Huang, Y.-T.; Wong, H.-F.; Yan, J.-L.; Wang, J.-J. Deep Learning–Based Brain Computed Tomography Image Classification with Hyperparameter Optimization through Transfer Learning for Stroke. *Diagnostics* **2022**, *12*, 807. [CrossRef] [PubMed]
174. Aho, K.; Harmsen, P.; Hatano, S.; Marquardsen, J.; Smirnov, V.E.; Strasser, T. Cerebrovascular disease in the community: Results of a WHO collaborative study. *Bull. World Health Organ.* **1980**, *58*, 113–130.
175. Sacco, R.L.; Kasner, S.E.; Broderick, J.P.; Caplan, L.R.; Connors, J.J.; Culebras, A.; Elkind, M.S.; George, M.G.; Hamdan, A.D.; Higashida, R.T.; et al. An Updated Definition of Stroke for the 21st Century: A statement for healthcare professionals from the American heart association/American stroke association. *Stroke* **2013**, *44*, 2064–2089. [CrossRef]
176. Iancu, A.; Buleu, F.; Chita, D.S.; Tutelca, A.; Tudor, R.; Brad, S. Early Hemorrhagic Transformation after Reperfusion Therapy in Patients with Acute Ischemic Stroke: Analysis of Risk Factors and Predictors. *Brain Sci.* **2023**, *13*, 840. [CrossRef] [PubMed]
177. Ho-Tin-Noé, B.; Desilles, J.-P.; Mazighi, M. Thrombus composition and thrombolysis resistance in stroke. *Res. Pr. Thromb. Haemost.* **2023**, *7*, 100178. [CrossRef] [PubMed]
178. Gurewich, V. The sequential, complementary and synergistic activation of fibrin-bound plasminogen by tissue plasminogen activator and pro-urokinase. *Fibrinolysis* **1989**, *3*, 59–66. [CrossRef]
179. van der Ende, N.A.M.; Roozenbeek, B.; Smagge, L.E.M.; Luijten, S.P.R.; Aerden, L.A.M.; Kraayeveld, P.; Wijngaard, I.R.v.D.; Nijeholt, G.J.L.; Hertog, H.M.D.; Flach, H.Z.; et al. Dual thrombolytic therapy with mutant pro-urokinase and small bolus alteplase for ischemic stroke (DUMAS): Study protocol for a multicenter randomized controlled phase II trial. *Trials* **2022**, *23*, 1–10. [CrossRef]
180. Gurewich, V. Activation of Fibrin-Bound Plasminogen by Pro-Urokinase and Its Complementariness with That by Tissue Plasminogen Activator. *Enzyme* **1988**, *40*, 97–108. [CrossRef]

181. Yang, Y.; Gu, B.; Xu, X.Y. In Silico Study of Different Thrombolytic Agents for Fibrinolysis in Acute Ischemic Stroke. *Pharmaceutics* **2023**, *15*, 797. [CrossRef]
182. Choi, H.-Y.; Cho, Y.; Kim, W.; Minn, Y.-K.; Kang, G.-H.; Jang, Y.-S.; Lee, Y.; Kim, J.-G.; Kim, J.; Cho, Y.; et al. Analysis of Mortality in Intracerebral Hemorrhage Patients with Hyperacute Ischemic Stroke Treated Using Thrombolytic Therapy: A Nationwide Population-based Cohort Study in South Korea. *J. Pers. Med.* **2022**, *12*, 1260. [CrossRef] [PubMed]
183. Nisar, T.; Hanumanthu, R.; Khandelwal, P. Symptomatic Intracerebral Hemorrhage after Intravenous Thrombolysis: Predictive Factors and Validation of Prediction Models. *J. Stroke Cerebrovasc. Dis.* **2019**, *28*, 104360. [CrossRef] [PubMed]
184. Aamir, R.; Fyffe, C.; Korin, N.; Lawrence, D.A.; Su, E.J.; Kanapathipillai, M. Heparin and Arginine Based Plasmin Nanoformulation for Ischemic Stroke Therapy. *Int. J. Mol. Sci.* **2021**, *22*, 11477. [CrossRef] [PubMed]
185. Knowland, D.; Arac, A.; Sekiguchi, K.J.; Hsu, M.; Lutz, S.E.; Perrino, J.; Steinberg, G.K.; Barres, B.A.; Nimmerjahn, A.; Agalliu, D. Stepwise Recruitment of Transcellular and Paracellular Pathways Underlies Blood-Brain Barrier Breakdown in Stroke. *Neuron* **2014**, *82*, 603–617. [CrossRef] [PubMed]
186. Shibata, K.; Hashimoto, T.; Miyazaki, T.; Miyazaki, A.; Nobe, K. Thrombolytic Therapy for Acute Ischemic Stroke: Past and Future. *Curr. Pharm. Des.* **2019**, *25*, 242–250. [CrossRef] [PubMed]
187. National Institute of Neurological Disorders and Stroke rt-PA Stroke Study Group. Tissue Plasminogen Activator for Acute Ischemic Stroke. *N. Engl. J. Med.* **1995**, *333*, 1581–1588. [CrossRef]
188. Lindley, R.I.; Wardlaw, J.M.; Whiteley, W.N.; Cohen, G.; Blackwell, L.; Murray, G.D.; Sandercock, P.A.; Baigent, C.; Chadwick, D.; Tyrrell, P.; et al. Alteplase for Acute Ischemic Stroke: Outcomes by clinically important subgroups in the Third International. *Stroke* **2015**, *46*, 746–756. [CrossRef]
189. Parsons, M.; Spratt, N.; Bivard, A.; Campbell, B.; Chung, K.; Miteff, F.; O'Brien, B.; Bladin, C.; McElduff, P.; Allen, C.; et al. A Randomized Trial of Tenecteplase versus Alteplase for Acute Ischemic Stroke. *N. Engl. J. Med.* **2012**, *366*, 1099–1107. [CrossRef]
190. Yamaguchi, T.; Mori, E.; Minematsu, K.; Nakagawara, J.; Hashi, K.; Saito, I.; Shinohara, Y. Alteplase at 0.6 mg/kg for Acute Ischemic Stroke Within 3 Hours of Onset: Japan Alteplase Clinical Trial (J-ACT). *Stroke* **2006**, *37*, 1810–1815. [CrossRef]
191. Campbell, B.C.; Mitchell, P.J.; Churilov, L.; Yassi, N.; Kleinig, T.J.; Dowling, R.J.; Yan, B.; Bush, S.J.; Dewey, H.M.; Thijs, V.; et al. Tenecteplase versus Alteplase before Thrombectomy for Ischemic Stroke. *N. Engl. J. Med.* **2018**, *378*, 1573–1582. [CrossRef]
192. Sommer, K.N.; Bhurwani, M.M.S.; Mokin, M.; Ionita, C.N. Evaluation of challenges and limitations of mechanical thrombectomy using 3D printed neurovascular phantoms. *Proc. SPIE Int. Soc. Opt. Eng.* **2021**, *11601*, 53–62. [CrossRef]
193. Gu, B.; Piebalgs, A.; Huang, Y.; Longstaff, C.; Hughes, A.D.; Chen, R.; Thom, S.A.; Xu, X.Y. Mathematical Modelling of Intravenous Thrombolysis in Acute Ischaemic stroke: Effects of Dose Regimens on Levels of Fibrinolytic Proteins and Clot Lysis Time. *Pharmaceutics* **2019**, *11*, 111. [CrossRef] [PubMed]
194. Endres, M.; Dirnagl, U.; Moskowitz, M.A. Chapter 2 The ischemic cascade and mediators of ischemic injury. In *Handbook of Clinical Neurology, in Stroke*; Elsevier: Amsterdam, The Netherlands, 2008; Volume 92, pp. 31–41. [CrossRef]
195. Tong, X.; Wang, Y.; Fiehler, J.; Bauer, C.T.; Jia, B.; Zhang, X.; Huo, X.; Luo, G.; Wang, A.; Pan, Y.; et al. Thrombectomy Versus Combined Thrombolysis and Thrombectomy in Patients With Acute Stroke: A Matched-Control Study. *Stroke* **2021**, *52*, 1589–1600. [CrossRef] [PubMed]
196. Rha, J.-H.; Saver, J.L. The Impact of Recanalization on Ischemic Stroke Outcome. *Stroke* **2007**, *38*, 967–973. [CrossRef]
197. Bhatia, R.; Hill, M.D.; Shobha, N.; Menon, B.; Bal, S.; Kochar, P.; Watson, T.; Goyal, M.; Demchuk, A.M. Low Rates of Acute Recanalization With Intravenous Recombinant Tissue Plasminogen Activator in Ischemic Stroke: Real-world experience and a call for action. *Stroke* **2010**, *41*, 2254–2258. [CrossRef]
198. González, R.G.; Furie, K.L.; Goldmacher, G.V.; Smith, W.S.; Kamalian, S.; Payabvash, S.; Harris, G.J.; Halpern, E.F.; Koroshetz, W.J.; Camargo, E.C.; et al. Good outcome rate of 35% in IV-tPA-treated patients with computed tomography angiography confirmed severe anterior circulation occlusive stroke. *Stroke* **2013**, *44*, 3109–3113. [CrossRef]
199. Iglesias-Rey, R.; Rodríguez-Yáñez, M.; Rodríguez-Castro, E.; Pumar, J.M.; Arias, S.; Santamaría, M.; López-Dequidt, I.; Hervella, P.; Correa-Paz, C.; Sobrino, T.; et al. Worse Outcome in Stroke Patients Treated with rt-PA Without Early Reperfusion: Associated Factors. *Transl. Stroke Res.* **2017**, *9*, 347–355. [CrossRef]
200. Alonso-Alonso, M.L.; Pérez-Mato, M.; Sampedro-Viana, A.; Correa-Paz, C.; Ávila-Gómez, P.; Sobrino, T.; Campos, F.; Castillo, J.; Iglesias-Rey, R.; Hervella, P. Fibrin-Targeted Nanoparticles for Finding, Visualizing and Characterizing Blood Clots in Acute Ischemic Stroke. *Pharmaceutics* **2022**, *14*, 2156. [CrossRef]
201. Adams, H.P.; Brott, T.G.; Furlan, A.J.; Gomez, C.R.; Grotta, J.; Helgason, C.M.; Kwiatkowski, T.; Lyden, P.D.; Marler, J.R.; Torner, J.; et al. Guidelines for Thrombolytic Therapy for Acute Stroke: A Supplement to the Guidelines for the Management of Patients With Acute Ischemic Stroke. A statement for healthcare professionals from a Special Writing Group of the Stroke Council, American Heart Association. *Circulation* **1996**, *94*, 1167–1174. [CrossRef]
202. Frank, D.; Zlotnik, A.; Boyko, M.; Gruenbaum, B.F. The Development of Novel Drug Treatments for Stroke Patients: A Review. *Int. J. Mol. Sci.* **2022**, *23*, 5796. [CrossRef]
203. Shah, P.P.; Szaflarski, J.P.; Allendorfer, J.; Hamilton, R.H. Induction of neuroplasticity and recovery in post-stroke aphasia by non-invasive brain stimulation. *Front. Hum. Neurosci.* **2013**, *7*, 888. [CrossRef] [PubMed]
204. Cichon, N.; Wlodarczyk, L.; Saluk-Bijak, J.; Bijak, M.; Redlicka, J.; Gorniak, L.; Miller, E. Novel Advances to Post-Stroke Aphasia Pharmacology and Rehabilitation. *J. Clin. Med.* **2021**, *10*, 3778. [CrossRef] [PubMed]

205. Chieffo, R.; Giatsidis, F.; Santangelo, R.; Alyagon, U.; Comola, M.; Zangen, A.; Comi, G.; Leocani, L. Repetitive Transcranial Magnetic Stimulation With H-Coil Coupled With Cycling for Improving Lower Limb Motor Function After Stroke: An Exploratory Study. *Neuromodul. Technol. Neural Interface* **2020**, *24*, 916–922. [CrossRef]
206. Kim, W.-S.; Kwon, B.S.; Gil Seo, H.; Park, J.; Paik, N.-J. Low-Frequency Repetitive Transcranial Magnetic Stimulation Over Contralesional Motor Cortex for Motor Recovery in Subacute Ischemic Stroke: A Randomized Sham-Controlled Trial. *Neurorehabilit. Neural Repair* **2020**, *34*, 856–867. [CrossRef]
207. Liu, Y.; Yin, M.; Luo, J.; Huang, L.; Zhang, S.; Pan, C.; Hu, X. Effects of transcranial magnetic stimulation on the performance of the activities of daily living and attention function after stroke: A randomized controlled trial. *Clin. Rehabil.* **2020**, *34*, 1465–1473. [CrossRef]
208. Zanardi, R.; Poletti, S.; Prestifilippo, D.; Attanasio, F.; Barbini, B.; Colombo, C. Transcranial direct current stimulation: A novel approach in the treatment of vascular depression. *Brain Stimul.* **2020**, *13*, 1559–1565. [CrossRef]
209. Power, J.D.; Schlaggar, B.L. Neural plasticity across the lifespan. *Wiley Interdiscip. Rev. Dev. Biol.* **2016**, *6*, e216. [CrossRef] [PubMed]
210. Alia, C.; Spalletti, C.; Lai, S.; Panarese, A.; Lamola, G.; Bertolucci, F.; Vallone, F.; Di Garbo, A.; Chisari, C.; Micera, S.; et al. Neuroplastic Changes Following Brain Ischemia and their Contribution to Stroke Recovery: Novel Approaches in Neurorehabilitation. *Front. Cell Neurosci.* **2017**, *11*, 76. [CrossRef]
211. Hatcher, J.F.; Dempsey, R.J. Citicoline: Neuroprotective mechanisms in cerebral ischemia. *J. Neurochem.* **2002**, *80*, 12–23. [CrossRef]
212. Lim, C.-M.; Kim, S.-W.; Park, J.-Y.; Kim, C.; Yoon, S.H.; Lee, J.-K. Fluoxetine affords robust neuroprotection in the postischemic brain via its anti-inflammatory effect. *J. Neurosci. Res.* **2009**, *87*, 1037–1045. [CrossRef]
213. Szelenberger, R.; Kostka, J.; Saluk-Bijak, J.; Miller, E. Pharmacological Interventions and Rehabilitation Approach for Enhancing Brain Self-repair and Stroke Recovery. *Curr. Neuropharmacol.* **2019**, *18*, 51–64. [CrossRef] [PubMed]
214. O'Collins, V.E.; Macleod, M.R.; Donnan, G.A.; Horky, L.L.; van der Worp, B.H.; Howells, D.W. 1026 Experimental treatments in acute stroke. *Ann. Neurol.* **2006**, *59*, 467–477. [CrossRef] [PubMed]
215. Goenka, L.; Satyanarayana, C.R.U.; George, M. Neuroprotective agents in Acute Ischemic Stroke—A Reality Check. *BioMed. Pharmacother.* **2018**, *109*, 2539–2547. [CrossRef] [PubMed]
216. Ford, G.A. Clinical pharmacological issues in the development of acute stroke therapies. *Br. J. Pharmacol.* **2008**, *153*, S112–S119. [CrossRef]
217. Frese, A.; Husstedt, I.; Ringelstein, E.; Evers, S. Pharmacologic Treatment of Central Post-Stroke Pain. *Clin. J. Pain* **2006**, *22*, 252–260. [CrossRef] [PubMed]
218. Greener, J.; Enderby, P.; Whurr, R. Pharmacological treatment for aphasia following stroke. *Cochrane Database Syst. Rev.* **2001**, *2001*, CD000424. [CrossRef]
219. Hackett, M.L.; Anderson, C.S.; House, A.O. Management of Depression After Stroke. *Stroke* **2005**, *36*, 1092–1097. [CrossRef]
220. Ghozy, S.; Reda, A.; Varney, J.; Elhawary, A.S.; Shah, J.; Murry, K.; Sobeeh, M.G.; Nayak, S.S.; Azzam, A.Y.; Brinjikji, W.; et al. Neuroprotection in Acute Ischemic Stroke: A Battle Against the Biology of Nature. *Front. Neurol.* **2022**, *13*, 870141. [CrossRef]
221. Cramer, S.C. Repairing the human brain after stroke. II. Restorative therapies. *Ann. Neurol.* **2008**, *63*, 549–560. [CrossRef]
222. Chollet, F.; Tardy, J.; Albucher, J.-F.; Thalamas, C.; Berard, E.; Lamy, C.; Bejot, Y.; Deltour, S.; Jaillard, A.; Niclot, P.; et al. Fluoxetine for motor recovery after acute ischaemic stroke (FLAME): A randomised placebo-controlled trial. *Lancet Neurol.* **2011**, *10*, 123–130. [CrossRef]
223. Chollet, F.; Albucher, J.-F. Strategies to Augment Recovery After Stroke. *Curr. Treat. Opt. Neurol.* **2012**, *14*, 531–540. [CrossRef] [PubMed]
224. Miyauchi, J.T.; Tsirka, S.E. Advances in immunotherapeutic research for glioma therapy. *J. Neurol.* **2017**, *265*, 741–756. [CrossRef] [PubMed]
225. Lattanzi, S.; Norata, D.; Divani, A.A.; Di Napoli, M.; Broggi, S.; Rocchi, C.; Ortega-Gutierrez, S.; Mansueto, G.; Silvestrini, M. Systemic Inflammatory Response Index and Futile Recanalization in Patients with Ischemic Stroke Undergoing Endovascular Treatment. *Brain Sci.* **2021**, *11*, 1164. [CrossRef] [PubMed]
226. Rabinstein, A.A. Update on Treatment of Acute Ischemic Stroke. *Contin. Lifelong Learn. Neurol.* **2020**, *26*, 268–286. [CrossRef]
227. Pagiola, I.; Chassin, O.; Gallas, S.; Riviere, M.S.; Legris, N.; Mihalea, C.; Caroff, J.; Ikka, L.; Chalumeau, V.; de Aguiar, G.B.; et al. Extending the stroke treatment window beyond DAWN in patients with very slow progressor type collaterals: How far can we go? *J. Cerebrovasc. Endovasc. Neurosurg.* **2021**, *23*, 354–358. [CrossRef]
228. de Castro-Afonso, L.H.; Abud, T.G.; Pontes-Neto, O.M.; Monsignore, L.M.; Nakiri, G.S.; Cougo-Pinto, P.T.; de Oliveira, L.; dos Santos, D.; Dias, F.A.; Fábio, S.C.R.; et al. Mechanical thrombectomy with solitaire stent retrieval for acute ischemic stroke in a Brazilian population. *Clinics* **2012**, *67*, 1379–1386. [CrossRef]
229. Gupta, V.; Parthasarathy, R. Mechanical Thrombectomy: Answering Unanswered. *Ann. Indian Acad. Neurol.* **2019**, *23*, 13–19. [CrossRef]
230. Derex, L.; Cho, T.-H. Mechanical thrombectomy in acute ischemic stroke. *Rev. Neurol.* **2017**, *173*, 106–113. [CrossRef]
231. Krishnan, R.; Mays, W.; Elijovich, L. Complications of Mechanical Thrombectomy in Acute Ischemic Stroke. *Neurology* **2021**, *97*, S115–S125. [CrossRef]
232. Kim, H.Y.; Bin Back, D.; Choi, B.-R.; Choi, D.-H.; Kwon, K.J. Rodent Models of Post-Stroke Dementia. *Int. J. Mol. Sci.* **2022**, *23*, 10750. [CrossRef]

233. Chin, Y.; Kishi, M.; Sekino, M.; Nakajo, F.; Abe, Y.; Terazono, Y.; Hiroyuki, O.; Kato, F.; Koizumi, S.; Gachet, C.; et al. Involvement of glial P2Y1 receptors in cognitive deficit after focal cerebral stroke in a rodent model. *J. Neuroinflamm.* **2013**, *10*, 95. [CrossRef] [PubMed]
234. Ward, R.; Valenzuela, J.P.; Li, W.; Dong, G.; Fagan, S.C.; Ergul, A. Poststroke cognitive impairment and hippocampal neurovascular remodeling: The impact of diabetes and sex. *Am. J. Physiol. Circ. Physiol.* **2018**, *315*, H1402–H1413. [CrossRef] [PubMed]
235. Lin, J.; Frontera, J.A. Decompressive Hemicraniectomy for Large Hemispheric Strokes. *Stroke* **2021**, *52*, 1500–1510. [CrossRef] [PubMed]
236. Scott, R.M.; Smith, E.R. Moyamoya Disease and Moyamoya Syndrome. *N. Engl. J. Med.* **2009**, *360*, 1226–1237. [CrossRef] [PubMed]
237. Canavero, I.; Vetrano, I.G.; Zedde, M.; Pascarella, R.; Gatti, L.; Acerbi, F.; Nava, S.; Ferroli, P.; Parati, E.A.; Bersano, A. Clinical Management of Moyamoya Patients. *J. Clin. Med.* **2021**, *10*, 3628. [CrossRef]
238. Fukui, M. Guidelines for the diagnosis and treatment of spontaneous occlusion of the circle of Willis ('moyamoya' disease). Research Committee on Spontaneous Occlusion of the Circle of Willis (Moyamoya Disease) of the Ministry of Health and Welfare, Japan. *Clin. Neurol. Neurosurg.* **1997**, *99* (Suppl. 2), S238–S240. [CrossRef]
239. Miyamoto, S.; Yoshimoto, T.; Hashimoto, N.; Okada, Y.; Tsuji, I.; Tominaga, T.; Nakagawara, J.; Takahashi, J.C.; Yamada, K.; Tomata, Y.; et al. Effects of Extracranial–Intracranial Bypass for Patients With Hemorrhagic Moyamoya Disease: Results of the Japan Adult Moyamoya Trial. *Stroke* **2014**, *45*, 1415–1421. [CrossRef]
240. Takahashi, J.C.; Funaki, T.; Houkin, K.; Inoue, T.; Ogasawara, K.; Nakagawara, J.; Kuroda, S.; Yamada, K.; Miyamoto, S.; Yoshimoto, T.; et al. Significance of the Hemorrhagic Site for Recurrent Bleeding: Prespecified Analysis in the Japan Adult Moyamoya Trial. *Stroke* **2016**, *47*, 37–43. [CrossRef]
241. Haynes, M.S.; Alder, K.D.; Bellamkonda, K.; Kuzomunhu, L.; Grauer, J.N. Incidence, predictors, and timing of post-operative stroke following elective total hip arthroplasty and total knee arthroplasty. *PLoS ONE* **2020**, *15*, e0239239. [CrossRef]
242. Somnuke, P.; Kattiyawittayakul, K.; Wongyingsinn, M.; Raksakietisak, M. Incidence, risk factors and outcomes of early post-operative stroke in a university-based, tertiary care hospital: A 7-year retrospective case-control study. *Medicine* **2023**, *102*, e34363. [CrossRef]
243. Powers, W.J.; Rabinstein, A.A.; Ackerson, T.; Adeoye, O.M.; Bambakidis, N.C.; Becker, K.; Biller, J.; Brown, M.; Demaerschalk, B.M.; Hoh, B.; et al. Guidelines for the Early Management of Patients With Acute Ischemic Stroke: 2019 Update to the 2018 Guidelines for the Early Management of Acute Ischemic Stroke: A Guideline for Healthcare Professionals From the American Heart Association/American Stroke Association. *Stroke* **2019**, *50*, e344–e418, Correction in *Stroke* **2019**, *50*, e440–e441. [CrossRef] [PubMed]
244. Turc, G.; Bhogal, P.; Fischer, U.; Khatri, P.; Lobotesis, K.; Mazighi, M.; Schellinger, P.D.; Toni, D.; de Vries, J.; White, P.; et al. European Stroke Organisation (ESO)—European Society for Minimally Invasive Neurological Therapy (ESMINT) Guidelines on Mechanical Thrombectomy in Acute Ischaemic StrokeEndorsed by Stroke Alliance for Europe (SAFE). *Eur. Stroke J.* **2019**, *4*, 6–12. [CrossRef] [PubMed]
245. Ansari, J.; Triay, R.; Kandregula, S.; Adeeb, N.; Cuellar, H.; Sharma, P. Endovascular Intervention in Acute Ischemic Stroke: History and Evolution. *Biomedicines* **2022**, *10*, 418. [CrossRef] [PubMed]
246. Roaldsen, M.B.; Lindekleiv, H.; Mathiesen, E.B. Intravenous thrombolytic treatment and endovascular thrombectomy for ischaemic wake-up stroke. *Cochrane Database Syst. Rev.* **2021**, *2021*, CD010995. [CrossRef]
247. Hankey, G.J. Stroke. *Lancet* **2016**, *389*, 641–654. [CrossRef]
248. Savitz, S.I.; Baron, J.-C.; Fisher, M.; Albers, G.W.; Arbe-Barnes, S.; Boltze, J.; Broderick, J.; Broschat, K.O.; Elkind, M.S.V.; En'wezoh, D.; et al. Stroke Treatment Academic Industry Roundtable X: Brain Cytoprotection Therapies in the Reperfusion Era. *Stroke* **2019**, *50*, 1026–1031. [CrossRef]
249. Fisher, M.; Savitz, S.I. Pharmacological brain cytoprotection in acute ischaemic stroke—Renewed hope in the reperfusion era. *Nat. Rev. Neurol.* **2022**, *18*, 193–202. [CrossRef]
250. Chen, F.; Qi, Z.; Luo, Y.; Hinchliffe, T.; Ding, G.; Xia, Y.; Ji, X. Non-pharmaceutical therapies for stroke: Mechanisms and clinical implications. *Prog. Neurobiol.* **2014**, *115*, 246–269. [CrossRef]
251. Wang, Y.; Meng, R.; Liu, G.; Cao, C.; Chen, F.; Jin, K.; Ji, X.; Cao, G. Intracranial atherosclerotic disease. *Neurobiol. Dis.* **2018**, *124*, 118–132. [CrossRef]
252. Molina, C.A.; Montaner, J.; Abilleira, S.; Ibarra, B.; Romero, F.; Arenillas, J.F.; Alvarez-Sabín, J. Timing of Spontaneous Recanalization and Risk of Hemorrhagic Transformation in Acute Cardioembolic Stroke. *Stroke* **2001**, *32*, 1079–1084. [CrossRef]
253. García-Cobos, R.; Frank-García, A.; Gutiérrez-Fernández, M.; Díez-Tejedor, E. Citicoline, use in cognitive decline: Vascular and degenerative. *J. Neurol. Sci.* **2010**, *299*, 188–192. [CrossRef] [PubMed]
254. Álvarez-Sabín, J.; Román, G.C. The Role of Citicoline in Neuroprotection and Neurorepair in Ischemic Stroke. *Brain Sci.* **2013**, *3*, 1395–1414. [CrossRef] [PubMed]
255. Stroke Unit Trialists' Collaboration. Collaborative systematic review of the randomised trials of organised inpatient (stroke unit) care after stroke. *BMJ* **1997**, *314*, 1151–1159. [CrossRef]
256. Campbell, B.C.V.; De Silva, D.A.; Macleod, M.R.; Coutts, S.B.; Schwamm, L.H.; Davis, S.M.; Donnan, G.A. Ischaemic stroke. *Nat. Rev. Dis. Prim.* **2019**, *5*, 70. [CrossRef]

257. Hsieh, H.-C.; Liao, R.-D.; Yang, T.-H.; Leong, C.-P.; Tso, H.-H.; Wu, J.-Y.; Huang, Y.-C. The clinical effect of Kinesio taping and modified constraint-induced movement therapy on upper extremity function and spasticity in patients with stroke: A randomized controlled pilot study. *Eur. J. Phys. Rehabil. Med.* **2021**, *57*, 511–519. [CrossRef] [PubMed]
258. Stoll, G.; Nieswandt, B. Thrombo-inflammation in acute ischaemic stroke—Implications for treatment. *Nat. Rev. Neurol.* **2019**, *15*, 473–481. [CrossRef] [PubMed]
259. Schnieder, M.; Bähr, M.; Kirsch, M.; Maier, I.; Behme, D.; Riedel, C.H.; Psychogios, M.-N.; Brehm, A.; Liman, J.; von Arnim, C.A.F. Analysis of Frailty in Geriatric Patients as a Prognostic Factor in Endovascular Treated Patients with Large Vessel Occlusion Strokes. *J. Clin. Med.* **2021**, *10*, 2171. [CrossRef]
260. Liu, X.; Dai, Q.; Ye, R.; Zi, W.; Liu, Y.; Wang, H.; Zhu, W.; Ma, M.; Yin, Q.; Li, M.; et al. Endovascular treatment versus standard medical treatment for vertebrobasilar artery occlusion (BEST): An open-label, randomised controlled trial. *Lancet Neurol.* **2020**, *19*, 115–122. [CrossRef]
261. Levy, E.I.; Siddiqui, A.H.; Crumlish, A.; Snyder, K.V.; Hauck, E.F.; Fiorella, D.J.; Hopkins, L.N.; Mocco, J. First Food and Drug Administration-Approved Prospective Trial of Primary Intracranial Stenting for Acute Stroke: SARIS (stent-assisted recanalization in acute ischemic. *Stroke* **2009**, *40*, 3552–3556. [CrossRef]
262. Hasan, T.F.; Hasan, H.; Kelley, R.E. Overview of Acute Ischemic Stroke Evaluation and Management. *Biomedicines* **2021**, *9*, 1486. [CrossRef]
263. Meisel, C.; Schwab, J.M.; Prass, K.; Meisel, A.; Dirnagl, U. Central nervous system injury-induced immune deficiency syndrome. *Nat. Rev. Neurosci.* **2005**, *6*, 775–786. [CrossRef] [PubMed]
264. Chamorro, A.; Urra, X.; Planas, A.M. Infection After Acute Ischemic Stroke: A manifestation of brain-induced immunodepression. *Stroke* **2007**, *38*, 1097–1103. [CrossRef] [PubMed]
265. Vogelgesang, A.; Grunwald, U.; Langner, S.; Jack, R.; Bröker, B.M.; Kessler, C.; Dressel, A. Analysis of Lymphocyte Subsets in Patients With Stroke and Their Influence on Infection After Stroke. *Stroke* **2008**, *39*, 237–241. [CrossRef]
266. Urra, X.; Cervera, Á.; Villamor, N.; Planas, A.; Chamorro, A. Harms and benefits of lymphocyte subpopulations in patients with acute stroke. *Neuroscience* **2009**, *158*, 1174–1183. [CrossRef] [PubMed]
267. Brait, V.H.; Arumugam, T.V.; Drummond, G.R.; Sobey, C.G. Importance of T Lymphocytes in Brain Injury, Immunodeficiency, and Recovery after Cerebral Ischemia. *J. Cereb. Blood Flow Metab.* **2012**, *32*, 598–611. [CrossRef]
268. Bi, Y.; Shen, J.; Chen, S.-C.; Chen, J.-X.; Xia, Y.-P. Prognostic value of neutrophil to lymphocyte ratio in acute ischemic stroke after reperfusion therapy. *Sci. Rep.* **2021**, *11*, 6177. [CrossRef]
269. Lattanzi, S.; Cagnetti, C.; Provinciali, L.; Silvestrini, M. Neutrophil-to-lymphocyte ratio and neurological deterioration following acute cerebral hemorrhage. *Oncotarget* **2017**, *8*, 57489–57494. [CrossRef]
270. de Mendivil, A.O.; Alcalá-Galiano, A.; Ochoa, M.; Salvador, E.; Millán, J.M. Brainstem Stroke: Anatomy, Clinical and Radiological Findings. *Semin. Ultrasound CT MRI* **2013**, *34*, 131–141. [CrossRef]
271. Haddad, J.J.; Saadé, N.E.; Safieh-Garabedian, B. Cytokines and neuro–immune–endocrine interactions: A role for the hypothalamic–pituitary–adrenal revolving axis. *J. Neuroimmunol.* **2002**, *133*, 1–19. [CrossRef]
272. Yeh, S.-J.; Huang, K.-Y.; Wang, T.-G.; Chen, Y.-C.; Chen, C.-H.; Tang, S.-C.; Tsai, L.-K.; Yip, P.-K.; Jeng, J.-S. Dysphagia screening decreases pneumonia in acute stroke patients admitted to the stroke intensive care unit. *J. Neurol. Sci.* **2011**, *306*, 38–41. [CrossRef]
273. Shim, R.; Wong, C.H.Y. Ischemia, Immunosuppression and Infection—Tackling the Predicaments of Post-Stroke Complications. *Int. J. Mol. Sci.* **2016**, *17*, 64. [CrossRef] [PubMed]
274. Suri, R.; Rodriguez-Porcel, F.; Donohue, K.; Jesse, E.; Lovera, L.; Dwivedi, A.K.; Espay, A.J. Post-stroke Movement Disorders: The Clinical, Neuroanatomic, and Demographic Portrait of 284 Published Cases. *J. Stroke Cerebrovasc. Dis.* **2018**, *27*, 2388–2397. [CrossRef]
275. Thieme, H.; Mehrholz, J.; Pohl, M.; Behrens, J.; Dohle, C.; Morkisch, N.; Borgetto, B.; Winter, J.; Hunter, S.M.; Sim, J.; et al. Mirror Therapy for Improving Motor Function After Stroke. *Stroke* **2013**, *44*, e1–e2. [CrossRef] [PubMed]
276. Lindenberg, R.; Renga, V.; Zhu, L.L.; Nair, D.; Schlaug, G. Bihemispheric brain stimulation facilitates motor recovery in chronic stroke patients. *Neurology* **2010**, *75*, 2176–2184. [CrossRef]
277. Lo, A.C.; Guarino, P.D.; Richards, L.G.; Haselkorn, J.K.; Wittenberg, G.F.; Federman, D.G.; Ringer, R.J.; Wagner, T.H.; Krebs, H.I.; Volpe, B.T.; et al. Robot-Assisted Therapy for Long-Term Upper-Limb Impairment after Stroke. *N. Engl. J. Med.* **2010**, *362*, 1772–1783. [CrossRef]
278. Wang, L.E.; Fink, G.R.; Diekhoff, S.; Rehme, A.K.; Eickhoff, S.B.; Grefkes, C. Noradrenergic enhancement improves motor network connectivity in stroke patients. *Ann. Neurol.* **2010**, *69*, 375–388. [CrossRef]
279. Acler, M.; Robol, E.; Fiaschi, A.; Manganotti, P. A double blind placebo RCT to investigate the effects of serotonergic modulation on brain excitability and motor recovery in stroke patients. *J. Neurol.* **2009**, *256*, 1152–1158. [CrossRef] [PubMed]
280. Pollock, A.; Baer, G.; Langhorne, P.; Pomeroy, V. Physiotherapy treatment approaches for the recovery of postural control and lower limb function following stroke: A systematic review. *Clin. Rehabil.* **2007**, *21*, 395–410. [CrossRef]
281. Sütbeyaz, S.; Yavuzer, G.; Sezer, N.; Koseoglu, B.F. Mirror Therapy Enhances Lower-Extremity Motor Recovery and Motor Functioning After Stroke: A Randomized Controlled Trial. *Arch. Phys. Med. Rehabil.* **2007**, *88*, 555–559. [CrossRef]
282. Doyle, S.; Bennett, S.; Fasoli, S.E.; McKenna, K.T. Interventions for sensory impairment in the upper limb after stroke. *Cochrane Database Syst. Rev.* **2010**, *2010*, CD006331. [CrossRef]

283. Almekhlafi, M.; Kuczynski, A.; Demchuk, A. Therapeutic hypothermia: Applications in adults with acute ischemic stroke. *Brain Circ.* **2019**, *5*, 43–54. [CrossRef] [PubMed]
284. Woodruff, T.M.; Thundyil, J.; Tang, S.-C.; Sobey, C.G.; Taylor, S.M.; Arumugam, T.V. Pathophysiology, treatment, and animal and cellular models of human ischemic stroke. *Mol. Neurodegener.* **2011**, *6*, 11. [CrossRef] [PubMed]
285. Lees, T.; Shad-Kaneez, F.; Simpson, A.M.; Nassif, N.T.; Lin, Y.; Lal, S. Heart Rate Variability as a Biomarker for Predicting Stroke, Post-stroke Complications and Functionality. *Biomark. Insights* **2018**, *13*, 1177271918786931. [CrossRef] [PubMed]
286. Hussain, I.; Park, S.-J. Quantitative Evaluation of Task-Induced Neurological Outcome after Stroke. *Brain Sci.* **2021**, *11*, 900. [CrossRef]
287. Hussain, I.; Park, S.-J. Prediction of Myoelectric Biomarkers in Post-Stroke Gait. *Sensors* **2021**, *21*, 5334. [CrossRef]
288. Onose, G.; Anghelescu, A.; Blendea, D.; Ciobanu, V.; Daia, C.; Firan, F.C.; Oprea, M.; Spinu, A.; Popescu, C.; Ionescu, A.; et al. Cellular and Molecular Targets for Non-Invasive, Non-Pharmacological Therapeutic/Rehabilitative Interventions in Acute Ischemic Stroke. *Int. J. Mol. Sci.* **2022**, *23*, 907. [CrossRef]
289. Saunders, D.H.; Sanderson, M.; Hayes, S.; Johnson, L.; Kramer, S.; Carter, D.D.; Jarvis, H.; Brazzelli, M.; Mead, G.E. Physical fitness training for stroke patients. *Cochrane Database Syst. Rev.* **2020**, *2020*, CD003316. [CrossRef]
290. Winstein, C.J.; Rose, D.K.; Tan, S.M.; Lewthwaite, R.; Chui, H.C.; Azen, S.P. A randomized controlled comparison of upper-extremity rehabilitation strategies in acute stroke: A pilot study of immediate and long-term outcomes. *Arch. Phys. Med. Rehabil.* **2004**, *85*, 620–628. [CrossRef]
291. Kendall, B.J.; Gothe, N.P. Effect of Aerobic Exercise Interventions on Mobility among Stroke Patients: A Systematic Review. *Am. J. Phys. Med. Rehabil.* **2016**, *95*, 214–224. [CrossRef]
292. Brunelli, S.; Iosa, M.; Fusco, F.R.; Pirri, C.; Di Giunta, C.; Foti, C.; Traballesi, M. Early body weight-supported overground walking training in patients with stroke in subacute phase compared to conventional physiotherapy: A randomized controlled pilot study. *Int. J. Rehabil. Res.* **2019**, *42*, 309–315. [CrossRef]
293. Shahid, J.; Kashif, A.; Shahid, M.K. A Comprehensive Review of Physical Therapy Interventions for Stroke Rehabilitation: Impairment-Based Approaches and Functional Goals. *Brain Sci.* **2023**, *13*, 717. [CrossRef] [PubMed]
294. Rost, N.S.; Brodtmann, A.; Pase, M.P.; van Veluw, S.J.; Biffi, A.; Duering, M.; Hinman, J.D.; Dichgans, M. Post-Stroke Cognitive Impairment and Dementia. *Circ. Res.* **2022**, *130*, 1252–1271. [CrossRef] [PubMed]
295. Huang, Y.-Y.; Chen, S.-D.; Leng, X.-Y.; Kuo, K.; Wang, Z.-T.; Cui, M.; Tan, L.; Wang, K.; Dong, Q.; Yu, J.-T. Post-Stroke Cognitive Impairment: Epidemiology, Risk Factors, and Management. *J. Alzheimer's Dis.* **2022**, *86*, 983–999. [CrossRef] [PubMed]
296. Kalaria, R.N.; Akinyemi, R.; Ihara, M. Stroke injury, cognitive impairment and vascular dementia. *Biochim. Biophys. Acta (BBA) Mol. Basis Dis.* **2016**, *1862*, 915–925. [CrossRef]
297. Hussain, A.; Lee, M.; Rana, J.; Virani, S.S. Epidemiology and risk factors for stroke in young individuals: Implications for prevention. *Curr. Opin. Cardiol.* **2021**, *36*, 565–571. [CrossRef]
298. Furie, K. Epidemiology and Primary Prevention of Stroke. *Contin. Lifelong Learn. Neurol.* **2020**, *26*, 260–267. [CrossRef]
299. Wu, S.; Wu, B.; Liu, M.; Chen, Z.; Wang, W.; Anderson, C.S.; Sandercock, P.; Wang, Y.; Huang, Y.; Cui, L.; et al. Stroke in China: Advances and challenges in epidemiology, prevention, and management. *Lancet Neurol.* **2019**, *18*, 394–405. [CrossRef]
300. Caprio, F.Z.; Sorond, F.A. Cerebrovascular Disease: Primary and Secondary Stroke Prevention. *Med. Clin. N. Am.* **2018**, *103*, 295–308. [CrossRef]
301. Feigin, V.L.; Krishnamurthi, R.V.; Parmar, P.; Norrving, B.; Mensah, G.A.; Bennett, D.A.; Barker-Collo, S.; Moran, A.E.; Sacco, R.L.; Truelsen, T.; et al. Update on the Global Burden of Ischemic and Hemorrhagic Stroke in 1990-2013: The GBD 2013 Study. *Neuroepidemiology* **2015**, *45*, 161–176. [CrossRef]
302. Pandian, J.D.; Gall, S.L.; Kate, M.P.; Silva, G.S.; Akinyemi, R.O.; Ovbiagele, B.I.; Lavados, P.M.; Gandhi, D.B.C.; Thrift, A.G. Prevention of stroke: A global perspective. *Lancet* **2018**, *392*, 1269–1278. [CrossRef]
303. Smith, A.C.; Bensink, M.; Armfield, N.; Stillman, J.; Caffery, L. Telemedicine and rural health care applications. *J. Postgrad. Med.* **2006**, *51*, 286–293.
304. Matusitz, J.; Breen, G.-M. Telemedicine: Its Effects on Health Communication. *Health Commun.* **2007**, *21*, 73–83. [CrossRef] [PubMed]
305. Baratloo, A.; Rahimpour, L.; Abushouk, A.I.; Safari, S.; Lee, C.W.; Abdalvand, A. Effects of Telestroke on Thrombolysis Times and Outcomes: A Meta-analysis. *Prehospital Emerg. Care* **2018**, *22*, 472–484. [CrossRef] [PubMed]
306. Tchero, H.; Teguo, M.T.; Lannuzel, A.; Rusch, E. Telerehabilitation for Stroke Survivors: Systematic Review and Meta-Analysis. *J. Med. Internet Res.* **2018**, *20*, e10867. [CrossRef]
307. Chae, S.H.; Kim, Y.; Lee, K.-S.; Park, H.-S. Development and Clinical Evaluation of a Web-Based Upper Limb Home Rehabilitation System Using a Smartwatch and Machine Learning Model for Chronic Stroke Survivors: Prospective Comparative Study. *JMIR mHealth uHealth* **2020**, *8*, e17216. [CrossRef]
308. Donkor, E.S. Stroke in the21stCentury: A Snapshot of the Burden, Epidemiology, and Quality of Life. *Stroke Res. Treat.* **2018**, *2018*, 3238165. [CrossRef]
309. Duncan, P.W.; Bushnell, C.; Sissine, M.; Coleman, S.; Lutz, B.J.; Johnson, A.M.; Radman, M.; Bettger, J.P.; Zorowitz, R.D.; Stein, J. Comprehensive Stroke Care and Outcomes: Time for a Paradigm Shift. *Stroke* **2021**, *52*, 385–393. [CrossRef]
310. Bonaventura, A.; Liberale, L.; Vecchié, A.; Casula, M.; Carbone, F.; Dallegri, F.; Montecucco, F. Update on Inflammatory Biomarkers and Treatments in Ischemic Stroke. *Int. J. Mol. Sci.* **2016**, *17*, 1967. [CrossRef]

311. Satani, N.; Savitz, S.I. Is Immunomodulation a Principal Mechanism Underlying How Cell-Based Therapies Enhance Stroke Recovery? *Neurotherapeutics* **2016**, *13*, 775–782. [CrossRef]
312. Azodi, S.; Jacobson, S. Cytokine Therapies in Neurological Disease. *Neurotherapeutics* **2016**, *13*, 555–561. [CrossRef]
313. Goyal, M.; Sutherland, G.R.; Lama, S.; Cimflova, P.; Kashani, N.; Mayank, A.; Psychogios, M.-N.; Spelle, L.; Costalat, V.; Sakai, N.; et al. Neurointerventional Robotics: Challenges and Opportunities. *Clin. Neuroradiol.* **2020**, *30*, 203–208. [CrossRef] [PubMed]
314. Patel, T.M.; Shah, S.C.; Pancholy, S.B. Long Distance Tele-Robotic-Assisted Percutaneous Coronary Intervention: A Report of First-in-Human Experience. *EClinicalMedicine* **2019**, *14*, 53–58. [CrossRef] [PubMed]
315. Zeleňák, K.; Krajina, A.; Meyer, L.; Fiehler, J.; ESMINT Artificial Intelligence and Robotics Ad hoc Committee; Behme, D.; Bulja, D.; Caroff, J.; Chotai, A.A.; Da Ros, V.; et al. How to Improve the Management of Acute Ischemic Stroke by Modern Technologies, Artificial Intelligence, and New Treatment Methods. *Life* **2021**, *11*, 488. [CrossRef]
316. Liu, T.; Fan, W.; Wu, C. A hybrid machine learning approach to cerebral stroke prediction based on imbalanced medical dataset. *Artif. Intell. Med.* **2019**, *101*, 101723. [CrossRef]
317. Kokkotis, C.; Giarmatzis, G.; Giannakou, E.; Moustakidis, S.; Tsatalas, T.; Tsiptsios, D.; Vadikolias, K.; Aggelousis, N. An Explainable Machine Learning Pipeline for Stroke Prediction on Imbalanced Data. *Diagnostics* **2022**, *12*, 2392. [CrossRef] [PubMed]
318. Poona, N.K.; van Niekerk, A.; Nadel, R.L.; Ismail, R. Random Forest (RF) Wrappers for Waveband Selection and Classification of Hyperspectral Data. *Appl. Spectrosc.* **2016**, *70*, 322–333. [CrossRef]
319. Yu, D.; Liu, Z.; Su, C.; Han, Y.; Duan, X.; Zhang, R.; Liu, X.; Yang, Y.; Xu, S. Copy number variation in plasma as a tool for lung cancer prediction using Extreme Gradient Boosting (XGBoost) classifier. *Thorac. Cancer* **2019**, *11*, 95–102. [CrossRef]
320. Gkantzios, A.; Kokkotis, C.; Tsiptsios, D.; Moustakidis, S.; Gkartzonika, E.; Avramidis, T.; Aggelousis, N.; Vadikolias, K. Evaluation of Blood Biomarkers and Parameters for the Prediction of Stroke Survivors' Functional Outcome upon Discharge Utilizing Explainable Machine Learning. *Diagnostics* **2023**, *13*, 532. [CrossRef]
321. Sarvestany, S.S.; Kwong, J.C.; Azhie, A.; Dong, V.; Cerocchi, O.; Ali, A.F.; Karnam, R.S.; Kuriry, H.; Shengir, M.; Candido, E.; et al. Development and validation of an ensemble machine learning framework for detection of all-cause advanced hepatic fibrosis: A retrospective cohort study. *Lancet Digit. Health* **2022**, *4*, e188–e199. [CrossRef]
322. Bos, J.M.; Attia, Z.I.; Albert, D.E.; Noseworthy, P.A.; Friedman, P.A.; Ackerman, M.J. Use of Artificial Intelligence and Deep Neural Networks in Evaluation of Patients With Electrocardiographically Concealed Long QT Syndrome From the Surface 12-Lead Electrocardiogram. *JAMA Cardiol.* **2021**, *6*, 532–538. [CrossRef]
323. Chandrabhatla, A.S.; Pomeraniec, I.J.; Ksendzovsky, A. Co-evolution of machine learning and digital technologies to improve monitoring of Parkinson's disease motor symptoms. *Npj Digit. Med.* **2022**, *5*, 32. [CrossRef] [PubMed]
324. Thompson, A.C.; Jammal, A.A.; Berchuck, S.I.; Mariottoni, E.B.; Medeiros, F.A. Assessment of a Segmentation-Free Deep Learning Algorithm for Diagnosing Glaucoma From Optical Coherence Tomography Scans. *JAMA Ophthalmol.* **2020**, *138*, 333–339. [CrossRef] [PubMed]
325. Sirsat, M.S.; Fermé, E.; Câmara, J. Machine Learning for Brain Stroke: A Review. *J. Stroke Cerebrovasc. Dis.* **2020**, *29*, 105162. [CrossRef] [PubMed]
326. Mainali, S.; Darsie, M.E.; Smetana, K.S. Machine Learning in Action: Stroke Diagnosis and Outcome Prediction. *Front. Neurol.* **2021**, *12*, 734345. [CrossRef]
327. Chavva, I.R.; Crawford, A.L.; Mazurek, M.H.; Yuen, M.M.; Prabhat, A.M.; Payabvash, S.; Sze, G.; Falcone, G.J.; Matouk, C.C.; de Havenon, A.; et al. Deep Learning Applications for Acute Stroke Management. *Ann. Neurol.* **2022**, *92*, 574–587. [CrossRef]
328. Zhu, S.; Gilbert, M.; Chetty, I.; Siddiqui, F. The 2021 landscape of FDA-approved artificial intelligence/machine learning-enabled medical devices: An analysis of the characteristics and intended use. *Int. J. Med. Inform.* **2022**, *165*, 104828. [CrossRef]
329. Chandrabhatla, A.S.; Kuo, E.A.; Sokolowski, J.D.; Kellogg, R.T.; Park, M.; Mastorakos, P. Artificial Intelligence and Machine Learning in the Diagnosis and Management of Stroke: A Narrative Review of United States Food and Drug Administration-Approved Technologies. *J. Clin. Med.* **2023**, *12*, 3755. [CrossRef]
330. Suri, J.S.; Agarwal, S.; Gupta, S.K.; Puvvula, A.; Viskovic, K.; Suri, N.; Alizad, A.; El-Baz, A.; Saba, L.; Fatemi, M.; et al. Systematic Review of Artificial Intelligence in Acute Respiratory Distress Syndrome for COVID-19 Lung Patients: A Biomedical Imaging Perspective. *IEEE J. Biomed. Health Inform.* **2021**, *25*, 4128–4139. [CrossRef]
331. Pal, D.; Reddy, P.B.; Roy, S. Attention UW-Net: A fully connected model for automatic segmentation and annotation of chest X-ray. *Comput. Biol. Med.* **2022**, *150*, 106083. [CrossRef]
332. Jain, P.K.; Dubey, A.; Saba, L.; Khanna, N.N.; Laird, J.R.; Nicolaides, A.; Fouda, M.M.; Suri, J.S.; Sharma, N. Attention-Based UNet Deep Learning Model for Plaque Segmentation in Carotid Ultrasound for Stroke Risk Stratification: An Artificial Intelligence Paradigm. *J. Cardiovasc. Dev. Dis.* **2022**, *9*, 326. [CrossRef]
333. Kaplan, P.W.; Rossetti, A.O. EEG Patterns and Imaging Correlations in Encephalopathy: Encephalopathy part II. *J. Clin. Neurophysiol.* **2011**, *28*, 233–251. [CrossRef]
334. Islam, M.S.; Hussain, I.; Rahman, M.; Park, S.J.; Hossain, A. Explainable Artificial Intelligence Model for Stroke Prediction Using EEG Signal. *Sensors* **2022**, *22*, 9859. [CrossRef] [PubMed]
335. Marsh, J.D.; Keyrouz, S.G. Stroke Prevention and Treatment. *J. Am. Coll. Cardiol.* **2010**, *56*, 683–691. [CrossRef] [PubMed]
336. Shrestha, S.; Coy, S.; Bekelis, K. Oral Antiplatelet and Anticoagulant Agents in the Prevention and Management of Ischemic Stroke. *Curr. Pharm. Des.* **2017**, *23*, 1377–1391. [CrossRef] [PubMed]

337. Hopkins, L.N. Blazing the Frontiers of Stroke Therapy. *Neurosurgery* **2019**, *85*, S1–S2. [CrossRef]
338. Saqqur, M.; Uchino, K.; Demchuk, A.M.; Molina, C.A.; Garami, Z.; Calleja, S.; Akhtar, N.; Orouk, F.O.; Salam, A.; Shuaib, A.; et al. Site of Arterial Occlusion Identified by Transcranial Doppler Predicts the Response to Intravenous Thrombolysis for Stroke. *Stroke* **2007**, *38*, 948–954. [CrossRef] [PubMed]
339. Sweid, A.; Hammoud, B.; Ramesh, S.; Wong, D.; Alexander, T.D.; Weinberg, J.H.; Deprince, M.; Dougherty, J.; Maamari, D.J.-M.; Tjoumakaris, S.; et al. Acute ischaemic stroke interventions: Large vessel occlusion and beyond. *Stroke Vasc. Neurol.* **2019**, *5*, 80–85. [CrossRef]
340. Ma, Z.; Ding, N.; Yu, Y.; Ma, Y.; Yuan, X.; Wang, Y.; Zhao, Y.; Luan, J.; Liu, J. Quantification of cerebral vascular perfusion density via optical coherence tomography based on locally adaptive regional growth. *Appl. Opt.* **2018**, *57*, 10117–10124. [CrossRef]
341. Yang, S.; Liu, K.; Ding, H.; Gao, H.; Zheng, X.; Ding, Z.; Xu, K.; Li, P. Longitudinal in vivo intrinsic optical imaging of cortical blood perfusion and tissue damage in focal photothrombosis stroke model. *J. Cereb. Blood Flow Metab.* **2018**, *39*, 1381–1393. [CrossRef]
342. Liu, K.; Yin, M.; Cai, Z. Research and application advances in rehabilitation assessment of stroke. *J. Zhejiang Univ. B* **2022**, *23*, 625–641. [CrossRef]
343. Kleinschnitz, C.; Fluri, F.; Schuhmann, M. Animal models of ischemic stroke and their application in clinical research. *Drug Des. Dev. Ther.* **2015**, *9*, 3445–3454. [CrossRef] [PubMed]
344. Dichgans, M.; Pulit, S.L.; Rosand, J. Stroke genetics: Discovery, biology, and clinical applications. *Lancet Neurol.* **2019**, *18*, 587–599. [CrossRef] [PubMed]
345. Herpich, F.; Rincon, F.M. Management of Acute Ischemic Stroke. *Crit. Care Med.* **2020**, *48*, 1654–1663. [CrossRef]
346. Chidambaram, S.B.; Rathipriya, A.G.; Mahalakshmi, A.M.; Sharma, S.; Hediyal, T.A.; Ray, B.; Sunanda, T.; Rungratanawanich, W.; Kashyap, R.S.; Qoronfleh, M.W.; et al. The Influence of Gut Dysbiosis in the Pathogenesis and Management of Ischemic Stroke. *Cells* **2022**, *11*, 1239. [CrossRef]
347. Kuriakose, D.; Xiao, Z. Pathophysiology and Treatment of Stroke: Present Status and Future Perspectives. *Int. J. Mol. Sci.* **2020**, *21*, 7609. [CrossRef] [PubMed]
348. Guruswamy, R.; ElAli, A. Complex Roles of Microglial Cells in Ischemic Stroke Pathobiology: New Insights and Future Directions. *Int. J. Mol. Sci.* **2017**, *18*, 496. [CrossRef]
349. Bonfanti, L.; Charvet, C.J. Brain Plasticity in Humans and Model Systems: Advances, Challenges, and Future Directions. *Int. J. Mol. Sci.* **2021**, *22*, 9358. [CrossRef]
350. Harpaz, D.; Eltzov, E.; Seet, R.C.; Marks, R.S.; Tok, A.I. Point-of-Care-Testing in Acute Stroke Management: An Unmet Need Ripe for Technological Harvest. *Biosensors* **2017**, *7*, 30. [CrossRef]
351. Ganesh, A.; Barber, P.A. The Cognitive Sequelae of Transient Ischemic Attacks—Recent Insights and Future Directions. *J. Clin. Med.* **2022**, *11*, 2637. [CrossRef]

MDPI
St. Alban-Anlage 66
4052 Basel
Switzerland
www.mdpi.com

MDPI Books Editorial Office
E-mail: books@mdpi.com
www.mdpi.com/books

www.ingramcontent.com/pod-product-compliance
Lightning Source LLC
Chambersburg PA
CBHW041431210926
51850CB00013B/3404